HANDBOOK OF METEOROLOGY

EDITED BY

F. A. BERRY, Jr.

*Captain, U.S.N., Officer-in-Charge, U.S. Navy
Weather Central, Washington, D.C.*

E. BOLLAY

Lieutenant Commander, U.S.N.R

NORMAN R. BEERS

New York   *London*
McGRAW-HILL BOOK COMPANY, INC.
1945

THE MAPLE PRESS COMPANY, YORK, PA.

PREFACE

The "Handbook of Meteorology" is designed to furnish the student and the professional meteorologist a convenient *text reference* for data, fundamental theory, and weather analysis and forecasting. The emphasis has been conscientiously placed on the scientific and engineering aspects of meteorology, rather than on current techniques. This emphasis is desirable at any time in a handbook, and particularly so at this time in view of recent and forthcoming changes in the methods of analysis and in the display of synoptic data (especially at upper levels).

The editors' intent has been that this handbook contain practical meteorological information. Their interpretation of what is "practical" is based on their collective experience and on their interpretation of the experience of returning students who have forecast the weather in the field under trying conditions. Accordingly the "Handbook of Meteorology" contains not only sections on fronts, air masses, upper air, etc., which are immediately suggested by its title, but also those fundamentals of mathematics and of the physical sciences on which meteorology is dependent. Some of these topics are frequently considered to be theoretical. When, however, they are written up in compact form and in meteorological language so as to be "available, useful, in practice or action," then they are rightly included in a book that is designed for daily use.

The practice of meteorology (by which is meant some direct or indirect aspect of weather forecasting) is like the other engineering sciences in that an answer to a problem is required in a specified time, often from meager data. Moreover, as is sometimes overlooked in the stress of war, "an engineer is one who can do at reasonable cost what almost any man can accomplish with unlimited resources at his disposal."* It is the editors' considered belief that the practice of meteorology is inevitably to become accepted as an engineering science.

The variety of problems that arises in meteorological practice is endlessly complicated by both space and time intervals. An aviator may stand by and demand to know precisely when he will have sufficient ceiling to take off. A constructor may save lives and money by knowing with some accuracy what will be the maximum rainfall over a large watershed. These and many other meteorological problems are not

* GILKEY, J. I., *Journal of Engineering*, January, 1945.

yet completely solved, though fine progress has been made with recent advances. Many meteorological problems are in fact not yet clearly stated as problems, nor can much be done to overcome present inadequacies until many men and women understand not only the basic principles of meteorology, but also the principal problems of at least one other activity, *e.g.*, aviation, agriculture, manufacture, transportation, distribution, communications, etc. The editors desire to acknowledge these problems even though the present edition of the handbook is limited in scope to aviation and hydrometeorology in this respect.

Perhaps greater cooperation is required among individuals and among nations in the practice of meteorology than in any other social effort of man. This is so, simply because even short-period forecasts often depend on synoptic data that are beyond the reach of the individual meteorologist or the individual nation. Problems involving longer periods and greater areas require even more data, from the surface and from upper levels of the atmosphere. Recognition of the cooperation required for successful weather services appeared at least a hundred years ago when John Ruskin wrote, "The meteorological Society, therefore, has been formed, not for a city, nor for a kingdom, but for the world."

The editors are deeply indebted to their contributors especially and to many others who have also cooperated generously to produce this handbook.

Mr. Charney wishes to acknowledge his indebtedness to Professor J. Kaplan for his introduction to the subject (Section IV), and for the kind encouragement he received in the preparation of the manuscript.

The editors assume full responsibility for the selection and arrangement of topics. Each section has been read by one or more critics, but individual authors were final judges as to how their contributions should be covered. Much desirable material was omitted for security reasons at the time of writing, but it is believed the integrated result of all that has been available provides the engineering science needed by a trained meteorologist in the practice of his profession. For the rules and regulations of the various meteorological services it is generally necessary to make reference to their individual publications. For the art of forecasting the weather according to the needs and desires of his audience, the meteorologist must defer to the association of his fellows and to experience.

THE EDITORS.

ANNAPOLIS, MARYLAND,
  September, 1945.

LIST OF CONTRIBUTORS

R. A. BAUMGARTNER, Lieut., U.S.N.R., Postgraduate School, United States Naval Academy, X.

NORMAN R. BEERS, Lieut., U.S.N.R., Postgraduate School, United States Naval Academy, I, V, X.

F. A. BERRY, JR., Capt., U.S.N., Staff Comair 7th Fleet, X.

E. BOLLAY, Lieut. Comdr., U.S.N.R., Fleet Weather Central, X.

JULE G. CHARNEY, Lecturer, Department of Meteorology, University of California, IV.

Civilian Staff, Institute of Tropical Meteorology, Rio Piedras, Puerto Rico, X.

FREDERICK A. FICKEN, Professor of Mathematics, University of Tennessee (on leave); Office of Field Service Consultant, Operations Research Group, U.S.N., II.

ALBERT S. FRY, Chief, Hydraulic Data Division, Tennessee Valley Authority, XIII.

H. G. HOUGHTON, Professor of Meteorology, Massachusetts Institute of Technology, III.

GEORGE R. JENKINS, Lieut., U.S.N.R., Postgraduate School, United States Naval Academy, IX, X.

H. LANDSBERG, Ph.D., Associate Professor of Meteorology, The University of Chicago, XII.

S. C. LOWELL, Lieut. Comdr., U.S.N.R., Postgraduate School, United States Naval Academy, III.

K. B. McEACHRON, Research Engineer, High Voltage Engineering Laboratory and Designing Engineer, Power Transformer Engineering Division, General Electric Company, Pittsfield, Mass.; Fellow, American Institute of Electrical Engineers, III.

J. F. O'CONNOR, Lieut., U.S.N.R., Postgraduate School, United States Naval Academy, X.

MILDRED B. OLIVER, Department of Meteorology, The University of Chicago, X.

VINCENT J. OLIVER, Department of Meteorology, The University of Chicago, X.

CARL-G. ROSSBY, Andrew MacLeish Distinguished Service Professorship and Chairman of the Department of Meteorology, The University of Chicago, VII.

P. L. SCHERESCHEWSKY, Formerly Chief Weather Service, French Armies, XI.

A. K. SHOWALTER, Meteorologist in Charge, Hydrometeorological Section, Office of Hydraulic Director, U.S. Weather Bureau, XIII.

H. J. STEWART, Daniel Guggenheim Aeronautical Laboratory, California Institute of Technology, VI.

H. U. SVERDRUP, Professor of Oceanography, University of California; Director, Scripps Institution of Oceanography, XIV.

LOUVAN E. WOOD, Meteorological Engineer, Friez Instrument Division, Bendix Aviation Corporation, VIII.

CONTENTS

For the detailed contents of any section consult the title page of that section. For the alphabetical index see pp. 1057 to 1068.

SECTION I

NUMERICAL AND GRAPHICAL DATA

By Norman R. Beers

and the Aerological Engineering Staff, Postgraduate School,
United States Naval Academy

CONTENTS

SECTION I

NUMERICAL AND GRAPHICAL DATA

By Norman R. Beers

and the Aerological Engineering Staff, Postgraduate School,
United States Naval Academy

Table 1.—Greek Prefixes

micro	1/1,000,000	deka	10
milli	1/1,000	hecto	100
centi	1/100	kilo	1,000
deci	1/10	mega	1,000,000

Table 2.—Greek Alphabet

A α or a	alpha	N ν	nu
B β	beta	Ξ ξ	xi
Γ γ	gamma	O o	omicron
Δ δ or ∂	delta	Π π	pi
E ϵ	epsilon	P ρ	rho
Z ζ	zeta	Σ σ	sigma
H η	eta	T τ	tau
Θ θ or ϑ	theta	Υ υ	upsilon
I ι	iota	Φ ϕ or φ	phi
K κ	kappa	X χ	chi
Λ λ	lambda	Ψ ψ	psi
M μ	mu	Ω ω	omega

MATHEMATICAL TABLES

TABLE 3.—RECIPROCALS OF NUMBERS*

N	0	1	2	3	4	5	6	7	8	9	Avg. diff.
1.00		.9990	.9980	.9970	.9960	.9950	.9940	.9930	.9921	.9911	
1	.9901	.9891	.9881	.9872	.9862	.9852	.9843	.9833	.9823	.9814	−10
2	.9804	.9794	.9785	.9775	.9766	.9756	.9747	.9737	.9728	.9718	
3	.9709	.9699	.9690	.9681	.9671	.9662	.9653	.9643	.9634	.9625	
4	.9615	.9606	.9597	.9588	.9579	.9569	.9560	.9551	.9542	.9533	−9
1.05	.9524	.9515	.9506	.9497	.9488	.9479	.9470	.9461	.9452	.9443	
6	.9434	.9425	.9416	.9407	.9398	.9390	.9381	.9372	.9363	.9355	
7	.9346	.9337	.9328	.9320	.9311	.9302	.9294	.9285	.9276	.9268	
8	.9259	.9251	.9242	.9234	.9225	.9217	.9208	.9200	.9191	.9183	−8
9	.9174	.9166	.9158	.9149	.9141	.9132	.9124	.9116	.9107	.9099	
1.10	.9091	.9083	.9074	.9066	.9058	.9050	.9042	.9033	.9025	.9017	
1	.9009	.9001	.8993	.8985	.8977	.8969	.8961	.8953	.8945	.8937	
2	.8929	.8921	.8913	.8905	.8897	.8889	.8881	.8873	.8865	.8857	
3	.8850	.8842	.8834	.8826	.8818	.8811	.8803	.8795	.8787	.8780	
4	.8772	.8764	.8757	.8749	.8741	.8734	.8726	.8718	.8711	.8703	
1.15	.8696	.8688	.8681	.8673	.8666	.8658	.8651	.8643	.8636	.8628	
6	.8621	.8613	.8606	.8598	.8591	.8584	.8576	.8569	.8562	.8554	−7
7	.8547	.8540	.8532	.8525	.8518	.8511	.8503	.8496	.8489	.8482	
8	.8475	.8467	.8460	.8453	.8446	.8439	.8432	.8425	.8418	.8410	
9	.8403	.8396	.8389	.8382	.8375	.8368	.8361	.8354	.8347	.8340	
1.20	.8333	.8326	.8319	.8313	.8306	.8299	.8292	.8285	.8278	.8271	
1	.8264	.8258	.8251	.8244	.8237	.8230	.8224	.8217	.8210	.8203	
2	.8197	.8190	.8183	.8177	.8170	.8163	.8157	.8150	.8143	.8137	
3	.8130	.8123	.8117	.8110	.8104	.8097	.8091	.8084	.8078	.8071	−6
4	.8065	.8058	.8052	.8045	.8039	.8032	.8026	.8019	.8013	.8006	
1.25	.8000	.7994	.7987	.7981	.7974	.7968	.7962	.7955	.7949	.7943	
6	.7937	.7930	.7924	.7918	.7911	.7905	.7899	.7893	.7886	.7880	
7	.7874	.7868	.7862	.7855	.7849	.7843	.7837	.7831	.7825	.7819	
8	.7812	.7806	.7800	.7794	.7788	.7782	.7776	.7770	.7764	.7758	
9	.7752	.7746	.7740	.7734	.7728	.7722	.7716	.7710	.7704	.7698	
1.30	.7692	.7686	.7680	.7675	.7669	.7663	.7657	.7651	.7645	.7639	
1	.7634	.7628	.7622	.7616	.7610	.7605	.7599	.7593	.7587	.7582	
2	.7576	.7570	.7564	.7559	.7553	.7547	.7541	.7536	.7530	.7524	
3	.7519	.7513	.7508	.7502	.7496	.7491	.7485	.7479	.7474	.7468	
4	.7463	.7457	.7452	.7446	.7440	.7435	.7429	.7424	.7418	.7413	
1.35	.7407	.7402	.7396	.7391	.7386	.7380	.7375	.7369	.7364	.7358	−5
6	.7353	.7348	.7342	.7337	.7331	.7326	.7321	.7315	.7310	.7305	
7	.7299	.7294	.7289	.7283	.7278	.7273	.7267	.7262	.7257	.7252	
8	.7246	.7241	.7236	.7231	.7225	.7220	.7215	.7210	.7205	.7199	
9	.7194	.7189	.7184	.7179	.7174	.7168	.7163	.7158	.7153	.7148	
1.40	.7143	.7138	.7133	.7128	.7123	.7117	.7112	.7107	.7102	.7097	
1	.7092	.7087	.7082	.7077	.7072	.7067	.7062	.7057	.7052	.7047	
2	.7042	.7037	.7032	.7027	.7022	.7018	.7013	.7008	.7003	.6998	
3	.6993	.6988	.6983	.6978	.6974	.6969	.6964	.6959	.6954	.6949	
4	.6944	.6940	.6935	.6930	.6925	.6920	.6916	.6911	.6906	.6901	
1.45	.6897	.6892	.6887	.6882	.6878	.6873	.6868	.6863	.6859	.6854	
6	.6849	.6845	.6840	.6835	.6831	.6826	.6821	.6817	.6812	.6807	
7	.6803	.6798	.6793	.6789	.6784	.6780	.6775	.6770	.6766	.6761	
8	.6757	.6752	.6748	.6743	.6739	.6734	.6729	.6725	.6720	.6716	
9	.6711	.6707	.6702	.6698	.6693	.6689	.6684	.6680	.6676	.6671	

$1/\pi = 0.318310$ $1/e = 0.367879$

Moving the decimal point in either direction in N requires moving it in the *opposite* direction in body of table (see p. 5).

* From Marks, "Mechanical Engineers' Handbook," 4th ed., McGraw-Hill, New York, 1941.

TABLE 3.—RECIPROCALS OF NUMBERS.*—(*Continued*)

N	0	1	2	3	4	5	6	7	8	9	Avg. diff.
1.50	.6667	.6662	.6658	.6653	.6649	.6645	.6640	.6636	.6631	.6627	−4
1	.6623	.6618	.6614	.6609	.6605	.6601	.6596	.6592	.6588	.6583	
2	.6579	.6575	.6570	.6566	.6562	.6557	.6553	.6549	.6545	.6540	
3	.6536	.6532	.6527	.6523	.6519	.6515	.6510	.6506	.6502	.6498	
4	.6494	.6489	.6485	.6481	.6477	.6472	.6468	.6464	.6460	.6456	
1.55	.6452	.6447	.6443	.6439	.6435	.6431	.6427	.6423	.6418	.6414	
6	.6410	.6406	.6402	.6398	.6394	.6390	.6386	.6382	.6378	.6373	
7	.6369	.6365	.6361	.6357	.6353	.6349	.6345	.6341	.6337	.6333	
8	.6329	.6325	.6321	.6317	.6313	.6309	.6305	.6301	.6297	.6293	
9	.6289	.6285	.6281	.6277	.6274	.6270	.6266	.6262	.6258	.6254	
1.60	.6250	.6246	.6242	.6238	.6234	.6231	.6227	.6223	.6219	.6215	
1	.6211	.6207	.6203	.6200	.6196	.6192	.6188	.6184	.6180	.6177	
2	.6173	.6169	.6165	.6161	.6158	.6154	.6150	.6146	.6143	.6139	
3	.6135	.6131	.6127	.6124	.6120	.6116	.6112	.6109	.6105	.6101	
4	.6098	.6094	.6090	.6086	.6083	.6079	.6075	.6072	.6068	.6064	
1.65	.6061	.6057	.6053	.6050	.6046	.6042	.6039	.6035	.6031	.6028	
6	.6024	.6020	.6017	.6013	.6010	.6006	.6002	.5999	.5995	.5992	
7	.5988	.5984	.5981	.5977	.5974	.5970	.5967	.5963	.5959	.5956	
8	.5952	.5949	.5945	.5942	.5938	.5935	.5931	.5928	.5924	.5921	
9	.5917	.5914	.5910	.5907	.5903	.5900	.5896	.5893	.5889	.5886	
1.70	.5882	.5879	.5875	.5872	.5869	.5865	.5862	.5858	.5855	.5851	−3
1	.5848	.5845	.5841	.5838	.5834	.5831	.5828	.5824	.5821	.5817	
2	.5814	.5811	.5807	.5804	.5800	.5797	.5794	.5790	.5787	.5784	
3	.5780	.5777	.5774	.5770	.5767	.5764	.5760	.5757	.5754	.5750	
4	.5747	.5744	.5741	.5737	.5734	.5731	.5727	.5724	.5721	.5718	
1.75	.5714	.5711	.5708	.5705	.5701	.5698	.5695	.5692	.5688	.5685	
6	.5682	.5679	.5675	.5672	.5669	.5666	.5663	.5659	.5656	.5653	
7	.5650	.5647	.5643	.5640	.5637	.5634	.5631	.5627	.5624	.5621	
8	.5618	.5615	.5612	.5609	.5605	.5602	.5599	.5596	.5593	.5590	
9	.5587	.5583	.5580	.5577	.5574	.5571	.5568	.5565	.5562	.5559	
1.80	.5556	.5552	.5549	.5546	.5543	.5540	.5537	.5534	.5531	.5528	
1	.5525	.5522	.5519	.5516	.5513	.5510	.5507	.5504	.5501	.5498	
2	.5495	.5491	.5488	.5485	.5482	.5479	.5476	.5473	.5470	.5467	
3	.5464	.5461	.5459	.5456	.5453	.5450	.5447	.5444	.5441	.5438	
4	.5435	.5432	.5429	.5426	.5423	.5420	.5417	.5414	.5411	.5408	
1.85	.5405	.5402	.5400	.5397	.5394	.5391	.5388	.5385	.5382	.5379	
6	.5376	.5373	.5371	.5368	.5365	.5362	.5359	.5356	.5353	.5350	
7	.5348	.5345	.5342	.5339	.5336	.5333	.5330	.5328	.5325	.5322	
8	.5319	.5316	.5313	.5311	.5308	.5305	.5302	.5299	.5297	.5294	
9	.5291	.5288	.5285	.5283	.5280	.5277	.5274	.5271	.5269	.5266	
1.90	.5263	.5260	.5258	.5255	.5252	.5249	.5247	.5244	.5241	.5238	
1	.5236	.5233	.5230	.5227	.5225	.5222	.5219	.5216	.5214	.5211	
2	.5208	.5206	.5203	.5200	.5198	.5195	.5192	.5189	.5187	.5184	
3	.5181	.5179	.5176	.5173	.5171	.5168	.5165	.5163	.5160	.5157	
4	.5155	.5152	.5149	.5147	.5144	.5141	.5139	.5136	.5133	.5131	
1.95	.5128	.5126	.5123	.5120	5118	.5115	.5112	.5110	.5107	.5105	
6	.5102	.5099	.5097	.5094	.5092	.5089	.5086	.5084	.5081	.5079	
8	.5076	.5074	.5071	.5068	.5066	.5063	.5061	.5058	.5056	.5053	−2
8	.5051	.5048	.5045	.5043	.5040	.5038	.5035	.5033	.5030	.5028	
9	.5025	.5023	.5020	.5018	.5015	.5013	.5010	.5008	.5005	.5003	

Moving the decimal point in either direction in *N* requires moving it in the *opposite* direction in body of table (see p. 5).

TABLE 3.—RECIPROCALS OF NUMBERS.*—(*Continued*)

N	0	1	2	3	4	5	6	7	8	9	Av. Diff.
2.0	.5000	.4975	.4950	.4926	.4902	.4878	.4854	.4831	.4808	.4785	− 24
1	.4762	.4739	.4717	.4695	.4673	.4651	.4630	.4608	.4587	.4566	− 21
2	.4545	.4525	.4505	.4484	.4464	.4444	.4425	.4405	.4386	.4367	− 20
3	.4348	.4329	.4310	.4292	.4274	.4255	.4237	.4219	.4202	.4184	− 18
4	.4167	.4149	.4132	.4115	.4098	.4082	.4065	.4049	.40ɔ2	.4016	− 17
2.5	.4000	.3984	.3968	.3953	.3937	.3922	.3906	.3891	.3876	.3861	− 15
6	.3846	.3831	.3817	.3802	.3788	.3774	.3759	.3745	.3731	.3717	− 14
7	.3704	.3690	.3676	.3663	.3650	.3636	.3623	.3610	.3597	.3584	− 13
8	.3571	.3559	.3546	.3534	.3521	.3509	.3497	.3484	.3472	.3460	− 12
9	.3448	.3436	.3425	.3413	.3401	.3390	.3378	.3367	.3356	.3344	− 12
3.0	.3333	.3322	.3311	.3300	.3289	.3279	.3268	.3257	.3247	.3236	− 11
1	.3226	.3215	.3205	.3195	.3185	.3175	.3165	.3155	.3145	.3135	− 10
2	.3125	.3115	.3106	.3096	.3086	.3077	.3067	.3058	.3049	.3040	− 10
3	.3030	.3021	.3012	.3003	.2994	.2985	.2976	.2967	.2959	.2950	− 9
4	.2941	.2933	.2924	.2915	.2907	.2899	.2890	.2882	.2874	.2865	− 8
3.5	.2857	.2849	.2841	.2833	.2825	.2817	.2809	.2801	.2793	.2786	− 8
6	.2778	.2770	.2762	.2755	.2747	.2740	.2732	.2725	.2717	.2710	− 8
7	.2703	.2695	.2688	.2681	.2674	.2667	.2660	.2653	.2646	.2639	− 7
8	.2632	.2625	.2618	.2611	.2604	.2597	.2591	.2584	.2577	.2571	− 7
9	.2564	.2558	.2551	.2545	.2538	.2532	.2525	.2519	.2513	.2506	− 6
4.0	.2500	.2494	.2488	.2481	.2475	.2469	.2463	.2457	.2451	.2445	− 6
1	.2439	.2433	.2427	.2421	.2415	.2410	.2404	.2398	.2392	.2387	− 6
2	.2381	.2375	.2370	.2364	.2358	.2353	.2347	.2342	.2336	.2331	− 6
3	.2326	.2320	.2315	.2309	.2304	.2299	.2294	.2288	.2283	.2278	− 5
4	.2273	.2268	.2262	.2257	.2252	.2247	.2242	.2237	.2232	.2227	− 5
4.5	.2222	.2217	.2212	.2208	.2203	.2198	.2193	.2188	.2183	.2179	− 5
6	.2174	.2169	.2165	.2160	.2155	.2151	.2146	.2141	.2137	.2132	− 5
7	.2128	.2123	.2119	.2114	.2110	.2105	.2101	.2096	.2092	.2088	− 4
8	.2083	.2079	.2075	.2070	.2066	.2062	.2058	.2053	.2049	.2045	− 4
9	.2041	.2037	.2033	.2028	.2024	.2020	.2016	.2012	.2008	.2004	− 4

$1/\pi = 0.31835$ $1/e = 0.367879$

Explanation of Table of Reciprocals (pp. 3–6). This table gives the values of $1/N$ for values of N from 1 to 10, correct to four figures. (Interpolated values may be in error by 1 in the fourth figure.)

To find the reciprocal of a number N **outside the range from 1 to 10,** note that moving the decimal point any number of places in either direction in column N is equivalent to moving it the same number of places in the **opposite direction** in the body of the table. For example:

$$\frac{1}{3.217} = 0.3109 \qquad \frac{1}{3217} = 0.000\,3109 \qquad \frac{1}{0.003217} = 310.9$$

Computations are frequently simplified by careful arrangement of the problem and recognition of adequate mathematical approximations. For example:

$$\frac{1}{1 \pm x} = 1 \mp x \qquad |x| << 1)$$

$$\frac{1}{(0.006944)^{0.5}} = \sqrt{144} = 12$$

TABLE 3.—RECIPROCALS OF NUMBERS.*—(Continued)

N	0	1	2	3	4	5	6	7	8	9	Avg. diff.
5.0	.2000	.1996	.1992	.1988	.1984	.1980	.1976	.1972	.1969	.1965	− 4
.1	.1961	.1957	.1953	.1949	.1946	.1942	.1938	.1934	.1931	.1927	
.2	.1923	.1919	.1916	.1912	.1908	.1905	.1901	.1898	.1894	.1890	
.3	.1887	.1883	.1880	.1876	.1873	.1869	.1866	.1862	.1859	.1855	
.4	.1852	.1848	.1845	.1842	.1838	.1835	.1832	.1828	.1825	.1821	− 3
5.5	.1818	.1815	.1812	.1808	.1805	.1802	.1799	.1795	.1792	.1789	
.6	.1786	.1783	.1779	.1776	.1773	.1770	.1767	.1764	.1761	.1757	
.7	.1754	.1751	.1748	.1745	.1742	.1739	.1736	.1733	.1730	.1727	
.8	.1724	.1721	.1718	.1715	.1712	.1709	.1706	.1704	.1701	.1698	
.9	.1695	.1692	.1689	.1686	.1684	.1681	.1678	.1675	.1672	.1669	
6.0	.1667	.1664	.1661	.1658	.1656	.1653	.1650	.1647	.1645	.1642	
.1	.1639	.1637	.1634	.1631	.1629	.1626	.1623	.1621	.1618	.1616	
.2	.1613	.1610	.1608	.1605	.1603	.1600	.1597	.1595	.1592	.1590	
.3	.1587	.1585	.1582	.1580	.1577	.1575	.1572	.1570	.1567	.1565	− 2
.4	.1563	.1560	.1558	.1555	.1553	.1550	.1548	.1546	.1543	.1541	
6.5	.1538	.1536	.1534	.1531	.1529	.1527	.1524	.1522	.1520	.1517	
.6	.1515	.1513	.1511	.1508	.1506	.1504	.1502	.1499	.1497	.1495	
.7	.1493	.1490	.1488	.1486	.1484	.1481	.1479	.1477	.1475	.1473	
.8	.1471	.1468	.1466	.1464	.1462	.1460	.1458	.1456	.1453	.1451	
.9	.1449	.1447	.1445	.1443	.1441	.1439	.1437	.1435	.1433	.1431	
7.0	.1429	.1427	.1425	.1422	.1420	.1418	.1416	.1414	.1412	.1410	
.1	.1408	.1406	.1404	.1403	.1401	.1399	.1397	.1395	.1393	.1391	
.2	.1389	.1387	.1385	.1383	.1381	.1379	.1377	.1376	.1374	.1372	
.3	.1370	.1368	.1366	.1364	.1362	.1361	.1359	.1357	.1355	.1353	
.4	.1351	.1350	.1348	.1346	.1344	.1342	.1340	.1339	.1337	.1335	
7.5	.1333	.1332	.1330	.1328	.1326	.1325	.1323	.1321	.1319	.1318	
.6	.1316	.1314	.1312	.1311	.1309	.1307	.1305	.1304	.1302	.1300	
.7	.1299	.1297	.1295	.1294	.1292	.1290	.1289	.1287	.1285	.1284	
.8	.1282	.1280	.1279	.1277	.1276	.1274	.1272	.1271	.1269	.1267	
.9	.1266	.1264	.1263	.1261	.1259	.1258	.1256	.1255	.1253	.1252	
8.0	.1250	.1248	.1247	.1245	.1244	.1242	.1241	.1239	.1238	.1236	
.1	.1235	.1233	.1232	.1230	.1229	.1227	.1225	.1224	.1222	.1221	
.2	.1220	.1218	.1217	.1215	.1214	.1212	.1211	.1209	.1208	.1206	
.3	.1205	.1203	.1202	.1200	.1199	.1198	.1196	.1195	.1193	.1192	
.4	.1190	.1189	.1188	.1186	.1185	.1183	.1182	.1181	.1179	.1178	− 1
8.5	.1176	.1175	.1174	.1172	.1171	.1170	.1168	.1167	.1166	.1164	
.6	.1163	.1161	.1160	.1159	.1157	.1156	.1155	.1153	.1152	.1151	
.7	.1149	.1148	.1147	.1145	.1144	.1143	.1142	.1140	.1139	.1138	
.8	.1136	.1135	.1134	.1133	.1131	.1130	.1129	.1127	.1126	.1125	
.9	.1124	.1122	.1121	.1120	.1119	.1117	.1116	.1115	.1114	.1112	
9.0	.1111	.1110	.1109	.1107	.1106	.1105	.1104	.1103	.1101	.1100	
.1	.1099	.1098	.1096	.1095	.1094	.1093	.1092	.1091	.1089	.1088	
.2	.1087	.1086	.1085	.1083	.1082	.1081	.1080	.1079	.1078	.1076	
.3	.1075	.1074	.1073	.1072	.1071	.1070	.1068	.1067	.1066	.1065	
.4	.1064	.1063	.1062	.1060	.1059	.1058	.1057	.1056	.1055	.1054	
9.5	.1053	.1052	.1050	.1049	.1048	.1047	.1046	.1045	.1044	.1043	
.6	.1042	.1041	.1040	.1038	.1037	.1036	.1035	.1034	.1033	.1032	
.7	.1031	.1030	.1029	.1028	.1027	.1026	.1025	.1024	.1022	.1021	
.8	.1020	.1019	.1018	.1017	.1016	.1015	.1014	.1013	.1012	.1011	
.9	.1010	.1009	.1008	.1007	.1006	.1005	.1004	.1003	.1002	.1001	

Moving the decimal point in either direction in N requires moving it in the *opposite* direction in body of table (see p. 5).

TABLE 4.—COMMON LOGARITHMS*
(special table)

Number	0	1	2	3	4	5	6	7	8	9	Avg. diff.
1.00	0.0000	0004	0009	0013	0017	0022	0026	0030	0035	0039	4
1.01	0043	0048	0052	0056	0060	0055	0059	0073	0077	0082	
1.02	0086	0090	0095	0099	0103	0107	0111	0116	0120	0124	
1.03	0128	0133	0137	0141	0145	0149	0154	0158	0162	0166	
1.04	0170	0175	0179	.0183	0187	0191	0195	0199	0204	0208	
1.05	0212	0216	0220	0224	0228	0233	0237	0241	0245	0249	
1.06	0253	0257	0261	0265	0269	0273	0278	0282	0286	0290	
1.07	0294	0298	0302	0306	0310	0314	0318	0322	0326	0330	
1.08	0334	0338	0342	0346	0350	0354	0358	0362	0366	0370	
1.09	0374	0378	0382	0386	0390	0394	0398	0402	0406	0410	
1.10	0.0414	0418	0422	0426	0430	0434	0438	0441	0445	0449	
1.11	0453	0457	0461	0465	0469	0473	0477	0481	0484	0488	
1.12	0492	0496	0500	0504	0508	0512	0515	0519	0523	0527	
1.13	0531	0535	0538	0542	0546	0550	0554	0558	0561	0565	
1.14	0569	0573	0577	0580	0584	0588	0592	0596	0599	0603	
1.15	0607	0611	0615	0618	0622	0626	0630	0633	0637	0641	
1.16	0645	0648	0652	0656	0660	0663	0667	0671	0674	0678	
1.17	0682	0686	0689	0693	0697	0700	0704	0708	0711	0715	
1.18	0719	0722	0726	0730	0734	0737	0741	0745	0748	0752	
1.19	0755	0759	0763	0766	0770	0774	0777	0781	0785	0788	
1.20	0.0792	0795	0799	0803	0806	0810	0813	0817	0821	0824	
1.21	0828	0831	0835	0839	0842	0846	0849	0853	0856	0860	
1.22	0864	0867	0871	0874	0878	0881	0885	0888	0892	0896	
1.23	0899	0903	0906	0910	0913	0917	0920	0924	0927	0931	
1.24	0934	0938	0941	0945	0948	0952	0955	0959	0962	0966	
1.25	0969	0973	0976	0980	0983	0986	0990	0993	0997	1000	3
1.26	1004	1007	1011	1014	1017	1021	1024	1028	1031	1035	
1.27	1038	1041	1045	1048	1052	1055	1059	1062	1065	1069	
1.28	1072	1075	1079	1082	1086	1089	1092	1096	1099	1103	
1.29	1106	1109	1113	1116	1119	1123	1126	1129	1133	1136	
1.30	0.1139	1143	1146	1149	1153	1156	1159	1163	1166	1169	
1.31	1173	1176	1179	1183	1186	1189	1193	1196	1199	1202	
1.32	1206	1209	1212	1216	1219	1222	1225	1229	1232	1235	
1.33	1239	1242	1245	1248	1252	1255	1258	1261	1265	1268	
1.34	1271	1274	1278	1281	1284	1287	1290	1294	1297	1300	
1.35	1303	1307	1310	1313	1316	1319	1323	1326	1329	1332	
1.36	1335	1339	1342	1345	1348	1351	1355	1358	1361	1364	
1.37	1367	1370	1374	1377	1380	1383	1386	1389	1392	1396	
1.38	1399	1402	1405	1408	1411	1414	1418	1421	1424	1427	
1.39	1430	1433	1436	1440	1443	1446	1449	1452	1455	1458	
1.40	0.1461	1464	1467	1471	1474	1477	1480	1483	1486	1489	
1.41	1492	1495	1498	1501	1504	1508	1511	1514	1517	1520	
1.42	1523	1526	1529	1532	1535	1538	1541	1544	1547	1550	
1.43	1553	1556	1559	1562	1565	1569	1572	1575	1578	1581	
1.44	1584	1587	1590	1593	1596	1599	1602	1605	1608	1611	
1.45	1614	1617	1620	1623	1626	1629	1632	1635	1638	1641	
1.46	1644	1647	1649	1652	1655	1658	1661	1664	1667	1670	
1.47	1673	1676	1679	1682	1685	1688	1691	1694	1697	1700	
1.48	1703	1706	1708	1711	1714	1717	1720	1723	1726	1729	
1.49	1732	1735	1738	1741	1744	1746	1749	1752	1755	1758	

Moving the decimal point n places to the right [or left] in the number requires adding +n [or −n] in the body of the table (see p. 9).

* From Marks, "Mechanical Engineers' Handbook," 4th ed., McGraw-Hill, New York, 1941.

TABLE 4.—COMMON LOGARITHMS.*—(*Continued*)
(*special table*)

Number	0	1	2	3	4	5	6	7	8	9	Avg. diff.
1.50	0.1761	1764	1767	1770	1772	1775	1778	1781	1784	1787	3
1.51	1790	1793	1796	1798	1801	1804	1807	1810	1813	1816	
1.52	1818	1821	1824	1827	1830	1833	1836	1838	1841	1844	
1.53	1847	1850	1853	1855	1858	1861	1864	1867	1870	1872	
1.54	1875	1878	1881	1884	1886	1889	1892	1895	1898	1901	
1.55	1903	1906	1909	1912	1915	1917	1920	1923	1926	1928	
1.56	1931	1934	1937	1940	1942	1945	1948	1951	1953	1956	
1.57	1959	1962	1965	1967	1970	1973	1976	1978	1981	1984	
1.58	1987	1989	1992	1995	1998	2000	2003	2006	2009	2011	
1.59	2014	2017	2019	2022	2025	2028	2030	2033	2036	2038	
1.60	0.2041	2044	2047	2049	2052	2055	2057	2060	2063	2066	
1.61	2068	2071	2074	2076	2079	2082	2084	2087	2090	2092	
1.62	2095	2098	2101	2103	2106	2109	2111	2114	2117	2119	
1.63	2122	2125	2127	2130	2133	2135	2138	2140	2143	2146	
1.64	2148	2151	2154	2156	2159	2162	2164	2167	2170	2172	
1.65	2175	2177	2180	2183	2185	2188	2191	2193	2196	2198	
1.66	2201	2204	2206	2209	2212	2214	2217	2219	2222	2225	
1.67	2227	2230	2232	2235	2238	2240	2243	2245	2248	2251	
1.68	2253	2256	2258	2261	2263	2266	2269	2271	2274	2276	
1.69	2279	2281	2284	2287	2289	2292	2294	2297	2299	2302	
1.70	0.2304	2307	2310	2312	2315	2317	2320	2322	2325	2327	2
1.71	2330	2333	2335	2338	2340	2343	2345	2348	2350	2353	
1.72	2355	2358	2360	2363	2365	2368	2370	2373	2375	2378	
1.73	2380	2383	2385	2388	2390	2393	2395	2398	2400	2403	
1.74	2405	2408	2410	2413	2415	2418	2420	2423	2425	2428	
1.75	2430	2433	2435	2438	2440	2443	2445	2448	2450	2453	
1.76	2455	2458	2460	2463	2465	2467	2470	2472	2475	2477	
1.77	2480	2482	2485	2487	2490	2492	2494	2497	2499	2502	
1.78	2504	2507	2509	2512	2514	2516	2519	2521	2524	2526	
1.79	2529	2531	2533	2536	2538	2541	2543	2545	2548	2550	
1.80	0.2553	2555	2558	2560	2562	2565	2567	2570	2572	2574	
1.81	2577	2579	2582	2584	2586	2589	2591	2594	2596	2598	
1.82	2601	2603	2605	2608	2610	2613	2615	2617	2620	2622	
1.83	2625	2627	2629	2632	2634	2636	2639	2641	2643	2646	
1.84	2648	2651	2653	2655	2658	2660	2662	2665	2667	2669	
1.85	2672	2674	2676	2679	2681	2683	2686	2688	2690	2693	
1.86	2695	2697	2700	2702	2704	2707	2709	2711	2714	2716	
1.87	2718	2721	2723	2725	2728	2730	2732	2735	2737	2739	
1.88	2742	2744	2746	2749	2751	2753	2755	2758	2760	2762	
1.89	2765	2767	2769	2772	2774	2776	2778	2781	2783	2785	
1.90	0.2788	2790	2792	2794	2797	2799	2801	2804	2806	2808	
1.91	2810	2813	2815	2817	2819	2822	2824	2826	2828	2831	
1.92	2833	2835	2838	2840	2842	2844	2847	2849	2851	2853	
1.93	2856	2858	2860	2862	2865	2867	2869	2871	2874	2876	
1.94	2878	2880	2882	2885	2887	2889	2891	2894	2896	2898	
1.95	2900	2903	2905	2907	2909	2911	2914	2916	2918	2920	
1.96	2923	2925	2927	2929	2931	2934	2936	2938	2940	2942	
1.97	2945	2947	2949	2951	2953	2956	2958	2960	2962	2964	
1.98	2967	2969	2971	2973	2975	2978	2980	2982	2984	2986	
1.99	2989	2991	2993	2995	2997	2999	3002	3004	3006	3008	

* From Marks, "Mechanical Engineers' Handbook," 4th ed., McGraw-Hill, New York, 1941.

TABLE 4.—COMMON LOGARITHMS.*—(*Continued*)
(*special table*)

Number	0	1	2	3	4	5	6	7	8	9	Avg. diff.
1.0	0.0000	0043	0086	0128	0170	0212	0253	0294	0334	0374	
1.1	0414	0453	0492	0531	0569	0607	0645	0682	0719	0755	
1.2	0792	0828	0864	0899	0934	0969	1004	1038	1072	1106	See pages 40–41
1.3	1139	1173	1206	1239	1271	1303	1335	1367	1399	1430	
1.4	1461	1492	1523	1553	1584	1614	1644	1673	1703	1732	
1.5	1761	1790	1818	1847	1875	1903	1931	1959	1987	2014	
1.6	2041	2068	2095	2122	2148	2175	2201	2227	2253	2279	
1.7	2304	2330	2355	2380	2405	2430	2455	2480	2504	2529	
1.8	2553	2577	2601	2625	2648	2672	2695	2718	2742	2765	
1.9	2788	2810	2833	2856	2878	2900	2923	2945	2967	2989	
2.0	0.3010	3032	3054	3075	3096	3118	3139	3160	3181	3201	21
2.1	3222	3243	3263	3284	3304	3324	3345	3365	3385	3404	20
2.2	3424	3444	3464	3483	3502	3522	3541	3560	3579	3598	19
2.3	3617	3636	3655	3674	3692	3711	3729	3747	3766	3784	18
2.4	3802	3820	3838	3856	3874	3892	3909	3927	3945	3962	17
2.5	3979	3997	4014	4031	4048	4065	4082	4099	4116	4133	17
2.6	4150	4166	4183	4200	4216	4232	4249	4265	4281	4298	16
2.7	4314	4330	4346	4362	4378	4393	4409	4425	4440	4456	16
2.8	4472	4487	4502	4518	4533	4548	4564	4579	4594	4609	15
2.9	4624	4639	4654	4669	4683	4698	4713	4728	4742	4757	15
3.0	0.4771	4786	4800	4814	4829	4843	4857	4871	4886	4900	14
3.1	4914	4928	4942	4955	4969	4983	4997	5011	5024	5038	14
3.2	5051	5065	5079	5092	5105	5119	5132	5145	5159	5172	13
3.3	5185	5198	5211	5224	5237	5250	5263	5276	5289	5302	13
3.4	5315	5328	5340	5353	5366	5378	5391	5403	5416	5428	13
3.5	5441	5453	5465	5478	5490	5502	5514	5527	5539	5551	12
3.6	5563	5575	5587	5599	5611	5623	5635	5647	5658	5670	12
3.7	5682	5694	5705	5717	5729	5740	5752	5763	5775	5786	12
3.8	5798	5809	5821	5832	5843	5855	5866	5877	5888	5899	11
3.9	5911	5922	5933	5944	5955	5966	5977	5988	5999	6010	11
4.0	0.6021	6031	6042	6053	6064	6075	6085	6096	6107	6117	11
4.1	6128	6138	6149	6160	6170	6180	6191	6201	6212	6222	10
4.2	6232	6243	6253	6263	6274	6284	6294	6304	6314	6325	10
4.3	6335	6345	6355	6365	6375	6385	6395	6405	6415	6425	10
4.4	6435	6444	6454	6464	6474	6484	6493	6503	6513	6522	10
4.5	6532	6542	6551	6561	6571	6580	6590	6599	6609	6618	10
4.6	6628	6637	6646	6656	6665	6675	6684	6693	6702	6712	10
4.7	6721	6730	6739	6749	6758	6767	6776	6785	6794	6803	9
4.8	6812	6821	6830	6839	6848	6857	6866	6875	6884	6893	9
4.9	6902	6911	6920	6928	6937	6946	6955	6964	6972	6981	9

$\log \pi = 0.4971$ $\log \pi/2 = 0.1961$ $\log \pi^2 = 0.9943$ $\log \sqrt{\pi} = 0.2486$
$\log e = 0.4343$ $\log (0.4343) = 0.6378 - 1$

Pages 9 and 10 give the common logarithms of numbers between 1 and 10, correct to four places. Moving the decimal point n places to the right [or left] in the number is equivalent to adding n [or $-n$] to the logarithm. Thus, $\log 0.017453 = 0.2419 - 2$, which may also be written $\bar{2}.2419$ or $8.2419 - 10$. See Sec. II for complete discussion of logarithms.

$$\log (ab) = \log a + \log b \qquad \log (a^N) = N \log a$$
$$\log \left(\frac{a}{b}\right) = \log a - \log b \qquad \log (\sqrt[N]{a}) = \frac{1}{N} \log a$$

*From Marks, "Mechanical Engineers' Handbook," 4th ed., McGraw-Hill, New York, 1941.

(*Continued on p. 10*)

TABLE 4.—COMMON LOGARITHMS.*—(*Continued*)
(*special table*)

Number	0	1	2	3	4	5	6	7	8	9	Avg. diff.
5.0	0.6990	6998	7007	7016	7024	7033	7042	7050	7059	7067	9
5.1	7076	7084	7093	7101	7110	7118	7126	7135	7143	7152	8
5.2	7160	7168	7177	7185	7193	7202	7210	7218	7226	7235	8
5.3	7243	7251	7259	7267	7275	7284	7292	7300	7308	7316	8
5.4	7324	7332	7340	7348	7356	7364	7372	7380	7388	7396	8
5.5	7404	7412	7419	7427	7435	7443	7451	7459	7466	7474	8
5.6	7482	7490	7497	7505	7513	7520	7528	7536	7543	7551	8
5.7	7559	7566	7574	7582	7589	7597	7604	7612	7619	7627	8
5.8	7634	7642	7649	7657	7664	7672	7679	7686	7694	7701	7
5.9	7709	7716	7723	7731	7738	7745	7752	7760	7767	7774	7
6.0	0.7782	7789	7796	7803	7810	7818	7825	7832	7839	7846	7
6.1	7853	7860	7868	7875	7882	7889	7896	7903	7910	7917	7
6.2	7924	7931	7938	7945	7952	7959	7966	7973	7980	7987	7
6.3	7993	8000	8007	8014	8021	8028	8035	8041	8048	8055	7
6.4	8062	8069	8075	8082	8089	8096	8102	8109	8116	8122	7
6.5	8129	8136	8142	8149	8156	8162	8169	8176	8182	8189	7
6.6	8195	8202	8209	8215	8222	8228	8235	8241	8248	8254	7
6.7	8261	8267	8274	8280	8287	8293	8299	8306	8312	8319	6
6.8	8325	8331	8338	8344	8351	8357	8363	8370	8376	8382	6
6.9	8388	8395	8401	8407	8414	8420	8426	8432	8439	8445	6
7.0	0.8451	8457	8463	8470	8476	8482	8488	8494	8500	8506	6
7.1	8513	8519	8525	8531	8537	8543	8549	8555	8561	8567	6
7.2	8573	8579	8585	8591	8597	8603	8609	8615	8621	8627	6
7.3	8633	8639	8645	8651	8657	8663	8669	8675	8681	8686	6
7.4	8692	8698	8704	8710	8716	8722	8727	8733	8739	8745	6
7.5	8751	8756	8762	8768	8774	8779	8785	8791	8797	8802	6
7.6	8808	8814	8820	8825	8831	8837	8842	8848	8854	8859	6
7.7	8865	8871	8876	8882	8887	8893	8899	8904	8910	8915	6
7.8	8921	8927	8932	8938	8943	8949	8954	8960	8965	8971	6
7.9	8976	8982	8987	8993	8998	9004	9009	9015	9020	9025	5
8.0	0.9031	9036	9042	9047	9053	9058	9063	9069	9074	9079	5
8.1	9085	9090	9096	9101	9106	9112	9117	9122	9128	9133	5
8.2	9138	9143	9149	9154	9159	9165	9170	9175	9180	9186	5
8.3	9191	9196	9201	9206	9212	9217	9222	9227	9232	9238	5
8.4	9243	9248	9253	9258	9263	9269	9274	9279	9284	9289	5
8.5	9294	9299	9304	9309	9315	9320	9325	9330	9335	9340	5
8.6	9345	9350	9355	9360	9365	9370	9375	9380	9385	9390	5
8.7	9395	9400	9405	9410	9415	9420	9425	9430	9435	9440	5
8.8	9445	9450	9455	9460	9465	9469	9474	9479	9484	9489	5
8.9	9494	9499	9504	9509	9513	9518	9523	9528	9533	9538	5
9.0	0.9542	9547	9552	9557	9562	9566	9571	9576	9581	9586	5
9.1	9590	9595	9600	9605	9609	9614	9619	9624	9628	9633	5
9.2	9638	9643	9647	9652	9657	9661	9666	9671	9675	9680	5
9.3	9685	9689	9694	9699	9703	9708	9713	9717	9722	9727	5
9.4	9731	9736	9741	9745	9750	9754	9759	9763	9768	9773	5
9.5	9777	9782	9786	9791	9795	9800	9805	9809	9814	9818	5
9.6	9823	9827	9832	9836	9841	9845	9850	9854	9859	9863	4
9.7	9868	9872	9877	9881	9886	9890	9894	9899	9903	9908	4
9.8	9912	9917	9921	9926	9930	9934	9939	9943	9948	9952	4
9.9	9956	9961	9965	9969	9974	9978	9983	9987	9991	9996	4

* From Marks, "Mechanical Engineers' Handbook," 4th ed., McGraw-Hill, New York, 1941

TABLE 5.—DEGREES AND MINUTES EXPRESSED IN RADIANS*
(See also pp. 12 and 22)

Degrees						Hundredths				Minutes	
1°	.0175	61°	1.0647	121°	2.1118	0°.01	.0002	0°.51	.0039	1′	.0003
2	.0349	2	1.0821	2	2.1293	2	.0003	2	.0091	2′	.0006
3	.0524	3	1.0996	3	2.1468	3	.0005	3	.0093	3′	.0009
4	.0698	4	1.1170	4	2.1642	4	.0007	4	.0094	4′	.0012
5°	.0873	65°	1.1345	125°	2.1817	.05	.0009	.55	.0096	5′	.0015
6	.1047	6	1.1519	6	2.1991	6	.0010	6	.0098	6′	.0017
7	.1222	7	1.1694	7	2.2166	7	.0012	7	.0099	7′	.0020
8	.1396	8	1.1868	8	2.2340	8	.0014	8	.0101	8′	.0023
9	.1571	9	1.2043	9	2.2515	9	.0016	9	.0103	9′	.0026
10°	.1745	70°	1.2217	130°	2.2689	0°.10	.0017	0°.60	.0105	10′	.0029
1	.1920	1	1.2392	1	2.2864	1	.0019	1	.0106	11′	.0032
2	.2094	2	1.2566	2	2.3038	2	.0021	2	.0108	12′	.0035
3	.2269	3	1.2741	3	2.3213	3	.0023	3	.0110	13′	.0038
4	.2443	4	1.2915	4	2.3387	4	.0024	4	.0112	14′	.0041
15°	.2618	75°	1.3090	135°	2.3562	.15	.0026	.65	.0113	15′	.0044
6	.2793	6	1.3265	6	2.3736	6	.0028	6	.0115	16′	.0047
7	.2967	7	1.3439	7	2.3911	7	.0030	7	.0117	17′	.0049
8	.3142	8	1.3614	8	2.4086	8	.0031	8	.0119	18′	.0052
9	.3316	9	1.3788	9	2.4260	9	.0033	9	.0120	19′	.0055
20°	.3491	80°	1.3963	140°	2.4435	0°.20	.0035	0°.70	.0122	20′	.0058
1	.3665	1	1.4137	1	2.4609	1	.0037	1	.0124	21′	.0061
2	.3840	2	1.4312	2	2.4784	2	.0038	2	.0126	22′	.0064
3	.4014	3	1.4486	3	2.4958	3	.0040	3	.0127	23′	.0067
4	.4189	4	1.4661	4	2.5133	4	.0042	4	.0129	24′	.0070
25°	.4363	85°	1.4835	145°	2.5307	.25	.0044	.75	.0131	25′	.0073
6	.4538	6	1.5010	6	2.5482	6	.0045	6	.0133	26′	.0076
7	.4712	7	1.5184	7	2.5656	7	.0047	7	.0134	27′	.0079
8	.4887	8	1.5359	8	2.5831	8	.0049	8	.0136	28′	.0081
9	.5061	9	1.5533	9	2.6005	9	.0051	9	.0138	29′	.0084
30°	.5236	90°	1.5708	150°	2.6180	0°.30	.0052	0°.80	.0140	30′	.0087
1	.5411	1	1.5882	1	2.6354	1	.0054	1	.0141	31′	.0090
2	.5585	2	1.6057	2	2.6529	2	.0056	2	.0143	32′	.0093
3	.5760	3	1.6232	3	2.6704	3	.0058	3	.0145	33′	.0096
4	.5934	4	1.6406	4	2.6878	4	.0059	4	.0147	34′	.0099
35°	.6109	95°	1.6581	155°	2.7053	35	.0061	.85	.0148	35′	.0102
6	.6283	6	1.6755	6	2.7227	6	.0063	6	.0150	36′	.0105
7	.6458	7	1.6930	7	2.7402	7	.0065	7	.0152	37′	.0108
8	.6632	8	1.7104	8	2.7576	8	.0066	8	.0154	38′	.0111
9	.6807	9	1.7279	9	2.7751	9	.0068	9	.0155	39′	.0113
40°	.6981	100°	1.7453	160°	2.7925	0°.40	.0070	0°.90	.0157	40′	.0116
1	.7156	1	1.7628	1	2.8100	1	.0072	1	.0159	41′	.0119
2	.7330	2	1.7802	2	2.8274	2	.0073	2	.0161	42′	.0122
3	.7505	3	1.7977	3	2.8449	3	.0075	3	.0162	43′	.0125
4	.7679	4	1.8151	4	2.8623	4	.0077	4	.0164	44′	.0128
45°	.7854	105°	1.8326	165°	2.8798	.45	.0079	.95	.0166	45′	.0131
6	.8029	6	1.8500	6	2.8972	6	.0080	6	.0168	46′	.0134
7	.8203	7	1.8675	7	2.9147	7	.0082	7	.0169	47′	.0137
8	.8378	8	1.8850	8	2.9322	8	.0084	8	.0171	48′	.0140
9	.8552	9	1.9024	9	2.9496	9	.0086	9	.0173	49′	.0143
50°	.8727	110°	1.9199	170°	2.9671	0°.50	.0087	1°.00	.0175	50′	.0145
1	.8901	1	1.9373	1	2.9845					51′	.0148
2	.9076	2	1.9548	2	3.0020					52′	.0151
3	.9250	3	1.9722	3	3.0194					53′	.0154
4	.9425	4	1.9897	4	3.0369					54′	.0157
55°	.9599	115°	2.0071	175°	3.0543					55′	.0160
6	.9774	6	2.0246	6	3.0718					56′	.0163
7	.9948	7	2.0420	7	3.0892					57′	.0166
8	1.0123	8	2.0595	8	3.1067					58′	.0169
9	1.0297	9	2.0769	9	3.1241					59′	.0172
60°	1.0472	120°	2.0944	180°	3.1416					60′	.0175

Arc 1° = 0.0174533   Arc 1′ = 0.000290888   Arc 1″ = 0.00000484814
1 radian = 57°.295780 = 57° 17′.7468 = 57° 17′ 44″ .806

* From Marks, "Mechanical Engineers' Handbook," 4th ed., McGraw-Hill, New York, 1941.

Table 6.—Radians Expressed in Degrees*

rad	deg	rad	deg	rad	deg	rad	deg	rad	deg
0.01	0°.57	.64	36°.67	1.27	72°.77	1.90	108°.86	2.53	144°.96
2	1°.15	.65	37°.24	8	73°.34	1	109°.43	4	145°.53
3	1°.72	6	37°.82	9	73°.91	2	110°.01	2.55	146°.10
4	2°.29	7	38°.39	1.30	74°.48	3	110°.58	6	146°.68
.05	2°.86	8	38°.96	1	75°.06	4	111°.15	7	147°.25
6	3°.44	9	39°.53	2	75°.63	1.95	111°.73	8	147°.82
7	4°.01	.70	40°.11	3	76°.20	6	112°.30	9	148°.40
8	4°.58	1	40°.68	4	76°.78	7	112°.87	2.60	148°.97
9	5°.16	2	41°.25	1.35	77°.35	8	113°.45	1	149°.54
.10	5°.73	3	41°.83	6	77°.92	9	114°.02	2	150°.11
1	6°.30	4	42°.40	7	78°.50	2.00	114°.59	3	150°.69
2	6°.88	.75	42°.97	8	79°.07	1	115°.16	4	151°.26
3	7°.45	6	43°.54	9	79°.64	2	115°.74	2.65	151°.83
4	8°.02	7	44°.12	1.40	80°.21	3	116°.31	6	152°.41
.15	8°.59	8	44°.69	1	80°.79	4	116°.88	7	152°.98
6	9°.17	9	45°.26	2	81°.36	2.05	117°.46	8	153°.55
7	9°.74	.80	45°.84	3	81°.93	6	118°.03	9	154°.13
8	10°.31	1	46°.41	4	82°.51	7	118°.60	2.70	154°.70
9	10°.89	2	46°.98	1.45	83°.08	8	119°.18	1	155°.27
.20	11°.46	3	47°.56	6	83°.65	9	119°.75	2	155°.84
1	12°.03	4	48°.13	7	84°.22	2.10	120°.32	3	156°.42
2	12°.61	.85	48°.70	8	84°.80	1	120°.89	4	156°.99
3	13°.18	6	49°.27	9	85°.37	2	121°.47	2.75	157°.56
4	13°.75	7	49°.85	1.50	85°.94	3	122°.04	6	158°.14
.25	14°.32	8	50°.42	1	86°.52	4	122°.61	7	158°.71
6	14°.90	9	50°.99	2	87°.09	2.15	123°.19	8	159°.28
7	15°.47	.90	51°.57	3	87°.66	6	123°.76	9	159°.86
8	16°.04	1	52°.14	4	88°.24	7	124°.33	2.80	160°.43
9	16°.62	2	52°.71	1.55	88°.81	8	124°.90	1	161°.00
.30	17°.19	3	53°.29	6	89°.38	9	125°.48	2	161°.57
1	17°.76	4	53°.86	7	89°.95	2.20	126°.05	3	162°.15
2	18°.33	.95	54°.43	8	90°.53	1	126°.62	4	162°.72
3	18°.91	6	55°.00	9	91°.10	2	127°.20	2.85	163°.29
4	19°.48	7	55°.58	1.60	91°.67	3	127°.77	6	163°.87
.35	20°.05	8	56°.15	1	92°.25	4	128°.34	7	164°.44
6	20°.63	9	56°.72	2	92°.82	2.25	128°.92	8	165°.01
7	21°.20	1.00	57°.30	3	93°.39	6	129°.49	9	165°.58
8	21°.77	1	57°.87	4	93°.97	7	130°.06	2.90	166°.16
9	22°.35	2	58°.44	1.65	94°.54	8	130°.63	1	166°.73
.40	22°.92	3	59°.01	6	95°.11	9	131°.21	2	167°.30
1	23°.49	4	59°.59	7	95°.68	2.30	131°.78	3	167°.88
2	24°.06	1.05	60°.16	8	96°.26	1	132°.35	4	168°.45
3	24°.64	6	60°.73	9	96°.83	2	132°.93	2.95	169°.02
4	25°.21	7	61°.31	1.70	97°.40	3	133°.50	6	169°.60
.45	25°.78	8	61°.88	1	97°.98	4	134°.07	7	170°.17
6	26°.36	9	62°.45	2	98°.55	2.35	134°.65	8	170°.74
7	26°.93	1.10	63°.03	3	99°.12	6	135°.22	9	171°.31
8	27°.50	1	63°.60	4	99°.69	7	135°.79	3.00	171°.89
9	28°.07	2	64°.17	1.75	100°.27	8	136°.36	1	172°.46
.50	28°.65	3	64°.74	6	100°.84	9	136°.94	2	173°.03
1	29°.22	4	65°.32	7	101°.41	2.40	137°.51	3	173°.61
2	29°.79	1.15	65°.89	8	101°.99	1	138°.08	4	174°.18
3	30°.37	6	66°.46	9	102°.56	2	138°.66	3.05	174°.75
4	30°.94	7	67°.04	1.80	103°.13	3	139°.23	6	175°.33
.55	31°.51	8	67°.61	1	103°.71	4	139°.80	7	175°.90
6	32°.09	9	68°.18	2	104°.28	2.45	140°.37	8	176°.47
7	32°.66	1.20	68°.75	3	104°.85	6	140°.95	9	177°.04
8	33°.23	1	69°.33	4	105°.42	7	141°.52	3.10	177°.62
9	33°.80	2	69°.90	1.85	106°.00	8	142°.09	1	178°.19
.60	34°.38	3	70°.47	6	106°.57	9	142°.67	2	178°.76
1	34°.95	4	71°.05	7	107°.14	2.50	143°.24	3	179°.34
2	35°.52	1.25	71°.62	8	107°.72	1	143°.81	4	179°.91
3	36°.10	6	72°.19	9	108°.29	2	144°.39	3.15	180°.48

Interpolation

.0002			0°.01
04			.02
06			.03
08			.05
.0010			0°.06
12			.07
14			.08
16			.09
18			.10
.0020			0°.11
22			.13
24			.14
26			.15
28			.16
.0030			0°.17
32			.18
34			.19
36			.21
38			.22
.0040			0°.23
42			.24
44			.25
46			.26
48			.28
.0050			0°.29
52			.30
54			.31
56			.32
58			.33
.0060			0°.34
62			.36
64			.37
66			.38
68			.39
.0070			0°.40
72			.41
74			.42
76			.44
78			.45
.0080			0°.46
82			.47
84			.48
86			.49
88			.50
.0090			0°.52
92			.53
94			.54
96			.55
98			.56

Multiples of π

1	3.1416	180°
2	6.2832	360°
3	9.4248	540°
4	12.5664	720°
5	15.7080	900°
6	18.8496	1080°
7	21.9911	1260°
8	25.1327	1440°
9	28.2743	1620°
10	31.4159	1800°

* From Marks, "Mechanical Engineers' Handbook," 4th ed., McGraw-Hill, New York, 1941.

TABLE 7.—NATURAL SINES AND COSINES*
Natural sines at intervals of 0°.1, or 6′

Deg.	°.0 =(0′)	°.1 (6′)	°.2 (12′)	°.3 (18′)	°.4 (24′)	°.5 (30′)	°.6 (36′)	°.7 (42′)	°.8 (48′)	°.9 (54′)			Avg. diff.
											0.0000	90°	
0°	0.0000	0017	0035	0052	0070	0087	0105	0122	0140	0157	0175	89	17
1	0175	0192	0209	0227	0244	0262	0279	0297	0314	0332	0349	88	17
2	0349	0366	0384	0401	0419	0436	0454	0471	0488	0506	0523	87	17
3	0523	0541	0558	0576	0593	0610	0628	0645	0663	0680	0698	86	17
4	0698	0715	0732	0750	0767	0785	0802	0819	0837	0854	0.0872	85	17
5	0.0872	0889	0906	0924	0941	0958	0976	0993	1011	1028	1045	84	17
6	1045	1063	1080	1097	1115	1132	1149	1167	1184	1201	1219	83	17
7	1219	1236	1253	1271	1288	1305	1323	1340	1357	1374	1392	82	17
8	1392	1409	1426	1444	1461	1478	1495	1513	1530	1547	1564	81	17
9	1564	1582	1599	1616	1633	1650	1668	1685	1702	1719	0.1736	80°	17
10°	0.1736	1754	1771	1788	1805	1822	1840	1857	1874	1891	1908	79	17
11	1908	1925	1942	1959	1977	1994	2011	2028	2045	2062	2079	78	17
12	2079	2096	2113	2130	2147	2164	2181	2198	2215	2233	2250	77	17
13	2250	2267	2284	2300	2317	2334	2351	2368	2385	2402	2419	76	17
14	2419	2436	2453	2470	2487	2504	2521	2538	2554	2571	0.2588	75	17
15	0.2588	2605	2622	2639	2656	2672	2689	2706	2723	2740	2756	74	17
16	2756	2773	2790	2807	2823	2840	2857	2874	2890	2907	2924	73	17
17	2924	2940	2957	2974	2990	3007	3024	3040	3057	3074	3090	72	17
18	3090	3107	3123	3140	3156	3173	3190	3206	3223	3239	3256	71	17
19	3256	3272	3289	3305	3322	3338	3355	3371	3387	3404	0.3420	70°	16
20°	0.3420	3437	3453	3469	3486	3502	3518	3535	3551	3567	3584	69	16
21	3584	3600	3616	3633	3649	3665	3681	3697	3714	3730	3746	68	16
22	3746	3762	3778	3795	3811	3827	3843	3859	3875	3891	3907	67	16
23	3907	3923	3939	3955	3971	3987	4003	4019	4035	4051	4067	66	16
24	4067	4083	4099	4115	4131	4147	4163	4179	4195	4210	0.4226	65	16
25	0.4226	4242	4258	4274	4289	4305	4321	4337	4352	4368	4384	64	16
26	4384	4399	4415	4431	4446	4462	4478	4493	4509	4524	4540	63	16
27	4540	4555	4571	4586	4602	4617	4633	4648	4664	4679	4695	62	16
28	4695	4710	4726	4741	4756	4772	4787	4802	4818	4833	4848	61	15
29	4848	4863	4879	4894	4909	4924	4939	4955	4970	4985	0.5000	60°	15
30°	0.5000	5015	5030	5045	5060	5075	5090	5105	5120	5135	5150	59	15
31	5150	5165	5180	5195	5210	5225	5240	5255	5270	5284	5299	58	15
32	5299	5314	5329	5344	5358	5373	5388	5402	5417	5432	5446	57	15
33	5446	5461	5476	5490	5505	5519	5534	5548	5563	5577	5592	56	15
34	5592	5606	5621	5635	5650	5664	5678	5693	5707	5721	0.5736	55	14
35	0.5736	5750	5764	5779	5793	5807	5821	5835	5850	5864	5878	54	14
36	5878	5892	5906	5920	5934	5948	5962	5976	5990	6004	6018	53	14
37	6018	6032	6046	6060	6074	6088	6101	6115	6129	6143	6157	52	14
38	6157	6170	6184	6198	6211	6225	6239	6252	6266	6280	6293	51	14
39	6293	6307	6320	6334	6347	6361	6374	6388	6401	6414	0.6428	50°	13
40°	0.6428	6441	6455	6468	6481	6494	6508	6521	6534	6547	6561	49	13
41	6561	6574	6587	6600	6613	6626	6639	6652	6665	6678	6691	48	13
42	6691	6704	6717	6730	6743	6756	6769	6782	6794	6807	6820	47	13
43	6820	6833	6845	6858	6871	6884	6896	6909	6921	6934	6947	46	13
44	6947	6959	6972	6984	6997	7009	7022	7034	7046	7059	0.7071	45°	12
45°	0.7071												

	°.9 =(54′)	°.8 (48′)	°.7 (42′)	°.6 (36′)	°.5 (30′)	°.4 (24′)	°.3 (18′)	°.2 (12′)	°.1 (6′)	°.0 (0′)		Deg.	

* From Marks, "Mechanical Engineers' Handbook," 4th ed., McGraw-Hill, New York, 1941.

(*Continued on p.* 14)

TABLE 7.—NATURAL SINES AND COSINES.*—(Continued)

Deg.	°.0 =(0′)	°.1 (6′)	°.2 (12′)	°.3 (18′)	°.4 (24′)	°.5 (30′)	°.6 (36′)	°.7 (42′)	°.8 (48′)	°.9 (54′)			Avg. diff.
											0.7071	45°	
45°	0.7071	7083	7096	7108	7120	7133	7145	7157	7169	7181	7193	44	12
46	7193	7206	7218	7230	7242	7254	7266	7278	7290	7302	7314	43	12
47	7314	7325	7337	7349	7361	7373	7385	7396	7408	7420	7431	42	12
48	7431	7443	7455	7466	7478	7490	7501	7513	7524	7536	7547	41	12
49	7547	7559	7570	7581	7593	7604	7615	7627	7638	7649	0.7660	40°	11
50°	0.7660	7672	7683	7694	7705	7716	7727	7738	7749	7760	7771	39	11
51	7771	7782	7793	7804	7815	7826	7837	7848	7859	7869	7880	38	11
52	7880	7891	7902	7912	7923	7934	7944	7955	7965	7976	7986	37	11
53	7986	7997	8007	8018	8028	8039	8049	8059	8070	8080	8090	36	10
54	8090	8100	8111	8121	8131	8141	8151	8161	8171	8181	0.8192	35	10
55	0.8192	8202	8211	8221	8231	8241	8251	8261	8271	8281	8290	34	10
56	8290	8300	8310	8320	8329	8339	8348	8358	8368	8377	8387	33	10
57	8387	8396	8406	8415	8425	8434	8443	8453	8462	8471	8480	32	9
58	8480	8490	8499	8508	8517	8526	8536	8545	8554	8563	8572	31	9
59	8572	8581	8590	8599	8607	8616	8625	8634	8643	8652	0.8660	30°	9
60°	0.8660	8669	8678	8686	8695	8704	8712	8721	8729	8738	8746	29	9
61	8746	8755	8763	8771	8780	8788	8796	8805	8813	8821	8829	28	8
62	8829	8838	8846	8854	8862	8870	8878	8886	8894	8902	8910	27	8
63	8910	8918	8926	8934	8942	8949	8957	8965	8973	8980	8988	26	8
64	8988	8996	9003	9011	9018	9026	9033	9041	9048	9056	0.9063	25	7
65	0.9063	9070	9078	9085	9092	9100	9107	9114	9121	9128	9135	24	7
66	9135	9143	9150	9157	9164	9171	9178	9184	9191	9198	9205	23	7
67	9205	9212	9219	9225	9232	9239	9245	9252	9259	9265	9272	22	7
68	9272	9278	9285	9291	9298	9304	9311	9317	9323	9330	9336	21	6
69	9336	9342	9348	9354	9361	9367	9373	9379	9385	9391	0.9397	20°	6
70°	0.9397	9403	9409	9415	9421	9426	9432	9438	9444	9449	9455	19	6
71	9455	9461	9466	9472	9478	9483	9489	9494	9500	9505	9511	18	6
72	9511	9516	9521	9527	9532	9537	9542	9548	9553	9558	9563	17	5
73	9563	9568	9573	9578	9583	9588	9593	9598	9603	9608	9613	16	5
74	9613	9617	9622	9627	9632	9636	9641	9646	9650	9655	0.9659	15	5
75	0.9659	9664	9668	9673	9677	9681	9686	9690	9694	9699	9703	14	4
76	9703	9707	9711	9715	9720	9724	9728	9732	9736	9740	9744	13	4
77	9744	9748	9751	9755	9759	9763	9767	9770	9774	9778	9781	12	4
78	9781	9785	9789	9792	9796	9799	9803	9806	9810	9813	9816	11	3
79	9816	9820	9823	9826	9829	9833	9836	9839	9842	9845	0.9848	10°	3
80°	0.9848	9851	9854	9857	9860	9863	9866	9869	9871	9874	9877	9	3
81	9877	9880	9882	9885	9888	9890	9893	9895	9898	9900	9903	8	3
82	9903	9905	9907	9910	9912	9914	9917	9919	9921	9923	9925	7	2
83	9925	9928	9930	9932	9934	9936	9938	9940	9942	9943	9945	6	2
84	9945	9947	9949	9951	9952	9954	9956	9957	9959	9960	0.9962	5	2
85	0.9962	9963	9965	9966	9968	9969	9971	9972	9973	9974	9976	4	1
86	9976	9977	9978	9979	9980	9981	9982	9983	9984	9985	9986	3	1
87	9986	9987	9988	9989	9990	9990	9991	9992	9993	9993	9994	2	1
88	9994	9995	9995	9996	9996	9997	9997	9997	9998	9998	0.9998	1	0
89	0.9998	9999	9999	9999	9999	0000	0000	0000	0000	0000	1.0000	0°	0
90°	1.0000												

	°.9 =(54′)	°.8 (48′)	°.7 (42′)	°.6 (36′)	°.5 (30′)	°.4 (24′)	°.3 (18′)	°.2 (12′)	°.1 (6′)	°.0 (0′)	Deg.

TABLE 8.—NATURAL TANGENTS AND COTANGENTS*
Natural tangents at intervals of 0°.1, or 6'

Deg.	°.0 =(0')	°.1 (6')	°.2 (12')	°.3 (18')	°.4 (24')	°.5 (30')	°.6 (36')	°.7 (42')	°.8 (48')	°.9 (54')			Diff. Av.
											0.0000	90°	
0°	0.0000	0017	0035	0052	0070	0087	0105	0122	0140	0157	0175	89	17
1	0175	0192	0209	0227	0244	0262	0279	0297	0314	0332	0349	88	17
2	0349	0367	0384	0402	0419	0437	0454	0472	0489	0507	0524	87	17
3	0524	0542	0559	0577	0594	0612	0629	0647	0664	0682	0699	86	18
4	0699	0717	0734	0752	0769	0787	0805	0822	0840	0857	0.0875	85	18
5	0.0875	0892	0910	0928	0945	0963	0981	0998	1016	1033	1051	84	18
6	1051	1069	1086	1104	1122	1139	1157	1175	1192	1210	1228	83	18
7	1228	1246	1263	1281	1299	1317	1334	1352	1370	1388	1405	82	18
8	1405	1423	1441	1459	1477	1495	1512	1530	1548	1566	1584	81	18
9	1584	1602	1620	1638	1655	1673	1691	1709	1727	1745	0.1763	80°	18
10°	0.1763	1781	1799	1817	1835	1853	1871	1890	1908	1926	1944	79	18
11	1944	1962	1980	1998	2016	2035	2053	2071	2039	2107	2126	78	18
12	2126	2144	2162	2180	2199	2217	2235	2254	2272	2290	2309	77	18
13	2309	2327	2345	2364	2382	2401	2419	2438	2456	2475	2493	76	18
14	2493	2512	2530	2549	2568	2586	2605	2623	2642	2661	0.2679	75	19
15	0.2679	2698	2717	2736	2754	2773	2792	2811	2830	2849	2867	74	19
16	2867	2886	2905	2924	2943	2962	2981	3000	3019	3038	3057	73	19
17	3057	3076	3096	3115	3134	3153	3172	3191	3211	3230	3249	72	19
18	3249	3269	3288	3307	3327	3346	3365	3385	3404	3424	3443	71	19
19	3443	3463	3482	3502	3522	3541	3561	3581	3600	3620	0.3640	70°	20
20°	0.3640	3659	3679	3699	3719	3739	3759	3779	3799	3819	3839	69	20
21	3839	3859	3879	3899	3919	3939	3959	3979	4000	4020	4040	68	20
22	4040	4061	4081	4101	4122	4142	4163	4183	4204	4224	4245	67	21
23	4245	4265	4286	4307	4327	4348	4369	4390	4411	4431	4452	66	21
24	4452	4473	4494	4515	4536	4557	4578	4599	4621	4642	0.4663	65	21
25	0.4663	4684	4706	4727	4748	4770	4791	4813	4834	4856	4877	64	21
26	4877	4899	4921	4942	4964	4986	5008	5029	5051	5073	5095	63	22
27	5095	5117	5139	5161	5184	5206	5228	5250	5272	5295	5317	62	22
28	5317	5340	5362	5384	5407	5430	5452	5475	5498	5520	5543	61	23
29	5543	5566	5589	5612	5635	5658	5681	5704	5727	5750	0.5774	60°	23
30°	0 5774	5797	5820	5844	5867	5890	5914	5938	5961	5985	6009	59	24
31	6009	6032	6056	6080	6104	6128	6152	6176	6200	6224	6249	58	24
32	6249	6273	6297	6322	6346	6371	6395	6420	6445	6469	6494	57	25
33	6494	6519	6544	6569	6594	6619	6644	6669	6694	6720	6745	56	25
34	6745	6771	6796	6822	6847	6873	6899	6924	6950	6976	0.7002	55	26
35	0.7002	7028	7054	7080	7107	7133	7159	7186	7212	7239	7265	54	26
36	7265	7292	7319	7346	7373	7400	7427	7454	7481	7508	7536	53	27
37	7536	7563	7590	7618	7646	7673	7701	7729	7757	7785	7813	52	28
38	7813	7841	7869	7898	7926	7954	7983	8012	8040	8069	8098	51	28
39	8098	8127	8156	8185	8214	8243	8273	8302	8332	8361	0.8391	50°	29
40°	0.8391	8421	8451	8481	8511	8541	8571	8601	8632	8662	8693	49	30
41	8693	8724	8754	8785	8816	8847	8878	8910	8941	8972	9004	48	31
42	9004	9036	9067	9099	9131	9163	9195	9228	9260	9293	9325	47	32
43	9325	9358	9391	9424	9457	9490	9523	9556	9590	9623	0.9657	46	33
44	0.9657	9691	9725	9759	9793	9827	9861	9896	9930	9965	1.0000	45°	34
45°	1.0000												

	°.9 =(54')	°.8 (48')	°.7 (42')	°.6 (36')	°.5 (30')	°.4 (24')	°.3 (18')	°.2 (12')	°.1 (6')	°.0 (0')	Deg.

* From Marks, "Mechanical Engineers' Handbook," 4th ed., McGraw-Hill, New York, 1941.

(Continued on p. 16)

TABLE 8.—NATURAL TANGENTS AND COTANGENTS.*—(*Continued*)

Deg.	°.0 =(0')	°.1 (6')	°.2 (12')	°.3 (18')	°.4 (24')	°.5 (30')	°.6 (36')	°.7 (42')	°.8 (48')	°.9 (54')		Avg. diff.
										1.0000	45°	
45°	1.0000	0035	0070	0105	0141	0176	0212	0247	0283	0319	0355	44	35
46	0355	0392	0428	0464	0501	0538	0575	0612	0649	0686	0724	43	37
47	0724	0761	0799	0837	0875	0913	0951	0990	1028	1067	1106	42	38
48	1106	1145	1184	1224	1263	1303	1343	1383	1423	1463	1504	41	40
49	1504	1544	1585	1626	1667	1708	1750	1792	1833	1875	1.1918	40°	41
50°	1.1918	1960	2002	2045	2088	2131	2174	2218	2261	2305	2349	39	43
51	2349	2393	2437	2482	2527	2572	2617	2662	2708	2753	2799	38	45
52	2799	2846	2892	2938	2985	3032	3079	3127	3175	3222	3270	37	47
53	3270	3319	3367	3416	3465	3514	3564	3613	3663	3713	3764	36	49
54	3764	3814	3865	3916	3968	4019	4071	4124	4176	4229	1.4281	35	52
55	1.4281	4335	4388	4442	4496	4550	4605	4659	4715	4770	4826	34	55
56	4826	4882	4938	4994	5051	5108	5166	5224	5282	5340	5399	33	57
57	5399	5458	5517	5577	5637	5697	5757	5818	5880	5941	6003	32	60
58	6003	6066	6128	6191	6255	6319	6383	6447	6512	6577	6643	31	64
59	1.6643	6709	6775	6842	6909	6977	7045	7113	7182	7251	1.7321	30°	67
60°	1.732	1.739	1.746	1.753	1.760	1.767	1.775	1.782	1.789	1.797	1.804	29	7
61	1.804	1.811	1.819	1.827	1.834	1.842	1.849	1.857	1.865	1.873	1.881	28	8
62	1.881	1.889	1.897	1.905	1.913	1.921	1.929	1.937	1.946	1.954	1.963	27	8
63	1.963	1.971	1.980	1.988	1.997	2.006	2.014	2.023	2.032	2.041	2.050	26	9
64	2.050	2.059	2.069	2.078	2.087	2.097	2.106	2.116	2.125	2.135	2.145	25	9
65	2.145	2.154	2.164	2.174	2.184	2.194	2.204	2.215	2.225	2.236	2.246	24	10
66	2.246	2.257	2.267	2.278	2.289	2.300	2.311	2.322	2.333	2.344	2.356	23	11
67	2.356	2.367	2.379	2.391	2.402	2.414	2.426	2.438	2.450	2.463	2.475	22	12
68	2.475	2.488	2.500	2.513	2.526	2.539	2.552	2.565	2.578	2.592	2.605	21	13
69	2.605	2.619	2.633	2.646	2.660	2.675	2.689	2.703	2.718	2.733	2.747	20°	14
70°	2.747	2.762	2.778	2.793	2.888	2.824	2.840	2.856	2.872	2.888	2.904	19	16
71	2.904	2.921	2.937	2.954	2.971	2.989	3.006	3.024	3.042	3.860	3.078	18	17
72	3.078	3.096	3.115	3.133	3.152	3.172	3.191	3.211	3.230	3.251	3.271	17	19
73	3.271	3.291	3.312	3.333	3.354	3.376	3.398	3.420	3.442	3.465	3.487	16	22
74	3.487	3.511	3.534	3.558	3.582	3.606	3.630	3.655	3.681	3.706	3.732	15	24
75	3.732	3.758	3.785	3.812	3.839	3.867	3.895	3.923	3.952	3.981	4.011	14	28
76	4.011	4.041	4.071	4.102	4.134	4.165	4.198	4.230	4.264	4.297	4.331	13	32
77	4.331	4.366	4.402	4.437	4.474	4.511	4.548	4.586	4.625	4.665	4.705	12	37
78	4.705	4.745	4.787	4.829	4.872	4.915	4.959	5.005	5.050	5.097	5.145	11	44
79	5.145	5.193	5.242	5.292	5.343	5.396	5.449	5.503	5.558	5.614	5.671	10°	53
80°	5.671	5.730	5.789	5.850	5.912	5.976	6.041	6.107	6.174	6.243	6.314	9	
81	6.314	6.386	6.460	6.535	6.612	6.691	6.772	6.855	6.940	7.026	7.115	8	
82	7.115	7.207	7.300	7.396	7.495	7.596	7.700	7.806	7.916	8.028	8.144	7	
83	8.144	8.264	8.386	8.513	8.643	8.777	8.915	9.058	9.205	9.357	9.514	6	
84	9.514	9.677	9.845	10.02	10.20	10.39	10.58	10.78	10.99	11.20	11.43	5	
85	11.43	11.66	11.91	12.16	12.43	12.71	13.00	13.30	13.62	13.95	14.30	4	
86	14.30	14.67	15.06	15.46	15.90	16.35	16.83	17.34	17.89	18.46	19.08	3	
87	19.08	19.74	20.45	21.20	22.02	22.90	23.86	24.90	26.03	27.27	28.64	2	
88	28.64	30.14	31.82	33.69	35.80	38.19	40.92	44.07	47.74	52.08	57.29	1	
89	57.29	63.66	71.62	81.85	95.49	114.6	143.2	191.0	286.5	573.0	∞	0°	
90°	∞												

	°.9 =(54')	°.8 (48')	°.7 (42')	°.6 (36')	°.5 (30')	°.4 (24')	°.3 (18')	°.2 (12')	°.1 (6')	°.0 (0')	Deg.

TABLE 9.—EXPONENTIALS*

$[e^n \text{ and } e^{-n}]$

n	e^n	Diff.	n	e^n	Diff.	n	e^n	n	e^{-n}	Diff.	n	e^{-n}	n	e^{-n}
0.00	1.000	10	0.50	1.649	16	1.0	2.718*	0.00	1.000	-10	0.50	.607	1.0	.368*
.01	1.010	10	.51	1.665	17	.1	3.004	.01	0.990	-10	.51	.600	.1	.333
.02	1.020	10	.52	1.682	17	.2	3.320	.02	.980	-10	.52	.595	.2	.301
.03	1.030	11	.53	1.699	17	.3	3.669	.03	.970	-9	.53	.589	.3	.273
.04	1.041	10	.54	1.716	17	.4	4.055	.04	.961	-10	.54	.583	.4	.247
0.05	1.051	11	0.55	1.733	18	1.5	4.482	0.05	.951	-9	0.55	.577	1.5	.223
.06	1.062	11	.56	1.751	17	.6	4.953	.06	.942	-10	.56	.571	.6	.202
.07	1.073	10	.57	1.768	18	.7	5.474	.07	.932	-9	.57	.566	.7	.183
.08	1.083	11	.58	1.786	18	.8	6.050	.08	.923	-9	.58	.560	.8	.165
.09	1.094	11	.59	1.804	18	.9	6.686	.09	.914	-9	.59	.554	.9	.150
0.10	1.105	11	0.60	1.822	18	2.0	7.389	0.10	.905	-9	0.60	.549	2.0	.135
.11	1.116	11	.61	1.840	19	.1	8.166	.11	.896	-9	.61	.543	.1	.122
.12	1.127	12	.62	1.859	19	.2	9.025	.12	.887	-9	.62	.538	.2	.111
.13	1.139	11	.63	1.878	18	.3	9.974	.13	.878	-9	.63	.533	.3	.100
.14	1.150	12	.64	1.896	20	.4	11.02	.14	.869	-8	.64	.527	.4	.0907
0.15	1.162	12	0.65	1.916	19	2.5	12.18	0.15	.861	-9	0.65	.522	2.5	.0821
.16	1.174	11	.66	1.935	19	.6	13.46	.16	.852	-8	.66	.517	.6	.0743
.17	1.185	12	.67	1.954	20	.7	14.88	.17	.844	-9	.67	.512	.7	.0672
.18	1.197	12	.68	1.974	20	.8	16.44	.18	.835	-8	.68	.507	.8	.0608
.19	1.209	12	.69	1.994	20	.9	18.17	.19	.827	-8	.69	.502	.9	.0550
0.20	1.221	13	0.70	2.014	20	3.0	20.09	0.20	.819	-8	0.70	.497	3.0	.0498
.21	1.234	12	.71	2.034	20	.1	22.20	.21	.811	-8	.71	.492	.1	.0450
.22	1.246	13	.72	2.054	21	.2	24.53	.22	.803	-8	.72	.487	.2	.0408
.23	1.259	12	.73	2.075	21	.3	27.11	.23	.795	-8	.73	.482	.3	.0369
.24	1.271	13	.74	2.096	21	.4	29.96	.24	.787	-8	.74	.477	.4	.0334
0.25	1.284	13	0.75	2.117	21	3.5	33.12	0.25	.779	-8	0.75	.472	3.5	.0302
.26	1.297	13	.76	2.138	22	.6	36.60	.26	.771	-8	.76	.468	.6	.0273
.27	1.310	13	.77	2.160	21	.7	40.45	.27	.763	-7	.77	.463	.7	.0247
.28	1.323	13	.78	2.181	22	.8	44.70	.28	.756	-8	.78	.458	.8	.0224
2.9	1.336	14	.79	2.203	23	.9	49.40	.29	.748	-7	.79	.454	.9	.0202
0.30	1.350	13	0.80	2.226	22	4.0	54.60	0.30	.741	-8	0.80	.449	4.0	.0183
.31	1.363	14	.81	2.248	22	.1	60.34	.31	.733	-7	.81	.445	.1	.0166
.32	1.377	14	.82	2.270	23	.2	66.69	.32	.726	-7	.82	.440	.2	.0150
.33	1.391	14	.83	2.293	23	.3	73.70	.33	.719	-7	.83	.436	.3	.0136
.34	1.405	14	.84	2.316	24	.4	81.45	.34	.712	-7	.84	.432	.4	.0123
0.35	1.419	14	0.85	2.340	23	4.5	90.02	0.35	.705	-7	0.85	.427	4.5	.0111
.36	1.433	15	.86	2.363	24	5.0	148.4	.36	.698	-7	.86	.423	5.0	.00674
.37	1.448	14	.87	2.387	24	6.0	403.4	.37	.691	-7	.87	.419	6.0	.00248
.38	1.462	15	.88	2.411	24	7.0	1097.	.38	.684	-7	.88	.415	7.0	.000912
.39	1.477	15	.89	2.435	25	8.0	2981.	.39	.677	-7	.89	.411	8.0	.000335
0.40	1.492	15	0.90	2.460	24	9.0	8103.	0.40	.670	-6	0.90	.407	9.0	.000123
.41	1.507	15	.91	2.484	25	10.0	22026.	.41	.664	-7	.91	.403	10.0	.000045
.42	1.522	15	.92	2.509	26	$\pi/2$	4.810	.42	.657	-6	.92	.399	$\pi/2$	.208
.43	1.537	16	.93	2.535	25	$2\pi/2$	23.14	.43	.651	-7	.93	.395	$2\pi/2$	.0432
.44	1.553	15	.94	2.560	26	$3\pi/2$	111.3	.44	.644	-6	.94	.391	$3\pi/2$	.00898
0.45	1.568	16	0.95	2.586	26	$4\pi/2$	535.5	0.45	.638	-7	0.95	.387	$4\pi/2$	.00187
.46	1.584	16	.96	2.612	26	$5\pi/2$	2576.	.46	.631	-6	.96	.383	$5\pi/2$	.000388
.47	1.600	16	.97	2.638	26	$6\pi/2$	12392.	.47	.625	-6	.97	.379	$6\pi/2$	.000081
.48	1.616	16	.98	2.664	27	$7\pi/2$	59610.	.48	.619	-6	.98	.375	$7\pi/2$	.000017
.49	1.632	17	.99	2.691	27	$8\pi/2$	286751.	.49	.613	-6	.99	.372	$8\pi/2$	.000003
0.50	1.649		1.00	2.718				0.50	0.607		1.00	.368		

NOTE: Do not interpolate in this column.

$e = 2.71828$ $1/e = 0.367879$ $\log_{10} e = 0.4343$ $1/(0.4343) = 2.3026$

$\log_{10}(0.4343) = \bar{1}.6378$ $\log_{10}(e^n) = n(0.4343)$

* From Marks, "Mechanical Engineers' Handbook," 4th ed., McGraw-Hill, New York, 1941.

TABLE 10.—HYPERBOLIC LOGARITHMS*

	n	$n\,(2.3026)$	$n\,(0.6974\text{--}3)$
These two pages give the natural (hyperbolic, or Napierian) logarithms ($\log_e$) of numbers between 1 and 10, correct to four places. Moving the decimal point n places to the right [or left] in the number is equivalent to adding n times 2.3026 [or n times $\bar{3}.6974$] to the logarithm. Base $e = 2.71828+$	1 2 3 4 5 6 7 8 9	2.3026 4.6052 6.9078 9.2103 11.5129 13.8155 16.1181 18.4207 20.7233	0.6974-3 0.3948-5 0.0922-7 0.7897-10 0.4871-12 0.1845-14 0.8819-17 0.5793-19 0.2767-21

Number	0	1	2	3	4	5	6	7	8	9	Avg. diff.
1.0	0.0000	0100	0198	0296	0392	0488	0583	0677	0770	0862	95
1.1	0053	1044	1133	1222	1310	1398	1484	1570	1655	1740	87
1.2	1823	1906	1989	2070	2151	2231	2311	2390	2469	2546	80
1.3	2624	2700	2776	2852	2927	3001	3075	3148	3221	3293	74
1.4	3365	3436	3507	3577	3646	3716	3784	3853	3920	3988	69
1.5	0.4055	4121	4187	4253	4318	4383	4447	4511	4574	4637	65
1.6	4700	4762	4824	4886	4947	5008	5068	5128	5188	5247	61
1.7	5306	5365	5423	5481	5539	5596	5653	5710	5766	5822	57
1.8	5878	5933	5988	6043	6098	6152	6206	6259	6313	6366	54
1.9	6419	6471	6523	6575	6627	6678	6729	6780	6831	6881	51
2.0	0.6931	6981	7031	7080	7129	7178	7227	7275	7324	7372	49
2.1	7419	7467	7514	7561	7608	7655	7701	7747	7793	7839	47
2.2	7885	7930	7975	8020	8065	8109	8154	8198	8242	8286	44
2.3	8329	8372	8416	8459	8502	8544	8587	8629	8671	8713	43
2.4	8755	8796	8838	8879	8920	8961	9002	9042	9083	9123	41
2.5	0.9163	9203	9243	9282	9322	9361	9400	9439	9478	9517	39
2.6	9555	9594	9632	9670	9703	9746	9783	9821	9858	9895	38
2.7	0.9933	9969	*0006	*0043	*0080	*0116	*0152	*0188	*0225	*0260	36
2.8	1.0296	0332	0367	0403	0438	0473	0508	0543	0578	0613	35
2.9	0647	0682	0716	0750	0784	0818	0852	0886	0919	0953	34
3.0	1.0986	1019	1053	1086	1119	1151	1184	1217	1249	1282	33
3.1	1314	1346	1378	1410	1442	1474	1506	1537	1569	1600	32
3.2	1632	1663	1694	1725	1756	1787	1817	1848	1878	1909	31
3.3	1939	1969	2000	2030	2060	2090	2119	2149	2179	2208	30
3.4	2238	2267	2296	2326	2355	2384	2413	2442	2470	2499	29
3.5	1.2528	2556	2585	2613	2641	2669	2698	2726	2754	2782	28
3.6	2809	2837	2865	2892	2920	2947	2975	3002	3029	3056	27
3.7	3083	3110	3137	3164	3191	3218	3244	3271	3297	3324	27
3.8	3350	3376	3403	3429	3455	3481	3507	3533	3558	3584	26
3.9	3610	3635	3661	3686	3712	3737	3762	3788	3813	3838	25
4.0	1.3863	3888	3913	3938	3962	3987	4012	4036	4061	4085	25
4.1	4110	4134	4159	4183	4207	4231	4255	4279	4303	4327	24
4.2	4351	4375	4398	4422	4446	4469	4493	4516	4540	4563	23
4.3	4586	4609	4633	4656	4679	4702	4725	4748	4770	4793	23
4.4	4816	4839	4861	4884	4907	4929	4951	4974	4996	5019	22
4.5	1.5041	5063	5085	5107	5129	5151	5173	5195	5217	5239	22
4.6	5261	5282	5304	5326	5347	5369	5390	5412	5433	5454	21
4.7	5476	5497	5518	5539	5560	5581	5602	5623	5644	5665	21
4.8	5686	5707	5728	5748	5769	5790	5810	5831	5851	5872	20
4.9	5892	5913	5933	5953	5974	5994	6014	6034	6054	6074	20

* From Marks, "Mechanical Engineers' Handbook," 4th ed., McGraw-Hill, New York, 1941.

$$\ln x = \log_e x = (2.3026)\log_{10} x \qquad \log_{10} x = (0.4343)\log_e x$$

where $\quad 2.3026 = \log_e 10$ and $0.4343 = \log_{10} e.$

TABLE 10.—HYPERBOLIC LOGARITHMS.*—(Continued)

Number	0	1	2	3	4	5	6	7	8	9	Avg. diff.
5.0	1.6094	6114	6134	6154	6174	6194	6214	6233	6253	6273	20
5.1	6292	6312	6332	6351	6371	6390	6409	6429	6448	6467	19
5.2	6487	6506	6525	6544	6563	6582	6601	6620	6639	6658	19
5.3	6677	6696	6715	6734	6752	6771	6790	6803	6827	6845	18
5.4	6864	6882	6901	6919	6938	6956	6974	6993	7011	7029	18
5.5	1.7047	7066	7084	7102	7120	7138	7156	7174	7192	7210	18
5.6	7228	7246	7263	7281	7299	7317	7334	7352	7370	7387	18
5.7	7405	7422	7440	7457	7475	7492	7509	7527	7544	7561	17
5.8	7579	7596	7613	7630	7647	7664	7681	7699	7716	7733	17
5.9	7750	7766	7783	7800	7817	7834	7851	7867	7884	7901	17
6.0	1.7918	7934	7951	7967	7984	8001	8017	8034	8050	8066	16
6.1	8083	8099	8116	8132	8148	8165	8181	8197	8213	8229	16
6.2	8245	8262	8278	8294	8310	8326	8342	8358	8374	8390	16
6.3	8405	8421	8437	8453	8469	8485	8500	8516	8532	8547	16
6.4	8563	8579	8594	8610	8625	8641	8656	8672	8687	8703	15
6.5	1.8718	8733	8749	8764	8779	8795	8810	8825	8840	8856	15
6.6	8871	8886	8901	8916	8931	8946	8961	8976	8991	9006	15
6.7	9021	9036	9051	9066	9081	9095	9110	9125	9140	9155	15
6.8	9169	9184	9199	9213	9228	9242	9257	9272	9286	9301	15
6.9	9315	9330	9344	9359	9373	9387	9402	9416	9430	9445	14
7.0	1.9459	9473	9488	9502	9516	9530	9544	9559	9573	9587	14
7.1	9601	9615	9629	9643	9657	9671	9685	9699	9713	9727	14
7.2	9741	9755	9769	9782	9796	9810	9824	9838	9851	9865	14
7.3	1.9879	9892	9906	9920	9933	9947	9961	9974	9988	*0001	13
7.4	2.0015	0028	0042	0055	0069	0082	0096	0109	0122	0136	13
7.5	2.0149	0162	0176	0189	0202	0215	0229	0242	0255	0268	13
7.6	0281	0295	0308	0321	0334	0347	0360	0373	0386	0399	13
7.7	0412	0425	0438	0451	0464	0477	0490	0503	0516	0528	13
7.8	0541	0554	0567	0580	0592	0605	0618	0631	0643	0656	13
7.9	0669	0681	0694	0707	0719	0732	0744	0757	0769	0782	12
8.0	2.0794	0807	0819	0832	0844	0857	0869	0882	0894	0906	12
8.1	0919	0931	0943	0956	0968	0980	0992	1005	1017	1029	12
8.2	1041	1054	1066	1078	1090	1102	1114	1126	1138	1150	12
8.3	1163	1175	1187	1199	1211	1223	1235	1247	1258	1270	12
8.4	1282	1294	1306	1318	1330	1342	1353	1365	1377	1389	12
8.5	2.1401	1412	1424	1436	1448	1459	1471	1483	1494	1506	12
8.6	1518	1529	1541	1552	1564	1576	1587	1599	1610	1622	12
8.7	1633	1645	1656	1668	1679	1691	1702	1713	1725	1736	11
8.8	1748	1759	1770	1782	1793	1804	1815	1827	1838	1849	11
8.9	1861	1872	1883	1894	1905	1917	1928	1939	1950	1961	11
9.0	2.1972	1983	1994	2006	2017	2028	2039	2050	2061	2072	11
9.1	2083	2094	2105	2116	2127	2138	2148	2159	2170	2181	11
9.2	2192	2203	2214	2225	2235	2246	2257	2268	2279	2289	11
9.3	2300	2311	2322	2332	2343	2354	2364	2375	2386	2396	11
9.4	2407	2418	2428	2439	2450	2460	2471	2481	2492	2502	11
9.5	2.2513	2523	2534	2544	2555	2565	2576	2586	2597	2607	10
9.6	2618	2628	2638	2649	2659	2670	2680	2690	2701	2711	10
9.7	2721	2732	2742	2752	2762	2773	2783	2793	2803	2814	10
9.8	2824	2834	2844	2854	2865	2875	2885	2895	2905	2915	10
9.9	2925	2935	2946	2956	2966	2976	2986	2996	3006	3016	10
10.0	2.3026										

Moving the decimal point n places to the right [or left] in the number requires adding n times 2.3026 [or n times $(0.6974 - 3)$] in the body of the table. See auxiliary table of multiples on top of the preceding page.

TABLE 11.—HYPERBOLIC SINES*
$$[\sinh x = \tfrac{1}{2}(e^x - e^{-x})]$$

x	0	1	2	3	4	5	6	7	8	9	Avg. diff
0.0	.0000	.0100	.0200	.0300	.0400	.0500	.0600	.0701	.0801	.0901	100
1	.1002	.1102	.1203	.1304	.1405	.1506	.1607	.1708	.1810	.1911	101
2	.2013	.2115	.2218	.2320	.2423	.2526	.2629	.2733	.2837	.2941	103
3	.3045	.3150	.3255	.3360	.3466	.3572	.3678	.3785	.3892	.4000	106
4	.4108	.4216	.4325	.4434	.4543	.4653	.4764	.4875	.4986	.5098	110
0.5	.5211	.5324	.5438	.5552	.5666	.5782	.5897	.6014	.6131	.6248	116
6	.6367	.6485	.6605	.6725	.6846	.6967	.7090	.7213	.7336	.7461	122
7	.7586	.7712	.7838	.7966	.8094	.8223	.8353	.8484	.8615	.8748	130
8	.8881	.9015	.9150	.9286	.9423	.9561	.9700	.9840	.9981	1.012	138
9	1.027	1.041	1.055	1.070	1.085	1.099	1.114	1.129	1.145	1.160	15
1.0	1.175	1.191	1.206	1.222	1.238	1.254	1.270	1.286	1.303	1.319	16
1	1.336	1.352	1.369	1.386	1.403	1.421	1.438	1.456	1.474	1.491	17
2	1.509	1.528	1.546	1.564	1.583	1.602	1.621	1.640	1.659	1.679	19
3	1.698	1.718	1.738	1.758	1.779	1.799	1.820	1.841	1.862	1.883	21
4	1.904	1.926	1.948	1.970	1.992	2.014	2.037	2.060	2.083	2.106	22
1.5	2.129	2.153	2.177	2.201	2.225	2.250	2.274	2.299	2.324	2.350	25
6	2.376	2.401	2.428	2.454	2.481	2.507	2.535	2.562	2.590	2.617	27
7	2.646	2.674	2.703	2.732	2.761	2.790	2.820	2.850	2.881	2.911	30
8	2.942	2.973	3.005	3.037	3.069	3.101	3.134	3.167	3.200	3.234	33
9	3.268	3.303	3.337	3.372	3.408	3.443	3.479	3.516	3.552	3.589	36
2.0	3.627	3.665	3.703	3.741	3.780	3.820	3.859	3.899	3.940	3.981	39
1	4.022	4.064	4.106	4.148	4.191	4.234	4.278	4.322	4.367	4.412	44
2	4.457	4.503	4.549	4.596	4.643	4.691	4.739	4.788	4.837	4.887	48
3	4.937	4.988	5.039	5.090	5.142	5.195	5.248	5.302	5.356	5.411	53
4	5.466	5.522	5.578	5.635	5.693	5.751	5.810	5.869	5.929	5.989	58
2.5	6.050	6.112	6.174	6.237	6.300	6.365	6.429	6.495	6.561	6.627	64
6	6.695	6.763	6.831	6.901	6.971	7.042	7.113	7.185	7.258	7.332	71
7	7.406	7.481	7.557	7.634	7.711	7.789	7.868	7.948	8.028	8.110	79
8	8.192	8.275	8.359	8.443	8.529	8.615	8.702	8.790	8.879	8.969	87
9	9.060	9.151	9.244	9.337	9.431	9.527	9.623	9.720	9.819	9.918	96
3.0	10.02	10.12	10.22	10.32	10.43	10.53	10.64	10.75	10.86	10.97	11
1	11.08	11.19	11.30	11.42	11.53	11.65	11.76	11.88	12.00	12.12	12
2	12.25	12.37	12.49	12.62	12.75	12.88	13.01	13.14	13.27	13.40	13
3	13.54	13.67	13.81	13.95	14.09	14.23	14.38	14.52	14.67	14.82	14
4	14.97	15.12	15.27	15.42	15.58	15.73	15.89	16.05	16.21	16.38	16
3.5	16.54	16.71	16.88	17.05	17.22	17.39	17.57	17.74	17.92	18.10	17
6	18.29	18.47	18.66	18.84	19.03	19.22	19.42	19.61	19.81	20.01	19
7	20.21	20.41	20.62	20.83	21.04	21.25	21.46	21.68	21.90	22.12	21
8	22.34	22.56	22.79	23.02	23.25	23.49	23.72	23.96	24.20	24.45	24
9	24.69	24.94	25.19	25.44	25.70	25.96	26.22	26.48	26.75	27.02	26
4.0	27.29	27.56	27.84	23.12	28.40	28.69	28.98	29.27	29.56	29.86	29
1	30.16	30.47	30.77	31.08	31.39	31.71	32.03	32.35	32.68	33.00	32
2	33.34	33.67	34.01	34.35	34.70	35.05	35.40	35.75	36.11	36.48	35
3	36.84	37.21	37.59	37.97	38.35	38.73	39.12	39.50	39.91	40.31	39
4	40.72	41.13	41.54	41.96	42.38	42.81	43.24	43.67	44.11	44.56	43
4.5	45.00	45.46	45.91	46.37	46.84	47.31	47.79	48.27	48.75	49.24	47
6	49.74	50.24	50.74	51.25	51.77	52.29	52.81	53.34	53.88	54.42	52
7	54.97	55.52	56.08	56.64	57.21	57.79	58.37	58.96	59.55	60.15	58
8	60.75	61.36	61.98	62.60	63.23	63.87	64.51	65.16	65.81	67.47	64
9	67.14	67.82	68.50	69.19	69.88	70.58	71.29	72.01	72.73	73.46	71
5.0	74.20										

If $x > 5$, $\sinh x = \tfrac{1}{2}(e^x)$ and $\log_{10} \sinh x = (0.4343)x + 0.6990 - 1$, correct to four significant figures.

* From Marks, "Mechanical Engineers' Handbook," 4th ed., McGraw-Hill, New York, 1941.

TABLE 12.—HYPERBOLIC COSINES*

$$[\cosh x = \tfrac{1}{2}(e^x + e^{-x})]$$

x	0	1	2	3	4	5	6	7	8	9	Avg. diff.
0.0	1.000	1.000	1.000	1.000	1.001	1.001	1.002	1.002	1.003	1.004	1
1	1.005	1.006	1.007	1.008	1.010	1.011	1.013	1.014	1.016	1.018	2
2	1.020	1.022	1.024	1.027	1.029	1.031	1.034	1.037	1.039	1.042	3
3	1.045	1.048	1.052	1.055	1.058	1.062	1.066	1.069	1.073	1.077	4
4	1.081	1.085	1.090	1.094	1.098	1.103	1.108	1.112	1.117	1.122	5
0.5	1.128	1.133	1.138	1.144	1.149	1.155	1.161	1.167	1.173	1.179	6
6	1.185	1.192	1.198	1.205	1.212	1.219	1.226	1.233	1.240	1.248	7
7	1.255	1.263	1.271	1.278	1.287	1.295	1.303	1.311	1.320	1.329	8
8	1.337	1.346	1.355	1.365	1.374	1.384	1.393	1.403	1.413	1.423	10
9	1.433	1.443	1.454	1.465	1.475	1.486	1.497	1.509	1.520	1.531	11
1.0	1.543	1.555	1.567	1.579	1.591	1.604	1.616	1.629	1.642	1.655	13
1	1.669	1.682	1.696	1.709	1.723	1.737	1.752	1.766	1.781	1.796	14
2	1.811	1.826	1.841	1.857	1.872	1.888	1.905	1.921	1.937	1.954	16
3	1.971	1.988	2.005	2.023	2.040	2.058	2.076	2.095	2.113	2.132	18
4	2.151	2.170	2.189	2.209	2.229	2.249	2.269	2.290	2.310	2.331	20
1.5	2.352	2.374	2.395	2.417	2.439	2.462	2.484	2.507	2.530	2.554	23
6	2.577	2.601	2.625	2.650	2.675	2.700	2.725	2.750	2.776	2.802	25
7	2.828	2.855	2.882	2.909	2.936	2.964	2.992	3.021	3.049	3.078	28
8	3.107	3.137	3.167	3.197	3.228	3.259	3.290	3.321	3.353	3.385	31
9	3.418	3.451	3.484	3.517	3.551	3.585	3.620	3.655	3.690	3.726	34
2.0	3.762	3.799	3.835	3.873	3.910	3.948	3.987	4.026	4.065	4.104	38
1	4.144	4.185	4.226	4.267	4.309	4.351	4.393	4.436	4.480	4.524	42
2	4.568	4.613	4.658	4.704	4.750	4.797	4.844	4.891	4.939	4.988	47
3	5.037	5.087	5.137	5.188	5.239	5.290	5.343	5.395	5.449	5.503	52
4	5.557	5.612	5.667	5.723	5.780	5.837	5.895	5.954	6.013	6.072	58
2.5	6.132	6.193	6.255	6.317	6.379	6.443	6.507	6.571	6.636	6.702	64
6	6.769	6.836	6.904	6.973	7.042	7.112	7.183	7.255	7.327	7.400	70
7	7.473	7.548	7.623	7.699	7.776	7.853	7.932	8.011	8.091	8.171	78
8	8.253	8.335	8.418	8.502	8.587	8.673	8.759	8.847	8.935	9.024	86
9	9.115	9.206	9.298	9.391	9.484	9.579	9.675	9.772	9.869	9.968	95
3.0	10.07	10.17	10.27	10.37	10.48	10.58	10.69	10.79	10.90	11.01	11
1	11.12	11.23	11.35	11.46	11.57	11.69	11.81	11.92	12.04	12.16	12
2	12.29	12.41	12.53	12.66	12.79	12.91	13.04	13.17	13.31	13.44	13
3	13.57	13.71	13.85	13.99	14.13	14.27	14.41	14.56	14.70	14.85	14
4	15.00	15.15	15.30	15.45	15.61	15.77	15.92	16.08	16.25	16.41	16
3.5	16.57	16.74	16.91	17.08	17.25	17.42	17.60	17.77	17.95	18.13	17
6	18.31	18.50	18.68	18.87	19.06	19.25	19.44	19.64	19.84	20.03	19
7	20.24	20.44	20.64	20.85	21.06	21.27	21.49	21.70	21.92	22.14	21
8	22.36	22.59	22.81	23.04	23.27	23.51	23.74	23.98	24.22	24.47	23
9	24.71	24.96	25.21	25.46	25.72	25.98	26.24	26.50	26.77	27.04	26
4.0	27.31	27.58	27.86	28.14	28.42	28.71	29.00	29.29	29.58	29.88	29
1	30.18	30.48	30.79	31.10	31.41	31.72	32.04	32.37	32.69	33.02	32
2	33.35	33.69	34.02	34.37	34.71	35.06	35.41	35.77	36.13	36.49	35
3	36.86	37.23	37.60	37.98	38.36	38.75	39.13	39.53	39.93	40.33	39
4	40.73	41.14	41.55	41.97	42.39	42.82	43.25	43.68	44.12	44.57	43
4.5	45.01	45.47	45.92	46.38	46.85	47.32	47.80	48.28	48.76	49.25	47
6	49.75	50.25	50 75	51.26	51.78	52.30	52.82	53.35	53.89	54.43	52
7	54.98	55.53	56.09	56.65	57.22	57.80	58.38	58.96	59.56	60.15	58
8	60.76	61.37	61.99	62.61	63.24	63.87	64.52	65.16	65.82	66.48	64
9	67.15	67.82	68.50	69.19	69.89	70.59	71.30	72.02	72 74	73.47	71
5.0	74.21										

If $x > 5$, $\cosh x = \tfrac{1}{2}(e^x)$ and $\log_{10} \cosh x = (0.4343)x + 0.6990 - 1$, correct to four significant figures.

* From Marks, "Mechanical Engineers' Handbook," 4th ed., McGraw-Hill, New York, 1941.

172640

TABLE 13.—DECIMAL EQUIVALENTS*

From minutes and seconds into decimal parts of a degree

Min	Decimal	Sec	Decimal
0'	0°.0000	0"	0°.0000
1	.0167	1	.0003
2	.0333	2	.0006
3	.05	3	.0003
4	.0667	4	.0011
5'	.0833	5"	.0014
6	.10	6	.0017
7	.1167	7	.0019
8	.1333	8	.0022
9	.15	9	.0025
10'	0°.1667	10"	0°.0028
1	.1833	1	.0031
2	.20	2	.0033
3	.2167	3	.0036
4	.2333	4	.0039
15'	.25	15"	.0042
6	.2667	6	.0044
7	.2833	7	.0047
8	.30	8	.005
9	.3167	9	.0053
20'	0°.3333	20"	0°.0056
1	.35	1	.0058
2	.3667	2	.0061
3	.3833	3	.0064
4	.40	4	.0067
25'	.4167	25"	.0069
6	.4333	6	.0072
7	.45	7	.0075
8	.4667	8	.0078
9	.4833	9	.0081
30'	0°.50	30"	0°.0083
1	.5167	1	.0086
2	.5333	2	.0089
3	.55	3	.0092
4	.5667	4	.0094
35'	.5833	35"	.0097
6	.60	6	.01
7	.6167	7	.0103
8	.6333	8	.0106
9	.65	9	.0108
40'	0°.6667	40"	0°.0111
1	.6833	1	.0114
2	.70	2	.0117
3	.7167	3	.0119
4	.7333	4	.0122
45'	.75	45"	.0125
6	.7667	6	.0128
7	.7833	7	.0131
8	.80	8	.0133
9	.8167	9	.0136
50'	0°.8333	50"	0°.0139
1	.85	1	.0142
2	.8667	2	.0144
3	.8833	3	.0147
4	.90	4	.015
55'	.9167	55"	.0153
6	.9333	6	.0156
7	.95	7	.0158
8	.9667	8	.0161
9	.9833	9	.0164
60'	1.00	60"	0°.0167

From decimal parts of a degree into minutes and seconds (exact values)

Decimal	Min/Sec	Decimal	Min/Sec
0°.00	0'	0°.50	30'
1	0' 36"	1	30' 36"
2	1' 12"	2	31' 12"
3	1' 48"	3	31' 48"
4	2' 24"	4	32' 24"
0°.05	3'	0°.55	33'
6	3' 36"	6	33' 36"
7	4' 12"	7	34' 12"
8	4' 48"	8	34' 48"
9	5' 24"	9	35' 24"
0°.10	6'	0°.60	36'
1	6' 36"	1	36' 36"
2	7' 12"	2	37' 12"
3	7' 48"	3	37' 48"
4	8' 24"	4	38' 24"
0°.15	9'	0°.65	39'
6	9' 36"	6	39' 36"
7	10' 12"	7	40' 12"
8	10' 48"	8	40' 48"
9	11' 24"	9	41' 24"
0°.20	12'	0°.70	42'
1	12' 36"	1	42' 36"
2	13' 12"	2	43' 12"
3	13' 48"	3	43' 48"
4	14' 24"	4	44' 24"
0°.25	15'	0°.75	45'
6	15' 36"	6	45' 36"
7	16' 12"	7	46' 12"
8	16' 48"	8	46' 48"
9	17' 24"	9	47' 24"
0°.30	18'	0°.80	48'
1	18' 36"	1	48' 36"
2	19' 12"	2	49' 12"
3	19' 48"	3	49' 48"
4	20' 24"	4	50' 24"
0°.35	21'	0°.85	51'
6	21' 36"	6	51' 36"
7	22' 12"	7	52' 12"
8	22' 48"	8	52' 48"
9	23' 24"	9	53' 24"
0°.40	24'	0°.90	54'
1	24' 36"	1	54' 36"
2	25' 12"	2	55' 12"
3	25' 48"	3	55' 48"
4	26' 24"	4	56' 24"
0°.45	27'	0°.95	57'
6	27' 36"	6	57' 36"
7	28' 12"	7	58' 12"
8	28' 48"	8	58' 48"
9	29' 24"	9	59' 24"
0°.50	30'	1°.00	60'

Decimal	Sec
0°.000	0".0
1	3".6
2	7".2
3	10".8
4	14".4
0°.005	18"
6	21".6
7	25".2
8	28".8
9	32".4
0°.010	36"

Common fractions

8ths	16ths	32nds	64ths	Exact decimal values
			1	.01 5625
		1	2	.03 125
			3	.04 6875
	1	2	4	.06 25
			5	.07 8125
		3	6	.09 375
			7	.10 9375
1	2	4	8	.12 5
			9	.14 0625
		5	10	.15 625
			11	.17 1875
	3	6	12	.18 75
			13	.20 3125
		7	14	.21 875
			15	.23 4375
2	4	8	16	.25
			17	.26 5625
		9	18	.28 125
			19	.29 6875
	5	10	20	.31 25
			21	.32 8125
		11	22	.34 375
			23	.35 9375
3	6	12	24	.37 5
			25	.39 0625
		13	26	.40 625
			27	.42 1875
	7	14	28	.43 75
			29	.45 3125
		15	30	.46 875
			31	.48 4375
4	8	16	32	.50
			33	.51 5625
		17	34	.53 125
			35	.54 6875
	9	18	36	.56 25
			37	.57 8125
		19	38	.59 375
			39	.60 9375
5	10	20	40	.62 5
			41	.64 0625
		21	42	.65 625
			43	.67 1875
	11	22	44	.68 75
			45	.70 3125
		23	46	.71 875
			47	.73 4375
6	12	24	48	.75
			49	.76 5625
		25	50	.78 125
			51	.79 6875
	13	26	52	.81 25
			53	.82 8125
		27	54	.84 375
			55	.85 9375
7	14	28	56	.87 5
			57	.89 0625
		29	58	.90 625
			59	.92 1875
	15	30	60	.93 75
			61	.95 3125
		31	62	.96 875
			63	.98 4375

* From Marks, "Mechanical Engineers' Handbook," 4th ed., McGraw-Hill, New York, 1941.

TABLE 14.—PROBABILITY INTEGRAL* $\dfrac{2}{\sqrt{\pi}} \displaystyle\int_0^x e^{-t^2}\, dt$

x	0	1	2	3	4	5	6	7	8	9
0.0	.00000	1128	2256	3384	4511	5637	6762	7886	9008	0128
.1	.11246	2362	3476	4587	5695	6800	7901	8999	0094	1184
.2	.22270	3352	4430	5502	6570	7633	8690	9742	0788	1828
.3	.32863	3891	4913	5928	6936	7938	8933	9921	0901	1874
.4	.42839	3797	4747	5689	6623	7548	8466	9375	0275	1167
.5	.52050	2924	3790	4646	5494	6332	7162	7982	8792	9594
.6	.60386	1168	1941	2705	3459	4203	4938	5663	6378	7084
.7	.67780	8467	9143	9810	0468	1116	1754	2382	3001	3610
.8	.74210	4800	5381	5952	6514	7067	7610	8144	8669	9184
.9	.79691	0188	0677	1156	1627	2089	2542	2987	3423	3851
1.0	.84270	4681	5084	5478	5865	6244	6614	6977	7333	7680
.1	.88020	8353	8679	8997	9308	9612	9910	0200	0484	0761
2	.91031	1296	1553	1805	2051	2290	2524	2751	2973	3190
.3	.93401	3606	3807	4002	4191	4376	4556	4731	4902	5067
.4	.95229	5385	5538	5686	5830	5969	6105	6237	6365	6490
.5	.96611	6728	6841	6952	7059	7162	7263	7360	7455	7546
.6	.97635	7721	7804	7884	7962	8038	8110	8181	8249	8315
.7	.98379	8441	85C0	8558	8613	8667	8719	8769	8817	8864
.8	.98909	8952	8994	9035	9074	9111	9147	9182	9216	9248
.9	.99279	9309	9338	9366	9392	9418	9443	9466	9489	9511
2.0	.99532	9552	9572	9591	9609	9626	9642	9658	9673	9688
.1	.99702	9715	9728	9741	9753	9764	9775	9785	9795	9805
.2	.99814	9822	9831	9839	9846	9854	9861	9867	9874	9880
.3	.99886	9891	9897	9902	9906	9911	9915	9920	9924	9928
.4	.99931	9935	9938	9941	9944	9947	9950	9952	9955	9957
.5	.99959	9961	9963	9965	9967	9969	9971	9972	9974	9975
.6	.99976	9978	9979	9980	9981	9982	9983	9984	9985	9986
.7	.99987	9987	9988	9989	9989	9990	9991	9991	9992	9992
.8	.99992	9993	9993	9994	9994	9994	9995	9995	9995	9996
.9	.99996	9996	9996	9997	9997	9997	9997	9997	9997	9998

* From Allen, "Six-place Tables," 6th ed., McGraw-Hill, New York, 1941.

TABLE 15.—VALUES OF SPECIAL CONSTANTS WITH THEIR COMMON LOGARITHMS*

Constant	Logarithm	Constant	Logarithm
$\pi = 3.14159\ldots\ldots$	.49715	$\dfrac{1}{\pi^3} = .032252\ldots\ldots$	$8.50855 - 10$
$2\pi = 6.28319\ldots\ldots$	.79818	$\dfrac{1}{\pi^4} = .010266\ldots\ldots$	$8.01140 - 10$
$4\pi = 12.56637\ldots\ldots$	1.09921	$\sqrt{\pi} = 1.7725\ldots\ldots$	.24857
$\dfrac{\pi}{6} = .52360\ldots\ldots$	$9.71900 - 10$	$\sqrt[3]{\pi} = 1.4646\ldots\ldots$	.16572
$\dfrac{\pi}{4} = .78540\ldots\ldots$	$9.89509 - 10$	†$e = 2.7183\ldots\ldots$	.43429
		$e^2 = 7.3891\ldots\ldots$	.86859
$\dfrac{\pi}{3} = 1.0472\ldots\ldots$	.02003	$e^3 = 20.086\ldots\ldots$	1.30288
$\dfrac{\pi}{2} = 1.5708\ldots\ldots$	.19612	$e^4 = 54.598\ldots\ldots$	1.73718
$\dfrac{4\pi}{3} = 4.1888\ldots\ldots$	.62209	$\dfrac{1}{e} = .36788\ldots\ldots$	$9.56571 - 10$
$\pi^2 = 9.8696\ldots\ldots$	.99430	$\dfrac{1}{e^2} = .13534\ldots\ldots$	$9.13141 - 10$
$\pi^3 = 31.006\ldots\ldots$	1.49145	$\dfrac{1}{e^3} = .049787\ldots\ldots$	$8.69712 - 10$
$\pi^4 = 97.409\ldots\ldots$	1.98860	$\dfrac{1}{e^4} = .018316\ldots\ldots$	$8.26282 - 10$
$\dfrac{1}{\pi} = .31831\ldots\ldots$	$9.50285 - 10$		
$\dfrac{1}{\pi^2} = .10132\ldots\ldots$	$9.00570 - 10$		

* From O'Rourke, "General Engineering Handbook," 2d ed., McGraw-Hill, New York, 1940.

† $\log_e 10 = \dfrac{1}{\log_{10} e} = 2.302585.$

TABLE 16.—THE 0.286 ($= R/c_p$) AND 3.5 ($= c_p/R$) POWERS OF NUMBERS FROM 1.0 TO 5.0*

N	$N^{0.286}$	$N^{3.5}$
1.00	1.00	1.00
1.01	1.0028	1.036
1.02	1.0057	1.072
1.03	1.0085	1.109
1.04	1.011	1.147
1.05	1.014	1.186
1.06	1.017	1.226
1.07	1.0195	1.267
1.08	1.022	1.309
1.09	1.025	1.352
1.10	1.028	1.396
1.11	1.0305	1.441
1.12	1.033	1.487
1.13	1.0355	1.534
1.14	1.038	1.582
1.15	1.0405	1.631
1.16	1.043	1.681
1.17	1.046	1.732
1.18	1.0485	1.784
1.19	1.051	1.838
1.20	1.0535	1.893
1.21	1.056	1.949
1.22	1.0585	2.006
1.23	1.061	2.065
1.24	1.0635	2.124
1.25	1.066	2.184
1.26	1.0685	2.245
1.27	1.071	2.308
1.28	1.073	2.373
1.29	1.0755	2.438
1.30	1.078	2.505
1.31	1.0805	2.573
1.32	1.0825	2.642
1.33	1.085	2.713
1.34	1.087	2.785
1.35	1.0895	2.858
1.36	1.092	2.933
1.37	1.094	3.009
1.38	1.0965	3.087
1.39	1.099	3.166
1.40	1.101	3.247
1.41	1.1035	3.329
1.42	1.1055	3.413
1.43	1.1075	3.498
1.44	1.110	3.584
1.45	1.112	3.672
1.46	1.1145	3.762
1.47	1.1165	3.854
1.48	1.1185	3.947
1.49	1.121	4.041

N	$N^{0.286}$	$N^{3.5}$
1.50	1.123	4.135
1.52	1.127	4.330
1.54	1.131	4.53
1.56	1.135	4.74
1.58	1.140	4.96
1.60	1.144	5.18
1.62	1.148	5.41
1.64	1.152	5.65
1.66	1.156	5.90
1.68	1.160	6.15

N	$N^{0.286}$	N	$N^{0.286}$
1.70	1.164	2.50	1.300
1.72	1.168	2.55	1.307
1.74	1.172	2.60	1.314
1.76	1.1755	2.65	1.321
1.78	1.179	2.70	1.328
1.80	1.183	2.75	1.335
1.82	1.187	2.80	1.342
1.84	1.1905	2.85	1.349
1.86	1.194	2.90	1.356
1.88	1.198	2.95	1.363
1.90	1.2015	3.00	1.369
1.92	1.205	3.1	1.382
1.94	1.2085	3.2	1.395
1.96	1.212	3.3	1.407
1.98	1.216	3.4	1.419
2.00	1.219	3.5	1.431
2.05	1.228	3.6	1.443
2.10	1.237	3.7	1.454
2.15	1.245	3.8	1.467
2.20	1.253	3.9	1.476
2.25	1.261	4.0	1.487
2.30	1.279	4.1	1.497
2.35	1.277	4.2	1.507
2.40	1.285	4.3	1.517
2.45	1.293	4.4	1.527
		4.5	1.537
		4.6	1.547
		4.7	1.556
		4.8	1.566
		4.9	1.575
		5.0	1.584

* Courtesy of P. J. Kiefer.

UNITS AND MEASURES

In scientific work of any kind, measurements of various quantities must be made. In order that the results of the measurement may be made useful to others, the *unit* of measure must be stated. Each civilized country maintains certain *standards* of measure against which the usual measuring devices are calibrated. These standards are defined in terms of easily reproducible quantities when possible. The standards of different countries are carefully checked among themselves.

Systems of Units. A choice must be made among available measures and a system of units constructed. The more common systems in use in meteorology are the *centimeter, gram, second,* the *meter, ton, second,* and the *foot, pound, second.* These are abbreviated to *cgs, mts,* and *fps,* respectively. In addition to these mechanical units, the meteorologist is also concerned with thermal units and, to a lesser extent, electrical units. See Bureau of Standards publications for precise definitions of the various units. Conversion tables among the various units are provided below. Nomograms and scales for conversion of frequently used quantities appear through the book. These may be used generally for conversions to about the same accuracy with which the usual meteorological element can be measured. If extreme accuracy is desired, recourse may be had to the tabulated values.

Meter. The fundamental unit of length is the meter. This is taken to be the distance between two specified lines on a platinum-iridium bar kept under atmospheric pressure and at zero degrees centigrade temperature. An alternative standard in case the bar should be damaged is expressed in terms of wave lengths of light under specified conditions of emission. One meter equals 100 centimeters.

Kilogram. The fundamental unit of mass is the kilogram. This is taken to be the mass of a certain cylinder of platinum-iridium. One kilogram equals 1,000 grams. One thousand kilograms equals one metric ton.

Second. The fundamental unit of time is the second. This is taken to be one 86,400th part of the mean solar day. Sixty seconds equals one minute. Sixty minutes equals one hour.

Dyne. The dyne is the cgs unit of force. It is the force required to accelerate a mass of one gram with an acceleration of one centimeter per second per second.

Millibar. The millibar is the meteorological unit of pressure. One millibar is 1,000 dynes per square centimeter.

Erg. The erg is the cgs unit of energy. It is the work done by a force of one dyne moving a distance of one centimeter. One joule is equal to 10,000,000 ergs.

*Ohm.** The unit of resistance shall be what is known as the *international ohm,* which is substantially equal to one thousand million units of resistance of the cgs system of electromagnetic units, and is represented by the resistance offered to an unvarying electric current by a column of mercury at the temperature of melting ice fourteen and four thousand five hundred and twenty-one ten-thousandths (14.4521) grams in mass, of a constant cross-sectional area, and of the length of one hundred and six and three-tenths (106.3) centimeters.

*Ampere.** The unit of current shall be what is known as the *international ampere,* which is one-tenth of the unit of current of the cgs system of electromagnetic units, and is the practical equivalent of the unvarying current, which, when passed through a solution of nitrate of silver in water in accordance with standard specifications, deposits silver at the rate of one thousand one hundred and eighteen millionths (0.001118) of a gram per second.

---

* From, Knowlton, "Standard Handbook for Electrical Engineers," 7th ed., McGraw-Hill, New York. 1941.

*Volt.** The unit of electromotive force shall be what is known as the *international volt*, which is the electromotive force that, steadily applied to a conductor whose resistance is one international ohm, will produce a current of an international ampere, and is practically equivalent to one thousand fourteen hundred and thirty-fourths (1,000/1,434) of the electromotive force between the poles or electrodes of the voltaic cell known as *Clark's cell*, at a temperature of fifteen degrees centigrade (15°C.), and prepared in the manner described in the standard specifications.

*Coulomb.** The unit of quantity shall be what is known as the *international coulomb*, which is the quantity of electricity transferred by a current of one international ampere in one second.

*Farad.** The unit of capacity shall be what is known as the *international farad*, which is the capacity of a condenser charged to a potential of one international volt by one international coulomb of electricity.

*Joule.** The unit of work shall be the *joule*, which is equal to ten million units of work in the cgs system, and which is practically equivalent to the energy expended in one second by an international ampere in an international ohm.

*Watt.** The unit of power shall be the *watt*, which is equal to ten million units of power in the cgs system, and which is practically equivalent to the work done at the rate of one joule per second. The *kilowatt* is 1,000 watts.

A *watt-hour* is 3,600 joules.

A *kilowatt-hour* is 3,600,000 joules or energy equivalent to work performed in one hour at the average rate of 1 kilowatt.

*Henry.** The unit of induction shall be the *henry*, which is the induction in a circuit when the electromotive force induced in this circuit is one international volt while the inducing current varies at the rate of one ampere per second.

Calorie. The calorie is a unit of heat energy. It is the heat required to raise one gram of water one degree centigrade under specified conditions (see Table 73). The calorie is here usually replaced by the smaller joule. One calorie equals 4.186 joules (approximately). The British thermal unit (Btu) is similarly the heat energy required to raise one pound of water one degree Fahrenheit under specified conditions.

Degree. A degree of temperature is rigorously defined in terms of the work done by ideal heat engines. See Sec. V for complete discussion. The relations among the more common scales of temperature are as follows:

$$\text{degrees centigrade} = \frac{5}{9}\ (\text{degrees Fahrenheit} - 32)$$

$$\text{degrees Kelvin} = \text{degrees centigrade} + 273.16$$

$$\text{degrees Rankine} = \text{degrees Fahrenheit} + 459.69$$

The Kelvin and Rankine scales are so-called "Absolute temperature scales."

* From Knowlton, "Standard Handbook for Electrical Engineers," 7th ed., McGraw Hill, New York, 1941.

TABLE 17.—MECHANICAL UNITS AND DIMENSIONS

Quantity	Dimensions	Meter, ton, second	N	Centimeter, gram, second
Length..............	L	1 meter (m)	10^2	1 centimeter (cm)
Mass...............	M	1 ton (1,000 kg)	10^6	1 gram
Time...............	T	1 second (sec)	1	1 second (sec)
Volume.............	L^3	1 m³	10^6	1 cubic centimeter (cm³)
Specific volume......	$M^{-1}L^3$	1 ton⁻¹ m³	1	1 gram⁻¹ cm³
Density.............	ML^{-3}	1 ton m⁻³	1	1 gram cm⁻³
Velocity.............	LT^{-1}	1 m sec⁻¹	10^2	1 cm sec⁻¹
Acceleration.........	LT^{-2}	1 m sec⁻²	10^2	1 cm sec⁻²
Force...............	MLT^{-2}	1 ton m sec⁻²	10^8	1 gram cm sec⁻² (1 dyne)
Pressure............	$ML^{-1}T^{-2}$	1 ton m⁻¹ sec⁻²	10^4	1 gram cm⁻¹ sec⁻² (barye)
Energy..............	ML^2T^{-2}	1 ton m² sec⁻² (1 kilojoule)	10^{10}	1 gram cm² sec⁻² (1 erg)
Specific energy.......	L^2T^{-2}	1 m² sec⁻² (1 kilojoule ton⁻¹)	10^4	1 cm² sec⁻² (1 erg gram⁻¹)

TABLE 18.—CONVERSION OF UNITS*

Length

1 inch = 25.40005 millimeters (mm)   1 millimeter = 0.0393700 inches (in.)
1 foot = 0.3048006 meters (m)   1 meter = 3.280833 feet = 39.3700 inches (in.)

1 mile = 1.609347 kilometers (km)   1 kilometer = 0.621370 miles
1 nautical mile = 6,080.27 feet (ft) (Bowditch)
1 nautical mile = 6,080.20 feet (ft) (Smithsonian)
1 mile (land) = 5,280 feet (ft)
1 μ (micron) = 10^{-4} cm
1 Å (Ångstrom unit) = 10^{-8} cm

Mass

1 avoirdupois pound = 0.4535924 kilogram (kg)
1 kilogram = 2.204622 avoirdupois pounds (lb)
1 avoirdupois ounce = 0.0283495 kilograms (kg)
1 metric ton = 1,000 kilograms = 10^6 grams

Velocity

1 mile per hour = $^{44}\!/_{30}$ feet per second (fps)
1 foot per second = $^{30}\!/_{44}$ miles per hour (mph)
1 mile per hour = 0.4470409 meters per second (mps)
1 knot = 1 nautical mile per hour = 1.6889444 feet per second (fps)
1 knot = 0.51479 meters per second (mps)
1 meter per second = 1.9424 knots = 2.236932 miles per hour (mph)

Pressure (Meteorological)

1 inch mercury (standard gravity and temperature) = 33.86395 millibars (mb)
1 millimeter mercury (standard gravity and temperature) = 1.33322387 millibars (mb)
1 millibar = 0.02952993 inches mercury = 0.75008 millimeters mercury
1 millibar = 1,000 dynes per square centimeter

Energy (Meteorological)

1 erg = 10^{-7} joule   1 joule = 10^7 ergs = 10^7 dyne cm
1 calorie (15°C) = 4.1858 joules
* Values from Smithsonian Tables.

TABLE 19.—CONTINENTAL MEASURES OF LENGTH WITH THEIR METRIC AND ENGLISH
EQUIVALENTS*

Measure	Metric equivalent	English equivalent
El (Netherlands)	1 m	3.2808 ft
Fathom, Swedish 6 ft	1.7814 m	5.8445 ft
Foot, Russian	0.30480 m	1 ft
Foot, Denmark	0.31385 m	1.0297 ft
Mile, German Sea	1.852 km	1.1508 statute miles
Mile, Swedish 36,000 feet	10.69 km	6.642 statute miles
Mile, Norwegian 36,000 feet	11.2986 km	7.02 statute miles
Mile, Netherlands (mijl)	1 km	0.6214 statute miles
Mile, Prussian (law of 1868)	7.500 km	4.660 statute miles
Mile, Danish	7.5324 km	4.6804 statute miles
Palm, Netherlands	0.1 m	0.3281 ft
Sagene (Russian)	2.1336 m	7 ft
Werst, or Versta (Russian) = 500 sashjene	1.0668 km	3.500 ft

* From Smithsonian Tables.

TABLE 20.—ASTRONOMICAL AND GEODETIC CONSTANTS*

Semiaxes of the earth's ellipsoid
$$a = 6.378388 \times 10^8 \quad \text{cm}$$
$$c = 6.356912 \times 10^8 \quad \text{cm}$$
$$\text{Flattening} = \frac{a - c}{a} = \frac{1}{297} = 0.00337$$

Surface of the earth = 5.101×10^{18} cm^2
Volume of the earth = 1.083×10^{27} cm^3
Quadrant of a meridian = 10,002.288 km
Quadrant of the equator = 10,019.148 km
Gravity at sea level in latitude φ
$$g = 978.049(1 + 0.0052884 \sin^2 \varphi - 0.0000059 \sin^2 2\varphi) \quad \text{cm/sec/sec}$$
Gravity at 45° = 980.6294 cm/sec/sec
Angular velocity of the earth's rotation, $\omega = 7.292115851 \times 10^{-5}$ rad/sec
Velocity of a point on the equator = 46,500 cm/sec
Centrifugal force at the equator divided by gravity at the equator
$$\frac{\omega^2 a}{g_e} = 0.003468 = \frac{1}{288}$$
Mean distance earth to sun = 149.5×10^6 km
Average velocity around sun = 29.8 km/sec
 * After Gutenberg.

TABLE 21.—U.S. CUSTOMARY WEIGHTS AND MEASURES
Length

12 inches = 1 foot
3 feet = 1 yard
5½ yards = 1 rod, pole, or perch
40 poles = 1 furlong
8 furlongs = 1 mile
5,280 feet = 1 mile
3 miles = 1 league
4 inches = 1 hand
9 inches = 1 span
6 feet = 1 fathom
6080.20 feet = 1 nautical mile
120 fathoms = 1 cable length
7.92 inches = 1 link
100 links = 1 chain
80 chains = 1 mile

TABLE. 21.—U.S. CUSTOMARY WEIGHTS AND MEASURES.—(*Continued*)

Area

160 square rods = 1 acre
640 acres = 1 square mile
1 circular mil = area of circle 0.001 inch in diameter

Volume

1 cord of wood = 128 cubic feet
1 perch of masonry = 16½ to 25 cubic feet
4 gills = 1 pint (liquid)
2 pints = 1 quart (liquid)
4 quarts = 1 gallon (liquid)
7.4805 gallons = 1 cubic foot
1 barrel petroleum oil, unrefined = 42 gallons (liquid), by trade custom
2 pints = 1 quart (dry)
8 quarts = 1 peck (dry)
4 pecks = 1 bushel(dry)
1 barrel = 7,056 cubic inches
1 register ton = 100 cubic feet
1 U.S. shipping ton = 40 cubic feet
1 British shipping ton = 42 cubic feet
1 board foot = 144 cubic inches

Weights

Avoirdupois weight:
  16 drams = 1 ounce
  16 ounces = 7,000 grains = 1 pound
  100 pounds = 1 central
  2,000 pounds = 1 short ton
  2,240 pounds = 1 long ton

In Great Britain:
  14 pounds = 1 stone
  2 stone = 1 quarter
  8 stone = 1 hundredweight (cwt)
  20 hundredweight = 1 ton

Troy Weight

24 grains = 1 scruple
3 scruples = 1 dram
8 drams = 1 ounce
12 ounces = 1 pound

TABLE 22.—ENERGY CONVERSION TABLE

	Ergs	Reciprocal	Gram-calories	Reciprocal
1 erg.	1	1	0.2390×10^{-7}	4.184×10^{7}
1 joule.	10^{7}	10^{-7}	0.2390	4.184
1 gram-wt-cm (g = 980.6).	980.6	1.0198×10^{-3}	0.2344×10^{-4}	4.267×10^{4}
1 gram-calorie.	4.184×10^{7}	0.2390×10^{-7}	1	1
1 kilogram-calorie.	4.184×10^{10}	0.2390×10^{-10}	1,000	10^{-3}
1 thermie or ton-calorie.	4.184×10^{13}	0.2390×10^{-13}	1,000,000	10^{-6}
1 foot-grain (g = 981.2).	1.938×10^{3}	0.5160×10^{-3}	0.4632×10^{-4}	2.159×10^{4}
1 foot-pound (g = 981.2)‡.	1.356×10^{7}	0.7371×10^{-7}	0.3242	3.084
1 foot-long-ton (g = 981.2).	3.039×10^{10}	0.3291×10^{-10}	0.7263×10^{3}	1.377×10^{-3}
1 foot-short-ton (g = 981.2).	2.713×10^{10}	0.3686×10^{-10}	0.6485×10^{3}	1.542×10^{-3}
1 British thermal unit.	1.055×10^{10}	0.9475×10^{-10}	0.2522×10^{3}	3.964×10^{-3}
1 watthour.	3.600×10^{10}	0.2778×10^{-10}	0.8604×10^{3}	1.162×10^{-3}
1 kilowatthour.	3.600×10^{13}	0.2778×10^{-13}	0.8604×10^{6}	1.162×10^{-6}
1 horsepower-hour* (746.1 whr).	2.686×10^{13}	0.3723×10^{-13}	0.6420×10^{6}	1.558×10^{-6}
1 horsepower-hour (745.7 whr, g = 980.6).	2.684×10^{13}	0.3725×10^{-13}	0.6416×10^{6}	1.559×10^{-6}
1 metric horsepower-hour (736 whr)†.	2.650×10^{13}	0.3774×10^{-13}	0.6333×10^{6}	1.579×10^{-6}
1 metric horsepower-hour (735.5 whr, g = 980.6).	2.648×10^{13}	0.3777×10^{-13}	0.6328×10^{6}	1.580×10^{-6}
1 kilojoule.	10^{10}	10^{-10}	239.0	4.184×10^{-3}

* For g = 981.2 for approximate latitude of London.
† For g = 981.3 for approximate latitude of Berlin.
  ‡ The local foot-pound varies between the equator and the poles, according to the local intensity of gravitation, between the limits 1.352 and 1.359 joules.
  From Knowlton, "Standard Handbook for Electrical Engineers," 7th ed., McGraw-Hill, New York, 1941.

TABLE 23.—INTENSITY OF PRESSURE (FORCE PER UNIT AREA)

	Barys or dynes per sq cm	Reciprocal	Grams per sq cm	Reciprocal	Standard atmospheres of 760 mm mercury	Reciprocal
1 baryc, or dyne per sq cm	1	1	1.0198×10^{-3}	$980.62*$	0.9869×10^{-6}	1.013×10^{6}
1 bar	10^{6}	10^{-6}	1.0198×10^{3}	0.9806×10^{-3}	0.9869	1.013
1 gr per sq cm or 1 cm water	$980.62*$	1.0198×10^{-3}	1	1	0.9678×10^{-3}	1.033×10^{3}
1 standard atmosphere, 760 mm mercury at 0 C	1.0132×10^{6}	0.9869×10^{-6}	1.033×10^{3}	0.9678×10^{-3}	1	1
1 lb avoird. per sq ft	478.8	0.2089×10^{-2}	0.4882	2.048	0.4725×10^{-3}	2.116×10^{3}
1 kg per sq meter	98.06	1.020×10^{-2}	0.1	10	0.9678×10^{-4}	1.033×10^{4}
1 kg per hectare	0.9806×10^{-2}	1.020×10^{2}	10^{-5}	10^{5}	0.9678×10^{-3}	1.033×10^{3}
1 mm pure mercury 0 C	1.333×10^{3}	0.7501×10^{-3}	1.359	0.7355	1.316×10^{-3}	0.760×10^{3}
1 in. pure mercury at 0 C	3.386×10^{4}	0.2953×10^{-4}	34.53	0.02896	0.03342	29.92
1 lb avoird. per sq in	0.6894×10^{5}	1.450×10^{-5}	70.31	0.01422	0.06804	14.70
1 long ton per sq ft	1.072×10^{6}	0.9324×10^{-6}	$1,094$	0.9144×10^{-3}	1.058	0.9448
1 short ton per sq ft	0.9575×10^{6}	1.044×10^{-6}	976.5	1.024×10^{-3}	0.9451	1.058
1 ft of water	2.989×10^{3}	0.3346×10^{-4}	30.48	0.03281	2.950×10^{-2}	33.90
1 in. of water	2.491×10^{3}	0.4015×10^{-3}	2.54	0.3937	2.458×10^{-3}	406.8

* The range of variation in sea-level gravitation intensity is from 978.03 at the equator to 983.22 at the poles.
From Knowlton, "Standard Handbook for Electrical Engineers," 7th ed., McGraw-Hill, New York, 1941.

TABLE 24.—POWER

	Watts	Reciprocal	Kilowatts	Reciprocal	Ergs per sec or abwatts	Reciprocal
1 British horsepower ($g = 980.6$)	745.7	1.341×10^{-3}	0.7457	1.341	7.457×10^{9}	1.341×10^{-10}
1 metric horsepower ($g = 980.6$)	735.5	1.360×10^{-3}	0.7355	1.360	7.355×10^{9}	1.360×10^{-10}
1 British thermal unit per hour	0.2931	3.412	0.2931×10^{-3}	3.412×10^{3}	0.2931×10^{7}	3.412×10^{-7}
1 gram-calorie per second	4.184	0.2390	4.184×10^{-3}	0.2390×10^{3}	4.184×10^{7}	0.2390×10^{-7}
1 gram-calorie per minute	6.973×10^{-2}	1.434×10^{2}	6.973×10^{-5}	1.434×10^{5}	6.973×10^{5}	0.1434×10^{-5}
1 gram-calorie per hour	1.162×10^{-3}	0.8606×10^{3}	1.162×10^{-6}	0.8606×10^{6}	1.162×10^{4}	0.8606×10^{-4}
1 kilogram-calorie per second	4.184×10^{3}	0.2390×10^{-3}	4.184	0.2390	4.184×10^{10}	0.2390×10^{-10}
1 kilogram-calorie per minute	6.973×10^{1}	0.1434×10^{-1}	6.973×10^{-2}	1.434×10^{2}	6.973×10^{8}	0.1434×10^{-8}
1 kilogram-calorie per hour	1.162	0.8606	1.162×10^{-3}	0.8606×10^{3}	1.162×10^{7}	0.8606×10^{-7}
1 water horsepower, 3,960 U.S. gallons per min lifted 1 ft	746.7	1.339×10^{-3}	0.7467	1.339	0.7467×10^{10}	1.339×10^{-10}
1 foot-pound per sec against standard g	1.356	0.7375	1.356×10^{-3}	737.5	1.356×10^{7}	0.7375×10^{-7}
1 foot-pound per min against standard g	0.02260	44.25	0.2260×10^{-4}	4.425×10^{4}	0.2260×10^{6}	4.425×10^{-6}
1 kilogram-meter per sec against standard g	9.806	1.020×10^{-1}	9.806×10^{-3}	1.020×10^{2}	9.806×10^{8}	1.020×10^{-8}

TABLE 25.—ANGULAR VELOCITY*

	Degrees per sec	Reciprocal	Grades per sec	Reciprocal	Radians per sec	Reciprocal	Revs. per sec	Reciprocal
1 degree per second	1	1	1.111	0.90	0.01745	57.30	0.002778	360
1 grade per second	0.90	1.111	1	1	0.01571	63.66	0.0025	400
1 radian per sec	57.30	0.01745	63.66	0.01571	1	1	0.1592	6.283
1 revolution per sec	360	0.002778	400	0.0025	6.283	0.1592	1	1
1 degree per minute	0.01667	60	0.01852	54	0.0002909	3,433	0.4630×10^{-4}	21,600
1 grade per minute	0.0150	66.667	0.01667	60	0.0002618	3,820	0.4167×10^{-4}	24,000
1 radian per minute	0.9549	1.047	1.061	0.9425	0.01667	60	0.002653	377.0
1 revolution per minute	6.0	0.1667	6.667	0.15	0.1047	9.551	0.01667	60
1 degree per hour	0.2778×10^{-3}	3,600	0.3086×10^{-3}	3,240	0.4848×10^{-5}	206,300	0.7716×10^{-6}	1.296×10^{6}
1 grade per hour	0.25×10^{-3}	4,000	0.2778×10^{-3}	3,600	0.4363×10^{-5}	229,200	0.6944×10^{-6}	1.44×10^{6}
1 radian per hour	0.01592	62.83	0.01768	56.55	0.2778×10^{-3}	3,600	0.4421×10^{-5}	2.262×10^{4}
1 revolution per hour	0.1	10	0.1111	9.00	0.1745×10^{-2}	573.1	0.2778×10^{-3}	3.6×10^{3}

* From Knowlton, "Standard Handbook for Electrical Engineers," 7th ed., McGraw-Hill, New York, 1941.

TABLE 26.—ELECTRICAL UNITS*

Symbol	Quantity	Equation	Practical unit	Value of practical units, electromagnetic units
I, i	Current............	$I = E/Z,\ I = Q/t$	Ampere.........	10^{-1}
Q, q	Quantity............	$Q = It$	Coulomb.........	10^{-1}
			Ampere-hour.....	360
$E, e,$	Electromotive force..	$E = W/Q$	Volt............	10^8
R, r	Resistance...........	$R = P/I^2 = E/I$	Ohm............	10^9
ρ	Resistivity..........	$\rho = RA/L$	Ohms per circular mil-foot........	
			Ohms per centimeter cube....	10^6
λ	Conductivity........	$\lambda = 1/\rho$	Mhos per unit volume...........	
C	Capacitance........	$C = Q/E$	Farad...........	10^{-9}
L	Inductance..........	$L = n\phi10^{-8}/I$	Henry..........	10^9
	Time constant.......	L/R	Second...........	1
			Henry per ohm...	1
T	Period or cycle......	$T = 1/f$	Second...........	1
f	Frequency...........	$f = 1/T$	Cycles per second	1
ω	Angular velocity.....	$\omega = 2\pi f$	....	......
X_L	Inductive reactance..	$X_L = 2\pi fL$	Ohm............	10^9
X_C	Capacitive reactance.	$X_C = 1/2\pi fC$	Ohm............	10^9
X, x	Reactance...........	$X = X_L - X_C$	Ohm............	10^9
Z, z	Impedance..........	$Z = \sqrt{R^2 + X^2}$	Ohm............	10^9
G, g	Conductance........	$G = R/Z^2$	Mho............	10^{-9}
B, b	Susceptance.........	$B = X/Z^2$	Mho............	10^{-9}
Y, y	Admittance..........	$Y = \sqrt{G^2 + L^2}$ $= 1/Z$	Mho............	10^{-9}
P	Electric power.......	$P = EI = I^2R$ $= EI\cos\theta$ (a-c)	Watt............ Watt............	10^7 10^7
W	Electric energy.......	$W = Pt$	Joule............ Watthour........ Kilowatt-hour....	10^7 36×10^9 36×10^{12}
	Power factor (p.f.)...	$\dfrac{EI\cos\theta}{EI} = \dfrac{\text{real } P}{\text{apparent } P}$		
	Reactive factor......	$\dfrac{EI\sin\theta}{EI} = \dfrac{\text{reactive } P}{\text{apparent } P}$		

n = number of turns; t = time in seconds; f = frequency; ϕ = flux.
* From Marks, "Mechanical Engineers' Handbook," 4th ed., McGraw-Hill, New York, 1941.

TABLE 27.—DEGREES FAHRENHEIT TO DEGREES CENTIGRADE

Fahren- heit	Centigrade									
	0	1	2	3	4	5	6	7	8	9
−110	−78.9	−79.4	−80.0	−80.6	−81.1	−81.7	−82.2	−82.8	−83.3	−83.9
−100	−73.3	−73.9	−74.4	−75.0	−75.6	−76.1	−76.7	−77.2	−77.8	−78.3
− 90	−67.8	−68.3	−68.9	−69.4	−70.0	−70.6	−71.1	−71.7	−72.2	−72.8
− 80	−62.2	−62.8	−63.3	−63.9	−64.4	−65.0	−65.6	−66.1	−66.7	−67.2
− 70	−56.7	−57.2	−57.8	−58.3	−58.9	−59.4	−60.0	−60.6	−61.1	−61.7
− 60	−51.1	−51.7	−52.2	−52.8	−53.3	−53.9	−54.4	−55.0	−55.6	−56.1
− 50	−45.6	−46.1	−46.7	−47.2	−47.8	−48.3	−48.9	−49.4	−50.0	−50.6
− 40	−40.0	−40.6	−41.1	−41.7	−42.2	−42.8	−43.3	−43.9	−44.4	−45.0
− 30	−34.4	−35.0	−35.6	−36.1	−36.7	−37.2	−37.8	−38.3	−38.9	−39.4
− 20	−28.9	−29.4	−30.0	−30.6	−31.1	−31.7	−32.2	−32.8	−33.3	−33.9
− 10	−23.3	−23.9	−24.4	−25.0	−25.6	−26.1	−26.7	−27.2	−27.8	−28.3
− 0	−17.8	−18.3	−18.9	−19.4	−20.0	−20.6	−21.1	−21.7	−22.2	−22.8
0	−17.8	−17.2	−16.7	−16.1	−15.6	−15.0	−14.4	−13.9	−13.3	−12.8
10	−12.2	−11.7	−11.1	−10.6	−10.0	− 9.4	− 8.9	− 8.3	− 7.8	− 7.2
20	− 6.7	− 6.1	− 5.6	− 5.0	− 4.4	− 3.9	− 3.3	− 2.8	− 2.2	− 1.7
30	− 1.1	− 0.6	0	0.6	1.1	1.7	2.2	2.8	3.3	3.9
40	4.4	5.0	5.6	6.1	6.7	7.2	7.8	8.3	8.9	9.4
50	10.0	10.6	11.1	11.7	12.2	12.8	13.3	13.9	14.4	15.0
60	15.6	16.1	16.7	17.2	17.8	18.3	18.9	19.4	20.0	20.6
70	21.1	21.7	22.2	22.8	23.3	23.9	24.4	25.0	25.6	26.1
80	26.7	27.2	27.8	28.3	28.9	29.4	30.0	30.6	31.1	31.7
90	32.2	32.8	33.3	33.9	34.4	35.0	35.6	36.1	36.7	37.2
100	37.8	38.3	38.9	39.4	40.0	40.6	41.1	41.7	42.2	42.8
110	43.3	43.9	44.4	45.0	45.6	46.1	46.7	47.2	47.8	48.3

PHYSICAL DATA

TABLE 28.—CHEMICAL ELEMENTS[1]

Element	Symbol	Atomic weight*	Valence	Element	Symbol	Atomic weight*	Valence
Aluminum	Al	26.97	3	Molybdenum	Mo	95.95	3, 4, 5, 6, 8
Antimony	Sb	121.76	3, 5	Neodymium	Nd	144.27	3
Argon[2]	A	39.94	0	Neon[2]	Ne	20.183	0
Arsenic[3]	As	74.91	3, 5	Nickel	Ni	58.69	2, 3, 4
Barium	Ba	137.36	2	Nitrogen[5]	N	14.008	3, 5
Beryllium	Be	9.02	2	Osmium	Os	190.2	2, 3, 4, 6, 8
Bismuth	Bi	209.00	3, 5	Oxygen[5]	O	16.000	2
Boron[3]	B	10.82	3	Palladium	Pd	106.7	2, 4
Bromine[4]	Br	79.916	1, 3, 5	Phosphorus[3]	P	31.02	3, 5
Cadmium	Cd	112.41	2	Platinum	Pt	195.23	2, 4
Cesium	Cs	132.91	1	Polonium	Po	(210)	2, 4
Calcium	Ca	40.08	2	Potassium	K	39.096	1
Carbon[3]	C	12.010	2, 4	Praseodymium	Pr	140.92	3
Cerium	Ce	140.13	3, 4	Protactinium	Pa	231	5
Chlorine[5]	Cl	35.457	1, 3, 5, 7	Radium	Ra	226.05	2
Chromium	Cr	52.01	2, 3, 6	Radon[9] (radium emanation)	Rn	222	0
Cobalt	Co	58.94	2, 3	Rhemium	Re	186.31	1, 4, 7
Columbium (Niobium)	Cb	92.91	2, 3, 4, 5	Rhodium	Rh	102.91	3, 4
Copper	Cu	63.57	1, 2	Rubidium	Rb	85.48	1
Dysprosium	Ds	162.46	3	Ruthenium	Ru	101.7	3, 4, 6, 8
Erbium	Er	167.2	3	Samarium	Sa	150.43	3
Europium	Eu	152.0	2, 3	Scandium	Sc	45.10	3
Fluorine[6]	F	19.00	1	Selenium[3]	Se	78.96	2, 4, 6
Gadolinium	Gd	156.9	3	Silicon[3]	Si	28.06	4
Gallium	Ga	69.72	2, 3	Silver	Ag	107.880	1
Germanium	Ge	72.60	2, 4	Sodium	Na	22.997	1
Gold	Au	197.2	1, 3	Strontium	Sr	87.63	2
Hafnium	Hf	178.6	4	Sulfur[3]	S	32.06	2, 4, 6
Helium[2]	He	4.003	0	Tantalum	Ta	130.88	4, 5
Holmium	Ho	163.5	3	Tellurium[3]	Te	127.61	2, 4, 6
Hydrogen[7]	H	1.0081	1	Terbium	Tb	159.2	3
Indium	In	114.76	1, 2, 3	Thallium	Tl	204.39	1, 3
Iodine[3]	I	126.92	1, 3, 5, 7	Thorium	Th	232.12	4
Iridium	Ir	193.1	2, 3, 4, 6	Thulium	Tm	169.4	3
Iron	Fe	55.84	2, 3	Tin	Sn	118.70	2, 4
Krypton[2]	Kr	83.7	0	Titanium	Ti	47.9	3, 4
Lanthanum	La	138.92	3	Tungsten	W	183.92	3, 4, 5, 6
Lead	Pb	207.21	2, 4	Uranium	U	238.07	4, 6, 8
Lithium[8]	Li	6.940	1	Vanadium	V	50.95	1, 2, 3, 4, 5
Lutecium	Lu	175.0	3	Xenon[2]	Xe	131.3	0
Magnesium	Mg	24.32	2	Ytterbium	Yb	173.04	2, 3
Manganese	Mn	54.93	2, 3, 4, 6, 7	Yttrium	Yt	88.92	3
Mercury	Hg	200.61	1, 2	Zinc	Zn	65.38	2
				Zirconium	Zr	91.22	4

* The atomic weights are based upon oxygen having a weight of 16.000 by definition.
[1] All the elements are metals, except as otherwise indicated.
[2] Inert gas.   [3] Metalloid.   [4] Liquid.   [5] Gas.
[6] Most active gas.   [7] Lightest gas.   [8] Lightest metal.   [9] Not placed.
From Marks, "Mechanical Engineers' Handbook," 4th ed., McGraw-Hill, New York, 1941.

TABLE 29.—APPROXIMATE DENSITIES AND SPECIFIC GRAVITIES*
Water at 39°F and normal atmospheric pressure taken as unity

Substance	Specific gravity	Density, lb/ft³	Substance	Specific gravity	Density, lb/ft³
Aluminum	2.55–2.80	165	Acid, nitric (91%)	1.50	94
Brass	8.60	535	Acid, sulfuric (87%)	1.80	112
Bronze	8.0	509	Chloroform	1.500	95
Copper	8.9	556	Ether	0.736	46
German silver	8.60	536	Oils, mineral	0.9	57
Gold	19.3	1,205	Oils, vegetable	0.95	58
Iron	7.7	485			
Lead	11.34	710	Ashes	0.7	42
Mercury (4°C)	13.5852 (gr/cm³)	847	Earth, dry, loose	1.2	76
			Lime	0.9	60
Nickel	8.9	537	Portland cement	3	196
Platinum	21.5	1,330	Rubble masonry	2	135
Silver	10.5	656	Sand, loose	1.5	98
Tin	7.3	459	Slags	1.5	98
Tungsten	19.2	1,200			
Zinc	7.0	440	Excavations in water:		
			Clay	1.28	80
Glass, crown	2.45–2.72	161	River mud	1.44	90
Crystal	2.90–3.00	184	Sand or gravel	0.96	60
Flint	3.2 –4.7	247			
Leather	0.86–1.02	59	Asbestos	2.1–2.8	153
Paper	0.70–1.15	58	Chalk	1.8–2.8	143
Rubber, caoutchouc	0.92–0.96	59	Clay	1.8–2.6	137
Rubber goods	1–2	94	Granite	2.6–2.7	165
Salt, granulated, piled	0.77	48	Limestone	2.1–2.86	155
Timber, ash	0.55	34	Sandstone	2.0–2.6	143
Oak	0.80	50	Shale	2.6–2.9	172
Pine	0.48	30			
Teak	0.99	62	Coal, hard	1.6	90
			Gasoline	0.70–0.75	45
Alcohol, ethyl	0.789	49	Petroleum	0.87	54
Acid, muriatic (40%)	1.20	75			

* Condensed from Marks "Mechanical Engineers' Handbook," 4th ed., McGraw-Hill, New York, 1941.

TABLE 30.—FREEZING POINTS OF LIQUIDS AT ATMOSPHERIC PRESSURE, IN DEGREES FAHRENHEIT*

Alcohol, methyl	−148	Linseed oil	− 4
Ammonia	−108	Mercury	− 38
Carbon bisulfide	−168	Turpentine	+ 14
Chloroform	− 82	Sulfuric acid	−105
Ether, ethyl	−180	Sea water	+ 28
Glycerin	− 40	Toluene	−149

* From O'Rourke, "General Engineering Handbook," 2d ed., McGraw-Hill, New York, 1940.

TABLE 31.—MELTING POINTS OF VARIOUS SOLIDS, IN DEGREES FAHRENHEIT*

Aluminum	1200	Lead	622
Brasses:		Mercury	−38
80 Cu + 20 Zn	1845	Nickel	2646
50 Cu + 50 Zn	1615	Paraffin	129
20 Cu + 80 Zn	1300	Phosphorus	111
Bronze, 90 Cu + 10 Sn	1840	Platinum	3190
Cast iron, gray	2500	Porcelain	2820
	(approx.)	Silver	1761
Cast iron, white	2000	Steel (about)	2400
	(approx.)	Tin	449
Copper	1981	Tin solder	275 to 350
Gold	1945	Wrought iron	2500
Iron	2795		(approx.)

* From O'Rourke, "General Engineering Handbook," 2d ed., McGraw-Hill, New York, 1940.

TABLE 32.—COEFFICIENTS OF LINEAR EXPANSION PER DEGREE FAHRENHEIT, ORDINARY ROOM TEMPERATURES*
The values given below should be divided by 10^6

Material	Coefficient	Material	Coefficient
Aluminum	12.8	Lead	16.2
Antimony	6.3	Masonry	2.5 to 5.0
Bismuth	7.4	Nickel	7.1
Brasses, bronzes	9.3 to 12.0	Paraffin wax	61
Carbon coke	3.0	Platinum	4.9
Carbon graphite	4.4	Porcelain	2
Cement and concrete	5.5 to 8.0	Rubber	43
Copper	9.2	Sandstone	3.9 to 6.7
Glass	3.3 to 5.0	Silver	10.5
Gold	7.9	Slate	3.3 to 5.5
Granite	4 to 5	Solder	13.4
Ice	28	Steel	5.6 to 7.3
Iridium	3.6	Tin	11.0
Iron	5.5 to 6.7	Wood, with grain	2 to 5
		Zinc	18.0

* From O'Rourke, "General Engineering Handbook," 2d ed., McGraw-Hill, New York, 1940.

TABLE 33.—COEFFICIENTS OF CUBICAL EXPANSION PER DEGREE FAHRENHEIT AT
ORDINARY ROOM TEMPERATURES*
The values given below should be divided by 10^3

Alcohol, ethyl...............	0.61	Petroleum, California........	0.43
Alcohol, methyl..............	0.80	Petroleum, Pennsylvania......	0.50
Calcium chloride ($CaCl_2$) solu-		Petroleum, Texas............	0.42
tion, 5 to 50 per cent.......	0.28	Turpentine.................	0.54
Glycerin....................	0.28	Water......................	0.115
Mercury....................	0.10	Ice........................	0.62

* Condensed from Marks, "Mechanical Engineers' Handbook," 4th ed., McGraw-Hill, New York, 1941.

TABLE 34.—SPECIFIC HEATS OF SOLIDS, AND LIQUIDS,* BTU PER POUND DEGREE
FAHRENHEIT

Material	Specific heat	Material	Specific heat	Material	Specific heat
Solids		Ice.............	0.500	Liquids	
Aluminum.......	0.218	Iron, pure.......	0.110	Acetic acid.....	0.510
Antimony........	0.052	Iron, cast.......	0.130	Alcohol, ethyl...	0.650
Asbestos.........	0.200	Iron, wrought...	0.110	Alcohol, methyl..	0.590
Ashes...........	0.200	Lead............	0.031	Ammonia (liq. 0°	
Bismuth.........	0.030	Marble..........	0.210	C).............	1.012
Brass...........	0.090	Mica...........	0.208	Benzine.........	0.450
Brickwork.......	0.220	Nickel..........	0.108	Chloroform (liq.	
Bronze..........	0.104	Paraffin.........	0.589	30°C).........	0.235
Carbon graphite..	0.160	Platinum........	0.032	Ether..........	0.540
Cement, portland	0.271	Porcelain.......	0.255	Gasoline........	0.530
Chalk..........	0.220	Rubber, hard....	0.339	Glycerin........	0.580
Charcoal........	0.200	Sand...........	0.190	Hydrochloric acid	0.600
Cinders.........	0.180	Silver..........	0.056	Kerosene........	0.470
Clay...........	0.190	Steel...........	0.120	Lead...........	0.030
Cobalt..........	0.103	Stone...........	about	Mercury........	0.030
Coal...........	0.240		0.200	Olive oil........	0.400
Copper.........	0.093	Tin............	0.054	Petroleum......	0.504
Cork...........	0.485	Tungsten........	0.034	Sea water......	0.940
Glass...........	0.160 to	Wood...........	0.450 to	Sulfuric acid....	0.336
	0.200		0.650	Turpentine......	0.420
Gold...........	0.032	Wool...........	0.393	Water (20°C)....	1.000
Gypsum........	0.259	Zinc...........	0.094		

* From O'Rourke, "General Engineering Handbook," 2d ed., McGraw-Hill, New York, 1940.

TABLE 35.—LATENT HEAT OF FUSION, BTU PER POUND*

Aluminum......	128.0	Lead..........	9.8	Zinc..........	46.8
Copper.........	78.0	Mercury.......	5.0	Glycerin.......	76.5
Iron, gray cast..	40.0	Nickel.........	133.0	Ice............	144.0
Iron, white cast.	60.0	Tin............	25.4	Sulfur.........	15.8

* Condensed from Marks, "Mechanical Engineers' Handbook," 4th ed., McGraw-Hill, New York, 1941.

TABLE 36.—LATENT HEAT OF VAPORIZATION AT ATMOSPHERIC PRESSURE, BTU PER POUND*

Water.........	970.2	Ethyl alcohol (98%)	406.0	Mercury....	122.2
Ammonia......	589.3	Sulfur dioxide.......	167.1	Sulfur.......	120.0
Methyl alcohol.	482.0	Gasoline...........	133.0–145.0	Kerosene....	105.0–110.0

* Condensed from Marks, "Mechanical Engineers' Handbook," 4th ed., McGraw-Hill, New York, 1941.

TABLE 37.—PROPERTIES OF GASES*

Gas	Chemical symbol	Number of atoms	Approx. molecular weight	Density relative to air	Specific heat† at 32°F, Btu/lb/°F		$k = \dfrac{C_p}{C_v}$	Gas constant R
					C_p	C_v		
Air...............	(Mech. mixture)		(29)	1.000	0.240	0.171	1.40	53.3
Carbon dioxide.....	CO_2	3	44	1.520	0.203	0.158	1.28	35.0
Carbon monoxide..	CO	2	28	0.968	0.248	0.177	1.40	55.3
Helium............	He	1	4	0.137	1.251	0.754	1.66	387
Hydrogen..........	H_2	2	2	0.070	3.14	2.16	1.46	762
Methane..........	CH_4	5	16	0.554	0.540	0.416	1.30	96.5
Nitrogen..........	N_2	2	28	0.970	0.248	0.177	1.40	55.3
Oxygen...........	O_2	2	32	1.105	0.217	0.155	1.40	48.2

* From O'Rourke, "General Engineering Handbook," 2d ed., McGraw-Hill, New York, 1940.
† For the variation of specific heat with temperature, see Table 38.

TABLE 38.—INSTANTANEOUS SPECIFIC HEATS OF GASES, IN BTU PER POUND DEGREE FAHRENHEIT*

$$C_p = a' + bT + cT^2$$
$$C_v = a + bT + cT^2$$

in which T is the absolute temperature in degrees Rankine

Gas	a'	a	b	c
Air...........................	0.239	0.1705	0	$4.138(10)^{-9}$
Carbon dioxide:				
$T < 2900$.................	0.1625	0.1175	$8.86(10)^{-5}$	$-1.36(10)^{-8}$
$T > 2900$.................	0.277	0.232	$9.55(10)^{-6}$	0
Carbon monoxide.............	0.2475	0.1765	0	$4.28(10)^{-9}$
Hydrogen....................	2.98	2.00	$3.31(10)^{-4}$	0
Methane....................	0.216	0.092	$6.59(10)^{-4}$	0
Nitrogen....................	0.2475	0.1765	0	$4.28(10)^{-9}$
Oxygen....................	0.216	0.154	0	$3.75(10)^{-9}$

* From O'Rourke, "General Engineering Handbook," 2d ed., McGraw-Hill, New York, 1940.

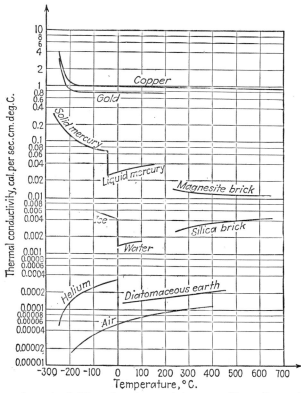

Fig. 1.—Thermal conductivities of various substances. (*From Zemansky, "Heat and Thermodynamics," McGraw-Hill, New York*, 1943.)

TABLE 39.—SPECIFIC THERMAL CONDUCTIVITIES OF METALS*

$$k = k_0 + at$$

Metal	k_0 Btu per hr per °F per sq ft for a thickness of 1 in.	a
Aluminum	1380	0.3
Brass (90-10)	700	0.6
Brass (70-30)	520	0.2
Cast iron	350	−0.15
Copper (pure)	2700	−0.24
Lead	240	−0.1
Nickel	400	−0.03
Silver	3000	−0.2
Steel (mild)	450	−0.2
Tin	450	−0.2
Wrought iron	430	−0.1
Zinc	790	−0.15

* From O'Rourke, "General Engineering Handbook," 2d ed., McGraw-Hill, New York, 1940.

TABLE 40.—SPECIFIC THERMAL CONDUCTIVITIES OF LIQUIDS*

$$k = k_0 + at$$

Liquid	k_0 Btu per hr per °F per sq ft for 1 in	a
Alcohol, ethyl	1.3	−0.0005
Alcohol, methyl	1.6	−0.001
Brine (25 % CaCl₂)	3.3	0.006
Brine (25 % NaCl)	2.7	0.006
Kerosene	1.1	−0.0008
Petroleum oil	1.0	−0.0002
Water	3.6	0.007

* From O'Rourke, "General Engineering Handbook," 2d ed., McGraw-Hill, New York, 1940.

TABLE 41.—SPECIFIC THERMAL CONDUCTIVITIES OF GASES AND VAPORS*

$$k = k_{32}\left(\frac{492 + C}{T + C}\right)\left(\frac{T}{492}\right)^{1.5}$$

Gas or vapor	k_{32} Btu per hr per °F per sq ft for a thickness of 1 in.	C
Air	0.155	225
Ammonia	0.139	
Hydrogen	1.1	169
Methyl chloride	0.058	
Nitrogen	0.157	205
Oxygen	0.161	259
Steam	0.1	1850

* From O'Rourke, "General Engineering Handbook," 2d ed., McGraw-Hill, New York, 1940.

TABLE 42.—RESISTIVITIES AND TEMPERATURE COEFFICIENTS OF METALS*
To obtain the resistivity in microhms for 1 cm³, divide by 6.01

Material	Resistivity at 0°C, circular mil-ft	Temp. coefficient per °C, at 20°C
Aluminum	17.1	0.00390
Copper, annealed	9.35	0.00393
Iron, pure	53.00	0.00600
Gold	12.36	0.00365
Lead	115.00	0.00390
Magnesium	30.00	0.00381
Mercury	564.00	0.00072
Nickel	41.60	0.00500
Platinum	66.00	0.00370
Silver	8.85	0.00400
Tantalum	87.60	0.00330
Tin	78.00	0.00365
Tungsten (hard drawn)	33.00	0.00320
Zinc	34.50	0.00400

* From O'Rourke, "General Engineering Handbook," 2d ed., McGraw-Hill, New York, 1940.

TABLE 43.—RESISTIVITIES AND TEMPERATURE COEFFICIENTS OF LOW-TEMPERATURE-
COEFFICIENT ALLOYS*

Material	Resistivity at 20°C, circular mil-ft	Temp. coefficient per °C, at 20°C
Copper-nickel......................	100 to 250	0.000005 to 0.0004
Copper-nickel-zinc (nickel silver).......	200 to 290	0.0002 to 0.00027
Iron-nickel........................	200 to 700	0.00034 to 0.001
Iron-nickel-chromium................	520 to 720	0.00016 to 0.00072
Copper-manganese-nickel.............	249 to 270	0.000025 to 3 × 10⁻⁵

* From O'Rourke, "General Engineering Handbook," 2d ed., McGraw-Hill, New York, 1940.

TABLE 44.—POTENTIAL GRADIENT AT VARIOUS ALTITUDES*

z, km............................	0	0.5	1.5	3	6	9
Gradient, volts/m................	130	50	30	20	10	5

* After Schweidler.

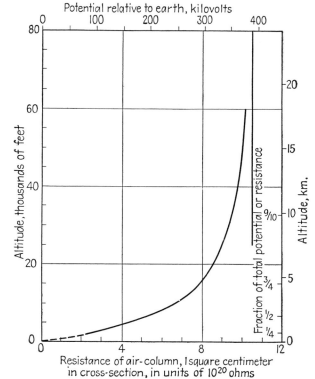

FIG. 2.—Potential and resistance between the earth and points in the atmosphere,
flight of Explorer II, Nov. 11, 1935. (*From Fleming, "Terrestrial Magnetism and Electricity," McGraw-Hill, New York, 1939.*)

TABLE 45.—CRITICAL DATA*

Substance	Critical temp., °C	Critical press., atm	Critical density, gram/cm³
Water	374	218	0.40
Sulfur dioxide	157	78	0.52
Carbon dioxide	31	73	0.46
Oxygen	−119	50	0.43
Nitrogen	−147	34	0.31
Hydrogen	−240	13	0.03
Helium	−268	2.3	0.07

* From Zemansky, "Heat and Thermodynamics," 2d ed., McGraw-Hill, New York, 1943.

TABLE 46.—TRIPLE-POINT DATA*

Substance	Centigrade temp.	Press., mm
Water	0.0098	4.579
Sulfur dioxide	− 72.7	16.3
Carbon dioxide	− 56.6	3880
Oxygen	−218.4	2.0
Argon	−189.2	512.2
Nitrogen	−209.8	96.4
Hydrogen	−259.1	51.4
Helium (lower triple point)	−271.0	38.65

* From Zemansky, "Heat and Thermodynamics," 2d ed., McGraw-Hill, New York, 1943.

TABLE 47.—CENTIGRADE TEMPERATURES OF FIXED POINTS*

Fixed points	Substance	Designation	Temp., °C
Standard	Ice	N.M.P. (ice point)	0.000
	Water	N.B.P. (steam point)	100.000
Basic	Oxygen	N.B.P. (oxygen point)	−182.97
	Sulfur	N.B.P. (sulfur point)	444.60
	Antimony	N.M.P. (antimony point)	630.50
	Silver	N.M.P. (silver point)	960.5
	Gold	N.M.P. (gold point)	1063.0
Secondary	Hydrogen	N.B.P.	−252.78
	Nitrogen	N.B.P.	−195.81
	Mercury	N.M.P.	− 38.86
	Sodium sulfate	Transition point	32.38
	Naphthalene	N.B.P.	217.96
	Tin	N.M.P.	231.85
	Benzophenone	N.B.P.	305.9
	Cadmium	N.M.P.	320.9
	Lead	N.M.P.	327.3
	Zinc	N.M.P.	419.5

* From Zemansky, "Heat and Thermodynamics," 2d ed., McGraw-Hill, New York, 1943.

TABLE 48.—VELOCITY OF SOUND IN VARIOUS GASES AT 0°C

For any given gas the velocity increases as the square root of the absolute temperature

Gas	Mps
Argon	308
Air	331
Helium	971
Nitrogen	378
Water vapor	401
Hydrogen	1,261
Carbon dioxide	258

TABLE 49.—VELOCITY OF SOUND IN PURE AND IN SEA WATER (35°/oo) (1,013 MB)*

Temp., °C	Pure water, mps	Salt water, mps
0	1,400	1,445
10	1,445	1,485
20	1,480	1,520
30	1,505	1,545

* After Sverdrup.

TABLE 50.—VELOCITY OF SOUND IN DRY AIR

Temp.	Velocity, mps
−70	286
−60	292
−50	299
−40	306
−30	312
−20	319
−10	325
0	331
10	337
20	342
30	348
40	354
50	360
60	366
70	371
80	376
90	382
100	386

TABLE 51.—COEFFICIENTS OF DIFFUSION*

	$Cm^2 Sec^{-1}$
Air vs. oxygen	0.178
Air vs. hydrogen	0.661
Air vs. water vapor	0.24
Carbon dioxide vs. oxygen	0.180
Carbon dioxide vs. hydrogen	0.56
Hydrogen vs. oxygen	0.72
Argon vs. helium	0.706

* After Linke.

TABLE 52.—HEAT CONDUCTIVITY AND COEFFICIENT OF VISCOSITY OF AIR*

Temp., °C	Conductivity, cal cm^{-1} sec^{-1} gram^{-1}	Viscosity, gram cm^{-1} sec^{-1}	Temp., °C	Conductivity, cal cm^{-1} sec^{-1} gram^{-1}	Viscosity, gram cm^{-1} sec^{-1}
−70	0.000 0422	0.000 172	10	0.000 0549	0.000 177
−50	0453	147	20	0565	181
−30	0485	152	30	0581	186
−10	0517	162	40	0597	191
0	0533	172			

* After Linke.

TABLE 53.—COEFFICIENTS OF DYNAMIC FRICTION

Glass on glass...................................... 0.40
Carbon on glass.................................... 0.18
Wood on wood...................................... 0.25 to 0.50
Lubricated metal surfaces.......................... 0.05 to 0.2
Steel on ice....................................... 0.02
Rubber on dry concrete............................. 0.5 to 0.8
Leather on metals.................................. 0.6   ..

TABLE 54.—COEFFICIENTS OF VISCOSITY

Substance	gram cm^{-1} sec^{-1}	Temp., °C
Air..............................	0.0181×10^{-2}	20
Hydrogen.........................	0.0097×10^{-2}	20
Mercury..........................	1.56×10^{-2}	20
Oil, machine.....................	$100\text{–}600 \times 10^{-2}$	20
Water............................	1.7938×10^{-2}	0
Water............................	1.0087×10^{-2}	20
Water............................	0.2839×10^{-2}	100

TABLE 55.—ELASTIC MODULI
Average values of elastic constants, mb

Material	Bulk modulus k	Modulus of rigidity n	Young's modulus y
Aluminum................	7.6×10^8	2.4×10^8	6.0×10^8
Copper...................	14.3×10^8	4.2×10^8	12.0×10^8
Steel....................	18.1×10^8	8.0×10^8	20.0×10^8
Lead.....................	5.0×10^8	0.54×10^8	1.6×10^8
India rubber.............		0.00016×10^8	0.05×10^8
Crown glass..............	5.0×10^8	2.9×10^8	7.1×10^8
Oak.....................			1.3×10^8

TABLE 56.—SURFACE TENSIONS OF LIQUIDS IN CONTACT WITH AIR AT SPECIFIED
TEMPERATURES

Substance	Dynes/cm	°C
Aluminum (molten)	520	750
Benzene	28.9	20
Carbon tetrachloride	26.8	20
Ether	17.0	20
Ethyl alcohol	22.3	20
Mercury	476	20
Soap solution	28	20
Water	72.75	20

TABLE 57.—BLACK-BODY RADIATION

$E = \sigma T^4$, where $\sigma = 0.826 \times 10^{-10}$ cal cm^{-2} min^{-1}

Temp., °C	0	1	2	3	4	5	6	7	8	9
−40	0.244	239	235	231	227	223	219	216	212	208
−30	288	283	279	274	270	265	261	256	252	248
−20	339	333	328	323	318	312	308	303	298	293
−10	395	389	384	378	372	366	360	355	349	344
− 0	459	452	446	439	432	426	420	414	407	401
+ 0	459	466	472	479	486	493	500	508	515	522
+10	530	537	545	553	560	568	576	584	592	601
+20	609	617	626	634	643	651	660	669	678	687
+30	696	706	715	724	734	743	753	762	772	782

$t =$	−50	−55	−60	−65	−70°C		
$E =$	0.204	0.187	0.170	0.155	0.140		
$t =$	40	45	50	55	60	65	70°C
$E =$	0.792	0.844	0.899	0.956	1.015	1.078	1.143

TABLE 58.—INTENSITY OF RADIATION FROM A BLACK BODY (UNIT ANGLE)*

In 10^2 cal cm^{-2} min^{-1} for $\Delta\lambda = 1\ \mu$

$1\ \mu = 10^{-4}$ cm

λ $\begin{array}{l}T=\\t=\end{array}$	313 40	293 20	273 0	253 −20	233 −60	213 −60	193°K −80°C
3 μ	0.017	0.006	0.002	0.000	0	0	0
4	0.181	0.083	0.034	0.012	0.004	0.001	0
5	0.580	0.311	0.152	0.066	0.025	0.008	0.002
6	1.069	0.636	0.322	0.176	0.078	0.030	0.009
7	1.469	0.941	0.564	0.312	0.156	0.068	0.025
8	1.706	1.155	0.738	0.440	0.238	0.116	0.049
9	1.792	1.265	0.849	0.535	0.310	0.163	0.076
10	1.765	1.289	0.899	0.593	0.365	0.205	0.102
11	1.669	1.252	0.902	0.618	0.397	0.234	0.124
12	1.537	1.179	0.872	0.615	0.409	0.253	0.141
13	1.314	1.088	0.822	0.594	0.408	0.261	0.152
14	1.246	0.989	0.761	0.563	0.396	0.261	0.158
16	0.980	0.798	0.632	0.484	0.355	0.245	0.158
18	0.765	0.634	0.514	0.404	0.305	0.219	0.148
20	0.597	0.502	0.414	0.332	0.257	0.190	0.133
22	0.468	0.408	0.333	0.271	0.214	0.162	0.117
24	0.371	0.319	0.269	0.222	0.178	0.137	0.101
26	0.296	0.256	0.218	0.182	0.148	0.116	0.087
28	0.238	0.208	0.178	0.150	0.123	0.098	0.075
30	0.192	0.170	0.146	0.124	0.103	0.083	0.064
34	0.131	0.116	0.101	0.087	0.073	0.060	0.047
40	0.077	0.069	0.061	0.051	0.045	0.038	0.031
50	0.036	0.033	0.029	0.026	0.022	0.019	0.016
60	0.019	0.017	0.016	0.014	0.012	0.011	0.009
70	0.011	0.010	0.009	0.008	0.007	0.006	0.005
80	0.007	0.006	0.006	0.005	0.004	0.004	0.003
100	0.003	0.003	0.002	0.002	0.002	0.002	0.002

* After Planck.

TABLE 59.—APPROXIMATE EMISSIVITIES OF VARIOUS SURFACES, AS COMPILED BY HOTTEL

Values at intermediate temperatures may be obtained by linear interpolation

Material	Temp. range, °C	Emissivity
Polished metals:		
Aluminum	250–600	0.039–0.057
Brass	250–400	0.033–0.037
Chromium	50–550	0.08–0.26
Copper	100	0.018
Iron	150–1000	0.05–0.37
Nickel	20–350	0.045–0.087
Zinc	250–350	0.045–0.053
Filaments:		
Molybdenum	750–2600	0.096–0.29
Platinum	30–1200	0.036–0.19
Tantalum	1300–3000	0.19–0.31
Tungsten	30–3300	0.032–0.35
Other materials:		
Asbestos	40–350	0.93–0.95
Ice (wet)	0	0.97
Lampblack	20–350	0.95
Rubber (gray)	25	0.86

TABLE 60.—BEAUFORT SCALE FOR WIND VELOCITIES

Winds correspond to those at 33 ft above the surface in an open position. Extension from force 11 in accordance with the Beaufort formula.

Velocity (mph) = 1.87 $\sqrt{F^3}$, where F is Beaufort number

Beaufort wind force	Description	Miles (statute) per hour	Knots	Mps*
0	Calm	Less than 1	Less than 1	Less than 0.4
1	Light air	1–3	1–3	0.4–1.5
2	Light breeze	4–7	4–6	1.6–3.3
3	Gentle breeze	8–12	7–10	3.4–5.4
4	Moderate breeze	13–18	11–16	5.5–7.9
5	Fresh breeze	19–24	17–21	8.0–10.7
6	Strong breeze	25–31	22–27	10.8–13.8
7	Moderate gale	32–38	28–33	13.9–17.1
8	Fresh gale	39–46	34–40	17.2–20.7
9	Strong gale	47–54	41–47	20.8–24.4
10	Whole gale	55–63	48–55	24.5–28.4
11	Storm	64–73	56–63	28.5–33.5
12	Hurricane	74–82	64–71	
13		83–92	72–80	
14		93–103	81–89	
15		104–114	90–99	
16		115–125	100–109	
17		126–136	110–118	

* *War Department Technical Manual*, TM1-235.

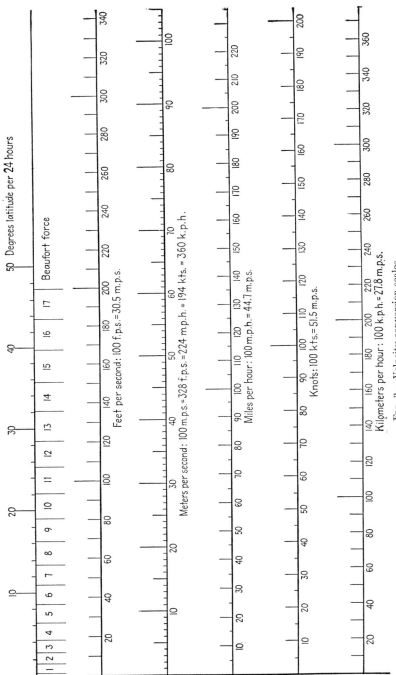

Fig. 3.—Velocity conversion scales.

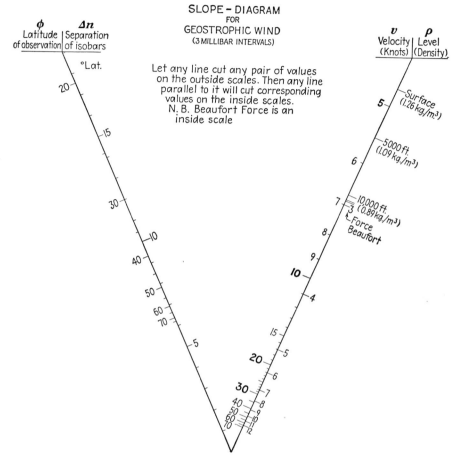

Fɪɢ. 4.—Slope diagram for geostrophic wind. Valid for any latitude, pressure gradient, and air density. (*Courtesy of J. Fulford.*)

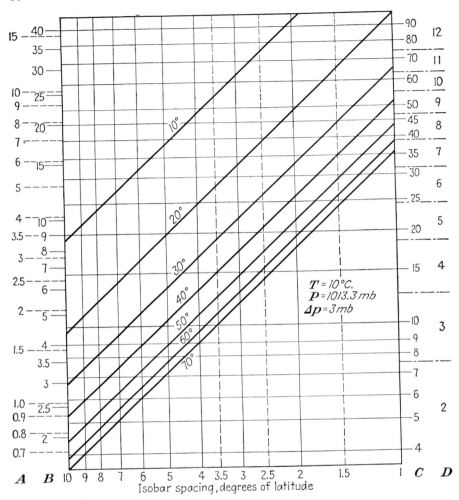

FIG. 5.—Key:
A, wind speed in degrees latitude per 12 hr
B, wind speed, mps
C, wind speed, mph
D, Beaufort number
Geostrophic wind scale for sea-level, 3-mb isobars. (*After University of Chicago.*)

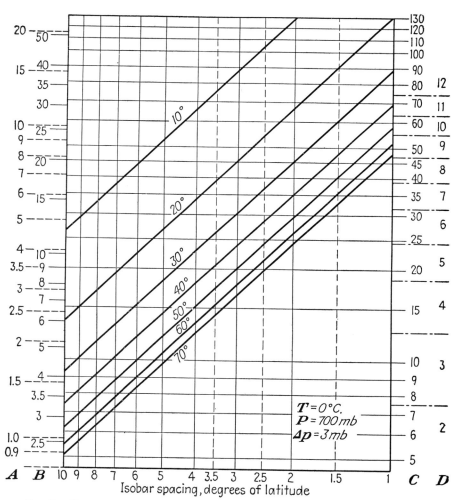

Fig. 6.—Key:

        A, wind speed in degrees latitude per 12 hr
        B, wind speed, mps
        C, wind speed, mph
        D, Beaufort number
Geostrophic wind scale for 10,000 ft-level, 3-mb isobars.   (*After University of Chicago.*)

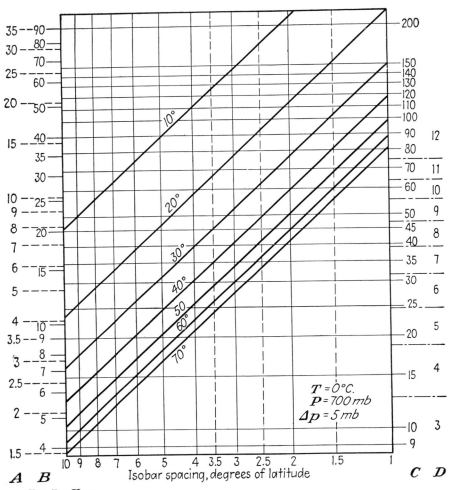

Fig. 7.—Key:

     A, wind speed in degrees latitude per 12 hr
     B, wind speed, mps
     C, wind speed, mph
     D, Beaufort number
Geostrophic wind scale for 10,000 ft-level, 5-mb isobars.   (*After University of Chicago.*)

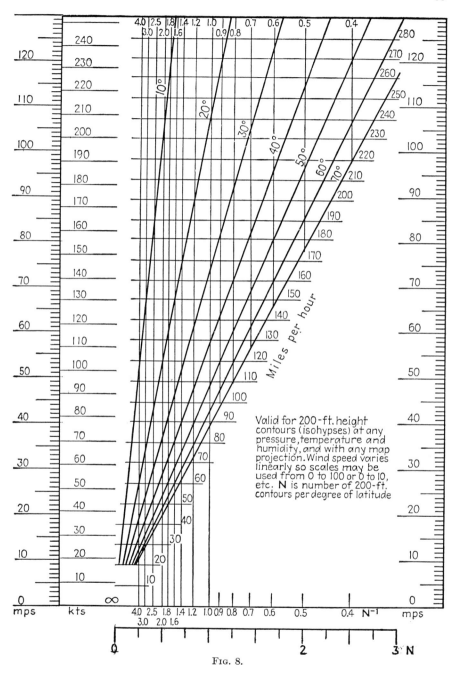

Valid for 200-ft. height contours (isohypses) at any pressure, temperature and humidity, and with any map projection. Wind speed varies linearly so scales may be used from 0 to 100 or 0 to 10, etc. N is number of 200-ft. contours per degree of latitude

Fig. 8.

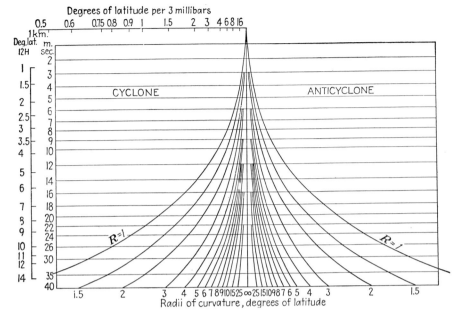

Degrees of latitude per 3 millibars

CYCLONE   ANTICYCLONE

Radii of curvature, degrees of latitude

GRADIENT WIND NOMOGRAM

A. Place straightedge so that it runs from latitude
   to observed wind speed at 1 km. (Horizontal lines)
B. Straightedge should cross curve representing radius of
   curvature of isobar in degrees of latitude
C. Upper scale gives pressure gradient in degrees of latitude per 3 mb.
D. Scales at left give wind speed and wind travel at 1 km.
E. Right half of nomogram gives two wind speeds for
   anticyclone. Lower value is used

$$2\omega v \sin\phi - \frac{1dp}{\rho dn} = \pm\frac{v^2}{r}$$

2 mps = 1 knot +3%
2 mps = 1 mph −10%

Latitude

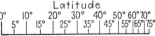

Fig. 9.—Nomogram for gradient wind, 3-mb isobars. (*Courtesy of D. F. Rex and H. V. Church.*)

Gradient-wind Isopleths

The families of isopleths in Figs. 11 to 26 give the gradient wind at every 5° latitude from 25 to 60°. The gradient-wind equation is

$$v_{gr} = \frac{\text{pressure gradient}}{\rho\lambda} \pm \frac{v_{gr}^2}{\lambda r}$$

where v_{gr} = gradient wind
   ρ = density of the air
   λ = Coriolis parameter ($2\omega \sin \varphi$)
   r = radius of curvature of the path
   $-$ = cyclonic curvature
   $+$ = anticyclonic curvature

The isopleths are valid for $\rho = 1.11$ kg per m³. The gradient wind varies *approximately* inversely as the density for a given pressure gradient, latitude, and radius of curvature.   The speed of the wind is shown in meters per second and by the appropriate Beaufort number.   Radius of curvature is given in degrees of latitude.   Pressure gradient is given in number of 3-mb isobars per degree of latitude.   A reciprocal scale is provided for convenient change from pressure gradient to isobar spacing.

On each chart, the geostrophic wind is given along the right-hand side of the figure ranged above the sign for an infinite radius of curvature.   This scale provides a ready estimate of the difference between the gradient and geostrophic wind.

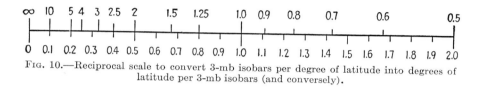

Fig. 10.—Reciprocal scale to convert 3-mb isobars per degree of latitude into degrees of latitude per 3-mb isobars (and conversely).

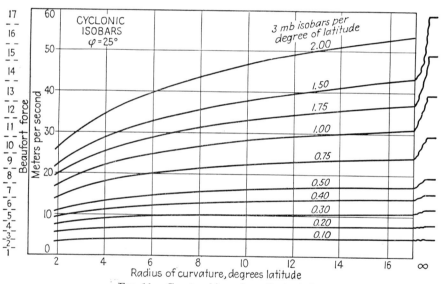

Fig. 11.—Geostrophic and gradient wind.

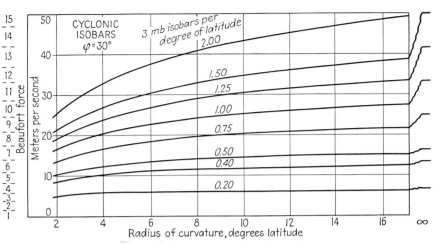

Fig. 12.—Geostrophic and gradient wind.

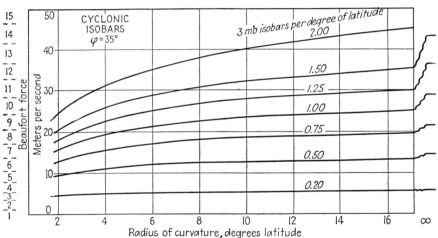

Fig. 13.—Geostrophic and gradient wind.

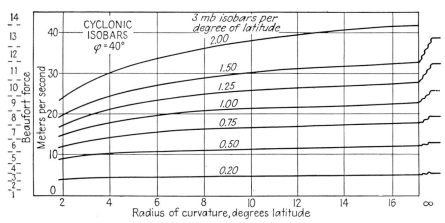

Fig. 14.—Geostrophic and gradient wind.

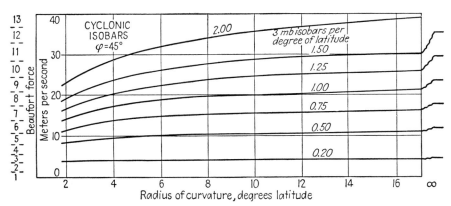

Fig. 15.—Geostrophic and gradient wind.

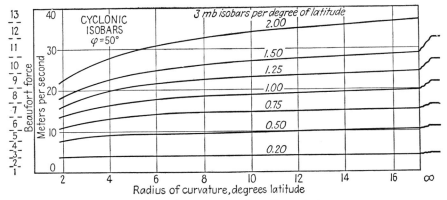

Fig. 16.—Geostrophic and gradient wind.

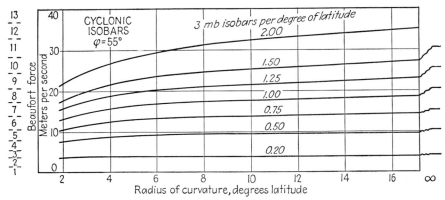

FIG. 17.—Geostrophic and gradient wind.

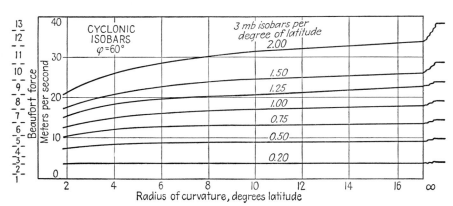

FIG. 18.—Geostrophic and gradient wind.

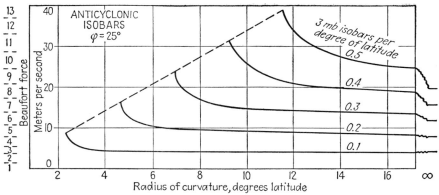

Fig. 19.—Geostrophic and gradient wind.

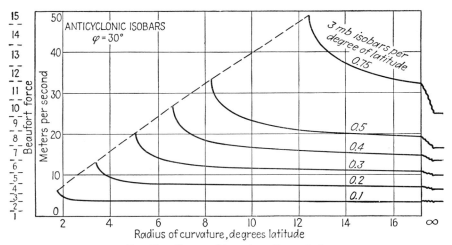

Fig. 20.—Geostrophic and gradient wind

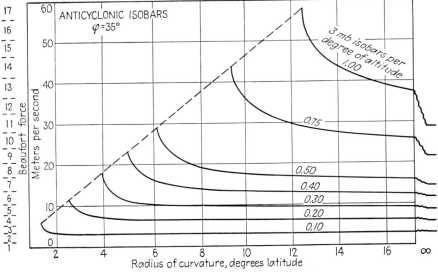

FIG. 21.—Geostrophic and gradient wind.

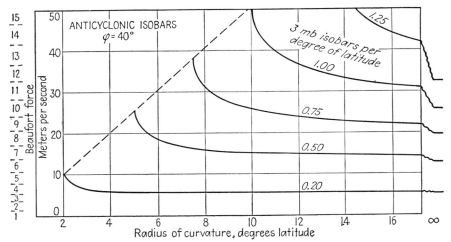

FIG. 22.—Geostrophic and gradient wind.

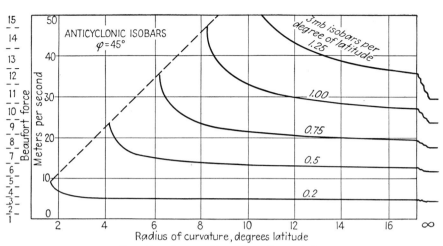

FIG. 23.—Geostrophic and gradient wind.

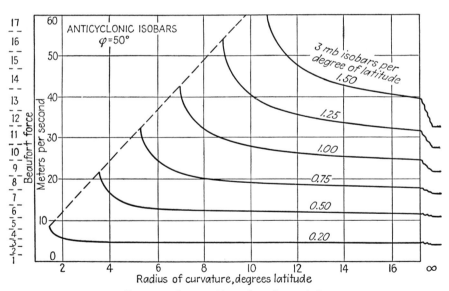

FIG. 24.—Geostrophic and gradient wind.

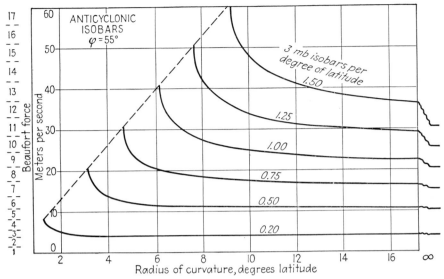

FIG. 25.—Geostrophic and gradient wind.

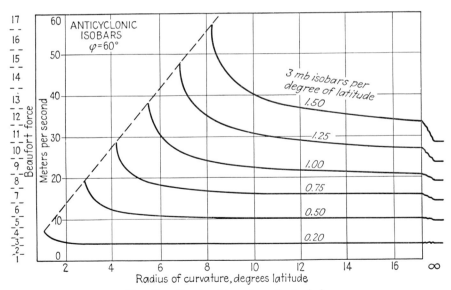

FIG. 26.—Geostrophic and gradient wind.

TABLE 61.—CORRECTIONS FOR INDICATED WIND VELOCITIES
Velocities indicated

By 3-cup anemometer, mph	By 4-cup anemometer, mph	By 4-cup anemometer with beaded cups, mph	By small airway anemometer, mph	Corrections, whole mph
0–16*	0–8	0–5	0–35	+1
17–26	9–12	6–13	35–57	0
27–35	13–16	14–20	(Corrections under	−1
36–44	17–20	21–27	higher velocities not	−2
45–52	21–24	28–34	yet determined; use	−3
53–61	25–28	35–41	zero)	−4
62–70	29–32	42–48		−5
71–79	33–36	49–55		−6
80–87	37–39	56–62		−7
88–96	40–43	63–69		−8
97–105	44–47	70–75		−9
106–114	48–51	76–82		−10
115–122	52–54	83–89		−11
123–132	55–58	90–96		−12
133–139	59–62	97–103		−13
140–149	63–65	104–110		−14
150–157	66–69	111–117		−15
158–166	70–73	118–124		−16
167–174	74–77	125–131		−17
175–184	78–80	132–138		−18
185–192	81–84	139–145		−19
193–200	85–88	146–152		−20
	89–91	153–158		−21
	92–95	159–165		−22
	96–99	166–171		−23
	100–103	172–178		−24
	104–106	179–185		−25
	107–110	186–192		−26
	111–114	193–200		−27
	115–117			−28
	118–121			−29

* Inconsequential variation from the rule for disposal of decimals disregarded as 2 and 4 mph
Reprint from Instructions No. 1, 1935.
NOTE: Corrections to be applied to wind velocities determined by anemometers. Correction to be added when the sign is plus and subtracted when the sign is minus. From *U.S. Weather Bur. Circ. N.*

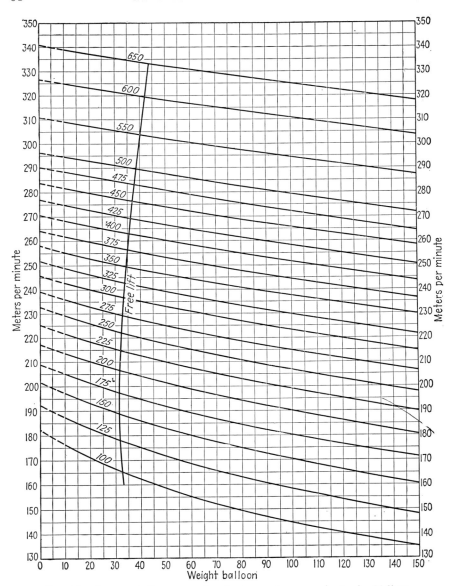

FIG. 27.—Average balloon ascensional rates, meters per minute, for Helium.

TABLE 62.—FREE LIFT FOR DEFINITE INFLATION*
Average ascensional rates given in meters per minute

Rates of ascent	140	160	180	200	220	240	260
Weight of balloons, grams	1	1	1	1	1	1	1
15	43.8	68.6	106.0	161.1	240.4		
16	44.8	69.8	107.3	162.6	242.1		
17	45.7	71.0	108.7	164.2	243.8		
18	46.7	72.2	110.0	165.7	245.5		
19	47.6	73.3	111.4	167.2	247.1		
20	48.5	74.4	112.7	168.7	248.7		
21	49.4	75.5	114.0	170.1	250.3		
22	50.3	76.5	115.3	171.6	251.9		
23	51.2	77.6	116.5	173.1	253.5		
24	52.0	78.7	117.8	174.5	255.1		
25	52.8	79.7	119.0	175.9	256.6		
26	53.6	80.7	120.2	177.3	258.2		
27	54.4	81.7	121.4	178.7	259.7		
28	55.2	82.7	122.6	180.0	261.2		
29	56.0	83.7	123.8	181.4	262.7		
30	56.8	84.7	125.0	182.7	264.2	376.9	
31	57.6	85.6	126.2	184.0	265.7	378.5	
32	58.3	86.6	127.3	185.4	267.2	380.1	
33	59.1	87.5	128.5	186.7	268.6	381.7	
34	59.8	88.5	129.6	188.0	270.1	383.3	
35	60.6	89.4	130.7	189.3	271.5	384.9	
36	61.3	90.3	131.8	190.6	273.0	386.5	
37	62.0	91.2	132.9	191.9	274.4	388.1	
38	62.7	92.1	134.0	193.1	275.8	389.6	
39	63.4	93.0	135.0	194.4	277.2	391.2	
40	64.1	93.9	136.1	195.6	278.6	392.7	547.0
41	64.8	94.8	137.1	196.8	280.0	394.2	548.6
42	65.5	95.6	138.2	198.1	281.4	395.8	550.3
43	66.2	96.5	139.2	199.3	282.8	397.3	551.9
44	66.9	97.3	140.3	200.5	284.2	398.8	553.5
45	67.6	98.2	141.3	201.7	285.5	400.3	555.1
46	68.2	99.0	142.3	202.9	286.9	401.8	556.7
47	68.9	99.9	143.3	204.1	288.2	403.3	558.3
48	69.5	100.7	144.3	205.3	289.6	404.8	559.9
49	70.2	101.6	145.4	206.5	290.9	406.2	561.5
50	70.8	102.4	146.4	207.6	292.2	407.7	563.1
51	71.4	103.2	147.4	208.8	293.5	409.2	564.7
52	72.1	104.0	148.3	209.0	294.8	410.6	566.3
53	72.7	104.8	149.3	211.0	296.1	412.1	567.8
54	73.4	105.6	150.2	212.2	297.4	413.5	569.4
55	74.0	106.3	151.2	213.3	298.7	414.9	570.9
56	74.6	107.1	152.1	214.4	300.0	416.3	572.4
57	75.2	107.9	153.1	215.6	301.3	417.7	574.0
58	75.8	108.7	154.0	216.7	302.6	419.1	575.5
59	76.4	109.4	155.0	217.8	303.8	420.5	577.0
60	77.0	110.2	155.9	218.9	305.1	421.9	578.6

* From *Aerographers' Manual*.

TABLE 63.—AVERAGE ASCENSIONAL RATES (m per min) FOR VARIOUS HELIUM FREE LIFTS (1) AND VARIOUS WEIGHT BALLOONS (W)

1	Average ascensional rates												
2,000	501.8				493.8		486.2		479.0		472.3		465.9
1,900	497.8				488.3		480.4		473.0		466.1		459.5
1,800	492.4		487.9		483.5		475.4		467.7		460.5		453.7
1,700	487.5		482.8		478.3		469.7		461.8		454.3		447.4
1,600	481.9		476.9		472.2		463.3		455.5		447.3		440.1
1,500	476.4		470.9		466.0		456.7		448.0		440.0		432.5
1,400	469.9		464.4		459.2		449.4		440.4		432.1		424.4
1,300	462.1		456.4		450.8		440.6		431.2		422.6		414.6
1,200	453.8		447.7		442.0		431.2		421.3		412.5		404.2
1,100	444.4	441.4	437.9	434.8	431.7	425.9	420.4	415.1	410.0	405.3	400.8	396.4	392.2
1,000	430.1	426.8	423.1	419.8	416.6	410.5	404.7	399.2	394.0	389.1	384.4	380.0	375.7
950	420.3	416.7	413.2	409.8	406.6	400.3	394.4	388.8	383.6	378.6	373.9	369.4	365.0
900	408.6	404.9	401.3	397.9	394.5	388.1	382.0	376.6	371.2	366.2	361.4	356.9	352.6
850	398.1	394.3	390.5	387.1	383.6	377.1	371.0	365.3	359.9	354.9	349.1	345.5	341.2
800	385.4	381.4	377.5	374.0	370.6	363.9	357.7	352.0	346.6	341.5	336.7	332.1	327.8
750	370.7	366.7	363.0	359.2	355.6	348.9	342.7	336.8	331.4	326.3	322.1	317.0	312.7
700	357.1	352.9	349.0	345.1	341.5	334.7	328.4	322.5	317.0	311.9	307.1	302.6	298.3
650	340.7	336.4	332.3	328.5	324.8	317.9	311.5	305.6	300.2	295.1	290.3	285.8	281.6
600	326.9	322.5	318.3	314.3	310.6	303.5	297.0	291.1	285.6	280.5	275.7	271.3	267.1
550	311.0	306.4	302.0	298.0	294.2	287.0	280.5	274.5	269.0	263.9	259.2	254.8	250.7
500	296.6	291.9	287.5	283.7	279.2	271.8	264.6	259.1	253.6	248.6	243.9	239.5	235.4
475	290.6	285.7	281.0	276.7	272.5	264.1	258.5	252.4	246.8	241.7	237.1	232.7	228.7
450	284.3	279.2	274.2	270.0	265.9	258.3	251.5	245.3	239.8	234.7	223.0	225.6	221.6
425	277.0	271.8	266.9	262.4	258.2	250.3	243.6	237.4	231.9	226.7	221.1	217.7	213.8
400	271.1	265.5	260.5	255.8	251.2	243.6	236.6	230.5	224.8	219.7	215.0	210.7	206.8
375	264.4	258.7	253.6	248.8	244.3	236.3	229.3	223.0	217.3	212.2	207.5	203.3	199.3
350	258.5	252.6	247.1	242.3	238.2	229.5	222.3	215.9	210.2	205.1	200.5	196.2	192.3
325	251.3	245.2	239.6	234.5	229.8	221.4	214.2	207.8	202.1	197.0	192.4	188.0	184.3
300	245.6	239.2	233.1	228..	223.0	214.6	207.1	200.7	194.9	189.8	185.2	181.0	177.1
275	238.5	231.7	225.6	219.1	213.1	206.3	198.8	192.2	186.5	181.2	176.8	172.7	168.5
250	231.5	224.3	217.9	212.1	207.0	197.9	190.3	183.7	178.0	172.9	168.4	164.3	160.5
225	226.1	217.0	210.2	204.2	198.8	189.6	181.9	175.2	169.5	164.4	159.9	155.9	152.2
200	217.3	208.9	210.7	195.4	189.9	180.3	172.4	165.8	160.1	155.1	150.3	146.7	143.1
175	209.1	200.0	192.3	185.7	179.9	171.1	162.2	155.6	149.9	149.5	140.6	136.8	133.3
150	201.8	191.7	183.5	173.6	170.4	160.2	152.2	145.6	140.0	135.1	130.9	127.1	123.8
125	192.5	181.1	172.0	166.6	158.2	148.0	140.1	133.4	128.0	123.4	119.3	115.7	112.6
100	(182.5)	169.4	158.2	151.4	144.7	134.5	126.5	120.2	115.0	110.5	106.7	103.4	(100.4)
W =	0	25	50	75	100	150	200	250	300	350	400	450	500

TABLE 64.—AVERAGE ASCENSIONAL RATES (m per min) FOR VARIOUS HYDROGEN FREE LIFTS (1) AND VARIOUS WEIGHT BALLOONS (*W*)

1	Average ascensional rates												
2,000	517.6				502.2		502.4		494.0		487.0		480.5
1,900	512.3				503.6		495.5		487.8		480.7		473.9
1,800	507.7		503.1		498.7		490.2		482.3		474.9		467.9
1,700	502.7		497.9		493.2		484.4		476.2		468.5		461.3
1,600	496.9		491.8		487.0		477.8		469.3		461.3		453.9
1,500	491.0		485.6		480.5		470.9		462.0		453.8		446.1
1,400	484.5		478.9		473.4		463.4		454.2		445.6		438.7
1,300	476.6		470.6		464.9		454.4		444.6		435.8		427.6
1,200	468.1		461.7		455.7		444.5		434.5		425.3		416.8
1,100	458.3	454.9	451.6	448.4	445.2	439.2	433.5	428.1	422.9	418.0	413.3	408.8	404.5
1,000	443.5	440.2	436.3	432.9	429.6	423.3	417.4	411.7	406.4	401.3	396.4	391.8	387.4
950	433.5	429.7	426.1	422.6	419.2	412.8	406.7	400.9	395.6	390.4	385.6	380.9	376.5
900	421.4	417.5	413.9	410.3	406.8	400.3	394.1	388.3	382.8	377.7	372.8	368.1	363.7
850	410.6	406.6	402.8	399.1	395.6	388.9	382.6	376.7	371.2	366.0	361.0	356.3	351.9
800	397.4	393.4	389.4	385.7	382.1	375.3	369.9	363.0	357.4	352.1	347.2	342.5	338.0
750	382.3	378.2	374.2	370.4	365.9	359.8	353.4	347.4	341.8	336.5	331.6	326.9	322.5
700	368.2	363.9	359.8	355.9	352.2	345.1	338.6	332.6	326.9	321.7	316.7	312.1	307.7
650	351.3	346.9	342.7	338.7	334.9	327.8	321.2	315.2	309.6	304.3	299.4	294.8	290.5
600	337.1	332.6	328.2	324.1	320.2	313.0	306.3	300.2	294.5	289.2	284.3	279.8	275.4
550	320.8	316.1	311.6	307.4	303.4	296.0	289.2	283.1	277.4	272.2	267.3	261.2	258.6
500	305.9	301.1	296.4	292.0	287.9	280.3	273.5	267.3	261.6	256.3	251.5	247.0	242.8
475	299.7	294.6	289.8	285.4	281.2	273.5	266.5	260.3	254.5	249.3	244.4	240.0	235.8
450	293.2	287.9	383.0	278.3	274.2	266.4	259.3	253.0	247.3	242.0	237.2	232.7	228.5
425	285.7	280.3	275.3	270.6	266.2	258.1	251.2	244.8	239.1	233.8	229.0	224.6	220.4
400	279.4	273.8	268.6	263.8	259.4	251.2	244.1	237.6	231.8	226.6	221.7	217.3	213.2
375	272.6	266.8	262.5	256.6	252.0	243.7	236.4	229.9	224.2	218.8	214.0	209.1	205.6
350	266.5	260.5	254.9	249.8	245.1	236.7	229.3	222.7	216.8	211.5	206.7	202.3	198.3
325	259.2	252.9	247.1	241.9	237.0	228.4	220.9	214.3	208.4	203.1	198.4	194.0	190.0
300	253.3	246.7	240.6	235.2	230.2	221.3	213.7	207.0	201.1	195.8	191.0	186.6	182.7
275	246.0	238.9	232.6	227.0	221.8	212.7	205.0	198.3	192.3	187.1	182.3	178.0	174.1
250	238.7	231.3	224.7	218.7	213.4	204.1	196.3	189.5	183.6	178.3	173.6	169.4	165.5
225	232.0	223.8	216.8	210.6	205.1	195.5	187.5	180.7	174.8	169.6	164.9	160.7	156.9
200	224.1	215.5	208.0	201.5	195.8	186.0	177.8	171.0	165.1	159.9	155.4	151.3	147.6
175	215.7	206.3	198.3	191.5	185.5	175.5	167.3	160.4	154.6	149.5	145.1	141.1	137.5
150	208.1	197.8	189.1	180.6	175.6	165.2	156.6	150.1	144.3	139.4	135.0	131.1	127.7
125	198.3	186.8	177.4	169.7	163.2	152.6	144.3	137.6	132.0	127.2	123.0	119.3	116.1
100	188.2	174.7	163.1	156.1	149.3	138.6	130.5	123.9	118.5	114.0	110.0	106.6	103.5
W =	0	25	50	75	100	150	200	250	300	350	400	450	500

NUMERICAL AND GRAPHICAL DATA

TABLE 65.—ALTITUDE TIMETABLES FOR VARIOUS RATES OF ASCENT

Ascensional rate in m per min

Min.	150	160	170	180	190	200	210	220	230	240	250	260	270
1	180	192	204	216	228	240	252	264	276	288	300	312	324
2	345	368	391	414	437	460	483	506	529	552	575	598	621
3	510	544	578	612	646	680	714	748	782	816	850	884	918
4	668	712	756	801	846	890	934	979	1,024	1,068	1,112	1,157	1,202
5	825	880	935	990	1,045	1,100	1,155	1,210	1,265	1,320	1,375	1,430	1,485
6	975	1,040	1,105	1,170	1,235	1,300	1,365	1,430	1,495	1,560	1,625	1,690	1,755
7	1,125	1,200	1,275	1,350	1,425	1,500	1,575	1,650	1,725	1,800	1,875	1,950	2,025
8	1,275	1,360	1,445	1,530	1,615	1,700	1,785	1,870	1,955	2,040	2,125	2,210	2,295
9	1,425	1,520	1,615	1,710	1,805	1,900	1,995	2,090	2,185	2,280	2,375	2,470	2,565
10	1,575	1,680	1,785	1,890	1,995	2,100	2,205	2,310	2,415	2,520	2,625	2,730	2,835
11	1,725	1,840	1,955	2,070	2,185	2,300	2,415	2,530	2,645	2,760	2,875	2,990	3,105
12	1,875	2,000	2,125	2,250	2,375	2,500	2,625	2,750	2,875	3,000	3,125	3,250	3,375
13	2,025	2,160	2,295	2,430	2,565	2,700	2,835	2,970	3,105	3,240	3,375	3,510	3,645
14	2,175	2,320	2,465	2,610	2,755	2,900	3,045	3,190	3,335	3,480	3,625	3,770	3,915
15	2,325	2,480	2,635	2,790	2,945	3,100	3,255	3,410	3,565	3,720	3,875	4,030	4,185
16	2,475	2,640	2,805	2,970	3,135	3,300	3,465	3,630	3,795	3,960	4,125	4,290	4,455
17	2,625	2,800	2,975	3,150	3,325	3,500	3,675	3,850	4,025	4,200	4,375	4,550	4,725
18	2,775	2,960	3,145	3,330	3,515	3,700	3,885	3,070	4,255	4,440	4,625	4,810	4,995
19	2,925	3,120	3,315	3,510	3,705	3,900	4,095	4,290	4,485	4,680	4,875	5,070	5,265
20	3,075	4,280	3,485	3,690	3,895	4,100	4,205	4,510	4,715	4,920	5,125	5,330	5,535
21	3,225	3,440	3,655	3,870	4,085	4,300	4,515	4,730	4,935	5,160	5,375	3,590	5,805
22	3,375	3,600	3,825	4,050	4,275	4,500	4,725	4,950	5,175	5,400	5,625	5,850	6,075
23	3,525	3,760	3,995	4,230	4,465	4,700	4,935	5,170	5,405	5,640	5,875	6,110	6,345
24	3,675	3,920	4,165	4,410	4,655	4,900	5,145	5,390	5,635	5,880	6,125	6,370	6,615
25	3,825	3,080	4,335	4,590	4,845	5,100	5,355	5,610	5,865	6,120	6,375	6,630	6,885
26	3,975	4,240	4,505	4,770	5,035	5,300	5,565	5,830	6,095	6,360	6,625	6,890	7,155
27	4,125	4,400	4,675	4,950	5,225	5,500	5,775	6,050	6,325	6,600	6,875	7,150	7,425
28	4,275	4,560	4,845	5,130	5,415	5,700	5,985	6,270	6,555	6,840	7,125	7,410	7,695
29	4,425	4,720	5,015	5,310	5,605	5,900	6,195	6,490	6,785	7,080	7,375	7,670	7,965
30	4,574	4,880	5,185	5,490	5,795	6,100	6,405	6,710	7,015	7,320	7,625	7,930	8,235
31	4,725	5,040	5,355	5,670	5,985	6,300	6,615	6,930	7,245	7,560	7,875	8,190	8,505
32	4,875	5,200	5,525	5,850	6,175	6,500	6,825	7,150	7,475	7,800	8,125	8,450	8,775
33	5,025	5,360	5,695	6,030	6,365	6,700	7,035	7,370	7,705	8,040	8,375	8,710	9,045
34	5,175	5,520	5,865	6,210	6,555	6,900	7,245	7,500	7,985	8,280	8,625	8,970	9,315
35	5,325	5,680	6,035	6,390	6,745	7,100	7,455	7,810	8,165	8,520	8,875	9,230	9,585

TABLE 66.—CEILING BALLOON ELEVATION
Free lift 40 grams with hydrogen or 45 grams with helium.   10-gram balloon.
Average rate of ascension after $1\frac{1}{2}$ min = 6 fps

Time interval, min*	Altitude, ft	Altitude, m
$\frac{1}{2}$	250	76.2
1	480	146
$1\frac{1}{2}$	670	204
2	850	259
$2\frac{1}{2}$	1,030	314
3	1,210	369
$3 + t$	$1,210 + 180t$	$369 + 55t$

* Time t is in half minutes.

TABLE 67.—WIND-VELOCITY PRESSURES
Air at 15°C and 760 mm mercury.

True Wind Speed	Velocity Pressure, lb/ft^2
10	0.26
20	1.02
30	2.30
40	4.09
50	6.39
60	9.21
70	12.53
80	16.36
90	20.71
100	25.57
110	30.94
120	36.82

TABLE 68.—THERMAL PROPERTIES OF WATER—SOLID, AND SATURATED LIQUID AND VAPOR PHASES +50° TO 40°C*

Values based on results of international research as tabulated by Keenan and Keyes, 1936, but with magnitudes in metric units and relative to zero values of enthalpy and entropy for ice at 200°K, or −73°C

Temp., °C	Saturated vapor pres., mb	Saturated vapor density, grams/m.³	Enthalpy, joules/gram			Entropy, joules/gram, °K		
			h_f	h_{fg}	h_g	s_f	s_{fg}	s_g
50	123.3	83.1	676	2,383	3,059	2.49	7.37	9.86
48	111.5	75.6	668	2,388	3,056	2.46	7.43	9.89
46	100.9	68.8	659	2,393	3,052	2.44	7.49	9.93
44	91.1	62.5	651	2,398	3,049	2.41	7.56	9.97
42	82.0	56.6	643	2,402	3,045	2.39	7.62	10.01
40	73.7	51.2	634	2,407	3,041	2.36	7.68	10.04
38	66.2	46.3	626	2,412	3,038	2.33	7.75	10.08
36	59.4	41.8	618	2,416	3,034	2.30	7.82	10.12
34	53.2	37.6	610	2,421	3,031	2.28	7.88	10.16
32	47.5	33.8	601	2,426	3,027	2.25	7.95	10.20
30	42.43	30.4	592	2,431	3,023	2.22	8.02	10.24
28	37.78	27.3	584	2,436	3,020	2.19	8.09	10.28
26	33.65	24.4	576	2,441	3,016	2.17	8.16	10.33
24	29.82	21.8	567	2,446	3,013	2.14	8.22	10.37
22	26.40	19.4	559	2,450	3,009	2.11	8.30	10.41
20	23.37	17.31	551	2,454	3,005	2.08	8.37	10.45
18	20.61	15.37	543	2,459	3,002	2.06	8.44	10.50
16	18.16	13.65	534	2,464	2,998	2.03	8.52	10.55
14	15.98	12.09	526	2,468	2,994	2.00	8.59	10.59
12	14.03	10.68	517	2,473	2,990	1.97	8.67	10.64
10	12.28	9.41	509	2,478	2,987	1.94	8.75	10.69
8	10.73	8.29	501	2,482	2,983	1.91	8.83	10.74
6	9.35	7.27	492	2,487	2,979	1.88	8.91	10.79
4	8.13	6.37	484	2,492	2,976	1.85	8.99	10.84
2	7.05	5.56	475	2,497	2,972	1.82	9.07	10.89
0	6.105	4.85	467	2,502	2,969	1.79	9.16	10.95

Temp., °C	Over water	Over ice	Over water	Over ice	h_i	h_{ig}	h_g	s_i	s_{ig}	s_g
0	6.105	6.105	4.85	4.85	134	2,835	2,969	0.57	10.38	10.95
−2	5.27	5.17	4.22	4.14	129	2,836	2,965	0.55	10.45	11.00
−4	4.54	4.37	3.66	3.53	125	2,836	2,961	0.54	10.53	11.07
−6	3.90	3.69	3.17	3.00	121	2,837	2,958	0.52	10.61	11.13
−8	3.34	3.10	2.74	2.54	117	2,837	2,954	0.51	10.69	11.20
−10	2.86	2.60	2.36	2.14	113	2,837	2,950	0.49	10.78	11.27
−12	2.44	2.18	2.03	1.81	109	2,838	2,947	0.48	10.86	11.34
−14	2.07	1.80	1.74	1.51	105	2,838	2,943	0.46	10.95	11.41
−16	1.75	1.51	1.48	1.28	101	2,838	2,939	0.44	11.04	11.48
−18	1.48	1.25	1.26	1.06	97	2,839	2,936	0.43	11.12	11.55
−20	1.24	1.04	1.07	0.892	93	2,839	2,932	0.41	11.21	11.62
−22		0.854		0.738	89	2,839	2,928	0.40	11.30	11.70
−24		0.702		0.612	85	2,839	2,924	0.38	11.39	11.77
−26		0.576		0.506	81	2,840	2,921	0.37	11.48	11.85
−28		0.468		0.414	77	2,840	2,917	0.35	11.58	11.93
−30		0.381		0.340	74	2,840	2,914	0.34	11.67	12.01
−32		0.310		0.279	70	2,840	2,910	0.32	11.77	12.09
−34		0.205		0.227	67	2,839	2,906	0.30	11.87	12.17
−36		0.202		0.185	63	2,839	2,902	0.29	11.97	12.26
−38		0.163		0.151	60	2,839	2,899	0.27	12.07	12.34
−40		0.131		0.122	56	2,839	2,895	0.26	12.17	12.43
−73					0	2,836	2,836	0.0	14.20	14.20

* KIEFER, PAUL J., *Monthly Weather Rev.*, November, 1941, vol. 69.

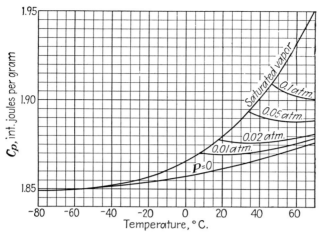

Fig. 28.—Variation of specific heat of water vapor (constant pressure) with temperature and pressure. (*After C. H. Meyers.*)

TABLE 69.—DEW-POINT TEMPERATURE (°F) VS. VAPOR PRESSURE (IN.)*

−30	0.007008	20	0.10274	70	0.7390
	0.007440		0.10779		0.7646
	0.007897		0.11306		0.7910
	0.008379		0.11857		0.8181
	0.008888		0.12432		0.8460
−25	0.009425	25	0.13033	75	0.8748
	0.009993		0.13660		0.9044
	0.010592		0.14314		0.9350
	0.011223		0.14997		0.9664
	0 011888		0.15710		0.9987
−20	0.012591	30	0.16453	80	1.0319
	0.013332		0.17228		1.0663
	0.014113		0.18036		1.1015
	0.014935		0.1878		1.1377
	0.015801		0.1955		1.1749
−15	0.016713	35	0.2035	85	1.2132
	0.017673		0.2118		1.2526
	0.018684		0.2203		1.2930
	0.019747		0.2292		1.3346
	0.020866		0.2382		1.3774
−10	0.022042	40	0.2477	90	1.4215
	0.023280		0.2575		1.4667
	0.024581		0.2676		1.5131
	0.025948		0.2781		1.5608
	0.027385		0.2890		1.6097
−5	0.028895	45	0.3003	95	1.6600
	0.030481		0.3119		1.7118
	0.032146		0.3239		1.7648
	0.033894		0.3362		1.8193
	0.035729		0.3491		1.8753
0	0.037655	50	0.3624	100	1.9327
	0.039675		0.3762		1.9917
	0.041794		0.3904		2.0521
	0.044017		0.4050		2.1141
	0.046348		0.4201		2.1778
5	0.048791	55	0.4357	105	2.2432
	0.051351		0.4518		2.3102
	0.054034		0.4684		2.3789
	0.056845		0.4856		2.4494
	0.059788		0.5033		2.5218
10	0.062870	60	0.5216	110	2.5959
	0.066097		0.5405		2.6719
	0.069474		0.5599		2.7498
	0.073009		0.5800		2.8297
	0.076708		0.6007		2.9115
15	0.080579	65	0.6220	115	2.9952
	0.084626		0.6440		3.0810
	0.088857		0.6667		3.1691
	0.093280		0.6901		3.2594
	0.097904		0.7142		3.3517
				120	3.4463

* From U.S. Weather Bureau Psychometric Slide Rule, Mar. 23, 1942.

NOTE: Data involving vapor pressures for temperature below 32°F are based on saturation vapor pressures with respect to ice.

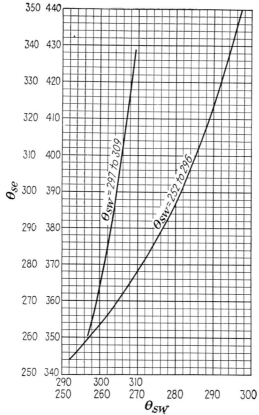

FIG. 29.—Pseudo-wet-bulb potential temperature vs pseudo-equivalent potential temperature.

TABLE 70.—PSEUDO-EQUIVALENT POTENTIAL TEMPERATURE VS. PSEUDO-WET-BULB POTENTIAL TEMPERATURE*

θ_{sw}	θ_{se}	θ_{sw}	θ_{se}
253	255.1	293	334.3
255	257.5	295	342.1
257	260.0	297	350.8
259	262.5	299	360.4
261	265.1	301	371.0
263	267.8	303	382.8
265	270.6	305	396.0
267	273.6	307	410.9
269	276.6	309	427.6
271	279.9	311	446.6
273	283.3	313	468.2
275	286.9		
277	290.7		
279	294.9		
281	299.3		
283	304.0		
285	309.1		
287	314.6		
289	320.6		
291	327.1		

* From *U.S. Weather Bur. Form* 1147-A.

TABLE 71.—PSEUDO-EQUIVALENT POTENTIAL TEMPERATURE*

$10^3 W_m$	θ_{se}/θ	Potential temperature									
		270	275	280	285	290	295	300	305	310	315
2	1.0009 exp (5/T)	275.4	280.5	285.6	290.7	295.8	300.9	306.0	311.1	316.2	321.3
3	1.0014 exp (7.5/T)	278.0	283.1	288.3	293.4	298.5	303.7	308.9	314.0	319.2	324.4
4	1.0018 exp (10/T)	280.5	285.7	291.0	296.1	301.2	306.5	311.7	316.9	322.0	327.2
6	1.0028 exp (15/T)	285.7	291.0	296.2	301.5	306.8	312.1	317.4	322.7	328.0	333.3
8	1.0037 exp (20/T)	291.1	296.4	301.7	307.0	312.3	317.7	323.1	328.5	333.9	339.3
10	1.0046 exp (25/T)	295.4	301.0	306.6	312.2	317.8	323.4	329.0	334.6	340.2	345.7
12	1.0055 exp (30/T)	300.5	306.2	311.9	317.6	323.3	329.0	333.4	340.4	346.1	351.7

* After Holmboe.

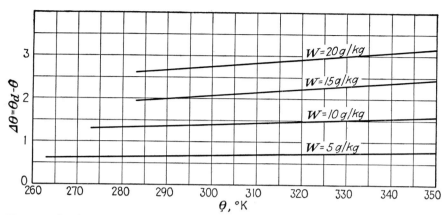

FIG. 30.—Partial potential temperature correction. Add the correction to θ to obtain θ_d.
$$\theta_d = \theta(1 + 0.461W).$$

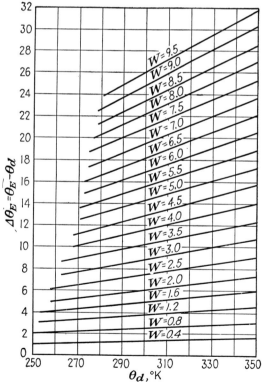

FIG. 31.—Rossby's equivalent potential temperature correction. Add the correction to θ_d to obtain θ_E.

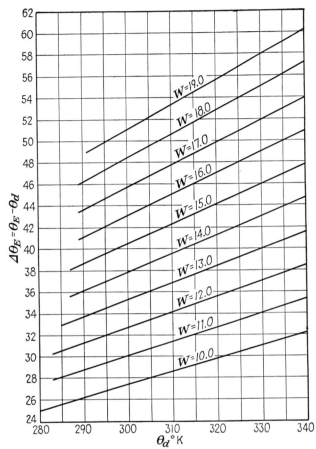

Fig. 32.—Rossby's equivalent potential temperature correction. Add the correction to θ_d to obtain θ_E.

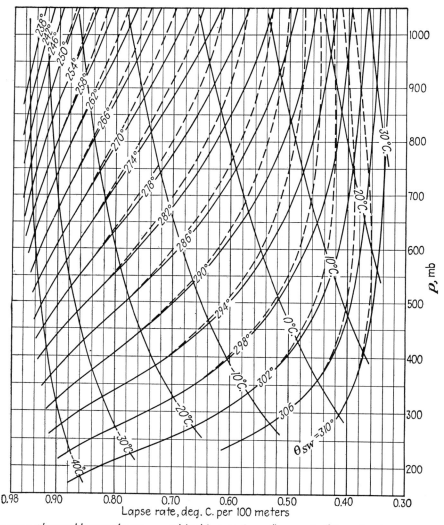

———, observed temperatures; — —, virtual temperatures; "over-water" throughout

FIG. 33.—Pseudoadiabatic lapse rates for various pressures and temperatures. (After P. J. Kiefer.)

TABLE 72.—SATURATION MIXING RATIOS, GRAM PER KG, FOR VARIOUS TEMPERATURES ($t°C$) AND PRESSURES p MB*

$t°C$ \ p mb	400	450	500	550	600	650	700	750	800	850	900	950	1,000	1,050
40				96.64	87.46	79.87	73.49	68.06	63.37	59.29	55.70	52.52	49.68	47.14
35		89.09	79.05	71.04	64.51	59.08	54.49	50.56	47.16	44.19	41.57	39.25	37.17	35.30
30	74.02	64.94	57.84	52.15	47.47	43.56	40.25	37.41	34.94	32.77	30.86	29.16	27.64	26.27
25	53.62	47.21	42.17	38.10	34.75	31.94	29.55	27.49	25.70	24.13	22.74	21.51	20.40	19.39
20	38.71	34.17	30.59	27.68	25.28	23.27	21.55	20.06	18.77	17.64	16.63	15.73	14.93	14.20
15	27.75	24.55	22.01	19.94	18.23	16.79	15.56	14.50	13.57	12.76	12.04	11.39	10.81	10.29
10	19.73	17.48	15.69	14.23	13.02	12.00	11.24	10.37	9.71	9.13	8.62	8.16	7.75	7.37
5	13.88	12.31	10.58	10.04	9.19	8.47	7.86	7.33	6.87	6.46	6.10	5.77	5.48	5.22
0	9.65	8.56	7.69	6.99	6.40	5.90	5.48	5.11	4.79	4.50	4.25	4.03	3.82	3.64
−5	6.64	5.90	5.30	4.82	4.41	4.07	3.78	3.53	3.30	3.11	2.93	2.78	2.64	2.51
−10	4.50	4.00	3.60	3.27	2.99	2.76	2.56	2.39	2.24	2.11	1.99	1.89	1.79	1.71
−15	3.00	2.67	2.40	2.18	2.00	1.85	1.71	1.60	1.50	1.41	1.33	1.26	1.20	1.14
−20	1.97	1.75	1.57	1.43	1.31	1.21	1.12	1.05	0.983	0.925	0.873	0.827	0.786	0.748
−25	1.26	1.12	1.01	0.918	0.841	0.776	0.721	0.673	0.631	0.594	0.561	0.531	0.504	0.480
−30	0.795	0.707	0.636	0.578	0.530	0.489	0.454	0.424	0.397	0.374	0.353	0.335	0.318	0.303
−35	0.489	0.435	0.391	0.356	0.326	0.301	0.280	0.261	0.245	0.230	0.217	0.206	0.196	0.186
−40	0.295	0.262	0.236	0.214	0.196	0.181	0.168	0.157	0.147	0.139	0.131	0.124	0.118	0.112
−45	0.173	0.154	0.138	0.126	0.115	0.106	0.099	0.092	0.086	0.081	0.077	0.073	0.069	0.066

* From *U.S. Weather Bur. Cir. N.*

NOTE.—Saturation with respect to liquid water assumed at temperatures below 0°C.

TABLE 73.—MEASUREMENTS BY OSBORNE, STIMSON, AND GINNINGS, NATIONAL BUREAU OF STANDARDS, 1939

Temp., °C	$\dfrac{\text{Abs. joules}}{\text{gram °C}}$	Temp., °C	$\dfrac{\text{Abs. joules}}{\text{gram °C}}$
0	4.2177	50	4.1807
5	4.2022	55	4.1824
10	4.1922	60	4.1844
15	**4.1858**	65	4.1868
20	4.1819	70	4.1896
25	4.1796	75	4.1928
30	4.1785	80	4.1964
35	4.1782	85	4.2005
40	4.1786	90	4.2051
45	4.1795	95	4.2103
50	4.1807	100	4.2160

TABLE 74.—WET-ADIABATIC LAPSE RATE (°C PER 100 GM*)
Condensation to liquid water

Temp., °C	Pressure, mb									
	1000	900	800	700	600	500	400	300	200	100
40	0.315	304	293	282	269	255	241	223	198	155
35	0.337	326	313	300	285	271	256	234	210	165
30	0.366	352	338	321	304	287	269	247	221	176
25	0.400	384	366	347	329	309	286	262	233	189
20	0.440	422	402	382	359	335	309	280	246	204
15	0.487	468	446	422	397	369	338	303	265	217
10	0.539	520	498	471	442	411	375	335	290	232
5	0.596	578	555	527	497	460	420	375	318	250
0	0.659	640	616	587	554	517	472	422	354	271
− 5	0.723	705	680	653	619	579	536	480	405	304
−10	0.780	762	741	717	687	650	603	543	461	343
−15	0.833	819	802	780	753	720	677	615	532	397
−20	0.879	867	853	835	815	785	746	692	609	465
−25	0.913	904	893	880	863	842	811	764	690	537
−30	0.941	936	929	920	907	890	866	829	765	629
−35	0.964	961	956	948	939	927	910	883	835	710
−40	0.978	976	972	968	962	954	942	922	886	800
−45	0.989	988	986	983	979	973	964	952	931	870
−50	0.995	994	993	991	988	984	979	971	956	909
−55	0.999	998	997	995	994	991	988	983	972	941
−60	1.001	1.001	1.000	1.000	0.999	0.998	0.996	0.992	0.985	0.966
−65	1.003	1.002	1.002	1.002	1.002	1.001	1.000	0.999	0.995	0.985

* After G. Stüve.

TABLE 75.—WET-ADIABATIC LAPSE RATE (°C PER 100 GM)*
Sublimation to ice or snow

Temp., °C	Pressure, mb									
	1000	900	800	700	600	500	400	300	200	100
− 0	0.600	578	554	525	493	457	413	363	306	236
− 5	0.680	658	634	606	573	535	490	432	362	271
−10	0.751	732	709	683	653	617	571	504	423	316
−15	0.818	802	782	760	732	679	652	591	504	371
−20	0.872	862	846	827	802	773	734	679	588	447
−25	0.914	905	893	880	862	839	809	761	683	535
−30	0.945	940	932	922	910	894	872	835	772	640
−35	0.970	965	960	953	945	935	921	896	850	743
−40	0.983	980	977	974	968	962	952	935	904	825
−45	0.992	991	989	986	983	979	973	964	944	891
−50	0.997	996	995	994	992	990	987	980	968	934
−55	1.000	0.999	999	998	997	996	995	991	983	961
−60	1.003	1.002	1.002	1.002	1.001	1.000	1.000	0.998	0.993	0.981
−65	1.004	1.004	1.003	1.003	1.003	1.003	1.002	1.001	1.000	0.994

* After G. Stüve.

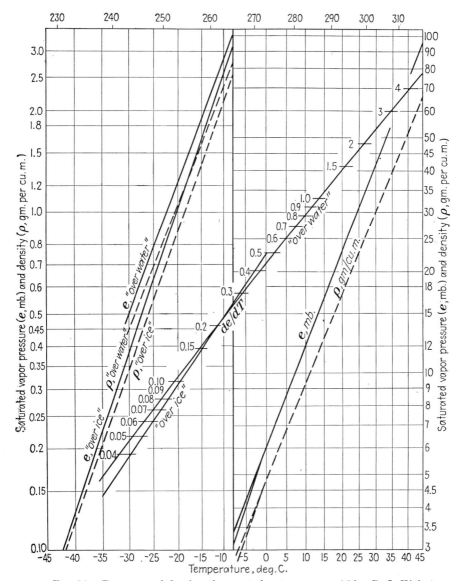

Fɪɢ. 34.—Pressure and density of saturated water vapor.   (*After P. J. Kiefer.*)

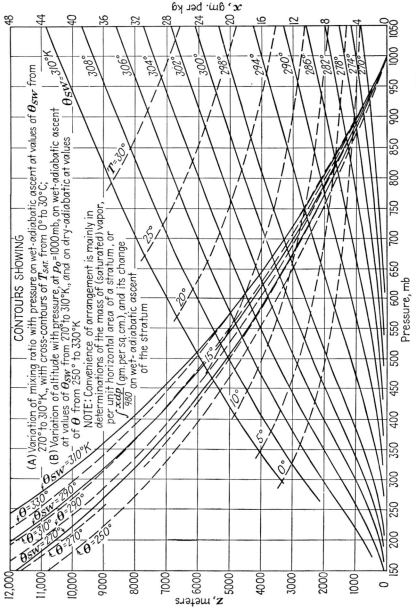

FIG. 35.—Chart for computing amount of precipitable water vapor in the atmosphere.   (*After P. J. Kiefer.*)

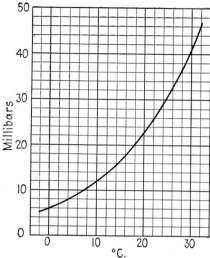
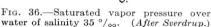

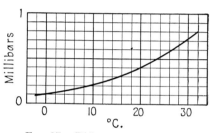

Fig. 36.—Saturated vapor pressure over water of salinity 35 °/₀₀. (*After Sverdrup.*)

Fig. 37.—Difference; saturated vapor pressure over pure water minus saturated vapor pressure over sea water (salinity 35 °/₀₀). (*After Sverdrup.*)

PRESSURE MEASUREMENT

A properly exposed mercurial barometer is used to determine the atmospheric pressure at a station. The direct result of the reading of the instrument, uncorrected for any errors, is known as the *observed reading*. Care must be used in reading the vernier as well as in setting it at the top of the meniscus. The observed reading may be corrected as outlined below to give the *actual pressure* at the ivory point of the barometer. The actual pressure is the "static pressure" of the atmosphere at the time and place of reading.

Pumping. During windy, gusty weather the barometer is subject to rapid and irregular oscillations. This phenomenon is known as pumping. Pumping is a reflection of actual changes in the static pressure, and the amount cannot, in general, be calculated. In strong winds, the fluctuations may amount to as much as 3 mb pressure. The term pumping is also applied to the more violent oscillations of the mercury caused by the movement of a ship at sea. These violent oscillations are damped by means of a glass tube with a constricted bore. Reasonably accurate readings may be obtained when the barometer is pumping by taking the mean of its highest and lowest positions during a period of 1 min.

Correction for Scale Errors, Capillarity, Etc. This is a mean difference between the readings of a given instrument and those of the standard barometer duly corrected. This quantity embraces all outstanding errors in the subdivision of the scale, or its total length; errors in the adjustment of the sighting edge to the zero line of the vernier; errors of capillarity, imperfect vacuum, etc. A card can be made showing the sum of these corrections. In a well-constructed barometer, the sum is about constant over the scale and rarely exceeds 0.20 mb.

Correction for Temperature. Changes in the temperature of its surroundings will produce changes in the length of the scale of the barometer and changes in the density of the mercury. Linear expansion of brass is 0.0000102 per °F. Cubical expansion of mercury is 0.0001010 per °F. Barometers reading in inches of mercury are manufactured to read true pressure (*i.e.*, zero temperature correction) at about 28.5°F. Barometers reading in millimeters of mercury are manufactured to read true pressure at 0°C. Barometers reading in millibars are, in England, generally adjusted to read true pressure at 12°C. The tables for correction usually give the correction for both scale and mercury density in one set of figures. An error is therefore introduced if the uncorrected reading of a mercurial barometer expressed in metric units is converted into English units, or vice versa, and a temperature correction afterward applied to the result. *The conversion of readings from one unit to another can then be made only after each reading has been fully corrected for temperature using the proper tables or graph of corrections.*

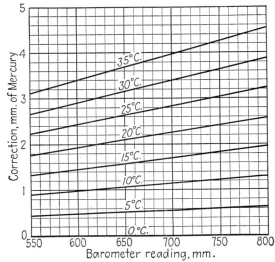

Fig. 38.—Correction of the barometer to standard temperature. Metric measure. Isopleths of temperature of attached thermometer. For temperatures above (below) 0°C the correction is to be subtracted (added). (*Plotted from Smithsonian Meteorological Tables.*)

Each barometer is supplied with an "attached thermometer." It is assumed that the mercury and scale are at one uniform temperature, the temperature of the attached thermometer. The correction necessary for various temperatures and pressures is shown in Figs. 38 and 39 and in Table 76. The curves and tables referred to are computed for brass scales.

Correction for Local Gravity. This correction breaks down into two parts: (1) the correction based on the variation of the force of gravity with latitude—*latitude term*, and (2) the correction based on the variation of gravity with altitude above sea level. These corrections appear because the *weight* of a given *mass* of mercury varies over the earth's surface. For the most precise work, the local value of gravity must be determined carefully. For all routine meteorological reports, it is sufficient to take local gravity at sea level to be given by

$$g_\varphi = 980.62(1 - 0.002640 \cos 2\varphi + 0.000007 \cos^2 2\varphi)$$

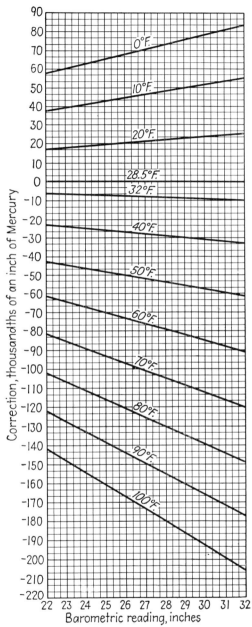

FIG. 39.—Correction of the barometer to standard temperature. English measure. Isopleths of temperature of attached thermometer. (*Plotted from Smithsonian Meteorological Tables.*)

The correction for altitude is

$$c \text{ (dynes)} = -0.0003086h \qquad \text{(m)}$$
$$c \text{ (dynes)} = -0.000094h \qquad \text{(ft)}$$

The height of the barometer in terms of the value of gravity is given by

$$h_{45,0} = h_{\varphi,z} \frac{g_{\varphi,z}}{g_{45,0}}$$

where $h_{45,0}$ = height of the barometer corrected to standard gravity

$h_{\varphi,z}$ = height of the barometer as read corrected for scale errors, etc.

$g_{\varphi,z}$ = value of gravity at the ivory point of the barometer

$g_{45,0}$ = standard value of gravity (980.62).

It will be noted that the correction for gravity varies with the pressure. At a shore station, it is customary to prepare a card of corrections combining gravity with the almost constant errors of scale, etc. For use on shipboard, a table of corrections corresponding to different latitudes is necessary. The correction for altitude is negligible on shipboard and at low-level shore stations.

Actual Pressure. The preceding discussion briefly illustrates the exceedingly complex corrections that are required to be made to a mercurial barometer if *very accurate* values of static pressure are to be deduced from the observed reading. For all routine work, it is adequate to proceed as shown in the below example.

Example. The observed reading of a mercurial barometer equipped with a brass scale is 30.08 in.; attached thermometer 58°F; latitude 38°45′N; elevation of barometer 100 ft above sea level.

Observed reading.....................	1,018.6 mb	30.08 in.
Scale, etc., error (from calibration card supplied by manufacturer or cognizant bureau)................................		−0.004 in.
Temperature correction (Fig. 39)........		−0.080 in.
Latitude correction (Fig. 42)............		−0.019 in.
Total correction (sum of above three)....		−0.103 in.
Actual pressure at ivory point..........	1,015.2 mb	29.98 in.

Reduced Pressures. The actual pressure at the ivory point of the barometer is generally of secondary interest. For meteorological purposes, several reduced pressures are employed. Reductions are commonly carried out for *station pressure,* which is a pressure corresponding to an adopted or station elevation. This may and generally does differ from the elevation of the barometer. *Sea-level barometric pressure* (M.S.L. pressure) is the theoretical pressure that would be exerted by the atmosphere at a station at a given time, *if that station were at sea level.* The M.S.L. pressure is intended to represent the barometric pressure at sea level under the prevailing meteorological conditions of temperature and station pressure. It is designed to give fairly smooth, consistent isobars on consecutive weather maps, so far as is practicable (*U.S. Weather Bureau Circular N,* 1941). The *altimeter setting* is a pressure, in inches, used for setting a pressure-scale-type sensitive altimeter in an airplane so that upon landing of the airplane at an airport the pointers of the instrument will indicate very closely the field elevation above sea level, provided that the instrument is functioning properly and is free from error and that the setting was determined by a properly equipped station near the time and place of landing and was furnished to the pilot just prior to landing.

When the difference in altitude between the barometer and the elevation to which it is desired to reduce the actual pressure (at the barometer) exceeds about 100 ft the calculations require some care. Definite procedures are established by the U.S. Weather Bureau for use in the United States. Similar procedures are established by the weather services of other countries. See for example Weather Bureau Form 1154-A, a *pressure-altitude chart* for reduction to sea level (1,050 to 800 mb). The procedures outlined below will give acceptable results for most stations.

Reduction to Sea Level. The M.S.L. pressure is defined above. When the observed reading is taken at a point in the atmosphere that is directly over the sea (*e.g.*, on shipboard) a very accurate value of the M.S.L. pressure may be obtained since the pressure, temperature, and humidity of the actual atmosphere between the barometer and the sea surface are measurable. In practice, for such low-level stations, an amount is added to the actual pressure. This amount or *reduction for elevation* is obtained by entry in Table 80.

At other than low-level stations, the temperature during the last 12 hr, the elevation, and the latitude are considered in the computation of the reduction for elevation. In the United States, tables prepared by the Weather Bureau for each station give the sea-level pressure for any set of values of actual pressure and average temperature during the last 12 hr. The average temperature used is one-half the sum of the temperature "now" and the temperature 12 hr ago. For very high elevation stations, a "plateau correction" is involved. Exact procedures used will vary between different weather services.

For intermediate level stations, the Admiralty Weather Manual gives the formula

$$\log p_0 = \log p + \frac{h}{221T}$$

where p_0 = M.S.L. pressure

p = actual pressure at station level

h = altitude of station, ft

T = mean absolute temperature of the air column between station and M.S.L. The above formula may be used when h is not greater than about 1,000 ft. The temperature T is usually taken to be the temperature at the station (dry-bulb temperature in shelter). The very high pressures reported over Siberia in winter and the low pressures reported over African plateaus in summer are due, at least in part, to the method of reduction to sea level.

Reduction to Sea Level. *Example.* The observed reading of the mercurial barometer on board a vessel in Guantanamo Bay area, latitude 20°N, longitude 75°10'W, is 29.95 in.; temperature of attached thermometer 74°F; barometer 60 ft above sea level.

Observed reading	1,014.2 mb	29.95 in.
Temperature correction (Fig. 39)		−0.12 in.
Latitude correction (Fig. 42)		−0.06 in.
Actual pressure		29.77 in.
Reduction for elevation (Table 80)		+0.06 in.
M.S.L. pressure	1,010.2 mb	29.83 in.

In the above example, the *corrections* for instrument error and for gravity due to altitude are considered negligible.

Gold Slide. Barometer corrections at sea may be very simply obtained by means of the gold slide. This is a device attached firmly to the barometer in the position usually occupied by the attached thermometer. It consists of a combined thermometer and sliding scale. The scale is adjusted for height above M.S.L. and for latitude. The combined correction and reduction to M.S.L. is read off the scale in line with the top of the mercury in the thermometer. The barometer is then read and the M.S.L. pressure obtained by applying the correction-reduction term from the scale to the observed reading. A complete description will be found in the Admiralty Weather Manual.

Aneroid Barometer. The aneroid barometer is described in Sec. VIII. The precision aneroid as used by the Navy is a very accurate instrument and is especially designed for shipboard use. A multiple-spring arrangement damps vibrations. Similarly constructed aneroids behave differently when subject to the same changes of pressure and temperature, so that *each instrument must be separately calibrated.* The instrument is subject to hysteresis lag and a secular error due to slow changes of the material of construction. This error is allowed for by periodic comparisons between the aneroid and a mercurial barometer. The response is almost immediate to changes in pressure. *No correction for temperature or latitude is necessary.* Actual pressures as read on the aneroid are reduced to any desired elevation precisely as actual pressures are determined by the mercurial barometer.

Altimeter Setting. The altimeter setting is a pressure in inches of mercury. It is the station pressure reduced to sea level in accordance with the U.S. Standard Atmosphere. See Sec. V for a complete discussion of the standard atmosphere. Tables and a nomogram showing variations of pressure, temperature, and elevation in the U.S. Standard atmosphere follow.

Caution: It should be noted that the altimeter setting does not depend on prevailing temperatures, as the M.S.L. pressure does depend on prevailing temperatures. It is important that altimeter settings not be confused with M.S.L. pressure. The former is used by aviation interests in setting barometric altimeters; the latter is used by meteorologists in drawing isobars on weather maps. In airways radio broadcasts, the altimeter setting is known as the *Kollsman number.*

Station Elevation. Station elevation by aviation interests is usually considered to be a point 10 ft above the mean elevation of the runway. Observed reading of the barometer is first *corrected* to give the actual pressure. The actual pressure is then *reduced* to give the pressure at the station elevation, *i.e.,* to give the station pressure. The station pressure is finally *reduced* to give altimeter setting (Kollsman number).

Altimeter Setting. *Example.* Mean elevation of runway = 30 ft. Station elevation = 10 ft above runway = 40 ft above M.S.L. Elevation of ivory point of barometer = 60 ft above M.S.L. Current actual pressure at the ivory point (after making corrections to the observed reading) = 29.825 in.

1. *Station pressure* is obtained by *adding* 0.001 in. for each foot the station elevation is *below* the ivory point, thus

$$\text{Station pressure} = (60 - 40) \times 0.001 + 29.825 = 29.845 \text{ in.}$$

2. *Altimeter setting* is obtained by entering Table 81 with the station elevation to find the amount of the reduction. Enter with 40 ft to find a reduction of 0.044 in. Then

$$\text{Altimeter setting} = 29.845 + 0.044 = 29.889 \text{ in.}$$

The above setting would be reported to the nearest 0.01 in.; *i.e.,* as 29.89 in.

NOTE: It is especially to be noted that established stations within the United States are furnished individually prepared tables which will be used in determining the altimeter setting. Entry is made with station pressure to the nearest 0.01 in. Altimeter setting is read directly without interpolation. For stations not equipped with such tables, the above procedure is adequate.

Pressure Altitude and Altimeter Setting. An alternate but equivalent procedure for determination of altimeter setting involves use of tables of altitude vs. pressure according to the U. S. Standard atmosphere. See for example Table 86a. To obtain altimeter setting proceed as follows: (1) Correct and reduce observed barometric reading to the actual pressure (p), 10 ft above the airport runway by methods given above.   (2) Enter Table 86a with pressure (p) in inches and read the corresponding elevation (z).   This elevation is known as the *pressure altitude*.   (3) Subtract the surveyed elevation of the airport runway from the pressure altitude.   The result in feet is known as the *pressure altitude variation*.   (4) Enter Table 86a with the pressure altitude variation and read the corresponding pressure.   This pressure is the *altimeter setting*.

Example.

1. Pressure (p), corrected and reduced to elevation 10 feet
   above runway.. 28.30 in.
2. Pressure altitude, z (Table 86a)........................ 1,533 ft
3. Pressure altitude variation when elevation of runway is
   500 ft ($z - h$).. 1,033 ft
4. Altimeter setting (Table 86a)........................... 28.82 in.

TABLE 76.—TEMPERATURE CORRECTION OF THE KEW PATTERN BAROMETER (MILLI-
BAR GRADUATIONS)

Corrections (mb) to be applied to the readings of Kew pattern barometers to reduce them to 285°K.

Attached thermometer, °C (add)	Barometer readings										Attached thermometer, °C (subtract)
	860	880	900	920	940	960	980	1000	1020	1040	
11	0.15	0.15	0.15	0.16	0.16	0.16	0.17	0.17	0.17	0.18	13
10	0.30	0.30	0.31	0.32	0.32	0.33	0.34	0.34	0.35	0.36	14
9	0.44	0.45	0.46	0.47	0.48	0.49	0.50	0.51	0.52	0.53	15
8	0.59	0.61	0.62	0.63	0.64	0.66	0.67	0.68	0.70	0.71	16
7	0.74	0.76	0.77	0.79	0.81	0.82	0.84	0.86	0.87	0.89	17
6	0.89	0.91	0.93	0.95	0.97	0.99	1.01	1.03	1.05	1.07	18
5	1.04	1.06	1.08	1.11	1.13	1.15	1.17	1.20	1.22	1.24	19
4	1.19	1.21	1.24	1.26	1.29	1.32	1.34	1.47	1.49	1.42	20
3	1.33	1.36	1.39	1.42	1.45	1.48	1.51	1.54	1.57	1.60	21
2	1.48	1.51	1.55	1.58	1.61	1.64	1.68	1.71	1.74	1.78	22
1	1.63	1.66	1.70	1.74	1.77	1.81	1.85	1.88	1.92	1.95	23
0	1.78	1.82	1.86	1.90	1.93	1.97	2.01	2.05	2.09	2.13	24
− 1	1.93	1.97	2.01	2.05	2.10	2.14	2.18	2.22	2.27	2.31	25
− 2	2.08	2.12	2.17	2.2	2.26	2.30	2.35	2.39	2.44	2.49	26
− 3	2.22	2.27	2.32	2.37	2.42	2.47	2.52	2.57	2.61	2.66	27
− 4	2.37	2.42	2.48	2.53	2.58	2.63	2.68	2.74	2.79	2.84	28
− 5	2.52	2.57	2.63	2.69	2.74	2.80	2.85	2.91	2.96	3.02	29
− 6	2.67	2.73	2.78	2.84	2.90	2.96	3.02	3.08	3.14	3.20	30
− 7	2.82	2.88	2.94	3.00	3.06	3.13	3.19	3.25	3.31	3.37	31
− 8	2.97	3.03	3.09	3.16	3.22	3.29	3.35	3.42	3.49	3.55	32
− 9	3.11	3.18	3.25	3.32	3.39	3.45	3.52	3.59	3.66	3.73	33
−10	3.26	3.33	3.40	3.48	3.55	3.62	3.69	3.76	3.83	3.91	34
−11	3.41	3.48	3.56	3.63	3.71	3.78	3.86	3.93	4.01	4.08	35
−12	3.56	3.63	3.71	3.79	3.87	3.95	4.03	4.10	4.18	4.26	36
−13	3.71	3.79	3.87	3.95	4.03	4.11	4.19	4.28	4.36	4.44	37
−14	3.86	3.94	4.02	4.11	4.19	4.28	4.36	4.45	4.53	4.62	38
−15	4.00	4.09	4.18	4.26	4.35	4.44	4.53	4.62	4.71	4.79	39
−16	4.15	4.24	4.33	4.42	4.51	4.61	4.70	4.79	4.88	4.97	40
−17	4.30	4.39	4.49	4.58	4.68	4.79	4.86	4.96	5.05	5.15	41
−18	4.45	4.54	4.64	4.74	4.84	4.93	5.03	5.13	5.23	5.33	42

TABLE 77.—NORMAL VALUE OF THE ACCELERATION OF GRAVITY AT SEA LEVEL (CM PER SEC^2)

Latitude, °	0°	2°	4°	6°	8°
80	983.06	983.12	983.16	983.19	983.21
70	982.61	982.72	982.82	982.91	982.99
60	981.92	982.07	982.22	982.36	982.49
50	981.07	981.25	981.42	981.59	981.76
40	980.17	980.35	980.53	980.71	980.89
30	979.33	979.49	979.65	979.82	980.00
20	978.64	978.76	978.89	979.03	979.18
10	978.19	978.26	978.34	978.43	978.53
0	978.04	978.04	978.06	978.10	978.14
(90° = 983.22)		(45° = 980.62)			

TABLE 78.—NORMAL VALUES OF $\left(\dfrac{g_\varphi}{9.8}\right)$ AT VARIOUS LATITUDES

Latitude, °	0°	1°	2°	3°	4°	5°	6°	7°	8°	9°
80	1.0031	1.0032	1.0032	1.0032	1.0032	1.0032	1.0033	1.0033	1.0033	1.0033
70	1.0027	1.0027	1.0028	1.0028	1.0029	1.0029	1.0030	1.0030	1.0031	1.0031
60	1.0020	1.0020	1.0021	1.0022	1.0023	1.0023	1.0024	1.0024	1.0025	1.0026
50	1.0011	1.0012	1.0013	1.0014	1.0014	1.0015	1.0016	1.0017	1.0018	1.0019
40	1.0002	1.0003	1.0004	1.0004	1.0005	1.0006	1.0007	1.0008	1.0009	1.0010
30	0.9993	0.9994	0.9995	0.9996	0.9996	0.9997	0.9998	0.9999	1.0000	1.0001
20	0.9986	0.9987	0.9987	0.9988	0.9989	0.9989	0.9990	0.9991	0.9992	0.9992
10	0.9982	0.9982	0.9982	0.9983	0.9983	0.9983	0.9984	0.9984	0.9985	0.9986
0	0.9980	0.9980	0.9980	0.9980	0.9980	0.9980	0.9981	0.9981	0.9981	0.9981

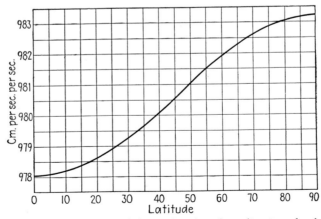

FIG. 40.—Normal value of the acceleration of gravity at sea level.

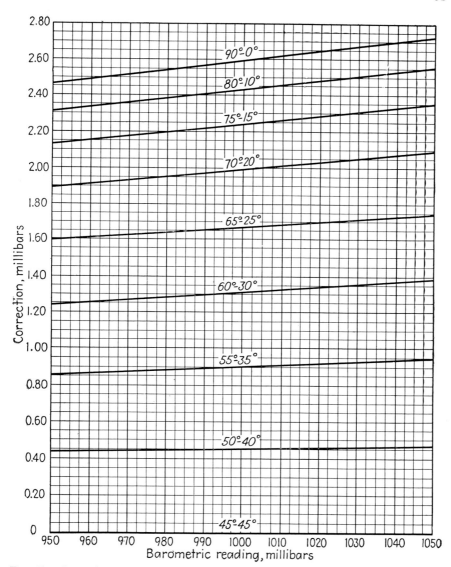

F𝖨𝗀. 41.—Correction of the barometer to standard latitude. Millibars. For latitudes above (below) 45° the values are to be added (subtracted).

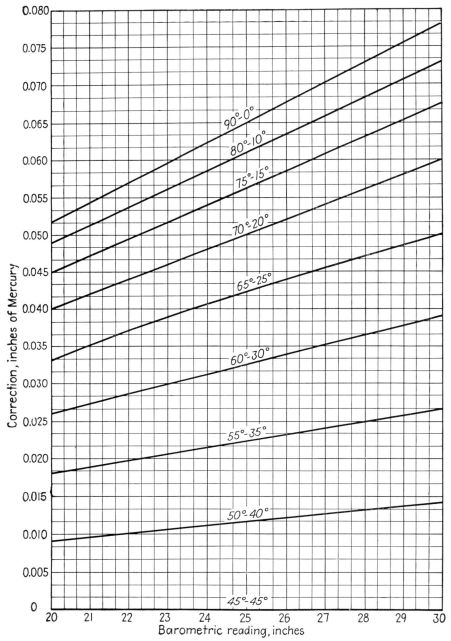

Fig. 42.—Correction of the barometer to standard latitude. English measure. For latitudes above (below) 45° the values are to be added (subtracted).

TABLE 79.—INFLUENCE OF GRAVITY ON BAROMETRIC OBSERVATIONS*
Correction for altitude to be subtracted

Height, ft	Reading of barometer, in.			
	18	22	26	30
2,000			0.003	0.004
4,000			0.006	
6,000		0.008	0.009	
8,000		0.011		
10,000	0.011	0.013		

* From *U.S. Weather Bur. Circ. F.*

TABLE 80.—REDUCTION OF BAROMETRIC READING TO MEAN SEA LEVEL
Reading 30 in.   The correction is always to be added

Height, ft	Temperature of air (dry bulb in shelter), °F				
	0	20	40	60	80
10	0.01	0.01	0.01	0.01	0.01
20	0.02	0.02	0.02	0.02	0.02
30	0.04	0.04	0.03	0.03	0.03
40	0.05	0.05	0.04	0.04	0.04
50	0.06	0.06	0.06	0.05	0.05
60	0.07	0.07	0.07	0.06	0.06
70	0.09	0.08	0.08	0.08	0.07
80	0.10	0.09	0.09	0.09	0.08
90	0.11	0.11	0.10	0.10	0.09
100	0.12	0.12	0.11	0.11	0.10

TABLE 81.—CORRECTION FACTORS TO BE ADDED TO STATION PRESSURE TO OBTAIN
ALTIMETER SETTING
For all pressures between 29.0 and 31.0 in. Hg

Altitude of Station Elevation above M.S.L., Ft	Correction in In. Hg
0	0.000
10	0.011
20	0.022
30	0.033
40	0.044
50	0.055
60	0.066
70	0.077
80	0.088
90	0.099
100	0.110
h	$0.0011h$

TABLE 82.—REDUCTION OF STATION PRESSURE TO M.S.L. PRESSURE*

$$\Delta p = k \cdot p, \qquad \text{where} \qquad k = (e^{\frac{zg}{RT}} - 1)$$

Tabulated values of k are valid for p in mb or in mm

Station elevation, m	Mean temperature, °C							
	-30	-20	-10	0	$+10$	$+20$	$+30$	$+40$
20	0.0028	027	026	025	024	023	023	022
40	0.0056	054	052	050	048	047	045	044
60	0.0085	081	079	076	073	070	068	066
80	0.0113	109	105	101	097	094	090	087
100	0.0141	136	131	126	121	117	113	109
120	0.0170	164	158	152	146	141	136	131
140	0.0199	191	184	177	170	165	159	154
160	0.0228	219	211	203	195	188	181	176
180	0.0257	247	237	228	219	212	204	199
200	0.0286	274	264	254	244	236	227	221
220	0.0315	302	291	280	269	260	250	224
240	0.0344	330	317	306	294	284	273	266
260	0.0373	357	344	331	319	308	296	289
280	0.0402	385	371	357	344	332	320	311
300	0.0431	413	397	383	369	356	344	334
320	0.0460	441	424	409	394	380	367	357
340	0.0490	470	451	435	419	404	391	379
360	0.0519	498	479	461	444	428	414	402
380	0.0548	526	506	487	470	453	437	424
400	0.0578	555	533	513	495	477	461	447
420	0.0608	584	560	539	520	502	485	470
440	0.0638	612	588	566	546	526	508	493
460	0.0668	641	615	592	571	551	532	515
480	0.0698	670	643	618	596	575	555	538
500	0.0728	698	671	645	622	600	579	561

* From Linke.

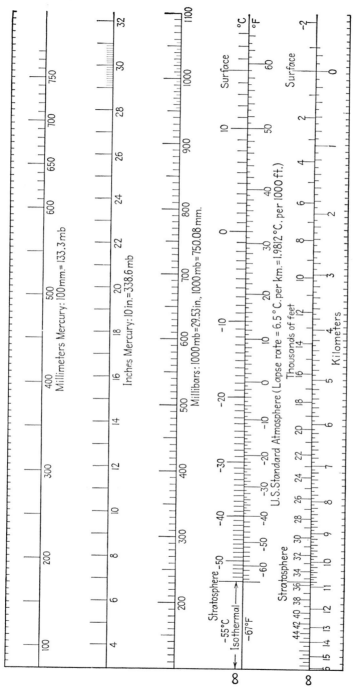

FIG. 43.—Barometric scale conversions and U.S. Standard atmosphere.

SCALES for changing INDICATED ALTITUDE to TRUE ALTITUDE
using Flight Temperatures

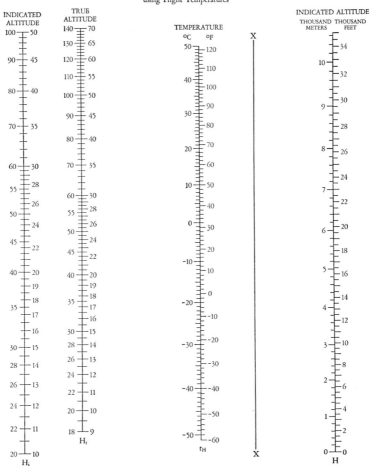

Fig. 44.—Indicated altitude to true altitude using flight temperatures. Put a straight-edge from the temperature scale t_H to the altitude (over sea level) on scale H and make a mark where it goes across line XX. Put the straightedge from this mark to the indicated altitude (over the ground) on scale H_i. Where it goes across H_t is the true altitude (over the ground). Use the right side of scale H_i with the right side of scale H_t and the left side of scale H_i with the left side of H_t. The numbers on these scales may be used for any altitude units; *e.g.*, 35 may be used for 350 m, 3,500 ft, 3,500 m, 35,000 ft, etc. (*Scales by permission.*)

SCALES for changing INDICATED ALTITUDE to TRUE ALTITUDE
using Mean Temperatures

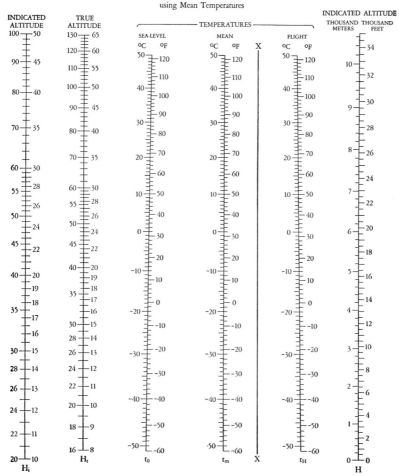

FIG. 45.—Indicated altitude to true altitude using mean temperatures. Put a straight-edge from the sea-level temperature on scale t_0 to the flight temperature on scale t_H. Make a mark where it goes across the mean temperature on scale t_m. Put the straightedge from this mark to the altitude on scale H, and make a mark where it goes across line XX. Put the straightedge from this mark to the indicated altitude on scale H_i. Where it goes across scale H_t is the true altitude. Proceed similarly with ground temperature and indicated altitude over the ground to obtain true altitude over the ground. Use the right (left) side of scale H_i with the right (left) side of scale H_t. The numbers on scales may be used with any altitude units. (*Scales by permission.*)

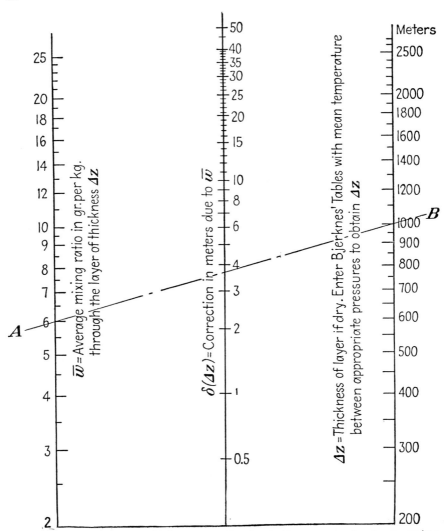

Fɪɢ. 46.—Nomogram for thickness correction due to the presence of water vapor in a stratum. Correction is always added. $\delta(\Delta z) = 0.61\bar{w}\,\Delta z$. *Rule:* To obtain correction lay down straightedge as shown above to find 3.7 m correction to be added because of mean mixing ratio equal to $\bar{w} = 6.0$ grams per kg in a layer 1,000 m thick if dry.

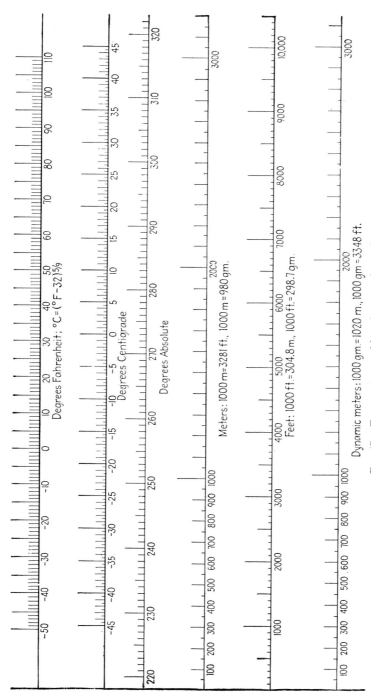

Fig. 47.—Temperature and length conversion scales.

TABLE 83.—INCHES OF STANDARD MERCURY INTO MILLIBARS

Inch	0	1	2	3	4	5	6	7	8	9
27.0	914.3	14.7	15.0	15.3	15.7	16.0	16.4	16.7	17.0	17.4
27.1	917.7	18.1	18.4	18.7	19.1	19.4	19.7	20.1	20.4	20.8
27.2	921.1	21.4	21.8	22.1	22.5	22.8	23.1	23.5	23.8	24.1
27.3	924.5	24.8	25.2	25.5	25.8	26.2	26.5	26.9	27.2	27.5
27.4	927.9	28.2	28.5	28.9	29.2	29.6	29.9	30.2	30.6	30.9
27.5	931.3	31.6	31.9	32.3	32.6	33.0	33.3	33.6	34.0	34.3
27.6	934.6	35.0	35.3	35.7	36.0	36.3	36.7	37.0	37.4	37.7
27.7	938.0	38.4	38.7	39.0	39.4	39.7	40.1	40.4	40.7	41.1
27.8	941.4	41.8	42.1	42.4	42.8	43.1	43.4	43.8	44.1	44.5
27.9	944.8	45.1	45.5	45.8	46.2	46.5	46.8	47.2	47.5	47.9
28.0	948.2	48.5	48.9	49.2	49.5	49.9	50.2	50.6	50.9	51.2
28.1	951.6	51.9	52.3	52.6	52.9	53.3	53.6	53.9	54.3	54.6
28.2	955.0	55.3	55.6	56.0	56.3	56.7	57.0	57.3	57.7	58.0
28.3	958.3	58.7	59.0	59.4	59.7	60.0	60.4	60.7	61.1	61.4
28.4	961.7	62.1	62.4	62.8	63.1	63.4	63.8	64.1	64.4	64.8
28.5	965.1	65.5	65.8	66.1	66.5	66.8	67.2	67.5	67.8	68.2
28.6	968.5	68.8	69.2	69.5	69.9	70.2	70.5	70.9	71.2	71.6
28.7	971.9	72.2	72.6	72.9	73.2	73.6	73.9	74.3	74.6	74.9
28.8	975.3	75.6	76.0	76.3	76.6	77.0	77.3	77.7	78.0	78.3
28.9	978.7	79.0	79.3	79.7	80.0	80.4	80.7	81.0	81.4	81.7
29.0	982.1	82.4	82.7	83.1	83.4	83.7	84.1	84.4	84.8	85.1
29.1	985.4	85.8	86.1	86.5	86.8	87.1	87.5	87.8	88.2	88.5
29.2	988.8	89.2	89.5	89.8	90.2	90.5	90.9	91.2	91.5	91.9
29.3	992.2	92.6	92.9	93.2	93.6	93.9	94.2	94.6	94.9	95.3
29.4	995.6	95.9	96.3	96.6	97.0	97.3	97.6	98.0	98.3	98.6
29.5	999.0	99.3	99.7	00.0	00.4	00.7	01.0	01.4	01.7	02.0
29.6	1002.4	02.7	03.1	03.4	03.7	04.1	04.4	04.7	05.1	05.4
29.7	1005.8	06.1	06.4	06.8	07.1	07.5	07.8	08.1	08.5	08.8
29.8	1009.1	09.5	09.8	10.2	10.5	10.8	11.2	11.5	11.9	12.2
29.9	1012.5	12.9	13.2	13.5	13.9	14.2	14.6	14.9	15.2	15.6
30.0	1015.9	16.3	16.6	16.9	17.3	17.6	18.0	18.3	18.6	19.0
30.1	1019.3	19.6	20.0	20.3	20.7	21.0	21.3	21.7	22.0	22.4
30.2	1022.7	23.0	23.4	23.7	24.0	24.4	24.7	25.1	25.4	25.7
30.3	1026.1	26.4	26.8	27.1	27.4	27.8	28.1	28.4	28.8	29.1
30.4	1029.5	29.8	30.1	30.5	30.8	31.2	31.5	31.8	32.2	32.5
30.5	1032.9	33.2	33.5	33.9	34.2	34.5	34.9	35.2	35.6	35.9
30.6	1036.2	36.6	36.9	37.3	37.6	37.9	38.3	38.6	38.9	39.3
30.7	1039.6	40.0	40.3	40.6	41.0	41.3	41.7	42.0	42.3	42.7
30.8	1043.0	43.3	43.7	44.0	44.4	44.7	45.0	45.4	45.7	46.1
30.9	1046.4	46.7	47.1	47.4	47.8	48.1	48.4	48.8	49.1	49.5
31.0	1049.8	50.1	50.5	50.8	51.1	51.5	51.8	52.2	52.5	52.8

Table 84.—The U.S. Standard Atmosphere

Press., mb	Elevation		Temp., °C	Press., mb	Elevation		Temp., °C
	Ft	M			Ft	M	
1,080	−1,780	−543	18.5	540	16,400	5,000	−17.5
1,070	−1,520	−463	18.0	530	16,860	5,140	−18.4
1,060	−1,260	−384	17.5	520	17,330	5,280	−19.3
1,050	−990	−302	17.0	510	17,800	5,430	−20.3
1,040	−730	−220	16.4	500	18,280	5,570	−21.2
1,030	−460	−140	15.9	490	18,770	5,720	−22.2
1,020	−180	−60	15.4	480	19,260	5,870	−23.2
1,013.3	**0,000**	**000**	**15.0**	470	19,770	6,030	−24.2
1,010	90	30	14.8	460	20,280	6,180	−25.2
1,000	370	110	14.3	450	20,800	6,340	−26.2
990	640	200	13.7	440	21,330	6,500	−27.3
980	930	280	13.2	430	21,870	6,670	−28.3
970	1,210	370	12.6	420	22,420	6,840	−29.4
960	1,490	450	12.1	410	22,990	7,010	−30.5
950	1,780	540	11.5	400	23,560	7,180	−31.7
940	2,060	630	10.9	390	24,150	7,360	−32.8
930	2,350	720	10.3	380	24,740	7,540	−34.0
920	2,650	810	9.8	370	25,350	7,730	−35.2
910	2,940	900	9.2	360	25,980	7,920	−36.5
900	3,240	990	8.6	350	26,610	8,110	−37.7
890	3,540	1,080	8.0	340	27,270	8,310	−39.0
880	3,850	1,170	7.4	330	27,940	8,520	−40.4
870	4,150	1,270	6.8	320	28,620	8,730	−41.7
860	4,460	1,360	6.2	310	29,320	8,940	−43.1
850	4,780	1,460	5.5	300	30,050	9,160	−44.5
840	5,100	1,550	4.9	290	30,790	9,390	−46.0
830	5,420	1,650	4.3	280	31,550	9,620	−47.5
820	5,740	1,750	3.6	270	32,330	9,860	−49.1
810	6,060	1,850	3.0	260	33,150	10,100	−50.7
800	6,390	1,950	2.3	250	33,980	10,360	−52.3
790	6,720	2,050	1.7	250	34,840	10,620	−54.0
780	7,060	2,150	1.0	**234**	**35,332**	**10,769**	**−55.0**
770	7,400	2,250	0.3	230	35,740	10,890	**−55.0**
760	7,740	2,360	−0.3	220	36,670	11,180	**−55.0**
750	8,080	2,460	−1.0	210	37,640	11,470	**−55.0**
740	8,440	2,570	−1.7	200	38,660	11,790	**−55.0**
730	8,790	2,680	−2.4	190	39,740	12,110	**−55.0**
720	9,150	2,790	−3.1	180	40,870	12,460	**−55.0**
710	9,510	2,900	−3.8	170	42,060	12,820	**−55.0**
700	9,880	3,010	−4.6	160	43,330	13,210	**−55.0**
690	10,250	3,120	−5.3	150	44,680	13,620	**−55.0**
680	10,620	3,240	−6.0	140	46,130	14,060	**−55.0**
670	11,000	3,350	−6.8	130	47,680	14,530	**−55.0**
660	11,380	3,470	−7.6	120	49,360	15,040	**−55.0**
650	11,770	3,590	−8.3	110	51,180	15,600	**−55.0**
640	12,160	3,710	−9.1	100	53,170	16,210	**−55.0**
630	12,570	3,830	−9.9				
620	12,970	3,950	−10.7				
610	13,380	4,080	−11.5				
600	13,790	4,200	−12.3				
590	14,220	4,330	−13.2				
580	14,640	4,460	−14.0				
570	15,070	4,590	−14.9				
560	15,510	4,730	−15.7				
550	15,950	4,860	−16.6				

TABLE 85.—PRESSURE, MB AT VARIOUS GEOMETRIC HEIGHTS ABOVE SEA LEVEL
According to the U.S. Standard Atmosphere*

Geometric height, hundreds of meters above sea level	0	1	2	3	4	5	6	7	8	9
−0	1,013	1,025	1,038	1,050						
+0	1,013	1,001	989	978	966	955	943	932	921	910
10	899	888	877	866	856	846	835	825	815	805
20	795	785	775	766	756	747	737	728	719	710
30	701	692	683	675	666	658	649	641	633	624
40	616	608	600	593	585	577	570	562	555	547
50	540	533	526	519	512	505	498	491	485	478
60	472	465	459	453	446	440	434	428	422	416
70	410	405	399	393	388	382	377	372	366	361
80	356	351	346	341	336	331	326	321	317	312
90	307	303	298	294	289	285	281	277	272	268
100	264	260	256	252	248	245	241	237	233	230
110	226	223	219	216	212	209	206	203	200	196
120	193	190	187	185	182	179	176	173	171	168
130	165	163	160	158	155	153	151	148	146	144
140	141	139	137	135	133	131	129	127	125	123
150	121	119	117	115	113	112	110	108	107	105
160	103	102	100	99	97	96	94	93	91	90
170	88	87	86	84	83	82	80	79	78	77
180	76	74	73	72	71	70	69	68	67	66
190	65	64	63	62	61	60	59	58	57	56
200	55.2									
210	47.2									
220	40.4									
230	34.5									
240	29.5									
250	25.2									
260	21.6									
270	18.4									
280	15.8									
290	13.5									
300	11.5									

* From *U.S. Weather Bur. Circ. N.*

TABLE 86.—ALTITUDES, PRESSURES, AND TEMPERATURES*
U.S. Standard Atmosphere

Ft	M	In. mercury	Mm mercury	Air temp.,°F	Air temp.,°C	Mean temp.,°F	Mean temp.,°C
−1,640.4	−500		806.2	+65.0	+18.3	+61.9	+16.6
−1,000	−304.8	31.02	787.9	62.6	17.0	60.8	16.0
0	0	29.92	760.0	59.0	15.0	59.0	15.0
+1,000	+304.8	28.86	732.9	55.4	13.0	57.2	14.0
1,640.4	500		716.0	53.1	11.7	56.1	13.4
2,000	609.6	27.82	706.6	51.8	11.0	55.4	13.0
3,000	914.4	26.81	681.1	48.4	9.1	53.6	12.0
3,280.8	1,000		674.1	47.3	8.5	53.1	11.7
4,000	1,219.2	25.84	656.3	44.8	7.1	51.8	11.0
4,921.2	1,500		634.2	41.4	5.2	50.2	10.1
5,000	1,524.0	24.89	632.3	41.2	5.1	50.0	10.0
6,000	1,828.8	23.98	609.0	37.6	3.1	48.2	9.0
6,561.7	2,000		596.2	35.6	2.0	47.1	8.4
7,000	2,133.6	23.09	586.4	34.0	+1.1	46.4	8.0
8,000	2,438.4	22.22	564.4	30.6	−0.8	44.6	7.0
8,202.1	2,500		560.1	29.8	−1.2	44.2	6.8
9,000	2,743.2	21.38	543.2	27.0	−2.8	42.8	6.0
9,842.5	3,000		525.8	23.9	−4.5	41.4	5.2
10,000	3,048.0	20.58	522.6	23.4	−4.8	41.0	5.0
11,000	3,352.8	19.79	502.6	19.8	−6.8	39.2	4.0
11,483	3,500		493.2	18.0	−7.8	38.3	3.5
12,000	3,657.6	19.03	483.3	16.2	−8.8	37.2	2.9
13,000	3,962.4	18.29	464.5	12.6	−10.8	35.4	1.9
13,123	4,000		462.2	12.2	−11.0	35.2	1.8
14,000	4,267.2	17.57	446.4	9.1	−12.7	33.6	0.9
14,764	4,500		432.9	7.6	−14.2	32.2	+0.1
15,000	4,572.0	16.88	428.8	5.5	−14.7	31.8	−0.1
16,000	4,876.8	16.21	411.8	1.9	−16.7	29.8	−1.2
16,404	5,000		405.1	+0.5	−17.5	29.2	−1.6
17,000	5,181.6	15.56	395.3	−1.7	−18.7	28.0	−2.2
18,000	5,486.4	14.94	379.4	−5.3	−20.7	26.2	−3.2
18,045	5,500		378.7	−5.4	−20.8	26.1	−3.3
19,000	5,791.2	14.33	364.0	−8.7	−22.6	25.7	−4.3
19,685	6,000		353.8	−11.2	−24.0	23.0	−5.0
20,000	6,096.0	13.75	349.1	−12.3	−24.6	22.5	−5.3
21,000	6,400.8	13.18	334.7	−15.9	−26.6	20.7	−6.3
21,325	6,500		330.2	−17.1	−27.3	19.9	−6.7
22,000	6,705.6	12.63	320.8	−19.5	−28.6	18.7	−7.4
22,966	7,000		307.8	−22.9	−30.5	16.9	−8.4
23,000	7,010.4	12.10	307.4	−23.1	−30.6	16.9	−8.4
24,000	7,315.2	11.59	294.4	−26.5	−32.5	14.9	−9.5
24,606	7,500		286.8	−28.7	−33.7	13.8	−10.1
25,000	7,620.0	11.10	281.9	−30.1	−34.5	13.1	−10.5
26,000	7,924.8	10.62	269.8	−33.7	−36.5	+11.1	−11.6
26,247	8,000		266.9	−34.6	−37.0	10.6	−11.9

TABLE 86.—ALTITUDES, PRESSURES, AND TEMPERATURES.*—(*Continued*)
U.S. Standard Atmosphere

Ft	M	In. mercury	Mm mercury	Air temp.,°F	Air temp.,°C	Mean temp.,°F	Mean temp.,°C
27,000	8,229.6	10.16	258.1	−37.3	−38.5	9.1	−12.7
27,887	8,500		248.1	−40.5	−40.3	7.5	−13.6
28,000	8,534.4	9.72	246.9	−40.9	−40.5	7.3	−13.7
29,000	8,839.2	9.29	236.0	−44.5	−42.5	5.4	−14.8
29,528	9,000		230.5	−47.0	−43.5	4.3	−15.4
30,000	9,144.0	8.88	225.6	−47.9	−44.4	3.4	−15.9
31,000	9,448.8	8.48	215.5	−53.5	−46.4	1.6	−16.9
31,168	9 500		213.8	−54.1	−46.7	+1.2	−17.1
32,000	9,753.6	8.10	205.8	−55.1	−48.4	−0.4	−18.0
32,808	10,000		198.2	−58.0	−50.0	−2.0	−18.9
33,000	10,058	7.73	196.4	−58.7	−50.4	−2.0	−19.1
34,000	10,363	7.38	187.4	−62.3	−52.4	−4.4	−20.2
34,449	10,500		183.4	−63.9	−53.3	−5.3	−20.7
35,000	10,668	7.04	178.7	−65.7	−54.3	−6.3	−21.3
36,000	10,973	6.71	170.4	−67.0	−55.0	−8.1	−22.3
36,089	11,000		169.7	−67.0	−55.0	−8.3	−22.4
37,000	11,278	6.39	162.4	−67.0	−55.0	−9.9	−23.3
37,730	11,500		156.9	−67.0	−55.0	−11.2	−24.0
38,000	11,582	6.10	154.9	−67.0	−55.0	−11.7	−24.3
39,000	11,887	5.81	147.6	−67.0	−55.0	−13.4	−25.2
39,370	12,000		145.0	−67.0	−55.0	−13.4	−25.2
40,000	12,192	5.54	140.7	−67.0	−55.0	−14.8	−26.0
41,000	12,497	5.28	134.2	−67.0	−55.0	−16.2	−26.8
41,010	12,500		134.1	−67.0	−55.0	−16.2	−26.8
42,000	12,802	5.04	127.9	−67.0	−55.0	−17.7	−27.6
42,651	13,000		124.0	−67.0	−55.0	−18.8	−28.2
43,000	13,106	4.80	122.0	−67.0	−55.0	−18.9	−28.3
44,000	13,411	4.58	116.3	−67.0	−55.0	−20.2	−29.0
44,291	13,500		114.7	−67.0	−55.0	−20.7	−29.2
45,000	13,716	4.36	110.8	−67.0	−55.0	−21.3	−29.6
45,932	14,000		106.0	−67.0	−55.0	−22.4	−30.2
46,000	14,021	4.16	105.7	−67.0	−55.0	−22.4	−30.2
47,000	14,326	3.97	100.7	−67.0	−55.0	−23.4	−30.8
47,572	14,500		98.0	−67.0	−55.0	−24.2	−31.2
48,000	14,630	3.781	96.05	−67.0	−55.0	−24.5	−31.4
49,000	14,935	3.605	91.57	−67.0	−55.0	−25.4	−31.9
49,212	15,000		90.6	−67.0	−55.0	−25.4	−31.9
50,000	15,240	3.436	87.30	−67.0	−55.0	−25.6	−32.0

* From Irvin, "Aircraft Instruments," 2d ed., McGraw-Hill, New York, 1944.

TABLE 86a.—ALTITUDE-PRESSURE TABLE (IN. HG)*
U.S. Standard Atmosphere

Pressure, in. Hg	Altitude, Ft.									
	0.00	0.01	0.02	0.03	0.04	0.05	0.06	0.07	0.08	0.09
21.0	9,471	9,458	9,446	9,434	9,422	9,409	9,397	9,385	9,372	9,360
21.1	9,348	9,336	9,323	9,311	9,299	9,287	9,274	9,262	9,250	9,238
21.2	9,225	9,213	9,201	9,189	9,176	9,164	9,152	9,140	9,128	9,116
21.3	9,103	9,091	9,079	9,067	9,055	9,043	9,030	9,018	9,006	8,994
21.4	8,982	8,970	8,958	8,946	8,933	8,921	8,909	8,897	8,885	8,873
21.5	8,861	8,849	8,837	8,825	8,813	8,801	8,789	8,776	8,764	8,752
21.6	8,740	8,728	8,716	8,704	8,692	8,680	8,668	8,656	8,644	8,632
21.7	8,620	8,608	8,596	8,584	8,572	8,560	8,548	8,536	8,524	8,512
21.8	8,500	8,489	8,477	8,465	8,453	8,441	8,429	8,417	8,405	8,393
21.9	8,381	8,369	8,357	8,346	8,334	8,322	8,310	8,298	8,286	8,274
22.0	8,262	8,250	8,239	8,227	8,215	8,203	8,191	8,179	8,168	8,156
22.1	8,144	8,132	8,120	8,109	8,097	8,085	8,073	8,061	8,050	8,038
22.2	8,026	8,014	8,003	7,991	7,979	7,967	7,956	7,944	7,932	7,920
22.3	7,909	7,897	7,885	7,873	7,862	7,850	7,838	7,827	7,815	7,803
22.4	7,791	7,780	7,768	7,756	7,745	7,733	7,721	7,710	7,698	7,686
22.5	7,675	7,663	7,652	7,640	7,628	7,617	7,605	7,593	7,582	7,570
22.6	7,559	7,547	7,535	7,524	7,512	7,501	7,489	7,478	7,466	7,454
22.7	7,443	7,431	7,420	7,408	7,397	7,385	7,374	7,362	7,350	7,339
22.8	7,327	7,316	7,304	7,293	7,281	7,270	7,258	7,247	7,235	7,224
22.9	7,212	7,201	7,189	7,178	7,167	7,155	7,144	7,132	7,121	7,109
23.0	7,098	7,086	7,075	7,064	7,052	7,041	7,029	7,018	7,006	6,995
23.1	6,984	6,972	6,961	6,949	6,938	6,927	6,915	6,904	6,893	6,881
23.2	6,870	6,858	6,847	6,836	6,824	6,813	6,802	6,790	6,779	6,768
23.3	6,756	6,745	6,734	6,722	6,711	6,700	6,688	6,677	6,666	6,655
23.4	6,643	6,632	6,621	6,610	6,598	6,587	6,576	6,564	6,553	6,542
23.5	6,531	6,519	6,508	6,497	6,486	6,475	6,463	6,452	6,441	6,430
23.6	6,418	6,407	6,396	6,385	6,374	6,363	6,351	6,340	6,329	6,318
23.7	6,307	6,296	6,284	6,273	6,262	6,251	6,240	6,229	6,218	6,206
23.8	6,195	6,184	6,173	6,162	6,151	6,140	6,129	6,118	6,106	6,095
23.9	6,084	6,073	6,062	6,051	6,040	6,029	6,018	6,007	5,996	5,985
24.0	5,974	5,962	5,951	5,940	5,929	5,918	5,907	5,896	5,885	5,874
24.1	5,863	5,852	5,841	5,830	5,819	5,808	5,797	5,786	5,775	5,764
24.2	5,753	5,742	5,731	5,720	5,709	5,698	5,687	5,676	5,666	5,655
24.3	5,644	5,633	5,622	5,611	5,600	5,589	5,578	5,567	5,555	5,545
24.4	5,534	5,524	5,513	5,502	5,491	5,480	5,469	5,458	5,447	5,436
24.5	5,425	5,415	5,404	5,393	5,382	5,371	5,360	5,350	5,339	5,328
24.6	5,317	5,306	5,295	5,285	5,274	5,263	5,252	5,241	5,230	5,220
24.7	5,209	5,198	5,187	5,176	5,166	5,155	5,144	5,133	5,123	5,112
24.8	5,101	5,090	5,080	5,069	5,058	5,047	5,037	5,026	5,015	5,004
24.9	4,994	4,983	4,972	4,961	4,951	4,940	4,929	4,919	4,908	4,897
25.0	4,886	4,876	4,865	4,854	4,844	4,833	4,822	4,812	4,801	4,790
25.1	4,780	4,769	4,758	4,748	4,737	4,726	4,716	4,705	4,695	4,684
25.2	4,673	4,663	4,652	4,642	4,631	4,620	4,610	4,599	4,588	4,578
25.3	4,567	4,557	4,546	4,536	4,525	4,514	4,504	4,493	4,483	4,472
25.4	4,462	4,451	4,440	4,430	4,419	4,409	4,398	4,388	4,377	4,367
25.5	4,356	4,346	4,335	4,325	4,314	4,304	4,293	4,283	4,272	4,262
25.6	4,251	4,241	4,230	4,220	4,209	4,199	4,188	4,178	4,167	4,157
25.7	4,146	4,136	4,125	4,115	4,105	4,094	4,084	4,073	4,063	4,052
25.8	4,042	4,032	4,021	4,011	4,000	3,990	3,980	3,969	3,959	3,948
25.9	3,938	3,928	3,917	3,907	3,896	3,886	3,876	3,865	3,855	3,845
26.0	3,834	3,824	3,814	3,803	3,793	3,782	3,772	3,762	3,751	3,741
26.1	3,731	3,720	3,710	3,700	3,689	3,679	3,669	3,659	3,648	3,638
26.2	3,628	3,617	3,607	3,597	3,586	3,576	3,566	3,556	3,545	3,535
26.3	3,525	3,515	3,504	3,494	3,484	3,474	3,463	3,453	3,443	3,433
26.4	3,422	3,412	3,402	3,392	3,382	3,371	3,361	3,351	3,341	3,331
26.5	3,320	3,310	3,300	3,290	3,279	3,269	3,259	3,249	3,239	3,229
26.6	3,218	3,208	3,198	3,188	3,178	3,168	3,157	3,147	3,137	3,127
26.7	3,117	3,107	3,097	3,086	3,076	3,066	3,056	3,046	3,036	3,026
26.8	3,016	3,005	2,995	2,985	2,975	2,965	2,955	2,945	2,935	2,925
26.9	2,915	2,905	2,895	2,884	2,874	2,864	2,854	2,844	2,834	2,824

* From Irvin, "Aircraft Instruments," 2d ed., McGraw-Hill, New York, 1944.

TABLE 86a.—ALTITUDE-PRESSURE TABLE (IN. HG).*—(*Continued*)
U.S. Standard Atmosphere

Pressure, in. Hg	Altitude, Ft.									
	0.00	0.01	0.02	0.03	0.04	0.05	0.06	0.07	0.08	0.09
27.0	2,814	2,804	2,794	2,784	2,774	2,764	2,754	2,744	2,734	2,724
27.1	2,714	2,704	2,694	2,684	2,674	2,664	2,654	2,644	2,634	2,624
27.2	2,614	2,604	2,594	2,584	2,574	2,564	2,554	2,544	2,534	2,524
27.3	2,514	2,504	2,494	2,484	2,474	2,464	2,454	2,444	2,434	2,425
27.4	2,415	2,405	2,395	2,385	2,375	2,365	2,355	2,345	2,335	2,325
27.5	2,315	2,306	2,296	2,286	2,276	2,266	2,256	2,246	2,236	2,226
27.6	2,217	2,207	2,197	2,187	2,177	2,167	2,158	2,148	2,138	2,128
27.7	2,118	2,108	2,098	2,098	2,079	2,069	2,059	2,049	2,040	2,030
27.8	2,020	2,010	2,000	1,990	1,981	1,971	1,961	1,951	1,942	1,932
27.9	1,922	1,912	1,902	1,893	1,883	1,873	1,863	1,854	1,844	1,834
28.0	1,824	1,814	1,805	1,795	1,785	1,776	1,766	1,756	1,746	1,737
28.1	1,727	1,717	1,707	1,698	1,688	1,678	1,668	1,659	1,649	1,639
28.2	1,630	1,620	1,610	1,601	1,591	1,581	1,572	1,562	1,552	1,542
28.3	1,533	1,523	1,513	1,504	1,494	1,484	1,475	1,465	1,456	1,446
28.4	1,436	1,427	1,417	1,407	1,398	1,388	1,378	1,369	1,359	1,350
28.5	1,340	1,330	1,321	1,311	1,302	1,292	1,282	1,273	1,263	1,254
28.6	1,244	1,234	1,225	1,215	1,206	1,196	1,186	1,177	1,167	1,158
28.7	1,148	1,139	1,129	1,120	1,110	1,100	1,091	1,081	1,072	1,062
28.8	1,053	1,043	1,034	1,024	1,015	1,005	995	986	976	967
28.9	957	948	938	929	919	910	900	891	881	872
29.0	863	853	844	834	825	815	806	796	787	777
29.1	768	758	749	739	730	721	711	702	692	683
29.2	673	664	655	645	636	626	617	607	598	589
29.3	579	570	560	551	542	532	523	514	504	495
29.4	485	476	467	457	448	439	429	420	410	401
29.5	392	382	373	364	354	345	336	326	318	308
29.6	298	289	280	270	261	252	242	233	224	215
29.7	205	196	187	177	168	159	149	140	131	122
29.8	112	103	94	85	75	66	57	47	38	29
29.9	20	10	+1	−8	−17	−26	−36	−45	−54	−63
30.0	−73	−82	−91	−100	−110	−119	−128	−137	−146	−156
30.1	−165	−174	−183	−192	−202	−211	−220	−229	−238	−248
30.2	−257	−266	−275	−284	−293	−303	−312	−321	−330	−339
30.3	−348	−358	−367	−376	−385	−394	−403	−412	−421	−431
30.4	−440	−449	−458	−467	−476	−485	−494	−504	−513	−522
30.5	−531	−540	−549	−558	−567	−576	−585	−594	−604	−613
30.6	−622	−631	−640	−649	−658	−667	−676	−685	−694	−703
30.7	−712	−721	−730	−740	−749	−758	−767	−776	−785	−794
30.8	−803	−812	−821	−830	−839	−848	−857	−866	−875	−884
30.9	−893	−902	−911	−920	−929	−938	−947	−956	−965	−974
31.0	−983	−992	−1,001	−1,010	−1,019	−1,028	−1,037	−1,046	−1,055	−1,064

* From Irvin, "Aircraft Instruments," 2d ed., McGraw-Hill, New York, 1944.

TABLE 87.—DISTANCE BETWEEN STANDARD ISOBARIC SURFACES IN DYNAMIC METERS*

Mean virtual temp. stratum, °C	0	1	2	3	4	5	6	7	8	9
100 mb										
−100	3442	3422	3402	3382	3362	3343	3323	3303	3283	3263
− 90	3641	3621	3601	3581	3561	3542	3522	3502	3482	3462
− 80	3840	3820	3800	3780	3760	3741	3721	3701	3681	3661
− 70	4039	4019	3999	3979	3959	3939	3920	3900	3880	3860
− 60	4238	4218	4198	4178	4158	4138	4119	4099	4079	4059
− 50	4437	4417	4397	4377	4357	4337	4318	4298	4278	4258
− 40	4636	4616	4596	4576	4556	4536	4516	4497	4477	4457
− 30	4835	4815	4795	4775	4755	4735	4715	4696	4676	4656
200 mb										
− 90	2130	2118	2107	2095	2083	2072	2060	2048	2037	2025
− 80	2246	2235	2223	2211	2200	2188	2176	2165	2153	2142
− 70	2363	2351	2339	2328	2316	2304	2293	2281	2270	2258
− 60	2479	2467	2456	2444	2432	2421	2409	2398	2386	2374
− 50	2595	2584	2572	2561	2549	2537	2526	2514	2502	2491
− 40	2712	2700	2689	2677	2665	2654	2642	2630	2619	2607
− 30	2828	2817	2805	2793	2782	2770	2758	2747	2735	2723
− 20	2945	2933	2921	2910	2898	2886	2875	2863	2851	2840
300 mb										
− 80	1594	1585	1577	1569	1561	1552	1544	1536	1528	1519
− 70	1676	1668	1660	1652	1643	1635	1627	1619	1610	1602
− 60	1759	1751	1742	1733	1726	1718	1709	1701	1693	1685
− 50	1841	1833	1825	1817	1808	1800	1792	1784	1775	1767
− 40	1924	1916	1908	1899	1891	1883	1874	1866	1858	1850
− 30	2007	1998	1990	1982	1974	1965	1957	1949	1941	1932
− 20	2089	2081	2073	2064	2056	2048	2040	2031	2023	2015
− 10	2172	2164	2155	2147	2139	2130	2122	2114	2106	2097
400 mb										
− 70	1300	1294	1287	1281	1275	1268	1262	1255	1249	1243
− 60	1364	1358	1351	1345	1339	1332	1326	1319	1313	1307
− 50	1428	1422	1416	1409	1403	1396	1390	1384	1377	1371
− 40	1492	1486	1480	1473	1467	1460	1454	1448	1441	1435
− 30	1556	1550	1544	1537	1531	1524	1518	1512	1505	1499
− 20	1621	1614	1608	1601	1595	1588	1582	1576	1569	1563
− 10	1685	1678	1672	1665	1659	1653	1646	1640	1633	1627
− 0	1749	1742	1736	1729	1723	1717	1710	1704	1697	1691
500 mb										
− 60	1115	1109	1104	1099	1094	1089	1083	1078	1073	1068
− 50	1167	1162	1157	1151	1146	1141	1136	1130	1125	1120
− 40	1219	1214	1209	1204	1198	1193	1188	1183	1178	1172
− 30	1272	1266	1261	1256	1251	1246	1240	1235	1230	1225
− 20	1324	1319	1314	1308	1303	1298	1293	1287	1282	1277
− 10	1376	1371	1366	1361	1355	1350	1345	1340	1335	1329
− 0	1429	1423	1418	1413	1408	1403	1397	1392	1387	1382
600 mb + 0	1429	1434	1439	1444	1450	1455	1460	1465	1471	1476

TABLE 87.—DISTANCE BETWEEN STANDARD ISOBARIC SURFACES IN DYNAMIC METERS. *
(Continued)

	Mean virtual temp. stratum, °C	0	1	2	3	4	5	6	7	8	9	
600 mb—												—600 mb
	− 50	987	982	978	973	969	965	960	956	951	947	
	− 40	1031	1027	1022	1018	1013	1009	1004	1000	996	991	
	− 30	1075	1071	1066	1062	1058	1053	1049	1044	1040	1035	
	− 20	1119	1115	1111	1106	1102	1097	1093	1088	1084	1080	
	− 10	1164	1159	1155	1150	1146	1142	1137	1133	1128	1124	
	− 0	1208	1204	1199	1195	1190	1186	1181	1177	1173	1168	
	+ 0	1208	1212	1217	1221	1226	1230	1235	1239	1243	1248	
	+ 10	1252	1257	1261	1265	1270	1274	1279	1283	1288	1292	
700 mb—												—700 mb
	− 40	893	889	885	882	878	874	870	866	862	859	
	− 30	931	928	924	920	916	912	908	905	901	897	
	− 20	970	966	962	958	954	951	947	943	939	935	
	− 10	1008	1004	1000	997	993	989	985	981	977	974	
	− 0	1046	1043	1039	1035	1031	1027	1023	1020	1016	1012	
	+ 0	1046	1050	1054	1058	1062	1066	1069	1073	1077	1081	
	+ 10	1085	1089	1092	1096	1100	1104	1108	1112	1115	1119	
	+ 20	1123	1127	1131	1135	1138	1142	1146	1150	1154	1158	
800 mb—												—800 mb
	− 40	788	784	781	778	774	771	767	764	761	757	
	− 30	822	818	815	811	808	805	801	798	795	791	
	− 20	855	852	849	845	842	838	835	832	828	825	
	− 10	889	886	882	879	876	872	869	866	862	859	
	− 0	923	920	916	913	909	906	903	899	896	893	
	+ 0	923	926	930	933	937	940	943	947	950	953	
	+ 10	957	960	964	967	970	974	977	980	984	987	
	+ 20	991	994	997	1001	1004	1008	1011	1014	1018	1021	
	+ 30	1024	1028	1031	1035	1038	1041	1045	1048	1051	1055	
900 mb—												—900 mb
	− 40	705	702	699	696	693	690	687	684	680	677	
	− 30	735	732	729	726	723	720	714	714	711	708	
	− 20	765	762	759	756	753	750	747	744	741	738	
	− 10	795	792	789	786	783	780	777	774	771	768	
	− 0	826	823	820	817	814	811	808	804	801	798	
	+ 0	826	829	832	835	838	841	844	847	850	853	
	+ 10	856	859	862	865	868	871	874	877	880	883	
	+ 20	886	889	892	895	898	901	904	907	910	913	
	+ 30	916	919	922	925	928	931	935	938	941	944	
	+ 40	947	950	953	956	959	962	965	968	971	974	
1000 mb—												—1000 mb
		0	1	2	3	4	5	6	7	8	9	

* After V. Bjerknes.

TABLE 88.—VIRTUAL TEMPERATURE DIFFERENCE AT SATURATION
$(T_v - T)_s$

Temp., °C	Pressure, mb									Temp., °C
	1,000	900	800	700	600	500	400	300	200	
	$T_v - T$									
−50	0.0	0.0	0.0	0.0	0.0	0.0	0.0	0.0	0.0	
−40	0.0	0.0	0.0	0.0	0.0	0.0	0.0	0.0	0.1	
−30	0.0	0.0	0.0	0.1	0.1	0.1	0.1	0.1	0.2	
−20	0.1	0.1	0.1	0.1	0.2	0.2	0.2	0.3		
−10	0.2	0.2	0.2	0.2	0.3	0.3	0.4			
− 5	0.3	0.3	0.3	0.4	0.4	0.5	0.6	900	1,000	
− 2	0.4	0.5	0.5	0.6	0.7	0.8		mb	mb	
− 1	0.5	0.6	0.7	0.8	0.9	1.1		9.9	8.8	40
0	0.6	0.6	0.7	0.8	0.9	1.2		9.3	8.3	39
1	0.6	0.7	0.8	0.9	1.0	1.3		8.8	7.9	38
2	0.7	0.8	0.8	1.0	1.1			8.3	7.4	37
3	0.7	0.8	0.9	1.0	1.2		800	7.8	7.0	36
4	0.8	0.9	1.0	1.1	1.3		mb	7.3	6.6	35
5	0.8	0.9	1.1	1.2	1.4		7.8	6.9	6.2	34
6	0.9	1.0	1.1	1.3	1.5		7.3	6.5	5.8	33
7	1.0	1.1	1.2	1.4	1.6		6.9	6.1	5.5	32
8	1.1	1.3	1.4	1.6	1.9		6.5	5.7	5.2	31
9	1.2	1.4	1.5	1.7	2.0		6.1	5.4	4.8	30
10	1.3	1.5	1.6	1.9			5.7	5.1	4.6	29
11	1.4	1.6	1.8	2.0			5.4	4.8	4.3	28
12	1.5	1.7	1.9	2.1			5.0	4.5	4.0	27
13	1.6	1.8	2.0	2.3		700	4.7	4.2	3.8	26
14	1.7	1.9	2.2	2.5		mb	4.5	4.0	3.6	25
15	1.8	2.0	2.3	2.6		4.8	4.2	3.7	3.3	24
16	2.0	2.2	2.5	2.8		4.5	3.9	3.5	3.1	23
17	2.1	2.3	2.6	3.0		4.2	3.7	3.3	2.9	22
18	2.3	2.5	2.8	3.2		3.9	3.4	3.0	2.7	21
19	2.4	2.7	3.0	3.4		3.7	3.2	2.9	2.6	20

The tabulated values may be multiplied by the relative humidity to give the actual virtual temperature difference.

TABLE 89.—PRESSURE INCREASE, MB PER 500 M THICKNESS*

Pressure at the lower level, mb	Mean virtual temperature															
	40°	38°	36°	34°	32°	30°	28°	26°	24°	22°	20°	18°	16°	14°	12°	10°
1,050	55.8	56.1	56.5	56.8	57.2	57.5	57.9	58.3	58.7	59.1	59.4	59.9	60.3	60.7	61.1	61.5
1,000	53.1	53.4	53.8	54.1	54.5	54.8	55.2	55.5	55.9	56.3	56.6	57.0	57.4	57.8	58.2	58.6
950	50.5	50.8	51.1	51.4	51.8	52.1	52.4	52.8	53.1	53.5	53.8	54.2	54.6	54.9	55.3	55.7
900	47.8	48.1	48.4	48.7	49.1	49.4	49.7	50.0	50.3	50.7	51.0	51.3	51.7	52.0	52.4	52.7
850						46.6	46.9	47.2	47.5	47.9	48.2	48.5	48.8	49.1	49.5	49.8
800						43.9	44.1	44.4	44.7	45.0	45.3	45.6	45.9	46.2	46.6	46.9
750											42.5	42.8	43.1	43.4	43.7	44.0
700											39.7	40.0	40.2	40.5	40.8	41.0

Pressure at the lower level, mb	10°	8°	6°	4°	2°	0°	−2°	−4°	−6°	−8°	−10°	−12°	−14°	−16°	−18°	−20°
1,050	61.5	62.0	62.4	62.8	63.6	63.7	64.2	64.6	65.1	65.6	66.1	66.6	67.0	67.5	68.0	68.6
1,000	58.6	59.0	59.4	59.8	60.3	60.7	61.1	61.5	62.0	62.4	62.9	63.4	63.8	64.3	64.8	65.3
950	55.7	56.1	56.4	56.8	57.2	57.6	58.0	58.4	58.9	59.3	59.8	60.2	60.6	61.1	61.6	62.0
900	52.7	53.1	53.5	53.8	54.2	54.6	55.0	55.4	55.8	56.2	56.6	57.0	57.4	57.9	58.3	58.8
850	49.8	50.2	50.5	50.8	51.2	51.5	51.9	52.3	52.7	53.0	53.4	53.8	54.2	54.6	55.0	55.5
800	46.9	47.2	47.5	47.8	48.2	48.5	48.8	49.2	49.6	49.9	50.3	50.6	51.0	51.4	51.8	52.2
750	44.0	44.3	44.6	44.9	45.2	45.5	45.8	46.1	46.5	46.8	47.1	47.5	47.8	48.2	48.6	48.9
700	41.0	41.3	41.6	41.9	42.2	42.4	42.7	43.0	43.4	43.7	44.0	44.3	44.6	45.0	45.3	45.6
650	38.1	38.4	38.6	38.9	39.2	39.4	39.7	40.0	40.3	40.6	40.9	41.2	41.5	41.8	42.1	42.4
600	35.2	35.4	35.7	35.9	36.2	36.4	36.6	36.9	37.2	37.5	37.7	38.0	38.3	38.6	38.8	39.1
550						33.3	33.6	33.8	34.1	34.3	34.6	34.8	35.1	35.3	35.6	35.9
500						30.3	30.5	30.7	31.0	31.2	31.4	31.6	31.9	32.1	32.4	32.6
450											28.3	28.5	28.7	28.9	29.1	29.3
400											25.1	25.3	25.5	25.7	25.9	26.1

Pressure at the lower level, mb	−20°	−22°	−24°	−26°	−28°	−30°	−32°	−34°	−36°	−38°	−40°	−42°	−44°	−46°	−48°	−50°
1,050	68.6	69.1	69.6	70.1	70.7	71.3	71.8	72.4	73.0	73.5	74.1					
1,000	65.3	65.8	66.3	66.8	67.3	67.9	68.4	69.0	69.5	70.0	70.6					
950	62.0	62.5	63.0	63.5	64.0	64.5	65.0	65.5	66.0	66.5	67.0					
900	58.8	59.2	59.6	60.1	60.6	61.1	61.5	62.0	62.5	63.0	63.5					
850	55.5	55.9	56.3	56.8	57.2	57.7	58.1	58.6	59.0	59.5	60.0					
800	52.2	52.6	53.0	53.4	53.8	54.3	54.7	55.1	55.5	56.0	56.4	56.8	57.3	57.7	58.2	58.7
750	48.9	49.3	49.7	50.1	50.5	50.9	51.2	51.7	52.1	52.5	52.9	53.3	53.7	54.1	54.6	55.1
700	45.6	46.0	46.4	46.7	47.1	47.5	47.8	48.2	48.6	49.0	49.3	49.7	50.1	50.5	51.0	51.4
650	42.4	42.7	43.0	43.4	43.7	44.1	44.4	44.8	45.1	45.5	45.8	46.2	46.5	46.9	47.3	47.8
600	39.1	39.4	39.7	40.0	40.4	40.7	41.0	41.3	41.6	42.0	42.3	42.6	43.0	43.3	43.7	44.1
550	35.9	36.1	36.4	36.7	37.0	37.3	37.5	37.8	38.1	38.5	38.7	39.1	39.5	39.7	40.1	40.5
500	32.6	32.8	33.1	33.3	33.6	33.9	34.1	34.4	34.7	35.0	35.2	35.5	35.8	36.1	36.5	36.8
450	29.3	29.6	29.8	30.0	30.2	30.5	30.7	31.0	31.2	31.5	31.7	32.0	32.2	32.5	32.8	33.1
400	26.1	26.3	26.5	26.7	26.9	27.1	27.3	27.6	27.8	28.0	28.2	28.4	28.7	28.9	29.2	29.4
350	22.8	23.0	23.2	23.4	23.6	23.7	23.9	24.1	24.3	24.5	24.7	24.9	25.1	25.3	25.5	25.8
300	19.6	19.7	19.9	20.0	20.2	20.3	20.5	20.7	20.8	21.0	21.2	21.3	21.5	21.7	21.9	22.1
250											17.6	17.8	17.9	18.1	18.2	18.4
200											14.0	14.2	14.3	14.4	14.5	14.7
150											10.5	10.6	10.7	10.8	10.9	11.0
100												7.0	7.1	7.2	7.3	7.4

Pressure at the lower level, mb	−50°	−52°	−54°	−56°	−58°	−60°	−62°	−64°	−66°	−68°	−70°	−72°	−74°	−76°	−78°	−80°
400	29.4	29.7	30.0	30.2	30.5	30.8										
350	25.8	26.0	26.2	26.4	26.7	27.0										
300	22.1	22.3	22.5	22.7	22.9	23.1	23.3	23.5	23.8	24.0	24.2	24.4	24.6	24.8	25.1	25.3
250	18.4	18.6	18.7	18.9	19.1	19.2	19.4	19.6	19.8	20.0	20.2	20.3	20.5	20.7	20.9	21.1
200	14.7	14.8	14.9	15.1	15.2	15.4	15.5	15.7	15.8	16.0	16.1	16.3	16.4	16.6	16.7	16.9
150	11.0	11.1	11.2	11.3	11.4	11.5	11.6	11.7	11.8	12.0	12.1	12.2	12.3	12.4	12.5	12.6
100	7.4	7.4	7.5	7.5	7.6	7.7	7.8	7.8	7.9	8.0	8.0	8.1	8.2	8.3	8.3	8.4

* After Cannegieter.

TABLE 90.—TABLE OF FACTORS TO CHANGE METERS INTO DYNAMIC METERS (GM)*
Use mean values from table.

φ	0°	10°	20°	30°	40°	50°	60°	70°	80°
0 m	2.19	2.18	2.14	2.07	1.98	1.89	1.81	1.74	1.69%
2,000	2.26	2.24	2.20	2.13	2.04	1.96	1.87	1.80	1.76
4,000	2.32	2.30	2.26	2.19	2.11	2.02	1.93	1.86	1.82
6,000	2.38	2.37	2.32	2.25	2.17	2.08	1.99	1.92	1.88
8,000	2.44	2.43	2.38	2.32	2.23	2.14	2.06	1.99	1.94
10,000	2.51	2.49	2.45	2.38	2.29	2.26	2.12	2.05	2.00
12,000	2.57	2.55	2.51	2.44	2.35	2.26	2.18	2.11	2.06
14,000	2.63	2.61	2.57	2.50	2.42	2.32	2.24	2.17	2.13
16,000	2.69	2.68	2.63	2.56	2.48	2.39	2.30	2.23	2.19
18,000	2.75	2.74	2.69	2.62	2.54	2.45	2.36	2.29	2.25
20,000	2.81	2.80	2.75	2.68	2.60	2.51	2.43	2.36	2.31
22,000	2.88	2.86	2.82	2.75	2.66	2.57	2.49	2.42	2.37
24,000	2.94	2.92	2.88	2.81	2.72	2.63	2.55	2.48	2.44
26,000	3.00	2.98	2.94	2.87	2.78	2.70	2.61	2.54	2.50
28,000	3.06	3.04	3.00	2.93	2.85	2.76	2.67	2.60	2.56

* After Linke.
Example. At 40° latitude, 14,000 m above sea level becomes

$$14,000 - \frac{1.98 + 2.42}{2} \times \frac{14,000}{100} = 13,692 \text{ gm.}$$

TABLE 91.—TABLE OF FACTORS TO CHANGE DYNAMIC METERS (GM) INTO METERS*
Use mean values from table.

Gm	$\varphi = 0°$	10°	20°	30°	40°	50°	60°	70°	80°
0 m	2.24	2.23	2.19	2.11	2.02	1.93	1.84	1.77	1.72
2,000	2.31	2.29	2.25	2.18	2.09	2.00	1.91	1.83	1.79
4,000	2.38	2.35	2.31	2.24	2.16	2.06	1.97	1.90	1.85
6,000	2.44	2.42	2.38	2.30	2.22	2.12	2.03	1.96	1.92
8,000	2.50	2.49	2.44	2.37	2.28	2.19	2.10	2.03	1.98
10,000	2.57	2.55	2.51	2.44	2.34	2.25	2.17	2.09	2.04
12,000	2.64	2.62	2.57	2.50	2.41	2.31	2.23	2.16	2.11
14,000	2.70	2.68	2.64	2.56	2.48	2.38	2.29	2.22	2.18
16,000	2.76	2.75	2.70	2.63	2.54	2.45	2.35	2.28	2.24
18,000	2.83	2.82	2.76	2.69	2.60	2.51	2.42	2.35	2.30
20,000	2.90	2.88	2.83	2.75	2.67	2.57	2.49	2.42	2.36
22,000	2.97	2.94	2.90	2.82	2.73	2.64	2.55	2.48	2.43
24,000	3.03	3.01	2.96	2.89	2.80	2.70	2.62	2.54	2.50
26,000	3.09	3.07	3.03	2.95	2.86	2.77	2.68	2.60	2.56
28,000	3.16	3.14	3.09	3.02	2.93	2.84	2.74	2.67	2.63

* After Linke.

TABLE 92.—DIMENSIONS OF EARTH, LANDS AND SEAS*

	Area, km²	Average depth, m	Average height, m	Volume, km³
Whole earth........	510,100,000			1,083,000,000,000
Whole ocean........	361,160,000	4,117		1,486,900,000
All land............	148,940,000			
Pacific Ocean.......	165,200,000	4,282		707,500,000
Atlantic Ocean......	82,400,000	3,926		323,600,000
Indian Ocean.......	73,500,000	3,963		291,000,000
Arctic Ocean........	14,090,000	1,205		17,000,000
Seas...............	25,970,000			
Asia...............	44,134,000		960	
Europe............	10,009,000		340	
Africa.............	29,834,000		750	
North America......	24,063,000		720	
South America......	17,788,000		590	
Australia..........	8,901,000		340	
Antarctica.........	14,169,000		2,200	

* From Gutenberg, "Internal Constitution of the Earth," McGraw-Hill, New York, 1939.

TABLE 93.—TEMPERATURE LAG (t DAYS)*

Depth		κ			
Ft	m	0.0031 dry soil	0.0049 damp soil	0.0064 crystal rocks	0.0133 sandstone
1	0.30	10.0	8.0	7.0	4.8
5	1.52	50.2	39.9	34.9	24.2
10	3.05	100.4	79.9	69.9	48.5
15	4.57	150.6	119.8	104.8	72.7
20	6.10	200.8	159.7	139.8	96.9
25	7.62	251.0	199.7	174.7	121.2
30	9.14	301.2	239.6	209.6	145.4
35	10.67	351.4	279.5	244.6	169.7

* From Gutenberg, "Internal Constitution of the Earth," McGraw-Hill, 1939.

TABLE 94.—APPROXIMATE TIME LAG IN THE TRANSMISSION OF SOLAR HEAT THROUGH WALLS AND ROOFS*

Material and Thickness of Wall or Roof	Time Lag, Hr.
2 in. pine..	1½
3 in. concrete + 1 in. cork..................................	2
4 in. gypsum...	2½
6 in. concrete..	3
22 in. brick and tile..	10
16 in. concrete + 1.5 in. cork..............................	19

* From A.S.H.V.E. Guide.

TABLE 95.—LENGTH OF 1° MERIDIAN AT DIFFERENT LATITUDES*

Latitude, °	Geographic Miles (1' of the Equator)
0	59.594
10	59.612
20	59.665
30	59.745
40	59.845
50	59.951
60	60.051
70	60.132
80	60.186
90	60.204

* From Smithsonian Tables.

TABLE 96.—LENGTH OF 1° OF THE PARALLEL AT DIFFERENT LATITUDES*

Latitude, °	Geographic Miles (1' of the Equator)
0	60.000
5	59.773
10	59.095
15	57.969
20	56.404
25	54.411
30	52.006
35	49.204
40	46.027
45	42.498
50	38.644
55	34.493
60	30.076
65	25.428
70	20.583
75	15.578
80	10.453
85	5.247
90	0.000

* From Smithsonian Tables.

TABLE 97.—MOLECULAR DENSITIES OF THE GASES OF THE ATMOSPHERE*

z, km	N_2	O_2	A	CO_2	He	Total
0	2.0×10^{19}	5.4×10^{18}	2.4×10^{17}	7.7×10^{15}	1.0×10^{14}	2.6×10^{19}
20	1.5×10^{18}	4.0×10^{17}	1.8×10^{16}	5.7×10^{14}	7.1×10^{12}	1.9×10^{18}
40	8.4×10^{16}	2.3×10^{16}	1.0×10^{15}	3.2×10^{13}	4.3×10^{11}	1.1×10^{17}
60	7.8×10^{15}	2.1×10^{15}	9.4×10^{13}	3.0×10^{12}	4.0×10^{10}	9.9×10^{15}
80	8.0×10^{14}	2.2×10^{14}	9.6×10^{12}	3.0×10^{11}	4.1×10^{9}	1.0×10^{15}
100	1.2×10^{14}	3.3×10^{13}	1.5×10^{12}	4.6×10^{10}	6.3×10^{8}	1.6×10^{14}
120	2.0×10^{13}	5.4×10^{12}	2.4×10^{11}	7.7×10^{9}	1.3×10^{8}	2.6×10^{13}
140	3.5×10^{12}	9.3×10^{11}	4.2×10^{10}	1.3×10^{9}	1.8×10^{7}	4.4×10^{12}
160	6.1×10^{11}	1.5×10^{11}	5.3×10^{9}	1.5×10^{8}	5.6×10^{6}	7.7×10^{11}
180	1.1×10^{11}	2.2×10^{10}	4.8×10^{8}	1.0×10^{7}	4.4×10^{6}	1.4×10^{11}
200	2.2×10^{10}	3.3×10^{9}	4.5×10^{7}	7.5×10^{5}	3.5×10^{6}	2.5×10^{10}
220	3.2×10^{9}	5.1×10^{8}	4.4×10^{6}	4.7×10^{4}	2.8×10^{6}	3.7×10^{9}
240	5.3×10^{8}	7.8×10^{7}	4.1×10^{5}	3.6×10^{3}	2.2×10^{6}	6.1×10^{8}
260	8.7×10^{7}	1.2×10^{7}	3.9×10^{4}	2.9×10^{2}	1.8×10^{6}	1.0×10^{8}
280	1.4×10^{7}	1.8×10^{6}	3.6×10^{3}	2.3×10	1.4×10^{6}	1.7×10^{7}
300	6.4×10^{6}	2.8×10^{5}	3.3×10^{2}	1.8	1.1×10^{6}	7.7×10^{6}
350	2.6×10^{4}	2.6×10^{3}			7.1×10^{5}	7.3×10^{5}
400					4.1×10^{5}	4.1×10^{5}

* From Fleming, "Terrestrial Magnetism and Electricity," McGraw-Hill, New York, 1939.

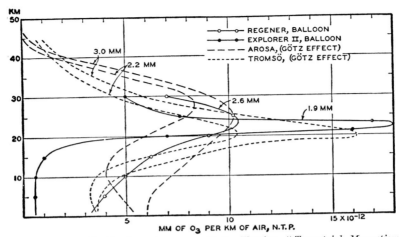

Fig. 48.—Vertical distribution of ozone. (*From Fleming, "Terrestrial Magnetism and Electricity," McGraw-Hill, New York, 1939.*)

TABLE 98.—CHARACTERISTICS OF FALLING RAINDROPS*

Drops			Number of drops per m²/sec								
Diameter		Volume, mm³	Rain "looking very ordinary"	Rain with breaks during which the sun shone	Beginning of a short fall like a thundershower	Sudden rain from a small cloud	Violent rain like a cloudburst, with some hail	Period of heaviest cloudburst	Period of less heavy cloudburst	Ending period of continuous fall
Mm	In.		1	2	3	4	5	6	7	8	9
0.5	0.019	0.065	1,000	1,600	129	60	0	100	514	679	7
1.0	0.039	0.524	200	120	100	280	50	1,300	423	524	233
1.5	0.059	1.77	140	60	73	160	50	500	359	347	113
2.0	0.079	4.19	140	200	100	20	150	200	138	295	46
2.5	0.098	8.18	0	0	29	20	0	0	156	205	7
3.0	0.118	14.1	0	0	57	0	200	0	138	81	0
3.5	0.138	22.4	0	0	0	0	0	0	0	28	32
4.0	0.157	33.5	0	0	0	0	50	0	0	20	39
4.5	0.177	47.7	0	0	0	0	0	200	101	0	0
5.0	0.196	65.4	0	0	0	0	0	0	0	0	25
Total...............			1,480	1,980	488	540	500	2,300	1,829	2,179	502
Rate of rainfall:											
Mm per min.........			0.06	0.07	0.10	0.04	0.31	0.72	0.57	0.38	0.25
In. per hr..........			0.14	0.16	0.23	0.10	0.74	1.69	1.35	0.89	0.60

* From Meinzer, "Hydrology," McGraw-Hill, New York, 1942.

TABLE 99.—VELOCITY OF FALLING RAINDROPS*

Diameter, mm	Maximum falling velocity, mps	
	Lenard	Laws
1.0	4.4	
2.0	5.9	6.6
3.0	7.0	8.0
4.0	7.7	8.8
5.0	7.9	9.2
5.5	8.0	9.3
6.5	7.8	

* From Meinzer, "Hydrology," McGraw-Hill, New York, 1942.

TABLE 100.—DEPTH OF WATER IN INCHES CORRESPONDING TO THE WEIGHT OF SNOW
(OR RAIN) COLLECTED IN AN 8-IN. GAUGE*
1 lb = 0.5507 in.

Weight, lb	0.00	0.01	0.02	0.03	0.04	0.05	0.06	0.07	0.08	0.09
0.0	0.00	0.01	0.01	0.02	0.02	0.03	0.03	0.04	0.04	0.05
0.1	0.06	0.06	0.07	0.07	0.08	0.08	0.09	0.09	0.10	0.10
0.2	0.11	0.12	0.12	0.13	0.13	0.14	0.14	0.15	0.15	0.16
0.3	0.17	0.17	0.18	0.18	0.19	0.19	0.19	0.20	0.21	0.22
0.4	0.22	0.23	0.23	0.24	0.24	0.25	0.25	0.26	0.26	0.27
0.5	0.28	0.28	0.29	0.29	0.30	0.30	0.31	0.31	0.32	0.33
0.6	0.34	0.34	0.34	0.35	0.35	0.36	0.36	0.37	0.38	0.38
0.7	0.39	0.39	0.40	0.40	0.41	0.41	0.42	0.43	0.43	0.44
0.8	0.44	0.45	0.45	0.46	0.46	0.47	0.47	0.48	0.49	0.49
0.9	0.50	0.50	0.51	0.51	0.52	0.52	0.53	0.54	0.54	0.55

* From Smithsonian Tables.

TABLE 101.—DEPTH OF WATER CORRESPONDING TO THE WEIGHT OF A CYLINDRICAL
SNOW CORE 2.655 IN. IN DIAMETER*
$\frac{1}{5}$ lb = 1 in.

Weight pounds.............	0.0	0.5	1.0	1.5	2.0	2.5	3.0
Inches, water..............	0.00	2.50	5.00	7.50	10.00	12.50	15.00

* From Smithsonian Tables.

TABLE 102.—TEMPERATURE, $T°K$, OF THE DAY ATMOSPHERE*

z, km	T, °K	z, km	T, °K
0	287	60	260
10	220	80	320
20	225	100	360
30	230	200	360
40	240	220	360

* After Maris.

TABLE 103.—CORIOLIS PARAMETER AND DERIVATIVE ($\lambda = 2\omega \sin \varphi$)

$$\frac{\partial \lambda}{\partial y} = \frac{(2\omega \cos \varphi)}{a}$$

Latitude, ° (φ)	$\lambda \times 10^4$ sec^{-1}	$\dfrac{\partial \lambda}{\partial y} \times 10^{13}$ cm^{-1} sec^{-1}
0	0.00	2.29
15	0.38	2.12
30	0.73	1.98
45	1.03	1.62
60	1.26	1.14
75	1.41	0.59
90	1.46	0.00

$a = 6.37 \times 10^8$ cm = radius of the earth
$\omega = 7.29 \times 10^{-5}$ sec^{-1} = angular velocity of the earth

TABLE 104.—HEIGHT OF HOMOGENEOUS ATMOSPHERE, METERS (AT 45° LATITUDE)

$$H_\varphi = \frac{g_\varphi}{g_{45}} \times H_{45}$$

Surface temp., °C	0	1	2	3	4	5	6	7	8	9
−30	7,113	7,084	7,054	7,025	6,996	6,966	6,937	6,908	6,878	6,849
−20	7,405	7,376	7,347	7,318	7,289	7,260	7,230	7,200	7,171	7,142
−10	7,689	7,668	7,639	7,610	7,581	7,552	7,522	7,492	7,463	7,434
− 0	7,991	7,962	7,932	7,903	7,874	7,844	7,815	7,786	7,756	7,727
+ 0	7,991	8,020	8,049	8,078	8,108	8,138	8,167	8,196	8,225	8,254
+10	8,283	8,312	8,341	8,370	8,400	8,430	8,459	8,488	8,517	8,546
+20	8,576	8,605	8,634	8,664	8,693	8,722	8,752	8,781	8,810	8,840
+30	8,869	8,898	8,927	8,956	8,986	9,016	9,045	9,074	9,103	9,132

TABLE 105.—DENSITY OF AIR, KG PER M^3

Pressure, mb	Virtual temp., °C											
	−70	−60	−50	−40	−30	−20	−10	0	10	20	30	40
100	0.172	0.164	0.156	0.150	0.143	0.138	0.132	0.128	0.123	0.119	0.115	0.111
200	0.343	0.327	0.312	0.299	0.287	0.275	0.265	0.255	0.246	0.238	0.230	0.223
300	0.514	0.491	0.468	0.449	0.430	0.413	0.397	0.383	0.369	0.357	0.345	0.334
400	0.686	0.654	0.625	0.598	0.573	0.550	0.530	0.510	0.492	0.475	0.460	0.446
500	0.858	0.818	0.781	0.748	0.717	0.689	0.662	0.648	0.615	0.594	0.575	0.556
600	1.030	0.981	0.937	0.897	0.860	0.826	0.795	0.766	0.738	0.713	0.689	0.668
700	1.202	1.146	1.095	1.047	1.004	0.965	0.927	0.894	0.862	0.833	0.805	0.779
800	1.374	1.310	1.250	1.197	1.146	1.102	1.059	1.020	0.986	0.952	0.920	0.891
900	1.544	1.472	1.406	1.345	1.290	1.239	1.192	1.148	1.108	1.071	1.035	1.002
1,000	1.715	1.635	1.562	1.495	1.434	1.376	1.325	1.276	1.230	1.190	1.150	1.113
1,100	1.887	1.801	1.720	1.645	1.578	1.515	1.459	1.405	1.354	1.308	1.265	1.225

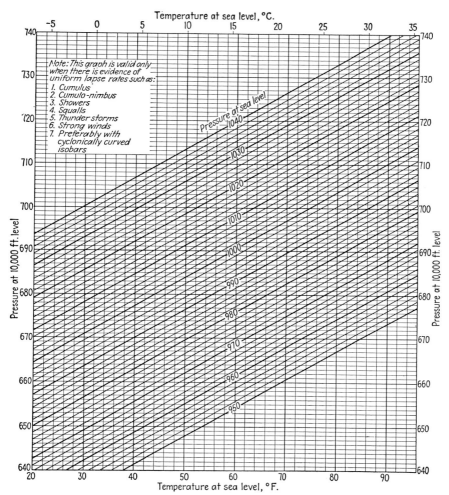

Fig. 49.—Graph for *approximately* extrapolating the pressure at the 10,000-ft level from surface observations. Computed on the assumption that the sounding lies along a wet adiabatic. Note the restrictions on use of the graph listed in upper left hand corner of figure.

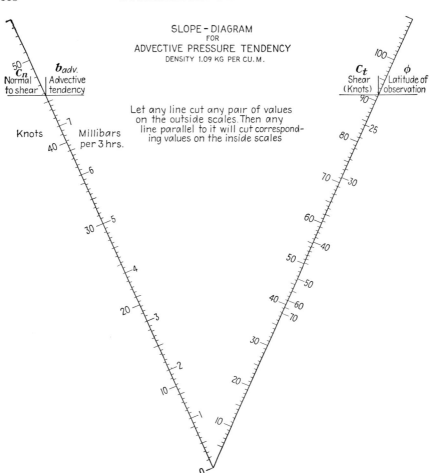

Fɪɢ. 50.—Slope diagram for advective pressure tendency.   (*Courtesy of J. Fulford.*)

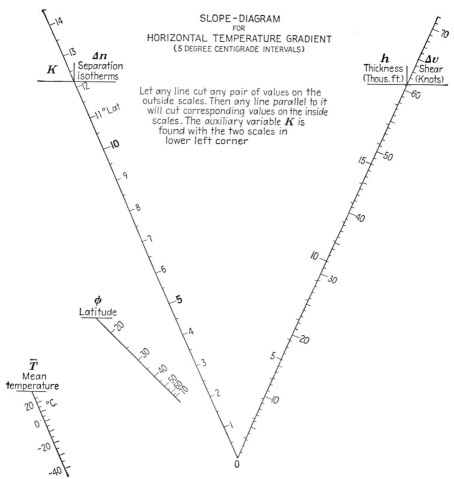

SLOPE - DIAGRAM
FOR
HORIZONTAL TEMPERATURE GRADIENT
(5 DEGREE CENTIGRADE INTERVALS)

Let any line cut any pair of values on the
outside scales. Then any line parallel to it
will cut corresponding values on the inside
scales. The auxiliary variable **K** is
found with the two scales in
lower left corner

Fɪɢ. 51.—Slope diagram for horizontal temperature gradient.   (*Courtesy of J. Fulford.*)

TABLE 106.—HEIGHT OF CLOUD OR CEILING, FT, LIGHT BEAM PROJECTED VERTICALLY*

Angle	tan	Base			Angle	tan	Base		
		500 ft h	1,000 ft h	1,500 ft h			500 ft h	1,000 ft h	1,500 ft h
5	0.08749	44	87	131	46	1.0355	518	1,036	1,554
6	0.10510	52	105	157	47	1.0724	536	1,072	1,608
7	0.12278	62	123	185	48	1.1106	556	1,111	1,667
8	0.14054	70	141	211	49	1.1504	575	1,150	1,725
9	0.15838	79	158	237	50	1.1918	596	1,192	1,788
10	0.17633	88	176	264	51	1.2349	618	1,235	1,853
11	0.19438	97	194	291	52	1.2799	640	1,280	1,920
12	0.21256	106	213	319	53	1.3270	664	1,327	1,991
13	0.23087	116	231	347	54	1.3764	688	1,376	2,064
14	0.24933	124	249	373	55	1.4281	714	1,428	2,142
15	0.26795	134	268	402	56	1.4826	742	1,483	2,225
16	0.28675	144	287	430	57	1.5399	770	1,540	2,310
17	0.30573	153	306	459	58	1.6002	800	1,600	2,400
18	0.32492	162	325	487	59	1.6643	832	1,664	2,496
19	0.34433	172	344	516	60	1.7321	866	1,732	2,598
20	0.36397	182	364	546	61	1.8040	902	1,804	2,706
21	0.38386	192	384	576	62	1.8807	940	1,881	2,821
22	0.40403	202	404	606	63	1.9626	982	1,963	2,945
23	0.42447	212	424	636	64	2.0503	1,025	2,050	3,075
24	0.44523	222	445	667	65	2.1445	1,072	2,144	3,216
25	0.46631	233	466	699	66	2.2460	1,123	2,246	3,369
26	0.48773	244	488	732	67	2.3559	1,178	2,356	3,534
27	0.50953	255	510	765	68	2.4751	1,238	2,475	3,173
28	0.53171	266	532	798	69	2.6051	1,302	2,605	3,907
29	0.55431	277	554	831	70	2.7475	1,374	2,748	4,122
30	0.57735	288	577	865	71	2.9042	1,452	2,904	4,356
31	0.60086	300	601	901	72	3.0777	1,539	3,078	4,617
32	0.62487	312	625	937	73	3.2709	1,636	3,271	4,907
33	0.64941	324	649	973	74	3.4874	1,744	3,487	5,231
34	0.67451	338	675	1,013	75	3.7321	1,866	3,732	5,598
35	0.70021	350	700	1,050	76	4.0108	2,006	4,011	6,017
36	0.72654	364	727	1,091	77	4.3315	2,166	4,332	6,498
37	0.75355	377	754	1,131	78	4.7046	2,352	4,705	7,057
38	0.78129	390	781	1,171	79	5.1446	2,572	5,145	7,717
39	0.80978	405	810	1,215	80	5.6713	2,836	5,671	8,507
40	0.83910	420	839	1,259	81	6.3138	3,157	6,314	9,471
41	0.86929	434	869	1,303	82	7.1154	3,558	7,115	10,673
42	0.90040	450	900	1,350	83	8.1443	4,072	8,144	12,276
43	0.93252	466	933	1,399	84	9.5144	4,757	9,514	14,211
44	0.96569	483	966	1.449	85	11.430	5,715	11,430	17,175
45	1.0000	500	1,000	1,500	86	14.301	7,150	14,301	21,441

* From *U.S. Weather Bur. Circ. N.*

Bibliography

Smithsonian Meteorological Tables.

Smithsonian Physical Tables.

Smithsonian Mathematical Tables.

National Bureau of Standards Mathematical Tables (MT1 to MT17).

International Critical Tables.

BOWDITCH, N.: "American Practical Navigator."

Aerographer's Manual, U.S. Navy.

War Department Technical Manual, "The Weather Observer," TM1-235.

Admiralty Weather Manual.

Instructions for Airway Meteorological Serivce, *U.S. Weather Bur. Circ. N.*

Measurement of Pressure, *U.S. Weather Bur. Circ. F.*

Pilot Balloons, *U.S. Weather Bur. Circ. O.*

U.S. Weather Bureau psychrometric tables.

Tables for Horizontal Distance of Pilot Balloons; 30 grams, *U.S. Weather Bur. Form* 1043.

Tables for Horizontal Distance of Pilot Balloons; 100 grams, *U.S. Weather Bur. Form* 10434A.

"Meteorologisches Taschenbuch," IV, Linke, Leipzig.

Hydrographic Office publications (various).

U.S. Weather Bureau publications (various).

Nautical Almanac.

Instructions for Modulated Audio Frequency Radiosonde Observations, *U.S Weather Bur., Circ. P.*, 5th ed., 1945.

SECTION II

METEOROLOGICAL MATHEMATICS
AND CALCULATIONS

By Frederick A. Ficken

CONTENTS

SECTION II

METEOROLOGICAL MATHEMATICS
AND CALCULATIONS

By Frederick A. Ficken

1. REAL NUMBERS

The applications of mathematics are based fundamentally on the familiar system of real numbers.

1.11. Rational Numbers. The numbers $\cdots, -3, -2, -1, 0, 1, 2, 3, \cdots$ are integers. Any two integers may be added, subtracted, or multiplied. If p and q are integers, and $q \neq 0$, then $r = p/q$ is a *rational number*. Any two rational numbers may be added, subtracted, multiplied, or divided, provided that the *divisor* is *not* zero. The rational numbers are real numbers.

1.12. Representation of Rationals on a Line. On a given line, let a point O and a different point A be chosen. Accept OA as a unit of distance. Let $r = p/q$ be any rational number, with the signs of p and q so adjusted that q is positive. Divide OA

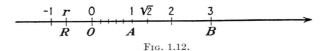

Fig. 1.12.

into q equal parts. From O measure off p of these parts, in the direction of A or in the opposite direction according as r is positive or negative, *e.g.*, $p = -3, q = 5, r = -3/5$ in the figure. For each rational number r, this construction yields precisely one point R on the line.

1.13. Real Numbers and Their Representation on a Line. It is easy to find, on the line of Fig. 1.12, many points not obtainable in this way from any rational number r, *e.g.*, lay off from O the diagonal (length $\sqrt{2}$) of a square whose side is OA. These gaps in the rationals are filled by *irrational* real numbers devised for the purpose by mathematicians. Every real number is rational or irrational. Thus the real numbers may be represented on the line of Fig. 1.12 in such a way that each real number r measures the directed distance from O of precisely one point R of the line, and, conversely, each point R of the line has its directed distance from O measured by precisely one real number r.

The directed distance *from* the point R *to* the point S is *always* $s - r$. In Fig. 1.12, the directed distance from A to B is $3 - 1 = 2$; that from B to A is $1 - 3 = -2$.

1.21. Order and Inequalities among Real Numbers. If $r - s$ is positive, we say that r is greater than s and write $r > s$ or $s < r$. If either $r > s$ or $r = s$, write $r \geqslant s$; similarly for $r \leqslant s$. In terms of Fig. 1.12, $r < s$ means that the displacement from R to S has the same direction as the displacement from O to A, *i.e.*, the directed distance from R to S is positive; if $s < r$, reverse the direction. To denote that r is not equal to s, write $r \neq s$. To denote that either $r = s$ or $r = -s$, write $r = \pm s$.

Properties of Inequalities. 1. If $a < b$ and $b < c$, then $a < c$. Special Case. If $a_1 \leqslant a_2 \leqslant \cdots \leqslant a_n$, then $a_1 = a_n$ if and only if $a_1 = a_2 = \cdots = a_n$; thus one inequality would require $a_1 < a_n$.

124

2. If a and b are any real numbers, then precisely one of the following three relations holds: $a > b$, $a = b$, $a < b$.

3. If $a < b$, then $a + c < b + c$.

4. a. If $a < b$ and $0 < r$, then $ar < br$, e.g., since $2 < 3$ and $0 < \frac{1}{6}$, $\frac{1}{3} < \frac{1}{2}$.

   b. If $a < b$ and $r < 0$, then $br < ar$, e.g., since $-2 < -1$ and $-3 < 0$, $3 < 6$.

The *arithmetic* (A), *geometric* (G), and *harmonic* (H) *means* of positive numbers $a_1, \cdots, a_n$ are given by

$$A = \frac{a_1 + \cdots + a_n}{n}, \qquad G = \sqrt[n]{a_1 \cdot a_2 \cdots \cdots a_n}, \qquad \frac{1}{H} = \frac{1}{n}\left(\frac{1}{a_1} + \cdots + \frac{1}{a_n}\right).$$

By means of Cauchy's inequality

$$(a_1 b_1 + \cdots + a_n b_n)^2 \leqslant (a_1{}^2 + \cdots + a_n{}^2)(b_1{}^2 + \cdots + b_n{}^2)$$

it can be shown that $H \leqslant G \leqslant A$, equality holding when and only when $a_1/b_1 = \cdots = a_n/b_n$. Cauchy's inequality is a consequence of Lagrange's identity

$$(a_1{}^2 + \cdots + a_n{}^2)(b_1{}^2 + \cdots + b_n{}^2) = (a_1 b_1 + \cdots + a_n b_n)^2 + [(a_1 b_2 - a_2 b_1)^2$$
$$+ \cdots + (a_1 b_n - a_n b_1)^2 + (a_2 b_3 - a_3 b_2)^2 + \cdots + (a_2 b_n - a_n b_2)^2 + \cdots +$$
$$(a_{n-1} b_n - a_n b_{n-1})^2]$$

holding for any a's and b's. The bracketed expression is never negative.

1.22. Absolute Value (Numerical Value, Modulus) of a Real Number. The absolute value $|r|$ of a real number r is equal to r if $r \geqslant 0$ and to $-r$ if $r < 0$. Thus $|2| = |-2| = 2$. On the line of Fig. 1.12, $|r|$ is the undirected distance between O and R.

Properties of $|r|$. 1. $|r| \geqslant 0$. 2. $|rs| = |r||s|$. 3. $||r| - |s|| \leqslant |r \pm s| \leqslant |r| + |s|$.

If a and b are fixed real numbers and x a variable real number, $|x - a| \leqslant b$ means that $-b \leqslant x - a \leqslant b$, or that $a - b \leqslant x \leqslant a + b$, and conversely. The totality of such values of x is said to form a closed *interval* with end points $a - b$ and $a + b$. The interval is open if $<$ replaces $\leqslant$ throughout.

1.23. Continuity of the Real Numbers. In contrast to the rational numbers, the real numbers are gapless. This important property may be formulated as follows: If $a_1 < a_2 < a_3 < \cdots$, but $a_i < A$ for $i = 1, 2, 3, \cdots$ and some *fixed* A, then there is a unique number $b \leqslant A$ such that $a_i \leqslant b$ for $i = 1, 2, 3, \cdots$ but $c < b$ implies that $c < a_i$ for some value of i. The number b is called the *least upper bound* or *supremum* of the family of numbers a_i, $i = 1, 2, 3, \cdots$.

1.24. Decimal Representation of the Real Numbers. It follows from 1.23 that, for example, the rational numbers 3, 3.1, 3.14, 3.141, 3.1415, 3.14159, $\cdots$ have associated with them a unique number b which is not less than any of them and not greater than any number that is not less than any of them. The number b, of course, could be π for this example. Any decimal $d_n d_{n-1} \cdots d_1 d_0 . d_{-1} d_{-2} d_{-3} \cdots$, where each d is a digit (one of the integers 0, 1, $\cdots$, 9) represents a real number, and every real number has such an expansion. If, from some position on, each digit is 9, the last digit that differs from 9 may be increased by 1. Thus $0.0279999 \cdots = 0.028$.

REMARK: The decimal representation for r will repeat if and only if r is rational (see 4.42).

1.3. Operations with the Real Numbers. The following rules hold. For definitions and proofs, see a text on the theory of functions of a real variable.

1. *Addition and Subtraction.*

$$a + b = b + a \qquad a + (b + c) = (a + b) + c$$
$$a + 0 = a \qquad a + (-a) = a - a = 0$$

2. *Multiplication and Division.*

$$ab = ba \qquad a(bc) = (ab)c$$
$$1 \cdot a = a \qquad a \cdot \frac{1}{a} = 1 \qquad \text{(if } a \neq 0)$$
$$a(b + c) = ab + ac$$
$$a(-b) = (-a)b = -ab \qquad (-a)(-b) = ab$$

Important: $a \cdot 0 = 0$; $ab = 0$ implies that either $a = 0$ or $b = 0$. *Division by zero is not defined.*

3. *Fractions.*

$$a \cdot \frac{1}{b} = \frac{a}{b} \qquad \frac{a}{b} \cdot \frac{c}{d} = \frac{ac}{bd}$$

$$\frac{a}{b} \div \frac{c}{d} = \frac{a}{b} \Big/ \frac{c}{d} = \frac{a}{b} \cdot \frac{d}{c} = \frac{ad}{bc} \qquad \text{(``Invert the divisor and multiply.'')}$$

$$\frac{a}{c} + \frac{b}{c} = \frac{a + b}{c} \qquad a + \frac{b}{c} = \frac{ac}{c} + \frac{b}{c} = \frac{ac + b}{c}$$

$$\frac{a}{b} + \frac{c}{d} = \frac{ad}{bd} + \frac{bc}{bd} = \frac{ad + bc}{bd}$$

$$-\frac{a}{b} = \frac{-a}{b} = \frac{a}{-b} \qquad \frac{-a}{-b} = -\left(-\frac{a}{b}\right) = \frac{a}{b}$$

4. *Exponents and Radicals.*

$$a^p \cdot a^q = a^{p+q} \qquad (a^p)^q = a^{pq}$$
$$a^p b^p = (ab)^p \qquad a^{-p} = \frac{1}{a^p}$$
$$a^0 = 1 \text{ (except possibly when } a = 0) \qquad a^{\frac{1}{p}} = \sqrt[p]{a} \qquad a^{\frac{q}{p}} = (a^q)^{\frac{1}{p}}$$

Example. $\quad 3^{-2} = \frac{1}{9} \qquad (-27)^{\frac{1}{3}} = \sqrt[3]{-1}\sqrt[3]{27} = -3$
$$(\frac{1}{16})^{-\frac{3}{2}} = 16^{\frac{3}{2}} = (16^3)^{\frac{1}{2}} = (4^6)^{\frac{1}{2}} = 4^{\frac{6}{2}} = 4^3 = 64$$

Caution: If $a < 0$, $a^{\frac{1}{p}}$ is not a real number unless p is an odd integer (see 5.7).

Radicals are generally best managed with fractional exponents, except possibly for square and cube roots. Square roots can occasionally be eliminated from a denominator (or numerator).

$$\frac{a}{b \pm \sqrt{c}} \cdot \frac{b \mp \sqrt{c}}{b \mp \sqrt{c}} = \frac{a(b \mp \sqrt{c})}{b^2 - c}$$

1.41. Constants, Variables, Functions. Typical problems in meteorology and other natural sciences seek relations between numbers that measure the magnitudes of physical quantities. For the purposes of a particular problem, a magnitude whose variation is known (or assumed) to be negligible is called a *constant*, *e.g.*, the mass of a shell during its flight, the density of an incompressible fluid. Other magnitudes are called *variables*. Generally, though by no means always, letters from the second half of the alphabet denote variables and those from the first half denote constants.

The possible values of a variable, which together form its *range*, are often thought to be continuously distributed in much the same way as points are distributed on a line. At the same time, the range is frequently limited, *e.g.*, relative humidity in per cent can be *any* number *between* 0 *and* 100.

Let x and y be two variables and suppose that, whenever a value for x is given, a value (or values) for y can be found. Under these circumstances, the (*dependent*) variable y is said to be a *function* of the (*independent*) variable x, and we write $y = f(x)$, saying "*y* equals *f* of *x*."

Caution: 1. It is *not* implied that there is necessarily any "causal" connection between x and y. 2. It is *not* necessary for y to have different values for different values of x. 3. $f(x)$ does *not* mean a number f times a number x; f may be thought of as an operation that yields y when applied to x.

Let $y = f(x)$ and suppose that the value a of x is given, so that a value, say, b, can be found for y. This is expressed by writing $b = f(a)$; thus $f(a)$ denotes the value found for y when $x = a$. $f(x)$ is *single-valued* or *multiple-valued* according as the rule f applied to a yields just one $f(a)$ for every a or yields more than one value for some a; unless otherwise stated, all functions are assumed to be single-valued.

The operation f for finding y when x is known commonly consists in (1) making a prescribed measurement, (2) calculating from a prescribed formula, or (3) estimating or measuring the ordinate on a given graph.

Example. 1. Temperature T (°F) is a function $f(t)$ of time t (hr):

t	6	7	8	9	10	11	12	13	14
T	35	36	36	38	41	45	50	51	50

so that $f(9) = 38$, $f(13) = 51$, etc.

2. $f(x) = 3x^2 - 5x + 2$, so that $f(-2) = 3(-2)^2 - 5(-2) + 2 = 24$, etc.

3. In place of the discrete observations of 1, a thermograph would record a smooth curve or graph (see Fig. 1.42), the abscissae and ordinates of whose points (t, T) give the times t and temperatures T simultaneously observed.

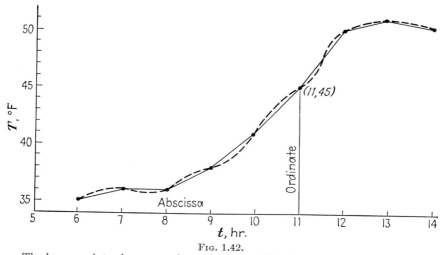

Fɪɢ. 1.42.

The heavy points alone come from the table. The broken line is put in to carry the eye along, and may or may not be an appropriate representation of the actual behavior of (t, T) between the heavy points. A thermograph might give the dotted curve. If a graph is expected later to accommodate further data, space for the additional points should be allowed at the outset.

A variable y is said to be a function $f(x_1, \cdots, x_n)$ of several variables x_1, x_2, $\cdots$, x_n if, whenever a value is given for each of the variables x_1, $\cdots$, x_n, a value for y can be found. The foregoing remarks apply.

1.42. Graphical Representation of $y = f(x)$. As described in 1.12 and 1.13, x and y may each be represented on a line, and $y = f(x)$ will give, for points a on the x-*axis* (line), points $f(a)$ on the y-axis. Place these axes at right angles in a plane; their point of intersection is called the *origin*. The axis of the independent (dependent) variable, here $x(y)$, is usually horizontal (vertical) with the *abscissa* x (*ordinate* y) increasing to the right (upward). The number pair (x,y) then corresponds uniquely to a point P found by measuring off the abscissa x (always the *first* of the number pair) to get a point on the x-axis, there erecting a perpendicular to the x-axis, and then measuring off on this perpendicular the ordinate y. Only the relevant parts of the axes usually appear. On them the unit of measure (scale) may be adjusted to permit an effective display of the data. The table (t,T) above gives the adjoining graph (Fig. 1.42).

1.5. Mathematical Induction. Given, for each positive integer n, a proposition P_n; to show that P_n is true for each sufficiently large value of n.

Step 1. Verify P_n for some specific value q of n.

Step 2. Prove that P_n (or, occasionally, P_q and P_{q+1} and $\cdots$ and P_n) implies P_{n+1} whenever $n \geqslant q$.

Conclusion. P_n is true whenever $n \geqslant q$.

Ordinarily $q = 0$, 1, 2, 3 or some other small integer. Steps 1 and 2 are both necessary. A proof proceeding in precisely this manner is a proof by *mathematical induction* (no connection with empirical induction). Such proofs occupy a basic position in mathematical literature.

Example. Let P_n be E_n: $1^2 + 2^2 + \cdots + n^2 = n(n + 1)(2n + 1)/6$.

Step 1. E_1 is $1^2 = 1(1 + 1)(2 + 1)/6$, which is true.

Step 2. Add $(n + 1)^2$ to each side of E_n and simplify the right member. Since the resulting equation is E_{n+1}, we see that E_n implies E_{n+1} whenever $n \geqslant 1$.

Conclusion. E_n is true for $n \geqslant 1$.

1.6. Necessary and Sufficient Conditions. Let P and Q be two propositions, and suppose that P *implies* Q (*i.e.*, if P is true, then Q is true). Then (the truth of) P is a *sufficient* condition for (the truth of) Q ("Q if P"), and (the truth of) Q is a *necessary* condition for (the truth of) P ("P only if Q"). Q is a necessary and sufficient condition for P ("Q if and only if P") thus means that P implies Q and Q implies P, and thus any proposition R which implies (is implied by) either implies (is implied by) the other; P and Q are then *equivalent* (logically). Let a, b, c be the sides of a triangle, A, B, C the opposite angles. In order that $a^2 = b^2 + c^2$, it is necessary, but not sufficient, that $a > b$; it is sufficient, but not necessary, that $a = 5$, $b = 4$, $c = 3$; it is necessary and sufficient that $A = 90$ deg. The *converse* of "If P, then Q" is "If Q, then P," and "P is equivalent to Q" may therefore be stated "P implies Q, and conversely."

2. COMPUTATION

2.1. Symbols of Aggregation. Parentheses ( ), brackets [ ], and braces { } group into a single term the entire expression they surround. The vinculum ‾‾‾‾‾‾ (now rare) and the fraction bar / or — group into a single term the entire expression below either symbol or above the bar. The sign appearing before an aggregated sum is to be distributed to each term of the aggregate expression.

Example. $2 + [3 - (5 - 7)] = 2 + [3 - (-2)] = 2 + [3 + 2] = 2 + 5 = 7$

$$-\{a + \overline{b - c}\} = -a - \overline{b - c} = -a - b - (-c) = c - a - b$$

$$-\frac{x - 2}{x - 3} = \frac{-x + 2}{x - 3} = \frac{x - 2}{-x + 3}$$

Caution: Exponents apply to precisely the terms indicated by the symbols; they are not to be distributed to the terms of an aggregated sum, *e.g.*, $2x^3$ is twice x^3, while $(2x)^3 = 2^3x^3 = 8x^3$. **Supercaution:** $\sqrt[p]{a^p \pm b^p} \neq a \pm b$ unless $p = 1$ (see 4.6).

By convention, signs of aggregation are frequently omitted when no confusion can arise, *e.g.*, $a + (bc) = a + bc$. Generally speaking, $+$ and $-$ separate more strongly (bind more weakly) than $\cdot$ (or $\times$) and $\div$, so that the latter operations should be performed first; thus $a + bc$ means "multiply b and c and add a to the result,"

$$e.g., \ a + b \div c = a + \frac{b}{c} = \frac{b + ac}{c} \neq \frac{a + b}{c} = \frac{a}{c} + \frac{b}{c} = a \div c + b \div c$$

Caution: The fraction bar binds the *entire* numerator into a single term *and* the *entire* denominator into a single term,

$$e.g., \ \frac{2 + 3}{2 - 8} = -\frac{5}{6}, \ but \ \frac{2 + 3}{2 - 8} \neq -\frac{3}{8}.$$

2.2. Change of Units. If (u) and (u') are two units for measuring the same physical quantity, a constant k will be known such that $1(u) = k(u')$. $1\,\mathrm{m} = 100\,\mathrm{cm} = 0.001\,\mathrm{km} = 3.281\,\mathrm{ft} = 0.0006214\,\mathrm{mile}$. Thus $q(u) = qk(u') = q'(u')$, where $q' = qk$.

Example. To change lb/in.² to kg/cm², given that $1\,\mathrm{lb} = 0.4536\,\mathrm{kg}$ and $1\,\mathrm{in.} = 2.540\,\mathrm{cm}$.

$$1.532 \ \frac{\mathrm{lb}}{\mathrm{in.}^2} = 1.532 \ \frac{0.4536\,\mathrm{kg}}{(2.540\,\mathrm{cm})^2} = \frac{1.532 \times 0.4536}{(2.540)^2} \ \frac{\mathrm{kg}}{\mathrm{cm}^2}$$

To change cm (to m) to ft: $21.03\,\mathrm{cm} = 21.03\,\mathrm{m}/100 = 0.2103\,\mathrm{m} = 0.2103\,(3.281\,\mathrm{ft}) = 0.2103 \times 3.281\,\mathrm{ft}$. Learn the process; the formula is too easily inverted in memory and in application.

2.31. Scientific Notation, Significant Figures, and Relative Error. In computing with real numbers, we express each number as a decimal that is broken off at a certain position, thus usually introducing an error. In expressing π for theoretical purposes, one writes $\pi = 3.1415$, giving the digits correctly as far as they go. This expression gives the left *end* point of the interval from 3.1415 to 3.1416 in which π lies. Although the last recorded digit (5) is correct, the error $\pi - 3.1415$ may (and here does) amount to almost as much as 0.0001, *i.e.*, to 1 in the last digit recorded. Practical notation, to which we adhere, is different. A decimal is to be broken off so as to give the *mid*point of an interval. The last digit may be incorrect, but the error is at most *half* a unit in the last position. Since $\pi = 3.141592+$, we write $\pi = 3.1416$ (to four decimal places), meaning that $3.14155 \leqslant \pi < 3.14165$. A decimal n expanded to q places ($q = 3$ for thousandths, $q = -2$ for hundreds, etc.) thus carries a notational error of $0.5 \cdot 10^{-q} = \Delta n$ (say "delta n").

Rule for Rounding Off: If the leftmost discarded digit is 5 or more, increase by 1 the last digit retained. To three decimal places, $\pi = 3.142$ ($\Delta \pi = 0.5 \cdot 10^{-3}$); to two, $\pi = 3.14$ ($\Delta \pi = 0.5 \cdot 10^{-2}$). **Caution:** Round off to the desired level in one step; to two places, $7.36495 = 7.36$, while repeated roundings off give the false result 7.37. Some authorities refine the above rule, when the leftmost discarded digit is 5 followed by zeros only, to make the last retained digit even: to 0 places, 31.50 and 32.50 both become 32.

Given n, there is a number n^*, $1 \leqslant n^* < 10$, and an integer p_n such that $n = n^* \cdot 10^{p_n}$, expressing n in *scientific notation*.

The factor 10^{p_n} merely locates the decimal point. *Rule:* Multiplying by 10 moves the decimal point one place to the *right;* multiplying by 10^{-1} (*i.e.*, $\div$ 10) moves the decimal point one place to the *left*.

The number n^* has no name but could well be called the significant part of n. Its f_n digits (if it is carried to $f_n - 1$ decimal places) are said to be the *significant figures* (or digits) of n. Thus $\Delta n^* = 0.5 \cdot 10^{-f_n+1}$ and $\Delta n = 0.5 \cdot 10^{p_n - f_n + 1}$.

Example. It is known of the gas constant R ergs per mole deg that $83145000 < R < 83155000$. Here $R^* = 8.315, f_R = 4, p_R = 7, R = 8.315 \times 10^7$. For the Boltzmann constant $k = 1.37 \times 10^{-16}$ erg per mole deg, $k^* = 1.37, f_k = 3, p_k = -16$. At 40°N latitude, to three significant figures, the acceleration of gravity is $g = 9.80 \times 10^2$ cm per sec². **Caution:** Be sure to include final significant ciphers in n^*; $g = 9.8 \times 10^2$ would give only two significant figures.

In speaking of *order of magnitude*, one says that n is "of the order of" 10^{p_n} (or of the order $d \times 10^{p_n}$, where d is the integral part of n^*), and Δn is of the order of $10^{p_n - f_n + 1}$.

The error Δn in a measurement, which experimental care seeks to minimize, can usually be given precisely to one significant figure, but rarely to more. If a measurement produces a decimal n', then we can say of the unknown "true" value n only that $n' - \Delta n \leqslant n \leqslant n' + \Delta n$; this is often expressed by writing $n = n' \pm \Delta n$, e.g., the (average) density of the earth is $\rho = 5.517 \pm 0.004$ grams per cm³, meaning $5.513 \leqslant \rho \leqslant 5.521$. Since it would be inconvenient to compute with such an expression as $n' \pm \Delta n$, one wishes to write $n = n'$. This raises the question of how to choose n' so that its notational accuracy may most closely represent the accuracy of the measurement. The most conservative usage (no usage is unanimous) puts $\Delta n' = 0.5 \cdot 10^{-q}$ if $0.5 \cdot 10^{-q-1} < \Delta n \leqslant 0.5 \cdot 10^{-q}$. However, the $(q + 1)$th decimal place of n' is often recorded in such a way as to place n' as close as possible to the midpoint of the shortest interval within which n is known to lie, *i.e.*, to give a best available estimate of the "correct" $(q + 1)$th place. Thus, to say that the density of the earth is $\rho' = 5.52$ conveys the erroneous impression that $5.515 \leqslant \rho < 5.525$, and the approximation 5.517 is ordinarily used.

The statement $n = n' \pm \Delta n$ is sometimes interpreted to mean that n has the average or mean value n' and the probable error Δn (see 14.24).

The (maximum) *relative error* in n is $\Delta n/n$. Thus

$$\frac{\Delta n}{n} = \frac{0.5 \cdot 10^{p_n - f_n + 1}}{n^* \cdot \quad 10^{p_n}} = \frac{1}{2 \cdot n^* \cdot 10^{f_n - 1}}$$

Note that $n^* \cdot 10^{f_n - 1}$ is always an integer consisting of precisely the significant figures of n, and that the sign of Δn is chosen, conventionally, so that $\Delta n/n > 0$.

Example. $\Delta R/R = 1/(2 \cdot 8{,}315)$, or one half part in $8{,}315$; $dk/k = 1/(2 \cdot 137)$, or one half part in 137.

When expressed as a percentage, $\Delta n/n$ is called *percentage error*. Thus the percentage error in k is slightly less than 0.4 per cent.

The fact that $1 \leqslant n^* < 10$ implies that $\Delta n/n \leqslant 1/(2 \cdot 10^{f_n - 1})$.

Thus a knowledge of the *number* of significant figures, disregarding their values, gives for the relative error a crude estimate that is frequently adequate for practical purposes. Conversely, suppose we know (1) that $\Delta n/n \leqslant 1/(2 \cdot n_1 \cdot 10^{g-1})$, where $1 \leqslant n_1 < 10$ and g is an integer not less than 1, and (2) that $\Delta n/n$ may exceed $1/(2 \cdot 10^g)$, so that the given estimate cannot be improved (except possibly by establishing it for a larger $n_1 < 10$). Two cases are possible. If $n^* \leqslant n_1$, then $\Delta n/n \leqslant 1/(2 \cdot n_1 \cdot 10^{g-1}) \leqslant 1/(2 \cdot n^* \cdot 10^{g-1})$, and we are justified in carrying n to $f_n = g$ significant figures. On the other hand, if $n_1 < n^*$, then our estimate for the relative error does not imply that $\Delta n/n \leqslant 1/(2 \cdot n^* \cdot 10^{g-1})$ but does imply that $\Delta n/n \leqslant 1/(2 \cdot n^* \cdot 10^{(g-1)-1})$; whence only $f_n = g - 1$ figures in n may be given with genuine assurance. It often happens, however, that the origin of n (*e.g.*, as a product or quotient) guarantees that its gth computed figure is a best available estimate of the correct gth figure and so may appropriately be recorded. This connection between $\Delta n/n$ and f_n accounts for the interest accorded to f_n in computation.

As a practical gauge of the reliability of a measurement, $\Delta n/n$ is almost always preferable to Δn. $\Delta n/n$ relates the size of the error to the size of the quantity measured, indicating, for example, that an error of 1 in. could possibly be tolerated in measuring a mile, but not in measuring 1 cm. Again, if a unit u' replaces u, $1(u) = k(u')$, then $\Delta n'/n' = \Delta n/n$; in any system of units within the limits of multiplicative error (see 2.32), $\Delta n/n$ is thus given by the same number, and hence n may be given to the same number of significant figures.

2.32. Significant Figures in Computation. A result computed from numerical data depends for its accuracy on the accuracy of the data and on the extent to which the particular computation preserves this accuracy. It is exceedingly laborious to carry along a large number of digits at each stage of a computation, particularly when tables are used. It is misleading to quote a result with an accuracy beyond that warranted by the data and their treatment. Both torpor and integrity thus suggest a preliminary estimate of the relative error of the result in terms of the relative errors in the data.

The computer must be aware of the accuracy of his data. A carefully written discussion will contain a statement of the error convention that is used in it. Unless the contrary is explicitly stated, conservative practical notation permits a computer to assume that the data he is given are so expressed that, if $n = n^* \cdot 10^{p_n}$ in scientific notation and n^* has f_n significant digits, then $0.5 \cdot 10^{p_n - f_n} < \Delta n \leq 0.5 \cdot 10^{p_n - f_n + 1}$.

In addition (positive terms), errors *add:* $\Delta(x + y + \cdots + z) = \Delta x + \Delta y + \cdots + \Delta z$. Also, the relative error in a sum cannot exceed the greatest relative error in the summands; it is often very much less, *e.g.*, $3,131.7 + 0.0002 = 3,131.7$, with a relative error less than $1/60,000$, although the relative error in 0.0002 is $\frac{1}{4}$.

Recommended Procedure. Settle each case on its merits according to the principle that a doubtful figure in a column renders the whole column doubtful. More precisely, select a summand s that is carried to the least number q of decimal places, *i.e.*, has the greatest error $\Delta s = 0.5 \cdot 10^{-q}$; round off each summand to q decimal places (or preferably one or two more, if they are available, to avoid influencing the last figure of the result by an accumulation of small errors); the result should be rounded off to q places, or possibly fewer, according as the sum of all the errors is not or is greater than Δs.

Example.

1.00957		1.0096
21.35	becomes	21.35
0.095412		0.0954
		22.46

Although $\Delta(x - y) = \Delta x + \Delta y$ (+ to get the *maximum* error on the right), there is no similar "rule" for subtraction. Thus $34,152 - 34,151$ could be anything between 0 and 2 and assuredly carries a large relative error. When possible, experiment and computation should be arranged beforehand so as to avoid anticipated subtractions of numbers close to each other.

In other cases, an estimate of Δn and $\Delta n/n$ is most easily found with the aid of calculus. The differential dn is an approximation to Δn whose reliability varies, in a rough way, inversely as the size of the second derivative. Thus $d(x^p) = px^{p-1} dx$ and $d(x^p)/x^p = p \, dx/x$, so that the error $\Delta(x^p)$ in x^p is approximately px^{p-1} times the error Δx in x, while the relative error $\Delta(x^p)/x^p$ is approximately p times the relative error $\Delta x/x$ in x.

Products and quotients (of positive quantities) are controlled by the approximate equation $\Delta(xy^{\pm 1})/xy^{\pm 1} = \Delta x/x + \Delta y/y$ (+ on right to get maximum relative error). By the inequalities 3 of 1.22, the relative error of the result is not greater than the sum (and not less than the difference) of the relative errors of the factors.

Rule for $xy^{\pm 1}$: Let f_x and f_y denote the number of significant figures in x and y. If $f_x < f_y$ (or $f_y < f_x$), round off y (or x, etc.) to y' with $f_{y'} = f_x + 1$. Carry xy' (or x/y') to $f_x + 1$ significant figures and then round it off to f_x figures. Ordinarily only the first $f_x - 1$ figures are assuredly correct, but the f_xth figure computed in this manner is the best available estimate of the correct f_xth figure.

Caution: When several multiplications or divisions are performed, relative errors *add;* the "rule" quickly becomes unreliable.

When only a certain number of significant figures is desired in the result, multiplication and division can be substantially abbreviated by the following arrangements of the work:

$$N = (4.3\,1\,8 \times 10^4)\,(8.7\,6\,2 \times 1\,0^{-2})$$

$$
\begin{array}{l}
4.3\,1\,8 \times 1\,0^4 \\
8.7\,6\,2 \times 1\,0^{-2} \\
\hline
3\,4\,5\,4\,4 \\
3\,0\,2\,3\,\cdot \\
2\,5\,9\,\cdot\,\cdot \\
9\,\cdot\,\cdot\,\cdot \\
\end{array}
$$

MULTIPLICATION

$$N = \overline{3\,7.8\,4} \times 10^2 = 3.7\,8\,4 \times 10^3$$

Use the digits of the multiplier from left to right. After finishing one line, start the right end of the succeeding line one place farther to the right. Round off, actually writing down for final addition only one more place than the number of places desired correct.

$$N = (3.7\,8\,4 \times 1\,0^3) \div (8.7\,6\,2 \times 1\,0^{-2}) = (3.7\,8\,4 \div 8.7\,6\,2) \times 1\,0^5$$

$$
\begin{array}{r}
8.7\,6\,2\,|\,3.7\,8\,4\,\lfloor.4 \\
3\,5\,0\,5 \\
8\,7\,6\,|\quad 2\,7\,9\,\lfloor 3 \\
2\,6\,3 \\
8\,8\,|\quad 1\,6\,\lfloor 1 \\
9 \\
9\,|\quad 7\,\lfloor 8 \\
\end{array}
$$

DIVISION

$$N = 0.4\,3\,1\,8 \times 1\,0^5 = 4.3\,1\,8 \times 1\,0^4$$

Round off at each step in divisor and its product by digits of quotient.

2.33. The Decimal Point. As indicated in the preceding examples, the decimal point is best located by means of scientific notation. The powers of 10 are easily managed, and there is little difficulty with products and quotients of numbers lying between 1 and 10.

Example. $0.008312 \times 21.35 = (8.312 \times 10^{-3})(2.135 \times 10) = (17.75) \times 10^{-2} = 1.775 \times 10^{-1}$. $11,300 \div 0.00215 = (1.13 \times 10^4) \div (2.15 \times 10^{-3}) = (1.13 \div 2.15) \times 10^7 = 0.526 \times 10^7 = 5.26 \times 10^6$.

2.4. Logarithms. If b is any positive number different from 1, and n is any positive number, then a number l can be computed such that $b^l = n$. l is called the logarithm of n to the base b: $l = \log_b n$. Being exponents, logarithms obey the rules for exponents.

$$\log_b pq = \log_b p + \log_b q \qquad\qquad \log_b p^q = q \log_b p$$
$$\log_b 1 = 0 \qquad\qquad\qquad\qquad \log_b b = 1 \text{ for any } b$$

If $b > 1$, $\log_b n > 0$ or < 0 according as $n > 1$ or < 1

Only two numbers are in practical use as bases for logarithms. Logarithms to the base 10 are called *common logarithms* and written log (no base indicated). Logarithms to the base $e = 2.71828 \cdots$ are called *natural* (Napierian, hyperbolic) *logarithms*, and written $\log_e$ or (here) ln (technical practice, not unanimous).

2.41. Common Logarithms. If $1 \leqslant n \leqslant 10$, then $0 \leqslant \log n \leqslant 1$. The logarithms of numbers between 1 and 10 have been tabulated. (For errors, interpolation, etc., see 2.5.) For example, log $4.15 = .6590$ (*i.e.*, $4.56 = 10^{.6590}$). If $n < 1$ or $10 < n$, use scientific notation: $n = n^* \cdot 10^{p_n}$, $1 \leqslant n^* < 10$, p_n an integer. Then log $n = \log n^* + \log 10^{p_n}$. Log n^* may be found from the tables and is called the *mantissa* of log n. Log $10^{p_n} = p_n$ is called the *characteristic* of log n. In going from log n to n, if log n is not between 0 and 1, write log $n = \log n^* + p_n$, where p_n is an integer and $0 \leqslant \log n^* < 1$. Then n^* may be found from the tables, and $n = n^* \cdot 10^{p_n}$. For example, log $45{,}600 = \log (4.56 \times 10^4) = .6590 + 4 = 4.6590$. If log $n = 4.6590 = .6590 + 4$, then $n^* = 4.56$ and $n = 4.56 \times 10^4 = 45{,}600$. If log $n = -2.3410 = .6590 - 3$, then $n^* = 4.56$ and $n = 4.56 \times 10^{-3}$. A negative characteristic is sometimes written before the mantissa with a superscript minus sign; such a sign applies *only* to the characteristic: log $(4.56 \times 10^{-3}) = .6590 - 3 = \overline{3}.6590$.

In the course of a computation, the mantissa may threaten to become negative, or a (negative) characteristic to become fractional; each emergency can and should be avoided, as indicated in the following examples:

Multiplication. $n = .000796 \times 8740 = (7.96 \times 10^{-4})(8.74 \times 10^3)$

log 7.96×10^{-4}	$.9009 - 4$
log 8.74×10^3	$.9415 + 3$
(Adding)	$\overline{1.8424 - 1}$
log $n = .8424$, $n = 6.957$	

Division. $n = 796 \div .00874 = (7.96 \times 10^2) \div (8.74 \times 10^{-3})$

log 7.96×10^2	$1.9009 + 1$
log 8.74×10^{-3}	$.9415 - 3$
(Subtracting)	$\overline{.9584 + 4}$
log $n = .9584 + 4$	
$n = 9.09 \times 10^4 = 90{,}900$	

Here the mantissa would be negative after subtraction if the precaution had not been taken of writing $1.9009 + 1$ instead of $.9009 + 2$.

Exponentiation. $n = (.00874)^{1/5}$. Here we wish to divide log $.00874 = .9415 - 3$ by 5, and the characteristic threatens to become the fraction $\frac{3}{5}$. Write log $.00874 = 2.9415 - 5$ (or $7.9415 - 10$ or $27.9415 - 30$, etc.), so that the characteristic is exactly divisible by 5. Then log $n = (2.9415 - 5)/5 = .5883 - 1$, $n = 3.88 \times 10^{-1} = .388$.

2.42. Natural Logarithms. Write $n = n^* \cdot e^{p_n}$, $1 \leq n^* < e$, p_n an integer, the latter determined by a table of integral powers of e. Mantissas range from 0 to 1 and are found in tables. Characteristics are interpreted as integral powers of e, *e.g.*, $n = 2.37 \times 3.51$, ln $n = \ln 2.37 + \ln 3.51 = .8629 + 1.2256 = 2.0885 = \ln 8.05$, $n = 8.05$.

It is often simpler to go over to logs for actual computation. This may be done by means of the equations log $x = .4342943 \cdot \ln x$, ln $x = 2.3025851 \cdot \log x$.

2.5. Tabular Work and Interpolation. Besides the tables given elsewhere in this volume, there are tables of squares, square roots, $\frac{3}{2}$ powers (which, used inversely, give $\frac{2}{3}$ powers), cubes, areas and volumes of spheres, unit conversions, etc. Generally speaking, any function whose values are needed frequently has been tabulated. Logarithms are available to 16 places, though empirical data justifying their use are rare.

Many excellent tables of the elementary functions have recently been produced by the Work Projects Administration (City of New York). Less common functions are tabulated in the "Funktionentafeln" of Jahnke and Emde, which also contains many useful graphs and a bibliography (to 1933) on numerical, graphical, and mechanical computation.

A table for the function $y = \log x$ appears in part as follows:

	0	1	2	3
7.0	8451	8457	8463	8470
7.1	8513	8519	8525	8531
7.2	8573	8579	8585	8591

Here, as in many tables, the absent decimal points are to be supplied.

Caution: Examine carefully a table whose use is contemplated for conventional notation of this sort, usually explained at the head of the table, *e.g.*, the initial 8 in the tabular entries might be given only once.

From the table, we see that $\log 7.11 = .8519$. This is understood to mean that, to four decimal places, *i.e.*, to within $.5 \times 10^{-4}$, the log of the (precise) number $7.11000 \cdots$ is $.8519$; it could, of course, be computed to any required accuracy. N.B. *Absolute* error and *decimal-place* accuracy are involved in tabular work. In practice, we are likely to know instead that, to two decimal places, $x = 7.11$. It is then absurd to compute $\log x$ to more than a certain number of places. It is shown in calculus that, for any function $y = f(x)$, the error Δy in y when x has the value a is approximately $dy = f'(a)\Delta x$, where $f'(a)$ is the value when $x = a$ of the derivative of y with respect to x. Roughly, when y is changing rapidly (slowly) per unit change in x, the tabulated (computed) value of y is properly given to fewer (more) places than the given value of x. For $\log x$, it can be shown that $dy = .43429\Delta x/x$, which decreases as x increases. In our example, putting $x = 7.11$, $\Delta x = .005$, we find $\Delta y = .3 \times 10^{-3}$, so that y may properly be given to three decimal places. The fourth place given in the table may be assumed to be chosen in accordance with the convention explained in 2.31.

Suppose now that, to three decimals, $x = 7.114$.

$$\log 7.110 = .8519$$
$$\log 7.114 = .85??$$
$$\log 7.120 = .8525$$

While x runs from 10 to 20 in the last two places, a jump of 10, $\log x$ runs from 19 to 25, a jump of 6. Assuming that the jump in y is uniformly distributed, each unit jump in x would contribute .1 of the total jump, or .6. Four jumps in x will contribute $4 \times .6$ or 2.4, giving $\log 7.114 = .85214 = .8521$. This particular process for finding values of $f(x)$ for values of x between those given in the table is called *linear interpolation*. Further illustration, from trigonometry:

$$\cos 51.70° = .6198$$
$$\cos 51.76° = .61??$$
$$\cos 51.80° = .6184$$

Here the cosine jumps *down* 14 while the angle jumps *up* 10. A jump of 6 *up* in the angle will hence give a jump of $6 \times 1.4 = 8$ *down* in the cosine, yielding cos 51.76° = .6190; alternatively, a jump of 4 *down* (from 51.80°) in the angle will give a jump of $4 \times 1.4 = 6$ *up* (from .6184) in the cosine, yielding the same result.

Inverse interpolation goes the other way, *e.g.*, cos 51.7? = .6195. Here the cosine jumps down 3; 2 jumps of 1.4 give a jump of 2.8, which is as close to 3 as an integral number of jumps of 1.4 can get. Hence cos 51.72° = .6195.

Recommendations. Master the process rather than try to learn a "rule." Be sure to follow the direction (increase or decrease) of the table. In most simple cases, interpolation can be performed mentally; this is expedited by the "tabular differences" and small tables of "proportional parts" which often accompany tables.

Thus the table of logarithms from which the above excerpt was taken amounts to a schema from which, by interpolation, a number given to three decimal places determines its logarithm to four places (last place best estimate), while a log given to four decimal places determines a number to three decimal places. This conveys the maximum consistent information in the minimum space, which is the object of the table. A table giving five-place logs for three-place values of x allows the determination, by interpolation, of the logs of numbers given to four places, and so on.

The assumption, in linear interpolation, that the jump in y is uniformly distributed over the jump in x usually introduces an error. In most tables encountered in elementary practice, except possibly at an extremity of the table, this error rarely amounts to more than 1 in the last decimal place to which the table is carried. Roughly speaking, this error will be large or small according as the curve $y = f(x)$ is bending rapidly or slowly at the point in question (see 8.4); unusually large tabular differences warn against reckless interpolation (*e.g.*, for log x near $x = 1$; for tan x near $x = 90°$). If investigation (perhaps via calculus) discloses that linear interpolation is unreliable, more refined formulas are available [Newton-Gregory (14.14)]; or the desired value may be computed directly, *e.g.*, with series. Various texts[1] should be consulted on these and allied questions.

2.6. Graphical and Mechanical Aids. Physical data may be presented for use in graphical form (see Sec. I).

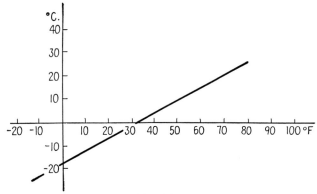

FIG. 2.6a.

Contour charts are useful in dealing with several variables. The isobars on a weather map are lines of equal pressure (level pressure curves, pressure contour lines). The familiar *pv*-diagram in thermodynamics usually carried curves along which temperature is constant. It often has also another family, the isentropes, along **any** one of which entropy is constant. For a contour chart used in fog prediction, see Sec. V. *Principle.* When the values of two variables are known, one or more values for the other may be read off the chart.

Computation from a formula may be permanently embodied in a graph, with accuracy to two and often three figures. In Fig. 2.6*b*, the graduations for the two variables have been entered on opposite sides of the same line. Since distances between graduations are proportional to differences between labels, these scales (for degrees centigrade and Fahrenheit) are called *uniform.*

Fig. 2.6*b*.

Always indicate, with units, the quantity whose values are given by the labels on a scale.

Nonuniform Scales. Even in case $y = f(x)$ is not linear, representation on one line is possible for a range of values of x within which no value of y occurs more than once. Using chosen values a of x, compute $f(a)$, getting a table of values. On the line, choose a fixed origin O and a fixed length m, called the *modulus* of the scale. Changing the sign of m changes the direction of the scale. Choose m so as to accommodate the desired range of values of $f(x)$ in the length available. Thus, if f ranges from -15 to 140, and 8 in. are available, take $m = 8/[140 - (-15)] = 0.05$ approximately, *i.e.*, 0.5 in. for a jump of 10 in $f(x)$. From O, measure off $mf(a)$, and assign the label a to the point so obtained. For $y = \log x$, with $m = 10$ cm, we get the logarithmic scale.

Fig. 2.6*c*.

Semilogarithmic graph paper has one uniform scale and one logarithmic, at right angles. If $y = ae^{bx}$, then $\log y = \log a + bx \log e$ and the graph on semilog paper is a straight line.

Logarithmic graph paper has two logarithmic scales at right angles. If $y = ax^b$, then $\log y = \log a + b \log x$, and the graph on log paper is a straight line.

Slide Rule. Two logarithmic scales slide along each other in such a way that segments (logs) may be added, and hence multiplications and divisions may be performed. Most rules also contain other scales for other computations. Depending on the size and quality of the rule and the part of the rule used, accuracy varies from three to five significant figures. Full directions come with each instrument. Though desirable, knowledge of logarithms is not necessary. Special slide rules may be constructed for special purposes.

Other Instruments. Besides the familiar adding machine, there are machines for multiplying and dividing (and extracting simple roots), solving systems of linear equations, solving an algebraic equation, finding the coefficients of a Fourier series (harmonic analyzer), measuring areas (*i.e.*, integrating), etc. Information may be secured from the manufacturers. Machines for the more elaborate problems are often expensive and require highly trained operators.

2.7. Alignment charts (nomograms or nomographic charts) are widely used. Scales are so graduated and so placed that a ruler or taut thread laid across the scales will cut them in a set of values that satisfy a given equation. The validity of a nomogram usually depends on similarity of triangles. To supplement the following remarks, see refs. 2 and 3. An interesting nomogram connecting barometric pressure, absolute and relative humidity, dew point, and several temperatures occurs in ref. 4.

Many common nomograms treat equations of the form $f(u) + g(v) = h(w)$. Draw two parallel lines λ, μ usually vertical, at opposite edges of the available space. On them construct the respective scales $x = lf(u)$ and $y = mg(v)$, choosing the moduli l and m so as to stay on the available lengths. Draw a line ν parallel to λ and μ and so placed that its respective distances from these lines are in the ratio l/m. Let the points P and Q, labeled $u = a$, $v = b$, be joined by a line α cutting ν in a point R. Find c such that $f(a) + g(b) = h(c)$, and give R the label c. Let $n = lm/(l + m)$. Using $nh(c)$ for the point P, put the scale $z = nh(w)$ on the line ν, and the chart is ready for use. For, let any line β cut λ, μ, and ν in points P', Q', and R', labeled a', b', and c'; by properly equating ratios of appropriate segments, and using the fact that $PR/RQ = l/m$, one may readily verify that $f(a') + g(b') = h(c')$, as required.

To treat an equation of the form $f(u)g(v) = h(w)$, take logarithms, and write $F(u) = \log f(u), \cdots, H(w) = \log h(w)$, getting $F(u) + G(v) = H(w)$. Then use the preceding process.

Example. $uv^{1.41} = w$ becomes $\log u + 1.41 \log v = \log w$. Of course, if $f(u) = u$ (or 10^u) then the scale for $F(u)$ reduces to a logarithmic scale (or a uniform scale).

Let the equation $f(u) + g(v) + h(w) = \phi(t)$ be given. First construct a nomogram for $z = f(u) + g(v)$, it being unnecessary to graduate the z-scale. Then construct a nomogram for $z + h(w) = \phi(t)$, using the z-scale and a conveniently placed w-scale as the "outside" parallel scales, with the t-scale properly placed between and parallel to them. In graduating the t-scale, let $u = a$, $v = b$, $w = c$, and find d so that $f(a) + g(b) + h(c) = \phi(d)$. The modulus on the t-scale turns out to be $lmn/(lm + mn + nl)$, if the u, v, w moduli are l, m, n. If two lines intersect at a point on the z-scale, then the u and v values picked out by one line and the w and t values picked out by the other give a quadruple of values that will satisfy the given equation. $f(u) + g(v) = h(w) + \phi(t)$ is handled similarly. Proceed similarly, step by step, for the sum of a larger number of functions.

Take logarithms to reduce $f(u)g(v)h(w) = \phi(t)$ or $f(u)g(v) = h(w)\phi(t)$ to the case just described.

A term of the form $1/f(u)$ may be replaced by $f_1(u) = 1/f(u)$.

In case several scales are being used at once, a key should be included indicating which pairs, triples, etc., are to be used together. The equation should always be given. In simple cases, this may suffice.

If it is inconvenient to introduce logarithmic scales in treating $f(u) = g(v)h(w)$, one may proceed as follows: At opposite ends of the available space, draw parallel $x = lf(u)$ and $y = mg(v)$ scales, oppositely directed. Draw a line ν joining the points $u = 0$ and $v = 0$ on these scales. On the y-scale, choose a convenient point Q distant β from $v = 0$. On the x-scale superpose temporarily the auxiliary scale $x' = \beta lh(w)/m$. To get the w-scale on the line ν, project this x'-scale, along with its w-labels, from the point Q on ν. This completes the nomogram, as may again be verified by means of similar triangles.

The equation $f(u)g(v) = h(w)\phi(t)$ may be written as $f(u)/\phi(t) = h(w)/g(v)$. This proportionality may be displayed by several different arrangements of the scales, and the result is sometimes called a *proportionality chart*. The moduli $m_u, \cdots, m_t$ must be proportional: $m_u/m_t = m_w/m_v$. Often the scales are superposed in pairs, u and w together and t and v together, the two lines making any convenient angle with each other.

For an equation of the form $f(u) + g(v) = h(w)/\phi(t)$, or $\phi(t)f(u) + \phi(t)g(v) = h(w)$, construct parallel scales $x = lf(u)$ and $y = lg(v)$ in the same or opposite directions according as $f(u)$ and $g(v)$ have opposite or like signs.   On the line of the u-scale, and with the same origin, put a scale $z = mh(w)$.   Draw the line, of length K, joining the origins of the u- and v-scales.   On this line, with $n = Km/l$, and origin at the origin of the u-scale, put a scale $T = n\phi(t)$, completing the nomogram.   If two parallel lines cut the diagram, labels u and v picked out by one and w and t picked out by the other will together satisfy the equation.

3. ELEMENTARY GEOMETRY

Conics are discussed in 7.31.

3.1. Definitions.   3.11. Circle.   As is the case with many names of curves and surfaces, *circle* can mean, according to context, either the circumference, or the part of the plane inside (or inside and on) the circumference.   Let O denote the center.   If P is on the circumference, the segment OP is a *radius;* its length r is "the" radius of the circle.   If A and B are on the circumference, the segment AB is a *chord* of C; a *diameter* if it contains O, as AOP.   A chord *subtends* (divides the circumference into) two *arcs*, a longer (shorter) of which is called the *major (minor)* arc. A diameter subtends a *semicircle*.   The area between a chord and either of its arcs is a *segment*.   A straight line that has two (one) points in common with a circle is a *secant (tangent)*, e.g., s (t).

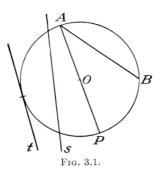

Fig. 3.1.

3.12. Angle.   Degrees, Radians.   Let l be a line through a point Q.   Discard one of the two halves into which Q divides l.   The residual half line (including Q) is called a *ray* issuing from Q.   Rays r and r' issuing from Q form an *angle* α whose *vertex* is Q and whose *sides* are r and r'.   Let C be a circle with center at Q.   Then α is a central angle in C. Place 360 equally spaced points on the circumference of C.   If r and r' pass through an adjacent pair of these points, the measure of α is 1 *degree* (1°).   The total angle at Q is 360° or 1 *revolution*.

In addition to the revolution and the degree, there is one other angular unit in common scientific use.   In this unit, the *radian*, the total angle at the center of the circle is 2π.   Since the circumference is $2\pi r$, 1 rad is a central angle whose sides have between them (*intercept*) a shorter arc equal in length to the radius of the circle. If s denotes the arc intercepted by an angle of α rad, then $s = r\alpha$. $1° = \pi/180$ rad $= 0.017453$ rad; 1 rad $= 180/\pi$ deg $= 57.2958° = 57°17.75'$.

If A, B, C lie in that order on a straight line, the angle ABC (*i.e.*, the angle with vertex B and sides passing through A and C) is a *straight angle* (180°).   The *left side* of an angle is that one which, by a clockwise rotation about the vertex of not more than 180°, can be made to coincide with the other (*right*) side.   A *right angle* is half a straight angle (90°); the sides of a right angle are *perpendicular* (orthogonal) to each other.   If $0 \leqslant \alpha < 90°$, α is *acute;* if $90° < \alpha < 180°$, α is *obtuse*.   If $\alpha + \beta = 90°$, α and β are *complements* of each other; if $\alpha + \beta = 180°$, α and β are *supplements* of each other.   If the vertex of an angle is on a circle, and its sides are secants, the angle is *inscribed* in the circle.

3.13. Polygons.   If points A_1, A_2, $\cdots$, A_n are given, and, for $i = 1$, $\cdots$, $n - 1$, the segment $A_i A_{i+1}$ is drawn, the result is a *broken line*.   If $A_n = A_1$, the broken line is a *polygon* with *vertices* A_1, $\cdots$, A_{n-1} and *sides* $A_1 A_2$, $\cdots$, $A_{n-1}A_1$, the sum of whose lengths is the *perimeter*.   A polygon with four sides is a *quadrilateral;* the sides $A_{12}A$ and A_3A_4 are *opposite*, as are the other two.   A *trapezoid (parallelogram)* is a quadrilateral with one pair (both pairs) of opposite sides parallel. (Two lines in a plane are *parallel* if they do not meet.) A *rhombus* is a parallelogram with a pair of adjacent sides equal. A *rectangle* is a parallelogram whose angles are right angles. A *square* is a rectangle with equal sides. If each vertex of a polygon lies on the circumference of a circle, the polygon is *inscribed* in the circle, which *circumscribes* the polygon.

3.14. Triangle. A triangle is a polygon with three sides. It is usual to letter the vertices with capital letters and the sides opposite the respective vertices with corresponding small letters. It is usually unambiguous to let $A, \cdots (a, \cdots)$ stand

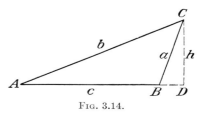

for a vertex (line), or the angle with that vertex (segment), or the measure of the angle (length of the side) according to context. A triangle is *isosceles* (*equilateral*) if two (all three) sides are equal; the odd side (isosceles case) is the *base*. If one angle is a right angle, the triangle is a *right* triangle; the side opposite the right angle is the *hypotenuse*, and the other sides are *legs*. A segment from a vertex to and perpendicular to the opposite

Fig. 3.14.

side is an *altitude* (h). A segment joining a vertex to the midpoint of the opposite side is a *median*. An angle between one side and the extension of an adjacent side (DBC) is an *exterior angle* of the triangle; the nonadjacent interior angles are its *opposite interior angles* (A and C). Two triangles whose angles are equal in pairs are *similar*.

3.15. Congruence. Two configurations are *congruent* (or *equal*) if it is possible to move either rigidly into a position coinciding precisely with that of the other. (Corresponding distances, angles, area, volumes, etc., of congruent configurations are equal.)

3.16. Reflections; Symmetry. If O is a given point, the *reflection* of a point

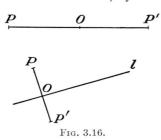

$P \neq O$ *in* O is the point $P' \neq P$ on the line OP such that $OP = OP'$. The reflection of P in a *line l* or *plane* π is the reflection of P in the point O of l or π nearest to P. A configuration that is transformed into itself by reflecting each of its points in a point O, a line l, or a plane π is said to be *symmetric* about O, l, or π. Thus a circle is symmetric about its center, and about any diameter.

3.2. Theorems and Formulas. 3.21. Circle. Through three distinct noncollinear points precisely one circle can be drawn. A line through a point P

Fig. 3.16.

on the circumference of a circle C with center O is tangent to C if and only if it is perpendicular to OP. If the lines RP and RQ are tangent at P and Q to the same circle, then the lengths RP and RQ are equal. A central angle is measured by the included arc. An inscribed angle is measured by half the included arc and hence is a right angle if and only if the included arc is a semicircle.

3.22. Angle. Two angles are equal (or supplementary) if their sides are either parallel or perpendicular right to right and left to left (or right to left and left to right). If l is a line, P a point on l, and Q a point not on l, then P is the point on l closest to Q, and the distance PQ is the distance between Q and l, if and only if QP is perpendicular to l. If the angle α has vertex O, a ray inside issuing from O bisects α if and only if each point of the ray is equidistant from the sides of O. P is on the line that bisects a segment AB perpendicularly if and only if $AP = PB$.

3.23. Polygons. Opposite sides and angles of a parallelogram are equal. Each diagonal of a parallelogram divides it into a pair of congruent triangles. The diagonals of a parallelogram bisect each other, perpendicularly if and only if the parallelogram is a rhombus.

3.24. Triangle. $a - b < c < a + b$ unless the triangle collapses (letters may be permuted). Ordered according to magnitude, a, b, c have the same order as A, B, C. A triangle is isosceles (equilateral) if and only if two angles (all angles) are equal. An exterior angle is the sum of the two opposite interior angles. The sum of the three angles of a triangle is 180°. $C = 90°$ if and only if $c^2 = a^2 + b^2$ (Pythagoras). The three altitudes meet in a point; so do the angle bisectors, the perpendicular bisectors of the sides, and the medians, the latter in a point two-thirds of the way from each vertex to the opposite side.

Two triangles are congruent if they have respectively equal two sides and the included angle, a side and the two adjacent angles, three sides, or, under certain conditions, two sides and the angle opposite one of them. From each of these sets of data, then, a triangle may be constructed (see 5.6).

If two triangles are similar, their sides, altitudes, medians, perimeters, etc., are proportional. Two lines are cut in proportional segments by a family of parallel lines. (One of two similar figures is an enlargement of the other, as in photography.)

3.25. Areas. *Circle* with radius r, diameter $d = 2r$, and circumference $C = 2\pi r = \pi d$ has area $\pi r^2 = \pi d^2/4 = c^2/(4\pi)$.

Circular sector with central angle $\alpha° = \beta$ rad and arc $s = r\beta = \pi r\alpha/180$ has area $rs/2 = r^2\beta/2 = \pi r^2\alpha/360$.

Circular Segment. Subtract triangle (vertex at center) from sector.

Triangle with altitude a' on side a and semiperimeter $s = (a + b + c)/2$ has area $aa'/2 = ab \sin C/2 = \sqrt{s(s - a)(s - b)(s - c)}$, where $\sin C = a'/b$.

Trapezoid with parallel sides b and b' distant h apart has area $h(b + b')/2$.

Parallelogram and *rectangle* thus have area bh, and a *square* of side s has area s^2.

3.3. Constructions (some easier with drawing equipment). *To construct the perpendicular bisector of a segment AB.* With a fixed radius $r > AB/2$, and A and B as centers, describe circular arcs intersecting in P and Q. PQ is the required line.

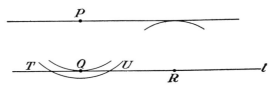

Fig. 3.3.

To construct through a given point P a parallel to a given line d.

Adjust compass so that, with P as center, the arc just touches l, at Q. With $R \neq Q$ on l as center and radius PQ, draw an arc on the same side of l as P; the tangent from P to this arc is the required parallel. Or drop a perpendicular PQ from P on l, and construct at P a perpendicular to PQ.

To drop a perpendicular from a given point P to a given line l. (1) PQ is the required line, where Q is the point in the preceding construction. (2) If greater precision is required, use a longer radius, the arc then cutting l in T and U. The perpendicular bisector of TU is the required line (PQ).

To erect a perpendicular at a given point P of a given line l. With P as center and any convenient radius, describe an arc cutting l in Q and R. The perpendicular bisector of QR is the required line.

To bisect an angle α *with vertex at O.* Lay off equal distances OA and OB on the sides of α. With A and B as centers and a fixed radius $r > AB/2$, describe arcs intersecting at C. OC is the required bisector.

To construct a circle through noncollinear points ABC, or to find the circle circumscribing a triangle ABC, or to find the center of a circular arc ABC (by taking C not on the line AB). The perpendicular bisectors of AB and BC intersect in the center O of the required circle. OA is its radius.

To draw a tangent to a given circle, center O, from an external point A. Let B be the midpoint of OA. The circle with center B and radius BO cuts the given circle in the points of tangency.

3.4. Three-dimensional Definitions and Theorems. Two nonparallel nonintersecting lines are *skew* (to each other). Two nonskew lines or a line and a point, or three noncollinear points determine a plane. Two planes are parallel or have a line l in common; in the latter case, the *dihedral angle* between the planes is the angle between two lines, one in each plane, each line perpendicular to l. Three planes that have a point but not a line in common form a *trihedral* angle. A solid bounded by planes is a *polyhedron;* of these the simplest is a *tetrahedron* (four planes). A solid

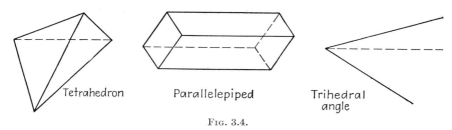

Tetrahedron Parallelepiped Trihedral angle

Fig. 3.4.

bounded by three pairs of parallel planes is a *parallelepiped*, which is *rectangular* if the planes intersect at right angles, and a *cube* if the edges are of equal length. If P is a point on a plane π, a line l through P is *normal* (perpendicular) to π if and only if it is perpendicular to every line in π through P. The *axis* of a circle is a line through its center normal to its plane.

Let C be a curve in a plane π and l a line in π not cutting C. The surface generated by revolving C about l as *axis* is a *surface of revolution,* and the solid bounded is a *solid of revolution.* Each point of C traces out a circle with l as axis.

Let C be a curve in a plane π and l' a line not in or parallel to π. Let a line l move so as to cut C and be parallel to l'. The surface S so formed is a *cylindrical surface* (or *cylinder*, if C is closed) with *generatrix* C, and the lines with which l coincides during its motion are the *generators* of S. If C is a polygon, S is a *prism.* If C is a circle (ellipse, parabola, hyperbola) S is *circular* (*elliptic, parabolic, hyperbolic*). If l' is normal to π, S is a *right* cylindrical surface. If C is closed, the interior of C is the *base* of S.

Let C be a curve in a plane π and P a point not in π. Let a line l move so as to contain P and cut C. The surface S so formed is a *conical surface* (or *cone*, if C is closed) with *generatrix* C and vertex P, and the lines with which l coincides during its motion are the *generators* of S. If C is a polygon, S is a *pyramid.* If C is a circle (ellipse, parabola, hyperbola), S is circular (elliptic, parabolic, hyperbolic). If C has a center O, and P is on the normal to π through O, S is a *right cone.* If C is closed, the interior of C is the *base* of S.

3.41. Areas and Volumes. A *prism* or *cylinder* contained between the plane π of its base and a parallel plane h units away from π has volume $Bh = Nl$ and lateral area pl, where l is the length of a generator, B and p are the area and perimeter of the base curve C, and N is the area of a section of the solid by a plane normal to the generators.

A *cone* (or *pyramid*) with vertex h perpendicular units from its base has volume $Bh/3$, where B is the area of the base.

Right circular cylinder, height h, base of radius r, has volume $\pi r^2 h$ and lateral area $2\pi rh$, hence total area $2\pi r(h + r)$.

Right circular cone, height h, base of radius r, has volume $V = \pi r^2 h/3$ and lateral area $A = \pi r \sqrt{r^2 + h^2}$, hence total area $\pi r(r + \sqrt{r^2 + h^2})$. In terms of the *semivertical angle* $\alpha = \arctan (r/h)$, $V = \frac{1}{3}\pi h^3 \tan^2 \alpha = \frac{1}{3}\pi r^3 \cot \alpha$ and $A = \pi r^2 \csc \alpha = \pi h^2 \tan \alpha \sec \alpha$. (For $\tan \alpha$, etc., see 5.)

Areas of *surfaces* and volumes of *solids* of *revolution* may often be found by Pappus' theorem: In a plane π, let a curve of length s (if closed, bounding an area a) lie entirely on one side of a line l. Revolve the curve about l. Let the centroid (see 10.6) of the curve (area) be distant $r_s(r_a)$ units from l, so that it moves on a circle of circumference $2\pi r_s(2\pi r_a)$. Then the area generated by the curve is $2\pi sr_s$, and the volume generated by the area is $2\pi ar_a$. However, unless the position of the centroids is obvious, by virtue of symmetry or otherwise, they will have to be located by integration, and it is then often easier to compute the area or volume directly by integration (see 9.1). In some instances the centroid and/or the volume may be found experimentally.

3.5. Sphere. A plane π that meets a sphere S either is *tangent* to the surface of S (*i.e.*, has just one point in common with it), or cuts it in a circle C, and every circle on S lies in just one cutting plane. The axis of C cuts S in two points that are the *poles* of C. If π contains the center O of S, π is a *diametral* plane; C is then a *great circle*, and any diameter of C is a *diameter* of S. Two points P, Q on the surface of S determine, along with O, a plane OPQ that cuts S in a great circle C containing P and Q; the minor arc PQ of C is the shortest route *on* S between P and Q. The part of S (surface or solid) between two parallel planes (one of which may be tangent) is a *zone* or *segment*.

Let O be a given point and T a given surface. Draw the ray OP for every point P on T. Let S be a sphere of radius 1 and center O. That *area* of the surface of S, every point of which lies in some ray OP, is the *solid angle* Ω *subtended at O by T*. The complete solid angle at O is 4π *steradians* = 1 *steregon*. (Units for Ω are rarely needed.)

3.51. Volumes and Areas. Let a sphere S have radius r. Then its volume is $4\pi r^3/3$ and its area is $4\pi r^2$.

A zone or segment between planes π and π', h units apart cutting out circles of radii a and a' ($a' = 0$ if π' is tangent to S), has volume $\pi h(3a^2 + 3a'^2 + h^2)/6$ and lateral area $2\pi rh$.

A spherical triangle bounded by the arcs of three great circles making angles A, B, C at the vertices has area $\pi r^2 E°/180°$, where the spherical excess $E° = A + B + C - 180°$.

3.6. The Earth. The earth is an oblate spheroid with an *equatorial radius* of 6,378.4 km, while the *polar radius* (distance from a center to a pole) is 6,359.9 km. This departure from sphericity is of little interest to meteorologists, who treat the earth as a sphere of radius 6,370 km. The area of the earth is 5.09951×10^8 km²; its volume 1.082841×10^{12} km³.

If P is a point on the surface and O the center of the earth, and N and S the north and south poles, then the semicircle NPS is the *meridian* of P. If this meridian cuts the equator in E, the angle EOP is the *latitude* φ of P. φ is positive or negative according as P is in the Northern or Southern Hemisphere. All points with the same latitude lie on a *parallel*. The value 981 cm per sec² used for the acceleration g of gravity at sea level is an approximation that is sufficiently exact for most meteorological purposes. More accurately, at $\varphi = 45°$, $g = 980.62$; elsewhere, $g = 980.62$ $(1 - 0.00259 \cos 2\varphi)$.

The angle at N between the meridian NPS of P and the meridian NGS of Greenwich is the *longitude* μ of P, positive or negative according as NPS is west or east of NGS.

The length of one minute of arc of a great circle is a *nautical mile*, approximately 6,080 ft or 1.1515 ordinary (statute) miles. If P' is φ', μ', then the angle $POP' = \psi$ is given by $\cos \psi = \sin \varphi \sin \varphi' + \cos \varphi \cos \varphi' \cos (\mu - \mu')$. Expressed in minutes, ψ gives the number of nautical miles in an arc of a great circle joining P and P'. A *knot* is a speed of 1 nautical mile per hr.

3.7. Maps. A map represents a portion of the surface of the earth on a plane. A map such that, at each point, the angle between two terrestrial directions equals the angle between the corresponding directions on the map is *conformal*. On such a map, at each point, the scale is the same in every direction. Shapes of sufficiently small terrestrial areas are displayed quite accurately, but sizes are often given very roughly. An *equal-area*, or *authalic*, map has the property that equal terrestrial areas are displayed by equal areas on the map. Most maps have one or the other of these properties. No map can have both. The cartographer balances the demands of these properties and others, of interest mainly in navigation, with a view to the uses to which his map will be put.

3.71. Mercator (conformal, excessively nonauthalic toward the poles). The earth is projected on a cylinder tangent to the earth at the equator, not in a geometrically simple manner, but so that each parallel of latitude has the same length on the map and so that the scale on each meridian is equal, at each point, to the scale on the parallel through the point. The cylinder is then rolled out onto a plane. This map is useful to navigators because a straight line on the map represents a *rhumb line* or *loxodrome*, i.e., a curve making a fixed angle with the meridians. The scale is good only very close to the equator.

3.72. Lambert Secant Cone (conformal, appreciably but not excessively nonauthalic). A portion of the earth is projected (again in a geometrically complicated manner) on a secant cone that contains two chosen *standard* parallels. For United States weather maps, 30 and 60° are standard; for Civil Aeronautic Authority maps, 33 and 45° are standard. On the standard parallels, the scale is exact. Between them, it is decreased by not more than about 1 per cent. Outside them, distortion increases more rapidly. Graphical measurement of even comparatively long distances is fairly reliable, unless the latitudes differ too greatly.

3.73. Authalic Maps. Many of these are in use, the commonest which depict the entire globe being similar in character to those of Mollweide or Aitoff. These maps represent the entire earth on an ellipse, with the major axis twice the minor axis. They are especially useful when, as in climatology, one wishes to indicate clearly the distribution over the earth's area of such items as rainfall, nebulosity, temperature, air currents, etc.

4. FORMAL ALGEBRA

Operations with algebraic expressions follow the rules of 1.3 for operations with real numbers.

4.1. Special Products and Factoring.

$$a(x + y) = ax + ay$$
$$(ax + by)(cx + dy) = acx^2 + (bc + ad)xy + bdy^2$$

1. $(ax \pm by)^2 = a^2x^2 \pm 2abxy + b^2y^2$
2. $(ax + by)(ax - by) = a^2x^2 - b^2y^2$ [*not* $(ax - by)^2$; see 1]

3. $(ax + b)(cx + d) = acx^2 + (bc + ad)x + bd$
4. $(x + b)(x + d) = x^2 + (b + d)x + bd$

$$(ax \pm by)(a^2x^2 \mp abxy + b^2y^2) = a^3x^3 \pm b^3y^3$$
$$(ax + by + cz)^2 = a^2x^2 + b^2y^2 + c^2z^2 + 2bcyz + 2cazx + 2abxy$$

Use either the upper or the lower ambiguous sign throughout. The expressions that are "multiplied out" on the right appear "factored" on the left. To factor $a^nx^n - b^ny^n$, n any positive integer, divide by $ax - by$. To factor $a^nx^n + b^ny^n$, n any *odd* integer, divide by $ax + by$; not factorable for *even* n.

4.2. Σ and Π Notation. It is often advantageous to abbreviate $a_1 + \cdots + a_n$ by writing $\sum_1^n a_i$. The notation means that the index i is to be given the values from 1 to n, inclusive, and the terms thus obtained are to be added up. Similarly,

$$\prod_1^n a_i = a_1 \cdot a_2 \cdot \cdots \cdot a_n$$

(less common). For example

$$\sum_1^3 a_ix_i^2 = a_1x_1^2 + a_2x_2^2 + a_3x_3^2, \qquad \prod_1^4 a_ib_i = a_1b_1a_2b_2a_3b_3a_4b_4.$$

The averages in 1.21 may be written

$$A = \frac{1}{n}\sum_1^n a_i, \qquad G = \left(\prod_1^n a_i\right)^{\frac{1}{n}}, \qquad H^{-1} = \frac{1}{n}\sum a_i^{-1}.$$

When there is only one index, $\sum_1^n a_i$ is sometimes written simply Σa_i, or, carelessly, Σa. If there are several indices, it is important to be quite clear about which are being summed: $y_i = \sum_{j=1}^n a_{ij}x_j$ means $y_i = a_{i1}x_1 + a_{i2}x_2 + \cdots + a_{in}x_n$. In work with tensors, indices occur as superscripts as well as subscripts; such indices are not to be confused with exponents.

4.3. Ratio and Proportion; Variation. The *ratio of x to y* is x/y. A *proportion* is an equality between two ratios: $x/y = a/b$, read "x is to y as a is to b." (Formerly written $x:y::a:b$; for extremes, means, antecedents, consequents, etc., and the Euclidean theory of proportion, see any elementary algebra or geometry. Direct treatment as equality of ratios recommended.) For example (Avogadro's law), at given temperature and pressure, the densities of two gases are proportional to their molecular weights: $(M_1/V_1)/(M_2/V_2) = m_1/m_2$, where M, V, and m denote mass, volume, and molecular weight, respectively.

If $a/b = b/c$, then $b^2 = ac$, and b is said to be a *mean proportional* between a and c.

The locutions "y varies as x," "y varies directly as x," "y is proportional to x," and the symbol $y \propto x$ all mean that the ratio of y to x is a constant $k \neq 0$, called the *constant of proportionality:* $y = kx$. "y varies inversely as x" means that y varies as the inverse of x, i.e., as $1/x$; $y = k/x$ or $xy = k$. "y varies (jointly) as x and t" means that y varies as the product xt; $y = kxt$. The constant of proportionality is determined from a knowledge of one particular set of associated values of the variables, e.g., Newton's law of gravitation states that the attractive force between two particles varies as their masses and inversely as the square of their distance apart: $F = kmm'/r^2$. Measuring m in grams, r in centimeters, and F in dynes, it may be observed experimentally that the force between two 1-kg masses placed 10 cm apart is 6.66×10^{-4} dynes. Hence $k = 6.66 \times 10^{-8}$ cm^3 per sec^2 gram, and $F = 6.66 \times 10^{-8} \, mm'/r^2$, from which the attraction between two masses in given relative position may be computed at once without further measurement.

4.41. Arithmetic Progressions. The sequence of numbers $a_1, a_2, \cdots, a_n, \cdots$ is an *arithmetic progression* (AP) when and only when $a_2 - a_1 = a_3 - a_2 = \cdots = a_n - a_{n-1} = \cdots = d$, called the *common difference*. An AP can thus be written $a, a + d, a + 2d, \cdots$. The nth term is $t_n = a + (n - 1)d$. The sum of the first n terms is $s_n = n[2a + (n - 1)d]/2 = n(a + t_n)/2$ (n times the average of the first and last terms). For example, $5, 1\frac{2}{3}, 1\frac{1}{3}, \cdots$ is an AP with $a = 5$, $d = -1/3$, $t_8 = 5 + 7(-1/3) = \frac{8}{3}$, $s_8 = 8(5 + \frac{8}{3})/2 = 9\frac{2}{3}$.

4.42. Geometric Progression. The sequence of numbers $a_1, a_2, \cdots, a_n, \cdots$ is a *geometric progression* (GP) if and only if $a_2/a_1 = a_3/a_2 = \cdots = a_n/a_{n-1} = \cdots = r$, called the *common ratio*. A GP can thus be written $a, ar, ar^2, \cdots$. The nth term is $t_n = ar^{n-1}$. The sum of the first n terms is $s_n = a(1 - r^n)/(1 - r) = (a - rt_n)/(1 - r)$ if $r \neq 1$. For example, $3, 3(1.04), 3(1.04)^2, \cdots$ is a GP with $a = 3$, $r = 1.04$, $t_5 = 3(1.04)^4$, $s_5 = 3[1 - (1.04)^5]/(1 - 1.04) = 16.26$.

As n increases, r^n does or does not get indefinitely small according as $|r| < 1$ or $|r| \geqslant 1$. In case $|r| < 1$, the series $a(1 + r + r^2 + \cdots)$ is said to converge, and to have the sum $a/(1 - r)$; if $|r| \geqslant 1$, the series diverges and is of no interest here. The rational number represented by a repeating decimal may be found by summing a GP. For example, $2.3\dot{5}8\dot{2} = 2.3582582582 \cdots = 2.3 + 582 \cdot 10^{-4} + 582 \cdot 10^{-7} + \cdots = 2.3 + 582 \cdot 10^{-4}(1 + 10^{-3} + 10^{-6} + \cdots) = 2.3 + 582 \cdot 10^{-4}/(1 - 10^{-3}) = 23,559/9,990$.

4.5. Permutations and Combinations. A *permutation* of n distinct objects a is an arrangement of the objects in a definite *order*: $a_1, a_2, \cdots, a_n$. There are $n(n - 1)(n - 2) \cdots \cdot 3 \cdot 2 \cdot 1 = n!$ (say "n factorial") permutations of the a's. If the a's are alike in groups, k_1 in a group, $\cdots$, k_p in a group, with $k_1 + \cdots + k_p = n$, then the number of distinct permutations is $n!/(k_1!k_2! \cdot \cdots \cdot k_p!)$. For example, 7 flags of different colors can be flown from a mast in $7!$ different orders; if 3 of the 7 flags are red, 2 white, and 2 blue, the number of orders (permutations) is $7!/3!2!2! = 2 \cdot 3 \cdot 5 \cdot 7$.

If, instead of using all n of the objects, we use only r of them, there will be $n!/(n - r)!$ permutations of the n distinct objects *taken r at a time*.

If we are not interested in the order in which the r objects are arranged, and agree that the same r objects shall constitute one *combination* (regardless of order), then we find that there are $_nC_r = C_r^n = C_{n.r} = \binom{n}{r} = n!/(n - r)!r!$ *combinations of n objects taken r at a time*. From the 7 different flags there can be drawn $7!/5!2!$ different pairs ($r = 2$). N.B. $C_r^n = C_{n-r}^n$. Also, $0! = 1$ (definition), so that $C_0^n = C_n^n = 1$. C_r^n is often written $n(n - 1) \cdot \cdots \cdot (n - r + 1)/r! = n(n - 1) \cdot \cdots \cdot (r + 1)/(n - r)!$.

If n is large, $n!$ may be approximated by means of *Stirling's inequality (formula)*: $\sqrt{2\pi}\, n^{n+\frac{1}{2}}e^{-n} < n! < \sqrt{2\pi}\, n^{n+\frac{1}{2}}e^{-n}(1 + 1/4n)$, where e is the natural base for logarithms (see 2.4).

4.6. Binomial Theorem. If n is a positive integer,

$$(a + b)^n = C_0^n a^n b^0 + C_1^n a^{n-1}b^1 + C_2^n a^{n-2}b^2 + C_3^n a^{n-3}b^3 + \cdots + C_{n-2}^n a^2 b^{n-2}$$
$$+ C_{n-1}^n a^1 b^{n-1} + C_n^n a^0 b^n = a^n + na^{n-1}b + \frac{n(n - 1)}{2!} a^{n-2}b^2$$
$$+ \frac{n(n - 1)(n - 2)}{3!} a^{n-3}b^3 + \cdots + b^n.$$

Example. $(x - y^{\frac{2}{3}})^5 = x^5 + 5\,x^4(-y^{\frac{2}{3}}) + \frac{5 \cdot 4}{1 \cdot 2} x^3(-y^{\frac{2}{3}})^2 + \frac{5 \cdot 4 \cdot 3}{1 \cdot 2 \cdot 3} x^2(-y^{\frac{2}{3}})^3$
$$+ \frac{5 \cdot 4 \cdot 3 \cdot 2}{1 \cdot 2 \cdot 3 \cdot 4} x(-y^{\frac{2}{3}})^4 + (-y^{\frac{2}{3}})^5 = x^5 - 5x^4 y^{\frac{2}{3}} + 10x^3 y^{\frac{4}{3}} - 10x^2 y^2 + 5xy^{\frac{8}{3}} - y^{\frac{10}{3}}.$$

For C_r^n, see 4.5. In case n is not a positive integer, the series does not terminate (see 8.74).

4.7. Determinants. The symbols

$$D_1 = a, \qquad D_2 = \begin{vmatrix} a & b \\ c & d \end{vmatrix}, \qquad D_3 = \begin{vmatrix} a & b & c \\ d & e & f \\ g & h & i \end{vmatrix}, \qquad \cdots ,$$

$$D_n = \begin{vmatrix} a_{11} & a_{12} & \cdots & a_{1n} \\ a_{21} & a_{22} & \cdots & a_{2n} \\ \cdots & \cdots & \cdots & \cdots \\ a_{n1} & a_{n2} & \cdots & a_{nn} \end{vmatrix} = |a_{ij}|, \cdots$$

are *determinants*. D_n is of *order* n. The horizontal (vertical) lines are called *rows* (*columns*); the individual entries are called *elements*. Each D denotes an expression formed in a specific way from its elements. When fully written out (expanded), D_n is a sum of $n!$ terms. Perhaps the most intelligible direct method of expansion is contained in the formula

$$\text{(by the } j\text{th column)} \sum_{i=1}^{n} (-1)^{i+j} a_{ij} A_{ij} = D_n = \sum_{i=1}^{n} (-1)^{i+j} a_{ji} A_{ji} \text{ (by the } j\text{th row)}$$

holding for any chosen *fixed* j, wherein A_{pq} denotes that D_{n-1} obtained by deleting from D_n the row and column in which a_{pq} occurs (for Σ, see 4.2). Thus $D_2 = (-1)^{1+2}b(c) + (-1)^{2+2}d(a) = ad - bc$. Using the left expression for D_n with $j = 2$ (*i.e.*, using the second column), we get

$$D_3 = -b \begin{vmatrix} d & f \\ g & i \end{vmatrix} + e \begin{vmatrix} a & c \\ g & i \end{vmatrix} - h \begin{vmatrix} a & c \\ d & f \end{vmatrix} = -b(di - gf) + e(ai - cg) - h(af - dc)$$

$$= aei + bfg + cdh - ceg - bdi - afh.$$

The formula for D_n reduces D_4 to D_3's, etc.

Elementary Properties. 1. $D = 0$ if every element in some one row (or column) vanishes. $D = 0$ if it has two rows (columns) with the elements of one proportional to (*i.e.*, a constant k times) the elements of the other.

2. $|a_{ij}| = |a_{ji}|$, *i.e.*, if the columns and rows of D' are the rows and columns of D, in the same order, than $D = D'$.

3. If D' consists of D with one pair of rows (or columns) interchanged, then $D' = -D$.

4. If D' consists of D with each element in any one row (or column) multiplied by the same constant k, then $D' = kD$. Thus a common factor of all the elements of a row (or column) may be brought out and written as a multiplier of D.

5. If D' arises from D by multiplying each element of some one row (or column) by the same constant k and adding it to some other row (or column), then $D' = D$.

For addition, multiplication, other means of expansion, etc., see a text. For application to linear equations, see 6.2.

4.8. Polynomials. A monomial $ax_1^{p_1} \cdot x_2^{p_2} \cdot \cdots \cdot x_k^{p_k}$ has *degree* p_i *in* x_i and total degree $p_1 + \cdots + p_k$. A function $f(x_1, \cdots, x_k) = \Sigma a_{p_1 p_2 \ldots p_k} x_1^{p_1} \cdot x_2^{p_2} \cdot \cdots \cdot x_k^{p_k}$, where the indices of summation p run independently from 0 to n, is a *polynomial* in $x_1, \cdots, x_n$ of degree equal to the total degree of the monomial of highest total degree. $x_1^4 x_2^2 x_3 x_4^3 + 3x_1^6 x_2 - 7x_1^3 x_2^2 x_3 x_4^2$ is a polynomial of degree 10 in $x_1, \cdots, x_4$. Thus $f(x) = a_n x^n + a_{n-1} x^{n-1} + \cdots + a_1 x + a_0$ with $a_n \neq 0$ is a polynomial of degree n in x; it is *linear* if $n = 1$, *quadratic* if $n = 2$, *cubic* if $n = 3$, and *quartic* if $n = 4$, etc.

To divide $f(x)$ by $x - c$, write down the coefficients $a_n, \cdots, a_0$, being sure to supply a zero for *each* missing coefficient, including a_0. Enter c (not $-c$) as divisor, bring down a_n, multiply a_n by c and *add* (algebraically) to a_{n-1}; multiply the result by c and add to a_{n-2}; $\cdots$; the end result is the *remainder* and is also the *value* $f(c)$ of the polynomial when $x = c$. The other figures appearing below the line are the coefficients of the *quotient*. This process is called *synthetic division* or [because the remainder is $f(c)$] *synthetic substitution*.

$$
\begin{array}{cccc|c}
a_n & a_{n-1} & a_1 & a_0 & \underline{c} \\
 & a_n c & & & \\
\hline
a_n & a_n c + a_{n-1} & & f(c) &
\end{array}
$$

To divide $f(x) = 3x^4 + 2x^2 - x$ by $x + 2$: N.B. $x + 2 = x - (-2)$, so $c = -2$

$$
\begin{array}{ccccc|c}
3 + 0 & + 2 & - 1 & + 0 & & \underline{-2} \\
 - 6 & + 12 & - 28 & + 58 & & \\
\hline
3 - 6 & + 14 & - 29 & + 58 & &
\end{array}
$$

Hence $3x^4 + 2x^2 - x = (3x^3 - 6x^2 + 14x - 29)(x + 2) + 58$. The quotient is $3x^3 - 6x^2 + 14x - 29$; the remainder is $58 = f(-2)$.

To divide $f(x)$ by $ax - b$, divide synthetically by $x = b/a$, getting $f(x) = q(x)$ $(x - b/a) + r = (ax - b)q(x)/a + r$, so that the quotient, on division by $ax - b$, is $q(x)/a$ and the remainder is $r = f(b/a)$.

Evidently $f(x) = (x - c)q(x)$ when and only when $f(c) = 0$ (*i.e.*, remainder 0); if $f(c) = 0$, c is a *root* or *solution* of the *algebraic equation* $f(x) = 0$. Every equation of degree $n \geqslant 1$ has at least one root c, which may be complex (see 5.7) ("fundamental theorem of algebra"). Proceeding to find roots of $q(x)$, etc., we finally get $f(x) = a_n(x - c_1)(x - c_2) \cdots (x - c_n)$, expressing $f(x)$ as a product of k linear factors. If $c_1, \cdots, c_l$, $l \leqslant n$ are the *distinct* c's, then $f(x) = a_n(x - c_1)^{m_1} \cdots \cdots (x - c_l)^{m_l}$, where $m_1 + \cdots + m_l = n$; m_i is the *multiplicity* of c_i ($i = 1, \cdots, l$). Hence, counting m times a root of multiplicity m, a polynomial of degree n has *exactly* n roots. If $f(x)$ has *real* coefficients, and $f(a + bi) = 0$, then $f(a - bi) = 0$; complex roots occur in conjugate pairs. If $c = a + bi$, the factors $x - (a + bi)$ and $x - (a - bi)$ may be multiplied together to get $x^2 - 2ax + a^2 + b^2$. Hence, if $f(x)$ has *real* coefficients, $f(x)$ may be expressed as a product of *real linear* and *quadratic* factors. Thus

$$
x^3 - 1 = (x - 1)\left(x - \frac{-1 + i\sqrt{3}}{2}\right)\left(x - \frac{-1 - i\sqrt{3}}{2}\right) = (x - 1)(x^2 + x + 1).
$$

An equation of *odd* degree with *real* coefficients has at least one *real* root. To solve an algebraic equation, see 6.3.

4.9. Partial Fractions. Let $R(x) = N(x)/D(x)$ be a quotient of two polynomials with real coefficients (a *rational function* in technical terminology), in which every factor common to N and D has been removed, and in which the indicated division has been carried out, if necessary, until the degree of N is *less* than the degree of D.

Suppose that, in factored form (see 4.8), $D(x) = \prod_{i=1}^{n_l} (a_i x + b_i)^{l_i} \prod_{k=1}^{n_q} (c_k x^2 + d_k x + e_k)^{q_k}$,

where the $a_i x + b_i$ are the n *distinct linear* factors of D, and the $c_k x^2 + d_k x + e_k$ are the *distinct* (irreducible) *quadratic* factors of D. Then, by methods indicated in the following example, *unique* constant A's, B's, and C's can be found such that

$$
R(x) = \sum_{i=1}^{n_l} \left[\sum_{j=1}^{l_i} \frac{A_{ij}}{(a_i x + b_i)^j} \right] + \sum_{k=1}^{n_q} \left[\sum_{m=1}^{q_k} \frac{B_{km} x + C_{km}}{(c_k x^2 + d_k x + e_k)^m} \right].
$$

The fractions on the right are *partial fractions* for $R(x)$. N.B. (Useful checks.)
$\sum_1^{n_l} l_i + 2 \sum_1^{n_q} q_k$ = degree of D. For each i, $A_{il_i} \neq 0$; for each k, either $B_{kq_k} \neq 0$
or $C_{kq_k} \neq 0$.

We are to find numbers $A, B, \cdots, F$ so that the sum of the partial expressions
on the right (written down in accordance with the preceding formula, $n_l = n_q = 1$,
$l_1 = 2 = q_1$) is equal to the expression $R(x)$ on the left;

$$\frac{x^4 - 8x^3 + 24x^2 - 14x + 7}{(2x-1)^2(x^2+2)^2} = \frac{A}{2x-1} + \frac{B}{(2x-1)^2} + \frac{Cx+D}{x^2+2} + \frac{Ex+F}{(x^2+2)^2}.$$

Clearing of fractions,

$$x^4 - 8x^3 + 24x^2 - 14x + 7 = A(2x-1)(x^2+2)^2 + B(x^2+2)^2 \\ + (Cx+D)(2x-1)^2(x^2+2) + (Ex+F)(2x-1)^2.$$

This is an identity (see 6.1). By expanding and equating coefficients, or by differentiating repeatedly and putting $x = 0$ (see 8.2, 8.7), we could find enough linear equations to determine the required constants (see 6.2). However, it is recommended that the following processes be tried first, using linear equations only as a last resort.

In the identity, put $x = \frac{1}{2}$ (chosen so as to make $2x - 1 = 0$). The result is $81/16 = 81B/16$, whence $B = 1$. Insert this value for B, transpose that term to the left, and divide each side by $2x - 1$. The result is

$$-4x^2 + 8x - 3 = A(x^2+2)^2 + (Cx+D)(2x-1)(x^2+2) + (Ex+F)(2x-1)$$

Put $x = \frac{1}{2}$ again, obtaining $A = 0$. Divide again by $2x - 1$, obtaining $-2x + 3 = (Cx+D)(x^2+2) + Ex + F$. If, now, either C or D were not to vanish, there would be a term either in x^3 or in x^2 on the right, while there is no such term on the left. Hence $C = D = 0$, and we have $-2x + 3 = Ex + F$. Equating coefficients (or putting $x = 0$, etc.), $E = -2, F = 3$. Hence

$$R(x) = \frac{1}{(2x-1)^2} - \frac{2x-3}{(x^2+2)^2}.$$

It is wise to verify the result before using it.

5. TRIGONOMETRY

5.1. Rectangular Cartesian Coordinates. As in 1.42, take horizontal (x) and vertical (y) axes intersecting at an origin O. Use now the *same* scale on each axis, and measure distances on *each* axis *from* O. Each point P of the plane then has uniquely associated with it a pair of numbers (x,y), which are called its rectangular cartesian *coordinates*. The axes divide the plane into four *quadrants*, in which x and y have the signs indicated in Fig. 5.1. (When $x > 0$ is measured to the right and $y > 0$ upward, as in the figure, the axes are "right-handed"; reversing the direction of *one* axis gives a "left-handed" pair, equally appropriate but rarely used.)

5.2. Angles. In trigonometry and more advanced work, an angle is understood to be the amount by which a ray (see 3.12) has been rotated from a certain *initial* ray in order to coincide with a certain *terminal* ray. In *standard position*, the vertex of an angle is at the origin and its initial ray is the positive x-axis (the *initial* direction). Rotation is counted positively in the counterclockwise direction (from initial direction toward positive y-axis) and negatively in the clockwise direction. Thus an angle of 745 deg means two counterclockwise revolutions (720 deg) and 25 deg more; $-7\pi/2$ rad means a clockwise revolution (2π) and $\frac{3}{4}$ revolution = $3\pi/2$ more.

5.3. The Trigonometric (Circular) Functions. Let θ be an angle in standard position and let $P \neq O$ be a point on the terminal ray. Drop a perpendicular from P

on the x-axis.   Find the coordinates (x,y) of P and the distance $OP = r = \sqrt{x^2 + y^2}$ $> 0.$ Then

$$\sin \theta = \frac{y}{r} \qquad \cos \theta = \frac{x}{r} \qquad \tan \theta = \frac{y}{x}$$

$$\csc \theta = \frac{r}{y} \qquad \sec \theta = \frac{r}{x} \qquad \cot \theta = \frac{x}{y}$$

By similar triangles, these ratios are independent of the position of P on the terminal ray and hence define functions of θ (see 1.41).   Sin θ has the sign of y, cos θ has the sign of x, and tan θ and cot θ have the same sign, which is $+$ or $-$ according as the signs of x and y agree or disagree; each of these signs is determined immediately by the quadrant in which the terminal ray of θ lies.

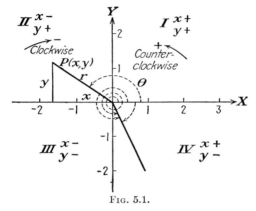

Fɪɢ. 5.1.

From the behavior of x and y when the sense of θ is reversed, we see that $\sin (-\theta) = -\sin \theta$, $\cos (-\theta) = \cos \theta$, etc.   Since $|x| \leqslant r$ and $|y| \leqslant r$, $|\sin \theta| \leqslant 1$, $|\cos \theta| \leqslant 1$, $|\sec \theta| \geqslant 1$ and $|\csc \theta| \geqslant 1$ always.   Further rotation of $k \cdot 360$ deg $= 2k\pi$ rad, k any *integer*, will give an angle with the same terminal ray as θ, hence the same ratios of x, y, r, hence the same values for any of the functions; *i.e.*, these functions are *periodic*, with period 2π (see graphs in 8.9).

The definitions fail whenever an x or y vanishes in a denominator.   For $\theta = \pi/2$, $x = 0$, $y > 0$, and tan θ is not defined.   However (see 8.13), $\tan \theta \to +\infty$ as $\theta \to \pi/2^-$, and $\tan \theta \to -\infty$ as $\theta \to \pi/2^+$.   By inspecting the quadrantal angles, the sides of an isosceles right triangle (side 1) or the sides of an equilateral triangle (side 2) with an altitude inserted, one arrives at the following table, where an entry $\pm \infty$ or $\mp \infty$ means

$\theta°$	$0°$	$30°$	$45°$	$60°$	$90°$	$180°$	$270°$	$360°$
θ rad	0	$\pi/6$	$\pi/4$	$\pi/3$	$\pi/2$	π	$3\pi/2$	2π
$\sin \theta$	0	$\frac{1}{2}$	$1/\sqrt{2}$	$\sqrt{3}/2$	1	0	-1	0
$\cos \theta$	1	$\sqrt{3}/2$	$1/\sqrt{2}$	$\frac{1}{2}$	0	-1	0	1
$\tan \theta$	0	$1/\sqrt{3}$	1	$\sqrt{3}$	$\pm \infty$	0	$\pm \infty$	0
$\cot \theta$	$\mp \infty$	$\sqrt{3}$	1	$1/\sqrt{3}$	0	$\mp \infty$	0	$\mp \infty$
$\sec \theta$	1	$2/\sqrt{3}$	$\sqrt{2}$	2	$\pm \infty$	-1	$\mp \infty$	1
$\csc \theta$	$\mp \infty$	2	$\sqrt{2}$	$2/\sqrt{3}$	1	$\pm \infty$	-1	$\mp \infty$

that the function has the limit indicated by the upper or lower sign, respectively, as θ tends to its value from below or above.

Tables refer to positive acute angles. To find values for other angles θ, find from

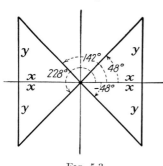

a diagram the positive acute angle θ' with *numerically* the same ratios of x, y, r; then the function of θ will have *numerically* the same value as for θ', and its sign will be determined by the quadrant in which θ lies.

Thus, for $142°$, $228°$, $-48°$, $-142°$, $312°$, $\theta' = 48°$; since $142°$ is in II, $\cos 142° = -\cos 48°$, $\tan 142° = -\tan 48°$; since $312°$ is in IV, $\cos 312° = \cos 48°$, $\sin 312° = -\sin 48°$; etc.

Going the other way, suppose $\tan \theta = -k$, where $k = \tan 48°$. Then $\theta = 142° + k \cdot 360°$ in II or $\theta = 312° + k \cdot 360°$ in IV, k *any integer.*

Fig. 5.3.

5.41. Identities among Functions of a Single Angle. An ambiguous sign before a radical is determined by the quadrant in which θ lies.

$$\sec \theta = \frac{1}{\cos \theta} \qquad \tan \theta = \frac{\sin \theta}{\cos \theta}$$

$$\csc \theta = \frac{1}{\sin \theta} \qquad \cot \theta = \frac{\cos \theta}{\sin \theta}$$

$$\cot \theta = \frac{1}{\tan \theta}$$

$$\sin^2 \theta + \cos^2 \theta = 1 \qquad \cos \theta = \pm \sqrt{1 - \sin^2 \theta}, \text{ etc.}$$
$$1 + \tan^2 \theta = \sec^2 \theta \qquad \sec \theta = \pm \sqrt{1 + \tan^2 \theta}, \text{ etc.}$$
$$1 + \cot^2 \theta = \csc^2 \theta \qquad \csc \theta = \pm \sqrt{1 + \cot^2 \theta}, \text{ etc.}$$

N.B. *Any* one angle may be substituted for *every* occurrence of θ in any identity; *e.g.*, $\sin^2 (2\pi\nu t) + \cos^2 (2\pi\nu t) = 1$.

5.42. Identities among Functions of Two Angles. Use *either* upper *or* lower of ambiguous signs on both sides.

$$\sin (\theta \pm \phi) = \sin \theta \cos \phi \pm \cos \theta \sin \phi$$
$$\cos (\theta \pm \phi) = \cos \theta \cos \phi \mp \sin \theta \sin \phi$$
$$\tan (\theta \pm \phi) = \frac{\tan \theta \pm \tan \phi}{1 \mp \tan \theta \tan \phi}$$
$$\sin \theta + \sin \phi = 2 \sin \tfrac{1}{2}(\theta + \phi) \cos \tfrac{1}{2}(\theta - \phi)$$
$$\sin \theta - \sin \phi = 2 \cos \tfrac{1}{2}(\theta + \phi) \sin \tfrac{1}{2}(\theta - \phi)$$
$$\cos \theta + \cos \phi = 2 \cos \tfrac{1}{2}(\theta + \phi) \cos \tfrac{1}{2}(\theta - \phi)$$
$$\cos \theta - \cos \phi = -2 \sin \tfrac{1}{2}(\theta + \phi) \sin \tfrac{1}{2}(\theta - \phi)$$

Use these formulas to learn that

$$\cos (\theta + 90°) = \cos \theta \cos 90° - \sin \theta \sin 90° = -\sin \theta$$
$$\sin (\pi - \theta) = \sin \pi \cos \theta - \cos \pi \sin \theta = \sin \theta$$
$$\sin A \cos B = \tfrac{1}{2}[\sin (A + B) + \sin (A - B)], \text{ etc.}$$

5.43. Multiple Angles.

$$\sin 2\theta = 2 \sin \theta \cos \theta$$
$$\cos 2\theta = \cos^2 \theta - \sin^2 \theta = 1 - 2 \sin^2 \theta = 2 \cos^2 \theta - 1$$
$$\sin 3\theta = 3 \sin \theta - 4 \sin^3 \theta$$

$$\cos 3\theta = 4 \cos^3 \theta - 3 \cos \theta$$
$$\sin 4\theta = 8 \cos^3 \theta \sin \theta - 4 \cos \theta \sin \theta$$
$$\cos 4\theta = 8 \cos^4 \theta - 8 \cos^2 \theta + 1$$

$$\sin \frac{\theta}{2} = \pm \sqrt{\frac{1 - \cos \theta}{2}} \qquad \cos \frac{\theta}{2} = \pm \sqrt{\frac{1 + \cos \theta}{2}}$$

$$\tan \frac{\theta}{2} = \frac{\sin \theta}{1 + \cos \theta} = \frac{1 - \cos \theta}{\sin \theta} = \pm \sqrt{\frac{1 - \cos \theta}{1 + \cos \theta}}$$

5.5. Inverse Functions. Trigonometric Equations. If $y = \sin x$, then $x = $ arc $\sin y = \sin^{-1} y$ (-1 *not* an exponent) is the *inverse sine* of y or *arc* (or angle) whose *sine* is y, usually encountered with x (y) as independent (dependent) variable: $y = $ arc $\sin x$. Similarly for arc $\cos x$, arc $\tan x$, etc. As indicated by their graphs (see 8.9), these functions are *multiple-valued*. Unless directed otherwise, especially in integral calculus, use the following *principal* values:

	Range of definition	Principal values
$y = $ arc $\sin x$	$|x| \leqslant 1$	$-90° = -\dfrac{\pi}{2} \leqslant y \leqslant \dfrac{\pi}{2} = 90°$
$y = $ arc $\cos x$	$|x| \leqslant 1$	$0° = 0 \leqslant y \leqslant \pi = 180°$
$y = $ arc $\tan x$	$-\infty < x < \infty$	$-90° = -\dfrac{\pi}{2} \leqslant y \leqslant \dfrac{\pi}{2} = 90°$
$y = $ arc $\cot x$	$-\infty < x < \infty$	$0° = 0 < y < \pi = 180°$
$y = $ arc $\sec x$	$1 \leqslant |x|$	$0° = 0 \leqslant y \leqslant \pi = 180°$
$y = $ arc $\csc x$	$1 \leqslant |x|$	$-90° = -\dfrac{\pi}{2} < y < \dfrac{\pi}{2} = 90°$

The last three functions occur rarely, being reducible to the first three as follows: arc $\cot x = \pi/2 - $ arc $\tan x$, arc $\sec x = $ arc $\cos 1/x$, arc $\csc x = $ arc $\sin 1/x$; also, arc $\cos x = \pi/2 - $ arc $\sin x$, arc $\csc x = \pi/2 - $ arc $\sec x$. These equations are valid for all x. If the principal value of one function is substituted in an equation, the computed value of the other function will be its principal value.

Caution: The relation arc $\cot x = $ arc $\tan (1/x)$ does not have this property when $x < 0$.

In solving equations, try first with identities to simplify as much as possible, *e.g.*, expressing all terms as functions of the same angle, preferably as the same function. Solve algebraically and seek inverse functions. Be sure to get *all* angles θ, $0 \leqslant \theta < 360° = 2\pi$ rad which satisfy the equation; then $(\theta + 360k)° = (\theta + 2k\pi)$ rad are also solutions for *any* integer k.

Example. $\sin^2 2\theta + 3 \cos^2 \theta = 3$ may be written $1 - \cos^2 2\theta + 3(\cos 2\theta + 1)/2 = 3$, or $(2 \cos 2\theta - 1)(\cos 2\theta - 1) = 0$, whence either $\cos 2\theta = \frac{1}{2}$ and $2\theta = $ arc $\cos \frac{1}{2} = 60°$, $300°$, $420°$, $660°$, or $\cos 2\theta = 1$ and $2\theta = $ arc $\cos 1 = 0°$, $360°$; finally, then, $\theta = 0°$, $30°$, $150°$, $180°$, $210°$, $330°$.

5.61. Right Triangles (Notation in 3.14). If $C = 90°$, then

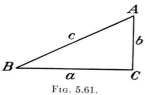

Fig. 5.61.

$$\sin B = \frac{b}{c} = \frac{\text{side opposite } B}{\text{hypotenuse}}$$

$$\cos B = \frac{a}{c} = \frac{\text{side adjacent } B}{\text{hypotenuse}}$$

$$\tan B = \frac{b}{a} = \frac{\text{side opposite } B}{\text{side adjacent } B}$$

and similarly for the other functions of B. Use these relations to find unknown from known parts, *e.g.*, $a = b \cot B$.

The (orthogonal, perpendicular) *projection* of c along the horizontal is $a = c \cos B$ $= c \sin A$; along the vertical $b = c \sin B = c \cos A$. Similarly, if α is an area in a plane π the projection of α on a plane π' making an angle θ with π is $\alpha' = \alpha \cos \theta$.

5.62. Oblique Triangles. For area and other information, see 3.25. As a guide, and to detect gross inconsistencies, draw a rough sketch to scale beforehand. When two angles are known, $A + B + C = 180°$ gives the third. Use logs except with the cosine law. As soon as four parts are known, the sine law may be used. When a check is desired, use any relation not used in the solution, or $c \cos B + b \cos C = a$, etc.

Formulas.

$$\frac{a}{\sin A} = \frac{b}{\sin B} = \frac{c}{\sin C} \qquad \text{(sine law)}$$

$$a^2 = b^2 + c^2 - 2bc \cos A \qquad \text{(similarly for } b \text{ and } B, c \text{ and } C; \text{ cosine law)}$$

$$\frac{a - b}{a + b} = \frac{\tan \frac{1}{2}(A - B)}{\tan \frac{1}{2}(A + B)} \qquad \text{(similarly for } b \text{ and } c, c \text{ and } a; \text{ tangent law)}$$

When using the tangent law, it is convenient to adjust the notation so that $A - B > 0$. N.B. $\tan \frac{1}{2}(A + B) = \cot \frac{1}{2}C$.

1. Given A, B, (hence C) and a: sine law.
2. Given a, b, C: tangent law with logs, or cosine law.
3. Given a, b, c: (*a*) If few significant figures are present, or a table of squares is available, use the cosine law without logs. (*b*) $\cos A = -1 + (b + c + a)(b + c - a)/(2bc)$. The second term on the right may be computed with logs; then subtract the 1 and find A in a table of *natural* functions. (*c*) For a method based on the "half-angle" formulas, see any text.
4. Given a, A, b (ambiguous case): sine law. If $b \sin A/a = \sin B > 1$, there is no solution. If $\sin B < 1$, then $B = B_1 < 90°$ or $B = B_2 > 90°$; there is a solution with B_1 if $A + B_1 < 180°$, and a solution with B_2 if $A + B_2 < 180°$.

5.7. Complex Numbers. There is a number, denoted by i (or $\sqrt{-1}$), with the property that $i^2 = -1$. On account of such unreal behavior, this number is said to

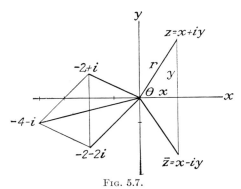

Fig. 5.7.

be *imaginary*. Long and regrettably sanctioned by usage, the adjective "imaginary" does not imply any intangible or illusory quality. If x and y are real numbers, then $z = x + iy$ is a *complex* number (in standard form) with *real part* $R(z) = x$ and *imaginary part* $I(z) = y$. [Some authors put $I(z) = iy$.] If $I(z) = 0$, the behavior of z is indistinguishable from that of the real number x, and z is said to be *purely real;* if $R(z) = 0$, z is *purely imaginary.*

The complex number $z = x + iy$ can be represented in a plane (*Argand diagram*) by the point whose rectangular cartesian coordinates (see 5.1) are (x,y). The *conjugate* of $z = x + iy$ is $\bar{z} = x - iy$; thus $\bar{z}$ and z have $(x,0)$ as their midpoint. The real number $|z| = r = \sqrt{x^2 + y^2}$ is called the *modulus* (or absolute value) of z. The angle $Ph\ z = \theta$ such that $\cos \theta = x/r$ and $\sin \theta = y/r$ is called the *phase* (or argument, or amplitude) of z. Thus $Ph\ z$ is multiple-valued; its *principal* value $ph\ z$ is so chosen that $-\pi < ph\ z \leqq \pi$. Hence $z = x + iy = r \cos \theta + ir \sin \theta$ (trigonometric form for z), and $\bar{z} = r \cos (-\theta) + ir \sin (-\theta) = r \cos \theta - ir \sin \theta$.

Equality. $z_1 = z_2$ if and only if $x_1 = x_2$ *and* $y_1 = y_2$.

Addition. $z_1 + z_2 = x_1 + x_2 + i(y_1 + y_2)$; add real and imaginary parts separately. Geometrically, $z_1 + z_2$ is joined to the origin by a line that is the diagonal of a parallelogram whose sides are z_1 and z_2. N.B. The segment from the origin to the point representing $z_2 - z_1$ is equal and parallel to the segment from z_1 to z_2.

Multiplication. $z_1 z_2 = (x_1 + iy_1)(x_2 + iy_2) = x_1 x_2 - y_1 y_2 + i(x_1 y_2 + y_1 x_2) = r_1 r_2 [\cos (\theta_1 + \theta_2) + i \sin (\theta_1 + \theta_2)]$; multiply moduli and add phases (sometimes getting nonprincipal phases for the product). Thus iz is z rotated positively through a right angle. To divide: $z_1/z_2 = (r_1/r_2)[\cos (\theta_1 - \theta_2) + i \sin (\theta_1 - \theta_2)]$.

Exponentiation. If m is an integer, $z^m = r^m(\cos m\theta + i \sin m\theta)$. If n is an integer, $z^{\frac{1}{n}} = r^{\frac{1}{n}}(\cos \phi_k + i \sin \phi_k)$, where $\phi_k = (\theta + 2k\pi)/n$ for $k = 0, 1, \cdots, n - 1$; the *principal* value of this multiple-valued function arises from the choice $k = 0$. Finally, $z^{\frac{m}{n}} = (z^{\frac{1}{n}})^m$, where $z^{\frac{m}{n}}$ is principal if and only if $z^{\frac{1}{n}}$ is. (If the exponent is irrational or complex, the situation is more elaborate.[5])

With these definitions and restrictions, the rules of operation stated in 1.3 hold for complex numbers. We note that complex numbers are not ordered as the reals were (see 1.21). However, $|z|$ has the properties enumerated in 1.22 for $|x|$; points z for which $|z - a| < k$ lie inside a circle with center a and radius k.

Rule for Computation: Express each complex number in the form $z = x + iy$ or $z = r(\cos \theta + i \sin \theta)$. Proceed as with real numbers, replacing i^2 by -1 whenever it occurs.

Caution: Preliminary expression in standard form is important. Thus $\sqrt{-3} \cdot \sqrt{-5} = i \sqrt{3} \cdot i \sqrt{5} = - \sqrt{15}$, not $\sqrt{(-3)(-5)} = \sqrt{15}$.

Special Rules: z is real (imaginary) if and only if $z = \bar{z}$ $\qquad (z = - \bar{z})$.

$$\overline{z_1 + z_2} = \bar{z}_1 + \bar{z}_2 \qquad \overline{z_1 z_2} = \bar{z}_1 \cdot \bar{z}_2 \qquad \overline{\dfrac{p}{zq}} = \dfrac{\bar{p}}{\bar{z}\bar{q}} \qquad |\bar{z}| = |z| \qquad |z|^2 = z\bar{z}$$

Hence, to find $1/z$, multiply above and below by $\bar{z}$ to get $1/z = \bar{z}/|z|^2$.

$$|z_1 z_2| = |z_1||z_2| \qquad \left|\dfrac{1}{z}\right| = \dfrac{1}{|z|} \qquad \left|\dfrac{p}{zq}\right| = |z|\dfrac{p}{q} \qquad |z_1 + z_2| \leqq |z_1| + |z_2|$$

$$\dfrac{1}{i} = -i = \dfrac{1}{i} \qquad i^3 = i^2 \cdot i = -i \qquad i^4 = i^2 \cdot i^2 = 1 \qquad i^5 = i^4 \cdot i = i, \cdots$$

$$i^{4k+1} = i \qquad i^{4k+2} = -1 \qquad i^{4k+3} = -i \qquad i^{4k} = 1, \text{ etc.}$$

Examples. *Algebraic to Trigonometric Form.* $5 - 8i = \sqrt{89} (5/\sqrt{89} - 8/\sqrt{89}\ i) = \sqrt{89} (\cos \theta + i \sin \theta)$, where $\sqrt{89} \cos \theta = 5$ and $\sqrt{89} \sin \theta = -8$, $-\pi < \theta \leqq \pi$ for principal value, so that $\theta = -58° = -1.0123$ rad. Thus $5 - 8i = \sqrt{89} [\cos (-58°) + i \sin (-58°)] = \sqrt{89} [\cos (-1.0123) + i \sin (-1.0123)]$.

Addition. $(2 + 3i) - (5 - 4i) = (2 - 5) + i(3 + 4) = -3 + 7i.$

Multiplication.   Algebraic form: $(2 + 3i)(-5 + i) = -10 + 3i^2 + 2i - 15i = -10 - 3 - 13i = -13 - 13i.$

Trigonometric form: $3(\cos 122° + i \sin 122°) \cdot \sqrt{2} \ (\cos 75° + i \sin 75°) = 3\sqrt{2} \ (\cos 197° + i \sin 197°) = 3\sqrt{2} \ [\cos (-163°) + i \sin (-163°)]$ (if principal phase of product is desired) $= 3\sqrt{2} \ (\cos 163° - i \sin 163°).$

Division.   Algebraic form: $\dfrac{2 + i}{3 - i} = \dfrac{2 + i}{3 - i} \cdot \dfrac{3 + i}{3 + i} = \dfrac{5 + 5i}{10} = \dfrac{1 + i}{2}.$

Trigonometric form: $\sqrt{5} \ [\cos (-18°) + i \sin (-18°)] \div \frac{1}{3}[\cos 37° + i \sin 37°] = 3\sqrt{5} \ [\cos (-55°) + i \sin (-55°)] = 3\sqrt{5} \ (\cos 55° - i \sin 55°).$

Exponentiation.   Algebraic form: Using binomial theorem (4.6), $(\sqrt{2} - 3i)^5 = (\sqrt{2})^5 + 5(\sqrt{2})^4(-3i) + 10(\sqrt{2})^3(-3i)^2 + 10(\sqrt{2})^2(-3i)^3 + 5(\sqrt{2})(-3i)^4 + (-3i)^5 = 4\sqrt{2} - 60i - 180\sqrt{2} i^2 - 540i^3 + 405\sqrt{2} i^4 - 243i^5 = 4\sqrt{2} - 60i - 180\sqrt{2} + 540i + 405\sqrt{2} - 243i = 229\sqrt{2} + 273i.$

Trigonometric form: $[3(\cos 112° + i \sin 112°)]^{5/2} = 3^{5/2}\left(\cos \dfrac{5 \cdot 112°}{2} + i \sin \dfrac{5 \cdot 112°}{2}\right) = 3^{5/2} (\cos 280° + i \sin 280°) = 3^{5/2} \ [\cos (-80°) + i \sin (-80°)]$ (if principal phase is desired).

It results from definitions adopted in the theory of functions of a complex variable, if t is a real number and e is the natural base of logarithms (see 2.4), that

$$e^{it} = \cos t + i \sin t \qquad \text{(Euler)}.$$

Hence $z = r(\cos \theta + i \sin \theta) = re^{i\theta}$ (exponential form for z), $z^{-1} = e^{-i\theta}/r$, $z_1 z_2 = r_1 r_2 e^{i(\theta_1 + \theta_2)}$. In particular, $e^{\frac{\pi i}{2}} = i$, $e^{\pi i} = -1$, $e^{\frac{3\pi i}{2}} = -i$, $e^{2\pi i} = 1$. Note that $|e^{i\theta}|^2 = \cos^2 \theta + \sin^2 \theta = 1$, so that $|e^{i\theta}| = 1$. Thus $e^{i\theta}$ always lies on the *unit circle*, which has its center at the origin and radius 1. Also, $e^{i\phi}z$ is z rotated counterclockwise about the origin through an angle ϕ; for this and other reasons, the exponential form of a complex number is often the most elegant means of dealing with rotational and periodic phenomena, *e.g.*, the problems arising in connection with alternating electric currents.

5.8. Hyperbolic Functions.   These functions (see the graphs in 8.9) are related to the hyperbola much as the circular (trigonometric) functions are related to the circle.

$$\sinh x = \frac{e^x - e^{-x}}{2} \qquad \cosh x = \frac{e^x + e^{-x}}{2}$$

$$\tanh x = \frac{\sinh x}{\cosh x} \qquad \coth x = \frac{\cosh x}{\sinh x}$$

$$\operatorname{sech} x = \frac{1}{\cosh x} \qquad \operatorname{cosech} x = \frac{1}{\sinh x}$$

They are connected with each other by the further relations

$$\sinh (-x) = -\sinh x \qquad \cosh (-x) = \cosh x$$
$$\cosh^2 x - \sinh^2 x = 1 \qquad \tanh^2 x + \operatorname{sech}^2 x = 1 \qquad \coth^2 x - \operatorname{cosech}^2 x = 1$$
$$\sinh (x \pm y) = \sinh x \cosh y \pm \cosh x \sinh y$$
$$\cosh (x \pm y) = \cosh x \cosh y \pm \sinh x \sinh y$$
$$\cosh 2x = \cosh^2 x + \sinh^2 x \qquad \sinh 2x = 2 \sinh x \cosh x$$

With properly generalized definitions

$$\cos ix = \cosh x \qquad \sin ix = i \sinh x$$
$$\cosh ix = \cos x \qquad \sinh ix = i \sin x$$

Since these functions are exponential, their inverse functions are logarithmic.

$$\text{arc sinh } x = \ln (x + \sqrt{x^2 + 1}) \qquad \text{arc cosh } x = \ln (x \pm \sqrt{x^2 - 1}) \qquad (|x| \geqslant 1)$$

$$\text{arc tanh } x = \frac{1}{2} \ln \left(\frac{1 + x}{1 - x} \right) \qquad (|x| < 1)$$

$$\text{arc coth } x = \frac{1}{2} \ln \left(\frac{x + 1}{x - 1} \right) \qquad (|x| > 1)$$

6. SOLUTION OF EQUATIONS

6.1. Identities and Equations. $f(x_1, \cdots, x_n) \equiv g(x_1, \cdots x_n)$ (*i.e., f identical with g*) means that $f(a_1, \cdots, a_n) = g(a_1, \cdots, a_n)$ for *every* set of numbers $a_1, \cdots, a_n$. Polynomials f and g are identical if and only if, whenever a term of f and a term of g are of equal degree in each of the variables, their coefficients are equal. If functions f and g are not identical, they may be equal for some values of the variables and unequal for others, or unequal for all; then $f = g$ is a (conditional) *equation*. A *solution* of an equation is a set of values of the variables which makes the equation true; if $f(a) = 0$, the solution $x = a$ is also said to be a *root* of the equation $f(x) = 0$ [or of the function $f(x)$] or a *zero* of $f(x)$.

Identity. $x^2 - 1 = (x - 1)(x + 1)$, true for *all* x.

Equation. $x^2 - 1 = 3$, true for $x = \pm 2$, otherwise false.

The practical problem for an identity (frequent in trigonometry) is to prove it; for an equation, to solve it. Processes given in texts for solving equations usually show merely that, *if* there *is* a solution, it *must* be one of the numbers yielded by the process; occasionally some or all of these numbers fail actually to satisfy the equation. When in doubt, check by substituting back into the equation.

On "solving" the equation $\dfrac{13}{5} - \dfrac{12}{x + 5} = \dfrac{18}{5x} - \dfrac{18}{x(x + 5)}$, we find that either $x = 1$ or $x = 0$. $x = 1$ satisfies the equation and is a solution, while $x = 0$ makes the equation meaningless. When writing down an equation, it is wise to note values of x that give rise to a vanishing denominator. In this example, merely writing the equation down requires (implicitly) that $x \neq 0$ and $x \neq -5$.

On "solving" the equation $x = \sqrt{x + 6}$, we find $x = 3$ or $x = -2$. The former checks, whereas the latter gives the absurdity $-2 = 2$ and so is not a solution. A glance at the equation shows that $x \geqslant -6$ for the right member to be real; since the right member is then $\geqslant 0$, the left member x must be $\geqslant 0$ for the equation to be satisfied. The "solution" -2, introduced by squaring the equation, is said to be *extraneous.*

Before attempting to solve an equation that is at all complicated, one should attempt to write it in several different forms, choosing then the one that seems most manageable. Trigonometric identities, logarithms, algebraic substitutions, etc., may lead to substantial simplification (see the trigonometric example in 5.5).

6.2. Linear Equations. If b and the a's are constants, the equation $a_1 x_1 + \cdots + a_n x_n = b$ is a *linear* equation. If the x's are cartesian coordinates, the equation represents a "hyperplane," which, if $n = 3$, is an ordinary plane in x_1-x_2-x_3-space and, if $n = 2$, is a straight line in the x_1-x_2-plane; hence the name *linear* (for geometry, see 3.4, 7.62). The equation can be solved to express any one of the variables that has a nonvanishing coefficient in terms of the others; if they are given, it may be computed.

We deal with a system S of e equations in v variables.

$$S \begin{cases} a_{11}x_1 + \cdots + a_{1v}x_v = b_1 \\ \cdots \cdots \cdots \cdots \cdots \cdots \cdots \\ a_{e1}x_1 + \cdots + a_{ev}x_v = b_e \end{cases}$$

S is *homogeneous* or *nonhomogeneous* according as $b_1 = \cdots = b_e = 0$ or not. A homogeneous system always has the trivial solution $x_1 = \cdots = x_v = 0$. Also, if $x_1, \cdots, x_v$ is a solution of a homogeneous system, then so is $tx_1, \cdots, tx_v$ for *any t*.

General Process. Write down first an equation in which x_1 has a nonvanishing coefficient. Divide this equation by that coefficient, getting an equation in which x_1 has coefficient 1. Use this equation to eliminate x_1 from the remaining $e - 1$ equations. Next, choose one of the latter equations in which x_2 or some other chosen variable has a nonvanishing coefficient, and, proceeding similarly, eliminate that variable from the remaining $e - 2$ equations. After eliminating x_r from the remaining $e - r$ equations, continue until either (1) the system T has been reached, or (2) all the equations have been used up ($r = e$). (The integer r is called the *rank* of S.)

$$T \begin{cases} x_1 + \alpha_{12}x_2 + \alpha_{13}x_3 + \cdots \qquad\qquad + \alpha_{1v}x_v = \beta_1 \\ x_2 + \alpha_{23}x_3 + \cdots \qquad\qquad + \alpha_{2v}x_v = \beta_2 \\ \cdots \cdots \cdots \cdots \cdots \cdots \cdots \cdots \cdots \cdots \cdots \\ x_r + \alpha_{rr+1}x_{r+1} + \cdots \qquad + \alpha_{rv}x_v = \beta_r \\ 0 \cdot x_{r+1} + \cdots \qquad + 0 \cdot x_v = \beta_{r+1} \\ \cdots \cdots \cdots \cdots \cdots \cdots \cdots \cdots \cdots \\ 0 \cdot x_k + \cdots + 0 \cdot x_v = \beta_e \end{cases}$$

In case 1, if (*a*) some one of $\beta_{r+1}, \cdots, \beta_e$ fails to vanish, the inconsistency $0 \neq 0$ has been reached, and T (and S) is *inconsistent* (has no solutions). However, if (*b*) $\beta_{r+1} = \cdots = \beta_e = 0$, then any solution of the first r equations will satisfy the remaining $e - r$ equations; the latter are said to depend on the former, and the situation coincides with case 2 in that all independent equations have been used up. In case 2, the equations may be solved, in turn, starting with the rth, for x_r, x_{r-1}, $\cdots$, x_2, x_1 in terms of $x_{r+1}, \cdots, x_v$. Values arbitrarily assigned to the latter determine unique values for the former. Any set of values thus obtained from T satisfies S. In case 1*b* or 2, then, S is *consistent* (has solutions).

Examples for $e = v = 3$.

$$S \begin{cases} x - y + 2z = 3 \\ 2x \qquad + z = 1 \\ 3x + 2y + z = 4 \end{cases} \qquad T \begin{cases} x - y + 2z = 3 \\ y - \tfrac{3}{2}z = -\tfrac{5}{2} \\ z = 3 \end{cases}$$

Hence $y = 2$ and $x = -1$. $r = 3$; case 2. Unique solution.
The planes represented by S have just the point $(-1,2,3)$ in common.

$$S \begin{cases} x - y + 2z = 3 \\ 2x \qquad + z = 1 \\ 3x - y + 3z = 4 \end{cases} \qquad T \begin{cases} x - y + 2z = 3 \\ y - \tfrac{3}{2}z = -\tfrac{5}{2} \\ 0 = 0 \end{cases}$$

$r = 2$; Case 1*b*. Infinitely many solutions.
Hence $y = \tfrac{3}{2}z - \tfrac{5}{2}$, $x = -\tfrac{1}{2}z + \tfrac{1}{2}$, with z arbitrary. These two planes intersect in a line, which lies in each of the three planes represented by S.

$$S \begin{cases} x - y + 2z = 3 \\ 2x \qquad + z = 1 \\ 3x - y + 3z = 5 \end{cases} \qquad T \begin{cases} x - y + 2z = 3 \\ y - \tfrac{3}{2}z = -\tfrac{5}{2} \\ 0 \neq \tfrac{1}{2} \end{cases}$$

$r = 2$; case 1*a*. No solutions.
The three planes represented by S intersect in pairs in three distinct parallel lines.

General Results. N.B. $r \leqslant e$ and $r \leqslant v$ naturally.

Homogeneous System. Since $\beta_1 = \cdots = \beta_e = 0$, case 1$a$ is impossible. If $r < v$, as happens naturally if $e < v$, $x_1, \cdots, x_r$ may be found from T in terms of $x_{r+1}, \cdots, x_v$. All choices of $x_{r+1}, \cdots, x_v$ *except* 0, $\cdots$, 0 give nontrivial solutions $x_1, \cdots, x_v$ of S. If $r = v$, there are no variables $x_{r+1}, \cdots, x_v$, and S has only the trivial solution.

Nonhomogeneous System. Case 1a, which requires $r < e$, means S inconsistent. If S is consistent, all solutions are found as in the homogeneous case; *every* choice of $x_{r+1}, \cdots, x_v$ gives a solution. If $r = v$, there is no choice and the solution is unique.

N.B. Before commencing the solution, the variables may be permuted so as to make a given one fall among $x_1, \cdots, x_r$ or among $x_{r+1}, \cdots, x_v$, as desired. The integer r is always the same, no matter how the computation is performed.

Special Case: $e = v$; Determinants. Let A denote the (vth order) determinant of the a's in S. For $i = 1, \cdots, v$, let B_i denote the determinant obtained by replacing the ith column of A by the column of the b's. *If $A \neq 0$, then ($r = v$ and) $x_i = B_i/A$ are the unique solutions of S (Cramer's rule).*

For the first example

$$A = \begin{vmatrix} 1 & -1 & 2 \\ 2 & 0 & 1 \\ 3 & 2 & 1 \end{vmatrix} = 5 \qquad B_x = \begin{vmatrix} 3 & -1 & 2 \\ 1 & 0 & 1 \\ 4 & 2 & 1 \end{vmatrix} = -5$$

$$B_y = \begin{vmatrix} 1 & 3 & 2 \\ 2 & 1 & 1 \\ 3 & 4 & 1 \end{vmatrix} = 10 \qquad B_z = \begin{vmatrix} 1 & -1 & 3 \\ 2 & 0 & 1 \\ 3 & 2 & 4 \end{vmatrix} = 15$$

so that $x = -5/+5 = -1$, $y = +10/+5 = 2$, $z = +15/+5 = 3$, as before. In the other two examples, $A = 0$.

If $A = 0$ and some $B_i \neq 0$, the equations are inconsistent. If $A = 0$ and each $B_i = 0$, use the general procedure.

In the important *homogeneous* case, where each $B_i = 0$ naturally, S will have a nontrivial solution if and only if $A = 0$. Using the A_{ji} defined in 4.7, $x_i = (-1)^{i+j}A_{ji}$ ($i = 1, \cdots, n$; any *fixed* j) will satisfy S. If these all vanish for any *one* j, they will vanish for every j, thus giving only the trivial solution. Then use the general process.

In simple cases, especially if there is some reason to conjecture consistency, the routine of the general process can be relaxed. In the first example, eliminate y at once by adding twice the first equation to the last (and divide out a 5), getting $x + z = 2$. Subtract this fourth equation from the second: $x = -1$. Substitute in the fourth to get $z = 3$, and then in the third to get $y = 2$. Alternative: Find $z = 1 - 2x$ from the second equation, substitute in the first and third, and eliminate either x or y from the resulting two equations in x and y.

Solutions should be checked; substitution of $x_1, \cdots, x_r$ expressed in terms of $x_{r+1}, \cdots, x_v$ should give identities. The work of solution can be arranged to provide a numerical check at each step (Doolittle method). Elaborate machines are available for really complicated cases (Mallock's machine). Graphical methods are useless beyond $v = 2$, and superfluous then, but geometrical ideas and terminology clarify the situation.

6.3. Algebraic Equations. If $f(x_1, \cdots, x_k)$ is a polynomial of degree n, $f(x_1, \cdots, x_k) = 0$ is an *algebraic* equation of degree n. Of main practical interest are the cases $k = 1$ and $k = 2$, to which the general case reduces when values are assigned to $k - 1$ or $k - 2$ of the variables.

6.31. $n = 2$; *quadratic equations.* N.B. $\sqrt{p}$, $p > 0$, *always* means the *positive* number whose square is p.

1. $k = 1$. The equation may be written $f(x) = ax^2 + bx + c = 0$ with $a \neq 0$. If $f(x)$ can be factored, set each factor $= 0$. Otherwise, write $f(x)$ in the form (*completing* the *square*)

$$a \left(x + \frac{b}{2a} \right)^2 - \frac{b^2 - 4ac}{4a}$$

whence
$$x = \frac{-b \pm \sqrt{b^2 - 4ac}}{2a}$$

These two solutions are real and distinct, real and equal, or conjugate complex, according as the *discriminant* $b^2 - 4ac$ is > 0, $= 0$, or < 0. The sum of the roots is $-b/a$; their product is c/a. The graph of $y = f(x)$ is a parabola with axis parallel to the y-axis. The parabola crosses the x-axis when $f(x) = 0$.

2. $k = 2$. The equation represents a *conic* (see 7.31). Treating the equation as a quadratic in one variable, with coefficients containing the other, it may be solved as in (1) to give one variable in terms of the other. Thus $2x^2 + 3xy - 5y^2 - x + 2y + 3 = 0$ may be written $-5y^2 + (3x + 2)y + 2x^2 - x + 3 = 0$, which is a quadratic in y with $a = -5$, $b = 3x + 2$, $c = 2x^2 - x + 3$. Applying the formula in (1), we find

$$y = \frac{-(3x + 2) \pm \sqrt{(3x + 2)^2 - 4(-5)(2x^2 - x + 3)}}{2(-5)}$$

or
$$y = \frac{3x + 2 \pm \sqrt{49x^2 - 8x + 64}}{10}$$

3. Simultaneous systems ($k = 2$). To solve a linear and a quadratic, solve the linear for one variable and substitute in the quadratic; solve this and substitute in the linear to find the other variable. (There are generally two solutions, which may coincide, or be complex.) To solve two quadratics, proceed similarly. (There are generally four solutions, some or all of which may coincide or be complex.) Occasionally the form of the equations permits one variable to be eliminated easily. If one of the equations is linear, it may help to square it; check to avoid extraneous solutions (see 6.1). For specific cases and examples, see an algebra text.

To solve graphically, plot the two curves and measure the abscissae and ordinates of their points of intersection.

6.32. $n > 2$, $k = 1$. Moderately complicated algebraic solutions are known for cubics and quartics; see a text on algebra or theory of equations. If $n \geqslant 5$, no algebraic solution exists. The content of 4.8 is useful here.

In the following, it is assumed, to simplify some of the statements, that $a_n > 0$; this may always be secured without changing the roots by multiplying the equation by -1, if necessary. Roots are counted with their multiplicities (see 4.8). As soon as one real root c has been found, one may divide out $x - c$ and proceed with the reduced equation $f(x)/(x - c) = q(x) = 0$; if a complex root $a + bi$ of an equation with *real* coefficients is found, divide out $x^2 - 2ax + a^2 + b^2$.

1. $f(x) = 0$ has between $x = a$ and $x = b$ an even (perhaps 0) or an odd number of roots according as $f(a)$ and $f(b)$ have the same or opposite signs. Thus $x^3 - 2 = 0$ has either 1 or 3 positive real roots, for $f(0) = -2 < 0$, while $f(x) > 0$ assuredly for large $x > 0$.

2. *Descartes's Rule.* Examine the sequence of coefficients a_n, a_{n-1}, $\cdots$, a_0. If two successive coefficients (ignoring missing ones) have different signs, a change or *variation* in sign is said to occur. *Rule:* If $f(x)$ has real coefficients, the number of *positive* real roots is equal to the number of variations in sign or less by a positive even integer; the number of *negative* roots of $f(x)$ is equal to the number of variations of

sign of $f(-x)$ or less by a positive even integer. Precisely one variation indicates the presence of exactly one root of the appropriate kind. In $f(x) = x^5 - 3x^2 + 1 = 0$, there are two variations of sign, hence two or zero positive roots; since $f(0) = 1$, $f(1) = -1$, and $f(x) > 0$ for large $x > 0$, there are two roots > 0. In $f(-x) = -x^5 - 3x^2 + 1$ there is one variation of sign, and hence $f(x) = 0$ has one negative root. Thus $f(x) = 0$ has one negative, two positive, and a pair of conjugate complex roots.

3. A *table of values* may be constructed by synthetic substitution (see 4.8). By using (1), we can locate roots at or between values of x used in constructing the table.

4. *Limits for the Roots.* If, when $p > 0$ is substituted synthetically in $f(x)$, each number on the third line of the synthetic computation is $\geqq 0$, then $f(x)$ has no positive roots $> p$, and p is an *upper limit* for the roots of $f(x)$. *Alternative Rule:* If the term of highest degree with negative coefficient is of degree r and the numerically greatest negative coefficient is a_q, then $1 + \sqrt[r]{-a_q}$ is an upper limit for the roots of $f(x)$. If p is an upper limit for roots of $f(-x)$, $-p$ is a lower limit for the negative roots of $f(x)$. [Note that it is necessary, if $f(x)$ is of odd degree, to multiply $f(-x)$ by -1 in order to make $a_n > 0$.]

5. *Rational Roots.* If the coefficients of $f(x)$ are *integers*, $x = p/q$, p and q *integers*, cannot satisfy $f(x) = 0$ unless p divides a_0 and q divides a_n. Hence the *possible* rational roots are those fractions whose numerators divide a_0 and denominators divide a_n; these may be listed and tested by synthetic division. Candidates can be eliminated by Descartes's rule, a knowledge of the limits, etc. If $f(x) = 6x^4 - 5x^3 + 27x^2 + 7x - 10$, the only possible rational roots are ± 1, ± 2, ± 5, ± 10, $\pm \frac{1}{2}$, $\pm \frac{5}{2}$, $\pm \frac{1}{3}$, $\pm \frac{2}{3}$, $\pm \frac{5}{3}$, $\pm \frac{10}{3}$, $\pm \frac{1}{6}$, $\pm \frac{5}{6}$. However, 1 is an upper and -1 a lower limit for the roots, and all candidates can therefore be discarded except those lying between -1 and $+1$, *i.e.*, $\pm \frac{1}{2}$, $\pm \frac{1}{3}$, $\pm \frac{2}{3}$, $\pm \frac{1}{6}$, $\pm \frac{5}{6}$. Descartes's rule reveals one negative root. Testing, we find $-2/3$ to be a root, and $f(x)/(3x + 2) = 2x^3 - 3x^2 + 11x - 5 = g(x)$. The only possible rational root of $g(x)$ is now $\frac{1}{2}$. This does turn out to be a root, and $f(x) = (3x + 2)(2x - 1)(x^2 - x + 5)$. Solving $x^2 - x + 5 = 0$, we find the other roots of $f(x)$ to be $(1 \pm i \sqrt{19})/2$.

6. $x^n = a$. If $a = Ae^{i\alpha}$, $A > 0$, the n solutions of this equation are $\sqrt[n]{A}\, e^{i\theta_p}$, where, for $p = 0, 1, 2, \cdots, n - 1$, $\theta_p = (\alpha + 2p\pi)/n$. See 5.7.

6.4. General Methods (for remarks on polar coordinates, see 7.5). **6.41. Graphical.** Compute or observe a table of values and draw a graph, using available information from calculus. Values of the abscissa for which the curve crosses the axis are roots of $f(x) = 0$. Often $f(x)$ can be expressed as $g(x) - h(x)$, g and h plotted on the same axes, and the abscissae of their points of intersection are then roots of $f(x) = 0$. Thus, to solve $x^5 - 5x^2 + 3x - 1 = 0$, plot $g(x) = x^5$ and $h(x) = 5x^2 - 3x + 1$. To solve $\sin 5x = \tan 2x$, plot $g(x) = \sin 5x$ and $h(x) = \tan 2x$. To solve simultaneously two equations $f(x,y) = 0$ and $g(x,y) = 0$, plot the two curves on the same axes; the abscissae and ordinates (x,y) of their points of intersection are the required solutions. The accuracy of a graph can be improved by enlarging the table of values and drawing a large-scale graph of a small portion of the curve, but this generally requires enough numerical information to make the graph superfluous. If pictorial display or a good *estimate* of the roots is desired, a graph is very useful.

6.42. Numerical Approximation. The process of trial and error may often be systematized, depending, of course, on the equation. For *Horner's method* of approximating *irrational* real roots of an algebraic equation, see an algebra text. Generally speaking, any aid in computation may be adopted to facilitate the solution of equations. There is a nomogram for solving quadratic equations; they can also be solved with a slide rule.

Tabular Interpolation. If the functions involved have been tabulated, an estimated root can be refined to the accuracy permitted by the table, *e.g.*, $10^x = 7 + x$, or $x = \log (x + 7)$. In a log table, taking $x = .8960$, we see that $\log (x + 7) = \log 7.896 = .8974 > x$, while, when $x = .8980$, $\log (x + 7) = \log 7.898 = .8975 < x$. Thus the root of the equation lies between .8960 and .8980, apparently closer to the latter. Trying $x = .8974$, we find $\log (x + 7) = \log 7.897 = .8974$ within the limits of tabular error, completing the solution.

6.43. Newton's Method (using derivatives; see 8.2). Let $y = f(x)$ have only one root between a and b, and let $f(a)$ and $f(b)$ have opposite signs. Let $f'(x)$ and $f''(x)$ not vanish for $a \leqslant x \leqslant b$. Choose that one of a, b whose sign is the same as that of $f''(x)$; for the upper (lower) curve, in the figure, choose a (b). (Success of the method is *assured* with this choice but conditional with the other.) Using a (similar process for b), draw the tangent t_1 to the curve at $[a, f(a)]$ and let a_1 be the point where this tangent meets the axis. Draw the tangent t_2 to the curve at the point $[a_1, f(a_1)]$, and let a_2 be the point where t_2 cuts the axis. Proceed until $f(a_n) = 0$ to the desired accuracy. In formulas: $a_1 = a - f(a)/f'(a)$, $a_2 = a_1 - f(a_1)/f'(a_1)$, $\cdots$, $a_{n+1} = a_n - f(a_n)/f'(a_n)$, $\cdots$. In solving $f(x) = x^4 + 0.053x^3 - 2x^2 - 3.12 = 0$, we find that $f(1) = -4.067$, $f(2) = 5.304$, while $f'(x) = 4x^3 + 0.159x^2 - 4x$ and $f''(x) = 12x^2 + 0.318x - 4$ are both > 0 for $1 \leqslant x \leqslant 2$. Take $x = 2$ to start. The first

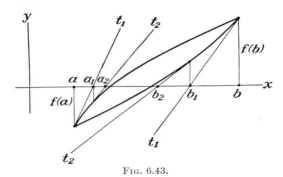

Fig. 6.43.

improvement gives $x = 2 - f(2)/f'(2) = 2 - 5.304/24.64 = 1.78$. The second improvement gives $x = 1.78 - f(1.78)/f'(1.78) = 1.78 - 0.93/15.94 = 1.723$. The third improvement gives $x = 1.723 - f(1.723)/f'(1.723) = 1.723 - 0.0269/12.32 = 1.7208$. Now $f(1.7208) = -0.0036$, so that, owing perhaps to rounding off in the computation, we have overshot the mark. However, $f(1.7209) = 0.0122$, so that, to four places, $x = 1.7208$ is the desired root. The comparatively slow convergence in this example is due to the large value of $f''(x)$ near the root. The method applies to any $f(x)$ that satisfies the requirements stated at the outset. For a simplification in the case of *polynomials*, based on the principle of Horner's method, see ref. 6, where other applications also are given.

A useful method for finding the least root of a power series has recently been improved (see ref. 7).

For solution by inverse interpolation, false position, a combination of Newton's method and false position which gives control of the error at each stage, approximation to complex roots, etc., see ref. 1, especially Chap. VI.

7. ANALYTIC GEOMETRY

When a coordinate system has been set up in the plane, or in space, it becomes possible to associate with each curve C an equation $y = f(x)$ [or $g(x,y) = 0$], and with each surface S an equation $z = f(x,y)$ [or $g(x,y,z) = 0$], in such a way that those and only those points P lie on C or S which have coordinates satisfying the appropriate equation. It is then possible to use algebraic methods to settle geometric questions, and the intuitive evidence of geometric configurations to illuminate algebraic problems. Thus, to find the coordinates of points of intersection, solve equations simultaneously; to find simultaneous solutions, graph and measure coordinates of points of intersection (see 6.31). For elementary definitions, see 3.1, 3.4.

In the following, (x,y) denote rectangular cartesian coordinates (see 5.1). The graph of an equation in these coordinates will be symmetric (see 3.16): about the x-axis (y-axis) if replacing y by $-y$ (x by $-x$) leaves the equation unaltered; about the origin if replacing x by $-x$ and y by $-y$ leaves the equation unaltered. Thus $x^2 = 2y$ (or $y^2 = 2x$) is symmetric about the y- (or x-) axis, but not about the x- (or y-) axis; $y = \sin x$ is symmetric about the origin but not about either axis.

7.1. Points and Lines. The *displacement from* $P_1(x_1,y_1)$ *to* $P_2(x_2,y_2)$ results from a horizontal displacement $x_2 - x_1$ and a vertical displacement $y_2 - y_1$. Hence the distance P_1P_2 is $\sqrt{(x_2 - x_1)^2 + (y_2 - y_1)^2}$. The direction of the displacement is indicated by the *inclination*, which is the angle α, $-90° < \alpha \leqslant 90°$ between the segment P_1P_2 and the direction of positive x-axis. If $-90° < \alpha < 90°$, the *slope* of the segment is $m = \tan \alpha = (y_2 - y_1)/(x_2 - x_1)$; if $\alpha = 90°$, then $x_2 = x_1$, the segment is vertical, m is nonexistent strictly, $m = \infty$ carelessly. The slope m of a line l is the slope of any segment on it. $m = 0$ if l is horizontal, $m > 0$ or $m < 0$ according as l rises or falls as x *increases*. If θ is the angle from l_1 to l_2, positively counterclockwise, then $\tan \theta = (m_2 - m_1)/(1 + m_1m_2)$. Hence two nonvertical lines are parallel if and only if $m_1 = m_2$, perpendicular if and only if $m_1m_2 + 1 = 0$.

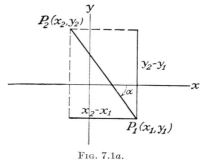

Fig. 7.1a.

The point P, which *divides* P_1P_2 *in the ratio* r_1/r_2, i.e., so that (directed distances) $P_1P/PP_2 = r_1/r_2$, has coordinates $x = (r_2x_1 + r_1x_2)/(r_1 + r_2)$ and $y = (r_2y_1 + r_1y_2)/(r_1 + r_2)$. For the *midpoint* $r_1 = r_2$ and $x = (x_1 + x_2)/2$, $y = (y_1 + y_2)/2$. P and $r_1/r_2 = \lambda$ vary together as follows (see 8.13): $\lambda = 0$ for $P = P_1$, $\lambda = 1$ for the midpoint, $\lambda \to +\infty$ as $P \to P_2$ inside the segment. $\lambda \to -1$ as P recedes indefinitely from P_1 and as P recedes indefinitely from P_2. $\lambda \to -\infty$ as $P \to P_2$ outside the segment.

The area of a triangle with vertices $P_i(x_i,y_i)\,(i = 1,2,3)$ is the numerical value of

$$\tfrac{1}{2}\begin{vmatrix} x_1 & y_1 & 1 \\ x_2 & y_2 & 1 \\ x_3 & y_3 & 1 \end{vmatrix} = \tfrac{1}{2}[x_1(y_2 - y_3) + x_2(y_3 - y_1) + x_3(y_1 - y_2)]$$

the sign of this expression being $+$ or $-$ according as P_1, P_2, P_3 occur in counterclockwise or clockwise order. The area K of a triangle with sides l_i: $A_ix + B_iy + C_i = 0$ $(i = 1,2,3)$ is the numerical value of

$$K = \frac{1}{2}\frac{\begin{vmatrix} A_1 & B_1 & C_1 \\ A_2 & B_2 & C_2 \\ A_3 & B_3 & C_3 \end{vmatrix}}{\begin{vmatrix} A_1 & B_1 \\ A_2 & B_2 \end{vmatrix} \cdot \begin{vmatrix} A_2 & B_2 \\ A_3 & B_3 \end{vmatrix} \cdot \begin{vmatrix} A_3 & B_3 \\ A_1 & B_1 \end{vmatrix}}$$

the sign of this expression being $+$ or $-$ according as l_1, l_2, l_3 occur in the triangle in counterclockwise or clockwise order (for determinants, see 4.7).

In the plane, with rectangular cartesian coordinates, straight lines and only straight lines have linear equations $Ax + By + C = 0$, not both $A = 0$ and $B = 0$. The *intercepts* are $a = -C/A$ on the x-axis and $b = -C/B$ on the y-axis. The slope is $m = -b/a = -A/B$. The *normal* (perpendicular) from the origin on l makes with the positive x-axis an angle θ such that $\cos \theta = A/\sqrt{A^2 + B^2}$, $\sin \theta = B/\sqrt{A^2 + B^2}$ and has length $p = -C/\sqrt{A^2 + B^2}$ measured positively *from* the *line* in the direction of the terminal ray of θ. Measured positively in that same direction, the perpendicu-

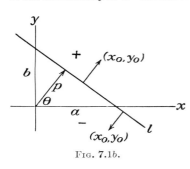

lar distance *from* l *to* the point (x_0,y_0) is $d = (Ax_0 + By_0 + C)/\sqrt{A^2 + B^2}$. The equations of the bisectors of the angle between l and l' are $\sqrt{A'^2 + B'^2} \ (Ax + By + C) = \pm \sqrt{A^2 + B^2} (A'x + B'y + C')$; one sign will give the interior bisector, the other the exterior, easily paired off with a rough diagram. The angle ϕ between the positive directions of the normals to l and l' is given by $\cos \phi = (AA' + BB')/ (\sqrt{A^2 + B^2} \sqrt{A'^2 + B'^2})$. The lines are perpendicular if and only if $AA' + BB' = 0$; parallel if and only if $A/A' = B/B'$, and coincident if and only if also $B/B' = C/C'$.

Fig. 7.1b.

Equations of Lines with Prescribed Properties.

Through (x_1,y_1) and (x_2,y_2)　　　$(y - y_1)(x_2 - x_1) = (y_2 - y_1)(x - x_1)$ or $\begin{vmatrix} x & y & 1 \\ x_1 & y_1 & 1 \\ x_2 & y_2 & 1 \end{vmatrix} = 0$

Through (x_1,y_1) with slope m　　　$y - y_1 = m(x - x_1)$

With slope m and y-intercept b　　　$y = mx + b$

With intercepts a and b　　　$\dfrac{x}{a} + \dfrac{y}{b} = 1$

Parallel to y-axis (no b)　　　$x = a$

Parallel to x-axis (no a)　　　$y = b$

With normal from origin in direction θ and of length p　　　$x \cos \theta + y \sin \theta = p$

Families of lines may be represented by allowing at least one of the constants in the equation to vary, so that each specific value it may have will give rise to a line. Thus $y = 3x + b$ represents the family of lines with slope 3, each parallel to the line $y = 3x$ and each with its y-intercept given by b.

7.2. Change of Rectangular Cartesian Coordinates. *Translation to Parallel Axes with New Origin.* The point P, whose coordinates are (x,y) when the origin is at O, has coordinates $(x',y') = (x - a, y - b)$ when the origin is shifted to a point O' whose (x,y) coordinates are (a,b). See the example in 7.31.

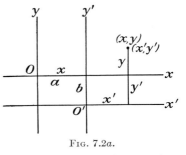

Fig. 7.2a.

Rotation of Axes with Fixed Origin. If the axes are rotated through an angle θ, new coordinates (x',y') of P are $x' = x \cos \theta + y \sin \theta$, $y' = -x \sin \theta + y \cos \theta$.

To get the equation in (x',y') corresponding to a given equation in (x,y), solve these equations to get $x = x' \cos \theta - y' \sin \theta$, $y = x' \sin \theta + y' \cos \theta$. Substitute for x and y in the given equation, and simplify. See the example in 7.31.

7.31. The Conic Sections. The second-degree equation $f(x,y) = Ax^2 + 2Bxy + Cy^2 + 2Dx + 2Ey + F = 0$, not all of A, B, C vanishing, represents a *conic section S* (curve in which a right circular cone is cut by the xy-plane). If $f(a,b) = 0$, the line *tangent* to S at (a,b) has the equation $Aax + B(ay + bx) + Cby + D(a + x) + E(b + y) + F = 0$.

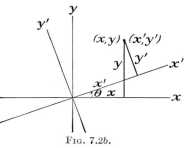

FIG. 7.2b.

Let
$$\Delta = \begin{vmatrix} A & B & D \\ B & C & E \\ D & E & F \end{vmatrix} \qquad \delta = \begin{vmatrix} A & B \\ B & C \end{vmatrix} \qquad \delta' = \begin{vmatrix} C & E \\ E & F \end{vmatrix}$$

If and only if the determinant (see 4.7) $\Delta = 0$, $f(x,y)$ can be written as the product of two linear functions (perhaps with complex coefficients), so that S "degenerates." If $\delta < 0$, S will consist of distinct intersecting straight lines. If $\delta > 0$, S will consist of a single point. If $\delta = 0$, S will consist of two lines that are parallel, coincident, or imaginary according as $\delta' < 0$, $\delta' = 0$, or $\delta' > 0$.

If $\Delta \neq 0$, the equation represents a *parabola* if $\delta = 0$, a *hyperbola* if $\delta < 0$, and an *ellipse* if $\delta > 0$; the ellipse is complex (no graph) or real according as the signs of A and Δ agree or disagree and, if real, is a *circle* if $A = C$ and $B = 0$. Thus, for the equation $32x^2 + 52xy - 7y^2 - 100x - 250y - 355 = 0$,

$$\Delta = \begin{vmatrix} 32 & 26 & -50 \\ 26 & -7 & -125 \\ -50 & -125 & -355 \end{vmatrix} = 162,000 > 0 \qquad \text{and} \qquad \delta = \begin{vmatrix} 32 & 26 \\ 26 & -7 \end{vmatrix} = -900 < 0,$$

so that the curve is a hyperbola.

The foregoing tests apply to *any* second-degree equation and may be applied before or after translation and/or rotation of axes. As in the example, the computation of Δ is often tedious. If this threatens to be the case, especially if the following "reduction" must be at least partly carried through, wait until later to test with Δ; the progress of the reduction may yield the necessary information without recourse to Δ. On the other hand, reference to Δ and the δ's may make an awkward reduction unnecessary.

If $B \neq 0$, the troublesome xy term may be removed by rotating the axes (see 7.2) through an angle θ such that $\tan 2\theta = 2B/(A - C)$, $\theta = 45°$ if $A = C$. Assuming this done, the equation is of the form $Ax^2 + Cy^2 + 2Dx + 2Ey + F = 0$, not both $A = 0$ and $C = 0$. If either A or $C = 0$, the equation represents a *parabola;* if A and C have opposite signs, it is a *hyperbola* that is rectangular if $A = -C$; if A and C have the same signs, it is an *ellipse* that is a *circle* if $A = C$.

In the example, we have $\tan 2\theta = 52/(32 + 7) = \frac{4}{3}$, whence $\cos 2\theta = \frac{3}{5}$, $\cos \theta = 2/\sqrt{5}$, $\sin \theta = 1/\sqrt{5}$, so that $x = (2x' - y')/\sqrt{5}$, $y = (x' + 2y')/\sqrt{5}$. Substituting and simplifying, we get $9x'^2 - 4y'^2 - 18\sqrt{5}\,x' - 16\sqrt{5}\,y' - 71 = 0$.

To simplify further, factor out the coefficients of the squared terms and complete the squares (see 6.31), getting equations as follows:

Parabola $(y - k)^2 = 4p(x - h)$ or $(x - h)^2 = 4p(y - k)$

Hyperbola $\dfrac{(x - h)^2}{a^2} - \dfrac{(y - k)^2}{b^2} = +1 \text{ or } -1$

Ellipse $\dfrac{(x - h)^2}{a^2} + \dfrac{(y - k)^2}{b^2} = +1 \text{ or } -1$

Thus, the example becomes $9(x'^2 - 2\sqrt{5}\,x' + 5) - 4(y'^2 + 4\sqrt{5}\,y' + 20) = 71 + 45 - 80 = 36$, or $(x' - \sqrt{5})^2/4 - (y' + 2\sqrt{5})^2/9 = 1$, with $h = \sqrt{5}$, $k = -2\sqrt{5}$. Translate the origin (see 7.2) to (h,k), finally getting the equations

Parabola $y^2 = 4px$ or $x^2 = 4py$

Hyperbola $\dfrac{x^2}{a^2} - \dfrac{y^2}{b^2} = +1 \text{ or } -1$

Ellipse $\dfrac{x^2}{a^2} + \dfrac{y^2}{b^2} = +1 \text{ or } -1$

Finally, putting $\bar{x} = x' - \sqrt{5}$, $\bar{y} = y' + 2\sqrt{5}$, the example becomes $\bar{x}^2/4 - \bar{y}^2/9 = 1$.

If the -1 occurs on the right in the case of the ellipse, the curve is complex and has no graph in the ordinary sense. If $a^2 = b^2$, the ellipse reduces to a *circle* of radius a; $x^2 + y^2 = a^2$ or $(x - h)^2 + (y - k)^2 = a^2$ according as the center is at the origin or at (h,k).

Theorem. If a point P moves in such a way that its distance to a fixed point F (*focus*) is in a constant ratio e (*eccentricity*) to its distance to a fixed line d (*directrix*), then P moves on a parabola if $e = 1$, an ellipse if $e < 1$, or a hyperbola if $e > 1$.

FIG. 7.31a.

(This e is entirely unrelated to the natural base for logarithms.)

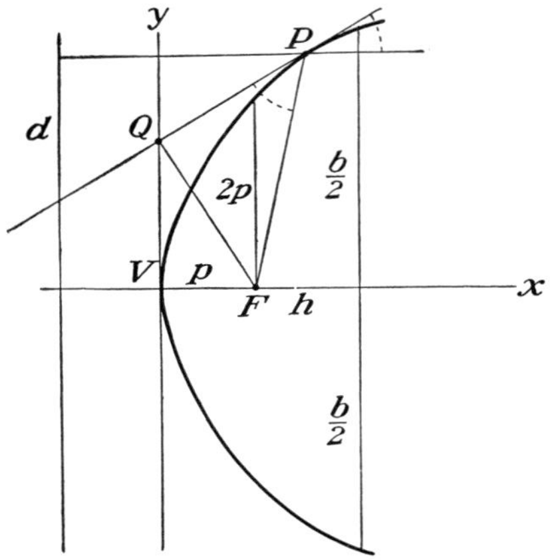

A line through the focus perpendicular to the directrix is evidently a line of symmetry (see 3.16) for the figure and is called an *axis* (geometric). A point V on the curve where the tangent to the curve is perpendicular to the *focal radius* FV is called a *vertex*. A point of symmetry is called a *center;* an ellipse has a center, and so does a hyperbola, but a parabola has none. The chord through a focus perpendicular to the axis is called a *latus rectum.*

Parabola. Place the axis of the parabola along the x-axis, with the origin midway between the focus and directrix, and let the focus be $(p,0)$, so that the equation of the directrix is $x = -p$. ($p > 0$ in Fig. 7.31a; if $p < 0$, the curve opens to the left.) Then the equation of the parabola is $y^2 = 4px$, and its vertex is at the origin. The latus rectum has length $4p$. The fact that $(p,2p)$ is on the parabola is useful in drawing a rough sketch quickly.

For the equation $x^2 = 4py$, the directrix is horizontal, the axis is along the y-axis, and the curve opens up or down according as $p > 0$ or $p < 0$. The equation $y = ax^2 + bx + c$ represents a parabola. To get geometric information, write $\left(x + \dfrac{b}{2a}\right)^2$

$= \dfrac{1}{a}\left(y + \dfrac{b^2 - 4ac}{4a}\right)$ and apply the theory: vertex at $(-b/2a,\ -(b^2 - 4ac)/4a)$, $p = 1/4a$, etc. The tangent at P makes equal angles with the focal radius PF and the ray through P parallel to the axis.   If the tangent at P cuts the tangent at the vertex in a point Q, then $PQF = 90°$.

A chord perpendicular to the axis cuts off a *segment* of base b equal to the length of the chord and *height* h equal to the distance from the chord to the vertex.   The area of a segment is $bh/3$.   For two segments of the same parabola, $b_1^2/b_2^2 = h_1/h_2$.

Ellipse and Hyperbola.   Each of these curves has two vertices V, V' on the axis, and their midpoint is the center.   Placing the origin there and the axis along the x-axis, the equations assume their standard forms.

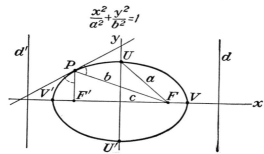

$$\frac{x^2}{a^2}+\frac{y^2}{b^2}=1$$

Fig. 7.31b.—Here $a > b$.   The foci and the major axis lie on the y-axis if and only if $b > a$. $b = a$ gives a circle.

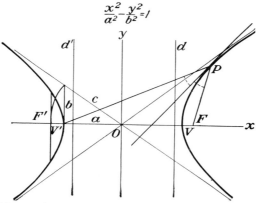

$$\frac{x^2}{a^2}-\frac{y^2}{b^2}=1$$

Fig. 7.31c.—The right member of the equation is -1 if and only if the foci and the transverse axis lie on the y-axis.   (See "conjugate hyperbolas" below.)

Let $c = \sqrt{a^2 - b^2}$ for the ellipse, $c = \sqrt{a^2 + b^2}$ for the hyperbola.   Each curve has the following: eccentricity $e = c/a$; two foci, at $(\pm c,0)$; two directrices, equations $x = \pm a^2/c = \pm a/e$; and two latera recta of length $2b^2/a$.   The tangent makes equal angles with the focal radii.

For the *ellipse*, $PF + PF' = 2a$; the *sum* of the focal radii is constant. VV', of length $2a$, is called the *major* axis; UU', of length $2b$, is the *minor* axis.   The area is πab.

For the *hyperbola*, $PF - PF' = 2a$ on one branch and $PF' - PF = 2a$ on the other; the *difference* of the focal radii is constant.   The segment VV', of length $2a$, is the *transverse* axis; a segment of length $2b$ on the y-axis, centered at O (and usually not drawn), is called the *conjugate* axis.   The asymptotes, to one of which P gets arbitrarily close as P recedes indefinitely on the hyperbola, have equations $x^2/a^2 - y^2/b^2 = 0$, *i.e.*, $y = \pm bx/a$. In our example, $a = 2$, so the transverse axis is 4; $b = 3$, so the conjugate axis is 6; the latera recta are $2 \cdot \frac{9}{2} = 9$; $c = \sqrt{4 + 9} = \sqrt{13}$; $e = \sqrt{13}/2$. In $(\bar{x},\bar{y})$ coordinates, the vertices are $(\pm 2,0)$, the foci are $(\pm\sqrt{13},0)$, the directrices are $\bar{x} = \pm 4/\sqrt{13}$, and the asymptotes are $\bar{y} = \pm 3\bar{x}/2$. To get this information back into (x',y') coordinates, put $\bar{x} = x' - \sqrt{5}$, $\bar{y} = y' + 2\sqrt{5}$; for (x,y), put $x' = (2x + y)/\sqrt{5}$, $y' = (-x + 2y)/\sqrt{5}$, *i.e.*, $\bar{x} = (2x + y - 5)/\sqrt{5}$, $\bar{y} = (-x + 2y + 10)/\sqrt{5}$. For example, the (x,y) equations of the asymptotes are thus $2(-x + 2y + 10) = \pm 3(2x + y - 5)$.

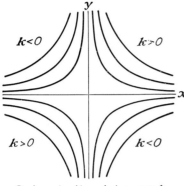

$k<0$ $k>0$

$k>0$ $k<0$

Rectangular Hyperbolas: $xy = k$

Fig. 7.31d.

If the transverse axis is vertical, the equation is $x^2/a^2 - y^2/b^2 = -1$, and the eccentricity is c/b. This particular hyperbola has the same asymptotes as $x^2/a^2 - y^2/b^2 = 1$, its foci are the same distance c from the center, and the transverse axis of either is the conjugate axis of the other. Two hyperbolas related in this way are *conjugate* hyperbolas.

In case $a = b$, the asymptotes are perpendicular, and the hyperbola is said to be *rectangular*. If the axes are rotated through 45 deg, so as to coincide with the asymptotes, the equation becomes $xy = k$.

Conversely, an equation of the form $xy - kx - hy - m = 0$ may be written in the form $(x - h)(y - k) = kh + m$. When the origin is translated to (h,k), the new equation is $xy = hk + m$, which represents a rectangular hyperbola.

7.32. Constructions Involving the Conics. 1. *Parabola*, given focus F, directrix d, and axis l (through F perpendicular to d) cutting d in D. The midpoint V of FD is the vertex of the parabola. If A is any point on the ray VF, erect a perpendicular p to l at A. The circle with center F and radius AD cuts p in two points of the parabola.

To construct a *tangent* at a point P, let the perpendicular through P to the axis cut the axis in A. Choose A' on the axis so that V bisects AA'. Then PA' is the tangent at P. Or bisect by b the angle AFP and construct through P a parallel to b.

To construct the *axis* l, and hence find the *vertex* V, when merely the curve is given, draw two parallel chords and bisect them in

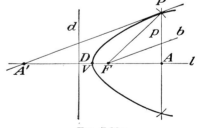

Fig. 7.32a.

M and M'. Then MM' is parallel to l. A chord perpendicular to MM' will have its midpoint A on l, which may thus be drawn through A parallel to MM'.

Continuing, to find the *focus* F, construct at V the line making arc tan 2 with l. This line cuts the curve in a point Q from which a perpendicular to l will cut l in F. The *directrix* is then perpendicular to l at a point D such that $DV = VF$.

2. *Ellipse*, given foci and "constant sum" $2a$. Take a string of length $2a$, fix its ends at foci, and move a pencil so as to keep the string taut. If major and minor semiaxes are given, draw circles with centers O and radii a and b. If these are cut in R and Q by a radius, perpendiculars to the axes through R and Q will intersect in a point P of the ellipse. Or mark the point P a units from one end on a straightedge of

length $a + b$. If its ends slide on the axes, P will move on the ellipse. To construct a tangent at a point P, bisect the (external) angle between the focal radii.

To find the *center*, when given merely the curve, draw the line MM' joining the midpoints of two parallel chords, and the line NN' joining the midpoints of two parallel chords not parallel to the first pair. Then MM' and NN' intersect in the center.

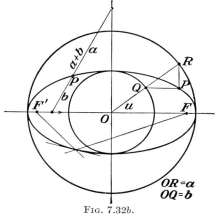

Given the center O, the largest inscribed (smallest circumscribed) circle (center at O, of course) will touch the ellipse at the ends of the *minor (major) axis*.

To find the *foci*, draw an arc with center at one end of the minor axis and radius equal to the major semiaxis. This arc will cut the major axis in the foci (because $c^2 = a^2 - b^2$).

3. *Hyperbola*, given foci F and F', hence their mid-point O, and the constant difference $2a$, hence the vertices V and V'. If A is on VF, circular arcs with center F and radius VA and center

Fig. 7.32b.

F' with radius $V'A$ will intersect in two points of the hyperbola.

7.4. Parametric Equations. Instead of investigating $y = f(x)$, or $F(x,y) = 0$ it is often advantageous to express both x and y in terms of a *parameter* (auxiliary variable) u. Points on the graph will be distinguished from each other by values of u, which thus amounts to a coordinate *on the curve*. The parameter may or may not have a useful geometric or physical interpretation.

Line joining $P_1(x_1,y_1)$ and $P_2(x_2,y_2)$: $x = x_1 + u(x_2 - x_1)$, $y = y_1 + u(y_2 - y_1)$. Here $u = 0$ gives P_1, $u = 1$ gives P_2, $u = \frac{1}{2}$ their mid-point, etc.

Line through (h,k), making an angle θ with positive x-axis: $x = h + u \cos \theta$, $y = k + u \sin \theta$. Here u denotes distance along the line from (h,k).

Circle, center (h,k), radius r: $x = h + r \cos u$, $y = k + r \sin u$; starting at $(h + r, k)$ when $u = 0$, P moves once around the circle as u runs from 0 to 2π.

Ellipse, center (h,k), semiaxes a and b: $x = h + a \cos u$, $y = k + b \sin u$. The angle u is indicated in the construction in Fig. 7.32b.

Hyperbola, center (h,k), transverse and conjugate axes a and b: $x = h + a \cosh u$, $y = k + b \sinh u$; or $x = h + a \sec v$, $y = k + b \tan v$.

Parabola: $x = v_0u \cos \alpha$, $y = v_0u \sin \alpha - gu^2/2$. [Equations of motion of a projectile fired when (time) $u = 0$ from the origin with initial velocity v_0, at an inclination of α to the horizontal, neglecting all forces except weight of projectile.]

The nonparametric equations are obtained by eliminating the parameter. Thus, for the ellipse, write $\cos u = (x - h)/a$, $\sin u = (y - k)/b$, square, and add.

7.5. Polar Coordinates. The distance r of a point P from the *pole* (origin) O and the angle θ made by the ray OP with the *initial ray* (positive x-axis), θ being positive in the counterclockwise sense, are *polar coordinates* (r,θ) of P. P also has polar coordinates $(r,\theta + 2k\pi)$, k any integer. It is also convenient to allow r to be negative, in which case a distance $|r|$ is measured off along the ray issuing from O in the direction opposite to that of the terminal ray of θ. Hence $P(r,\theta)$ also has coordinates $[-r,\theta + (2k + 1)\pi]$, k any integer.

If (x,y) are rectangular cartesian coordinates with the positive x-axis coinciding with the initial ray, and θ positive in the direction from the positive x- to the positive

y-axis, and if (r,θ) are any polar coordinates of a point P, then $x = r \cos \theta$, $y = r \sin \theta$. Conversely, $r = \sqrt{x^2 + y^2}$ and θ such that $\cos \theta = x/r$, $\sin \theta = y/r$ are polar coordinates of P.

Equations. 1. *Line* whose normal from the pole makes an angle α with the initial direction and is p units long: $r \cos (\theta - \alpha) = p$.

2. *Circle* with center at the origin and radius a: $r = a$. With center at (r_1,θ_1) and radius a: $r^2 + r_1{}^2 - 2rr_1 \cos (\theta - \theta_1) = a^2$.

3. *Ray* issuing from origin making angle α with initial ray: $\theta = \alpha$.

4. *Conic* with focus at pole and directrix distant p units from pole and perpendicular to initial direction: $r = ep/(1 + e \cos \theta)$, where e is eccentricity (see 7.31). $2ep$, often denoted by $2l$, is the length of the *latus rectum*.

In plotting curves $r = f(\theta)$, some care is needed with negative values of r. In particular, an equation $r^2 = f(\theta)$ leads to $r = \pm \sqrt{f(\theta)}$. Thus, when $f(\theta) < 0$, r is

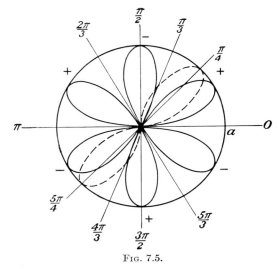

FIG. 7.5.

imaginary (and is not graphed); when $f(\theta) > 0$, the positive value of r gives rise to a point on the terminal line of θ, and the negative value will give a second point, on the terminal line reversed. The curve traced out by the negative square root may or may not coincide (perhaps in a different order) with the curve traced out by the positive square root.

The six-looped curve is the graph of $r^2 = a^2 \sin 3\theta$. The loops labeled $+$ $(-)$ arise from the positive (negative) square root. The two-looped (dotted) curve is the graph of $r^2 = a^2 \sin 2\theta$. Each loop is traced out once by the positive and once by the negative square root.

Similarly, care is needed in finding points of intersection. This may be done graphically if an estimate to a small number of significant figures will be satisfactory and care is exercised to choose among the coordinates of a point a pair that actually satisfies both equations. In solving $r = f(\theta)$ and $r = g(\theta)$ analytically, there will be four possibilities to consider: $f(\theta + 2k\pi) = g(\theta + 2l\pi)$, $f(\theta + (2k + 1)\pi) = -g(\theta + 2l\pi)$, $f(\theta + 2k\pi) = -g(\theta + (2k + 1)\pi)$, and $-f(\theta + (2k + 1)\pi) = -g(\theta + (2k + 1)\pi)$. If α satisfies any one of these equations (with k_0 in f and l_0 in g), e.g., the last but one, then $[f(\alpha + 2k_0\pi),\alpha]$ and $\{-g[\alpha + (2k_0 + 1)\pi],\alpha\}$ are coordinates of the same point, which lies on both curves.

7.6. Points, Lines, and Planes in Space. 7.61. Coordinates. Equations and Surfaces. Let three mutually orthogonal (perpendicular) planes intersect at a point O. Use two of the lines so formed as x- and y-axes for a system of rectangular cartesian coordinates (see 5.1) in their plane. On the third line, the z-axis, use the same unit of measure as on the other axes, and measure positively away from O in such a way that, viewed *from* the *positive* z-axis, the rotation in the xy-plane from the positive x-axis toward the positive y-axis is counterclockwise. Each point of space then has associated with it a unique ordered triple of numbers (x,y,z) which are its *rectangular cartesian coordinates*. Each coordinate measures the perpendicular distance to the plane (*coordinate plane*) of the other two. This is a *right-handed system* of axes. Reversing the direction of an odd (even) number of axes gives a left-handed (right-handed) system. The coordinate planes divide space into eight octants, each of which is characterized by a definite set of signs for the coordinates; only the $(+,+,+)$ octant has a name; it is called the *first octant*.

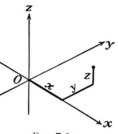

Fig. 7.6a.

The equation $x = a$ defines a plane parallel to the yz-plane; similarly for $y = b$ and $z = c$. Two such equations define a line parallel to the axis of the variable not mentioned. Three such equations fix a point.

Given a system of rectangular cartesian coordinates, let the coordinates $(x,y,0)$ in the xy-plane be replaced by polar coordinates $(\rho,\phi,0)$ (see 7.5). Then each point in space has coordinates (ρ,ϕ,z) which are its *cylindrical* coordinates. Surfaces $\rho = a$ are right circular cylinders with axis the z-axis and radius a. Surfaces $\phi = \alpha$ are half planes with the z-axis as the free edge. Surfaces $z = c$ are planes parallel to the xy-plane. Curves $\rho = a$, $\phi = \alpha$, z variable are lines parallel to the z-axis. Curves $\rho = a$, $z = c$, ϕ variable are circles. Curves $\phi = \alpha$, $z = a$, r variable are rays issuing from points on the z-axis and running out indefinitely parallel to the $\rho\phi$-plane. (ρ,ϕ,z) and (x,y,z) are related by the equations $\rho = \sqrt{x^2 + y^2}$, $\cos \phi = x/\rho$, $\sin \phi = y/\rho$, and $x = \rho \cos \phi$, $y = \rho \sin \phi$.

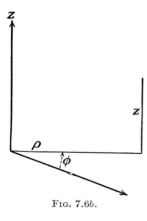

Fig. 7.6b.

Cylindrical coordinates are useful in studying configurations symmetric about a line (*axial symmetry*). In dealing with a surface of revolution, set $\phi = 0$, say, and plot the curve $\rho = f(z)$ which it is desired to revolve. Letting ϕ vary has the effect of rotating this curve about the z-axis, so that the equation $\rho = f(z)$, interpreted in cylindrical coordinates, is the equation of the required surface. Conversely, any equation $\rho = f(z)$ represents a surface of revolution with the z-axis as axis of revolution.

Let an *axis* (ray) l issue from a *pole* (point) O. In the plane π perpendicular to l, choose an arbitrary initial ray m issuing from O. Then the *spherical coordinates* of P are (r, θ, φ), where $r(0 \leqslant r < + \infty)$ is the distance OP, $\theta(0 \leqslant \theta \leqslant 2\pi)$ is the angle, measured positively when counterclockwise as viewed from l, from m to the ray n which is the projection of the ray OP on the plane π, and $\varphi \left(-\dfrac{\pi}{2} \leqslant \varphi \leqslant \dfrac{\pi}{2} \right)$ is the angle from n to the ray OP, measured positively when l and OP are on the same side of the plane π. The equation $r = a$ defines a sphere with center at the pole and radius a; $a = 1$ gives the *unit sphere*. The equation $\varphi = \alpha$ defines a half cone with vertex O, axis l

and semivertical angle $\frac{\pi}{2} - \alpha$; $\varphi = 0$ gives the plane π. The equation $\theta = \beta$ gives a half plane with the line containing l as the free edge. $r = a$ and $\varphi = \alpha$ taken together define a circle with center on l and radius $a \cos \alpha$ lying on the sphere $r = a$, around which θ varies. $r = a$ and $\theta = \beta$ taken together give a semicircle of radius a in the

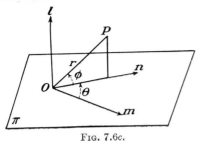

FIG. 7.6c.

half plane $\theta = \beta$, around which φ varies. $\varphi = \alpha$ and $\theta = \beta$ taken together give a ray issuing from O along which r varies.

If the origin of (x, y, z) is the pole and the xy-plane is the plane π and the positive x-axis is the line m, then $x = r \cos \varphi \cos \theta$, $y = r \cos \varphi \sin \theta$, and $z = r \sin \varphi$, while $r = \sqrt{x^2 + y^2 + z^2}$, $\varphi = $ arc sin z/r (principal value), and as in plane polar coordinates, θ is so chosen between 0 and 2π that $\cos \theta = x/\sqrt{x^2 + y^2}$, and

$$\sin \theta = y/\sqrt{x^2 + y^2}.$$

If O is at the center of the earth, with the axis l passing through the North Pole, and m is coplanar with Greenwich and the North Pole, then the plane π is the equatorial plane, θ gives east longitude, and φ gives latitude.

NOTE: Most nonmeteorological books use φ to denote the angle here denoted by θ and, instead of the angle here denoted by φ, use the angle (from 0 to π) from the ray l to the ray OP, calling this angle θ.

7.62. Lines and Planes. Let $P_1(x_1,y_1,z_1)$ and $P_2(x_2,y_2,z_2)$ be two points. The displacement from P_1 to P_2 results from displacements $x_2 - x_1$ in the direction of the positive x-axis, $y_2 - y_1$ along the positive y-axis, and $z_2 - z_1$ along the positive z-axis. These directed segments are called the *projections* or *components* of the directed segment P_1P_2. The *distance* P_1P_2 is evidently $d = \sqrt{(x_2 - x_1)^2 + (y_2 - y_1)^2 + (z_2 - z_1)^2}$. The angles α, β, γ between 0 and 180 deg measured from the positive x-, y-, z-axes are the *direction angles* of the segment P_1P_2, or of any line containing this segment, and their cosines are the *direction cosines* of the segment or line. Further, $x_2 - x_1 = d \cos \alpha$, $y_2 - y_1 = d \cos \beta$, and $z_2 - z_1 = d \cos \gamma$ are *direction numbers* of the segment or line; if $d = 1$ these numbers reduce to the cosines. Using the expression for d, we see that $\cos^2 \alpha + \cos^2 \beta + \cos^2 \gamma = 1$. Any three numbers λ, μ, ν, not all zero, are direction numbers for any one (l) of a family of parallel lines. If $\rho = \sqrt{\lambda^2 + \mu^2 + \nu^2}$, then $\cos \alpha = \lambda/\rho$, $\cos \beta = \mu/\rho$, and $\cos \gamma = \nu/\rho$ are direction

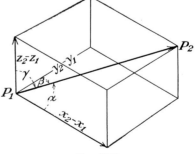

FIG. 7.62.

cosines and α, β, γ direction angles of l. If t is any real number $\neq 0$, $t\lambda$, $t\mu$, and $t\nu$ will also be direction numbers of l; if $t < 0$ the angles α, β, γ are carried into their supplements, and the direction of l is reversed. The vanishing of a direction number means that the line is parallel to the plane of the other two variables.

$P(x,y,z)$ will lie on the line l through P_1 and P_2 if and only if $x - x_1, y - y_1, z - z_1$ are direction numbers for l, *i.e.*, if and only if there is a number t such that $x - x_1 = t(x_2 - x_1)$, $y - y_1 = t(y_2 - y_1)$, and $z - z_1 = t(z_2 - z_1)$.

$$x = x_1 + t(x_2 - x_1)$$
$$y = y_1 + t(y_2 - y_1) \Bigg\} \qquad (1)$$
$$z = z_1 + t(z_2 - z_1)$$

$$\frac{x - x_1}{x_2 - x_1} = \frac{y - y_1}{y_2 - y_1} = \frac{z - z_1}{z_2 - z_1} \qquad (= t) \qquad (2)$$

$$\frac{x - x_1}{\lambda} = \frac{y - y_1}{\mu} = \frac{z - z_1}{\nu} \qquad (= t) \qquad (3)$$

(1) are *parametric* equations for l. Equations (2) and (3), sometimes said to be "symmetric," do not contain the parameter and are satisfied by points on the line from P_1 (2) through P_2 and (3) in a direction with numbers λ, μ, ν. (The latter could be replaced by $\cos \alpha$, $\cos \beta$, $\cos \gamma$.) The angle θ between the directions of l_1 and l_2 determined by their cosines has $\cos \theta = \cos \alpha_1 \cos \alpha_2 + \cos \beta_1 \cos \beta_2 + \cos \gamma_1 \cos \gamma_2$. ($l_1$ and l_2 need not intersect.) For perpendicularity, $\cos \theta = 0$.

Planes and only planes have linear equations $Ax + By + Cz + D = 0$ with not all of A, B, C vanishing. If one of A, B, C vanishes, the plane is parallel to the axis of the corresponding variable. The *intercepts* are $a = -D/A$, $b = -D/B$, and $c = -D/C$ on the x-, y-, z-axes, respectively. The numbers A, B, C are direction numbers of any line *normal* (perpendicular) to the plane. Measured positively in the direction whose numbers are A, B, C, the distance *from* the plane *to* $P_0(x_0,y_0,z_0)$ is $(Ax_0 + By_0 + Cz_0 + D)/\sqrt{A^2 + B^2 + C^2}$. The angle θ between two planes is the angle between their normals. The planes are perpendicular if and only if $AA' + BB' + CC' = 0$; parallel if and only if $A/A' = B/B' = C/C'$, and coincident if and only if also $C/C' = D/D'$.

Equations of Planes. Through (x_1,y_1,z_1) with normal in direction with numbers λ, μ, ν

$$\lambda(x - x_1) + \mu(y - y_1) + \nu(z - z_1) = 0$$

Distant p units from origin with normal in direction with angles α, β, γ

$$x \cos \alpha + y \cos \beta + z \cos \gamma = p$$

With intercepts a, b, c
$$\frac{x}{a} + \frac{y}{b} + \frac{z}{c} = 1$$

Through P_1, P_2, P_3
$$\begin{vmatrix} x & y & z & 1 \\ x_1 & y_1 & z_1 & 1 \\ x_2 & y_2 & z_2 & 1 \\ x_3 & y_3 & z_3 & 1 \end{vmatrix} = 0$$

To find where a line cuts a plane, substitute from its equations into the equation of the plane.

Two planes determine a line. Its direction numbers are $BC' - B'C$, $CA' - A'C$, $AB' - A'B$. To write down equations for the line, the coordinates of a point on it are needed. To find these, assign a numerical value to one of the variables in the equations for the planes and solve for the other two. *Alternative Procedure:* Manipulate the equations for the planes so that they may be put into form (3) above.

Example.
$$\begin{cases} 2x - 3y + 4z + 5 = 0 \\ 3x + 4y - 5z + 6 = 0 \end{cases}$$
$$-17y + 22z + 3 = 0$$
$$22x + y + 49 = 0$$
$$\frac{x + {}^{49}\!/_{22}}{-22} = y = \frac{z + {}^{3}\!/_{22}}{{}^{17}\!/_{22}}$$

Subtract double the second equation from treble the first. Similarly, eliminate z. Solving each resulting equation for y, we get equations of the form (3), which say that the line passes through $(-49/22, 0, -3/22)$ and has direction numbers -22, 1, ${}^{17}\!/_{22}$.

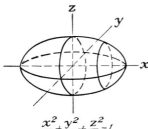

$$\frac{x^2}{a^2}+\frac{y^2}{b^2}+\frac{z^2}{c^2}=1$$

Ellipsoid

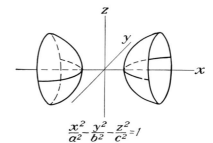

$$\frac{x^2}{a^2}-\frac{y^2}{b^2}-\frac{z^2}{c^2}=1$$

Two sheeted (parted) Hyperboloid

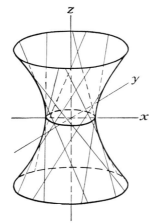

$$\frac{x^2}{a^2}+\frac{y^2}{b^2}-\frac{z^2}{c^2}=1$$

One sheeted (unparted) Hyperboloid
(Ruled surface)

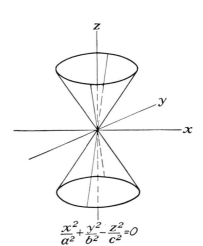

$$\frac{x^2}{a^2}+\frac{y^2}{b^2}-\frac{z^2}{c^2}=0$$

Cone

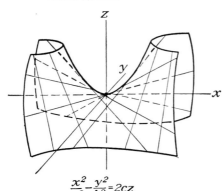

$$\frac{x^2}{a^2}-\frac{y^2}{b^2}=2cz$$

Hyperbolic paraboloid, saddle surface
(Ruled surface)

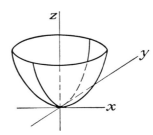

$$\frac{x^2}{a^2}+\frac{y^2}{b^2}=2cz$$

Elliptic paraboloid

Fig. 7.7.

All the foregoing and related problems are clarified and simplified by the methods of elementary vector analysis (see 12.3).

7.7. Quadric Surfaces. The second-degree equation $f(x,y,z) = Ax^2 + By^2 + Cz^2 + 2Dxy + 2Exz + 2Fyz + 2Lx + 2My + 2Nz + K = 0$ represents a *quadric surface* (or *conicoid*) Q. If $f(a,b,c) = 0$, the *tangent plane* to Q at (a,b,c) is

$$Aax + Bby + Ccz + D(ay + bx) + E(az + cx) + F(bz + cy) + L(a + x)$$
$$+ M(b + y) + N(c + z) + K = 0$$

and the *normal* line at (a,b,c) has direction numbers

$$Aa + Db + Ec + L \qquad Da + Bb + Fc + M \qquad Ea + Fb + Cc + N$$

For classification of quadrics, transformations of coordinates, and the reduction of the general second-degree equation, similar to the treatment of conics in 7.31, see a text.

A determinant d may be formed for Q precisely as Δ was formed for the conic S in 7.31. If $d = 0$, Q degenerates into a pair of planes or a cone or a cylinder. In the latter case, coordinates can be so chosen that one of the variables is missing; the generators are parallel to the axis of that variable; and, in the plane of the two variables appearing in the equation, the equation represents the generatrix. The cone and the possible results of the reduction in case $d \neq 0$ are the following:

A *ruled surface* contains a family of straight lines. Each ruled quadric Q contains two families of rulings, and through each point of Q there passes one ruling of each family. Each ruling cuts every ruling of the other family but none of its own family.

Special Cases. If any two of a, b, c are equal, the ellipsoid or hyperboloid is an *ellipsoid* or *hyperboloid* of *revolution*. An ellipsoid of revolution is a *spheroid*, *prolate* if the odd axis is longest, *oblate* if the odd axis is shortest. If $a = b = c$, the ellipsoid is a sphere. If $a = b$, the elliptic paraboloid becomes a *paraboloid* of *revolution*, while the hyperbolic paraboloid becomes rectangular. If $a = b$, the (elliptic) cone becomes *circular*.

8. DIFFERENTIAL CALCULUS

8.1. Limits and Continuity. 8.11. Limits of Variables. A numerical variable v is said to be *bounded* or *unbounded above* (or *below*) according as there is or is not a number M such that $v \leqslant M$ (or $M \leqslant v$) throughout the entire course of the variation of v. If v is bounded both above and below (*i.e.*, $|v| \leqslant M$ for some $M \geqslant 0$), one says simply that v is *bounded*.

A variable v taking on the values 2, $\frac{3}{2}$, $\frac{5}{4}$, $\frac{9}{8}$, $\frac{17}{16}$, $\cdots$ can be made to differ from 1 by as little as may be required by following its variation far enough. This statement is abbreviated by writing $v \to 1$ or $\lim v = 1$ and saying that "v tends to 1," "v approaches 1," or "the *limit* of v is 1." If v is any variable and c any constant, $v \to c$ thus means that, given any positive number ϵ (however small), there is a definite stage in the variation of v *beyond* which each value of v, without exception, has the property that $|v - c| < \epsilon$. If v represents a physical quantity, in whose measurement an error of ϵ can be tolerated, then v is physically indistinguishable from c as soon as $|v - c| < \epsilon$. Note that if, from some stage on, $v = c$, then $v \to c$, but that $v \to c$ does not imply that v ever actually *has* the value c. The basic concept here roughly formulated kept its mystery until the nineteenth century, and can be handled confidently only after practice.

The terminology is extended to deal with a variable taking on such values as 1, 2, 4, 8, 16, $\cdots$ or -1, -2, -4, -8, $\cdots$. Such a variable v is said to "increase (decrease) without limit," or to "become positively (negatively) infinite." Abbreviating, we write $v \to +\infty$ (or $-\infty$) and say that v "tends to" or "approaches" or "has the limit" $+\infty$ (or $-\infty$). These abbreviated notations and locutions do not imply that the symbol ∞ or the word "infinity" denotes a number; attempts to use ∞ as a number can only produce nonsense.

If each of the assertions $v \to +\infty$, $v \to -\infty$, and $v \to c$ (including $v = c$) is false (for every c), v is said to *oscillate*—finitely or infinitely according as v is or is not bounded.

Theorems on Limits. $\lim (u + v + \cdots + w) = \lim u + \lim v + \cdots + \lim w$, $\lim (u \cdot v \cdot \cdots \cdot w) = (\lim u) \cdot (\lim v) \cdots \cdots (\lim w)$, $\lim (u/z) = (\lim u)/(\lim z)$, *provided* that the necessary limits exist and $\lim z \neq 0$. For "indeterminate forms," *e.g.*, $\lim (u/z)$ when $u \to 0$ and $z \to 0$, see 8.5.

8.12. Regions.   Limits of Variable Points.   A connected portion R of space, or of a surface S (perhaps plane), or of a curve C (perhaps straight) that has been singled out by means of a system of boundary surfaces, or curves on S, or points (at most two) on C is called a *region.* (On a straight line, an interval is a region.) Points of

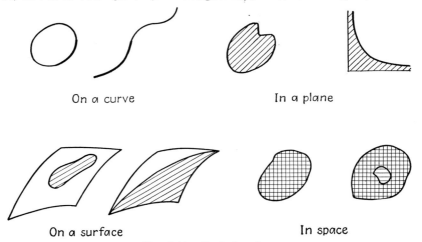

On a curve   In a plane

On a surface   In space

Fig. 8.12.—Typical regions.

space, or of S, or of C are thus either interior to R, or exterior to R, or on the boundary of R. (Note that the exterior may be disconnected and so not a region.)

A region R is *bounded* if it is contained in a sphere. The diameter $\|R\|$ of R is the least distance not exceeded by the distance between any two points of R. The volume, area, or length of R is $|R|$.

A point having a certain property is *isolated* if it is interior to a region of which no other point has the property.

Now let A be a variable point, B a fixed point. $A \to B$ or $\lim A = B$ means that, given any region R (however small $\|R\|$ may be) to which B is interior, there is a definite stage in the variation of A *beyond* which A is interior to R. If a coordinate system is given covering some fixed region R_0 to which B is interior, then $A \to B$ if and only if the (variable) coordinates of A approach the (fixed) coordinates of B.

8.13. Limits of Functions.   Let $f(P)$ be a function which, for each point P in a region R, defines a single real number $f(P)$. Let Q be a fixed point in R. We are interested in the behavior of $f(P)$ as $P \to Q$. This behavior will depend, in general, on the positions taken by P in its approach to Q. If $f(P) \to l$ (l constant) for *every* way in which Q can approach P, then we write $f(P) \to l$ as $P \to Q$ (no qualification).

It often happens that we wish to allow P to approach Q from one side of a surface through Q, or through points not on a curve C through Q (in space), or from one side of a curve through Q on a surface, or from one side of Q on a curve. In order for lim $f(P)$ to exist, all these qualified limits must exist and have the same value. Only in the case of a function of a single real variable is there a standard notation: $f(a+) =$ lim $f(u)$ as $u \to a$, $u > a$; $f(a-) =$ lim $f(u)$ as $u \to a$, $u < a$. Thus $f(u) \to l$ as $u \to a$ means that $f(a-) = l = f(a+)$.

8.14. Continuity of Functions. $f(P)$ is said to be *continuous* at Q if and only if: (1) $f(Q) = a$ exists; (2) $f(P) \to l$ as $P \to Q$; (3) $a = l$. $f(P)$ is continuous in a region if it is continuous at every point of the region. Roughly, if $y = f(x)$ is continuous, its graph will have no breaks and can be drawn without lifting the pencil from the paper.

Suppose that, at a point Q, $f(P) \to l$ as $P \to Q$, but $f(Q) \neq l$. Or suppose that, at each point Q of a curve C on a surface S or in space, $f(P) \to l$ (depending on Q) as $P \to Q$ through points not on C, but $f(Q) \neq l$ for some Q on C. Or suppose that,

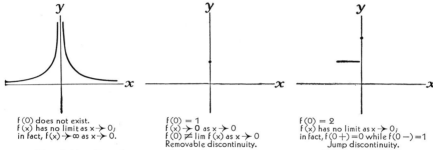

f(0) does not exist. / f(x) has no limit as x → 0; / in fact, f(x) → ∞ as x → 0.

f(0) = 1 / f(x) → 0 as x → 0 / f(0) ≠ lim f(x) as x → 0 / Removable discontinuity.

f(0) = 2 / f(x) has no limit as x → 0; / in fact, f(0+) = 0 while f(0−) = 1 / Jump discontinuity.

FIG. 8.14.—Functions $y = f(x)$ discontinuous at $x = 0$, continuous elsewhere.

at each point Q of a surface S in space, $f(P) \to l$ (depending on Q) as $P \to Q$ through points not on S, but that $f(Q) \neq l$ for some Q on S. If $f(P)$ is otherwise continuous, these are said to be *removable* discontinuities. By defining $F(Q) = l$ at these points, $F(P) = f(P)$ elsewhere, we get a continuous function F that coincides with f except at these points.

Let Q denote an arbitrary point of a curve C, or of a curve C^* on a surface S, or of a surface S^* in space, and let P denote a point that varies on C or on S or in space. Let $f(P)$ be continuous in some region containing Q on C, or C^* on S, or S^* in space *except* that $f(P) \to l$ or $f(P) \to l'$ according as $P \to Q$ on C from one side or the other of Q, or on S from one side or the other of C^*, or in space from one side or the other of S^*, *and* that, in the latter two cases l and l' are continuous functions of Q on C^* and S^*, respectively. Then $f(P)$ is said to have a *jump* discontinuity at Q on C, or along C^* on S, or on S^* in space.

$f(P)$ is said to be *sectionally continuous* in a region R if every discontinuity is either removable or a jump, and these discontinuities are confined to a *finite* number of: points if R is on a curve C; of points and curves if R is on a surface S; and of points, curves, and surfaces if R is in space.

The temperature of the atmosphere is usually sectionally continuous, while the pressure is continuous but its derivatives may be only sectionally continuous. On the surface of a cold or warm front, the temperature and the derivatives of the pressure may have jump discontinuities.

8.2. Differentiation. Let $y = f(x)$ be given and $y = f(a)$ when $x = a$. If x changes by a (positive or negative) amount $h = \Delta x$ ("delta x," *increment* in x), then y

will take on the value $f(a + h) = f(a) + k = f(a) + \Delta y$, where $k = \Delta y$ (increment in y) is the *change in y corresponding to the change h in x*. The ratio $k/h = \Delta y/\Delta x = [f(a + h) - f(a)]/h$, called the *difference quotient* or *increment ratio*, may be interpreted (1) as the *slope* of the *secant* line PQ, (2) as the *average rate of increase of y per unit increase in x* over the interval a, $a + h$.

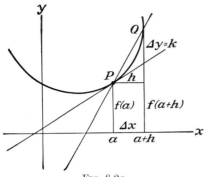

FIG. 8.2a.

If h is now restricted to (numerically) smaller and smaller values (with $h = 0$ excluded, of course), the point Q is then restricted [if the curve is continuous at P (see 8.14)] to a smaller and smaller neighborhood of P on the curve. The direction of the secant line then represents more and more closely the direction of the curve at P. Also, k/h gives the average rate of increase (of y per unit increase in x) over smaller and smaller intervals a, $a + h$ and thus represents more and more closely what happens *at $x = a$*. This suggests that k/h may have a limit as $h \to 0$ (see 8.13). If this limit exists, $f(x)$ is said to be *differentiable* when $x = a$. The limit is denoted by $f'(a)$ and is called the *derivative of $f(x)$ when $x = a$*. The process of finding the derivative is called *differentiation*.

In case $f'(a)$ exists, the secant PQ approaches the tangent at P as $h \to 0$, so that $f'(a)$ measures the *slope of the line tangent to the curve at P*. Also, $f'(a)$ is the *instantaneous* rate of increase of y per unit increase in x when $x = a$.

In order for $f'(a)$ to exist, $f(x)$ *must* be continuous when $x = a$. Continuity does not, however, imply differentiability. Thus, if $f(x) = |x|$ then, at $x = a = 0$, $k/h = +1$ or -1 according as $h > 0$ or $h < 0$, and cannot have a limit as $h \to 0$. If $\lim k/h$ as $h \to +0$ (or -0) exists (see 8.13), $f(x)$ is said to have a derivative from the right (or left); at $x = 0$, $|x|$ has a derivative from the right (left) equal to $+1$ (-1). In order for $f'(a)$ to exist, the derivatives from the right and from the left must both exist and be equal.

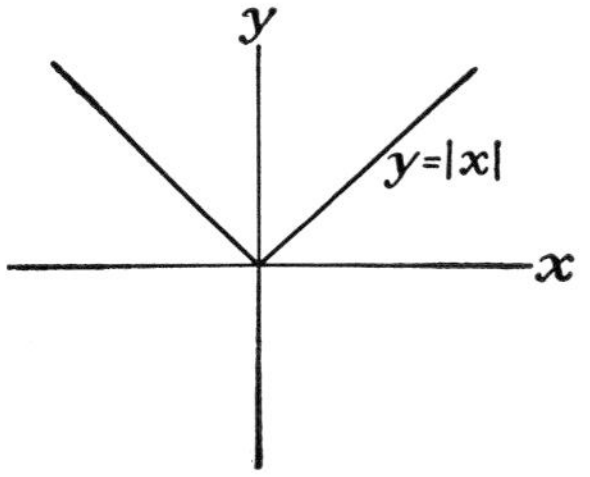

In case $f'(a)$ exists for every value a of x in an interval, the function $f'(x)$ is said to be the *derivative* of $f(x)$.

Alternative Notation. $f'(x) = y' = D_x f(x) = D_x y = dy/dx = df(x)/dx$. The derivative of $f'(x)$ is the *second*

FIG. 8.2b.

derivative of $f(x)$ and is denoted by $f''(x) = y'' = d^2y/dx^2$, etc.; the derivative of $f^{(n-1)}(x)$ is the *nth derivative* of $f(x)$ and is denoted by $f^{(n)}(x) = y^{(n)} = d^ny/dx^n$, etc.

8.21. General theorems on derivatives of $f(x)$, $g(x)$, $h(x)$, $\cdots$

$$(f + g + \cdots + h)' = f' + g' + \cdots + h'$$

$$(f \cdot g \cdots \cdots h)' = f' \cdot g \cdots \cdots h + f \cdot g' \cdots \cdots h + \cdots + f \cdot g \cdots \cdots h'$$

$$\frac{(f \cdot g \cdots \cdots h)'}{(f \cdot g \cdots \cdots h)} = \frac{f'}{f} + \frac{g'}{g} + \cdots + \frac{h'}{h} \qquad \text{(logarithmic differentiation)}$$

$$(f \cdot g)^{(n)} = f^{(n)} + C_1^n f^{(n-1)} g' + C_2^n f^{(n-2)} g'' + \cdots + C_{n-1}^n f' g^{(n-1)} + g^{(n)}$$

$$\text{(\textit{Leibnitz' rule; for} } C_k^n, \text{ see 4.5)}$$

$$\left(\frac{f}{g}\right)' = \frac{gf' - fg'}{g^2}$$

If y is a function of u and u is a function of x, then y is a function of x and $dy/dx = (dy/du)(du/dx)$. (*Function of a function; extremely important.*)

If y is a function of x and $dy/dx \neq 0$, then x is a function of y and $dx/dy = 1/(dy/dx)$. (*Inverse function.*)

If y is a function of u and x is a function of u and $dx/du \neq 0$, then y is a function of x and $dy/dx = (dy/du)/(dx/du)$. (*Parametric equations.*) N.B. Derivatives with respect to a parameter, especially one denoting time, are frequently indicated by superscript dots: $dx/du = \dot{x}$, $d^2x/du^2 = \ddot{x}$; thus $y' = \dot{y}/\dot{x}$ in the theorem.

8.22. Derivatives of Special Functions. "Angles" are in *radian* measure. If degrees are used, *either* convert to radians *or* multiply the stated derivatives of the trigonometric functions by $\pi/180 = 0.017453$, $(\sin x°)' = 0.017453 \cos x°$, and the stated derivatives of their inverses by 57.2957, $[(\text{arc sin } x)°]' = 57.2957/\sqrt{1 - x^2}$. Use the stated $(\text{arc sin } x)'$ or its negative according as arc sin x, $(|x| < 1)$, is in quadrant IV or I or in II or III. Reduce arc cos x, arc cot x, arc sec x, arc csc x as in 5.5. Note restrictions on inverse hyperbolic functions in 5.8.

Notation. c is any constant, a is any positive constant, $x < 0$ and $a \neq 1$ (unless $x = 1$) for $\log_a x$ to exist, $e = 2.71828 \cdots$ the natural base for logarithms (see 2.4), $\ln x = \log_e x$.

$$\frac{dc}{dx} = 0 \qquad\qquad \frac{d(cu)}{dx} = c\frac{du}{dx}$$

$$\frac{dx}{dx} = 1 \qquad\qquad \frac{d(u^c)}{dx} = cu^{c-1}\frac{du}{dx}$$

$$\frac{d(a^u)}{dx} = a^u \ln a \frac{du}{dx} \qquad\qquad \frac{d(e^u)}{dx} = e^u \frac{du}{dx}$$

$$\frac{d(\log_a u)}{dx} = \frac{\log_a e}{u}\frac{du}{dx} \qquad\qquad \frac{d(\ln u)}{dx} = \frac{1}{u}\frac{du}{dx}$$

$$\frac{d(\sin u)}{dx} = \cos u \frac{du}{dx} \qquad\qquad \frac{d(\cos u)}{dx} = -\sin u \frac{du}{dx}$$

$$\frac{d(\tan u)}{dx} = \sec^2 u \frac{du}{dx} \qquad\qquad \frac{d(\cot u)}{dx} = -\csc^2 u \frac{du}{dx}$$

$$\frac{d(\sec u)}{dx} = \sec u \tan u \frac{du}{dx} \qquad\qquad \frac{d(\csc u)}{dx} = -\csc u \cot u \frac{du}{dx}$$

$$\frac{d(\text{arc sin } u)}{dx} = \frac{1}{\sqrt{1 - u^2}}\frac{du}{dx} \qquad\qquad \frac{d(\text{arc tan } u)}{dx} = \frac{1}{(1 + u^2)}\frac{du}{dx}$$

$$\frac{d(\sinh u)}{dx} = \cosh u \frac{du}{dx} \qquad\qquad \frac{d(\cosh u)}{dx} = \sinh u \frac{du}{dx}$$

$$\frac{d(\tanh u)}{dx} = \text{sech}^2 u \frac{du}{dx} \qquad\qquad \frac{d(\text{arc sinh } u)}{dx} = \frac{1}{\sqrt{1 + u^2}}\frac{du}{dx}$$

$$\frac{d(\text{arc cosh } u)}{dx} = \pm\frac{1}{\sqrt{u^2 - 1}}\frac{du}{dx} \qquad\qquad \frac{d(\text{arc tanh } u)}{dx} = \frac{1}{(1 - u^2)}\frac{du}{dx}$$

Technical Suggestions. To handle complicated products, quotients, and exponents, take ln before differentiation. Thus, if $y = (1 - x^2)^x$, $\ln y = x \ln (1 - x^2)$, $y'/y = \ln (1 - x^2) - 2x^2/(1 - x^2)$, $y' = (1 - x^2)^x \ln (1 - x^2) - 2x^2(1 - x^2)^{x-1}$.

If y is defined *implicitly* by an equation of the form $f(x,y) = 0$, differentiate each side of the equation, *treating y as a function of x;* solve the result for y'. Thus, if $x^3 + y^3 + 3xy - 3 = 0$, $3x^2 + 3y^2y' + 3y + 3xy' = 0$, whence $y' = -(x^2 + y)/(y^2 + x)$.

To find an approximate value of a derivative from a graph, draw the tangent and find its slope; or, more accurately (if the graph itself is accurate enough to warrant the trouble), set a mirror normal to the curve (in such a way that the curve continues smoothly into the mirror) and if the normal has slope m the curve has slope $-1/m$. Only rough ideas can be formed easily from tables (especially empirical tables); $\Delta y/\Delta x$ in such a case will be a crude approximation to $f'(x)$, better if Δx is small. If more accurate information must be obtained and cannot be found otherwise, see ref. **1**, especially pars. 35ff.

8.3. Tangent; Differentials; Approximation; Polar Coordinates. The *tangent line* to $y = f(x)$ at $[a,f(a)]$ is $y - f(a) = f'(a)(x - a)$, if $f'(a)$ exists. The *normal line* (perpendicular to tangent) is $f'(a)[y - f(a)] = -(x - a)$. The angle between two curves is defined to be the angle between their tangent (or normal) lines (see **7.1**).

The equation of the tangent is often written $dy = f'(a)dx$, in terms of the *coordinates* $dy = y - f(a)$ and $dx = x - a$ *relative to* $[a,f(a)]$ of a point (x,y) on the tangent line. The variable dx (or dy) is called the *differential of x* (or y). Differentials may be introduced at any point $[a,f(a)]$ where $f'(a)$ exists. Since it reduces, when $x = a$, to $dy = f'(a)dx$, an equation $dy = f'(x)dx$ expresses compactly the relation between dy and dx at any point $[x,f(x)]$ on the curve $y = f(x)$. Also, if dx and dy refer to the point $[a,f(a)]$, and $dx \neq 0$, then $dy \div dx = f'(a)$; on allowing a to vary, we have, at *each* point, $dy \div dx = f'(x)$, which sanctions the previous notation

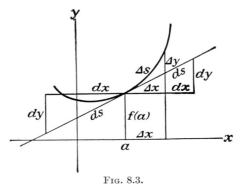

FIG. 8.3.

$dy/dx = f'(x)$. Any formula for differentiation may be written immediately as an equation between differentials. Thus $d(x^c) = cx^{c-1}\,dx$, $d(\sin x) = \cos x\,dx$, etc.

Suppose that we know $f(a)$, and that x is given an increment Δx so as to have the value $a + \Delta x$. It sometimes is but more often is not feasible to compute $f(a + \Delta x)$ directly from the definition of $f(x)$. The increment $\Delta y = f(a + \Delta x) - f(a)$ can be *approximated* with the value dy obtained by taking $dx = \Delta x$ in the relation $dy = f'(a)dx$. This is sometimes written $\Delta y \approx dy$. If desired, $f(a + \Delta x) = f(a) + \Delta y \approx f(a) + dy$.

The *error* $\Delta y - dy$ committed in writing $\Delta y = dy$ may often be estimated as follows: Let $\left|f''(x)\right| \leqslant M$ for x in the interval a, $a + \Delta x$. Then $\left|\Delta y - dy\right| \leqslant M(\Delta x)^2/2$.

Examples. To find the value of $\cos 32°$. Using the convenient neighboring value of $30°$ for a in the theory, we have $\cos 30° = \sqrt{3}/2 = .8660$, $\sin 30° = \frac{1}{2}$, $d(\cos x) = -\sin x\,dx$, and $dx = \Delta x = 2° = .03490$ (radians), so that $d(\cos x) = (-\frac{1}{2})(.03490) = -.0175$. Hence $\Delta y \approx -.0175$ and $\cos 32° \approx .8485$. To estimate the error, $(\cos x)'' = -\cos x$; for $30° \leqslant x \leqslant 32°$, $\left|(\cos x)''\right| \leqslant \cos 30° = .8660$, so that $\left|\Delta y - dy\right| \leqslant .8660(.0349)^2/2 = .00053$ (to five places). (Tables give $\cos 32° = .8480$ to four places.)

8.31. Polar Coordinates. As in the cartesian case, $f'(\alpha)$ computed from $r = f(\theta)$ when $\theta = \alpha$ may be interpreted as the rate of increase of r per unit increase in θ when

$\theta = \alpha$. However, it is not $\Delta\theta$ but the circular arc $\overset{\frown}{PQ} = r\,\Delta\theta$ that helps "displace" $P[f(\alpha),\alpha]$ to $R[f(\alpha + \Delta\theta),\alpha + \Delta\theta]$. To interpret $f'(\alpha)$ geometrically, we let the point $T[f(\alpha) + dr,\alpha + d\theta]$ move toward P along the tangent at P and can show that, as $d\theta \to 0$, the ratio $dr/(r\,d\theta) \to \cot\psi$. Thus $\cot\psi = f'(\alpha)/f(\alpha)$, and $f'(\alpha)$ is connected geometrically with the inclination of the tangent at $[f(\alpha),\alpha]$ to the radius vector (and not directly with its inclination to the initial direction, as in the cartesian case).

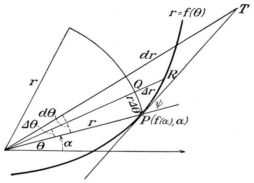

Fig. 8.31.

8.4. Maxima and Minima. Inflections. From the interpretation of $f'(a)$ as the instantaneous rate of increase of y per unit increase in x, it follows that $f'(a) > 0$, < 0, $= 0$ according as $f(x)$ is *increasing, decreasing,* or *stationary* (neither increasing nor decreasing) as x *increases* through the value $x = a$. The inclination and the slope of the tangent will be > 0, < 0, or $= 0$ in the respective cases.

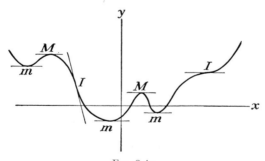

Fig. 8.4a.

$f(a)$ is a (relative) *maximum* M if, whenever x is close enough to a, $f(a) > f(x)$. $f(a)$ is a (relative) *minimum* m if, whenever x is close enough to a, $f(a) < f(x)$. As in the figure, an m may exceed an M. Maxima and minima are called *extreme values*. At an extreme value (except possibly at an end of an interval), $f(x)$ *must* be stationary; if $f'(a)$ exists, $f'(a) = 0$. Hence, to find stationary points, set $f'(x) = 0$ and solve for x. Moreover, if $f'(a) = 0$ *and* $f'(x)$ changes sign as x increases through a, then $f(a)$ *will* be extreme: m if first $f'(x) < 0$ (f decreasing), then $f'(a) = 0$ (f stationary), and then $f'(x) > 0$ (f increasing); and M if first $f'(x) > 0$ (f increasing), then $f'(a) = 0$ (f stationary), and then $f'(x) < 0$ (f decreasing) (*sign test*). If $f'(a) = 0$, $f(a)$ is thus

M or m according as $f'(x)$ is decreasing or increasing, *i.e.*, M if $f''(a) < 0$, m if $f''(a) > 0$ (*second derivative test;* indecisive if $f''(a) = 0$).

Use whichever test seems easier. The sign test may work when $f''(a)$ fails to exist or vanishes. Note that the tests give only *sufficient* conditions. An extreme value may exist where $f'(x)$ does not. Also, M or m may occur at the ends of an interval to which x is restricted, even though $f'(x) \neq 0$ there (as at M', where $f'(b)$ exists from the left and is positive). For a more general derivative test, valid when any finite number of derivatives vanish, see 8.73.

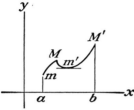

Fig. 8.4b.—"Tests" reveal only m'.

$f'(x)$ is increasing, decreasing, or stationary, as x increases through $x = a$, according as $f''(a) > 0$, < 0, or $= 0$. In the respective cases, the tangent line will be below the curve and rotating counterclockwise; or above the curve and rotating clockwise; or above and rotating clockwise (or below and rotating counterclockwise) to the left of $x = a$ and below and rotating counterclockwise (or above and rotating clockwise) to the right. When viewed from below, the curve is said to be *convex*, to be *concave*, or to have an *inflection point* (I on the figures) in the respective cases. To find inflection points, put $f''(x) = 0$ and solve for x; inflection occurs if $f''(x)$ changes sign as x increases through $x = a$. Thus $f'(x)$ has an extreme value at an inflection. Also, the curve crosses its tangent there. For curvature see 10.21.

8.5. The Mean-value Theorem. Indeterminate Forms. *Theorem.* If $f(x)$ is continuous for $a \leqslant x \leqslant b$ and $f'(x)$ exists for $a < x < b$, then there is a number c, $a < c < b$ such that $f(b) - f(a) = (b - a)f'(c)$.

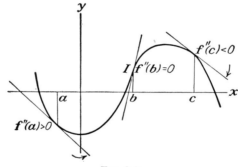

Fig. 8.4c.

It follows that, if $f'(x)$ vanishes identically, $f(x)$ is a constant. (No other *proof* is known for this important fact.)

Generalized Theorem. If $f(x)$ and $g(x)$ are continuous for $a \leqslant x \leqslant b$, and, for $a < x < b$, $f'(x)$ and $g'(x)$ exist and $g'(x)$ does not vanish, then there is a number c, $a < c < b$, such that $[f(b) - f(a)]/[g(b) - g(a)] = f'(c)/g'(c)$.

Suppose now that $f(a) = g(a) = 0$. For $x = a$, the quotient $f(x)/g(x)$ then assumes the meaningless or *indeterminate form* $\%$. Nevertheless, $f(x)/g(x)$ may have a limit (see 8.13) as $x \to a$. In the computation of du/dv, for example, $\Delta u/\Delta v$ becomes meaningless if 0 is substituted for Δu and for Δv, although $\lim (\Delta u/\Delta v)$ may exist as $\Delta v \to 0$ and $\Delta u \to 0$ with Δv. Putting $f(a) = g(a) = 0$ and $b = a + h$ in the generalized theorem, we have $f(a + h)/g(a + h) = f'(c)/g'(c)$, where $a < c < a + h$. If $h \to 0$, then $a + h \to a$ and $c \to a$, and the left member will have a limit (or become positively or negatively infinite) if and only if the right member does the same. For example, if $f(x) = x^n - a^n$, $g(x) = x - a$, $f(a + h)/g(a + h) = [(a + h)^n - a^n]/[(a + h) - a] = nc^{n-1}/1$, where $a < c < a + h$. Hence $f(x)/g(x) \rightarrow na^{n-1}$ as $x \rightarrow a$. If it happens that $f'(a) = g'(a) = 0$, we apply the same process to $f'(x)/g'(x)$, learn that $f(a + h)/g(a + h) = f''(c_1)/g''(c_1)$, where $a < c_1 < c$, and seek a limit (perhaps $+ \infty$ or $- \infty$) of $f''(x)/g''(x)$ as $x \rightarrow a$. Continue with higher derivatives, if necessary, until a decision is reached. Note that the process gives essentially *one-sided* limits, from the right or left according as $h > 0$ or $h < 0$; these may differ. If this process seems objectionable, series expansions (see 8.7) may be helpful.

To find $\lim [(x - \sin x)/x^3]$ as $x \rightarrow 0$. Since $x - \sin x \rightarrow 0$ and $x^3 \rightarrow 0$ as $x \rightarrow 0$, this form is indeterminate, so we find $(x - \sin x)' = 1 - \cos x$ and $(x^3)' = 3x^2$ and seek their limits. Since $1 - \cos x \rightarrow 0$ and $3x^2 \rightarrow 0$ as $x \rightarrow 0$, we find $(1 - \cos x)' = \sin x$ and $(3x^2)' = 6x$ and seek their limits. Since $\sin x \rightarrow 0$ and $6x \rightarrow 0$ as $x \rightarrow 0$, we find $(\sin x)' = \cos x$ and $(6x)' = 6$ and seek their limits. Now $\cos x \rightarrow 1$ and $6 \rightarrow 6$ as $x \rightarrow 0$, so that $[(x - \sin x)/x^3] \rightarrow \frac{1}{6}$ as $x \rightarrow 0$. Using series, $x - \sin x = (x^3/6) - (x^5/120) + \cdots$, so that $(x - \sin x)/x^3 = \frac{1}{6} - (x^2/120) + \cdots \rightarrow \frac{1}{6}$ as $x \rightarrow 0$.

Applications. 1. If $f(x) \rightarrow 0$ and $g(x) \rightarrow \pm \infty$ as $x \rightarrow a$, then $\lim [f(x) \cdot g(x)] = \lim \{f(x)/[1/g(x)]\}$. Since $1/g(x) \rightarrow 0$ as $x \rightarrow a$, the latter limit can be sought by the foregoing method.

2. If $f(x) \rightarrow \pm \infty$ and $g(x) \rightarrow \mp \infty$, then $f(x) + g(x)$ threatens to become $\pm \infty \mp \infty$. Try some transformation to reduce to $\frac{0}{0}$ or to case 1 (and then to $\frac{0}{0}$). Thus $f(x) = 1/\ln x \rightarrow \pm \infty$ and $1/(1 - x) \rightarrow \mp \infty$ according as $x \rightarrow 1^+$ or $x \rightarrow 1^-$. We seek $\lim [f(x) + g(x)]$ as $x \rightarrow 1+$. $1/\ln x + 1/(1 - x) = (\ln x + 1 - x)/[(1 - x)\ln x]$. If $F(x) = \ln x + 1 - x$ and $G(x) = (1 - x) \ln x$, then $F(x) \rightarrow 0$ and $G(x) \rightarrow 0$ as $x \rightarrow 1+$, and the general method may be applied. $F'(x) = (1/x) - 1$ and $G'(x) = -\ln x + (1/x) - 1$, both of which approach 0, but $F''(x) = -1/x^2$ and $G''(x) = -1/x - 1/x^2$, so that $F''(x)/G''(x) \rightarrow \frac{1}{2}$ and hence $f(x) + g(x) \rightarrow \frac{1}{2}$ as $x \rightarrow 1+$.

Technical Note. $F'(x)/G'(x) = (-1 + 1/x)/(-1 + 1/x - \ln x) = (1 - x)/(1 - x - x \ln x)$; the latter form is easier to manage: simplify the quotient and differentiate (*separately*) the numerator and the denominator of the simplified quotient.

3. $[f(x)]^{g(x)}$ is indeterminate in case (a) $f(x) \rightarrow 0$ and $g(x) \rightarrow 0$ or (b) $f(x) \rightarrow \pm \infty$ and $g(x) \rightarrow 0$, or (c) $f(x) \rightarrow 1$ and $g(x) \rightarrow \pm \infty$. Take logarithms, getting $g(x) \ln f(x)$, which is of form 1, etc. To find $\lim x^x$ as $x \rightarrow 0$: $y = x^x$, $\ln y = x \ln x = \ln x/(1/x)$, which has the same limit as $(1/x)/(-1/x^2) = -x \rightarrow 0$ as $x \rightarrow 0$; since $\ln y \rightarrow 0$, $y \rightarrow 1$ as $x \rightarrow 0$.

8.6. Series of Constant Terms. Let $a_0, a_1, a_2, \cdots$ be constants and define the *partial sum* $A_n = a_0 + \cdots + a_n$. The tentative symbol $\Sigma a_n = a_0 + a_1 + \cdots$ (*i.e.*, $n = 0, 1, 2, \cdots$ in the summation; see 4.2) has meaning if and only if $A_n \rightarrow A$ as $n \rightarrow \infty$ (see 8.11, 8.13) with $- \infty < A < \infty$; the *series* Σa_n is then said to *converge*, and $A = \Sigma a_n$ is its *sum*, e.g., $\pi = 3.1415 \cdots = 3 + 1 \cdot 10^{-1} + 4 \cdot 10^{-2} + 1 \cdot 10^{-3} + 5 \cdot 10^{-4} + \cdots$. Otherwise the series *diverges* and the symbol Σa_n is rejected (here) as meaningless.

Convergence Tests. 1. If Σa_n converges, then $a_n \rightarrow 0$ (but *not* conversely).

2. If $p_n \geqslant 0$ and $P_n = p_0 + p_1 + \cdots + p_n$, then either $P_n \rightarrow + \infty$ as $n \rightarrow \infty$ or $P_n \leqslant M$ for all n, Σp_n converges and $\Sigma p_n \leqslant M$.

3. Let $p_n \geqslant 0$. If Σp_n converges, and $|a_n| \leqslant p_n$, then Σa_n converges. If Σp_n diverges and $a_n \geqslant p_n$, then Σa_n diverges. (*Comparison test.*)

4. If $|a_{n+1}/a_n| \rightarrow k$ as $n \rightarrow \infty$ then: if $k < 1$, Σa_n converges; if $k > 1$, Σa_n diverges; if $k = 1$ the test fails. (*Ratio test.*)

5. If Σa_n alternates (in sign, *i.e.*, $\pm a_n = (-1)^n |a_n|$), and if $a_n \rightarrow 0$ *steadily* (or *monotonically*) (*i.e.*, $|a_n| \geqslant |a_{n+1}|$), then Σa_n converges, $|A_n - A| \leqslant |a_{n+1}|$, and A lies between a_0 and $a_0 - a_1$. (*Alternating series.*)

These crude tests are useful. When they fail, exceedingly delicate tests may be found in the literature. Note that convergence or divergence is determined by a_n for large n, *i.e.*, by the "tail" of the series. A test is thus decisive whenever its conditions are met for *almost all n*, meaning for all n exceeding some particular value. The early terms are essential if the value of the sum is needed.

If the terms a_n are themselves sums, the a_n must be computed before Σa_n is investigated, unless the series $\Sigma a_n'$ obtained by dropping the parentheses converges. Thus $(1 - 1) + (1 - 1) + \cdots = 0$, but $1 - 1 + 1 - 1 + 1 - + \cdots$ diverges; however, $(1 - \frac{1}{2}) + (\frac{1}{4} - \frac{1}{8} + \frac{1}{16}) - \frac{1}{32} + \cdots = 1 - \frac{1}{2} + \frac{1}{4} - \frac{1}{8} \cdots$ because the latter series converges. In a convergent series, then, parentheses may be introduced at pleasure.

If $\Sigma a_n = A$ and $\Sigma b_n = B$, then $\Sigma(a_n \pm b_n) = A \pm B$ and, if k is any real number, $\Sigma(ka_n) = k\Sigma a_n = kA$.

If Σa_n converges, it is said to converge *absolutely* or *conditionally* according as $\Sigma|a_n|$ converges or diverges. (Note that Σa_n converges if $\Sigma|a_n|$ does, but not conversely.) If Σa_n converges absolutely, then its terms may be rearranged in any order without affecting the sum A (false if convergence is merely conditional), *e.g.*, add first all the positive terms, then all the negative terms, and take the difference.

Let Σa_n and Σb_n be given, and form Σc_n by formal multiplication: $c_n = a_0 b_n + a_1 b_{n-1} + \cdots + a_n b_0$. If all these series converge, to A, B, and C, respectively, then $AB = C$. If one of Σa_n and Σb_n converges absolutely and the other converges, then Σc_n converges to $C = AB$, where $A = \Sigma a_n$ and $B = \Sigma b_n$.

8.7. Power Series. Taylor and Maclaurin Series. Applications. A series $\Sigma a_n(x - c)^n$ is a *power series* (powers of $x - c$). There is a number R, $0 \leqslant R \leqslant +\infty$ (series useless unless $R > 0$), called the *radius of convergence*, such that the series converges if $|x - c| < R$ (*convergence interval J*), diverges if $|x - c| > R$, and special study is needed when $|x - c| = R$. If $|a_{n+1}/a_n| \to k$ as $n \to \infty$, then $R = 1/k$. For each x in J, the series converges absolutely. Also, convergence is *uniform* (see the literature) over any closed interval (see 1.22) entirely inside J.

Inside J, $f(x) = \Sigma a_n(x - c)^n$ is continuous. The series obtained by differentiating or integrating (see 8.2, 9.2) this series term by term also has convergence interval J, and their sums are $f'(x)$ and $\int f(x)dx$ respectively. If $g(x) = \Sigma b_n(x - c)^n$ is a series with convergence interval J', and if K is a shorter of J and J', then, inside K, $f(x)$ and $g(x)$ may be added, subtracted, multiplied, and (if $b_0 \neq 0$) divided formally as if they were polynomials, and the resulting series converges inside K (except in case of division, when, if there are in K values of x for which $g(x) = 0$, the new radius of convergence may be smaller than that of K, though still *positive*). Also, $f(x) = g(x)$ for all x inside K, if and only if $a_n = b_n$ for all n. Finally, let $h(y) = \Sigma d_k(y - l)^k$, convergence radius R', and suppose we wish $F(x) = h[f(x)]$; $F(x) = \Sigma d_k[\Sigma a_n(x - c)^n - l]^k$ if $|\Sigma a_n(x - c)^n| - l| < R'$, and, for all x with this property, the series for $F(x)$ may be obtained by expanding and collecting according to powers of $x - c$. *Briefly:* Inside its convergence interval, a power series may be treated substantially as a polynomial.

8.71. Taylor's Theorem. If $f(x)$ and its derivatives up to order $n + 1$ are continuous for $|x - c| < r$, then

$$f(x) = f(c) + f'(c)(x - c) + \frac{f''(c)(x - c)^2}{2!} + \cdots + \frac{f^{(n)}(c)(x - c)^n}{n!} + R_n(x)$$

where $$R_n(x) = \frac{1}{n!}\int_c^x f^{(n+1)}(t)(x - t)^n dt = \frac{1}{n!}f^{(n+1)}(\xi)(x - c)^n$$

with ξ between c and x. (For integration, see 9.1.) If $f(x)$ has derivatives of all orders, and if, for $|x - c| < R$, $R_n(x) \to 0$ as $n \to \infty$, then the power series $\Sigma f^{(n)}(c)$

$(x - c)^n/n!$ has convergence interval $|x - c| < R'$ for some $R' \geqslant R$, in which its sum is $f(x)$.

On putting $x = c + h$, one gets

$$f(c + h) = f(c) + hf'(c) + \cdots + \frac{h^n f^{(n)}(c)}{n!} + R_n(h)$$

where

$$R_n(h) = \frac{1}{n!}\int_0^h f^{(n+1)}(c + u)(h - u)^n\, du = \frac{h^n}{n!} f^{(n+1)}(c + \eta)$$

with η between 0 and h.

8.72. Maclaurin's theorem is Taylor's theorem in the particular case $c = 0$. A polynomial is a Maclaurin series with only a finite number of terms.

8.73. Maxima and Minima. To determine whether or not $f(c)$ is an extreme value of $f(x)$, expand $f(x)$ in powers of $(x - c)$: $f(x) = \Sigma f^{(n)}(c)(x - c)^n/n!$. Let $f^{(k)}(c)$, $k \geqslant 2$, be the first nonvanishing derivative of $f(x)$ when $x = c$. Then x can be taken so close to c that $f(x) - f(c)$ has the same sign as $f^{(k)}(c)(x - c)^k/k!$. If k is odd, this expression changes sign as x increases through c, and $[c,f(c)]$ is a point of inflection (with a horizontal tangent). If k is even, $f(c)$ is a maximum or a minimum according as $f^{(k)}(c) < 0$ or $f^{(k)}(c) > 0$.

8.74. Series and Convergence Intervals for Special Functions. To get series for $f(-x)$, $f(x^2)$, etc., substitute in series for $f(x)$ and change convergence interval appropriately.

$$(1 + x)^p = 1 + px + p\frac{(p - 1)}{2} x^2 + \cdots + \frac{p(p - 1)\cdots(p - n + 1)}{n!} x^n$$
$$+ \cdots \text{ (binomial series) any real } p \qquad |x| < 1$$

$$\frac{1}{1 \pm x} = 1 \mp x + x^2 \mp x^3 + \cdots + (\mp)x^n + \cdots \text{ (geometric series)} \qquad |x| < 1$$

$$\sqrt{1 \pm x} = 1 \pm \frac{1}{2} x - \frac{1 \cdot 1}{2 \cdot 4} x^2 \pm \frac{1 \cdot 1 \cdot 3}{2 \cdot 4 \cdot 6} x^3 - \frac{1 \cdot 1 \cdot 3 \cdot 5}{2 \cdot 4 \cdot 6 \cdot 8} x^4 \pm \cdots \qquad |x| < 1$$

$$\frac{1}{\sqrt{1 \pm x}} = 1 \mp \frac{1}{2} x + \frac{1 \cdot 3}{2 \cdot 4} x^2 \mp \frac{1 \cdot 3 \cdot 5}{2 \cdot 4 \cdot 6} x^3 + \frac{1 \cdot 3 \cdot 5 \cdot 7}{2 \cdot 4 \cdot 6 \cdot 8} x^4 \mp \cdots \qquad |x| < 1$$

$$\ln(1 + x) = x - \frac{x^2}{2} + \frac{x^3}{3} + \cdots + (-1)^{n+1}\frac{x^n}{n} + \cdots \qquad |x| < 1$$

$$\ln\frac{1 + x}{1 - x} = 2\left(x + \frac{x^3}{3} + \frac{x^5}{5} + \cdots + \frac{x^{2n+1}}{2n + 1} + \cdots\right) \qquad |x| < 1$$

$$\ln\frac{x + 1}{x - 1} = 2\left[\frac{1}{x} + \frac{1}{3}\left(\frac{1}{x}\right)^3 + \frac{1}{5}\left(\frac{1}{x}\right)^5 + \cdots + \frac{1}{2n + 1}\left(\frac{1}{x}\right)^{2n+1} + \cdots\right] \qquad |x| > 1$$

$$\ln x = \frac{x - 1}{x} + \frac{1}{2}\left(\frac{x - 1}{x}\right)^2 + \frac{1}{3}\left(\frac{x - 1}{x}\right)^3 + \cdots + \frac{1}{n}\left(\frac{x - 1}{x}\right)^n + \cdots \qquad x > \frac{1}{2}$$

$$\ln x = 2\left[\left(\frac{x - 1}{x + 1}\right) + \frac{1}{3}\left(\frac{x - 1}{x + 1}\right)^3 + \frac{1}{5}\left(\frac{x - 1}{x + 1}\right)^5\right.$$
$$\left. + \cdots + \frac{1}{2n + 1}\left(\frac{x - 1}{x + 1}\right)^{2n+1} + \cdots\right] \qquad x > 0$$

$$e^x = 1 + x + \frac{x^2}{2!} + \frac{x^3}{3!} + \cdots + \frac{x^n}{n!} + \cdots \qquad |x| < \infty$$

$$e^{-x^2} = 1 - x^2 + \frac{x^4}{2!} - \frac{x^6}{3!} + \cdots + (-1)^n\frac{x^{2n}}{n!} + \cdots \qquad |x| < \infty$$

$$\cosh x = 1 + \frac{x^2}{2!} + \frac{x^4}{4!} + \cdots + \frac{x^{2n}}{(2n)!} + \cdots \qquad |x| < \infty$$

$$\sinh x = x + \frac{x^3}{3!} + \frac{x^5}{5!} + \cdots + \frac{x^{2n+1}}{(2n + 1)!} + \cdots \qquad |x| < \infty$$

$$\cos x = 1 - \frac{x^2}{2!} + \frac{x^4}{4!} - + \cdots + (-1)^n \frac{x^{2n}}{(2n)!} + \cdots \qquad |x| < \infty$$

$$\sin x = x - \frac{x^3}{3!} + \frac{x^5}{5!} - + \cdots + (-1)^n \frac{x^{2n+1}}{(2n+1)!} + \cdots \qquad |x| < \infty$$

$$\text{arc } \sin x = x + \frac{1}{2}\frac{x^3}{3} + \frac{1 \cdot 3}{2 \cdot 4}\frac{x^5}{5} + \cdots + \frac{(2n)!}{(2^n n!)^2} \cdot \frac{x^{2n+1}}{2n+1} + \cdots \qquad |x| < 1$$

$$\text{arc } \tan x = x - \frac{x^3}{3} + \frac{x^5}{5} - + \cdots + (-1)^n \frac{x^{2n+1}}{2n+1} + \cdots \qquad |x| < 1$$

$$\text{arc } \tan x = \frac{\pi}{2} - \frac{1}{x} + \frac{1}{3x^3} - \frac{1}{5x^5} + \cdots + (-1)^{n+1} \frac{1}{(2n+1)x^{2n+1}} + \cdots \qquad |x| > 1$$

$$\text{arc } \sinh x = x - \frac{1}{2} \cdot \frac{x^3}{3} + \frac{1 \cdot 3}{2 \cdot 4}\frac{x^5}{5} - + \cdots + (-1)^n \frac{(2n)!}{(2^n n!)^2}\frac{x^{2n+1}}{(2n+1)} + \cdots \qquad |x| < 1$$

$$\text{arc } \tanh x = x + \frac{x^3}{3} + \frac{x^5}{5} + \cdots + \frac{x^{2n+1}}{2n+1} + \cdots \qquad |x| < 1$$

8.75. Note on Applications. Useful theoretical results are often obtained, especially from the binomial, exponential, and logarithmic series, by taking advantage of known relations of magnitude to ignore terms of order higher than x^2 or even all but first-order terms. The following specific examples are adapted from ref. 8.

In cooling air from temperature T to wet-bulb temperature T_w, the expression for the gain in entropy has a factor $f = (T - T_w)/T_w - \ln(T/T_w)$. But $\ln(T/T_w) = \ln[1 + (T - T_w)/T_w] = [(T - T_w)/T_w] - \frac{1}{2}[(T - T_w)/T_w]^2 + \cdots$. Now $T - T_w$ is positive and small compared with T_w. Since the series alternates (see 8.6, test 5) the error committed in dropping terms of order 3 and more is at most as large as the first term dropped. If this error is negligible compared with the terms kept, then f becomes $\frac{1}{2}[(T - T_w)/T_w]^2$ to a satisfactory degree of approximation.

The fact that x is so small that its higher powers are negligible is often expressed by writing $1 >> x$ or $x << 1$. In this example, $(T - T_w)/T_w << 1$, or $T - T_w << T_w$.

Since measurements of T carry a possible error of as much as 1°C, it is not only superfluous but misleading to carry the series far enough to ensure a much smaller computed error. Physical limitations often indicate in this way appropriate standards of accuracy in the use of series.

The work done in a certain vaporization cycle is $dw = (L + dL)dT/[A(T + dT)]$, where L is the latent heat of vaporization, T is temperature, and $1/A$ is the mechanical equivalent of heat. Keeping $dT << T$, the binomial theorem gives $(T + dT)^{-1} = T^{-1}(1 + dT/T)^{-1} = T^{-1}(1 - dT/T)$ to the first order. Since $dL << 1$, the product $dL\,dT$ may be neglected, and $dW = L\,dT/(AT)$ to the first order.

The relation between the pressures p_1 at the top and p_2 at the bottom of a layer of air of thickness h_1, lapse rate α_1, and temperature T_2 at the bottom is given by $h_1 = T_2[1 - (p_1/p_2)^{R\alpha_1/g}]/\alpha_1$, where g is the acceleration of gravity and R is the gas constant. Using the exponential series

$$\left(\frac{p_1}{p_2}\right)^{R\alpha_1/g} = e^{R\alpha_1 \ln(p_1/p_2)/g} = 1 + \frac{R\alpha_1 \ln(p_1/p_2)}{g} + \frac{R^2\alpha_1^2 (\ln(p_1/p_2))^2}{2g^2} + \cdots.$$

In the lower troposphere, and for layers thin enough to ensure that $p_2 < ep_1$, $|\ln(p_1/p_2)| < 1$ and $R\alpha_1 << g$. Hence to the first order, $(p_1/p_2)^{R\alpha_1/g} = 1 + R\alpha_1 \ln(p_1/p_2)/g$ and $h_1 = T_2 R \ln(p_2/p_1)/g$.

8.8. Computation with (Maclaurin) Series (Taylor series essentially the same, with appropriate trivial changes). Let $S_n(x) = \sum_0^n a_k x^k$, with $a_k = f^{(k)}(0)/k!$. Then $|f(x) - S_n(x)| = |R_n(x)|$ (see 8.71–2). Hence $|f(x) - S_n(x)| \leqslant M|x|^n/n!$, if $|f^{(n+1)}(t)| \leqslant M$ for $0 < t < x$ (or $-x < t < 0$, according to whether we wish to compute with

$x > 0$ or $x < 0$). This estimates the error when $f(x)$ is replaced by an *exact* value of $S_n(x)$. Thus $|\sin x - \sum_0^n (-1)^k x^{2k+1}/(2k + 1)!| = |x^{2n+1} \sin \eta/(2n + 1)!|$ with $0 < \eta < x$. Since $|\sin \eta| \leqslant 1$, the error is at most $|x|^{2n+1}/(2n + 1)!$, which decreases with n, rapidly if $|x| \leqslant 1$.

General Recommendation. Choose a series for which $R_n(x)$ decreases [$S_n(x)$ converges] rapidly with n when x is in the range where the computation must be performed. Thus, to compute $\ln \frac{3}{2}$, either $\ln (1 + x) = x - x^2/2 + x^3/3 + \cdots$ with $x = \frac{1}{2}$ or $\ln [(1 + x)/(1 - x)] = 2(x + x^3/3 + x^5/5 + \cdots)$ with $x = \frac{1}{5}$ could be used, but the latter converges much more rapidly.

Special Cases. If after the proposed substitution of a value a for x, the resulting series of constant terms will *alternate*, use the information in 8.6, test 5. If the terms of the resulting series are *positive*, it may be advisable to compute the (sum of the) remainder directly or perhaps find a number which it cannot exceed by comparing the terms with those of a conveniently chosen geometric series (see 8.74, 4.42).

Having decided how many terms can safely be neglected, we wish next to substitute a for x and compute $S_n(a)$ "exactly" (so as to have an error no larger than $|R_n(a)|$). But a will usually be given as an approximate decimal. It is quite possible for the error thus introduced in computing $S_n(a)$ to exceed $|R_n(a)|$, which then becomes illusory as an estimate of the error in $f(a)$, and the following question arises: What accuracy in a corresponds to what accuracy in $S_n(a)$? The answer cannot be given unconditionally but depends on the special circumstances. It is *usually* enough to have x given to two or three more decimal places than are required in $S_n(a)$. When in doubt, keep track of the errors in each term; if their most unfavorable total is below the allowable error, success is assured; if not, more places of x must be used, and it may help to use more terms of the series, thus decreasing the error from that source. Methods are available for improving the convergence of series. For this, and on all questions connected with series, see ref. 9.

Example. To compute $\ln 2$, use the series for $y = \ln [(1 + x)/(1 - x)]$ with $x = \frac{1}{3}$. Thus $\ln 2 = 2\Sigma a_n$, where $a_n = [(2n + 1) \cdot 3^{2n+1}]^{-1}$. Estimation of error via derivatives: $y^{(2k+2)} = (2k + 1)![(1 - x)^{-(2k+2)} + (-1)^{(2k+1)}(1 + x)^{-(2k+2)}]$, whose maximum for $0 \leqslant x \leqslant \frac{1}{3}$ occurs when $x = \frac{1}{3}$, and so is $(2k + 1)![(\frac{3}{2})^{2k+2} - (\frac{3}{4})^{2k+2}]$, whence we may take $M = (2k + 1)!(\frac{3}{2})^{2k+2}$ and $|R_{2k+1}(\frac{1}{3})| < M(\frac{1}{3}^{2k+1})$ $(1/(2k + 1)!)$ or $|R_{2k+1}| < \frac{3}{2}^{2k+2}$(for $k = 4$, $|R_{2k+1}| < \frac{3}{1024} < 0.003$). It is easier and more revealing in this case to examine Σa_n directly.

$$0 < r_n = \frac{1}{(2n + 3)3^{2n+3}} + \frac{1}{(2n + 5)3^{2n+5}} + \cdots < \frac{1}{(2n + 3)3^{2n+3}}$$
$$\left(1 + \frac{1}{9} + \frac{1}{9^2} + \cdots\right) = \frac{1}{(2n + 3)3^{2n+3}} \cdot \frac{9}{8} = \frac{1}{8(2n + 3)3^{2n+1}}.$$

(If we wish, the last expression is less than $a_n/8$.) Hence, for $n = 4$, $r_n < 1/(8 \cdot 11 \cdot 3^9) < 3^{-13} < 0.6 \times 10^{-6}$ so that we can hope to get five decimal places and quite a good idea of the sixth.

$$a_0 = .3333333^+$$
$$a_1 = .0123457^-$$
$$a_2 = .0008230^+$$
$$a_3 = .0000653^+$$
$$a_4 = .0000056^+$$
$$a_5 = .0000005^+$$
$$\overline{.3465734}$$

Writing **7** places for each term, and noting that the result may be excessive by $\frac{5}{2}$ units in the last place, or in defect by $\frac{1}{2}$, we get $0.3465733 < \frac{1}{2} \ln 2 < 0.3465737$, or, on account of R_n, $.3465733 < \frac{1}{2}\ln 2 < .3465743$, whence $.6931466 < \ln 2 < .6931486$. To five places, then, $\ln 2 = .69315$; to six places, $\ln 2 = .693147$ (very likely).

8.9. Graphs. These graphs are for qualitative use only, being too small and rough for interpolation. On the graphs of the inverse circular and hyperbolic functions, the principal values, if any, correspond to points lying on the heavy branches of the curves.

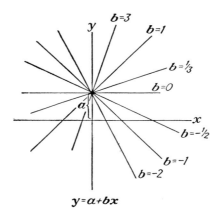

$$y = a + bx$$

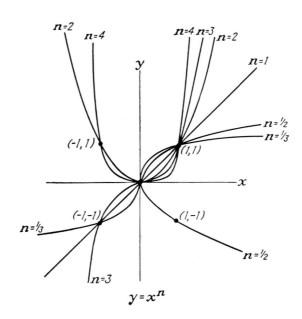

$$y = x^n$$

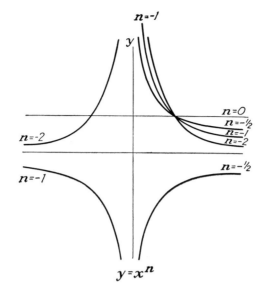

$y = x^n$

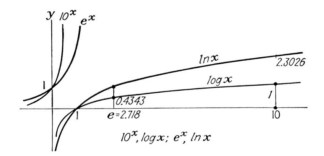

$10^x, \log x; \; e^x, \ln x$

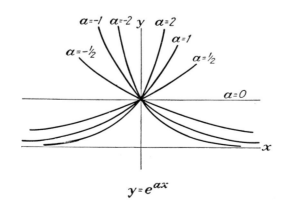

$y = e^{ax}$

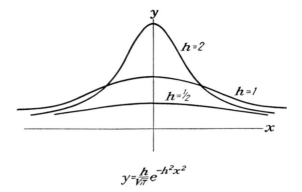

$$y = \frac{h}{\sqrt{\pi}} e^{-h^2 x^2}$$

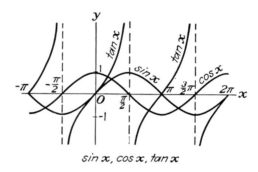

$\sin x, \cos x, \tan x$

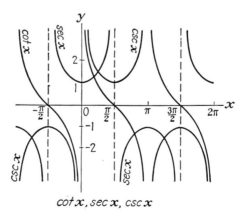

$\cot x, \sec x, \csc x$

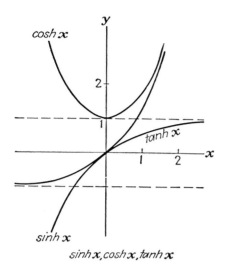

$\sinh x, \cosh x, \tanh x$

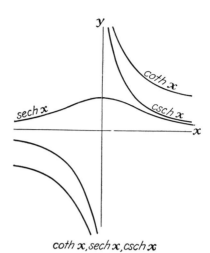

$\coth x, \operatorname{sech} x, \operatorname{csch} x$

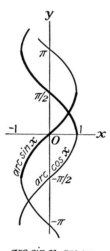

$\arcsin x, \arccos x$

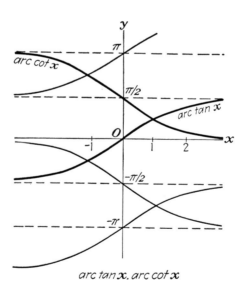

$\arctan x, \operatorname{arc} \cot x$

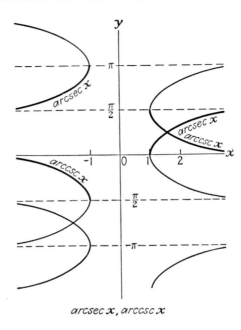

arcsec x, arccsc x

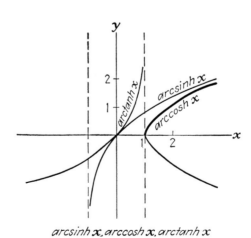

arcsinh x, arccosh x, arctanh x

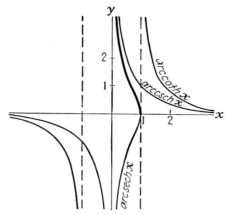

arccoth x, arcsech x, arccsch x

9. INTEGRAL CALCULUS

9.1. Definite Integrals. Let $y = f(x)$ be given for $a \leqslant x \leqslant b$ (interval I). Subdivide I by n points $a = x_0 < x_1 < \cdots < x_{n-1} < x_n = b$ into subintervals I_i: $x_{i-1} \leqslant x \leqslant x_i$, of length $\Delta_i x = x_i - x_{i-1}$. Let ξ_i be any value of x in I_i. Then $f(\xi_i)\Delta_i x$ is the area of a rectangle with base $\Delta_i x$ and height $f(\xi_i)$, and this rectangle is an approximation [if $f(x) \geqslant 0$ for x in I_i] to the area between the curve, I_i, and the ordinates $f(x_i)$ and $f(x_{i+1})$. The sum $\sum = \sum_1^n f(\xi_i)\Delta_i x$ is an approximation [if $f(x) \geqslant 0$ for x in I] to the area bounded by the curve, I, and the ordinates $f(a)$ and $f(b)$.

Let the number n of intervals I_i be increased in such a way that the length $||\Delta x||$ of the *longest* I_i tends to zero. If Σ approaches a limit l which is independent of the particular sequence of subdivisions used and of the choice, at each stage, of the numbers ξ_i in each I_i, then $f(x)$ is said to be *integrable* over the *range* I, the limit l is called the *definite integral* of the *integrand* $f(x)$ from the *lower limit* (of integration) a to the *upper limit* (of integration) b, and is denoted by

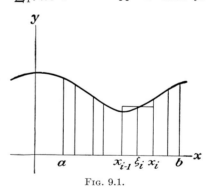

$$\int_a^b f(x)dx = \lim \sum_1^n f(\xi_i)\Delta_i x \text{ as } n \to \infty \text{ and}$$

FIG. 9.1.

$||\Delta x|| \to 0$. $f(x)$ is integrable over I if $f(x)$ is bounded in I and has only a finite number of discontinuities there; in particular, $f(x)$ is integrable over I if it is sectionally continuous over I.

Note that the integral may be written down in terms of any variable whatever:
$$\int_a^b f(x)dx = \int_a^b f(t)dt = \int_a^b f(\theta)d\theta = \int_a^b f(p)dp, \text{ etc.}$$

If $f(x) \geqslant 0$ $[f(x) \leqslant 0]$ over I, then $\int_a^b f(x)dx \geqslant 0$ $(\leqslant 0)$. In the former (latter) case, the *area* bounded by the curve I and the ordinates $f(a)$ and $f(b)$ is $\int_a^b f(x)dx$ $\left\{\int_a^b [-f(x)]dx\right\}$.

Caution: In computing areas by integration, it is highly important to watch for changes in the sign of the integrand; thus the area between the sine curve and the interval $0 \leqslant x \leqslant 2\pi$ is $\int_0^\pi \sin x \, dx + \int_\pi^{2\pi} (-\sin x) dx = 4$, while $\int_0^{2\pi} \sin x \, dx = 0$.

In polar coordinates, the area bounded by a curve and radii $r_1 = f(\theta_1)$ and $r_2 = f(\theta_2)$ is $\frac{1}{2} \int_{\theta_1}^{\theta_2} r^2 \, d\theta$. The limits are to be chosen so that, as θ increases from θ_1 to θ_2, $r = f(\theta)$ passes just once over the required area.

Volume of a Solid with Known Cross Section. Suppose that a solid has the property that the area of every cross section by a plane perpendicular to a line l can be computed (or is known). Taking l as x-axis, express this area A as a function $A(x)$. Suppose that, as x runs from a to b, the solid is covered just once. Then the volume is $\int_a^b A(x) dx$.

Volume of a Solid of Revolution. Suppose that the solid is generated by rotating $y = f(x)$ about the x-axis. Then the cross sections by planes perpendicular to the x-axis have area $A(x) = \pi y^2$, and the volume is $\pi \int_a^b [f(x)]^2 \, dx$.

Area of a Solid of Revolution. The (curved) lateral area of the solid is $2\pi \int_a^b y \sqrt{1 + y'^2} \, dx$, where y' denotes the derivative of y with respect to x (see 8.2).

9.11. Theorems on Definite Integrals.

$$\int_a^a f(x) dx = 0 \qquad\qquad \int_a^b f(x) dx = -\int_b^a f(x) dx$$

$$\int_a^b f(x) dx + \int_b^c f(x) dx = \int_a^c f(x) dx \qquad \int_a^b cf(x) dx = c \int_a^b f(x) dx$$

$$\int_a^b [f(x) + g(x)] dx = \int_a^b f(x) dx + \int_a^b g(x) dx$$

$$\left[\int_a^b f(x) g(x) dx \right]^2 \leqslant \int_a^b [f(x)]^2 \, dx \int_a^b [g(x)]^2 \, dx \text{ (Schwarz)}$$

Mean-value Theorem. The *mean value* $\bar{y} = \overline{f(x)}$ of y between a and b is $\left[\int_a^b f(x) dx \right] / (b - a)$. If $m \leqslant f(x) \leqslant M$ for $a \leqslant x \leqslant b$, then $m \leqslant \bar{y} \leqslant M$. In fact, if f is continuous, $\bar{y} = f(\xi)$ for some ξ, $a < \xi < b$.

Integration by Substitution. If $x = g(t)$, $a = g(c)$, $b = g(d)$, and, for $c \leqslant t \leqslant d$, $g'(t)$ is continuous and does not vanish, then $\int_a^b f(x) dx = \int_c^d f[g(t)] g'(t) dt$.

Integration by Parts. $\int_a^b f(x) g'(x) dx = f(b)g(b) - f(a)g(a) - \int_a^b f'(x) g(x) dx$.

Differentiation of an Integral. 1. If $F(x) = \int_a^b f(x,t) dt$, then $F'(x) = \int_a^b f_x(x,t) dt$, provided that $f_x(x,t)$ (see 10.1) is continuous.

2. If $F(x) = \int_x^b f(t) dt$, then $F'(x) = -f(x)$ at each point of continuity of $f(x)$.

3. If $F(x) = \int_a^x f(t) dt$, then $F'(x) = f(x)$ at each point of continuity of $f(x)$.

4. If $F(x) = \int_{g(x)}^{h(x)} f(x, t) dt$, then

$$F'(x) = \int_{g(x)}^{h(x)} f_x(x, t) dt + f[x, h(x)] h'(x) - f[x, g(x)] g'(x)$$

Fundamental Theorem of Integral Calculus. If $f(x)$ is continuous for $a \leqslant x \leqslant b$, and if $F'(x) = f(x)$, then, for $a \leqslant x \leqslant b$, $F(x) = F(a) + \int_a^x f(t) dt$. If $f(x)$ has a finite number of points of discontinuity but is bounded, apply this theorem to intervals of continuity between successive points of discontinuity.

9.12. Extensions (sometimes called *improper integrals*). 1. *Infinite Range.* If $\int_a^b f(x)dx$ has a limit l as $b \to +\infty$ (or as $a \to -\infty$) we write $l = \int_a^\infty f(x)dx$ (or $\int_{-\infty}^b f(x)dx$); otherwise the latter symbols are meaningless. Further, $\int_{-\infty}^\infty f(x)dx = \int_{-\infty}^a f(x)dx + \int_a^\infty f(x)dx$ for any a, $-\infty < a < \infty$ *provided that each* of the integrals on the right exists. Criteria for the existence of $\lim \int_a^b f(x)dx$ as $b \to \infty$ have been developed similar to those for series.

Caution: For convergence, it is not sufficient that $f(x) \to 0$ as $x \to \infty$.

The easiest general process for determining convergence is comparison with a function of known behavior. Thus, if $c > 0$, $\int_c^\infty x^\alpha\, dx$ converges or diverges according as $\alpha < -1$ or $\alpha \geqq -1$. Hence $\int_c^\infty f(x)dx$ will converge if $|f(x)| \leqq Mx^\alpha$, $\alpha < -1$, and diverge if $|f(x)| \geqq Mx^\alpha$, $\alpha \geqq -1$, in each case for some constant $M > 0$ and all sufficiently large values of x. This test can be applied if a β can be found such that $x^\beta f(x) \to l$ as $x \to \infty$.

2. $f(x) \to +\infty$ (*or* $-\infty$) *as* $x \to a$. Such discontinuities are usually isolated and may be assumed to be at an end point $x = a$ or $x = b$ of the range. If $\int_{a+\epsilon}^b f(x)dx \to l$ (or $\int_a^{b-\epsilon} f(x)dx \to l$) as $\epsilon \to 0$, $\epsilon > 0$, we write $l = \int_a^b f(x)dx$; otherwise this symbol is meaningless. Convergence or divergence can often be determined by comparison with $\int_a^b (x-a)^\alpha\, dx \left[\text{or } \int_a^b (b-x)^\alpha\, dx \right]$, which converge if $\alpha > -1$ and diverge if $\alpha \leqq -1$.

9.2. Indefinite Integrals. The fundamental theorem in 9.11 reduces the computation of $\int_a^b f(x)dx$ to the problem of finding a function $F(x)$ such that $F'(x) = f(x)$ over an interval in which $f(x)$ is continuous. Any such function is called an *antiderivative,* a *primitive function,* or an *indefinite intergal* of $f(x)$ and is denoted by $\int f(x)dx$ with no limits of integration.

Note: If $F(x)$ is an antiderivative of $f(x)$, so is $F(x) + C$ for *any* constant C.

TABLE OF INTEGRALS

1. $\displaystyle\int (du + \cdots + dw) = u + \cdots + w + C$

2. $\displaystyle\int c\, du = c \int du = cu + c$

3. $\displaystyle\int \left(\frac{du}{dv}\right) dv = \int du = u + C$

4. $\displaystyle\int f[g(t)]g'(t)dt = \int f(u)du \qquad \text{if } u = g(t)$

5. $\displaystyle\int u\, dv = uv - \int v\, du$

6. $\displaystyle\int u^n\, du = \frac{u^{n+1}}{n+1} + C \qquad n \neq -1$

7. $\displaystyle\int \frac{du}{u} = \ln|u| + C = \ln|u| + \ln c = \ln|cu|$

8. $\displaystyle\int (a+bu)^n\, du = \frac{(a+bu)^{n+1}}{(n+1)b} + C \qquad n \neq -1$

9. $\int \dfrac{du}{a + bu} = \dfrac{1}{b} \ln |a + bu| + C$

10. $\int \dfrac{du}{a + bu^2} = \dfrac{1}{\sqrt{ab}} \arctan \sqrt{\dfrac{b}{a}}\, u + C \qquad ab > 0$

11. $\int \dfrac{du}{a + bu^2} = \dfrac{1}{\sqrt{-ab}} \operatorname{arc\,tanh} \sqrt{-\dfrac{b}{a}}\, u + C \qquad ab < 0$

$\qquad = \dfrac{1}{2\sqrt{-ab}} \ln \dfrac{a + u\sqrt{-ab}}{a - u\sqrt{-ab}} + C \qquad ab < 0$

12. $\int \sqrt{a + bu^2}\, du = \dfrac{u}{2} \sqrt{a + bu^2} + \dfrac{a}{2\sqrt{b}} \ln (u\sqrt{b} + \sqrt{a + bu^2}) + C \qquad b > 0$

13. $\int \sqrt{a + bu^2}\, du = \dfrac{u}{2} \sqrt{a + bu^2} + \dfrac{a}{2\sqrt{-b}} \arcsin \sqrt{-\dfrac{b}{a}}\, u + C \qquad b < 0$

14. $\int \dfrac{du}{\sqrt{a + bu^2}} = \dfrac{1}{\sqrt{b}} \ln (u\sqrt{b} + \sqrt{a + bu^2}) + C \qquad b > 0$

15. $\int \dfrac{du}{\sqrt{a + bu^2}} = \dfrac{1}{\sqrt{-b}} \arcsin \sqrt{-\dfrac{b}{a}}\, u + C \qquad b < 0$

16. $\int \dfrac{\sqrt{a + bu^2}}{u}\, du = \sqrt{a + bu^2} + \sqrt{a} \ln \dfrac{\sqrt{a + bu^2} - \sqrt{a}}{u} + C \qquad a > 0$

17. $\int \dfrac{\sqrt{a + bu^2}}{u}\, du = \sqrt{a + bu^2} - \sqrt{-a} \arctan \dfrac{\sqrt{a + bu^2}}{\sqrt{-a}} + C \qquad a < 0$

18. $\int \dfrac{du}{u\sqrt{a + bu^2}} = \dfrac{1}{\sqrt{a}} \ln \dfrac{\sqrt{a + bu^2} - \sqrt{a}}{u} + C \qquad a > 0$

19. $\int \dfrac{du}{u\sqrt{a + bu^2}} = \dfrac{1}{\sqrt{-a}} \operatorname{arc\,sec} \sqrt{-\dfrac{b}{a}}\, u + C \qquad a < 0$

20. $\int a^u\, du = \dfrac{a^u}{\ln a} + C$

21. $\int e^u\, du = e^u + C$

22. $\int \ln u\, du = u \ln u - u + C$

23. $\int \sin u\, du = -\cos u + C$

24. $\int \cos u\, du = \sin u + C$

25. $\int \tan u\, du = -\ln |\cos u| + C = \ln |\sec u| + C$

26. $\int \cot u\, du = \ln |\sin u| + C$

27. $\int \sec u\, du = \ln |\sec u + \tan u| + C$

28. $\int \csc u\, du = \ln |\csc u - \cot u| + C$

29. $\int \arcsin u\, du = u \arcsin u + \sqrt{1 - u^2} + C$

30. $\int \arctan u\, du = u \arctan u - \tfrac{1}{2} \ln (1 + u^2) + C$

31. $\int \operatorname{arc\,sec} u\, du = u \operatorname{arc\,sec} u - \ln |u + \sqrt{u^2 - 1}| + C$

32. $\displaystyle\int \sin^2 u \, du = \frac{u}{2} - \frac{1}{4} \sin 2u + C$

33. $\displaystyle\int \cos^2 u \, du = \frac{u}{2} + \frac{1}{4} \sin 2u + C$

34. $\displaystyle\int \sec^2 u \, du = \tan u + C$

35. $\displaystyle\int \sec^3 u \, du = \frac{1}{2}(\sec u \tan u + \ln |\sec u + \tan u|) + C$

36. $\displaystyle\int \csc^2 u \, du = -\cot u + C$

37. $\displaystyle\int \sec u \tan u \, du = \sec u + C$

38. $\displaystyle\int \csc u \cot u \, du = -\csc u + C$

39. $\displaystyle\int \frac{du}{1 + \cos u} = \tan \frac{u}{2} + C$

40. $\displaystyle\int \frac{du}{a + b \cos u} = \frac{2}{\sqrt{a^2 - b^2}} \arctan \frac{\sqrt{a^2 - b^2} \tan \dfrac{u}{2}}{a + b} + C \qquad b^2 < a^2$

41. $\displaystyle\int \frac{du}{a + b \cos u} = \frac{1}{\sqrt{b^2 - a^2}} \ln \frac{\sqrt{b^2 - a^2} \tan (u/2) + a + b}{\sqrt{b^2 - a^2} \tan (u/2) - a - b} + C \qquad a^2 < b^2$

42. $\displaystyle\int \frac{du}{a^2 \cos^2 u + b^2 \sin^2 u} = \frac{1}{ab} \arctan \left(\frac{b \tan u}{a}\right) + C$

43. $\displaystyle\int \sinh u \, du = \cosh u + C$

44. $\displaystyle\int \cosh u \, du = \sinh u + C$

45. $\displaystyle\int \tanh u \, du = \ln \cosh u + C$

46. $\displaystyle\int \coth u \, du = \ln \sinh u + C$

47. $\displaystyle\int \operatorname{sech} u \, du = 2 \arctan e^u + C$

48. $\displaystyle\int \operatorname{csch} u \, du = \ln \tanh \frac{u}{2} + C$

49. $\displaystyle\int \operatorname{arc\,sinh} u \, du = u \operatorname{arc\,sinh} u - \sqrt{1 + u^2} + C$

50. $\displaystyle\int \operatorname{arc\,tanh} u \, du = u \operatorname{arc\,tanh} u + \frac{1}{2} \ln(1 - u^2) + C$

51. $\displaystyle\int \sinh^2 u \, du = \frac{1}{4} \sinh 2u - \frac{1}{2}u + C$

52. $\displaystyle\int \cosh^2 u \, du = \frac{1}{4} \sinh 2u + \frac{1}{2}u + C$

53. $\displaystyle\int \operatorname{sech}^2 u \, du = \tanh u + C$

54. $\displaystyle\int \operatorname{csch}^2 u \, du = \coth u + C$

55. $\displaystyle\int e^{au} \sin bu \, du = \frac{e^{au}(a \sin bu - b \cos bu)}{a^2 + b^2} + C$

56. $\displaystyle\int e^{au} \cos bu \, du = \frac{e^{au}(b \sin bu + a \cos bu)}{a^2 + b^2} + C$

57. $\displaystyle\int_0^\pi \sin au \cos bu\, du = \int_0^\pi \cos au \cos bu\, du$

$$= \int_0^\pi \sin au \sin bu\, du = 0 \qquad \text{unless } a = b$$

58. $\displaystyle\int_0^\pi \sin^2 au\, du = \int_0^\pi \cos^2 au\, du = \frac{\pi}{2}$

59. $\displaystyle\int_0^\infty \frac{\sin au}{u}\, du = \frac{\pi}{2},\ 0,\ -\frac{\pi}{2} \qquad \text{according as } a > 0,\ a = 0,\ \text{or } a < 0$

60. $\displaystyle\int_0^\infty e^{-a^2u^2}\, du = \frac{\sqrt{\pi}}{2a}$

Reduction Formulas

Many integrals can be reduced to the foregoing by one or more integrations by parts (No. 5).

Example. Compare the first of the following formulas, with $p = 2$, $q = 1$, $r = 7$:

$$\int u^2(a+bu)^7\, du = \frac{u^2(a+bu)^8}{8b} - \frac{1}{4b}\int u(a+bu)^8\, du$$

$$= \frac{u^2(a+bu)^8}{8b} - \frac{u(a+bu)^9}{36b^2} + \frac{1}{36b^2}\int (a+bu)^9\, du$$

$$= \frac{(a+bu)^8}{4b}\left(\frac{u^2}{2} - \frac{u(a+bu)}{9b} + \frac{(a+bu)^2}{90b^2}\right) + C$$

Applied to a general expression, one integration by parts gives a *reduction formula*, which can be used to step the exponents up or down until one of the preceding formulas applies.

Typical Reduction Formulas. Note that a formula loses meaning if one of the denominators vanishes; but then some other formula applies. In No. 68, if $p + q = 0$, the integrand is $\int \tan^q u\, du$ or $\int \cot^q u\, du$ and responds to 69 or its analogue for the cotangent.

61. $\displaystyle\int u^p(a+bu^q)^r\, du = \frac{u^{p-q+1}(a+bu^q)^{r+1}}{bq(r+1)} - \frac{p-q+1}{bq(r+1)}\int u^{p-q}(a+bu^q)^{r+1}\, du$

62. $\displaystyle\int u^p(a+bu^q)^r\, du = \frac{u^{p+1}(a+bu^q)^r}{p+1} - \frac{rqb}{p+1}\int u^{p+q}(a+bu^q)^{r-1}\, du$

63. $\displaystyle\int \frac{du}{(a+bu^2)^p} = \frac{1}{2a(p-1)}\frac{u}{(a+bu^2)^{p-1}} + \frac{2p-3}{2a(p-1)}\int \frac{du}{(a+bu^2)^{p-1}}$

64. $\displaystyle\int u^n a^{bu}\, du = \frac{u^n a^{bu}}{b\ln a} - \frac{n}{b\ln a}\int u^{n-1}a^{bu}\, du$

65. $\displaystyle\int u^n(\ln u)^p\, du = \frac{u^{n+1}(\ln u)^p}{n+1} - \frac{p}{n+1}\int u^n(\ln u)^{p-1}\, du$

66. $\displaystyle\int \sin^n u\, du = -\frac{\sin^{n-1}u\cos u}{n} + \frac{n-1}{n}\int \sin^{n-2}u\, du$

67. $\displaystyle\int \frac{du}{\sin^n u} = -\frac{\cos u}{(n-1)\sin^{n-1}u} + \frac{n-2}{n-1}\int \frac{du}{\sin^{n-2}u}$

68. $\displaystyle\int \cos^p u \sin^q u\, du = \frac{\cos^{p-1}u\sin^{q+1}u}{p+q} + \frac{p-1}{p+q}\int \cos^{p-2}u\sin^q u\, du$

69. $\displaystyle\int \tan^n u\, du = \frac{\tan^{n-1}u}{n-1} - \int \tan^{n-2}u\, du$

70. $\displaystyle\int u^n \cos au\, du = \frac{u^n \sin au}{a} - \frac{n}{a}\int u^{n-1}\sin au\, du$

Tactical Notes. 1. Integration can *always* be checked: $d[\int f(x)dx]/dx = f(x)$.

2. If none of these formulas is immediately applicable, try a change of form or a substitution; in trigonometric cases, an identity may help. Do not forget, when putting $x = g(t)$, also to put $dx = g'(t)dt$.

3. If $f(x)$ involves $Ax^2 + Bx + C$, complete the square (see 6.31) to get $a + bu^2$. Then use 10 to 19, perhaps after first using 61 to 63. Or, possibly with an adjustment of sign of the whole integrand, $a + bu^2$ may be regarded, according to the signs of a and b, as one of $\alpha^2 + \beta^2 u^2$, $\alpha^2 - \beta^2 u^2$, $\beta^2 u^2 - \alpha^2$; in the respective cases, try the substitutions $\beta u = \alpha \tan \theta$, $\alpha \sin \theta$, $\alpha \sec \theta$, with $\alpha > 0$ and $\beta > 0$.

4. If $f(x) = P(x)/Q(x)$, with P and Q polynomials (see 4.8), first divide until the degree of the numerator $P^*(x)$ is *less* than the degree of $Q(x)$. Then find partial fractions for $P^*(x)/Q(x)$ (see 4.9).

5. The integral table of B. O. Pierce (Ginn) will be found adequate for almost all work. Libraries often have the massive tables of Bierens de Haan.

9.3. Approximations for $\displaystyle\int_a^b f(x)dx$. In graphical methods, plot $y = f(x)$ and locate the area A associated with the integral.

1. Count the squares of A. Very rough.

2. Use material of known weight per unit area; cut out A and weigh it. Rough without thoroughly homogeneous material and good balances.

3. Use a planimeter or integraph; directions are given with each instrument.

4. Compute an approximating sum, as in the definition, first drawing in the upper horizontal sides of the approximating rectangles so as to exclude from each as nearly as possible as much of what belongs as is included of what does not belong.

5. Compute the areas $[f(x_{i+1}) + f(x_i)]\Delta_i x/2$ of the trapezoids $[x_i, f(x_i)]$, $(x_i, 0)$, $(x_{i+1}, 0)$, $[x_{i+1}, f(x_{i+1})]$ and add; simplify the computation by using equal subdivisions if possible.

6. *Simpson's Rule.* Use an *even* number of *equal* subdivisions: x_0, $x_0 + \Delta x$, $x_0 + 2\,\Delta x$, $\cdots$, $x_0 + 2m\,\Delta x$. Then the approximation is $\Delta x/3$ times the sum $f(x_0) + f(x_{2m}) + 2[f(x_2) + f(x_4) + \cdots + f(x_{2m-2})] + 4[f(x_1) + f(x_3) + \cdots + f(x_{2m-1})]$. The error is numerically not greater than $M(b - a)(\Delta x)^4/180$, where M is the numerically largest value of $f^{(4)}(x)$ for $a \leqslant x \leqslant b$. If $2m + 1$ subdivisions are given, use the rule for $2m$ of them, and approximate the extra area by 4 or 5.

7. Subdivide the range by introducing points $a = x_0 < x_1 < \cdots < x_n = b$. Starting at $P_0(a,0)$, draw a line with slope $f(a)$ cutting $x = x_1$ in $P_1(x_1, \eta_1)$; through P_1 draw a line with slope $f(x_1)$ cutting $x = x_2$ in $P_2(x_2, \eta_2)$; $\cdots$; draw through P_{n-1} a line with slope $f(x_{n-1})$, cutting $x = b$ in $P_n(b, \eta_n)$. Then η_n is an approximation to $\displaystyle\int_a^b f(x)dx$ which may be improved by increasing n, and often also by using the number $f[(x_i + x_{i+1})/2]$ instead of $f(x_i)$ for the slope of the line leaving P_i. Note that the curve drawn is an approximate graph of $F(x) = \displaystyle\int_a^x f(x)dx$.

8. Expand $f(x)$ in series, if possible, and integrate term by term.

9. Replace $f(x)$ by an approximation $\phi(x)$ good for x between a and b and such that $\phi(x)$ can be integrated.

9.4. Fourier Series. A series $T(x) = \frac{1}{2}a_0 + \sum_1^\infty (a_n \cos nx + b_n \sin nx)$ is called a *trigonometric series*. $T(x)$ is the *Fourier series* $F(x)$ of $f(x)$ in the interval $-\pi < x < \pi$, and we write $f(x) \sim F(x)$, if $\pi a_n = \displaystyle\int_{-\pi}^\pi f(x) \cos nx\, dx$ $(n = 0,1,2, \cdots)$ and $\pi b_n = \displaystyle\int_{-\pi}^\pi f(x) \sin nx\, dx$ $(n = 1,2,3, \cdots)$. If any $T(x)$ represents $f(x)$, its coefficients must be those of $F(x)$; proof is based on 57 and 58 of the Integral Tables (page 196).

Theorems. If $f(x)$ and $f'(x)$ are both sectionally continuous in $-\pi \leqslant x \leqslant \pi$, then $F(x) = [f(x+) + f(x-)]/2$; in particular, $F(x) = f(x)$ at every point where $f(x)$ is continuous.

If $f(x)$ is continuous and $f'(x)$ is sectionally continuous in $-\pi \leqslant x \leqslant \pi$, then the convergence of $F(x)$ is uniform and absolute, and wherever $f''(x)$ exists, $f'(x) = F'(x)$ (differentiated term by term).

If $f(x)$ is sectionally continuous in $-\pi \leqslant x \leqslant \pi$, and $f(x) \sim F(x)$, then $\int_{-\pi}^{x} f(t)dt = \int_{-\pi}^{x} F(t)dt$ (integrated term by term).

REMARKS: 1. If $f(-x) = f(x)$, $b_n = 0$ for all n, and $F(x)$ is a "cosine series"; if $f(-x) = -f(x)$, $a_n = 0$ for all n, and $F(x)$ is a "sine series."

2. If $f(x)$ is to be treated over $a \leqslant x \leqslant b$, put $f(x) = \phi(t)$, where $x = (b-a)t/(2\pi) + (b-a)/2$, and treat $\phi(t)$ for $-\pi < t < \pi$.

3. If $f'(-\pi+)$ and $f'(\pi-)$ both exist (see 8.13), then $F(\pi) = F(-\pi) = [f(-\pi+) + f(\pi-)]/2$.

4. $F(x + 2\pi) = F(x)$, so that the sum of a Fourier series is *periodic*, with period 2π. Hence $f(x) = F(x)$ outside the interval $-\pi < x < \pi$ implies that $f(x)$ is periodic.

5. The coefficients may be approximated (see 9.3). An elaborate machine (harmonic analyzer) is available for finding them.

6. $F_n(x)$ is often treated as an approximation to $f(x)$ "in the mean," i.e., the "error" is taken to be the mean error $\int_{-\pi}^{\pi} |f(x) - F_n(x)|^2 \, dx = \int_{-\pi}^{\pi} |f(x)|^2 \, dx - \pi[\tfrac{1}{2}a_0^2 + \sum_1^n (a_k^2 + b_k^2)]$, where $F_n(x) = \tfrac{1}{2}a_0 + \sum_1^n (a_k \cos kx + b_k \sin kx)$. When this type of approximation is used, convergence as $n \to \infty$ for particular values of x is of only secondary interest. As soon as an n has been found that makes the mean error satisfactorily small, $F_n(x)$ is used as an approximation to $f(x)$; the relative behavior of $F_n(x)$ and $f(x)$ may then be quite irregular at particular points.

Example. To find $F(x)$ for $f(x) = x + x^2$ in $-1 < x < 1$. Put $x = t/\pi$, $\phi(t) = f(t/\pi) = t/\pi + t^2/\pi^2$ and seek $\Phi(t)$ for $\phi(t)$ in $-\pi < t < \pi$.

$$a_0 = \frac{1}{\pi} \int_{-\pi}^{\pi} \left(\frac{t}{\pi} + \frac{t^2}{\pi^2} \right) dt = \frac{2}{3}.$$

If $n > 0$

$$a_n = \frac{1}{\pi^2} \int_{-\pi}^{\pi} \left(t + \frac{t^2}{\pi} \right) \cos nt \, dt = \frac{4 \cos n\pi}{\pi^2 n^2} = \frac{4(-1)^n}{\pi^2 n^2}$$

$$b_n = \frac{1}{\pi^2} \int_{-\pi}^{\pi} \left(t + \frac{t^2}{\pi} \right) \sin nt \, dt = \frac{2(-1)^{n+1}}{\pi n}$$

$$\Phi(t) = \frac{1}{3} + \sum_1^{\infty} \left[\frac{4(-1)^n}{\pi^2 n^2} \cos nt + \frac{2(-1)^{n+1}}{\pi n} \sin nt \right]$$

$$F(x) = \frac{1}{3} + \frac{2}{\pi} \sum_1^{\infty} \frac{(-1)^n}{n} \left(\frac{2}{\pi n} \cos n\pi x - \sin n\pi x \right)$$

Special series for $-\pi < x < \pi$:

$$x = 2 \left(\sin x - \frac{\sin 2x}{2} + \frac{\sin 3x}{3} - \frac{\sin 4x}{4} + - \cdots \right)$$

$$x^2 = \frac{\pi^2}{3} - 4 \left(\cos x - \frac{\cos 2x}{2^2} + \frac{\cos 3x}{3^2} - + \cdots \right)$$

The following series are valid in $0 < x < \pi$:

$$1 = \frac{4}{\pi} \left(\sin x + \frac{\sin 3x}{3} + \frac{\sin 5x}{5} + \cdots \right)$$

$$x = \frac{\pi}{2} - \frac{4}{\pi} \left(\cos x + \frac{\cos 3x}{3^2} + \frac{\cos 5x}{5^2} + \cdots \right)$$

$$x^2 = \frac{2}{\pi}\left[(\pi^2 - 4)\sin x - \pi^2 \frac{\sin 2x}{2} + \left(\pi^2 - \frac{4}{3^2}\right)\frac{\sin 3x}{3} - \pi^2 \frac{\sin 4x}{4}\right.$$

$$\left. + \left(\pi^2 - \frac{4}{5^2}\right)\frac{\sin 5x}{5} - + \cdots\right]$$

Note that the right member gives the left member with sign changed in $-\pi < x < 0$.

9.41. Fourier Integral. If $f(x)$ is not periodic, then it can be represented by a Fourier series only inside some chosen interval. It can be represented for all x by a *Fourier integral*, by virtue of the theorem: If $f(x)$ and $f'(x)$ are sectionally continuous in every finite interval, and $\int_{-\infty}^{\infty} |f(x)|dx$ exists (converges), then, for $-\infty < x < +\infty$

$$\frac{1}{2}[f(x +) + f(x -)] = \frac{1}{\pi}\int_0^\infty \left\{\int_{-\infty}^\infty f(u)\cos[t(u - x)]du\right\}dt$$

10. FUNCTIONS OF SEVERAL VARIABLES

10.1. Differentiation. If $z = F(x,y)$ is the equation of a surface S, and a value b is assigned to y, then $z = F(x,b) = g(x)$. $g'(x)$ is called the *partial derivative* of z

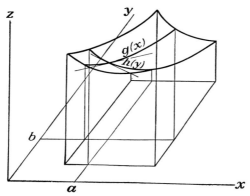

Fɪɢ. 10.1.

with respect to x and is denoted by $\partial z/\partial x$, $\partial F/\partial x$, z_x, F_x (or $F_x{}'$). At (a,b) $z_x = F_x(a,b)$ $= \lim [F(a + h,b) - F(a,b)]/h$ as $h \to 0$ measures the rate of increase of z per unit increase in x, when y has the value b, as x increases through the value $x = a$; also z_x is the slope at $x = a$ of the curve $z = g(x)$ in the plane $y = b$. $\partial z/\partial y$ is defined, denoted, and interpreted similarly. If $w = f(x,y,z)$, and fixed values b and c are assigned to y and z, $w = f(x,b,c) = g(x)$, and $g'(x)$ is the partial derivative $w_x = \partial w/\partial x$ (etc.) of w with respect to x. w_x measures the rate of increase of w per unit increase in x, when $y = b$ and $z = c$, as x increases through the value $x = a$, but the geometric interpretation would require four dimensions.

Example. If z measures kilometers above the surface of the earth, the adiabatic *lapse rate* of temperature $T°C$ for moist unsaturated air is $\partial T/\partial z = -9.8$, which says that T decreases 9.8°C per km of climb.

To compute partial derivatives, treat as *constants* all except the variable of differentiation. If $w = x^2y^3 - y\sin xz^2$, $w_x = 2xy^3 - z^2y\cos xz^2$, $w_y = 3x^2y^2 - \sin xz^2$, $w_z = -2xyz\cos xz^2$. Higher partial derivatives are simply partial derivatives of partial derivatives. In the example, $\partial^2 w/\partial x\,\partial y = (w_x)_y = w_{xy} = 6xy^2 - z^2\cos xz^2$ and $\partial^2 w/\partial y\,\partial x = (w_y)_x = w_{yx} = 6xy^2 - z^2\cos xz^2$. Note that $w_{xy} = w_{yx}$. Whenever (as usually happens in practice) a repeated partial derivative is continuous, regardless of the order in which the differentiations are performed, the result is the same regardless of the order of differentiation; the differentiations may then be carried out in any order.

The plane tangent to S at (a,b,c) is

$$\text{for } f(x,y,z) = k \qquad f_x(x - a) + f_y(y - b) + f_z(z - c) = 0$$
$$\text{for } z = F(x,y) \qquad F_x(x - a) + F_y(y - b) - (z - c) = 0$$

Accordingly, the normal line has direction numbers f_x, f_y, f_z, or $F_x, F_y, -1$; the derivatives are evaluated at (a,b,c). Notation occasionally met for $z = F(x,y)$: $p = z_x$, $q = z_y$, $r = z_{xx}$, $s = z_{xy}(= z_{yx})$, $t = z_{yy}$. N.B. The surfaces $f(x,y,z) = k$ [or the curves $F(x,y) = k$ in the xy-plane] are called *level surfaces* of f (or *level curves* or *contour lines* of F); in either case, the curves that are perpendicular (*orthogonal trajectories*) to the level surfaces (or curves) are called *streamlines* or *lines of flow*. Thus a streamline is tangent, at each of its points, to the line normal to the level surface (or curve in the plane) passing through the point.

When $z = F(x,y)$, differentials $dx = x - a$, $dy = y - b$, $dz = z - c$ are defined at each point (a,b,c) in such a way that $P(a + dx, b + dy, c + dz)$ lies in the plane tangent to $z = f(x,y)$ at $[a,b,f(a,b)]$; thus $dz = F_x \, dx + F_y \, dy$, and this expression is called the *total differential* of z. Similarly, $w = f(x,y,z)$ has total differential $dw = f_x \, dx + f_y \, dy + f_z \, dz$; although the full geometric interpretation needs four dimensions, we note that $dw = 0$ for P on the plane tangent at (a,b,c) to $f(x,y,z) = f(a,b,c)$. If increments Δx, $\cdots$ are given to the independent variables, then $f(x, \cdots)$ suffers an increment Δf. By setting $dx = \Delta x$, $\cdots$, a value for df is obtained that is a useful approximation to Δf(see 8.3).

As (x,y,z) moves through (a,b,c) in a direction whose cosines (see 7.62) are l, m, n, the rate of increase of $f(x,y,z)$ per unit increase in distance s measured in this direction is $df/ds = lf_x + mf_y + nf_z$; this is the *directional derivative* of f in the direction l, m, n. In the plane case, $dF/ds = lF_x + mF_y = F_x \cos \phi + F_y \sin \phi$, where the direction l, m makes an angle ϕ with the positive x-axis. More generally, if x, y, and z are functions of a parameter u, so that (x,y,z) moves along a curve (see 7.4, 8.2), then $df/du = \dot{f} = f_x \dot{x} + f_y \dot{y} + f_z \dot{z}$; similarly in the plane.

If x, y, z depend on u and v, then $f(x,y,z)$ is a function of u and v; $f_u = f_x x_u + f_y y_u + f_z z_u$; similarly for f_v.

In case there are many possible independent variables, it may be desirable to indicate explicitly in the notation by a subscript those which are held constant. In thermodynamics, $p = -(\partial U/\partial V)_S$ thus means that the pressure p is the negative of the partial derivative of the internal energy U with respect to the volume V, entropy S being held constant.

Again, in two total differentials of the same function with respect to the same variables, coefficients of the differentials of the same variables are equal: $dF = A \, dx + B \, dy = C \, dx + D \, dy$ implies $A = C = F_x$ and $B = D = F_y$. Using the thermodynamic variables above, along with (absolute) temperature T, $dS = (\partial S/\partial U)_V \, dU + (\partial S/\partial V)_U \, dV = (dU + p \, dV)/T$ thus implies that $(\partial S/\partial U)_V = 1/T$ and $(\partial S/\partial V)_U = p/T$. For a full treatment of thermodynamic variables, see ref. 10.

10.11. Implicit Functions. $f(x,y,z) = k$ defines z *implicitly* as a function of x and y, and the equation may often be solved to get $z = F(x,y)$. In fact, if $f(a,b,c) = k$ and $f_z = (a,b,c) \neq 0$, and f and its derivatives are continuous, then there is a certain region R of the xy-plane containing (a,b) such that $z = F(x,y)$ is a single-valued function of x and y for (x,y) in R and $c = F(a,b)$. The curves on the surface $f(x,y,z) = k$ along which $f_z = 0$ often decompose the surface into pieces, each of which can be represented in the form $z = F(x,y)$. Thus $x^2 + y^2 + z^2 = k$, $k > 0$ represents a sphere, and $f_z = 2z = 0$ on the xy-plane. The hemisphere above (or below) that plane is given by $z = \sqrt{k - (x^2 + y^2)}$ [or $z = -\sqrt{k - (x^2 + y^2)}$].

To compute z_x (or z_y) from $f(x,y,z) = k$ without first solving for z, differentiate each side of the equation with respect to x (or y), treating z as a function of x wherever it appears: $f_x + f_z z_x = 0$, whence $z_x = -f_x/f_z$ (or $z_y = -f_y/f_z$).

Precisely similar remarks apply to curves given by $F(x,y) = k$ in the plane. Also, $F_x + F_y y' = 0$, whence $y' = -F_x/F_y$ (see 8.22). Continue to get higher derivatives. Thus $f_{xx} + 2f_{xz}z_x + f_{zz}(z_x)^2 + f_z z_{xx} = 0$ permits the computation of z_{xx} (wherever $f_z \neq 0$), etc.

10.12. Change of Coordinates. Suppose that $u = \xi(x,y)$ and $v = \eta(x,y)$ are two functions with continuous derivatives such that for each (x,y) in a region R of the xy-plane there is one and only one point (u,v) in a region R' of the uv-plane and that the *Jacobian* or *functional determinant* (see 4.7)

$$J = \frac{\partial(u,v)}{\partial(x,y)} = \begin{vmatrix} u_x & u_y \\ v_x & v_y \end{vmatrix}$$

does not vanish in R. Then (u,v) may be used as coordinates for the region R. The equations connecting (x,y) and (u,v) are said to define a *transformation of coordinates*.

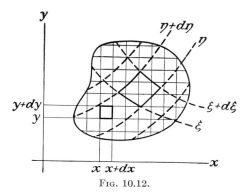

FIG. 10.12.

Under similar conditions, $u = \xi(x,y,z)$, $v = \eta(x,y,z)$, $w = \zeta(x,y,z)$ introduce coordinates (u,v,w) ranging over a region V' in uvw-space for a region V of xyz-space

$$J = \frac{\partial(u,v,w)}{\partial(x,y,z)} = \begin{vmatrix} u_x & u_y & u_z \\ v_x & v_y & v_z \\ w_x & w_y & w_z \end{vmatrix}.$$

The *inverse* transformation replaces coordinates (u,v) by coordinates (x,y) [or (u,v,w) by (x,y,z)]. The Jacobian of the inverse transformation is the reciprocal of the Jacobian of the transformation. Also, $\dfrac{\partial(x,y)}{\partial(u,v)} \cdot \dfrac{\partial(u,v)}{\partial(\lambda,\mu)} = \dfrac{\partial(x,y)}{\partial(\lambda,\mu)}$; similarly for three variables. For coordinates previously mentioned (see 7.61)

$$\frac{\partial(x,y)}{\partial(r,\theta)} = r \qquad \frac{\partial(x,y,z)}{\partial(\rho,\theta,z)} = \rho \qquad \frac{\partial(x,y,z)}{\partial(r,\theta,\phi)} = r^2 \cos \varphi$$

To find the expression for a function after a change of coordinates, substitute from the equations of transformation: $f(u,v) = f[\xi(x,y), \eta(x,y)] = F(x,y)$, etc. To find derivatives, use the rules outlined above: $\partial f/\partial u = F_x x_u + F_y y_u$, etc.

REMARK: In order that there may be a *functional relation* $\phi(\xi,\eta) = 0$ between $\xi(x,y)$ and $\eta(x,y)$, it is necessary and sufficient that $\partial(\xi,\eta)/\partial(x,y)$ vanish *identically*. Thus $\xi = x - y$ and $\eta = e^{x-y}$ have Jacobian zero and are connected by the relation $\eta = e^\xi$.

The foregoing applies with purely formal changes to k functions of k variables.

10.13. Mean-value Theorem. If $f(x,y,z)$ has derivatives throughout the box $a < x < a + h,\ b < y < b + k,\ c < z < c + l$, then there is a number θ, $0 < \theta < 1$, such that $\Delta f = f(a + h, b + k, c + l) - f(a,b,c) = hf_x + kf_y + lf_z$, with the derivatives evaluated at $(a + \theta h, b + \theta k, c + \theta l)$. Similarly for $F(x,y)$ in the plane.

10.14. Taylor's Theorem. Under conditions similar to those of 8.71, $f(a + h, b + k, c + l) = f(a,b,c) + (hf_x + kf_y + lf_z) + Q(h,k,l) + \cdots + R_n$, where $2Q(h,k,l) = h^2 f_{xx} + k^2 f_{yy} + l^2 f_{zz} + 2hk f_{xy} + 2hl f_{xz} + 2kl f_{yz}$ and the derivatives are evaluated at (a,b,c). Similarly for $F(x,y)$ in the plane.

10.15. Extreme Values. Definitions similar to those in 8.4. f must be stationary: $f_x = f_y = f_z = 0$. The solutions of these equations are points (a,b,c) where f *may* have an extreme value. h,k,l can be taken so small that the sign of Δf is the sign of $Q(h,k,l)$. For $f(a,b,c)$ to be a *minimum*, Q must be *positive* for *all* h, k, l not all zero. Then the *quadratic form* Q is said to be *positive definite*. Necessary and sufficient conditions that Q be positive definite are that

$$\begin{vmatrix} f_{xx} & f_{xy} & f_{xz} \\ f_{yx} & f_{yy} & f_{yz} \\ f_{zx} & f_{zy} & f_{zz} \end{vmatrix} > 0, \qquad \begin{vmatrix} f_{xx} & f_{xy} \\ f_{yx} & f_{yy} \end{vmatrix} > 0, \text{ and } f_{xx} > 0.$$

(For determinants, see 4.7.) To find maxima, seek minima of $-f$, so that $-Q$ must be positive definite.

For $F(x,y)$ to be a minimum, $F_x = F_y = 0$ and $Q = h^2 F_{xx} + 2hk F_{xy} + k^2 F_{yy}$ must be positive definite, for which the necessary and sufficient conditions are that $F_{xx}F_{yy} - (F_{xy})^2 > 0$ and $F_{xx} > 0$. Maximum of F is minimum of $-F$.

10.21. Curves and Surfaces. A *curve* C may be given by equations $f_1(x,y,z) = k_1$, $f_2(x,y,z) = k_2$ (intersection of two surfaces). The tangent line then has direction numbers $f_{1y}f_{2z} - f_{1z}f_{2y}, f_{1z}f_{2x} - f_{1x}f_{2z}, f_{1x}f_{2y} - f_{1y}f_{2x}$. It is usually preferable to obtain a parametric representation $x = \xi(u)$, $y = \eta(u)$, $z = \zeta(u)$, in which u ranges over a region of a u-(parameter) axis (see 7.4). For curves in the xy-plane, $\zeta(u) = 0$ identically; other representations: $F(x,y) = k$, $y = f(x)$.

If s denotes length measured along C from some fixed point, then, in x, y, z- or r, θ, ϕ,- or ρ, ϕ, z-coordinates, respectively

$$ds^2 = dx^2 + dy^2 + dz^2$$
$$ds^2 = dr^2 + r^2 \cos^2 \varphi d\theta^2 + r^2 d\varphi^2$$
$$ds^2 = d\rho^2 + \rho^2 d\varphi^2 + dz^2$$

If the independent variable is u, divide by du^2 to get $(ds/du)^2$. Special cases in the plane:

$$\left(\frac{ds}{dx}\right)^2 = 1 + \left(\frac{dy}{dx}\right)^2 \qquad \left(\frac{ds}{d\theta}\right)^2 = r^2 + \left(\frac{dr}{d\theta}\right)^2$$

Use the positive or negative square root according as s increases or decreases as u increases. Then the arc length from the point u_0 to the point u is $s = \int_{u_0}^u (ds/du)du$.

When s is used as parameter, in cartesian coordinates, $1 = \dot{x}^2 + \dot{y}^2 + \dot{z}^2$, and $\dot{x}$, $\dot{y}$, and $\dot{z}$ may be interpreted as the direction cosines of the line T tangent at P to the curve. It can be shown that $\ddot{x}$, $\ddot{y}$, and $\ddot{z}$ are direction numbers of a line N perpendicular to T at P. N is called the *principal normal*. The number $\kappa = (\ddot{x}^2 + \ddot{y}^2$

$+\ ^{-2})^{1/2}$ is called the *curvature* of C at P and $\rho = 1/\kappa$ is called the *radius of curvature*. A circle of radius ρ in the plane of T and N (*osculating plane*) with its center on N distant ρ units from P on the side from which the curve appears concave is called the *circle of curvature* or the *osculating circle*. κ measures the rate at which the tangent line is rotating per unit change in s. For a plane curve

$$y = f(x) \qquad \kappa = \frac{y''}{(1 + y'^2)^{3/2}}$$

$$x = \xi(u),\ y = \eta(u) \qquad \kappa = \frac{\dot{x}\ddot{y} - \dot{y}\ddot{x}}{(\dot{x}^2 + \dot{y}^2)^{3/2}}$$

$$r = g(\theta) \qquad \kappa = \frac{2r'^2 - rr'' + r^2}{(r'^2 + r^2)^{3/2}}$$

In the first case, if y'^2 is numerically small, $\kappa = y''$ (approximately). At a point of inflection, $\kappa = 0$ and the circle of curvature degenerates to a straight line.

For a plane curve given by $F(x,y) = k$ (constant), at a point where $F_y \neq 0$, $y' = -F_x/F_y$, $(ds/dx)^2 = 1 + (F_x/F_y)^2$, $\kappa = \pm(F_{xx}F_y^2 - 2F_{xy}F_xF_y + F_{yy}F_x^2)/(F_x^2 + F_y^2)^{3/2}$, with the sign opposite to that of F_y.

10.22. Surfaces S arise with roughly equal frequencies in the forms

I. $f(x,y,z) = k$ (const)

II. $z = F(x,y)$

III. $x = \xi(u,v) \qquad y = \eta(u,v) \qquad z = \zeta(u,v)$

In II (or III) the point (x,y) [or (u,v)] ranges over a region R of the xy-plane (or uv-plane). For the normal line (hence tangent plane) for I and II, see 10.1. In case III, it can be shown that the direction cosines (see 7.62) of one normal are given by

$$\Delta \cos \alpha = \frac{\partial(y,z)}{\partial(u,v)} \qquad \Delta \cos \beta = \frac{\partial(z,x)}{\partial(u,v)} \qquad \Delta \cos \gamma = \frac{\partial(x,y)}{\partial(u,v)}$$

where $\Delta = \sqrt{EG - F^2}$, in which $E = x_u^2 + y_u^2 + z_u^2$, $F = x_ux_v + y_uy_v + z_uz_v$, and $G = x_v^2 + y_v^2 + z_v^2$. The normal in the opposite direction has the sign of each cosine changed. For a curve on S given by a relation between u and v, $ds^2 = E\,du^2 + 2F\,du\,dv + G\,dv^2$; to get ds for I and II compute dz and substitute in $ds^2 = dx^2 + dy^2 + dz^2$.

It can also be shown that the area da of a small portion of the surface, bounded by $u = u_0$, $u = u_0 + du$, $v = v_0$, $v = v_0 + dv$, is given to a satisfactory degree of approximation by $\Delta\,du\,dv$, where Δ is evaluated for u_0, v_0. Thus $\Delta \cos \alpha\,du\,dv$, etc., are the projections of da on the yz-plane, etc.

In case II, u and v are replaced by x and y, $E = 1 + z_x^2$, $F = z_xz_y$, $G = 1 + z_y^2$, $\Delta = \sqrt{1 + z_x^2 + z_y^2}$, etc. In case I, if $f_z \neq 0$, $E = 1 + f_x^2/f_z^2$, $F = f_xf_y/f_z^2$, $G = 1 + f_y^2/f_z^2$, $\Delta = \sqrt{f_x^2 + f_y^2 + f_z^2}/|f_z|$, etc.

In the important special case of a sphere (radius b), at whose center spherical coordinates r, ϕ, θ have been set up, $da = b^2 \cos \varphi\,d\theta\,d\phi$.

10.31. Multiple (Double, Triple, and Repeated) Integrals. Let the equation $z = f(x,y)$ define a surface S. [If f is multiple-valued, decompose S into pieces on each of which f is single-valued (see 10.11).] Let R be the projection of S on the xy-plane. Subdivide R into n small regions R_i ($i = 1, \cdots, n$), let Δ_ia be the area of R_i, and let (ξ_i, η_i) be a point in R_i. Then $f(\xi_i, \eta_i)\Delta_ia$ is the volume of a prism P_i of base Δ_ia and height $f(\xi_i, \eta_i)$. If $f(x,y) \geqslant 0$ for (x,y) in R_i, this volume is an approximation to the volume of the solid whose base and sides are those of P_i but whose "roof" is that piece S_i of S which projects into R_i, whereas the roof of P_i is a piece of the plane $z = f(\xi_i, \eta_i)$. The sum $\sum = \sum_1^n f(\xi_i, \eta_i) \Delta_i a$ is an approximation [if $f(x,y) \geqslant 0$ for (x,y) in R] to the "volume" between S and R. Let the number n of small regions R_i into which R is divided increase indefinitely, in such a way that the diameter $||\Delta a||$ of that region R_i with *longest* diam-

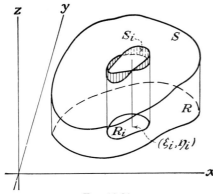

eter tends to 0 (see 8.12). If Σ approaches a limit l which is independent of the way in which the process of subdivision of R into R_i took place and of the choice, at each stage, of the points (ξ_i, η_i) in R_i, then $f(x,y)$ is said to be integrable over the *range* R, the limit l is called the (definite) *double integral* of the integrand $f(x,y)$ over R and is denoted by

$$\iint_R f(x,y)\,da = \lim \sum_1^n f(\xi_i, \eta_i)\Delta_i a \quad \text{as}$$

$n \to \infty$ and $||\Delta a|| \to 0$. $f(x,y)$ is integrable over R if it is bounded in R and continuous except at points that are isolated (see 8.12) or lie on a finite number N

Fig. 10.31.

of continuous curves $\phi_i(x,y) = k_i$ ($i = 1, \cdots, N$) such that ϕ_{ix} and ϕ_{iy} are sectionally continuous.

If $f(x,y) \geqslant 0$ [$f(x,y) \leqslant 0$] over R, then $\displaystyle\int\!\!\int_R f(x,y)\,da \geqslant 0$ ($\leqslant 0$). In the former

(latter) case the *volume* between the surface and R is equal to the double integral of f (or $-f$). As in the computation of areas, changes in the sign of f must be noted carefully. If $f(x,y) = 1$, then $\sum_1^n \Delta_i a = |R|$ and the double integral gives the area $|R|$ of R.

In precisely the same manner, one defines the *triple* integral $\displaystyle\int\!\!\int\!\!\int_V f(x,y,z)\,dv$

$= \lim \sum_1^n f(\xi_i,\eta_i,\zeta_i)\Delta_i V$ as $n \to \infty$ and $||\Delta V|| \to 0$ where V is the region inside a closed surface that is subdivided into n small regions V_i of volume $\Delta_i v$, (ξ_i,η_i,ζ_i) is a point of V_i and $||\Delta v||$ is the diameter of that V_i with the greatest diameter. The integral exists, *i.e.*, f is integrable over V, if f is bounded in V and continuous except at points that are isolated or lie on a finite number of continuous curves or surfaces given by functions with derivatives that are sectionally continuous. If $f(x,y,z) = 1$, the integral reduces to $\sum_1^n \Delta_i v = |V|$, the volume of V.

10.311. Extensions. 1. In case it is desired to extend a double or triple integral over a region R or V that is not confined by its boundary to a finite portion of the plane or of space, insert auxiliary boundary curves or surfaces so as to have a region R' or V' to which the above definitions apply. Compute the integrals over R' or V' and let the auxiliary boundaries recede indefinitely. Then $\displaystyle\int\!\!\int_R f(x,y)\,da = \lim$

$\displaystyle\int\!\!\int_{R'} f(x,y)\,da$ and $\displaystyle\int\!\!\int\!\!\int_V f(x,y,z)\,dv = \lim \int\!\!\int\!\!\int_{V'} f(x,y,z)\,dv$, provided that the

limits on the right do not depend on the manner in which the auxiliary boundaries disappear. Even for a continuous integrand, the proviso often requires rather delicate study, usually involving an interchange of the order of integration (see 10.32) with at least one limit ∞. The question can sometimes be settled by comparison with the following integrals: If R or V does not contain the origin, then $\displaystyle\iint\limits_{R} (x^2 + y^2)^{\frac{\alpha}{2}}\, da$

is convergent if $\alpha < -2$ and divergent if $\alpha \geqslant -2$, while $\displaystyle\iiint\limits_{V} (x^2 + y^2 + z^2)^{\frac{\alpha}{2}}\, dv$

is convergent if $\alpha < -3$ and divergent if $\alpha \geqslant -3$.

2. In case the integrand becomes infinite at isolated points Q, or along curves C, or over surfaces S in R or V, insert auxiliary boundary curves or surfaces so as to exclude these discontinuities, getting a region R' or V' to which the ordinary definitions apply. Proceed as in extension 1, deflating each auxiliary boundary so that it coincides, in the limit, with the point, curve, or surface whose discontinuities it excludes. The integrals are defined as in extension 1 with the same provisional condition. Useful for comparison: If (a,b) is in R or (a,b,c) in V, then $\displaystyle\iint\limits_{R} [(x-a)^2 + (y-b)^2]^{\frac{\alpha}{2}}\, da$

converges if $\alpha > -2$ and diverges if $\alpha \leqslant -2$, while $\displaystyle\iiint\limits_{V} [(x-a)^2 + (y-b)^2$

$+ (z-c)^2]^{\frac{\alpha}{2}}\, dv$ converges if $\alpha > -3$ and diverges if $\alpha \leqslant -3$.

10.312. Theorems. The foregoing integrals, as well as the line and surface integrals to follow, share most of the properties stated in 9.11 for $\displaystyle\int_a^b f(x)dx$. Once and for all:

The integral of $f + g$ is the integral of f plus the integral of g.

The integral of cf, c constant, is c times the integral of f.

If the range of integration is made up of two (or a finite number of) partial ranges with no (pair having a) common region, the integral over the whole range is the sum of the integrals over the partial ranges.

The mean value $\bar{f}$ of f over a range R [curve (or line), surface (or plane), volume] of measure (length, area, volume) $|R|$ is the integral of f throughout R divided by $|R|$. If, throughout R, $m \leqslant f \leqslant M$, then $m \leqslant \bar{f} \leqslant M$, and if f is continuous, $\bar{f} = f(P)$ for some point P in R.

10.32. Repeated (or Iterated) Integrals. The diagram depicts a surface S with equation $z = f(x,y)$, where f is continuous over the region R. Assign to y the value η, getting $z = f(x,\eta) = g(x)$ for $\alpha(\eta) \leqslant x \leqslant \beta(\eta)$, where (α,η) is the point T and (β,η) is the point U. The cross-sectional plane area $TUVW$ is then $h(\eta) = \displaystyle\int_{\alpha(\eta)}^{\beta(\eta)} g(x)dx$ $= \displaystyle\int_{a(\eta)}^{\beta(\eta)} f(x,\eta)dx$. Similarly, the cross-sectional plane area $ABCD$ is $k(\xi) = \displaystyle\int_{\gamma(\xi)}^{\delta(\xi)} f(\xi,y)dy$. Applying the process for finding the volume of a solid with known cross-sectional area (see 9.1), we find that the volume of the solid above R and below S [assuming $f(x,y) \geqslant 0$ over R for convenience] is

$$\int_a^b k(x)dx = \int_a^b \left[\int_{\gamma(x)}^{\delta(x)} f(x,y)dy \right] dx = \int_a^b \int_{\gamma(x)}^{\delta(x)} f(x,y)dy\, dx = \int_c^d h(y)dy$$
$$= \int_c^d \left[\int_{\alpha(y)}^{\beta(y)} f(x,y)dx \right] dy = \int_c^d \int_{\alpha(y)}^{\beta(y)} f(x,y)dx\, dy$$

These are called *repeated* or *iterated* integrals. Note the order: work *first* with the *inside* differential and limits. Treat discontinuities in accordance with earlier methods [see 10.311(2)]. If parallels to the axes meet the boundary of R in more than two points, break R up into smaller regions whose boundaries are met in at most two points. The three expressions for the volume must be equal.

$$\iint\limits_{R} f(x,y)da = \int_{a}^{b} \int_{\gamma(x)}^{\delta(x)} f(x,y)dy\ dx = \int_{c}^{d} \int_{\alpha(y)}^{\beta(y)} f(x,y)dx\ dy$$

Thus a double integral may be computed by repeated single integrations.

In a similar manner, one finds an expression for a triple integral as a repeated integral.

$$\iiint\limits_{V} f(x,y,z)dv = \int_{a}^{b} \int_{\gamma(x)}^{\delta(x)} \int_{\lambda(x,y)}^{\mu(x,y)} f(x,y,z)dz\ dy\ dx,$$

the limits of integration being so chosen as precisely to cover the field V of integration. Similarly, one can evaluate the triple integral by integrating $f(x,y,z)dy\ dz\ dx$, or $f(x,y,z)dx\ dz\ dy$, etc. (six possible orders), in each case choosing the limits so as precisely to cover V.

When R or V extends "to infinity" [first extension (10.311)], the repeated integrals can often be shown to be equal by virtue of appropriate *uniformity* of convergence. For details, see a text.

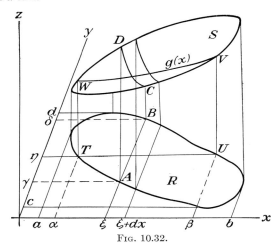

Fig. 10.32.

When, in any particular problem, it is known that the double or triple integral exists, then the repeated integrals are equal regardless of order of integration, and the order may often be chosen so as materially to simplify the calculation. Many computations may be conducted at pleasure with single, double, or triple integrals; tentative steps often disclose the simplest attack. In any repeated integral, an integration with *constant* limits may be delayed as long as may be desired; if $\mu(x,y)$ and $\lambda(x,y)$ are constants in the above triple integral, the integration with respect to z may be carried out at any convenient time. If *all* limits are constants, *any* order may be used.

NOTE: When it is desired to acknowledge the variables in which $f(x,y)$ or $f(x,y,z)$ is expressed, the double or triple integral is often written $\iint\limits_{R} f(x,y)dx\ dy$ or $\iiint\limits_{V} f(x,y,z)dx\ dy\ dz$ without thereby committing the writer to a specific order of integration and the attendant limits.

10.33. Areas of Surfaces as Double Integrals. If the projection of S: $z = F(x,y)$ on the xy-plane covers a region R, then (see 10.22) $\Delta\, du\, dv = \sqrt{1 + z_x{}^2 + z_y{}^2}\, dx\, dy$, since $u = x$, $v = y$. The area of S is $\displaystyle\int\int_R \sqrt{1 + z_x{}^2 + z_y{}^2}\, dx\, dy$. If S has equation

$f(x,y,z) = k$, and $f_z \neq 0$ over R, the area is $\displaystyle\int\int_R (\sqrt{f_x{}^2 + f_y{}^2 + f_z{}^2}/|f_z|)dx\, dy$.

10.4. Multiple Integrals in Other Coordinates. In the iterated integrals of 10.32, the da or dv of the double or triple integrals were replaced by $dx\, dy$ or $dx\, dy\, dz$, which may be regarded as the area and volume of boxes (see Fig. 10.12) with sides parallel to the coordinate axes and dimensions $dx \times dy$ and $dx \times dy \times dz$. Suppose now that coordinates (ξ,η) or (ξ,η,ζ) are introduced.

$$x = \phi(\xi,\eta), \qquad y = \psi(\xi,\eta) \qquad \text{or} \qquad x = \phi(\xi,\eta,\zeta), \qquad y = \psi(\xi,\eta,\zeta), \qquad z = \chi(\xi,\eta,\zeta)$$

In order to express the multiple integrals as iterated integrals, it now becomes necessary to use "boxes" with curved sides.

$$\xi,\ \eta,\ \xi + d\xi,\ \eta + d\eta \qquad \text{or} \qquad \xi,\ \eta,\ \zeta,\ \xi + d\xi,\ \eta + d\eta,\ \zeta + d\zeta$$

It can be shown that

$$\int\int_R f(x,y)dx\, dy = \int\int_{R'} f[\phi(\xi,\eta),\ \psi(\xi,\eta)]\frac{\partial(\phi,\psi)}{\partial(\xi,\eta)}\, d\xi\, d\eta$$

and $$\int\int\int_V f(x,y,z)dx\, dy\, dz = \int\int\int_{V'} f[\phi(\xi,\eta,\zeta),\ \psi(\xi,\eta,\zeta),\ \chi(\xi,\eta,\zeta)]\frac{\partial(\phi,\psi,\chi)}{\partial(\xi,\eta,\zeta)}\, d\xi\, d\eta\, d\zeta,$$

where $\partial(\phi,\psi)/\partial(\xi,\eta)$ or $\partial(\phi,\psi,\chi)/\partial(\xi,\eta,\zeta)$ is the Jacobian (J) (determinant of partial derivatives see 10.12), and is assumed not to vanish for (ξ,η) in R' or (ξ,η,ζ) in V'. Note that, at (ξ,η) or (ξ,η,ζ), $J > 0$ or $J < 0$ according as the two ξ, η or three ξ, η, ζ positive "axes" agree or disagree with the x, y- or x, y, z-axes in respect of being right-handed or left-handed (see 7.61). A point where J vanishes is a *singular* point of the transformation; detailed study of each singular point is usually required. The following are *special cases:* Polar coordinates

$$\int\int_R f(x,y)dx\, dy = \int\int_{R'} f(r\cos\theta,\ r\sin\theta)r\, dr\, d\theta$$

Cylindrical coordinates

$$\int\int\int_V f(x,y,z)dx\, dy\, dz = \int\int\int_{V'} f(\rho\cos\phi,\ \rho\sin\phi,\ z)\rho\, d\rho\, d\phi\, dz$$

Spherical coordinates

$$\int\int\int_V f(x,y,z)dx\, dy\, dz = \int\int\int_{V'} f(r\cos\varphi\cos\theta,\ r\cos\varphi\sin\theta,\ r\sin\varphi)r^2\cos\varphi\, dr\, d\theta\, d\phi$$

Calculations can often be simplified by using the coordinate system best adapted to the problem; in particular, a convenient location of the origin should be sought.

The equations of a transformation of coordinates may be interpreted instead as equations of *point transformations* mapping two planes or spatial regions on each other, in such a way that the point with coordinates (ξ,η) in R' is mapped on the point

(x,y) in R, etc. Point transformations are said to be (locally) area- or volume-preserving or authalic in case boxes with sufficiently small equal corresponding dimensions have areas or volumes whose ratios tend to 1 as the longest dimension tends to zero. By inspecting the above formulas for the special case in which $f \equiv 1$, one sees that the transformation will have this property if its Jacobian is identically 1 (see 3.7).

10.5. Line and Surface Integrals. Let $f(x,y,z)$ be defined in a region V in which C is a curve given by $y = \phi(x)$, $z = \psi(x)$. If $P_1 [x_1, \phi(x_1), \psi(x_1)]$ and $P_2 [x_2, \phi(x_2), \psi(x_2)]$ are two points on C, the integral $\int_{C\, P_1}^{P_2} f(x,y,z)dx = \int_{x_1}^{x_2} f[x, \phi(x), \psi(x)]dx$ is called the

line integral or *curvilinear integral* of f along C from P_1 to P_2; similarly for $\int_{C\, P_1}^{P_2} F(x,y)dx$

$= \int_{x_1}^{x_2} F[x, \phi(x)]dx$ along $y = \phi(x)$ in the plane, and similarly for line integrals with respect to y and z. If the curve has parametric representation $x = \xi(u)$, $y = \eta(u)$,

$z = \zeta(u)$, $\int_{C\, P_1}^{P_2} f(x,y,z)dx = \int_{u_1}^{u_2} f[\xi(u), \eta(u), \zeta(u)] \dot{x}\, du$; similarly for the other integrals.

Reversing the direction of C changes the sign of the integral: $\int_{C\, P_1}^{P_2} = - \int_{C\, P_2}^{P_1}$.

If arc length s is used as a parameter, then for example

$$\int_{C\, P_1}^{P_2} f(x,y,z)dy = \int_{s_1}^{s_2} f(\xi(s), \eta(s), \zeta(s))\, \dot{y}\, ds$$

and $\dot{y} = \cos \beta$ if the tangent to C has the direction angles α, β, γ. Thus $\int_{C\, P_1}^{P_2} f\, dx$

$+ g\, dy + h\, dz = \int_{s_1}^{s_2} (f \cos \alpha + g \cos \beta + h \cos \gamma)ds$. The symbol $\oint$ is often used to indicate that C is closed.

Plane Area by Line Integral. If C is a closed curve in the xy-plane, bounding a region R whose area is A, then $A = \frac{1}{2} \int_C x\, dy - y\, dx$, C being described in the positive direction, *i.e.*, so that R is always to the left.

Entropy. If T, Q, and S_A denote absolute temperature, heat, and entropy of a state represented in the familiar pv-diagram by a point A, then $S_B - S_A = \int_{C\, A}^{B} dQ/T$,

where C is any reversible path from A to B.

If $z = \phi(x,y)$ is a surface S in V, $\iint_S f(x,y,z)dx\, dy = \iint_R f[x,y,\phi(x,y)]dx\, dy$,

where R is the projection of S on the xy-plane, is the *surface integral* of f over S; similarly for integrals over surfaces $x = \phi(y,z)$ or $y = \phi(z,x)$. If the rectangle dx, dy is the projection of a small portion of the surface, of area da, then $da \cos \gamma = dx\, dy$, $0 \leqslant \cos \gamma \leqslant 1$, and $\iint_S f(x,y,z)dx\, dy = \iint_S f(x,y,z) \cos \gamma\, da$. If the surface is expressed by parametric equations (see 10.22), then $\cos \gamma$ and da may be replaced in the integral by $|\partial(x,y)/\partial(u,v)|/\Delta$ and $\Delta\, du\, dv$, so that

$$\iint_S f(x,y,z)dx\, dy = \iint_{R'} f[\xi(u,v), \eta(u,v), \zeta(u,v)]|\partial(x,y)/\partial(u,v)|\, du\, dv$$

where R' is the region in the uv-plane corresponding to S. The *area* of S is

$$\iint_S da = \iint_{R'} \sqrt{EG - F^2}\, du\, dv \text{ (see 10.22).}$$ If it is desired, as it often is, to take

account of the direction of the normal, remove the absolute value sign and watch the sign of the integrand; changing the direction of the normal changes the sign of the integral.

10.51. In the following theorems, P, Q, and R are functions of x, y, and z which, along with their derivatives, have sufficient continuity (usually present in practice) throughout a sufficiently large region to ensure that the necessary integrals exist.

Suppose that we have a surface S whose complete boundary is C (perhaps several curves). Let one side of S be (arbitrarily) called positive, and let l, m, n denote the normal directed into space from the positive side of the surface. Let C be described (in each of its parts) in such a way that an observer walking along C on the positive side of S always has S to his left.

Stokes' Theorem.

$$\int_C P\, dx + Q\, dy + R\, dz = \iint_S [l(R_y - Q_z) + m(P_z - R_x) + n(Q_x - P_y)]da$$

Suppose that we have a bounded region V of space whose complete boundary S is composed of closed surfaces and that l, m, n are the cosines of the normal directed outward (*away* from S).

Divergence Theorem (Gauss).

$$\iint_S (Pl + Qm + Rn)da = \iiint_V (P_x + Q_y + R_z)dv$$

N.B. 1. In case the integrand of the triple integral has discontinuities, exclude them from V by surfaces, which contribute further surface integrals on the left. When these auxiliary surfaces are deflated, contributions to the surface integral sometimes remain.

2. To apply this theorem to a region extending out indefinitely, introduce a large surface (*e.g.*, a sphere), apply the theorem to the region so bounded, and then let the auxiliary surface recede indefinitely. In fortunate cases, the integral over it has a limit (usually zero), and the volume integral also converges.

These remarkable theorems underlie many applications of mathematics. They are often expressed in vector form (see 12.52, 12.53, 12.7 where the meaning of "divergence" is noted and Green's theorems are stated). Technically, their use permits line and surface, and surface and volume integrals to be transformed into one another. In the plane, they have essentially the same content: If C is the complete

boundary of R, $\int_C P\, dx + Q\, dy = \iint_R (Q_x - P_y)dx\, dy$.

Under certain restrictions on the region involved, the following statements are equivalent:

1. $\int_C^{\,B}_{\,A} P\, dx + Q\, dy + R\, dz$ depends only on the positions of A and B and not on the curve C.

2. $\oint_C P\, dx + Q\, dy + R\, dz = 0$ for *every closed* curve C.

3. $R_y = Q_z$ and $P_z = R_x$ and $Q_z = P_y$ identically.

4. There is a function $\phi(x,y,z)$, namely, $\displaystyle\int_C^{(x,y,z)} P\,dx + Q\,dy + R\,dz$ (A fixed, any

curve C) such that $\phi_x = P$ and $\phi_y = Q$ and $\phi_z = R$ and ϕ is single-valued.

When any one of these statements is true, the expression $P\,dx + Q\,dy + R\,dz$ is said to be an *exact* or *complete* differential. In fact, the expression is then the total differential $d\phi$ of $\phi(x,y,z)$ (see 10.1) and P, Q, R are direction numbers of the normals to the surfaces $\phi = $ constant.

Under certain restrictions on the region involved, the following statements are equivalent:

1. $\displaystyle\iint_S (Pl + Qm + Rn)da$ depends only on the boundary C of S and not on the

surface bounded by C.

2. $\displaystyle\iint_S (Pl + Qm + Rn)da = 0$ for *every closed* surface S.

3. $P_x + Q_y + R_z = 0$ identically.

4. There are functions A, B, and C such that $P = C_y - B_z$ and $Q = A_z - C_x$ and $R = B_x - A_y$.

10.6. Applications; Formulas. In addition to the lengths, areas, and volumes for which formulas have already been found, many physical quantities such as mass, centroid, moment of inertia, work, pressure, attraction, etc., may be expressed as definite integrals.

Densities occur frequently: mass (or electric or magnetic charge) per unit volume, mass per unit length, force per unit mass, force per unit area, etc.

Illustrations of General Procedure. To find $\rho(x,y,z) = $ mass per unit volume *at* $P(x,y,z)$, let R be a region containing P, find the mass M contained in R, compute the *average* mass $M/|R|$ of R per unit volume, and take the limit as $||R|| \rightarrow 0$, P always in R (for $|R|$ and $||R||$, see 8.12). To find force $F(x,y,z)$ per unit area in a certain direction at $P(x,y,z)$, let R be a plane region containing P and perpendicular to the given direction, find the total force F exerted on R normally (*i.e.*, in the given direction), compute the *average* force $F/|R|$ per unit area, and take the limit as $||R|| \rightarrow 0$, P always in R. (Densities may be thought of as curvilinear, superficial, or solid generalized derivatives.)

Typical Application. Suppose we wish to know by how much (dp) the pressure p at height h (measured vertically upward) exceeds the pressure $p + dp$ at height $h + dh$. Evidently $dp = -g\,dm$, where dm is the mass of air in a prism with area 1 cm² as base and height dh. By the definition of ρ, $dm = \rho\,dh$ (approximately), so that $dp = -g\rho\,dh$.

In the following, we use dR to denote dv, da, or ds according as the region R over which P varies is solid, on a surface, or on a curve. Also, we use one integral sign, corresponding to R on a curve; use two for da, three for dv, if more explicitness is desired. To handle any specific problem, limits of integration are to be provided according to directions in 10.32. Specializations to various coordinate systems and representations follow the lines of 10.4. For discrete distributions, replace integrals by sums.

Mass m distributed over R with density ρ: $m = \displaystyle\int_R \rho(P)dR$, with P in dR. N.B. ρ is constant if the distribution is *homogeneous*, or *uniform* (*i.e.*, $|R_1| = |R_2|$ implies $m_1 = m_2$).

Center of Mass $\bar{P}(\bar{x},\bar{y},\bar{z})$: $m\bar{x} = \int_R \rho(x,y,z)x \, dR$.   Similarly for $\bar{y}$ and $\bar{z}$.   If $\rho \equiv 1$, in fact, if ρ is constant, $\bar{P}$ is associated with R as a geometric configuration rather than necessarily with a distribution of matter.   It is then often called the *centroid* (terminology not standard).   For distributions of experimental size, $\bar{P}$ coincides with the *center of gravity* as defined in physics.   In case the distribution is uniform, and R has a line or plane of symmetry, $\bar{P}$ lies in that line or plane.

Moment of Inertia I of the distribution about a line l from which P has distance $d(P)$: $I = \int_R \rho(P)[d(P)]^2 \, dR$, with P in dR.   *Radius of gyration* λ of the distribution about l is defined by $m\lambda^2 = I$.   In case $\rho \equiv 1$, one speaks of the moment of inertia of the configuration.

Attraction K, per unit mass, exerted by the distribution at a point Q in the direction of a ray k issuing from Q: $K = \int_R \dfrac{\rho(P) \cos \theta}{r^2} \, dR$, with P in dR, where r denotes the distance PQ and θ is the angle $(0 \leqslant \theta \leqslant \pi)$ between k and the ray PQ.

Total Force F exerted normally on an area S by a normal force of magnitude σ per unit area: $F = \int_S \sigma(P)da$, with P in da.   If the fluid pressure against a confining surface at a point is p, the total force exerted on a small area da of the surface is $p \, da$; if this small area is displaced through a distance ds along the normal, the work done by the fluid is approximately $p \, da \, ds = p \, dv$, where $dv = da \, ds$ is the volume of a small prism whose base is (approximately) da, and whose height is ds.

Work W done by a force F in a displacement from Q_1 to Q_2 along a curve C: $W = \int_{C}^{Q_2} F_t(P)ds$, with P in ds, where F_t denotes the projection of F at P along the tangent to C at P, directed from P to Q.

Mass of fluid ϕ leaving a surface S, per unit time, if the density at P is $\rho(P)$, the velocity of the particle at P has magnitude $V(P)$ and direction making an angle $\theta(P)$ with the *outward* normal to S at P: $\phi = \int_S \rho(P)V(P) \cos \theta(P)da$, with P in da.

Typical Specializations (of These and Other Integrals).   1. Mass of a circular disk of radius c whose density is $a + br$, a and b constants, r the distance from the center. In polar coordinates, $m = \int_0^{2\pi} \int_0^c (a + br)r \, dr \, d\theta = \pi c^2(3a + 2bc)/3$.

2. Centroid of a uniform (constant mass σ per unit area) hemispherical shell of radius a, so that $m = 2\pi a^2\sigma$.   Using spherical coordinates, with the θ-plane as equatorial plane, $\bar{z}$ to denote the height of the centroid above this plane, and $z = a \sin \varphi$ in the integrand, $2\pi a^2\sigma\bar{z} = \int_0^{\frac{\pi}{2}} \int_0^{2\pi} \sigma a \sin \varphi \, a^2 \cos \varphi \, d\theta \, d\varphi = \pi \sigma a^3 \int_0^{\frac{\pi}{2}} \sin 2\varphi \, d\varphi = \pi\sigma a^3$, whence $\bar{z} = a/2$.

3. The moment of inertia of the area inside the curve $r^2 = a^2 \cos 2\theta$ about a line containing the initial ray is

$$4 \int_0^{\frac{\pi}{4}} \int_0^{a\sqrt{\cos 2\theta}} (r \sin \theta)^2 \, r \, dr \, d\theta = a^4 \int_0^{\frac{\pi}{4}} \sin^2 \theta \cos^2 2\theta \, d\theta$$

$$= \frac{a^4}{2} \int_0^{\frac{\pi}{4}} (\cos^2 2\theta - \cos^3 2\theta)d\theta$$

$$= \frac{a^4}{2} \int_0^{\frac{\pi}{4}} \left[\frac{1}{2}(1 + \cos 4\theta) - (1 - \sin^2 2\theta) \cos 2\theta\right]d\theta = a^4\left(\frac{\pi}{16} - \frac{1}{6}\right)$$

4. The (gravitational) potential energy ϕ per unit mass at a point P is the work required to bring a unit mass to the point from a standard position (of zero energy). In terrestrial problems, the standard position for given latitude and longitude is usually mean sea level, from which h is measured vertically upward. Hence $\phi = \int_0^h g\,du$, since the work is independent of the path, which may be taken vertically. If $g_0 = 980.6$ cm per sec^2 is the acceleration of gravity when $h = 0$, then, at height h, $g = g_0(1 - 2h/R_0)$ (approximately), where $R_0 = 6{,}370$ km is the mean radius of the earth. Hence $\phi = g_0 \int_0^h (1 - 2u/R_0)du = g_0 h(1 - h/R_0)$, which is commonly called the *geopotential*.

5. In studying turbulence (see ref. 8, Chap. IX, especially par. 52), it becomes necessary to find the amount of horizontal momentum transferred through a horizontal surface S by perturbations u' horizontally and w' vertically of a steady horizontal velocity u. Taking the x-axis in the direction of u, the y-axis horizontal and perpendicular to u, and the z-axis vertical, the horizontal momentum M of unit volume is $\rho(u + u')$, where ρ is the density of the air. The amount of this momentum transferred per unit time through a horizontal area $da = dx\,dy$ is thus $Mw'\,da$, so that the total momentum flowing across S is $\iint_S \rho(u + u')w'\,dx\,dy$ per unit time.

11. DIFFERENTIAL EQUATIONS

11.1. Definitions. An equation involving differentials or derivatives is a *differential equation* (*DE*), *ordinary* or *partial* according as there is just one or more than one independent variable. The *order* is the order of the derivative of highest order, and the *degree* is the power to which that derivative is raised. A *solution* is a function that satisfies the equation. To verify that an alleged solution actually is a solution, find the necessary derivatives and substitute in the *DE*. An ordinary *DE* of order n will usually have a solution (sometimes called the *general solution*) containing n arbitrary or disposable constants. These constants may often be chosen to make the *particular solution* so obtained satisfy desirable initial or boundary conditions.

Caution. 1. Some *DE* have solutions (some of which are called *singular*) not obtainable in this manner from the "general" solution.

2. Some *DE* have no solutions. Unless a specific solution can be found, it is often difficult to establish conditions for the existence and/or uniqueness of solutions. We assume in the following methods of finding solutions that the functions that occur have properties adequate for the validity of the processes used. See texts for details and for other methods, *e.g.*, ref. 11.

11.2. Equations of the First Order and First Degree. 1. $y' = f(x)$. Then $y = \int f(x)dx + C$ (see 9.1 to 9.2). More precisely, $y = b + \int_a^x f(t)dt$ has $y' = f(x)$ and $y = b$ when $x = a$. For example, to find a curve whose slope is $y' = 6x^2 - 5\cos x$ and which has the abscissa 3 when $x = \pi$: $y = 2x^3 - 5\sin x + C$, $3 = 2\pi^3 - 5\sin \pi + C$, $C = 3 - 2\pi^3$, $y = 3 + 2(x^3 - \pi^3) - 5\sin x$.

2. *Variables Separate.* $f(x)dx + g(y)dy = 0$. *Solution.* $\int f(x)dx + \int g(y)dy = C$. In other cases, try to separate variables by algebraic manipulation; if this succeeds, variables are *separable*.

3. *Homogeneous.* $y' = f(x,y)$ where $f(tx,ty) = t^k f(x,y)$ for some k. Substitute vx for y or vy for x (transforming the equation, of course). In the result, the variables will be separable. Solve and replace v by y/x or x/y.

4. *Exact.* $P(x,y)dx + Q(x,y)dy = 0$ with $P_y = Q_x$ so that there is a function z such that $dz = P\,dx + Q\,dy$ (see 10.1, 10.51). *Solution.* $C = \int_{(a,b)}^{(x,y)} P\,dx + Q\,dy$

along any curve from any chosen (a,b) to (x,y) (see 10.5). More directly, let $z = \int P(x,y)dx + \phi(y)$, treating y in the integration as constant, ϕ arbitrary; compute $z_y = \phi'(y) + \int P_y\,dx$ and set it equal to Q, getting a *DE* for ϕ from which x disappears (since $P_y = Q_x$). Solve for ϕ, and substitute in the expression for z. Then $z = C$ is a solution of the original *DE*. Or, let $z = \int Q(x,y)dy + \psi(x)$, etc., e.g., $(\sin y - y \sin x - e^{2x})dx + (x \cos y + \cos x)dy = 0$. Since $P_y = \cos y - \sin x = Q_y$, this is exact. Let $z = \int (x \cos y + \cos x)dy + \psi(x) = x \sin y + y \cos x + \psi(x)$. Then $z_x = \sin y - y \sin x + \psi'(x)$ must equal $P = \sin y - y \sin x - e^{2x}$. This will be true if $\psi'(x) = -e^{2x}$, $\psi(x) = -e^{2x}/2$. Hence solutions of the *DE* are $x \sin y + y \cos x - e^{2x}/2 = C$.

5. *Integrating Factor.* $P(x,y)dx + Q(x,y)dy = 0$, $P_y \neq Q_x$. It is sometimes possible to multiply the *DE* by a factor $f(x,y)$ that will make the result exact (see 4). If algebraic manipulations and trial factors fail, see a text for hints.

6. *Linear.* $y' + P(x)y = Q(x)$. Multiply by the integrating factor $f(x) = e^{\int P(x)dx}$. Then the solution is y where $yf(x) = \int Q(x)f(x)dx$, e.g., in the theory of the atmosphere one meets the *DE* $dp = -g\rho\,dz$, where p denotes pressure, g the acceleration of gravity, ρ density, and z height. If the air is dry, and the vertical lapse rate of temperature T is a constant α, so that $T = T_0 - \alpha z$, then, by using the equation of state $p = \rho RT$ (R the gas constant), the *DE* may be written in the form $dp/dz + g[R(T_0 - \alpha z)]^{-1}p = 0$. Here $Q(z) = 0$, $P(z) = g[R(T_0 - \alpha z)]^{-1}$, and the integrating factor $f(z)$ is $(T_0 - \alpha z)^{\frac{-g}{R\alpha}}$. Hence the solution is $(T_0 - \alpha z)^{\frac{-g}{R\alpha}} p = C$. If $p = p_0$ when $z = 0$, then $C = p_0 T_0^{\frac{-g}{R\alpha}}$, and the solution may be written $p = p_0(1 - \alpha z/T_0)^{\frac{g}{R\alpha}}$.

11.3 Equations of First Order and Higher Degree. 1. Solve for y' and try to integrate the resulting *DE*.

2. Solve for y (or x) in terms of x (or y) and y', and differentiate with respect to x (or y). Solve this new *DE* for x (or y) as a function of y'. Along with the given *DE*, we then have parametric equations $x = f(y')$, $y = g(y')$ for a solution in terms of the parameter y'. It may or may not be possible, or desirable, to obtain a relation between x and y by eliminating the parameter y', e.g., $y'^2 + y' = e^y$, or $y = \ln [y'(1 + y')]$. Differentiation gives

$$y' = \frac{1}{y'}\frac{dy'}{dx} + \frac{1}{1 + y'}\frac{dy'}{dx}, \quad \text{or} \quad \frac{dx}{dy'} = \frac{1}{y'^2} + \frac{1}{y'(1 + y')}$$

Integrating with respect to y', we find $x = -1/y' + \ln [y'/(1 + y')] + C$, so that x and y can be found when y' is given such that $y'(1 + y') > 0$.

Special Case (Clairaut). $y = y'x + \phi(y')$ has the solutions $y = cx + \phi(c)$ for each constant c.

11.4. Special Equations of Second and Higher Order. 1. If y does not appear explicitly, write the *DE* as a first-order *DE* with y' and x as dependent and independent variables. Solve to get y', and then solve the resulting first-order *DE* to get y.

2. If x does not appear explicitly, use y' and y as dependent and independent variable [putting $y'(dy'/dy)$ for y''] and proceed as in 1.

3. $y'' = f(y)$ (common in mechanics). Multiply each side by $2y'$ and integrate to get y'. Then integrate again (variables separable).

4. $d^ny/dx^n = f(x)$. Integrate n times with respect to x.

11.51. Linear Equations. A *DE* of the form

$$f_n(x)y^{(n)} + f_{n-1}(x)y^{(n-1)} + \cdots + f_1(x)y' + f_0(x)y = g(x)$$

where $y^{(k)} = d^ky/dx^k$, is said to be linear. Solutions are sought over an interval in which $f_n(x)$ does not vanish. The *DE* is *homogeneous* (*HE*) if $g(x) = 0$; with each *DE* there is associated an *HE* obtained by replacing $g(x)$ in the *DE* by 0.

If $y_1, \cdots, y_k$ satisfy the *HE*, and $c_1, \cdots, c_k$ are constants, then $y = c_1y_1 + \cdots + c_ky_k$ satisfies the *HE*. If $y_1, \cdots, y_n$ satisfy the *HE*, and if $c_1y_1 + \cdots + c_ny_n = 0$ for each x implies that $c_1 = \cdots = c_n = 0$ (*i.e.*, $y_1, \cdots, y_n$ are linearly independent), then every solution of the *HE* may be written in the form $y = c_1y_1 + \cdots + c_ny_n$ with properly chosen c's, and $y_c = c_1y_1 + \cdots + c_ny_n$ with arbitrary c's is called the *complementary function* of the *DE*.

If $g(x) \neq 0$ and y^* is any specific solution, called a *particular integral*, of the *DE*, then the most general solution of the *DE* is $y = y^* + y_c$.

1. *Reduction of Order.* Suppose that a solution y_1 of the *HE* has been found. Let $y = y_1z$, and substitute in the *HE*. The result will be a new *HE* of order $n - 1$ involving z and its first $n - 1$ derivatives with respect to x. Solve this, perhaps by finding a solution z_1, and putting $z = z_1w$, and again reducing the order. Finally, restore y by substitutions $\cdots$, $w = z/z_1$, $z = y/y_1$.

2. y^* *When y_c Is Known.* If $y_c = c_1y_1 + \cdots + c_ny_n$, let $y = g_1(x)y_1 + g_2(x)y_2 + \cdots + g_n(x)y_n$.

$$y_1g_1' + y_2g_2' + \cdots + y_ng_n' = 0$$
$$y_1'g_1' + \cdots + y_n'g_n' = 0$$
$$\cdots \cdots \cdots \cdots \cdots \cdots$$
$$y_1^{(n-2)}g_1' + \cdots + y_n^{(n-2)}g_n' = 0$$
$$y_1^{(n-1)}g_1' + \cdots + y_n^{(n-1)}g_n' = g(x)/f_n(x)$$

Set up the foregoing system of equations. Solve these n linear algebraic equations (see 6.2) for the n unknowns $g_1', \cdots, g_n'$, integrate each result to get $g_1, \cdots, g_n$, and substitute these in the expression for y; the result is $y = y^* + y_c$. (*Variation of parameters.*)

3. To find y_c when $f_n(x), \cdots, f_0(x)$ are constants $a_0, \cdots, a_n$, substitute $y = Ae^{mx}$ in the *HE* as a trial solution, getting the *characteristic* equation $a_nm^n + a_{n-1}m^{n-1} + \cdots + a_1m + a_0 = 0$. Solve this algebraic equation (see 6.3), getting the *distinct* roots $m = m_1, \cdots, m_p$, some or all of which may be complex. Then, for $h = 1, \cdots, p$, if m_h is a root of multiplicity μ_h, the function $y_h = (c_{h0} + c_{h1}x + \cdots + c_{h\mu_h-1}x^{\mu_h-1})e^{m_hx}$ is a solution, and $y_c = y_1 + \cdots + y_p$ is the complementary function.

If the a's are real, complex roots occur in conjugate pairs: $m_h = \bar{m}_j = \alpha_h + i\beta_h = \alpha_h - i\beta_h$. In this case (using 5.7), $y_h + y_j$ may be written in the form

$$e^{\alpha_hx}[(d_{h0} + d_{h1}x + \cdots + d_{h\mu_h-1}x^{\mu_h-1})\cos\beta_h + (d'_{h0} + d'_{h1}x + \cdots + d'_{h\mu_h-1}x^{\mu_h-1})\sin\beta_hx],$$

where the d's and d'''s may be but are not necessarily complex.

Example. Let the characteristic equation be $(m^2 + 2m + 4)^2(m - 2) = 0$, $m_1 = 2$, $m_2 = -1 + i\sqrt{3}$ ($\mu_2 = 2$), $m_3 = -1 - i\sqrt{3}$ ($\mu_3 = 2$). Then $y_1 = c_1e^{2x}$, $y_2 = (c_{20} + c_{21}x)e^{(-1+i\sqrt{3})x}$, $y_3 = (c_{30} + c_{31}x)e^{(-1-i\sqrt{3})x}$, $y_c = y_1 + y_2 + y_3$; or write $y_2 + y_3$ in the form $e^{-x}[(d_{20} + d_{21}x)\cos\sqrt{3}\,x + (d_{20}' + d_{21}'x)\sin\sqrt{3}\,x]$.

4. In case the f's are not constant, y_c is not so easily found. If, in particular, $f_k(x) = a_kx^k$ for $k = 0, 1, \cdots, n$, the substitution $x = e^t$ reduces the *HE* to one in which the derivatives do have constant coefficients. Solve the new *HE* by the method of 3 and put $t = \ln x$ in the result.

5. To find the particular integral y^* when the f's are constant a's (more general methods including the Laplace transform, in texts): (a) Let $g(x) = Ax^pe^{qx}$, p a positive integer or zero, q any constant (real or complex). If q is a root of multiplicity k of the characteristic equation (see 3), so that $k = 0$ if q is not a root, substitute in the *DE* a trial expression of the form $y = (b_0 + b_1x + \cdots + b_px^p)x^ke^{qx}$, and determine

particular values of the b's so as to satisfy it. With these specific b's, $y^* = (b_0 + b_1 x + \cdots + b_p x^p) x^k e^{qx}$.

b. Special Case. If $g(x) = A e^{rx} \cos sx$ or $A e^{rx} \sin sx$, r and s real, try $y = e^{rx}(B_1 \cos sx + B_2 \sin sx)$, and determine values for B_1 and B_2 by requiring that y satisfy the *DE*. Or, put $\cos sx = (e^{isx} + e^{-isx})/2$, $\sin sx = (e^{isx} - e^{-isx})/(2i)$ and proceed as in (a).

c. If $g(x) = g_1(x) + \cdots + g_l(x)$, replace $g(x)$ in the *DE* by $g_h(x)$, getting l equations DE_h for $h = 1, \cdots, l$. Find y_h^* for each DE_h. Then y^* for the original *DE* is $y_1^* + \cdots + y_l^*$.

6. If two or more *DE* with several dependent variables have the same independent variable, try by differentiation and algebraic manipulation to get a *DE* involving only one dependent variable. Solve this and substitute in the other *DE*, which will then have one less dependent variable. Repeat.

11.52. Second-order Linear Equations with Constant Coefficients; Vibrations. Let x denote distance measured from a fixed point on a line, and let t denote time. A particle of mass m moving on the line subject to a force proportional to the displacement x will move in such a way that $m\ddot{x} + kx = 0$, where $k > 0$ or $k < 0$ according as the force is directed from the displaced position x toward or away from the origin, *i.e.*, according as the force is *restoring* or *disturbing*. We assume $k > 0$. Then the solution of the *DE* is

$$x = a \cos \omega_0 t + b \sin \omega_0 t = c \cos (\omega_0 t - \gamma)$$

where $\omega_0 = \sqrt{k/m}$ and $a, b, c = \sqrt{a^2 + b^2}$ and $\gamma = \tan^{-1}(b/a)$ are arbitrary constants determined by the values of x and $\dot{x}$ for some specific value of t, *e.g.*, $t = 0$. Thus the particle *vibrates* (*simple harmonic motion*) about its position of equilibrium at the origin with a *frequency* $\nu_0 = \omega_0/2\pi$ or angular frequency ω_0, *i.e.*, making ν_0 complete oscillations per unit time, or ω_0 complete oscillations per 2π units of time. The *amplitude* c is the maximum of $|x|$ and is attained at instants separated from each other by time intervals of length $1/2\nu_0$. The time $1/\nu_0$ required for a complete oscillation is called the *period* of the oscillation.

In case the motion is opposed by resistance proportional to the velocity, *i.e.*, *damped*, the *DE* is $m\ddot{x} + 2\beta\dot{x} + kx = 0$, with $\beta > 0$. In case $\beta^2 > km$ (strong damping), or $\beta^2 = km$ (critical damping), the particle will gradually approach the origin without oscillation; these cases do not interest us here. If $\beta^2 < km$, the solution of the *DE* is

$$x = e^{-\frac{\beta}{m}t}(a \cos \omega_1 t + b \sin \omega_1 t) = c e^{-\frac{\beta}{m}t} \cos (\omega_1 t - \gamma)$$

with $\omega_1^2 = (km - \beta^2)/m^2 = \omega_0^2 - \beta^2/m^2$ and the previous meanings of the other constants. Thus the amplitude $c e^{-\frac{\beta t}{m}}$ (for successive values of t giving maximum $|x|$) is depreciated as t increases by the *damping factor* $e^{-\frac{\beta t}{m}}$ and ultimately becomes imperceptible.

Since no external force has been applied, the foregoing vibrations are said to be *characteristic*, or *natural*, or *free*, and ω_0 (or ω_1) is the *characteristic* or *natural frequency* of the free undamped (or damped) vibrations. In case a further force of magnitude $f(t)$ is applied along the line of motion, the *DE* becomes $m\ddot{x} + 2\beta\dot{x} + kx = f(t)$. The case in which $f(t) = F \cos \omega t$ (or $F \sin \omega t$), F constant, is of special interest; the applied force varies periodically, having a frequency $\nu = \omega/2\pi$. In this case the solution of the *DE* is

$$x = c e^{-\frac{\beta t}{m}} \cos (\omega_1 t - \gamma) + \frac{F \cos (\omega t - \delta)}{\sqrt{m^2(\omega_0^2 - \omega^2)^2 + 4\beta^2\omega^2}}$$

where $\omega_0 = \sqrt{k/m}$ is the natural angular frequency of the undamped system, $\omega_1{}^2 = \omega_0{}^2 - \beta^2/m^2$, and $\tan \delta = 2\beta\omega/[m(\omega_0{}^2 - \omega^2)]$. On account of its disappearance as t increases, the vibration given by the complementary function (first term) is called a *transient*. After the transient becomes imperceptible, a *steady state* is reached in which only the oscillations given by the particular integral (second term) occur. These have the frequency ω of the impressed force and are called *forced* oscillations. For a system with given ω_0, the amplitude of the forced vibrations is greatest if $\omega = \omega_0$, so that forced and undamped free vibrations have the same frequency and are then said to be in resonance. The presence or absence of resonance often accounts (though not always respectively) for the success or failure of mechanical and electrical equipment. Trouble threatens when $\beta = 0$ and $\omega = \omega_0$.

11.6. Series Solutions (see 8.7). Especially when other methods fail, and especially when the behavior of y is important for values of x near some fixed $x = a$, it may be helpful to seek a series $f(x) = (x - a)^p \sum_0^\infty c_k (x - a)^k$, $c_0 \neq 0$ which will satisfy the *DE*. Write the *DE* so that each function of x (except y) is expressible as a series of powers of $x - a$. Even if this is not possible, there *may* still be a series solution, perhaps in *negative* powers of $x - a$; see the literature. Compute the derivatives of $f(x)$, writing $f(x) = \sum_0^\infty c_k(x - a)^{k+p}$ and differentiating term by term. Substitute them in the *DE*, which will then state, after arrangement according to powers of $x - a$, that two power series are equal. If this is to be true, coefficients of equal powers of $x - a$ must be equal. By equating these coefficients, find values for p and the c's in $f(x)$. When these values are substituted in $f(x)$, the resulting series, if it converges, has a sum y which satisfies the equation.

REMARKS: 1. In the linear case, this method may be used to find y_c and/or y^*. N.B. 0 is the power series each of whose coefficients vanishes.

2. Ordinarily, if the *DE* is of order n, n values of p will appear to be suitable. Ordinarily, one (possibly more) of the early c's (*e.g.*, c_0, since $c_0 \neq 0$) will turn out to be unrestricted, thus amounting to an arbitrary constant of integration, while later c's will be determined by earlier c's and a value of p. In fortunate cases, each value of p will give rise to a solution. In less fortunate cases (common if values of p differ by an integer), only one value of p will work; for procedure in such cases, see the literature (*e.g.* ref. 12).

3. In most of the linear cases that arise in practice, if each coefficient has a series in powers of $x - a$ and if each of these series converges for $|x - a| < R$, then the series obtained in the above manners (even in the less fortunate cases) converge for $|x - a| < R$ and satisfy the *DE*.

Example. $xy'' + y' + xy = 0$ (Bessel's equation of order zero). This is linear and homogeneous, and the coefficients are already expressed as powers of x; we therefore take $a = 0$ and seek p and c's such that $f(x) = x^p \sum_0^\infty c_k x^k = \sum_0^\infty c_k x^{k+p}$ will satisfy the *DE*.

$$f(x) = c_0 x^p + c_1 x^{p+1} + c_2 x^{p+2} + c_3 x^{p+3} + \cdots + c_k x^{p+k} + \cdots$$
$$f'(x) = pc_0 x^{p-1} + (p + 1)c_1 x^p + (p + 2)c_2 x^{p+1} + \cdots + (p + k)c_k x^{p+k-1} + \cdots$$
$$f''(x) = p(p - 1)c_0 x^{p-2} + (p + 1)pc_1 x^{p-1}$$
$$+ \cdots + (p + k)(p + k - 1)c_k x^{p+k-2} + \cdots$$

Multiply $f(x)$ and $f''(x)$ by x, add, and equate coefficients of x^{p-1}, x^p, x^{p+1}, $\cdots$ to zero.

$$p(p - 1)c_0 + pc_0 = 0 \qquad\qquad p^2 c_0 = 0$$
$$(p + 1)pc_1 + (p + 1)c_1 = 0 \qquad\qquad (p + 1)^2 c_1 = 0$$
$$(p + 2)(p + 1)c_2 + (p + 2)c_2 + c_0 = 0 \qquad (p + 2)^2 c_2 = -c_0$$
$$(p + 3)(p + 2)c_3 + (p + 3)c_3 + c_1 = 0 \qquad (p + 3)^2 c_3 = -c_1$$
$$\cdots \cdots \cdots \cdots \cdots \cdots \cdots \cdots$$
$$(p + k)(p + k - 1)c_k + (p + k)c_k + c_{k-2} = 0 \qquad (p + k)^2 c_k = -c_{k-2}$$
$$\cdots \cdots \cdots \cdots \cdots \cdots \cdots \cdots$$

By the first equation, since $c_0 \neq 0$, $p = 0$ (twice). Since $p = 0$, the second equation shows that $c_1 = 0$. It then follows from the last equation that $c_k = 0$ for each odd k. The third and the last equations show that, if $k = 2h$ is even, $c_{2h} = (-1)^h c_0 / [2^2 \cdot 4^2 \cdot \cdots \cdot (2h)^2] = (-1)^h c_0 / (2^h h!)^2$. Hence

$$f(x) = c_0 \left(1 - \frac{x^2}{(2^1 1!)^2} + \frac{x^4}{(2^2 2!)^2} + \cdots + \frac{(-1)^h x^{2h}}{(2^h h!)^2} + \cdots \right)$$

is a solution of the *DE*, if it converges (as it does, for all x). However, the *DE* is of the second order, so we expect to find another solution. This is a less fortunate case; by methods found in texts, we obtain another solution.

$$\phi(x) = f(x) \ln x + C \left(\frac{x^2}{(2^1 1!)^2} - \left(1 + \frac{1}{2} \right) \frac{x^4}{(2^2 2!)^2} + \left(1 + \frac{1}{2} + \frac{1}{3} \right) \frac{x^6}{(2^3 3!)^2} \right.$$
$$\left. - \cdots + \cdots + (-1)^{h+1} \left(1 + \frac{1}{2} + \cdots + \frac{1}{h} \right) \frac{x^{2h}}{(2^h h!)^2} + \cdots \right)$$

where C is an arbitrary constant. Note however that, as often happens in less fortunate cases, $\phi(x)$ is of no use *at $x = 0$* unless $c_0 = 0$, since otherwise $|f(x) \ln x| \to \infty$ as $x \to 0+$.

11.7. Graphical and Numerical Solutions. Geometrically, an equation $f(x,y,y') = 0$ determines a value of y', hence a *direction*, at each point $P(x,y)$. Further, an equation $f(x,y,y',y'') = 0$ determines a value of y'', hence a *curvature* for each direction (given by y') at each point $P(x,y)$. For first-order equations, then, it is often helpful to construct a *line-element* diagram, i.e., at (a,b) draw a short line with slope y' computed from the *DE* when $x = a$ and $y = b$.

As the adjoining example indicates, such a diagram often reveals useful qualitative information about the solutions. We see that the solutions of $x - yy' = 0$ form a family of equilateral hyperbolas (see 7.31) whose equation turns out to be $x^2 - y^2 = C$. Note the indeterminacy of y' when $x = y = 0$. Such *singular points* of a *DE* require special care.

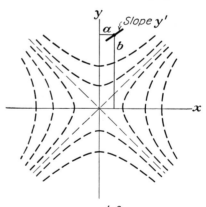

$x\text{-}yy'=0$
Line element diagram
FIG. 11.7.

11.71. To solve $y' = f(x,y)$ *graphically*, beginning at $P_0(a,b)$, chosen so that y will have the value b when $x = a$, draw from P_0 a line l_0 of slope $m_0 = f(a,b)$. On l_0 close to P_0, choose $P_1(a_1,b_1)$. From P_1 draw a line l_1 of slope $m_1 = f(a_1,b_1)$. On l_1 close to P_1, choose $P_2(a_2,b_2)$. From P_2 draw l_2 with slope $m_2 = f(a_2,b_2)$. Continue as far as may be desired. The broken line thus obtained is an approximation to a solution that usually deviates farther from the "true" solution through P_0 the farther one gets away from P_0. The approximation may be improved by taking P_i closer to P_{i-1}, thus requiring more steps to cover the desired range of x, and by using for m_i an average value such as $[f(a_i,b_i) + f(a_{i+1},b_{i+1})]/2$ or $\frac{1}{4}[f(a_{i-1},b_{i-1}) + 2f(a_i,b_i) + f(a_{i+1},b_{i+1})]$.

11.72. To solve $y' = f(x,y)$ approximately by *series*, agreeing that $y = b$ when $x = a$, find y'', y''', $\cdots$ from the *DE* ($y'' = f_x + f_y y' = f_x + f_y f$, etc.) and substitute $x = a$, $y = b$, thus getting the values $y_a^{(k)}$ of the derivatives when $x = a$. Then y is represented approximately by the first few terms of the power series $y = b +$

$\sum_1^\infty y_a^{(k)}(x-a)^k/k!$. Especially if the series shows signs of converging rapidly, values of y computed from its first few terms with x close to a are fairly reliable, may be found quickly, and form a basis for more refined methods of approximation. In principle, of course, this process may be continued to any required degree of accuracy, but in practice the difficulties connected with finding the higher derivatives are often prohibitive, and it may happen that the series converges (if at all) too slowly for practical purposes unless x is very close to a. It is possible, of course, after obtaining a point $(a+h, b+k)$, to begin all over again using this point in place of (a,b), get a point $(a+h+h_1, b+k+k_1)$, and so on.

11.73. Numerical Approximation by Picard's Method. We are given the *DE* $y' = f(x,y)$ and desire a solution such that $y = b$ when $x = a$. A line-element diagram or some other evidence may suggest that y is a function of some specific form y_0, with $y_0 = b$ when $x = a$; if no such suggestion is reliably apparent, put $y_0 = b$ (constant). The first approximation is $y_1 = b + \int_a^x f[t,y_0(t)]dt$. The second is $y_2 = b + \int_a^x f[t,y_1(t)]dt$. The third is $y_3 = b + \int_a^x f[t,y_2(t)]dt$, and so on. If x is kept close enough to a, it can be shown that as $n \to \infty$ the functions $y_n(x)$ tend (uniformly) to a function y that satisfies the *DE*. If the integrations offer difficulty, they may be performed approximately (see 9.3).

Numerous developments of the foregoing methods are available (see refs. 13 and 14).

11.8. Linear Second-order Partial Differential Equations (*PDE*). We give only general remarks on three of these equations. For notation and expressions for $\nabla^2\psi$ in various coordinate systems, see 12.52, 12.6.

Laplace's equation	$\nabla^2\psi = 0$
Heat equation	$\nabla^2\psi = a^2\psi_t$
Wave equation	$\nabla^2\psi = a^2\psi_{tt}$

"General" solutions are often inaccessible and rarely useful. Instead, one seeks particular solutions that satisfy the other conditions of the problem at hand. A useful attack is to express ψ tentatively as a *product* of functions each of which depends on only *one* of the independent variables. If an independent variable is missing in the equation, take the corresponding function to be identically 1.

In rectangular coordinates	$\psi = X(x)Y(y)Z(z)T(t)$
In cylindrical coordinates	$\psi = R(\rho)\Theta(\theta)Z(z)T(t)$
In spherical coordinates	$\psi = R(r)\Theta(\theta)\Phi(\phi)T(t)$

Example. Laplace's equation in polar coordinates in two dimensions is $\rho(\rho\psi_\rho)_\rho + \psi_{\theta\theta} = 0$. $\psi = R(\rho)\Theta(\theta)$ will satisfy the equation if and only if

$$\rho(\rho R')'\Theta + R\Theta'' = 0 \qquad \text{or} \qquad \frac{\rho(\rho R')'}{R} + \frac{\Theta''}{\Theta} = 0$$

In the latter equation, the first term is a function of ρ alone, while the second is a function of θ alone. The equation can thus be satisfied only if each term is a constant, and these constants are numerically equal and of opposite sign. Hence we may write

$$\rho(\rho R')' = \lambda R \qquad \Theta'' = -\lambda\Theta$$

If $\lambda \neq 0$, the solutions of the first are $R = a\rho^{\sqrt{\lambda}} + b\rho^{-\sqrt{\lambda}}$; those of the second are $\Theta = ce^{i\sqrt{\lambda}\theta} + de^{-i\sqrt{\lambda}\theta}$. If $\lambda = 0$, the solutions of the first are $R = a\ln\rho + b$; those of the second are $\Theta = c\theta + d$.

In a particular problem, one usually finds that only certain specific values of λ are permissible if $\psi = R\Theta$ is to satisfy the initial or boundary conditions of the problem. These values are called *eigenvalues* (proper values, characteristic values) of λ. Corresponding to each eigenvalue there will be eigenfunctions R_λ and Θ_λ such that $\psi_\lambda = R_\lambda \Theta_\lambda$ satisfies the *PDE*. Having found these solutions ψ_λ, one seeks constants c_λ such that $\psi = \sum_\lambda c_\lambda \psi_\lambda$ not only satisfies the *PDE* but meets any other conditions imposed by the problem.

In other coordinate systems and for other dimensions, the analysis is similar. See ref. 10, Chap. VII, and ref. 15.

The solution $\psi(x,t)$, for $t > 0$, of the heat equation for an infinite bar, with initial temperature $\psi(x,0) = f(x)$, is sometimes useful in the form

$$\psi(x,t) = \frac{a}{2\sqrt{\pi t}} \int_{-\infty}^{\infty} f(\xi) e^{-\frac{a^2(\xi-x)^2}{4t}} d\xi = \frac{1}{\sqrt{\pi}} \int_{-\infty}^{\infty} f\left(x + \frac{2}{a}\sqrt{t}\,X\right) e^{-X^2} dX$$

where $X = a(\xi - x)/(2\sqrt{t})$.

12. VECTOR ANALYSIS

12.1. Definitions. Vectors are used to represent displacements, velocities, accelerations, momenta, forces, etc. For practical purposes, a vector is best conceived as a directed line segment or arrow leading from an *initial point P* to a *terminal point Q*, and it is then often called the *position vector* of Q relative to the origin P. The *direction* of the vector is the direction from P to Q, and the distance PQ is the *magnitude* or length of the vector.

Notation. Greek letters denote numerical functions and Latin, or Roman, small letters denote numbers, both often called *scalars* to emphasize their nondirectedness. Vectors are denoted by boldface type. The magnitude of the vector $\mathbf{V}$ is denoted by $|\mathbf{V}|$.

The vector $t\mathbf{V} = \mathbf{V}t$, $t \neq 0$ is defined to have the same initial point as $\mathbf{V}$, the same or the opposite direction according as $t > 0$ or $t < 0$, and $|t\mathbf{V}| = |t||\mathbf{V}|$. It is convenient to recognize, at each point P, the vector of zero length (initial and terminal point P) and undefined direction as the *zero vector* at P, and to denote this vector by $\mathbf{0}$. $\mathbf{V} = \mathbf{0}$ if and only if $|\mathbf{V}| = 0$. Complete the above definition by putting $0\mathbf{V} = \mathbf{0}$. Then any vector $\mathbf{V}$ may be *multiplied* by any real number t. The *sum* $\mathbf{U} = \mathbf{V} + \mathbf{W}$ of $\mathbf{V}$ and $\mathbf{W}$ is defined to have the direction and length of the diagonal of the parallelogram with $\mathbf{V}$ and $\mathbf{W}$ as adjacent sides. These operations have the properties

$$\mathbf{V} + \mathbf{W} = \mathbf{W} + \mathbf{V} \qquad \mathbf{U} + (\mathbf{V} + \mathbf{W}) = (\mathbf{U} + \mathbf{V}) + \mathbf{W} \qquad \mathbf{V} + \mathbf{0} = \mathbf{V}$$
$$\mathbf{V} + (-1)\mathbf{V} = \mathbf{0} \qquad ||\mathbf{V}| - |\mathbf{W}|| \leqslant |\mathbf{V} + \mathbf{W}| \leqslant |\mathbf{V}| + |\mathbf{W}|$$
$$t\mathbf{V} + u\mathbf{V} = (t + u)\mathbf{V} \qquad t(\mathbf{V} + \mathbf{W}) = t\mathbf{V} + t\mathbf{W} \qquad |t\mathbf{V}| = |t||\mathbf{V}|$$

REMARKS: 1. Define $-\mathbf{V}$ to be $(-1)\mathbf{V}$; then $\mathbf{V} - \mathbf{V} = \mathbf{0}$.

2. $\mathbf{V}/|\mathbf{V}|$ is a vector of length 1 or a *unit* vector with the same origin and direction as $\mathbf{V}$.

3. $\mathbf{W} - \mathbf{V}$ is equal in length and parallel to the vector *from* the terminal point of $\mathbf{V}$ *to* the terminal point of $\mathbf{W}$.

4. (All vectors with common origin O.) Two (or three) vectors $\mathbf{U}$ and $\mathbf{V}$ (and $\mathbf{W}$) are *dependent* or *independent* according as their terminal points are or are not collinear (or coplanar) with the point O. $\mathbf{U}$ and $\mathbf{V}$ (and $\mathbf{W}$) are dependent or independent according as $a\mathbf{U} + b\mathbf{V} = \mathbf{0}$ (or $a\mathbf{U} + b\mathbf{V} + c\mathbf{W} = \mathbf{0}$) does not or does imply that $a = b = 0$ (or $a = b = c = 0$). If $\mathbf{U}$ and $\mathbf{V}$ (and $\mathbf{W}$) are independent, then every vector $\mathbf{X} \neq \mathbf{0}$ in the plane of $\mathbf{U}$ and $\mathbf{V}$ (or in space) can be expressed in one and only one way as $\mathbf{X} = a\mathbf{U} + b\mathbf{V}$ (or $\mathbf{X} = a\mathbf{U} + b\mathbf{V} + c\mathbf{W}$) with not all the numbers a, b (and c) vanishing.

The *scalar* (or dot) *product* $\mathbf{U} \cdot \mathbf{V}$ is a *number* equal to $|\mathbf{U}||\mathbf{V}| \cos \theta$, where θ is the angle between $\mathbf{U}$ and $\mathbf{V}$. The expression $\mathbf{A} \cdot \mathbf{A} = |\mathbf{A}|^2$ is often written $\mathbf{A}^2$. $\mathbf{U} \cdot \mathbf{V} = 0$ if and only if $|\mathbf{U}| = 0$, or $|\mathbf{V}| = 0$, or $\cos \theta = 0$ and $\mathbf{U}$ and $\mathbf{V}$ are perpendicular.

Properties.

$$\mathbf{U} \cdot \mathbf{V} = \mathbf{V} \cdot \mathbf{U} \qquad \mathbf{U} \cdot (\mathbf{V} + \mathbf{W}) = \mathbf{U} \cdot \mathbf{V} + \mathbf{U} \cdot \mathbf{W}$$

The *vector* (or cross) *product* $\mathbf{W} = \mathbf{U} \times \mathbf{V}$ is the *vector* such that: (1) $\mathbf{W}$ is perpendicular to $\mathbf{U}$ and to $\mathbf{V}$ (and hence to their plane); (2) $\mathbf{W}$ is so directed that, when viewed from its terminal point, the rotation *from* $\mathbf{U}$ *to* $\mathbf{V}$ is positive (counterclockwise); (3) $|\mathbf{W}| = |\mathbf{U}||\mathbf{V}| \sin \theta$. Thus $|\mathbf{W}|$ is the area of the parallelogram with $\mathbf{U}$ and $\mathbf{V}$ as adjacent sides. $\mathbf{W} = 0$ if and only if $|\mathbf{U}| = 0$, or $|\mathbf{V}| = 0$, or $\sin \theta = 0$ and $\mathbf{U}$ and $\mathbf{V}$ are parallel. *N.B.* $\mathbf{A} \times \mathbf{A} = 0$.

Properties. $\quad \mathbf{U} \times \mathbf{V} = -\mathbf{V} \times \mathbf{U} \qquad \mathbf{U} \times (\mathbf{V} + \mathbf{W}) = \mathbf{U} \times \mathbf{V} + \mathbf{U} \times \mathbf{W}$

The *scalar triple product* $\mathbf{U} \cdot \mathbf{V} \times \mathbf{W}$, meaning $\mathbf{U} \cdot (\mathbf{V} \times \mathbf{W})$, is equal to $+$ or $-$ the volume of the parallelepiped with $\mathbf{U}$, $\mathbf{V}$, $\mathbf{W}$ as coinitial edges according as $\mathbf{U}$ is on the same side of the plane of $\mathbf{V}$ and $\mathbf{W}$ as $\mathbf{V} \times \mathbf{W}$ or on the opposite side. Thus $\mathbf{U}$, $\mathbf{V}$, and $\mathbf{W}$ are coplanar if and only if $\mathbf{U} \cdot \mathbf{V} \times \mathbf{W} = 0$. In other words, $\mathbf{U} \cdot \mathbf{V} \times \mathbf{W} = 0$ or $\neq 0$ according as $a\mathbf{U} + b\mathbf{V} + c\mathbf{W} = 0$ for numbers a, b, c not all zero, or not, *i.e.*, according as $\mathbf{U}$, $\mathbf{V}$, and $\mathbf{W}$ are dependent or independent.

Properties.

$$\mathbf{U} \cdot \mathbf{V} \times \mathbf{W} = \mathbf{V} \cdot \mathbf{W} \times \mathbf{U} = \mathbf{W} \cdot \mathbf{U} \times \mathbf{V} = -\mathbf{V} \cdot \mathbf{U} \times \mathbf{W} = -\mathbf{U} \cdot \mathbf{W} \times \mathbf{V}$$
$$= -\mathbf{W} \cdot \mathbf{V} \times \mathbf{U}$$

Dot and cross may be interchanged: $\mathbf{U} \cdot \mathbf{V} \times \mathbf{W} = \mathbf{U} \times \mathbf{V} \cdot \mathbf{W}$, etc.

The *vector triple product* $\mathbf{U} \times (\mathbf{V} \times \mathbf{W})$ is evaluated thus:

$$\mathbf{U} \times (\mathbf{V} \times \mathbf{W}) = (\mathbf{U} \cdot \mathbf{W})\mathbf{V} - (\mathbf{U} \cdot \mathbf{V})\mathbf{W} \qquad (\mathbf{U} \times \mathbf{V}) \times \mathbf{W} = (\mathbf{U} \cdot \mathbf{W})\mathbf{V} - (\mathbf{V} \cdot \mathbf{W})\mathbf{U}$$

Note that, although parentheses are unnecessary on the right but may be used if desired, they are essential on the left. $\mathbf{U} \cdot \mathbf{W}\mathbf{V}$ could only mean the number $\mathbf{U} \cdot \mathbf{W}$ multiplying the vector $\mathbf{V}$. The mere juxtaposition of two symbols for vectors, such as $\mathbf{W}\mathbf{V}$, has no meaning; a dot or cross must be interposed.

$$(\mathbf{U} \times \mathbf{V}) \cdot (\mathbf{W} \times \mathbf{X}) = (\mathbf{U} \cdot \mathbf{W})(\mathbf{V} \cdot \mathbf{X}) - (\mathbf{U} \cdot \mathbf{X})(\mathbf{V} \cdot \mathbf{W})$$

Expand $(\mathbf{U} \times \mathbf{V}) \times (\mathbf{W} \times \mathbf{X})$ by applying twice the rule for a vector triple product; several (equal) results possible.

Caution: *Never* try to divide by a vector. If a vector equation must be solved, try to multiply (dot or cross) by a vector in such a way as to isolate the number or vector sought, or to give it a numerical coefficient by which division will be possible.

In many problems, only the direction and magnitude of a vector are relevant, not its initial point. It is then usual not to distinguish between two parallel vectors of the same sense and length but to give the name *free vector* to the totality of all such vectors (one at each point). The foregoing definitions apply without change to free vectors.

In other circumstances (often with forces), a vector may be restricted to a line or a plane. Such vectors are sometimes called *bound, or sliding,* vectors.

12.2. Components. Let a fixed point O be the origin of a rectangular cartesian coordinate system. The vector $\mathbf{R}$ from O to a point $R(x,y,z)$ is the position vector of the point relative to O. Let $\mathbf{i}$, $\mathbf{j}$, and $\mathbf{k}$ denote, respectively, the (unit) position vectors of the points with coordinates $(1,0,0)$, $(0,1,0)$, and $(0,0,1)$. Then $\mathbf{R} = x\mathbf{i} + y\mathbf{j} + z\mathbf{k}$ (vector addition). The vectors $x\mathbf{i}$, $y\mathbf{j}$, and $z\mathbf{k}$, or, more commonly, the numbers x, y, z are called the *components* of $\mathbf{R}$ along $\mathbf{i}$, $\mathbf{j}$, and $\mathbf{k}$.

$$|\mathbf{R}| = \sqrt{x^2 + y^2 + z^2} = r \text{ (standard notation)} \qquad t\mathbf{R} = tx\mathbf{i} + ty\mathbf{j} + tz\mathbf{k}$$

$$\mathbf{R}_1 + \mathbf{R}_2 = (x_1 + x_2)\mathbf{i} + (y_1 + y_2)\mathbf{j} + (z_1 + z_2)\mathbf{k} \qquad \mathbf{R}_1 \cdot \mathbf{R}_2 = x_1 x_2 + y_1 y_2 + z_1 z_2$$

$$\mathbf{R}_1 \times \mathbf{R}_2 = (y_1 z_2 - z_1 y_2)\mathbf{i} + (z_1 x_2 - x_1 z_2)\mathbf{j} + (x_1 y_2 - y_1 x_2)\mathbf{k} = \begin{vmatrix} \mathbf{i} & \mathbf{j} & \mathbf{k} \\ x_1 & y_1 & z_1 \\ x_2 & y_2 & z_2 \end{vmatrix}$$

$$\mathbf{R}_1 \cdot \mathbf{R}_2 \times \mathbf{R}_3 = \begin{vmatrix} x_1 & y_1 & z_1 \\ x_2 & y_2 & z_2 \\ x_3 & y_3 & z_3 \end{vmatrix} \qquad \mathbf{i}^2 = \mathbf{j}^2 = \mathbf{k}^2 = 1 \qquad \mathbf{i} \cdot \mathbf{j} = \mathbf{j} \cdot \mathbf{k} = \mathbf{k} \cdot \mathbf{i} = 0$$

$$\mathbf{i} \times \mathbf{j} = \mathbf{k} = -\mathbf{j} \times \mathbf{i} \qquad \mathbf{j} \times \mathbf{k} = \mathbf{i} = -\mathbf{k} \times \mathbf{j} \qquad \mathbf{k} \times \mathbf{i} = \mathbf{j} = -\mathbf{i} \times \mathbf{k}$$

12.3. Geometric Applications in three dimensions; can be specialized to two. Here $\mathbf{R}$ is the variable ("running") position vector of a point that moves about on a plane or line, provided that it satisfies the indicated equation. Some of the equations and formulas given in 7.62 arise naturally by expressing the following in components (coordinates):

Line through $\mathbf{R}_0$ parallel to $\mathbf{V}$ $\qquad \mathbf{R} = \mathbf{R}_0 + t\mathbf{V}$

N.B. If $\mathbf{V}$ is a unit vector, t measures the directed distance along the line from $\mathbf{R}_0$ to $\mathbf{R}$

Line through $\mathbf{R}_0$ and $\mathbf{R}_1$ $\qquad \mathbf{R} = \mathbf{R}_0 + t(\mathbf{R}_1 - \mathbf{R}_0)$

Plane through $\mathbf{R}_0$ with normal $\mathbf{N}$ $\qquad (\mathbf{R} - \mathbf{R}_0) \cdot \mathbf{N} = 0$

Distance *from* $(\mathbf{R} - \mathbf{R}_0) \cdot \mathbf{N} = 0$ *to* $\mathbf{R}_1$, measured positively in the direction of $\mathbf{N}$

$$(\mathbf{R}_1 - \mathbf{R}_0) \cdot \mathbf{N}/|\mathbf{N}|$$

Plane through $\mathbf{R}_0$, $\mathbf{R}_1$, and $\mathbf{R}_2$

$$\mathbf{R} = \mathbf{R}_0 + t(\mathbf{R}_1 - \mathbf{R}_0) + u(\mathbf{R}_2 - \mathbf{R}_0)$$

or

$$(\mathbf{R} - \mathbf{R}_0) \cdot (\mathbf{R}_0 - \mathbf{R}_1) \times (\mathbf{R}_1 - \mathbf{R}_2) = 0$$

$\mathbf{R} = \mathbf{R}_0 + t\mathbf{V}$ cuts $(\mathbf{R} - \mathbf{R}_1) \cdot \mathbf{N} = 0$ when $(\mathbf{R}_0 + t\mathbf{V} - \mathbf{R}_1) \cdot \mathbf{N} = 0$, *i.e.*, for $t = (\mathbf{R}_1 - \mathbf{R}_0) \cdot \mathbf{N}/\mathbf{V} \cdot \mathbf{N}$ $\qquad (\mathbf{V} \cdot \mathbf{N} \neq 0)$

The line of intersection of $(\mathbf{R} - \mathbf{R}_0) \cdot \mathbf{N}_0 = 0$ and $(\mathbf{R} - \mathbf{R}_0) \cdot \mathbf{N}_1 = 0$ is parallel to $\mathbf{N}_0 \times \mathbf{N}_1$.

12.4. Vector Functions of a Single Variable. If, for each value of a numerical variable t, a point $U(t)$ is specified along with a definite vector $\mathbf{W}$ at $U(t)$, we say that $\mathbf{W}$ is a function of t and write $\mathbf{W} = \mathbf{W}(t)$. In many important special cases, $U(t)$ is a fixed point while the terminal point of $\mathbf{W}$ varies, the length and direction of $\mathbf{W}$ depending on it. A variable vector $\mathbf{V}$ has a constant vector $\mathbf{A}$ as limit: lim $\mathbf{V} = \mathbf{A}$, or $\mathbf{V} \to \mathbf{A}$, if and only if the initial and terminal points of $\mathbf{V}$ approach, respectively, the initial and terminal points of $\mathbf{A}$. $\mathbf{V} \to \mathbf{A}$ if and only if $|\mathbf{V}| \to |\mathbf{A}|$ and, unless $|\mathbf{A}| = 0$, the direction of $\mathbf{V}$ approaches that of $\mathbf{A}$. Continuity is defined as for numerical functions (see 8.14).

If t changes by an amount Δt, we construct at $U(t)$ a vector $\mathbf{W} + \Delta\mathbf{W}$ equal in length and parallel in direction to $\mathbf{W}(t + \Delta t)$, which is at $U(t + \Delta t)$. Then seek the limit of the vector $\Delta\mathbf{W}/\Delta t$ as $\Delta t \to 0$. If it exists, lim $\Delta\mathbf{W}/\Delta t = \dot{\mathbf{W}}(t) = \mathbf{W}'(t) = d\mathbf{W}/dt$ is the *derivative* of $\mathbf{W}$ with respect to t. If $U(t)$ is fixed, $\dot{\mathbf{W}}$ is tangent to the path γ of the terminal point of $\mathbf{W}$. If s denotes arc length along γ, $|d\mathbf{W}/ds| = 1$. Thus $\mathbf{T} = d\mathbf{W}/ds$ is the *unit* tangent to γ at $\mathbf{W}(t)$. $d\mathbf{T}/ds = d^2\mathbf{W}/ds^2 = \kappa\mathbf{N}$, where κ is the curvature and $\mathbf{N}$ is the unit principal normal (see 10.21) to γ at $\mathbf{W}(t)$. The vector $\mathbf{B} = \mathbf{T} \times \mathbf{N}$ is the *binormal* to γ at $\mathbf{W}(t)$. The arc rate of rotation of the binormal is called the *torsion* τ of γ. It can be shown that $d\mathbf{B}/ds = -\tau\mathbf{N}$ and $d\mathbf{N}/ds = \tau\mathbf{B} - \kappa\mathbf{T}$. These (Frenet) formulas are more complicated if arc length is not the parameter.

Properties of Derivatives.

$$(\mathbf{V} + \mathbf{W})' = \mathbf{V}' + \mathbf{W}' \qquad [\phi(t)\mathbf{V}]' = \phi'\mathbf{V} + \phi\mathbf{V}'$$

If t is a function of u, $d\mathbf{V}/du = (d\mathbf{V}/dt)(dt/du)$.

If $\mathbf{V} = v_1(t)\mathbf{i} + v_2(t)\mathbf{j} + v_3(t)\mathbf{k}$, then $\mathbf{V}' = v_1'\mathbf{i} + v_2'\mathbf{j} + v_3'\mathbf{k}$.

$|\mathbf{V}|$ is constant if and only if $\mathbf{V} \cdot \mathbf{V}' = 0$ for all t.

$\mathbf{V}$ has constant direction if and only if $\mathbf{V} \times \mathbf{V}' = 0$ for all t.

$$(\mathbf{U} \cdot \mathbf{V})' = \mathbf{U}' \cdot \mathbf{V} + \mathbf{U} \cdot \mathbf{V}' \qquad (\mathbf{U} \times \mathbf{V})' = \mathbf{U}' \times \mathbf{V} + \mathbf{U} \times \mathbf{V}'$$
$$(\mathbf{U} \cdot \mathbf{V} \times \mathbf{W})' = \mathbf{U}' \cdot \mathbf{V} \times \mathbf{W} + \mathbf{U} \cdot \mathbf{V}' \times \mathbf{W} + \mathbf{U} \cdot \mathbf{V} \times \mathbf{W}'$$
$$[\mathbf{U} \times (\mathbf{V} \times \mathbf{W})]' = \mathbf{U}' \times (\mathbf{V} \times \mathbf{W}) + \mathbf{U} \times (\mathbf{V}' \times \mathbf{W}) + \mathbf{U} \times (\mathbf{V} \times \mathbf{W}')$$

12.5. Vector Fields. If a vector $\mathbf{V} = \mathbf{V}(x,y,z) = \mathbf{V}(\mathbf{R}) = v_1\mathbf{i} + v_2\mathbf{j} + v_3\mathbf{k}$ is defined at each point $\mathbf{R} = x\mathbf{i} + y\mathbf{j} + z\mathbf{k}$ of a region (see 8.12) of space, the family of these vectors forms a vector *field*. If $\mathbf{V}$ is constant, the family of these vectors is indistinguishable from a free vector. The components v_1, v_2, v_3 of $\mathbf{V}$ are best conceived as related to vectors $\mathbf{i}, \mathbf{j}, \mathbf{k}$ at the point R rather than at the origin O from which $\mathbf{R}$ issues, although the $\mathbf{i}, \mathbf{j}, \mathbf{k}$ at R and those at O are congruent. The initial attention needed to provide $\mathbf{i}, \mathbf{j}, \mathbf{k}$ at R habitually will be amply repaid by extended experience, especially since there is no alternative when curvilinear coordinates are used.

12.51. Gradient. If $\phi(x,y,z) = \phi(\mathbf{R})$ is a function of x, y, z, then $\phi_x\mathbf{i} + \phi_y\mathbf{j} + \phi_z\mathbf{k}$, where $\phi_x = \partial\phi/\partial x$, etc., is a vector field called the *gradient* of ϕ and written grad ϕ or $\nabla\phi$ (say "del ϕ"). The total differential of ϕ is $d\phi = \nabla\phi \cdot d\mathbf{R}$, where $d\mathbf{R} = dx\,\mathbf{i} + dy\,\mathbf{j} + dz\,\mathbf{k}$. If $\mathbf{U}$ is a unit vector, the directional derivative (see 10.1) of ϕ in the direction of $\mathbf{U}$ is $d\phi/ds = \nabla\phi \cdot \mathbf{U}$. The direction of $\nabla\phi$ is the direction in which ϕ *increases* most rapidly, and $|\nabla\phi|$ is equal to the rate of increase of ϕ in that direction. At each point $\mathbf{R}_0$, $\nabla\phi$ is normal to the level surface $\phi(\mathbf{R}) = \phi(\mathbf{R}_0)$.

12.52. Flux. Divergence. Let Σ be a surface inside a region in which a vector field $\mathbf{V}$ is defined, $\mathbf{N}$ the *unit* normal to Σ. Then the surface integral $\iint \mathbf{V} \cdot \mathbf{N}\,da$ over Σ (see 10.5) is the *flux* f of $\mathbf{V}$ across Σ. $\mathbf{V} \cdot \mathbf{N}$ is often called the *normal component* of $\mathbf{V}$, $|\mathbf{V} - \mathbf{V} \cdot \mathbf{N}\mathbf{N}|$ its *tangential component;* they are sometimes denoted by V_n and V_t. If $\mathbf{V}$ represents at each point the velocity of a particle of a moving fluid (density 1 everywhere and always) instantaneously situated at that point, then f measures the rate of total flow across Σ out of the region away from which $\mathbf{N}$ points into the region into which $\mathbf{N}$ points.

Now let Σ be *closed*, and $\mathbf{N}$ directed *outward*. Then f will measure, in the same physical context, the net rate of decrease of the amount of fluid inside Σ, or the rate of expansion of the volume instantaneously occupied by this fluid. The flux f divided by the volume Δv enclosed by Σ gives the average rate of expansion $f/\Delta v$ of the volume per unit volume. Now let P be a point inside Σ, and let Σ shrink toward P in such a way that $\|\Sigma\| \to 0$ (see 8.12). If $f/\Delta v \to l$ regardless of other details of the manner in which Σ shrinks toward P, then this number l, obtainable at each point P from the field $\mathbf{V}$ is called, on account of the interpretation just given, the *divergence* of $\mathbf{V}$, written div $\mathbf{V} = \nabla \cdot \mathbf{V}$ (say "del dot V"). If $\mathbf{V} = v_1\mathbf{i} + v_2\mathbf{j} + v_3\mathbf{k}$, and the derivatives v_{1x}, v_{2y}, v_{3z} are continuous, then the divergence exists and

$$\nabla \cdot \mathbf{V} = v_{1x} + v_{2y} + v_{3z} = \frac{\partial v_1}{\partial x} + \frac{\partial v_2}{\partial y} + \frac{\partial v_3}{\partial z}$$

The *divergence theorem* (see 10.51) says that $\iint \mathbf{V} \cdot \mathbf{N}\,da = \iiint \nabla \cdot \mathbf{V}\,dv$, the surface integral extending over a closed surface Σ whose outward normal is $\mathbf{N}$ and the volume integral extending over the interior of Σ. A field for which $\nabla \cdot \mathbf{V} = 0$ everywhere is called *solenoidal* or *source-free* (see 10.51). div grad $\phi = \nabla \cdot \nabla \phi = \nabla^2 \phi$ ("del squared ϕ") is called the Laplacian of ϕ; if $\nabla^2 \phi = 0$ throughout a region, ϕ is said to be *harmonic* in that region.

Other Theorems.

$$\iiint_V \nabla \phi \, dv = \iint_\Sigma \phi \mathbf{N} \, da \qquad\qquad \iiint_V \nabla \times \mathbf{W} \, dv = \iint_\Sigma \mathbf{N} \times \mathbf{W} \, da$$

12.53. Circulation. Curl. Let γ be a closed curve inside a region in which a vector field $\mathbf{V}$ is defined, s the arc length of γ, and $\mathbf{T}$ its unit tangent. Then $c = \int_\gamma \mathbf{V} \cdot \mathbf{T} \, ds = \int_\gamma \mathbf{V} \cdot d\mathbf{R}$ (see 10.5) is the *circulation* of $\mathbf{V}$ around γ. If $\mathbf{V}$ represents a force experienced by a particle of unit mass placed (on γ) at the initial point of $\mathbf{V}$, then c measures the work done by $\mathbf{V}$ as that particle is carried once around γ.

Now let P be a fixed point, $\mathbf{N}$ a unit vector at that point, π a plane through P normal to $\mathbf{N}$, and γ a closed nonself-crossing curve in π bounding an area Δa (containing P) and so directed that an observer standing on the side of π into which $\mathbf{N}$ points may walk along γ and keep Δa at his left. Compute c for γ and the field $\mathbf{V}$, divide by Δa, and seek a limit as γ shrinks on π toward P in such a way that $\|\gamma\| \to 0$. If $\lim c/\Delta a$ exists regardless of other details of the manner in which γ shrinks toward P, then the number l obtained in this way from the field $\mathbf{V}$, at each point P and for each direction N is the component along $\mathbf{N}$ of a vector called the *curl* or rotation or vorticity of $\mathbf{V}$, written curl $\mathbf{V} = \text{rot } \mathbf{V} = \nabla \times \mathbf{V}$ (say "del cross V"). If $\mathbf{V} = v_1\mathbf{i} + v_2\mathbf{j} + v_3\mathbf{k}$, and the necessary derivatives are continuous, then the curl exists and

$$\nabla \times \mathbf{V} = (v_{3y} - v_{2z})\mathbf{i} + (v_{1z} - v_{3x})\mathbf{j} + (v_{2x} - v_{1y})\mathbf{k}$$

Stokes' theorem (see 10.51) says that $\int_\gamma \mathbf{V} \cdot d\mathbf{R} = \iint \mathbf{N} \cdot \nabla \times \mathbf{V} \, da$, the surface integral being extended over a surface Σ whose complete boundary is γ, with γ and $\mathbf{N}$ having the proper relative directions. A field for which $\nabla \times \mathbf{V} = 0$ is called *irrotational* or *vortex-free* or *lamellar* (see 10.51). If $\mathbf{V}$ represents the velocity of a moving fluid, $\nabla \times \mathbf{V}$ is intimately connected with the vortices of the fluid.

Other Theorems.

$$\int_\gamma \phi \, d\mathbf{R} = \iint_\Sigma \mathbf{N} \times \nabla\phi \, da \qquad\qquad \int_\gamma d\mathbf{R} \times \mathbf{V} = \iint_\Sigma (\mathbf{N} \times \nabla) \times \mathbf{V} \, da$$

12.54. Partial derivatives $\mathbf{V}_x = \partial \mathbf{V}/\partial x = \lim\{[\mathbf{V}(x + \Delta x, y, z) - \mathbf{V}(x,y,z)]/\Delta x\}$ as $\Delta x \to 0$. Similarly for $\mathbf{V}_y$, and $\mathbf{V}_z$ (see 10.1, 12.4). If $\mathbf{V} = v_1\mathbf{i} + v_2\mathbf{j} + v_3\mathbf{k}$, then $\mathbf{V}_x = v_{1x}\mathbf{i} + v_{2x}\mathbf{j} + v_{3x}\mathbf{k}$, etc. The vector $\mathbf{V}_x$ measures the rate of change of $\mathbf{V}$ per unit change in x. If $\mathbf{W} = w_1\mathbf{i} + w_2\mathbf{j} + w_3\mathbf{k}$, then $\mathbf{W} \cdot \nabla \mathbf{V} = (\mathbf{W} \cdot \nabla)\mathbf{V} = w_1\mathbf{V}_x + w_2\mathbf{V}_y + w_3\mathbf{V}_z$ (definition). If $|\mathbf{W}| = 1$, then $\mathbf{W} \cdot \nabla \mathbf{V}$ is the *directional derivative* $d\mathbf{V}/ds$ of $\mathbf{V}$ in the direction of $\mathbf{W}$, *i.e.*, the rate of change of $\mathbf{V}$ per unit change in distance s measured in that direction. In computing $\mathbf{W} \cdot \nabla \mathbf{V}$, *first* form the operator $\mathbf{W} \cdot \nabla = w_1\partial/\partial x + w_2\partial/\partial y + w_3\partial/\partial z$, and *then* apply this operator directly to $\mathbf{V}$. Similarly

$$(\nabla \cdot \nabla)\mathbf{V} = \nabla^2\mathbf{V} = (\partial^2/\partial x^2 + \partial^2/\partial y^2 + \partial^2/\partial z^2)\mathbf{V} = \mathbf{V}_{xx} + \mathbf{V}_{yy} + \mathbf{V}_{zz}$$

Caution. $\nabla \mathbf{V}$, without dot or cross, has no meaning save as a second-order tensor (see the literature).

$N.B.\ \nabla \cdot \mathbf{V} = \mathbf{i} \cdot \mathbf{V}_x + \mathbf{j} \cdot \mathbf{V}_y + \mathbf{k} \cdot \mathbf{V}_z$ and $\nabla \times \mathbf{V} = \mathbf{i} \times \mathbf{V}_x + \mathbf{j} \times \mathbf{V}_y + \mathbf{k} \times \mathbf{V}_z$

Properties of the Foregoing Operators.

$$\nabla(\phi + \psi) = \nabla\phi + \nabla\psi \qquad\qquad \nabla(\phi\psi) = \phi\nabla\psi + \psi\nabla\phi$$
$$\nabla \cdot (\mathbf{U} + \mathbf{V}) = \nabla \cdot \mathbf{U} + \nabla \cdot \mathbf{V} \qquad \nabla \times (\mathbf{U} + \mathbf{V}) = \nabla \times \mathbf{U} + \nabla \times \mathbf{V}$$
$$\nabla \cdot (\phi\mathbf{V}) = \nabla\phi \cdot \mathbf{V} + \phi\nabla \cdot \mathbf{V} \qquad \nabla \times (\phi\mathbf{V}) = \nabla\phi \times \mathbf{V} + \phi\nabla \times \mathbf{V}$$
$$\nabla \times \nabla\phi = 0 \qquad\qquad \nabla \cdot \nabla \times \mathbf{V} = 0$$
$$\nabla \cdot (\mathbf{U} \times \mathbf{V}) = \mathbf{V} \cdot \nabla \times \mathbf{U} - \mathbf{U} \cdot \nabla \times \mathbf{V} \qquad \nabla \times (\nabla \times \mathbf{V}) = \nabla(\nabla \cdot \mathbf{V}) - \nabla^2\mathbf{V}$$

$$\nabla(\mathbf{U} \cdot \mathbf{V}) = \mathbf{U} \cdot \nabla\mathbf{V} + \mathbf{V} \cdot \nabla\mathbf{U} + \mathbf{U} \times (\nabla \times \mathbf{V}) + \mathbf{V} \times (\nabla \times \mathbf{U})$$
$$\nabla \times (\mathbf{U} \times \mathbf{V}) = \mathbf{V} \cdot \nabla\mathbf{U} - \mathbf{U} \cdot \nabla\mathbf{V} + \mathbf{U}\nabla \cdot \mathbf{V} - \mathbf{V}\nabla \cdot \mathbf{U}$$

If $\mathbf{R} = x\mathbf{i} + y\mathbf{j} + z\mathbf{k}$ and $r = |\mathbf{R}|$, then $\nabla \times \mathbf{R} = 0$, $\nabla \cdot \mathbf{R} = 3$, $\nabla r = \mathbf{R}/r$, $\nabla^2(1/r) = 0$, and $\mathbf{V} \cdot \nabla\mathbf{R} = \mathbf{V}$ for any $\mathbf{V}$.

12.6. Curvilinear Coordinates. Let x, y, z be replaced by new coordinate variables, ξ, η, ζ (see 10.12). Assume that ξ, η, ζ form an *orthogonal* system, *i.e.*, at each point the three coordinate curves are mutually perpendicular. (For more complicated cases, see any text on tensor analysis.) If $\mathbf{R} = x\mathbf{i} + y\mathbf{j} + z\mathbf{k}$, then the derivatives $\mathbf{R}_\xi$, $\mathbf{R}_\eta$, $\mathbf{R}_\zeta$ are vectors tangent to the coordinate curves. Let $\mathbf{i}_1 = \mathbf{R}_\xi/|\mathbf{R}_\xi|$, $\mathbf{i}_2 = \mathbf{R}_\eta/|\mathbf{R}_\eta|$, $\mathbf{i}_3 = \mathbf{R}_\zeta/|\mathbf{R}_\zeta|$ be unit vectors tangent to the coordinate curves. These vectors are also normal to the coordinate surfaces: $\mathbf{i}_1 = \nabla\xi/|\nabla\xi|$, $\cdots$, $\mathbf{i}_3 = \nabla\zeta/|\nabla\zeta|$. It is usual to introduce functions $h_1 = |\mathbf{R}_\xi| = 1/|\nabla\xi|$, $h_2 = |\mathbf{R}_\eta| = 1/|\nabla\eta|$, and $h_3 = |\mathbf{R}_\zeta| = 1/|\nabla\zeta|$. Then, symbolically

$$\nabla = \frac{\mathbf{i}_1}{h_1}\frac{\partial}{\partial\xi} + \frac{\mathbf{i}_2}{h_2}\frac{\partial}{\partial\eta} + \frac{\mathbf{i}_3}{h_3}\frac{\partial}{\partial\zeta}$$

Let ϕ, $\mathbf{V} = v_1\mathbf{i}_1 + v_2\mathbf{i}_2 + v_3\mathbf{i}_3$, and $\mathbf{W}$ be expressed in terms of ξ, η, ζ. Then

$$\nabla\phi = \frac{\mathbf{i}_1}{h_1}\phi_\xi + \frac{\mathbf{i}_2}{h_2}\phi_\eta + \frac{\mathbf{i}_3}{h_3}\phi_\zeta = \phi_\xi\nabla\xi + \phi_\eta\nabla\eta + \phi_\zeta\nabla\zeta$$

$$\mathbf{W} \cdot \nabla\mathbf{V} = \frac{w_1}{h_1}\mathbf{V}_\xi + \frac{w_2}{h_2}\mathbf{V}_\eta + \frac{w_3}{h_3}\mathbf{V}_\zeta$$

$$\nabla \cdot \mathbf{V} = \frac{1}{h_1 h_2 h_3}[(h_2 h_3 v_1)_\xi + (h_3 h_1 v_2)_\eta + (h_1 h_2 v_3)_\zeta]$$

$$\nabla^2\phi = \frac{1}{h_1 h_2 h_3}\left[\left(\frac{h_2 h_3}{h_1}\phi_\xi\right)_\xi + \left(\frac{h_3 h_1}{h_2}\phi_\eta\right)_\eta + \left(\frac{h_1 h_2}{h_3}\phi_\zeta\right)_\zeta\right]$$

$$\nabla \times \mathbf{V} = [(h_3 v_3)_\eta - (h_2 v_2)_\zeta]\frac{\mathbf{i}_1}{h_2 h_3} + [(h_1 v_1)_\zeta - (h_3 v_3)_\xi]\frac{\mathbf{i}_2}{h_3 h_1} + [(h_2 v_2)_\xi - (h_1 v_1)_\eta]\frac{\mathbf{i}_3}{h_1 h_2}$$

In cylindrical coordinates ρ, ϕ, z, $h_1 = 1$, $h_2 = \rho$, $h_3 = 1$

and $$\nabla\psi = \mathbf{i}_1\psi_\rho + \frac{\mathbf{i}_2}{\rho}\psi_\phi + \mathbf{i}_3\psi_z$$

$$\nabla \cdot \mathbf{V} = \frac{1}{\rho}(\rho v_1)_\rho + \frac{1}{\rho}v_{2\phi} + v_{3z}$$

$$\nabla^2\psi = \frac{1}{\rho}(\rho\psi_\rho)_\rho + \frac{1}{\rho^2}\psi_{\phi\phi} + \psi_{zz}$$

$$\nabla \times \mathbf{V} = \left(\frac{1}{\rho}v_{3\phi} - v_{2z}\right)\mathbf{i}_1 + (v_{1z} - v_{3\rho})\mathbf{i}_2 + \left(v_{2\rho} - \frac{1}{\rho}v_{1\phi}\right)\mathbf{i}_3$$

In spherical coordinates r, θ, φ, $h_1 = 1$, $h_2 = r \cos \varphi$, $h_3 = r$ and

$$\nabla \psi = \mathbf{i}_1 \psi_r + \frac{\mathbf{i}_2}{r \cos \varphi} \psi_\theta + \frac{\mathbf{i}_3}{r} \psi_\varphi$$

$$\nabla \cdot \mathbf{V} = \frac{1}{r^2}(r^2 v_1)_r + \frac{1}{r \cos \varphi} v_{2\theta} + \frac{1}{r \cos \varphi}(\cos \varphi \, v_3)_\varphi$$

$$\nabla^2 \psi = \frac{1}{r^2}(r^2 \psi_r)_r + \frac{1}{r^2 \cos^2 \varphi} \psi_{\theta\theta} + \frac{1}{r^2 \cos \varphi}(\cos \varphi \, \psi_\varphi)_\varphi$$

$$\nabla \times \mathbf{V} = [v_{3\theta} - (\cos \varphi \, v_2)_\varphi] \frac{\mathbf{i}_1}{r \cos \varphi}$$

$$+ [v_{1\varphi} - (rv_3)_r] \frac{\mathbf{i}_2}{r}$$

$$+ [(r \cos \varphi \, v_2)_r - v_{1\theta}] \frac{\mathbf{i}_3}{r \cos \varphi}$$

12.7. Green's Theorems. *Harmonic Functions.* We are given a closed surface Σ over which and throughout whose interior I surface and volume integrals are to be extended. In the divergence theorem, put $\mathbf{V} = \phi \nabla \psi$. The result is

$$\iiint \nabla \phi \cdot \nabla \psi \, dv = \iint \phi \mathbf{N} \cdot \nabla \psi \, da - \iiint \phi \, \nabla^2 \psi \, dv$$

Interchange ϕ and ψ and subtract, getting

$$\iint (\phi \mathbf{N} \cdot \nabla \psi - \psi \mathbf{N} \cdot \nabla \phi) da = \iiint (\phi \nabla^2 \psi - \psi \nabla^2 \phi) dv$$

These are *Green's theorems.* N.B. $\mathbf{N} \cdot \nabla \phi$ is often called the *normal derivative* of ϕ and written $\partial \phi / \partial n$.

In the first of these theorems, take $\phi = 1$ and suppose that $\nabla^2 \psi = 0$ throughout some region containing Σ, *i.e.*, ψ is *harmonic.* The result is $\iint \mathbf{N} \cdot \nabla \psi \, da = \iint \partial \psi / \partial n \, da = 0$. Thus the flux of the gradient of a harmonic function across any closed surface is zero.

Let r denote distance measured from a fixed point P. Putting $\phi = 1/r$ in the second theorem above, one can show that

$$\iint \left[\frac{1}{r} \frac{\partial \psi}{\partial n} - \psi \frac{\partial}{\partial n}\left(\frac{1}{r}\right) \right] da - \iiint \frac{\nabla^2 \psi}{r} dv = \begin{cases} 0 \\ \text{or} \\ 4\pi\psi(P) \end{cases}$$

according as P is outside or inside Σ. This formula expresses the value of ψ at P in terms of the values of ψ and $\partial \psi / \partial n$ on the boundary and of $\nabla^2 \psi$ in I. If ψ is *harmonic*, then the volume integral disappears and *only* the boundary values are needed. In fact, if Σ is a *sphere* with center at P, $\psi(P)$ is the average of the values of ψ on Σ. Unless a harmonic function ψ is constant, it cannot have a greatest or least value in I. Finally, these results can be used to show that there is essentially just one function ψ harmonic in I and having on Σ either given values of ψ or given values of $\partial \psi / \partial n$. The foregoing theorems are of great practical importance in connection, for example, with the distribution of electric charge.

13. MECHANICS

13.1. Kinematics of a Particle. A moving point or particle is located by its position vector $\mathbf{R} = x\mathbf{i} + y\mathbf{j} + z\mathbf{k}$ referred to a specific origin O at which a specific *frame* $\mathbf{i}$, $\mathbf{j}$, $\mathbf{k}$ of mutually orthogonal unit reference vectors has been installed. (For vectors, see 12.) The motion of O and of the frame is assumed known. The velocity and acceleration of (the terminal point of) $\mathbf{R}$ as measured or observed or computed in this frame will be called its *apparent* velocity and acceleration.

Time will be denoted by t, arc length along the path or trajectory of $\mathbf{R}$ by s, and derivatives with respect to t by dots. The *unit* tangent, normal, and binormal to the path are denoted by $\mathbf{T}$, $\mathbf{N}$, and $\mathbf{B}$. The apparent velocity vector $\mathbf{V}$ and apparent acceleration vector $\mathbf{A}$ are conceived with their initial points at the position $\mathbf{R}$ of the particle. To get cartesian equations, write down from the stated equations a separate equation for each component.

$\mathbf{V} = \dot{\mathbf{R}} = \dot{x}\mathbf{i} + \dot{y}\mathbf{j} + \dot{z}\mathbf{k}$. The *speed* is $\dot{s} = |\mathbf{V}| = v = (\dot{x}^2 + \dot{y}^2 + \dot{z}^2)^{\frac{1}{2}}$, and $\mathbf{V} = v\mathbf{T} = \dot{s}d\mathbf{R}/ds$.

$\mathbf{A} = \dot{\mathbf{V}} = \ddot{\mathbf{R}} = \ddot{x}\mathbf{i} + \ddot{y}\mathbf{j} + \ddot{z}\mathbf{k} = \dot{v}\mathbf{T} + v^2 \dfrac{d\mathbf{T}}{ds} = \dot{v}\mathbf{T} + \kappa v^2\mathbf{N}$, where κ is the curvature of the path (see 10.21, 12.4). The terms $\dot{v}\mathbf{T}$ and $\kappa v^2\mathbf{N}$ are often called, respectively, the *tangential* and *normal* accelerations. They are the components of $\mathbf{A}$ referred to the (moving) frame constituted by the instantaneous positions of $\mathbf{T}$, $\mathbf{N}$, and $\mathbf{B}$. Since the component along $\mathbf{B}$ vanishes, the particle is moving instantaneously in the plane of $\mathbf{T}$ and $\mathbf{N}$ (osculating plane of the path), in which both $\mathbf{V}$ and $\mathbf{A}$ lie.

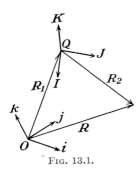

Fig. 13.1.

Suppose now that, at the specified origin O, we have a frame $\mathbf{I}$, $\mathbf{J}$, $\mathbf{K}$, which is moving rigidly. Viewed from the i-j-k frame, this motion will appear to be a turning about the point O. At each instant there will be a vector $\boldsymbol{\Omega}$, called the *angular velocity* vector, about which the I-J-K frame will be instantaneously rotating with angular speed $|\boldsymbol{\Omega}|$. If $\mathbf{U}$ (origin O) is any vector attached rigidly to the I-J-K frame, its apparent rate of change can be shown to be $\dot{\mathbf{U}} = \boldsymbol{\Omega} \times \mathbf{U}$. To emphasize its source in the movement of the I-J-K frame, $\boldsymbol{\Omega} \times \mathbf{U}$ is called the *drag* derivative of $\mathbf{U}$. For example (see 12.4), the angular velocity of the T-N-B frame at points of the path is $\boldsymbol{\Omega} = \kappa\mathbf{B} + \tau\mathbf{T}$, and the Frenet formulas give the drag derivatives of $\mathbf{T}$, $\mathbf{N}$, and $\mathbf{B}$.

If $\mathbf{U} = u_1\mathbf{I} + u_2\mathbf{J} + u_3\mathbf{K}$ is not rigidly attached to the frame, its components u_1, u_2, u_3 will also change. Thus

$$\dot{\mathbf{U}} = \dot{u}_1\mathbf{I} + \dot{u}_2\mathbf{J} + \dot{u}_3\mathbf{K} + u_1\dot{\mathbf{I}} + u_2\dot{\mathbf{J}} + u_3\dot{\mathbf{K}} = \dot{u}_1\mathbf{I} + \dot{u}_2\mathbf{J} + \dot{u}_3\mathbf{K} + \boldsymbol{\Omega} \times \mathbf{U}$$

since $\dot{\mathbf{I}} = \boldsymbol{\Omega} \times \mathbf{I}$, etc. The sum of the first three terms is precisely the rate of change of $\mathbf{U}$ as measured by an observer in the I-J-K frame. This is often called the *relative* derivative and is here denoted by $\dfrac{d'\mathbf{U}}{dt}$. (The notation $\partial\mathbf{U}/\partial t$ is sometimes used but requires great care.)

Result. $\dot{\mathbf{U}} = \dfrac{d'\mathbf{U}}{dt} + \boldsymbol{\Omega} \times \mathbf{U}$: apparent derivative is relative derivative plus drag derivative.

Suppose, finally, that we have a moving point $Q = \mathbf{R}_1$ and an I-J-K frame at Q moving with instantaneous angular velocity $\boldsymbol{\Omega}$. The apparent velocity and acceleration of $\mathbf{R} = \mathbf{R}_1 + \mathbf{R}_2$ are the following (since, in computing $\mathbf{A}$, $\boldsymbol{\Omega} \times \boldsymbol{\Omega} = 0$):

$$\mathbf{V} = \dot{\mathbf{R}} = \dot{\mathbf{R}}_1 + \dot{\mathbf{R}}_2 = \dot{\mathbf{R}}_1 + \frac{d'\mathbf{R}_2}{dt} + \boldsymbol{\Omega} \times \mathbf{R}_2$$

$$\mathbf{A} = \dot{\mathbf{V}} = \ddot{\mathbf{R}}_1 + \frac{d'^2\mathbf{R}_2}{dt^2} + \boldsymbol{\Omega} \times \frac{d'\mathbf{R}_2}{dt} + \left(\frac{d'\boldsymbol{\Omega}}{dt} + \boldsymbol{\Omega} \times \boldsymbol{\Omega}\right) \times \mathbf{R}_2 + \boldsymbol{\Omega} \times \left(\frac{d'\mathbf{R}_2}{dt} + \boldsymbol{\Omega} \times \mathbf{R}_2\right)$$

$$\mathbf{A} = \ddot{\mathbf{R}}_1 + \frac{d'^2\mathbf{R}_2}{dt^2} + 2\boldsymbol{\Omega} \times \frac{d'\mathbf{R}_2}{dt} + \frac{d'\boldsymbol{\Omega}}{dt} \times \mathbf{R}_2 + \boldsymbol{\Omega} \times (\boldsymbol{\Omega} \times \mathbf{R}_2)$$

The term $2\boldsymbol{\Omega} \times d'\mathbf{R}_2/dt$ is called the *Coriolis acceleration*.

Fortunately, it is rarely necessary to use this entire expression for **A**. For instance, if Q and the **I-J-K** frame are fixed in and moving with a rigid body, and $\mathbf{R}_2$ leads always to the same point of the body, then $d'\mathbf{R}_2/dt$ and $d'^2\mathbf{R}_2/dt^2$ both vanish. Again, if $Q = O$, $\mathbf{R}_1$ vanishes. If the component of its angular velocity arising from orbital revolution about the sun is neglected, which may nearly always be done, the earth has constant angular velocity of magnitude $\omega = 2\pi/(24 \times 3{,}600) = 7.292 \times 10^{-5}$ rad per sec about its axis.

In special coordinate systems, it is often advantageous to use a frame determined in a natural way at each point O by the coordinate system.

1. Cylindrical coordinate in space and, with $z = 0$, polar coordinates in the plane· **I** *radial* (direction of increasing ρ), **J** *transverse* (direction of increasing θ), **K** vertical

$$\mathbf{R} = \rho\mathbf{I} + z\mathbf{K} \qquad\qquad \mathbf{\Omega} = \dot\theta\mathbf{K}$$
$$\mathbf{V} = \dot\rho\mathbf{I} + \rho\dot\theta\mathbf{J} + \dot z\mathbf{K} \qquad \mathbf{A} = (\ddot\rho - \rho\dot\theta^2)\mathbf{I} + (\rho\ddot\theta + 2\dot\rho\dot\theta)\mathbf{J} + \ddot z\mathbf{K}$$

2. Spherical coordinates in space, **I, J, K**, in directions of increasing r, θ, ϕ, respectively:

$$\mathbf{R} = r\mathbf{I}$$
$$\mathbf{V} = \dot r\mathbf{I} + r\dot\theta \cos \varphi\mathbf{J} - r\dot\phi\mathbf{K}$$
$$\mathbf{\Omega} = \dot\theta \sin \varphi\mathbf{I} + \dot\phi\mathbf{J} + \dot\theta \cos \varphi\mathbf{K}$$
$$\mathbf{A} = (\ddot r - r\dot\phi^2 - r\dot\theta^2 \cos^2 \varphi)\mathbf{I}$$
$$+ (r\ddot\theta \cos \varphi + 2\dot r\dot\theta \cos \varphi)\mathbf{J}$$
$$+ (-r\ddot\phi - 2\dot r\dot\phi + r\dot\theta^2 \sin \varphi \cos \varphi)\mathbf{K}$$

3. If the origin is a point fixed on the surface of the earth at latitude φ, **I** eastward, **J** northward, and **K** vertical, $\mathbf{R} = x\mathbf{I} + y\mathbf{J} + z\mathbf{K}$, and ω denotes the magnitude of the earth's angular velocity, then $\mathbf{\Omega} = \omega \cos \varphi\mathbf{J} + \omega \sin \varphi\mathbf{K}$ and

$$\mathbf{V} = \frac{d'\mathbf{R}}{dt} + \mathbf{\Omega} \times \mathbf{R} = (\dot x + z\omega \cos \varphi - y\omega \sin \varphi)\mathbf{I} + (\dot y + x\omega \sin \varphi)\mathbf{J} + (\dot z - x\omega \cos \varphi)\mathbf{K}$$

$$\mathbf{A} = \frac{d'\mathbf{V}}{dt} + 2\mathbf{\Omega} \times \mathbf{V} = (\ddot x - 2\dot y\omega \sin \varphi + 2\dot z\omega \cos \varphi - x\omega^2)\mathbf{I}$$

$$+ (\ddot y + 2\dot x\omega \sin \varphi + \omega^2 \sin \varphi \, (z \cos \varphi - y \sin \varphi))\mathbf{J}$$

$$+ (\ddot z - 2\dot x\omega \cos \varphi - \omega^2 \cos \varphi \, (z \cos \varphi - y \sin \varphi))\mathbf{K}$$

The terms of **A** involving ω^2 are much too small to be of practical interest, being of the order of the local variation of g and considerably smaller than various accidental items of which the equations take no account. For practical purposes, then

$$\mathbf{A} = [\ddot x + 2\omega(\dot z \cos \varphi - \dot y \sin \varphi)]\mathbf{I} + (\ddot y + 2\dot x\omega \sin \varphi)\mathbf{J} + (\ddot z - 2\dot x\omega \cos \varphi)\mathbf{K}$$

13.2. Kinetics of a Particle. The vector $\mathbf{M} = m\mathbf{V}$ is the *momentum* of a particle of mass m moving with apparent velocity **V**. If the resultant force acting on the particle is **F**, Newton's law for the motion of the particle is $\dot{\mathbf{M}} = \mathbf{F} = m\mathbf{A}$.

Example. Let gravity be the only force acting. Use the frame of example 3 above, so that $\mathbf{F} = -mg\mathbf{K}$. Then the equation of motion is

$$[\ddot x + 2\omega(\dot z \cos \varphi - \dot y \sin \varphi)]\mathbf{I} + (\ddot y + 2\dot x\omega \sin \varphi)\mathbf{J} + (\ddot z - 2\dot x\omega \cos \varphi)\mathbf{K} = -g\mathbf{K}$$

From the last two components we find that

$$\dot y = b - 2\omega x \sin \varphi \qquad \dot z = c - gt + 2\omega x \cos \varphi$$

where b and c are constants of integration. When these values are used in the equation for $\ddot{x}$, the result is

$$\ddot{x} + 4\omega^2 x = 2\omega(b \sin \varphi - c \cos \varphi) + 2\omega g t \cos \varphi$$

whence
$$x = A \cos 2\omega t + B \sin 2\omega t + \frac{1}{2\omega}[b \sin \varphi + (gt - c) \cos \varphi]$$

where A and B are constants of integration. Using this value for x, y and z may be found by integrating the expressions for $\dot{y}$ and $\dot{z}$.

The *impulse* of a force acting during the period from $t = t_0$ to $t = t_1$ is $\mathbf{I} = \mathbf{M}_1 - \mathbf{M}_0 = \int_{t_0}^{t_1} \mathbf{F}\, dt$. If the duration of the action of a force is negligible, the force is said to be impulsive, and $\mathbf{I}$ is defined to be the change produced in $\mathbf{M}$.

The *moment of momentum* (angular momentum) about a point O, referred to which the position vector of the particle is $\mathbf{R}$, is $\mathbf{H} = \mathbf{R} \times \mathbf{M}$, which is the *moment* of $\mathbf{M}$ about O. Hence, since $\dot{\mathbf{R}} \times \mathbf{V} = 0$, $\dot{\mathbf{H}} = \mathbf{R} \times \dot{\mathbf{M}} = \mathbf{R} \times \mathbf{F} = \mathbf{L}$, where $\mathbf{L} = \mathbf{R} \times \mathbf{F}$ is the moment of $\mathbf{F}$ about O. To get the angular momentum $\mathbf{M}_l$ about a line l, find $\mathbf{H}$ for any point O on the line and resolve along a unit vector $\mathbf{U}$ parallel to the line. Similarly for the moment $\mathbf{F}_l$ of a force $\mathbf{F}$ about the line l. Thus $\mathbf{M}_l = \mathbf{U} \cdot \mathbf{H} = \mathbf{U} \cdot \mathbf{R} \times \mathbf{M}$ and $\mathbf{F}_l = \mathbf{U} \cdot \mathbf{R} \times \mathbf{F}$, whence $\dot{\mathbf{M}}_l = \mathbf{F}_l$.

Work. Energy. The scalar $k = m\mathbf{V}^2/2$ is the *kinetic energy* of the particle. The *work* done by $\mathbf{F}$ on the particle in a displacement from P to Q, along a curve γ, with unit tangent $\mathbf{T}$, is a scalar whose value is computed by the curvilinear integral (see 10.5, 10.6)

$$W(P,Q;\gamma) = \int_{P\gamma}^{Q} \mathbf{F} \cdot \mathbf{T}\, ds,$$

where s measures arc length along γ. When multiplied scalarly by $\mathbf{V}\, dt$, the equation of motion becomes $m\mathbf{V} \cdot \dot{\mathbf{V}}\, dt = \mathbf{F} \cdot \mathbf{V}\, dt = \mathbf{F} \cdot \mathbf{T}\, ds$. Integrating from P to Q along γ, we find $k_Q - k_P = W(P,Q;\gamma)$; the change in k is equal to the work done. If the force $\mathbf{F}$ is an irrotational or conservative field, as the gravitational field is, the integral of $\mathbf{F}$ is independent of γ and $W(P,Q;\gamma) = \phi(P) - \phi(Q)$, where the scalar ϕ is defined, except possibly for an added constant, by $\phi(Q) = -W(O,Q;\gamma)$ for some fixed point O and any path γ from O to Q and is called the *potential energy*. In this case, $k_Q - k_P = \phi(P) - \phi(Q)$, whence $k_P + \phi(P) = k_Q + \phi(Q)$, so that the sum of kinetic and potential energies is *constant* (*conservation of energy, energy equation*). In this case, $\mathbf{F} = -\nabla\phi$ (see 10.51, 12.51).

The change in k caused by an impulsive force with impulse $\mathbf{I}$ is $k_1 - k_0 = \mathbf{I} \cdot (\mathbf{V}_1 + \mathbf{V}_0)/2$.

13.3. System of Particles. For $i = 1, \cdots, n$ (range of all summations), let a particle of mass m_i have position vector $\mathbf{R}_i$, apparent velocity $\mathbf{V}_i$, apparent acceleration $\mathbf{A}_i$, linear momentum $\mathbf{M}_i$, and angular momentum $\mathbf{H}_i = \mathbf{R}_i \times \mathbf{M}_i$ about O, and be acted on by a resultant force $\mathbf{F}_i$ with moment $\mathbf{L}_i = \mathbf{R}_i \times \mathbf{F}_i$ about O.

The momentum of the system is $\mathbf{M} = \Sigma\mathbf{M}_i = \Sigma m_i\mathbf{V}_i$. If a point $\bar{R}$, the center of mass (see 10.6), is defined by the equation $m\bar{\mathbf{R}} = \Sigma m_i\mathbf{R}_i$, where $m = \Sigma m_i$ is the total mass of the system, then $\Sigma m_i\mathbf{V}_i = \Sigma m_i\dot{\mathbf{R}}_i = m\dot{\bar{\mathbf{R}}}$, and $\mathbf{M} = m\dot{\bar{\mathbf{R}}}$. By adding the equations $m_i\dot{\mathbf{V}}_i = \mathbf{F}_i$ and putting $\mathbf{F} = \Sigma\mathbf{F}_i$, we find $\dot{\mathbf{M}} = \mathbf{F} = m\ddot{\bar{\mathbf{R}}}$. Thus $\bar{R}$ moves as if the mass of the whole system were concentrated there and had applied to it all the forces $\mathbf{F}_i$. In particle, *if* $\mathbf{F} = 0$, $\dot{\mathbf{M}} = 0$ so that the linear momentum of the system is constant (*conservation of linear momentum*).

The moment of momentum or angular momentum of the system about O is $\mathbf{H} = \Sigma \mathbf{H}_i = \Sigma \mathbf{R}_i \times m_i \mathbf{V}_i$. Since $\dot{\mathbf{R}}_i \times \mathbf{V}_i = 0$, $\dot{\mathbf{H}} = \Sigma \mathbf{R}_i \times m_i \dot{\mathbf{V}}_i = \Sigma \mathbf{R}_i \times \mathbf{F}_i = \Sigma \mathbf{L}_i = \mathbf{L}$, where $\mathbf{L} = \Sigma \mathbf{L}_i$ denotes the resultant moment about O of all the forces $\mathbf{F}_i$ (*not* the "moment of the resultant"). In particular, *if* $\mathbf{L} = 0$, the angular momentum of the system about O is constant (*conservation of angular momentum*). Similar theorems follow for the moment of momentum about any point and about any line l (see 13.2).

For a continuous distribution, each summation in the preceding statements is replaced by an integral.

13.4. Fluid Kinematics. Suppose a particle (or point) with position vector $\mathbf{R}$ is moving with velocity $\mathbf{V}$. Let $\phi(x,y,z,t)$ be a scalar function, *e.g.*, density, which varies from point to point at a given instant and from time to time at a given point. We are interested in the rate of change of ϕ per unit change in time computed at each instant at the position R of the moving particle. This derivative is called the *particle derivative*, and is denoted by $D\phi/Dt$. At time t, let the particle be at $\mathbf{R}$, and at $t + \Delta t$ let it be at $\mathbf{R} + \Delta \mathbf{R}$. The new value of ϕ is the value at $\mathbf{R} + \Delta \mathbf{R}$ and $t + \Delta t$. The change in ϕ may thus be attained in two steps: (1) a change $\Delta_s \phi$ to the value of ϕ at $\mathbf{R} + \Delta \mathbf{R}$ and time t and (2) a change $\Delta_t \phi$ from the value $\phi + \Delta_s \phi$ to the value at $\mathbf{R} + \Delta \mathbf{R}$ and time $t + \Delta t$. Since t is fixed in the computation of $\Delta_s \phi$, and $\Delta \mathbf{R} = \mathbf{V} \Delta t$ approximately, it follows that $\Delta_s \phi = \mathbf{V} \cdot \nabla \phi \, \Delta t$ approximately. Since the point $\mathbf{R} + \Delta \mathbf{R}$ is fixed during the computation of $\Delta_t \phi$, it follows that $\Delta_t \phi = (\partial \phi/\partial t)\Delta t$ approximately, where the partial derivative is here computed at the point $\mathbf{R} + \Delta \mathbf{R}$. On dividing $\Delta \phi = \Delta_s \phi + \Delta_t \phi$ by Δt and letting $\Delta t \to 0$, we see that

$$\frac{D\phi}{Dt} = \frac{\partial \phi}{\partial t} + \mathbf{V} \cdot \nabla \phi = \frac{\partial \phi}{\partial t} + v_1 \phi_x + v_2 \phi_y + v_3 \phi_z$$

Similarly, if $\mathbf{W}$ is a vector

$$\frac{D\mathbf{W}}{Dt} = \frac{\partial \mathbf{W}}{\partial t} + \mathbf{V} \cdot \nabla \mathbf{W} = \frac{\partial \mathbf{W}}{\partial t} + v_1 \mathbf{W}_x + v_2 \mathbf{W}_y + v_3 \mathbf{W}_z$$

(see 12.54). In particular, the acceleration of the particle is $D\mathbf{V}/Dt = \partial \mathbf{V}/\partial t + \mathbf{V} \cdot \nabla \mathbf{V}$. Note that $\partial \phi/\partial t$, $\partial \mathbf{W}/\partial t$, and $\partial \mathbf{V}/\partial t$ are computed at the point occupied at time t.

Considered as a "rigid body," a small portion of a moving fluid has approximately the velocity $\mathbf{V}$ of any one of its particles and angular velocity $\mathbf{\Omega} = \frac{1}{2}\nabla \times \mathbf{V}$. See also 12.52, 12.53.

13.41. Equation of Continuity. Let Σ be a closed surface fixed in space, I its interior. If ρ denotes density (mass per unit volume), the total mass inside Σ is $\iiint \rho \, dv$, with the integral extended over I. The rate of increase of this mass is the partial derivative of this integral with respect to t. Under conditions on the continuity of ρ which are usually assumed to hold in practice, this derivative is $\iiint (\partial \rho/\partial t) \, dv$. On the other hand, the increase in mass is the total amount flowing in across Σ, *i.e.*, $-\iint \rho \mathbf{V} \cdot \mathbf{N} \, da$, extended over Σ, where $\mathbf{N}$ is the outward normal to Σ (see 12.52). By the divergence theorem $-\iint \rho \mathbf{V} \cdot \mathbf{N} \, da = -\iiint \nabla \cdot (\rho \mathbf{V})dv$. Hence $\iiint [\partial \rho/\partial t + \nabla \cdot (\rho \mathbf{V})]dv = 0$, if no fluid is created or destroyed in I. Since this must hold for *all* closed surfaces Σ, it follows that the integrand vanishes:

$$\frac{\partial \rho}{\partial t} + \nabla \cdot (\rho \mathbf{V}) = \frac{\partial \rho}{\partial t} + \mathbf{V} \cdot \nabla \rho + \rho \nabla \cdot \mathbf{V} = \frac{D\rho}{Dt} + \rho \nabla \cdot \mathbf{V} = 0$$

This is the *equation of continuity*. If the fluid is *incompressible* (ρ constant everywhere and always), the equation becomes $\nabla \cdot \mathbf{V} = 0$. In the case of the wind, the assumption that no creation or destruction takes place seems justified. However, if one were studying the motion of water vapor in the air, then evaporation or condensation might occur at certain points. Such points, where the fluid is "created" or "destroyed," are called *sources* and *sinks* in the general theory. Their presence requires additional terms in the equation of continuity.

13.5. Fluid Kinetics. Let dv be a small element of volume of a fluid, within which there is accordingly a mass $dm = \rho\, dv$. The forces operating on dm are conventionally divided into two classes: (1) *body* forces, which are considered to act on each separate particle of dm and to have intensities proportional to dm and (2) *surface* forces, which are considered to be exerted on the boundaries of dm by contiguous matter and to have intensities proportional to the area da of the portion of the boundary acted on. A fluid is *ideal* or *viscous* (frictional) according as all surface forces do or do not act normal to da. The viscous effect of microscopic friction between particles of a fluid or of a plastic solid has this difference from that observed between a book and a table: any force, however weak, will sufficiently overcome viscosity to produce fluid motion or plastic distortion, while no force weaker than a certain minimum will accelerate the book from its static position on the table.

13.51. Ideal Fluid. The surface force on da in this case is independent of the direction of the normal to da, is called *pressure*, and is denoted by p (units of force per unit area, counted positively when exerted outward). Hence the total force exerted by surrounding matter on the surface of dm is $-\iint p\mathbf{N}\, da$ extended over the surface bounding dm. Since dm is small, this is very nearly $-\nabla p\, dv$ (see 10.6, 12.52). Denoting by $\mathbf{F}$ the body forces per unit mass, we thus have the equation of motion

$$dm\, \frac{D\mathbf{V}}{Dt} = \mathbf{F}\, dm - \nabla p\, dv$$

$$\rho\, \frac{D\mathbf{V}}{Dt} = \rho\, \frac{\partial \mathbf{V}}{\partial t} + \rho \mathbf{V} \cdot \nabla \mathbf{V} = \rho \mathbf{F} - \nabla p$$

$$\frac{\partial \mathbf{V}}{\partial t} + \mathbf{V} \cdot \nabla \mathbf{V} = \mathbf{F} - \frac{1}{\rho} \nabla p$$

or, since

$$\tfrac{1}{2}\nabla \mathbf{V}^2 = \mathbf{V} \cdot \nabla \mathbf{V} + \mathbf{V} \times (\nabla \times \mathbf{V}),$$

$$\frac{\partial \mathbf{V}}{\partial t} - \mathbf{V} \times (\nabla \times \mathbf{V}) = \mathbf{F} - \frac{1}{\rho} \nabla p - \frac{1}{2} \nabla \mathbf{V}^2$$

The flow is *steady* when, at each point of space, the velocity $\mathbf{V}$ of the particle instantaneously at that point is independent of the time. Then $\partial \mathbf{V}/\partial t = 0$ and the equation of motion becomes

$$\mathbf{V} \cdot \nabla \mathbf{V} = \mathbf{F} - \frac{1}{\rho} \nabla p$$

Only in the steady state are the trajectories of the particles the same as the family of curves to which, at each instant, the vectors of the field $\mathbf{V}$ are tangent.

If the force $\mathbf{F}$ is conservative, $\mathbf{F} = -\nabla \phi$, the equation becomes

$$\mathbf{V} \cdot \nabla \mathbf{V} = - \left(\nabla \phi + \frac{1}{\rho} \nabla p \right)$$

The equation of state connecting p, ρ, and the temperature may sometimes be used, especially in a region of constant temperature, to transform the term $(\nabla p)/\rho$.

If the velocity field $\mathbf{V}$ is irrotational, use the second form for the equation of motion, in which $\nabla \times \mathbf{V}$ now vanishes.

$$\frac{\partial \mathbf{V}}{\partial t} + \frac{1}{2} \nabla \mathbf{V}^2 = \mathbf{F} - \frac{\nabla p}{\rho}$$

In the steady case with conservative **F**

$$\frac{1}{2}\nabla V^2 + \nabla\phi + \frac{1}{\rho}\nabla p = 0$$

whence, by integration along a streamline

$$\frac{1}{2}V^2 + \phi + \int \frac{dp}{\rho} = C$$

with different constants, usually, on different streamlines. Especially in the incompressible case, when ρ is constant so that $\int dp/\rho = p/\rho$, this is called *Bernoulli's equation*. In the important special case in which $\phi = gz$, this equation becomes

$$z + \frac{p}{\rho g} + \frac{V^2}{2g} = C$$

In the case of the wind, above the viscous boundary layer, a frame **I-J-K** at a fixed point on the surface of the earth is used (see 13.1). Then the relative derivative DV/Dt must be augmented by the drag derivative $\mathbf{\Omega} \times \mathbf{V}$, where $\mathbf{\Omega}$ is the angular velocity of the earth. Also, $\mathbf{F} = -\nabla\phi$ with $\phi = gz$, so that $\mathbf{F} = -g\mathbf{K}$. The equation of motion now is

$$\frac{\partial \mathbf{V}}{\partial t} + \mathbf{V} \cdot \nabla\mathbf{V} + \mathbf{\Omega} \times \mathbf{V} = -g\mathbf{K} - \frac{1}{\rho}\nabla p$$

Using the expressions for the components of acceleration found in 13.1, we may write the equations

$$\ddot{x} + 2\omega(\dot{z}\cos\varphi - \dot{y}\sin\varphi) = -\frac{1}{\rho}\frac{\partial p}{\partial x}$$

$$\ddot{y} + 2\omega\,\dot{x}\sin\varphi = -\frac{1}{\rho}\frac{\partial p}{\partial y}$$

$$\ddot{z} - 2\omega\dot{x}\cos\varphi = -\frac{1}{\rho}\frac{\partial p}{\partial z} - g$$

13.52. Viscous Winds. It is not easy to take account, in a practically useful way, and with full generality, of the effects of viscosity as they are encountered near the surface of the earth. The general equations contain the divergence of the velocity, which is ignored here. Although the coefficient of viscosity varies with temperature and altitude, the value $\mu = 1.71 \times 10^{-4}$ gram per cm per sec is ordinarily used. The equations of motion are then augmented by an acceleration $\mu(\nabla^2\mathbf{V})/\rho$ on the right. Finally, it is usually sufficient to keep only the term $(\mu/\rho)\partial^2\mathbf{V}/\partial z^2$. With these assumptions, the equations of motion are

$$\ddot{x} + 2\omega(\dot{z}\cos\varphi - \dot{y}\sin\varphi) = -\frac{1}{\rho}\frac{\partial p}{\partial x} + \frac{\mu}{\rho}\frac{\partial^2\dot{x}}{\partial z^2}$$

$$\ddot{y} + 2\omega\dot{x}\sin\varphi = -\frac{1}{\rho}\frac{\partial p}{\partial y} + \frac{\mu}{\rho}\frac{\partial^2\dot{y}}{\partial z^2}$$

$$\ddot{z} - 2\omega\dot{x}\cos\varphi = -\frac{1}{\rho}\frac{\partial p}{\partial y} - g$$

Turbulent motion is still more complicated, involving eddy viscosity whose effect is much stronger than that of the above molecular viscosity μ. For details, see ref. 16 and Sec. VI.

14. EMPIRICAL DATA

14.1. Finding a Formula. It is often advantageous to find a formula $y = f(x)$ which "fits" a given table of values, in the sense that, if $y = b$ when $x = a$ in the table, then the value $f(a)$ of y is b.

14.11. Lagrange Formula. If $(a_i, b_i)(i = 0, \cdots, n)$ are $n + 1$ tabulated pairs of corresponding values, with all the a_i distinct, then

$$f(x) = b_0 \frac{(x - a_1)(x - a_2) \cdots (x - a_n)}{(a_0 - a_1)(a_0 - a_2) \cdots (a_0 - a_n)} + b_1 \frac{(x - a_0)(x - a_2) \cdots (x - a_n)}{(a_1 - a_0)(a_1 - a_2) \cdots (a_1 - a_n)}$$
$$+ \cdots + b_n \frac{(x - a_0)(x - a_1) \cdots (x - a_{n-1})}{(a_n - a_0)(a_n - a_1) \cdots (a_n - a_{n-1})}$$

is the unique polynomial of degree n such that $f(a_i) = b_i$ exactly.

Ordinarily, one wishes a polynomial of lower degree or some other more advantageous expression and must accordingly be content with a formula that, for some or all of the a_i, gives values $f(a_i)$ for y which are satisfactory approximations, in some sense or other, to the b_i.

14.12. Rectification of Data. Plot the data and join the points by a smooth curve. From the appearance of this tentative curve, and from information regarding the physical source of the data, an experienced observer may be able to conjecture the nature of an appropriate function $y = f(x)$. Fortunately, many of the functions that fit empirical data can be expressed, by a change of variables, in linear form. From the given table, a new table may be computed for the new variables, and the data in the new table are then said to be *rectified*.

Actual computation of a new table may often be avoided by plotting on logarithmic or semilogarithmic paper (see 2.6). Again, tests using the given data, perhaps suitably modified, may suffice to reject a proposed relation between x and y. In case 6b below, for example, the equation $\Delta \log y = b \Delta \log x$ tells us to compute from the given (x, y) table a table $(\log x, \log y)$, form the successive differences $\Delta \log x = \log x_{i+1} - \log x_i$ and $\Delta \log y$, and then test to see whether or not $\Delta \log y$ is proportional to $\Delta \log x$. If so, the constant b of proportionality is the exponent of x in $f(x) = ax^b$. On the other hand, suppose that we have found four corresponding pairs in the $(\log x, \log y)$ table, formed the differences, and tested for proportionality with intolerably bad failure. We would then not bother with further expansion of the $(\log x, \log y)$ table, but would look for some other function than ax^b.

14.13. Determination of Constants. The data in the rectified table (u, v) will satisfy approximately an equation of the form $v = \alpha + \beta u$. Three methods are in common use for determining values of α and β.

1. *Graphical.* Plot the (u, v) table of values, and draw a line that seems to "fit" these points as closely as possible. Determine α and β either graphically by measuring the v-intercept α and computing the slope β from appropriate measurements (see 8.9, $y = a + bx$), or algebraically by substituting into $v = \alpha + \beta u$ the coordinates of two points *on* the line and solving for α and β. This method is very ready but quite rough.

2. *Averages.* Divide the (u, v) table into two parts with roughly equal numbers of pairs in the two parts. If $(u_i, v_i)(i = 1, \cdots, p_1)$ is one part, compute $U_1 = \left(\sum_1^{p_1} u_i\right)/p_1$, $V_1 = \left(\sum_1^{p_1} v_i\right)/p_1$, and U_2 and V_2 similarly from the other part. Then the line through (U_1, V_1) and (U_2, V_2) is the required line. This method is usually a satisfactory compromise between economy and reliability.

3. *Least Squares.* Under certain assumptions, it can be shown that the values of α and β that are most probably correct are those so chosen as to make $\sum_1^n (v_{i0} - v_{ic})^2$

Function	Rectified form	Tabular test	Linear graph
1. $y = a + bx$	$y = a + bx$	$\Delta y = b \Delta x$	y vs. x
2. $y = a + \dfrac{b}{x}$	$y = a + b\left(\dfrac{1}{x}\right)$	$\Delta y = b \Delta \left(\dfrac{1}{x}\right)$	y vs. $\dfrac{1}{x}$
3. $\dfrac{1}{y} = a + \dfrac{b}{x}$ or $y = \dfrac{x}{ax+b}$	$\dfrac{1}{y} = a + b\left(\dfrac{1}{x}\right)$	$\Delta \left(\dfrac{1}{y}\right) = b \Delta \left(\dfrac{1}{x}\right)$	$\dfrac{1}{y}$ vs. $\dfrac{1}{x}$
4a.* $\dfrac{1}{y-c} = a + \dfrac{b}{x}$	$\dfrac{1}{y-c} = a + b\left(\dfrac{1}{x}\right)$	$\Delta \left(\dfrac{1}{y-c}\right) = b \Delta \left(\dfrac{1}{x}\right)$	$\dfrac{1}{y-c}$ vs. $\dfrac{1}{x}$
4b. $y = \dfrac{x}{ax+b} + c$	$\dfrac{x-x_k}{y-y_k} = ax_k + b + \dfrac{a}{b}(ax_k+b)x$†	$\dfrac{x-x_k}{y-y_k} = \dfrac{a(ax_k+b)}{b}\Delta x$	$\dfrac{x-x_k}{y-y_k}$ vs. x
5.* $y = a + bx + cx^2$	$\dfrac{y-y_k}{x-x_k} = b + c(x - x_k)$†	$\dfrac{y-y_k}{x-x_k} = c\,\Delta(x - x_k)$	$\dfrac{y-y_k}{x-x_k}$ vs. $x - x_k$
6.* $y = ax^b$	a. $y = a(x^b)$ b. $\log y = \log a + b \log x$	$\Delta y = a\,\Delta(x^b)$ $\Delta \log y = b \Delta \log x$	$\log y$ vs. $\log x$ or y vs. x on logarithmic paper y vs. x^b
7.* $y = a + cx^b$	a. $y = a + c(x^b)$ b. $\log (y - a) = \log c + b \log x$	$\Delta y = c\,\Delta(x^b)$ $\Delta \log (y - a) = b \Delta \log x$	$\log (y - a)$ vs. $\log x$ or $y - a$ vs. x on logarithmic paper y vs. x^b
8. $y = ac^{bx}$	$\log y = \log a + bx \log c$ $\quad = \log a + bx \log e$	$\Delta \log y = b \log e\, \Delta x = \log c\, \Delta x$	$\log y$ vs. x or y vs. x on semilogarithmic paper
9.* $y = a + ce^{bx}$	$\log (y - a) = \log c + bx \log e$	$\Delta \log (y - a) = b \log e\, \Delta x$	$\log (y - a)$ vs. x or $y - a$ vs. x on semilogarithmic paper
10. $y = axe^{bx} = axc^x$	$\log \dfrac{y}{x} = \log a + bx \log e$ $\quad = \log a + x \log c$	$\Delta \log \dfrac{y}{x} = b \log e\, \Delta x$ $\quad = \log c\, \Delta x$	$\log \dfrac{y}{x}$ vs. x or $\dfrac{y}{x}$ vs. x on semilogarithmic paper

*Use 4a if y is known or easily found to have the value c when $x = 0$. Otherwise use 4b, when, as in case 5, the linear equation of the rectified graph will give all the necessary constants. Use 6a or b and 7a or b according as the value of b is or is not known. In case 7, if the value a of y when $x = 0$ is not known or easily found, choose values x_1, x_2, and $x_3 = \sqrt{x_1 x_2}$, preferably appearing in the table, and find the corresponding values y_1, y_2, y_3, interpolating graphically or otherwise (see 14.13) for y_3 unless x_3 is in the table; then $a = (y_1 y_2 - y_3^2)/(y_1 + y_2 - 2y_3)$. In case 9, with x_1, x_2, $x_3 = \frac{1}{2}(x_1 + x_2)$, y_1, y_2, and y_3 chosen similarly, $a = (y_1 y_2 - y_3^2)/(y_1 + y_2 - 2y_3)$ again.

† (x_k, y_k) is a point chosen from the given table.

a minimum, where $v_{ic} = \alpha + \beta u_{i0}$ and (u_{i0}, v_{i0}) are the n observed pairs of values in the rectified table. To find these values of α and β, write down the n equations $v_{i0} = \alpha + \beta u_{i0}$ and add them, and write down the n equations $u_{i0}v_{i0} = \alpha u_{i0} + \beta u_{i0}^2$ and add them. The resulting *normal equations*

$$\Sigma v_{i0} = n\alpha + \beta \Sigma u_{i0} \qquad \Sigma u_{i0}v_{i0} = \alpha \Sigma u_{i0} + \beta \Sigma u_{i0}^2$$

may be solved for α and β. This method is undoubtedly the most reliable, but by far the most laborious. Its use is unwarranted except when quite accurate data are given and quite an accurate result is desired.

If the data (x_i, y_i) do not submit to rectification, other treatment becomes necessary. One may start with a function $f(x)$ suggested by the data as an approximation, compute new data (x_i, y_i^*), where $y_i^* = y_i - f(x_i)$, and try to rectify the new data. Proceeding, obtain successive approximations until a satisfactory fit results. Or, especially if the phenomenon is roughly periodic, try to use a few terms of a Fourier series (see 9.4); an example is treated in ref. 8, par. 62. See also ref. 17. On this point, and for a wealth of details and examples, see ref. 2. For fitting the normal curve, see 14.25.

14.14. Interpolation. Suppose the data consist of *equally spaced* values of x: $a, a + h, a + 2h, \cdots$ and corresponding values of $y: f(a), f(a + h), \cdots$, and that

x	y	Δ	Δ^2	$\cdots$
a	$f(a)$			
$a + h$	$f(a + h)$	$\Delta f(a)$	$\Delta^2 f(a)$	
$a + 2h$	$f(a + 2h)$	$\Delta f(a + h)$	$\Delta^2 f(a + h)$	$\Delta^3 f(a)$
$a + 3h$	$f(a + 3h)$	$\Delta f(a + 2h)$	$\Delta^2 f(a + 2h)$	$\Delta^3 f(a + h)$
$a + 4h$	$f(a + 4h)$	$\Delta f(a + 3h)$	$\Delta^2 f(a + 3h)$	$\Delta^3 f(a + 2h)$
$\cdots$	$\cdots$	$\Delta f(a + 4h)$	$\cdots$	$\Delta^3 f(a + 3h)$
$\cdots$	$\cdots$	$\cdots$	$\cdots$	$\cdots$

we wish a value $f(a + th)$ of y for the value $a + th$, $0 < t < 1$, of x. First construct a *difference table*, as follows: Let $\Delta f(x) = f(x + h) - f(x)$, and, generally, $\Delta^k f(x + mh) = \Delta^{k-1} f[x + (m + 1)h] - \Delta^{k-1} f(x + mh)$. Then (*Newton-Gregory formula*)

$$f(a + th) = f(a) + t\,\Delta f(a) + \frac{t(t - 1)}{2!}\,\Delta^2 f(a) + \frac{t(t - 1)(t - 2)}{3!}\,\Delta^3 f(a) + \cdots$$

continuing as long as substantial contributions accrue [usually not beyond $\Delta^3 f(a)$]. Values of the coefficients $t(t - 1)/2, \cdots$ for convenient values of t are included in many collections of tables. If $\Delta^{k+1} f(a) = \Delta^{k+2} f(a) = \cdots = 0$, or if these differences are neglected, the series reduces to a polynomial of degree k in t. For $k = 1$, we have linear interpolation (see 2.5). Conversely, if $f(x)$ is a polynomial, $\Delta^{k+1} f(x) = 0$ for any x. Hence a necessary and sufficient condition that a polynomial of degree k exist which passes through all the points occurring in a given table is that each of the computed differences Δ^{k+1} should vanish. For the many refinements of the process of interpolation, see ref. 1.

Missing values in a table can occasionally be furnished on an experimental basis. See, for example, ref. 17, pages 69–70.

The process of extrapolation, *i.e.*, of extending a table or even a graph beyond the greatest or least values of its independent variable, is generally highly unreliable unless the equation is known, and it should not be undertaken without well-grounded assumptions.   This is true particularly of data obtained experimentally, because the end values in the table are often observed under the least reliable conditions.

14.2. Statistics.   Except in its rudimentary descriptive aspects, the statistical treatment of empirical data presents essentially difficult and complicated problems on which progress is continual.   To supplement the following general remarks see current journals and ref. 18; ref. 19 for modern theory; ref. 1 for graduation (smoothing), search for periodicities, etc.; ref. 20 for measurements; and ref. 17 (pages 53–76) and Sec. XII for climatic data.

14.21. Frequency Tables.   Histograms.   Suppose that n independent and equally reliable observations $x_1, \cdots, x_n$ of a physical quantity are available, *e.g.*, monthly precipitations for a period of years, temperatures at scattered points of a gas, repeated estimates of the reading of a barometer at a given time.

Unless n is quite small, one will construct a frequency table.   If M is a greatest of the x's and m a least, $M - m$ is called the *range* of the data.   On an x-axis choose a convenient interval containing m and M.   Divide this interval conveniently into k

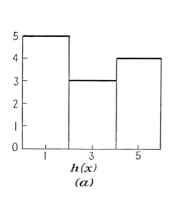

$h(x)$

(a)

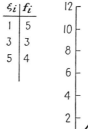

ξ_i	f_i
1	5
3	3
5	4

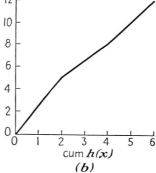

cum $h(x)$

(b)

Fig. 14.21.

smaller intervals of equal length δ, where $\frac{1}{2}\delta$ is not less than and often equal to the allowable error in the specification of the x's.   If temperatures are given to the nearest degree, for instance, then $\delta \geqslant 1$ deg.   Let ξ_i be the value of x at the mid-point of the ith interval, $i = 1, \cdots, k$.   Define $f_i = f(\xi_i)$ to be the number of the given x's that fall between $\xi_i - \frac{1}{2}\delta$ and $\xi_i + \frac{1}{2}\delta$, *i.e.*, in the ith *class*.   Of course $\sum_1^k f_i = n$. The integer f_i states the *frequency* and the fraction f_i/n states the *relative frequency* with which the observed x's fall in the ith class.   The table (ξ_i, f_i) is a *frequency table*.

Define $h(x)$ for any x to have the value f_i if x is in the ith class, and $h(x) = 0$ otherwise, with the exception that various conventions are used, according to convenience, regarding the points $\xi_i \pm \frac{1}{2}\delta$ at the ends of the class intervals, *e.g.*, let $h(\xi_i + \frac{1}{2}\delta) = \frac{1}{2}(f_i + f_{i+1})$.   The graph of $h(x)$ is a *histogram* of the given data and is often used to present a frequency table pictorially.   Again, one often encounters a *cumulative* histogram or *ogive*, which is a graph of $\operatorname{cum}h(x) = 1/\delta \int_{-\infty}^{x} h(t)dt$.   Thus $\delta \operatorname{cum}h(a)$ is the *area* between $h(x)$ and the x-axis and to the left of the ordinate $h(a)$, so that $\operatorname{cum}h(a)$ is the frequency with which x is observed to be not greater than a. For large enough values of x, $\operatorname{cum}h(x) = n$.   Sometimes f_i/n is used instead of f_i in the definition of $h(x)$; then $\operatorname{cum}h(x) = 1$ for large enough x.

Example. The $n = 12$ x's: 0.70, 0.89, 1.23, 5.44, 2.31, 5.88, 4.32, 1.19, 2.38, 3.77, 4.59, 1.75 might represent monthly rainfall, in inches, at some locality. Classified with $\xi_1 = 1$, $\xi_2 = 3$, $\xi_3 = 5$, so that $\delta = 2$, the data would give the table, histogram, and cumulative histogram displayed in Fig. 14.21.

14.22. Continuous Distributions. If the accuracy of measurements can be improved, δ can be decreased, but more intervals will then be necessary. The histogram will have more and narrower steps and may become less and less appreciably distinct from a smooth curve. For many theoretical and practical reasons, one studies curves $y = \phi(x)$, usually so adjusted that $\int_{-\infty}^{\infty} \phi(x)dx = 1$, which may be treated as histograms idealized in this manner and may be used to approximate histograms arising in practice. Just as the area of that part of the histogram between the ordinates $h(a)$ and $h(b)$ gives, when divided by $n\delta$, the relative frequency with which the observed x actually has fallen between a and b, so the area $\int_{a}^{b} \phi(x)dx$ gives the *probability* that an x observed at random will fall between a and b. The function $\phi(x)$ is said to be a continuous *distribution function*, even though it may have jump discontinuities, to emphasize the fact that a random observation may have any one of a continuous set of values rather than being necessarily restricted to a discrete set like the ξ_i ($i = 1, \cdots, k$) above. There are, of course, theoretical probability distributions for discrete variables, *e.g.*, in games of chance. Accordingly, one speaks of a distribution in the discrete case too, meaning a frequency table or a histogram.

14.23. Definitions. The following quantities may be computed from a frequency table (ξ_i, f_i) or its histogram. Summations are for $i = 1, \cdots, k$. If data are not classified into a frequency table, put $k = n$ and, for each i, $\xi_i = x_i$ and $f_i = 1$. Classification introduces errors that are negligible unless the data are quite precise or the classification quite crude; for adjustments to account for these errors in some cases, see Sheppard's corrections in texts. If a continuous distribution function is given, replace the summations by appropriately chosen integrations.

Modes are those ξ_i with a greatest f_i, *i.e.*, with the highest ordinate on the histogram, so that a mode occurs more frequently than any nonmodal observation.

The *median* is that x for which $\text{cumh}(x) = n/2$; observations occur with equal frequency above and below the median. Those x for which $\text{cumh}(x) = n/4$ or $3n/4$ are called the first and third *quartiles*.

The arithmetic *mean* is $\bar{x} = (1/n)\Sigma f_i\xi_i$. It may be interpreted as the abscissa of the centroid of the distribution.

The ith *deviation* from the mean, or *residual*, or (less happily) *error*, is $\xi_i - \bar{x}$. By the definition of $\bar{x}$, $\Sigma f_i(\xi_i - \bar{x}) = 0$ always.

The *mean deviation* is $(1/n)\Sigma f_i|\xi_i - \bar{x}|$.

The *variance* or *dispersion* is $\sigma^2 = (1/n)\Sigma f_i(\xi_i - \bar{x})^2 = (1/n)\Sigma f_i\xi_i^2 - \bar{x}^2$. The positive square root σ of the variance is the *standard deviation*, or the *root mean square deviation*. σ will be small or large according as the distribution is or is not heavily concentrated in the vicinity of $\bar{x}$. *Tchebycheff's inequality:* For any distribution for which σ^2 exists, and for any positive number λ, the probability that $|x - \bar{x}| > \lambda\sigma$ is not greater than $1/\lambda^2$.

The variable $X_i = (\xi_i - \bar{x})/\sigma$ is said to be *standardized*. The distribution of the X's, expressed in the frequency table (X_i, f_i), has mean 0 and variance 1.

The *skewness* of the distribution is $\alpha_3 = (1/n)\Sigma f_i X_i^3$. For a distribution symmetric about $\bar{x}$, $\alpha_3 = 0$. In a rough way, α_3 measures the departure of the distribution from symmetry about the mean. If a distribution has just one mode x^*, then $(\bar{x} - x^*)/\sigma$ is sometimes used instead of α_3 as a measure of skewness.

The *kurtosis* is $\alpha_4 = (1/n)\Sigma f_i X_i^4$, whose size, like that of σ, indicates the dispersion of the ξ's from $\bar{x}$.

For the distribution of the previous example, $(\xi_i, f_i) = (1,5), (3,3), (5,4)$: the mode is $\xi_1 = 1$; the median is $2\frac{2}{3}$; the mean is $\frac{1}{12}(5 \cdot 1 + 3 \cdot 3 + 4 \cdot 5) = 1\frac{7}{6}$; the deviations from the mean are $-1\frac{1}{6}, \frac{1}{6}, 1\frac{3}{6}$; the mean deviation is $\frac{1}{12}(5 \cdot 1\frac{1}{6} + 3 \cdot \frac{1}{6} + 4 \cdot 1\frac{3}{6}) = \frac{55}{36}$; the variance is $\frac{1}{12}(5(1\frac{1}{6})^2 + 3(\frac{1}{6})^2 + 4(1\frac{3}{6})^2) = \frac{107}{36}$; the standard deviation is $\sqrt{107}/6 = 1.72$; the standardized X_i are $-1\frac{1}{6}/(\sqrt{107}/6)$ $= -11/\sqrt{107}, 1/\sqrt{107}, 13/\sqrt{107}$;

$$\alpha_3 = \frac{1}{12}\left(5 \cdot \left(\frac{-11}{\sqrt{107}}\right)^3 + 3\left(\frac{1}{\sqrt{107}}\right)^3 + 4\left(\frac{13}{\sqrt{107}}\right)^3\right) = 0.16$$

14.24. The Normal Curve. The continuous distribution function of greatest practical importance, on which, incidentally, the method of least squares is based, is the *normal* function $y = he^{-h^2(x-\bar{x})^2}/\sqrt{\pi}$, where the constant h is called the *modulus of precision* (see 8.9). In case $\bar{x} = 0$ and $h = 1$, the cumulated function $erf(x) = 2/\sqrt{\pi}\int_0^x e^{-t^2}\,dt$ is called the *error function*. This function is tabulated. To find the probability that a random observation of a normally distributed variable will lie between a and b, find

$$\int_a^b y\,dx = \frac{1}{\sqrt{\pi}}\int_{h(a-\bar{x})}^{h(b-\bar{x})} e^{-t^2}\,dt = \frac{1}{2}\left[erf\left[h(b-\bar{x})\right] \pm erf\left[h(a-\bar{x})\right]\right]$$

Use the tabulated values of *erf* and the plus or minus sign according as a and b are or are not separated by $\bar{x}$. In particular, the probability that a random observation will deviate from $\bar{x}$ by as much as k (on either side) is $\int_{\bar{x}-k}^{\bar{x}+k} y\,dx = 2\int_{\bar{x}}^{\bar{x}+k} y\,dx = erf(hk)$. Note that these probabilities give theoretical *relative* frequencies; to get theoretical frequencies, multiply them by the number n of observations in the case at hand.

The normal distribution has mode, median, and mean $\bar{x}$, mean deviation $\dfrac{1}{\sqrt{\pi}}$

$$\int_{-\infty}^{\infty} h|x - \bar{x}|e^{-h^2(x-\bar{x})^2}\,dx = \frac{1}{h\sqrt{\pi}}, \sigma^2 = 1/(2h^2), \sigma = 1/(h\sqrt{2}),\text{ standardized variable}$$

$X = (x - \bar{x})/(h\sqrt{2})$, $\alpha_3 = 0$, $\alpha_4 = 3$. The *probable error* or *quartile* q is that deviation which is equally likely to be exceeded or not; for a normal distribution, $erf(q) = \frac{1}{2}$ and $q = 0.67449\sigma$. By virtue of the connection between h and σ, the normal distribution is often written $y = \dfrac{1}{\sigma\sqrt{2\pi}} e^{-\frac{(x-\bar{x})^2}{2\sigma^2}}$. The area under this curve lies, approximately: 68 per cent between $\bar{x} - \sigma$ and $\bar{x} + \sigma$; 96 per cent between $\bar{x} - 2\sigma$ and $\bar{x} + 2\sigma$; 100 per cent between $\bar{x} - 3\sigma$ and $\bar{x} + 3\sigma$; actually less than 0.3 per cent beyond $\bar{x} \pm 3\sigma$.

14.25. Fitting the Normal Curve. Many empirical distributions (ξ_i, f_i) can be satisfactorily approximated by the normal curve. If this is possible, use $\bar{x}$ and σ computed from the data as the $\bar{x}$ and σ in the equation. To test for normality, draw a histogram and approximate it by a smooth curve, which should resemble the normal curve fairly closely. The relative frequencies of deviations from the mean not exceeding σ, 2σ, and 3σ should be approximately those given above the normal curve. The skewness and kurtosis should not differ too greatly from 0 and 3, respectively. For the more delicate χ^2 test, see texts.

14.26. Precision of Measurements. In case $x_1, \cdots, x_n$ are (ideally) simultaneous observations of the same quantity, *e.g.*, readings of a barometer, it is commonly assumed that the *accidental* or *random* errors in the observations are distributed normally about the true value. On this assumption, the most probable true value is the arithmetic mean $\bar{x}$ computed from the observations. The precision of the measurement is indicated by the standard deviation σ or the probable error 0.67449σ computed from the data, more exactly, by $\sigma' = \sigma \sqrt{n/(n-1)}$ or $0.67449\sigma'$. In order to exclude satisfactorily the effect of accidental errors, it is rarely economical to take more than 10 or 12 readings before striking an average.

In case one of the readings x_i is suspiciously far out of line, compute $\bar{x}$ with x_i included in the data, and use the normal curve to find the probability of getting a deviation as large as $|x_i - \bar{x}|$. If this probability is less than $1/2n$, discard x_i (Chauvenet's criterion). This is ordinarily justified if $|x_i - \bar{x}| > 3.1\sigma$, or approximately 4.5 probable errors.

14.27. Correlation (for details see ref. 18). Suppose that n independent and equally reliable *pairs* of observations $(x_1,y_1), \cdots , (x_n,y_n)$ of a pair of physical quantities are available, e.g., height and girth of n men, monthly precipitation and average temperature at a given locality for a period of years. The points (x_i,y_i) located in the plane will form a *scatter diagram*. By classifying the x's and the y's separately, as in 14.21, one gets a cross-classification into kK *classes*, with centers at $(\xi_i,\eta_I)(i = 1, \cdots ,k; I = 1, \cdots ,K)$, and a *frequency table* (ξ_i,η_I,f_{iI}). Of course, $n = \Sigma f_{iI}$ summed for $i = 1, \cdots , k$ and $I = 1, \cdots , K$ independently. The frequency table is commonly displayed as a double-entry table, with columns labeled ξ_i, rows labeled η_I, and the entry f_{iI} in the ξ_i, η_I position in the table. The histogram is now a family of rectangular prisms constructed and interpreted in a natural manner.

Continuous distribution functions $\phi(x,y)$ are also encountered. The probability that $a < x < b$ and $c < y < d$ simultaneously is then $\int_a^b \int_c^d \phi(x,y)dy\, dx$. In dealing with continuous distributions, replace summations in the following remarks by appropriate integrations.

By fixing one of the variables and summing over the other, one gets the *marginal distributions*. Thus the marginal table for the ξ's is (ξ_i,f_i), where $f_i = \sum_1^K f_{iI}$ over I, while that for the η's is (η_I,f_I), where $f_I = \sum_1^k f_{iI}$ over i. Of course, $\Sigma f_i = n = \Sigma f_I$. One finds from the marginal distributions, separately for each variable: $\bar{x}$, $\bar{y}$, deviations $\xi_i - \bar{x}$ and $\eta_I - \bar{y}$, σ_x, σ_y, and standardized variables $X_i = (\xi_i - \bar{x})/\sigma_x$ and $Y_I = (\eta_I - \bar{y})/\sigma_y$. If the data have not been classified, i and I both run from 1 to $n = k = K$.

It is usually desired, when a value of one of the variables is known, to form an idea of the mean and standard deviation of the corresponding distribution of the other. Giving the point (ξ_i,η_I) the weight f_{iI}, fit a line to the table (ξ_i,η_I) by least squares (see 14.13). The result is

$$y - \bar{y} = r\frac{\sigma_y}{\sigma_x}(x - \bar{x}), \text{ i.e., } Y = rX \qquad \text{or} \qquad x - \bar{x} = r\frac{\sigma_x}{\sigma_y}(y - \bar{y}), \text{ i.e., } X = rY$$

according as the sum of the squared vertical or horizontal deviations from the line is required to be a minimum. These lines are the *regression lines* of y on x and of x on y, respectively, and could equally well be found by fitting the column means or the row means. The coefficient

$$r = \frac{1}{n}\sum f_{iI}X_iY_I = \frac{1}{n\sigma_x\sigma_y}\sum f_{iI}(\xi_i - \bar{x})(\eta_I - \bar{y}) = \frac{1}{\sigma_x\sigma_y}\left(\frac{1}{n}\sum f_{iI}\xi_i\eta_I - \bar{x}\bar{y}\right)$$

summed for i and I running independently over their ranges, is called the product-moment *correlation coefficient*. The number $(1/n)\Sigma f_{iI}(\xi_i - \bar{x})(\eta_I - \bar{y})$ is sometimes called the *covariance* of x and y. The standard deviation of r is $(1 - r^2)/\sqrt{n}$.

Suppose now that a value a of x is given. Substituting in the regression equation gives $y = b_a = \bar{y} + r\sigma_y(a - \bar{x})/\sigma_x$. The number b_a is then an estimated value of y when x is known to have the value a. The best estimate, however (see 14.26), is the mean $\bar{y}_a$ of the distribution of those y's for which $x = a$. If, for each a, $b_a = \bar{y}_a$, regression is said to be *linear*. Unless regression is approximately linear, b_a may not be a useful estimate of y, and r should be employed only with caution. Instead of computing for each a the standard deviation of y, given $x = a$, about the estimated value b_a, it is usual to use for all x the *standard error of estimate* $S_y = \sigma_y \sqrt{1 - r^2}$, the minimum of whose square is sought in finding the regression line. Given $x = a$, the theory thus estimates $y = b_a$ with a standard deviation S_y.

The regression line of x on y may be used similarly to estimate x when y is given. Then $S_x = \sigma_x \sqrt{1 - r^2}$.

It can be shown that $-1 \leqslant r \leqslant 1$ always. If $r = 0$, the regression lines are the horizontal line $y = \bar{y}$ and the vertical line $x = \bar{x}$, while $S_y = \sigma_y$ and $S_x = \sigma_x$. In this case, knowledge of either variable does not improve an estimate of the other. At the other extreme, when $r = \pm 1$, the two lines coincide, $S_y = S_x = 0$, and all the observations lie on the line. Between the extremes, especially for normal distributions, and depending on the value of r, it is possible to make rather definite statements regarding the probabilities for one variable when the other is known.

The commonest continuous distribution in two variables is the normal distribution

$$\phi(x,y) = \frac{1}{2\pi\sigma_x\sigma_y \sqrt{1 - r^2}} e^{-\frac{1}{2(1-r^2)} \left(\frac{x^2}{\sigma_x^2} - \frac{2rxy}{\sigma_x\sigma_y} + \frac{y^2}{\sigma_y^2} \right)}$$

Regression is linear.

It is not safe to conclude from a numerically large value of r that there is any causal connection between x and y. Thus the rainfalls in Minneapolis and St. Paul doubtless have r nearly 1 but are common effects of the same complex of causes, rather than either being the cause of the other.

Bibliography

1. WHITTAKER, E. T., and G. ROBINSON: "The Calculus of Observations," Blackie, London, 1932.
2. LIPKA, J.: "Graphical and Mechanical Computation," Wiley, New York, 1918.
3. DAVIS, D. S.: "Empirical Equations and Nomography," McGraw-Hill, New York, 1943.
4. VAN VOORHIS, M. G.: "How to Make Alignment Charts," McGraw-Hill, New York, 1937.
5. HARDY, G. H.: "Pure Mathematics," Chap. X, Cambridge, London, 1933.
6. DICKSON, L. E.: "First Course in the Theory of Equations," p. 95, Wiley, New York, 1922.
7. *Am. Math. Monthly*, **40**: 462–465 (1942).
8. HEWSON, E. W., and R. W. LONGLEY: "Meteorology: Theoretical and Applied," Part I, Wiley, New York, 1944.
9. KNOPP, K.: "Theory and Application of Infinite Series," Blackie, London, 1928.
10. MARGENAU, H., and G. M. MURPHY: "Mathematics of Physics and Chemistry," Chap. I, Van Nostrand, New York, 1943.
11. AGNEW, R. P.: "Differential Equations," McGraw-Hill, New York, 1942.
12. PIAGGIO, H. T. H.: "Differential Equations," Chaps. IX and X, G. Bell, London, 1931.
13. SHERWOOD, T. K., and C. E. REED: "Applied Mathematics in Chemical Engineering," pp. 112–140, McGraw-Hill, New York, 1939.
14. *Bull. Nat. Research Council*, 1933.
15. CHURCHILL, R. V.: "Fourier Series and Boundary Value Problems," McGraw-Hill, New York, 1941.

16. HAURWITZ, B.: "Dynamic Meteorology," Chap. X, McGraw-Hill, New York, 1941.
17. LANDSBERG, H.: "Physical Climatology," par. 24, The Pennsylvania State College, 1941.
18. KENNEY, J. F.: "Mathematics of Statistics," Van Nostrand, New York, 1939.
19. WILKS, S. S.: "Mathematical Statistics," Princeton University Press, Princeton, N. J., 1943.
20. WORTHING, A. G., and J. GEFFNER: "Treatment of Experimental Data," Wiley, New York, 1943.

SECTION III

PHYSICS OF ATMOSPHERIC PHENOMENA

By H. G. Houghton, S. C. Lowell, and K. B. McEachron

CONTENTS

PHYSICS OF ATMOSPHERIC PHENOMENA

By H. G. Houghton, S. C. Lowell, and K. B. McEachron

VISIBILITY

By H. G. Houghton

Introduction. The term *visibility* as commonly used in aviation and in meteorology means the maximum horizontal distance at which prominent objects can be seen or can be recognized. Because the word visibility has a different meaning in popular usage, Bennett[1] and also Middleton[2] have suggested the more specific term *visual range* for the meteorological element. However, since this is written for meteorologists, the word visibility rather than visual range will be used. The definition of visibility as adopted by the International Meteorological Organization[3] specifies it as the distance at which a prominent object such as a house or a tree can be recognized. This is a difficult definition to follow in practice, since the identification of an object depends to a considerable extent on familiarity and on the nature of the object. It is also quite unsatisfactory from a theoretical point of view. For these reasons and to conform with usual American practice, the visibility will be defined here as the maximum distance at which an object can be seen against its background.

Although Bergeron recognized that visibility was at times a useful characteristic property for the identification of air masses, only very limited use of it in this way has been made. This is due in part to the effects of local sources of atmospheric pollution, nonuniformity of visibility observations, and the modifying effects of condensation and precipitation. The reverse problem of forecasting the visibility from a knowledge of the synoptic conditions and local factors is of considerable importance. A more careful and systematic use of visibility as a meteorological element might lead to a better understanding of the forecasting problem.

The visibility is of major importance in the operation of aircraft. With the perfection of blind-flying techniques, visibility has ceased to be so critical along the route but has become more important at terminals because of the poorer flying conditions permitted by the instrumental developments. Although instrument approach and landing systems are being perfected, such landings are still more hazardous and require much more time to execute than a normal contact approach and landing. Conditions of poor visibility and low ceiling often result in serious congestion and delay in the air above a busy air terminal. For aviation purposes, it is the lower visibilities, under 3 miles, say, that are the most important. (A visibility of 3 miles or less requires instrument flight procedures.) Such conditions are usually due to fog, precipitation, dense haze, or smoke. The variation of the visibility with elevation is also of importance, as will be pointed out below.

Daytime Visibility. Simple observations reveal that, the more distant an object is from the observer, the brighter it appears. If the objects are viewed against the horizon sky, a distance will finally be reached at which the brightness of the object so closely approaches that of the sky background that the object becomes invisible. This limiting distance is the visibility. The increase in the apparent brightness of the object with the distance from the observer is due to the sun- and skylight scattered back toward the observer by the atmospheric suspensoids in the optical path. The scattering particles, in order of increasing size, range from the air molecules through condensation nuclei, smoke, dust, and fog to precipitation elements. Particles that are smaller than the wave length of light, such as air molecules, condensation nuclei, and smoke, scatter blue light more strongly than red. This accounts for the blue appearance of distant objects often observed when there is light haze or light smoke present. Particles larger than the wave length of light, such as dust, fog, and precipitation, scatter nonselectively. Since most poor visibility is due to these larger particles, the scattered light is white except when the substance from which the particles are formed is colored, as in dust storms and certain smokes. The angular distribution of the scattered light is dependent on the size of the particles and on the spatial distribution of the incident light. It might appear from the above brief discussion of scattering that a theoretical treatment of the visibility as a scattering problem would be extremely difficult. Most of the complications can be avoided by an ingenious treatment due to Koschmieder.[4]

Consider a black object viewed against the horizon sky through an atmosphere having a scattering coefficient k. (The scattering coefficient is the fraction of the incident light scattered per unit distance along the beam.) Further, let the flux density due to sunlight and skylight be E_0, and assume that this is constant along the optical path. This implies that the sky is either perfectly clear or uniformly clouded. As will be shown later, this assumption does not greatly affect the generality of the result.

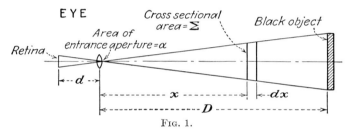

Fig. 1.

Figure 1 is a schematic representation of a black object at a distance D from the eye. The first problem is to compute the apparent brightness of the black object due to the light scattered into the eye from the cone subtended by the object. For the present purpose, the brightness B may be represented by the flux density produced on the retina of the eye. Consider an elementary slice of the volume of length dx and cross section Σ. Since the scattering coefficient is k and the incident radiation is E_0, the total quantity of light scattered by the elementary volume is $kE_0\Sigma\,dx$. The scattered light per unit solid angle scattered toward the eye is some fraction c of the total. The factor c is a function of the angular distribution of the sun- and skylight and the size of the scattering particles, and is assumed to be independent of x. Thus the scattered light per unit solid angle in the direction of the eye from the elementary volume is $dI = ckE_0\Sigma\,dx$.

Owing to the scattering loss between the elementary volume and the eye, only a fraction e^{-kx} of this will reach the eye. If the entrance aperture of the eye has an area α, the solid angle subtended by the eye is α/x^2. Hence the total scattered flux entering the eye from the elementary volume is

$$dF = dI\,\frac{\alpha}{x^2}\,e^{-kx}$$

$$= cE_0\,\Sigma\alpha\,\frac{k}{x^2}\,e^{-kx}\,dx$$

If d is the focal length of the eye, the area of the image of the object on the retina is, by simple geometric optics, $\Sigma d^2/x^2$. The flux density on the retina, or the brightness, is then dF divided by the area of the image, or

$$dB_0 = cE_0 \frac{\alpha}{d^2} ke^{-kz}\, dx$$

The total apparent brightness of the black object is obtained by integration from $x = 0$ to $x = D$, which gives

$$B_0 = cE_0 \frac{\alpha}{d^2}(1 - e^{-kd}) \qquad (1)$$

Now, as pointed out by Koschmieder,[4] the brightness of the sky immediately adjacent to the black object must be given by the same expression integrated to $x = \infty$ since the sky brightness is produced in the same manner as the apparent brightness of the black object. This readily gives

$$B_b = cE_0 \frac{\alpha}{d^2} \qquad (2)$$

In the absence of color contrast, an object is visible as long as the ratio of the difference between the brightness of object and background and the background brightness exceeds a certain threshold value. In symbols, the limiting condition is

$$\frac{B_b - B_0}{B_b} \geqq \epsilon \qquad (3)$$

where ϵ is the brightness contrast threshold of the eye. For the average eye under daylight conditions, ϵ has a value of 0.01 to 0.02. Returning to the expressions for the apparent brightness of a black object and of the horizon sky, it is clear that the visibility will be equal to the distance D_m at which the brightness contrast is equal to the threshold value ϵ.

$$\frac{B_b - B_0}{B_b} = 1 - (1 - e^{-kD_m}) = \epsilon$$

or

$$e^{-kD_m} = \epsilon$$

$$D_m = \frac{1}{k}\ln\frac{1}{\epsilon} \qquad (4)$$

where D_m is the visibility of a black object against the horizon sky. Since the lines of sight to the object and to the sky background are adjacent and since the brightness of the object is so nearly equal to that of the background, variations in E_0, c, and k with x would have compensating effects on B_0 and B_b. It is therefore believed that Eq. (4) should be generally true for all atmospheric conditions. It is to be noted that the visibility is independent of the nature of the atmospheric suspensoids and also of the sun and sky illumination. Taking ϵ as a constant, the visibility is inversely proportional to the scattering coefficient, which is a physical property of the atmosphere. Actually ϵ is subject to some variations that will be discussed below.

If the object is not black but has a diffuse reflection coefficient A, the brightness of the object is the sum of the brightness due to the scattered light in the optical path and the brightness of the object itself. A general expression can be derived only for the case of a uniformly clouded sky, since otherwise the brightness of the object is a function of its orientation with respect to the sun. The visibility of an object of reflection coefficient A seen against the horizon sky is[2]

$$D_m = \frac{1}{k}\ln\left(\frac{1 - A/2}{\epsilon}\right) \qquad (5)$$

It will be noted that if $A = 0$ (black object), this expression reduces to Eq. (4). For a perfectly white object ($A = 1$) and $\epsilon = 0.02$, the ratio of the visibility of the white object to that of a black object for the same scattering coefficient is about 0.82.

Since perfectly white objects are seldom found, the difference would normally be less than this. If the sun is not obscured, the above result will not hold, particularly if the object is a specular reflector or if the sun is behind the observer. If the object is not dark-colored (albedo > 0.25), it should not be viewed with the sun behind the plane of the observer if large errors are to be avoided. It is often stated that colored objects should not be used as visibility marks. However, if the objects do not have a large albedo, most of the apparent brightness at the visual range will be due to the scattered light, and the apparent color of the object will not be much affected by its actual color.

The only background considered in the above developments has been the horizon sky. A similar expression may be derived for the visibility of an object of reflectivity A_0 viewed against a background of reflectivity A_b where the background and object are at the same distance from the observer. It must be assumed that the sky is uniformly clouded. The result is[2]

$$D_m = \frac{1}{k} \ln \left[\frac{1}{\epsilon} \left(\frac{|A_0 - A_b|}{2} + \epsilon - \frac{\epsilon A_0}{2} \right) \right] \tag{6}$$

If there is a reasonable difference between A_0 and A_b, the terms ϵ and $\epsilon A_0/2$ may be neglected. The visibility depends not only on the scattering coefficient, but also on the difference between the reflection coefficients of the object and its background. (Note that $|A_0 - A_b|$ is a numerical difference and is never negative.) Reflection coefficients vary widely for different objects and backgrounds, and their values are not commonly available.

Of the three formulas for the visibility given above, Eq. (4) is the simplest, is valid for all weather conditions, and expresses the visibility only in terms of a physical property of the atmosphere and a constant of the eye. For these reasons, it seems that the visibility of a black object against the horizon sky should be taken as the standard visibility, although it is recognized that this standard cannot always be attained in practice.

The threshold brightness contrast ratio ϵ has been considered as a constant above, but it is subject to considerable variation under certain conditions. The value of ϵ varies from one person to another, but independent visibility estimates made by trained observers using the same marks seldom show significant differences. The value of ϵ increases quite rapidly when the general illumination level falls below a certain minimum, which is, however, well below usual daylight values. The apparent value of ϵ also increases rapidly as the angle subtended by the object at the eye becomes less than about 2 deg. The variation of ϵ with the angle subtended by the object is shown in the table below, in which the value of ϵ is taken as unity for an angle of 3 deg.

Subtended angle, deg.	0.25	0.5	1.0	2.0	3.0	5.0
Relative value of ϵ	8.6	4.0	2.0	1.2	1.0	1.0

This effect is due to light diffracted by the edges of the object into the line of sight. The table shows that small objects such as chimneys, flagpoles, etc., are not desirable markers for the estimation of the visibility. The apparent value of ϵ may also be decreased by the use of objects that subtend angles greater than about 5 deg. Objects of excessive size partly shade the optical path, thus decreasing the scattered light. Since small particles scatter mostly in a forward direction, this effect is quite important. A source of light in the field of view such as the sun also produces an apparent increase in ϵ. This usually can be avoided by a proper choice of direction, but, if not, the object should be viewed through a blackened tube that will exclude the light from the "dazzle source." Finally, it also appears that the apparent value of ϵ increases when the visibility is very poor, as in dense fog. It has been found that, when the visibility is about 300 m, the apparent value of ϵ is about doubled; and, when the visibility is about 50 m, ϵ is approximately quadrupled. In considering the effect of variations of ϵ on the visibility, it should be remembered that the visibility is proportional to $\ln 1/\epsilon$ and hence that a twofold increase in ϵ will correspond to a decrease of only about 15 per cent in the visibility.

The scattering coefficient in the lower atmosphere is usually determined by the suspended particulate matter such as dust, smoke, fog, etc. In the presence of fog, precipitation, or smoke, the visibility may be a function of the direction of sight. Present practice is to report a visibility such that equal or greater values exist over at least half of the horizon. If the visibility in any quadrant is significantly different from the average visibility, it is reported separately with the direction indicated. This is particularly important in the case of terminal visibilities of less than 3 miles. Variability in the visibility should also be reported if the visibility is poor.

Owing to the normal stable stratification of the atmosphere and the fact that most of the atmospheric suspensoids are introduced near the ground, the scattering

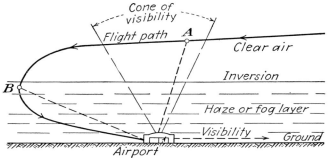

Fig. 2.

coefficient usually decreases with elevation. This is often marked when an inversion is present that keeps the smoke, dust, etc., in the layer between the ground and the inversion.

The visibility as transmitted to the pilot of an approaching airplane is normally based on observations made at the ground level in a horizontal direction. This information is directly applicable only during the latter part of the landing approach and often does not tell the pilot the maximum distance from which he can see the terminal. This is due both to the decrease of the scattering coefficient with elevation and to the terrestrial background of the objects viewed by the pilot. Normally these effects act in opposition, but only rarely do they exactly compensate for each other. No general rule can be given, although it is evident that, in the case of a well-stirred atmosphere (approximately adiabatic lapse rate), the effect of the terrestrial background will be more important than the change of scattering coefficient with elevation, and the visibility from the airplane will be less than that observed at the ground, assuming the ground observer uses marks seen against the sky. The reverse case of a very stable stratification or inversion condition with thick smoke, haze, or ground fog may cause an even greater discrepancy with the pilot being able to see much farther than the ground observer. This condition often confuses inexperienced pilots, because the terminal may be easily visible from overhead but be lost to sight in the landing approach. Such a case is illustrated in Fig. 2. It is assumed here that the horizontal visibility below the inversion is constant and greater than the height of the inversion. The visibility above the inversion is assumed very large compared with that below it. From a point such as A, the terminal will be visible since the path through the haze of fog layer is considerably less than the visibility. At some point such as B, the terminal will disappear because of the increased length of the path in the haze. (This limiting path length within the haze layer is less than the visibility because of the effect of the terrestrial background.) From all points outside the "cone of visibility" determined by the limiting point B, the terminal will be invisible.

In an effort to eliminate some of the difficulties discussed above, there may be a tendency to use visibility markers viewed against a terrestrial background, or markers located at as high an elevation as possible. This is not desirable in view of the effect of changing atmospheric conditions. If it is necessary to interpret the visibility as determined at the ground, the pilot is usually in a better position to do so than the observer. Also, the horizontal visibility at the ground applies to the most critical part of the landing procedure. It is much more important to follow standard methods for determining the visibility so that the information furnished the pilot will be comparable at all terminals.

Visibility at Night. The visibility of nonluminous objects at night follows the same laws as the visibility during the daytime, but the illumination at night is so low that the value of ϵ is large and of indeterminable value. It has been estimated that the visibility in full moonlight is about 20 per cent of the visibility in daylight in the same atmosphere.

The visibility at night is usually taken to mean the maximum distance at which a light is visible. The flux density at a distance x from a point source of candle power I_0 in an atmosphere having a scattering coefficient k is given by

$$E = \frac{I_0 e^{-kx}}{x^2} \tag{7}$$

The light will be visible as long as the flux density at the eye exceeds the threshold value E_0. Thus the visibility of a light is

$$D_m = \frac{1}{k}\left(\ln \frac{I_0}{E_0} - 2 \ln D_m \right) \tag{8}$$

This can be solved for D_m by "cut and try" or from charts if I_0 and E_0 are known. The value of E_0 is dependent on the background illumination and the degree of dark adaptation of the eye. For absolute darkness and complete dark adaptation, $E_0 \approx 10^{-13}$ lumens per cm^2. (One lumen is the luminous flux in one solid angle from a source of one candle power.) Langmuir and Westendorp[5] found the following empirical relationship between E_0 and the background brightness B:

$$E_0 = 3.5 \times 10^{-9} \sqrt{B}$$

For full moonlight, this gives a value of E_0 of about 2×10^{-11} lumens per cm^2. It is probable that the value of E_0 is subject to much more variation under usual conditions than ϵ, the threshold brightness contrast ratio.

The visibility of a light is also dependent on its candle power, which may vary from less than 100 to several million candle power in the beam of a beacon.

By combining Eqs. (4) and (8), it is possible to express the visibility of a black object against the horizon sky in daytime D_d, in terms of the visibility D_m of a light of candle power I_0 at night, in an atmosphere of the same scattering coefficient. The result is

$$D_d = \frac{D_n \log \epsilon}{\log E_0 - \log I_0 + \log D_n} \tag{9}$$

The uncertainty as to the proper value of E_0 makes this expression unsatisfactory for computing D_d as a function of D_n. M. G. Bennett[1] has carried out a series of experimental determinations of the relationship, and his results are given in Table 1.

TABLE 1[1]

Daytime visibility	Visual range of lights at night in the same atmosphere			
	1 cp	10^2 cp	10^4 cp	10^6 cp
27 yd	38 yd	53 yd	70 yd	87 yd
55	69	111	139	174
110	125	204	269	347
220	226	371	525	689
550	451	829	1,240	1,690
1,100	738	1,490	1$\frac{1}{3}$ mi.	2 mi.
1$\frac{1}{4}$ mi.	1,140	1$\frac{1}{2}$ mi.	2$\frac{1}{2}$ mi.	3$\frac{2}{3}$
2$\frac{1}{2}$ mi.	1,650 yd	2$\frac{1}{4}$	4$\frac{2}{3}$	7
4$\frac{1}{3}$	1$\frac{1}{4}$ mi.	3$\frac{3}{4}$	7$\frac{1}{2}$	12
6$\frac{1}{4}$	1$\frac{1}{3}$	4$\frac{3}{4}$	10	16
12$\frac{1}{2}$	1$\frac{1}{2}$	7$\frac{1}{2}$	18	30
18	1$\frac{2}{3}$	9	24	44
31	1$\frac{3}{4}$	11	34	67

[1] The values above the horizontal line are experimental; those below are extrapolated.

By inserting the data of Table 1 in Eq. (8), it is found that the value of E_0 is about 3×10^{-11} lumens per cm^2 if ϵ is taken as 0.01. However, there is a range of at least 10 to 1 in the computed values of E_0, and a change in ϵ to 0.02 reduces the computed value of E_0 by a factor of 100. These results suggest the desirability of more experimental investigations of the type carried out by Bennett.

Table 1 clearly indicates the great dependence of the nighttime visibility on the candle power of the light observed. For visibilities under 1,000 yd, almost any light that might ordinarily be used will give a visibility considerably greater than the corresponding daytime visibility. This at least partly explains the frequently reported increase in visibility from late afternoon to evening under conditions of poor visibility. To make nighttime visibilities directly comparable with daytime visibilities, it is evident that the candle power of the lights used as marks should increase with their distance from the point of observation. It is not usually feasible to provide a special chain of lights for this purpose. Alternatively, the equivalent visibility of a light of known candle power may be determined from Table 1. For example, if a 100-cp light is just visible at 200 yd, the equivalent daytime visibility is 110 yd. Airport boundary and obstruction lights, street lights, etc., are usually about 100 cp, and they may be used for visibilities up to about 4 miles, although the visibility will be overestimated in the lower ranges. Beacons and lighthouses may have the equivalent of from one to several million candle power in their beams, and the visibility of such lights may be several times the daytime visibility. Most instructions to observers warn against the use of such lights as visibility markers. However, if no other light is available at the proper distance, beacons must be used. With the aid of Bennett's table, some correction can be made. It should also be remembered that an airman is interested first in how far he can see the airway and airport beacons, and later in how far he can see the landing and obstruction lights. These two visibilities may differ by a factor of 2 or more, as is apparent from the table. The decrease of the scattering coefficient with altitude may be even more marked at night than during the day because of the greater stability. This will always make the visibility from the air greater than from the ground, because at night the background is not a determining factor as it is in the daytime.

It has been found that flashing lights are not picked up more readily than steady lights. The visibility of a flashing light is determined by its mean candle power, *i.e.*, by its actual candle power multiplied by the ratio of the time on to the time of a complete cycle. Under laboratory conditions, red lights are picked up better than any other color of equal candle power, but in practice the difference appears to be negligible.

The visibility of point sources is much greater than that of a diffuse extended source. It appears that the threshold flux density for large diffuse sources is of the order of 10,000 times that for a point source.

Instrumental Methods. At most stations, it will be found impossible or impractical to locate a sufficient number of visibility marks that satisfy the requirements discussed above. In many cases, relatively large distances exist between suitable marks. Under such conditions, the observer is instructed to estimate visibilities intermediate between the marks by noting the sharpness with which the nearer mark stands out and its apparent color. It is extremely difficult to make reliable estimates of the visibility in this fashion.

Local differences of this kind in the visibility reports from neighboring stations lead to inconsistencies that are very confusing to pilots. One possible solution to this problem is the adoption of a standard instrumental procedure for the determination of the visibility. The significant factor that determines the visibility is the scattering coefficient. If this is known, the visibility can be computed from Eqs. (4) and (8) if standard values of the threshold brightness contrast ratio ϵ and of the threshold flux density E_0 are adopted. The scattering coefficient can be determined by measuring the transmission of light over a path of known length. Instruments for this purpose are expensive and must be placed with a fixed orientation, with the result that the indication may not always be representative. It appears that a visual-comparison photometer can be designed that will measure the ratio of the apparent brightness of a black object and the horizon sky. The scattering coefficient could then be readily computed.

Several empirical visibility meters have been developed in Europe, in which ground-glass disks of increasing roughness are interposed between the eye and an object until the latter disappears. These instruments do not determine the scattering coefficient, but they may be calibrated empirically to read the visibility. Although they have been found to be useful in Europe, such instruments have not been used in this country.

Neither of the instruments mentioned above is suitable for measuring the visibility of lights at night. However, a comparison photometer with a self-contained reference light source has been constructed for this purpose,[6] and this might be combined with the photometer for daylight observations to produce a universal instrument.

Forecasting of the Visibility. The forecasting of the visibility is largely a local problem. The most important condition to forecast is poor visibility (3 miles or less). The reduction of visibility below 3 miles may be due to fog, smoke, dust, exceptionally dense haze, or precipitation (especially snow and heavy rain). The forecasting of precipitation and of fog is too broad a problem to be considered here, and reference should be made to the next part of this Section and to Sec. X for a discussion of these subjects.   Haze is thought to consist of a suspension of very small particles which may be either solid or liquid.   The distinction between haze and smoke or dust is not always very clear.   The concentration of suspensoids is dependent on the past history of the air (particularly its trajectory and the amount of precipitation it has yielded) and on its vertical stability.   Since air masses reaching a station with a given wind direction often have similar life histories, there is usually a fairly good correlation between wind direction and visibility.   The effects of local sources of pollution such as smoke and dust are often marked.   The reduction in visibility in such cases is dependent upon the amount of smoke discharged, the state of the ground (in case of dust), the wind direction, wind velocity, and the stability of the atmosphere. Stability, particularly when an inversion is present, tends to confine the smoke or dust to the layer near the ground.   With a given amount of smoke being discharged, the greater the wind velocity the smaller is the concentration of smoke in the air. Also, high wind velocities cause vertical mixing which distributes the suspensoids through a deeper layer and hence improves the horizontal visibility at the ground. For given ground conditions, the amount of dust added to the air increases with the wind velocity.   Because of the effect of vertical stability, the visibility tends to be a minimum in early morning and a maximum in the early afternoon.

As will be appreciated from the above discussion, local factors are of major importance in determining the visibility, and this is equally true of fog and, to some extent, of precipitation.   A careful statistical study of visibility at a given station, based on an understanding of the physical principles involved, will generally greatly improve the visibility forecasts.

Summary of Suggestions and Rules for the Estimation of the Visibility.   For convenience, the more important points that bear on the practical aspects of the estimation of the visibility are summarized below.   A more complete discussion of each item has already been given in the preceding pages.

1. Objects used for daytime visibility marks should be silhouetted against the horizon sky and should be as dark as possible, *i.e.*, very light colors or glossy surfaces should be avoided.

2. All objects used as daytime visibility marks should subtend an angle of at least 0.5 deg and preferably not more than 5 deg at the eye.

3. Visibility marks should not have greatly different elevations, because of the variation of the scattering coefficient with altitude.

4. Even if completely black, the objects should be so oriented that the sun is not in the field of view at observation time.   If the object is not of low albedo (0.3 or less), the sun should preferably not be behind the plane of the observer.

5. If possible, separate series of marks should be available in more than one azimuth so that the variation of visibility with direction can be determined.

6. At night, low candle power noncollimated lights should be used for the shorter distances (*e.g.*, boundary lights, obstruction lights, streetlights, etc.).   Beacons and lighthouses if used at all should be selected only for the most distant marks.

7. Lights should not be used if they must be viewed against a background of general illumination such as will be produced by a city.

8. Estimates of the visibility at night should not be made until the eyes become dark-adapted.   (At least several minutes—30 min may be required for complete dark-adaptation.)

9. For details as to the method of coding and reporting the visibility, the observer should be familiar with Circular N^7 or its equivalent.

Bibliography

1. BENNETT, M. G.: *Quart. J. Roy. Meteorolog. Soc.*, **61**: 179–188 (1935).
2. MIDDLETON, W. E. K.: "Visibility in Meteorology," 2d ed., The University of Toronto Press, 1941.

3. Organisation Météorologique Internationale, Conférence des directeurs à Varsovie, 1935, Vol. 1 (O.M.I. No. 29).
4. Koschmieder, H.: *Beitr. Phys. freien Atmos.*, **12** : 33–53, 171–181 (1924).
5. Langmuir, I. and W. F. Westendorp: *Physics*, 1 : 273–317 (1931).
6. Middleton, W. E. K.: *Gerlands Beitr. Geophys.*, **44** : 358–375 (1935).
7. Instructions for Airway Meteorological Service, *U.S. Weather Bur. Circ. N*, 5th ed., 1941.

CONDENSATION AND PRECIPITATION

By S. C. Lowell

Saturation and Supersaturation. Air is defined to be *saturated* when the water vapor it contains would be in equilibrium with a plane surface of pure water at the same temperature as the air, *i.e.*, when the rates of transition of water molecules from the liquid to the gaseous phase and from the gaseous to the liquid phase are equal. The partial pressure of the water vapor in the air is then called the *saturation vapor pressure* corresponding to the given temperature, or conversely we may speak of the *saturation temperature* corresponding to a given vapor pressure. The relation between saturation vapor pressure and temperature is shown in Fig. 34, page 81, and in Table 68, page 70. The *relative humidity* of moist air is defined to be the ratio of the existing vapor pressure at the given temperature to the saturation vapor pressure at the same temperature. If the air contains more water vapor than is required to saturate it at the given temperature (*i.e.*, the relative humidity is greater than 100 per cent), then we say that it is *supersaturated*. Supersaturation may be produced in many ways, *e.g.*, by mixing of two air masses of different temperature, both of which are at or near saturation, by radiational and contact cooling, and, of particular importance in the atmosphere, by adiabatic expansion.

The saturation vapor pressure was defined above for equilibrium with a plane surface of pure water. In a water cloud, the liquid water is in the form of spherical droplets, which, as we shall see later, are solutions of electrolytes and may in addition carry free electrical charges, all these factors influencing the equilibrium vapor pressure over the water droplet. We shall consider the effects separately.

The equilibrium vapor pressure e_r' over a drop of pure water of radius r is greater than the saturation vapor pressure e' defined above, as was shown by Lord Kelvin,[1] who derived the following relation:

$$\rho R T \ln \frac{e_r'}{e'} = \frac{2\sigma}{r} \tag{1}$$

where ρ is the density of the water, R and T are the gas constant and absolute temperature of the vapor, and σ is the surface tension of the water drop. Equation (1) shows that at a given temperature the equilibrium vapor pressure increases as the droplet radius decreases. This means that evaporation can take place from droplets of very small radius into air whose relative humidity is greater than 100 per cent.

If the water contains an electrolyte in solution, the equilibrium vapor pressure over the surface of the solution is reduced according to Raoult's law, the ratio of the equilibrium vapor pressure e_s' to the saturation vapor pressure e' being equal to the molar concentration of the water in the solution, *i.e.*,

$$\frac{e_s'}{e'} = \frac{m}{m + M} \tag{2}$$

where $m/(m + M)$ is the molar fraction of water contained in the total mass $m + M$ of water and electrolyte. But for a water drop, $m + M$ is proportional to the volume, hence to r^3, and we may write

$$\frac{e_s'}{e'} = \frac{km}{r^3} \tag{3}$$

where k is a constant of proportionality, and m is the mass of water in the drop.

252

If the water drop carries a free charge E then, as was shown by J. J. Thomson,[2] Eq. (1) must be modified as follows:

$$\rho RT \ln \frac{e_r'}{e'} = \frac{2\sigma}{r} - \frac{E^2}{8\pi r^4} \qquad (4)$$

Actually the correction is of little importance, since very minute droplets seldom carry more than one electronic charge, and the first term on the right-hand side far outbalances the second.

Of greater importance in the atmosphere in its effect upon equilibrium vapor pressures is the coexistence of water in all three phases, solid, liquid, and vapor. As a glance at Table 68, page 70 shows, the saturation vapor pressure over water at temperatures below 0°C is appreciably greater than the saturation vapor pressure over ice at the same temperature. Air that is saturated with respect to the water

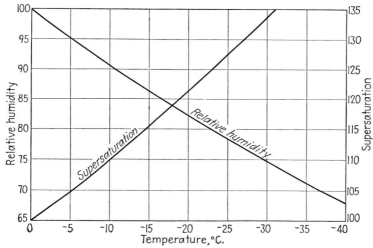

Fig. 1.—Relative humidity corresponding to saturation over an ice surface (scale on left) and supersaturation over ice surface corresponding to 100 per cent relative humidity (scale on right).

droplets will be supersaturated with respect to the ice particles. Hence, if ice crystals and water droplets coexist, in a cloud, say, there will be a vapor-pressure gradient such that water molecules will evaporate from the liquid-drop surfaces and sublime on the ice-crystal surfaces. The relative humidity corresponding to saturation over an ice surface is shown in Fig. 1. The importance of this in connection with the occurrence of ice-crystal clouds such as cirrus is immediately evident. For example, at a temperature of $-20°C$ the humidity need only be about 83 per cent for the air to be saturated with respect to an ice surface. Figure 1 also shows the supersaturation that exists over an ice surface at different temperatures when the relative humidity is 100 per cent. For example, at a temperature of $-10°C$ the vapor pressure at a relative humidity of 100 per cent is 10 per cent greater than the equilibrium vapor pressure over an ice surface, and there is a considerable tendency for water vapor to sublime on any ice crystals that may be present. The importance of this effect on the colloidal instability of water-ice clouds will be brought out later.

There is an additional supersaturation process that may occasionally be important. If a cold drop of water is brought into juxtaposition to a drop at a higher temperature, by turbulent mixing, say, then, since the equilibrium vapor pressure over the cold drop is lower than that over the warmer drop, there will be a tendency for water molecules to evaporate from the warm drop, lowering its temperature, and to condense on the cold drop, raising its temperature. The process is similar to that occurring between liquid drops and ice particles. It is interesting to compare the magnitude of the two effects. The dashed curve of Fig. 2 shows the difference between the saturation vapor pressures over water and over ice as a function of the air temperature. The curve shows a maximum difference of 0.27 mb at about $-12°C$. Curve A shows the temperature difference between adjacent water droplets that would cause a saturation-vapor-pressure difference of the same magnitude as the maximum difference between water and ice, *i.e.*, 0.27 mb. Curve B shows the temperature difference between adjacent water droplets that would give a difference in saturation pressure equal to one-half the maximum difference between water and ice, *i.e.*, 0.13 mb. Curve C shows the temperature corresponding to one-fifth the maximum water-ice effect, *i.e.*, 0.05 mb. For very low air temperatures, impossibly high temperature differences between adjacent water drops would be required in order to get vapor-pressure gradients of a magnitude comparable to the water-ice effect, but this is by

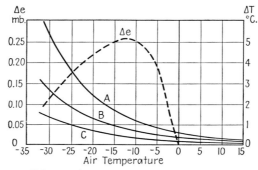

Fig. 2.—Temperature difference between neighboring cloud elements as a factor in producing supersaturation. (*Courtesy of S. Petterssen.*)

no means true at higher temperatures, where a fraction of a degree centigrade may suffice.

Condensation and Sublimation. The process of transition from the invisible-vapor state to the visible-liquid state is called condensation. At temperatures below 0°C, water vapor may pass directly to the solid state, in which case the process is called sublimation. Both processes are accompanied by the release of considerable quantities of heat.

Condensation and sublimation are not direct consequences of a state of saturation or even slight supersaturation. Laboratory experiments performed on air that has been carefully washed and filtered through cotton gauze to remove all foreign particles present in it show that a supersaturation of at least 420 per cent is required to cause condensation. Cloud-chamber photographs show that in this case the condensation takes place first on the negative ions. With a further increase in supersaturation (to about 600 per cent), the positive ions act as the *condensation nuclei*, and at 790 per cent all ions are active as nuclei. With air that has not been freed of the suspended solid and liquid particles, it is found that condensation may even commence before saturation is reached. Thus there seems definite evidence that the condensing (or sublimation) of water vapor in the atmosphere requires the presence of a nucleus for each droplet (or ice crystal) that is formed.

Most of the foreign particles in the air seem to be of terrestrial origin and find their way into the air by numerous processes. The breaking of surf or waves, for instance, causes small droplets to be carried into the air from the ocean surface. When the droplet evaporates, a residue of salts (chiefly chlorides) is left suspended in the air and is carried to higher levels of the atmosphere. Sandstorms or gusty, turbulent winds inject billions of small particles into the air. Combustion processes furnish finely divided carbon and, more important, sulfur dioxide, which ultimately becomes sulfuric acid owing to the oxidizing action of sunlight [$SO_2 + (O) \rightarrow SO_3$; $SO_3 + H_2O \rightarrow H_2SO_4$]. These foreign particles vary greatly in size, ranging from 10^{-3} to approximately 10 μ in radius. Their number also varies widely, from several million per cubic centimeter for heavily polluted air to only a few per cubic centimeter in mountainous regions or at high levels where the air is very pure. Landsberg[3] has shown that the nuclei count undergoes wide changes due to the varying conditions of air mass, relative humidity, trajectory, wind, electrical potential, and radioactivity, and that definite diurnal and annual variations are indicated.

The foreign particles existing in the atmosphere may be divided roughly into three categories. Hygroscopic particles, such as sodium chloride and sulfuric acid, make up one large group. Generally speaking they act as condensation nuclei, even at temperatures well below 0°C, probably because, although they are of nuclear dimensions, they exist as strong salt solutions with a consequent lowering of their freezing point. The condensation nuclei may begin to gather moisture at humidities as low as 70 per cent. This accounts, in large part, for the extremely hazy appearance of the sky before a rain. The increasing relative humidity allows the hygroscopic particles in the air to start absorbing moisture and increase in size. The air figuratively "rains haze" before any actual precipitation occurs.

Minute crystalline particles, such as silica or quartz grains, are thought to act as *sublimation nuclei*, according to Wegener.[4] This theory is in part substantiated by the fact that melted firn ice, *i.e.*, the partly compacted granular snow at the upper end of a glacier, will not conduct electricity and hence contains no free ions, whereas rain water will conduct electricity, though its conductivity is rather low. Since this snow is formed entirely by sublimation processes, it is at least evidence that sublimation does not take place on the usual hygroscopic condensation nuclei. The hexagonal form of the silica crystals favors the formation of ice crystals, which are known to have the same structure.

Amorphous, nonhygroscopic particles, such as soot or ordinary dust, are classified as *neutral particles*. They may act as condensation nuclei but probably rarely do so, owing to the superabundance of hygroscopic particles in the air, sufficient at all times for the condensation process. The condensation nuclei are several thousand times as numerous in the atmosphere as the sublimation nuclei. In fact, at times there may be a definite deficiency of the latter.

The action of hygroscopic nuclei in initiating condensation may be seen from Fig. 3, which with some modifications follows H. Köhler[5] and H.G. Houghton.[6] The dashed curve gives the relative humidity at which condensation will occur on nonhygroscopic uncharged spherical particles as a function of their radii. The solid curves are for nuclei of sodium chloride of dry weight as shown. As the figure shows, condensation may commence on the hygroscopic nuclei at comparatively low relative humidities, but these small nuclei cannot grow to cloud-droplet size until the relative humidity becomes large enough to cause condensation on a neutral particle of about the same radius. This might lead one to suppose that neutral particles would be nearly as effective for condensation nuclei as the hygroscopic particles, but present evidence indicates that this is not so, owing perhaps to the fact that the hygroscopic particles are in the form of liquid droplets even below saturation, a shape that seems more conducive to condensation than the irregular shape of dust particles. Since the larger nuclei are most hygroscopic, condensation will take place first on them, and the smaller nuclei, which as the figure shows require higher relative humidities, will become active only if there are not enough of the larger nuclei present. In spite of this seeming preference for large nuclei, condensation generally takes place on nuclei having a considerable mass spectrum.

As condensation proceeds on the droplets, the effects of the radius of curvature, electric charge, and dissolved hygroscopic substance on the equilibrium vapor pressure over the droplet diminish and, for average-sized nuclei (radii 10^{-6} to 10^{-5} cm), are no longer important by the time the droplet radius exceeds 10^{-4} cm. H. G. Houghton[7]

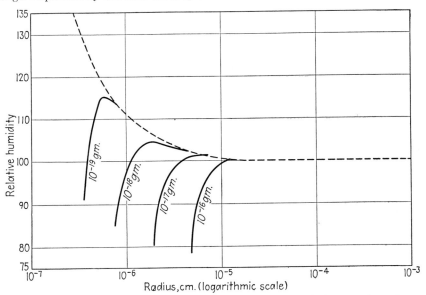

Fig. 3.—Relative humidity (or supersaturation) needed to induce condensation on spherical nuclei as a function of their radii. The dashed curve is for neutral particles. The solid curves apply to nuclei of sodium chloride of dry weights as indicated. Computed for a temperature of 0°C and a saturation vapor pressure of 6.105 mb. (*After H. G. Houghton.*)

derived an expression for the growth of such droplets from which a number of conclusions may be drawn. His equation states that

$$a^2 = A^2 + 8k(D - D_0)t \qquad (5)$$

where a = drop diameter at time t, A = initial drop diameter, k = diffusion coefficient of water vapor in air, and $(D - D_0)$ = difference between the water-vapor density in the atmosphere and at the drop surface. Equation (5) shows that for the drop diameter to increase $(D - D_0)$ must be positive, or in other words the atmosphere must be supersaturated with respect to the droplet surface. The equation also shows that aging tends to make the droplets uniform in size. A computation of Prof. Houghton's, for example, on two droplets initially 0.2 and 2.0 μ in diameter showed that the larger would be only 10.2 μ when the smaller had grown to 10.0 μ.

By means of the condensation process just described, it is possible to get droplets ranging in diameter up to 200 μ. This size of droplet corresponds roughly to large cloud particles or small drizzle particles. It is not possible to get large drops solely by the process of condensation.

Diem[8] has shown that each cloud type has a definite range of radii and that the mass spectra for the different types differ markedly. Stratocumulus cloud has a range of radii from about 2 to 20 μ with a maximum at 6 μ. Nimbostratus cloud contains drops with radii between 2 μ and a diffuse upper limit of 35 μ. There are two maxima: one wide and flat from 4 to 11 μ and a secondary maximum for the larger drops at about 22 μ. Stratocumulus may have a drop spectrum similar to nimbostratus, usually occurring when light snow falls into the stratocumulus. Altostratus cloud has a range from 2 to 25 μ with a maximum at 9.5 μ. No recent information is available for the cumulus-type clouds.

Sublimation Forms. When ice forms in the free air because of sublimation processes it may do so in a great variety of geometrical forms, all included under the collective classification of snow. This polymorphism, giving rise as it does to thousands of exquisite natural patterns, has been the object of much study. W. A. Bentley,[9] for example, made over 4,800 photomicrographs of different snow crystals. Nearly all the forms are transparent, and they owe their white appearance en masse to the light they reflect from their brilliant facets. The size of snow crystals, which ranges from $\frac{1}{100}$ to $\frac{1}{2}$ in. in diameter, depends on the temperature at which they form and the type and thickness of the cloud through which they fall. Large flake formation occurs at warmer temperatures, while the smaller, more solid type of crystal is produced at colder temperatures.

The fact that the water molecule contains three atoms favors the formation of triangular or hexagonal crystalline forms, of which the columnar and tabular forms predominate. The columnar forms comprise the hexagonal columns and the long, thin, needle-shaped columns (ice spicules), while the tabular forms include all those, either solid or branching, which form on a thin tabular plane. When the temperature is very low, solid triangular or hexagonal plates and tiny columns are most prevalent. According to Bentley,[10] the crystal types may be arranged in the following order with respect to the proportion of the total snowfall which they furnish: branching tabular forms, granular forms, plates with branching exteriors, plate forms, columns, needle forms, and compound forms.

Hail, another type of ice that occurs in the atmosphere, is formed through a quite different process. Strong, rising convective currents, as, for example, in a cumulonimbus cloud, cause intense supersaturation resulting in raindrops that are carried aloft and freeze in the cooling air. These frozen drops are called *hail*. The hailstones may fall after reaching a certain level and descend through a region of the cloud containing supercooled water that freezes on the hailstone or leaves a coating of water much as ice forms on aircraft.

Repetitions of the ascending and descending motion result in a concentric structure of clear and opaque ice and the possible formation of very large hailstones. Bilham and Relf[11] arrived at a theoretical upper limit for hailstone weight of about 1.5 lb on the basis of aerodynamic theory, but weights of over 2 lb have been recorded; these very large stones possibly having been formed by the cementing together of two hailstones. Schumann[12] has discussed the thermodynamics of hailstone formation.

Soft hail is not really hail at all but is a form of snow, consisting of pellets of closely packed ice crystals. It breaks with a splash upon striking a hard surface, whereas true hail neither rebounds as much nor disintegrates as easily. As a rule, true hail is characteristic of the violent summer thunderstorms, while soft hail usually accompanies the less severe winter or spring storms.

A. Wegener[13] has made a very convenient classification of the different sublimation forms on the basis of the temperature and the relative saturation with respect to an ice or water surface. His scheme is shown in Fig. 4, where the abscissa is the temperature of the sublimation product and the ordinate is the vapor pressure. In the schematic phase diagram, T is the triple point (0.0098°C, 6.105 mb), *i.e.*, the intersection of the evaporation curve BTD with the sublimation curve AT and the fusion curve CT. If the existing vapor pressure is

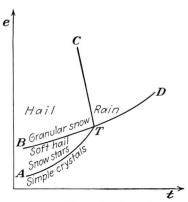

Fig. 4.—Schematic phase diagram showing sublimation forms. (*After A. Wegener and K. Wegener.*)

equal to the equilibrium vapor pressure over ice or only very slightly greater, then the simple halo-producing crystals are formed. If the vapor pressure is somewhat greater than in the preceding case, then, because more moisture is available, the corners of the crystals branch out, yielding the diversiform snow stars. When the supersaturation (with respect to an ice surface) is still further increased, the branching out proceeds in all directions, leading to the spherical crystals of soft hail. When the vapor pressure becomes equal to or surpasses the equilibrium vapor pressure over supercooled drops, then supercooled water droplets may coexist with the still growing soft hail, and there is a possibility of glaze deposition on the soft hail crystals, leading to the formation of what Bergeron calls *granular snow*. If the supersaturation with respect to a supercooled water surface becomes very great, true hail may be formed.

Cloud Elements and Precipitation Elements. It was pointed out earlier that condensation processes alone cannot produce droplets over 200 μ in diameter, *i.e.*, about the size of small drizzle particles. These might fall out of a cloud, but they would probably evaporate before reaching the earth, though that of course would depend on the height of the cloud and the relative saturation of the air below the cloud. According to Findeisen,[14] the falling distance that a drop can cover in unsaturated air is proportional to the fourth power of the radius. For this reason, clouds that contain small droplets have sharp edges, while those containing larger cloud elements have very ill-defined edges and trailing wisps or virga. Since, in general, ice crystals are of larger size than water droplets and have a consequent greater rate of fall, the edges of ice clouds (see page 259) are much more diffuse than are the edges of water clouds. Süring[15] has recommended that the dividing line between cloud elements and precipitation elements be made at a radius of 10^{-2} cm (100 μ). The radius of most raindrops is considerably larger than this limit, however, and may be as large as 0.55 cm. Bernard[16] has made an attempt to correlate the distribution in size of raindrops with precipitation intensity.

Colloidal Nature of Clouds. It was pointed out by A. Schmauss[17,18] as early as 1919 that clouds could be regarded as colloids, or more particularly aerosols, because they are composed of very large numbers of finely divided liquid and solid particles suspended in air. This aspect of condensation phenomena has received much attention in recent years and has been developed and extended by T. Bergeron,[19] W. Findeisen,[14] and others[4,20] to form the basis of our modern theories of condensation and precipitation.

Colloidal Instability in Clouds. A cloud, like all colloidal sols, has a constant tendency toward self-destruction because of the forces that act to promote coalescence of the cloud droplets. This tendency to coalesce is called colloidal instability and, since it eventually leads to large droplets that cannot be supported against the attraction of gravity, is manifested by precipitation. Coalescence, with a consequent decrease in the degree of dispersion of the water droplets, may be a result of one or more of the following factors:

1. Nonuniform charge or lack of charge on the cloud elements.
2. Variation in size of the water droplets.
3. Temperature difference between adjoining cloud elements.
4. Turbulent motion or relative motion of cloud elements due to gravitational sedimentation.
5. Coexistence of all three phases of water, solid, liquid, and vapor.

The effect of 1 and 4 is to facilitate the fusion of droplets and increase the number of collisions between droplets, while, as was shown previously in considering supersaturation phenomena, 2, 3, and 5 create vapor pressure gradients causing diffusion transport of water from the smaller to the larger droplets, from the warmer to the colder droplets, or from the liquid phase to the ice phase.

Precipitation from Pure Water Clouds. The cloud droplets may grow in size by condensation and coalescence. Condensation processes are insufficient to produce large droplets, and Bergeron[19] is of the opinion that colloidal instability due to any of the first four factors above will not ordinarily result in precipitation of greater intensity than drizzle or wet fog. Later investigations by H. G. Houghton and W. H. Radford[21] cast some doubt upon this conclusion, however, and show that, if cloud particles of unequal size occur at the same level, the effect of gravitational sedimentation may cause large raindrops to be formed. According to Houghton,[6] the size of the drop that may be formed by this process is a function of the liquid-water content of the cloud, the depth of the cloud layer, the vertical convective velocity, and the size of the cloud particles. A temperature difference between adjoining cloud droplets brought about by intense turbulent mixing or by uneven illumination of the cloud top (Reynolds effect) might, as Petterssen[22] has shown, conceivably upset the colloidal stability of the cloud. As Fig. 2 shows, this temperature difference would not have to be large at temperatures above 0°C to produce an appreciable vapor-pressure gradient between cloud particles.

Precipitation from Pure-ice Clouds. Owing to the smaller number of sublimation nuclei that are available, the number of cloud elements will be smaller in a pure-ice cloud, but for the same reason they will probably be of larger size and grow faster than the pure-water-cloud elements. Also, as Houghton[6] points out, ice crystals in the form of skeletal snowflakes with their large surface-to-volume ratio and smaller fall velocities may reach a larger weight by sublimation than is possible for water droplets. With continuous adiabatic cooling, large-sized soft hail pellets may even be formed. Since these particles can fall through a rather thick unsaturated layer without completely evaporating, the probability of precipitation is usually greater from ice clouds than from water clouds.

Conditions are particularly favorable for coalescence of snow crystals in an ice cloud, if the temperature is not too low, because of the form of the crystals, their large surface areas, and zigzag paths of fall. It is believed that when two crystals collide the increased pressure causes the ice to melt along their interface. The melting reduces the pressure, and the two crystals freeze together. Large snowflakes may form in this way. Conditions are most favorable with thawing snow. Rain is formed by melting of the falling snowflakes, if the temperature below the cloud is greater than 0°C, and is thus seen to be a special case of snow formation.

Ice clouds occur quite frequently in the atmosphere. In fact, among layer-type clouds, they occur more often than pure-water clouds. They are usually found at intermediate levels, as, for example, the "typical altostratus," in the precipitation shield of the advancing warm front. The precipitation intensity associated with them is usually light to moderate.

Precipitation from Water-ice Clouds.   The coexistence of all three phases of water, *i.e.*, solid, liquid, and vapor, will produce colloidal instability in a cloud. As proposed by Bergeron[19] and extended by Findeisen,[14] this is deemed to be the most important effect in initiating precipitation and the only one capable of producing heavy precipitation. It was pointed out earlier that the equilibrium vapor pressure is considerably lower over ice than it is over water. Therefore, if supercooled drops and ice crystals coexist in a cloud, there will be a diffusion transport of water vapor from the drops to the ice crystals and a consequent rapid growth of the ice crystals at the expense of the water drops, which may entirely evaporate. As these crystals get larger, they will fall through the lower layers of the cloud, which may still consist of supercooled drops, and they will grow still further by sublimation. This process alone would not account for large raindrops (formed from melted snow), according to Houghton,[6] but during the fall a large number of collisions will take place between the falling crystals and the cloud elements, and to a minor extent between the precipitation elements themselves.

Precipitation from Stratus-type Clouds.   In Fig. 5, which follows Findeisen,[14,23] various kinds of precipitation-producing stratus-type clouds are shown. The cloud

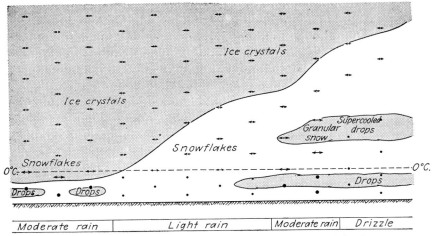

Fig. 5.—A nimbostratus-cloud system.   (*After W. Findeisen.*)

system is considered to be moving from left to right with an upglide motion. The 0°C isotherm is shown as a horizontal dashed line approximately 1,500 ft above the surface.

Two pure-water stratus-type clouds are shown at the extreme right, the upper one, since it is above the freezing level, consisting of supercooled drops. Light drizzle falls from both clouds, but the drops from the upper cloud evaporate before reaching the ground, so that only the lower cloud yields actual precipitation. Above the water clouds is an ice cloud expanding downward, *e.g.*, an altostratus that has thickened to include a part of the 0°C isotherm. It has a few fractostratus containing water droplets below it.   Small ice crystals fall from the higher portions of the cloud (in the upper right corner of the figure), but being small they evaporate before reaching the water clouds.   From the thicker part of the ice cloud, larger ice crystals fall, and these reach the water cloud consisting of supercooled drops.   There they grow rapidly by sublimation and also pick up a glaze coating forming granular snow particles, which fall out at a faster rate.   When they pass below the freezing level, they melt, forming large raindrops which together with the drizzle from the low cloud give moderate rain.   During this process, the supercooled stratus rapidly dissolves, while the lower stratus disappears more gradually.   After the water clouds have passed, the intensity of the rain diminishes but increases again as the base of the ice cloud lowers because of the increased size of the snowflakes.   The fractostratus clouds are unimportant for the formation of precipitation but are chiefly caused by convection currents in the nearly saturated air below the nimbostratus cloud.

Altocumulus clouds consisting of supercooled water droplets are often found along the upper boundary of the altostratus.   They are formed of course by convectional motion but arise because the continual falling out of ice crystals from the upper part of the altostratus exhausts the limited supply of sublimation nuclei.   Since sublimation is no longer possible, condensation may take place on the condensation nuclei, which are always available in sufficient numbers.   Usually the altocumulus layer formed in this way is only 200 to 300 m thick, but occasionally it may attain a thickness of 1,000 m.

Precipitation from Cumulus-type Clouds.   The several stages in the growth and decay of cumulus clouds are shown in Fig. 6, which follows Findeisen.[14,23]

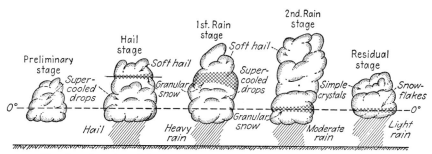

Fig. 6.—Growth and decay of cumuliform clouds.   (*After W. Findeisen.*)

In its *preliminary stage*, the cumulus cloud consists entirely of water drops, both above and below the 0°C level.

As the cumulus develops upward and the temperature decreases in the upper part of the cloud, the supersaturation with respect to an ice surface finally becomes so great (see Fig. 1) that sublimation begins on the sublimation nuclei.   This generally takes place in the neighborhood of −18°C, therefore about 3,000 m above the 0° level (indicated by the dotted line in stage II of Fig. 6).   Above this boundary, a rapid transition from supercooled droplets to ice crystals takes place, and concomitantly many collisions between water droplets and ice crystals occur, forming soft hail and granular snow.   When these particles become too heavy to be supported by the ascending air currents in the cloud, they penetrate the supercooled-water cloud lying below the critical level and grow rapidly in size forming hailstones.   The greater the distance between the critical level and the 0° isotherm and the more intense the vertical currents, the greater will be the size of the resulting hailstones.   If these particles are large enough, they will not melt after falling below the 0° isotherm and will reach the ground as hailstones.   This sequence of events, when it occurs, constitutes the *hail stage*.

Following closely upon the hail stage is a period of very intense rainfall called the *first rain stage*.   During this stage, the supercooled-water cloud becomes transformed into an ice cloud, owing to the gradual sedimentation of the heavier ice particles into it from the upper part of the cloud.   This creates a deep zone in which ice crystals and supercooled droplets coexist and in which the ice crystals grow rapidly at the expense of the droplets.   This mixing zone works gradually downward until it reaches the 0° level in the cloud.   During this process, the undercooled water cloud has been gradually used up, and with its completion the major part of the precipitation activity of the cumulonimbus is ended.

In the *second rain stage*, water droplets below the 0° isotherm are carried above it by the vertical currents and brought next to falling ice particles, now much smaller than before, however.   The size of the raindrops therefore decreases and with it the precipitation intensity.

During the *residual stage* of the cumulonimbus, the ascending currents disappear, and subsidence may even take place.   This results in a change of part of the snow-flakes and soft hail to simple ice crystals by partial evaporation.   There are still a few crystals falling out, however, which will cohere in the vicinity of the 0° level to form snowflakes and yield a slight after-rain.

The crucial factors in determining the character of the various stages are the duration and intensity of the convective current that builds the cumulonimbus. With intense heating, as in the tropics, this current may last for a long time with considerable magnitude, and the hail stage and first rain stage will last for several hours. If the ascending current is weak or of short duration, on the other hand, no hail is likely to fall, and the intense precipitation will have but a brief span.   If subsidence sets in prematurely, the cumulonimbus may shrink and disappear after only a mild shower.

Bibliography

1. KELVIN (SIR WILLIAM THOMSON), LORD: *Proc. Roy. Soc. Edinburgh*, February, 1870.
2. THOMSON, J. J., and G. P. THOMSON: "Conduction of Electricity through Gases," 2 vols., Macmillan, New York, 1928–1933.
3. LANDSBERG, H.: Atmospheric Condensation Nuclei, in English in *Ergebnisse der kosmischen Physik*, **3**: 155–252 (1938).
4. WEGENER, A.: "Thermodynamik," Barth, Leipzig, 1924.
5. KÖHLER, H.: *Meddel. Statens Met.-Hydrograf. Anstalt*, **2** (5), 1925; *Geofysiske Publ.*, **2** (1 and 6), 1921.
6. HOUGHTON, H. G.: *Bull. Am. Meteorolog. Soc.*, Vol. 19, No. 4, 1938.
7. HOUGHTON, H. G.: *Physics*, **4**: 419 (1933).
8. DIEM, M.: *Ann. Hydrog.*, **70**: 142–150 (1942).
9. BENTLEY, W. A., and W. J. HUMPHREYS: "Snow Crystals," McGraw-Hill, New York, 1937.
10. BENTLEY, W. A.: "Snow," in Encyclopaedia Britannica, 14th ed.
11. BILHAM, E. G., and E. F. RELF: *Quart. J. Roy. Meteorolog. Soc.*, **63**: 149 (1937).
12. SCHUMANN, T.E.W.: *Quart. J. Roy. Meteorolog. Soc.*, **64**: 11 (1938).
13. WEGENER, A., and K. WEGENER: "Physik der Atmosphäre," Barth, Leipzig, 1935.
14. FINDEISEN, W.: *Meteorolog. Z.*, 55, April, 1938.
15. SÜRING, R.: "Leitfaden der Meteorologie," p. 125, Leipzig, 1927.
16. BERNARD, M.: (Physics of the Earth, IX) "Hydrology," p. 41, McGraw-Hill, New York, 1942.
17. SCHMAUSS, A.: *Meteorolog. Z.*, 16, 1919; 1, 1920.

18. SCHMAUSS, A., and A. WIGAND: "Die Atmosphäre als Kolloid," F. Vieweg, Brunswick, 1929.
19. BERGERON, T.: *Mémoire Assoc. Mét. de l'U.G.G.I.*, Lisbon, 1933.
20. PEPPLER, W.: *Beitr. Phys. freien Atmos.*, **23**: 275 (1936).
21. HOUGHTON, H. G., and W. H. RADFORD: *M.I.T. and Woods Hole Ocean. Inst., Papers Met. and Phys. Oceanog.*, **6** (4), 1938.
22. PETTERSSEN, S.: "Weather Analysis and Forecasting," p. 46, McGraw-Hill, New York, 1940.
23. FINDEISEN, W.: *Z. angew. Meteorolog.*, **55** (7), 1938.

LIGHTNING AND LIGHTNING PROTECTION

By K. B. McEachron

THEORY OF LIGHTNING

1. Origin of Thundercloud Charges. The exact process is not known through which a cloud builds up electrical charges to such a magnitude as to cause a lightning stroke. Several theories[1-3] have been advanced to account for the accumulation of these charges, but the problem is complex and not subject to easy proof or determination through the construction of models. It is desirable to point out here that certain meteorological conditions are necessary to produce a thunderstorm.

Lightning storms seem to have certain characteristics[4] that are peculiar to a given storm. As an illustration, some storms consist mostly of cloud-to-cloud strokes, while in other storms cloud-to-ground strokes predominate. Furthermore, it has been shown[4] that some storms produce strokes having a higher percentage of successive current peaks within the stroke than other storms. These habits are apparently due to differences in the location of centers of charge within the cloud; the successive-current-peak type (multiple stroke) being the result of the discharge of various centers in sequence.

In the theory of the process of production of the charged cloud, a separation of charges must take place. Considerable data[5,6] indicate that the ground end of at least 95 per cent of cloud-to-transmission-line strokes are positive. Thus, it seems certain that for most strokes the lower surface or cloud base is negatively charged, at least over most of its area. There is evidence that regions of positive charge can exist in the base of the cloud.[7,44] That the more remote or upper portions of the cloud are positively charged is evidenced by oscillograms,[8] which show a change from negative to positive after some time has elapsed. The action of convection currents in breaking up drops of water that have coalesced and raising these droplets to the upper portions of the cloud with a subsequent reuniting and falling is regarded as an important part of the mechanism of separating electrical charges in the cloud. It is quite obvious that the air currents must do the work of separating the charges and keeping them separated within the cloud. When these convection currents die out, the charges will come together, and the cloud returns to approximately its original condition.

Mechanism of Lightning Discharge

2. Conditions existing on the earth, prior to a stroke between cloud and earth, must be recognized in order that one may be familiar with the stroke process. This is necessary for a proper understanding of the protective measures taken and may help to explain why protection is not always attained.

As a negative charge builds up in the cloud base, a corresponding positive charge appears in the earth underneath by induction, the negative earth charges being driven away from under the negative cloud. Thus, as a storm cloud is carried along above the surface of the earth, its counterpart of opposite polarity is carried along the surface of the earth underneath. These positive charges, being attracted by the negative charges in the cloud, move up transmission towers, along overhead ground wires, up church steeples, flagpoles, trees, and other conducting or semiconducting objects.

264

When the gradient is high, a corona discharge, or St. Elmo's fire, may be seen. Thus it is clear that, with a cloud of considerable height, say 5,000 ft, a very large area of the earth is involved underneath.

3. Stroke Mechanism. Through the use of photographs taken of lightning strokes, where the lens has been moved rapidly with respect to the photographic film, Schonland[9,10] found that the lightning stroke is initiated by a streamer from the cloud which progresses toward the earth in a series of steps. The time taken for this initiating step leader to reach the earth was in some cases of the order of 0.01 sec. When it reached the earth, a "return stroke" took place, progressing from the earth to the cloud. At times, this return stroke did not get all the way back to the cloud.

That more than one discharge can take place through the same path between cloud and ground has been known for many years. In 1905, Larson[11] published a photo-

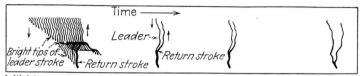

STILL CAMERA
PHOTOGRAPH
OF STROKE

Initial downward stepped leader   Subsequent discharges with downward continuous leaders

HIGH SPEED BOYS CAMERA PHOTOGRAPH OF THE SAME STROKE

CLOUD TO EARTH

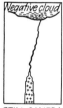

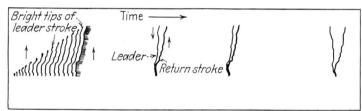

STILL CAMERA
PHOTOGRAPH
OF STROKE

Initial upward stepped leader   Subsequent discharges with downward continuous leaders

HIGH SPEED BOYS CAMERA PHOTOGRAPH OF THE SAME STROKE

CLOUD TO TALL CONDUCTING STRUCTURE

FIG. 1.—Schematic diagram of mechanism of lightning discharge.[8]

graph of 40 such discharges occurring in a stroke that persisted for 0.624 sec, as shown by the moving camera he used.

Schonland's work showed that discharges, or current peaks that followed the path determined by the step-leader discharge, were in general preceded by a downward leader called a *continuous leader*. When this made contact with the earth, a return discharge occurred from the earth just as in the case of the initial step leader. The velocity of the step leader is of the order of 50 m per μsec with time intervals of about 100 μsec between steps. On the other hand, the continuous leaders have a velocity of about 1 to 23 m per μsec. The return stroke from the earth moves up the channel with a velocity of from about 20 to 140 m per μsec, which on the average is considerably less than half the speed of light.

McEachron[8] has shown that at least 80 per cent of the strokes to the Empire State Building are initiated by the building, rather than by the cloud. Apparently the stress at the top of the building, due to its great height (1,275 ft to top of lightning-collecting system) and its configuration, is such that a step leader begins at the building and progresses toward the cloud. The velocities are of the same order of magnitude that Schonland found for the downward step leader. No return stroke was found from the cloud, but instead a continuing current flowed, in one case as long as 0.625 sec. Following the initial step leader, successive discharges may occur, and these, where leaders have been photographed, are always downward followed by the return discharge from the building. It appears, therefore, that the mechanism involved in the discharges, once the path is established, is the same whether to high or low structures or to open country (see Fig. 1). The propagation velocities of both the continuous leader and the return discharge are of the same order whether the stroke was initiated by the cloud or by a tall object.

Unpublished data seem to bear out the expectation that, as the heights of structures become less, the earth-initiated stroke becomes less frequent. Possibly it could occur from some structure having a height of the order of 100 ft or more located at the top of a high rocky pinnacle (such as the figure of Christ on Corcovado in Rio de Janeiro, which is known to have been struck several times), but in general structures of ordinary height will not initiate lightning strokes.

Many photographs[8,10] taken both in this country and abroad show continuing illumination between successive current peaks as shown by the moving lens or film camera or as seen sometimes when the path is blown along by the wind. Oscillo-

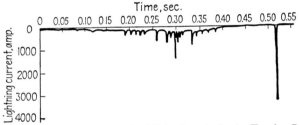

FIG. 2.—Crater-lamp oscillograph record of lightning stroke to Empire State Building, Aug. 11, 1937.

grams (of which Fig. 2 is an example) taken by McEachron in 1937, show how closely the change in density of such a photograph can be correlated with current in the stroke (except of course where the film is overexposed, which is not likely to be the case with the continuing part of the stroke). Thus, the photographs, now that they have been checked by oscillograms, become good evidence to indicate that a stroke of lightning is in reality a direct-current arc between the cloud and the earth with superimposed current peaks that may occur in almost any manner, depending upon the construction of the cloud and the location of concentrations of charge.

It seems certain now that the direction of branching is also the direction of propagation of the initial step leader; still photographs that show branching will therefore reveal information on direction of propagation. As a rule, branching[8,10] occurs only at the time of the initial discharge; succeeding discharges, if present, seldom show any branching.

It also seems clear that direction of propagation is determined by the configuration of the electrodes rather than by polarity as was thought some years ago to be the case.

As the stress in the cloud becomes higher and higher, such a condition is finally reached that a leader starts on its way to the earth, picking the best path from instant to instant, which is the reason why the discharge does not follow the electrostatic field. As it progresses, it in effect lays down a negative[12] space charge in the area surrounding the path and carries with it a high concentration of charge in the progressing end of the leader. The leader is connected to the cloud at all times, since it requires a supply of charges from the cloud to keep it going. From the point of view of the electrostatic field, it is as though a conductor had been let down from the cloud, which as it progresses will cause a continual redistribution of charge on the earth underneath. As the leader comes closer and closer, charges[13] in the earth move in to the point underneath the downcoming leader until the stress on the earth's surface may become so great that the air is broken down and streamers form from the earth's surface. Figure 3, a drawing of a photograph taken on a beach in New Jersey, shows the existence of the streamers. It should be noticed that the earth streamers are branched in an upward direction and that there are downward streamers on the stroke itself. When the step leader formed the downward streamers, there was no connecting stroke to the ground, but the potential gradient was sufficiently high so that three upward streamers formed and the stroke made contact with the longest of the three.

The return-discharge phenomenon means that the positive earth charge* is moving up the channel created by the downward-moving leader. A field exists between these charges and those in the space surrounding the channel, and also in the cloud and in any branches that may have been formed by the initial step leader. The movement of these charges up into the channel constitutes the flow of current, and here, apparently, is the reason why there was no return discharge following the upward leader from the Empire State Building, although there was one from the earth following the downward step leader. The mobility of charges in the cloud is low, while the mobility of charges in the earth is very much greater. Thus, when the leader reached the earth, charges could move quickly up into its channel, giving

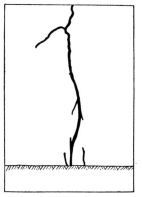

FIG. 3.—Lightning stroke to beach showing upward streamers.

rise to a relatively high current peak. The cloud, however, is not capable of supplying charges so quickly, and therefore the current peak cannot develop; instead a slow discharge results.

Although there is yet no oscillographic proof, there is considerable evidence that strokes to the earth, and particularly to regions of low resistance of considerable area, begin with a current peak. It is suggested by McEachron and McMorris[13] that the movement of charges in to the point to be struck, prior to actual contact, represents the wave front, and the tail is determined by the movement of charges into the channel after contact.

Goodlet,[12] however, is of the opinion that "as the leader stroke moves toward its goal, point discharge and displacement currents will flow, but the current in the object struck is probably small until the start of the main or return stroke."

A consideration of the preceding discussion will disclose the fact that, in high-resistance areas, charges may have to come from considerable distances, with the result that considerable voltage is required to get the charges to the point struck. This means that, in high-resistance areas, although the stroke current may be less, the area involved will be much greater than if the ground resistance were low. In

---

* When reference is made to the movement of positive charges along a conductor, it is to be understood that this refers to the movement of the electric field and the positive ends of the field terminating on the conductor. Thus the word *charges* is used to signify the terminal points of the electric field, either at conductors or at other bodies. It is also to be remembered that the movement of a positive charge from left to right is equivalent to the movement of a negative charge in the opposite direction.

areas of high-resistance soil, the presence of extensive buried conductors, such as cables or pipe lines, will influence the availability of charge, which would result in easier formation of streamers from the earth above such conductors, thus tending to direct the stroke to the earth above such conductors. This would not be expected to have any appreciable influence in determining the path of the stroke until the step leader was only a few hundred feet from the buried conductor. It has been suggested that underground areas of low conductivity might have a similar effect.[36]

Electrical Characteristics of Lightning

4. Voltage. The potential between cloud and ground just prior to a stroke of lightning has been variously estimated as from 80,000,000 to 1,000,000,000 volts.[14,15,37] These values mean little to the protection engineer, since it is the potential due to lightning that appears on the transmission line or distribution circuit that is of importance to him. If the current in a wave traveling along a line is known, a satisfactory value of voltage is obtained by multiplying the current by the surge impedance Z of the conductor, which is frequently taken as 500 ohms.

The potential that can appear upon apparatus is limited only by either protective devices or flashover of insulating structures, such as line insulators, plus the impulse strength of wood or other insulating support, including the effect of ground resistance. Direct strokes must be given consideration on all circuits carried overhead and on underground circuits connected to overhead conductors. Increasing the insulation between conductors and between conductors and ground is not of itself an effective means of preventing flashover due to lightning. Some reduction can be obtained by this means, and on low-voltage circuits the spacing may be made great enough so that system current does not ordinarily follow the lightning flashover, thus reducing the number of interruptions to service.[16] However, increasing the insulation allows higher voltage traveling waves to reach stations.

The potential allowed by line insulation is determined by flashover values resulting from laboratory tests. The flashover potential is dependent upon the shape of the wave applied, as well as upon its crest value. As the time to flashover becomes shorter, the flashover voltage becomes higher and may be of the general order of twice the 60-cycle crest flashover when the rate of voltage rise is approximately 1,000 kv per μsec, and flashover is on the front of the wave.

Lightning may cause voltages in conductors by induction as well as by direct stroke to the conductor. The electrostatic field from a charged cloud induces in the earth beneath charges of the opposite sign, and these will be found also on transmission-line conductors within the cloud field. Charges, having the same sign as the cloud, will be driven off to remote parts of the line, or over the surface of insulators to ground as a slow leakage. Since any charge bound on the conductors will be released as the field is reduced in strength and will travel away from under the cloud along the conductor at the speed of light, the potential developed as the result of a sudden change of cloud field will depend not only on the change of field gradient but on the rate of change

$$V = agh \tag{1}$$

where V = induced voltage

a = a factor less than unity, dependent upon rate of change of the cloud field, and the distribution of bound charge

g = actual gradient in volts per foot where line is located

h = height of line, ft

Peek[15] gives a maximum value of 100 kv per ft gradient, while Norinder[17] has measured values of ag up to 80 kv per ft.

In a traveling wave, it is necessary that the energy be equally divided between the electromagnetic and the electrostatic fields. When the released bound charge has fully developed into a traveling wave in a conductor of sufficient length, it will be found that the voltage of the traveling wave is half of that induced, since the electromagnetic field is developed at the expense of the electrostatic.

In general, experience seems to show, in light of the available data, that lines insulated for a system voltage of more than 69 kv will not be troubled much by induced voltages; but, as the insulation becomes less with lower voltage circuits, the voltages induced by lightning are of greater importance.

Direct measurements of induced voltages on circuits are not available, but it is known that with a negative cloud base the polarity of the traveling wave, owing to the release of the bound charge, would be positive. Some indication of the magnitude of voltages induced in this manner may be obtained from the results of field studies of currents through distribution[18] and station-type[19,38] lightning arresters. In the general problem of protection[39] of structures from the effects of lightning, the usual

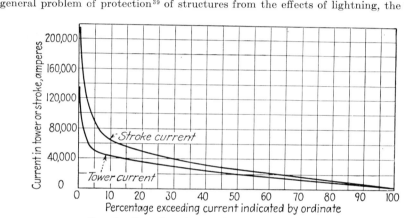

Fig. 4.—Data from 734 strokes (Lewis and Foust).[20]

lightning-rod system, though effective for direct-stroke protection, may require modification if protection from induced effects is to be obtained. Some form of Faraday cage may be the solution, particularly if small sparks are to be avoided, as in the case of explosive dusts or gases.

5. Total Stroke Current—Duration—Charge. The most comprehensive data relating to current in the stroke have been collected by Lewis and Foust[20] in cooperation with electrical-supply companies. These measurements were made with the magnetic link[21,22] arranged to read current in transmission towers. In some cases, current was measured in all four legs. In other cases, measurements were made in only one leg and extrapolated. The stroke currents were obtained by adding together currents in all towers that seemed to be involved in a particular stroke. It is to be noted that 50 per cent of the stroke currents are in excess of 23,000 amp (Fig. 4).

Experience curves are given showing lightning currents through distribution[23] (Fig. 5) and station lightning arresters[19] (Fig. 6). The original data show, among other characteristics, a much greater percentage of positive records for distribution circuits than were recorded on transmission lines. It seems likely that many of these are the result of the release of bound charges with negative cloud base. In general, the results obtained represent records of surges that have traveled to the arrester through a considerable length of conductor (which is much longer in the case of the station arresters) and have become modified because of travel. Thus these data give information as to lightning-arrester currents to be expected, but as for the stroke itself they indicate only that the stroke current was of greater magnitude, undoubtedly being in many cases more than twice the current values given for arresters.

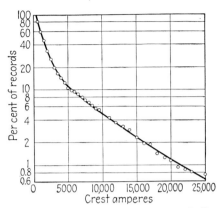

Fig. 5.—Lightning currents through distribution arresters (McEachron and Mc-Morris).[23] Of 1,608 records obtained, 1,011 were negative and 597 were positive.

Walter[24] suggested in 1905 that the continuing illumination observed on moving-camera photographs of multiple lightning strokes was a continuing current. It remained for McEachron[8] to show in 1939 that the continuing illumination did represent current, through the use of rotating-lens or film cameras of the Boys[25] type in conjunction with oscillographs, installed in the top of the Empire State Building, one of which was capable of recording over 1 sec of time.

Broadly speaking, the lightning stroke consists of one or more current peaks superimposed on a more or less continuous

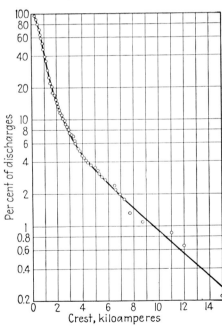

Fig. 6.—Lightning currents through station arresters (Gross and McMorris).[19] Of 459 records obtained, 401 were negative and 58 were positive.

current flow. This is illustrated by the record[40] shown in Fig. 7 taken with cathode-ray and magnetic-type oscillographs. For strokes to the earth itself or to structures of low height, it is believed the stroke would be of the downward-leader type initiated with a current peak. The stroke recorded in Fig. 7 was probably of the upward-step-leader type.

It is important to recognize the existence of the continuing type of discharge, since it is no doubt the cause of most of the thermal effects of lightning, such as fires, holes

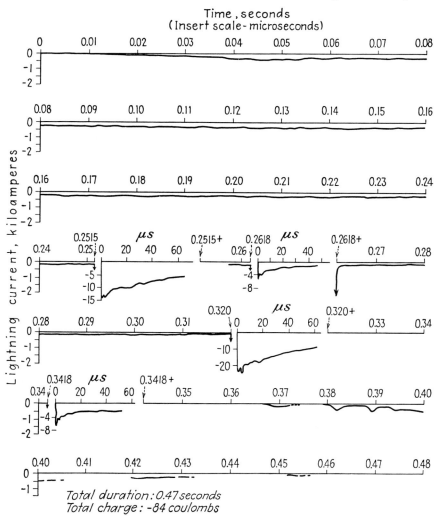

Fig. 7.—Composite replotted from low-speed and high-speed oscillograms showing four current peaks of stroke 3 to the Empire State Building, June 26, 1940 (McEachron).[40]

burned in the skin of airplanes, burning of cable sheaths and conductors, and is responsible in part for the blowing of fuses on distribution circuits.[26]

The duration of 75 strokes determined by the magnetic oscillograph or the Boys camera is given in Fig. 8. These results show that 50 per cent of the strokes had durations of 0.35 sec or longer. The maximum duration recorded was 1.5 sec.

The total charge, including current peaks for 49 strokes to the Empire State Building, is given in Fig. 9. The curve indicates that 50 per cent of the strokes to the building had a charge of 25 coulombs or more, although the average is 37. The result agrees well with the values suggested by Bruce and Golde.[37] The maximum charge recorded in Fig. 9 is 165 coulombs, although McEachron and Hagenguth[41] report a

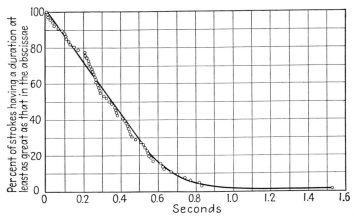

Fig. 8.—Stroke duration as a function of the frequency of occurrence (McEachron)[8,40] recorded by Boys camera or oscillograph. Based on 75 strokes to the Empire State Building, 1935–1940.

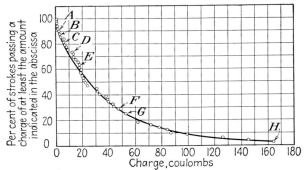

Fig. 9.—Frequency of occurrence of stroke charge based largely on low-speed oscillographic data. Total charge of lightning strokes recorded by low-speed oscillographs, Empire State Building, 1937–1940. Based on 49 strokes.

Coulombs in Two-polarity Strokes

A	B	C	D	E	F	G	H
− 1.4	2.0	5.4	12.0	8.2	30	36	160.0
+ 1.2	1.1	1.8	2.3	9.9	17	17	2.6

Other strokes negative charge only (McEachron).[40]

possible maximum charge of 240 coulombs based on data from holes burned in a thin metal sphere on top of the WSM broadcasting tower in Nashville, Tenn.

Polarity. Cloud-to-ground strokes seem to be predominantly from a negative cloud base. The oscillograms of 49 strokes to the Empire State Building taken during 1937–1940 show that 41 were entirely negative, the remaining eight being partly positive and partly negative. There is no certain evidence that any of the eight

TABLE 1.—DATA ON STROKE CURRENTS

(1) Voltage of system on which measurements were made, kv	(2) Range of tower currents, amp	(3) Range of stroke currents, amp	(4) Per cent under values given in col. (2) or (3)	(5) Per cent negative, all strokes	(6) Authority reference No.
6–220	100,000 30,000 20,000		100 90 82	87	6
66, 115, 154		130,000 100,000 60,000	100 90	97	45
150	40,000	70,000	100	70	46
69–220		150,000 30,000 2,000	99 50	93	20
Induced on aerial by direct strokes, measured with cathode-ray oscillograph		120,000 60,000 60,000	100 80% of all positive 70% of all negative		30
Strokes to tall structures, measured by fulchronograph		160,000 10,000 5,000 1,000	100 80 50 7		28
Lightning to Empire State Building, measured oscillo-graphically		58,000 14,000 7,000	100 80 50	84	40

TABLE 2.—COULOMBS

(1) Per cent at least equal to col. (2)	(2) Coulombs	(3) Authority reference No.
2 20 50 80	165 62 25 5	8
20 50 }measured with antenna 70	1.6 0.6 0.3	30
Measured charge of cloud	4–60 5–200 Average 30 }	28 47

TABLE 3.—CHARACTERISTICS OF CURRENT WAVE SHAPE IN DIRECT STROKES

Front and duration, time in μsec			How determined	Authority reference No.
Wave front	Time to half value	Total duration		
2–6		50 or more	Based on rate of rise and propagation data	12
0–10		Up to 60	Loop coupled to cathode-ray oscillograph	30
3.5–10		23 to 45	Cathode-ray oscillograph to high vertical conductor	27
0–6	6–150	100–300	Cathode-ray oscillograph to Empire State Building	40
	Varied from 20 to 90	50 to 20,000	Fulchronograph records to structures 530 to 585 ft in height	28

Rate of current rise, amp/μsec

Average	Maximum		
20,000	40,000	Klydonograph and loop to tower	46
6,000	40,000	Cathode-ray oscillograph to high vertical conductor	27
1,700–13,000	36,000	Empire State Building	40
	30,000	Loop coupled to cathode ray oscillograph	30
174–6,000 amp 43,000 amp/μsec (highest reported)		Range for currents to tall structures Transmission-line fulchronograph data	28

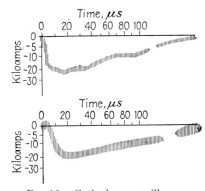

FIG. 10.—Cathode-ray oscillograms of a lightning stroke to the Empire State Building, Aug. 8, 1938. Two discharges; 89,200 μsec between discharges (McEachron).[29]

began positive. It is evident, therefore, that current reversals do take place, but apparently they are the exception rather than the rule.

6. Peak Current Wave Shapes and Charge. Very few data are available showing wave shapes of lightning currents measured directly. Some Russian data[27] are given in the table together with results from the Empire State investigation, and fulchronograph records from strokes to buildings and chimneys ranging from 530 to 585 ft in height.[23]

In 1938, McEachron[29] secured some oscillograms of successive current peaks in strokes to the Empire State Building. In one case, as many as 12 current peaks were measured in 0.28 sec. The wave shapes of all were similar. Three of them reached a crest current of 5,000 amp. In another case two peaks, 89,200 μsec apart, were secured and show considerable similarity in wave shape (see Fig. 10).

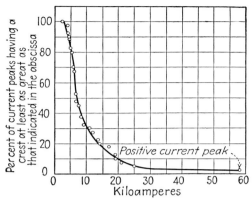

Fig. 11.—Frequency of occurrence of current peaks of various magnitudes determined oscillographically. Amplitude of current peaks recorded by high-speed cathode-ray oscillograph on the Empire State Building. Results based on 13 strokes (McEachron).[40]

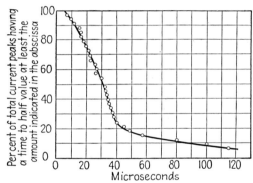

Fig. 12.—Duration of current peaks measured to half value as a function of frequency of occurrence. Based on 33 oscillographic records. Duration of current peaks recorded by high-speed cathode-ray oscillograph on the Empire State Building, 1938 and 1939. Results based on 11 strokes (McEachron).[40]

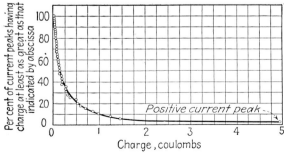

Fig. 13.—Charge of current peaks as a function of frequency of occurrence. Coulomb values based on time to half value of crest current. Charge in current peaks recorded by high-speed cathode-ray oscillograph. Charge passed until current decays to one half of crest. Empire State Building, 1938 and 1940. Results based on 13 strokes (McEachron).[40]

It cannot be said that lightning has a particular wave shape; it can be expressed only between limits and in terms of the frequency of occurrence, as indicated in the curves and tables.

Rate of Rise. It should be pointed out that, for transmission lines, rates indicated in the table apply only at the stricken point, and then not after a line-to-ground arc-over. The rates of rise on a transmission line will be only half the rates indicated (unless the stroke occurs at the very end of the line), and these will decrease rapidly with travel.

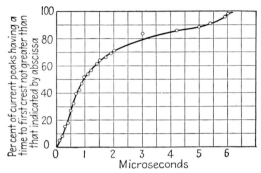

Fig. 14.—Time to first crest of current peaks as a function of frequency of occurrence. Recorded by high-speed cathode-ray oscillograph. Empire State Building, 1938 and 1940. Results based on 13 strokes (McEachron).[40]

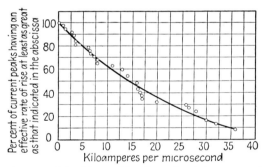

Fig. 15.—Effective rate of rise (slope of line through points on wave front at 10 per cent and 90 per cent of first crest) of current peaks of lightning strokes as a function of frequency of occurrence. Recorded by high-speed cathode-ray oscillograph. Empire State Building, 1938 and 1940. Results based on 11 strokes (McEachron).[40]

In Figs. 11 to 15 are plotted data obtained from cathode-ray oscillograms of current peaks in strokes to the Empire State Building. The building itself may have some modifying effect on these characteristics as presented, but it is believed that once the path between cloud and the building is established the succeeding current peaks will not be much influenced by the building. These data may or may not be representative of the first current peak of the down step leader to structures of ordinary height. However, because of the paucity of data on wave shape, these results are believed to be useful until more complete data have become available.

A more detailed description of these data may be obtained from the original paper,[40] but it is worthy of note that the current peak with the greatest magnitude, 58,000 amp, was positive in polarity and had a charge to half-current value of 4.9 coulombs, which is more than three times larger than obtained from any other current peak recorded oscillographically.

TABLE 4.—COULOMBS. INDIVIDUAL CURRENT PEAKS

Peak current range	No. of peaks	Coulomb range	Authority reference No.
2,100–5,000	8	0.19–0.28	29
5,000–21,000	28	0.28–1.47	29
56,000 (positive)	1	4.9	29
17,000–31,000	4	0.12–1.5	27
21,000	4	4–60	28

7. Effects on Objects Struck. Nonconductors are often shattered by lightning, while conductors may be burned or even vaporized. The high current peak may shatter trees or poles without setting fire to them, while a succession of current peaks with continuing current may well cause a fire. High-current short-time discharges passed through a No. 14 rubber-covered wire can eliminate the wire but leave the rubber apparently undamaged. Such cases have been reported from the field, and similar results have been obtained in the laboratory.[42,43] Hollow or flat conductors are often crushed by high lightning currents, owing to both thermal and magnetic effects.

The pressure effects developed inside trees or chimneys, or inside dwellings because of the pressure in the spark often blow such structures apart violently. The effects upon electrical apparatus are too complicated to be discussed here. It seems likely, in view of the rather large number of coulombs found in lightning discharges, that some overhead-conductor burning thought to have been due to power current may be due to lightning.

Methods are available for satisfactorily protecting most equipment and structures from the effects of lightning. In general, such protection is best undertaken only by those skilled in the art.

8. Lightning to Aircraft. Lightning storms represent a hazard to aircraft,[39] not only on account of turbulence, which affects the pilot's ability to handle the craft, or static, which may interfere with radio communications or cause other trouble, but on account of lightning itself. With lighter-than-air craft, special precautions must be taken to prevent the formation of sparks that might cause fire. This is particularly hazardous if inflammable gases are present. The principle of the Faraday cage may be used with interconnection of metal parts to prevent the formation of sparks resulting from the flow of current through the structure due to changes in electrostatic field or perhaps to direct strokes involving the structure.

Captive balloons should have the cable and winch well grounded. If personnel are to be located close to such cable and winch during times when lightning currents may be flowing to the earth through such cables, the use of a grounded metal mat to which the winch is connected and on which the personnel are located would add materially to their safety even though the cable and winch are well grounded. If such a ground mat is not available, a buried cable surrounding the winch and operating personnel and grounded with two or more ground rods spaced along the buried conductor would operate to reduce the potential gradient along the surface of the earth close to the winch in the event of a stroke to the balloon or cable. Of course, the winch should be independently grounded and interconnected with such a buried conductor. Telephone or other wiring entering the grounded area should have grounded conductors bonded to the grounded mat or cable, and the live conductors should be connected to the common ground through protective gaps or arresters.

Airplanes are struck by lightning occasionally. Holes burned in the metal skin indicate that the most frequent path of current flow due to lightning is from wing tip to wing tip and from nose to tail. Apparently cloud-to-cloud discharges are usually responsible for such burns. Planes do not seem to be involved in cloud-to-ground strokes very frequently, which may help to explain a relative predominance of thermal effects compared with mechanical or explosive effects that are encountered frequently in connection with structures on the earth's surface. The compensation of magnetic compasses may be upset because of the magnetic effects of lightning stroke currents on magnetic material in the airplane, such as engine parts. Pilots may be temporarily blinded by lightning flashes (particularly at night) that may interfere with reading of instruments immediately after the flash. Radio antennas may be damaged, particularly if they are of the trailing type. Metal propellers have been pitted from the flow of lightning current.

All-metal airplanes appear to stand up under lightning conditions quite satisfactorily since the occupants and the apparatus within the metal skin are within an almost perfect Faraday cage. However, some satisfactory form of lightning protection for radio equipment should be provided. Airplanes with nonmetal wings and fuselage are less likely to become a part of the path of a lightning stroke, but, unless some successful form of conduction for the lightning current is provided, such planes would be hazardous to operate in lightning storms on account of the possibilities of mechanical and thermal effects.

PROTECTION OF ORDINARY BUILDINGS AND MISCELLANEOUS PROPERTY INCLUDING HANGARS FOR AIRCRAFT

9. Considerations for Protection. Whether or not protection is justified will depend upon the value and nature of the building and its contents, the frequency of occurrence of lightning storms, the degree of shielding offered by other structures, as well as the availability of firefighting apparatus. In some states, a reduction in insurance rates can be obtained as a result of protecting buildings against lightning. The hazard to human beings is practically eliminated if they are within a properly protected structure.

10. Operation of a lightning-rod system depends upon the principle of intercepting the lightning stroke before it reaches the structure to be protected and discharging the lightning current to ground through a sufficiently low resistance so that dangerous voltages do not result on account of the IR drop. Tall trees cannot be depended upon to protect near-by structures; if they are provided with lightning rods, however, they may act as other similar tall grounded structures to protect near-by objects within the cone of protection.

11. In placing a lightning-rod system on any structure, *air terminals* should be provided for the purpose of keeping the earth end of the lightning arc away from the building, and also to provide a cone of protection so that the building itself is completely protected. Air terminals should be spaced at intervals not exceeding 25 ft and should not be less than 10 in. in height. "National Bureau of Standards Handbook H21" should be consulted for details.

12. Roof and down conductors should be coursed in such a way as to join each air terminal to all the rest. At least two paths to the ground should be provided. "Dead ends" should not exceed 16 ft.

13. Roof and down conductors may be made of copper, copper-clad steel, galvanized steel, or a metal alloy that is as resistant to corrosion as copper. Copper cables should weigh not less than 187.5 lb per 1,000 ft. Thus, a round solid conductor of at least ¼ in. in diameter is required. If steel or galvanized-steel conductors are used, they must have a net weight of steel of not less than 320 lb per 1,000 ft.

14. Metal-roofed buildings should be treated as any other building from the point of view of lightning rods and down conductors, unless the metal sheets of which the roof is composed are made electrically continuous by means of suitable bonding or interlocking. Such a roof should be provided with air terminals to receive the lightning arc and should be connected to the earth at, at least, two points, preferably on opposite corners.

15. Grounds are very important, connections being made to water pipes where available. In some cases, it will be necessary to provide an extensive buried system of wires to obtain a suitably low ground resistance. The treatment of grounds depends, of course, upon the character of the soil and its conductivity.

16. Bonding and Grounding of Metal Bodies. Metal bodies extending through the structure to be protected, or wholly exterior to the structure, should be bonded to the nearest lightning conductor and under some conditions may require an additional separate grounding. Metal bodies situated wholly within the structure, and those projecting through the sides of the building below the second floor, which come within 6 ft of a lightning conductor or a metal body connected thereto, should be bonded to the nearest lightning conductor and, if of considerable length, should be grounded at the lowest or farthest extremity.

17. Protection of Hangars. The same general principles apply as for ordinary buildings, but special attention is required because of the great height and area involved. If the structure is built with a steel frame and all parts are securely bonded together, it is necessary only to provide suitable air terminals extending through the roof of the building and to give suitable attention to the grounding of the steel frame. Where air terminals are spaced 25 ft or less on roof ridges, or flat surfaces, the height of the terminal should not be less than 4 ft 10 in. For each additional foot of separation above 25 ft, the air terminal should be increased in height not less than 2 in. "National Bureau of Standards Handbook H21" should be consulted for details.

Flat roofs should be divided into rectangles having sides not exceeding 50 ft in length by drawing lines parallel to the edge of the roof and air terminals erected at the intersection of these lines.

With reference to grounding, if a water-pipe system enters the structure, the building frame should be bonded to it at the point of entrance, and in addition artificial grounds should be provided for the steel columns or roof trusses at not less than half of the footings and distributed as uniformly about the building perimeter as possible. Where satisfactory grounds cannot be obtained from driven pipes, the building may be surrounded by a buried conductor that is interconnected to the building steel at convenient points to carry out the general principle of securing grounds at not less than half of the footings. Exterior metal bodies, such as roof flashings or down spouts, should be bonded to the lightning conductor itself. Interior metallic bodies should be independently grounded and if within 10 ft of a lightning conductor should be bonded to it.

18. Protection of Areas. In general, areas may be most satisfactorily protected by the use of masts with interconnected wires from their tops all suitably grounded to an interconnected ground system.[31] This arrangement gives a subdivision of the lightning currents down any one mast and provides for further subdivisions of currents in the ground. The area to be protected should be wholly within a cone of protection that will have a base radius of approximately two times the height of the axis of the cone. For a more detailed discussion of the cone of protection, reference should be made to "National Bureau of Standards Handbook H21."

19. Protection of Structures Containing Inflammable or Explosive Materials.
In this connection, extra precautions must be taken to prevent all sparks, no matter
how minute, from occurring within the protected area. Since these sparks can result
from induction, it is essential that all metal objects be bonded together and frequently connected to ground. It may sometimes be desirable to build a complete
Faraday cage, which of course represents the most certain form of protection. When
areas are particularly large, overhead ground wires or grounded mesh may be used.
Before undertaking the protection of such structures, one should consult those experienced in this field to be sure that all the proper precautions are taken.

The same method discussed under the general heading "protection of areas"
may be used for buildings, namely, the use of masts with or without interconnection
of overhead wires. In general, if conditions permit, the use of overhead ground wires
is desirable rather than masts alone since it reduces the current per support and will
give better protection in the area between the masts involved.[48]

20. Static Electricity. Protection of the contents of a building against the effects
of static electricity resulting from changes in electrostatic fields due to sources external
to the building may be accomplished through the use of a Faraday cage. Many steel-
frame buildings are inherently self-protecting unless explosive dusts are present or
some special process is being carried on in which the elimination of small sparks is
necessary. An all-metal structure with no openings would, of course, eliminate static
induced from outside fields; as the number and size of openings increase, the protection
obviously decreases. However, static may be produced inside such a structure
because of operations going on within that involve moving belts or friction, or because
of many other causes. In general, it can be stated that the separation of any non-
conducting material is likely to produce static. It can also be produced in many
other ways.

The control of static requires special knowledge and cannot be successfully treated
here. However, in certain industries, the use of high humidity over 60 per cent with
a suitable mixture of carbon dioxide gas is successful in combating static. The
general principle of grounding objects that either may come in contact with each
other or are separated by a small gap is probably the most satisfactory general method.
Use of conductive flooring and conductive shoes and the elimination of wool or silk
clothing for personnel helps in prevention of explosions due to static in ordnance
plants and hospitals. For more detailed information with reference to static electricity and its control, reference should be made to current literature on the subject.[32,33,34,35]

Bibliography

1. SIMPSON, G. C.: Electricity of Rain and Its Origin in Thunderstorms, *Phil. Trans.
   Roy. Soc. (London)*, A, **209**: 379–413 (1909).
2. WILSON, C. T. R.: Some Thundercloud Problems, *J. Franklin Inst.*, **208**: 1
   (July, 1929).
3. GEITEL, H.: Zur Frage nach dem Ursprunge der Niederschlags Elektrizität,
   Physik. Z., **17**: 455, (1916).
4. McEACHRON, K. B.: Multiple Lightning Strokes, *Trans. A.I.E.E.*, **53**: 1633–1637
   (1934).
5. LEWIS, W. W., and C. M. FOUST: Lightning Surges on Transmission Lines—
   Natural Lightning, *Gen. Elec. Rev.*, 543–555 (1936).
6. GRÜNEWALD, H.: Research on Operating Disturbances Due to Lightning and on
   Protection of Overhead Lines against Lightning, *Paper* 323, Conférence Internationale des Grands Réseaux Électriques, 1939.
7. JENSEN, J. C.: The Branching of Lightning and the Polarity of Thunderclouds,
   J. Franklin Inst., **216**: 707 (1933).
8. McEACHRON, K. B.: Lightning to the Empire State Building; *J. Franklin Inst.*,
   227: 149–217 (1939).

9. SCHONLAND, B. F. J., and H. COLLENS: Progressive Lightning, *Proc. Roy. Soc. (London)*, A (849) 654–674 (1934).
10. SCHONLAND, B. F. J., D. J. MALAN, and H. COLLENS: Progressive Lightning, II, *Proc. Roy. Soc. (London)*, A, **152** (A877): 595–625 (1935).
11. LARSON, A.: Photographing Lightning with a Moving Camera, *Smithsonian Inst. Ann. Rept.* 119 (1905).
12. GOODLET, B. L.: Lightning, *J.I.E.E.*, **81**: 487 (1937); **82**: 494 (1938).
13. MCEACHRON, K. B., and W. A. MCMORRIS: The Lightning Stroke—Mechanism of Discharge, *Gen. Elec. Rev.*, 487, 1936.
14. WILSON, C. T. R.: Investigation on Lightning Discharges and on the Electric Field of Thunderstorms, *Phil. Trans. Roy. Soc. (London)*, A, **221**: 73 (1920).
15. PEEK, F. W., JR.: Lightning, *Trans. A.I.E.E.*, **50** (3): 1077 (1931).
16. ANDREWS, F. E.: Performance of Wood Insulation in Transmission Lines, *Elec. World*, Apr. 25, 1931.
17. NORINDER, HARALD: On the Nature of Lightning Discharges, *J. Franklin Inst.*, **218**: 73 (1934).
18. MCEACHRON, K. B., and W. A. MCMORRIS: Discharge Currents in Distribution Arresters, *Trans. A.I.E.E.*, **54**: 1395–1399 (1935).
19. GROSS, I. W., and W. A. MCMORRIS: Lightning Currents in Arresters at Stations, *Trans. A.I.E.E.*, **59**, 1940.
20. LEWIS, W. W., and C. M. FOUST: Lightning Investigations on Transmission Lines, VII, *Trans. A.I.E.E.*, **59**, 1940.
21. FOUST, C. M., and H. P. KUEHNI: The Surge Crest Ammeter, *Gen. Elec. Rev.*, **35**: 644–648 (1932).
22. FOUST, C. M., and G. F. GARDNER: A New Surge Crest Ammeter, *Gen. Elec. Rev.*, **37**: 324–327 (1934).
23. MCEACHRON, K. B., and W. A. MCMORRIS: Discharge Currents in Distribution Arresters, II, *Trans. A.I.E.E.*, **57**, 1938.
24. WALTER, B.: Über das Nachleuchten der Luft bei Blitzschlagen, *Ann. Physik*, **18**: 863 (1905).
25. HAGENGUTH, J. H.: Lightning Recording Instruments, *Gen. Elec. Rev.*, 1940.
26. BERGVALL, R. C., and E. BECK: Lightning and Lightning Protection on Distribution Systems, *Trans. A.I.E.E.*, **59**, 1940.
27. SOKOLNIKOW, J. C.: Elektrichestvo, **58** (2) 1; abstract, *ETZ*, **58**, 32 (Aug. 12, 1937).
28. WAGNER, C. F., G. D. MCCANN, and E. BECK: Field Investigations of Lightning, *Trans. A.I.E.E.*, **60**: 1222–1229 (1941).
29. MCEACHRON, K. B.: Wave Shapes of Successive Lightning Current Peaks, *Elec. World*, **113** (6): 428 (1940).
30. NORINDER, H.: Lightning Currents and Their Variations, *J. Franklin Inst.*, **220** (1): 69–92 (1935).
31. WAGNER, C. F., G. D. MCCANN and C. M. LEAR: Shielding of Substations, *Trans. A.I.E.E.*, **61**: 96–100 (1942).
32. Static Electricity, National Fire Protection Association, Boston, Mass., 1941.
33. Static Electricity in Nature and Industry, *Bur. Mines Bull.* 368.
34. Static Electricity, *Bur. Standards Circ.* C438.
35. Combustible Anesthetics and Operating Room Explosions, National Fire Protection Association, Boston, Mass.
36. MATHIAS, ÉMILE: Eléctricité Atmosphérique. La Théorie de Dauzère sur la conductibilité de l'air dans les régions esposées à la foudre. *Compt. rend.*, **201**: 314–317, July 29, 1935.
37. BRUCE, C. E. R., and R. H. GOLDE: The Lightning Discharge, *J.I.E.E.*, **88**: 487 (December, 1941).
38. GROSS, I. W., G. D. MCCANN, and E. BECK: Field Investigation of the Characteristics of Lightning Currents Discharged by Arresters, *Trans. A.I.E.E.*, **61**: 266–271 (1942).
39. "Code for Protection against Lightning," Parts I, II, III, Bureau of Standards Handbook H21 (1937).

40. McEachron, K. B.: Lightning to the Empire State Building, II, *Trans. A.I.E.E.*, **60**: 885–890 (1941).
41. McEachron, K. B., and J. H. Hagenguth: Effect of Lightning on Thin Metal Surfaces, *Trans. A.I.E.E.*, **61**: 559–564 (1942).
42. McEachron, K. B., and J. L. Thomason: Testing with High Impulse Currents, *Gen. Elec. Rev.*, **38** (3) 126 (1935).
43. Bellaschi, P. L.: Lightning Currents in Field and Laboratory, *Trans. A.I.E.E.*, **54**, 1935.
44. Simpson, G. C.: The Distribution of Electricity in Thunder Clouds, *Proc. Royal Soc. (London)*, August, 1937.
45. Rokkaku, H.: Lightning on Transmission Lines; *Paper* 321, Conférence Internationale des Grands Réseaux Électriques, 1939.
46. Berger, K.: Résultats des mesures effectuées au cours des orages de 1934/1935, *Assoc. Suisse des Élec. Bul.*, **27**: 145–163 (1936).
47. Workman, E. J., and R. E. Holzer: The Origin and Types of Discharges in Thunderstorms, report presented at meeting of International Geophysical Union, September, 1939.
48. Lightning Protection, *Ordnance Department Safety Bulletin* 70, Chicago, Ill., June, 1943.

SECTION IV

RADIATION

By J. Charney

CONTENTS

SECTION IV

RADIATION

By J. Charney

INTRODUCTION AND DEFINITIONS

Definition of Radiation. All matter not at the absolute zero of temperature sends out energy into the surrounding space in the form of electromagnetic waves. The propagation of this energy as well as the energy itself is called radiation. Radiation is easily distinguished from other forms of heat transfer, such as conduction and convection, by its speed of propagation, which equals that of light, and by the fact that no intervening material medium is required for its transmission.

The phenomena embraced by the term radiation can be classified according to wave length. At the lower end of the known spectrum lies cosmic radiation with wave lengths as low as 10^{-14} cm, while at the upper end lies radiation from power lines, with wave lengths measured in kilometers. In between are found the gamma rays, X rays, ultraviolet radiation, visible light, infrared radiation, and radio waves. The radiation of interest to the meteorologist is that from the sun, the earth, and the atmosphere and lies within the ultraviolet, visible, and infrared spectral regions. The wave-length unit customarily employed in these regions is the micron (μ), which is equal to 10^{-4} cm. In this unit, the wave lengths of the appreciable solar and terrestrial radiation fall between 0.15 and 120 μ.

Absorption and Emission of Radiation in Gases. The nature of the radiation emitted by a body depends on its physical state. Emission by a gas will differ characteristically from emission by a liquid or solid, and the same is true of absorption. To understand the differences, it is necessary to consider separately the emissive and absorptive processes of gases, and of liquids and solids.

According to the atomic theory, a gas is a collection of molecules, each of which independently emits radiant energy at the expense of its internal energy. Each atom in the molecule is visualized as a miniature solar system, consisting of light negatively charged electrons rotating in different orbits about a heavy positively charged nucleus. The kinetic and potential energies of the electrons together constitute the internal energy of the atom. On the basis of the electromagnetic theory, the physicists of the nineteenth century attempted to predict the wave lengths and intensities of the radiation emitted by the rotating electrons in the atom. Their utter failure in this respect, and also their failure to predict the emission spectrum of the so-called "black body," resulted in radical revision of the foundations of the classical mechanics of the atom.

The two fundamental hypotheses of the new theory are: (1) an atom can exist only in certain definite energy states, E_1, E_2, $\cdots$ etc., whose values are determined in accordance with certain quantization rules, and (2) the frequency of the radiation emitted by an atom is given by the relation

$$\nu = \frac{E_m - E_n}{h} \tag{1}$$

where E_m and E_n are the respective energies in their initial and final states, and h is a

284

universal constant. The frequency ν is related to the wave length λ by the formula

$$\lambda\nu = c \tag{2}$$

where c is the velocity of light. Since the permissible energy values E_m and E_n belong to a discrete set, the emission spectrum of a gas will consist of a discontinuous set of frequencies. Absorption consists of the reverse process. The absorbed frequency will correspond to an energy transition from a lower to a higher state, so that the sign of the right-hand side of Eq. (1) will be changed.

The emission spectrum of a gas can be measured by passing radiation from the gas through a narrow slit and then through a prism or grating. The image of the slit, when focused on the retina of the eye or on a photographic plate, is found to consist of many separate bright lines against a dark background, each corresponding to a different frequency. The absorption spectrum can be obtained by sending light from a source emitting a continuous range of frequencies through the gas and then through the prism or grating, and will consist of dark lines against a bright background.

The absorption spectrum of a monatomic gas is somewhat simpler in appearance than that of a gas whose molecules consist of more than one atom. The former consists of well-defined lines while the latter seems to be made up of a series of bands, especially when viewed by an instrument with low dispersive power. Closer inspection of the bands, however, reveals that they are composed of many fine lines. This difference can be explained under the assumption that the rotational and vibrational energies of the atoms in a molecule as well as the electronic energies of the atoms can take only discrete values. Denoting these energies by E_{rot}, E_{vib}, E_{elec}, the second fundamental hypothesis gives for the frequency emitted by a molecule

$$\nu = \frac{(E_{elec} + E_{vib} + E_{rot})' - (E_{elec} + E_{rot} + E_{vib})}{h} \tag{3}$$

where primed quantities refer to the initial state and unprimed quantities refer to the final state of the molecule. With each change in the electronic configurations of the atoms in the molecule are associated independent changes in the vibrational and rotational energy levels. Thus, while a given electronic transition in an *atom* results in a single frequency emitted or absorbed, a corresponding change of electronic energy in a *molecule* is associated with many frequencies. The lines in a single band are associated with fixed electronic and vibrational transitions, but with varying rotational changes.

Influence of Pressure and Temperature on Absorption. No mention was made in the preceding pages of the influence of neighboring molecules on emission or absorption by a molecule of a gas. Actually the energy levels of a molecule will be somewhat altered by the electric forces exerted by contiguous molecules. When the molecule undergoes a transition from one energy state to another, the resulting frequency will differ slightly from its undisturbed value. The result is a broadening of the spectral lines in proportion to the degree of influence of the molecules of a gas upon one another. An increase in the rate of impact should therefore be expected to increase the width of the spectral lines and so permit the absorption of a greater quantity of radiation. Since the rate of impact is directly proportional to pressure and inversely proportional to the square root of temperature, absorption will be increased by higher pressures and lower temperatures.

Emission and Absorption by Solids and Liquids. If the considerations of the last paragraph are applied to a molecule in a solid or liquid, it should be expected that the enormously greater proximity of the neighboring molecules would so disturb the energy levels of an emitting or absorbing molecule that the spectral lines would merge with one another to produce a continuous spectrum. This explains why all solids and liquids emit and absorb radiation over a continuous range of wave lengths, but with intensities that vary with wave length. The total quantity of radiation increases with temperature, as does also the frequency at which the maximum energy is radiated. Thus a body that is being heated first becomes a dull red, then yellow, and eventually white. A more precise statement of the laws governing radiation from gases, liquids, and solids will be given in the following pages.

The Calculus of Radiation. 1. *Definition of Flux.* The flux of radiation is the total quantity of radiant energy traversing a surface from one side to another per unit area per unit time. Thus if Δv is the energy passing through a surface element $\Delta\sigma$ in the time Δt, the flux F is the limit of the quotient $\Delta v/\Delta\sigma \, \Delta t$ as both $\Delta\sigma$ and Δt tend toward zero. The unit of radiative energy most commonly employed in meteorological calculations is the 15°C gram-calorie, and the flux is usually expressed in gram-calories per square centimeter per minute.

2. *Definition of Intensity.* When beams of radiation traverse a surface in all directions, it becomes necessary to define a measure of the quantity of radiation flowing in a given direction. Let $\Delta\sigma$ be a small surface element containing the point P. Let N be the normal to $\Delta\sigma$ at P, and let L be a line through P making an angle θ with

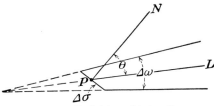

FIG. 1.—Definition of intensity.

N. Around L construct an elementary cone of solid angle $\Delta\omega$, intersecting the periphery of $\Delta\sigma$, thus producing a volume in the form of a truncated cone abutting on $\Delta\sigma$. Let Δv be the energy transmitted through $\Delta\sigma$ in the time Δt within the above volume (see Fig. 1). The limit of the expression $\Delta v/\Delta t \, \Delta\omega \, \Delta\sigma \cos\theta$ as Δt, $\Delta\omega$, and $\Delta\sigma$ approach zero, keeping P and L fixed, is called the intensity I of the field of radiation at P and in the direction L. As so defined, the intensity in a certain direction is the flux per unit solid angle across a surface normal to this direction.

It should be remarked that the above definition of intensity breaks down in the case of a parallel beam; for in any direction other than that of the beam the intensity is zero, while in the direction of the beam the quotient $\Delta v/\Delta t \, \Delta\omega \, \Delta\sigma \cos\theta$ becomes infinite since $\Delta v/\Delta t \, \Delta\sigma \cos\theta$ approaches a finite limit as $\Delta\omega$ approaches zero.

3. *Relation between Flux and Intensity.* Since a unit area whose normal is inclined at an angle θ with the direction L projects into $\cos\theta$ units on a plane normal to L, the flow of energy through this unit area per unit solid angle along L is $I \cos\theta$. Hence the flux confined in the differential solid angle $d\omega$ is $I \cos\theta \, d\omega$. The total flux through a given surface is therefore equal to the integral

$$F = \int I \cos\theta \, d\omega \tag{4}$$

taken over the half sphere of which the surface is in the diametral plane. The net flux would be obtained by integration over the entire sphere. Introducing the azimuthal sphere coordinate ϕ, Eq. (4) becomes

$$F = \int_0^{2\pi} d\phi \int_0^{\frac{\pi}{2}} I \cos\theta \sin\theta \, d\theta \tag{5}$$

This expression for the flux cannot be integrated until I is known as a function of θ and ϕ. In the event that I is independent of direction, the field of radiation is said to be isotropic. In this case, the integration may be performed with the result

$$F = \pi I \tag{6}$$

or *the flux through an arbitrary surface in an isotropic radiation field is π times the intensity.*

4. *Lambert's Law.* If we denote the intensity of radiation emitted from unit surface of a body in a direction making an angle θ with the normal to the surface by I_θ, Lambert's law states that I_θ is proportional to cos θ, or

$$I_\theta = I_0 \cos \theta \tag{7}$$

where I_0 is the value of I_θ at normal emergence. But if I is the total intensity of radiation flowing away from the surface in the direction θ, we know from the definition of I that $I_\theta = I \cos \theta$. Introducing this expression into Eq. (7), we obtain

$$I = I_0 = \text{const} \tag{7'}$$

or *the intensity of the radiation emitted by a body is independent of direction.* In accordance with Lambert's law, an incandescent sphere when viewed from a distance appears to be a uniformly illuminated disk. Lambert's law is in general exact only for a perfectly absorbing body (a so-called "black" body); it holds for others only to the extent to which they approximate a black body. The slight darkening of the outer portion of the sun's disk can be attributed to the imperfect absorbing power of the gases in its atmosphere.

5. *Specific or Monochromatic Intensity.* Since the emissive and absorptive properties of most materials depend on wave length, it is useful to define a specific or monochromatic intensity as the intensity referred to unit wave-length interval. Thus $I_\lambda = dI/d\lambda$. The specific intensity referred to unit frequency interval is similarly defined by the relation $I_\nu = dI/d\nu$. From Eq. (2), it easily follows that

$$I_\lambda = -\frac{c}{\lambda^2} I_\nu \tag{8}$$

Ordinarily the term *intensity* will be used in place of specific intensity where this can be done without ambiguity.

Black-body Radiation, Kirchhoff's Law. A relationship between the absorptive and emissive properties of a body will now be established by the following idealized experiment. Consider an evacuated enclosure surrounded on all sides by walls that completely absorb radiation of all wave lengths. If the whole system is adiabatically isolated from its surroundings, the temperature of the walls will become constant, and the intensity of the radiation field within the enclosure will become everywhere the same and independent of direction. In this state of radiative equilibrium, the intensity of emission from the walls must equal the constant value of the intensity within the enclosure. Thus the intensity of emission from a part of the wall will be independent of its composition and also of the direction. A body that completely absorbs radiation of all wave lengths is called a *black body.* We have therefore proved the theorem: *The intensity of the radiation emitted by a black body is independent of its composition and of the direction of emission.* It will be noticed that the second part of the statement is equivalent to Eq. (7'). It follows that the intensity of emission from a black body can only be a function of temperature and wave length. The validity of this law in no way depends on the particular condition of radiative equilibrium that we have assumed in its derivation, since it concerns only the rate of emission of a body, a quantity that is not influenced by the surrounding radiation field.

Now place a slab of absorbing material within the enclosure and let it come to temperature equilibrium with its surroundings. Let a stream of radiation, moving in an arbitrary direction, strike the slab. A certain fraction a_λ of the incident black-body intensity E_λ will be absorbed and a fraction $1 - a_\lambda$ transmitted. Let e_λ denote the intensity of emission from the slab in the given direction. Since the intensity of radiation in the enclosure is everywhere the same, the intensity of the incident beam must equal that of the emergent beam, which consists of the transmitted black-body intensity plus the emitted intensity. We therefore have

$$E_\lambda = (1 - a_\lambda)E_\lambda + e_\lambda \qquad (9)$$

or

$$\frac{e_\lambda}{a_\lambda} = E_\lambda \qquad (9')$$

This is Kirchhoff's law as applied to the emission from a slab and may be stated: *The ratio of the intensity of emission to the fractional absorption of a slab is equal to the black-body intensity at the same wave length and temperature.* Reflection and scattering have not been considered, but it can be shown that the theorem remains true when they are taken into account.

In the above form, Kirchhoff's law is adapted to the treatment of radiative transfer in a gas where emission and absorption take place *within* the body. An analogous form can be derived from the relation between emission and absorption *at the surface* of a liquid or solid, which reflects but does not transmit radiation. For this purpose, we replace a section of the wall of our enclosure by a reflecting material and again allow the system to come to temperature equilibrium. As before, the intensity of radiation from the walls and in the enclosure must equal that of a black body. The intensity of a beam of radiation issuing from the reflecting part of the wall will consist of two parts; one part will be the intensity of emission e_λ' of the wall itself, and, if the reflecting material is perfectly smooth, the other part will be the reflected intensity of the beam that comes from a direction symmetrical to that of the given beam with respect to the normal to the surface. If a_λ' is the fraction of the incident radiation absorbed by the surface, then $1 - a_\lambda'$ is the fraction reflected, and the total intensity of the beam coming from the wall is therefore equal to $e_\lambda' + (1 - a_\lambda')E_\lambda$; and this quantity must be equal to E_λ, the intensity of radiation from within the enclosure traveling in the opposite direction. We have, therefore

$$E_\lambda = (1 - a_\lambda')E_\lambda + e_\lambda' \qquad (10)$$

or

$$\frac{e_\lambda'}{a_\lambda'} = E_\lambda \qquad (10')$$

It is an immediate consequence of Kirchhoff's law that the intensity emitted by a body can never exceed the black-body intensity and can equal it only in the spectral regions where the body is opaque. The earth's surface, for example, is nearly opaque to long-wave atmospheric radiation while it reflects some short-wave solar radiation. It follows that it will emit black-body radiation in the former and less than black-body radiation in the latter region. A body that absorbs a constant fraction of the incident radiation at all wave lengths is called a *gray* body. According to Kirchhoff's law, a gray body will emit a constant fraction of the black-body intensity.

Planck's Radiation Law. We have shown that the intensity of emission from a black body depends only on its temperature and the wave length. The exact form of the relationship was first derived by Planck and may be written

$$E_\lambda = \frac{2hc^2}{\lambda^5} \frac{1}{e^{\frac{hc}{\lambda kT}} - 1} \qquad (11)$$

where h is known as Planck's constant and has the value 6.55×10^{-27} erg sec, c is the velocity of light 3×10^{10} cm per sec, and k, Boltzmann's constant, is equal to 1.37×10^{-16} erg per deg. T is the absolute temperature of the body.

The black-body intensity per unit frequency becomes, with the aid of Eqs. (2) and (8)

$$E_\nu = -\frac{2'\nu^3}{c^2}\frac{1}{e\frac{h\nu}{kT} - 1} \tag{12}$$

Figure 2 shows the black-body emission curves drawn for the temperatures 200 and 300°K.

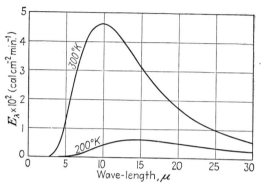

Fig. 2.—Black-body emission spectra for $\Delta\lambda = 1\mu$.

Wien's Displacement Law. Color Temperature of the Sun. From Fig. 2, it can be seen that the wave length corresponding to the maximum intensity decreases as the temperature increases. The quantitative expression of this rule can be derived from Planck's general law by the ordinary calculus rule for obtaining the maximum value of a function. Setting the derivative with respect to λ of the right-hand side of Eq. (11) equal to zero, we find that *the maximal value of λ is inversely proportional to the absolute temperature T.* Or

$$\lambda_m = \frac{a}{T} \tag{13}$$

where a has the value 0.288 cm deg.

The maximum intensity of solar radiation is in the blue-green at $0.475\ \mu$. On the assumption that the sun is a perfect black body, Wien's law gives for its temperature the value 6090°K, which is known as the *color temperature* of the sun.

Stefan-Boltzmann's Law. The total intensity of black-body emission can be obtained by integrating Eq. (11) or (12) between the limits 0 and ∞. Thus

$$E = -\int_0^\infty E_\nu\, d\nu = \int_0^\infty E_\lambda\, d\lambda \tag{14}$$
$$= \int_0^\infty \frac{2h}{c^2}\frac{\nu^3\, d\nu}{e^{\frac{h\nu}{kT}} - 1}$$
$$= \frac{2k^4T^4}{c^2h^3}\int_0^\infty \frac{x^3\, dx}{e^x - 1}$$

The last integral can be shown to have the value $\pi^5/15$. Hence

$$E = \frac{2\pi^4k^4}{15c^2h^3}T^4 = \frac{\sigma}{\pi}T^4 \tag{14'}$$

where $\sigma = 2\pi^5k^4/15c^2h^3$. This is the Stefan-Boltzmann law, according to which *the total energy emitted by a black body varies as the fourth power of the absolute temperature.*

From the nature of the coordinates in Fig. 2 and from the first line in Eq. (14), it follows that the area underneath each curve in the figure represents the total black-body intensity emitted at its corresponding temperature.

It was proved on page 287 that the intensity of emission from a black surface is independent of direction. The flux from the surface is therefore given by Eq. (6)

$$F_b = \pi E = \sigma T^4 \tag{15}$$

The constant σ has the value 5.70×10^{-5} erg cm^{-2} sec^{-1} deg^{-4}, or 0.817×10^{-10} cal cm^{-2} min^{-1} deg^{-4}.

Beer's Law of Absorption. When a beam of monochromatic radiation is transmitted through an infinitesimal distance dl in an absorbing medium, a certain fraction of the intensity will be absorbed, and it is assumed that this fraction is independent of the intensity but proportional to the density ρ of the absorbing substance and to the distance dl. It follows that

$$\frac{dI_\lambda}{I_\lambda} = -k_\lambda \rho \, dl \tag{16}$$

The proportionality constant k_λ is known as the *coefficient of absorption* of the medium. Integration of Eq. (16) gives

$$I_\lambda = I_{\lambda 0} e^{-k_\lambda \int_0^l \rho \, dl} \tag{16'}$$

where $I_{\lambda 0}$ is the initial intensity and l the total distance traversed by the beam. The integral $\int_0^l \rho \, dl$ represents the mass of absorbing material in a column of unit cross section extending the distance l. It is called the *optical path length* and is denoted by m. Substituting this symbol for $\int_0^l \rho \, dl$ in Eq. (16), we obtain the formula

$$I_\lambda = I_{\lambda 0} e^{-k_\lambda m} \tag{16''}$$

which is known as *Beer's law.*

We have tacitly supposed that k_λ depends only on wave length; but it was shown above that the absorbing power of a substance is a function of pressure and temperature as well. In the event that these quantities vary along the path of the beam k_λ will also vary, and Eq. (16') would be modified by placing k_λ under the integral sign in the exponential term.

In dealing with the transmission of radiation through the atmosphere, it is usually assumed that the absorbing materials are uniformly stratified in a horizontal direction. The density ρ will then depend only on the vertical coordinate z, and the path length m can be written as $\int_{z_0}^{z_1} \rho \, dz \sec \theta$ where θ is the angle between the direction of the beam and the vertical. The quantity $\int_{z_0}^{z_1} \rho \, dz$ is called the *optical depth* or *optical thickness* of the layer between the heights z_0 and z_1 and is equal to the number of grams of absorbing material in a vertical column 1 sq cm in cross-sectional area, extending from z_0 to z_1.

The Equation of Radiative Transfer. Consider a beam of monochromatic radiation that traverses the distance dl in an absorbing medium. By Eq. (16), the change in intensity due to absorption is $-k_\lambda \rho \, dlI_\lambda$, and the fractional absorption is $k_\lambda \rho \, dl$. By Kirchhoff's law [Eq. (9')], the intensity emitted in the direction of the beam is $k_\lambda \rho \, dlE_\lambda$. Therefore the net change in intensity will be

$$dI_\lambda = -k_\lambda \rho \, dlI_\lambda + k_\lambda \rho \, dlE_\lambda \tag{17}$$

and, introducing the change of variable $dm = \rho \, dl$, we obtain the equation of radiative transfer

$$\frac{dI_\lambda}{dm} = -k_\lambda(I_\lambda - E_\lambda) \tag{17'}$$

By means of this equation, it is theoretically possible to calculate the intensity of radiation at any point in the atmosphere, provided that the distribution of the absorbing substances and the coefficients of absorption are known.

Calculation of Flux from a Horizontal Layer. The central problem in the theory of radiative heat transfer in the atmosphere is the calculation of the flux emitted by a horizontal layer from one of its faces. As will be seen, all questions relating to the heating and cooling of the atmosphere by radiation are ultimately referred to the determination of flux. Let us consider a layer, bounded by horizontal planes of infinite extent, in which the absorbing substances are uniformly stratified in the horizontal direction, and first calculate the flux at the base of the layer that originates in an infinitesimal sheet of optical thickness du. If the intensity emitted by the sheet in the direction θ (see Fig. 3) is dI_λ and the optical depth measured upward from the base

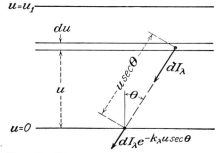

Fig. 3.—Transmitted intensity from an infinitesimal layer.

of the layer is u, the intensity transmitted to the base of the layer will be $dI_\lambda e^{-k_\lambda u \sec \theta}$, by Beer's law. Introducing this expression for the transmitted intensity into Eq. (5), we obtain for the flux of radiation at the base of the layer coming from the infinitesimal sheet

$$dF_\lambda = \int_0^{2\pi} d\phi \int_0^{\frac{\pi}{2}} dI_\lambda e^{-k_\lambda u \sec \theta} \sin \theta \cos \theta \, d\theta \tag{18}$$

But from Kirchhoff's law [Eq. (9')] the intensity emitted by a slab in a given direction is equal to the black-body intensity multiplied by the fractional absorption in the same direction. In the present case, the fractional absorption in the infinitesimal sheet is $k_\lambda \, du \sec \theta$. Therefore $dI_\lambda = k_\lambda \, du \sec \theta E_\lambda$, and Eq. 18 becomes, after integrating with respect to ϕ

$$dF_\lambda = 2\pi E_\lambda \, du \int_0^{\frac{\pi}{2}} k_\lambda e^{-k_\lambda u \sec \theta} \sin \theta \, d\theta \tag{19}$$

If the integral $2 \int_0^{\frac{\pi}{2}} e^{-k_\lambda u \sec \theta} \sin \theta \cos \theta \, d\theta$ is denoted by τ_f the expression for dF_λ reduces to

$$dF_\lambda = \pi E_\lambda \frac{d\tau_f}{du} du \tag{20}$$

The total downward flux F at the base of the layer is obtained by integrating Eq. (20), from $u = 0$ to $u = u_1$, where u_1 is the optical depth of the whole layer, and over the whole range of wave lengths, from $\lambda = 0$ to $\lambda = \infty$. Performing this integration and substituting the monochromatic black-body flux $(F_\lambda)_b$ for πE_λ, in which we are justified by Eq. (6), we obtain

$$F = \int_0^\infty d\lambda \int_0^{u_1} (F_\lambda)_b \frac{d\tau_f}{du} \, du \qquad (21)$$

In actual computations of flux in the atmosphere, a graphical method is used for the evaluation of the above integral. This method will be discussed below.

SOLAR RADIATION

Nature of Solar Radiation. Observations conducted over a period of years indicate that the solar radiation does not change appreciably from year to year and varies only with latitude and season. A measure of the quantity of the incoming solar radiation is the *solar constant*, which is defined as the flux of solar radiation at the outer boundary of the earth's atmosphere received on a surface normal to the sun's direction at the earth's mean distance from the sun. The solar constant has the value 1.94 cal cm^{-2} min^{-1}.

The spectral distribution of solar radiation closely approximates that of a black body. On the assumption that the sun is perfectly black, one can compute by means of Boltzmann's law the temperature it should have in order for the flux at the outer limit of the earth's atmosphere to equal the solar constant. This quantity is known as the *effective temperature* of the sun and is equal to 5760°K. The *effective temperature* differs from the *color temperature* of the sun defined on page 289. Selective absorption in the sun's atmosphere reduces the total radiation but leaves relatively unchanged the wave length corresponding to the maximum intensity.

It follows from Planck's law that a black body at the temperature of the sun will radiate upward of 99 per cent of its energy between the wave lengths 0.15 and 4 μ. Roughly one-half of this radiation will lie in the visible region of the spectrum between 0.38 and 0.77 μ, and the remainder in the invisible ultraviolet and infrared regions.

Geographical and Seasonal Distribution of Solar Radiation. In the absence of an atmosphere, the flux of solar radiation reaching a point on the earth's surface depends only on the zenith angle θ of the sun and the sun's distance from the earth. If I denotes the intensity of solar radiation, the flux is given by Eq. (4).

$$F = \int I \cos \theta \, d\omega \qquad (22)$$

extended over the solid angle $\Delta\omega$, subtended by the sun. Within this small angle, I may be regarded as constant. Therefore

$$F = I \cos \theta \, \Delta\omega = I \cos \theta \frac{\pi a^2}{r^2} \qquad (23)$$

where a is the radius of the sun and r its distance from the earth. When $\theta = 0$ and $r = r_m$, the mean distance of the sun from the earth, the above expression reduces to the solar constant. Thus

$$S = \pi \frac{a^2}{r_m^2} I \qquad (24)$$

We note that the intensity is a much larger quantity than the solar constant. Substituting Eq. (24) into Eq. (23), we obtain

$$F = S \cos \theta \frac{r_m^2}{r^2} \qquad (25)$$

and if no great accuracy is required, the factor r_m^2/r^2 can be taken as unity. We then derive the familiar cosine law for solar radiation, namely

$$F = S \cos \theta \qquad (25')$$

The total energy delivered per unit area on the earth's surface in the course of a **day** is obtained by integrating Eq. (25) over the period during which the sun is above the horizon. The evaluation of this integral is an astronomical problem, and only the result is reproduced here. If ϕ is the latitude of the place, δ the solar declination, and H the hour angle between sunrise and noon, or noon and sunset, the total flux is

$$Q = \int F \, dt = \frac{24}{\pi} \frac{r_m{}^2}{r^2} S \sin \phi \sin \delta (H - \tan H) \qquad (26)$$

where

$$\cos H = -\tan \phi \tan d \qquad (27)$$

The solar declination can be found for any time of year from a nautical almanac.

Figure 4 was computed by means of Eq. 26 and shows the dependence of the daily flux of solar radiation in cal cm^{-2} day^{-1} on latitude and season. The horizontal coordinate is given as time of year in the lower margin and solar declination in the upper margin; the vertical coordinate is the latitude. It is noteworthy that, owing to the length of day at the summer solstice of the Northern Hemisphere, the North

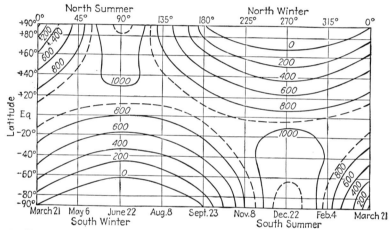

FIG. 4.—Daily insolation in cal cm^{-2} day^{-1} received at the earth's surface in the absence of an atmosphere. (*After Milankovitch's computations.*)

Pole receives the maximum daily total insolation, and the equator receives a minimum. At the winter solstice, these conditions are reversed; indeed latitudes above 68° receive no radiation at all. The total radiation received by the Southern Hemisphere during the southern summer is larger than the amount received by the Northern Hemisphere during its summer, for in the Southern Hemisphere the earth is closer to the sun.

Because of the presence of the earth's atmosphere, the solar radiation received at the surface of the earth is somewhat less than is shown in Fig. 4. The depletion of energy in the solar beam is greatest at high latitudes where, because of the large zenith angle, its path through the atmosphere is longest. For this reason, the region of maximum insolation at the earth's surface is no longer found at the pole in summer but is shifted to about 30° latitude.

Depletion of Solar Radiation. 1. *Absorption in the Atmosphere.* A study of the solar spectrum reveals the presence of numerous fine absorption lines and bands. These consist in part of the well-known Fraunhofer lines and are produced by absorption by the gases and vapors in the sun's atmosphere. The remaining lines are due to absorption by the gases in the earth's atmosphere.

One of the most striking characteristics of the observed solar spectrum is the abrupt termination of the short wave-length end at 0.29 μ. This is caused primarily by ozone in the earth's atmosphere and represents a loss of over 5 per cent of the original solar intensity. Ozone is distributed throughout the atmosphere up to heights of more than 50 km with the greatest concentration occurring between 25 and 30 km. The high temperatures of the upper atmosphere are due in part to ozone absorption.

With the exception of ozone, the significant absorption of solar radiation by the gaseous components of the atmosphere can be entirely attributed to water vapor. Oxygen has several absorption bands, but these are so narrow that they represent a very minute loss of solar energy. The remaining gases either do not absorb at all or absorb negligible quantities.

Our quantitative knowledge of the absorption spectrum of water vapor in the short-wave region is mainly due to Fowle.[1] His results are represented in Fig. 5 by a curve giving the relationship between the fractional absorption and the optical path length. Also included in the figure is a similar curve computed from a large number of measurements of the absorption of solar radiation in the atmosphere by Kimball[2] and Hoelper.[3] Fowle's observations were taken in the laboratory and do not take into account the variation of atmospheric absorption with pressure and temperature. If the proper pressure and temperature corrections were made, it is probable that the two curves would be brought into close agreement.

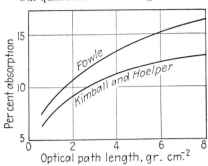

Fig. 5.—Absorption of solar radiation by water vapor according to Fowle and to Kimball and Hoelper.

The nongaseous components of the atmosphere such as dust, smoke, and salt particles must be included as absorbing agents. Their effect is highly variable but should in most cases be quite small.

2. Heating of Atmosphere by Absorption of Solar Radiation. From Fowle's measurements, Mügge and Möller[4] derived an empirical formula from which one obtains the heating of the atmosphere by solar radiation. If $(\Delta T)_3$ denotes the temperature rise in 3 hr, u is the optical depth of the atmosphere water vapor above a given level, and θ is the zenith angle, then the formula

$$(\Delta T)_3 = 0.039(u \sec \theta)^{-0.70} \tag{28}$$

gives the rate of change in temperature at that level. Since the altitude of the sun does not remain constant, Eq. (28) must be integrated over the course of the sun's altitude during the day to obtain the total heating. From extensive computations by means of Eq. (28), Tanck[5] found that the absorption of solar radiation produces a heating of from 0.3 to 0.6°C per day according to the amount of water vapor.

3. Scattering and Diffuse Reflection. According to the electromagnetic theory of light, a small charged particle placed in the path of a beam of light will experience an alternating electric force. If the inertia of the particle is sufficiently small, it will oscillate with the same frequency as that of the imposed wave and will therefore emit electromagnetic radiation of the same frequency.

The electronic charges in a molecule are acted upon in this way, and the emitted radiation is scattered or reflected. Scattering occurs when the size of the aggregation in which the molecules are found is of the order of the wave length of the incident radiation or less and is characterized by the dependence on wave length, long waves being less effectively scattered than short ones.

If I is the intensity of the scattered light in a direction making an angle θ with the incident beam and $I_{\frac{\pi}{2}}$ is the intensity in the normal direction, it can be shown that the following law holds for scattering by very small particles:

$$I = I_{\frac{\pi}{2}}(1 + \cos^2 \theta) \tag{29}$$

Thus the amount of scattering is greatest in the direction of the incident beam and in the opposite direction, and least in the normal direction.

A coefficient of scattering may be defined in the same manner as was the coefficient of absorption. In a medium such as pure dry air, having an index of refraction μ, and containing n particles per cubic centimeter, the coefficient of scattering s_λ is very nearly equal to

$$s_\lambda = \frac{32\pi^3}{3n\lambda^4}(\mu - 1)^2 \tag{30}$$

This expression conforms to the well-known law discovered by Lord Rayleigh, which states that *scattering is inversely proportional to the fourth power of the wave length of the incident radiation.* If it is desired to compute the depletion of the intensity of a beam of light, the quantity $k_\lambda + s_\lambda$ should be used in place of k_λ in Beer's law.

In the atmosphere, sunlight is scattered by the molecules of dry air and water vapor, and also by very small solid impurities. Among the more obvious optical phenomena produced by scattering may be mentioned the blue color of the sky and the red color of the sun and clouds at sunrise or sunset. The sky is blue because it is made visible by scattered light, which, according to Rayleigh's law is rich in the shorter wave lengths; whereas the sun and clouds are seen by direct light since a preponderance of the short waves have been removed during its long passage through the atmosphere.

When the diameters of the scattering particles exceed the wave length of the incident light, the coefficient of scattering is no longer inversely proportional to the fourth power but is inversely proportional to a smaller power of the wave length; and, when the particles become sufficiently large, the dispersal of radiation is equally effective for all wave lengths and is called *diffuse reflection.* Large dust and haze particles, water droplets, and ice crystals reflect rather than scatter light; and, since diffuse reflection is nonselective, all wave lengths are reflected equally. Thus, contrary to popular belief, there is no particular color of light that will penetrate a fog most effectively.

When large solid particles are present in the atmosphere, the sky is suffused with a whitish tinge due to diffusely reflected light. The depth of the blue color of the sky can therefore be regarded as a measure of the amount of impurities in the atmosphere.

4. *The Total Depletion of Solar Radiation.* As described above, the incoming solar radiation suffers depletion in the following ways: (1) absorption by ozone in the upper atmosphere; (2) scattering by dry air; (3) absorption, scattering, and diffuse reflection by suspended solid particles; and (4) absorption and scattering by water vapor. We have seen that the solar intensity is reduced about 5 per cent by ozone absorption. Kimball[6] has computed the effect of the factors (2), (3), and (4) for several localities from observed intensities of solar radiation. In Table 1 is shown the percentage of extinction of direct radiation from a zenith sun caused by each of the above factors at three representative stations. The total depletion given in the table does not represent the percentage that fails to reach the earth, for some of the scattered

and reflected radiation eventually arrives at the ground as diffuse radiation.   Kimball estimates that on a clear day about 50 per cent of the scattered and diffusely reflected light is returned to the ground in this way.

TABLE 1.—PERCENTAGE DEPLETION OF DIRECT RADIATION FROM A ZENITH SUN BY THE CONSTITUENTS OF THE ATMOSPHERE AT REPRESENTATIVE STATIONS*

	Dry air	Water vapor	Solid impurities	Total
Washington, D.C.				
Summer......................	9	17	10	36
Winter......................	9	8	9	26
Apia, Samoa				
Summer......................	9	21	5	35
Winter......................	9	20	8	37
Mount Whitney, California				
Summer......................	6	2	3	11

* After Kimball.

Reflection of Solar Radiation by the Earth's Surface and by Clouds.   Albedo of the Earth.   With the single exception of a snow surface, the surface of the earth is a poor reflector of solar radiation.   Fresh snow reflects 80 to 85 per cent of the incident radiation, while old snow may reflect as little as 40 per cent.   Grass reflects 10 to 33 per cent, rock 12 to 15 per cent, dry earth 14 per cent, and wet earth 8 to 9 per cent.   Virtually no radiation is reflected by a smooth water surface when the sun is within 40 deg of the zenith; the reflection coefficient then increases from 2 per cent at a zenith distance of 43 deg to 40 per cent at 85 deg.

The percentage reflection from a cloud deck was given by Aldrich[7] as 78 per cent. His measurements were made on low stratus clouds from 180 to 500 m thick.   It might be expected that a larger proportion of solar radiation would be reflected by clouds of greater thickness, since the opportunities for internal multiple reflection would be increased.   A theoretical investigation conducted by Hewson[8] indicates a rather wide variation of reflectivity with mean drop size in the cloud, cloud thickness, and density.   The variation of the reflection coefficient with cloud thickness for clouds containing 1.0 gram m^{-3} of liquid water and with an average drop diameter of 10^{-3} cm is plotted in Fig. 6.   No accurate measurements of the absorption of solar radiation by a cloud have yet been made.   Although it is known that the absorption coefficients of liquid water are very small in the short-wave region of the spectrum, the path length of a ray passing through a cloud may be so magnified by internal reflection that appreciable absorption might occur.   Hewson has calculated the fractional absorption in a cloud as a function of thickness, and these results are also shown in Fig. 6.

The *albedo* of the earth is a quantity used to measure the total reflecting power of the earth plus its atmosphere.   It is defined as the fraction of the incoming solar radiation returned to space by scattering and reflection in the atmosphere and by reflection at clouds and at the earth's surface.   It represents the unused fraction of the incoming solar energy; the part that is absorbed neither in the atmosphere nor at the earth.   The mean albedo can be determined astronomically by comparing the intensities of reflected light from the moon when part is illuminated by direct sunlight and part by reflected light from the earth.   Aldrich determined its value indirectly by assuming that, on the average, 8 per cent of the incoming radiation is reflected from the earth's surface, 9 per cent reflected and scattered from a cloudless sky, and 78 per cent reflected by clouds, and that the mean cloud cover of the earth is 52 per

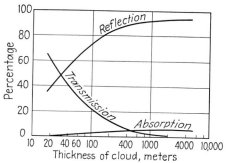

Fig. 6.—Percentage reflection, absorption, and transmission of solar radiation by clouds. (*After Hewson.*)

cent.   Under these assumptions, the albedo is 0.43, a value that agrees fairly well with the astronomical calculations.

Ångström[9] gives the formula

$$A = 0.17 + 0.53C \tag{31}$$

for the relation between albedo and cloud amount, where A is the albedo and C the number of tenths of sky covered by clouds.

TERRESTRIAL RADIATION

Black radiation at terrestrial or atmospheric temperatures, say, between 200 and 330°K is practically all contained within the limits 4 and 120 μ; whereas, at the effective temperature of the sun, about 6000°K, the bulk of the radiation is contained within the limits 0.15 and 4 μ.   Because of the mutual exclusiveness of the two regions, it is customary in meteorology to refer to the former as the long-wave region and the latter as the short-wave region.

The various substances composing the earth's surface radiate very nearly as black bodies.   This is true even for surfaces that are highly reflecting in the visible and short wave regions of the spectrum, such as a snow surface.   What is important is that snow, together with other terrestrial bodies, is an excellent absorber in the long-wave infrared region.   By Kirchhoff's law it must therefore emit black radiation in this region, and as a glance at Fig. 2 shows, all but an infinitesimal portion of the radiation from a black body at terrestrial temperatures is contained in the infrared wave lengths.

Besides the earth, a cloud of sufficient thickness (about 50 m) may also be regarded as a black radiator.   Because of the extremely large values of the absorption coefficients of liquid water in the long-wave region of the spectrum, it is estimated that a cloud or dense fog about 50 m thick will absorb practically all long-wave radiation and will therefore also radiate as a black body.

Absorption of Long-wave Radiation in the Atmosphere.   It was stated on page 294 that the atmospheric absorption of solar radiation is slight, being mainly due to water vapor but also in part to ozone and oxygen. The atmosphere is less pervious to terrestrial radiation, owing mainly to the greatly increased absorbing capacity of water vapor in the long-wave regions of the spectrum. In addition, carbon dioxide has a narrow but intense absorption band centered at 14.7 μ and extending from 12 to 16.3 μ, and ozone has weaker bands centered at 7 and 10 μ. The absorption spectrum of water vapor is shown in Fig. 7. Marked absorption is shown between a wide range of wave lengths, extending from about 4.5 to beyond 80 μ. The values of k_λ in the figure were measured by Hettner[10] in 1918. Later determinations indicate that Hettner's values are too large but that the principal characteristics are correct.

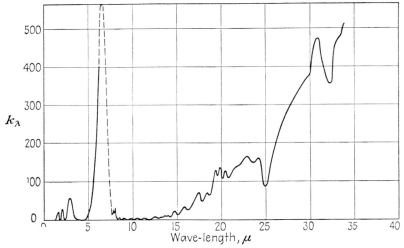

FIG. 7.—Absorption spectrum of water vapor.

Simpson's Computation of Long-wave Radiation in the Atmosphere. In order to avoid having to deal with the highly variable coefficients of absorption of water vapor, early investigators of heat transfer by long-wave radiation treated the atmosphere as a gray body. The lack of success that attended their efforts led Simpson[11] to propose a method that takes into account the variability of the water-vapor spectrum and is yet quite simple in application. Simpson regards the atmosphere as being composed of horizontal layers with optical depths of 0.03 gram of water vapor and 0.06 gram of carbon dioxide per square centimeter. Using Hettner's data except in the region 9 to 12 μ, where according to Fowle[1] but contrary to Hettner no absorption takes place, and taking the carbon dioxide band at 14.7 μ into account, he was able to reduce the main features of atmospheric absorption of long-wave radiation by such a layer to

    1. Effectively complete absorption from 5.5 to 7 μ, and from 14 μ upward.
    2. Effectively complete transparency from 8.5 to 11 μ and below 4 μ.
    3. Incomplete absorption from 7 to 8.5 μ and 11 to 14 μ.

On the basis of the Simpson wave-length categories, it is possible to give a quantitative estimate of the upward flux at any level. For this purpose, we divide the atmosphere below this level into layers containing 0.03 gram of water vapor and 0.06 gram of carbon dioxide per square centimeter. If Δp is the variation in pressure through one of the layers expressed in millibars and q the mean specific humidity in grams per gram, then

$$\Delta p = \frac{0.03 \times 10^{-3}}{q} g \tag{32}$$

for example, if $q = 0.01$, we obtain 2.9 mb for the difference in pressure, which corresponds to a difference in elevation of about 100 ft in the lower atmosphere. Consider now the upward flux at the level in question in each of the three categories. Since each layer is opaque to radiation in 1, it follows by Kirchhoff's law that the radiation in this category will have originated in the first layer below the given level and can be computed from the mean temperature of the layer, a quantity that will not deviate much from the temperature of the given level. The radiation in 2 must have originated at the ground and can be calculated from the surface temperature. Of the radiation in 3 we cannot make any equally direct statement. It will be intermediate in amount between the radiation of this kind at the temperature of the earth and that at the temperature of the level, and we cannot be too far away if we take the mean value as a first approximation. Figure 8 illustrates the method used in computing the

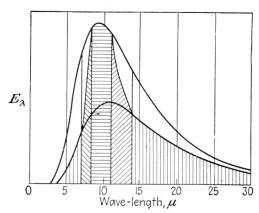

Fig. 8.—Illustration of Simpson's method.

flux. The upper curve represents black-body radiation from the ground, and the lower curve represents black-body radiation at the temperature of the level. The flux in the first category originates just below the level and is represented by the vertically hatched region from 5.5 to 7 μ and above 14 μ. The flux in the second category comes from the ground and is represented by the horizontally hatched area. The diagonally hatched region from 7 to $8\frac{1}{2}$ μ and from 11 to 14 μ represents the flux in the third category. The sum of the shaded areas represents the upward flux of radiation.

The Generalized Transmission Function. At the time that Simpson introduced his simplifications, too little was known of the absorption spectrum of water vapor to warrant any more complicated treatment of long-wave radiative transfer. However, recent determinations of water-vapor absorption have made it feasible to evaluate the flux by means of the exact expression given in Eq. 21.

Since the rapid variation of the absorption coefficients from line to line in the water-vapor spectrum makes the direct integration of Eq. (21) a practical impossibility, a method has been devised by Elsasser[12] for replacing the absorption coefficient k_λ by a more tractable function. The absorption coefficient is involved in the function $e^{-k_\lambda u}$, which appears in the expression for τ_f. This function, according to Beer's law, represents the fractional transmission of a beam of monochromatic radiation through the optical path u. In order to simplify the calculations, Elsasser replaces the function $e^{-k_\lambda u}$ in the following way: The whole wave-length interval in which water vapor absorbs is divided into a number of small subintervals, and the average value of $e^{-k_\lambda u}$ is determined for each of these intervals. We thus define the *generalized transmission function* $\tau_1(u, \lambda)$ by the equation

$$\tau_I = (e^{-k_\lambda u})_{av} \tag{33}$$

For details concerning the determination of the function τ_I, the reader is referred to Elsasser's original paper. It suffices for the present discussion to know that $\tau_I(u, \lambda)$ can be represented by the function

$$\tau_I = 1 - \phi\left(\sqrt{\frac{l_\lambda u}{2}}\right) \tag{34}$$

where ϕ is the probability integral $\phi(x) = \int_0^x e^{-s^2}\, ds$, and l_λ is the function of wave length that replaces the absorption coefficient k_λ, and τ_1 is called the *generalized absorption coefficient*. Its variation with wave length is shown in Fig. 9. Since l_λ is much less variable than k_λ, it lends itself more easily to computation.

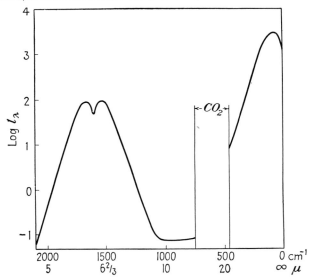

Fig. 9.—The generalized absorption coefficient.

It was explained qualitatively above how absorption depends on pressure and temperature. It has been established both theoretically and empirically that l_λ is approximately proportional to the square root of the pressure and inversely proportional to the fourth root of the temperature. The temperature variation is small and can be neglected. We may therefore write

$$l_\lambda = (l_\lambda)_s \sqrt{\frac{p}{p_s}} \tag{35}$$

where p_s is a standard pressure, which will be taken as 1,000 mb, and $(l_\lambda)_s$ is the value of the generalized absorption coefficient at this pressure. As indicated previously, to allow for the variation of k_λ with pressure, the expression $k_\lambda u$ must be replaced by $\int k_\lambda \, du$. Similarly, it can be shown in the present case that the term $l_\lambda u$ must be replaced in τ_I by $\int l_\lambda \, du = (l_\lambda)_s \int \sqrt{p/p_s} \, du$. We therefore define the *corrected optical depth* u' by the relationship

$$u' = \int \sqrt{\frac{p}{p_s}} \, du \tag{36}$$

The expression for τ_1 then becomes

$$\tau_I = 1 - \phi \left[\sqrt{\frac{(l_\lambda)_s u'}{2}} \right] \tag{34'}$$

ELSASSER RADIATION CHART

1. *Construction of the Chart.* The computation of the flux of radiation from a layer has been reduced on page 292 to the evaluation of the integral

$$F = \int_0^\infty d\lambda \int_0^{u_1} (F_\lambda)_b \frac{d\tau_f}{du} \, du \tag{21}$$

By substitution of τ_I for $e^{-k_\lambda u}$, the function τ_f becomes

$$\tau_f = \int_0^{\frac{\pi}{2}} \tau_I[(l_\lambda)_s u' \sec \theta] \sin \theta \cos \theta \, d\theta \tag{37}$$

This integration cannot be performed directly, since $(l_\lambda)_s$ is known only as an empirical function of λ. We resort to a graphical method devised by Mügge and Möller.[4] Introducing the function

$$Q(u',T) = \int_0^\infty \frac{d(F_\lambda)_b}{dT} \tau_f \, d\lambda \tag{38}$$

and integrating the right-hand side of Eq. (21) by parts we obtain

$$F = \int_0^{T_0} Q[u'(T_0),T]dT + \int_{T_0}^{T_1} Q[u'(T),T]dT + \int_{T_1}^0 Q[u_1'(T_1),T]dT \tag{39}$$

where T_0 is the temperature at the base of the layer and T_1 the temperature at the top. The expression $u'(T)$ represents the relation between the corrected optical depth measured from the base of the layer to an intermediate level and the temperature at that level. By definition $u'(T_0) = 0$,and $u'(T_1) = u'$, the optical thickness of the entire layer.

The three integrals in Eq. (39) together define a closed path in the QT-plane, and the flux is equal to the area enclosed by the path. In order to emphasize that portion of the chart which lies in the region of meteorological interest, the coordinates Q and T are changed by means of the area-preserving transformation

$$x = aT^2 \qquad y = \frac{Q(u',T)}{2aT} \tag{40}$$

From the nature of the above transformation, it can be shown that the moisture isopleths $u' = $ const and the isotherms $T = $ const have the following properties:

    a. The isotherms are vertical lines.
    b. The isopleth for $u' = \infty$ is a horizontal line.
    c. All the moisture isopleths intersect in a point at which the absolute temperature is zero.
    d. u increases downward along any isotherm from $u' = 0$ to $u' = \infty$. See Fig. 10.

2. *Representation of the Flux Due to Carbon Dioxide in the Radiation Chart.* Although water vapor is the most important absorber of infrared radiation in the atmosphere, the effect of carbon dioxide is appreciable and cannot be ignored. Carbon

dioxide absorption is concentrated chiefly in a rather narrow region of the spectrum where it is very intense. We assume, as an approximation, that the carbon dioxide in an infinitesimally thin layer of air absorbs all radiation within this region and therefore, by Kirchhoff's law, emits black-body radiation corresponding to its tem-

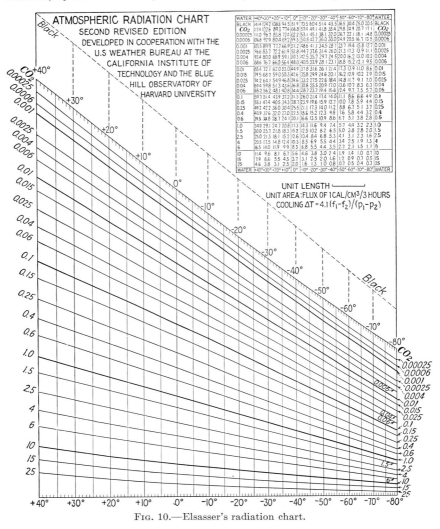

Fig. 10.—Elsasser's radiation chart.

perature. If λ_0 and λ_1 are the wave-length limits of the region of intense carbon dioxide absorption, we have

$$F_{CO_2} = \int_{\lambda_0}^{\lambda_1} (F_\lambda)_b \, d\lambda$$

Since F_{CO_2} is a function of temperature only, we may express it in the radiation chart by drawing an isopleth above the moisture isopleths such that the wedge-shaped area bounded by this isopleth, the $u = 0$ isopleth, and the $T = T_0$ isotherm is equal to the carbon dioxide flux at the temperature T_0. In the chart, this isopleth is denoted *black* for reasons that will shortly become evident. In order to avoid the duplication of flux, water vapor is taken to be transparent in the interval λ_0 to λ_1.

3. *The Optical Depth.* The optical depth or optical thickness of a horizontally stratified layer was defined as the integral $\int_{z_0}^{z_1} \rho \, dz$, where ρ is the density of the absorbing material and z_0, z_1 the vertical coordinates of the horizontal bounding planes. In the present case, ρ is equal to the water-vapor density ρ_w, which is given by $q\rho_a$, where q is the specific humidity and ρ_a the air density. Substituting this expression, we obtain

$$u = \int_{z_0}^{z_1} q\rho_a \, dz \qquad \text{or} \qquad du = q\rho_a \, dz \qquad (41)$$

and if ρ_a is eliminated by means of the hydrostatic equation $dp = -g\rho_a \, dz$, du becomes

$$du = -\frac{1}{g} q \, dp \qquad (42)$$

Introducing this expression into Eq. (36), we obtain for the corrected optical thickness

$$u' = -\frac{1}{g} \int_{p_0}^{p_1} q \, \sqrt{p/p_s} \, dp \qquad (43)$$

The prime will be omitted from u in the following discussion with the understanding that by u the *corrected* optical depth is always meant.

4. *Graphical Determination of Flux from a Layer.* In order to calculate the flux from the base of a layer, it is necessary to measure the area bounded by the paths of integration defined by the integrals in Eq. (39). Along the first path, represented by the line OB in Fig. 11, $u = 0$, while T increases from 0 to T_0, its value at the base of the layer. The second path, BC, is the plot of corresponding u and T values through the layer beginning with $u = 0$, $T = T_0$ at the base and extending to $u = u_1$, $T = T_1$ at the top. Along the third path CO, $u = u_1$, and T decreases from T_1 to 0. The area $OBCO$ therefore represents the water-vapor flux for all wave lengths with the exception of the region of carbon dioxide absorption. The carbon dioxide flux is given by the area OAB, since by our assumptions the carbon dioxide radiation originates immediately above the base in an infinitesimal layer whose temperature is T_0. The total flux from the base is therefore represented by the area $OABCO$.

5. *Black-body Flux.* It follows from Kirchhoff's law that the flux emitted by an isothermal layer of infinite optical depth is the same as that of a black body, since the fractional absorption in such a layer is unity. On the radiation chart, the flux from an infinite layer of water vapor at temperature T_0 is represented by the area $OBDO$, for the three paths of integration are, respectively, the $u = 0$ isopleth, the $T = T_0$ isotherm, and the $u = \infty$ isopleth. To this area must be added the area $OABO$ representing the black carbon dioxide flux. Thus total black-body emission at temperature T_0 is represented by the triangle $OADO$ in the diagram.

Applications of the Radiation Chart. The radiation chart can be applied wherever it is desired to calculate heat exchange by long-wave radiation in the atmosphere, the only requirement being an upper-air sounding that gives simultaneous values of temperature, pressure, and humidity. The following examples will illustrate its use:

1. *Net Flux at the Ground.* To calculate the downward flux at the ground, the atmosphere may be considered to be a layer extending from the ground up to a point above which the moisture content is negligible (usually 300 to 200 mb is sufficiently high). The downward flux is represented by an area such as $OABCO$ in Fig. 11, and the net flux is represented by the area $OCBDO$.

2. Net Flux at an Intermediate Level. At an upper level L in the atmosphere, the downward flux is determined as in the preceding case, except that T_0 in Fig. 11 should be taken as the temperature at L instead of the ground temperature. To obtain the upward flux, u is measured downward from L. T increases with u along the curve *BE* until the ground is reached; the ground is then replaced by an isothermal layer of infinite optical depth so that, as u increases to ∞, T_G, the ground temperature, remains constant. The upward flux will be represented by the area *OABEFO*, and the net flux by the area *OCBEFO*.

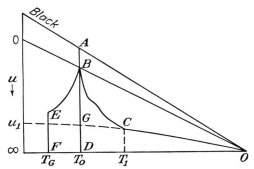

Fig. 11.—Representation of flux on the radiation chart.

3. Net Flux at a Cloud Base or Cloud Deck. The flux at the base or deck of a cloud can be determined in an analogous manner. Since a cloud of sufficient thickness may be regarded as a black body, the problem of calculating the net flux at the top of the cloud is the same as the problem of calculating the net flux at the ground. The area *OCBDO* represents the net flux at a cloud deck, provided that T_0 is the temperature at the deck. The upward flux at the base is found just as for any intermediate level and is represented by the area *OABEFO*. The downward flux is black-body flux and is given by the area *OABDO*. The net flux is therefore represented by the area *BEFDB*.

4. Effect of an Inversion. The net flux at the ground in the case of an inversion is illustrated in Fig. 12. The flux is the algebraic sum of the signed areas.

5. Measurement of Areas on the Radiation Chart. The measurement of areas on the chart is facilitated by a table that gives the areas of the wedge-shaped figures bounded by the $u = \infty$ isopleth, $a\,u = $ const isopleth, and an isotherm. For instance, in order to calculate the area *OCBDO* in Fig. 11, we read from the table the wedge-shaped area bounded by the

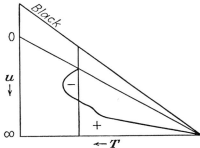

Fig. 12.—Effect of an inversion.

$u = \infty$ isopleth, the $u = u_1$ isopleth, and the $T = T_0$ isotherm. This leaves only the small area *BGCB*, which may be measured by means of a planimeter or by dividing it into simple geometrical figures whose areas are easy to evaluate.

6. Determination of the $u(T)$ Relationship. Expressing p in millibars, q in grams per kilogram, u in grams per square centimeter, and writing the integral expression for u [Eq. (43)] as a summation of small terms, we obtain

$$u = -\frac{1}{1,000}\sum q_i \sqrt{\frac{p_i}{1,000}}\,\Delta_i p \tag{44}$$

where q_i is the mean specific humidity and p_i the mean pressure in the pressure interval $\Delta_i p$. The summation extends upward or downward from the level at which the flux is being computed according as the downward or the upward flux is desired. The method of calculation of the $u(T)$ relationship is illustrated by Table 2 taken from an actual sounding.

TABLE 2.—COMPUTATION OF OPTICAL DEPTH

p, mb	T, °C	q, gram/kg	q_i	$\Delta_i p$	$\sqrt{\dfrac{p_i}{1,000}}$	$\Delta_i u$	u for reference level at		
							974	605	279
974	16	5.6					0	1.22	1.35
			5.9	14	0.98	0.08			
960	19	6.2					0.08	1.14	1.27
			5.9	38	0.97	0.22			
922	19	5.5					0.30	0.92	1.05
			4.5	80	0.93	0.34			
842	11	3.5					0.64	0.58	0.71
			3.5	52	0.91	0.17			
790	8	3.5					0.81	0.41	0.54
			3.5	48	0.88	0.15			
742	4	3.5					0.96	0.26	0.39
			2.9	62	0.83	0.15			
680	2	2.3					1.11	0.11	0.24
			1.9	75	0.80	0.11			
605	−3	1.4					1.22	0	0.13
			1.2	69	0.76	0.06			
536	−9	1.0					1.28	0.06	0.07
			0.9	54	0.71	0.03			
482	−15	0.7					1.31	0.09	0.04
			0.5	82	0.66	0.03			
400	−25	0.3					1.34	0.12	0.01
			0.2	78	0.60	0.01			
322	−37	0.1					1.35	0.13	0
			0.1	43	0.55	0.00			
279	−44	0.1					1.35	0.13	0

SYNOPTIC APPLICATIONS

Radiational Cooling of the Free Atmosphere. The change in the mean temperature of an atmospheric layer can be computed from the values F_2 and F_1 of the net flux at the base and top of the layer. The net loss of heat $F_2 - F_1$ must be at the expense of the cooling of the layer. If ΔT is the change in mean temperature, and p_1 and p_2 denote the pressures at the base and top

$$F_2 - F_1 = -c\rho\,\Delta T\left(\frac{p_2 - p_1}{g}\right) \tag{45}$$

Expressing the pressures in millibars and the flux in cal cm^{-2} 3 hr^{-1}, we derive the following numerical formula for the cooling in 3 hr.

$$(\Delta T)_{3hr} = -4.1\frac{F_2 - F_1}{p_2 - p_1} \tag{46}$$

The applications of the above formula to the free atmosphere show that long-wave radiation produces everywhere a net cooling of from 1 to 3°C per day. The cooling is greatest in regions of high specific humidity. Thus equatorial regions experience more radiational cooling than polar regions. A further consequence of the differential cooling is the tendency for an increase in stability in a given air mass due to the more rapid cooling of the more moist lower layers.

The heating of the atmosphere by the direct absorption of solar radiation during the day is probably not greater than 0.6°C per day and therefore does not compensate for the cooling by long-wave radiation.

Formation of Polar Air Masses by Radiative Cooling. The variation in the rate of radiational cooling in the atmosphere due to differences in humidity and temperature is normally not sufficient to create marked air-mass discontinuities. Moreover, the radiative cooling must in the main be compensated by turbulent convection and by the release of the heat of condensation when precipitation occurs, for we know that the atmosphere does not undergo permanent cooling. But, according to Wexler,[13] an important exception is found in the formation of polar air masses. He has investigated the modification that occurs when a maritime polar air mass arrives in the winter over a snow-covered continent and there stagnates. In this case, the compensating influence of conduction and condensation is slight. The heat transfer is due primarily to radiation. Initially the snow surface sends more radiation into space than it receives from the atmosphere, the outward flux being represented by an area such as *OCBDO* in Fig. 11. Since the direct absorption of sunlight is negligible because of the low altitude of the sun and also because of the high reflective power of a snow surface, the snow surface will cool until it no longer sends off more radiation than it receives. But this state of radiative equilibrium cannot persist. The net flux vanishes only at the ground, while at higher levels it is upward. Hence the air above the ground will continue to cool. As the air cools, it sends down less radiation to the ground, and the ground cools further. The cooling of the atmosphere, which is initially confined to a thin layer adjacent to the surface, is gradually extended upward by radiative cooling until a near isothermal layer develops from the top of the ground inversion to a height where the temperature becomes the same as that of the air before cooling. The successive stages in the process are shown in Fig. 13.

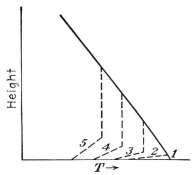

FIG. 13.—Successive stages in the formation of a polar continental air mass by radiative cooling.

Radiative Heating and Cooling of Clouds. The net flux at a cloud deck is the difference between the upward black-body flux from the cloud and the downward flux from the atmosphere. Since the downward radiation originates at lower temperatures and besides equals only a fraction of the black-body radiation corresponding to these temperatures, it will be exceeded by the upward flux. And since the net flux in the interior of the cloud must be zero, the top of a cloud will lose heat by radiation. At the base of a cloud, the downward black flux is exceeded by the flux from below, and the base will therefore be heated. The radiation chart areas representing the net outgoing flux from the top of a cloud and the net incoming flux at the base are shown in Fig. 14. The temperatures at the base and top are, respectively, T_1 and T_2; the ground temperature is T_G.

The radiational cooling of the top and heating of the base cause the cloud to become less stable. This effect is considered to be a factor contributing to the release of nocturnal thunderstorms in maritime tropical air masses.

Sky Radiation. The downward flux of long-wave radiation from the atmosphere is a factor of extreme importance in the forecasting of diurnal temperature variations, nocturnal fogs, and nocturnal frosts. We shall first consider the downward flux from a cloudless sky.

1. *Flux from a Clear Sky.* In the absence of clouds, the downward flux depends on the variation of specific humidity and temperature in the atmosphere. The nature of the dependency is clearly shown by the radiation chart diagram (Fig. 14). Several formulas have been proposed for its determination, but they all suffer from the defect that the flux is evaluated in terms of surface temperature and humidity

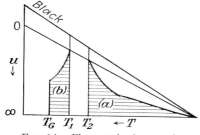

Fig. 14.—Flux at the base and top of an overcast. Area (a) represents the net outgoing flux at the top; area (b) the net incoming flux at the base.

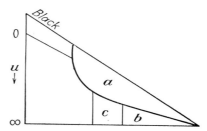

Fig. 15.—Dependence of sky radiation on cloud height.

without taking into account the moisture-temperature distribution with height. We should therefore expect nothing better than a rough approximation.

Ångström[14] has proposed the following formula for the downward flux R:

$$R = \sigma T^4(A - B10^{-\gamma e}) \qquad (47)$$

T is the surface air temperature and e is the partial pressure of water vapor at the surface. If e is expressed in millibars, the constants A, B, and γ have the following empirical values:

$$A = 0.81 \qquad B = 0.24 \qquad \gamma = 0.052$$

A different formula suggested by Brunt[15] is

$$R = \sigma T^4(a + b\sqrt{e}) \qquad (48)$$

The best mean values of the constants a and b obtained from a series of independent determinations are

$$a = 0.44 \qquad b = 0.08$$

Attempts have been made to give theoretical justifications for the above formulas, but that they are essentially empirical can be seen if we take $e = 0$; then the ratio $R/\sigma T^4$ for the atmosphere equals $A - B$ or a, depending on which equation is used. This would indicate that a dry atmosphere radiates about one-half as much as the ground. Since the only other appreciable source of long-wave radiation in the atmosphere besides water vapor is carbon dioxide, it would follow that the emission of atmospheric carbon dioxide is nearly 0.5 the black-body flux from the ground, whereas actual measurements give not more than 0.2 for this ratio. The above formulas are nevertheless valuable, for they depend only on surface observations of temperature and humidity. But, where aerological soundings are available, the radiation chart is decidedly preferable.

2. *Flux from a Cloudy Sky.* A layer of clouds acts as a black body and greatly increases the downward sky radiation. Figure 15 shows the downward flux from an overcast sky as represented on the Elsasser radiation chart. The area a represents the flux from a clear sky, $a + b$ the flux from a high overcast sky, and $a + b + c$ the flux from a low overcast sky. It can thus be seen that the downward flux decreases rapidly with increasing cloud height.

Ångström and Asklöf[17] have given a modification of the formula for sky radiation which takes into account the effect of clouds. If R_0 is the downward flux from a clear sky as computed by Eq. (47), the flux from an overcast sky is written

$$R = \sigma T^4 - \lambda(\sigma T^4 - R_0) \tag{49}$$

where λ is a factor depending on the height of the cloud base. Table 3 was computed by Ångström and Asklöf from observational data. From theoretical con-

TABLE 3

Cloud height, km	1.5	3	7
λ	0.14	0.25	0.80

siderations, Phillips[18] calculated a different set of values, given in Table 4. For low clouds, there is quite good agreement between Tables 3 and 4. The discrepancy in the values of λ for high clouds is attributed by Phillips to the fact that the overcast

TABLE 4

Cloud height, km	2	5	8
λ	0.17	0.38	0.45

observed by Asklöf during the night may sometimes not have been sufficiently thick to be black and may even have had occasional breaks.

Brunt[19] has derived a formula for the nocturnal decrease in surface temperature when the net outward flux from the ground is known. If ρ_1, c_1, and k_1 are, respectively, the density, specific heat, and specific conductivity of the ground, T the temperature at sunset, t the time after sunset, and R_n the net flux of long-wave radiation at the ground, the decrease in temperature is given by the equation

$$\Delta T = \frac{2}{\sqrt{\pi}} \frac{R_N}{\rho_1 c_1 \sqrt{k_1}} \sqrt{t} \tag{50}$$

The chief difficulty in the way of using the above formula lies in the uncertainty of the quantity $\rho_1 c_1 \sqrt{k_1}$. The addition of 20 per cent of water to dry soil will increase this quantity fivefold.

Brunt's equation shows that the change in temperature for a specified net flux is inversely proportional to $\rho_1 c_1 \sqrt{k}$. This factor therefore indicates the nature of the dependence of the nocturnal variation of temperature upon the type of ground. Thus, for example, $\rho_1 c_1 \sqrt{k_1}$ is nine times as large for the average soil as for snow, and we should expect the temperature variation over a snow surface to be many times that over a soil surface.

A fundamental shortcoming in Brunt's formula is its failure to take into account the effect of turbulent convection from the atmosphere into the ground. It should not, therefore, be expected to apply on windy nights.

The Terrestrial Heat Balance. The total flux of solar radiation received by the earth is equal to its cross-sectional area multiplied by the solar constant. To obtain the mean flux over the earth's surface, we divide by its area. Hence, if r is the radius of the earth, the mean flux becomes

$$\frac{\pi r^2 \times 1.94}{4\pi r^2} = 0.485 \qquad \text{cal cm}^{-2}\text{ min}^{-1}$$

Of this amount, it has been estimated by Bauer and Phillips[20] that 33 per cent is returned to space by direct reflection at clouds and at the surface of the earth, and 10 per cent by diffuse reflection and scattering. Thus a total of 43 per cent is returned to space without absorption. This figure has been defined as the earth's mean albedo (see page 297). Of the remaining 57 per cent, 14 per cent is absorbed by the atmosphere and 43 per cent by the earth.

Since the mean temperatures of the atmosphere do not change perceptibly from year to year, a balance must exist between heat income and outgo. The atmosphere receives heat by the following processes: (1) absorption of short-wave solar radiation, (2) absorption of long-wave terrestrial radiation, (3) condensation of water vapor, and (4) conduction from the earth. The amount under item (1) has already been taken as 14 per cent of the mean incoming solar radiation. The atmospheric absorption of radiation from the earth, under item (2), is estimated to be 0.533 cal cm^{-2} min^{-1} or 112 per cent of the mean incoming solar flux. Wüst[21] estimates the total precipitation over the whole earth to be 390,000 km^3 per year, which corresponds to an average gain of 0.086 gram-cal cm^{-2} min^{-1} or 18 per cent released as latent heat of condensation under item (3). Finally, Sverdrup[21] has given a value of 0.010 cal cm^{-2} min^{-1} or 2 per cent for conduction into the atmosphere under item (4).

The atmosphere loses heat by radiation downward to the earth and upward to space. The mean downward flux is estimated to be 0.466 cal cm^{-2} min^{-1}, or 96 per cent of the mean incoming flux, and the outward flux to space is estimated at 0.243 cal cm^{-2} min^{-1} or 50 per cent.

On the basis of the above values, the heat budget of the atmosphere can be summarized in Table 5.

TABLE 5.—HEAT BUDGET OF THE ATMOSPHERE, IN PER CENT

Receives		Loses	
Absorption of solar radiation	14	Radiation to earth	96
Absorption of terrestrial radiation	112	Radiation to space	50
Condensation of water vapor	18		146
Conduction from earth	2		
	146		

The mean temperature of the earth as well as that of the atmosphere does not change perceptibly from year to year. Hence the earth and atmosphere as a whole must receive as much energy from the sun as they send off to space by long-wave radiation. Since 57 per cent of the mean incoming solar radiation is absorbed and only 50 per cent radiated back into space by the atmosphere, 7 per cent must come directly from the earth. If we add the 7 per cent of the earth's outgoing flux of radiation transmitted through the atmosphere to the 112 per cent absorbed by the atmosphere, we obtain a total of 119 per cent of the mean incoming radiation for the quantity emitted by the earth. It may at first seem paradoxical that the earth emits more radiation than is received outside the atmosphere from the sun, but the apparent contradiction is resolved if we take into account the fact that the earth receives atmospheric as well as the transmitted solar radiation.

Geographical Distribution of the Outgoing Radiation. Although a radiational heat balance exists for the earth and atmosphere as a whole, there is a rather marked discrepancy between income and outgo at particular latitudes. The latitudinal variation of incoming radiation was described in the table on page 292. In order to calculate the corresponding variation in the outgoing long-wave radiation by means of the Elsasser chart, it is necessary to know the mean distribution of the meteorological variables p, T, and f throughout the atmosphere. Since these data are not now available, recourse is had to Simpson's method[22] outlined on page 298. Simpson[22] assumes that the stratosphere contains at least 0.03 gram per cm² of water vapor and 0.06 gram of carbon dioxide. On this basis, the outgoing radiation in the first of his categories will originate in the stratosphere, the radiation in the second category will originate either at the earth or at a cloud, and the radiation in the third category will

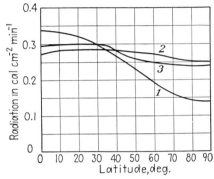

Fig. 16.—Latitudinal variation of yearly incoming and outgoing radiation.

be midway between black-body radiation at the temperature of the stratosphere and black-body radiation from the earth or a cloud. The total flux can be calculated if the latitudinal variation of the mean temperature of the earth and the mean cloud temperature are known. Simpson took the latter to be 261°K for all latitudes. He also assumed a mean cloud amount of five-tenths for all latitudes, so that his final figure for the total radiation is the mean of the figures for clear and cloudy skies. Figure 16 shows the variation of incoming and outgoing radiation with latitude. Curve 1 represents the incoming radiation, and curve 2 the outgoing radiation as computed by Simpson. His calculations gave a mean value of 0.271 cal cm⁻² min⁻¹ for the outgoing flux, which compares favorably with the value 0.276 determined by multiplying the solar constant by the albedo.

The Simpson wave-length categories were based upon measurements of water-vapor absorption that have since been revised. He overestimated the absorptive power of 0.03 gram per cm² of water vapor, and thicker layers should have been used. Moreover, it now appears that the stratosphere does not even contain 0.03 gram per cm² of water vapor. The radiation in the first category will therefore have originated partly in the troposphere and consequently at higher temperatures. Despite these criticisms, the essential correctness of Simpson's approach is borne out by later investigations. The work of Bauer and Phillips[23] may be mentioned in this connection. They assumed as given the distribution of temperature and moisture with height and calculated the flux by dividing the water-vapor spectrum into three regions in each of which the coefficient of absorption was given a constant value. The outgoing flux was then calculated by integrating the equation of radiative transfer derived on page 291.   Their results are represented in Fig. 16 by curve 3.   It will be seen that both Simpson's and Bauer and Phillips' calculations give an excess of incoming over outgoing radiation from 0 to about 30 or 35° and a defect from there to 90°.

In order to create a net balance between incoming and outgoing radiation, there must be a meridional transport of heat from lower to higher latitudes.   The energy may be transported by atmospheric or ocean currents as sensible heat, or as latent heat of evaporation.   It does not follow that the atmospheric winds are entirely responsible for the heat transport, but it is quite certain that any theory of the general circulation must take into account a considerable poleward transport of heat by winds.

Bibliography

1. FOWLE, F. E.: *Smithsonian Inst. Misc. Collections*, **68** (8), 1917.
2. KIMBALL, H. H.: *Monthly Weather Rev.*, **58**: 43 (1930).
3. HOELPER, O.: *Meteorolog. Z.*, **54**: 458 (1937).
4. MÜGGE, R., and F. MÖLLER: *Z. Geophys.*, **8**: 53 (1932).
5. TANCK, H. J.: *Ann. Hydrog.*, **6**: 47 (1940).
6. KIMBALL, H. H.: *Monthly Weather Rev.*, **55**: 168 (1927).
7. ALDRICH, L. B.: *Smithsonian Inst. Misc. Collections*, **69** (10), 1919.
8. HEWSON, E. W.: *Quart. J. Roy. Meteorolog. Soc.*, **69**: 47 (1943).
9. ÅNGSTRÖM, A.: *Gerlands Beitr. Geophys.*, **15**: 1 (1936).
10. HETTNER, G.: *Ann. Physik*, **55**: 476 (1918).
11. SIMPSON, G. C.: *Mem. Roy. Meteorolog. Soc.*, **2** (6), 1928.
12. ELSASSER, W. M.: *Quart. J. Roy. Meteorolog. Soc.*, 66 suppl. 41, 1940; see also *Harvard Meteorolog. Study* 6, 1942 (contains an excellent bibliography).
13. WEXLER, H.: *Monthly Weather Rev.*, **64**: 122 (1936).
14. ÅNGSTRÖM, A.: *Smithsonian Inst. Misc. Collections*, **65** (3), 1915; see also *Gerlands Beitr. Geophys.*, **21**: 145 (1929).
15. BRUNT, D.: *Quart. J. Roy. Meteorolog. Soc.*, **58**: 389 (1932).
16. ÅNGSTRÖM, A.: *Gerlands Beitr. Geophys.*, **21**: 145 (1929).
17. ASKLÖF: *Geog. Ann.*, **2**: 253 (1920).
18. PHILLIPS, H.: *Gerlands. Beitr. Geophys.*, **56**: 229 (1940).
19. BRUNT, D.: "Physical and Dynamical Meteorology," 2d ed., p. 139, Cambridge, London, 1941.
20. BAUER, F., and H. PHILLIPS: *Gerlands Beitr. Geophys.*, **45**: 82 (1935); **47**: 218 (1936).
21. SVERDRUP, H. U.: "Oceanography for Meteorologists," p. 5, Prentice-Hall, New York, 1942.
22. SIMPSON, G. C.: *Mem. Roy. Meteorolog. Soc.*, **3** (21), 1928.
23. BAUER, F., and H. PHILLIPS: *Gerlands Beitr. Geophys.*, **45**: 82 (1935).

METEOROLOGICAL THERMODYNAMICS AND ATMOSPHERIC STATICS

By Norman R. Beers

CONTENTS

SECTION V

METEOROLOGICAL THERMODYNAMICS AND ATMOSPHERIC STATICS

By Norman R. Beers

INTRODUCTION AND FIRST LAW

Thermodynamics in General. The science of thermodynamics is fundamentally the logical structure that follows a statement of the first and the second laws of thermodynamics. The first law of thermodynamics is a special statement of the law of the conservation of energy. The second law concerns itself with the question of the direction in which a process involving heat transfer takes place. Neither of these laws can be fully understood or safely applied until a precise understanding of certain fundamentals is obtained. The first articles of this section will give the necessary background for the subject; the balance of the section will be devoted to the applications of thermodynamics to meteorological practice.

Thermodynamics studies the end results (initial and final) of energy transitions from one form to another and of energy transmission (or transfer) from one system to another when the change is specified to occur in some particular fashion or process. The laws of thermodynamics coupled with the numerical results of a few experiments enable us to solve many useful problems in all the sciences. The more refined and sophisticated techniques of kinetic theory and statistical mechanics, especially in their quantum versions, add an understanding of the basic mechanism whereby these processes of nature occur. This is an understanding with which thermodynamics is not immediately concerned.

The type of question to which thermodynamics gives the answer, in meteorology, for example, is illustrated by the following: What is the final temperature of a mass of 10 kg of saturated air that is heated isobarically at 1 atmosphere of pressure by the addition of 1 kw-hr of energy as heat if the initial temperature was 10°C and if there is no change in water-vapor content during the process? We note that the statement of the problem specifies the process and the initial *state* (or condition) of the air completely. Nothing is stated as to the fundamental manner or mechanism of the heat transfer; *i.e.*, the answer is the same whether the heat is transferred by radiation or conduction or convection. Such a specification as the above is typical.

When we wish to apply thermodynamics to some particular type of problem, we must confine our attention to some particular part of the world with its attendant quantities, machines, etc. Such an isolated set of objects and concepts is known as a *thermodynamic system*, or simply as a *system*, *i.e.*, a system is that part of the universe about which we decide to think throughout a portion of an investigation. Thermodynamics is concerned with systems in *equilibrium states*, *i.e.*, states in which the system is in mechanical, thermal, and chemical equilibrium.

A system is said to be in an equilibrium state when it is completely determined (or described) by one or more variables or "thermodynamic coordinates," independent of the time. It is the evidence of experience that any finite system may be completely specified by a finite number of thermodynamic coordinates or *properties*. It is also found by experience that the properties needed to specify a state are in general functionally related. The equation or functional relation among the properties of a system is known as the *equation of state.*

It may be noted that the term *state* is not used in thermodynamics as it is in elementary physics to denote changes of form of a substance, *e.g.*, water changing to ice or steam. For such changes in thermodynamics, the expression *change of phase* is used. The substance is then said to be in the solid, liquid, or vapor *phase, i.e.*, ice, liquid water, or water vapor.

An illustration will illuminate the generalizations of the preceding paragraphs. Consider the earth and its atmosphere. This we may take as a *system.* The thermodynamic *state* of the atmosphere is known when we know the *properties* of pressure and temperature, and the relative humidity at every point. We say the system is in *thermodynamic equilibrium* when the properties are constant with time (or almost constant). If no clouds are present, the atmosphere is in the gaseous (oxygen, nitrogen, etc.) and vapor (water-vapor) phase. Clouds are small droplets of liquid water or ice (or both).

It should finally be noted among these general statements that thermodynamics is concerned only with continuous media; *i.e.*, it is assumed that all elements of the systems with which we operate may be discussed using the ordinary processes of calculus. Dryden[1] observes in a recent review of the statistical theory of turbulence, "The volume is small in comparison with the dimensions of interest in the flow but large enough to include many molecules. A cube of size 0.001 mm, containing at atmospheric pressure about 2.7×10^7 molecules, satisfies this condition."

Thermodynamics and Meteorology. The earth's atmosphere may be compared with the working substance of a huge thermodynamic engine. Energy is received by the atmosphere ultimately from the sun. This energy heats the surface of the earth and the air adjacent thereto unevenly owing to the variable nature of the surface and to unequal and nonsimultaneous receipt. Thus the polar regions in winter receive no direct sunlight while equatorial regions have an almost 12-hr day. Land surfaces and the air over deserts are heated well in excess of 100°F. at midday, while the oceans and adjacent atmosphere have a maximum temperature of around 85°F.

This unequal heating carried on over the entire earth coupled with the rotation of the earth and the attendant Coriolis acceleration lies at the foundation of the large-scale, or *general, circulations* of the atmosphere. On a smaller scale, local variations in heating are responsible for the monsoons. On a still smaller scale, these local variations cause land, sea, and valley breezes and local storms of the thunderstorm type.

We shall see that thermodynamics aids in the understanding of how and why the large-scale phenomena are as they are observed to be in the atmosphere. The science of thermodynamics will, in fact, probably be in a front place in the finally accepted theory of these phenomena. Such a theory does not exist at present. Thermodynamics does permit us to make many useful calculations and predictions for a generally local situation.

The essence of the difference between meteorology and most sciences is the fact that meteorology studies a system that cannot be controlled by man. No adequate laboratory has ever been devised to simulate the atmosphere itself for study of the weather. Laboratory methods are useful for certain small-scale determinations; but in general the meteorologist must search for idealized processes that he considers "most likely" to occur, in lieu of controlled experiment in the laboratory. If the first selected ideal process fails to produce results that are in fair correlation with observations, he simply drops that concept and searches for another.

A ready acceptance of what we shall call physical approximations to the truth (*e.g.*, that isentropic processes may occur in nature) necessarily typifies the meteorologist. This situation makes him the more ready to accept mathematical approximations.   Yet neither physical nor mathematical approximations are invoked except when necessary to solve otherwise intractable problems.   *The essential item is to understand the limitations of the approximations used.*

Work.   Work is the product of force and the distance through which the force has moved.   This implies that the force has moved some material body (or mass point), *i.e.*, that work is done on a body by a force that moves it, or by a body that moves against a restraining force.   The general mathematical expression for work is simply

$$\text{Work} = W = \int \mathbf{F} \cdot d\mathbf{s} \tag{1}$$

where $\mathbf{F}$ is the vector representing the force and $d\mathbf{s}$ is the vector representing the element of displacement.

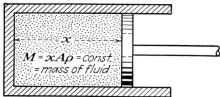

FIG. 1.—Fluid performs work on a piston during expansion

Work Done by an Expanding Fluid. Consider a cylinder fitted with a piston and filled with a nonviscous compressible fluid.   If the piston is permitted to move a distance dx, the work done by the fluid as it expands is $pA\,dx$, where p is the pressure in the fluid and A the cross-sectional area of the piston. The mass of fluid in the cylinder remains constant; $M = Ax\rho$.   Therefore, $M\,dv = A\,dx$, and the work done per unit mass of fluid contained in the cylinder is

$$dw = p\,dv \tag{2}$$

where $v = \dfrac{1}{\rho}$ is the specific volume of the fluid (see Fig. 1).

We may consider the atmosphere of the earth to be divided up by a series of planes into elements of volume, each of which is large enough to contain many molecules of air and small enough so that pressure and specific volume are constant throughout each element.   Pressure and specific volume may of course vary as we go from one element of volume to the next.   Each small element of volume may expand and in expanding do work on its surroundings; or it may be compressed and have work done on itself by its surroundings.   The amount of work done per unit mass will be $p\,dv$ by Eq. (2).   The amount of work done by the entire atmosphere would be found by integrating $\rho p\,dv$ throughout the atmosphere.

Energy.   Planck refers to energy briefly as the faculty to produce external effects. This general concept includes the chemical, radiation, kinetic, mechanical potential, radioactive, etc., energies of our daily experience.   More precise definitions of the energies involved in meteorological thermodynamics follow.

The *kinetic energy* of a mass M moving with velocity $\mathbf{u}$ is equal to $Mu^2/2$ in Newtonian mechanics.   The quantity $Mu^2/2$ is equal numerically to the work required to accelerate the mass M from zero velocity up to velocity $\mathbf{u}$.

$$\text{K.E.} = \int \mathbf{F} \cdot d\mathbf{s} = \int M\frac{d\mathbf{u}}{dt} \cdot d\mathbf{s} = M \int_0^u \mathbf{u} \cdot d\mathbf{u} = \frac{Mu^2}{2} \tag{3}$$

We note that the kinetic energy is measured relative to the origin from which the velocity is measured.   The kinetic energy of a bomb loaded in and carried by an aircraft in flight, for example, is zero relative to the aircraft but is $Mu^2/2$ relative to the ground when u is the ground speed of the aircraft.

The *potential energy* of a mass M in the earth's gravitational field is numerically equal to the work required to raise the mass to its present position against the gravitational force acting on it. We have

$$\text{P.E.} = \int_{z_0}^{z} Mg \, dz = Mg(z - z_0) \tag{4}$$

where z_0 is the position from which the lifting began and the expression on the right implies that for elevations that concern us in meteorology the acceleration of gravity is independent of position. This amounts to a mathematical approximation, but one that is near enough to the truth so that the small variations caused by change of g with latitude and elevation are negligible for present purposes. We note again that this energy is measured relative to an origin. We generally take mean sea level as the point of zero potential energy so that the energy of a mass is written simply as Mgz.

The *flow energy* of a fluid per unit mass of fluid is the product pv, where p is the absolute pressure and v the specific volume of the fluid at the point considered. The concept of flow energy (which was introduced by the engineers) is demonstrated as follows: Consider a region of space R which contains fluid. Let a part of the region be surrounded by a surface S. Consider an element of that surface ΔA in area. Let the pressure at the center of ΔA be p. Imagine a small

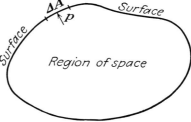

FIG. 2.—Flow work.

amount of fluid to be forced past ΔA. The amount of fluid forced out will be $\Delta A \, \Delta x$ where Δx is the linear displacement of fluid normal to ΔA. The mass of fluid forced out will be

$$\Delta M = \Delta A \, \Delta x \rho = \frac{\Delta A \, \Delta x}{v}$$

where ρ is the density of the fluid at the point and v its specific volume. The mechanical work done by the pressure in moving unit mass of fluid is

$$\frac{\Delta W}{\Delta M} = \frac{p \, \Delta A \, \Delta x}{\rho \, \Delta A \, \Delta x} = pv \tag{5}$$

We note that the faculty of the fluid to do mechanical work in moving unit mass of fluid is given by the product pv (see Fig. 2).

The *internal energy* of a system is the sum total of energies inside the boundaries of the system. This must include in general the mechanical, chemical, electrical, etc., energies of all fundamental particles of the system. The term for internal energy must not, however, count again those energies already given an accounting. Thus we have already itemized the kinetic energy of organized velocity, the mechanical potential energy, and the flow energy; these energies will not be counted as internal energies in meteorological thermodynamics. Nor will we need to consider chemical or electrical energies. The release of electrical energy in a thunderstorm is of course tremendous; but its analysis is not feasible here or necessary for the usual weather processes in the atmosphere. The remaining item of internal energy important for meteorology is the energy due to the disorganized motions of the particles (atoms and molecules) of the atmosphere. These random velocities of the particles are functions of the temperature of the system, as is shown in kinetic theory. It will be shown later that for certain systems (*e.g.*, a perfect or ideal gas) the internal energy is a function of temperature alone.   We designate internal energy by the letter E and internal energy per unit mass of fluid (*i.e., specific internal energy* by the letter e).

Temperature and Heat.   We are aware of what we refer to as the *temperature* of a material body as a fact of everyday experience.   Initially we classify bodies as *hot* or *cold* according to how they feel.   We say the hot body has a higher temperature than the cold.   We observe further that, if a hot body is brought close to a cold body, the two bodies will after a period of time appear equally hot (or cold).   This qualitative concept of temperature must be replaced by a quantitative concept later, but in the early analysis it is useful as it stands.

It has been indicated above that the internal energy of a system is a function of the temperature of the system.   Then if the temperature changes the internal energy may also change.   The mechanism through which the internal energy has changed is studied in physics; this mechanism (or these mechanisms) is not the subject of thermodynamics.   Thermodynamics is concerned, however, with the end results of the change.   The energy that is transferred from one body to another by virtue of a temperature difference between the bodies will be referred to exclusively as *heat*.   Once the energy has been transferred, it will be known as kinetic energy, potential energy, internal energy, chemical energy, etc.; it is no longer *heat*.   *In analogy: the precipitation from the atmosphere may be referred to as rain when it is falling.   After it fills a lake (or a water tumbler), it is no longer rain; it is then water.*

The important concept to bear in mind is that for thermodynamic analyses *heat is energy in transmission* because of a temperature gradient.   The concept of energy in transmission (or transfer) also applies to mechanical work.   Thus work may be done on a body or by a body, as heat may be added to or taken away from a system.   The net result in either case is that the energy content (more precisely internal energy) may be changed.

Units of heat may be transformed into units of work by using the appropriate numerical factor.   For example, one calorie of heat is approximately equal to 4.186 joules of mechanical work.   Either the calorie or the joule is a unit of energy.

The First Law of Thermodynamics.   The first law of thermodynamics is a statement of the law of conservation of energy for a thermodynamic system.   As such it carries the weight of the law of conservation of energy and is offered without "proof," being a conclusion reached on the evidence of experience.   In words, for a thermodynamic system, *heat added to the system between initial and final states equals the increase in internal energy of the system plus the work done by the system.*   For small changes in the thermodynamic properties, this statement is expressed algebraically by the equation

$$dq' = de + dw' \qquad (6)$$

The primes on the small amounts of heat (q) and of work (w) indicate that these quantities are not uniquely determined by a particular change of state.   The internal energy is by its very definition dependent solely upon the thermodynamic coordinates; it is itself a coordinate and a function of state.   The energy in transfer between the system and its surroundings during a state change, however, depends upon the process chosen in going from the initial to the final state.   Thus heat and work are not functions of state.   Mathematically, we say that dw' and dq' are not perfect differentials in Eq. (6).   The primes are not written generally; but it is necessary to note that *heat and work depend on the path in all cases.*

The general statement of Eq. (6) will take a more special form when the system is specified.   The work term may include shaft work of a turbine, for example.   Or dw' may designate only the work done by an expanding fluid.   In the above equation and hereafter, all energy terms are written with the joule as a unit.   Small letters (e, q,

w, etc.) designate joules per gram of fluid while capitals designate energy for the entire system in joules (E, Q, W, etc.). The first law in its differential form for the earth's atmosphere is by the above argument and by Eq. (2) easily seen to be

$$dq' = de + p\,dv \tag{7}$$

where the only external work done is done by an expansion of the nonviscous fluid acting against an applied pressure.

Mechanical Equilibrium in the Atmosphere. The equations of motion for a non-viscous fluid on a nonrotating earth are in their Eulerian form

$$\frac{du}{dt} = X - \frac{1}{\rho}\frac{\partial p}{\partial x} = \frac{\partial u}{\partial t} + \mathbf{v}\cdot\nabla u$$

$$\frac{dv}{dt} = Y - \frac{1}{\rho}\frac{\partial p}{\partial y} = \frac{\partial v}{\partial t} + \mathbf{v}\cdot\nabla v \tag{8}$$

$$\frac{dw}{dt} = Z - \frac{1}{\rho}\frac{\partial p}{\partial z} = \frac{\partial w}{\partial t} + \mathbf{v}\cdot\nabla w$$

where the vector velocity of the fluid at a point (x,y,z) at time t is given by $\mathbf{v} = (u,v,w)$. The vector (X,Y,Z) represents the extraneous forces acting at the point. The pressure is given by p and the density of the fluid by ρ. For the case where the extraneous forces have a potential Ω, Eqs. (8) give an energy relation at once. We multiply the equations in order by the scalars u, v, and w and add to get

$$\frac{d}{dt}\left(\frac{u^2 + v^2 + w^2}{2} + \Omega\right) + \frac{1}{\rho}\mathbf{v}\cdot\nabla p - \frac{\partial\Omega}{\partial t} = 0 \tag{9}$$

When the potential is steady ($\partial\Omega/\partial t = 0$), as it is, for example, in the earth's gravitational field, we have

$$\frac{d}{dt}\left(\frac{u^2 + v^2 + w^2}{2} + \Omega\right) + \frac{1}{\rho}\mathbf{v}\cdot\nabla p = 0 \tag{10}$$

where $-\nabla p$ is the pressure gradient. If we assume further that the pressure is steady ($\partial p/\partial t = 0$) and choose the axis of z vertically upward, we have

$$\frac{d}{dt}\left(\frac{u^2}{2} + gz\right) + v\frac{dp}{dt} = 0 \tag{11}$$

where u is now the total velocity, $g\,dz = d\Omega$, and $v = \dfrac{1}{\rho}$. Equation (11) reduces to Bernoulli's equation on integration, provided that the specific volume v is a constant. In meteorology, the change in v during flow is generally too great to be neglected, and Eq. (11) must be used as it stands. It is easily shown that Eq. (11) is also valid for a rotating earth, since the terms involving Coriolis acceleration cancel out.

It has been assumed in deriving Eq. (11) that the fluid of the system is nonviscous. If viscosity is to be taken into account, additional terms must be added to the Eqs. (8). In general the expanded forms of (8) are nonintegrable. To obtain usable results, it is assumed that the above equation as it stands will apply to the earth's atmosphere. The results of observation indicate that the assumption is warranted to a good approximation for all usual circumstances. As the equation is applied in thermodynamics, to smooth steady flow, it is an excellent approximation.

Equation (11) may be integrated to quadrature immediately. We find

$$\frac{u_1{}^2 - u_2{}^2}{2} + g(z_1 - z_2) + \int_2^1 v\,dp = 0 \tag{12}$$

where subscripts 1 and 2 refer to any two points along the streamline of flow. We note that streamlines coincide with trajectories since it has been assumed that the flow is steady. The first two terms of Eq. (12) are sometimes known as the "mechanical effects" evident in the fluid. The mechanical effects, *i.e.*, the sum of the change in kinetic and potential energies along the streamline, may obviously be determined between two points when the specific volume is known as a function of pressure along the streamline.

Alternative Derivation of First Law for Atmosphere. It is useful to demonstrate the truth of Eq. (7) for the atmosphere in another way. We consider the sum of the energies shown previously to exist in any flowing fluid, *i.e.*, the sum of the kinetic, potential, internal, and flow energies. Adding these, we have at a point 1, $u_1^2/2 + gz_1 + e_1 + p_1v_1$. At a second point 2 on the streamline, we have

$$\frac{u_2^2}{2} + gz_2 + e_2 + p_2v_2 + {}_1q_2 \qquad (13)$$

where ${}_1q_2$ is the heat added between the two points. By the law of the conservation of energy, these two values must be equal. We have then from the above in differential form

$$u\,du + g\,dz + de + d(pv) = dq' \qquad (14)$$

Eliminating the terms contained in Eq. (11) expressed in differential form, we find $dq' = de + p\,dv$, which is the first law as expressed by Eq. (7). The function $e + pv = h$, known as the *enthalpy*, will be useful in later developments.

It is important to note that Eq. (14) is valid for all steady flow, *i.e.*, it holds good even when the fluid is viscous.

SECOND LAW, TEMPERATURE, AND ENTROPY

The Second Law of Thermodynamics. The second law of thermodynamics may be stated in any one of a variety of ways according to the point of view it is wished to emphasize. The law deals with the question of the direction in which a process involving heat transfer takes place. Early statements emphasized concepts pertaining to heat flow in ideal engines. The emphasis today in theoretical thermodynamics is on the concept of entropy. We quote several statements of the law as summarized by Zemansky.[2]

1. Kelvin-Planck Statement.—It is impossible to construct an engine which, operating in a cycle, will produce no effect other than the extraction of heat from a reservoir, and the performance of an equivalent amount of work.
2. Clausius Statement.—It is impossible to construct a device which, operating in a cycle, will produce no effect other than the transference of heat from a cooler to a hotter body.
3. The Entropy Principle.—As a result of natural processes, the entropy of the universe is increasing.
4. The Principle of the Degradation of Energy.—As a result of natural processes, energy is becoming unavailable for work.
5. The Probability Principle.—As a result of natural processes, the disorder of the universe is increasing.

The second law like the first is accepted on the evidence of experience. The statements of Kelvin, Planck, and Clausius are essentially negative; but they are more immediately useful for our purposes than are either of the other statements of the law. It may be shown that the above statements of the law are equivalent.

Definitions Useful for Second Law. It is necessary to investigate some theoretical matters before making any applications of the second law. A quantitative definition of temperature is essential. This will be based on an analysis of a few simple idealized processes that are known collectively as a Carnot cycle. We consider a system composed of a working substance (*e.g.*, the steam in a reciprocating engine) and other items such as pulleys, pistons, cylinders, reservoirs, etc. If the system is taken through a series of changes such as expansions, additions of heat, etc., and is finally brought back to its initial state, we say the system has executed a complete cycle. The Carnot cycle is a series of two isothermal and two adiabatic changes.

An *isothermal process* is a state change in which the temperature of the system remains constant. We note that our fundamental qualitative concept of temperature as "hotness" or "coldness" is adequate for an understanding of a constant temperature. An *adiabatic process* is a state change in which no heat is added to (or taken from) the system. A *reservoir* is a system of such nature that it may be used as a *source* or *sink* of heat without sensibly altering its temperature. A true reservoir does not exist in nature, though the oceans, say, approximate a reservoir very well for short periods of time. Thus the temperature of the ocean water at a depth of 1 m is almost constant from 0000h to 1200h of the same day.

An *insulator* is defined to be a substance that will not transmit heat. A true insulator does not exist in nature, but various substances transmit so little heat in comparison with that transmitted by others for the same temperature gradient that for practical purposes they may be considered to be insulators. Air, cork, wool, etc., are good practical insulators. A substance that transmits heat perfectly is known as a *conductor*. Perfect conductors do not exist in nature, but they are approximated by silver and copper.

The *working substance* of a system is any matter (gaseous, fluid, or solid) which may be used to receive energy either as heat or work and to store it in the form of internal energy only to return the energy later either as heat or work. It is the evidence of experience that an equation of state exists among the coordinates needed to specify the state of the substance, though for our initial analysis it is not necessary that the equation of state be known. It is assumed that working substances exist which may be put through reversible cycles.

A *reversible process* or state change is a process that is assumed to take place ideally. It is conceived to occur so that it might actually be reversed anywhere along the process path in such a way that *both the system and its surroundings return to their original condition.* An *irreversible process,* on the other hand, is one such that the system cannot be brought back to its original condition without requiring a conversion or degradation of some external energy. All actual processes in nature are more or less irreversible. Ideal processes such as the slow compression of a gas or the extension of a steel spring may be made to approach reversibility as nearly as we please by careful control of the surroundings. In meteorology, the nonturbulent expansion of an ascending parcel of air through its environment may be considered reversible, provided the ascent occurs quickly enough so that a negligible quantity of heat is transferred. Any general textbook of thermodynamics devotes considerable space to a discussion of reversibility and irreversibility.

Carnot Cycle. The Carnot cycle consists of two adiabatic and two isothermal processes executed by any thermodynamic system whatever. The processes are assumed to be carried out as follows:

a—b, with the system in thermodynamic equilibrium at temperature T_1, a reversible adiabatic is performed until the system is in thermodynamic equilibrium at a higher temperature T_2.

b—c, the system is maintained in thermodynamic equilibrium at T_2 while a reversible isothermal is performed during which the system absorbs heat Q_2.

c—d, a reversible adiabatic is performed so that the temperature drops from T_2 to T_1.

d—a, the system is maintained in thermodynamic equilibrium at T_1 while a reversible isothermal is performed during which the system rejects heat Q_1.

Efficiency of Carnot Cycle. The efficiency of any heat engine is defined to be

$$\eta = \frac{\text{useful work done per cycle}}{\text{heat added per cycle}} \tag{15}$$

The Carnot cycle consists of reversible ideal processes. Therefore there are no losses. The useful work done during the cycle is then $W = Q_2 - Q_1$. The heat rejected to the cold reservoir at T_1 is not "lost." But it has been made unavailable for useful work in the present system. In other words, there has been a "degradation" of energy. It is a corollary of the second law that Q_1 is a positive number for any temperature of the sink at T_1 other than the absolute zero of temperature. In terms of the heats Q_2 and Q_1, the efficiency of the cycle is then

$$\eta = \frac{Q_2 - Q_1}{Q_2} = \frac{W}{Q_2} \tag{16}$$

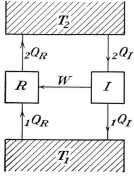

Fig. 3.—Reversible and irreversible engines working between reservoirs.

Corollaries of the Second Law. Keenan[3] has collected several corollaries of the second law in very useful form. The first of these are generally referred to as Carnot's propositions. The proofs are here paraphrased from the proofs given by Keenan; the corollaries are quoted by permission.

I. *It is impossible to construct an engine to work between two heat reservoirs, each having a fixed and uniform temperature, which will exceed in efficiency a reversible engine working between the same reservoirs.* PROOF: Assume the contrary to be true, letting an engine I which is not reversible have an efficiency η_I that is greater than the efficiency η_R of the reversible engine R, working between the same temperatures T_2 and T_1; $T_2 > T_1$. Let I draw enough heat from the hot reservoir to produce work W exactly equal to that required to operate the reversible engine R as a "heat pump," *i.e.*, reversibly. The flow of energy is indicated in Fig. 3. The efficiencies of the two engines are

$$\eta_I = \frac{{}_2Q_I - {}_1Q_I}{{}_2Q_I} = \frac{W}{{}_2Q_I} \qquad \eta_R = \frac{{}_2Q_R - {}_1Q_R}{{}_2Q_R} = \frac{W}{{}_2Q_R} \tag{17}$$

where $\eta_I > \eta_R$ by hypothesis. This requires that ${}_2Q_R > {}_2Q_I$ and ${}_1Q_R > {}_1Q_I$. But this states that heat is transferred in net amount greater than zero from the cold to the hot reservoir without other external effects. Since this conclusion contradicts the second law as stated by Clausius, our original assumption cannot be true. The corollary is proved.

II. *All reversible engines have the same efficiency when working between the same two reservoirs.* PROOF: The proof is like that in I above, *i.e.*, it may be shown by assuming the contrary that a conclusion is reached that violates the second law. We conclude that the corollary is true.

III. *A temperature scale may be defined which is independent of the nature of the thermodynamic substance.* PROOF: Consider two Carnot engines operating in tandem. Let the first absorb heat Q_3 from a reservoir at T_3 and reject heat Q_2 to a reservoir at T_2. The second engine is made to absorb the heat Q_2 from the reservoir at T_2 and finally to reject heat Q_1 to a reservoir at T_1. The scheme is shown in Fig. 4. The device may also be considered as a single Carnot engine working between reservoirs

T_3 and T_1.   We have $T_3 > T_2 > T_1$.   From Eq. (16), we write

$$\frac{Q_2}{Q_3} = 1 - {}_3\eta_2 = f(T_2, T_3)$$

$$\frac{Q_1}{Q_3} = 1 - {}_3\eta_1 = f(T_1, T_3)$$

$$\frac{Q_1}{Q_2} = 1 - {}_2\eta_1 = f(T_1, T_2)$$

where the function f is unknown save that it must be a function of the temperatures alone, since the efficiencies (corollary II) are functions of the temperature alone.   We introduce the identity

$$\frac{Q_1}{Q_2} \equiv \frac{Q_1/Q_3}{Q_2/Q_3}$$

whence

$$f(T_1, T_2) = \frac{f(T_1, T_3)}{f(T_2, T_3)} = \frac{\psi(T_1)}{\psi(T_2)}$$

and

$$\frac{Q_1}{Q_2} = f(T_1, T_2) = \frac{\psi(T_1)}{\psi(T_2)}$$

The first and third terms of the above equation state that the ratio of the heats in a reversible engine is equal to the ratio of some function of the temperatures between

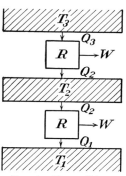

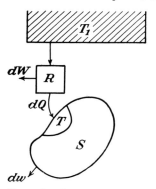

FIG. 4.—Reversible engines operating in tandem.

FIG. 5.—Inequality of Clausius.

which the engine operates.   The ratio may then be taken as the ratio of the temperatures themselves, when the temperatures are measured on the proper scale.   This scale is known as the *Kelvin temperature scale* or as the thermodynamic temperature scale.   Two temperatures T_1 and T_2 on the Kelvin scale are then given by the ratio of the heats added to and rejected to and by a Carnot cycle operating between the temperatures.

$$\frac{T_1}{T_2} = \frac{Q_1}{Q_2} \tag{18}$$

Since the ratio of the heats is independent of the working substance of the cycle, so also is the ratio of the temperatures when measured on the Kelvin scale.   The Kelvin scale is then unique.   Temperatures measured on the Kelvin scale will always be represented by T (or $°K$).

IV. *The Inequality of Clausius: Whenever a system executes a complete cycle, the integral of dQ/T around the cycle is less than zero, or in the limit is equal to zero.*   PROOF:

Consider a system S (see Fig. 5) that executes a cycle while work $\oint dw$ and heat $\oint dQ$ cross its boundaries. Let each element of heat dQ be supplied reversibly through a reversible engine R which in turn receives heat from a reservoir at T_1. By the first law we have $\oint dw = \oint dQ$ since $\oint dE = 0$. From the definition of the temperature scale, $(dW + dQ)/T_1 = dQ/T$ or $dW = [(T_1 - T)/T]dQ$ where T is the temperature at which dQ is supplied to the system and dW is the element of work done by the reversible engine. The total work done by the systems composed of S and R is

$$\Sigma W = \oint dw + \oint dW$$

whence on substituting from the above

$$\Sigma W = \oint dQ + \oint \frac{T_1 - T}{T} dQ = T_1 \oint \frac{dQ}{T}$$

But if ΣW is positive, the combined system of S and R constitutes a perpetual-motion machine of the second kind, and this violates the Kelvin-Planck statement of the second law. We also have that $T_1 > 0$. Therefore we must conclude that

$$\oint \frac{dQ}{T} \leqslant 0 \tag{19}$$

which proves the corollary.

V. *The cyclic integral of dQ/T is equal to zero for any reversible cycle. Consequently the integral of dQ/T for any reversible process between two states is a property of the system.* PROOF: Consider a system R' which executes a reversible cycle (see Fig. 6). Let a second system R'' surround R' and be affected by work and heat that pass the boundary of R'. R' and R'' together thus form an isolated system. Any heat that flows from R'' must enter R', and conversely, so that $dQ' = -dQ''$. For heat to flow from R'' to R', it is necessary that $T'' > T'$. We therefore have for any flow of heat $dQ'/T' \geqslant -dQ''/T''$. Now

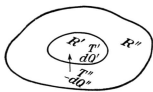

FIG. 6.—System to illustrate entropy as a property.

let both R'' and R' execute complete cycles. From the above, we have

$$\oint \frac{dQ'}{T'} + \oint \frac{dQ''}{T''} \geqslant 0 \tag{20}$$

But by Eq. (19) each integral is nonpositive. Therefore we must have each integral separately equal to zero. We may then write for any reversible cycle the general result

$$\oint_{\text{rev}} \frac{dQ}{T} = 0 \tag{21}$$

Since Eq. (21) is the necessary and sufficient condition that the integrand be a perfect differential, the integral of dQ/T is independent of the path of integration. Therefore $\int_i^f dQ/T$ is a thermodynamic property, since its value depends only on the initial and final states i and f of the system, provided that the heat is added reversibly. The property so defined is known as the entropy. In differential form, we have

$$dS = \frac{dQ}{T}\Big]_{\text{rev}} \tag{22}$$

where the letter S (or s for unit mass) will hereafter designate the entropy.

VI. *The entropy of an isolated system increases or in the limit remains constant.*
PROOF: Let any system go from an initial state to a final state by any reversible process
and return from final state to the initial state by any other reversible process.   Now
repeat the cycle letting the system follow the first reversible path to the final state
and then return to the initial state by some irreversible process (see Fig. 7).   By
corollary IV, we have

$$\int_{iR}^{f} \frac{dQ}{T} + \int_{fI}^{i} \frac{dQ}{T} \leqslant 0$$

and by V

$$\int_{iR}^{f} \frac{dQ}{T} + \int_{fR}^{i} \frac{dQ}{T} = 0$$

whence

$$\int_{fI}^{i} \frac{dQ}{T} - \int_{fR}^{i} \frac{dQ}{T} \leqslant 0$$

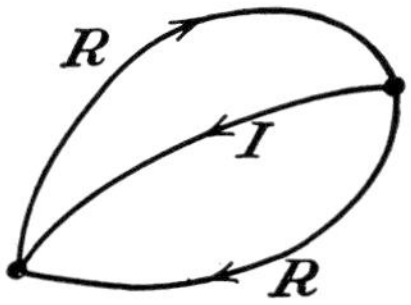

Fig. 7.—Reversible
paths *R*. Irreversible
path *I*.

On considering differentials and recalling Eq. (22), we see that

$$dS = \frac{dQ}{T}\bigg]_{\text{rev}} \geqslant \frac{dQ}{T} \tag{23}$$

For an isolated system $dQ = 0$.   Therefore

$$dS_{\text{isol}} \geqslant 0 \tag{24}$$

or the entropy of an isolated system cannot decrease.   The corollary is proved.
*When the entropy of a system is constant for a process ($dS \equiv 0$), the process is said to be
isentropic.*
Alternative Approach to Second Law.   An alternative and in some ways preferable
approach to a statement of the second law is given by Slater.[4]   He states the second
law as a postulate in the form

$$dS \geqslant \frac{dQ}{T} \tag{25}$$

where S denotes the entropy.   The equality sign is to be used when the process is
reversible and the inequality sign when the process is irreversible.   Slater remarks,
"We shall see that definite consequences can be drawn from it, and they prove to be
always in agreement with experiment."   This is an entirely adequate reason for
acceptance of the law without the labor of analyzing Carnot cycles and other general
systems by which the same conclusion was reached in Eq. (23).   It is useful, however,
for one who approaches the subject for the first time to start with somewhat less
abstract beginnings than the entropy directly.
Entropy as a Property.   The entropy as defined in Eq. (22) was shown in corollary
V to be a thermodynamic property.   The change in entropy of a system that has been
changed from state 1 to another state 2 may then be calculated along any reversible
path.   The value so calculated will be the change in entropy for all possible paths
connecting the same end states, whether reversible or irreversible.   On combining
the first and second laws, we have

$$dS = \frac{dE + dW'}{T}$$

for reversible processes.   By using this equation, one may determine the change in
entropy in terms of the change of internal energy and of the work done by the system.
It is clear from its definition that the entropy like internal energy, kinetic energy,
potential energy, etc., is measured from some datum or base state.

EQUATION OF STATE AND THERMODYNAMIC COORDINATES

Equation of State. Thermodynamics cannot of itself tell us what the equation of state of a system will be or how many variables it will involve. These questions rest with kinetic theory and statistical mechanics and with the evidence of experiment. For the systems with which we shall be concerned in meteorology, those sciences predict and experiment verifies that three thermodynamic coordinates (*i.e.*, properties) as a minimum are required. These three are related by an equation of state, determined by the system, which means that only two of the coordinates are independent. The most commonly used coordinates are **pressure** (p), **temperature** (T), and **specific volume** (v).

The atmosphere is found to obey the perfect gas law $(pv = RT)$ nearly enough for all practical purposes, so long as the air is not saturated. The gas constant R contains the relative humidity of the air as a parameter. It is unfortunate for the analysis, but true, that the amount of water vapor in the air cannot be determined as a function of the thermodynamic coordinates. When the amount of water vapor changes (as it does owing to mixing of air masses, vertical motions, and horizontal path traveled by the air mass) we must take account of the change by using different values of the constant R in the calculations. There is, in essence, a new thermodynamic system involved. The manner in which R changes with humidity is studied under "mixtures" (page 349). We emphasize here only that for any particular value of the humidity (less than saturation) the air does behave almost like a perfect gas. Its equation of state is then pv/T = constant.

Thermodynamic Coordinates. It is clear mathematically that any chosen function of thermodynamic coordinates will itself be a new coordinate. New functions (*i.e.*, coordinates) may be defined at will by any one of several methods, *e.g.*, (1) algebraic manipulation—since p is a coordinate, p^2 is another coordinate; (2) specification of a process—entropy is discovered to be a coordinate after defining it to be $dS = dQ/T$ where the heat is added reversibly; and (3) discovery of some combination of coordinates that eases the analysis—*enthalpy* (see the following).

The introduction of new coordinates is limited to those which ease and clarify the analysis. Sometimes the new coordinate is named (*e.g.*, enthalpy) and used as an integral part of the argument. At other times the new function is always written in terms of the old. Thus in meteorology $p^{0.286}$ is used as a coordinate on the adiabatic chart, but it bears no name and is always thought of in terms of pressure. The most frequently used thermodynamic coordinates are listed here.

1. p = pressure; one of the fundamental and directly observable coordinates.
2. T = temperature; one of the fundamental and directly observable coordinates.
3. v = specific volume (or its inverse density; $\rho = 1/v$).
4. e = internal energy per unit mass.
5. s = entropy per unit mass.
6. h = enthalpy per unit mass = $e + pv$ *by definition.*
7. ψ = Helmholtz free energy = $e - Ts$ *by definition.*
8. g = Gibbs free energy = $h - Ts$ *by definition.*
9. b = Availability = $h - T_0 s$ *by definition* (T_0 is a fixed and constant T).

Thermodynamic Formulas. Of all the coordinates that may be found, in whatever fashion, only two will be independent variables. This is the evidence of experience that is stated mathematically when we say an equation of state exists among the variables. It is then merely a matter of manipulation to express any one coordinate as a function of any other two. The form of the function (the equation of state) need not be known explicitly to obtain useful relations among the coordinates, since many of the physically observable items will be expressed in terms of derivatives. Meteorology does not need many of the formal relations that are available. A few of the more important are derived and listed below. For a more complete listing, see Bridgman.[5]

Pressure. The pressure in a fluid is the limit of the ratio of the normal force on an area to the area as the area decreases in size without limit. Thus $p = \lim\limits_{\Delta A \to 0} (\Delta F / \Delta A)$.

If the stress across the area is wholly normal for all orientations of the area (*i.e.*, no tangential stresses), it is easy to show that the limit above is the same whatever the orientation of the area; *i.e.*, the pressure in the fluid is the same in all directions. This result is descriptive of what are known as *perfect fluids*. Nature does not provide us with a true perfect fluid; but the air approximates to perfect conditions, provided that the flow is slow and nonturbulent. The static pressure in the atmosphere may in general be easily and accurately measured by a mercurial or aneroid barometer (see Sec. VIII on Meteorological Instruments).

Temperature. The temperature has been precisely defined in terms of the work done by reversible engines [Eq. (18)]. In principle, it may be measured as accurately as we please by means of a gas thermometer. In practice, it is measured by various types of thermometers that have been calibrated at fixed points against standards.

Specific Volume. The specific volume of a fluid is the reciprocal of its density. The density is the limit of the ratio of the mass of fluid in an element of volume to the volume as the volume decreases without limit. Thus $\rho = 1/v = \lim\limits_{\Delta \tau \to 0} (\Delta M / \Delta \tau)$. The specific volume of a confined fluid may be measured as accurately as we please by weighing the container both with and without the fluid and taking the ratio of the volume of the container to the mass of the fluid. In meteorological practice, the specific volume (or the density) is found from the equation of state after the pressure, temperature, and humidity have been measured.

Internal Energy. It has been deduced from thermodynamic considerations that the internal energy is a thermodynamic coordinate. Then if the change in internal energy between any two states is found by considering any process that connects the states, this value of the internal energy change will be the same for any other process connecting the states. Suppose heat energy is added to a system whose specific volume is constrained to remain constant. From the first law, we have (for a system in which $dw = p\, dv$)

$$dq]_v = de \tag{26}$$

since $p\, dv = 0$. Since the volume does not change and we have postulated a change in state, the temperature must change. The specific heat at constant volume is defined by the relation

$$c_v = \lim_{\Delta T \to 0} \left(\frac{\Delta q}{\Delta T}\right)_v \tag{27}$$

whence with the above $de = c_v\, dT$ and

$$e_2 - e_1 = \int_1^2 c_v\, dT \tag{28}$$

We note that the internal energy is specified numerically only when the base state from which it is measured is known. In meteorology, we may take the base state at a temperature of 200°K and a pressure of 1,000 mb. This base state is here chosen so that, at all pressures and temperatures found in the earth's atmosphere, the internal energy will be reckoned as positive. The specific heat as defined above is not necessarily constant. The magnitude of c_v is to be determined by experiment.

Entropy. Entropy was defined by Eq. (22). Suppose heat is added reversibly to a system. From Eq. (22), we have on integrating between states 1 and 2

$$s_2 - s_1 = \int_1^2 \frac{dq}{T}\Big]_{\text{rev}} \tag{29}$$

or on combining the first and second laws *for a system that can do work only by fluid expansion*

$$s_2 - s_1 = \int_1^2 \left(c_v \frac{dT}{T} + p \frac{dv}{T}\right)_{\text{rev}} \tag{30}$$

The result computed from Eq. (30) will be the change in entropy between the two states regardless of what process or combination of processes connects the two states (reversible or irreversible).

From Eq. (23), we have for irreversible processes that $T\,ds > de + p\,dv$ or

$$p\,dv < T\,ds - de \tag{31}$$

Since $p\,dv$ stands for the work done by the fluid, we see that the useful work that may be derived for a given change in state is less for an irreversible process than it would have been for a reversible process.

The entropy, like the internal energy, is specified numerically only when a base state is specified. The base state is again taken to be at 200°K and 1,000 mb. Since only changes of entropy have any significance in the following, the base state is arbitrary.

Enthalpy. The enthalpy is $h = e + pv$ by definition. In differential form, we have $dh = de + p\,dv + v\,dp$. Substituting into the combined first and second laws, we have

$$T\,ds \geqq dh - v\,dp = dq \tag{32}$$

where the equality holds for reversible processes and the inequality holds for irreversible processes. Suppose heat is added to a system whose pressure is constrained to remain constant. Then $v\,dp = 0$, and we have $dh = dq]_p$. The specific heat at constant pressure is defined by

$$c_p = \mathop{L}_{\Delta T \to 0}\left(\frac{\Delta q}{\Delta T}\right)_p \tag{33}$$

whence with the above $dh = c_p\,dT$ or

$$h_2 - h_1 = \int_1^2 c_p\,dT \tag{34}$$

There is no suggestion that c_p as defined above is a constant. Like c_v, it is a magnitude that must be determined by experiment. The enthalpy, like internal energy and entropy, is measured from the arbitrarily selected base state of 200°K and 1,000 mb.

Helmholtz Free Energy. The Helmholtz free energy is $\psi = e - Ts$ by definition. Using the combined first and second laws with this definition, we find

$$-d\psi \geqq s\,dT + p\,dv \tag{35}$$

This relation states that, for isothermal processes, the work done is not greater than the decrease in ψ and also that ψ is constant for reversible isothermal processes when no work is done. For an irreversible isothermal process, ψ decreases when no work is done.

Gibbs Free Energy. The Gibbs free energy is $g = h - Ts$ by definition. Using the combined first and second laws with this definition, we find

$$dg \leqq v\,dp - s\,dT \tag{36}$$

Then for isobaric isothermal processes, g is either constant (reversible) or decreasing (irreversible). The function g is important because of the many physical processes that take place isobarically and isothermally, *e.g.*, change of phase (melting of solids and the vaporization of liquids).

Availability. The availability is $b = h - T_0 s$ by definition. T_0 is a fixed value of the Kelvin temperature. It is the lowest temperature available for the rejection of energy as heat, and it is assumed that all heat is rejected at that temperature by the system (see Keenan[3]). Using the combined first and second laws with the above definition, we find

$$db \leqslant (T - T_0)ds \qquad (37)$$

Thus for isobaric reversible processes, $db = (T - T_0)ds$, and for isobaric irreversible processes, $db = (T - T_0)ds$. We note that the availability is less for the irreversible process than it is for the reversible process.

Maxwell Relations. The most important differential relations for reversible processes may be written

$$
\begin{aligned}
de &= -p\,dv + T\,ds \\
dh &= v\,dp + T\,ds \\
d\psi &= -p\,dv - s\,dT \\
dg &= v\,dp - s\,dT
\end{aligned}
\qquad (38)
$$

Since e, h, ψ, and g are all thermodynamic coordinates, we may write from the above at once

$$
\begin{aligned}
-p &= \left(\frac{\partial e}{\partial v}\right)_s & T &= \left(\frac{\partial e}{\partial s}\right)_v \\
v &= \left(\frac{\partial h}{\partial p}\right)_s & T &= \left(\frac{\partial h}{\partial s}\right)_p \\
-p &= \left(\frac{\partial \psi}{\partial v}\right)_T & -s &= \left(\frac{\partial \psi}{\partial T}\right)_v \\
v &= \left(\frac{\partial g}{\partial p}\right)_T & -s &= \left(\frac{\partial g}{\partial T}\right)_p
\end{aligned}
\qquad (39)
$$

on recalling the usual equation for the differential increase of a function of two independent variables.

If we assume further that the coordinates have continuous second partial derivatives, the order of differentiation is immaterial, and we have from the above on differentiation

$$
\begin{aligned}
-\left(\frac{\partial p}{\partial s}\right)_v &= \left(\frac{\partial T}{\partial v}\right)_s & \left(\frac{\partial p}{\partial T}\right)_v &= \left(\frac{\partial s}{\partial v}\right)_T \\
\left(\frac{\partial v}{\partial s}\right)_p &= \left(\frac{\partial T}{\partial p}\right)_s & -\left(\frac{\partial v}{\partial T}\right)_p &= \left(\frac{\partial s}{\partial p}\right)_T
\end{aligned}
\qquad (40)
$$

The four equations in (40) are known as the *Maxwell relations*.

SPECIFIC HEATS

Specific Heats as Observables. It will be evident from the preceding equations that internal energy, entropy, and enthalpy may be evaluated (given a base state) when the specific heats and the equation of state are known. The specific heat generally is defined by the limit

$$c = \underset{\Delta T \to 0}{L} \left(\frac{\Delta q}{\Delta T}\right) \qquad (41)$$

where Δq is the heat added per unit mass of substance and ΔT is the increase in temperature. According to this definition, the specific heat will have different values according to the constraint imposed as the heat is added. The two ways of adding heat and the resulting specific heats that are most useful are (1) isobaric processes and the specific heat at constant pressure, and (2) isosteric processes and the resulting specific heat at constant volume (or density). The specific heats are sometimes referred to 1 mole of the substance, in which case they are properly named *molar heats*. In meteorology, it is customary to refer always to 1 gram of the substance. The limits given by Eq. (41) according to whether the heat is added at constant pressure or constant volume are then correctly termed *specific heats*.

The specific heats are determined by measurement of the heat required to bring about certain changes in the state of a gram of substance. For a few substances, the values may be computed by simple kinetic theory and statistical mechanics. The results of theory and experiment match closely in most cases. It is emphasized here only that the values of the specific heats cannot be determined by thermodynamic theory. They are essentially functions of the constitution of the substance and must be determined by observation or a theory of matter.

General Relations between the Specific Heats. Some general relations between the specific heats at constant pressure and at constant volume may be deduced from thermodynamic theory. These results are useful in simplifying the theory and in providing a check on the experimental results obtained. From Eq. (41) and the first law of thermodynamics ($dq = de + p \, dv = dh - v \, dp$), we have

$$c_v = \left(\frac{\partial q}{\partial T}\right)_v = \left(\frac{\partial e}{\partial T}\right)_v \tag{42}$$

$$c_p = \left(\frac{\partial q}{\partial T}\right)_p = \left(\frac{\partial h}{\partial T}\right)_p \tag{43}$$

for the specific heats at constant volume and at constant pressure, respectively. When the heat is added reversibly, we have in addition

$$c_v = T \left(\frac{\partial s}{\partial T}\right)_v \tag{42a}$$

$$c_p = T \left(\frac{\partial s}{\partial T}\right)_p \tag{43a}$$

It may be shown by straightforward application of the results of the previous article that the ratio of the specific heats is given by

$$\frac{c_p}{c_v} = \frac{(\partial p/\partial v)_s}{(\partial p/\partial v)_T} \tag{44}$$

(see Joos[6] or Page[7]). Similarly, it may be shown that the difference of the specific heats is given by

$$c_p - c_v = -T \frac{(\partial v/\partial T)_p{}^2}{(\partial v/\partial p)_T} \tag{45}$$

We note from Eq. (45) that c_p is always greater than c_v, since $(\partial v/\partial p)_T$ is negative (a conclusion based on observation). From Eq. (44), it is noted that the ratio of the specific heats is equal to the ratio of the isentropic and isothermal slopes on a pressure-volume diagram. Also, since $c_p > c_v$, the isentropes are always steeper than the isotherms through any given point on the pressure-volume diagram.

Using Eqs. (44) and (45), one can find the values of c_p and c_v for any given substance by mechanical measurements alone. This result follows from the well-known result that the velocity of sound in a gas is given by

$$u^2 = \frac{pc_p}{\rho c_v} \tag{46}$$

Since u may be measured with reasonable accuracy by simple apparatus (*e.g.*, a Kundt's tube) and the density ρ may be measured directly, the ratio c_p/c_v is determinable without any heat measurements. Also one may determine the value of $c_p - c_v$ mechanically from the right-hand side of Eq. (45). This observation is not simple to make, but the values of the derivatives can be found and thence the difference of the specific heats without any measurement of the heat added. Using the results for c_p/c_v and $c_p - c_v$, one may eliminate and find the values of the specific heats separately. If the equation of state is known, the difference of the specific heats is easily found by direct differentiation (*e.g.*, when $pv = RT$, we find $c_p - c_v = R$).

Specific Heats as Functions. The specific heats will in general be functions of the thermodynamic coordinates. Using the Maxwell relations with the definitions of the specific heats, we find by direct differentiation that

$$\left(\frac{\partial c_v}{\partial v}\right)_T = T\left(\frac{\partial^2 p}{\partial T^2}\right)_v \tag{47}$$

$$\left(\frac{\partial c_p}{\partial p}\right)_T = -T\left(\frac{\partial^2 v}{\partial T^2}\right)_p \tag{48}$$

It is noted that, if the equation of state is known and c_v as a function of temperature at some particular volume, or c_p at some particular pressure, the specific heats may be calculated for all states. In the case of a perfect gas, Eqs. (47) and (48) reduce to

$$\left(\frac{\partial c_v}{\partial v}\right)_T = 0 \qquad \left(\frac{\partial c_p}{\partial p}\right)_T = 0 \tag{49}$$

Equation (49) shows that c_p and c_v are functions of temperature alone for any perfect gas. It was noted above that $c_p - c_v$ is constant for a perfect gas. This result does not imply that c_p and c_v are constant separately, however.

Specific Heats from Kinetic Theory. It is easily shown in elementary classical kinetic theory that the specific heats of the simple gases are absolute constants. These constants differ for different gases according to the ultimate structure postulated to exist in the gas. The value of c_p for a gas composed of spherical molecules, for example, turns out to be $5R/2$ where R is the gas constant. c_p for dumbbell-shaped molecules on the same classical theory is $7R/2$. The ratios of the specific heats of these gases are 5/3 and 7/5, respectively. These values correspond closely to the values as determined by experiment on appropriate gases.

Specific Heats of Solids and Liquids. The specific heat of a solid or liquid will also depend upon how the heat is added. Yet for many solids and liquids the expansion of the substance is so small on the addition of heat that the work done by the expansion $\int p\,dv$ may be neglected in comparison with the change in internal energy. A considerable literature has been developed on the specific heats of both solids and liquids on statistical mechanics and kinetic theory grounds. These investigations are beyond the scope of this book (see Slater,[4] Fowler,[8] etc.). For practical purposes in meteorology, we may always neglect the work done in expansion and state that any solid or liquid with which we are concerned has only one value of the specific heat no matter how the heat is added. This specific heat will be denoted by c_i for ice and c_w for liquid water. Neither c_i nor c_w is strictly constant; both will be functions of the temperature (of the ice or water).

Calorimetry. An excellent résumé of the entire subject of calorimetry is given by Saha and Srivastava.[9] The authors discuss the important experimental methods and results of the last 50 years. Methods are discussed under the headings of (1) mixtures,

(2) cooling, (3) change of state, and (4) electrical. Experimental methods for determination of specific heats are discussed under (1) measurement of c_p, (2) measurement of c_v, (3) determination of c_p/c_v. Other sources of information are Glazebrook;[10] Partington and Shilling;[11] "Handbuch der Experimentalphysik," Vol. 8, Part I; and especially the International Critical Tables.

The present generally accepted values for the properties of water and its vapor are those tabulated by Keenan and Keyes.[12] Kiefer[13] has recomputed values from this source and tabulated them in degrees centigrade and millibars of pressure for the convenience of the meteorologist. The tables of Kiefer are reproduced in Sec. I. Further tables are reproduced from U.S. Weather Bureau data.

THERMOMETRY AND THERMOMETRIC SUBSTANCES

Temperature. The concept of temperature was originally conceived on the basis of feeling. A body was judged hot or cold according to how it felt to the touch. As accurate measurements were beginning to be made, it was observed that as a body became "hotter" certain physical changes took place. Gases expanded, for example, or, if constrained to remain in the same volume, increased their pressure against the walls of the containing vessel. These physical changes were used as criteria for the establishment of arbitrary scales of "hotness." Thus temperature came to mean the amount of expansion of a column of mercury or of a volume of gas. The centigrade scale labels the temperature of the boiling point of water under 1 atm of pressure as 100 degrees centigrade (100°C), and the freezing point of water as zero degrees centigrade (0°C). The Fahrenheit scale labels the same temperatures as 212 and 32°F, respectively. The numerical relation between the two scales is then

$$\frac{°C}{°F - 32} = \frac{100}{180} \tag{50}$$

Thermometric Substances and the International Temperature Scale. It is clear that temperature as measured by the expansions, say, of different substances will not necessarily agree for values between or beyond the two points (freezing and boiling) at which they are made to agree. Experiment shows that the departure from agreement is too great to be ignored. In 1927, the Seventh General Conference of Weights and Measures representing 31 nations unanimously adopted a temperature standard. The following are excerpts from the report by Burgess[14] on the decisions reached by the conference:

1. The thermodynamic centigrade scale, on which the temperature of melting ice, and the temperature of condensing water vapor, both under the pressure of 1 standard atmosphere, are numbered 0 degrees and 100 degrees, respectively, is recognized as the fundamental scale to which all temperature measurements should ultimately be referable.
2. The experimental difficulties incident to the practical realization of the thermodynamic scale have made it expedient to adopt for international use a practical scale designated as the international temperature scale. This scale conforms with the thermodynamic scale as closely as is possible with present knowledge, and is designed to be definite, conveniently and accurately reproducible, and to provide means for uniquely determining any temperature within the range of the scale, thus promoting uniformity in numerical statements of temperature.
3. Temperatures on the international scale will ordinarily be designated as "°C," but may be designated as "°C (Int.)" if it is desired to emphasize the fact that the scale is being used.
4. The international temperature scale is based upon a number of fixed and reproducible equilibrium temperatures to which numerical values are assigned, and upon

the indications of interpolation instruments calibrated according to a specified procedure at the fixed temperatures.

5. The basic fixed points and the numerical values assigned them for the pressure of 1 standard atmosphere are given in the following table, together with formulas which represent the temperature (t_p) as a function of vapor pressure (p) over the range of 680 to 780 mm of mercury.

6. Basic fixed points of the international temperature scale:

a. Temperature of equilibrium between liquid and gaseous oxygen at the pressure of 1 standard atmosphere (oxygen point), $-182.97°C$.

$$t_p = t_{760} + 0.0126(p - 760) - 0.0000065(p - 760)^2$$

b. Temperature of equilibrium between ice and air-saturated water at normal atmospheric pressure (ice point), $0.000°C$.

c. Temperature of equilibrium between liquid water and its vapor at the pressure of 1 standard atmosphere (steam point), $100.000°C$

$$t_p = t_{760} + 0.0367(p - 760) - 0.000023(p - 760)^2$$

d. Temperature of equilibrium between liquid sulphur and its vapor at the pressure of 1 standard atmosphere (sulphur point), $444.60°C$

$$t_p = t_{760} + 0.0909(p - 760) - 0.000048(p - 760)^2$$

e. Temperature of equilibrium between solid silver and liquid silver at normal atmospheric pressure (silver point), $960.5°C$

f. Temperature of equilibrium between solid gold and liquid gold at normal atmospheric pressure (gold point), $1063°C$

Standard atmospheric pressure is defined as the pressure due to a column of mercury 760 mm high having a mass of 13.5951 g/cm^3, subject to a gravitational acceleration of 980.665 cm/sec^2, and is equal to $1,013,250$ dynes/cm^2. It is an essential feature of a practical scale of temperature that definite numerical values shall be assigned to such fixed points as are chosen. It should be noted, however, that the last decimal place given for each of the values in the table is significant only as regards the degree of reproducibility of that fixed point on the international temperature scale. It is not to be understood that the values are necessarily known on the thermodynamic centigrade scale to the corresponding degree of accuracy.

7. The means available for interpolation lead to a division of the scale into four parts.

a. From the ice point to $660°C$ the temperature t is deduced from the resistance R_t of a standard platinum resistance thermometer by means of the formula

$$R_t = R_0(1 + At + Bt^2)$$

The constants R_0, A, and B of this formula are to be determined by calibration at the ice, steam, and sulphur points, respectively. The purity and physical condition of the platinum of which the thermometer is made should be such that the ratio R_t/R_0 shall not be less than 1.390 for $t = 100$ degrees and 2.645 for $t = 444.6$ degrees.

b. From -190 degrees to the ice point, the temperature t is deduced from the resistance R_t of a standard platinum resistance thermometer by means of the formula

$$R_t = R_0[1 + At + Bt^2 + C(t - 100)t^3]$$

The constants R_0, A, and B are to be determined as specified above, and the additional constant C is determined by calibration at the oxygen point. The standard thermometer for use below 0 degrees must, in addition, have a ratio R_t/R_0 less than 0.250 for $t = -183$ degrees.

c. From $660°C$ to the gold point, the temperature t is deduced from the electromotive force e of a standard platinum vs. platinum-rhodium thermocouple, one junction of which is kept at a constant temperature of $0°C$ while the other is at the temperature t defined by the formula

$$e = a + bt + ct^2$$

The constants a, b, c are to be determined by calibration at the freezing point of antimony, and at the silver and gold points.

d. Above the gold point the temperature t is determined by means of the ratio of the intensity J_2 of monochromatic visible radiation of wave length λ cm, emitted by a black body at the temperature t_2, to the intensity J_1 of radiation of the same wave length emitted by a black body at the gold point, by means of the formula

$$\ln \frac{J_2}{J_1} = \frac{c_2}{\lambda} \left[\frac{1}{1,336} - \frac{1}{t + 273} \right]$$

The constant c_2 is taken as 1.432 cm degrees. The equation is valid if $\lambda(t + 273)$ is less than 0.3 cm degrees.

The measurement of surface temperatures in meteorology is generally made with mercury-in-glass thermometers. Very low temperatures are measured with spirit-in-glass thermometers. The thermometer may be calibrated to read in centigrade or Fahrenheit degrees. Temperatures are reported in degrees Fahrenheit at the surface and in degrees centigrade for upper-air readings. See Sec. VIII on Meteorological Instruments.

Gas as a Thermometric Substance. Gas thermometers are not used in routine meteorological work. An analysis of the gas thermometer is necessary, however, to establish the identity of the Kelvin scale and a temperature that can be measured in practice. Extensive experiments in all countries have established the fact that, at high enough temperatures and low enough pressures, the equation of state of any gas may be written

$$pv = R\theta \tag{51}$$

where p is the absolute pressure, v the specific volume, and θ the "absolute temperature" of the gas. R is a constant for any particular gas. Equation (51) is taken as the definition of the absolute temperature. Gas thermometers may be constructed to measure temperature on the basis of Eq. (51), using either constant volume or constant-pressure instruments. By extrapolation from measurements made at low temperatures, it is found that numerically

$$\theta = t + 273.2 \tag{52}$$

where t is the temperature on the centigrade scale.

The Perfect Gas. The perfect (or ideal) gas will be defined to be a gas having two precise characteristics, (1) that for all values of the thermodynamic coordinates its equation of state is $pv = R\theta$ exactly, and (2) that its internal energy is a function of temperature alone. Both these conditions are shown to be true in elementary kinetic theory for a simple gas in a force-free field. Thermodynamics cannot prove either relation, since it is not concerned with the ultimate structure of matter. The question of how closely actual gases follow the definition of a perfect gas is fully discussed in ref. 15.

For a perfect gas as defined above, the first law becomes

$$dq = c_v\, d\theta + p\, dv = de + p\, dv \tag{53}$$

where the specific heat at constant volume is given by

$$c_v = \left(\frac{\partial q}{\partial \theta} \right)_v = \frac{de}{d\theta} \tag{54}$$

since the internal energy depends on temperature alone. We also have the useful differential relation

$$p \, dv + v \, dp = R \, d\theta \qquad (55)$$

Kelvin Temperature Scale and Kelvin Degree. The Kelvin temperature scale was defined by Eq. (18). The Kelvin degree is established as follows: Let Q_n be the heat absorbed from a reservoir at T_n (the temperature on the Kelvin scale) by a reversible engine and Q_{n-1} be the heat rejected to a reservoir at T_{n-1} by the same engine. Consider a number of engines operating in tandem between reservoirs at progressively lower temperatures, each engine absorbing the heat rejected by the engine preceding it in line and rejecting heat to the engine that follows. Arrange the temperatures of successive reservoirs so that each engine does the same amount of work W. By the first law of thermodynamics, the work done by each engine is equal to $W = Q_n - Q_{n-1} = Q_{n-1} - Q_{n-2} = Q_{n-2} - Q_{n-3} = \cdots$. By the second law and the definition of the Kelvin temperatures we have $Q_n/Q_{n-1} = T_n/T_{n-1}$, $Q_{n-1}/Q_{n-2} = T_{n-1}/T_{n-2}$, $\cdots$. From these two relations we see that $Q_n(1 - T_{n-1}/T_n) = Q_{n-1}$ $(1 - T_{n-2}/T_{n-1}) = Q_{n-2}(1 - T_{n-3}/T_{n-2}) = \cdots$ and $T_n - T_{n-1} = T_{n-1} - T_{n-2} = T_{n-2} - T_{n-3} = \cdots$. We see then that equal intervals on the Kelvin scale are the intervals between which reversible engines will do the same work, when each engine absorbs the heat rejected by the engine ahead in line. It will be recalled that the Kelvin scale is unique, since the efficiency of all reversible engines operating between the same temperatures is the same for any working substance.

Now imagine 100 reversible engines operating between the temperatures of the steam point and the ice point. These engines define 100 equal temperature intervals when the engines are arranged as indicated above. Each such interval is called 1 degree on the Kelvin scale. We have $T_{\text{steam}} - T_{\text{ice}} = 100$. Temperatures on the Kelvin scale are designated by capital T or by °K. The thermodynamic centigrade scale has temperature intervals equal to the intervals on the Kelvin scale. On the centigrade scale, however, the steam and ice points are arbitrarily selected to be 100 and 0°, respectively. The relation between the centigrade and Kelvin scales is $T = t + 273.2$, as will be shown below.

Absolute Zero of Temperature. The concept of an absolute zero of temperature arises naturally from the Kelvin scale. Imagine a sequence of engines operating reversibly as before, with each performing an equal amount of work. Since $Q_{n-i}/Q_{n-(i+1)} = T_{n-i}/T_{n-(i+1)}$ and $T_{n-(i+1)} < T_{n-i}$ we see that $Q_{n-(i+1)} < Q_{n-i}$. But $Q_{n-i} - Q_{n-(i+1)} = \text{const}$ for all engines by hypothesis. It is evident that, if Q_n is finite (as it must be for T_n finite), then eventually a point will be reached at which zero heat is rejected. The temperature at which this occurs is designated as the absolute zero of temperature on the Kelvin scale. It will be noted that this concept of an absolute zero of temperature does not involve the ultimate structure of the working substance.

Kelvin Temperature Scale and Perfect Gas Thermometer. It is shown here that the temperature indicated by a gas thermometer (when the gas is perfect) is identical with the Kelvin temperature. Consider a thermodynamic system composed of the following items: (1) a cylinder fitted with a frictionless piston, both cylinder and piston being perfect insulators, (2) a cylinder head that may be removed or attached to the cylinder at will without the loss of work, heat, or gas—also perfectly insulating, (3) two reservoirs at temperatures θ_1 and θ_2 as measured on the absolute scale ($\theta_2 > \theta_1$), (4) a perfect gas that fills the cylinder, and (5) an extraneous force that may be applied at will to the piston. The system is sketched in Fig. 8. The cycle through which the gas passes is shown to pressure vs. specific volume coordinates in Fig. 9. The Carnot cycle is analyzed as follows:

a—b. With the insulating head in place and the system in thermodynamic equilibrium at θ_1, the extraneous force is applied and the gas is reversibly and adiabatically compressed until its temperature is θ_2. The first law requires that $_ae_b = -\int_a^b p\,dv = \int_a^b c_v\,d\theta$ since $dq = de + p\,dv = 0$.

b—c. The insulating head is removed and the reservoir at θ_2 is brought into contact with the working substance without heat or work. With the system in thermodynamic equilibrium at θ_2, the gas is permitted to expand reversibly and isothermally until a point is reached where $v_c > v_b$; the amount of the expansion is arbitrary. Since the process is isothermal, $dT = de = 0$, and, on introducing the equation of state, we have the heat added from the reservoir to be given by $_bq_c = \int_b^c p\,dv = R\theta_2 \ln (v_c/v_b)$. This quantity is also the work done by the gas during its expansion.

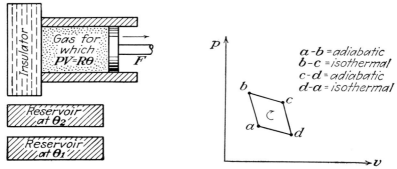

FIG. 8.—System to illustrate Carnot cycle.   FIG. 9.—Carnot cycle to *pv*-coordinates.

c—d. The reservoir at θ_2 is removed from the cylinder without heat or work, and the insulating head is restored. With the system in thermodynamic equilibrium at θ_2, the gas is permitted to expand reversibly and adiabatically. We have $dq = 0$, and the internal energy and temperature will decrease since the gas does work against the piston. The expansion is continued until the system is in thermodynamic equilibrium at θ_1. We have $_ce_d = -\int_c^d p\,dv = \int_c^d c_v\,d\theta$.

d—a. The insulating head is removed and the reservoir at θ_1 is brought to the cylinder without heat or work. The extraneous force is applied, and the gas is compressed reversibly and isothermally until the original specific volume is attained. The system is then in equilibrium at θ_1 and v_a, and the cycle has been completed. We have $_dq_a = \int_d^a p\,dv = R\theta_1 \ln (v_a/v_d)$.

For the adiabatic processes *a—b* and *c—d*, we have $dq = 0$. Then, on combining the equation of state with the first law, we find $c_v\,d\theta/\theta = -p\,dv/\theta = -R\,dv/v$. We have then

$$\int_a^b c_v \frac{d\theta}{\theta} = -R \ln \left(\frac{v_b}{v_a}\right) \qquad \text{and} \qquad \int_d^c c_v \frac{d\theta}{\theta} = R \ln \left(\frac{v_d}{v_c}\right)$$

The integrals above are equal, however, since c_v for a perfect gas is independent of specific volume or pressure and since the integrations are carried out between equal temperature intervals. Then $v_b/v_a = v_c/v_d$. We introduce this equation into the above results for the ratio $_dq_a/_bq_c$ and obtain at once

$$\frac{\theta_1}{\theta_2} = \frac{_dq_a}{_bq_c} = \frac{T_1}{T_2} \tag{56}$$

where the ratio on the right is the ratio of the Kelvin temperatures. This is equal to $_dq_a/_bq_c$ because the cycle is reversible. We see that the ratio of any two temperatures as measured on the Kelvin scale is the same as the ratio of the same two temperatures as measured on the gas scale (*i.e.*, the absolute scale). The two scales are further made to agree at the ice point and the steam point by setting $T_{st} - T_{ice} = \theta_{st} - \theta_{ice} = 100$ arbitrarily. It follows immediately that for all temperatures

$$\theta = T = t + 273.2 \tag{57}$$

where the right-hand side of Eq. (57) results from experiment using gas thermometers and extrapolating data down to the absolute zero of temperature.

Departures of the international scale from the Kelvin scale are too small to be of significance in routine meteorological work. Keenan[3] quotes results based on measurements made by Beattie that show less than 0.05°C difference between temperatures measured on the two scales. He concludes that no distinction need be made between the scales for any work of less than the most precise accuracy.

PROPERTIES OF PURE SUBSTANCES

State and Phase Changes; Descriptive. Consider a homogeneous thermodynamic system at a state 1 determined by the coordinates pvT. If one or more of the coordinates are changed, compatible with the equation of state, and if the system remains homogeneous in the same phase it was in at state 1, then the system is said to have undergone a simple change of state. A *phase* is any physically *homogeneous* aspect of a system. A simple illustration of change of state is the compression of a gas. Mathematically we may say that a system undergoes a change of state when all the thermodynamic coordinates and their first and second derivatives are continuous.

It is a matter of common knowledge, however, that familiar substances may be found under normal conditions in solid, liquid, or vapor form. Moreover these different phases may coexist in equilibrium. A system composed of two or more phases is referred to as *heterogeneous*. When a portion of the substance of the system that is in one phase (*e.g.*, ice) is transformed into another phase (*e.g.*, water), we say that a change of phase has occurred. There may be a change of phase in a system without a change of state of either component. Generally as the coordinates change there may be simultaneous phase and state changes.

Pressure-volume Diagram. Introduce about 1 gram of water into a closed evacuated container. Let the volume of the container be about 2 l. and the temperature of the water be just under 100°C. It is observed that, if the container and its contents are kept at the initial temperature of the water, all the water *evaporates* (*vaporizes*) into the space available. The system is homogeneous in its vapor phase. The state is represented by point 1 on the pv-diagram of Fig. 10. Now compress the vapor slowly and isothermally. The pressure will rise continuously for a time. At a certain value of the volume, it is observed that simultaneously (1) vapor condenses into liquid water, and (2) the rate of increase of pressure is zero $[(\partial p/\partial v)_T = 0]$. This is point 2 in the figure. The pressure at this point is named the *saturation vapor pressure* (sometimes the maximum vapor pressure). The vapor in the container is referred to as *saturated vapor*. In its initial condition at state 1, the vapor was unsaturated, or *superheated*.

Continue the compression past 2 isothermally. It is observed that the process is also isobaric. Condensation continues until nothing is left in the container except liquid water. This state is at point 3 in the figure. Throughout the isothermal, isobaric compression from 2 to 3, the system was heterogeneous in two phases (water vapor and liquid water). During the process, the phase of water vapor is changing to the phase of liquid water. There is no change of state in either component since temperature and pressure are not changed. At point 3 the system is again homogeneous, now in the liquid phase, and we refer to the state as *saturated liquid*.

Continue the compression past 3 isothermally. The pressure rises abruptly. The phase remains unchanged at any pressures found in the atmosphere. It may be remarked, however, that if the pressure is large enough the liquid may undergo a phase change into a form of solid (see Slater[4] for the several forms of ice found at high pressures).

According to the temperature at which the above experiment is performed, one obtains a different curve on the *pv*-diagram. Several of the isotherms for water are shown in Fig. 10. The locus of the points 2, 2′, etc., is known as the *saturated-vapor*

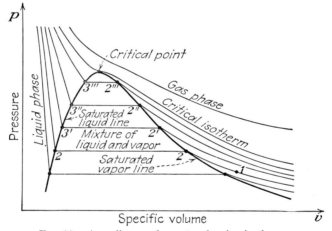

Fig. 10.—A *pv*-diagram for water showing isotherms.

line. The locus of the points 3, 3′, etc., is known as the *saturated-liquid line* for temperatures above 0°C and as the *saturated-ice line* for temperatures below 0°C (see triple-point data below). If the temperature of the experiment is below 0°C, the vapor does not condense into liquid water but sublimates into the solid phase (ice). If, on the other hand, the temperature of the experiment is greater than 374°C, it is observed that the system remains homogeneous no matter how high the pressure may rise. The "vapor" then behaves like a perfect gas in that only changes of state are possible with increase in pressure isothermally.

Critical Temperature. The lowest temperature for which the water vapor cannot be compressed into a liquid is known as the critical temperature of water. If the densities of both liquid and vapor are measured and plotted as functions of temperature, the temperature where the two curves meet is the critical temperature. In Fig. 10, the intersection of the saturated-vapor and the saturated-liquid lines is known as the *critical point*. In *pvt*-space, this point occurs for water at $T = 647°$K, $p = 218$ atm, and $v = 2.50$ cm³ per gram. Since all atmospheric temperatures are well below the critical temperature of water, all three phases of water exist normally in the atmosphere. The critical temperatures of nitrogen, oxygen, and argon (the main constituents of dry air) are $-147°$C, $-119°$C, and $-122°$C, respectively, so that these elements do not appear in equilibrium in their liquid or solid phases in the atmosphere.

Triple Point. Consider another ideal experiment. With the apparatus of the above system, keep the volume of the container constant and change the temperature by the addition of heat. At high enough temperatures, it is found that the space is filled with superheated vapor. If the temperature is lowered, a point will be reached at which some of the vapor begins to condense into liquid water. The system is in equilibrium with two phases present. If the temperature is lowered still further, the liquid will begin to freeze into ice and some of the vapor will sublimate into ice directly. The system is in equilibrium with three phases present. It is found that the temperature cannot be lowered to less than 0°C until sufficient heat has transferred from the system to freeze all the liquid water. When the container holds nothing but ice (and vapor), the temperature may be lowered still further. The system is now in equilibrium with two phases present, ice and vapor. The point in which a substance may exist in all three phases simultaneously is known as the triple point. For water, the data are accurately $t = 0.0098°C$ and $p = 6.105$ mb. (see Zemansky[2]). In meteorology, it is sufficient to take $t = 0°C$, as has been done in the above discussion. The triple point on the pv-diagram is of course not a point but a line, for the condition may occur for any value of the specific volume between the saturated-liquid line and the saturated-vapor line.

Pressure-temperature Diagram. Another representation that is useful is found when the saturation vapor pressure is plotted against temperature. From the pv-diagram, it is clear that there is a definite value of the temperature corresponding to each value of the saturation vapor pressure (and conversely). These values are found at the points 2, 2', etc. Thus we can plot a single curve on pT-coordinates representing saturation vapor pressure against temperature. The general shape of the curve is shown in Fig. 11. We observe that for temperatures above 0°C the pressure increases uniformly with temperature. At the triple point, the slope of the curve is discontinuous. Below 0°C, the saturation vapor

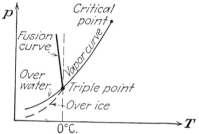

FIG. 11.—A pT-diagram for water at saturation.

pressure has one of two values according to whether the vapor is in equilibrium with ice or with supercooled water. The saturation vapor pressure is less over ice than it is over water.

Supercooled water is liquid water that exists at a temperature of less than 0°C. In the laboratory, the state is unstable. Introduction of a small crystal of ice, or even a mechanical shock of the container, will cause the supercooled water to change into the solid phase. In the upper atmosphere, however, there is ample evidence that supercooled water (at least in the form of small droplets of liquid water) may persist down to −40°C. This evidence is provided by observations of clouds that contain liquid-water droplets at this low temperature. Liquid-water droplets are not common below −10°C, but they do sometimes occur. Our observations must then determine the saturation vapor pressure for a system in which vapor condenses into liquid water (rather than sublimates into ice) even at temperatures below 0°C. The pressure is called the saturation vapor pressure "over water" or "over ice" according to whether it is determined for the vapor in equilibrium with supercooled water or with ice.

One may construct three-dimensional pvT-surfaces illustrating the properties of water described above. Or one may draw isopleths in the plane of any two thermodynamic coordinates, *e.g.*, Fig. 10, which shows isotherms in the pv-plane. The choice of diagram for illustration of the facts depends upon what one wishes to emphasize.

A substance that has no free surface and that fills all the space available in a container is called a *gas* if its temperature exceeds the critical temperature for the substance. If its temperature is less than the critical temperature, it is called a *vapor*. Thus in the atmosphere we refer to water vapor and to the gases nitrogen, oxygen, etc.

Temperature-entropy Diagram. Consider finally state and phase changes that occur isobarically and the construction of a temperature-entropy diagram. Introduce into an evacuated cylinder a small piece of ice at less than 0°C. Let the pressure applied to the system be maintained constant by a weight acting on a frictionless piston in the cylinder. Add heat to the system. The ice will increase in temperature. If the heat is added reversibly, the entropy increases in amount $ds = dq/T$. If the pressure is greater than a few millibars, the heating continues to increase the temperature of the ice at a rate proportional to the rate at which heat is transferred into the system. This increase in temperature continues until 0°C is reached. The specific heat of the ice (c_i) is almost constant. The increase in entropy during the warming from T_1 to 273°K is given by $\Delta s = \int c_i \, dT/T = c_i \ln (273/T_1)$.

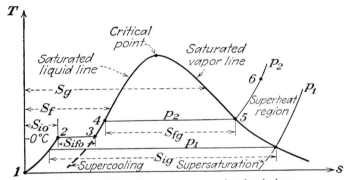

Fig. 12.—A Ts-diagram for water showing isobars.

We may continue to add heat, but no increase in temperature will be observed until all the ice is melted into liquid water. The heat required to melt unit mass of the solid (*i.e.*, to change its phase from solid to liquid) is called the *heat of fusion*. During this phase change, the increase in entropy will be $\Delta s = L_i/273$, where L_i is the latent heat of fusion.

The temperature of the water is now increased by causing more heat to enter the system. The temperature may be increased up to the temperature T_2, which is the temperature corresponding to the saturation vapor pressure equal to the pressure of the experiment. This is a state change of liquid water. The change in entropy of the system will be $\Delta s = c_w \ln (T_2/273)$, where c_w is the specific heat of the water. c_w is almost constant.

If more heat crosses the boundary of the system, the liquid water vaporizes at a constant temperature (and pressure). This is then a pure change of phase. The amount of heat necessary to vaporize unit mass of the liquid is called the *latent heat of vaporization*. The change in entropy will be $\Delta s = L/T_2$ where L is the latent heat of vaporization and T_2 is the temperature at which the pressure of the experiment is the saturation vapor pressure of the water.

When all the liquid has been vaporized, further addition of heat will produce a state change in the vapor phase. The increase in entropy for this state change is given by Eq. (30) to be $\Delta s = \int c_v \, dT/T + p \, dv/T$. This equation must take account of the increase in entropy due to work done by the fluid as well as that due to an increase in temperature.

According to the value of the pressure at which the experiment is performed, we get different curves on the Ts-diagram. Several such isobars are shown for water in Fig. 12. Since the melting point of ice at all atmospheric pressures is almost exactly 0°C, and since the heat of fusion is almost a constant, the saturated-ice line and the saturated-liquid line are almost identical with the family of isobars up to the point where vaporization begins. It should be noted, however, that strictly speaking the saturated-ice and saturated-liquid lines are the loci of the points in Ts-space where the solid begins to sublimate or the liquid to vaporize, respectively. On an enlarged scale, we should have a family of isobars in the neighborhood of the ice and liquid lines which, to any usual scale, fade into the loci, as shown in Fig. 13.

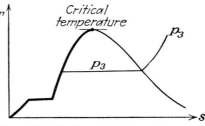

FIG. 13.—Family of isobars shown on enlarged scale by thickened line.

Just as it is possible to cool the liquid below 0°C without the appearance of the solid phase, so is it possible to cool vapor below the temperature corresponding to the saturation vapor pressure of the substance. The vapor is then known as *super-saturated vapor*. This state is also unstable but not uncommon in the atmosphere. Whether or not condensation begins as the vapor is cooled to the saturation point depends upon whether or not impurities are present. This is discussed further on pages 254–257. The regions of supercooled liquid and supersaturated vapor are noted on the Ts-diagram of Fig. 12.

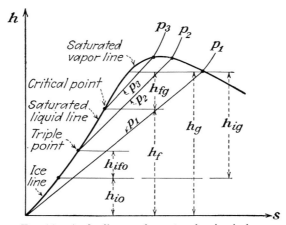

FIG. 14.—An hs-diagram for water showing isobars.

An enthalpy-entropy diagram may also be constructed on the basis of the above experiment. This diagram for water is shown in Fig. 14. The method of construction is as indicated in the figure.

Conclusions from Experiment. A number of conclusions may be deduced from the numerical results of the experiments described above. These are, for meteorological purposes:

1. The specific heats of ice and of liquid water are constants.
2. The melting (freezing) of ice (water) occurs isothermally at a temperature that is independent of the pressure of the experiment. The heat of fusion is a constant.
3. Given a constant pressure, the sublimation of ice or the vaporization or condensation of liquid water occurs at the temperature for which the pressure of the experiment is the saturation vapor pressure over ice or over water, respectively. The heat of vaporization decreases with increase in the pressure of the experiment until it is zero at the critical point. The heat of vaporization is equal to the heat of condensation.
4. Under constant pressure conditions in thermodynamic equilibrium

    a. Adding heat to water produces a pure change of state.
    b. Adding heat to ice produces a pure change of state.
    c. Adding heat to ice and water mixed at 0°C produces a pure change of phase.
    d. Adding heat to liquid water and water vapor mixed at saturation pressure produces a pure change of phase.
    e. Adding heat to vapor produces a pure change of state.

Not all the above statements are exact. They are, however, excellent approximations to the truth for the calculation of thermodynamic coordinates of water in its various phases for all practical purposes in meteorology.

Clapeyron's Equation. An important thermodynamic relation for the change of phase may be easily deduced from Maxwell's relations and the second law of thermodynamics. For any reversible process, we have $q = \int T \, ds$ for the heat added to a system in terms of the change in entropy of the system. For an isothermal process such as the vaporization of water under its saturation vapor pressure, we have $ds = (\partial s / \partial v)_T \, dv$ where entropy s is expressed as a function of the two properties specific volume v and temperature $T[s = s(v,T)]$. The heat required to vaporize the liquid is then $L = \int T(\partial e / \partial T)_v \, dv$ where we have replaced $(\partial s / \partial v)_T$ by $(\partial e / \partial T)_v$ using one of the Maxwell relations [Eq. (40)]. The saturation vapor pressure is designated by e in accordance with meteorological practice (rather than by p). The saturation vapor pressure is a function of temperature alone so that the above equation may be integrated at once to give

$$L = T \frac{de}{dT} (v_g - v_f) \tag{58}$$

where L is the latent heat of vaporization, v_g is the specific volume of the vapor, and v_f is the specific volume of the liquid. The latent heat of vaporization is also equal to the change in enthalpy, since by the first law $dq = dh$ when the process is isobaric. It is $h_{fg} \equiv L$ that is tabulated for the latent heat of vaporization.

Equation (58) is known as the *Clapeyron equation*. It is usually written as

$$\frac{de}{dT} = \frac{L}{T(v_g - v_f)} \tag{59}$$

and, since $v_g >> v_f$ for water, we have to an excellent approximation

$$\frac{de}{dT} = \frac{L}{Tv_g} \tag{60}$$

The usefulness of the above result occurs in supplying a means of checking experimental data. It also will occur in numerous meteorological formulas as a simplifying factor. If the equation of state and the latent heat of vaporization are known functions so that L and v may be expressed as functions of temperature, Eq. (60) may be integrated to give the saturation vapor pressure as a function of temperature (see below).

Numerical Results for Water. Generally accepted data for the thermodynamic properties of water are those tabulated by Keenan and Keyes.[12] Their tables are given in engineering units and thus are not convenient for direct use in meteorology. From these data, Kiefer[13] has computed the values and tabulated in centigrade degrees and joules of energy. Kiefer's very useful table is reproduced in Sec. I. Harrison[16] published earlier data of a more limited nature which agree well with Kiefer's table. Goodman[17] has computed the saturation vapor pressures from equations given by Washburn[18] and made some useful tabulations. Goodman states that his results are in agreement with the values published by Goff.[19] The general agreement among all published data is so good that Kiefer's table (which has the most convenient arrangement for meteorological work) is adopted as standard in this book. See Sec. I for various tables and graphs.

Using the data of Keenan and Keyes, Kiefer concludes that the saturation vapor pressure over water in the range $-20°C \leqslant t \leqslant 50°C$ may be accurately represented by the equation

$$\ln \frac{e_s}{6.105} = 25.22 \frac{(T - 273)}{T} - 5.31 \ln \frac{T}{273} \tag{61}$$

where e_s is the saturation vapor pressure in millibars. A formula due to Washburn is accepted for the saturation vapor pressure over ice. He gives

$$\ln \frac{e_s}{6.105} = 4{,}332 \left(\frac{1}{273} - \frac{1}{T} \right) + 2.31 \ln \frac{T}{273} \tag{62}$$

for the range $-40°C \leqslant t \leqslant 0°C$. T is the Kelvin temperature ($T = t + 273$). It is sometimes convenient to use the formula given by Tetens[20]

$$e_s = 6.11 \times 10^{\frac{at}{b+t}} \tag{63}$$

where $t = °C$ and the constants a and b have the following values:

1. Over water $a = 7.5$ $b = 237.3$
2. Over ice $a = 9.5$ $b = 265.5$

From the tabulated values of temperature, saturation vapor pressure, and saturation vapor density it is found that

$$\frac{e_s v}{T} = \text{const} = 0.461 \qquad \text{joules/gram °K} \tag{64}$$

where e_s is the pressure in centibars, v the specific volume in cubic meters per kilogram, and T the Kelvin temperature. It may be shown that the above ratio is also true for superheated vapor at pressures up to an atmosphere when the superheat is high enough. The obvious conclusion is that the vapor at the temperatures and pressures encountered in the atmosphere behaves like a perfect gas. This conclusion is valid so long as the vapor is not in thermodynamic equilibrium with a free surface of liquid water in a container. See Keenan and Keyes[12] for the properties of superheated vapor at high temperatures.

Enthalpies and entropies are tabulated in terms of joules per gram and of joules per gram degree Kelvin relative to the arbitrary base point of 200°K and 1,000 mb pressure. This base is chosen so that negative values are avoided for atmospheric temperatures and pressures. From the first law of thermodynamics, the enthalpy is in general given by the heat required to bring the system from the base state to the present state. Thus $dh = dq]_p$. The following notation is used:

h_i = enthalpy of ice at $t°C$ $t \leqslant 0$
h_{ig} = enthalpy of sublimation $t \leqslant 0$
$h_g = h_i + h_{ig}$ = enthalpy of saturated vapor $t \leqslant 0$
$h_{if0} = L_i$ = latent heat of fusion = enthalpy of fusion $t = 0$
h_f = enthalpy of liquid water $t \geqslant 0$
$h_{fg} = L$ = latent heat of vaporization = enthalpy of vaporization $t \geqslant 0$
$h_g = h_f + h_{fg}$ = enthalpy of saturated vapor $t \geqslant 0$
s_i = entropy of ice $t \leqslant 0$
s_{ig} = entropy of sublimation $t \leqslant 0$
$s_g = s_i + s_{ig}$ = entropy of saturated vapor $t \leqslant 0$
s_f = entropy of liquid water $t \geqslant 0$
s_{if0} = entropy of fusion $t = 0$
s_{fg} = entropy of vaporization $t \geqslant 0$
s_g = entropy of saturated vapor $t \geqslant 0$

The empirical equations for these quantities from Kiefer's tabulated data are as follows:

$$h_i = 134 + 1.835t \quad \text{joule/gram} \quad t \leqslant 0 \tag{65}$$

$$h_{ig} = 2{,}838(\pm 3) \quad \text{joule/gram} \quad t \leqslant 0 \tag{66}$$

$$h_g = 2{,}838 + 1.835t \quad \text{joule/gram} \quad t \leqslant 0 \tag{67}$$

$$L_i = 333 \quad \text{joule/gram} \quad t = 0 \tag{68}$$

$$h_f = 467 + 4.19t \quad \text{joule/gram} \quad t \geqslant 0 \tag{69}$$

$$h_{fg} = L = 2{,}502 - 2.38t \quad \text{joule/gram} \quad t \geqslant 0 \tag{70}$$

$$h_g = 2{,}969 + 1.81t \quad \text{joule/gram} \quad t \geqslant 0 \tag{71}$$

$$s_i = 0.57 - 1.835 \left[\ln \left(\frac{273}{T} \right) \right] \quad \text{joule/gram °K} \quad t \leqslant 0 \tag{72}$$

$$s_{ig} = \frac{2{,}838(\pm 3)}{T} \quad \text{joule/gram °K} \quad t \leqslant 0 \tag{73}$$

$$s_g = s_i + s_{ig} \quad t \leqslant 0 \tag{74}$$

$$s_{if0} = \frac{333}{273} = 1.22 \quad \text{joule/gram °K} \quad t = 0 \tag{75}$$

$$s_f = 1.79 + 4.19 \ln \frac{T}{273} \quad \text{joule/gram °K} \quad t \geqslant 0 \tag{76}$$

$$s_{fg} = \frac{2{,}502 - 2.38t}{t + 273} \quad \text{joule/gram °K} \quad t \geqslant 0 \tag{77}$$

$$s_g = s_f + s_{fg} \quad \text{joule/gram°K} \quad t \geqslant 0 \tag{78}$$

It should be noted that the above equations provide (as they must) for the enthalpy (and entropy) of saturated vapor to have the same value whether the liquid has vaporized at zero degrees or the ice has sublimated. A similar statement holds for the condensation or sublimation of vapor at zero degrees. Whenever accurate values of any entropy or enthalpy are desired for saturated conditions, one should refer to the tables in Sec. I. This is particularly true when it is desired to know the entropy or enthalpy change for a considerable range in temperature, for the specific heats concerned are not strictly constant even over the limited range of the table.

The internal consistency of the data is illustrated by a simple application of Clapeyron's equation. If we substitute the linear expression for h_{fg} into Eq. (60) and replace v_g by using Eq. (64), we find

$$\frac{de_s}{dT} = \frac{(3{,}152 - 2.38T)}{0.461T^2} e_s \tag{79}$$

whence on integrating

$$\ln \frac{e_s}{6.105} = 25 \frac{T - 273}{T} - 5.2 \ln \frac{T}{273} \tag{80}$$

which compares favorably with Eq. (61).

Direct observations of the value of the specific heats of water vapor in its super-heated state, for temperatures and pressures found in the atmosphere, are not available. Insofar as the vapor may be considered to be a perfect gas at and near to its saturation temperature and pressure, these observations are not necessary. The enthalpy of a perfect gas is a function of its temperature alone; therefore the enthalpy of superheated vapor is equal to the enthalpy of saturated vapor at the same temperature. The enthalpy of superheated vapor is then given by Eq. (71), and the specific heat at constant pressure is $c_p' = 1.81$ joules per gram °K. Goodman[17] quotes a value of 0.45 Btu per lb °F, which corresponds to 1.88 joules per gram °K. Haurwitz[21] quotes 0.466, corresponding to 1.95 joules per gram °K. Holmboe in a private communication quotes 1.911 joules per gram °K. C. H. Meyers (Bureau of Standards) in a private communication shows that c_p' varies from 1.85 to 1.88 joules per gram °K as the temperature and pressure vary between -60 and 60°C and 0.00 and 0.02 atmospheres. The divergence in these quoted values emphasizes the importance of going to the tables when accurate values of enthalpy or specific heat are required. For routine purposes, the variation of the specific heat may be neglected, and a mean value may be adopted for calculations. It will appear later that fairly large changes in the value of c_p' will occasion only small changes in the meteorological elements calculated. This is true because of the small amount (by weight) of water present in the atmosphere. In the following pages, a value of $c_p' = 1.81$ joules per gram °K is used in all numerical work.

PROPERTIES OF GASES

Substances that are in the vapor phase at a temperature in excess of the critical temperature are referred to as gases. At temperatures less than the critical temperature, the substance, while in the vapor phase, will be referred to simply as vapor. Thus the so-called "permanent" gases of the atmosphere (nitrogen, oxygen, argon, etc.) having critical temperatures well below any temperature in the atmosphere are correctly known as *gases*. The vapor of water, carbon dioxide, sulfur dioxide, etc., having critical temperatures above temperatures usually found in the atmosphere are correctly known as *vapors* for meteorological purposes. See Sec. I for critical-point data of various atmospheric gases.

Equation of State. More than 56 different equations of state have been proposed as functions relating the thermodynamic coordinates for actual gases. Of these the simplest, and for most purposes adequately correct, equation is the perfect gas equation of state ($pv/T = $ const). Data for the function pv/T are available in the International Critical Tables (Vol. 3) for various gases and wide ranges of temperatures. See also the more recent "Temperature, Its Measurement and Control in Science and Industry," ref. 15. Examination of the data shows that, for all the elements in the atmosphere in their vapor (or gaseous) phase, we may take $pv/T = $ const as a good approximation for all meteorological purposes (save, of course, research). The results of the data for pure substances may be summarized in the formula

$$pv = \frac{pV}{m} = \frac{BT}{m} = RT \tag{81}$$

where the symbols have the following meanings:

B = universal gas constant = 8.314 joules/mol °K

m = molecular weight of the substance

R = gas constant for the particular gas considered = B/m

V = volume of one molecular weight of the gas, m³

p = pressure, cb (1 cb = 10 mb)

v = specific volume, m^3/kg $\left(v = \dfrac{1}{\rho}\right)$

T = temperature on the Kelvin scale (*i.e.*, absolute scale)

Internal Energy. We have the experimental evidence of the Joule-Thomson experiment (see Zemansky[2]) that the internal energy of actual gases is dependent on temperature alone, to about the same order of accuracy that the perfect gas equation of state holds for actual gases. This experiment shows that, as a gas expands through a porous plug from a higher to a lower pressure adiabatically, the temperature of the gas remains the same. Since no organized work is done by such an expansion, it follows from an application of the first law of thermodynamics that the internal energy is a function of temperature alone. We write then for the internal energy of the gases and vapors of the atmosphere that

$$e = \text{internal energy} = e(T) \tag{82}$$

It may be remarked that, although we find the above conclusion to hold with sufficient accuracy for most meteorological purposes, for certain industrial processes the conclusion is invalid. In the Linde process for the liquefaction of air, for example, the final stages are possible because of the cooling effect of such an adiabatic expansion. *This fact is an excellent example of the necessity we have to examine all approximations (whether mathematical or physical) with the utmost care.*

We may on the other hand deduce the result that the internal energy of a perfect gas is a function of temperature alone when we admit that the temperature T in the gas equation is the thermodynamic temperature, for we have by the first and second laws that $de = T\,ds - p\,dv$, whence

$$\left(\frac{\partial e}{\partial v}\right)_T = T\left(\frac{\partial s}{\partial v}\right)_T - p = T\left(\frac{\partial p}{\partial T}\right)_v - p$$

where the right-hand side follows from Maxwell's relations in Eq. (40). On differentiating the equation of state, we find $\left(\dfrac{\partial p}{\partial T}\right)_v = \dfrac{R}{v}$ whence

$$\left(\frac{\partial e}{\partial v}\right)_T = \frac{RT}{v} - p = p - p = 0 \tag{83}$$

or the internal energy is indeed a function of the temperature alone. The above is a reversal of the argument used in the article on thermometry; there it was postulated that the internal energy was a function of the temperature alone and, on that basis, it was shown that the temperature appearing in the perfect gas equation of state was necessarily the thermodynamic temperature.

Enthalpy. Since we have $pv = RT$ and $e = e(T)$, then the enthalpy, which is $e + pv$ by definition, is equal to $e + RT$, *i.e.*, the enthalpy of a perfect gas is also a function of the temperature alone.

Difference of the Specific Heats. If we substitute the equation of state ($pv = RT$) into Eq. (45) for the difference of the specific heats, we find that

$$c_p - c_v = R \tag{84}$$

or the difference is a positive constant. Although c_p and c_v are not necessarily constant independently, they are functions of the temperature alone, as shown by Eq. (49).

Thermodynamic Properties. From the observable quantities of pressure, temperature, and specific heats, the other thermodynamic properties may be calculated. A base point, or datum zero, is adopted of 200°K and 1,000 mb pressure so that enthalpy, internal energy, and entropy will always have positive values in meteorological calculations. The base point is otherwise arbitrary, since all that is here desired is to know the changes in the properties between states. The formulas follow:

1. Specific volume (or its reciprocal, density) is calculated from the equation of state after p and T are measured

$$v = \frac{RT}{p} \qquad m^3/\text{kg} \tag{85}$$

2. Internal energy is given by the integration of Eq. (28) with c_v a constant by assumption (or of course a mean value)

$$e = c_v(T - 200) \qquad \text{joules/gram} \tag{86}$$

3. Enthalpy is given by integration of Eq. (34) with c_p a constant by assumption (or a mean value)

$$h = c_p(T - 200) \qquad \text{joules/gram} \tag{87}$$

4. Entropy is given by integration of Eq. (30) with both c_p and c_v as constants (or mean values)

$$s = c_p \ln \frac{T}{200} - R \ln \frac{p}{1,000} \qquad \text{joules/gram °K} \tag{88}$$

where the equation of state has been introduced to perform the integration. p is in millibars.

State Changes of Gases. The only changes that can occur in a gas will be state changes, *i.e.*, there can be no changes of phase, no matter how the thermodynamic properties change, because a gas is a substance in the vapor phase at a higher temperature than the critical temperature. The system then remains homogeneous in one phase for all changes in the coordinates. The equation of state is a function of three variables, any two of which are independent. It is useful to study the possible variations in two of the coordinates when one is constrained to remain constant.

A general way of applying the desired constraint is known as the *polytropic relation*. This was introduced by Emden in his researches on the atmospheres of the stars (see Milne[22] and Eddington[23]). It is assumed that the amount of heat dq that is added to a system is given by

$$dq = \xi \, dT \tag{89}$$

where ξ is a parameter. Physically ξ is simply a specific heat. If Eq. (89) is introduced into the differential expression for the first law of thermodynamics and integrated with c_p, c_v, and ξ as constants, using the equation of state $pv = RT$, we find

$$\xi \, dT = c_p \, dT - RT \frac{dp}{p} = T \, ds \tag{90}$$

whence
$$T = \text{const } p^{\frac{c_p - c_v}{c_p - \xi}} \tag{91}$$

Equation (91) is known as the *generalized Poisson equation*. Different state changes take place along what are called *process lines* (or isopleths) according to the value of the parameter ξ. The defined processes with which we are usually concerned are set out in Table 1.

TABLE 1

Value of ξ	Heat	Process requires	Name of process
$\xi = c_p$	$dq = c_p\,dT$	$p = $ const	Isobaric
$\xi = c_v$	$dq = c_v\,dT$	$v = $ const	Isosteric
$\xi = 0$	$dq = 0$ (Rev)	$s = $ const	Isentropic
$\xi = \pm\,\infty$	$(dT = 0)$	$\left\{\begin{array}{l} e = \text{const} \\ h = \text{const} \\ T = \text{const} \end{array}\right\}$	Isothermal

The *isentropic* process is of great interest in meteorology. This is sometimes referred to as the reversible adiabatic, or somewhat loosely as the adiabatic. With $\xi = 0$, Eq. (91) reduces to Poisson's equation

$$T = \text{const } p^{\frac{R}{c_p}} \tag{92}$$

where $R = c_p - c_v$. Equation (92) defines a family of process lines according to the value of the constant used in calculation. Each line (or *isentrope*) represents state changes possible with a constant value of *entropy*. The lines are also called *adiabatics*, (or *adiabats*) since the constraint calls for $dq = 0$. The constraint of constant temperature defines a family of lines known as *isotherms*.

Potential Temperature. Isopleths of constant entropy for dry air are almost universally known in meteorology as lines of *constant potential temperature*, or as dry adiabatics. The potential temperature θ is defined by the equation

$$\theta = T\left(\frac{1,000}{p}\right)^{\frac{R}{c_p}} \tag{93}$$

Physically we see from Poisson's equation and Eq. (93) that the potential temperatur is simply the temperature that a mass of dry air would attain if it were brought from its initial state to a pressure of 1,000 mb by an isentropic state change. The relation between entropy and potential temperature is easily found by differentiating Eq. (93) logarithmically and substituting into $ds = c_p\,dT/T - R\,dp/p$. We find $ds = c_p\,d\theta/\theta$ or, on integrating

$$s = c_p \ln\,\theta + \text{const} \tag{94}$$

Although the potential temperature is much more used in meteorology than is entropy both in explanation and in thermodynamic diagrams, yet the entropy is an invaluable concept in the development of the subject and in its understanding. In particular, entropy is useful in discussing changes of phase of water in the atmosphere. The relation in Eq. (94) may be used to change from potential temperature to entropy, or vice versa.

Mixtures of Gases. A mixture is a system whose component parts are completely homogeneous and in which no chemical reactions have occurred or are occurring. Thus dry air is a mixture of several gases. Humid air is a mixture of dry air and water vapor. The fundamental law concerning the mixtures of gases is the Gibbs-Dalton law, which states: (1) *the total pressure exerted by a mixture of perfect gases is equal to the sum of the partial pressures of each of the components when each component occupies* *by itself the volume of the mixture at the temperature of the mixture, and* (2) *the internal energy and the entropy of the mixture are equal to the sum of the internal energies and entropies of the component parts of the mixture when each component occupies by itself the volume of the mixture at the temperature of the mixture.*

The Gibbs-Dalton law is applied to determine the thermodynamic coordinates of a mixture of M_1 grams of component 1 with M_2 grams of component 2. Let m_1 and m_2 be the molecular weights of 1 and 2, respectively. By part 1 of the law

$$p = p_1 + p_2 \tag{95}$$

where p is the total pressure of the mixture of the two gases at temperature T contained in a volume V. The total mass of the mixture is $M_1 + M_2$. From the equation of state for perfect gases, we have

$$p = BT \left(\frac{\rho_1}{m_1} + \frac{\rho_2}{m_2} \right) \tag{96}$$

where ρ_1 and ρ_2 are the densities of the component gases. But $\rho_1 = M_1/V$ and $\rho_2 = M_2/V$. Also the density of the mixture is $\rho = (M_1 + M_2)/V$. Then

$$\rho_1 = \frac{M_1 \rho}{M_1 + M_2} \qquad \rho_2 = \frac{M_2 \rho}{M_1 + M_2} \tag{97}$$

and on combining Eqs. (96) and (97)

$$p = BT\rho \left[\frac{(M_1/m_1) + (M_2/m_2)}{M_1 + M_2} \right] \tag{98}$$

It follows from Eq. (98) that the mixture of perfect gases behaves mathematically like another perfect gas whose molecular weight and gas constant are given by

$$m = \frac{M_1 + M_2}{(M_1/m_1) + (M_2/m_2)} \tag{99}$$

$$R = \frac{M_1 R_1 + M_2 R_2}{M_1 + M_2} \tag{100}$$

where $R_1 = B/m_1$ and $R_2 = B/m_2$. B is the universal gas constant.

The utility of Eq. (98) is obvious. For any mixture of perfect gases, we have to deal with the perfect gas equation in the form $pv = RT$, where the gas constant R appears as a parameter of the system. R is given by Eq. (100) to be a mean value of the gas constants of the component gases. The above argument for a two-component mixture may be repeated similarly for a mixture of any number of components. It is thus easily shown that, for a mixture of n perfect gases

$$
\begin{aligned}
pv &= RT \\
R &= \frac{\Sigma M_i R_i}{\Sigma M_i} \\
v &= \frac{1}{\rho} \\
p &= \Sigma p_i \\
\rho &= \Sigma \rho_i
\end{aligned}
\tag{101}
$$

The internal energy and the entropy of the mixture are given by part 2 of the Gibbs-Dalton law to be

$$e = \frac{\Sigma M_i e_i}{\Sigma M_i} = \frac{\Sigma M_i c_{vi}(T - 200)}{\Sigma M_i} \tag{102}$$

$$s = \frac{\Sigma M_i s_i}{\Sigma M_i} \tag{103}$$

Following a simple transformation from Eqs. (101) and (102), the enthalpy of the mixture is found to be

$$h = e + pv = \frac{\Sigma M_i h_i}{\Sigma M_i} = \frac{\Sigma M_i c_{pi}(T - 200)}{\Sigma M_i} \tag{104}$$

The specific heats of the mixture per unit mass of mixture are given by

$$c_v = \frac{de}{dT} = \frac{\Sigma M_i c_{vi}}{\Sigma M_i} \tag{102'}$$

$$c_p = \frac{dh}{dT} = \frac{\Sigma M_i c_{pi}}{\Sigma M_i} \tag{104'}$$

from Eqs. (102) and (104).

State Changes of Mixtures of Gases.   A mixture of gases will be such that each component gas is at the temperature of the mixture.   Thus any change in temperature of the gas, so long as thermodynamic equilibrium is maintained, will produce the same change in temperature of each component.   This simple conclusion is not valid for the other properties.

Let p_i be the partial pressure of component i.   Then

$$p_i = R_i T \rho_i = \frac{R_i M_i T}{V} \tag{105}$$

where V is the total volume of the mixture.   Let M_i remain constant.   By differentiation of Eq. (105), we see that

$$\frac{dp_i}{p_i} = \frac{dT}{T} - \frac{dV}{V} = \frac{dp}{p} \tag{106}$$

where the right-hand side of Eq. (106) is obtained by differentiation of the equation of state for the mixture $(pv = RT)$ on the assumption that $R = $ const, *i.e.*, all the M_i are assumed to remain constant.   *The conclusion is reached in Eq. (106) that the percentage increase of pressure of each component and of the mixture is the same, provided that there is no change in the relative magnitudes of the components present (i.e., none has condensed out, for example).*

It is useful to investigate the change in entropy of the component gases of a mixture when the mixture itself is constrained to isentropic state changes.   Consider a binary mixture of M_1 grams of gas 1 and M_2 grams of gas 2.   Let R and R' and c_p and c_p', respectively, be the gas constants and the specific heats at constant pressure of the two gases.   For brevity, write $x = M_2/M_1$.   The entropy of the mixture is

$$M_1 s_1 + M_2 s_2 = \text{const} \tag{107}$$
or
$$s_1 + x s_2 = \text{const} \tag{108}$$

where s_1 and s_2 are the specific entropies of the components 1 and 2.   On differentiating while holding x constant, we have

$$ds_1 + x\, ds_2 = 0 \tag{109}$$

or, in terms of pressures and temperature

$$\frac{c_p\, dT}{T} - \frac{R\, dp_1}{p_1} + x\left(\frac{c_p'\, dT}{T} - \frac{R'\, dp_2}{p_2}\right) = 0 \tag{110}$$

By rearrangement and noting that the relative increments of the partial pressures are equal and equal to the relative increment of the total pressure, by Eq. (106) we have

$$\frac{dT}{T} = \frac{R + xR'}{c_p + c_{p'}} \frac{dp_1}{p_1} \tag{111}$$

The change in entropy of component 1 is given by the left-hand member of Eq. (110), whence, substituting dT/T from Eq. (111), we find on rearrangement

$$ds_1 = x \left[\frac{(R'/c_{p'}) - (R/c_p)}{(1/c_{p'}) + (x/c_p)} \right] \frac{dp_1}{p_1} \tag{112}$$

It is concluded that the entropy of component 1 changes during an isentropic state change for the mixture unless $R'/c_{p'} = R/c_p$. By Eq. (109), it follows that the entropy of component 2 will also change.

THERMODYNAMIC PROPERTIES OF THE EARTH'S ATMOSPHERE

Composition of the Atmosphere. Table 2 gives the composition of the earth's atmosphere. Above 20 km, the composition may be somewhat changed. Above 100 km approximately the lighter elements must predominate (Fleming,[25] and Chapman and Milne[26]). For the purpose of a thermodynamic investigation of the atmosphere for meteorological use, the distribution given in Table 2 may be assumed to hold at all elevations.

TABLE 2*

	Mass, %	Molecular weight
Nitrogen	75.51	28.016
Oxygen	23.15	32.000
Argon	1.28	39.944
Carbon dioxide	Variable	44.020
Water vapor	Variable	18.016
Liquid water	Variable	18.016
Solid water (ice or snow)	Variable	18.016
Other gases (neon, helium, radon, etc.)	Trace	
Impurities (dust, salt, smoke, etc.)	Trace	

* After Paneth.[24]

Vertical Structure of the Atmosphere. The atmosphere is a gaseous envelope that extends radially outward from the earth's surface for many kilometers. *The fundamental assumption is made that the substance of the atmosphere may be considered to be a thermodynamic system.* It is a matter of observation that the lower layers of the atmosphere are generally not isothermal. On the contrary, there is generally a gradient of temperature or lapse rate directed vertically upward. This temperature gradient will produce heat flow from lower to higher layers by conduction. There is also a nonzero flux of heat transfer by radiation across the layers of the atmosphere. The heat flow from both conduction and radiation is so small, however, that to a first approximation it is neglected and the atmosphere is assumed to be in thermodynamic equilibrium, *i.e.*, it is assumed that each element of mass is controlled in its thermodynamic coordinates by an equation of state. To an excellent approximation, this equation of state is given by $pv/T = $ const.

The lower layers of the atmosphere are in modified convective equilibrium, the modification being largely due to radiation. This means, in other words, that the major amount of the heat transfer in the lower layers is done by radiation. Convective (and/or advective) currents in the air carry warmer or colder masses of air from place to place and thus transfer or transport the energy of the mass. This transport is not called heat transfer in the language of thermodynamics. Thus, when a parcel of air is moved quickly and without turbulence from a region of high pressure to a region of low pressure (as in a vertical ascent), the heat transfer to the parcel is assumed to be zero. This is so because the gradients of temperature in the atmosphere are too small to transfer appreciable amounts of heat in the short time required for such a transport of air. The above transport is assumed to be isentropic. It follows from Poisson's equation that ascending air will be cooled and descending air warmed. The final result of such motions is that the temperature decreases with elevation above the earth. The temperature gradient cannot be found from Poisson's equation directly. One must take into account the effect of advective currents of air all at different potential temperatures (or entropies). When the vertical motions are sufficiently intense (as in a layer that is thoroughly mixed or kept stirred over a rough terrain), the gradient of temperature will closely approximate the gradient as determined from Poisson's equation (see pages 348, 373).

Above the tropopause, the atmosphere is in conductive equilibrium. Here the mass of air is relatively free from vertical movements. Radiation also plays a minor role. As equilibrium is established, we should expect isothermal conditions to prevail. That such a region does actually exist is well shown by observations taken from the records of radiosondes. At extreme altitudes, where the atmosphere is so rarefied that conduction no longer plays an important role, radiation must become the predominating influence. At these levels, the atmosphere is said to be in radiative equilibrium (see Milne[22]).

The lower layers of the atmosphere are also the "reservoir" for varying amounts of water vapor and liquid or solid water. Water is evaporated from the oceans and the earth's vegetation and carried aloft by motions of the air. When it ascends sufficiently far (provided that the amount is not changed by mixing), the water vapor must finally condense into liquid water or sublimate into snow. It is in these lower layers that all the weather occurs. But, although the atmosphere above the tropopause cannot hold enough water (because of the low temperatures present) to be a region of weather, there is ample evidence that the weather of the lower layers is not unaffected by what occurs at higher levels. This is so because the pressures existing at high levels must play a role in the movements of the air at lower levels and thus in the weather resulting from those movements (see Sec. X).

Terminology. The following terms are used in the analysis:

1. *Dry air* is a mixture of nitrogen, oxygen, argon, and a trace of carbon dioxide in the proportions given in Table 2.
2. *Humid or moist air* is dry air plus water vapor in amount less than saturation.
3. *Saturated air* is dry air plus water vapor at saturation vapor pressure.
4. *Wet air (or iced air)* is dry air plus saturated vapor and liquid water (or ice or snow).
5. *Supersaturated air* is dry air plus water vapor at greater than saturation vapor pressure.
6. *Supercooled air* is dry air plus saturated vapor and supercooled liquid water.

Dry, humid, saturated, and wet air are in stable thermodynamic equilibrium. Supersaturated and supercooled air are in metastable thermodynamic equilibrium. A suitable quantity and type of impurities present will prevent the occurrence of the metastable states. The impurities present do not affect the thermodynamic analysis; but they must be considered in the role of catalysts when the mechanism of condensation is studied. The other gases of the atmosphere (trace of neon, argon, etc.) are present in such small quantity that they do not affect the thermodynamic analysis. The single important parameter of the atmosphere, considered as a thermodynamic system, is the amount of water vapor present. Its composition in other respects may be considered constant for present purposes.

Dry Air. According to Table 2 and to Eq. (101), the gas constant for dry air is

$$R = \frac{M_N R_N + M_O R_O + M_A R_A}{M_N + M_O + M_A} = 0.287 \qquad \text{joule/gram °C} \tag{113}$$

This result comes from the observational fact that dry air is a mixture of several gases. We note that, since R/c_p is almost the same for nitrogen and for oxygen (about 0.286), a state change that is isentropic for the dry air will also be isentropic for the nitrogen and the oxygen separately by Eq. (112). The specific heat at constant pressure of dry air is 1.003 joules per gram °K. The equation for dry adiabatics (potential temperature lines) is then

$$\theta = T \left(\frac{1,000}{p} \right)^{0.286} \tag{114}$$

from Eq. (93). Any of the thermodynamic properties of dry air may be found from the appropriate equations, which follow from the equation of state

$$pv = 0.287T \tag{115}$$

Humid Air. We recall that for atmospheric temperatures and pressures we have $ev/T = 0.461$ for water vapor. This is the equation of state of the water vapor based on experimental results. It may be noted that the gas constant for water vapor from this result (*i.e.*, 0.461 joule per gram °K) is nearly equal to the ratio of the universal gas constant to the molecular weight of water, *i.e.*, $8.314/18.016 = 0.462$. It is then strongly indicated that the water vapor may be assumed to behave like a perfect gas. This assumption is made subject to the provision that the vapor is not in equilibrium with a flat surface of liquid water.

We then assume that humid air is a mixture of dry air plus water vapor in amount less than that required to saturate the space at the temperature of the space. Since both the vapor and the dry air behave like perfect gases, the equations developed for mixtures of perfect gases will apply to the humid air. We have then

$$pv = \frac{M_a R_a + M_v R_v}{M_a + M_v} T \tag{116}$$

where p, v, and T are the pressure, specific volume, and temperature of the humid air. M_a and M_v are the masses of dry air and water vapor, respectively. R_a and R_v are the gas constants for dry air and for water vapor, respectively.

A variety of different (but dependent) parameters have been introduced to designate the quantity of water vapor present. Some of these are defined below:

1. *Mixing ratio w* is defined by the relation

$$w = \frac{M_v}{M_a} \tag{117}$$

2. *Specific humidity q* is defined by

$$q = \frac{M_v}{M_v + M_a} \tag{118}$$

3. *Relative humidity f* is defined by

$$f = \frac{e}{e_s} \tag{119}$$

where e is the partial pressure of the vapor in the humid air and e_s is the saturation vapor pressure corresponding to the temperature of the mixture.

The dependence of the parameters upon each other is easily shown. From the above equations, we have

$$q = \frac{w}{1 + w} < w \tag{120}$$

The vapor pressure and total pressure in a space of volume V at temperature T are given by

$$e = \rho_v R_v T = \frac{M_v B T}{V m_v}$$
$$p = \rho R T = \frac{M_v + M_a}{V m} B T \tag{121}$$

respectively, where the symbols have their usual meanings. Take the ratio of these pressures and introduce the definition of w to find

$$w = \frac{(m_v/m_a)e}{p - e} = \frac{\epsilon e}{p - e} \tag{122}$$

or numerically, since $m_v/m_a = 0.622$ where m_v and m_a are the molecular weights of the water vapor and dry air, respectively, we have

$$w = \frac{0.622e}{p - e} \tag{123}$$

for the relation between the mixing ratio and the partial pressure of the water vapor. Note that the total pressure p enters into the equation. Equation (123) may be written in terms of relative humidity, pressure, and temperature (implicitly) by substituting Eq. (119). We find

$$w = \frac{\epsilon f e_s}{p - f e_s} \tag{124}$$

and recall that e_s is a function of temperature alone. The value of the *saturation mixing ratio* is defined to be the maximum value w may have at a specified temperature and pressure. This is given by $f = 100$ per cent to be

$$w_s = \frac{\epsilon e_s}{p - e_s} \tag{125}$$

Another useful parameter for designating the water vapor present in humid air is the *virtual temperature*. As we have seen, the equation of state for the mixture of dry air and water vapor is $pv = RT$ where the gas constant for the mixture is given by

$$R = \frac{M_a R_a + M_v R_v}{M_a + M_v} = \frac{R_a + w R_v}{1 + w}$$

The equation may also be written $pv = R_a T_v$, where T_v is a number which when multiplied by the gas constant for dry air gives the magnitude RT of the actual humid air. Then from the above we have

$$T_v = \frac{R}{R_a} T \tag{126}$$

Equation (126) is taken as the definition of the virtual temperature. It may be noted that T_v is not an actual temperature in that it could be measured by a thermometer. But T_v does have the dimensions of temperature, and it does give a measure of the humidity of the air.   Using the above equations we find that, in terms of the mixing ratio w

$$T_v = T \left[\frac{1 + w(R_v/R_a)}{1 + w} \right] \tag{127}$$

Since $R_v/R_a = m_a/m_v = 28.97/18.02 = 1.61$, we have numerically

$$T_v = T \left[\frac{1 + 1.61w}{1 + w} \right] \doteq T(1 + 0.61w) \tag{128}$$

where the approximation on the right-hand side is good whenever squared powers and higher of w may be neglected in comparison with unity.   Since the mixing ratio is known from the properties of water vapor to be less than 0.03 in almost all cases, the approximation indicated is valid for all except the most accurate calculations.   The difference between virtual temperature and temperature is always positive.   This difference will be greater for greater values of relative humidity (or mixing ratio). The difference is greatest when $f = 100$ per cent or when $w = w_s$.   Since w_s depends on temperature and pressure alone, the difference between temperature and virtual temperature at saturation depends upon temperature and pressure alone.   We have

$$(T_v - T)_s = 0.61 w_s T \tag{129}$$

for the difference at saturation.   For values of the relative humidity other than 100 per cent, we have approximately

$$T_v - T = (T_v - T)_s f \tag{130}$$

where f is the relative humidity.   This approximation is deduced from the approximate form of Eq. (123)

$$w = \frac{\epsilon e}{p - e} \doteq \frac{0.622e}{p - e} \doteq \frac{0.622 f e_s}{p} \doteq f w_s \tag{131}$$

since $e << p$.

It follows from the equation of state that *humid air is lighter than dry air*, i.e., given two samples of air at pressure p and temperature T, one humid the other dry, the ratio of their densities is less than one.   The pressure in either gas is given by

$$p = \rho R T = \rho_a R_a T$$

whence, substituting in the value for the gas constant R in terms of R_a and the mixing ratio of the humid air, we have

$$\rho = \frac{\rho_a R_a}{R} = \frac{\rho_a (1 + w)}{1 + 1.61w} < \rho_a \tag{132}$$

The specific heats of humid air are found from Eqs. (102′) and (104′) to be

$$c_p = \frac{M_a c_{pa} + M_v c_{pv}}{M_a + M_v} \tag{133}$$

$$c_v = \frac{M_a c_{va} + M_v c_{vv}}{M_a + M_v} \tag{134}$$

where c_{pa} and c_{pv} are the specific heats at constant pressure of dry air and of water vapor while c_{va} and c_{vv} are the specific heats of dry air and of water vapor at constant volume.   In terms of the mixing ratio, the above are

$$c_p = \frac{c_{pa} + wc_{pv}}{1 + w}$$
$$c_v = \frac{c_{va} + wc_{vv}}{1 + w} \qquad (135)$$

Equations (133) to (135) give the specific heats per unit mass of humid air. The specific heats per unit mass of dry air are defined similarly save that now only M_a appears in the denominators. In this case

$$c_p = c_{pa} + wc_{pv}$$
$$c_v = c_{va} + wc_{vv} \qquad (136)$$

The specific heats at constant pressure are the most useful of the above equations. Numerically we have, using $c_{pa} = 1.003$ joules per gram $°K$ and $c_{pv} = 1.81$ joules per gram $°K$

$$c_p = 1.003 + 0.80w \qquad \text{joules/gram } °K \text{ (per unit mass humid air)}$$
$$c_p = 1.003 + 1.80w \qquad \text{joules/gram } °K \text{ (per unit mass dry air)}$$

to a very good approximation.

State Changes in Dry or Humid Air. Given a mass of humid air ($f < 100$ per cent) we have, from the preceding equations

1. Isobaric state changes (constant pressure)

$$dq'\Big]_p = dh = c_p \, dT$$

2. Isosteric state changes (constant density)

$$dq'\Big]_v = de = c_v \, dT$$

3. Isentropic state changes (entropy, or potential temperature, constant)

$$ds = \frac{c_p \, dT}{T} - \frac{R \, dp}{p} = 0$$

or

$$\frac{T_2}{T_1} = \left(\frac{p_2}{p_1}\right)^{\left(\frac{R_a + wR_v}{c_{pa} + wc_{pv}}\right)} \qquad (137)$$

when the mixing ratio w is constant.

Neglecting squared and higher powers of w, we find numerically from Eq. (137) that

$$\frac{T_2}{T_1} = \left(\frac{p_2}{p_1}\right)^{0.286(1-0.2w)} \qquad (138)$$

In meteorological applications, a further approximation is generally introduced by neglecting the mixing-ratio term in the exponent of Eq. (138) altogether. In other words, it is assumed that sufficient accuracy is obtained by writing

$$\left(\frac{R}{c_p}\right)_{\text{humid air}} = \left(\frac{R}{c_p}\right)_{\text{dry air}} \qquad (139)$$

The error introduced by the approximation indicated in Eq. (139) is small. Its upper limit may be estimated as follows: Using the generally accepted value of $R_a/c_{pa} = 0.288$ (instead of the value 0.286 which better fits the magnitudes 0.287 for R_a and 1.003 for c_{pa}) and $R_v/c_{pv} = 0.461/1.81 = 0.255$ for water vapor, we find from Eq. (112) that

$$ds_a = -\frac{0.06w\,dp}{p} \qquad \text{joules/gram °K} \tag{140}$$

Assuming that w could have the very high mean value of 10 grams per kg for an expansion from 1,000 to 400 mb, we have, on integrating, that $|\Delta s| < 0.0005$ for the error in entropy. The error in potential temperature is about 0.2°C since $c_p\,\Delta\theta = \theta\,\Delta s$ by Eq. (94). It may then be safely assumed for calculations to meteorological accuracy that a state change that is isentropic for the mixture of dry air and water vapor with relative humidity less than 100 per cent will also be isentropic for the dry air component. The state change must, of course, be also isentropic for the water-vapor component.

Saturated and Wet Air. A mass of humid air may be brought to the saturation point in a variety of ways. We recall that saturated air is dry air plus water vapor at saturation vapor pressure for the temperature of the air. Saturation may be attained by (1) cooling the humid mixture down to the temperature for which the actual value of the mixing ratio is equal to the saturation mixing ratio, or (2) increasing the value of the mixing ratio until it attains the saturation value for the initial temperature. A combination of such simple processes may also occur. Several paths by which saturation may be reached are so important in nature that the saturation temperature attained is given a special name (*e.g.*, dew-point temperature). See pages 381, 390 for a discussion of meteorological temperatures.

Equations (116) to (140) inclusive may be applied to all states of humid air. After the air reaches saturation, the thermodynamic system is no longer a homogeneous mixture of gases for all state changes. If the air is warmed, or if some of the vapor is removed, then the air is again humid and the previous equations apply. But if the air is cooled, or if additional water is put into the space, then either the air is supersaturated or it is wet. Whether or not the air takes on the metastable state of supersaturation depends upon the condensation nuclei present. If correct numbers and types of hygroscopic nuclei are present, the vapor may condense at even less than 100 per cent relative humidity. If neither hygroscopic nor condensation nuclei are present, supersaturation is the rule. The question of the mechanism of condensation is not one for thermodynamics to answer (see pages 254, 257).

What may be termed the normal state of affairs is that, as cooling of a mass of humid air continues past the saturation temperature, either one of two things occurs, (1) the vapor condenses and remains in the space, or (2) the vapor condenses and is precipitated from the space. In nature the most likely occurrence is a combination of these two simple events. In the absence of specific information on the relative proportions of liquid water remaining and precipitating, one must determine the state and phase changes for the two possible events noted above. These are referred to, respectively, as *reversible adiabatic* and *pseudoadiabatic* changes. The important thing to notice is that in either case the air at saturation is in such a state that any further cooling will introduce a nonhomogeneous aspect, *i.e.*, both water vapor and liquid water will be present, even though the latter may be precipitated immediately after condensation. If at any later point the direction of the heat flow is changed, the pseudoadiabatic process is immediately homogeneous with only water vapor present in the dry air, since the liquid water has been precipitated. But the reversible adiabatic is homogeneous only after the liquid water present has been vaporized; up to this point the two phases of vapor and liquid water are present and the system is nonhomogeneous.

If saturation is reached only after the temperature has been lowered below 0°C, the normal situation is that water vapor sublimates directly into snow. If a nonhomogeneous system of liquid water and water vapor is cooled to 0°C, the system will either undergo another phase change with the water freezing into ice at 0°C, or only a state change will occur with supercooled air resulting. The latter result is not uncommon in the atmosphere, with the supercooling sometimes persisting well below 0°C.

The value of the mixing ratio that is computed and used in practice for temperatures below 0°C in the free atmosphere is generally taken from the saturation vapor pressure over water, even down to temperatures at which ice or snow must predominate over liquid water. In discussing this question, Goodman[17] states:*

Because there are two possible values of vapor pressure, there are theoretically two values for the specific humidity of saturated air at all temperatures below 32°F. One value of specific humidity corresponds to the vapor pressure over subcooled water and the other to the vapor pressure over ice. Theoretically, therefore, the specific humidity of saturated air at temperatures below 32°F would depend on whether the air was in contact with subcooled water or with ice. Actually, however, the specific humidity of saturated air below 32°F has been found to correspond to the vapor pressure over subcooled water whether the air is actually in contact with subcooled water or with ice. All of the weight of the experimental evidence, as quoted by Keyes and Smith[36] and by Ewell,[37,38] indicates that the moisture content of cold air in equilibrium with ice corresponds to the vapor pressure over subcooled water and not to the vapor pressure over ice. As Keyes and Smith[36] state, 'the water content of air in equilibrium with ice does not correspond to what would be expected on the basis of Dalton's law from the known vapor pressures of ice, but rather to the pressure of subcooled water.'

State and Phase Changes in Saturated Air. I. *Reversible Adiabatic (or Isentropic) Changes.* Consider a mass of saturated air at pressure p, temperature T, and mixing ratio $w_1 = \xi$. It is desired to determine the function relating temperature and pressure when the mass is cooled isentropically   The following symbols are used:

T = temperature
p = total pressure
e_s = partial pressure of water vapor
s_a = entropy of unit mass dry air
s_f = entropy of unit mass liquid water
s_{fg} = entropy of vaporization of water
s_g = entropy of unit mass of water vapor
M_a = mass of dry air present
M_g = mass of water vapor present
M_f = mass of liquid water present
c = specific heat of liquid water
c_{pa} = specific heat of dry air
R_a = gas constant of dry air
w = mixing ratio $= \dfrac{M_g}{M_a}$
y = "mixing ratio" for liquid water $= \dfrac{M_f}{M_a}$
$w + y = \xi$ = constant
L = latent heat of vaporization

Two assumptions are made, (1) that the mass of dry air cools isentropically, and (2) that the liquid water remains in the space considered.

The entropy of the system is given by

$$M_a s_a + M_g s_g + M_f s_f = \text{const} \tag{141}$$

* Reprinted with permission.

or in terms of w and y by

$$s_a + ws_g + ys_f = \text{const} \tag{142}$$

On differentiation we have

$$ds_a + w(ds_f + ds_{fg}) + y\,ds_f + (s_f + s_{fg})dw + s_f\,dy = 0 \tag{143}$$

or, on recalling that $s_g = s_f + s_{fg}$ and $w + y = \text{const}$, we have

$$ds_a + d(ws_{fg}) + (w + y)ds_f = 0 \tag{144}$$

Equation (144) may be integrated at once to give

$$s_a \Big]_1^2 + \left(\frac{wL}{T}\right)_2 - \left(\frac{wL}{T}\right)_1 + c\xi \ln\left(\frac{T_2}{T_1}\right) = 0 \tag{145}$$

If the small term involving ξ is neglected, we arrive at the approximation used by Rossby

$$c_{pa} \ln T - R_a \ln (p - e_s) + \frac{wL}{T} = \text{const} \tag{146}$$

since the entropy change of the dry air component is given by

$$s_a \Big]_1^2 = c_{pa} \ln\left(\frac{T_2}{T_1}\right) - R_a \ln \frac{p_2 - e_{s2}}{p_1 - e_{s1}} \tag{147}$$

It will be noted that Eqs. (145) and (146) give temperature as a function of pressure (or conversely) since e_s, w, ξ, and L are known as functions of T. The air remains saturated always, and w is therefore equal to w_s. To calculate the isentrope, one takes a starting point wherever desired and computes the constant in Eq. (146). Then, on assuming some other temperature, the corresponding pressure on the isentrope may be calculated. The wet isentropes (or adiabatics) determined by the above equations are also called the pseudo-wet-bulb potential temperature lines (see pages 387, 388).

II. *Pseudoadiabatic Changes.* If it is assumed that the water vapor condenses in the space under consideration (thus giving up heat to the dry air component and increasing its entropy) but is then immediately precipitated out, the analysis requires a small change. The preceding equations hold down to Eq. (144). As the liquid water does not remain in the space to add its entropy, we have in lieu of Eq. (145)

$$s_a \Big]_1^2 + \left(\frac{wL}{T}\right)_2 - \left(\frac{wL}{T}\right)_1 + c\int_1^2 w\frac{dT}{T} = 0 \tag{148}$$

Since $w \leqslant \xi$, we may again neglect the last term. For greater accuracy one may take mean values of the saturation mixing ratio at the two temperatures for which the computation is being made and then, by graphical methods, find the changes in pressure corresponding to particular temperatures. It will be noted that, to the approximation indicated in Eq. (146), the pseudoadiabatic and reversible adiabatic isopleths are identical lines.

III. *Hail Stage.* It is of some interest to set down the equation for the hail stage. It is assumed that isentropic cooling has proceeded from some initial point down to 0°C and that freezing occurs at that temperature. Add the following symbols to those used previously:

M_i = mass of ice present
s_i = entropy of unit mass of ice at 0°C

s_{if0} = entropy of fusion per unit mass at 0°C

$$z = \frac{M_i}{M_a} = \text{``mixing ratio'' for the solid (ice)}$$

$w + y + z = w_1$ = const = mixing ratio when the air is saturated and $y = z = 0$

w_2 = mixing ratio when the temperature reaches 0°C and $z = 0$

w_3 = mixing ratio when all the liquid water present has frozen ($y = 0$)

It is assumed that freezing occurs isentropically (for the entire mixture) and isothermally. The total entropy of the system is

$$s_a + w s_g + y s_f + z s_i$$

per unit mass of dry air. The entropies s_g, s_f, and s_i are each constant since the temperature remains at 0°C throughout the process. We recall that $s_g = s_i + s_{ig} = s_f + s_{fg}$ at 0°C. Also $s_f - s_i = s_{if0}$. On differentiating the above expression, we have after a simple reduction

$$ds_a + s_{if0}\,dy + s_{ig}\,dw = 0 \tag{149}$$

Equation (149) may be integrated between state 3 where the freezing is complete and state 2 where the freezing has commenced. We find

$$s_a\Big]_2^3 + s_{if0}(y_3 - y_2) + s_{ig}(w_3 - w_2) = 0$$

But $y_3 = 0$ and $y_2 = (w_1 - w_2)$. Also $s_a\Big]_2^3 = -R_a \ln (p - e)_3/(p - e)_2$. This gives finally

$$R_a \ln \frac{(p - e)_2}{(p - e)_3} = s_{if0}(w_1 - w_2) + s_{ig}(w_2 - w_3)$$

or in slightly different form

$$R_a \ln \frac{(p - e)_2}{(p - e)_3} = s_{if0}(w_1 - w_3) + s_{fg}(w_2 - w_3) \tag{150}$$

Since $w_1 > w_2$ and $e_2 = e_3 = 6.105$ mb, Eq. (150) may be written numerically

$$0.287 \ln \frac{p_2 - 6.1}{p_3 - 6.1} = 1.22(w_1 - w_3) + 9.16 \left(\frac{6.1}{p_2 - 6.1} - \frac{6.1}{p_3 - 6.1} \right) 0.622 \tag{151}$$

On selecting any initial point, it is found after calculation that the pressure p_3 after the isothermal, isentropic freezing is less than the pressure when the freezing commenced. In the atmosphere, this implies an isothermal layer of finite thickness in which the entrained water freezes as it ascends. Note that $p_2 - p_3$ is small in comparison with either p_2 or p_3 in any actual case. On expanding the term containing the logarithm in Eq. (151), it is found approximately that

$$p_2 - p_3 = \frac{s_{if0}p_3(w_1 - w_3)}{R_a} \tag{152}$$

Consider a numerical example. Maritime tropical air with an initial mixing ratio of 20 grams per kg ascends to the 0°C isotherm where it freezes at about 500 mb. The thickness of the isothermal layer will be approximately

$$\Delta p = \frac{(20 - 8) \times 500 \times 1.22}{1,000 \times 0.287} \doteq 25 \text{ mb} \tag{153}$$

It is noted that, even if all the liquid water is carried aloft, the isothermal layer is relatively thin. If some of the water precipitates out (as it undoubtedly will in nature), the isothermal layer will be appreciably thinner.

THERMODYNAMIC DIAGRAMS AND CALCULATIONS

Accuracy of Observations. The thermodynamic properties of the atmosphere and the parameter indicating the water-vapor content cannot be measured in the upper air as accurately as one might wish. In general, pressure and temperature are transmitted in synoptic messages to millibars and whole degrees centigrade, respectively. Mixing ratios are transmitted to tenths of grams per kilogram of dry air, but the last figure is generally in doubt. In making routine calculations from such reports, it is correct to introduce mathematical approximations that simplify the arithmetic so long as the error introduced by the approximation is less than the error of observation. To illustrate: 1. The mixing ratio is accurately given by $0.622e/(p - e)$ where p is the total pressure as read by a barometer and e is the partial pressure of the water vapor present. Since $e << p$ it is sufficiently accurate for most purposes to write

$$w = \frac{0.622e}{p} \tag{154}$$

2. The gas constant is given by Eq. (116) for humid air. In terms of the mixing ratio

$$R = R_a \frac{1 + (R_v/R_a)w}{1 + w} = R_a \frac{1 + 1.61w}{1 + w}$$

Since w is small in comparison with unity, it is sufficiently accurate for most purposes to expand the denominator by the binomial theorem and neglect squared and higher powers of w to obtain

$$R = R_a(1 + 0.61w) \tag{155}$$

Thermodynamic Data and Their Use. Thermodynamic data are collected and used in one or more of several ways.

I. *Tables.* Section I contains a selection of tables giving data useful in meteorology. When one wishes to find a related physical quantity as accurately as the observations will allow, one should refer to the appropriate table. The saturation vapor pressure over water, for example, is given in Table 68 for various temperatures as accurately as it is known by experiment. When a value for the pressure is desired that corresponds to a temperature falling between tabulated values, one interpolates between tabulated values. Tables are sometimes prepared in full so that interpolation is not required. See, for example, Table 87. This table of V. Bjerknes gives distances between standard isobaric surfaces at various mean virtual temperatures. Interpolation is neither necessary nor desirable in general, for the accuracy of observations is such that interpolation is not warranted.

II. *Charts and Graphs.* Observed data may be plotted on cross-section paper, and a smooth curve may be drawn connecting the observed points. If the scale and grid of the paper are well chosen, the graph may be as accurate as is desired. The graphical method is nearly always the most convenient way of recording data. For extended lists of observations, the graph is likely to become unwieldy if the scale is large enough to give much accuracy. Mistakes due to the human factor are also possible in reading the graph. In spite of some disadvantages, however, graphical analysis offers so much speed and convenience that it is widely used in all sciences.

III. *Equations.* Frequently a desired physical quantity may be easily expressed in terms of the elementary functions. Thus we have simple algebraic equations for mixing ratio and the gas constant in terms of pressures and w, respectively. It is sometimes desirable to substitute into these equations and solve for accurate numerical values of the unknowns. In other cases, the analytical expressions involve more complicated functions. The equations frequently do not admit of an explicit solution for one variable in terms of the known quantities. The numerical results must then be obtained by a process of "trial and error" or by a graphical analysis.

IV. *Thermodynamic Diagrams.* A thermodynamic diagram is a chart or graph on which isopleths of various thermodynamic coordinates are plotted against other thermodynamic coordinates, or functions of those coordinates. To restrict the definition even further, it is sometimes stated that the thermodynamic diagram must be one on which areas represent energy. Several of the various diagrams in general use will be discussed below. Any one of them is found to be of great value in routine meteorological work.

Vertical Structure of the Atmosphere and Origin of Thermodynamic Diagrams. The fundamental relation used in the development of all thermodynamic diagrams is the equation of mechanical equilibrium. In differential form, we have, from Eq. (11)

$$u \, du + g \, dz + v \, dp = 0 \tag{156}$$

When the unit mass of atmosphere to which this equation applies is stationary, we have the well-known equation of hydrostatic equilibrium

$$g \, dz + v \, dp = 0 \tag{157}$$

or

$$\frac{dp}{dz} = -g\rho \tag{158}$$

where ρ is the density of the atmosphere and the axis of z is vertically up.

Equation (157) may be integrated to quadrature immediately. We have

$$(z_2 - z_1)g = -\int_1^2 v \, dp \tag{159}$$

for the linear distance between the points 1 and 2 in the atmosphere. Introducing the equation of state for perfect gases, this becomes

$$z_2 - z_1 = -\frac{1}{g} \int_1^2 RT \frac{dp}{p} \tag{160}$$

It follows that, if the temperature and gas constant (which depends on the mixing ratio) are known as functions of pressure, one may integrate and thus determine the linear distance between isobaric surfaces. Temperature and pressure are chosen as fundamental variables because they are directly measurable, whereas the specific volume cannot be measured directly in the free atmosphere. It is a matter of personal choice what functions of the temperature and pressure are selected for coordinates of the diagram. It is sometimes helpful to choose functions that will make areas represent energy. In the pseudoadiabatic diagram, the functions are chosen to make the dry adiabatics straight lines.

Geopotential. The geopotential is defined to be the potential energy of unit mass in the earth's gravitational field. We take the acceleration due to gravity to be numerically 980 and effectively constant for meteorological thermodynamics. The unit of geopotential is the dynamic meter, which is defined to be 10^5 erg of energy. Thus in the earth's gravitational field the geopotential of a gram of mass at an elevation of z m is

$$\Phi = 980 \times 100z \quad \text{ergs/gram} \tag{161}$$

where the elevation z is measured above sea level, *i.e.*, the potential energy is taken zero at sea level.   In dynamic meters (gm), the geopotential is

$$= 0.98z \qquad \text{gm} \tag{162}$$

when z is in meters.   The dynamic meter is dimensionally energy per unit mass; but, since the acceleration of gravity is assumed constant, this quantity may be converted into linear distance above sea level in meters by division by **0.98**.

Pressure-volume Diagram.   From Eqs. (159) and (162), it is clear that, if the specific volume were plotted as a function of pressure between two points 1 and 2 in the atmosphere, then the geopotential between the two points would be represented by the area indicated on the right-hand side of Eq. (159).   Given a sounding in the atmosphere showing the specific volume as a function of pressure, one could then planimeter the area indicated in Fig. 15a and find the linear distance between any two points along the sounding.   The *pv*- (or Clapeyron) diagram is little used in meteorology because the specific volume is not directly measurable.

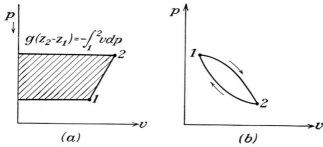

Fig. 15.—Areas as energy and *pv*-diagram.

Closed Areas as Energies on Thermodynamic Diagrams.   Consider the representation of any cyclic process on the *pv*-diagram.   Let some particular cycle be shown by the path 1-2-1 in a clockwise direction in Fig. 15*b*.   The total work done by the thermodynamic substance through the cycle is given by the line integral; work = $\oint p\, dv$.   From the first law of thermodynamics, we have $dq = de + p\, dv = dh - v\, dp$. From this we have on integrating around a cycle $\oint dq = \oint p\, dv = -\oint v\, dp$ since de and dh are perfect differentials.   But either of the integrals on the right-hand side of this equation gives the area enclosed by the curve that represents the cyclic process on the *pv*-diagram.   Therefore the area is equal to (1) the net amount of heat transfer during the cycle and also (2) the net work done on or by the substance.   This conclusion is equivalent to the usual expression that "areas on the thermodynamic diagram represent energies."   The scale of the energy per unit of area depends only upon the scales to which pressure and volume are laid down.

Equal-area Transformations.   The fact that energies transferred in a cyclic process and linear distances between points representing different states in the atmosphere can be determined by the measurement of an area on the *pv*-diagram suggests that one investigate equal-area transformations, *i.e.*, determine what other coordinates can be used on graphs so that an area of A cm², say, which lies inside some particular closed curve on the original diagram will remain exactly A cm² when the bounding curve is plotted on the new diagram.   The area may change its shape entirely, but the restriction is that the magnitude of the area remain the same.   The problem is essentially one of mapping.   We think of an area A on the *pv*-diagram being transferred to an *xy*-diagram by imagining the bounding curve transferred point by point. The coordinates x and y are so far unspecified functions of p and v. It may be shown that the areas are transformed equally when

$$\frac{\partial x}{\partial p}\frac{\partial y}{\partial v} - \frac{\partial x}{\partial v}\frac{\partial y}{\partial p} = 1 \qquad (163)$$

Equation (163) may be solved in general. It is sufficient for our purposes to note here that, if the x-coordinate, for example, is specified as a function of p and v, then the y-coordinate must satisfy Eq. (163) in order to give an equal area transformation. If we select RT as the abscissa of a thermodynamic diagram and want areas to represent energies, then from Eq. (163), and with $pv = RT$, we find that the ordinate must be a function satisfying

$$v\frac{\partial y}{\partial v} - p\frac{\partial y}{\partial p} = 1$$

This equation is satisfied if $y = -\ln p$. Other functions will also satisfy the equation. Any diagram constructed with any of these functions as an ordinate scale will therefore have areas representing energies just as does the pv-diagram.

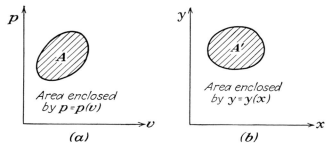

Fig. 16.—Transformation of coordinates. $x = x(p,v)$; $y = y(p,v)$.

The coordinates used on a number of thermodynamic diagrams are listed in Table 3 with indication of whether or not areas represent energies. The area-preserv-

TABLE 3

Chart name	Ordinate	Abscissa	Areas are preserving
Pseudoadiabatic (or Stüve)................	$-p^{R/c_p}$	RT	No
Emagram..............................	$-\ln p$	RT	Yes
Refsdal...............................	$-RT \ln p$	$\ln RT$	Yes
Temperature-entropy (or Tephigram)......	$-T$	$c_p \ln \theta$ (or s)	Yes

ing quality of each of the last three diagrams named above is easily shown by substituting into Eq. (163). Of all the diagrams listed, the adiabatic (or pseudoadiabatic) is most widely used in this country. The Refsdal is used widely abroad and has been chosen to illustrate the book of Petterssen.[33] The Emagram, which is similar to the pseudoadiabatic diagram, is at present gaining a following. The temperature-entropy or $T\theta$-diagram published by Kiefer has many advantages and is also becoming more widely used. The $T\theta$-diagram is a modification of the older Tephigram of Shaw.

Pseudoadiabatic Diagram. The pseudoadiabatic diagram will be used for illustrative purposes in this book. The construction of the chart is given in detail below.

Much that is said with respect to its construction will apply equally well to the construction of any other thermodynamic diagram, *i.e.*, all the diagrams are obtained from some transformation of coordinates on the pv-diagram.

To lay out an adiabatic chart, first draw a basic grid of lines with temperature measured along the horizontal axis, increasing to the right. The scale is linear and runs from about -40 to $40°C$. Isobars are laid off by labeling the base line as 1,050 mb and drawing in straight lines parallel to the base line at distances from other isobars given by the difference between the respective pressures to the R/c_p power. The scheme is illustrated in Fig. 17. It may be shown that the linear distance between isobars of equal isobaric spacing increases with decrease in pressure.

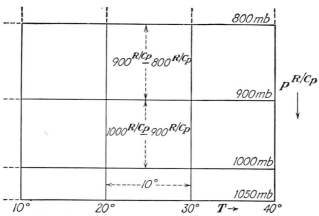

FIG. 17.—Construction of pseudoadiabatic diagram.

If the coordinates of the pseudoadiabatic diagram are substituted into the left-hand side of Eq. (163), we find

$$\frac{\partial x}{\partial p}\frac{\partial y}{\partial v} - \frac{\partial x}{\partial v}\frac{\partial y}{\partial p} = \frac{R}{c_p}\,p^{\frac{R}{c_p}} \tag{164}$$

But this term is the Jacobian of the transformation from pv to the new coordinates. For a small area at pressure p on the pv-diagram then, the transformation gives another area on the pseudoadiabatic chart. The ratio of the areas is the right-hand side of Eq. (164). Since the area on the pv-diagram represents energy, then the energy represented by an area on the pseudoadiabatic diagram is equal to that area divided by the right-hand side of Eq. (164). We have then

$$\text{Energy} = \text{area on pseudoadiabatic diagram} \times K \tag{165}$$

where the factor K is

$$K = \frac{c_p}{R}\cdot\frac{1}{p^{\frac{R}{c_p}}} = 0.937\,\frac{\theta}{T}$$

Values of K are tabulated below. We note that the factor is unity at about 800 mb, greater than unity for $p < 800$ and less than unity for $p > 800$. Thus the areas on the pseudoadiabatic chart may be taken to represent energies at 800 mb; the areas underestimate the energy for $p < 800$, and the areas overestimate the energy for $p > 800$.

TABLE 4

Pressure, mb	Factor K
1,000	0.94
900	0.97
800	1.00
700	1.04
600	1.08
500	1.14
400	1.22

It is important to note that the diagram may be constructed so that areas will represent energies at any one particular pressure desired, by the insertion of an appropriate factor. It is not possible to make any one pseudoadiabatic diagram be

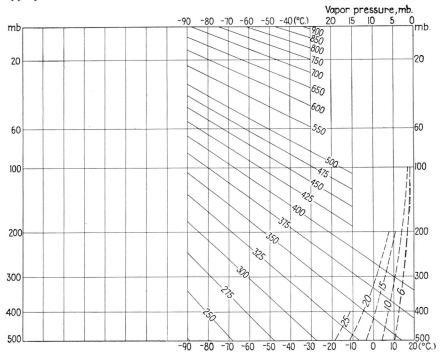

FIG. 18a.—Pseudoadiabatic diagram for upper levels.

area preserving at all pressures. The area factor is independent of the temperature (since θ/T is a function of pressure alone).

Isopleths of entropy (or potential-temperature lines) for dry or humid air are plotted from Poisson's equation

$$T = p^{\frac{R}{c_p}} \times \text{const}$$

These lines are correct for humid air to the approximation noted in Eq. (139). Since the ordinate of the diagram is $p^{\frac{R}{c_p}}$, the potential-temperature isopleths are straight lines. The slope of each line is different and is given by $(100)^{\frac{R}{c_p}}/\theta$ in terms of the potential temperature θ. If the diagram extended to zero degrees and zero pressure, the lines would spread out radially from the upper left-hand corner of the paper (as drawn).

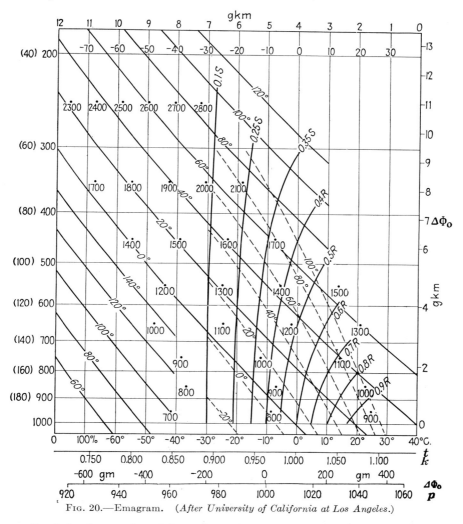

Fig. 20.—Emagram. (*After University of California at Los Angeles.*)

Isopleths of saturation mixing ratio are drawn using the equation

$$w_s = \frac{0.622e_s}{p - e_s} \tag{125}'$$

where p is the total pressure (the pressure shown on the diagram), and e_s is the saturation vapor pressure. Since e_s is a function of temperature alone, the above equation determines pressure as a function of temperature for any value of the parameter w_s. The saturation-mixing-ratio lines are not straight but curve to the region of lesser temperature as pressure is decreased.

Adiabatics for saturated air are computed from Eq. (148). The computation is made for the saturation vapor pressure over water even down to $-40°C$. These isopleths are also curved. Physically it is clear that the temperature cannot fall off so rapidly when liquid water is condensed from a mass being cooled adiabatically as it does from a mass that remains humid. The wet adiabatics then rise more steeply than do the dry adiabatics when viewed from the bottom of the chart. The temperature fall with decrease in pressure is smaller numerically along the wet adiabatic, or it has a smaller lapse rate.

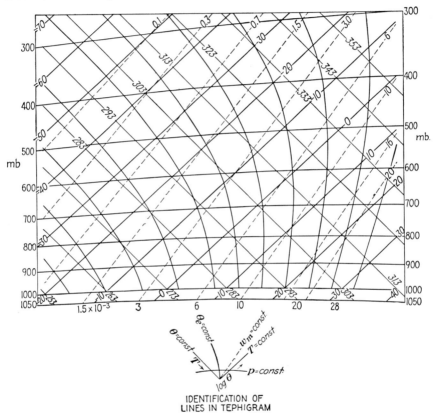

IDENTIFICATION OF
LINES IN TEPHIGRAM

Fig. 21.—Tephigram.   (*After U.C.L.A.*)

It is sometimes useful to have the virtual-temperature difference shown on the thermodynamic diagram. We recall that $T_v - T$ at saturation is a function of mixing ratio and temperature alone. But the mixing ratio at saturation is a function of temperature and pressure; then $T_v - T$ at saturation is a function of temperature and pressure. Any one of several schemes may be used to find $T_v - T$. It is not customary to construct isopleths. One form of the adiabatic chart has small vertical hatches on the standard isobars, where the horizontal spread of the hatch marks is equal to $T_v - T$. The Emagram shows the difference by numbers ranged along the isobars. On some forms of the Emagram, numbers ranging along the isobars (every 100 mb) show the virtual-temperature correction for thickness at saturation in dynamic meters directly. All such schemes can show the difference only at saturation. The actual value is taken to be the saturated difference times the relative humidity [Eq. (130)]. Simple nomograms may also be constructed to show the virtual-temperature difference. Since the values of w_s are shown on the pseudoadiabatic diagram, it is a simple mental calculation to get a good approximation to $T_v - T$ by using the equation $T_v - T = 0.61w_sfT$.

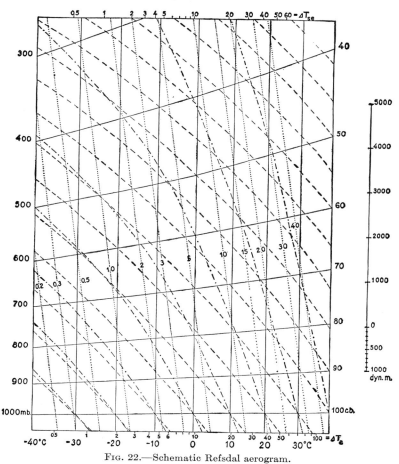

Fig. 22.—Schematic Refsdal aerogram.

Other Thermodynamic Diagrams. The most generally used thermodynamic diagrams are shown in skeleton form in Figs. 18 to 22. Since the basic grid of coordinates is different in each case, the various isopleths appear with different slopes and curvatures. Straight lines on one chart will be curved on another, etc. Areas on all the charts except the adiabatic chart are equal save for the possibility of a constant factor. Choice among the diagrams is personal when all are equally available. No one contains any information that cannot be made a part of any other. Some operations may be essentially simpler (or more clear) on one of the diagrams. When such operations have to be performed frequently, one naturally elects to use the diagram best suited to his purposes. For illustrations, we shall use either the pseudoadiabatic diagram or the $T\theta$-diagram of Kiefer. The latter is more convenient when operations involving entropy are derived and used.

Soundings above 400 *Mb.* It is the current practice of the U.S. Weather Bureau to put the results of radiosonde flights up to 400 mb on the teletype sequence some hours in advance of the results above that level. This has the advantage of getting the information early. The practice should not lead, however, to a neglect of the results at higher levels.

The standard pseudoadiabatic diagram (Fig. 18) runs to only 400 mb. Results above 400 mb are plotted on the supplementary diagram shown in Fig. 18a. The other thermodynamic diagrams in general use (Tephigram, Emagram, etc.) carry isobars above 400 mb. In any case, one may start over again at the bottom isobar

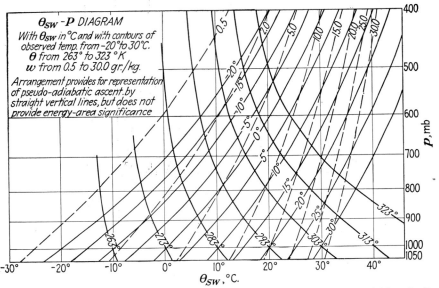

FIG. 22a.—A thermodynamic diagram with straight-line wet adiabats. (*After J. R. Barkley and R. Palmer.*)

by relabeling the various isopleths. This may be done on any chart whose distance between isobars is proportional to the logarithm of the ratio of the appropriate pressures; for, ln $^{10}\!\!/_5$ = ln $^4\!\!/_2$ = ln $^6\!\!/_3$, etc. The potential-temperature lines (dry adiabatics) and isotherms must also be relabeled (see Fig. 20).

Thermodynamic Diagrams in Meteorology. The thermodynamic diagram is the fundamental working tool of the meteorologist for all investigations of the atmosphere above the very surface layers. Radiosondes or aerographs are sent aloft, and from their records the meteorologist picks off the values of pressure, temperature, and humidity at all significant levels. These levels are plotted on the thermodynamic diagram by entering with pressure and temperature. The parameter used to indicate humidity is plotted alongside the point to give a more nearly complete picture of the state of the air. The points so located are connected by straight lines, and the resulting series of lines (or the *sounding*) gives a graphical picture over the station of the atmosphere at the time of the flight. The sounding as plotted on the thermodynamic diagram determines the following useful information:

1. Air mass or air masses over the station (see pages 609, 676).
2. Stability of the air (see pages 402, 693).
3. Fronts (see pages 647–649).
4. Subsidence inversions (see pages 734, 758).
5. Elevations of significant points and standard isobaric surfaces (see pages 373–379).

One must distinguish clearly between isopleths on the thermodynamic diagram and the sounding that is representative of the atmosphere at some particular time and place. Isopleths are the curves resulting from imposed mathematical constraints. The sounding is a series of lines connecting thermodynamic coordinates representative of the atmosphere at a definite time and place. Under certain conditions, the sounding may be coincident with some isopleth. The usual situation, however, is that the sounding will fall somewhere between isotherms and isentropes; *i.e.*, the physical processes in the atmosphere are not such as to make the entire layer of air assume a state of either conductive (whence isothermal) or adiabatic (whence isentropic) equilibrium. *Any particular sounding will show evidence of many simultaneous processes in the atmosphere.*

PRESSURE, TEMPERATURE, AND ALTITUDE IN THE ATMOSPHERE

It is important for many practical purposes to be able to find the altitude at which certain strata of air may exist in the atmosphere, *e.g.*, inversions, isothermal layers, moist layers, etc. For any distribution of temperature, pressure, and water vapor whatever in the atmosphere, the equation of mechanical equilibrium in a static condition is given by Eq. (158). The axis of z is vertically up; g is the acceleration of gravity; ρ is the density of the air. Substituting in the equation of state for a perfect gas, we find from Eq. (158)

$$\frac{dp}{dz} = -\frac{gp}{RT} \tag{166}$$

where R is the gas constant at the point considered. R is a function of the water-vapor content. If the temperature and water-vapor content are known as functions of pressure, then Eq. (166) may be integrated and the linear distance between any two pressures may be determined. The problem does not admit of a unique solution in general, because the temperature and water-vapor content of the air do not follow any known law. To a first approximation, the temperature may be assumed to decrease linearly with altitude. This simplified model is treated below. The general case must be solved with the aid of a thermodynamic diagram after temperatures and mixing ratios are determined by indirect observation (radiosonde or aerograph).

Lapse Rate of Temperature. The lapse rate of temperature in the atmosphere is defined by the equation

$$\gamma = -\frac{dT}{dz} \tag{167}$$

where γ is the lapse rate. Strictly, one should write $\gamma = -\partial T/\partial z$; but the notation of Eq. (167) is preferred here to emphasize the fact that the atmosphere is assumed quiescent enough so that (approximately) temperature is a function of elevation alone. In words, *lapse rate* is the space rate of decrease of temperature upward through the atmosphere at some particular time. A careful distinction should be made between lapse rate and the rate of decrease of temperature in an individual parcel as it ascends, even though the two may be *numerically* the same. On combining Eqs. (136) and (167), we have

$$\gamma = \frac{gp}{RT}\frac{dT}{dp} \qquad (168)$$

Thus the lapse rate is determined whenever temperature and R are known as functions of pressure. To a first approximation, R may be taken to be constant and equal to the gas constant for dry air ($R_a = 0.287$ joule/gram °K).

Isothermal Atmosphere. Suppose the atmosphere is at constant temperature T_0 between two pressures p_1 and p_2. This situation frequently exists for thin strata. Then from the preceding equations

$$\gamma = 0$$
$$z_2 - z_1 = \frac{RT_0}{g} \times \ln\frac{p_1}{p_2} \qquad (169)$$
$$p_2 = p_1 e^{-\frac{g}{RT_0}(z_2 - z_1)}$$

Equations (169) give either the linear distance between two pressures or the difference in pressure between two points at elevations of z_1 and z_2, respectively. Numerically, when T is in degrees Kelvin and z in meters, we have

$$z_2 - z_1 = 29.3 T_0 \ln\frac{p_1}{p_2} \qquad \text{m}$$
$$z_2 - z_1 = 67.4 T_0 \log_{10}\frac{p_1}{p_2} \qquad \text{m} \qquad (170)$$
$$z_2 - z_1 = 221.1 T_0 \log_{10}\frac{p_1}{p_2} \qquad \text{ft}$$

where as a first approximation the gas constant for dry air is used.

Homogeneous Atmosphere. It is sometimes useful to consider an atmosphere in which the density is constant with elevation. Such an atmosphere is said to be *homogeneous.* Since the density ρ_0 is constant in the equation of hydrostatic equilibrium, one may integrate immediately to obtain

$$p = p_0 - g\rho_0 z$$

where p_0 is the surface pressure, ρ_0 is the constant density, and g is the mean value of the acceleration of gravity. z-axis is up. Assuming that the material of the atmosphere is a perfect gas, one obtains by substitution

$$T = T_0 - \frac{gz}{R}$$

where T_0 is the surface temperature. The lapse rate through the homogeneous atmosphere is then given by

$$-\frac{dT}{dz} = \gamma = \frac{g}{R} = 3.42 \qquad \text{°C}/100 \text{ m}$$

on writing $g = 980$ cm per sec². This lapse rate is known as the *autoconvective lapse rate.*

Unlike the actual atmosphere, the homogeneous atmosphere has a definite thickness. The top of the atmosphere occurs where $p = 0$. Then from the above

$$H = \frac{p_0}{g\rho_0} = \frac{RT_0}{g}$$

where H is the thickness. Note that H will vary with latitude owing to changes in both g (average between surface and top of atmosphere) and T_0. See tables in Sec. I.

Adiabatic Equilibrium. When the atmosphere is in isentropic (or adiabatic) equilibrium

$$ds = \frac{dq}{T} = c_p \frac{dT}{T} - R \frac{dp}{p} = 0 \qquad (171)$$

and on combining this with Eq. (168), which is true for any gaseous atmosphere

$$\gamma_d = -\frac{dT}{dz} = \frac{g}{c_p} \qquad (172)$$

Numerically this becomes, for the earth's atmosphere

$$\begin{aligned} \gamma_d &= 0.98°C/100 \text{ m} = 1°C/100 \text{ gm} \\ \gamma_d &= 5.4°F/1{,}000 \text{ ft} \end{aligned} \qquad (173)$$

Since c_p varies only a little with the mixing ratio w, then γ_d is also a good approximation for humid air. The value γ_d is referred to as the *dry-adiabatic lapse rate*. The linear distance between two points of temperature T_1 and T_2 in an atmosphere that has a lapse rate equal to γ_d will be

$$\begin{aligned} z_2 - z_1 &= 102(T_1 - T_2) \qquad \text{m} \\ z_2 - z_1 &= 340(T_1 - T_2) \qquad \text{ft} \end{aligned} \qquad (174)$$

when the temperatures are measured on the Kelvin or centigrade scale.

International Standard Atmosphere. The I.C.A.N. (International Commission for Air Navigation) atmosphere according to which some altimeters are calibrated is defined as follows:

1. The air is dry, and its chemical composition is the same at all altitudes.
2. The value of gravity is uniform and equal to 980.62 cm per sec per sec.
3. The temperature and pressure at M.S.L. (mean sea level) are 15°C and 1,013.2 mb.
4. At any altitude z (meters) measured above M.S.L. and between 0 and 11,000 m the temperature of the air is equal to $t = 15 - 0.0065z$ °C.
5. For altitudes above 11,000 m, the temperature of the air is constant and equal to $-56.5°C$.

U.S. Standard Atmosphere. The U.S. Standard atmosphere is similar to the I.C.A.N. atmosphere. In the former

Gravity $= 980.665$
M.S.L. pressure $= 760$ mm $= 1{,}013.25$ mb
M.S.L. temperature $= 15°C = 288°K$
$\gamma =$ lapse rate $= 0.65°C$ per 100 m up to 10,769 m $= 35{,}332$ ft (*i.e.*, up to 234 mb)
$\gamma =$ lapse rate $= 0.00$ at and above tropopause
Tropopause and stratosphere temperature $= -55°C = 218°K$

Pressure and Elevation in Standard Atmosphere. In an atmosphere with a lapse rate of $\beta =$ const, and a surface temperature of T_0, the relation between pressure and elevation is found by substituting the expression for temperature into Eq. (166) and integrating. We find immediately when $\beta \neq 0$

$$p = p_0 \left(1 - \frac{\beta z}{T_0}\right)^{\frac{g}{R\beta}} \qquad (175)$$

where z is the elevation above the point where temperature and pressure are T_0 and p_0, respectively. Numerically, for the I.C.A.N. atmosphere

$$p = 1{,}013.2 \left(1 - \frac{0.0065z}{288}\right)^{5.2568} \tag{176}$$

or, on solving for altitude in terms of pressure

$$z = 44{,}308 \left[1 - \left(\frac{p}{1{,}013.2}\right)^{0.19023}\right] \tag{177}$$

where z is meters and p is in millibars. The constants appearing above will have slightly different values when computed for the U.S. Standard atmosphere.

Sea-level Pressure from Aircraft in Flight. Equation (175) may be inverted and used to determine the sea-level pressure below an aircraft in flight at a known elevation. The method depends upon the actual elevation having been measured accurately by some device other than an aneroid barometer (a barometric altimeter). Let the pressure, temperature, and elevation be p, T, and z. If the lapse rate in the atmosphere below the aircraft is constant and equal to β, the pressure at sea level is

$$p_0 = p \left(1 + \frac{\beta z}{T}\right)^{\frac{1}{0.292\beta}} \tag{178}$$

In the absence of better information, one may assume the standard atmosphere to exist below the aircraft. In practice, it will be found that a better value of the lapse rate may be estimated from previous flights over similar conditions.

Fortunately for the utility of the above method, the change in p_0 is relatively small with lapse rate. Then if one's estimate of the lapse rate is considerably off the true value, Eq. (178) may still be used to find the sea-level pressure to good approximation. To show this, differentiate Eq. (178).

$$\frac{dp_0}{p_0} = \frac{d\beta}{0.292\beta^2}\left[\frac{x}{1+x} - \ln(1+x)\right] \tag{179}$$

where for simplicity $x = \beta z/100T$. On expanding the term inside the bracket and making a rearrangement, it is found that

$$-\frac{dp_0}{p_0} = \frac{z^2 d\beta}{5{,}840T^2}\left(1 - \frac{4}{3}\frac{\beta z}{100T} + \frac{6}{4}\frac{\beta^2 z^2}{10^4 T^2} - \cdots\right) \tag{180}$$

where β is in degrees Centigrade per 100 m and z is in meters.

Example. Suppose the observations made at the aircraft show that $T = 15°C$ and $z = 1{,}000$ m. Suppose the best estimate of β is $0.85°C$ per 100 m. On substitution into the above, it is found that $dp_0/p_0 = -0.002d\beta$. Thus the lapse rate may be off as much as 0.20 °C per 100 m, and still the error in surface pressure is only 0.40 mb (since the order of magnitude of surface pressure is 1,000 mb always).

Correction to Altimeter for Surface Temperature. The altimeter is calibrated to read true elevation only when the temperature at the surface is 15°C. If the actual surface temperature is not equal to 15°C, a correction must be applied to the reading of the altimeter. In general, this correction will be greater the greater the indicated altitude and the greater the difference between surface temperature and 15°C. The correction is determined as follows: Let z be the true elevation and z' the indicated elevation. When the altimeter indicates z', the actual pressure is

$$p = 1{,}013.2\left(1 - \frac{0.0065z'}{288}\right)^{5.256} \tag{181}$$

but when the lapse rate is constant and equal to 0.65° per 100 m, the pressure at z m (true) is

$$p = 1{,}013.2 \left(1 - \frac{0.0065z}{T_0}\right)^{5.256} \tag{182}$$

where T_0 is the surface temperature. The pressure given by Eqs. (181) and (182) is of course the same. Equating the two expressions, we find after a simple rearrangement

$$z = z'\left(1 + \frac{\Delta T}{288}\right) \tag{183}$$

where ΔT is the difference between the surface temperature and 15°C. Thus $\Delta T = T_0 - 288$ or $\Delta T = t_0 - 15$. *When the surface temperature is more than 15°, the altimeter will read too low, and the correction is to be added. When the surface temperature is lower than 15°, the altimeter will read too high, and the correction is to be subtracted.*

If the lapse rate in the atmosphere is constant but not equal to 0.65, then Eq. (182) will contain the actual lapse rate β. Expanding the right-hand sides of Eqs. (181) and (175) by the binomial theorem, it is easy to show that Eq. (183) still gives the correction for the elevation to a first approximation. The approximation made is to neglect squared and higher powers of the quantity $\beta z/T_0$, which appears in the right-hand side of Eq. (182) when β is inserted. The approximation is good up to 5,000 m.

If the aircraft is flying above a frontal or subsidence inversion, the above simple correction may be in error.

Correction to Altimeter for Surface Pressure. The indicated altitude of any altimeter will, of course, be in considerable error unless the instrument has been adjusted to the correct "altimeter setting." The value of the altimeter setting will vary with the synoptic situation. If the altimeter is to be trusted, the pilot must obtain his correct setting either by radio or by estimate from the weather map. The latter choice is taken only when circumstances prevent the use of radio. **See the instructions supplied with the altimeter used.** See also Sec. I and Blackburn's "Basic Air Navigation." *

Navigational Computer. Dalton[35] describes a navigational computer that is now in wide use. This device provides a quick means by which an aircraft pilot may determine his correct altitude from the altitude indicated by the barometric altimeter. Entry on the computer is made with temperature at the aircraft and pressure altitude. Pressure altitude is the reading of the altimeter when it was set to read zero at the standard M.S.L. pressure. If the ground temperature is also known, a better correction is made by entry with mean-temperature and mean-pressure altitude.

Dalton arrives at his formula by assuming a constant lapse rate (not necessarily standard) and expanding the altimeter equation. He obtains the result

$$\frac{H}{Z} = \frac{\tfrac{1}{2}(T_1 + T_0)}{288 - \beta_s[(Z_1 + Z_0)/2]}$$

where H = "true thickness of layer" (*i.e.*, true altitude)

Z = indicated altitude = $Z_1 - Z_0$

Z_1 = pressure altitude at top of layer

Z_0 = pressure altitude at bottom of layer

T_0 = actual temperature at bottom of layer, °K

T_1 = actual temperature at top of layer, °K

β_s = standard lapse rate (0.65°C per 100 m)

By considering the various errors that may enter (surface temperature off 15°C, nonstandard lapse rate, and nonstandard surface pressure), Dalton shows that the

* McGraw-Hill, New York, 1944.

above formula gives an error of less than 0.11 per cent even for extreme departures from the standard atmosphere. *These conclusions are valid so long as the actual lapse rate is constant. The formula takes no account of frontal or other inversions. Even with inversions present, the error is generally not large.*

Kollsman Scales. See Sec. I for a reproduction of Kollsman scales used to change indicated altitude to true altitude.

Lapse Rate in Saturated Atmosphere. When the atmosphere is saturated, the lapse rate is not constant, even when it is in wet-adiabatic equilibrium. The functional relation between pressure and temperature for an atmosphere in wet-adiabatic equilibrium is given by Eq. (148). In differential form,

$$c_p \frac{dT}{T} - R_a \frac{d(p-e)}{p-e} + d\left(\frac{wL}{T}\right) + cw \frac{dT}{T} = 0 \tag{184}$$

where T = temperature
p = pressure
L = latent heat of vaporization
e = saturation vapor pressure of water at T
R_a = gas constant for dry air (0.287 joule/gram °K)
c_p = specific heat at constant pressure for dry air (1.003 joule/gram °K)
w = saturation mixing ratio at T and p grams/gram
c = specific heat of liquid water (4.19 joules/gram °K)

The value of the derivative dT/dp may be found by straightforward differentiation in Eq. (184). It is found that

$$\frac{dT}{dp} = \frac{\frac{R_a}{p-e} + \frac{0.622Le}{T(p-e)^2}}{\left[\frac{c_p + wc}{T} - \frac{wL}{T^2}\right] + \left\{\frac{w}{T}\frac{dL}{dT} + \frac{de}{dT}\left[\frac{R_a}{p-e} + \frac{L(0.622+w)}{T(p-e)}\right]\right\}} \tag{185}$$

On making the approximations that $R_a = R$ and $w = 0.622e/p$ and substituting Eq. (185) into Eq. (168), it is found that the lapse rate for an atmosphere in wet-adiabatic equilibrium is

$$\gamma_m = \gamma_d \frac{p + \frac{0.622Le}{R_aT}}{p + \frac{0.622}{c_p}\left[e\left(c + \frac{dL}{dT}\right) + \left(\frac{R_aT}{0.622} + L + \frac{Le}{p}\right)\frac{de}{dT}\right]} \tag{186}$$

If the further approximation is made that the relatively small terms Le/p and $R_aT/0.622$ are neglected, Brunt's[28] result is obtained

$$\gamma_m = \gamma_d \frac{p + \frac{0.622Le}{R_aT}}{p + \frac{0.622}{c_p}\left[e\left(c + \frac{dL}{dT}\right) + L\frac{de}{dT}\right]} \tag{187}$$

If we approximate still further by neglecting $e[c + (dL/dT)]$, we obtain the result given by Haurwitz[21]

$$\gamma_m = \gamma_d \frac{p + \frac{0.622Le}{R_aT}}{p + \frac{0.622L}{c_p}\frac{de}{dT}} \tag{188}$$

If Clapeyron's equation from Eq. (60) is written in the form

$$\frac{de}{dT} = \frac{Le}{R_vT^2} \qquad R_v = 0.461 \text{ joule/gram °K} \tag{189}$$

and substituted into Eq. (188), it is found that

$$\gamma_m = \gamma_d \frac{p + \dfrac{0.622 Le}{R_a T}}{p + \dfrac{0.622}{c_p} \dfrac{L^2 e}{R_v T^2}} \tag{190}$$

Since $L/T \gg 1$, we see that $\gamma_m < \gamma_d$. Brunt has published a chart giving the values of γ_m. Another form of the results due to Kiefer is given in Sec. I. Equation (190) is perhaps the simplest for calculation if required.

Lapse Rate in Polytropic Atmosphere. If the atmosphere is in polytropic equilibrium, one has from Eqs. (168) and (91) that

$$\gamma = \frac{g}{c_p - \xi}$$

and it is noted that $\gamma > \gamma_d$ when $0 < \xi < c_p$, as is usually the case in nature when the air is heated from below.

Pressure-height for an Actual Sounding. When the record of a radiosonde or aerograph flight is plotted on a thermodynamic diagram, it is found, in general, that

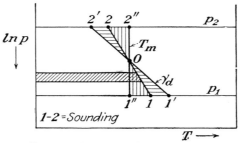

Fig. 23.—Area method on Emagram.

the sounding does not follow either an isotherm, a dry adiabatic, or a wet adiabatic. Relatively thin isothermal layers will often be found. Under certain conditions, *inversions* of temperature will be present (increase of temperature with elevation). Frequently rather thick layers will be in either dry- or wet-adiabatic equilibrium. *The usual situation is that the sounding will be composed of a combination of all possible types of thermodynamic equilibrium.* This result is due to radiation and to advective and convective currents of air aloft. There is then no simple analytical expression for temperature as a function of pressure, and graphical analysis must be used. *The information required from the sounding consists of the location above sea level of the strata of air that may be of significance to the forecaster.* A level is said to be "significant" when pressure, temperature, or humidity suffer abrupt change there.

I. *Dry Air.* Let points 1 and 2 in Fig. 23 represent two significant levels from an actual sounding. Assume that the air is dry. Let the curve connecting the two points represent the state of the atmosphere between the two points. *In practice, a straight line is drawn between significant points.* The distance between the two levels in dynamic meters is given by Eq. (159) and the gas law to be

$$g(z_2 - z_1) = \int_2^1 v \, dp = \int_2^1 RT \, d(\ln p) \tag{191}$$

We note that $RT \, d(\ln p)$ is an element of area on the thermodynamic diagram (see Fig. 23). Then, by the above equation, $g(z_2 - z_1)$ is equal to the area to the left of the sounding, provided that the temperature scale extends down to 0°K. A simple geometrical device enables us to get a numerical value for this area. Draw the line 1'-2' coincident with a dry adiabatic so that the areas 1-0-1' and 2-0-2' are equal. Then clearly $g(z_2 - z_1)$ is also equal to the area to the left of the dry adiabatic 1'-2'. This dry adiabatic will be referred to as the *mean dry adiabatic*.

For any dry-adiabatic line on the diagram, we have $c_p \, dT/T = R \, dp/p$. Substituting this equation into Eq. (191) gives

$$g(z_2 - z_1) = \int c_p \, dT = c_p(T_1' - T_2') \tag{192}$$

where T_1' and T_2' are the temperatures where the mean dry adiabatic intersects the original pressure levels. Numerically we have

$$g(z_2 - z_1) = 100.3(T_1' - T_2') \qquad gm \tag{193}$$

for the distance between p_1 and p_2.

An alternate method of the distance determination involves the drawing of a mean isotherm. In Fig. 23, let 1''-2'' be the isotherm through the point 0 which was determined by drawing the mean adiabatic. Recalling how the mean dry adiabatic was drawn, it is a simple geometrical exercise to show that area 2-0-2'' is equal to 1-0-1''. Then the area to the left of the isotherm is also equal to $g(z_2 - z_1)$. Therefore

$$g(z_2 - z_1) = RT_m \ln \frac{p_1}{p_2} \tag{194}$$

where T_m is the isotherm drawn to make areas 2-0-2'' and 1-0-1'' equal. Numerically we have

$$g(z_2 - z_1) = 28.7 T_m \ln \frac{p_1}{p_2} \qquad gm \tag{195}$$

II. *Humid Air.* When the air is humid, account must be taken of the water vapor present. Equation (191) now becomes

$$g(z_2 - z_1) = \int_2^1 RT \frac{dp}{p} = \int_2^1 R_a T_v \frac{dp}{p} \tag{196}$$

where R is the gas constant for the humid air. The virtual temperature has been introduced by the equation $RT = R_a T_v$. *The above fundamental relation then shows that the distance between the two levels of interest, in dynamic meters, is now given by the area to the left of the curve of virtual temperature.* If the air had been dry the gas constant would have been the gas constant for dry air, the virtual temperature would have been identical with the temperature, and Eq. (196) would have been identical with Eq. (191).

Once the virtual-temperature curve has been constructed for the sounding, one is concerned only with the area to the left of the new curve. This area is given by either one of the following equations, and each gives numerically the distance in dynamic meters between the two levels.

$$g(z_2 - z_1) = R_a T_{vm} \ln \frac{p_1}{p_2} = 28.7 T_{vm} \ln \frac{p_1}{p_2} \qquad gm \tag{197}$$

$$g(z_2 - z_1) = c_p(T_{v1}' - T_{v2}') = 100.3(T_{v1}' - T_{v2}') \qquad gm \tag{198}$$

T_{vm} is the mean virtual temperature. T_{v1}' and T_{v2}' are the temperatures at which the dry-adiabatic "mean" to the virtual-temperature curve cuts the isobars. Refer to Fig. 23 and think of the points 1 and 2 representing virtual temperatures rather than real temperatures.

The virtual-temperature curve is constructed by using Eq. (128); or by using Eq. (130) with the distances laid off for $(T_v - T)_s$ on the adiabatic chart; or by using a nomogram (see Fig. 19); or the "distance corrections" may be determined directly using the nomogram built into the Emagram. Any of these devices originates ultimately with the equation $RT = R_a T_v$ [Eq. (126)].

III. *Wet Air.* Suppose each small element of the atmosphere contains droplets of liquid water (as in a colloidal suspension). The difference in pressure between top and bottom of an element of thickness dz is $dp = -g\rho \, dz$ where ρ_a, ρ_v, and ρ_f are the densities of the dry air, water vapor, and liquid-water droplets, respectively. The total density is the sum of these three densities. The distance between two points in the atmosphere in static equilibrium is

$$g(z_2 - z_1) = \int_2^1 \frac{dp}{\rho_a + \rho_v + \rho_f} = \int_2^1 \frac{dp}{(p/RT) + \rho_f} < \int_2^1 RT \frac{dp}{p}$$

It may be noted that the distance as computed by the use of the preceding formulas is somewhat greater than the actual distance in the case where liquid-water droplets are present. But since $\rho_f \ll \rho_a$, the error is very small. Moreover the quantity ρ_f is not known in terms of the thermodynamic coordinates, nor is it easily observable. *Equation (197) or (198) is then used to calculate the linear distance between levels, even when saturation is present with some liquid water droplets.*

Note on the Assumption of Static Equilibrium. It is true that the equations developed above are not strictly accurate unless the velocity of flow is the same at points 1 and 2 [see Eq. (156)]. If numerical values are inserted, however, it is evident that the error introduced by assuming static equilibrium is entirely negligible in this application.

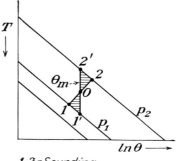

1-2 =*Sounding*
Area *1-O-1'*=*Area 2-O-2'*

Fig. 24.—Area method on $T\theta$-diagram.

Geopotential on the Temperature-entropy Diagram. The expression for change in geopotential (*i.e.*, altitude) on the $T\theta$-diagram or on the Tephigram takes an especially simple form. Using the general equation $g \, dz + v \, dp = 0$, the differential expression for potential temperature $d\theta/\theta = dT/T - (R/c_p)dp/p$, and the equation of state for perfect gases, it is found that

$$g \, dz = c_p T \frac{d\theta}{\theta} - c_p \, dT \tag{199}$$

Along a dry adiabatic, $d\theta = 0$, and changes in geopotential are then simply measured along a vertical straight line. The process for finding the change in geopotential between two pressure levels on a $T\theta$-chart is similar to that used when the sounding is plotted on the adiabatic diagram or Emagram. It may be shown that the change in geopotential is given by Eq. (198) where as before T_{v1}' and T_{v2}' are the temperatures where the mean dry adiabatic cuts the isobars p_1 and p_2. The areas to be equalized are shown by hatching in Fig. 24.

Recommended Procedures. I. *Mean Dry-adiabatic Method.* Plot the sounding. Using Eq. (129) or any other desired device, determine the virtual-temperature difference at each significant point at saturation. Multiply by the relative humidity to get $T_v - T$. Mark T_v at each significant level by a small check. Draw the dry adiabatic that equalizes areas between the T_v-curve, the isobars whose distance apart it is desired to know, and itself (see Fig. 23). The difference in elevation is given by Eq. (198). The horizontal spread between the points T_{v1}' and T_{v2}' in degrees multiplied by 100.3 gives the linear distance between the significant levels in dynamic meters.

The scheme may be inverted and used to locate the pressure at some particular level, say 10,000 *ft above sea level, accurately enough for most purposes.* This involves knowing the elevation of the station above which the sounding was taken. First equalize areas by a mean adiabatic up to a pressure that is close to that desired (*e.g.*, 850 mb for 5,000 ft or 700 mb for 10,000 ft). Then measure to the left from the intersection of the mean dry adiabatic and the surface isobar a distance in degrees equal to the elevation desired less the elevation of the station in hundreds of dynamic meters. The isotherm attained will intersect the adiabatic at the desired isobar and give the desired pressure plus or minus a millibar or two. If the pressure is desired more accurately, equalize areas a second time, now using as top isobar the pressure found by the first trial. Successive trials will converge to a unique pressure very rapidly.

II. *Mean Isotherm Method.* Plot the sounding and locate check marks for virtual temperature as before. The distance between the isobars is given by Eq. (197) in dynamic meters. Distances between standard isobars may be looked up in the appropriate tables (see Sec. I). Fletcher and Graham[27] have developed a nomogram to solve Eq. (197). Bellamy has a circular slide rule to solve it (University of Chicago Press). Again, the equation is easily solved in any particular case using tables or a slide rule.

III. *Mean Mixing Ratio.* Plot the sounding. Plot the mixing ratio on a separate curve to the left. Estimate the mean mixing ratio in the layer. The linear thickness is given by

$$z_2 - z_1 = 29.3 \int_2^1 T(1 + 0.61w) \frac{dp}{p} \qquad \text{m}$$

since $T_v = T(1 + 0.61w)$. This equation may be integrated to give

$$z_2 - z_1 = 29.3 T_m \ln \frac{p_1}{p_2} (1 + 0.61w_m)$$

where T_m is the mean temperature in the layer (not virtual temperature) and w_m is the mean mixing ratio in the layer. The above may be written as

$$z_2 - z_1 = (z_2 - z_1)' + \delta(\Delta z) \qquad (200)$$

where the first term is the thickness the layer would have if it were dry and the second term is the correction to take account of the water vapor present. The second term may be calculated with tables or easily determined by use of a nomogram (see Sec. I, Fig. 46). The first term may be calculated or obtained from the tables in Sec. I (these tables are now entered using mean temperature as an argument as if it were a virtual temperature). The above method has the slight advantage that virtual-temperature differences for the sounding do not have to be determined.

Note on Pseudoadiabatic Diagram. Any of the above procedures may be used on any thermodynamic diagram on which areas represent energies. The methods are also used on the pseudoadiabatic diagram, although differences in elevation must then be considered as approximations; since, as was shown earlier, the pseudoadiabatic diagram cannot be made area preserving at more than one pressure. The errors added are negligible, provided that the pressure intervals over which the procedures are carried out are not too great. Comparison between the results determined on different diagrams (pseudoadiabatic and $T\theta$) shows that it is safe to use intervals of about 300 mb.

Lapse Rate for Any Sounding. The average lapse rate in a layer may be obtained easily when the thickness of the layer and the decrease in temperature are known. We have

$$\gamma_{av} = -\frac{T_1 - T_2}{z_1 - z_2} \tag{201}$$

where the linear distance $z_1 - z_2$ is to be determined by one of the above methods. In principle, the lapse rate at a point may also be determined. This is scarcely worth while on an actual sounding unless reports of temperature, pressure, and mixing ratio have been received continuously. The idea may be illustrated by showing how to find graphically the lapse rate at a point along a wet adiabatic. From Eq. (168), we have

$$\gamma = -\frac{dT}{dz} = \frac{gp}{RT}\frac{dT}{dp} = \frac{g}{RT}\frac{dT}{d(\ln p)} \tag{202}$$

On the Emagram, linear distances along the vertical and horizontal are given by $-dy = d(\ln p)$ and $dx = R\,dT$. Therefore

$$\gamma = \frac{g}{R^2 T}\frac{dx}{dy} = \frac{g/R^2 T}{\tan \alpha} \tag{203}$$

where $\tan \alpha$ is the slope of the tangent to the wet adiabatic. This slope may be measured graphically and the lapse rate computed. Or a simple nomogram may be constructed to give γ as a function of T and α. Kiefer has made a similar analysis on his $T\theta$-diagram and constructed the nomogram (Fig. 19).

Lapse Rate of Potential Temperature. On combining the definition of the potential temperature, the equation of state for gases, and the equation of hydrostatic equilibrium, it is easy to show that the lapse rate of potential temperature is given by

$$\frac{d\theta}{dz} = \frac{\theta}{T}(\gamma_d - \gamma) \tag{204}$$

where θ = potential temperature, T = temperature, γ_d = the dry-adiabatic lapse rate, and γ = the actual lapse rate in the atmosphere.

METEOROLOGICAL TEMPERATURES

Meteorologists have found it useful to introduce more than a dozen different "temperatures" in the course of their practical and theoretical work. These are not different temperatures in the sense of being measured on different temperature scales (*e.g.*, centigrade or Fahrenheit). Only one is a temperature, in that it is a number corresponding to what would be read on a thermometer placed in the air under discussion. The others are different temperatures to which the air may be brought by one or more specific processes. For convenience, these meteorological temperatures may be classified as follows:

1. Temperature $= T$. The actual temperature of a material mass of air as measured by a thermometer ($T = t + 273$).

2. Saturation Temperatures.

Dew-point temperature $= T_d$
Condensation-level temperature $= T_c$
Wet-bulb temperature $= T_w$
Pseudo-wet-bulb temperature $= T_{sw}$

3. Potential Temperatures.

Potential temperature $= \theta$

Partial potential temperature $= \theta_d$

Pseudo-wet-bulb potential temperature $= \theta_{sw}$

Wet-bulb potential temperature $= \theta_w$

Pseudo-equivalent potential temperature $= \theta_{se}$

Equivalent potential temperature $= \theta_e$

4. Dry Temperatures.

Virtual temperature $= T_v$

Equivalent temperature $= T_e$

Pseudo-equivalent temperature $= T_{se}$

A given mass of air at some particular pressure, temperature, and mixing ratio (p, T, w) will attain any one of the above temperatures when subjected to the proper defining process(es). The different saturation temperatures are attained by various ways of bringing the sample of air to saturation. The potential temperatures are attained when the sample of air is brought to the 1,000-mb pressure level in some particular way. The dry temperatures are attained when the sample of air has all its moisture removed in some particular way. Each of the commonly used meteorological temperatures is discussed below. *It should be noted that other temperatures may be defined.* The dew-point, condensation-level, wet-bulb, and pseudo-wet-bulb temperatures do not, for example, exhaust the processes by which a sample of air may be brought to saturation. The prime purpose of the introduction of most of the different temperatures is to indicate the water-vapor content of the air.

Dew-point Temperature. Let a parcel of air initially at p, T, w be cooled isobarically by some physical process (*e.g.*, radiation) while the mixing ratio remains constant. The temperature attained at saturation is named the dew-point temperature and designated by T_d. We have for the process

$$
\begin{aligned}
p &= \text{total pressure is constant} \\
w &= \text{mixing ratio is constant} \\
e &= \text{vapor pressure is constant} \\
Q &= \text{heat lost by parcel is positive} \\
T_d &< T
\end{aligned}
\tag{205}
$$

Since w and e are constant, T_d is the temperature for which e is equal to the saturation vapor pressure. To find T_d from a table of saturation vapor pressures, enter the table with the value of e given and read the corresponding temperature. This is T_d. To find T_d on a thermodynamic diagram, pass along the isobar at p until the w_s line corresponding to w is reached. T_d is shown as point 1 in Fig. 25. Point O is the initial air. Note that if the parcel is permitted to ascend along the dry adiabatic for a short distance and then is cooled isobarically to w_s, the dew-point temperature is changed very little in comparison with the change in the original temperature of the air. This is so because the saturation-mixing-ratio line is almost vertical. The saturation-mixing-ratio line is sometimes called the *dew-point line*.

In nature, the dew-point temperature is attained on calm clear nights when radiative cooling produces dew or frost. In the observatory, it may be measured directly by cooling a polished surface of metal (say a can inside which is a mixture of ice and water) until condensation from the surrounding air begins. The dew-point temperature is particularly useful in synoptic meteorology and is regularly transmitted on the U.S. Weather Bureau teletype as a part of the surface report. Its value is usually determined from tables in which entry is made with the wet-bulb temperature. Use may also be made of the Psychrometric Slide Rule of the U.S. Weather Bureau.

Condensation-level Temperature. Let a parcel of air initially at p, T, w ascend dry-adiabatically with mixing ratio constant until saturation is attained. The temperature at saturation is named the condensation-level temperature and designated by T_c. We have for the process

$$p = \text{total pressure decreases}$$
$$w = \text{mixing ratio is constant}$$
$$e = \text{vapor pressure decreases}$$
$$s = \text{entropy is constant } (\Delta Q = 0)$$
$$T_c < T_d < T$$

T_c may be calculated by observing that with the above constraints

$$\frac{de}{e} = \frac{dp}{p}$$

$$T = p^{\frac{R}{c_p}} \times \text{const} \tag{206}$$

$$T_c = T \left(\frac{e_c}{e}\right)^{\frac{R}{c_p}}$$

for the process. Also $e_c = e_s$ at the temperature T_c. This calculation may be made by "trial and error" or graphically. The graphical solution is of course immediately

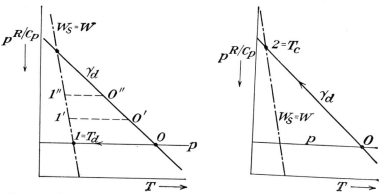

FIG. 25.—Dew-point temperature, T_d.   FIG. 26.—Condensation-level temperature, T_c.

available on the thermodynamic diagram, as shown in Fig. 26. Point O is the initial air. Point 2 is at the condensation-level temperature where the path O-2 is a dry adiabatic. This point is the intersection of the dry adiabatic through O and the $w_s = w$ line. The isobar through point 2 is known as the *condensation-level pressure* and designated by p_c.

In nature, the condensation-level temperature is approximated by the temperature at the base of clouds when the air below the clouds is thoroughly stirred. The temperature T_c is sometimes called the *cloud-level temperature* for this reason. T_c is not usually measured in the observatory, although it could be by causing a sample of air to be expanded adiabatically in a container until condensation began.

Wet-bulb Temperature. Let a stream of air at p, T, w flow past the bulb of a thermometer that is covered with a wet wick. Water will be evaporated from the wick by the humid air flowing past. The bulb of the thermometer will be cooled by the evaporation. When an equilibrium condition is reached, the following energy equation holds:

$$(T - T_w)(c_p + wc_p') = (w' - w)L_w \qquad (207)$$

where T = temperature of the approaching air

T_w = temperature of the leaving air = wet-bulb temperature

w = mixing ratio of the approaching air

w' = mixing ratio of the leaving air

c_p = specific heat at constant pressure of dry air

c_p' = specific heat at constant pressure of water vapor

L_w = latent heat of vaporization at the wet-bulb temperature

When the bulb is kept wet and well ventilated by a good stream of air (velocity about 10 mps), the equilibrium condition will be that the leaving air is saturated and

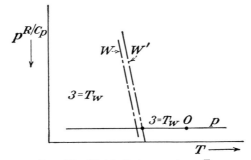

Fig. 27.—Wet-bulb temperature, T_w.

that the temperature of the bulb is equal to the temperature of the leaving air. Then w' is determined by T_w and may be calculated from

$$w' = \frac{0.622e_w}{p - e_w} \qquad (208)$$

where p is the pressure at the time of observation and $e_w = e_s$ at T_w. The calculation from Eqs. (207) and (208) may be made by trial and error or graphically. The graphical solution is simple on Kiefer's multipressure hygrometric chart (see page 391). T_w cannot be determined on the thermodynamic diagram directly.

Tables are available whereby the observed wet-bulb depression (*i.e.*, $T - T_w$) gives the relative humidity and the dew-point temperature of the air without calculation. A different set of tables will be required for different pressures. The data may also be given in graphical form. T_w is usually measured with a sling psychrometer (see Sec. VIII on Meteorological Instruments).

Since the process is isobaric, the wet-bulb temperature may be indicated on the thermodynamic diagram, as shown by point 3 in Fig. 27. Since $w' > w$, $T_w > T_d$. In nature, the wet-bulb temperature will be closely approximated when falling rain evaporates into warmer air beneath, thereby cooling it to saturation at the same time that its mixing ratio increases from its original value of w to w'. The heat required for vaporization is assumed to be supplied by the air.

Pseudo-wet-bulb Temperature. The pseudo-wet-bulb temperature is defined by a series of processes that may be indicated on the thermodynamic diagram. Let a parcel of air initially at p, T, w ascend dry-adiabatically to T_c. Then follow down a wet adiabatic until the original pressure is reached. The intersection of the wet adiabatic through T_c and the original isobar determines the pseudo-wet-bulb temperature, and it is designated by T_{sw}. See point 4 in Fig. 28. Physically the process is more nearly real if it is reversed. Thus let air be at point 4 and saturated. Let it ascend along the wet adiabatic with immediate precipitation of condensed liquid water until the mixing ratio has reduced to w. Then let the air descend along a dry adiabatic until the original pressure is attained. The parcel completes the process at

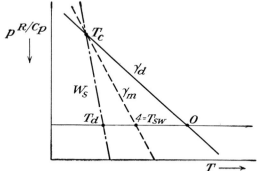

FIG. 28.—Pseudo-wet-bulb temperature, T_{sw}.

p, T, w. The temperature from which it started is named the pseudo-wet-bulb temperature.

T_{sw} may be calculated by inserting T_c and the condensation-level pressure p_c with the original pressure p into the equation for the wet adiabatic as given in Eq. (148). But the calculation is both laborious and unnecessary, since a graphical solution is at hand in the thermodynamic diagram.

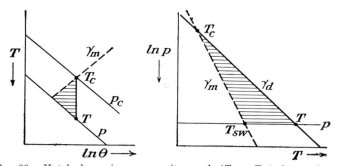

FIG. 29.—Hatched area in energy units equals $(T_w - T_{sw})$ times a factor.

The usefulness of T_{sw} lies in the fact that it is approximately equal to the wet-bulb temperature. Following Bleeker,[32] it may be shown that, to an excellent approximation

$$T_w - T_{sw} = \frac{A}{(c_p + 0.622L^2 e_{sw}/pR_v T_w T_{sw})} > 0 \qquad (209)$$

where e_{sw} is the saturation vapor pressure at T_{sw}, and A is the energy in joules per gram represented by the area hatched in Fig. 29. The other symbols have their usual meanings. If numerical values are inserted on the right-hand side of Eq.

(209), it is found that the difference between T_w and T_{sw} is positive and usually less than 0.5°C. Except for the most accurate work, it is adequate to take T_{sw} as determined graphically on the thermodynamic diagram to be equal numerically to the wet-bulb temperature T_w.

Potential Temperature. The potential temperature was defined by Eq. (93); $\theta = T(1{,}000/p)^{R/c_p}$. The potential temperature is found on the thermodynamic diagram by following a dry adiabatic from the p, T of the parcel to the 1,000-mb isobar. θ is shown as point 5 in Fig. 30. The potential temperature of any parcel of air that is located along the dry adiabatic through 5 is the same value θ. The dry adiabatic itself is then labeled by a value of θ equal to the temperature at the intersection of the dry adiabatic and the 1,000-mb isobar. The potential temperature has been found useful as the coordinate in some thermodynamic diagrams. In nature, its importance arises from the fact that parcels of air that ascend or descend in the atmosphere without heat transfer will always have the same potential temperature

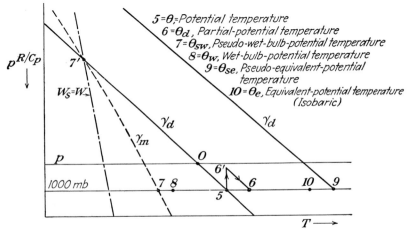

Fig. 30.—Potential temperatures.

so long as condensation is not reached. The potential-temperature lines are also lines of constant entropy. For any value of θ, the entropy s is given by $s = c_p \ln \theta + \text{const.}$

Partial Potential Temperature. The partial potential temperature as defined by Rossby[29] is

$$\theta_d = T \left(\frac{1{,}000}{p_d} \right)^{\frac{R}{c_p}} \tag{210}$$

where p_d is the partial pressure of the dry air $(p_d = p - e)$. In terms of θ, it is easy to show that

$$\theta_d = \theta \left(1 + \frac{e}{p_d} \right)^{\frac{R}{c_p}} \tag{211}$$

To a very good approximation, it is found on expanding that

$$\theta_d = \theta \left(1 + \frac{Re}{c_p p_d} \right) = \theta(1 + 0.461w) \tag{212}$$

where w is the mixing ratio of the parcel of air for which θ_d is desired. The potential temperature of the air may be read directly from the thermodynamic diagram. Then the partial potential temperature may be computed using Eq. (212). Tables are also available from which the correction term $0.461w\theta$ may be obtained by entering with θ and w. On the thermodynamic diagram, θ_d may be found by ascending isothermally from θ a distance equal to the vapor pressure, and then following the new θ line back to 1,000 mb. The path is shown in Fig. 30 as a triangle 5-6'-6. The point 6 is θ_d. The value of θ_d may be taken equal to θ only when w is very small.

Pseudo-wet-bulb Potential Temperature. The pseudo-wet-bulb potential temperature is defined to be the temperature at which the wet adiabatic intersects the 1,000-mb isobar. The equation defining θ_{sw} is, from Eq. (146)

$$c_p \ln \frac{T_1}{\theta_{sw}} - R_a \ln \frac{p_1 - e_1}{1,000 - e_2} + \frac{w_1 L_1}{T_1} - \frac{w_2 L_2}{\theta_{sw}} = 0 \tag{213}$$

where the state 1 is anywhere along the wet adiabatic and 2 is at 1,000 mb. The vapor pressures and mixing ratios in Eq. (213) are saturation values. The symbols have their usual meanings. Any parcel of air at saturation is said to have the pseudo-wet-bulb potential temperature of the wet adiabatic passing through the point p, T locating the parcel on the thermodynamic diagram.

Consider a parcel of air at saturation anywhere along the θ_{sw} line. Let the parcel descend to some p, T dry-adiabatically. Since this process is reversible (*i.e.*, in ascent the air would go back up the same dry adiabatic), we see that the parcel of air has the same θ_{sw} it had originally. Similarly, if the parcel ascends adiabatically, it will continue on the same θ_{sw} line. Clearly θ_{sw} for any parcel of air is unchanged by any adiabatic process. If a parcel of air at p, T, w is considered (not at saturation), its θ_{sw} is found by following up the dry adiabatic through the point until the condensation-level temperature is reached. Thus, for any parcel of air, its pseudo-wet-bulb potential temperature is found by determining the wet adiabatic through the condensation-level temperature and pressure.

The pseudo-wet-bulb potential temperature, θ_{sw}, is shown by point 7 in Fig. 30. The usefulness of θ_{sw} is in (1) identification of air masses and (2) stability investigations when layers of air are lifted as by a front or orographically.

Wet-bulb Potential Temperature. The wet-bulb potential temperature is the temperature at the intersection of the 1,000-mb isobar and the wet adiabatic through the wet-bulb temperature. Since we have noted previously that $T_w > T_{sw}$, it follows that $\theta_w > \theta_{sw}$. θ_w is shown as point 8 on Fig. 30. Since the wet-bulb and pseudo-wet-bulb temperatures are approximately the same numerically, the same may be said of their potential temperatures. The approximation is so good that except for the most accurate work the two temperatures may be taken to be equal numerically.

Pseudo-equivalent Potential Temperature. The pseudo-equivalent potential temperature is defined to be the temperature a parcel of air will attain if it ascends pseudoadiabatically until all its water vapor has been condensed out and then descends along a dry adiabatic to the 1,000-mb isobar. The temperature is designated by θ_{se}. From Eq. (213), we have

$$c_p \ln \frac{T'}{\theta_{sw}} - R_a \ln \frac{p'}{1,000 - e} - \frac{wL}{\theta_{sw}} = 0 \tag{214}$$

where T' and p' are the temperature and pressure at the level where the parcel has lost all its vapor. Since the parcel descends dry-adiabatically, we have

$$\theta_{se} = T' \left(\frac{1,000}{p'}\right)^{\frac{R}{c_p}} \tag{215}$$

On elimination between the two equations it is found that

$$\theta_{se} = \theta_{sw} \exp \left(\frac{wL}{\theta_{sw}c_p} + \frac{R_a e}{1{,}000 c_p} \right)$$

or, approximately, on expanding in series

$$\theta_{se} = \theta_{sw} + \frac{wL}{c_p} \tag{216}$$

where w and L are the saturation mixing ratio and latent heat of vaporization, respectively, at θ_{sw}. c_p is the specific heat of dry air at constant pressure.

It may be noted that θ_{sw} and θ_{se} are each uniquely determined by the same wet adiabatic. The former is the temperature at the intersection of the wet adiabatic with the 1,000-mb isobar. The latter is the temperature at the intersection of the 1,000-mb isobar with that dry adiabatic which is asymptotic to the wet adiabatic.

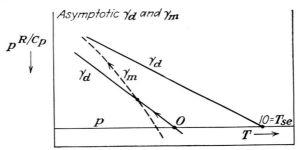

Fig. 31.—Pseudo-equivalent temperature, T_{se}.

Since θ_{sw} is easier to determine graphically on the pseudoadiabatic diagram, it is often preferred. On the other hand, θ_{se} is more useful in some theoretical equations so far as simplicity is concerned. The pseudoequivalent potential temperature is shown as point 9 on Fig. 30.

The relation between θ_{se} and the partial potential temperature is given by Rossby.[21] This is

$$\theta_{se} = \theta_d \exp \left(\frac{Lw}{T_c c_p} \right) \tag{217}$$

where exp means the base of the natural logarithms. The other terms are

θ_d = partial potential temperature
T_c = condensation-level temperature of the parcel
L = latent heat of vaporization at T_c
w = mixing ratio of the parcel
c_p = specific heat of dry air at constant pressure

Equations (216) and (217) provide relatively easy means of finding θ_{se} in terms of quantities that may be found directly from the pseudoadiabatic diagram.

Pseudo-equivalent Temperature. The pseudo-equivalent temperature is defined to be the temperature a parcel of air will attain if it ascends pseudoadiabatically until it loses all its vapor, and then descends dry-adiabatically to its original pressure level. The process is illustrated in Fig. 31. The pseudo-equivalent temperature is shown by point 10. From the definition and the equations for the adiabatics, it may be shown following Rossby[29] that approximately

$$T_{se} = T \exp \left(\frac{wL}{T_c c_p} \right) \tag{218}$$

where T_{se} = pseudo-equivalent temperature
 $\qquad T$ = initial temperature of the air
 $\qquad T_c$ = condensation-level temperature
 $\qquad w$ = initial mixing ratio of the air
 $\qquad L$ = latent heat of vaporization at T
 $\qquad c_p$ = specific heat of dry air at constant pressure

Equivalent Temperature. The equivalent temperature according to the original definition of Robitzsch is the temperature that the humid air would attain if all its water vapor were condensed out at constant pressure and the heat released (latent heat of vaporization) used to warm the air. As Rossby[29] has pointed out, it is impossible to visualize this process physically. It is preferred here to define the equivalent temperature by the equation (see Petterssen[33])

$$T_e = T_w + \frac{w'L}{c_p} \qquad (219)$$

where T_e = equivalent temperature
 $\qquad T_w$ = wet-bulb temperature
 $\qquad w'$ = mixing ratio at saturation at T_w
 $\qquad L$ = latent hear of vaporization at T_w
 $\qquad c_p$ = specific heat of dry air at constant pressure

The equivalent temperature may be calculated from Eq. (219) or determined from a nomogram (see Kiefer's hygrometric chart for one form). T_e as defined by Eq. (219) is sometimes known as the *isobaric equivalent temperature.*

Following Bleeker,[32] it may be shown that the relation between the equivalent temperature and the pseudo-equivalent temperature is given to an excellent approximation by

$$T_{se} - T_e = \frac{A'}{c_p} \qquad (220)$$

where A' is the energy in joules per gram represented by the area shown by hatching in Fig. 32. *It will be noted by reference to any particular case and with the use of any thermodynamic diagram that $T_{se} - T_e$ is a quantity too large to be neglected, i.e., we may not use the equivalent temperature for the pseudo-equivalent temperature, or conversely.*

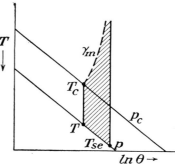

FIG. 32.—Hatched area proportional to $(T_{se} - T_e)$.

The order of magnitude of the difference between T_{se} and T_e is shown most easily by subtracting T_e as given in Eq. (219) from T_{se} as given in Eq. (218). On expanding the latter in series, it is found that

$$T_{se} - T_e = T\left(1 + \frac{wL}{T_c c_p}\right) - T_w - \frac{w'L}{c_p} \qquad (221)$$

On substituting for $T - T_w$ from Eq. (207), it is found approximately

$$T_{se} - T_e \doteq \frac{T - T_c}{T_c}\frac{wL}{c_p} > 0 \qquad (222)$$

where it is assumed L is the same for both T_c and T_w. T, T_c, and w are the temperature, condensation-level temperature, and mixing ratio, respectively, for the original air for which the difference $T_{se} - T_e$ is desired.

Rossby's Equivalent Temperature. Rossby[29] defines the equivalent temperature to be that temperature attained in the pseudoadiabatic cycle when the parcel returns to its original *partial* pressure, not to its original total pressure. Making this adjustment it is seen that the equivalent temperature of Rossby is *exactly*

$$T_E = T \exp. \left(\frac{wL}{T_c c_p}\right) \quad \text{and} \quad \theta_E = \theta_d \exp \left(\frac{wL}{T_c c_p}\right) \tag{217}'$$

This expression when expanded becomes to an excellent approximation equal to the right-hand side of Eq. (219). [To see this, replace T_w in terms of temperature and mixing ratios from Eq. (207).] Thus Rossby's equivalent temperature is almost exactly equal to Robitzsch's isobaric equivalent temperature. The small difference between them will be given by the right-hand side of Eq. (222). *The isobaric equivalent temperature is never greater than Rossby's equivalent temperature.*

Equivalent Potential Temperature. The equivalent potential temperature is the temperature attained by a parcel of air that follows the dry adiabatic through the equivalent temperature to the 1,000-mb isobar. Since $T_{se} > T_e$, the same may be said for their potential temperatures. θ_e is shown by point 10 in Fig. 30. The equivalent temperatures are now little used in comparison with the pseudo-equivalent temperatures.

Virtual Temperature. The virtual temperature T_v was introduced by Eq. (126) and used in geopotential determinations on the thermodynamic diagrams. It is unnecessary to think of any physical processes by which a parcel of air may be changed over into another state where p, T, w go into p, T_v, 0. The defining equation $T_v R_a = TR$ is sufficient to describe it and its practical use sufficient to justify it. One may of course think of the parcel of humid air as having been replaced by a parcel of dry air at the higher temperature. The virtual temperature is given to an excellent approximation by $T_v = T(1 + 0.61w)$ where T is the actual temperature, and w is the mixing ratio of the air in grams per gram.

Virtual Potential Temperature and Potential Virtual Temperature. Montgomery and Spilhaus[34] define the virtual-potential temperature by

$$\theta_v = T_v \left(\frac{1,000}{p}\right)^{\frac{R}{c_p}}$$

and the potential virtual temperature by

$$_v\theta = T_v \left(\frac{1,000}{p}\right)^{\frac{R_a}{c_{pa}}}$$

where the symbols have their usual meanings. These temperatures were introduced in a paper suggesting various modifications in upper-air analysis. In particular, it was suggested that isentropic analysis should be performed on one of these surfaces rather than on a surface of constant potential temperature.

MISCELLANEOUS THERMODYNAMIC CHARTS AND PROCESSES

Rossby Diagram. The Rossby diagram[29] is constructed with mixing ratio on a horizontal linear scale and partial potential temperature on a vertical logarithmic scale (see Fig. 33). On the diagram are drawn isopleths of equivalent potential temperature according to Eq. (217)'. This temperature is called *equivalent potential temperature* after Rossby. Also shown on the diagram are isopleths of condensation-level temperature and condensation-level partial pressure (T_c and p_{cd}, respectively). It is easy to see that each of these quantities is uniquely determined by θ_d and w (see any thermodynamic diagram).

The Rossby diagram is particularly useful in the study of air masses, especially in such studies as pertain to their identification. Since the basic coordinates are θ_d and w, we see that a thoroughly stirred humid air mass will be represented by a point on the diagram. This is so because neither the partial potential temperature nor the mixing ratio is changed by dry-adiabatic processes. When saturation is reached, however, the mixing ratio and the partial potential temperature will both change for any further ascent. Thus a very thin stratum of air in which w changes rapidly will occupy a line on the diagram, while a thick layer of stirred humid air will occupy only a point. Curves drawn on the diagram for well-defined air masses are known as *characteristic curves* of those air masses. The diagram is also very useful in distinguish-

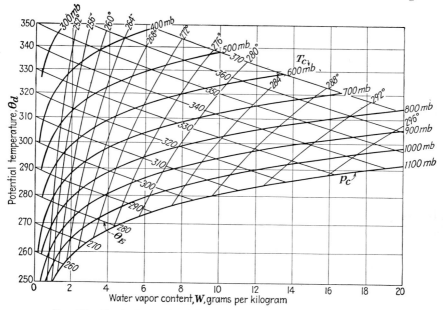

Fig. 33.—Equivalent potential-temperature diagram. (*After Rossby.*)

ing different types of inversions in the upper air. Thus a frontal inversion in which the mixing ratio may increase with elevation will stand out from a subsidence inversion in which the mixing ratio decreases rapidly with elevation.

Multipressure Hygrometric Chart. Kiefer's multipressure hygrometric chart contains facilities from which certain elements may be more readily obtained than from the thermodynamic diagram (see Fig. 34). The right-hand side of the diagram is drawn to basic coordinates of temperature and vapor pressure, both on a linear scale. The 100 per cent relative-humidity line shows the saturation vapor pressure at the temperature desired. Relative-humidity lines of less than 100 per cent are drawn using the equation $e = fe_s$ where f is relative humidity and e_s is the saturation vapor pressure. Dry-adiabatic lines for the *vapor pressure* are drawn using Poisson's equation

$$\frac{T_1}{T_2} = \left(\frac{e_1}{e_2}\right)^{\frac{R}{c_p}} \qquad \frac{R}{c_p} = 0.286 \tag{223}$$

These lines slope down and to the right on the diagram.

The wet-bulb temperature may be determined graphically for any pressure because of the following relation:

$$\frac{T - T_w}{e' - e} = \frac{0.622L}{p} \tag{224}$$

which is obtained by replacing w and w' in terms of e and e' in the right-hand side of Eq. (207) and making the usual approximations. Thus the ratio of the change in temperature (from original air temperature to wet-bulb temperature) to the change in vapor pressure is a function of total pressure and temperature (implicitly through the latent heat of vaporization L). The sloping lines of the rosettes on the chart are drawn from Eq. (224) or its equivalent, according to the value of L at the temperature from which the lines radiate.

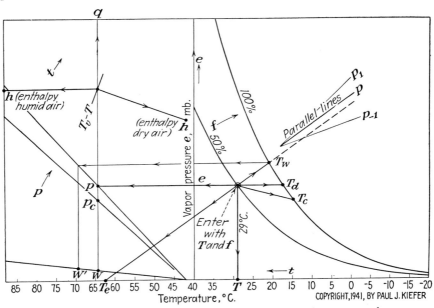

Fig. 35.—Skeleton multipressure hygrometric chart showing manner of use.

The left-hand side of the chart has scales for mixing ratio and specific humidity. These are determined by entering with vapor pressure and crossing over to the appropriate pressure line. The pressure lines on the left-hand side are drawn from the equation $w = 0.622e/(p - e)$ where the symbols have their usual meaning. At the top of the left-hand side of the chart is a nomogram for the determination of the virtual-temperature difference. This nomogram may be constructed using the relation $T_v - T = 0.61wT$. The enthalpy of the humid air and the enthalpy of dry air at p, T, w are shown by appropriate scales [$h = c_p(t + 73) + wL$]. *The manner of determination of the desired elements is shown in the skeleton diagram of Fig. 35.*

Amount of Precipitable Water. The total amount of precipitable water vapor in a column of air may be determined graphically when a sounding is available. It is also useful to investigate the total precipitable water vapor in a saturated column (at a particular value of θ_{sw}) from the surface to great heights. Solot[30] has given an analysis that is straightforward and as accurate as one may wish. He proceeds as follows:

The total mass of water vapor in a small element of humid air of 1 cm² cross section and dz thickness is $\Delta m_v = w\rho_a\,dz$. Introducing the equation of hydrostatic equilibrium and making changes of units as necessary, it is found that the linear thickness of liquid water that would be precipitated if all were precipitated is

$$0.0004 \int_1^2 w\,dp \quad \text{in.} \tag{225}$$

$$\tfrac{1}{980} \int_1^2 w\,dp \quad \text{cm} \tag{226}$$

if the column extends from p_1 to p_2 in millibars. *The mixing ratio w is to be expressed in grams per kilogram.* The value of w in an actual sounding is taken from the radio-sonde or aerograph record. If one wants the upper limit to the amount of precipitable water, one picks off w_s for the appropriate θ_{sw} from a thermodynamic diagram. One may of course construct isopleths on a special chart plotting w_s against p for various values of θ_{sw} (see Sec. I). The thickness of water in inches or centimeters is determined by numerical integration of either of the above equations.

Rates of Precipitation. Fulks[31] has shown that

$$r = \frac{780ab}{T} - \frac{2,666e}{T^2} \tag{227}$$

where r = mm/hr of precipitation from a 100-m thick saturated layer having an ascensional rate of 1 mps
a = wet adiabatic lapse rate, °C/100 m
$b = \dfrac{de}{dT}$, mb/°C
T = absolute temperature
e = saturation vapor pressure at T

Equation (227) is derived on the fundamental assumption that the weight of dry air in the ascending layer remains constant. Thus, taking a layer Δh in thickness with an average density ρ_a, the water vapor present is $w\rho_a\,\Delta h$. The rate of precipitation is numerically equal to the rate of change of water vapor present in the element (provided that precipitation is instantaneous with condensation). Then $r = \rho_a\,\Delta h(dw/dt)$. On making certain minor approximations and introducing the relations for mixing ratio in terms of pressure, etc., Eq. (227) results.

Isopleths calculated from Eq. (227) are plotted on the Emagram for $0.3 \leqslant r \leqslant 1.0$ (see Fig. 20). When the rate of precipitation is desired from layers of thickness other than 100 m, or with different rates of ascension, the value of r as read from the diagram is multiplied by the actual velocity in meters per second and thickness in hundreds of meters. Or one picks a mean value of r for each 100 m thick layer and adds the results to get the rate of precipitation from the total thickness.

The precipitation rate in any actual case must clearly take account of the divergence of the flow. Progress has been made in the solution of particular problems by private investigators. Each case must be solved according to the structure (frontal, orographic, or otherwise) that causes the air to ascend. Turbulence will evidently complicate the problem still further.

Isobaric Cooling of Saturated Air (Formation of Fog). Suppose a parcel of air at saturation loses heat to its surroundings. Condensation will begin immediately except in the rare instance that no condensation nuclei are present. The energy equation for the process assumed isobaric is

$$dq = c_p\,dT + d(wL) \tag{228}$$

If we approximate by assuming the latent heat of vaporization to be constant and use Clapeyron's equation, $de/dT = Le/R_vT^2$, in terms of finite differences, we obtain

$$\Delta q = \left(c_p + \frac{L^2\epsilon e}{pR_vT^2} \right) \Delta T \tag{229}$$

and

$$\Delta q = \left(\frac{c_pR_vT^2}{Le} + \frac{L\epsilon}{p} \right) \Delta e \tag{230}$$

where Δq = heat lost from parcel, joules/gram

c_p = specific heat of the humid air at constant pressure

L = latent heat of vaporization

$\epsilon = \dfrac{R_a}{R_v} = 0.622$

e = saturation vapor pressure at T, mb

T = absolute temperature (mean between initial and final)

p = total pressure, mb

R_v = gas constant for water vapor = 0.461   joules/gram °K

Using Eq. (229), one may obtain a good approximation to the temperature drop accompanying a loss of heat Δq; using Eq. (230) one may obtain a good approximation to the change in vapor pressure accompanying a loss of heat Δq. *If the rate at which heat is lost* (dq/dt) *is known from other considerations* (*observation and radiation equations, for example*), *the rate at which temperature will decrease may be determined.* The equations are so insensitive to small changes in T that the initial temperature may be used inside the brackets without introducing appreciable error.

If none of the condensate is precipitated out of the air, the water droplets remaining will eventually cause a reduction in visibility and fog. The concentration of droplets in grams per cubic meter will be given by $\Delta w\rho$, where Δw is the condensate in grams per kilogram and ρ the density of the air in kilograms per cubic meter. Introducing the equation of state and Clapeyron's equation, it is found to a good approximation that

$$\text{Concentration of droplets in grams/m}^3 = \frac{470Le}{T^3} \Delta T \tag{231}$$

where ΔT is the cooling below saturation, L is the latent heat of vaporization, and e and T are average values for the saturation vapor pressure in millibars and the temperature in degrees Kelvin between the initial and final states. These average values may be estimated accurately enough since the result is insensitive to small changes in T (in the denominator), and since e is a smooth function of T. Combining Eqs. (231) and (229), expressed as a time rate of heat loss, one may predict the time of onset of radiation fog, provided that one knows (1) the droplet concentration required to produce fog, and (2) the rate of heat loss dq/dt. These items depend entirely upon local conditions.

Fog-prediction diagrams may be constructed using the above results. Isopleths are drawn on the multipressure hygrometric chart or on other charts for different droplet concentrations. The isopleths may be calculated from Eq. (231) or some equivalent equation. A useful form is given by Petterssen[33] (see Fig. 36). Other forms of fog-prediction diagram may be constructed for particular stations when the rate of cooling is known under certain synoptic situations (see Fig. 37). Such diagrams are particularly useful in the forecasting of radiation fog. *It must be emphasized that such a diagram prepared for one station will not* (*generally speaking*) *work for any other station where radiation rates are different.*

It is interesting to note that the end result of cooling below saturation is the same whether the condensation occurs uniformly as the temperature is lowered or "all at once" after a state of supersaturation is reached. In the latter case, the change in temperature down to the state of greatest supersaturation is given by $c_p \, \Delta T = \Delta q$. When condensation occurs, an amount of heat $(w_i - w_f)L$ is used in warming the air. With reference to Fig. 38, where the notation is self-explanatory, we have

$$\frac{T_f - T'}{e_i - e_f} = \frac{0.622L}{p} \tag{232}$$

But Eq. (232) gives the slope of the pressure lines in the rosettes on Kiefer's multi-pressure hygrometric chart. Therefore, by using this chart, one may find the final

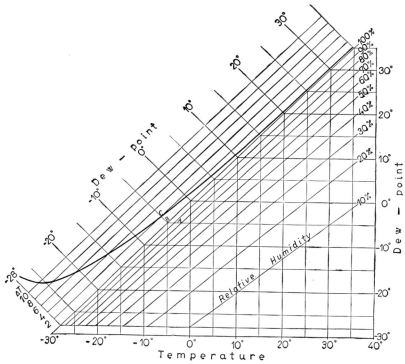

Fig. 36.—Fog-prediction diagram. (*From Petterssen, " Weather Analysis and Forecasting."*)

temperature graphically as indicated in Fig. 38. This procedure is based on the assumption that the heat transfer is known.

By an obvious extension of the above one may compute the amount of condensate from an engine exhaust in the free atmosphere when the engine characteristics are known. Such computations permit the designation of flight levels where "condensation trails" will not form.

Mixing of Two or More Air Masses. Consider for simplicity the mixing of two different masses of air. Let M, T, w, θ, and p with subscript 1 or 2 refer to mass, temperature, mixing ratio, potential temperature, and pressure of the first and second air, respectively. Assume that the mixing ratios are such that condensation due to the mixing will not occur. Let the two parcels of air that are to be mixed be represented by points 1 and 2 on the thermodynamic diagram of Fig. 39. Let each mass

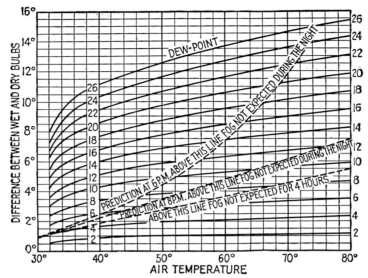

FIG. 37.—Taylor's fog-prediction diagram for Kew on calm, clear nights. The "fog-prediction lines" may conveniently be plotted on Fig. 36. (*From Petterssen, "Weather Analysis and Forecasting."*)

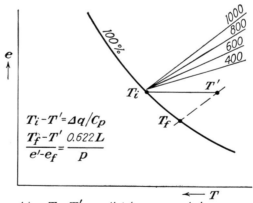

$$T_i - T' = \Delta q / C_p$$

$$\frac{T_f - T'}{e' - e_f} = \frac{0.622 L}{p}$$

Line $T_f - T'$ parallel to appropriate pressure line in rosette

FIG. 38.—Temperature attained after supersaturation and subsequent condensation to saturation.

come to an arbitrary pressure p' along a dry adiabatic. The new temperatures will be given by

$$T_1' = T_1 \left(\frac{p'}{p_1}\right)^{\frac{R}{c_p}} \qquad T_2' = T_2 \left(\frac{p'}{p_2}\right)^{\frac{R}{c_p}}$$

Now let the masses mix isobarically. The system is assumed to be isolated from its surroundings so that the heat given up by one mass in cooling is absorbed by the other in warming. Then $Q_1 = M_1 c_{p1}(T - T_1') = Q_2 = M_2 c_{p2}(T_2' - T)$ where T is the final temperature of the mixture. If the masses have the same constitution so that $c_{p1} = c_{p2}$, we obtain by rearrangement

$$T = \frac{M_1 T_1' + M_2 T_2'}{M_1 + M_2}$$

or

$$T = \frac{M_1 T_1 \left(\frac{p'}{p_1}\right)^{\frac{R}{c_p}} + M_2 T_2 \left(\frac{p'}{p_2}\right)^{\frac{R}{c_p}}}{M_1 + M_2} \tag{233}$$

The final potential temperature is obtained by noting that

$$\theta = T \left(\frac{1,000}{p'}\right)^{\frac{R}{c_p}} \qquad \theta_1 = T_1 \left(\frac{1,000}{p_1}\right)^{\frac{R}{c_p}} \qquad \theta_2 = T_2 \left(\frac{1,000}{p_2}\right)^{\frac{R}{c_p}}$$

whence with the above

$$\theta = \frac{M_1 \theta_1 + M_2 \theta_2}{M_1 + M_2} \tag{234}$$

Since the entropy of the mixture is given by $s = c_p \ln \theta + \text{const}$ per unit mass, it may be noted that $s \neq (M_1 s_1 + M_2 s_2)/(M_1 + M_2)$ where s_1 and s_2 are the specific entropies of the original masses. It may be shown by expansion of the terms in convergent

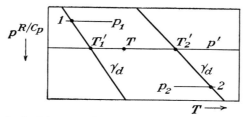

Fig. 39.—Result of mixing two air masses shown on a pseudoadiabatic diagram.

series that $s > (M_1 s_1 + M_2 s_2)/(M_1 + M_2)$. This result in words is that the specific entropy of the total mass after mixing is greater than the average specific entropy of the total masses before mixing. Such a result is expected, of course, because the mixing is essentially a nonreversible (and hence nonisentropic) process.

It may similarly be shown that the final mixing ratio is given by

$$w = \frac{M_1 w_1 + M_2 w_2}{M_1 + M_2} \tag{235}$$

provided that condensation does not occur. Introducing the relation between the mixing ratio and the vapor pressure, it is found that approximately

$$e = \frac{M_1 e_1 + M_2 e_2}{M_1 + M_2} \tag{236}$$

for the vapor pressure in the final mixture.

The vapor pressure as given by Eq. (236) may be in excess of the saturation vapor pressure for the final temperature as given by Eq. (233), even when neither one of the masses was originally at saturation. Thus in Fig. 40 the straight line connecting the two points representing the initial conditions on the hygrometric chart may have a portion of its length falling at humidities greater than 100 per cent. But it is recalled that Eq. (233) is valid only when condensation does not occur.   When account of the condensing vapor is taken, one finds a final temperature that is higher than that given by Eq. (233).   This final temperature will be found on the hygrometric chart at the intersection of the 100 per cent humidity curve with the extension of the wet-bulb line in the nearest rosette (at the appropriate pressure) running through the point T,e given by solving Eqs. (233) and (236).   See T_f in Fig. 40.

It appears at first thought that the mixing of two masses neither of which is at saturation might give a resulting mass with fog present.   If the amount of condensate is investigated, however, it is found that the greatest amount possible is not sufficient to produce fog unaided.   *The increase in relative humidity due to the masses mixing may,*

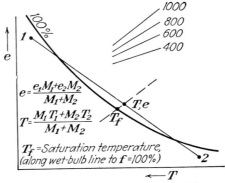

FIG. 40.—Result of mixing two air masses shown on a multipressure hygrometric chart.

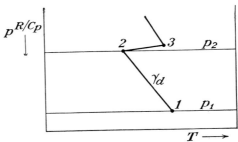

FIG. 41.—Thoroughly stirred stratum between p_1 and p_2.

on the other hand, be such that a relatively small amount of radiation or other cooling will produce fog.   These facts follow from the relatively small curvature of the saturation-vapor-pressure curve.

Thorough Stirring of a Bounded Stratum.   Consider a stratum of air bounded by pressures p_1 and p_2.   By bounded is meant that there is no interchange of properties between this stratum and the rest of the atmosphere, a situation that occurs frequently when there is a strong inversion aloft with fair winds under the inversion (see Fig. 41). When two elements of mass are mixed, their potential temperature becomes $\theta = (\theta_1 \, dM_1 + \theta_2 \, dM_2)/(dM_1 + dM_2)$ where θ_1 and θ_2 were their initial temperatures (provided that condensation does not occur).   When the entire stratum is thoroughly stirred (*i.e.*, mixed), its potential temperature is constant and equal to

$$\theta = \frac{\int \theta \, dM}{M} = \frac{\int \theta \, dp}{p_1 - p_2} \qquad (237)$$

Similarly, the mixing ratio attains a uniform value of

$$w = \frac{\int w \, dM}{M} = \frac{\int w \, dp}{p_1 - p_2} \tag{238}$$

throughout the layer.   The conditions may be represented by the equations

$$\frac{\partial \theta}{\partial z} = \frac{\partial w}{\partial z} = 0 \qquad p_1 < p < p_2 \tag{239}$$

The relative humidity increases upward through the layer.   This conclusion is almost obvious.   One may demonstrate the fact rigorously by taking the derivative of the relative humidity $(f = e/e_s)$, and introducing the facts that $w = $ const and $de/dT = Le/R_v T^2$ by Clapeyron's equation.   It is found that

$$\frac{1}{f} \frac{df}{dz} = \frac{\gamma_d}{T} \left[\frac{L}{R_v T} - \frac{c_p}{R_a} \right] > 0 \tag{240}$$

The above results are valid so long as there is no condensation.   Equation (240) also shows that, when condensation is reached in a previously unsaturated layer, the relative humidity becomes 100 per cent first in the topmost part of the layer.   Such a result would occur, for example, if a thoroughly stirred layer were cooled (either from the top or from the bottom).   If the cooling continues, the saturated layer will build down from the top of the stirred stratum.   Eventually stratus clouds will form, building from the top down.

Rossby[29] has treated in detail the case in which thorough mixing occurs in a bounded layer that is initially cold enough (for the amount of water vapor present) to cause saturation and subsequent condensation at the top.   He concludes that "thoroughly stirred strata, regardless of their degree of saturation, necessarily are characterized by a fairly constant equivalent potential temperature (θ_E)."   The conditions of Eq. (239) may then be replaced by the more general ones

$$\frac{\partial \theta_{se}}{\partial z} = \frac{\partial \theta_{sw}}{\partial z} = 0 \tag{241}$$

for a thoroughly stirred layer, regardless of the water vapor present.   *It may be concluded conversely that, if Eqs. (241) do not hold true in a stratum, then the layer is not thoroughly stirred.*   The most common sounding will show $\partial \theta_{sw}/\partial z > 0$ even when the humidities are reported at 100 per cent by the radiosonde or aerograph, just as in most soundings in humid air $\partial \theta/\partial z > 0$.   *In other words, the phenomenon of complete thorough stirring is not evident in the usual air mass.*

It may be noted that, when a stratum of air is thoroughly stirred, the addition of heat at any one level will warm the entire stratum almost uniformly.   Thus the potential temperature will increase by a finite amount, $\Delta \theta$ at all levels.   Similarly the loss of heat will cool the entire stratum uniformly.   Thus one must take into account the thickness of the thoroughly stirred layer in forecasting surface temperatures based on expected diurnal surface heating (or cooling).   *The thicker the layer the smaller will be the diurnal change in temperature (provided that the layer remains thoroughly stirred).*

Cloud Height above Thoroughly Stirred Stratum.   When there is sufficient moisture in the lower layers to cause condensation after thorough stirring, stratus-type clouds will appear at the top of the layer under the inversion.   The sounding under the clouds will be dry-adiabatic.   The height of the base of the clouds may be estimated when the surface temperature and surface dew-point temperature are known.   The approximate formula is

$$\text{Height of base of clouds} = 120(T_0 - T_{d0}) \qquad \text{°C, m}$$
$$\text{Height of base of clouds} = 200(T_0 - T_{d0}) \qquad \text{°F, ft}$$

where T_0 and T_{d0} are surface temperature and dew-point temperature, respectively. The difference $T_0 - T_{d0}$ is generally known as *spread*.

The above formulas are derived on the assumption that the sounding is dry-adiabatic and that the base of the clouds is at the level where the temperature is equal to the dew-point temperature. The mean temperature in the layer is assumed to be 280°K; if the mean temperature is different, the constants in the above equations will be slightly different.

Synoptically the height of the base of the clouds will be given by the above formulas to sufficient accuracy for most purposes when either (1) stratus appear at the top of a well-mixed layer, the mixing being due to moderate to strong winds under a pronounced inversion, or (2) cumulus appear above a layer through which the sounding is dry-adiabatic (in a warm sector or behind the cold front with moderate to strong winds).

Conservative Properties and Air-mass Identification. Any attribute or property of an air mass that changes relatively little from day to day as the air continues in its path is said to be *conservative*. No property as yet discovered is strictly constant for an air mass, nor is one likely to exist. Not only do heating, condensation, and evaporation change the thermodynamic properties of the air, but turbulent exchange of the very particles of the air will completely alter the original constitution of the air mass in time. Yet certain properties are so nearly constant through short periods of time and for the processes most likely to occur in nature that it is worth while to consider the various elements and the conditions under which they may be expected to be conservative. When a property is found to be almost conservative, it will be called *quasi-conservative*.

Dry-adiabatic temperature changes bear the constraint that no heat is gained or lost by the individual parcels of air, and that no water vapor is gained or lost. Thus for all possible movements (under this constraint), the parcel of air considered will remain on a potential-temperature surface in space (on a potential-temperature isopleth on a thermodynamic diagram). Moist-adiabatic (or wet-adiabatic) temperature changes bear the constraint that the only heat available for vaporization must be supplied by the air and that when condensation occurs heat is added to the air, *i.e.*, for the thermodynamic system as a whole (dry air plus water vapor and/or liquid water), the processes are adiabatic and the heat transfer is zero. Then, for all possible movements (under this constraint), the parcel of air considered while saturated will remain on a pseudo-wet-bulb-potential-temperature surface in space (and on a wet adiabatic on the thermodynamic diagram). *In brief, any adiabatic movements of the air in space will appear on the adiabatics when shown on the thermodynamic diagram.*

Nonadiabatic temperature changes bear the constraint that heat may be added to or lost by the parcels of air under consideration as they move about in space. The heat may be transferred by radiation or conduction. In such a case, $dq \neq 0$ in the first law of thermodynamics. The parcel will not remain on a potential-temperature surface in space or on a potential-temperature isopleth (either wet or dry) on the thermodynamic diagram. If heat is added so that the temperature may rise, the water vapor may remain constant or even increase. If heat is lost and the parcel cooled below saturation, water vapor will condense out, and the mixing ratio will decrease.

The process of evaporation from falling rain is assumed to bear the constraint that the heat of vaporization is supplied by the air. In such a case, for example, the wet-bulb temperatures will remain constant throughout the process because it coincides with their defining process. Similarly the equivalent temperatures will be constant for evaporation from falling rain.

The important results are summarized in Table 5 following Petterssen. Each of the conclusions may be reached by combining the definitions of the appropriate process(es) with the definitions of the element under consideration.

TABLE 5*

Humidity element	Conservative with respect to			
	Dry-adiabatic temperature changes	Moist-adiabatic temperature changes	Nonadiabatic temperature changes	Evaporation from falling rain
Relative humidity..............	No	Yes	No	No
Mixing ratio..................	Yes	No	Yes	No
Specific humidity..............	Yes	No	Yes	No
Dew-point temperature..........	Quasi-conservative	No	Yes	No
Wet-bulb temperature..........	No	No	No	Yes
Equivalent temperature..........	No	No	No	Yes
Pseudo-wet-bulb temperature.....	No	No	No	Quasi-conservative
Pseudo-equivalent temperature...	No	No	No	Quasi-conservative
Wet-bulb potential temperature...	Quasi-conservative	Quasi-conservative	No	Yes
Equivalent potential temperature.	Quasi-conservative	Quasi-conservative	No	Yes
Pseudo-wet-bulb potential temperature.....................	Yes	Yes	No	Quasi-conservative
Pseudo-equivalent potential temperature.....................	Yes	Yes	No	Quasi-conservative
Potential temperature...........	Yes	No	No	No

* From Petterssen.[33]

Kiefer has introduced a new element known as the *sigma function*, which has certain merits in air-mass analysis.[13] By definition, the sigma function is

$$\sigma = \Phi + c_p(t + 73) + wL \qquad (242)$$

where Φ = geopotential, joules/gram

$\quad t$ = °C

$\quad w$ = mixing ratio, grams/gram

$\quad L$ = latent heat of vaporization, joules/gram

$\quad c_p$ = specific heat of dry air at constant pressure

This may be written as $\sigma = \Psi + wL$ where Ψ is the stream function by definition (also called the *acceleration potential;* see Montgomery and Spilhaus).[34] Note that the units chosen for "sigma" are joules per gram, rather than mega-ergs per gram as transmitted by the U.S. Weather Bureau for stream function.

The definition of the sigma function arises out of the following considerations: The equation of mechanical equilibrium is combined with the equation for the wet adiabatics. We have, to a very good approximation

$$c_p \frac{dT}{T} - R_a \frac{dp_d}{p_d} + d\,\frac{wL}{T} = 0$$

and

$$g\,dz + v\,dp + u\,du = 0$$

whence on combining we find to a good approximation

$$c_p\,dT + g\,dz + u\,du + d(wL) = 0$$

or

$$\sigma + \frac{u^2}{2} = \text{const} \qquad (243)$$

The sigma function is tabulated in joules per gram from a base point of 200°K and sea level. In terms of the units chosen, the contribution of velocity in Eq. (243) is so small that it may be neglected for air-mass analysis.

TABLE 6.—SIGMA FUNCTION AT 1 KM IN SOURCE REGION

Air mass	T°C	w, grams/kg	θ_{sw}°K	σ, joules/gram
A, winter.................	−30	0.2	253	54
A, summer...............	0	4.0	278	94
cP, winter...............	−20	0.5	259	64
cP, summer..............	15	6.0	284	114
mP, winter..............	0	4.0	278	94
mP, summer.............	10	6.0	273	108
cT, summer.............	15	9.0	290	132
mT, summer.............	25	15.0	297	145

It is observed from its definition that the sigma function for an air mass may be expected to show about the same degree of conservatism that is shown by the pseudo-potential temperatures. In particular, it will be conservative for adiabatic processes and for evaporation from falling rain. The sigma function will not be conservative for nonadiabatic processes. From this point of view, the sigma function has no advantage over the proper potential temperatures. On the other hand, it does give, in joules per gram, the actual present energy of the parcel, and its changes from day to day may then be correlated with the heating that has occurred (at least for stream-line flow).

Stability and Instability. Some aspects of the mechanical stability of the atmosphere are considered here. See Sec. X for a more detailed analysis from a different point of view. Assume the atmosphere to be in a steady state with a known temperature-height curve (obtained from a radiosonde flight). Consider the effects of making a small virtual displacement of a "test parcel" of the atmosphere. Two assumptions are made about the displacement: (1) that the test parcel follows an adiabatic isopleth on the thermodynamic diagram, and (2) that the displacement introduces no change in the environment. *These assumptions will be excellent approximations to truth for small displacements.*

If the result of the displacement is such that forces are generated that tend to increase the displacement, the parcel is said to be in unstable equilibrium. If the result of the displacement is such that forces are generated that tend to restore the parcel to its original position, the parcel is said to have been in stable equilibrium in its environment. And finally, if the displacement generates no forces on the parcel, it is said to have been in neutral (or indifferent) equilibrium in its environment.

From an energy point of view, the parcel is in unstable, stable, or neutral equilibrium according as the displacement increases, decreases, or leaves unchanged its kinetic energy. Thus consider two parallel streams of flow in the atmosphere. Let the isobaric surfaces coincide with the geopotential surfaces initially. The equation of mechanical equilibrium may be written from Eq. (11) as

$$u_1 \, du_1 + v_1 \, dp_1 + g \, dz_1 = 0$$
$$u_2 \, du_2 + v_2 \, dp_2 + g \, dz_2 = 0 \tag{244}$$

where subscripts 1 and 2 refer to adjacent streams. Since the isobaric surfaces are assumed to coincide with the geopotential surfaces, we have $p_1 = p_2$ when $z_1 = z_2$. On subtracting the first from the second equation above, we have then

$$d \left(\frac{u_2{}^2 - u_1{}^2}{2} \right) + (v_2 - v_1)dp = 0 \tag{245}$$

for any pressure p. The variation in kinetic energy due to the displacement may be written from the above and the equation of state to be

$$\delta(\text{K.E.}) = R_a(T_{v1} - T_{v2}) \frac{dp}{p} \tag{246}$$

where T_{v1} and T_{v2} are the virtual temperatures in the two streams. Equation (246) is a general result applicable whether either both or only one of the streams is in motion relative to the earth.

The usual criteria for stability are derived from Eq. (246) by assuming one of the streams (say 2) to be at rest relative to the earth while the other is displaced slightly from rest. The left-hand side of Eq. (246) then gives the kinetic energy attained by the parcel due to the displacement. In terms of linear displacement in the z-direction (up), we have

$$-\delta(\text{K.E.}) = R_a(T_{v1} - T_{v2}) \frac{g \, dz}{p R_2 T_2} \tag{247}$$

From this equation, it is clear that the atmosphere at the point considered is in unstable, stable, or neutral equilibrium according as the right-hand side of Eq. (247) is negative, positive, or zero. The result may be expressed in terms of the actual lapse rate in the atmosphere and of the adiabatic lapse rate by writing

$$R_1 T_1 = R_1(T_0 - \gamma_d \, dz)$$
$$R_2 T_2 = R_2(T_0 - \gamma \, dz) \tag{248}$$

for dry or humid air (replace γ_d by γ_m for wet air). From the above, the final criteria for stability are

$$T_0(R_1 - R_2) + (\gamma R_2 - \gamma_d R_1) > 0 \qquad \text{unstable}$$
$$T_0(R_1 - R_2) + (\gamma R_2 - \gamma_d R_1) < 0 \qquad \text{stable} \tag{249}$$
$$T_0(R_1 - R_2) + (\gamma R_2 - \gamma_d R_1) = 0 \qquad \text{neutral}$$

where T_0 = temperature at the point considered

R_1 = gas constant at z in the environment (the same as the gas constant in the ascending parcel)

R_2 = gas constant at $z + dz$ in the environment

γ = actual lapse rate in the environment at z

γ_d = lapse rate along the adiabatic (replace γ_d by γ_m if the air is saturated)

When the variation of the water vapor with elevation is neglected, as is usually done, we have from the above that $R_1 = R_2$ and

$$\gamma > \gamma_d(\text{or } \gamma_m) \quad \text{unstable}$$
$$\gamma < \gamma_d(\text{or } \gamma_m) \quad \text{stable} \qquad\qquad (250)$$
$$\gamma = \gamma_d(\text{or } \gamma_m) \quad \text{neutral}$$

according to whether the air is humid (or dry) or saturated, respectively. From the definitions of the potential temperatures, it follows that the parcel is in unstable, stable, or neutral equilibrium according as the derivatives $\partial\theta/\partial z$ and $\partial\theta_{sw}/\partial z$ are nega-

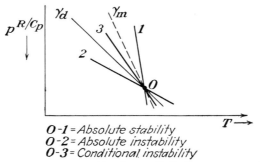

O-1 = *Absolute stability*
O-2 = *Absolute instability*
O-3 = *Conditional instability*

Fig. 42.—Classification of parcel stability for small displacement according to lapse rate.

tive, positive, or zero for humid (or dry) or saturated air, respectively. For perturbations that push the parcel down, the temperature change will follow along a dry adiabatic in any case (humid or saturated) so long as the atmosphere has no liquid water present to vaporize.

The above criteria are accurate but of limited use, since they apply strictly only for small displacements of a parcel. *If it is assumed that the parcel continues to follow*

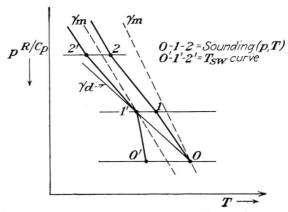

O-1-2 = *Sounding (p,T)*
O'-1'-2' = T_{SW} *curve*

Fig. 43.—Conditional instability of the stable type for a parcel.

an adiabatic isopleth for finite displacements, and that the environment is still unchanged, the results may be extended. It is easy to see that the right-hand side of Eq. (246) represents the area on the Emagram that lies between the adiabatic path of a parcel and the original sounding in the environment. This quantity will also approximately represent the corresponding area on a pseudoadiabatic diagram. For upward displacements, the energy is positive or negative according as the environment is colder or warmer than the ascending parcel. Positive area contributes to instability, and negative area contributes to stability. The stability of the atmosphere must be investigated for every significant level along the sounding. Three general types of stability are recognized for this case where finite displacements are permitted: (1) absolute instability, (2) absolute stability, and (3) conditional instability. Conditional instability is divided into the subclasses of (*a*) stable type, (*b*) pseudo-latent instability, and (*c*) real latent instability. Soundings exhibiting each type of stability are given in Figs. 42 to 45. A summary of the criteria is given in Table 7 from Petterssen.[33]

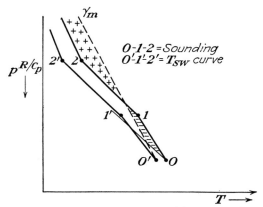

FIG. 44.—Conditional instability of the real latent type for a parcel.

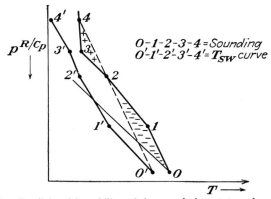

FIG. 45.—Conditional instability of the pseudo-latent type for a parcel.

It may be noted that the assumptions made in deriving the above criteria are too far from the truth to make the conclusions certain in borderline cases. When the area method indicates strong stability or strong instability, the criteria are generally valid. Otherwise the more refined methods provided by the circulation theorem should be used (see Sec. X).

Ascent or Descent of a Layer and Effect on Lapse Rate (Convective Stability). Consider the effects on the lapse rate in a stratum of air that is lifted bodily (as by a front or orographically). Two cases are considered.

I. The entire layer is and remains unsaturated. The quantity $(1/\theta)/(\partial\theta/\partial p)$ is invariant with respect to the motion, since θ is constant at every point (the lifting is

TABLE 7.—CRITERIA OF STABILITY AND INSTABILITY DEDUCED ON THE ASSUMPTION OF AN UNDISTURBED ENVIRONMENT (THE PARCEL METHOD)

	Nonsaturated air: displaced parcel not becoming saturated	Saturated air: displaced parcel remaining saturated
Absolute stability....	$\gamma < \gamma_d$ or $\dfrac{\partial \theta}{\partial z} > 0$	$\gamma < \gamma_m$ or $\dfrac{\partial \theta_{se}}{\partial z} > 0$ or $-\dfrac{\partial T_{sw}}{\partial z} < \gamma_m$ or $\dfrac{\partial \theta_{sw}}{\partial z} > 0$
Indifferent state......	$\gamma = \gamma_d$ or $\dfrac{\partial \theta}{\partial z} = 0$	$\gamma = \gamma_m$ or $\dfrac{\partial \theta_{se}}{\partial z} = 0$ or $-\dfrac{\partial T_{sw}}{\partial z} = \gamma_m$ or $\dfrac{\partial \theta_{sw}}{\partial z} = 0$
Absolute instability..	$\gamma > \gamma_d$ or $\dfrac{\partial \theta}{\partial z} < 0$	$\gamma > \gamma_m$ or $\dfrac{\partial \theta_{se}}{\partial z} < 0$ or $-\dfrac{\partial T_{sw}}{\partial z} > \gamma_m$ or $\dfrac{\partial \theta_{sw}}{\partial z} < 0$

	Nonsaturated air: displaced parcel becoming saturated
Absolute stability....	$\gamma < \gamma_m$
Conditional instability:	
Stable type........	$\gamma_m < \gamma < \gamma_d$ and also no potential pseudo-wet-bulb temperature line intersecting the ascent curve
Pseudo-latent......	$\gamma_m < \gamma < \gamma_d$ and also some of the potential pseudo-wet bulb temperature lines intersecting the ascent curve, and the positive area smaller than the negative area
Real latent........	$\gamma_m < \gamma < \gamma_d$ and also some of the potential pseudo-wet-bulb temperature lines intersecting the ascent curve, and the positive area larger than the negative area
Absolute instability..	$\gamma > \gamma_d$ or $\dfrac{\partial \theta}{\partial z} < 0$

assumed to occur adiabatically), and Δp remains the same provided that there is zero horizontal divergence. Then, using Eq. (204) it is found that

$$\frac{1}{\theta} \frac{\partial \theta}{\partial p} = \frac{1}{\theta} \frac{\partial \theta}{\partial z} \frac{\partial z}{\partial p} = \frac{\gamma - \gamma_d}{g\rho T} \tag{251}$$

or

$$\frac{\gamma - \gamma_d}{p} = \text{const} \tag{252}$$

In other words, the difference between the lapse rate and the dry-adiabatic lapse rate is proportional to the pressure (whether the layer is lifted or lowered). Ascent will cause the actual lapse rate to approach γ_d, and descent will cause it to depart from γ_d. If account is taken of horizontal divergence when it is nonzero, it is found that similarly

$$\frac{\gamma - \gamma_d}{A p} = \text{const} \tag{253}$$

where A is the cross-sectional area of the layer. This area will change with the movement of the stratum.

The above result may be shown with ease on any thermodynamic diagram. Suppose both top and bottom points representing a stratum of air are lifted along an adiabatic, and assume that the straight line connecting the two points continues to represent the sounding through the lifted layer. The above conclusions follow qualitatively, as seen in Fig. 46.

II. Part of the layer becomes saturated owing to lifting. This case does not bear a simple mathematical treatment as does case I. An investigation may be made on a thermodynamic diagram, however. Consider the sounding to be known through a given layer. It is found that if the layer is dry above and moist below, lifting will produce instability. If the layer is wet above and dry below, lifting tends to produce

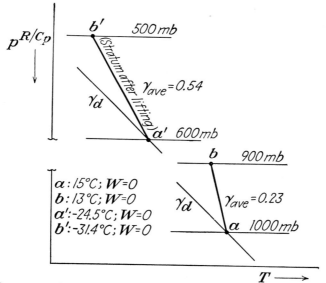

Fig. 46.—Dry stratum before and after lifting with zero divergence. Lapse rate increases as stratum ascends.

stability. The two cases are clearly shown in Figs. 47 and 48. It is assumed in both illustrations that the horizontal divergence of flow is zero.

Considering infinitely thin layers, it may be shown by use of the thermodynamic diagram that, for lifting with zero divergence, we have

$$\frac{\partial \theta_{sw}}{\partial z} < 0 \qquad \text{unstable}$$

$$\frac{\partial \theta_{sw}}{\partial z} > 0 \qquad \text{stable}$$

$$\frac{\partial \theta_{sw}}{\partial z} = 0 \qquad \text{neutral}$$

The derivatives may be evaluated approximately between any two levels in the sounding. The entire layer is said to be convectively unstable, stable, or neutral according to the value of the derivative.

If the amount of divergence is known (positive or negative) from considerations of the flow pattern, its effect may be estimated on the thermodynamic diagram. Thus assume that a layer 100 mb in thickness is raised from a lower level to a higher while 20 per cent of the mass of air between top and bottom flows out in the horizontal plane. After the lifting is completed, a spread of only 80 mb in pressure will fix the top and bottom of the original layer. Assuming that the original "top" of layer ascends adiabatically to 80 mb above the final position of the original "bottom," one obtains an approximate picture on the diagram of what the sounding would be after lifting. *This procedure is approximate at best when applied to thick layers. Since the actual amount of divergence is rarely known accurately, the method must be applied with caution.*

General Aspects of Viscosity. The results of this section are valid for a fluid that is mechanically and thermodynamically *perfect*. The fundamental assumptions have been made that (1) fluid friction is negligibly small, and (2) the fluid is capable

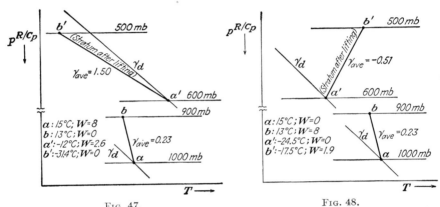

Fig. 47.

Fig. 48.

Fig. 47.—Bottom of layer becoming saturated while top remains dry during lifting with zero divergence. Stratum becomes unstable.

Fig. 48.—Top of layer becoming saturated while bottom remains dry during lifting with zero divergence. Stratum stability increases.

of executing reversible thermodynamic cycles. These assumptions were introduced mathematically by writing

I. Equation of mechanical equilibrium

$$g \, dz + u \, du + v \, dp = 0$$

II. First and second laws of thermodynamics

$$dq]_{\text{rev}} = T \, ds = de + p \, dv = dh - v \, dp$$

III. Equation of state

$$pv = RT$$

Equations I, II, III are excellent first approximations to reality in the atmosphere under very general conditions. They may be considered valid for (1) laminar flow, (2) heat exchange by continuous orderly processes, and (3) the temperatures and pressures encountered in the troposphere. When these conditions are not present, the fundamental equations must be modified. In a severe thunderstorm, for example, it is unlikely that any of the above three equations is applicable. Then conditions are such that the actual viscosity of the atmosphere cannot be neglected.

The direction of change introduced by departure from the above ideal conditions is given by general theory. The entropy of the system must increase (considered to be isolated). Friction effects will reduce the available energy. The quantitative results of an increase in entropy and of turbulence remain to be given.   The general case is perhaps impossible of treatment.   Special cases have been solved for simple boundary conditions.   These special cases do not concern us particularly in a meteorological analysis and are not discussed here.

Bibliography

1. DRYDEN, H. L.: *Quart. Appl. Math.*, **1**: 7 (1943).
2. ZEMANSKY, M. W.: "Heat and Thermodynamics," 2d ed., McGraw-Hill, New York, 1943.
3. KEENAN, J. H.: "Thermodynamics," Wiley, New York, 1941.
4. SLATER, J. C.: "Introduction to Chemical Physics," McGraw-Hill, New York, 1939.
5. BRIDGMAN, P. W.: "A Condensed Collection of Thermodynamic Formulas," Harvard University Press, Cambridge, Mass.
6. JOOS, G.: "Theoretical Physics," Blackie, Glasgow, 1934.
7. PAGE, L.: "Introduction to Theoretical Physics," London, Van Nostrand, New York, 1928.
8. FOWLER, R. H.: "Statistical Mechanics," Cambridge, 1936.
9. SAHA, M. N., and B. N. SRIVASTAVA: "A Treatise on Heat," Allahabad, 1935.
10. GLASEBROOK: "A Dictionary of Applied Physics," vol. 1.
11. PARTINGTON and SHILLING: "Specific Heats of Gases."
12. KEENAN, J. H., and KEYES, G. K.: "Thermodynamic Properties of Steam," Wiley, New York, 1936.
13. KIEFER, P. J.: "The Thermodynamic Properties of Water Vapor," *Monthly Weather Rev.*, **69** (November, 1941).
14. BURGESS: *Bur. Standards J. Research*, **1** (1928).
15. "Temperature, Its Measurement and Control in Science and Industry," Reinhold, New York, 1941.
16. HARRISON, L. P.: *Monthly Weather Rev.*, **62**: 247 (1934).
17. GOODMAN, W.: "Air Conditioning Analysis," Macmillan, New York, 1943.
18. WASHBURN, E. W.: *Monthly Weather Rev.*, **52**: 488 (1924).
19. GOFF, J. A.: *Heating, Piping, and Air Conditioning*, February, 1942.
20. TETANS, O: *Z. Geophys.*, **6**: 297 (1930).
21. HAURWITZ, B.: "Dynamic Meteorology," McGraw-Hill, New York, 1941.
22. MILNE, E. A.: "Thermodynamics of the Stars (Handbuch der Astrophysik)," Springer, Berlin, 1933.
23. EDDINGTON, A. S.: "Internal Constitution of the Stars," Cambridge, 1926.
24. PANETH, F. A.: *Quart. J. Roy. Meteorolog. Soc.*, **65**: 304 (1939).
25. FLEMING, J. A.: "Terrestial Magnetism and Electricity," McGraw-Hill, New York, 1939.
26. CHAPMAN, S., and E. A. MILNE: *Quart J. Roy. Meteorolog. Soc.*, **46**: 357 (1928).
27. FLETCHER, R. D., and R. D. GRAHAM: *Bull. Am. Meteorolog. Soc.*, September, 1943.
28. BRUNT, D.: "Physical and Dynamical Meteorology," Cambridge, 1939.
29. ROSSBY, C.-G.: Thermodynamics Applied to Air Mass Analysis, *M.I.T. Papers* 1932.
30. SOLOT, S. B.: *Monthly Weather Rev.*, **67**: 100 (1939).
31. FULKS, J. R.: *Monthly Weather Rev.*, October, 1935.
32. BLEEKER, W.: *Quart. J. Roy. Meteorolog. Soc.*, **65**: 282 (1939).
33. PETTERSSEN, S.: "Weather Analysis and Forecasting," McGraw-Hill, New York, 1940.
34. MONTGOMERY, R. B., and A. F. SPILHAUS: *J. Aeronaut. Science*, **8**: 276–283 (1941).
35. DALTON, P.: *J. Aeronaut. Science*, **4**: 154 (1937).
36. KEYES, F. G., and SMITH, L. B.: *Refrigerating Eng.*, **27** (March, 1934).
37. EWELL, A. W.: *Refrigerating Eng.*, **27** (March, 1934).
38. EWELL, A. W.: *Refrigerating Eng.*, **35** (March, 1938).

SECTION VI

KINEMATICS AND DYNAMICS OF FLUID FLOW

By H. J. Stewart

CONTENTS

SECTION VI

KINEMATICS AND DYNAMICS OF FLUID FLOW

By H. J. STEWART

1. Introduction. The science of meteorology is essentially a branch of fluid mechanics; however, it is in certain respects one of the most difficult branches. From an analytical standpoint, this is caused by the fact that those phases of fluid mechanics such as heat transfer, turbulence, and the stability of fluid motions which are of the utmost importance in any comprehensive theory of the general circulation of the atmosphere are phases for which the basic mathematical techniques either are not yet available or are extremely complex. In addition, the experimental data for the arctic regions and the stratosphere are rather meager; so there is as yet no general agreement as to the experimental facts of the motions of the atmosphere in these regions. In contrast to the majority of the physical sciences, the investigation of controlled laboratory experiments is seldom possible in meteorological problems, for the large scale of the atmospheric motions is often an essential feature. For these reasons, there does not exist at the present time any complete and coherent theory of the general atmospheric circulation. One of the earliest attempts at the development of such a theory was made in 1888 by Helmholz,[1] who to a certain extent anticipated the modern polar-front theories, which were formulated primarily by the Norwegian school of meteorologists headed by V. Bjerknes during the First World War.[2-4] The impossibility of an axially symmetric meridional circulation was pointed out by Exner[5] and Jeffreys.[6] The effect of the instability of the shearing zones near the belts of westerlies in determining the resulting cellular motion of the atmosphere has been discussed by Stewart.[7,8] The role of high-level isentropic mixing has been discussed by Rossby,[9] who has also discussed the effects of wave disturbances in the westerlies.[10]

Since these investigations have dealt primarily with special features of the motion of the atmosphere, it is often difficult to separate cause and effect in applying them to the larger problem. It is obvious that the development of a more complete theory of the general atmospheric circulation would be of great importance, not only for theoretical purposes, but particularly for the development of improved weather-forecasting techniques.

Although the general theory is still not complete, most of the details of the atmospheric motions are well understood and can be investigated successfully by means of the standard techniques of fluid mechanics. Some of the simpler problems will be discussed by purely kinematical methods; however, a complete consideration of the forces involved is necessary in most cases.

The essential property that characterizes a fluid is the fact that a fluid will not withstand any shear stress, however small it may be, without yielding to it. It is obvious that a fluid in motion may experience shear stresses, and under these conditions the fluid elements are continuously deformed. In many cases, the influence of these shearing stresses is small, and satisfactory approximations may be obtained without considering them. It will be assumed throughout that the atmosphere may be treated as a fluid continuum.

There are two general methods of describing fluid motions. These methods are generally referred to as the *Lagrangian* and the *Eulerian* methods, although both methods were used by Euler. In the Lagrangian method, the trajectories of the fluid elements are described. If the original coordinates of a fluid element are (a,b,c) and its coordinates at any other time t are (x,y,z), then the fluid motion is described in the Lagrangian method by the following equations:

$$x = x(a,b,c,t)$$
$$y = y(a,b,c,t) \tag{1.1}$$
$$z = z(a,b,c,t)$$

For a given fluid element, a, b, and c are fixed, and Eq. (1.1) obviously describes the trajectory of the fluid element. The velocities and accelerations of the element can be obtained by partial differentiations with respect to time. Similarly, the fluid characteristics pressure, temperature, and density are described in terms of the element coordinates and the time.

In the Eulerian system, the velocity field is described in terms of the spatial coordinates and time. If (u,v,w) are the velocity components in the (x,y,z) directions, respectively, in a system of rectangular coordinates, then the velocity field is described by the following system of equations:

$$u = u(x,y,z,t)$$
$$v = v(x,y,z,t) \tag{1.2}$$
$$w = w(x,y,z,t)$$

Similar expressions are also used for the pressure, temperature, and density fields. The Lagrangian equations can be considered as integrals of the Eulerian equations. The only modern writers who have used the Lagrangian method extensively are V. Bjerknes and his collaborators.[11] For the solution of most problems, the Eulerian method is somewhat easier, and for that reason it is almost always used. The Eulerian method will be used throughout the present work.

It is often necessary to consider the variation with time of some property of a given fluid element. In the Lagrangian notation, this is given by the partial derivative with respect to time; however, it is more complex in the Eulerian system, for in general all four coordinates of a given fluid element change with time. If a particle at time t is at the point (x,y,z), a short time dt later it will be at the point $(x + u\,dt, y + v\,dt, z + w\,dt)$. The change in any quantity $F(x,y,z,t)$ is thus

$$dF = \left(\frac{\partial F}{\partial t} + u\frac{\partial F}{\partial x} + v\frac{\partial F}{\partial y} + w\frac{\partial F}{\partial z} \right) dt \tag{1.3}$$

The quantity in parentheses is thus the time rate of change of F for the fluid element. Because of its frequent recurrence, it is given the special notation DF/Dt, which was first introduced by Stokes. Thus

$$\frac{DF}{Dt} = \frac{\partial F}{\partial t} + u\frac{\partial F}{\partial x} + v\frac{\partial F}{\partial y} + w\frac{\partial F}{\partial z} \tag{1.4}$$

2. Frontogenesis and Frontolysis. In general, the characteristics of the atmosphere vary rather slowly from point to point; however, a large portion of the extratropical weather is intimately associated with the existence of narrow zones where the characteristics vary rapidly. These transition zones are generally thought of as surfaces of discontinuity or frontal surfaces which separate two dissimilar air masses. One of the fundamental atmospheric processes that must be discussed is the process by which such atmospheric discontinuities are produced. This process is called *frontogenesis.* Negative frontogenesis, *i.e.,* the destruction of atmospheric discontinuities, is called *frontolysis.* Kinematical studies of this problem have been made by Bergeron,[12] Bjerknes,[11] and Petterssen.[13] The present study will follow the method outlined by Petterssen.

Since the problem involves the change with time of the gradients of fluid properties, it is necessary to consider in some detail the deformation of the elements in a fluid flow. The nature of the flow in the neighborhood of a point will thus be investigated. In order to simplify the analysis, the velocity field in a plane surface will be considered rather than the more general three-dimensional field. If u and v are the x- and y-components of the velocity at the point (x,y) as shown in Fig. 1, these velocity fields may be expanded in a Taylor series about the origin

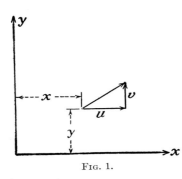

$$u = u + \left(\frac{\partial u}{\partial x}\right)_0 x + \left(\frac{\partial u}{\partial y}\right)_0 y + \cdots$$

$$v = v_0 + \left(\frac{\partial u}{\partial x}\right)_0 x + \left(\frac{\partial v}{\partial y}\right)_0 y + \cdots \qquad (2.1)$$

Fig. 1.

and for small enough values of x and y, the higher order terms may be neglected.

The significance of some of the various velocity derivatives may be shown by considering their change from one coordinate system to another. Let the x, y-coordi-

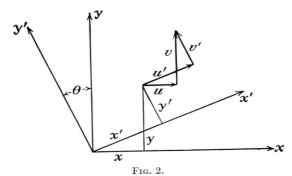

Fig. 2.

nate system be rotated through an angle θ to form the coordinate system x', y', as in Fig. 2. Then

$$u = u' \cos \theta - v' \sin \theta$$
$$v = u' \sin \theta + v' \cos \theta \qquad (2.2)$$

and $\qquad x' = x \cos \theta + y \sin \theta \qquad y' = -x \sin \theta + y \cos \theta \qquad (2.3)$

From this

$$\frac{\partial v}{\partial x} = \cos \theta \left(\sin \theta \frac{\partial u'}{\partial x'} + \cos \theta \frac{\partial v'}{\partial x'} \right) - \sin \theta \left(\sin \theta \frac{\partial u'}{\partial y'} + \cos \theta \frac{\partial v'}{\partial y'} \right) \qquad (2.4)$$

and $\qquad \dfrac{\partial u}{\partial y} = \sin \theta \left(\cos \theta \dfrac{\partial u'}{\partial x'} - \sin \theta \dfrac{\partial v'}{\partial x'} \right) + \cos \theta \left(\cos \theta \dfrac{\partial u'}{\partial y'} - \sin \theta \dfrac{\partial v'}{\partial y'} \right) \qquad (2.5)$

If these expressions are subtracted, it is seen that

$$\frac{\partial v}{\partial x} - \frac{\partial u}{\partial y} = \frac{\partial v'}{\partial x'} - \frac{\partial u'}{\partial y'} \qquad (2.6)$$

The quantity

$$2c = \frac{\partial v}{\partial x} - \frac{\partial u}{\partial y} \qquad (2.7)$$

is thus seen to be independent of the orientation of the coordinate axes and must represent some physical characteristic of the fluid motion. Its significance can be determined from a study of the motion of two infinitesimal lines of fluid particles which at the initial instant lie along the x- and y-coordinate axes, as shown in Fig. 3. An infinitesimal time dt later, the position of these particles is also shown. It may be seen that each row is subjected to a translation, a rotation, and a stretching. The

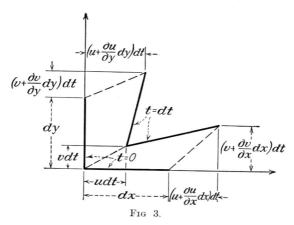

FIG 3.

elongation of the lines of particles is a second-order effect and may be neglected. From this, the angular velocity ω_1 of the particles originally on the x-axis is

$$\omega_1 = \frac{\partial v}{\partial x} \qquad (2.8)$$

Similarly the angular velocity ω_2 of the particles originally on the y axis is

$$\omega_2 = -\frac{\partial u}{\partial y} \qquad (2.9)$$

Counterclockwise rotations have been assumed to be positive. From this

$$c = \tfrac{1}{2}(\omega_1 + \omega_2) \qquad (2.10)$$

and c is seen to be the average rate of rotation of the fluid element.

Similarly

$$\frac{\partial u}{\partial x} = \cos\theta \left(\cos\theta\, \frac{\partial u'}{\partial x'} - \sin\theta\, \frac{\partial v'}{\partial x'}\right) - \sin\theta \left(\cos\theta\, \frac{\partial u'}{\partial y'} - \sin\theta\, \frac{\partial v'}{\partial y'}\right) \qquad (2.11)$$

and

$$\frac{\partial v}{\partial y} = \sin\theta \left(\sin\theta\, \frac{\partial u'}{\partial x'} + \cos\theta\, \frac{\partial v'}{\partial x'}\right) + \cos\theta \left(\sin\theta\, \frac{\partial u'}{\partial y'} + \cos\theta\, \frac{\partial v'}{\partial y'}\right) \qquad (2.12)$$

If these are added it is seen that

$$\frac{\partial u}{\partial x} + \frac{\partial v}{\partial y} = \frac{\partial u'}{\partial x'} + \frac{\partial v'}{\partial y'} \qquad (2.13)$$

The quantity

$$2b = \frac{\partial u}{\partial x} + \frac{\partial v}{\partial y} \tag{2.14}$$

is also seen to be independent of the orientation of the coordinate system.   If $(\partial u/\partial x)$ $+ (\partial v/\partial y)$ is integrated over the area S enclosed by a contour C (see Fig. 4), then, by Green's theorem for the plane

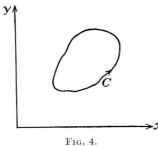

$$\iint_S \left(\frac{\partial u}{\partial x} + \frac{\partial v}{\partial y} \right) dS = \oint_C (u \, dy - v \, dx) \tag{2.15}$$

The right-hand integral is the net volume of fluid (per unit thickness in the z-direction) that flows out across the contour C in a unit time.   The quantity $(\partial u/\partial x) + (\partial v/\partial y)$ can thus be interpreted as the source strength per unit of area in the plane.   The constant b will be called the *coefficient of horizontal divergence.*

Fig. 4.

If Eqs. (2.4) and (2.5) are added, it is seen that

$$\frac{\partial v}{\partial x} + \frac{\partial u}{\partial y} = \cos 2\theta \left(\frac{\partial v'}{\partial x'} + \frac{\partial u'}{\partial y'} \right) + \sin 2\theta \left(\frac{\partial u'}{\partial x'} - \frac{\partial v'}{\partial y'} \right) \tag{2.16}$$

and if Eqs. (2.11) and (2.12) are subtracted, it seen that

$$\frac{\partial u}{\partial x} - \frac{\partial v}{\partial y} = \cos 2\theta \left(\frac{\partial u'}{\partial x'} - \frac{\partial v'}{\partial y'} \right) - \sin 2\theta \left(\frac{\partial v'}{\partial x'} + \frac{\partial u'}{\partial y'} \right) \tag{2.17}$$

From Eq. (2.16), the axes can always be placed so that $(\partial v/\partial x) + (\partial u/\partial y) = 0$ and $(\partial u/\partial x) - (\partial v/\partial y) > 0$.   Such axes will be called *principal axes.* For principal axes, the field of motion may be represented by three coefficients

$$
\begin{aligned}
a &= \frac{1}{2} \left(\frac{\partial u}{\partial x} - \frac{\partial v}{\partial y} \right) \\
b &= \frac{1}{2} \left(\frac{\partial u}{\partial x} + \frac{\partial v}{\partial y} \right) \\
c &= \frac{1}{2} \left(\frac{\partial v}{\partial x} - \frac{\partial u}{\partial y} \right)
\end{aligned}
\tag{2.18}
$$

Equation (2.1) may then be rewritten as

$$
\begin{aligned}
u &= u_0 + ax + bx - cy \\
v &= v_0 - ay + by + cx
\end{aligned}
\tag{2.19}
$$

The field of motion has thus been split into four component parts that can be studied separately; however, it is more convenient to discuss them in terms of the fluid streamlines.   A streamline is a line for which the tangent at any point gives the direction of the fluid motion at that point.   If $\mathbf{u}$ is the velocity vector at a point and $d\mathbf{r}$ is an element of the streamline through that point

$$\mathbf{u} \times d\mathbf{r} = 0 \tag{2.20}$$

This may be considered as a set of differential equations defining the streamlines for any given velocity field.   For motion in a plane where z is constant, this may be written as

$$\frac{dx}{u} = \frac{dy}{v} \tag{2.21}$$

or

$$0 = v\,dx - u\,dy \tag{2.22}$$

It should be noted that the streamlines are not the paths of the fluid particles (often called *streak lines* or *trajectories*) unless the motion is steady.

The first component of the field of motion, $u = u_0$ and $v = v_0$, is obviously a rectilinear field of motion and needs no further discussion. For the next component, $u = ax$ and $v = -ay$, the equation of the streamlines is

$$0 = -a(y\,dx + x\,dy) \tag{2.23}$$

or

$$xy = C_1 \tag{2.24}$$

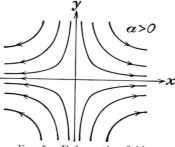

FIG. 5.—Deformation field.

where C_1 is a constant. The streamlines are thus rectangular hyperbolas. The streamlines for $a > 0$ are shown in Fig. 5. This field of motion is called a *deformation field*, and the x-axis is called the *deformation axis*. It must be remembered that the x-axis was chosen so that

$$\frac{\partial v}{\partial x} + \frac{\partial u}{\partial y} = 0 \qquad \text{and} \qquad a = \frac{1}{2}\left(\frac{\partial u}{\partial x} - \frac{\partial v}{\partial y}\right) > 0$$

For the third component, $u = bx$ and $v = by$, the equation of the streamlines is

$$0 = b(y\,dx - x\,dy) \tag{2.25}$$

or

$$\frac{y}{x} = C_2 \tag{2.26}$$

The streamlines are radial lines from the origin and are shown in Fig. 6 for $b > 0$. Such a flow with horizontal divergence is generally possible only if there is a vertical

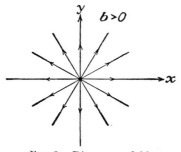

FIG. 6.—Divergence field.

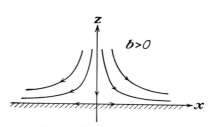

FIG. 7.—Vertical subsidence.

subsidence, and the flow in the xz-plane must be of the type shown in Fig. 7. If b is negative, the direction of flow is everywhere reversed, and the flow shows horizontal convergence.

The equation for the streamlines for the fourth component, $u = -cy$ and $v = cx$, is

$$0 = c(x\,dx + y\,dy) \tag{2.27}$$

or

$$x^2 + y^2 = C_3 \tag{2.28}$$

The streamlines are circles about the origin, and the motion shown in Fig. 8 for $c > 0$ corresponds to a rotation of the fluid about the origin with an angular velocity equal to c.

These results show that the field of motion in the neighborhood of a point may be considered as a uniform translation plus a deformation plus a divergence plus a rotation. This general result may easily be extended to three-dimensional motions.

The magnitude of the horizontal temperature gradient is a convenient measure of the rate of variation of an air mass' characteristics. In order for the tendency toward the generation of a front (or frontogenesis) to exist, it is necessary that the fluid motion

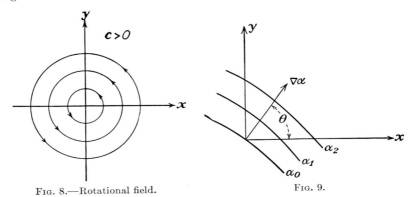

FIG. 8.—Rotational field.   FIG. 9.

be such that these horizontal gradients increase in magnitude with time. For simplicity, we shall follow Petterssen and consider, instead of the temperature, any conservative property $\alpha = \alpha(x,y,t)$. The function

$$F = \frac{D}{Dt} |\nabla\alpha| \tag{2.29}$$

is a measure of the tendency toward frontogenesis. If F is positive, frontogenesis exists; if F is negative, frontolysis exists. F is called the *frontogenetical function*.

Now

$$2|\nabla\alpha| \frac{D}{Dt} |\nabla\alpha| = \frac{D}{Dt} (\nabla\alpha)^2 \tag{2.30}$$

Since α is a conservative property

$$\frac{D\alpha}{Dt} = 0 \tag{2.31}$$

From the gradient of Eq. (2.31), it is seen that

$$\frac{\partial}{\partial t} \nabla\alpha + \mathbf{v} \cdot \nabla \nabla\alpha + \nabla\alpha \cdot \nabla\mathbf{v} = 0 \tag{2.32}$$

or

$$\frac{\partial}{\partial t} (\nabla\alpha)^2 + \mathbf{v} \cdot \nabla(\nabla\alpha)^2 = -2(\nabla\alpha) \cdot (\nabla\alpha \cdot \nabla\mathbf{v}) \tag{2.33}$$

Thus

$$F = -\frac{\nabla\alpha}{|\nabla\alpha|} \cdot (\nabla\alpha \cdot \nabla\mathbf{v}) \tag{2.34}$$

Since

$$\nabla\alpha = |\nabla\alpha|(\mathbf{i} \cos\theta + \mathbf{j} \sin\theta) \tag{2.35}$$

where θ is the angle from the x-axis measured counterclockwise to the vector $\nabla\alpha$ as in Fig. 9, the expression for the frontogenetical function is

$$F = -|\nabla\alpha| \left[\cos^2\theta \, \frac{\partial u}{\partial x} + \sin^2\theta \, \frac{\partial v}{\partial y} + \sin\theta\cos\theta \left(\frac{\partial u}{\partial y} + \frac{\partial v}{\partial x} \right) \right] \qquad (2.36)$$

If the x- and y-axes are principal axes, the velocity derivatives may be expressed in terms of the coefficients a, b, and c of Eq. (2.18), and

$$F = -|\nabla\alpha|(a\cos 2\theta + b) \qquad (2.37)$$

It is apparent from this that only the deformation and horizontal divergence fields are important for problems of frontogenesis. Although it is also obvious from physical considerations, Eq. (2.37) shows that horizontal convergence, $b < 0$, produces frontogenesis; however, the horizontal divergence or convergence in the atmosphere is generally very small; and a very good approximation can be obtained by neglecting the coefficient b. Since both $|\nabla\alpha|$ and a are positive, the frontogenetical function, with $b = 0$, will be positive if $45\ \mathrm{deg} < |\theta| < 135\ \mathrm{deg}$. This means that the lines of constant α must make an angle of less than 45 deg with the x-axis (the deformation axis) in order for frontogenesis to exist. Horizontal convergence increases this angle; and horizontal divergence decreases it. For a given deformation field, the frontogenetical effect is obviously strongest if the lines of constant α are parallel to the deformation axis ($|\theta| = 90$ deg).

A positive value of the frontogenetical function is certainly a necessary condition for the formation of a front; however, it is obviously not a sufficient condition. If a front is to form, it is most likely to form in the region where F is the largest. A more complete discussion of this problem is given in Petterssen's original paper.

The strongest frontogenetical region in the Northern Hemisphere is in the Pacific Ocean off the eastern Siberian coast. In this area, there is in the winter season a very strong temperature gradient between the cold land mass and the warm ocean; and the isotherms are therefore practically parallel to the coast line. In addition, the outflow of air from the continental anticyclone into the maritime air mass produces a strong deformation zone with its axis also practically parallel to the coast line. Similar conditions and another strong frontogenetical zone exist off the eastern coast of North America. It is the frontal systems created in these regions that are primarily responsible for the weather in western North America and in Europe.

3. Kinematical Analysis of the Pressure Field. The classical method of weather forecasting and one that has been used ever since the beginnings of technical meteorology involves the consideration of the past history of any given weather element or storm and, through extrapolation, the prediction of its future position and intensity. Although this qualitative idea is very old and has been widely used, the formal rigorous treatment of this method of forecasting has been given only fairly recently, primarily by Petterssen,[14] who discussed the problem of the prediction of changes in the surface pressure field by purely kinematical methods and without reference to the atmospheric forces involved.

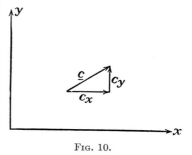

Fig. 10.

In order to develop these extrapolation methods, it is first necessary to calculate the time rate of change of any given quantity with respect to a moving point. Suppose a point is moving with a velocity **c** with c_x and c_y being the velocity components in the x- and y-directions, respectively, as in **Fig. 10,** then the time rate of change of any function $F = F(x,y,t)$ for this point is

$$\frac{\delta F}{\delta t} = \frac{\partial F}{\partial t} + \mathbf{c} \cdot \nabla F \tag{3.1}$$

The second time derivative with respect to this moving point is then

$$\frac{\delta^2 F}{\delta t^2} = \left(\frac{\partial}{\partial t} + \mathbf{c} \cdot \nabla \right) \frac{\delta F}{\delta t} \tag{3.2}$$

or $$\frac{\delta^2 F}{\delta t^2} = \frac{\partial^2 F}{\partial t^2} + 2\mathbf{c} \cdot \nabla \frac{\partial F}{\partial t} + \mathbf{c} \cdot \left(\mathbf{c} \cdot \nabla \nabla F \right) + \left(\frac{\partial \mathbf{c}}{\partial t} + \mathbf{c} \cdot \nabla \mathbf{c} \right) \cdot \nabla F \tag{3.3}$$

Now the acceleration of the moving point is

$$\mathbf{A} = \frac{\delta \mathbf{c}}{\delta t} \tag{3.4}$$

and Eq. (3.3) may therefore be, rewritten as

$$\frac{\delta^2 F}{\delta t^2} = \frac{\partial^2 F}{\partial t^2} + 2\mathbf{c} \cdot \nabla \frac{\partial F}{\partial t} + \mathbf{c} \cdot (\mathbf{c} \cdot \nabla \nabla F) + \mathbf{A} \cdot \nabla F \tag{3.5}$$

In a similar manner, expressions for computing all the higher order derivatives may be obtained.

As an application of Petterssen's method, these formulas will be used to compute the velocity of an isobar. Suppose the point P on the p_0 isobar in Fig. 11 is restricted to motions such that it remains on both the x-axis and the p_0 isobar. For this point, $c_y = 0$, and

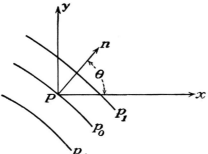

$$\frac{\delta p}{\delta t} = 0 = \frac{\partial p}{\partial t} + c_x \frac{\partial p}{\partial x} \tag{3.6}$$

The velocity with which the point moves along the x-axis (and thus the velocity of the isobar) is

$$c_x = - \frac{\partial p / \partial t}{\partial p / \partial x} \tag{3.7}$$

Fig. 11.

If the point P had been restricted to motion along the intersection of the p_0 isobar and the y-axis, then the velocity of the point would have been

$$c_y = - \frac{\partial p / \partial t}{\partial p / \partial y} \tag{3.8}$$

Similarly if P had been restricted to motion along the normal to the isobar, its velocity in the normal direction would have been

$$c_n = - \frac{\partial p / \partial t}{\partial p / \partial n} \tag{3.9}$$

Since $$\frac{\partial p}{\partial n} = \frac{\partial p}{\partial x} \sec \theta = \frac{\partial p}{\partial y} \csc \theta \tag{3.10}$$

the velocities have the relations

$$c_n = c_x \cos \theta = c_y \sin \theta \tag{3.11}$$

The significance of these relations may be seen easily by considering the displacements of the point P in a short time Δt. These are shown in Fig. 12. By Eq. (3.11), it can be seen that the three points P', P'', and P''' all lie on a straight line that is parallel to the isobar p_0 at the point P. The isobar p_0 has thus been subjected to a displacement parallel to itself. Since an isobar is a line and since the point on the line $P'P'''$ that the point P goes into is a matter of no consequence, all three equations (3.7), (3.8), and (3.9) are thus equivalent. As a matter of convenience, it is generally best to compute directly the displacement of an isobar in the direction normal to itself.

Equation (3.9) thus provides a direct method of estimating the rate at which the surface-pressure chart is changing; however, the forecaster is primarily interested in the behavior of special features of the pressure chart such as high- or low-pressure centers or fronts rather than the behavior of the separate isobars. Special formulas can easily be developed for treating these special features directly. A pressure center

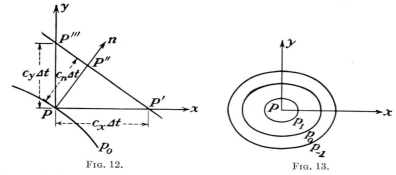

FIG. 12.   FIG. 13.

is characterized by the fact that at the center $\nabla p = 0$. This is true for either a high, a low, or a saddle point. If a point P is restricted to move with a pressure center, as shown in Fig. 13, it is characterized by the fact that

$$\frac{\delta}{\delta t}\, \nabla p = 0 \tag{3.12}$$

If this is expanded in terms of its components, it is seen that

$$0 = \frac{\partial^2 p}{\partial x\, \partial t} + c_x \frac{\partial^2 p}{\partial x^2} + c_y \frac{\partial^2 p}{\partial x\, \partial y}$$

and

$$0 = \frac{\partial^2 p}{\partial y\, \partial t} + c_x \frac{\partial^2 p}{\partial x\, \partial y} + c_y \frac{\partial^2 p}{\partial y^2} \tag{3.13}$$

These form a set of two simultaneous linear equations that may be solved simultaneously for the velocity components c_x and c_y.

These expressions may be considerably simplified through the proper orientation of the axes at the pressure center. Since the pressure is continuous, it may be expanded in a Taylor series about the origin

$$p = p_0 + A x^2 + B xy + C y^2 + \cdots \tag{3.14}$$

and for small enough values of x and y the higher order terms may be neglected. The isobars near the origin are thus seen to be plane quadratic curves (*i.e.*, ellipses or hyperbolas). If the x- and y-axes are oriented along the principal axes of these infinitesimal ellipses or hyperbolas, the coefficient B in Eq. (3.14) must then vanish, and for these principal axes

$$\frac{\partial^2 p}{\partial x\, \partial y} = 0 \tag{3.15}$$

at the origin.   Equation (3.13) may then be solved for the velocity of a pressure center
as

$$c_x = - \frac{\partial^2 p/\partial x\, \partial t}{\partial^2 p/\partial x^2}$$

$$c_y = - \frac{\partial^2 p/\partial y\, \partial t}{\partial^2 p/\partial y^2} \tag{3.16}$$

and

In order to use these simplified expressions, principal axes as shown in Fig. 13 must be
used.

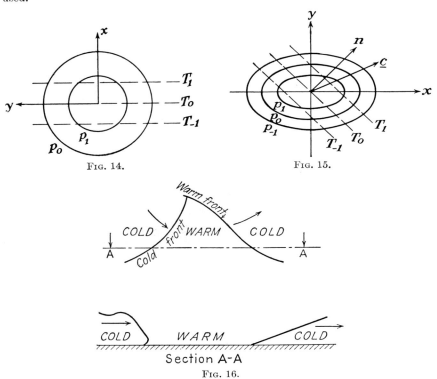

FIG. 14.   FIG. 15.

Section A-A

FIG. 16.

For a circular high, any axis is a principal axis, and it is convenient to take the
x-axis perpendicular to the isallobar ($T = \partial p/\partial t = $ const) through the pressure center
as in Fig. 14.   For this case, $\partial^2 p/\partial y\, \partial t$ and c_y both vanish, and since $\partial^2 p/\partial x\, \partial t > 0$
and $\partial^2 p/\partial x^2 < 0$, it is seen that c_x is positive.   The circular high thus moves in the
direction of increasing pressure tendency.   For an elliptic high with the x-axis the
long axis of the ellipse, the velocity is no longer along the normal to the isallobars as
$|\partial^2 p/\partial x^2| < |\partial^2 p/\partial y^2|$ but is rotated toward the long axis of the high, as shown in
Fig. 15.   Similar interpretations for Eq. (3.16) may easily be worked out for a low
pressure center or for a saddle point.

It is a little more difficult to determine the property of the pressure field that
characterizes a front.   If a section (see Fig. 16) is taken through a normal frontal
system, it is apparent that there must be a discontinuity in the pressure tendency at
a front that is moving, since the surface pressure at a point is a measure of the mass of

air in a unit column above the point, and the tendency at a point near a front is thus primarily a measure of the rate of change of thickness of the layer of cold air over the point. A front is thus characterized by discontinuities in the pressure derivatives; however, the pressure itself must be continuous, for the fluid elements would otherwise be subjected to infinitely large accelerations. This last fact can be used to develop a formula for computing the velocity of a front. Let points 1 and 2 be neighboring points, as shown in Fig. 17, separated only by the front at the point P, and with point 1 being in the warm air mass. Then if P is restricted to the intersection of the x-axis and the front

$$\frac{\delta}{\delta t}(p_2 - p_1) = 0 = \frac{\partial}{\partial t}(p_2 - p_1) + c_x \frac{\partial}{\partial x}(p_2 - p_1) \qquad (3.17)$$

The velocity of the front is thus

$$c_x = -\frac{(\partial p_2/\partial t) - (\partial p_1/\partial t)}{(\partial p_2/\partial x) - (\partial p_1/\partial x)} \qquad (3.18)$$

If the x-axis is set parallel to the warm sector isobars, the term $\partial p_1/\partial x = 0$, and the calculation is somewhat simplified. It may be noted that, if the point P is not at a front, both numerator and denominator of Eq. (3.18) vanish, and, as one might expect, the velocity is indeterminate; however, the tendencies and pressure gradients are discontinuous at a front, and definite results are therefore obtained for this case.

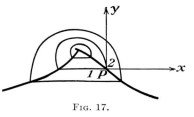

FIG. 17.

In these three examples, Eq. (3.1) has been used to compute the instantaneous velocity of an isobar, a pressure center, and a front. Similarly, Eq. (3.5) may be used to compute the instantaneous accelerations of these systems. This will be worked out explicitly for the acceleration of an isobar. Suppose the point P of Fig. 11 is restricted to motion along the intersection of the x-axis and the p_0 isobar. Then

$$\frac{\delta^2 p}{\delta t^2} = 0 = \frac{\partial^2 p}{\partial t^2} + 2c_x \frac{\partial^2 p}{\partial x\, \partial t} + c_x{}^2 \frac{\partial^2 p}{\partial x^2} + A_x \frac{\partial p}{\partial x} \qquad (3.19)$$

for this point, and its acceleration in the x-direction is thus

$$A_x = -\frac{\dfrac{\partial^2 p}{\partial t^2} + 2c_x \dfrac{\partial^2 p}{\partial x\, \partial t} + c_x{}^2 \dfrac{\partial^2 p}{\partial x^2}}{\dfrac{\partial p}{\partial x}} \qquad (3.20)$$

where the velocity c_x is given by Eq. (3.7). Similar expressions could be worked out for the accelerations of fronts or of pressure centers; however, it is apparent that they are quite complicated.

The application of these results to the problem of forecasting displacements may be illustrated by considering the displacement of an isobar. If $D = D(t)$ is the displacement of the point P of Fig. 11 along the x-axis, then D may be expanded in a Taylor series

$$D = \left(\frac{\delta D}{\delta t}\right)_0 t + \left(\frac{\delta^2 D}{\delta t^2}\right)_0 \frac{t^2}{2!} + \cdots \qquad (3.21)$$

where $(\delta D/\delta t)_0 = c_x$ and $(\delta^2 D/\delta t^2)_0 = A_x$. There are thus two obvious limitations on the use of these methods to compute the displacements. The first involves the accuracy with which any given term may be computed from the available experimental data, and the second arises from the error involved in terminating the series of Eq. (3.21) at any given point.

The primary factor involved in the experimental error in determining the various terms is the calculation of the time derivatives. Let us consider the barogram (see Fig. 18) at a station for which there is a warm-front passage at the time t_f. The actual rate of pressure change $(\partial p/\partial t)$ is not reported; the actual pressure change for the previous 3-hr period is given. For example, at time t_1, the pressure change Δp_1 would be reported, and at time t_2, the change Δp_2 would be given. By inspection, Δp_1 is a fairly good approximation to $\partial p/\partial t$ half a period earlier. On the other hand, Δp_2 is a very poor approximation to $\partial p/\partial t$ at time t_2; in fact, it is even of the wrong sign. Since the pressure differences are reported to a tenth of a millibar pressure, the minimum probable error in $\partial p/\partial t$ is about ± 0.1 mb per 3 hr except just behind a

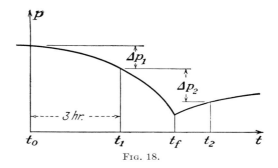

F IG . 18.

front where for all practical purposes $\partial p/\partial t$ is not reported. Since $\partial^2 p/\partial t^2$ must be computed by taking the difference in consecutive pressure changes, its probable error is about ± 0.2 mb per (3 hr)^2.

The pressure at a fixed point may be written as

$$p = p_0 + \left(\frac{\partial p}{\partial t}\right)_0 t + \frac{1}{2}\left(\frac{\partial^2 p}{\partial t^2}\right)_0 t^2 + \cdots \tag{3.22}$$

It is thus seen that the probable error in computing the pressure after 24 hr due to the term in $\partial p/\partial t$ is about ± 0.8 mb. The error due to the next term, however, is about ± 6.4 mb. This indicates that the velocity term may be computed fairly accurately; the acceleration term will probably be of the correct sign, but its accuracy is rather poor, and higher order terms cannot be computed. The use of this formula corresponds to the problem of single-barometer forecasting. At a fixed point, weather is roughly periodic with the cycle from fair weather to storm to fair weather taking a period that is generally of the order of magnitude of 1 week. This indicates that the higher order terms cannot be neglected for periods of much more than 1 day. Since they cannot be computed with any reasonable accuracy, this (single-barometer) method of forecasting is useful primarily for very short range forecasts with the upper limit being about 1 day.

The diurnal variation of any quantity has, of course, a period of 1 day; and the higher order derivatives of this diurnal variation are therefore comparatively large. It is thus generally necessary to subtract out the normal diurnal variation of any quantity, in particular the temperature for which the diurnal variation is often large. before attempting to extrapolate any quantity for a period of 1 or more days.

If the weather at a point that moves with the pressure system, as in Eq. (3.21) or in corresponding formulas for computing the displacements of pressure centers or fronts, is considered, the higher order derivatives are very much decreased in magnitude, and the period for which the formulas are useful is considerably extended. It is generally found that the use of only the first, or velocity, term in computing displacements of pressure systems gives satisfactory results for 6- or 12-hr forecasts. If the correction due to the acceleration term is included, satisfactory results can generally be obtained for 24-hr forecasts. In his original paper, Petterssen[14] shows a number of examples, and the agreement between the predicted and forecast 24-hr displacements of fronts, wedge lines, and pressure centers is generally satisfactory even for quite complex systems.

4. The Dynamical Principles. The problems of frontogenesis and of extrapolating the surface pressure field are purely geometrical, and kinematical methods were sufficient for their solution. In order to study the atmosphere as a three-dimensional fluid motion, it is necessary to use the more powerful methods of fluid dynamics and to study the forces involved. Since a fluid is a material system, all the principles cf mechanics can be used in the solution of these problems. The principles of conservation of mass, conservation of momentum (Newton's laws of motion), and conservation of energy will be used. The thermal equation of state of the fluid provides an additional relation. These three scalar and one vector equations can then be solved for the velocity vector and the pressure, temperature, and density of the fluid, subject to suitable boundary conditions. It is first necessary to express these principles in the proper Eulerian notation.

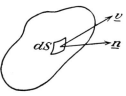

Fig. 19.

Consider an arbitrary fixed surface S in a fluid flow as in Fig. 19, with $\mathbf{n}$ being a unit vector that is an external normal, $\mathbf{v}$ being the fluid-velocity vector, and ρ being the fluid density. The rate of mass flow out through the surface S is

$$\iint_S \mathbf{n} \cdot \mathbf{v}\rho \, dS \tag{4.1}$$

On the other hand, the rate of decrease of mass within the surface is

$$-\iiint_\tau \frac{\partial \rho}{\partial t} \, d\tau \tag{4.2}$$

where τ is the volume enclosed by the surface S. By the principle of the conservation of mass, the decrease of mass within the surface must be caused by a flux of mass through the surface, and therefore

$$\iiint_\tau \frac{\partial \rho}{\partial t} \, d\tau + \iint_S \mathbf{n} \cdot \mathbf{v}\rho \, dS = 0 \tag{4.3}$$

If the surface integral is transformed to a volume integral by Green's theorem, it is seen that

$$\iiint_\tau \left[\frac{\partial \rho}{\partial t} + \nabla \cdot (\rho \mathbf{v}) \right] d\tau = 0 \tag{4.4}$$

Since the surface S was any arbitrarily chosen surface, Eq. (4.4) can be true in every case only if the integrand vanishes, or

$$\frac{\partial \rho}{\partial t} + \nabla \cdot (\rho \mathbf{v}) = 0 \tag{4.5}$$

This is called the *continuity equation* and is the mathematical representation of the principle of conservation of mass in a fluid flow. By differentiating the product $\rho \mathbf{v}$, the equation can also be written in the equivalent form

$$\frac{1}{\rho}\frac{D\rho}{Dt} + \nabla \cdot \mathbf{v} = 0 \tag{4.6}$$

From this, it is apparent that the continuity equation has an extremely simple form for an incompressible fluid for which $D\rho/Dt = 0$. It is then simply

$$\nabla \cdot \mathbf{v} = 0 \tag{4.7}$$

Before writing out the equations of motion, it is necessary to consider the state of stress in a fluid. In general, a fluid element will experience both normal and tangential

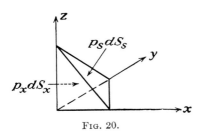

FIG. 20.

stresses; however, in many cases, the influence of the tangential stresses is very small. For this reason, the consideration of the effects of the tangential stresses will be delayed until later. For the present it will be supposed that the stresses in a fluid are normal to the surface of the fluid elements. In this case, the stress tensor is much simplified. Consider the infinitesimal tetrahedron $OABC$ shown in Fig. 20. Let p_s be the normal stress at the center of the slant face that has an area dS_s. Also let p_x be the normal stress at the center of the x face of area dS_x. Compression stresses are considered positive. The net force in the x direction is

$$p_x \, dS - p_s \, dS_s \cos \theta \tag{4.8}$$

plus higher order terms that arise from the variation of the stresses over their faces where θ is the angle between the x-axis and the external normal to the slant face. Now

$$dS_x = \cos \theta \, dS_s \tag{4.9}$$

and the expression (4.8) can therefore be written as

$$(p_x - p_s)dS_x \tag{4.10}$$

If the linear dimension of the fluid element is considered a first-order infinitesimal, the mass of the element is then a third-order infinitesimal. Since it is certainly necessary that the acceleration of the fluid element be finite, the net force acting on the fluid element must also be, at most, a third-order infinitesimal. The second-order term of expression (4.10) must thus be identically zero, or

$$p_x = p_s \tag{4.11}$$

in the limit as the size of the element approaches zero. Similar results can be obtained for the normal stresses on the y and z faces of the element. It is thus seen that the normal stress at a point must be the same in all directions if there are no shear stresses. A single parameter, the pressure, thus completely specifies the state of stress in a nonviscous fluid. This result is generally called *Pascal's law*.

In order to express Newton's laws of motion in the appropriate Eulerian form, consider a surface S (see Fig. 21) that encloses a particular mass of fluid of volume v.

As the fluid moves, this surface is carried with it and may be deformed, but it always encloses the same mass of fluid.   The only forces acting on this body of fluid are the normal pressure force and a body force, **F** per unit of mass, which is due to gravity. The possibility of other body forces due to electric or magnetic fields, etc., and of viscous shear stresses will be neglected here.   If **a** is the acceleration vector at any point, Newton's laws of motion require that

$$\iiint_\tau \rho\mathbf{a}\, d\tau = -\iint_S \mathbf{n} p\, dS + \iiint_\tau \rho\mathbf{F}\, d\tau \qquad (4.12)$$

since the mass of any element of volume is $\rho\, d\tau$.   The two terms on the right are the pressure and gravity forces acting on the body.   If the surface integral is transformed to a volume integral by Green's theorem, it is seen that

$$\iiint_\tau (\rho\mathbf{a} + \nabla p - \rho\mathbf{F})d\tau = 0 \qquad (4.13)$$

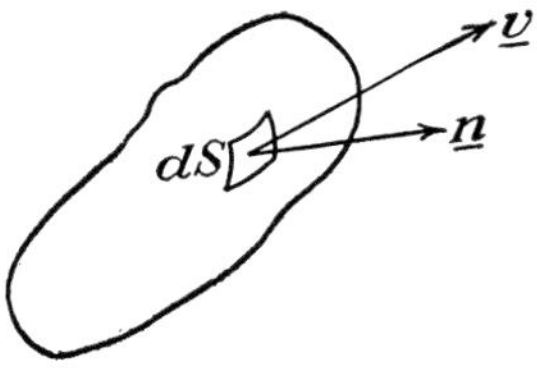

Since this relation must hold for any given mass of fluid, the integrand must vanish, or

$$\mathbf{a} = -\frac{1}{\rho}\nabla p + \mathbf{F} \qquad (4.14)$$

FIG. 21.

If the fluid-velocity vector **v** is measured with respect to a stationary coordinate system, the acceleration vector is obviously

$$\mathbf{a} = \frac{D\mathbf{v}}{Dt} \qquad (4.15)$$

so

$$\frac{D\mathbf{v}}{Dt} = -\frac{1}{\rho}\nabla p + \mathbf{F} \qquad (4.16)$$

These are the Eulerian equations of motion for a stationary coordinate system.

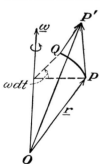

FIG. 22.

For meteorological purposes, it is most convenient to consider velocities relative to the surface of the rotating earth, and it is thus necessary to use the acceleration as measured in these rotating coordinates.   For this purpose, it is convenient to consider two coordinate systems, one fixed and the other rotating with the earth.   The origin of the two coordinate systems may be taken on the axis of rotation of the earth.   The angular velocity of the earth may be considered as constant and is a positive righthand rotation with respect to the north polar axis. The magnitude of the angular velocity of the earth ω is

$$\omega = \frac{2\pi}{\text{sidereal day}} = 7.292 \times 10^{-5}\ \text{sec}^{-1} \qquad (4.17)$$

Suppose a point P (see Fig. 22) moves to the point P' in time dt, as observed in the fixed coordinate system.   Let **r** be the vector distance from the common origin of the two coordinate systems.   Then

$$\overrightarrow{PP'} = d\mathbf{r} \qquad (4.18)$$

Now in time dt the point P, as observed in the rotating system, would have shifted to the point Q where

$$\overrightarrow{PQ} = \omega \times \mathbf{r}\, dt \tag{4.19}$$

The apparent change in $\mathbf{r}$, $d\mathbf{r}_a$, as observed in the rotating system, is thus $\overrightarrow{QP'}$, or

$$d\mathbf{r} = d\mathbf{r}_a + \omega \times \mathbf{r}\, dt \tag{4.20}$$

Thus
$$\frac{d\mathbf{r}}{dt} = \left(\frac{d\mathbf{r}}{dt}\right)_a + \omega \times \mathbf{r} \tag{4.21}$$

It is apparent that similar results hold for any vector quantity that can be represented by a directed line element, and thus for any vector quantity $\mathbf{A}$

$$\frac{d\mathbf{A}}{dt} = \left(\frac{d\mathbf{A}}{dt}\right)_a + \omega \times \mathbf{A} \tag{4.22}$$

where the subscript a means the rate of change as observed in the rotating system.

If $\mathbf{q}$ is the absolute velocity of the point P and $\mathbf{v}$ is its velocity relative to the surface of the earth

$$\mathbf{q} = \mathbf{v} + \omega \times \mathbf{r} \tag{4.23}$$

by Eq. (4.21). Also $\mathbf{a} = \dfrac{d\mathbf{q}}{dt}$, and therefore

$$\mathbf{a} = \left(\frac{d\mathbf{v}}{dt}\right)_a + 2\omega \times \mathbf{v} + \omega \times (\omega \times \mathbf{r}) \tag{4.24}$$

The angular velocity has been treated as constant in this differentiation. Now the first term on the right of Eq. (4.24) is obviously the *apparent or relative acceleration;* the second term is called the *Coriolis acceleration;* and the third term is the *centripetal acceleration.*

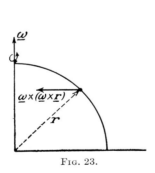

Fig. 23.

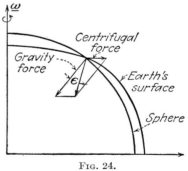

Fig. 24.

Consider a particle that is at rest on the surface of a spherical earth. Then, since $\mathbf{v} = 0$, it is accelerated toward the polar axis as in Fig. 23; and, in order to keep this mass stationary with respect to the rotating earth, a force sufficient to produce this acceleration must be applied. The gravitational force of mass attraction can supply the normal force; however, it cannot supply the tangential component. If the earth were a sphere, then, since even the earth's crust is slightly fluid, this tangential force would not be supplied, and the surface elements would slide toward the equator. Instead of a real centripetal acceleration, it is convenient to reverse its sign and to consider it as an apparent force, the centrifugal force. The earth's surface yields until the resultant of the centrifugal and gravity forces is normal to the surface, as shown in Fig. 24. By this process, the normal to the surface is rotated through the angle ϵ, which is quite small, having a maximum value of $700''$ at $45°$ latitude, either north or south. The resultant of these two forces that acts normal to the earth's surface is the apparent force of gravity. The apparent acceleration due to gravity g is

$$g = 980.621(1 - 0.00264 \cos 2\varphi) \qquad \text{cm/sec}^2 \qquad (4.25)$$

where φ is the latitude angle.

Since the centripetal acceleration is included in the apparent acceleration due to gravity, the Eulerian equations of motion for a coordinate system that is fixed with respect to the earth's surface are

$$\frac{D\mathbf{v}}{Dt} + 2\omega \times \mathbf{v} = -\frac{1}{\rho}\nabla p + \mathbf{F} \qquad (4.26)$$

For a perfect gas, the thermal equation of state is

$$p = \rho R T \qquad (4.27)$$

where $R = 8.3136 \times 10^7$ ergs per mol $\cdot$ °K is the universal gas constant and T is the absolute temperature (see Sec. V). For most purposes, the atmosphere may be treated as a perfect gas; although, in certain problems involving evaporation or condensation of water, this approximation is not valid.

The energy equation for a fluid can conveniently be worked out by considering the motion of a particular volume of fluid, as in Fig. 21. If the principle of conservation of energy is applied to the mass of fluid within the surface S, it is seen that

$$\iiint_\tau \rho \frac{Dq}{Dt}\, d\tau = \iiint_\tau \rho \frac{D}{Dt}\left(e + V + \frac{1}{2}\mathbf{v}\cdot\mathbf{v}\right) d\tau + \iint_S \mathbf{n}\cdot \mathbf{v}p\, dS \quad (4.28)$$

where q is the heat added per unit of mass, e is the internal energy per unit of mass,* and V is the gravitational potential energy per unit of mass. The volume integral on the right-hand side represents the part of the added energy that remains within the fluid mass, and the surface integral represents the energy transferred outside by work against the normal pressure stresses. For a viscous fluid, an additional surface integral must be added that gives the work done against viscous stresses. If the surface integral is transformed to a volume integral by Green's theorem, it is seen that

$$\iiint_\tau \left[\rho \frac{D}{Dt}\left(q - e - V - \frac{1}{2}\mathbf{v}\cdot\mathbf{v}\right) - \nabla\cdot(p\mathbf{v})\right] d\tau = 0 \qquad (4.29)$$

Since the fluid mass was arbitrarily chosen, the integrand must vanish, and therefore

$$\frac{Dq}{Dt} = \frac{D}{Dt}\left(e + V + \frac{1}{2}\mathbf{v}\cdot\mathbf{v}\right) + \frac{1}{\rho}\nabla\cdot(p\mathbf{v}) \qquad (4.30)$$

This may be taken as the energy equation for a nonviscous fluid; however, a much more familiar form may be obtained by combining this with the continuity equation and the equations of motion. If the scalar product of the velocity vector and Eq. (4.26) is computed, it is seen that

$$\mathbf{v}\cdot\frac{D\mathbf{v}}{Dt} = -\frac{1}{\rho}\mathbf{v}\cdot\nabla p + \mathbf{v}\cdot\mathbf{F} \qquad (4.31)$$

Since $\mathbf{F} = -\nabla V$ and $\partial V/\partial t = 0$, this can be written as

$$0 = \frac{D}{Dt}\left(V + \frac{1}{2}\mathbf{v}\cdot\mathbf{v}\right) + \frac{1}{\rho}\mathbf{v}\cdot\nabla p \qquad (4.32)$$

---

* It is assumed that mechanical energy units are used to express q and e.

If this is subtracted from Eq. (4.30), it is seen that

$$\frac{Dq}{Dt} = \frac{De}{Dt} + \frac{p}{\rho} \nabla \cdot \mathbf{v} \qquad (4.33)$$

From the continuity equation

$$\frac{1}{\rho} \nabla \cdot \mathbf{v} = -\frac{1}{\rho^2} \frac{D\rho}{Dt} = \frac{D}{Dt}\left(\frac{1}{\rho}\right) \qquad (4.34)$$

so that the energy equation can finally be written in its most familiar form as

$$\frac{Dq}{Dt} = \frac{De}{Dt} + p\frac{D}{Dt}\left(\frac{1}{\rho}\right) \qquad (4.35)$$

For a viscous fluid, additional terms must be added to the right-hand side of Eqs. (4.35), (4.33), and (4.30) to take into account the energy transferred by the viscous stresses.[15,16]

In the free atmosphere, heat is added to the fluid elements by radiation and conduction, and both are rather slow processes. For many fluid motions, this heat transfer may be neglected and $Dq/Dt = 0$. In this case, the temperature may be eliminated between the energy equation and the equation of state, and it is seen that the fluid elements follow an isentropic or reversible adiabatic process that may be written

$$\frac{D}{Dt}(p\rho^{-k}) = 0 \qquad (4.36)$$

where $k = c_p/c_v = 1.40$ for air.

A simple integral of the equations of motion can be obtained for the case of steady-state motion by considering the s component of Eq. (4.26), which is parallel to the streamline at any given point, as in Fig. 25. This is

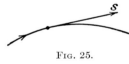

Fɪɢ. 25.

$$v\frac{\partial v}{\partial s} = -\frac{1}{\rho}\frac{\partial p}{\partial s} - \frac{\partial V}{\partial s} \qquad (4.37)$$

where v is the magnitude of the velocity vector at this point. This can be integrated if the density is a function only of the pressure to give

$$\frac{1}{2}v^2 + V + \int \frac{dp}{\rho} = \text{const. along the streamline} \qquad (4.38)$$

For a fluid of constant density, this gives the well-known Bernoulli equation

$$\frac{1}{2}v^2 + V + \frac{p}{\rho} = \text{const along a streamline} \qquad (4.39)$$

For a perfect gas that is carrying out the isentropic process of Eq. (4.36), this gives

$$\tfrac{1}{2}v^2 + V + c_pT = \text{const along a streamline} \qquad (4.40)$$

This may be called a generalized form of Bernoulli's equation.

The generalized Bernoulli equation (4.40) shows that, unless there are large changes in elevation, the temperature variation along a streamline for a fluid that moves adiabatically is quite small; in fact, a change in speed of air from 0 to 30 mps produces a temperature change of 0.4°C if there is no change in V. With the help of the equation of state, the adiabatic equation can be written as

$$\frac{d\rho}{\rho} = \frac{1}{k-1}\frac{dT}{T} \tag{4.41}$$

For this case, where the variation of the temperature is of the order of 0.4/300, it is seen that the variation in density is about ⅓ per cent for normal atmospheric conditions. This shows that, if the heat transfer between fluid elements is not important and if the changes in speed or elevation along the streamline are not large, the fluid may be considered as incompressible with a high degree of accuracy.

5. Circulation and Vorticity. If $d\mathbf{r}$ is an element of a contour C, as in Fig. 26, the circulation around the contour is defined as

$$\Gamma = \oint_C \mathbf{v} \cdot d\mathbf{r} \tag{5.1}$$

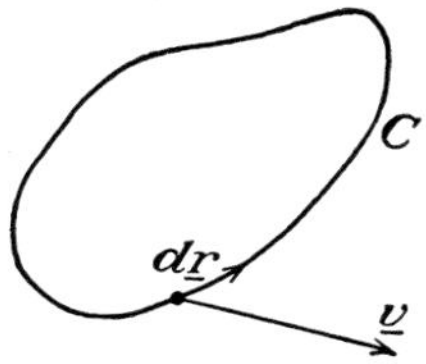

By Stokes' theorem, this may also be written as

$$\Gamma = \int\int_S \mathbf{n} \cdot (\nabla \times \mathbf{v})dS \tag{5.2}$$

Fig. 26.

where S is any surface that caps the contour C and $\mathbf{n}$ is a unit normal to the surface on the positive side as given by the right-hand rule. The quantity

$$\mathbf{\Omega} = \nabla \times \mathbf{v} \tag{5.3}$$

is called the *vorticity*. The interpretation of the vorticity vector is most easily obtained by considering the motion of a solid body. For such a body rotating with an angular velocity $\boldsymbol{\omega}$ about an axis, the velocity at any point is

$$\mathbf{v} = \mathbf{v}_o + \boldsymbol{\omega} \times \mathbf{r} \tag{5.4}$$

where $\mathbf{r}$ is the distance from the origin which is on the axis of rotation and $\mathbf{v}_o$ is the velocity of the origin. From this

$$\nabla \times \mathbf{v} = \nabla \times (\boldsymbol{\omega} \times \mathbf{r}) = 2\boldsymbol{\omega} \tag{5.5}$$

For a solid body's motion, the vorticity vector is thus seen to be twice the angular velocity vector. A similar interpretation can be obtained by expanding the vorticity in terms of the velocity components in a right-hand cartesian coordinate system. If the unit vectors in the x-, y-, and z-directions are $\mathbf{i}$, $\mathbf{j}$, and $\mathbf{k}$ respectively, and the velocity components are u, v, and w, the vorticity is

$$\mathbf{\Omega} = \mathbf{i}\left(\frac{\partial w}{\partial y} - \frac{\partial v}{\partial z}\right) + \mathbf{j}\left(\frac{\partial u}{\partial z} - \frac{\partial w}{\partial x}\right) + \mathbf{k}\left(\frac{\partial v}{\partial x} - \frac{\partial u}{\partial y}\right) \tag{5.6}$$

From Eqs. (2.7) and (2.10), the z-component of the vorticity is seen to be twice the angular velocity about the z-axis. Similar interpretations can also be made for the other components.

An important theorem can be obtained by considering the rate of change of circulation about a contour that is composed of a particular set of fluid elements. As the fluid elements move, the contour is carried with them. Since both $\mathbf{v}$ and $d\mathbf{r}$ vary

$$\frac{D\Gamma}{Dt} = \oint_C \frac{D\mathbf{v}}{Dt} \cdot d\mathbf{r} + \oint_C \mathbf{v} \cdot \frac{D}{Dt}(d\mathbf{r}) \tag{5.7}$$

Now $(D/Dt)(d\mathbf{r}) = d\mathbf{v}$; so

$$\oint_C \mathbf{v} \cdot \frac{D}{Dt}(d\mathbf{r}) = \oint_C d\left(\frac{v^2}{2}\right) = 0 \tag{5.8}$$

In the first integral, the apparent acceleration can be expressed by means of the equations of motion. Thus, for a nonviscous fluid

$$\frac{D\Gamma}{Dt} = -2 \oint_C \boldsymbol{\omega} \times \mathbf{v} \cdot d\mathbf{r} - \oint_C \frac{1}{\rho} (\nabla p) \cdot d\mathbf{r} + \oint_C \mathbf{F} \cdot d\mathbf{r} \qquad (5.9)$$

Since $\mathbf{F} = -\nabla V$ where V is the gravitational potential that is a single-valued function

$$\oint_C \mathbf{F} \cdot d\mathbf{r} = -\oint_C dV = 0 \qquad (5.10)$$

Also

$$\oint_C \frac{1}{\rho} (\nabla p) \cdot d\mathbf{r} = \oint_C \frac{dp}{\rho} \qquad (5.11)$$

Now the vector area of a surface capping the contour is

$$\mathbf{A} = \frac{1}{2} \oint_C \mathbf{r} \times d\mathbf{r} \qquad (5.12)$$

where $\mathbf{r}$ is the radius vector from some origin, and thus

$$\frac{D\mathbf{A}}{Dt} = \frac{1}{2} \oint_C \mathbf{v} \times d\mathbf{r} + \frac{1}{2} \oint_C \mathbf{r} \times d\mathbf{v} \qquad (5.13)$$

If the second term is integrated by parts, it is seen that

$$\frac{D\mathbf{A}}{Dt} = \oint_C \mathbf{v} \times d\mathbf{r} \qquad (5.14)$$

Since the dot and the cross can be interchanged in a triple scalar product and since $\boldsymbol{\omega}$ is a constant, Eq. (5.9) becomes

$$\frac{D}{Dt} (\Gamma + 2\boldsymbol{\omega} \cdot \mathbf{A}) = -\oint_C \frac{dp}{\rho} \qquad (5.15)$$

This result is somewhat simplified if one considers, instead of the contour C, its projection C' on the equatorial plane. Then, if the integration is carried around C' in a counterclockwise manner as viewed from the North Pole, Eq. (5.15) may be written as

$$\frac{D}{Dt} (\Gamma + 2\omega A) = -\oint_C \frac{dp}{\rho} \qquad (5.16)$$

where A is the area enclosed by the projected contour C'. This is known as the *Bjerknes circulation theorem.*[17] This theorem is extremely useful for the qualitative interpretation of motions that are produced by differential heating. It must be remembered that viscous stresses were neglected in this development.

From Eq. (5.1), it is apparent that the circulation around a contour may be considered as being the length of the contour multiplied by the mean tangential velocity. If one considers a square contour 1 km on a side, then the circumference is 4 km and the maximum projected area is 1 km². It is then seen that the $2\omega A$ term corresponds to a mean tangential velocity that is at most 4 cm per sec. The term $2\omega A$, which gives the effect of the rotation of the earth, is thus seen to be negligibly small unless very large contours or very small velocity gradients are considered.

The phenomena of the ocean breeze that is generally found on any warm summer afternoon near a coast line is one that can be discussed very simply in terms of the Bjerknes circulation theorem. Suppose that an air mass is stationary over a coast line and that the land surface is warmer than the sea surface as a result of radiation heating, as in Fig. 27. In this section the isobars, p_0 and p_{-1}, would be nearly horizontal; however, the isosteric surfaces s_0 and $s_1(s = 1/\rho)$ would be inclined as shown, owing to the differential heating at the surface. If one considers the contour $ABCDA$

$$\oint_C s \, dp = -(p_0 - p_{-1})(s_1 - s_0) \tag{5.17}$$

Thus

$$\frac{D}{Dt}(\Gamma + 2\omega A) = (p_0 - p_{-1})(s_1 - s_0) \tag{5.18}$$

Since the height to which the differential diurnal heating is important is small, the influence of the Coriolis term may be neglected, and it is seen that the circulation around the contour $ABCDA$ increases with time. This effect obviously produces an onshore breeze at the surface and an outward drift aloft.

At night, the radiation cooling of the land is greater than that of the sea; and late at night the horizontal temperature gradient is therefore normally reversed. This produces the offshore breeze, which reaches its maximum about dawn.

It may be noted that the circulation that is developed tends to rotate the isosteric surfaces toward the isobaric surfaces in the direction that places the lighter fluid above the heavier fluid. The circulation theorem is not too well suited for the actual calculation of velocities, since it really is a differential equation specifying the rate of change of vorticity, and this would be rather diffi-

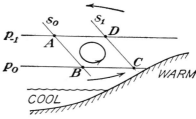

FIG. 27.

cult to integrate. It is, however, of great value for qualitative discussion of small-scale motions produced by differential heating such as the land and sea breezes discussed above or mountain and valley breezes.

In the previous example, circulations developed because the isosteric surfaces cut across the isobaric surfaces to form contours like $ABCDA$. These contours are called *solenoids* by Bjerknes. It would seem that, if the isobaric surfaces are also isosteric surfaces, *i.e.*, if $\rho = \rho(p)$, then no solenoids would be formed and no circulations would develop. Consider

$$P = \int_{p_0}^{p} \frac{dp}{\rho} \tag{5.19}$$

where p_0 is some standard pressure. Then P is a single-valued function of the pressure. Furthermore

$$dP = \frac{dp}{\rho} \tag{5.20}$$

so that

$$\oint_C \frac{dp}{\rho} = \oint_C dP = 0 \tag{5.21}$$

If the density is not a function of the pressure alone, then Eq. (5.20) is no longer true, and thus Eq. (5.21) is also no longer true.

The effects of the Coriolis term in the Bjerknes circulation theorem can be seen by considering its application to large-scale motions on the earth. In order to use the theorem as stated in Eq. (5.16), it is necessary to integrate around the contours so that the projection of the contour in the equatorial plane is described in a counter-clockwise manner as viewed from the North Pole; consequently the circulation for a contour that encloses a cyclone is positive for both the Northern and Southern Hemispheres. Similarly a contour that encloses an anticyclone has a negative circulation. If a Northern Hemisphere anticyclone moves southward, its projected area A diminishes; the circulation must therefore increase algebraically, and the intensity of the anticyclone must approach zero. Similarly a southward displacement of a cyclone causes its intensification. Northward displacements reverse these effects. Only the Coriolis term in the circulation theorem was considered in reaching these conclusions; the equally important pressure integral on the right-hand side and the effects of viscous terms superimpose additional effects.

An important application of the circulation theorem is to an infinitesimal contour C in an isentropic surface at latitud φ. In such a surface, the pressure and density are connected by the adiabatic equation

$$p\rho^{-k} = B \tag{5.22}$$

where B is a constant. Also the circulation around the contour is $\zeta \, dS$ where ζ is the component of the vorticity normal to the isentropic surface and dS is the area enclosed by the contour. Since the isentropic surfaces in the atmosphere are nearly horizontal, the area of the projection on the equatorial plane is $\sin \varphi \, dS$. Thus

$$\frac{D}{Dt}[(\zeta + 2\omega \sin \varphi)dS] = 0 \tag{5.23}$$

The heat-transfer processes in the free atmosphere are slow enough so that the fluid elements follow closely the adiabatic equation, Eq. (4.36). This means that the fluid elements remain in the same isentropic surfaces.

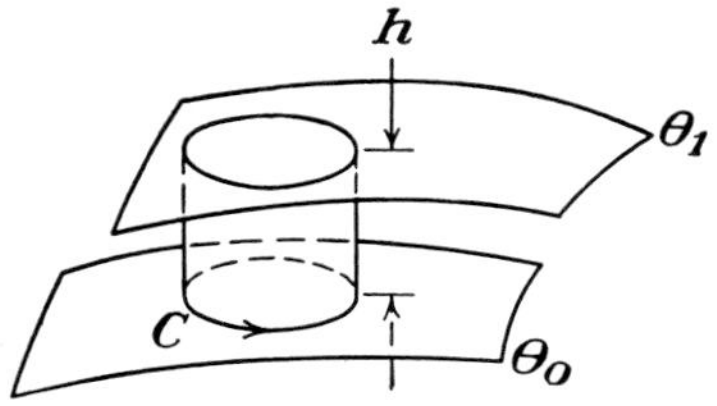

FIG. 28.

If the contour C is in the isentropic surface for which the potential temperature is θ_0, consider the small cylinder of fluid bounded by the contour C and extending from the θ_0 to the θ_1 isentropic surface as in Fig. 28. The mass of this element is

$$M = \rho h \, dS \tag{5.24}$$

Since the pressure change Δp between the surfaces θ_1 and θ_0 is the weight of a unit column of the fluid between these surfaces (neglecting vertical accelerations)

$$M = \frac{1}{g} \Delta p \, dS \tag{5.25}$$

As the mass of the fluid element is conserved, the area dS is thus seen to be inversely proportional to the pressure differential Δp, and Eq. (5.23) becomes

$$\frac{D}{Dt}\left(\frac{\zeta + 2\omega \sin \varphi}{\Delta p}\right) = 0 \tag{5.26}$$

The bracketed term is thus seen to be a conservative quantity as long as the motion is isentropic. If the fluid element is carried to a standard latitude φ_s and then stretched (or compressed) to a standard pressure differential Δp_s, the vorticity ζ_s for that condition is given by

$$\frac{\zeta_s + 2\omega \sin \varphi_s}{\Delta p_s} = \frac{\zeta + 2\omega \sin \varphi}{\Delta p} \tag{5.27}$$

The quantity ζ_s, called the *potential vorticity*, is also a conservative quantity, and it has been used by Starr and Neiburger[18] to trace the movements of fluid elements in an isentropic surface. Because of the inaccuracies in the observations of the winds aloft, there are, of course, fairly large experimental errors involved in the determination of the potential vorticity.

6. Atmospheric Wind Systems. There are several general methods by which the atmospheric winds may be investigated. The simplest method is to specify a velocity field and to calculate from this the corresponding pressure field. Since the atmospheric winds are on the average nearly horizontal, the restriction to horizontal motion will be made. Furthermore it must be remembered that, since the viscous stresses are being neglected, the pressure-velocity fields computed here do not apply in the lowest levels of the atmosphere (roughly the first kilometer) where viscous stresses are important.

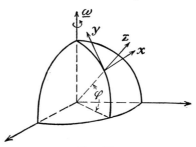

Fig. 29.

If a fixed coordinate system on the earth's surface is used in which the z-axis is normal to the earth's surface, the x-axis points east, and the y-axis points north, as in Fig. 29, then the equations of motion are [from Eq. (4.26)]

$$\frac{Du}{Dt} + 2\omega(w \cos \varphi - v \sin \varphi) = -\frac{1}{\rho}\frac{\partial p}{\partial x}$$

$$\frac{Dv}{Dt} + 2\omega u \sin \varphi = -\frac{1}{\rho}\frac{\partial p}{\partial y} \qquad (6.1)$$

$$\frac{Dw}{Dt} - 2\omega u \cos \varphi = -\frac{1}{\rho}\frac{\partial p}{\partial z} - g$$

where φ is the latitude angle. The small difference between the latitude angle and the angle between the z-axis and the equatorial plane has been neglected. If only horizontal motion is considered, w is zero, and the first two equations of motion are

$$\frac{Du}{Dt} - \lambda v = -\frac{1}{\rho}\frac{\partial p}{\partial x}$$

$$\frac{Dv}{Dt} + \lambda u = -\frac{1}{\rho}\frac{\partial p}{\partial y} \qquad (6.2)$$

where
$$\lambda = 2\omega \sin \varphi \qquad (6.3)$$

It may be noted that, with the restriction to horizontal motion, the x-axis may be arbitrarily oriented in the horizontal plane without changing the form of Eq. (6.2). If the second of these equations is multiplied through by $i = \sqrt{-1}$ and the two are added, a single equation

$$\frac{D}{Dt}(u + iv) + i\lambda(u + iv) = -\frac{1}{\rho}\left(\frac{\partial p}{\partial x} + i\frac{\partial p}{\partial y}\right) \qquad (6.4)$$

for the horizontal wind vector, $u + iv$, is obtained. It is often simpler to use this single equation involving the complex wind vector than to use the two equations involving the wind components.

The simplest possible wind system is the case in which the wind blows in straight parallel horizontal lines with a constant velocity.* The acceleration of the air parti-

---

* Strictly, some curvature of the path of the air particles is necessary because of the curvature of the earth's surface; however, that factor is neglected here.

cles relative to the earth's surface is then zero, and

$$u + iv = \frac{i}{\lambda\rho}\left(\frac{\partial p}{\partial x} + i\frac{\partial p}{\partial y}\right) \tag{6.5}$$

The velocity vector is thus seen to be perpendicular to the horizontal pressure gradient, or the isobars are parallel to the streamlines. Since the orientation of the axes is arbitrary, they may be placed so that v is zero and u positive. For this case, the isobars are parallel to the x-axis, and

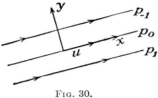

Fig. 30.

$$u = -\frac{1}{\lambda\rho}\frac{\partial p}{\partial y} \tag{6.6}$$

The flow and pressure patterns for the Northern Hemisphere are shown in Fig. 30.

In middle latitudes, the relative acceleration of the air is generally small when compared with the Coriolis acceleration. Equation (6.5) or (6.6) can thus be considered as a first approximation to the equilibrium wind for any given pressure distribution. This first approximation is called the *geostrophic wind*, and Eq. (6.5) or (6.6) is called the *geostrophic wind equation*.

It is frequently observed that, as an air mass moves along, its speed is nearly constant while its direction changes greatly. For a steady flow with no tangential acceleration

$$\frac{D}{Dt}(u + iv) = i\frac{u^2}{r} \tag{6.7}$$

where the axes have been placed so that $v = 0$ and $u > 0$ and r is the radius of curvature of the trajectory as shown in Fig. 31. The radius of curvature is considered positive when the curvature is positive as in Fig. 31. For the fluid element at the origin

$$i\frac{u^2}{r} + i\lambda u = -\frac{1}{\rho}\left(\frac{\partial p}{\partial x} + i\frac{\partial p}{\partial y}\right) \tag{6.8}$$

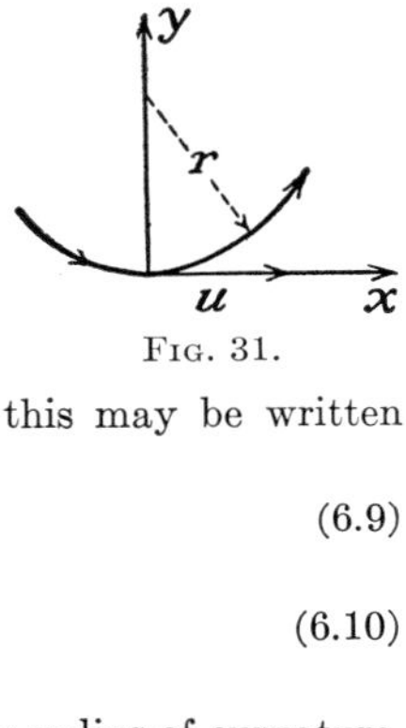

Fig. 31.

From this $\partial p/\partial x = 0$; and the isobars are therefore parallel to the streamlines in this case also. If the pressure gradient is replaced by the geostrophic wind $u_g = -(1/\lambda\rho)\,(\partial p/\partial y)$, this may be written

$$\frac{u^2}{r} + \lambda(u - u_g) = 0 \tag{6.9}$$

and

$$u = \frac{r\lambda}{2}\left(-1 \pm \sqrt{1 + \frac{4u_g}{r\lambda}}\right) \tag{6.10}$$

Since the wind velocity must approach u_g for large values of the radius of curvature, the positive sign must be used; therefore

$$u = \frac{r\lambda}{2}\left(-1 + \sqrt{1 + \frac{4u_g}{r\lambda}}\right) \tag{6.11}$$

The significance of the other root, which corresponds to the negative sign, will be discussed later. The wind velocity calculated from Eq. (6.11) is called the *gradient wind*.

In the Northern Hemisphere $\lambda > 0$, and the gradient wind is therefore less than the geostrophic wind if r is positive (*i.e.*, cyclonic curvature). This is shown by

Eq. (6.9). Similarly, the gradient wind is greater than the geostrophic wind if r is negative (*i.e.*, anticyclonic curvature).

Additional light is shed on this problem by considering the wind motion for the case in which there are no external applied forces. For this case, the geostrophic wind is zero, and Eq. (6.9) can be written

$$r_c = -\frac{u}{\lambda} \tag{6.12}$$

where r_c is the radius of curvature for this critical case. For this case, equilibrium is possible in the Northern Hemisphere only if r_c is negative. At 45°N latitude, $\lambda = 1.03 \times 10^{-4} \sec^{-1}$; therefore $r_c = -97$ km if u is 10 mps. For this radius of curvature, the centripetal and Coriolis accelerations are equal and opposite. From this, it is seen that in the Northern Hemisphere there are two types of motion with a clockwise rotation. In the first type, the magnitude of the radius of curvature is greater than that of r_c, and the Coriolis acceleration is larger than the centripetal acceleration. In order to produce this acceleration, the pressure must decrease to the left of the velocity, or $(\partial p/\partial y) < 0$ in Eq. (6.8). In the second type, the magnitude of the radius of curvature is less than that of r_c, and the Coriolis acceleration is smaller than the centripetal acceleration. For this case, the pressure must increase to the left of the velocity, or $(\partial p/\partial y) > 0$ in Eq. (6.8).

If the air is flowing in closed paths and the equilibrium is of the first type discussed above, the pressure must be a maximum in the center. This is the type of equilibrium found in all the large-scale systems in the Northern Hemisphere that rotate in a clockwise manner. Since these systems with high pressure in their center are called *anticyclones*, clockwise rotation in the Northern Hemisphere is often referred to as *anticyclonic rotation*. The equilibrium velocity for this case is the gradient wind, which is given in Eq. (6.11).

It is of considerable interest to note that, for anticyclonic motion, the quantity in parentheses in Eq. (6.11) varies from -1 for $u_g/\lambda r = -\frac{1}{4}$ to 0 for $u_g = 0$ or $r = \infty$. Thus the minimum value of u/r is

$$\left(\frac{u}{r}\right)_{\min} = -\frac{1}{2}\lambda = -\omega \sin \varphi \tag{6.13}$$

and this is equal but of opposite sign to the rate of rotation of the earth about an axis normal to the surface. This shows that, although an anticyclone rotates in a clockwise sense relative to the earth's surface, its absolute rotation is in the opposite sense.

For very small scale systems, equilibrium of the second type with low pressure in the center of a clockwise-rotating system is possible; however, such systems are of very little importance from the viewpoint of the meteorologist because of their small scale. The neglected root of Eq. (6.10) that corresponds to the negative sign gives the equilibrium velocity for this case.

If a system in the Northern Hemisphere rotates in a counterclockwise sense, it must, by Eq. (6.8), have a low pressure in its center. Since this is the type of motion observed in the middle-latitude cyclones, counterclockwise rotation is often referred to as *cyclonic rotation*. For cyclonic rotation, the equilibrium velocity is also given by Eq. (6.11).

It may be noted that, for both cyclonic and anticyclonic motion, the fluid moves along the isobars with the low pressure on the left. This is in accordance with the rule first enunciated by the Dutch meteorologist Buys Ballot. In middle latitudes, the observed winds have large enough radii of curvature so that the centripetal term in Eq. (6.8) is generally small compared with the Coriolis term. This means that the geostrophic wind is generally a fairly good approximation for these latitudes.   In the equatorial regions, the Coriolis parameter is very small; therefore the geostrophic wind is not a satisfactory approximation in these regions.   Furthermore, the small value of the Coriolis parameter makes the anticyclonic type of equilibrium with high pressure in the center of the system impossible in equatorial regions.

In the Southern Hemisphere, the Coriolis parameter λ is negative.   All the conclusions reached above for the Northern Hemisphere can be repeated, but the direction of the rotation must be reversed.   Thus, in the Southern Hemisphere, cyclones rotate in a clockwise sense; anticyclones rotate in a counterclockwise sense; and the wind blows with low pressure on its right.

The conditions under which the gradient-wind equation is valid are that (1) there are no viscous forces, (2) there are no tangential accelerations, and (3) the motion is steady.   The first of these is satisfied everywhere except close to the earth's surface; and the second is also generally nearly satisfied; however, in many situations the motion is far from steady.   Some indication of the effect of a changing pressure field can be seen by considering an initially flat pressure field that builds up to a steady gradient. Suppose

$$-\frac{1}{\rho}\left(\frac{\partial p}{\partial x} + i\frac{\partial p}{\partial y}\right) = i\lambda u_g(1 - e^{-kt}) \tag{6.14}$$

where u_g is the final steady-state geostrophic wind and k is a constant, and suppose that the fluid is initially stationary.   Then, from Eq. (6.4)

$$\frac{D}{Dt}(u + iv) + i\lambda(u + iv) = i\lambda u_g(1 - e^{-kt}) \tag{6.15}$$

and $u + iv = 0$ for $t = 0$.   The solution of this equation that satisfies this boundary condition is

$$u + iv = u_g\left(1 - \frac{i\lambda}{i\lambda - k}e^{-kt}\right) + u_g\frac{k}{i\lambda - k}e^{-i\lambda t} \tag{6.16}$$

If the wind vector at each instant were the geostrophic wind corresponding to the pressure gradient, it would be $u_s + iv_s$ where

$$u_s + iv_s = u_g(1 - e^{-kt}) \tag{6.17}$$

This might be called the "stationary" wind velocity.   The difference between these two

$$(u + iv) - (u_s + iv_s) = u_g\frac{k}{i\lambda - k}(e^{-i\lambda t} - e^{-kt}) \tag{6.18}$$

is the component caused by the changing pressure gradient.   This shows an oscillating component with a period $2\pi/\lambda$ that does not damp out and a second component that does damp out.   At 30°N latitude, this period $2\pi/\lambda$ is 1 day.   If $k = \lambda$ so that the pressure gradient would reach 95 per cent of its final value in 12 hr, the magnitude of the oscillating component would be $0.71u_g$.   From this, it is apparent that the wind velocity in a changing pressure field may vary considerably from the wind velocity for stationary equilibrium.

A different method of approach to this problem was given by Brunt and Douglas.[19] Equation (6.4) may be written as

$$u + iv = \frac{i}{\lambda\rho}\left(\frac{\partial p}{\partial x} + i\frac{\partial p}{\partial y}\right) + \frac{i}{\lambda}\frac{D}{Dt}(u + iv) \tag{6.19}$$

From this $$\frac{\partial}{\partial t}(u + iv) = \frac{i}{\lambda\rho}\left(\frac{\partial^2 p}{\partial x\,\partial t} + i\frac{\partial^2 p}{\partial y\,\partial t}\right) + \frac{i}{\lambda}\frac{\partial}{\partial t}\frac{D}{Dt}(u + iv) \tag{6.20}$$

if the variation of density is neglected. If this result is substituted in the last term of Eq. (6.19), it becomes

$$u + iv = \frac{i}{\lambda\rho}\left(\frac{\partial p}{\partial x} + i\,\frac{\partial p}{\partial y}\right) - \frac{1}{\lambda^2\rho}\left(\frac{\partial^2 p}{\partial x\,\partial t} + i\,\frac{\partial^2 p}{\partial y\,\partial t}\right) + \frac{i}{\lambda}\left(u\,\frac{\partial}{\partial x} + v\,\frac{\partial}{\partial y}\right)(u + iv)$$

$$- \frac{1}{\lambda^2}\frac{\partial}{\partial t}\frac{D}{Dt}(u + iv) \quad (6.21)$$

Brunt and Douglas suggest that the last two terms may be neglected in most cases, and their final result is that

$$u + iv = \frac{i}{\lambda\rho}\left(\frac{\partial p}{\partial x} + i\,\frac{\partial p}{\partial y}\right) - \frac{1}{\lambda^2\rho}\left(\frac{\partial^2 p}{\partial x\,\partial t} + i\,\frac{\partial^2 p}{\partial y\,\partial t}\right) \quad (6.22)$$

The first term is the geostrophic wind, and the second is thus the component caused by the changing pressure field. This component blows perpendicular to the isallobars (lines of constant pressure tendency, $\partial p/\partial t$) and is called the *isallobaric component* by Brunt and Douglas. It is obvious that this is only an approximation, since the isallobaric wind will in general have a component normal to the isobars; and for this condition the neglected terms in Eq. (6.21) cannot vanish except in very special circumstances. The isallobaric wind component apparently does, however, give an idea of the order of magnitude of the deviation of the wind from the geostrophic (or gradient) wind.

A similar method of approach is to consider the geostrophic wind as a first approximation to the wind velocity. This corresponds to neglecting the last term in Eq. (6.19). Then this first approximation can be used to evaluate the neglected term and to obtain in this manner a second approximation. This may be considered as a first step in an iterative process for determining the wind velocity for a given pressure field. This method was indicated by Van Mieghem;[20] however, no discussion of the convergence of this method has been given.

It is often of interest to determine the variation of the wind with height. For the cases in which the geostrophic wind is a satisfactory approximation, a very simple solution to this problem can be given; however, it is convenient to consider first the law of variation of pressure in the vertical. For horizontal motion, the equation for vertical equilibrium [see Eq. (6.1)] is

$$-2\omega u\,\cos\,\varphi = -\frac{1}{\rho}\frac{\partial p}{\partial z} - g \quad (6.23)$$

For any normal velocity, the left-hand term is extremely small. For example, if $u = 10^3$ cm per sec and $\varphi = 30°$N, $2\omega u\,\cos\,\varphi = 0.1$ cm² per sec. Since this term is only about one ten-thousandth of the acceleration due to gravity, its influence on the vertical pressure gradient is negligible, and, with sufficient accuracy

$$\frac{\partial p}{\partial z} = -g\rho \quad (6.24)$$

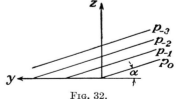

Fig. 32.

This is the well-known hydrostatic equation.

A very simple explanation of the variation of wind with height can be obtained by considering the slopes of the isobaric surfaces. If at a given point the x-axis is taken parallel to the wind at that point, as in Eq. (6.6), the isobars in the yz-plane are as shown in Fig. 32. The slope of the isobaric surface through the origin is

$$\tan\,\alpha = \frac{\partial p/\partial y}{\partial p/\partial z} \quad (6.25)$$

From Eqs. (6.24) and (6.6), this is

$$\tan \alpha = \frac{\lambda}{g} u \qquad (6.26)$$

where u is the geostrophic wind at the origin. The geostrophic wind is thus a measure of the steepness of the isobaric surface. From Eq. (6.24), the distance between consecutive surfaces is determined by the density; therefore, if the surfaces p_0 and p_{-1} are to be parallel, each isobar must also be an isosteric and thus an isothermal line. For this condition with parallel isobaric surfaces, the geostrophic wind would not vary with height. This condition was first given by Dines.[21] If the temperature increased to the right along the p_0 isobar, then the surface p_{-1} would be steeper than the surface p_0, and the geostrophic wind would increase with elevation. Conversely, if the temperature decreased to the right along the p_0 isobar, the geostrophic wind would decrease with elevation.

Analytic formulas for computing the variation of the geostrophic wind with elevation will now be developed. If the density is eliminated from Eq. (6.5) by the equation of state, the geostrophic-wind equation may be written

$$\frac{(u+iv)}{T} = i\frac{R}{\lambda}\left(\frac{\partial}{\partial x}\log p + i\frac{\partial}{\partial y}\log p\right) \qquad (6.27)$$

and the hydrostatic equation may be written

$$\frac{\partial}{\partial z}\log p = -\frac{g}{RT} \qquad (6.28)$$

From Eq. (6.27)

$$\frac{\partial}{\partial z}\left(\frac{u+iv}{T}\right) = \frac{iR}{\lambda}\left(\frac{\partial^2}{\partial x \, \partial z}\log p + i\frac{\partial^2}{\partial y \, \partial z}\log p\right) \qquad (6.29)$$

If the pressure is eliminated by means of the hydrostatic equation, this becomes

$$\frac{\partial}{\partial z}\left(\frac{u+iv}{T}\right) = \frac{ig}{\lambda T^2}\left(\frac{\partial T}{\partial x} + i\frac{\partial T}{\partial y}\right) \qquad (6.30)$$

If this is integrated with respect to z from a level z_0 to a higher level z, it is seen that

$$u+iv = (u_0+iv_0)\frac{T}{T_0} + i\frac{g}{\lambda}T\int_{z_0}^{z}\frac{1}{T^2}\left(\frac{\partial T}{\partial x} + i\frac{\partial T}{\partial y}\right)dz \qquad (6.31)$$

where the subscript 0 refers to conditions at the height z_0. The geostrophic wind thus has one component that is proportional to the absolute temperature and a second that depends on the horizontal temperature gradient. The second component is called the *thermal wind*.

If the temperature does not vary with height, Eq. (6.31) may be evaluated very simply, for the horizontal temperature gradient is in this case also independent of height. Thus

$$u+iv = u_0 + iv_0 + \frac{igz}{\lambda T_0}\left(\frac{\partial T}{\partial x} + i\frac{\partial T}{\partial y}\right) \qquad (6.32)$$

The thermal wind is seen to blow parallel to the isotherms, and, in the Northern Hemisphere, it moves with the low temperature on the left. It is thus seen that, as the altitude increases, the wind changes so as to approach the direction of the horizontal isotherms. For more complex temperature distributions, the evaluation of the thermal wind would be more difficult; however, the interpretation would, of course, be similar.

In middle latitudes and at fairly low elevations, the ratio $(\Delta T/\Delta p)$ is much greater for horizontal north-south variations than for vertical variations. This means that the velocity of the westerlies must increase with elevation. At high altitudes, in the lower regions of the stratosphere, the temperature gradient along the isobaric surfaces is reversed, and the westerly wind then diminishes with height. The maximum velocity of the westerlies is normally found near the base of the stratosphere.

These results also show an important difference between the migratory anticyclones of the middle latitudes and the semipermanent anticyclones such as the Azores or Pacific highs. The former are cold-air outbreaks from the polar regions; consequently, their thermal-wind component is opposed to the surface geostrophic (or gradient) wind. These systems are thus quite shallow. The semipermanent anticyclones are characterized by warm subsiding air in the core, and the wind therefore increases with elevation up to great heights in these systems.

7. Surfaces of Discontinuity. Since much of the weather of the middle latitudes is associated with the polar front, the equilibrium of the air masses in the neighborhood of these atmospheric discontinuity surfaces is a problem of considerable meteorological importance. Helmholtz[1] gave the earliest demonstration that such a discontinuity surface would be in equilibrium, provided that certain discontinuities exist in the density and velocity fields at the frontal surface. There are, of course, no real discontinuities of this sort in the atmosphere; however, the fluid-property gradients are so large in the frontal zones and these zones are narrow enough so that their thickness is practically negligible in any synoptic problem.

From the equations of motion, it is apparent that the atmospheric pressure field must be a continuous function, for there would otherwise be regions of infinite acceleration in the fluid. This fact can be used to establish the conditions for equilibrium of a frontal surface. The problem is, of course, very complicated in general; however, very simple solutions can be given for a few cases. Suppose a frontal surface separates two fluid masses, as in Fig. 33. Then, since the pressure must be a continuous function, the pressure derivatives in the direction parallel to the frontal surface must be the same in both air masses. Thus

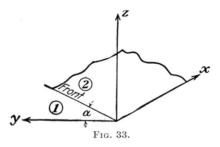

Fig. 33.

$$\frac{\partial p_1}{\partial y} \cos \alpha + \frac{\partial p_1}{\partial z} \sin \alpha = \frac{\partial p_2}{\partial y} \cos \alpha + \frac{\partial p_2}{\partial z} \sin \alpha \qquad (7.1)$$

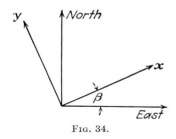

Fig. 34.

where the subscript 1 refers to the left-hand fluid and the subscript 2 refers to the right-hand fluid. From this

$$\tan \alpha = -\frac{(\partial p_1/\partial y) - (\partial p_2/\partial y)}{(\partial p_1/\partial z) - (\partial p_2/\partial z)} \qquad (7.2)$$

A similar expression could be written for the slope of the frontal surface in the x-direction.

The simplest case is that of a plane front with the air in each air mass moving in straight horizontal paths parallel to the front. In this case, the front traces a straight line in any horizontal plane, and it is convenient to take the x-axis along the front so that

$v = w = 0$. If the x-axis makes an angle β with the eastward direction, as in Fig. 34, the equations of motion are [from Eq. (4.26)]

$$\frac{Du}{Dt} + 2\omega(w \cos \varphi \cos \beta - v \sin \varphi) = -\frac{1}{\rho}\frac{\partial p}{\partial x}$$

$$\frac{Dv}{Dt} + 2\omega(-w \cos \varphi \sin \beta + u \sin \varphi) = -\frac{1}{\rho}\frac{\partial p}{\partial y} \qquad (7.3)$$

$$\frac{Dw}{Dt} - 2\omega \cos \varphi(u \cos \beta - v \sin \beta) = -\frac{1}{\rho}\frac{\partial p}{\partial z} - g$$

Since the relative accelerations are zero, the pressure gradients are

$$\frac{\partial p}{\partial x} = 0$$

$$\frac{\partial p}{\partial y} = -2\omega\rho u \sin \varphi \qquad (7.4)$$

$$\frac{\partial p}{\partial z} = -g\rho + 2\omega\rho u \cos \varphi \cos \beta$$

The condition that the frontal intersection in the xz-plane be horizontal is that

$$\frac{\partial p_1}{\partial x} - \frac{\partial p_2}{\partial x} = 0 \qquad (7.5)$$

This condition is obviously satisfied.   The slope of the front in the yz-plane is, from Eq. (7.2)

$$\tan \alpha = -\frac{2\omega \sin \varphi(\rho_1 u_1 - \rho_2 u_2)}{g(\rho_1 - \rho_2) - 2\omega \cos \varphi \cos \beta(\rho_1 u_1 - \rho_2 u_2)} \qquad (7.6)$$

It is of interest to consider the special case in which both air masses have the same density but different velocities.   For this case

$$\tan \alpha = \tan \varphi \sec \beta \qquad (7.7)$$

Thus, for an east-west front, $\alpha = \varphi$, and for a north-south front, $\alpha = \pi/2$. These are special cases of a more general result that is obtained by noting that, in these coordinates

$$\boldsymbol{\omega} = \omega(\cos \varphi \sin \beta \mathbf{i} + \cos \varphi \cos \beta \mathbf{j} + \sin \varphi \mathbf{k}) \qquad (7.8)$$

and

$$\mathbf{n} = -\sin \alpha \mathbf{j} + \cos \alpha \mathbf{k} \qquad (7.9)$$

where $\mathbf{n}$ is a unit normal to the frontal surface.   From these results, it is seen that

$$\boldsymbol{\omega} \cdot \mathbf{n} = \omega(-\cos \varphi \cos \beta \sin \alpha + \sin \varphi \cos \alpha) \qquad (7.10)$$

By Eq. (7.7), the right-hand side of Eq. (7.10) is zero, so that $\mathbf{n}$ is perpendicular to $\boldsymbol{\omega}$. This proves that such a frontal surface having the same density on both sides must be parallel to the polar axis.

A second special case that can be considered is the case in which both air masses have the same velocity but have different densities.   If $u = u_1 = u_2$, the slope of the front for this case is

$$\tan \alpha = -\frac{2\omega u \sin \varphi}{g - 2\omega u \cos \varphi \cos \beta} \qquad (7.11)$$

The second term in the denominator can be neglected, since, for $u = 10^3$ cm per sec, the product $\omega u = 7.29 \times 10^{-2}$ cm per sec². The second term is thus less than a thousandth of g for any possible wind velocity.   Furthermore, since ωu is so much smaller than g, the slope of the front shown by Eq. (7.11) is very small, and the front is practically horizontal.

For the special case of equal densities, the first term in the denominator of Eq. (7.6) vanished, and only the second term was left; however, it is easily seen that, if the temperature difference between the air masses is as much as 1°C, the first term is many hundreds of time as large as the second. A satisfactory approximation for all normal cases can thus be obtained by neglecting the second term. Thus, with sufficient accuracy

$$\tan \alpha = - \frac{\lambda}{g} \frac{(\rho_1 u_1 - \rho_2 u_2)}{(\rho_1 - \rho_2)} \tag{7.12}$$

Since the temperatures, but not the densities, are directly observed, it is convenient to express this result in terms of the temperatures. Since the pressure at neighboring points separated by the front must be the same, $\rho_1 T_1 = \rho_2 T_2$, and

$$\tan \alpha = - \frac{\lambda}{g} \frac{T_2 u_1 - T_1 u_2}{T_2 - T_1} \tag{7.13}$$

Let $$T_2 = T_1 + \Delta T$$

and $$u_2 = u_1 + \Delta u \tag{7.14}$$

Then $$\tan \alpha = \frac{\lambda T_1}{g} \left(\frac{\Delta u}{\Delta T} - \frac{u_1}{T_1} \right) \tag{7.15}$$

Furthermore, the second term in the bracket is normally very much smaller than the first, since $\Delta T / T$ is of the order of 0.05 while Δu is generally of the same order as u_1. The final approximate formula for the slope of the frontal surface is thus

$$\tan \alpha = \frac{\lambda T_1}{g} \frac{\Delta u}{\Delta T} \tag{7.16}$$

At latitude 30°N with $T_1 = 280°$K, $\Delta T = 10°$K and $\Delta u = 10^3$ cm per sec, this gives $\tan \alpha = 0.0021$. The slope of frontal surfaces may thus be expected to be of this general order of magnitude.

If the x-axis is placed along the front with the cold air on the left, ΔT is always positive. Thus if Δu is also positive (in the Northern Hemisphere), $\tan \alpha$ is small and positive, and α is a small positive angle. For this case, the warm air would lie on top of the cold air. This is the condition normally found at the polar front. For the other case, with Δu negative, the cold air would lie on top of the warm air, for α would be slightly less than π. These formulas have considered only the possibility of static equilibrium; the stability of the motions was not considered. It will later be shown that this condition with the cold air on top of the warm is unstable and would soon break down, and the air masses would mix. The condition with warm air on top of the cold will be seen to be comparatively stable.

This indicates that the front between a westerly current and an easterly current to the south (in the Northern Hemisphere) is unstable. This is undoubtedly one of the reasons why the sharp and well-defined frontal systems that are generally found on the polar side of the belt of westerlies are not also observed on the equatorial side between the westerlies and the trade winds.

These formulas were developed on the assumption that the acceleration of the fluid elements relative to the earth's surface was negligible when compared with the Coriolis acceleration. This approximation and Eq. (7.16) are satisfactory for the nascent cyclone and also in general for the warm front of a fully developed cyclone, for the horizontal velocity component parallel to the front is normally much larger than the upglide velocity components in these cases. On the other hand, the velocities of a cold front are more often nearly in a plane perpendicular to the surface frontal line. For this case, the Coriolis accelerations are generally negligible in comparison with the relative accelerations, and the pressure gradients are entirely different from those predicted by Eq. (7.4). A complete solution of this cold-front problem cannot be given; however, von Kármán[22] has shown for the case in which the motion is in the plane perpendicular to the surface frontal line that the initial slope, near the ground, must be about 60 deg. A complete solution of this problem has not yet been found.

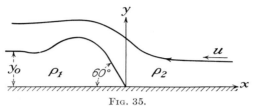

Fig. 35.

If a coordinate system is used in which the cold air mass is stationary, the flow is a steady-state flow with the warm air mass approaching and flowing up and over the cold air mass, as in Fig. 35. In each fluid, the density will be treated as a constant. In the stationary cold mass, the pressure is

$$p_1 = p_0 - g\rho_1 y \qquad (7.17)$$

where p_0 is the surface pressure in the cold mass. Since the motion is steady, the Bernoulli equation, Eq. (4.39), can be used to compute the pressure, and the Bernoulli constant will be the same for every streamline if it is assumed that the warm air mass for large values of x has a constant velocity U. Thus

$$p_2 = -\tfrac{1}{2}\rho_2(u^2 + v^2) - g\rho_2 y + C \qquad (7.18)$$

At the origin, where the velocity is zero, the pressure must be p_0; thus $C = p_0$; and

$$p_2 = p_0 - g\rho_2 y - \tfrac{1}{2}\rho_2(u^2 + v^2) \qquad (7.19)$$

Since the pressure must be continuous across the frontal surface, the boundary condition there is that $p_1 = p_2$, or

$$y = \frac{\rho_2}{\rho_1 - \rho_2} \times \frac{u^2 + v^2}{2g} \qquad (7.20)$$

Since the fluid has no upper boundary, the speed of the warm air far downstream may be expected to be again U, and the depth of the cold mass there, y_0, is therefore

$$y_0 = \frac{\rho_2}{\rho_1 - \rho_2} \times \frac{U^2}{2g} \qquad (7.21)$$

Since the fluid flow is two-dimensional, the continuity equation is

$$\frac{\partial u}{\partial x} - \frac{\partial v}{\partial y} = 0 \qquad (7.22)$$

This shows that

$$d\psi = -v\,dx + u\,dy \qquad (7.23)$$

must be a perfect differential. The significance of the function ψ is easily seen; for, in the xy-plane on the lines for which ψ is constant, $d\psi = 0$; or

$$0 = -v\,dx + u\,dy \qquad (7.24)$$

from Eq. (7.23).   The lines of constant ψ are thus streamlines [see Eq. (2.22)], and ψ is for this reason called the *stream function*.   The velocity components may be obtained from the stream function by differentiating, and, from Eq. (7.23)

$$u = \frac{\partial \psi}{\partial y}$$

and

$$v = -\frac{\partial \psi}{\partial x}$$

(7.25)

It may be noted that the velocity component in any given direction is the derivative of the stream function 90 deg to the left of that direction.

Fig. 36.—"The Black Blizzard" of Apr. 14, 1935, near Lamar, Colo.   (*Photograph from U.S. Soil Conservation Service.*)

Far ahead of the front, if viscous effects are neglected, the warm air moves with a uniform velocity; its vorticity here is therefore zero.   Since the density is assumed to be constant and the scale of the motion is small enough so that the Coriolis terms may be neglected, Bjerknes' circulation theorem shows that the vorticity will remain zero. Thus

$$\frac{\partial v}{\partial x} - \frac{\partial u}{\partial y} = 0$$

(7.26)

If the velocity components are eliminated by use of the stream function, it is seen that

$$\frac{\partial^2 \psi}{\partial x^2} + \frac{\partial^2 \psi}{\partial y^2} = 0$$

(7.27)

For this problem, it is more convenient to use polar coordinates $r = \sqrt{x^2 + y^2}$ and $\theta = \tan^{-1}(y/x)$.   In these coordinates, the equation for the stream function is

$$\frac{\partial^2 \psi}{\partial r^2} + \frac{1}{r}\frac{\partial \psi}{\partial r} + \frac{1}{r^2}\frac{\partial^2 \psi}{\partial \theta^2} = 0$$

(7.28)

A solution of this equation is

$$\psi = Cr^n \sin n\theta$$

(7.29)

where C and n are constants.   It may be noted that the streamline $\psi = 0$, which extends along the line $\theta = 0$, is continued past the origin on the line $\theta = \pi/n$.   The radial velocity at any point is

$$u_r = -\frac{1}{r}\frac{\partial \psi}{\partial \theta} = -nCr^{n-1}\cos n\theta \qquad (7.30)$$

and, along the streamline $\theta = \pi/n$, there is no tangential velocity component; and

$$u_r = nCr^{n-1} \qquad (7.31)$$

If the flow near the origin is to obey the boundary condition, Eq. (7.20), it is apparent that the velocity must vary like the square root of the distance from the origin. From Eq. (7.30), this is possible, provided that $n = \frac{3}{2}$ and π/n is thus $2\pi/3$.   The cold front thus makes an angle of $\pi/3$ or 60 deg with the surface, at least near the origin, as shown in Fig. 35.   Frictional effects on a moving cold air mass cause the nose to be rounded off near the ground, but this steep frontal surface is quite characteristic of those cold fronts in which the fluid flow is nearly two-dimensional.   Figure 36 is a photograph of a dust storm that shows this typical shape.   This very steep surface is also responsible for the very narrow band of storm activity associated with a line squall.

Of these two special solutions, the first was characterized by a horizontal flow parallel to the frontal surface with negligible relative accelerations, and the second was characterized by a two-dimensional flow in a plane perpendicular to the surface frontal line and large relative accelerations.   In the first case, the front was very flat; and, in the latter case, the front was quite steep.   The atmospheric fronts are intermediate types with the warm fronts generally closely resembling the first type and the cold fronts resembling the second.

8. General Effects of Viscosity.   In all the previous calculations, it was assumed that the shear stresses acting in the fluid were negligibly small.   In the upper levels of the atmosphere, this is generally a satisfactory approximation; however, the shear stresses are of great importance in the layer, about 1 km deep, just above the earth's surface.   The existence of these shear stresses is the phenomenon of viscosity or fluid friction.   In the present application, the fluid will be treated as incompressible.   The more general case in which compressibility is considered is discussed by Lamb.[15]

The simplest fluid flow in which the viscous stresses are of primary importance is that of a fluid between two parallel walls a distance h apart with one wall stationary and the other moving parallel to it with a constant velocity U.   For this case, it is observed that the fluid elements next to the boundaries move at the same velocity as the boundaries.   This condition of no slipping past solid boundaries is characteristic of all viscous-fluid motions except motions of a gas at very low pressures where the mean free path of the molecules is of the same order as the other geometric distances involved.[*]

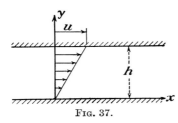

Fig. 37.

If the velocity U and spacing h are not too large, the fluid velocity between the parallel walls is found to vary linearly with the distance from the fixed wall, and the velocity distribution is as shown in Fig. 37.   It is also found that a shearing stress τ which tends to prevent the relative motion is exerted on both walls.   This shearing stress is found to be proportional to the ratio U/h, or

$$\tau = \mu\frac{U}{h} \qquad (8.1)$$

---

[*] A critical discussion of this boundary condition is given in S. Goldstein, "Modern Developments in Fluid Dynamics," Oxford, London, 1938.

where μ is the constant of proportionality. The coefficient μ is called the *viscosity coefficient*.

This experiment would be rather impractical to duplicate in practice; however, it is closely approximated by the flow between two concentric cylinders, one fixed and the other rotating, as long as the radii of the cylinders are large compared to the distance separating them. The latter type of flow is generally called *Couette flow*.

The viscosity coefficient of air is found experimentally to be almost independent of the air density (Maxwell's law) and to increase with temperature. An empirical formula[23] for this case is

$$\mu = 0.0001702(1 + 0.00329t + 0.000007t^2) \qquad \text{dyne sec/cm}^2 \qquad (8.2)$$

where t is the temperature in degrees centigrade. For liquids, the viscosity coefficient is found to decrease with temperature. For water

$$\mu = \frac{0.01779}{1 + 0.03368t + 0.00022t^2} \qquad \text{dyne sec/cm}^2 \qquad (8.3)$$

This result was obtained by Helmholtz[24] from Poiseuille's observations.

The shear stress on any plane in the fluid that is parallel to the wall must be the same as that on the walls in the fluid flow considered above, and this can be expressed in a more useful manner as

$$\tau = \mu \frac{du}{dy} \qquad (8.4)$$

since the velocity profile is linear. This expression was introduced by Newton and is generally called the *Newtonian friction law*. This simple law can be used to calculate the shear stress in any case for which the fluid flows in parallel sheets. As an example, it will be used to calculate the flow in a cylindrical tube.

Consider a circular cylinder of radius a with x being the distance along the axis and r being the distance from the axis, as in Fig. 38. Since the flow is in sheets parallel to the axis, the pressure in any plane perpendicular to the axis, such as sections 1 or 2 in Fig. 38, must be constant. For steady flow through this tube, the fluid cylinder of radius r must have no net force acting on it in the x-direction. The pressure forces

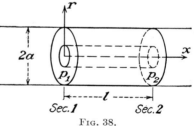

Fig. 38.

on the ends of the cylinder and the viscous stresses on the side must thus balance, or

$$\mu \frac{du}{dr} \cdot 2\pi rl + (p_1 - p_2)\pi r^2 = 0 \qquad (8.5)$$

Gravity forces have, of course, been omitted. From this

$$\frac{du}{dr} = -\frac{p_1 - p_2}{2\mu l} r \qquad (8.6)$$

The velocity must vanish at the wall where $r = a$. Thus

$$u = +\frac{p_1 - p_2}{2\mu l} \int_r^a r \, dr \qquad (8.7)$$

and
$$u = \frac{p_1 - p_2}{4\mu l} (a^2 - r^2) \qquad (8.8)$$

The velocity distribution is thus seen to be parabolic, is proportional to the pressure gradient along the axis of the tube, and is inversely proportional to the viscosity of the fluid. For this case, the volume of fluid flowing through the tube in a unit time is

$$V = \int_0^a 2\pi r u \, dr \qquad (8.9)$$

or

$$V = \frac{\pi(p_1 - p_2)}{8\pi l} a^4 \qquad (8.10)$$

This fourth-power law for the fluid flow was accurately verified in an extensive series of tests by Poiseuille, and this flow is usually called *Poiseuille flow*. Since the rate of fluid flow through a tube and the pressure gradient along it can be accurately measured, this gives a very good experimental method of determining the viscosity of a fluid.

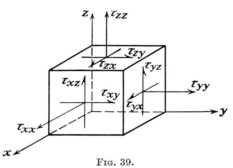

Fig. 39.

In order to investigate more complicated three-dimensional flows, it is necessary to generalize the simple friction law given by Eq. (8.4). The general notation for the viscous stresses on the surfaces of a fluid element is shown in Fig. 39. On the reverse side of the element, the direction of positive stresses is reversed. It may be noted that tension stresses are taken as positive. The double-subscript notation is used, with the first subscript giving the face on which the stress acts and the second subscript giving the direction in which the stress is acting. Thus τ_{xy} is a stress in the x face which is directed in the y-direction.

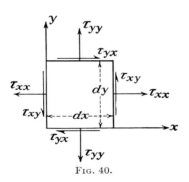

Fig. 40.

These nine stress components are not all independent. This can easily be seen by considering the moments acting on an infinitesimal element, as in Fig. 40. The moment tending to rotate the element about the z-axis is thus $(\tau_{xy} - \tau_{yx})dx \, dy \, dz$ plus higher order terms. This moment is a third-order infinitesimal unless $\tau_{xy} = \tau_{yx}$. On the other hand, the moment of inertia of the element about this axis is a fifth-order infinitesimal. The angular acceleration of the element would therefore be infinite unless $\tau_{xy} = \tau_{yx}$. Since the angular accelerations of the fluid elements must be finite

and similarly

and

$$\begin{aligned} \tau_{xy} &= \tau_{yx} \\ \tau_{yz} &= \tau_{zy} \\ \tau_{zx} &= \tau_{xz} \end{aligned} \qquad (8.11)$$

By this result, the number of independent viscous stress components is reduced to six. It is of interest to note that, because of this result, the order in which the subscripts are written is a matter of no consequence.

The elementary friction law, Eq. (8.4), shows that in this case the viscous stress is a linear function of the rate of distortion of the fluid elements. If a similar law is to hold for three-dimensional flows, then

$$\tau_{xy} = a_{11} \frac{\partial u}{\partial x} + a_{12} \frac{\partial u}{\partial y} + a_{13} \frac{\partial u}{\partial z} + a_{14} \frac{\partial v}{\partial x} + \cdots + a_{19} \frac{\partial w}{\partial z}$$

$$\tau_{xz} = a_{21} \frac{\partial u}{\partial x} + a_{22} \frac{\partial u}{\partial y} + \cdots \qquad\qquad + a_{28} \frac{\partial w}{\partial y} + a_{29} \frac{\partial w}{\partial z} \qquad (8.12)$$

etc.

where a_{11}, a_{12}, etc., are viscosity coefficients. Similar expressions could be written for the other stress components. Now these viscosity coefficients must be the same for any orientation of the coordinate system, and for parallel flows Eq. (8.12) must reduce to Eq. (8.4). For example, if $v = w = 0$, then $a_{12} = \mu$. By this means and by a rotation of the coordinate system, all the viscosity coefficients can be evaluated,

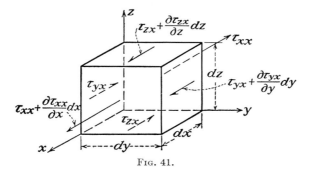

Fig. 41.

and it is found that the viscous stresses are given by the following expressions for an incompressible fluid:

$$\tau_{xy} = \mu \left(\frac{\partial u}{\partial y} + \frac{\partial v}{\partial x} \right) = \tau_{yx}$$

$$\tau_{yz} = \mu \left(\frac{\partial v}{\partial z} + \frac{\partial w}{\partial y} \right) = \tau_{zy}$$

$$\tau_{zx} = \mu \left(\frac{\partial w}{\partial x} + \frac{\partial u}{\partial z} \right) = \tau_{xz} \qquad (8.13)$$

$$\tau_{xx} = 2\mu \frac{\partial u}{\partial x}$$

$$\tau_{yy} = 2\mu \frac{\partial v}{\partial y}$$

$$\tau_{zz} = 2\mu \frac{\partial w}{\partial z}$$

The algebraic details of this calculation are rather involved and will not be duplicated here. The detailed calculations are given by Lamb.[15]

The equations of motion for a viscous fluid can be obtained by correcting the dynamical equation, Eq. (4.14) or (4.26), for the viscous stresses. Consider an element as in Fig. 41 in which all the stress components acting in the x-direction are shown. From this, it is apparent that the net force in the x-direction per unit of volume due to these viscous stresses is

$$\frac{\partial}{\partial x}\tau_{xx} + \frac{\partial}{\partial y}\tau_{yx} + \frac{\partial}{\partial z}\tau_{zx}$$

Corresponding expressions for the forces in the y- and z-directions can be written from symmetry. The equations of motion corresponding to Eq. (4.14) are then as follows:

$$
\begin{aligned}
a_x &= -\frac{1}{\rho}\frac{\partial p}{\partial x} + \frac{1}{\rho}\left(\frac{\partial}{\partial x}\tau_{xx} + \frac{\partial}{\partial y}\tau_{yx} + \frac{\partial}{\partial z}\tau_{zx}\right) + F_x \\
a_y &= -\frac{1}{\rho}\frac{\partial p}{\partial y} + \frac{1}{\rho}\left(\frac{\partial}{\partial x}\tau_{xy} + \frac{\partial}{\partial y}\tau_{yy} + \frac{\partial}{\partial z}\tau_{zy}\right) + F_y \\
a_z &= -\frac{1}{\rho}\frac{\partial p}{\partial z} + \frac{1}{\rho}\left(\frac{\partial}{\partial x}\tau_{xz} + \frac{\partial}{\partial y}\tau_{yz} + \frac{\partial}{\partial z}\tau_{zz}\right) + F_z
\end{aligned}
\tag{8.14}
$$

If the shear stresses are expressed in terms of the velocity components by Eq. (8.13), the equations of motion can be conveniently expressed in the vector form

$$\mathbf{a} = -\frac{1}{\rho}\nabla p + \mathbf{F} + \frac{\mu}{\rho}\nabla^2 \mathbf{v} \tag{8.15}$$

The variation of the viscosity coefficient has been neglected. These dynamical equations were first obtained by Navier.[25] The method of development is due to Stokes,[26] and Eqs. (8.15) are called the *Navier-Stokes equations.* The ratio $\nu = \mu/\rho$, which occurs in the Navier-Stokes equations is called the *kinematic viscosity coefficient.*

An important feature of the Navier-Stokes equations is seen if they are written in a dimensionless form. If in any given flow a characteristic velocity and length are U and d, respectively, dimensionless variables may be introduced by writing $\mathbf{v} = U\mathbf{v}'$, $t = (d/U)t'$, $p = \rho U^2 p'$, $\mathbf{a} = (U^2/d)\mathbf{a}'$, $x = dx'$, $y = dy'$, and $z = dz'$. The primed quantities are then the dimensionless variables. The Navier-Stokes equations then become (if the gravity forces are omitted)

$$\mathbf{a}' = -\nabla p' + \frac{1}{R}\nabla^2 \mathbf{v}' \tag{8.16}$$

where $R = \rho U d/\mu$, and the distance d has been taken as the unit of length. R is called the *Reynolds number* after Osborn Reynolds, who first pointed out the importance of this dimensionless ratio.[27] It is apparent from Eq. (8.16) that, if two flows have geometrically similar boundary conditions, then Eq. (8.16) will be the same and the flows will be geometrically similar, provided that the Reynolds number is the same for the two cases. The Reynolds number is thus a hydrodynamical measure of the scale of a fluid motion.

It would be possible to develop special solutions of the Navier-Stokes equations; however, these solutions are of little meteorological interest; for the atmospheric motions are almost invariably complicated by the phenomenon of turbulence. This phenomenon was first clearly discussed by Reynolds, who investigated the flow in a tube by injecting streamers of dye into the fluid. At low speeds, the fluid flowed smoothly in parallel layers, and this type of flow is for this reason called *laminar flow.* At high speeds, however, the flow was not smooth but turbulent with the fluid being thoroughly mixed. The latter type of flow is called *turbulent flow.* The critical Reynolds number (based on the mean velocity in the tube and the tube diameter) above which the flow was turbulent was found to be about 2,000. By taking great care to see that the fluid entered the pipe smoothly and with no critical disturbances, Ekman[28] was able to obtain laminar flows with Reynolds numbers as high as 24,000.

For laminar flow, the velocity distribution and shearing stresses are those given by Eqs. (8.8) and (8.4); however, in turbulent flow the velocity gradients in the center of the tube are much smaller and those near the wall are much larger than those for laminar flow with the same total discharge. Typical velocity profiles are shown in Fig. 42. The shearing stresses on the wall are also much higher for the turbulent flow. It thus appears that the Navier-Stokes equations will provide the proper solution for the case of laminar motion but not for turbulent motion. The same result is found in all other cases. Unfortunately, the scale of atmospheric motions (and thus their Reynolds numbers) is large enough so that these motions are almost invariably turbulent.

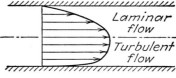

FIG. 42.

As an explanation of this difficulty, Reynolds suggested that the Navier-Stokes equations still apply, provided that they are applied to the instantaneous and rapidly varying turbulent velocities rather than to the mean velocity that is observed by normal instruments. This conjecture appears to be completely verified, although the general theory of turbulent motion is at present not complete. If the Navier-Stokes equations are to be used for turbulent-flow problems, it is thus necessary to express them in terms of the observed mean velocity components instead of the instantaneous velocities. This will be carried out in detail for the equation for equilibrium in the x-direction.

From Eq. (4.26)

$$\rho a_x = \rho\,\frac{\partial u}{\partial t} + \rho u\,\frac{\partial u}{\partial x} + \rho v\,\frac{\partial u}{\partial y} + \rho w\,\frac{\partial u}{\partial z} + 2\rho(\omega_y w - \omega_z v) \tag{8.17}$$

The continuity equation can be written as

$$0 = u\,\frac{\partial \rho}{\partial t} + u\,\frac{\partial}{\partial x}\,(\rho u) + u\,\frac{\partial}{\partial y}\,(\rho v) + u\,\frac{\partial}{\partial z}\,(\rho w) \tag{8.18}$$

If these are added, it is seen that

$$\rho a_x = \frac{\partial}{\partial t}\,(\rho u) + \frac{\partial}{\partial x}\,(\rho u^2) + \frac{\partial}{\partial y}\,(\rho u v) + \frac{\partial}{\partial z}\,(\rho u w) + 2\rho(\omega_y w - \omega_z v) \tag{8.19}$$

Let the velocity components and the pressure be written as

$$\begin{aligned} u &= \bar{u} + u' \\ v &= \bar{v} + v' \\ w &= \bar{w} + w' \\ p &= \bar{p} + p' \end{aligned} \tag{8.20}$$

where $\bar{u}$, $\bar{v}$, $\bar{w}$, and $\bar{p}$ are the mean values over a period δ which is of the order of a few hundredths of a second so that these means are the quantities normally observed. The primed quantities are thus the turbulent fluctuations which average out to zero in the time δ. These turbulent components can be measured directly only by means of instruments, such as the hot-wire anemometer,[29] that react extremely rapidly to velocity fluctuations. For the large-scale turbulent motions occurring in the atmosphere, the appropriate value for δ may be much increased. Now take the mean value of Eq. (8.19) over the period δ. The averaging process of taking the mean and the processes of differentiation can obviously be interchanged. Now

$$\overline{\rho u} = \overline{\rho\bar{u} + \rho u'} = \overline{\rho\bar{u}} + \overline{\rho u'} = \rho\bar{u} \tag{8.21}$$

Similarly

$$\overline{\rho u^2} = \rho \bar{u} \bar{u} + \overline{\rho u' u'}$$
$$\overline{\rho u v} = \rho \bar{u} \bar{v} + \overline{\rho u' v'} \qquad (8.22)$$
$$\overline{\rho u w} = \rho \bar{u} \bar{w} + \overline{\rho u' w'}$$

It should be noted that the density has been treated as constant. For the terms that are linear in the velocity components, it is seen that the instantaneous velocity is merely replaced by its mean value in this averaging process. On the other hand, the terms in which the velocity is quadratic have additional terms introduced that depend on the mean value of products of the turbulent components. The mean value of Eq. (8.19) is thus

$$\overline{\rho a_x} = \frac{\partial}{\partial t}(\rho \bar{u}) + \frac{\partial}{\partial x}(\rho \bar{u} \bar{u} + \overline{\rho u' u'}) + \frac{\partial}{\partial y}(\rho \bar{u} \bar{v} + \overline{\rho u' v'}) + \frac{\partial}{\partial z}(\rho \bar{u} \bar{w} + \overline{\rho u' w'})$$
$$+ 2\rho(\omega_y \bar{w} - \omega_z \bar{v}) \quad (8.23)$$

Now the continuity equation is linear in the velocity components, and its form is therefore unchanged by the averaging process. It can thus be multiplied through by $\bar{u}$, as in Eq. (8.18), and then be subtracted from Eq. (8.23). The final result is

$$\overline{\rho a_x} = \rho \frac{\partial \bar{u}}{\partial t} + \rho \bar{u} \frac{\partial \bar{u}}{\partial x} + \rho \bar{v} \frac{\partial \bar{u}}{\partial y} + \rho \bar{w} \frac{\partial \bar{u}}{\partial z} + 2\rho(\omega_y \bar{w} - \omega_z \bar{v}) + \frac{\partial}{\partial x}(\overline{\rho u' u'}) + \frac{\partial}{\partial y}(\overline{\rho u' v'})$$
$$+ \frac{\partial}{\partial z}(\overline{\rho u' w'}) \quad (8.24)$$

The effect of the turbulent velocity components appears only in the last three terms. Since the viscous stresses are linear in the velocity components, no additional terms are introduced on the right-hand side of Eq. (8.14). The equations of motion in terms of the mean velocities can thus be written as follows:

$$\frac{\partial \bar{u}}{\partial t} + \bar{u} \frac{\partial \bar{u}}{\partial x} + \bar{v} \frac{\partial \bar{u}}{\partial y} + \bar{w} \frac{\partial \bar{u}}{\partial z} + 2(\omega_y \bar{w} - \omega_z \bar{v}) = -\frac{1}{\rho} \frac{\partial \bar{p}}{\partial x} + F_x$$
$$+ \frac{1}{\rho} \left[\frac{\partial}{\partial x}(\bar{\tau}_{xx} - \overline{\rho u' u'}) + \frac{\partial}{\partial y}(\bar{\tau}_{yx} - \overline{\rho u' v'}) + \frac{\partial}{\partial z}(\bar{\tau}_{zx} - \overline{\rho u' w'}) \right]$$

$$\frac{\partial \bar{v}}{\partial t} + \bar{u} \frac{\partial \bar{v}}{\partial x} + \bar{v} \frac{\partial \bar{v}}{\partial y} + \bar{w} \frac{\partial \bar{v}}{\partial z} + 2(\omega_z \bar{u} - \omega_x \bar{w}) = -\frac{1}{\rho} \frac{\partial \bar{p}}{\partial y} + F_y$$
$$+ \frac{1}{\rho} \left[\frac{\partial}{\partial x}(\bar{\tau}_{xy} - \overline{\rho u' v'}) + \frac{\partial}{\partial y}(\bar{\tau}_{yy} - \overline{\rho v' v'}) + \frac{\partial}{\partial z}(\bar{\tau}_{zy} - \overline{\rho v' w'}) \right] \qquad (8.25)$$

$$\frac{\partial \bar{w}}{\partial t} + \bar{u} \frac{\partial \bar{w}}{\partial x} + \bar{v} \frac{\partial \bar{w}}{\partial y} + \bar{w} \frac{\partial \bar{w}}{\partial z} + 2(\omega_x \bar{v} - \omega_y \bar{u}) = -\frac{1}{\rho} \frac{\partial \bar{p}}{\partial z} + F_z$$
$$+ \frac{1}{\rho} \left[\frac{\partial}{\partial x}(\bar{\tau}_{xz} - \overline{\rho u' w'}) + \frac{\partial}{\partial y}(\bar{\tau}_{yz} - \overline{\rho v' w'}) + \frac{\partial}{\partial z}(\bar{\tau}_{zz} - \overline{\rho w' w'}) \right]$$

In these equations, $\bar{\tau}_{xx} = 2\mu(\partial \bar{u}/\partial x)$, etc., as in Eq. (8.13).

It is to be noted that these dynamical equations written in terms of the mean velocities are of exactly the same form as the Navier-Stokes equations except for the additional terms involving the mean values of the products of the turbulent velocities which appear on the right-hand side. These additional terms appear in the equations as though they were stresses, and they are for this reason generally called *apparent stresses*. It is these apparent stresses that cause the marked differences in the velocity distributions for laminar and turbulent flows. For atmospheric motions, it is generally found that the apparent stresses $-\rho \overline{u' v'}$, etc., are many thousands of times as large as the laminar type stresses, $\overline{\tau_{xy}}$, etc., and the laminar-type stresses can be neglected except in a very few cases.

Reynolds' introduction of the apparent stresses shows clearly the reason for the differences in laminar and turbulent flows; however, it does not permit the solution of turbulent-flow problems; for the apparent stresses appear as six new unknown quantities. The critical problem in turbulent motion is to express these apparent stresses in terms of the mean velocity components, and no general solution of this problem has yet been obtained. Since the turbulent velocities at any given point generally are caused by conditions some distance away, the complete theory will undoubtedly involve integral relations; however, many extremely useful results have been obtained by more or less intuitive methods that involve only differential relations.

If the phenomenon of viscosity in a laminar flow is considered from a molecular viewpoint, the viscous stresses appear as a transfer of momentum due to the random motion of the molecules.[30] It thus seems plausible to consider the turbulent motions as a macroscopic parallel to the random motion of the molecules. The stresses in a turbulent flow could then be given by a formula similar to the Newtonian friction law [Eq. (8.4) or the more general Eq. (8.13)]. For example, one could write, corresponding to Eq. (8.4)

$$\tau_{xy} = K\rho \frac{\partial \bar{u}}{\partial y} \tag{8.26}$$

where τ_{xy} is the total stress and K is an effective kinematic viscosity coefficient or, as it is more generally called, an exchange coefficient. This method was first used by Boussinesq.[31] This assumption, of course, gives no information regarding the magnitude of the exchange coefficient, and this must be determined by experiment for any given type of problem. If the turbulence was not isotropic, as in a stably stratified atmosphere, it would be necessary to use a different exchange coefficient for mixing in the various directions. This has been done in the case of high-level isentropic mixing in the atmosphere by Grimminger[32] and Rossby.[9]

The exchange coefficient can also be used to calculate the turbulent transport and diffusion not only of momentum but also of dust, moisture, heat, or any other property. The application of this principle to meteorological problems has been made by Wilhelm Schmidt[33] and many others. An excellent survey of the effects of turbulence on the motions of the atmosphere was given recently by Lettau.[34]

9. The Wind Structure and Turbulence near the Earth's Surface. In the lower levels of the atmosphere where the viscous stresses are significant, the wind may be considered as horizontal, and in addition the variation of the viscous stresses in the horizontal directions is so small that only the vertical variations need be considered. If it is assumed that the wind in any given horizontal plane is a uniform rectilinear flow, then the acceleration of the fluid elements relative to the earth's surface is also negligible. This is generally a satisfactory approximation. The equations of motion for this case are

$$\begin{aligned} -\rho\lambda\bar{v} &= -\frac{\partial \bar{p}}{\partial x} + \frac{\partial}{\partial z}\tau_{xz} \\ \rho\lambda\bar{u} &= -\frac{\partial \bar{p}}{\partial y} + \frac{\partial}{\partial z}\tau_{yz} \end{aligned} \tag{9.1}$$

where τ_{xz} and τ_{yz} are the total viscous stresses. In this case, it is again convenient to express these as a single complex equation

$$i\lambda\rho(\bar{u} + i\bar{v}) = -\left(\frac{\partial \bar{p}}{\partial x} + i\frac{\partial \bar{p}}{\partial y}\right) + \frac{\partial}{\partial z}(\tau_{xz} + i\tau_{zy}) \tag{9.2}$$

In terms of the exchange coefficient K

$$\tau_{zx} + i\tau_{zy} = K\rho \frac{\partial}{\partial z}(\bar{u} + i\bar{v}) \tag{9.3}$$

If the exchange coefficient is treated as a constant, for a first approximation, Eq. (9.2) can be written

$$i\lambda\rho(\bar{u} + i\bar{v}) = -\left(\frac{\partial \bar{p}}{\partial x} + i\frac{\partial \bar{p}}{\partial y}\right) + K\rho\frac{\partial^2}{\partial z^2}(\bar{u} + i\bar{v}) \tag{9.4}$$

Suppose the x-axis is placed along the surface isobar, so that the geostrophic wind is u_g where u_g is positive. For the comparatively shallow layer in which the viscous effects are important, the horizontal pressure gradient and the density and thus u_g can be considered as constant. Equation (9.4) then becomes

$$\frac{\partial^2}{\partial z^2}(\bar{u} + i\bar{v} - u_g) - i\frac{\lambda}{K}(\bar{u} + i\bar{v} - u_g) = 0 \tag{9.5}$$

This is to be solved with the boundary conditions that $\bar{u} + i\bar{v} = 0$ at the surface, $z = 0$, and that $\bar{u} + i\bar{v} = u_g$ at high levels. The general solution of Eq. (9.5) is

$$\bar{u} + i\bar{v} - u_g = C_1 e^{-(1+i)az} + C_2 e^{(1+i)az} \tag{9.6}$$

where

$$a = \sqrt{\frac{\lambda}{2K}} \tag{9.7}$$

From the boundary condition, it is seen that $C_2 = 0$ and $C_1 = -u_g$; therefore

$$\bar{u} + i\bar{v} = u_g[1 - e^{-(1+i)az}] \tag{9.8}$$

If the velocity vectors are plotted for various elevations, it is seen that their end points lie on an equiangular spiral that has the geostrophic wind for its limit point, as in Fig. 43. Since this solution was first obtained by Ekman, who applied it to the ocean currents produced by surface stresses,[35] this figure is called the *Ekman spiral*. Its application to atmospheric motions was first made by Åkerblom.[36]

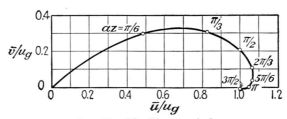

Fig. 43.—The Ekman spiral.

The results of this solution show that the viscous stresses cause the wind to blow across the isobars into the low-pressure zone. The maximum angle with which the wind blows across the isobars is 45 deg, and this occurs at the surface. In addition, it is seen that the deviation of the wind from the geostrophic wind and the viscous stresses diminish upward in an exponential manner.

If the Ekman spiral is used to analyze experimentally obtained wind distributions, the magnitudes of the exchange coefficient and the viscous stresses can be obtained. The most convenient experimental quantity for this purpose is the height H at which the wind blows parallel to the isobars. From Eq. (9.8), this occurs for $aH = \pi$; therefore

$$K = \frac{\lambda}{2}\left(\frac{H}{\pi}\right)^2 \tag{9.9}$$

From Eq. (9.3), the horizontal-shear-stress vector at any level is

$$\tau_{zx} + i\tau_{zy} = K\rho u_g(1 + i)ae^{-(1+i)az} \tag{9.10}$$

and the magnitude of the shearing stress at the surface τ_0 is therefore

$$\tau_0 = \frac{\lambda\rho H u_g}{\pi\sqrt{2}} \tag{9.11}$$

Observations of the height H for various geostrophic winds were made by Dobson,[37] and the results are shown in Table 1.

TABLE 1

u_g, cm/sec	460	910	1,560
H, cm	6×10^4	8×10^4	9×10^4
ρK, grams/cm sec	23	43	54
K, cm²/sec	2.0×10^4	3.7×10^4	4.7×10^4
τ_0, dynes/cm²	0.8	2.2	4.2

In these calculations, it was assumed that $\lambda = 1.14 \times 10^{-4}$ sec^{-1} and that $\rho = 1.15 \times 10^{-3}$ gram per cm³. These results give at least the order of magnitude of K and τ_0. It is of interest to note that the value of $\rho K/\mu$ is of the order of 10^5; therefore the stresses due to turbulent mixing are overwhelmingly large when compared with the laminar-type stresses.

Although the Ekman-spiral solution gives the general features of the wind in the friction layer, it can, of course, give no information on the variation of the exchange coefficient with altitude. This information can be obtained from the analysis of sounding-balloon data. The shear stresses can be computed by Eq. (9.1), and the exchange coefficients can then be computed by Eq. (9.2). Calculations of this sort have been made by Mildner.[38] His observations of the exchange coefficient which were obtained as the mean of 28 balloon runs on an October day near Leipzig show the results in Table 2.

TABLE 2

Height, m	80	135	190	240	295	405	460	510
ρK, grams/cm sec	125	270	310	500	246	117	70	70

This shows a linear variation of the exchange coefficient with altitude in the first 250 m and then a rapid decrease to a more or less constant value aloft.

If an atmosphere is initially stable, turbulent mixing tends to produce a condition of neutral stability throughout the layer that is mixed. This produces a very stable layer just at the top of the turbulent layer, as shown in Fig. 44, in which the potential temperature is plotted against altitude. This very stable layer would suppress the turbulence and slow down the growth of the turbulent layer. Such a condition may have caused the great decrease in the exchange coefficient observed by Mildner between 240 and 295 m, for the air mass in which his observations were made had a fairly stable lapse rate of about 6.5°C per km.

The most obvious weakness of the Ekman-spiral solution is the assumption that the exchange coefficient is the same at all levels. This is particularly so in the first few meters from the surface where the exchange coefficient must become very small. A solution that is valid near the surface can be obtained by means of the "mixing-length" hypothesis, which was introduced by Prandtl.[39] Let us consider a layer next to the surface that is thin enough so that the shear stress may be considered as constant. In this ground layer, place the x-axis parallel to the wind. The shear stress in this layer τ_{xz} is thus the shear stress at the ground τ_0, or

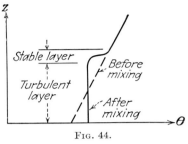

FIG. 44.

$$\tau_0 = -\rho \overline{u'w'} \tag{9.12}$$

where the laminar-type stresses have been neglected. In accordance with the exchange-coefficient hypothesis, it seems reasonable to suppose that u' is proportional to $\partial \bar{u}/\partial z$, and if the layer of air is neutrally stable, u' and w' should be proportional. Equation (9.12) can thus be written as

$$\tau_0 = \rho l^2 \left(\frac{\partial \bar{u}}{\partial z}\right)^2 \tag{9.13}$$

where l^2 is the constant of proportionality. The quantity l has the dimensions of a length and is called the *mixing length*. In terms of the mixing length, the exchange coefficient is given by

$$K = l^2 \left|\frac{\partial \bar{u}}{\partial z}\right| \tag{9.14}$$

where the magnitude of $\partial \bar{u}/\partial z$ is used, since both K and l^2 are essentially positive. The introduction of the mixing length does not in itself solve the turbulent-flow problem, for it merely replaces one unknown quality, the exchange coefficient, by another, the mixing length; however, it is generally found to be much easier to make suitable estimates of the mixing length than of the exchange coefficient.

An elegant method of estimating the mixing length in any given turbulent flow is provided by the principle of *mechanical similarity*, which was proposed by von Kármán.[40] He suggested that, in any fully developed turbulent flow, the velocity field in the neighborhood of any two given points must be geometrically similar but may have a different scale. Since this scale must be independent of the magnitude of the velocity, the simplest length that can provide such a scale is $\left|\dfrac{\partial \bar{u}/\partial z}{\partial^2 \bar{u}/\partial z^2}\right|$. Other lengths such as $\left|\dfrac{\partial^2 \bar{u}/\partial z^2}{\partial^3 \bar{u}/\partial z^3}\right|$ could also be formed from the velocity derivatives; but, if the similarity principle holds, they must all be proportional to the first; and the mixing length also must be proportional to this scale length. Thus

$$l = k \left|\frac{\partial u/\partial z}{\partial^2 \bar{u}/\partial z^2}\right| \tag{9.15}$$

where k is a universal constant.

By means of this expression for the mixing length, Eq. (9.13) can be written as

$$\left(\frac{\partial \bar{u}}{\partial z}\right)^{-2} \frac{\partial^2 \bar{u}}{\partial z^2} = -\frac{k}{\sqrt{\tau_0/\rho}} \tag{9.16}$$

The negative square root was used, since $\partial^2\bar{u}/\partial z^2$ is negative in the case under consideration. This may be integrated to obtain

$$\left(\frac{\partial\bar{u}}{\partial z}\right)^{-1} = \frac{k}{\sqrt{\tau_0/\rho}}\,(z + z_0) \tag{9.17}$$

where z_0 is an integration constant and z is supposed to be small enough so that the variation of τ_0 is negligible. From this

$$\frac{\partial\bar{u}}{\partial z} = \frac{1}{k}\sqrt{\frac{\tau_0}{\rho}}\frac{1}{z + z_0} \tag{9.18}$$

If this result is compared with Eq. (9.13), it is seen that

$$l = k(z + z_0) \tag{9.19}$$

Since z is measured from the surface of the earth, kz_0 is thus seen to be the mixing length at the surface; and z_0 must be a measure of the roughness of the surface and is called the *roughness coefficient*. If the surface is quite smooth, there may be a very thin layer with laminar flow next to the surface, and Eq. (9.19) cannot be assumed to hold down to the surface but will hold only to the outer limit of the laminar sublayer. The effects of this laminar sublayer have been considered by von Kármán.[41] Such laminar sublayers exist in the atmosphere at smooth-water surfaces, and Rossby[42] has applied von Kármán's results in his study of the momentum transfer at the sea surface.

If the possibility of the existence of a laminar sublayer is neglected, then z_0 is a measure of the roughness, and Eq. (9.18) holds down to the surface where $\bar{u} = 0$. If Eq. (9.18) is integrated from $z = 0$ to a height z, it is thus seen that

$$\bar{u} = \frac{1}{k}\sqrt{\frac{\tau_0}{\rho}}\log\left(1 + \frac{z}{z_0}\right) \tag{9.20}$$

This result was given by Prandtl.[43] For this case, the exchange coefficient is given by

$$K = k\sqrt{\frac{\tau_0}{\rho}}\,(z + z_0) \tag{9.21}$$

i.e., the exchange coefficient varies linearly with the distance from the surface. This is in agreement with Mildner's observations. This logarithmic velocity distribution is found to apply to the flow near any rough wall.

TABLE 3

Surface condition	z_0, cm	Observer
Smooth lawn	0.55	Hellmann[45]
Open fields	3.2	Shaw[46]
Sea surface (swells—no breakers)	4.0	Wüst[47]

This solution contains two arbitrary constants, k and z_0, which can be adjusted to give the best results. The best experimental data show that $k = 0.40$. As a result of channel tests with sand grains for the surface-roughness elements, Prandtl suggests that

$$z_0 = \frac{\epsilon}{30} \tag{9.22}$$

where ϵ is the actual height of the roughness element.   This is in rough agreement with the results found from the analysis of meteorological observations.   Several values of the roughness coefficient as reduced by Rossby and Montgomery[44] from observations by various observers are given in Table 3.

Similar roughness coefficients have been given by Paeschke.[48]   His results are given in Table 4.

TABLE 4

Surface Condition	z_0, cm
Smooth snow	0.5
Fallow field	2.1
Low grass	3.2
High grass	3.9
Wheat field	4.5

If Eq. (9.20) is solved for the shear stress τ_0, it is seen that

$$\tau_0 = \rho \left\{ \frac{k}{\log \left[1 + (z/z_0)\right]} \right\}^2 \bar{u}^2 \tag{9.23}$$

so that the shear stress varies as the square of the wind velocity at any fixed level in the logarithmic layer.   This is in agreement with an empirical law

$$\tau_0 = \rho \gamma^2 \bar{u}^2 \tag{9.24}$$

which had been proposed by G. I. Taylor[49] as a result of his analysis of Dobson's[37] observations.   Taylor's results showed $\gamma = 0.05$.   As the wind observations were made at a height of 30 m, this corresponds to a roughness coefficient of $z_0 = 1.0$ cm.

The logarithmic wind distribution fits the observed wind distribution in the layer next to the surface with a high degree of accuracy as long as the air is in a condition of neutral stability.   For very light winds, the turbulent mixing is frequently too weak to establish a neutrally stable layer.   The gravitational stability in this case reduces the intensity of vertical mixing, and thus the wind will increase faster with altitude for a given shear stress.   A dimensionless quantity that measures this effect is the Richardson number $\left[\dfrac{(g/\theta)(\partial\theta/\partial z)}{(\partial\bar{u}/\partial z)^2}\right]$ where θ is the potential temperature.   The first theoretical investigation of this problem was made by Exner;[50] however, his results were not satisfactory.   Since then this problem has been treated more successfully by Rossby and Montgomery[44] and by Sverdrup.[51]

Although the logarithmic velocity distribution fits the observed wind close to the surface very well, it contains the shear stress τ_0 as an undetermined parameter.   A more complete theory can be built up by combining the logarithmic solution with a wind spiral.   If there is to be continuity in the wind velocity and in the shear, the boundary condition at the bottom of the spiral regime is that

$$\bar{u} + i\bar{v} = C \frac{\partial}{\partial z} (\bar{u} + i\bar{v}) \tag{9.25}$$

where C is a real constant.   This condition is required since the wind velocity and shear in the logarithmic layer are in the same direction.   The other boundary condition is that the wind must approach the geostrophic wind at high levels.

If the exchange coefficient is treated as constant throughout the wind spiral, Eqs. (9.5) and (9.6) still apply.   Since $\bar{u} + i\bar{v} - u_g = 0$ for large values of z, $C_2 = 0$. If the wind at the bottom of the wind spiral ($z = 0$) is

$$(\bar{u} + i\bar{v})_0 = C_0 e^{i\alpha} \tag{9.26}$$

so that α is the angle at which the wind crosses the isobars at the ground, then

$$\bar{u} + i\bar{v} = u_g + (C_0 e^{i\alpha} - u_g)e^{-(1+i)az} \qquad (9.27)$$

If this is substituted in the boundary condition (9.25), it is seen that

$$C_0 = u_g(\cos\alpha - \sin\alpha) \qquad (9.28)$$

and that

$$C = \frac{1}{2a}(\cot\alpha - 1) \qquad (9.29)$$

With this value of C_0, Eq. (9.27) can be written

$$\bar{u} + i\bar{v} = u_g + \sqrt{2}u_g \sin\alpha \; e^{-(1+i)az + i\left(\alpha + \frac{3\pi}{4}\right)} \qquad (9.30)$$

The shear stress at the bottom of this layer is

$$\tau_{zx} + i\tau_{zy} = K\rho \frac{\partial}{\partial z}(\bar{u} + i\bar{v})|_{z=0} = K\rho \frac{C_0}{C} e^{i\alpha} \qquad (9.31)$$

The magnitude of this shear stress must equal the shear stress at the ground; therefore

$$\tau_0 = \rho u_g \sin\alpha \sqrt{2\lambda K} \qquad (9.32)$$

This solution of the wind-spiral equation was given by G. I. Taylor.[49]

If this wind spiral is to be fitted together with the logarithmic solution to obtain a complete solution, then the shear stress τ_0 of Eq. (9.32) must be used through the lower layer. If the wind is to be continuous

$$u_g(\cos\alpha - \sin\alpha) = \frac{1}{k}\sqrt{\frac{\tau_0}{\rho}}\log\frac{h + z_0}{z_0} \qquad (9.33)$$

where h is the depth of the lower layer. The exchange coefficient must also be continuous; therefore

$$K = k\sqrt{\frac{\tau_0}{\rho}}(h + z_0) \qquad (9.34)$$

Since the wind velocity, the shearing stress, and the exchange coefficient are continuous, the wind shear is also continuous. If τ_0 and h are eliminated between Eqs. (9 32) to (9.34), an expression is obtained that determines the angle α. This is

$$\frac{k\sqrt{u_g}(\cos\alpha - \sin\alpha)}{(2\lambda K)^{\frac{1}{4}}\sqrt{\sin\alpha}} = \log\frac{K^{\frac{3}{4}}}{kz_0\sqrt{u_g}\sin\alpha\,(2\lambda)^{\frac{1}{4}}} \qquad (9.35)$$

This solution can be used in the same manner as the Ekman spiral to analyze observed wind distributions. For this purpose, Dobson's observations will again be used. It is first necessary to determine the theoretical height at which the wind is in the gradient direction. Let this height H be written

$$H = h + H' \qquad (9.36)$$

where h is the thickness of the logarithmic layer and H' is the thickness of the spiral layer. From Eq. (9.30), the cross wind $\bar{v}$ will vanish at the height H' if

$$H' = \left(\alpha + \frac{3\pi}{4}\right)\sqrt{\frac{2K}{\lambda}} \qquad (9.37)$$

Also, from Eqs. (9.34) and (9.32)

$$h + z_0 = \frac{K^{3/4}}{k \sqrt{u_g \sin \alpha} \,(2\lambda)^{1/4}} \tag{9.38}$$

Since z_0 is negligibly small when compared with H, the sum of these two gives H. In the calculations, it is preferable to use dimensionless forms. Let

$$N_1 = \frac{2\lambda K}{u_g{}^2} \tag{9.39}$$

so that N_1 is a dimensionless exchange coefficient. Then

$$\frac{\lambda H}{u_g} = \left(\alpha + \frac{3\pi}{4}\right) N_1{}^{1/2} + \frac{N_1{}^{3/4}}{2k\sqrt{\sin\alpha}} \tag{9.40}$$

A second dimensionless parameter is

$$N = \frac{u_g}{\lambda z_0} \tag{9.41}$$

In terms of these two dimensionless parameters, the equation for α becomes

$$\frac{k(\cos\alpha - \sin\alpha)}{N_1{}^{1/4}\sqrt{\sin\alpha}} = \log\frac{NN_1{}^{3/4}}{2k\sqrt{\sin\alpha}} \tag{9.42}$$

If N and $\lambda H/u_g$ are given, Eqs. (9.42) and (9.40) can be solved simultaneously for the values of α and N_1. Then, from Eq. (9.32), the shear stress can be found, for

$$\tau_0 = \rho u^2{}_g \sin\alpha \, N_1{}^{1/2} \tag{9.43}$$

The depth of the logarithmic layer is given by

$$\frac{h}{H} = \left[1 + 2k\left(\alpha + \frac{3\pi}{4}\right)\sqrt{\sin\alpha}\, N_1{}^{-1/4}\right]^{-1} \tag{9.44}$$

In reducing Dobson's data, it was assumed that $\lambda = 1.14 \times 10^{-4}\,\text{sec}^{-1}$, $z_0 = 1.0$ cm and $\rho = 1.15 \times 10^{-3}$ gram per cm³. The details are given in Table 5.

TABLE 5

u_g, cm/sec	460	910	1560
H, m	600	800	900
$\alpha°$(observed)	13	$21\frac{1}{2}$	20
N	4.03×10^6	7.98×10^6	13.7×10^6
$\dfrac{\lambda H}{u_g}$	0.01555	0.01003	0.00658
N_1	2.93×10^{-5}	1.24×10^{-5}	5.49×10^{-6}
$\alpha°$ (computed)	13	17	21
$\dfrac{h}{H}$	0.068	0.043	0.037
$N_1 N^{3/4}$	2.37×10^5	2.80×10^5	2.79×10^5
τ_0, dynes/cm²	0.32	1.04	2.35
K, cm²/sec	2.72×10^4	4.51×10^4	5.38×10^4

It may be observed that this corrected theory predicts the angle that the wind blows across the isobars with a fairly high degree of accuracy. The thickness of the logarithmic layer is seen to be about 40 m. The shearing stress at the ground is seen to be somewhat smaller than that computed by the simple Ekman spiral, and the exchange coefficient is somewhat larger.   It may further be observed that the exchange coefficient varies roughly with the square root of the velocity, and the shear stress varies roughly with the three-halves power of the velocity.

This solution to the problem of the wind distribution in the friction layer is completely determined by u_g, λ, z_0, ρ, and K.   From these quantities may be formed the two dimensionless parameters N and N_1. Of these two, the first, N, is composed of u_g, λ, and z_0, which are all external parameters that determine the wind distribution. On the other hand, N_1 contains K, the exchange coefficient. Since the exchange coefficient is a measure of the turbulent mixing, it is an internal parameter determined by the fluid motion.   For this reason, one would expect to find a functional relationship of the form

$$N_1 = N_1(N) \tag{9.45}$$

From the analysis of Dobson's data, it appears that this may roughly be written as

$$N_1 = 2.6 \times 10^5 N^{-3/2} \tag{9.46}$$

Unfortunately, this formula cannot be considered as generally applicable, for there are several additional factors that have not been considered.   Probably the most important of these is the effect of gravitational stability.   Next is the fact that the surface roughness is far from homogeneous; and, although the roughness in the immediate neighborhood determines the wind distribution in the logarithmic layer, the turbulent mixing in the wind spiral must be determined by the mean roughness a considerable distance upwind.

In this wind-spiral solution, the exchange coefficient was considered as constant. Since the wind shear was found to decrease exponentially, this solution involves a mixing length that increases exponentially with altitude.   A solution of the wind spiral that avoids this difficulty has been given by Rossby,[52] who used a generalization of von Kármán's similarity principle to compute the mixing length.   In a later paper, Rossby and Montgomery[44] combined this wind spiral with a logarithmic layer near the ground in order to obtain a complete solution.   The general results of their analysis are similar to the somewhat simpler theory given here, which uses G. I. Taylor's solution of the wind spiral.

10. Diffusion of Properties by Eddies. It was pointed out that the laminar stresses, if viewed on a molecular scale, arise from the transport of momentum by the random motion of the molecules.   In exactly the same manner, the apparent stresses are caused by the turbulent transport of momentum.   Consider a surface element of area dS in the xy-plane, as in Fig. 45. Then the volume of fluid carried up through this surface by the turbulent motions in time dt is $w'\,dS\,dt$. Now this fluid has a momentum in the x-direction of an amount ρu per unit volume.   Thus the amount of x momentum transported in the z-direction is $\rho u w'\,dS\,dt$. If this is averaged over the time δ, it is thus seen that

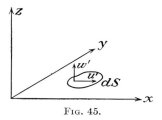

FIG. 45.

$\overline{\rho u' w'}$ is the net turbulent transport of x momentum in the z-direction per unit time and area.   This will be written as

$$T_z(\rho\bar{u}) = \overline{\rho u' w'} = -\tau_{zx} \tag{10.1}$$

Similar interpretations can be placed upon the other apparent shear stresses.

In order to simplify the following discussion, let us consider a motion with $\bar{v} = \bar{w} = 0$ and $\bar{u} = \bar{u}(z)$, as in Fig. 46. If it is supposed that the turbulent transfer arises from eddies of fluid that are initially in equilibrium with their surroundings and are then moved across the flow to some new position where they mix, then, in any given plane, $z = \text{const}$

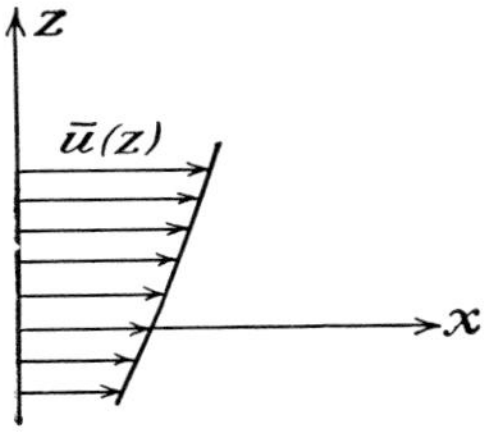

FIG. 46.

$$u' = -l' \frac{\partial \bar{u}}{\partial z} \tag{10.2}$$

where l' is the distance below the plane where the turbulent eddy was formed. This is, of course, only the first term of a Taylor series. Substituting this result in Eq. (10.1), it is seen that

$$T_z(\rho \bar{u}) = -\tau_{zx} = -\rho \overline{w'l'} \frac{\partial \bar{u}}{\partial z} \tag{10.3}$$

In other words, the exchange coefficient K is

$$K = \overline{w'l'} \tag{10.4}$$

The turbulent transport of any other fluid property may be discussed in a similar manner. In order to make the problem more specific, the transport of moisture will be considered. Let q be the specific humidity, so that ρq is the mass of water per unit volume. Then

$$q = \bar{q} + q' \tag{10.5}$$

where q' is the turbulent fluctuation of the specific humidity and $\bar{q}$ is its mean value. Then, again, the volume of fluid carried up through the surface dS' of Fig. 10.1 in time dt is $w' \, dS \, dt$, and this fluid has moisture per unit volume equal to ρq. Thus $\rho q w' \, dS \, dt$ is the turbulent transport of moisture through this element in time dt. If this is averaged over the time δ, it is seen that

$$T_z(\rho q) = \overline{\rho w' q'} \tag{10.6}$$

Also, if $\bar{q} = \bar{q}(z)$, we can assume

$$q' = -l' \frac{\partial \bar{q}}{\partial z} \tag{10.7}$$

so that

$$T_z(\rho q) = -K\rho \frac{\partial \bar{q}}{\partial z} \tag{10.8}$$

where K is given by Eq. (10.4).

If the turbulent motion is isotropic, the exchange coefficient is the same for the transport in any direction. On the other hand, if the turbulent motion is not isotropic as is the case in a fluid with a stable stratification, it would be necessary to use a different exchange coefficient for each direction in which the turbulent transport was computed.[*] As the mixing length l' depends on the rate at which the fluid proper-

---

[*] This idea may be formulated more precisely by use of tensor quantities. In the notation of cartesian tensors,[13] the shear stress τ_{ij} is a second-order tensor; and if it is assumed that the shear stress is a linear-tensor function of the fluid-deformation tensor $\partial \bar{u}_i / \partial x_j$, then

$$\tau_{ij} = \rho K_{ijlm} \frac{\partial \bar{u}_l}{\partial x_m} \tag{A}$$

where the exchange coefficients K_{ijlm} are the components of a fourth-degree tensor that has 3^4, or 81, components. Since $\tau_{ij} = \tau_{ji}$, these are reduced to 54 distinct components. If the turbulent mixing is isotropic and the fluid is incompressible, this may be further reduced to a single exchange coefficient K, and

$$\tau_{ij} = \rho K \left(\frac{\partial \bar{u}_i}{\partial x_j} + \frac{\partial \bar{u}_j}{\partial x_i} \right) \tag{B}$$

This corresponds exactly to the laminar stresses as given by Eq. (8.13). For nonisotropic turbulent ties are transferred to and from the turbulent eddies, it might also be expected that the exchange coefficients would vary from one property to another. This effect has been noticed; however, Sverdrup's observations[54] indicated that exchange coefficients for heat and momentum were practically the same in the atmosphere. On the other hand, he found considerable difference in the exchange coefficients of different properties in the ocean, where the stability is fairly high.

This method of developing the formulas for the turbulent stresses [Eq. (10.3)] and the exchange coefficient [Eq. (10.4)] is due to G. I. Taylor.[55] Since it involves the assumption that momentum is transported unchanged by the turbulent eddies through the distance l', these theories are generally classed as momentum-transfer theories. The assumption that momentum is transferred in this manner assumes that the turbulent-pressure fluctuations have no net effect on the momentum transfer. Taylor has shown that, if a motion (both the mean and the turbulent components) is strictly two-dimensional, the momentum is not conserved in the eddy-transfer process but the vorticity is. Taylor was thus led to develop his vorticity-transfer

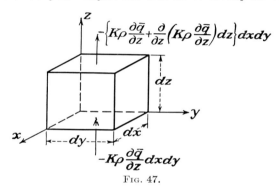

$$-\left\{K\rho\frac{\partial \bar{q}}{\partial z}+\frac{\partial}{\partial z}\left(K\rho\frac{\partial \bar{q}}{\partial z}\right)dz\right\}dxdy$$

$$-K\rho\frac{\partial \bar{q}}{\partial z}dxdy$$

Fɪɢ. 47.

theory. It can easily be shown that the vorticity-transfer theory cannot apply to turbulent flow near a wall, and for this case the momentum-transfer theory which gives the logarithmic velocity distribution gives a satisfactory solution. For certain other cases such as the flow in the wake behind a long rod, the vorticity-transfer theory can be applied quite successfully.

The turbulent transport as given by Eq. (10.8) can be used to develop the equation for turbulent diffusion. For this purpose, the laminar-diffusion terms will be omitted in order to simplify the calculations. Consider a volume element as in Fig. 47. Then the turbulent transport of water per unit of time into the element from below is

---

mixing, the reduction from 54 exchange coefficients to one is no longer possible, although many may be shown to vanish.

In comparison with this result, the turbulent transport of a scalar quantity such as the moisture $\rho \bar{q}$, is a first-order tensor T_i (ρq), which may be assumed to be a linear-tensor function of the specific-humidity gradient $\partial \bar{q}/\partial x_i$. Thus

$$T_i(\rho q) = -\rho K_{ij}\frac{\partial \bar{q}}{\partial x_j} \qquad (C)$$

For this case, the exchange tensor K_{ij} is a second-order tensor having 3^2, or 9, components. If the turbulent mixing is isotropic, the exchange tensor is again reduced to a single exchange coefficient K where

$$T_i(\rho q) = -\rho K \frac{\partial \bar{q}}{\partial x_i} \qquad (D)$$

This is the tensor form equivalent to Eq. (10.8). For the nonisotropic case, the nonvanishing components of K_{ij} may be identified with certain of the nonvanishing components of K_{ijlm}.

$-K\rho(\partial\bar{q}/\partial z)dx\,dy$. Similarly, the transport out of the top is $-\{K\rho(\partial\bar{q}/\partial z) + (\partial/\partial z)$ $[K\rho(\partial\bar{q}/\partial z)]dz\}\,dx\,dy$. Thus $\dfrac{\partial}{\partial z}\left(K\rho\dfrac{\partial\bar{q}}{\partial z}\right)$ is the net transport per unit volume and time into this volume element due to the turbulent mixing in the z-direction. Similar expressions give the effects due to mixing in the x- and y-directions. This net influx is of course equal to the rate of increase of water within the element, or

$$\rho\frac{D\bar{q}}{Dt} = \frac{\partial}{\partial x}\left(\rho K\frac{\partial\bar{q}}{\partial x}\right) + \frac{\partial}{\partial y}\left(\rho K\frac{\partial\bar{q}}{\partial y}\right) + \frac{\partial}{\partial z}\left(\rho K\frac{\partial\bar{q}}{\partial z}\right) \tag{10.9}$$

This is the turbulent-diffusion equation for isotropic turbulent mixing. In most cases, the density may be considered constant and eliminated from the equation. Since

$$\rho\frac{D\bar{q}}{Dt} = \frac{\partial}{\partial t}\left(\rho\bar{q}\right) + \nabla\cdot(\rho\bar{q}\mathbf{\bar{v}}) - \bar{q}\left[\frac{\partial\rho}{\partial t} + \nabla\cdot(\rho\mathbf{\bar{v}})\right] \tag{10.10}$$

and the bracketed term is identically zero by the equation of continuity, the diffusion equation, Eq. (10.9), can be written as

$$\frac{\partial}{\partial t}\left(\rho\bar{q}\right) + \nabla\cdot(\rho\bar{q}\mathbf{\bar{v}} - \rho K\,\nabla\bar{q}) = 0 \tag{10.11}$$

In this form the advective transport $\rho\bar{q}\mathbf{\bar{v}}$ can be more easily compared with the turbulent transport $-\rho K\,\nabla\bar{q}$.

These results will first be used to discuss the problem of evaporation into a steady wind. In order to simplify the calculations, it will be assumed that the motion is two-dimensional in the xz-plane, and furthermore the variation of the wind velocity, density, and exchange coefficient with elevation will be neglected. These assumptions completely neglect the wind structure of the lower atmosphere; therefore only qualitative results can thus be obtained. Since the horizontal turbulent transport is normally very small compared with the horizontal advective transport, the horizontal-turbulent-transport terms will be neglected. With these approximations, the diffusion equation becomes

$$\frac{\partial^2\bar{q}}{\partial z^2} - \frac{\bar{u}}{K}\frac{\partial\bar{q}}{\partial x} = 0 \tag{10.12}$$

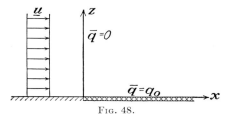

FIG. 48.

Suppose the air is initially dry so that $\bar{q} = 0$, and at $x = 0$ it passes over a moist surface where the saturation humidity is constant and equal to q_0. Since the horizontal turbulent transport is neglected, the air must still be dry for the line $x = 0$. The boundary conditions that the humidity must satisfy are thus $\bar{q} = 0$ for $x = 0$, and $\bar{q} = q_0$ for $z = 0$. The solution is to cover the quadrant in which both x and z are positive, as shown in Fig. 48.

This problem can be solved by the similarity method. Assume $\bar{q} = \bar{q}(\eta)$ where $\eta = bzx^\alpha$ and b and α are constants. Then

$$\frac{\partial\bar{q}}{\partial x} = \alpha\frac{\eta}{x}\frac{d\bar{q}}{d\eta} \tag{10.13}$$

and

$$\frac{\partial^2\bar{q}}{\partial z^2} = b^2x^{2\alpha}\frac{d^2\bar{q}}{d\eta^2} \tag{10.14}$$

From this

$$\frac{d^2\bar{q}}{d\eta^2} - \frac{\alpha\bar{u}}{Kb^2x^{2\alpha-1}}\eta\frac{d\bar{q}}{d\eta} = 0 \tag{10.15}$$

It may be observed that, if $\alpha = -\frac{1}{2}$, this reduces to a function of the single variable η. Furthermore, let $b^2 = \bar{u}/4K$; so

$$\eta = z \sqrt{\frac{\bar{u}}{4Kx}} \tag{10.16}$$

Then Eq. (10.15) becomes

$$\frac{d^2\bar{q}}{d\eta^2} + 2\eta \frac{d\bar{q}}{d\eta} = 0 \tag{10.17}$$

The boundary conditions are $\bar{q} = 0$ for $\eta = \infty$ and $\bar{q} = q_0$ for $\eta = 0$. Now an integrating factor for Eq. (10.17) is e^{η^2}; therefore, this may be integrated to

$$\frac{d\bar{q}}{d\eta} = Ce^{-\eta^2} \tag{10.18}$$

where C is a constant of integration. The integral of this that makes $\bar{q}$ vanish at $\eta = \infty$ is

$$\bar{q} = -C \int_\eta^\infty e^{-t^2}\, dt \tag{10.19}$$

From the other boundary condition, it is seen that

$$q_0 = -C \int_0^\infty e^{-t^2}\, dt = -C \frac{\sqrt{\pi}}{2} \tag{10.20}$$

Thus

$$\bar{q} = q_0 \frac{2}{\sqrt{\pi}} \int_\eta^\infty e^{-t^2}\, dt \tag{10.21}$$

This result may be conveniently expressed in terms of the error function $\Phi(x)$, which is tabulated in many convenient references.[56] (See also Sec. I.) Since

$$\Phi(x) = \frac{2}{\sqrt{\pi}} \int_0^x e^{-t^2}\, dt \tag{10.22}$$

Eq. (10.21) may be written

$$\bar{q} = q_0 \left[1 - \Phi(z \sqrt{\bar{u}/4Kx})\right] \tag{10.23}$$

This solution of the turbulent-diffusion problem was given by Jeffreys.[57] In this approximation, the specific humidity is constant along the curves for which η is constant, *i.e.*, along any parabola of the form $z^2 = kx$.

This solution may be used to estimate the rate of evaporation from the surface. This is

$$T_z(\rho q)|_{z=0} = -K\rho \frac{\partial \bar{q}}{\partial z}\bigg|_{z=0} \tag{10.24}$$

By the use of Eqs. (10.20), (10.18), and (10.16), this is

$$T_z(\rho q)|_{z=0} = \rho q_0 \sqrt{\frac{\bar{u}K}{\pi x}} \tag{10.25}$$

The total mass M of water evaporated per unit time in a strip of unit width in the y-direction and of length l in the x-direction is

$$M = \int_0^l T_z(\rho q)|_{z=0}\, dx \tag{10.26}$$

or

$$M = 2\rho q_0 \sqrt{\frac{\bar{u}Kl}{\pi}} \tag{10.27}$$

Since this solution treats the wind velocity and the exchange coefficient as being independent of the elevation, it seriously overestimates the total evaporation from large bodies of water. A more complete solution of this problem that considers some of the effects of the variation of the wind velocity and exchange coefficient with elevation has been given by Sutton.[58] More recent investigations of the evaporation from land and water surfaces which consider the linear variation of the exchange coefficient for small elevations have been given by Thornthwaite and Holtzman[59] and by Montgomery.[60] W. Schmidt[61] has considered the problem of evaporation from a moist surface of finite width. He used the same approximations as Jeffreys and obtained similar results.

A similar analysis can be used to discuss the problem of the heating (or cooling) of an air mass that has moved onto a warm (or cold) surface. The turbulent-diffusion equation [Eq. (10.9)] cannot be applied directly to the temperature field, for the temperature is not conserved during the eddy-mixing process. As the turbulent eddy is displaced from equilibrium, its temperature changes at the dry-adiabatic lapse rate; consequently, the heat transfer in the vertical is

$$T_z(Q) = -K\rho c_p \left(\frac{\partial T}{\partial z} + \Gamma\right) \tag{10.28}$$

where Γ is the dry-adiabatic lapse rate of $9.86°C$ per km. This shows that the turbulent transport of heat in a stable atmosphere is downward, and this heat transport tends to produce an air mass of neutral stability. The heat-diffusion equation is thus

$$\rho c_p \frac{DT}{Dt} = \frac{\partial}{\partial z}\left[K\rho c_p \left(\frac{\partial T}{\partial z} + \Gamma\right)\right] \tag{10.29}$$

if only the vertical turbulent transport of heat is considered. If the variation of $K\rho$ with elevation is neglected, this can be further simplified to

$$\frac{DT}{Dt} = K \frac{\partial^2 T}{\partial z^2} \tag{10.30}$$

Suppose an air mass has an initial temperature distribution given by

$$T = T_0 - \gamma z$$

where γ is the constant lapse rate. At time $t = 0$, this air mass arrives over a surface for which the temperature is T_1. Then, if horizontal variations in temperature are neglected and $K\rho$ is assumed to be independent of the elevation

$$\frac{\partial T}{\partial t} = K \frac{\partial^2 T}{\partial z^2} \tag{10.31}$$

In order to calculate the final temperature distribution, it is preferable to calculate the change in temperature ΔT where

$$\Delta T = T - (T_0 - \gamma z) \tag{10.32}$$

From Eq. (10.31)

$$\frac{\partial}{\partial t}(\Delta T) = K \frac{\partial^2}{\partial z^2}(\Delta T) \tag{10.33}$$

and the boundary conditions are $\Delta T = 0$ for $t = 0$ and $\Delta T = T_1 - T_0$ for $z = 0$. This is mathematically identical with the previously discussed moisture-diffusion problem, with the following quantities corresponding:

Heat diffusion...................................	ΔT	$T_1 - T_0$	t	z
Moisture diffusion..............................	$\bar{q}$	q_0	$x/\bar{u}$	z

By means of these substitutions, the final solution can be written immediately as

$$T = T_0 - \gamma z + (T_1 - T_0) \left[1 - \Phi \left(\frac{z}{\sqrt{4Kt}} \right) \right] \qquad (10.34)$$

This solution was given by Taylor.[55] The term in brackets in this equation is unity for $z = 0$ and diminishes to a value of 0.1 for $z/\sqrt{4Kt} = 1.2$. Thus the effect of the change in surface temperature is very small above this level. Taylor assumes that for all practical purposes this term has no effect above $z/\sqrt{4Kt} = 1$; therefore the height to which the effect of the surface temperature change has penetrated is

$$z^2 = 4Kt \qquad (10.35)$$

Taylor's observations of the virtual-temperature distribution over the Grand Banks of Newfoundland showed a marked inversion near the surface and neutrally stable air aloft. From the height of the inversion and the length of time the air mass had been over the cold sea surface, Taylor was able to estimate the magnitude of the exchange coefficient. His results are as follows:

Wind (Beaufort scale)	1	2	3
K, cm^2/sec	1,000	2,000	3,000

It may be noted that these values of the exchange coefficient are very much smaller than those computed from Dobson's data. The difference (a factor of 10) may be ascribed to stability in the surface inversion.

Further calculations using Eq. (10.34) have been made by Schwerdtfeger,[62] who used the equation to estimate the rate of heating of an air mass that is over a warm surface.

Another interesting application of the heat-diffusion equation is the problem of the diurnal temperature variation in the atmosphere. Suppose the air mass has a mean lapse rate γ and that the mean temperature distribution is therefore given by Eq. (10.30). Then, if the variation of the exchange coefficient in the vertical direction is neglected, the diurnal variation is determined by Eq. (10.33). The boundary conditions are as follows:

(a) At the ground ($z = 0$)

$$\Delta T = A \cos \nu t$$

(b) Very high ($z = \infty$)

$$\Delta T = 0$$

Note that the diurnal variation of the temperature at the surface has been approximated by a single cosine term of amplitude A and frequency $\nu = 2\pi$ (days)$^{-1}$. This could be considered as one term of a Fourier series; however, the first term by itself fits the surface conditions fairly well. Note that the time is zero at the maximum surface temperature condition, which generally occurs in the middle of the afternoon. Since the boundary condition (a) may be written in an exponential form and Eq. (10.33) is linear with constant coefficients, the solution must also be of an exponential form. It may easily be verified that a solution of Eq. (10.33) that fits these boundary conditions is

$$\Delta T = Ae^{-az} \cos (\nu t - az) \qquad (10.36)$$

where

$$a = \sqrt{\frac{\nu}{2K}} \qquad (10.37)$$

From this result, it appears that the amplitude of the diurnal variation of the temperature diminishes exponentially with elevation and that the time of the maximum temperature varies linearly with elevation. The exchange coefficient can be estimated from either the amplitude or the phase changes with elevation.

Observations of the diurnal temperature variation on the Eiffel Tower have been given by Schmidt.[33] His results are given in Table 6.

TABLE 6

Height, m	Amplitude, °C	Time of maximum (P.M.)
1.8	3.00	2:30
123.1	2.09	4:00
196.7	1.72	4:30
301.8	1.29	4:30

The exchange coefficient as computed from the first two amplitudes is 4.1×10^4 cm² per sec. The exchange coefficient computed from the last two is 4.9×10^4 cm² per sec. It may be observed that these exchange coefficients are of the same order of magnitude as those computed from Dobson's wind observations.

The assumption that $K\rho$ is independent of the elevation must be very poor in the layer closest to the surface in which the exchange coefficient must vary linearly. This indicates that Schmidt's simplified theory given above must underestimate the amplitude and phase change in the first layer and overestimate them aloft. This conclusion is supported by Schmidt's experimental data.

Haurwitz[63] has considered the diurnal variation of temperature in an atmosphere in which the exchange coefficient varies linearly with height. For this case, the solution can be expressed in terms of Bessel functions.

An interesting application of the theory of turbulent mixing for nonisotropic turbulence has been made by Grimminger.[64] Parr[65] had suggested that, although the gravitational stability must suppress the vertical mixing in the ocean, the mixing in the surfaces of constant density might be on a large scale. This conjecture is apparently verified. Rossby[66] suggested that similar large-scale lateral mixing may take place in the isentropic surfaces in the upper atmosphere. Grimminger applied this idea to the study of the lateral spread of the lines of constant specific humidity in these isentropic surfaces. He neglected vertical mixing, and from his experimental data he found the coefficient of lateral mixing to be between $K = 10^9$ cm² per sec and 10^{10} cm² per sec. Since this is of the order of 10^5 times as large as the vertical exchange coefficient and since the vertical-property gradients are about 10^2 times as large as the horizontal gradients, the horizontal mixing is apparently much more important than the vertical mixing. Following this result, Rossby[67] has discussed the possible effects of this lateral mixing on the general circulation of the atmosphere.

Another interesting and unusual application of the theory of turbulent diffusion has been made by Defant[68] and Lettau[34] and others. In this application, the migratory cyclones and anticyclones of the middle latitudes are treated as turbulent eddies in the general circulation of the atmosphere. For these eddies, the exchange coefficient may be computed directly by Eq. (10.4). For this case, K is found to be of the order of 5×10^{10} cm² per sec. This result was used to compute the meridional transport of heat in the earth's atmosphere.

For all the problems considered here, the turbulent transport was computed by making certain assumptions regarding the nature of the exchange coefficient. Many attempts have been made to develop a more complete theory of turbulent motion by statistical methods that would eliminate the necessity for these simplifying assumptions. The most important developments in these theories to date have been given by Taylor[69] and by von Kármán.[70] An excellent survey of the status of the statistical theories of turbulence has been given recently by Dryden.[71]

11. Energy Changes in Atmospheric Wind Systems. It has been shown that the viscous effects create surface stresses that tend to destroy the atmospheric motion, and the magnitude of this shear stress at the surface is about 1 dyne per cm^2. Since the mass of air in a unit column extending from the surface to the outer limit of the atmosphere is about 1 kg per cm^2, a stress of this magnitude must have rather large dissipative effects. These effects will now be studied in a little more detail.

Suppose the x-axis is taken parallel to the isobars, then the equation for equilibrium in the y-direction [see Eq. (9.1)] is

$$-\lambda \rho \bar{v} = \frac{\partial}{\partial z} (\tau_{zx}) \tag{11.1}$$

If this is integrated with respect to z, it is seen that

$$\int_0^z \rho \lambda \bar{v} \, dz = -\tau_{zx} \Big|_0^z \tag{11.2}$$

At high levels, the shear stress must vanish and at the ground $\tau_{zx} = \tau_0 \cos \alpha$. Thus

$$M = \int_0^\infty \rho \bar{v} \, dz = \frac{1}{\lambda} \tau_0 \cos \alpha \tag{11.3}$$

The integral is the total mass transport across the isobars. [It was assumed in Eq. (11.1) that the isobars were parallel at all levels in the friction layer.] This mass flows across the isobars into the low-pressure region. Since the pressure at any point is a measure of the mass of air above that point, the mass transport must tend to equalize the surface-pressure differential. For moderate winds, $\tau_0 \cos \alpha$ may be taken as about 1 dyne per cm^2 and for middle latitudes λ is about 10^{-4} sec^{-1}; therefore the mass transported across the isobar is about 10^4 grams per cm sec.

Suppose the isobar surrounds a circular low of radius R. Then the total transport inward is $2\pi R M$. Since the area is πR^2, the mean rate of pressure rise within this isobar is

$$\frac{\partial p}{\partial t} = g \frac{2\pi R M}{\pi R^2} = \frac{2gM}{R} \tag{11.4}$$

If $M = 10^4$ grams per cm sec and $R = 400$ km, then

$$\frac{\partial p}{\partial t} = \frac{1}{2} \text{ dyne/cm}^2 \text{ sec} = 1.8 \text{ mb/hr} \tag{11.5}$$

This rate of pressure increase in the low is high enough so that, if it were the only factor, almost any atmospheric low-pressure system would be destroyed in a period of about 1 day. Since the atmospheric cyclones last for much longer periods, there must be equally strong regenerative processes taking place in the levels above the friction layer.

Similar conclusions may be obtained from a consideration of the rate of dissipation of the atmospheric kinetic energy by the viscous stresses. Consider a fluid element, as shown in Fig. 49, which is of thickness dy in the y-direction.   Then the work done per unit time on the fluid below this element by the shearing stress on its bottom surface is $\bar{u}\tau_{zx}\,dx\,dy$.   On the other hand, the work done at the upper surface by the upper fluid is $\left[\bar{u}\tau_{zx} + \dfrac{\partial}{\partial z}(\bar{u}\tau_{zx})dz\right]dx\,dy$; therefore the net work done on the element per unit volume and time by the viscous stresses is $(\partial/\partial z)(\bar{u}\tau_{zx})$.   Now

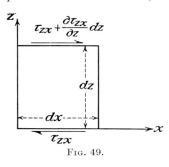

FIG. 49.

$$\frac{\partial}{\partial z}(\bar{u}\tau_{zx}) = \bar{u}\frac{\partial}{\partial z}\tau_{zx} + \tau_{zx}\frac{\partial\bar{u}}{\partial z} \qquad (11.6)$$

Since $(\partial/\partial z)\tau_{zx}$ is the net force acting on the element per unit of volume, the first term on the right-hand side represents the work done by the viscous stresses in accelerating the fluid element.   This is kinetic energy transferred from the rest of the fluid into the element.   Since the total work done on the element is done at the expense of the kinetic energy of the surrounding fluid, the difference $\tau_{zx}(\partial\bar{u}/\partial z)$ must represent the kinetic energy dissipated within the element and turned into heat.   Each of the other eight viscous stresses contributes a similar term to the dissipation function Φ, which gives the rate of dissipation of kinetic energy per unit volume.   Thus

$$\Phi = \tau_{xx}\frac{\partial\bar{u}}{\partial x} + \tau_{xy}\frac{\partial\bar{v}}{\partial x} + \tau_{xz}\frac{\partial\bar{w}}{\partial x} + \tau_{yx}\frac{\partial\bar{u}}{\partial y} + \tau_{yy}\frac{\partial\bar{v}}{\partial y} + \tau_{yz}\frac{\partial\bar{w}}{\partial y} + \tau_{zx}\frac{\partial\bar{u}}{\partial z} + \tau_{zy}\frac{\partial\bar{v}}{\partial z}$$
$$+ \tau_{zz}\frac{\partial\bar{w}}{\partial z} \qquad (11.7)$$

If the fluid motion is horizontal so that $\bar{w} = 0$ and if the horizontal variations of the velocity are neglected, the dissipation function is simply

$$\Phi = \tau_{zx}\frac{\partial\bar{u}}{\partial z} + \tau_{zy}\frac{\partial\bar{v}}{\partial z} \qquad (11.8)$$

The viscous dissipation in a unit column of the atmosphere is thus

$$D = \int_0^\infty \left(\tau_{zx}\frac{\partial\bar{u}}{\partial z} + \tau_{zy}\frac{\partial\bar{v}}{\partial z}\right)dz \qquad (11.9)$$

If this is integrated by parts, it is seen that

$$D = \left(\bar{u}\tau_{zx} + \bar{v}\tau_{zy}\right)_0^\infty - \int_0^\infty \left(\bar{u}\frac{\partial}{\partial z}\tau_{zx} + \bar{v}\frac{\partial}{\partial z}\tau_{zy}\right)dz \qquad (11.10)$$

The integrated part vanishes since the wind velocity vanishes at the surface and the shear stresses vanish aloft.   If the x-axis is taken parallel to the isobars, the equations of motion for the friction layer [see Eq. (9.1)] may be written

$$\frac{\partial}{\partial z}\tau_{zx} = -\lambda\rho\bar{v}$$

$$\frac{\partial}{\partial z}\tau_{zy} = \lambda\rho(\bar{u} - u_g) \qquad (11.11)$$

where $u_g = -(1/\lambda\rho)(\partial p/\partial y)$ is the geostrophic wind above the friction layer.   From this

$$\bar{u}\frac{\partial}{\partial z}\tau_{zx} + \bar{v}\frac{\partial}{\partial z}\tau_{zy} = -\lambda\rho u_g\bar{v} \qquad (11.12)$$

The viscous dissipation is thus

$$D = \lambda u_g \int_0^\infty \rho \bar{v} \, dz \tag{11.13}$$

By Eq. (11.3), this may finally be written

$$D = u_g \tau_0 \cos \alpha \tag{11.14}$$

A surface shear stress of 1 dyne per cm² corresponds to a geostrophic wind of about 10^3 cm per sec. The dissipation for this case is about 10^3 ergs per cm² sec. Since the mass in a unit column of air is about 10^3 grams, the total kinetic energy for this case is about $\frac{1}{2} \times 10^3 (10^3)^2 = 5 \times 10^8$ ergs per cm². In an hour the total dissipation would be 3.6×10^6 ergs per cm², or about 1 per cent of the total kinetic energy. From this, it appears that the kinetic energy of the winds must be completely replaced every 100 hr.

It is of interest to note that about half this viscous dissipation occurs in the lowest 20 or 30 m of the atmosphere. The dissipation in such a column of height z is

$$D_1 = \int_0^z \left(\tau_{zx} \frac{\partial \bar{u}}{\partial z} + \tau_{zy} \frac{\partial \bar{v}}{\partial z} \right) dz \tag{11.15}$$

If the x-axis is taken parallel to the wind close to the ground, then $\tau_{zx} = \tau_0$ and is constant for this shallow layer. Furthermore $\tau_{zy} = 0$; therefore

$$D_1 = \tau_0 \bar{u} \tag{11.16}$$

where $\bar{u}$ is the wind velocity at the height z. Since the wind velocity reaches half the geostrophic velocity at an elevation of from 20 to 30 m, half the total dissipation of kinetic energy occurs below this level.

In order to discuss the manner in which the energy of the winds is replenished, it is necessary to discuss the distribution of energy in the atmosphere. This energy exists as gravitational potential energy, internal energy, and kinetic energy. A very simple relationship exists between the internal and gravitational potential energy in the atmosphere. In a unit column, the potential energy is

$$P = \int_0^\infty z \rho g \, dz \tag{11.17}$$

By the hydrostatic equation, this may be written as

$$P = - \int_0^\infty z \frac{\partial p}{\partial z} \, dz \tag{11.18}$$

If this is integrated by parts, it is seen that

$$P = - pz \Big|_0^\infty + \int_0^\infty p \, dz \tag{11.19}$$

The integrated part vanishes at both limits; and, if the effects of moisture are neglected, $p = \rho R T$; therefore

$$P = R \int_0^\infty \rho T \, dz \tag{11.20}$$

If the effects of moisture are neglected, the internal energy* in this column is

$$E = c_v \int_0^\infty \rho T \, dz \tag{11.21}$$

* It is assumed that the internal energy and the specific heat coefficients are expressed in mechanical energy units.

Since $c_p = c_v + R$, the sum of the internal and potential energies is

$$P + E = c_p \int_0^\infty \rho T \, dz \qquad (11.22)$$

or

$$P + E = kE \qquad (11.23)$$

where k is the ratio of the specific heats c_p/c_v. For air, $k = 1.40$. In the atmosphere, particularly in the tropical regions, a considerable portion of the internal energy is contained as the latent heat of vaporization of water. This of course complicates the discussion and slightly changes the result. The importance of the water vapor in this problem has been discussed by Normand.[72] If the modification due to the water vapor is neglected, this result shows that, of any seven units of energy added to the internal and potential energy of the atmosphere, five units go into the internal energy and two units go into the potential energy.

Consider a fluid mass within a closed system, *i.e.*, a system with fixed insulating walls. For this case, there can be no energy transfer, either heat or mechanical, through the walls. The principle of the conservation of energy thus requires that

$$\frac{\partial}{\partial t} (K + P + E) = 0 \qquad (11.24)$$

where K is the total kinetic energy, P is the total gravitational potential energy, and E is the total internal energy. If this is integrated, it is seen that

$$\Delta K = -\Delta(P + E) \qquad (11.25)$$

If this closed system extends for the whole depth of the atmosphere, this can be written

$$\Delta K = -k \, \Delta E \qquad (11.26)$$

This shows that, in such a closed system, any increase of kinetic energy can be produced only by decreasing the internal and potential energies; furthermore, the kinetic energy dissipated by viscous stresses at once reappears as increases in the internal and potential energy.

Margules[73] first applied these formulas to the problem of estimating the amount of kinetic energy produced by any given change in a hydrodynamic system. Before considering such calculations in detail, the expression for the internal energy of an adiabatic layer will first be obtained. For such a layer extending between the levels z_1 and z_2

$$E = c_v \int_{z_1}^{z_2} \rho T \, dz \qquad (11.27)$$

By the hydrostatic equation, this is

$$E = -\frac{c_v}{g} \int_{p_1}^{p_2} T \, dp \qquad (11.28)$$

For an adiabatic layer, $Tp^{\frac{(1-k)}{k}}$ is constant; therefore

$$T \, dp = \frac{k}{2k - 1} d(Tp) \qquad (11.29)$$

Thus

$$E = \frac{c_p}{g(2k - 1)} (p_1 T_1 - p_2 T_2) \qquad (11.30)$$

Consider a unit column in the atmosphere with two adiabatic layers next to the surface. Then suppose the lower two layers overturn without mixing, as indicated in Fig. 50. Each layer may be assumed to retain its initial potential temperature, θ_1 for the original lower layer and θ_2 for the original upper layer. As the pressure at any point is the weight of the air above that level, the ground pressure p_g is unchanged as is the pressure at the top of the two layers p_t. Similarly if p_m is the pressure at the middle surface initially, $p_t + p_g - p_m$ is the pressure at the middle surface in the final state. The temperature at any point is given by the potential temperature for

$$T = \theta \left(\frac{p}{p_0}\right)^{\frac{k-1}{k}} \qquad (11.31)$$

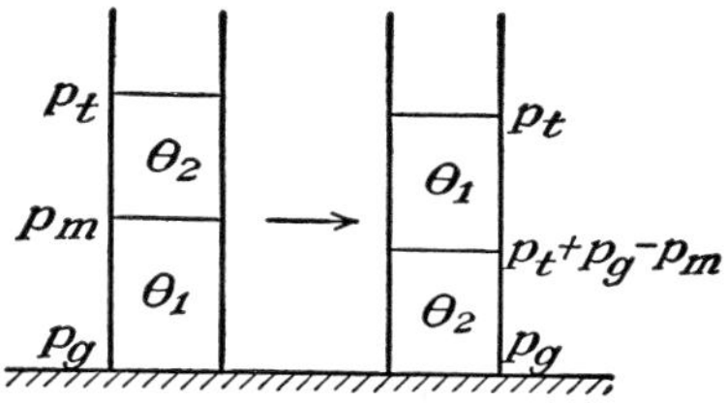

FIG. 50.

where p_0 is the standard pressure of 1,000 mb. By Eq. (11.30), the initial internal energy of the two layers is

$$E_i = \frac{c_p}{g(2k-1)} \left[\theta_1 p_g \left(\frac{p_g}{p_0}\right)^{\frac{k-1}{k}} - (\theta_1 - \theta_2) p_m \left(\frac{p_m}{p_0}\right)^{\frac{k-1}{k}} \right.$$
$$\left. - \theta_2 p_t \left(\frac{p_t}{p_0}\right)^{\frac{k-1}{k}} \right] \qquad (11.32)$$

Similarly, the internal energy of the two layers in the final state is

$$E_f = \frac{c_p}{g(2k-1)} \left[\theta_2 p_g \left(\frac{p_g}{p_0}\right)^{\frac{k-1}{k}} + (\theta_1 - \theta_2)(p_g + p_t - p_m) \left(\frac{p_g + p_t - p_m}{p_0}\right)^{\frac{k-1}{k}} \right.$$
$$\left. - \theta_1 p_t \left(\frac{p_t}{p_0}\right)^{\frac{k-1}{k}} \right] \qquad (11.33)$$

From this the change in internal energy is

$$\Delta E = E_f - E_i = -\frac{c_p(\theta_1 - \theta_2)}{g(2k-1)} \left[p_g \left(\frac{p_g}{p_0}\right)^{\frac{k-1}{k}} + p_t \left(\frac{p_t}{p_0}\right)^{\frac{k-1}{k}} - p_m \left(\frac{p_m}{p_0}\right)^{\frac{k-1}{k}} \right.$$
$$\left. - (p_g + p_t - p_m) \left(\frac{p_g + p_t - p_m}{p_0}\right)^{\frac{k-1}{k}} \right] \qquad (11.34)$$

The bracketed quantity is the difference of nearly equal quantities and is rather difficult to evaluate. As long as the pressures p_g and p_t are not very much different from p_0, a very simple approximate form can be used, for

$$\left(\frac{p}{p_0}\right)^{\frac{k-1}{k}} = \left(1 + \frac{p - p_0}{p_0}\right)^{\frac{k-1}{k}} = 1 + \frac{k-1}{k} \left(\frac{p - p_0}{p_0}\right) + \cdots \qquad (11.35)$$

If the two adiabatic layers are rather shallow, the higher order terms may be omitted, and Eq. (11.34) becomes

$$\Delta E = -\frac{2c_p(k-1)}{gk(2k-1)} (\theta_1 - \theta_2) \frac{(p_g - p_m)(p_m - p_t)}{p_0} \qquad (11.36)$$

From this, it is seen that the internal energy is decreased if the initial condition was unstable so that $\theta_2 < \theta_1$. By Eq. (11.26), the kinetic energy that could be released by this overturning is

$$\Delta K = \frac{2c_p(k-1)}{g(2k-1)} (\theta_1 - \theta_2) \frac{(p_g - p_m)(p_m - p_t)}{p_0} \qquad (11.37)$$

It should be noted that there is no change in the internal energy of the air above the p_t level.   The mass of air in the two adiabatic layers is

$$M = \frac{p_g - p_t}{g} \tag{11.38}$$

If the available kinetic energy were to be evenly distributed through this mass, it would correspond to a uniform velocity c where $\Delta K = \frac{1}{2} M c^2$ or

$$c^2 = 4 c_p \frac{k-1}{2k-1} (\theta_1 - \theta_2) \frac{(p_g - p_m)(p_m - p_t)}{p_0 (p_g - p_t)} \tag{11.39}$$

If the adiabatic layers are both 100 mb thick and the potential temperature difference is 10°C, the speed c is 21.1 mps.   For 200-mb layers with 10°C potential temperature difference, the speed c is 29.8 mps.

Margules carried out similar calculations using the exact formula, Eq. (11.34), for two adiabatic layers, each 2,000 m thick, with a 3°C potential temperature difference.   For this case, $c = 15$ mps.   Margules also considered the case of two adiabatic air masses that initially lie side by side and in the final state lie with the lighter ones above the heavier.   The horizontal extent of the two air masses was the same. For 3,000 m thick layers with a 5°C potential temperature difference, the speed c is 12.2 mps.   If the potential temperature difference is increased to 10°C, $c = 17.3$ mps.

These calculations show that the vertical motions that are observed during the development and occlusion process for a normal cyclone are a large enough energy source to explain the observed winds that accompany these storms.   These calculations, of course, show only that the energy is available; they do not show why the energy is not immediately dissipated through small-scale turbulent motions instead of being used to produce the large-scale motions of the winds.

Similar calculations giving the energy available in a single layer of air having an unstable lapse rate have been given by Littwin.[74]   The effects of moisture in a single layer of air with a dry-adiabatic lapse rate has also been considered by Littwin.[75]

12. Perturbation Theory.   It has been shown that dissimilar air masses could be in equilibrium with a sloping discontinuity surface separating them.   It was also seen that this horizontal juxtaposition of dissimilar air masses represents a certain potential energy source that might be transformed into kinetic energy of motion. It thus seems necessary to discuss the stability of these discontinuity surfaces in order to determine whether or in what manner this energy of mass distribution can be realized.   For this purpose, the linearized theory of small disturbances (or, briefly, the perturbation theory) is the appropriate mathematical tool.   Complete discussions of this problem have been made by V. Bjerknes and his collaborators.[11]   In this investigation, the effects of viscosity and of compressibility will be neglected.

The equations of motion are then

$$\frac{\partial \mathbf{v}}{\partial t} + \mathbf{v} \cdot \nabla \mathbf{v} + 2\boldsymbol{\omega} \times \mathbf{v} = -\frac{1}{\rho} \nabla p + \mathbf{F} \tag{12.1}$$

and the equation of continuity is

$$\nabla \cdot \mathbf{v} = 0 \tag{12.2}$$

Suppose these equations, together with suitable boundary conditions, are satisfied by a mean velocity field $\mathbf{V}$ and a mean pressure field P.   Suppose further that a small disturbance is superimposed on this mean field so that

$$\begin{aligned} \mathbf{v} &= \mathbf{V} + \mathbf{v}' \\ p &= P + p' \end{aligned} \tag{12.3}$$

where $\mathbf{v}'$ and p' are the velocity and pressure disturbances. The equations of motion become

$$\frac{\partial \mathbf{V}}{\partial t} + \mathbf{V} \cdot \nabla \mathbf{V} + 2\boldsymbol{\omega} \times \mathbf{V} + \frac{\partial \mathbf{v}'}{\partial t} + \mathbf{v}' \cdot \nabla \mathbf{v}' + 2\boldsymbol{\omega} \times \mathbf{v}' + \mathbf{V} \cdot \nabla \mathbf{v}' + \mathbf{v}' \cdot \nabla \mathbf{V}$$

$$= -\frac{1}{\rho} \nabla P - \frac{1}{\rho} \nabla p' + \mathbf{F} \quad (12.4)$$

Similarly, the equation of continuity is

$$\nabla \cdot \mathbf{V} + \nabla \cdot \mathbf{v}' = 0 \qquad (12.5)$$

Now the mean flow satisfies the equations of motion and continuity, and therefore the terms involving only the mean terms drop out. The equations for the disturbances are thus

$$\frac{\partial \mathbf{v}'}{\partial t} + \mathbf{v}' \cdot \nabla \mathbf{v}' + \mathbf{V} \cdot \nabla \mathbf{v}' + \mathbf{v}' \cdot \nabla \mathbf{V} + 2\boldsymbol{\omega} \times \mathbf{v}' = -\frac{1}{\rho} \nabla p' \qquad (12.6)$$

and

$$\nabla \cdot \mathbf{v}' = 0 \qquad (12.7)$$

From this, it is seen that the continuity equation has the same form for the disturbance terms as for the mean flow. If the disturbance velocity is assumed to be small when compared with the mean velocity, every term of Eq. (12.6) is a first-order term except for $\mathbf{v}' \cdot \nabla \mathbf{v}'$, which is of the second order and may be neglected for sufficiently small disturbances. If this term is omitted, then the equations of motion become

$$\frac{\partial \mathbf{v}'}{\partial t} + \mathbf{V} \cdot \nabla \mathbf{v}' + \mathbf{v}' \cdot \nabla \mathbf{V} + 2\boldsymbol{\omega} \times \mathbf{v}' = -\frac{1}{\rho} \nabla p' \qquad (12.8)$$

These linearized equations of motion, which are valid for sufficiently small disturbances, are called the *perturbation equations*. For the special case to be considered here, in which the disturbances in a uniform rectilinear flow in the x-direction so that $\mathbf{V} = U\mathbf{i}$ are considered, Eq. (12.8) takes the simpler form

$$\frac{\partial \mathbf{v}'}{\partial t} + U \frac{\partial \mathbf{v}'}{\partial x} + 2\boldsymbol{\omega} \times \mathbf{v}' = -\frac{1}{\rho} \nabla p' \qquad (12.9)$$

Let us consider, as an example of the application of the perturbation equations the two-dimensional problem of gravitational wave motion in a single layer of fluid. It will be assumed that the fluid is initially stationary so that $\mathbf{V} = 0$ and that the motion is on a small enough scale so that the Coriolis terms may be neglected. Suppose the mean depth of the layer is h and the wave amplitude is η, as shown in Fig. 51. The origin of the coordinate system is taken at the bottom of the layer. If the perturbation velocity components in the x- and z-directions are u and w, respectively, the perturbation equations for this case are

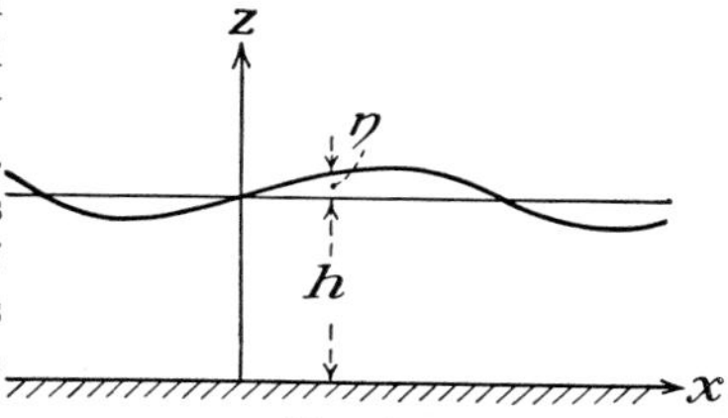

FIG. 51.

$$\frac{\partial u}{\partial t} + \frac{1}{\rho} \frac{\partial p'}{\partial x} = 0$$

$$\frac{\partial w}{\partial t} + \frac{1}{\rho} \frac{\partial p'}{\partial z} = 0 \qquad (12.10)$$

$$\frac{\partial u}{\partial x} + \frac{\partial w}{\partial z} = 0$$

Since these equations are linear with constant coefficients, the solution can be of an exponential form. An exponential solution that corresponds to a wave progressing in the positive x-direction with a velocity c ($c \neq 0$) is

$$u = A e^{\lambda z + ik(x-ct)}$$
$$w = B e^{\lambda z + ik(x-ct)}$$
$$p' = \rho C e^{\lambda z + ik(x-ct)}$$

(12.11)

where A, B, and C are complex constants and only the real part of these expressions are to be used. These will be a solution of the perturbation equations if

$$-ikcA + ikC = 0$$
$$-ikcB + \lambda C = 0$$
$$ikA + \lambda B = 0$$

(12.12)

From these, it is seen that

$$A = \frac{1}{c} C$$
$$B = \frac{\lambda}{ikc} C$$
$$\lambda = \pm k$$

(12.13)

Two solutions are thus found, one for $\lambda = k$ and the other for $\lambda = -k$.

Suppose A_1, B_1, and C_1 are the constants corresponding to the first solution $\lambda = k$, and A_2, B_2, and C_2 are the constants corresponding to the second solution. Since the equations are linear, the sum of these two solutions is a more general solution. For this case

$$w = (B_1 e^{kz} + B_2 e^{-kz}) e^{ik(x-ct)}$$

(12.14)

At the bottom of the layer, the vertical velocity must be zero, and this can be satisfied if $B_1 = -B_2$. From Eq. (12.13), it is seen that $A_1 = A_2$ and $C_1 = C_2$. In addition to this condition, the motion must also satisfy the boundary condition that the pressure is constant on the top surface where $z = h + \eta$. For the equilibrium condition

$$P = -g\rho(z - h) + p_0$$

(12.15)

if the pressure is taken as p_0 at the free surface. The total pressure is thus

$$p = -g\rho(z - h) + \rho C_1(e^{kz} + e^{-kz}) e^{ik(x-ct)} + p_0$$

(12.16)

This pressure will be p_0 at the free surface if

$$0 = -g\rho\eta + C_1\rho[e^{k(h+\eta)} + e^{-k(h+\eta)}] e^{ik(x-ct)}$$

(12.17)

Since the perturbation must be small, A, B, and C and η are all small quantities. If only first-order terms are retained, Eq. (12.17) becomes

$$\eta = \frac{2}{g} C_1 \cosh (kh) e^{ik(x-ct)}$$

(12.18)

The wave amplitude must also satisfy the purely kinematic condition at the wave surface that

$$\frac{\partial \eta}{\partial t} = w$$

(12.19)

If the second-order terms are again omitted, this may be written as

$$-2\frac{ikc}{g} C_1 \cosh (kh)e^{ik(x-ct)} = \frac{2C_1}{ic} \sinh (kh)e^{ik(x-ct)} \tag{12.20}$$

or

$$c^2 = \frac{g}{k} \tanh (kh) \tag{12.21}$$

This formula determines the speed of propagation of the wave c in terms of g, h, and the wave length $L = 2\pi/k$. The wave velocity has two possible values, one positive and one negative. This indicates that a wave disturbance of a given wave length will travel in either the positive or negative x-directions with the same speed.

These results may be summarized in a somewhat more simple manner by writing

$$\eta = ae^{ik(x-ct)} \tag{12.22}$$

so that the maximum wave height is given by the magnitude of a. From Eq. (12.18)

$$a = \frac{2}{g} C_1 \cosh (kh) \tag{12.23}$$

The final solution for the perturbations is as follows:

$$
\begin{aligned}
u &= \frac{ag \cosh (kz)}{c \cosh (kh)} e^{ik(x-ct)} \\
w &= -i\frac{ag \sinh (kz)}{c \cosh (kh)} e^{ik(x-ct)} \\
\frac{p'}{\rho} &= ag \frac{\cosh (kz)}{\cosh (kh)} e^{ik(x-ct)}
\end{aligned}
\tag{12.24}
$$

It is of interest to consider the special case for which the wave length is long compared to the depth. For this case, $kh = 2\pi h/L$ is a small quantity, and $\tanh (kh)$ is approximately kh. This is within 1 per cent of accurate if $L/h > 40$. For this case, the wave velocity equation is simply

$$c^2 = gh \tag{12.25}$$

so that disturbances of all wave lengths travel at the same speed. For this approximation, Eq. (12.24) shows that the velocity u is constant in any given vertical column and the vertical velocity w is a linear function of the distance from the bottom.

A second special case is that of wave motion in very deep water, where the wave length is small compared with the depth. For this case, $\tanh (kh) = 1$. This is within 1 per cent of accurate as long as $L/h < 0.4$. The waves in the open ocean thus come in this class. For this case

$$c^2 = \frac{g}{k} \tag{12.26}$$

and it is seen that long waves travel faster than short waves.

These results were computed on the assumption that the waves were of the progressive type and were moving with a wave velocity c; however, the results may be combined so as to obtain the theory for standing waves. Since the equations are linear, solutions may be superimposed. Consider one wave train for which the wave surface is given by Eq. (12.22), and superimpose upon this a second wave train having the same shape at $t = 0$ but traveling in the opposite direction. The combined wave surface is given by

$$\eta = ae^{ik(x-ct)} + ae^{ik(x+ct)} \tag{12.27}$$

or

$$\eta = 2a \cos (kct)e^{ikx} \tag{12.28}$$

This is a standing wave of twice the amplitude, the same wave length, and of a frequency kc. The velocity and pressure perturbations for the standing wave may be obtained by superposition from Eq. (12.24).

If one considers the wave motion in a fluid having two layers of different densities and a free surface at the top, the problem is complicated by the fact that there may be wave disturbances in the inner surface separating the two fluids. This problem has been considered by Ekman,[76] who used it to explain the phenomenon of "dead water" encountered in the Norwegian fiords, where there is a layer of fresh water over salt water.

After this preliminary discussion of gravity waves, we shall now consider the problem of wave disturbances in a frontal surface. The polar front will be approximated by a plane frontal surface with an east-west intersection with any horizontal plane and with a slope upward toward the north at an angle α. Take the x-axis in the frontal surface toward the east, the y-axis in the frontal surface toward the north, and the z-axis normal to the surface, as shown in Fig. 52. In this coordinate system

$$\boldsymbol{\omega} = \omega[\cos(\varphi - \alpha)\,\mathbf{j} + \sin(\varphi - \alpha)\,\mathbf{k}] \tag{12.29}$$

and
$$\mathbf{F} = -g(\sin\alpha\,\mathbf{j} + \cos\alpha\,\mathbf{k}) \tag{12.30}$$

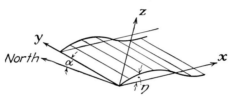

Fig. 52.

Suppose that in each air mass the velocity is horizontal, constant and parallel to the frontal surface so that $\mathbf{V} = U\mathbf{i}$. The flow pattern is then the same as that discussed on pages 441–443.

For the undisturbed state, the equations of motion give

$$\frac{\partial p}{\partial x} = 0$$

$$\frac{\partial p}{\partial y} = -\rho(2U\omega_z - F_y) \tag{12.31}$$

$$\frac{\partial p}{\partial z} = \rho(2U\omega_y + F_z)$$

The pressure in the undisturbed state can thus be written as

$$P = p_0 - \rho(2U\omega_z - F_y)y + \rho(2U\omega_y + F_z)z \tag{12.32}$$

If unprimed quantities are used to refer to the lower fluid and primed ones for the upper, the condition that the pressure must be continuous along the frontal surface $z = 0$ is that

$$\rho(2U\omega_z - F_y) = \rho'(2U'\omega_z - F_y) \tag{12.33}$$

This equation determines the slope of the frontal surface, for it may be written as [see Eqs. (12.29) and (12.30)]

$$\rho[2U\omega\sin(\varphi - \alpha) + g\sin\alpha] = \rho'[2U'\omega\sin(\varphi - \alpha) + g\sin\alpha] \tag{12.34}$$

From this
$$\tan\alpha = -\frac{2\omega\sin\varphi(\rho U - \rho' U')}{g(\rho - \rho') - 2\omega\cos\varphi(\rho U - \rho' U')} \tag{12.35}$$

This is the same formula for the slope of the frontal surface as that obtained from Eq. (7.6) with $\beta = 0$.

If the perturbation velocity components are u, v, and w, the perturbation equations for this case are as follows:

$$\frac{\partial u}{\partial t} + U \frac{\partial u}{\partial x} + 2w\omega_y - 2v\omega_z + \frac{1}{\rho}\frac{\partial p'}{\partial x} = 0$$

$$\frac{\partial v}{\partial t} + U \frac{\partial v}{\partial x} + 2u\omega_z + \frac{1}{\rho}\frac{\partial p'}{\partial y} = 0$$

$$\frac{\partial w}{\partial t} + U \frac{\partial w}{\partial x} - 2u\omega_y + \frac{1}{\rho}\frac{\partial p'}{\partial z} = 0 \qquad (12.36)$$

$$\frac{\partial u}{\partial x} + \frac{\partial v}{\partial y} + \frac{\partial w}{\partial z} = 0$$

An exponential solution of these equations which corresponds to a wave traveling in the x-direction with a velocity c is given by

$$u = A e^{\lambda z + ik(x-ct)}$$

$$v = B e^{\lambda z + ik(x-ct)}$$

$$w = C e^{\lambda z + ik(x-ct)} \qquad (12.37)$$

$$\frac{p'}{\rho} = D e^{\lambda z + ik(x-ct)}$$

It may be noted that the perturbation is assumed to be independent of the y-coordinate. Now Eq. (12.37) will be a solution of the perturbation equations, provided that

$$-ik(c - U)A - 2\omega_z B + 2\omega_y C + ikD = 0$$

$$2\omega_z A - ik(c - U)B = 0$$

$$-2\omega_y A - ik(c - U)C + \lambda D = 0 \qquad (12.38)$$

$$ikA + \lambda C = 0$$

From the last three of these equations, it is seen that

$$B = -\frac{i2\omega_z}{k(c - U)} A$$

$$C = -\frac{ik}{\lambda} A \qquad (12.39)$$

$$D = \left[\frac{2\omega_y}{\lambda} + \frac{k^2}{\lambda^2}(c - U)^2\right] A$$

The first equation of (12.38) will then be satisfied if

$$\frac{k^2}{\lambda^2} = 1 - \frac{4\omega_z^2}{k^2(c - U)^2} \qquad (12.40)$$

For the present, let us consider c as being real. Then, if $c - U$ is small enough, λ must be pure imaginary. This corresponds to perturbations that are periodic in the z-direction. If the fluid is of infinite extent on either side of the frontal surface, such periodic perturbations would require infinite kinetic energy and may be eliminated from consideration. This means that the frequency of the wave motion as seen by an observer moving with the air mass must be greater than $2\omega_z$ or

$$|k(c - U)| > 2\omega_z \qquad (12.41)$$

With this restriction, it is seen that

$$\lambda = \frac{\pm k}{\{1 - [4\omega_z^2/k^2(c - U)^2]\}^{1/2}} \tag{12.42}$$

If the motion is such that the Coriolis terms may be neglected, this reduces to $\lambda = \pm k$, as in Eq. (12.13). Since the amplitude of the frontal disturbance must vanish at large distances from the front, the λ for the lower fluid is

$$\lambda = k \left[1 - \frac{4\omega_z^2}{k^2(c - U)^2} \right]^{-1/2} \tag{12.43}$$

where k is considered as positive. Similarly

$$\lambda' = -k \left[1 - \frac{4\omega_z^2}{k^2(c - U')^2} \right]^{-1/2} \tag{12.44}$$

where primed quantities refer to the upper fluid and the principal value of the square root is used.

If the wave surface is given by

$$\eta = ae^{ik(x - ct)} \tag{12.45}$$

then the kinematic relations relating this amplitude with the velocity fields are that

$$\frac{\partial \eta}{\partial t} + U \frac{\partial \eta}{\partial x} = w|_{z=\eta}$$
$$\frac{\partial \eta}{\partial t} + U' \frac{\partial \eta}{\partial x} = w'|_{z=\eta} \tag{12.46}$$

Thus, if second-order quantities are neglected

$$-ik(c - U)a = C$$
$$-ik(c - U')a = C' \tag{12.47}$$

The other boundary condition that the fluid motion must satisfy is that the pressure must be continuous across the front, or

$$(P + p')_{z=\eta} = (P + p')'_{z=\eta} \tag{12.48}$$

By means of Eqs. (12.32), (12.33), (12.37), and (12.45), this can be written

$$\rho[(2U\omega_y + F_z)a + D] = \rho'[(2U'\omega_y + F_z)a + D'] \tag{12.49}$$

From Eqs. (12.39) and (12.47)

$$D = (c - U) \left[2\omega_y + \frac{k^2}{\lambda} (c - U) \right] a$$
$$D' = (c - U') \left[2\omega_y + \frac{k^2}{\lambda'} (c - U') \right] a \tag{12.50}$$

By means of these relations, the pressure equation that determines the wave velocity can be written as

$$\rho \left[(F_z + 2c\omega_y) + \frac{k^2}{\lambda} (c - U)^2 \right] = \rho' \left[(F_z + 2c\omega_y) + \frac{k^2}{\lambda'} (c - U')^2 \right] \tag{12.51}$$

Because the solutions of this equation are very complex, it is advisable to consider first some special cases.

Suppose the motion is such that the Coriolis accelerations are negligible. This is the case for disturbances of a fairly short wave length. For this case, $\alpha = 0$, $F_z = -g$,

$\lambda = k$, and $\lambda' = -k$.   The only restriction on c is that it cannot equal U or U'.   The wave-velocity equation is thus

$$\rho[-g + k(c - U)^2] = \rho'[-g - k(c - U')^2] \tag{12.52}$$

This is a quadratic equation for the wave velocity, and its solution is

$$c = \frac{\rho U + \rho' U'}{\rho + \rho'} \pm \left[\frac{g(\rho - \rho')}{k(\rho + \rho')} - \frac{\rho \rho'}{(\rho + \rho')^2}(U - U')^2 \right]^{\frac12} \tag{12.53}$$

If the density of the upper fluid ρ' is taken as zero, this shows that the wave velocity is

$$c = U \pm \sqrt{\frac{g}{k}} \tag{12.54}$$

so that the wave velocity relative to the moving fluid is given by the same expression as Eq. (12.26).   For the wave velocity in the frontal surface, the first term on the right-hand side of Eq. (12.53) is a mean velocity of the two fluid masses; the wave motion therefore has two equal velocities, one to the right and one to the left, relative to this mean velocity of the fluid masses.   This first term is called the *convective component* of the wave velocity, and the second is called the *dynamic component*. This solution apparently gives the wave velocity as a continuous function of the wave length $L = 2\pi/k$; however, it should be noted from Eq. (12.38) or (12.39) that the solution breaks down for $c = U$ or U'.

Consider the case for which both fluids are stationary and $U = U' = 0$, then Eq. (12.53) becomes

$$c = \pm \sqrt{\frac{g(\rho - \rho')}{k(\rho + \rho')}} \tag{12.55}$$

By comparison with Eq. (12.26), it is seen that, as the density of the upper fluid increases from zero, the wave velocity continuously decreases and becomes zero for $\rho = \rho'$.   For $\rho < \rho'$, *i.e.*, for the case with the heavier fluid on top, the wave velocity becomes an imaginary quantity.   From Eq. (12.45), it is seen that, if the imaginary part of c is positive, the wave amplitude increases with time.   The two imaginary solutions thus correspond to one wave that increases in amplitude and one wave that damps out.   For this case, the heavier fluid is on top, and the wave disturbance converts this potential energy of mass distribution into the wave motion, which increases in amplitude.   Such a wave, which increases in amplitude by converting other forms of energy into the kinetic energy of the wave motion, is called an *unstable wave*.   For this particular case, waves of any wave length are unstable.

From Eq. (12.53), it is seen that, if there is a velocity difference between the two fluids so that $U - U' \neq 0$, complex values of c and thus unstable waves can exist even if the lighter fluid is on top, provided that the wave length is short enough. The critical wave length that separates stable and unstable waves is

$$L_{\mathrm{crit}} = \frac{2\pi}{g} \frac{\rho \rho'}{(\rho^2 - \rho'^2)}(U - U')^2 \tag{12.56}$$

If $\rho'/\rho = 0.95$, which corresponds to an atmospheric temperature difference of 15°C if the upper air mass is at 300°K, and if $U - U' = 10$ mps, the critical wave length is 625 m.   For the unstable shorter waves, the wave energy is obtained from the kinetic energy of the motion of the fluid masses.   This instability of the short waves causes some mixing at any atmospheric frontal surface, and a narrow mixing zone between the two air masses is therefore created.

Stable waves of the type determined by Eq. (12.53) are frequently observed at the top or bottom of cloud strata.   The application of this formula to the study of such cloud waves has been made by Wegener.[77] A more complete theory taking into account the compressibility of the air has been applied to this problem by Haurwitz.[78]

If the wave length is very long, the Coriolis acceleration terms may not be neglected. If the Coriolis terms are included, the frontal surface is inclined: a certain potential energy of mass distribution therefore exists even though the lighter fluid is on top; for the heavier fluid may flow under and force the lighter fluid aloft. If the Coriolis terms are included, the wave velocity is given by Eq. (12.51).

Let us consider next the case for which the front is purely a velocity discontinuity so that $\rho = \rho'$. For this case, $\alpha = \varphi$, $\omega_z = 0$, $\omega_y = \omega$, $\lambda = k$, and $\lambda' = -k$. The wave-velocity equation is simply

$$(c - U)^2 = -(c - U')^2 \tag{12.57}$$

or
$$c = \frac{U + U'}{2} \pm i\,\frac{U - U'}{2} \tag{12.58}$$

It appears that for this case the Coriolis acceleration has no effect, and the instability due to the shearing motion of the air masses is observed. Since this instability exists for disturbances of all wave lengths, frontal systems of this type with the same density in both fluid masses must be expected to develop very wide mixing zones, owing to the unstable wave motion.

For this case, the Coriolis term involving ω_y was seen to have no effect. This is generally true; for this component occurs only in the term $F_z + 2c\omega_y$; and F_z is by far the largest term unless c is very large. However, if c were to be much larger than, say, 300 mps, the compressibility effects must also be considered. Thus, over the entire range for which Eq. (12.51) is applicable, ω_y may be neglected and the wave-velocity equation is

$$\rho\left[F_z + \frac{k^2}{\lambda}(c - U)^2\right] = \rho'\left[F_z + \frac{k^2}{\lambda'}(c - U')^2\right] \tag{12.59}$$

Let us consider next the special case for which the velocity is uniform across the front and $U = U'$. For this case, α is practically zero; therefore $F_z = -g$ and $\omega_z = \omega \sin \varphi$. By means of Eqs. (12.43) and (12.44), the wave-velocity equation [Eq. (12.59)] becomes

$$(c - U)^2\left[1 - \frac{4\omega_z^2}{k^2(c - U)^2}\right]^{\frac{1}{2}} = \frac{g(\rho - \rho')}{k(\rho + \rho')} \tag{12.60}$$

Since the square root is positive, it is seen that the wave is stable if $\rho > \rho'$ and unstable if $\rho < \rho'$. This is true for all wave lengths just as in the case for which the Coriolis terms were neglected.

The complete discussion of Eq. (12.59) for the case with both velocity and density discontinuities has been considered in some detail by Bjerknes et al.[79] Owing to the complexity of the problem, the calculations will not be repeated here. They find that for short wave lengths the wave velocity is given very closely by Eq. (12.53); however, certain additional wave velocities very close to the critical velocities $c = U$ or U' are found. For short wave lengths, these roots are not significant; however, for very long waves having a wave length of the order of 1,000 km, these additional wave velocities become complex. These very long wave disturbances are thus unstable if the Coriolis terms are included.

The existence of these unstable long waves is the basis of the wave theory of the origin of the polar-front cyclones that was proposed by V. Bjerknes and his collaborators. According to this theory, the unstable long-wave disturbances in the polar front cause the potential energy due to the sloping discontinuity surface to be converted into the kinetic energy of the wave motion. As this wave motion becomes more intense, it is observed on the synoptic scale as a young cyclone. Since the wave theory developed above assumed that the disturbance was small, it can be applied only to the initial stages of the development. The further development of the cyclone and the occlusion process cannot be discussed by this means. Some considerations of the development of finite waves that can be applied to the occlusion process have been made by Rosenhead.[80]

V. Bjerknes *et al.*[11] have also considered some of the effects of compressibility and of the restrictions due to the earth's surface on the polar-front waves.

13. Atmospheric Tidal Motions. After discussing the long cyclone waves in the polar front, it seems appropriate to consider next those even larger scale phenomena which involve motions of the entire atmosphere. Since half the total mass of the atmosphere occurs in the first 6 km from the surface, the atmosphere is an extremely shallow layer of fluid if it is viewed on a planetary scale. For any disturbance that involves a considerable portion of the atmosphere, the wave length must be very long in comparison with the depth of the fluid. In the theory of gravity waves in an incompressible fluid, it was seen that for such long waves the motion was primarily horizontal, with the ratio of the horizontal velocities to the vertical velocities being of the order of the wave length divided by the fluid depth. It will be seen that this same conclusion holds for the large-scale motions of the atmosphere where the effects of compressibility must be considered. It thus appears that the vertical accelerations may be neglected for these large-scale motions. Because the first class of fluid motions of this general nature to be investigated was the oceanic tides, these large-scale motions are generally called *tidal motions*.

Let us first consider the problem of small disturbances in a stationary atmosphere. Let z be the normal distance from the earth's surface, and let the density ρ_0 and the pressure p_0 in this stationary atmosphere be functions of z only. The pressure and density in the undisturbed condition are related by the hydrostatic equation

$$\frac{\partial p_0}{\partial z} = -g\rho_0 \tag{13.1}$$

In the disturbed condition, the velocity is $\mathbf{v}$ with the components (u,v,w); the density is $\rho = \rho_0 + \rho'$; and the pressure is $p = p_0 + p'$.

The equations of motion as linearized for small disturbances are

$$\frac{\partial \mathbf{v}}{\partial t} + 2\boldsymbol{\omega} \times \mathbf{v} = -\frac{1}{\rho_0}\nabla p' + \frac{\rho'}{\rho_0}\mathbf{F} \tag{13.2}$$

where $\mathbf{F}$ is the gravitational force vector or $\mathbf{F} = -g\mathbf{k}$ where $\mathbf{k}$ is a unit vector normal to the surface. Since ρ_0 is not a function of the time, this may also be written as

$$\frac{\partial^2 \mathbf{v}}{\partial t^2} + 2\boldsymbol{\omega} \times \frac{\partial \mathbf{v}}{\partial t} = -\frac{1}{\rho_0}\nabla \frac{\partial p'}{\partial t} + \frac{\mathbf{F}}{\rho_0}\frac{\partial \rho'}{\partial t} \tag{13.3}$$

If the equation of continuity is similarly linearized, it becomes

$$\frac{\partial \rho'}{\partial t} = -\rho_0\left(\nabla \cdot \mathbf{v} + \frac{w}{\rho_0}\frac{\partial \rho_0}{\partial z}\right) \tag{13.4}$$

We shall assume that for the disturbed motion the density is a function only of the pressure and that

$$\frac{Dp}{D\rho} = c^2 \tag{13.5}$$

so that c is the rate of propagation of a small pressure pulse. This may also be written

$$\frac{Dp}{Dt} = c^2 \frac{D\rho}{Dt} = -c^2 \rho \, \nabla \cdot \mathbf{v} \tag{13.6}$$

If this expression is also linearized, it is seen that

$$\frac{\partial p'}{\partial t} + w \frac{\partial p_0}{\partial z} = -c^2 \rho_0 \, \nabla \cdot \mathbf{v} \tag{13.7}$$

By means of Eq. (13.1), this may also be written as

$$\frac{\partial p'}{\partial t} = \rho_0 (gw - c^2 \, \nabla \cdot \mathbf{v}) \tag{13.8}$$

If the expressions for $\partial p'/\partial t$ and $\partial \rho'/\partial t$ are substituted in Eq. (13.3), it is seen that

$$\frac{\partial^2 \mathbf{v}}{\partial t^2} + 2\boldsymbol{\omega} \times \frac{\partial \mathbf{v}}{\partial t} = \nabla(c^2 \, \nabla \cdot \mathbf{v} - gw) + \mathbf{k} \, \nabla \cdot \mathbf{v} \left(g + \frac{c^2}{\rho_0} \frac{\partial \rho_0}{\partial z} \right) \tag{13.9}$$

These are the perturbation equations for small disturbances in a stationary atmosphere.

For certain cases, these equations may be simplified. The first case occurs if the atmosphere is in convective equilibrium, and the expansions due to the disturbances are also adiabatic. For this case

$$p = B\rho^\gamma \tag{13.10}$$

where B is a constant and $\gamma = c_p/c_v$. For this case

$$\frac{Dp}{D\rho} = \gamma B \rho^{\gamma-1} = \gamma \frac{p}{\rho} \tag{13.11}$$

Thus

$$\frac{Dp}{D\rho} = c^2 = \gamma RT \tag{13.12}$$

Since the density is a function only of the pressure

$$\frac{\partial p_0}{\partial z} = \frac{Dp}{D\rho} \frac{\partial \rho_0}{\partial z} \tag{13.13}$$

By Eqs. (13.1) and (13.12), this can be written

$$0 = \frac{c^2}{\rho_0} \frac{\partial \rho_0}{\partial z} + g \tag{13.14}$$

therefore the last term in Eq. (13.9) is identically zero.

This result is also obtained for an isothermal atmosphere that is in radiative equilibrium if the disturbances are slow enough so that they also follow an isothermal law. For this case

$$p = \rho RT \tag{13.15}$$

where T is constant; therefore

$$\frac{Dp}{D\rho} = c^2 = RT \tag{13.16}$$

Since the density is again a function of the pressure alone, Eqs. (13.13) and (13.14) apply for this case too. For either of these special cases, the perturbation equations are thus

$$\frac{\partial^2 \mathbf{v}}{\partial t^2} + 2\boldsymbol{\omega} \times \frac{\partial \mathbf{v}}{\partial t} = \nabla(c^2 \, \nabla \cdot \mathbf{v} - gw) \tag{13.17}$$

For the isothermal atmosphere, c^2 is constant; but, for the adiabatic atmosphere, c^2 varies with the altitude.

The perturbation equations neglecting the Coriolis terms have been discussed by Lamb,[15,81] and his treatment will be followed here. If the Coriolis terms are neglected, the perturbation equations for either the adiabatic or isothermal case are

$$\frac{\partial^2 \mathbf{v}}{\partial t^2} = \nabla(c^2 \nabla \cdot \mathbf{v} - gw) \tag{13.18}$$

This equation may be solved most readily in terms of the velocity potential function. The density is a function only of the pressure for both cases; therefore the Bjerknes circulation theorem, with the Coriolis term omitted, shows that the circulation around any arbitrary contour formed from a given set of fluid elements will remain constant. Since this circulation is zero for the undisturbed state, it must remain zero. Since the circulation around any contour is simply the flux of vorticity through the contour, the circulation around any arbitrary contour can be zero only if the vorticity is everywhere zero. The fluid motion for these perturbations is thus irrotational, or

$$\nabla \times \mathbf{v} = 0 \tag{13.19}$$

It is apparent that this condition satisfies Eq. (13.18). Since the curl of the velocity vector is zero, the velocity vector itself must be the gradient of some scalar function, or

$$\mathbf{v} = \nabla\varphi \tag{13.20}$$

The scalar function φ is called the *velocity potential*. In terms of the velocity potential, Eq. (13.18) becomes

$$\nabla \left(\frac{\partial^2 \varphi}{\partial t^2} - c^2 \nabla^2 \varphi + g \frac{\partial \varphi}{\partial z} \right) = 0 \tag{13.21}$$

The bracketed quantity must thus be independent of the spatial coordinates. Since an arbitrary function of time may be added to the velocity potential without changing the values of the velocity components, this may be integrated, and

$$\frac{\partial^2 \varphi}{\partial t^2} = c^2 \nabla^2 \varphi - g \frac{\partial \varphi}{\partial z} \tag{13.22}$$

In an adiabatic atmosphere, the temperature decreases upward at a constant rate If the subscript s denotes surface values, the conditions in the equilibrium state are given by

$$T_0 = T_s - \frac{g}{c_p} z \tag{13.23}$$

The atmosphere for this case has a definite outer limit at the point where T_0 vanishes. The depth of the atmosphere h is

$$h = \frac{c_p}{g} T_s \tag{13.24}$$

The temperature can thus be written as

$$T_0 = \frac{g}{c_p} (h - z) \tag{13.25}$$

By Eq. (13.12)

$$c^2 = \frac{Rg}{c_v} (h - z) \tag{13.26}$$

It is also convenient to introduce at this point the depth H of the equivalent homogeneous atmosphere where

$$H = \frac{p_s}{g\rho_s} = \frac{R}{g} T_s \qquad (13.27)$$

By Eq. (13.24), this may also be written

$$H = \frac{R}{c_p} h \qquad (13.28)$$

Since the adiabatic atmosphere has a definite upper limit, it has certain mathematical advantages over the isothermal atmosphere. Let us investigate the horizontal propagation of a wave in the adiabatic atmosphere. Suppose

$$\varphi = P(z)e^{i(kx-\sigma t)} \qquad (13.29)$$

This will satisfy Eq. (13.22) if

$$0 = c^2 \frac{d^2 P}{dz^2} - g \frac{dP}{dz} + (\sigma^2 - k^2 c^2)P \qquad (13.30)$$

In order to solve this equation, it is convenient to introduce $\xi = h - z$ and $m = c_v/R$ Then

$$0 = \xi \frac{d^2 P}{d\xi^2} + m \frac{dP}{d\xi} + \left(\frac{m\sigma^2}{gk} - k\xi\right) kP \qquad (13.31)$$

This may be further simplified by putting

$$P = e^{-k\xi}Q(\xi) \qquad (13.32)$$

Then
$$\xi \frac{d^2 Q}{d\xi^2} + (m - 2k\xi) \frac{dQ}{d\xi} - 2k\alpha Q = 0 \qquad (13.33)$$

where
$$2\alpha = m\left(1 - \frac{\sigma^2}{gk}\right) \qquad (13.34)$$

Equation (13.33) is the standard form for the confluent hypergeometric function[82]; therefore its solution is

$$Q = A_1 F_1(\alpha;m;2k\xi) + B(2k\xi)^{1-m} {}_1F_1(\alpha - m - 1; 2 - m; 2k\xi) \qquad (13.35)$$

where the confluent hypergeometric function is

$$_1F_1(\alpha;m;2k\xi) = 1 + \frac{\alpha}{1 \cdot m}(2k\xi) + \frac{\alpha(\alpha + 1)}{1 \cdot 2 \cdot m(m + 1)}(2k\xi)^2 + \cdots \qquad (13.36)$$

and A and B are arbitrary constants. If Eq. (13.6) is linearized, it is seen that

$$\frac{Dp}{Dt} = -c^2 \rho_0 \nabla^2 \varphi = -\rho_0 \left(\frac{\partial^2 \varphi}{\partial t^2} - g \frac{\partial \varphi}{\partial \xi}\right) \qquad (13.37)$$

At the outer limit of the atmosphere, this must vanish. Now ρ_0 varies as ξ^m, and near the outer limit, $\partial\varphi/\partial\xi$ varies as ξ^{-m} unless $B = 0$. Since Dp/Dt would thus approach a finite limit at the outer edge of the atmosphere, it follows that $B = 0$, and the velocity potential is

$$\varphi = Ac^{i(kx-\sigma t)-k\xi} {}_1F_1(\alpha;m;2k\xi) \qquad (13.38)$$

For any given wave length, the permissible values of α and thus of σ are determined from the other boundary condition that the vertical velocity must vanish at the surface, or $\partial \varphi / \partial \xi = 0$ for $\xi = h$.   Thus

$$\frac{2\alpha}{m}\left[1 + \frac{\alpha + 1}{1 \cdot (m+1)}\,(2kh) + \frac{(\alpha+1)(\alpha+2)}{1 \cdot 2(m+1)(m+2)}\,(2kh)^2 + \cdots\right]$$

$$= \left[1 + \frac{\alpha}{1 \cdot m}\,(2kh) + \frac{\alpha(\alpha+1)}{1 \cdot 2 \cdot m(m+1)}\,(2kh)^2 + \cdots\right] \quad (13.39)$$

This determines α as a function of kh.

Our chief interest is in waves that are long compared with the depth so that kh is small.   For very small values of kh, the first approximation to the solution of Eq. (13.39) is $2\alpha/m = 1$.   A second approximation is given by

$$\frac{2\alpha}{m} = \frac{1 + kh}{1 + (m + 2/m + 1)kh} = 1 - \frac{kh}{m+1} + 0(kh)^2 \quad (13.40)$$

From Eq. (13.34), it is seen that

$$\frac{\sigma}{k} = \sqrt{\frac{gh}{m+1}} \quad (13.41)$$

Since $m + 1 = c_p/R$, this can also be written as

$$\frac{\sigma}{k} = \sqrt{gH} \quad (13.42)$$

where H is the depth of the equivalent homogeneous atmosphere as given by Eq. (13.28).   For a surface temperature of 15°C, $H = 27{,}640$ ft and $\sigma/k = 943$ ft per sec. Since σ/k is the speed with which the wave moves, it is seen that the speed of propagation of the long waves in an adiabatic atmosphere is the same as the speed of propagation in the equivalent homogeneous atmosphere.

For these long waves, the velocity potential of Eq. (13.38) can be written as

$$\varphi = A e^{i(kx-\sigma t)}\left[1 + \frac{k^2}{m+1}\left(\frac{1}{2}\,\xi^2 - \xi h\right) + \cdots\right] \quad (13.43)$$

The remaining terms are of higher order in k and may be neglected.   The velocity components are thus

$$\begin{aligned} u &= ikA e^{i(kx-\sigma t)} \\ v &= 0 \\ w &= \frac{k^2 A z}{m+1}\,e^{i(kx-\sigma t)} \end{aligned} \quad (13.44)$$

Thus the horizontal velocity components are independent of the elevation, and the vertical velocity varies linearly with the elevation.   Furthermore $w/u = -ikz/(m+1)$, and this is of the same order as kh since $z \leqslant h$.   We see also from Eq. (13.22) that

$$c^2 \nabla \cdot \mathbf{v} - gw = \frac{\partial^2 \varphi}{\partial t^2} \quad (13.45)$$

so

$$c^2 \nabla \cdot \mathbf{v} - gw = - \frac{A k^2 h}{m+1}\,e^{i(kx-\sigma t)} \quad (13.46)$$

where the terms of order k^4 or higher have been neglected.   This quantity is also seen to be practically independent of the elevation for long waves.

In an isothermal atmosphere with isothermal expansions, Eq. (13.22) may also be applied.   Since c^2 is constant in this case, it is possible to find solutions that **are** independent of z and represent purely horizontal motions. For example, Eq. (13.22) is satisfied by

$$\varphi = A e^{ik(x-ct)} \tag{13.47}$$

For this case, the wave velocity is independent of the wave length and is given by $c = \sqrt{RT}$ or $c = \sqrt{gH}$ just as in Eq. (13.42). Furthermore, $v = w = 0$; and

$$u = ikA e^{ik(x-ct)} \tag{13.48}$$

The disturbance velocity is again independent of the altitude.

In an isothermal atmosphere with adiabatic expansions, it is not possible to use the velocity potential, for the density is no longer a function of the pressure alone. For this case

$$g + \frac{c^2}{\rho_0} \frac{\partial \rho_0}{\partial z} = -(\gamma - 1)g \tag{13.49}$$

If the Coriolis terms are omitted, Eq. (13.9) then becomes

$$\frac{\partial^2 \mathbf{v}}{\partial t^2} = \nabla(c^2 \nabla \cdot \mathbf{v} - gw) - \mathbf{k}(\gamma - 1)g \nabla \cdot \mathbf{v} \tag{13.50}$$

Since c^2 is constant, it is possible to find a solution with purely horizontal motions. It is easily verified that a solution of Eq. (13.50) is given by $v = w = 0$ and

$$u = A e^{ik(x-ct)+(\gamma-1)\frac{gz}{c^2}} \tag{13.51}$$

This is a wave propagated with the velocity $c = \sqrt{\gamma RT}$. For air, $\gamma = 1.40$; therefore the wave velocity for this case may be written $c = 1.18 \sqrt{gH}$. For a surface temperature of 15°C, this makes $c = 1{,}115$ ft per sec. It may be noted that for this case the horizontal velocity increases with the elevation; however, the momentum $\rho_0 u$ and the pressure variation Dp/Dt both diminish with altitude.

Lamb also calculated the speed of propagation of long waves in an atmosphere having a lapse rate of half the dry-adiabatic lapse rate with adiabatic disturbances. For this case, he found that the speed of propagation of long waves is $1.07 \sqrt{gH}$. For a surface temperature of 15°C, this is 1,010 ft per sec.

G. I. Taylor[83] has computed in the same manner the velocity of propagation of long waves in an atmosphere in which the temperature falls linearly from 283°K at the ground to 220°K at a height of 13 km and is constant above that level. For this case, he finds the velocity of propagation to be 1,024 ft per sec. This is very close to the value of 1,046 ft per sec observed from the air waves caused by the explosion of Krakatao in 1883.

This problem has been treated more recently by Pekeris.[84] He further refined Taylor's calculations by including the Coriolis terms and by considering the temperature maximum that occurs at about 60 km in the high levels of the stratosphere.

These same methods can be used to discuss the large-scale oscillations of the earth's atmosphere as a whole. If the effects of rotation are neglected, Eq. (13.22) can be used to discuss the isothermal oscillations of an isothermal atmosphere. In polar coordinates (r,θ,ϕ) with θ being the angle from the north polar axis and ϕ being the meridional angle, Eq. (13.22) may be written

$$g \frac{\partial \varphi}{\partial r} + \frac{\partial^2 \varphi}{\partial t^2} = c^2 \left[\frac{\partial^2 \varphi}{\partial r^2} + \frac{2}{r} \frac{\partial \varphi}{\partial r} + \frac{1}{r^2 \sin \theta} \frac{\partial}{\partial \theta} \left(\sin \theta \frac{\partial \varphi}{\partial \theta} \right) + \frac{1}{r^2 \sin^2 \theta} \frac{\partial^2 \varphi}{\partial \phi^2} \right] \tag{13.52}$$

If we assume that the motion is primarily horizontal so that $\partial \varphi / \partial r = 0$, this becomes

$$\frac{1}{\sin \theta} \frac{\partial}{\partial \theta} \left(\sin \theta \frac{\partial \varphi}{\partial \theta} \right) + \frac{1}{\sin^2 \theta} \frac{\partial^2 \varphi}{\partial \phi^2} - \frac{r^2}{c^2} \frac{\partial^2 \varphi}{\partial t^2} = 0 \tag{13.53}$$

Since the depth of the atmosphere is small compared with a, the radius of the earth r may be replaced by a in this equation. Assume that

$$\varphi = P(\theta)e^{i(\sigma t - m\phi)} \tag{13.54}$$

This represents a wave disturbance traveling along the latitude circles with an angular velocity σ/m. Since the velocity field must be single-valued, m must be an integer. With these assumptions, Eq. (13.53) may be written as

$$\frac{1}{\sin \theta} \frac{d}{d\theta} \left(\sin \theta \frac{dP}{d\theta} \right) + \left[n(n+1) - \frac{m^2}{\sin^2 \theta} \right] P = 0 \tag{13.55}$$

where

$$n(n+1) = \frac{\sigma^2 a^2}{c^2} \tag{13.56}$$

This is practically the standard form of the equation for the associated Legendre functions,[82] and it is well known that this equation has a solution that is finite over the range $0 \leqslant \theta \leqslant \pi$ only if n is an integer. The natural frequencies of an isothermal atmosphere with isothermal expansions on a nonrotating sphere is thus given by Eq. (13.56) with integer values of n. The periods of the natural oscillations are thus

$$\frac{2\pi}{\sigma} = \frac{2\pi}{\sqrt{n(n+1)}} \frac{a}{c} \tag{13.57}$$

If we take $a = 4,000$ miles and $c = 943$ ft per sec (15°C), the two longest periods are 27.6 hr and 16.0 hr, corresponding to $n = 1$ and $n = 2$, respectively.

If the effects of the earth's rotation are to be considered, Eq. (13.17) rather than Eq. (13.22) must be used. If the vertical velocities are neglected and r is set equal to a, Eq. (13.17) becomes, in polar coordinates

$$\frac{\partial^2 u}{\partial t^2} - 2\omega \frac{\partial v}{\partial t} \cos \theta = \frac{1}{a \sin \theta} \frac{\partial P}{\partial \phi}$$
$$\frac{\partial^2 v}{\partial t^2} + 2\omega \frac{\partial u}{\partial t} \cos \theta = - \frac{1}{a} \frac{\partial P}{\partial \theta} \tag{13.58}$$

where

$$P = \frac{c^2}{a \sin \theta} \left[\frac{\partial u}{\partial \phi} - \frac{\partial}{\partial \theta} (v \sin \theta) \right] \tag{13.59}$$

and where u and v are the velocity components to the east and north, respectively. In an isothermal atmosphere, c^2 is constant; therefore the solutions of these equations show the same velocities at all elevations. These same equations arise in the theory of the free oscillations of a liquid ocean of constant depth on a rotating sphere. The calculations of the periods of the free oscillations of an isothermal atmosphere were carried out in this manner by Margules.[85] He assumed, as in Eq. (13.54), that the solutions were of the form of a function of θ multiplied by $e^{i(\sigma t - m\phi)}$. He used a temperature of 0°C and a propagation velocity $c = 2.84 \times 10^4$ cm per sec. For the zonal type of oscillations ($m = 0$), he found the three longest periods to be 20.44, 12.28, and 9.59 hr. For $m = 1$, the three longest periods of waves traveling toward the east were found to be 13.87, 9.22, and 6.63 hr. For $m = -1$, the waves traveling to the west were found to have periods of 36.57, 10.22, and 6.77 hr. For $m = \pm 2$, he found the longest periods to be 11.94 and 18.42 hr for waves traveling east and west, respectively.

These free oscillations of the earth's atmosphere are of great interest in the study of the regular oscillations of the surface barometric pressure due to tidal motions of the atmosphere. It is observed that the most regular barometric oscillations have solar diurnal and semidiurnal periods. On the other hand, the corresponding lunar tides are almost unobserved. For example, the semidiurnal solar tide is found to have an amplitude of 0.0375 in. of mercury at the equator, while Chapman[86] finds the semidiurnal lunar tide to have an amplitude of 0.00036 in. of mercury at Greenwich. It has been suggested that the solar semidiurnal tide is amplified by a near resonance with one of the atmosphere's natural frequencies. In this connection, it is of interest to note that, according to Margules' calculations, the oscillation of the semidiurnal type that travels to the east with $m = 2$ has a period of 11.94 hr. Since the solar tide has a period of 12 hr, a very near resonance is indicated. It has been suggested by Kelvin,[87] that these oscillations may be due to the daily temperature variation; however, it would be expected that, if this were the cause, the solar diurnal tide rather than the solar semidiurnal tide would be the larger. This question has been discussed more completely by Chapman.[88]

It is unlikely that the near resonance indicated by Margules' calculations is a satisfactory solution of this problem as Margules assumed the disturbances to be isothermal. For disturbances of as short a period as 12 hr, it would be expected that the disturbances would be more nearly of an adiabatic nature. Pekeris[84] showed that, in an atmosphere with a temperature maximum in the upper stratosphere as is indicated by the propagation of sound waves through these regions, one of the modes of oscillation for adiabatic disturbances might be expected to have a period of about 12 hr. A rigorous theoretical treatment of this problem, however, is not yet possible, since the required experimental observations of the temperature in the high stratosphere are not yet available.

The types of oscillations that have been considered thus far have been small disturbances in an otherwise stationary atmosphere. Actually the atmosphere has a strong mean meridional velocity field composed primarily of westerly and trade-wind belts. These mean motions might be expected to modify long-wave disturbances in the atmosphere. It is reported by Taylor[83] that some of the irregularities in the Krakatao explosion wave could be interpreted as the effects of the trade-wind belts. On the other hand, the atmospheric wind speeds are only a small fraction of the speed of propagation of these long-wave disturbances; therefore the effects of the mean speeds might be expected to be small.

The longest period of any free oscillation of the atmosphere of the type considered thus far is of the order of 1 day, and these types of motion apparently have no great influence on the weather phenomena occurring in the lower levels of the atmosphere. Another class of long-wave oscillatory motions apparently has large influences on the surface weather. It was observed that the isobars in 10-km level charts generally are of a sinusoidal character. Rossby[89] suggested that these sinusoidal isobars represent wave disturbances in the belt of westerlies.

In his discussion of these waves, Rossby assumed the disturbance to be purely horizontal and the fluid density to be constant. With these assumptions, the horizontal divergence of the disturbance velocity field vanishes. He also assumed that the mean motion of the westerlies could be represented by a uniform rectilinear flow of velocity U. If we consider an infinitesimal horizontal contour of area dS in an incompressible fluid, the Bjerknes circulation theorem shows that

$$\frac{D}{Dt}[(\zeta + \lambda)dS] = 0 \qquad (13.60)$$

where ζ is the vorticity component normal to the surface and $\lambda = 2\omega \cos\theta$ is the Coriolis parameter and θ is the colatitude. If there is no horizontal divergence, dS cannot vary; therefore

$$\frac{D}{Dt}(\zeta + \lambda) = 0 \tag{13.61}$$

If the x-axis points east and the y-axis points north, the velocity components are $U + u$ and v where u and v are the disturbance velocities. Now

$$\frac{D\lambda}{Dt} = -2\omega \sin\theta \frac{D\theta}{Dt} = \frac{2\omega \sin\theta}{a} v = \beta v \tag{13.62}$$

where a is the radius of the earth. If the disturbance is assumed to be small and the curvature of the earth's surface is neglected, Eq. (13.61) can be written

$$\left(\frac{\partial}{\partial t} + U\frac{\partial}{\partial x}\right)\left(\frac{\partial v}{\partial x} - \frac{\partial u}{\partial y}\right) + \beta v = 0 \tag{13.63}$$

Since the equation of continuity for this case is simply

$$\frac{\partial u}{\partial x} + \frac{\partial v}{\partial y} = 0 \tag{13.64}$$

it is again possible to introduce the stream function ψ where

$$u = \frac{\partial \psi}{\partial y}$$
$$v = -\frac{\partial \psi}{\partial x} \tag{13.65}$$

The stream function satisfies the continuity equation identically. It will also satisfy Eq. (13.63) if

$$\left(\frac{\partial}{\partial t} + U\frac{\partial}{\partial x}\right)\left(\frac{\partial^2 \psi}{\partial x^2} + \frac{\partial^2 \psi}{\partial y^2}\right) + \beta\frac{\partial \psi}{\partial x} = 0 \tag{13.66}$$

The variation of β will be neglected. If it is assumed that the disturbance is independent of the y-coordinate, then

$$\psi = Ae^{ik(x-ct)} \tag{13.67}$$

will be a solution if

$$c = U - \frac{\beta}{k^2} \tag{13.68}$$

It may further be noted that a standing wave for which $c = 0$ is possible if

$$k = \sqrt{\frac{\beta}{U}} \tag{13.69}$$

The wave length for which $c = 0$ is 6,050 km at 45° latitude if $U = 15$ mps.

The observed long-wave disturbances in the atmosphere are apparently almost stationary standing waves of this type caused by geographical disturbances. It has been pointed out by Elliot[90] and by others that the assumption that there is no horizontal divergence is not quite verified; however, the horizontal divergence is generally small.

This analysis can easily be extended to take into account the fact that the atmosphere surrounds a spherical earth. To simplify the analysis, only stationary disturbances will be considered. Equation (13.61) is still valid where ζ is the radial component of the vorticity or

$$\zeta = \frac{1}{a \sin \theta} \left[\frac{\partial}{\partial \theta} (u \sin \theta) + \frac{\partial v}{\partial \phi} \right] \tag{13.70}$$

where u and v are again the east and north components of the velocity, respectively. Supposing that the mean velocity of the atmosphere is along the latitude circles and is a function only of the latitude, we can write

$$\begin{aligned} u &= U + u' \\ v &= v' \end{aligned} \tag{13.71}$$

where the primed quantities refer to the disturbance. Similarly

$$\zeta = Z + \zeta' \tag{13.72}$$

where

$$Z = \frac{1}{a \sin \theta} \left[\frac{\partial}{\partial \theta} (U \sin \theta) \right]$$

and

$$\zeta' = \frac{1}{a \sin \theta} \left[\frac{\partial}{\partial \theta} (u' \sin \theta) + \frac{\partial v'}{\partial \phi} \right] \tag{13.73}$$

If the disturbance is small, Eq. (13.61) may be linearized, and it becomes for steady-state disturbances

$$\frac{U}{a \sin \theta} \frac{\partial \zeta'}{\partial \phi} + \frac{1}{a} \left(2\omega \sin \theta - \frac{\partial Z}{\partial \theta} \right) v' = 0 \tag{13.74}$$

For this case, the equation of continuity is

$$\frac{\partial u'}{\partial \phi} - \frac{\partial}{\partial \theta} (v' \sin \theta) = 0 \tag{13.75}$$

For this case, the stream function is defined by

$$\frac{\partial \psi}{\partial \theta} = u'$$

$$\frac{\partial \psi}{\partial \phi} = v' \sin \theta \tag{13.76}$$

The continuity equation is thus satisfied identically. If the stream function is introduced into Eq. (13.74), it can be integrated once and becomes

$$\zeta' + \frac{1}{U} \left(2\omega \sin \theta - \frac{\partial Z}{\partial \theta} \right) \psi = F(\theta) \tag{13.77}$$

where $F(\theta)$ is an arbitrary function of θ. A particular integral of Eq. (13.77) involving $F(\theta) \neq 0$ shows motions parallel to the latitude circles. Since these solutions are of no interest, $F(\theta)$ can be set equal to zero. If ζ' is expressed in terms of the stream function, Eq. (13.77) becomes

$$\frac{1}{\sin \theta} \frac{\partial}{\partial \theta} \left(\sin \theta \frac{\partial \psi}{\partial \theta} \right) + \frac{1}{\sin^2 \theta} \frac{\partial^2 \psi}{\partial \phi^2} + \frac{a}{U} \left(2\omega \sin \theta - \frac{\partial Z}{\partial \theta} \right) \psi = 0 \tag{13.78}$$

This equation determines the disturbances for any given mean motion.

Except for the equatorial regions, the mean velocity is represented fairly well by

$$U = U_0 \sin \theta \tag{13.79}$$

where U_0 is a constant. The wave disturbances for this mean velocity field have been discussed by Haurwitz.[91] For this case, Eq. (13.78) becomes

$$\frac{1}{\sin\theta}\frac{\partial}{\partial\theta}\left(\sin\theta\frac{\partial\psi}{\partial\theta}\right) + \frac{1}{\sin^2\theta}\frac{\partial^2\psi}{\partial\phi^2} + 2\left(1 + \frac{\omega a}{U_0}\right)\psi = 0 \qquad (13.80)$$

The stream function is thus seen, as in Eq. (13.53), to be a surface harmonic of order n where

$$n(n+1) = 2\left(1 + \frac{\omega a}{U_0}\right) \qquad (13.81)$$

Both u' and v' must vanish at the poles, and the harmonic must have a period of 2π in ϕ. This restricts ψ to the form

$$\psi = \sum_m A_m P_n^m(\cos\theta)\,{\cos\atop\sin}(m\phi) \qquad (13.82)$$

where m is an integer greater than unity and n, also an integer, is given by Eq. (13.81). Since n must also be an integer, it is seen that oscillations of this type with no horizontal divergence anywhere can exist only for a discrete set of mean velocities. These velocities are shown in Table 7.

TABLE 7

n	U_0, mps	$\frac{1}{\sqrt{2}}U_0$, mps
2	231.5	163.6
3	92.6	65.4
4	51.4	36.3
5	33.1	23.4
6	23.2	16.4
7	17.1	12.1

The last column, $U_0/\sqrt{2}$, is the velocity at 45° latitude, in the middle of the westerlies. It may be noted that the velocities corresponding to $n = 5, 6$, and 7 cover the range of velocities normally encountered in the atmosphere. For intermediate values of the mean velocity, wave disturbances of this type are no longer possible; however, more complex types involving horizontal divergence are possible. If the restriction to steady-state motion is relaxed, similar families of moving waves may be found. The application of the theory of this type of wave motion to forecasting problems has been studied by Rossby,[89] Haurwitz,[91,92] Starr,[93] and others.

All the types of oscillatory motion studied thus far have had periods of only a few days. Since the weather varies widely from one year to the next, it would seem that the atmosphere must possess natural modes of oscillation having periods of the order of magnitude of, but different from, a year; for the nonseasonal variation of the only external parameter, the solar-energy input, is apparently extremely small. One type of motion that has shown extremely long periods has been discussed by Stewart[7,8] who investigated some of the properties of the subtropical high-pressure systems.

On the equatorial side of the belt of westerly winds, in both Northern and Southern Hemispheres, there is a strong shearing zone having anticyclonic vorticity. Now such shearing motions are generally unstable and must be expected to break up, and the vorticity can be expected either to diffuse throughout the fluid or to collect in stable vorticity concentrations, provided that such stable configurations exist. For example, the vorticity in the boundary layer of a fluid flowing in a channel is diffused throughout the fluid, whereas the vorticity in the wake behind a bluff body develops into the well-known von Kármán "vortex street" for a wide range of the Reynolds number. Stewart suggested that the subtropical high-pressure systems are dynamically stable vorticity concentrations similar to the "vortex street." In order to discuss this question, it is necessary to investigate the stability of such vorticity configurations.

It seems that the horizontal field of motion is of primary importance in determining the motion of these large-scale systems; therefore it is assumed that the atmosphere can be treated as a single layer of fluid of constant density with the vertical velocities being of small importance so that the pressure can be determined from the hydrostatic equation. It is also assumed that the apparent acceleration is negligible when compared with the Coriolis acceleration. In addition, the effects of the variation of the Coriolis parameter are neglected. This means that the fluid motions considered are those taking place on a rotating disk rather than on a rotating sphere.

By the hydrostatic equation, the pressure at any point is

$$p = g\rho(h - z) \tag{13.83}$$

where h is the depth of the atmosphere. The circulation theorem for this case shows that

$$\frac{D}{Dt}\left(\frac{\zeta + 2\omega}{h}\right) = 0 \tag{13.84}$$

where ω is the angular velocity of the disk. If the motion could have started from rest with a uniform depth h_0, this may be integrated and becomes

$$\zeta = \frac{\partial v}{\partial x} - \frac{\partial u}{\partial y} = \frac{2\omega}{h_0}(h - h_0) \tag{13.85}$$

If the velocity components are eliminated from this equation by means of the geostrophic-wind equations

$$-2\omega v = -g\frac{\partial h}{\partial x} \qquad 2\omega u = -g\frac{\partial h}{\partial y} \tag{13.86}$$

an expression determining the depth of the atmosphere (*i.e.*, the sea-level pressure) is obtained. This is

$$\frac{\partial^2 h}{\partial x^2} + \frac{\partial^2 h}{\partial y^2} - \frac{4\omega^2}{gh_0}(h - h_0) = 0 \tag{13.87}$$

This equation can be further simplified by the introduction of dimensionless variables, $X = 2\omega x/\sqrt{gh_0}$, $Y = 2\omega y/\sqrt{gh_0}$ and $\eta = (h - h_0)/h_0$. With these new variables, Eq. (13.87) becomes

$$\frac{\partial^2 \eta}{\partial X^2} + \frac{\partial^2 \eta}{\partial Y^2} - \eta = 0 \tag{13.88}$$

In terms of the dimensionless depth and dimensionless velocities defined by $U = u/\sqrt{gh_0}$ and $V = v/\sqrt{gh_0}$, the geostrophic-wind equation can be rewritten as

$$\frac{\partial \eta}{\partial X} = V \qquad \frac{\partial \eta}{\partial Y} = -U \tag{13.89}$$

The only solution of Eq. (13.88) that vanishes at infinity and represents flow in circles about the origin and thus corresponds to a simple vortex is

$$\eta = \alpha K_0(r) \tag{13.90}$$

where α is an arbitrary constant, $r = \sqrt{X^2 + Y^2}$ and $K_0(r)$ is a modified Bessel function[94] of the second kind. If α is positive, the motion is anticyclonic; if α is negative, the motion is cyclonic. All the motions considered in the present investigation are built up through superposition of vortices of this type. From Eq. (13.89), this vortex has a dimensionless tangential velocity u_θ given by

$$u_\theta = \frac{\partial \eta}{\partial r} = -\alpha K_1(r) \tag{13.91}$$

The geostrophic-wind equations used in the above development can be shown to be valid unless $r < < 1$.

Based on a homogeneous atmosphere having a mean sea-level pressure and density of 1.013×10^6 dynes per cm; and 1.22×10^{-3} gram per cm³, respectively, the same as the standard atmosphere, the characteristic velocities and distances used above to produce dimensionless variables are $\sqrt{gh_0} = 2.87 \times 10^4$ cm per sec and $\sqrt{gh_0}/2\omega = 1.97 \times 10^8$ cm. At a distance of 2,000 km from the center of the Pacific or Azores high-pressure systems, the characteristic velocity is of the order of 10 mps. Since $K_1(1) = 0.602$, this indicates that these anticyclones have a strength such that α is approximately 0.06.

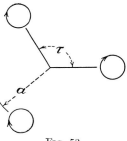

If the interaction between the Northern and Southern Hemispheres is neglected, the ring of subtropical anticyclones can be roughly represented by N equal anticyclones of the type given by Eq. (13.90), which are placed on a ring of radius a and spaced at equal angles $\tau = 2\pi/N$, as shown in Fig. 53. In the Northern Hemisphere, the Pacific and Azores high-pressure regions are well defined. They are about 120° longitude apart. There is some evidence of a third such system over India and equidistant from the other two; however, this evidence is far from conclusive, owing to the low-level interference from the monsoon. In the Southern Hemisphere, there are also three such systems. The best model is thus obtained with $N = 3$.

Fig. 53.

The surface deflection for such a system in its equilibrium state is

$$\eta = \alpha \sum_{n=1}^{N} K_0[a^2 + r^2 - 2ar \cos (\theta - n\tau)]^{\frac{1}{2}} \tag{13.92}$$

From Eq. (13.89), the dimensionless velocities in the radial and tangential directions u_r and u_θ, respectively, are given by

$$u_r = -\frac{1}{r}\frac{\partial \eta}{\partial \theta} \qquad u_\theta = \frac{\partial \eta}{\partial r} \tag{13.93}$$

The velocity of any vortex is the velocity at that point due to all the remaining vortices. From the second of the expressions of Eq. (13.93), the system shown in Fig. 53 is seen to have a dimensionless angular velocity Ω given by

$$\Omega = -\frac{\alpha}{a} \sum_{n=1}^{N-1} K_1 \left(2a \sin \frac{1}{2}n\tau\right) \sin \left(\frac{1}{2}n\tau\right) \tag{13.94}$$

In Table 8 are given the values of Ω and of a $\Omega \sqrt{gh_0}$, the linear velocity of the vortices, for $\alpha = 0.06$, $N = 3$, and $a = 3.0$ and 3.5. The values of $a = 3.0$ and 3.5 correspond to the ring of subtropical anticyclones being placed at latitude 37 and 29°, respectively. From this result, it is seen that such a vortex system would have a slow precession to the west. In the atmosphere, there is also a region of distributed cyclonic vorticity to the north of the westerly winds. It is easily seen that this cyclonic vorticity tends to produce an eastward displacement of the subtropical highs. It appears that these two displacements cancel one another so that the systems are practically stationary, a condition that is undoubtedly also imposed by the thermodynamic and topographical factors. Since the westward drift shown in Table 8 is very small, no attempt will be made to correct the vortex system shown in Fig. 53 in order to take into account the polar cyclonic vorticity.

TABLE 8.—ANGULAR VELOCITIES OF VORTEX SYSTEMS, $\alpha = 0.06$, $N = 3$

a	Ω	$a\Omega \sqrt{gh_0}$
3.0	−0.000112	−9.6 cm/sec
3.5	−0.000037	−3.7 cm/sec

If the nth vortex is displaced by a distance Δr_n in the radial direction and by an angle $\Delta\theta_n$ in the tangential direction, the surface deflection in the displaced condition is

$$\eta = \alpha \sum_{n=1}^{N} K_0[(a + \Delta r_n)^2 + r^2 - 2(a + \Delta r_n)r \cos (\theta - n\tau - \Delta\theta_n)]^{\frac{1}{2}} \quad (13.95)$$

The velocities of the vortices may be calculated as before by Eq. (13.93). If the displacements are small, the changes in velocity of the Nth vortex from the equilibrium value indicated by Eq. (13.94) may be written as

$$\Delta u_r = \frac{d}{dt}(\Delta r_N) = \frac{\alpha}{2} \sum_{n=1}^{N-1} \Delta r_n K_0(R_n) \sin n\tau$$
$$+ \alpha \sum_{n=1}^{N-1} a(\Delta\theta_n - \Delta\theta_N) \left[\frac{k_1(R_n)}{R_n} + K_0(R_n) \cos^2 \frac{1}{2} n\tau\right] \quad (13.96)$$

$$\Delta u_\theta = a \frac{d}{dt}(\Delta\theta_N) + \Omega \Delta r_N = \alpha \Delta r_N \sum_{n=1}^{N-1} \left[K_0(R_n) \sin^2 \frac{1}{2} n\tau - \frac{K_1(R_n)}{R_n} \cos n\tau\right]$$
$$+ \alpha \sum_{n=1}^{N-1} \left\{\Delta r_n \left[K_0(R_n) \sin^2 \frac{1}{2} n\tau + \frac{K_1(R_n)}{R_n}\right] + \frac{1}{2} a \Delta\theta_n K_0(R_n) \sin n\tau\right\}$$

where
$$R_n = 2a \sin \tfrac{1}{2} n\tau$$

It should be noted that the t in Eqs. (13.96) is a dimensionless time. If the actual time is t^*, then

$$t = 2\omega t^* \quad (13.97)$$

Expressions similar to Eq. (13.96) for the velocities of the other vortices could be be written from symmetry. These would form a set of simultaneous differential equations for the displacements.

If the N equations in each of the two sets indicated in Eqs. (13.96) are added, it is seen that

$$\frac{d}{dt}\left(\sum_{n=1}^{N} \Delta r_n\right) = 0$$

$$\frac{d}{dt}\left(\sum_{n=1}^{N} \Delta\theta_n\right) = \alpha \left(\sum_{n=1}^{N} \Delta r_n\right) \quad (13.98)$$
$$\left\{\sum_{n=1}^{N-1} \left[2K_0(R_n) \sin^2 \frac{1}{2} n\tau + \frac{1}{a} K_1(R_n) \sin \frac{1}{2} n\tau\right]\right\}$$

From this, it is seen that, if a mean value of a is chosen so that $\sum_{n=1}^{N} \Delta r_n = 0$ initially, then from Eqs. (13.98) both $\sum_{n=1}^{N} \Delta r_n$ and $\sum_{n=1}^{N} \Delta\theta_n$ will remain constant. These results correspond to similar equations for two-dimensional line vortices that state that the impulse of a system having no external forces remains constant.[15]

The disturbed motion of the vortex system as described in Eqs. (13.96) can best be discussed by considering the normal modes of oscillation. From the symmetry conditions, the displacements in each normal mode must be of the form $\Delta r_n = \Delta r_N e^{in\varphi}$ and $\Delta\theta_n = \Delta\theta_N e^{in\varphi}$ where φ characterizes the normal mode and is a member of the series $2\pi/N$, $4\pi/N$, $\cdots$, $2\pi(N-1)/N$, 2π. With this notation, Eqs. (13.96) can be written as

$$\frac{1}{\alpha}\frac{d}{dt}(\Delta r_N) = A \ \Delta r_N - Ba \ \Delta\theta_N \qquad \frac{a}{\alpha}\frac{d}{dt}(\Delta\theta_N) = C \ \Delta r_N + Aa \ \Delta\theta_N \qquad (13.99)$$

where $A = \tfrac{1}{2} \sum_{n=1}^{N-1} K_0(R_n)e^{in\varphi}\sin n\tau$

$$B = \sum_{n=1}^{N-1} (1 - e^{in\varphi})\left[\frac{K_1(R_n)}{R_n} + K_0(R_n)\cos^2\frac{1}{2}n\tau\right] \qquad (13.100)$$

$$C = \sum_{n=1}^{N-1}\left[K_0(R_n)(1+e^{in\varphi})\sin^2\frac{1}{2}n\tau + \frac{K_1(R_n)}{R_n}(1 - 2\cos n\tau + e^{in\varphi})\right]$$

If the amplitudes Δr_N and $\Delta\theta_N$ are assumed to vary like e^{ipt} and p is the dimensionless normal frequency, then, from Eq. (13.99),

$$p = \alpha(-iA \pm \sqrt{BC}) \qquad (13.101)$$

From Eq. (13.100), it can be seen that, for $\varphi = 2\pi$, $A = B = 0$, and there are thus two zero normal frequencies. These two zero frequencies are those shown in Eq. (13.98). It can also be seen that, for the specified values of φ, A is always a purely imaginary quantity; B is always real and not less than zero. From Eq. (13.101), the condition that the frequencies be real is that C be real and nonnegative. Complex frequencies, of course, characterize systems in which the amplitudes increase with time and are thus unstable. Now C is always real and is always positive for $N < 7$. If $N = 7$, C is always positive if $a > 71$. For $N > 7$, C is negative for one or more of the given values of φ. A value of $a > 71$ corresponds either to disturbances of such great wave lengths or to motions of such a shallow layer of air in the earth's atmosphere that it is probably of no significance. The vortex system is thus stable if N is less than or equal to six.

The frequencies and modes of oscillation will be discussed in some detail for the cases where $N = 3$ and $a = 3.0$ and 3.5. From Eq. (13.97), it may be seen that the period of an oscillation is given by

$$T = \left|\frac{1}{2p}\right| \qquad \text{sidereal days} \qquad (13.102)$$

The normal mode of oscillation, from Eq. (13.99), is given by

$$\frac{a \ \Delta\theta_N}{\Delta r_N} = \frac{a \ \Delta\theta_n}{\Delta r_n} = \frac{A - (ip/\alpha)}{B} \qquad (13.103)$$

The results for $a = 3.0$ and 3.5 are given in Tables 9 and 10, respectively. The two normal modes thus show a short-period (2,000 to 5,000 days) and a long-period oscillation (70,000 to 250,000 days). The path of the vortex is in each case an ellipse. For the short-period oscillation, the vortices are east of their mean position when traveling south and west when traveling north.   For the long-period oscillation, the sense of the rotation is reversed.

TABLE 9.—NORMAL MODES OF OSCILLATION FOR $N = 3$, $a = 3.0$, $\alpha = 0.06$

φ	$\dfrac{2\pi}{3}$		$\dfrac{4\pi}{3}$	
$-iA$	0.002248		−0.002248	
B	0.00414		0.00414	
C	0.001182		0.001182	
p	0.000268	0.000002	−0.000268	−0.000002
T days	1,865	2.5×10^5	1,865	2.5×10^5
$\dfrac{a\,\Delta\theta_N}{\Delta r_N}$	$-0.53i$	$0.53i$	$0.53i$	$-0.53i$

TABLE 10.—NORMAL MODES OF OSCILLATION FOR $N = 3$, $a = 3.5$, $\alpha = 0.06$

φ	$\dfrac{2\pi}{3}$		$\dfrac{4\pi}{3}$	
$-iA$	0.000859		−0.000859	
B	0.001467		0.001467	
C	0.000375		0.000375	
p	0.000096	0.000007	−0.000096	−0.000007
T days	5,200	7.1×10^4	5,200	7.1×10^4
$\dfrac{a\,\Delta\theta_N}{\Delta r_N}$	$-0.51i$	$0.51i$	$0.51i$	$-0.51i$

These calculations cannot be considered as a quantitative theory of the oscillations of the semipermanent high-pressure systems; however, the essential features of the calculation, vorticity concentrations at distances of roughly 10,000 km, are found in the atmosphere; therefore motions of this type must exist in the atmosphere.   It might be expected that the effects of coupling between the Northern and Southern Hemispheres and of any cyclonic vorticity concentrations on the polar sides of the westerly winds would be to decrease the period of the shortest oscillation and to introduce additional natural frequencies.   These results are of considerable interest particularly in view of the climatological evidence of atmospheric periods of the order of from 10 to 20 years.

Bibliography

1. VON HELMHOLTZ, H.: Sitzb. Akad. Wiss. Berlin, **501** (1873).
2. BJERKNES, J.: Geofysiske Publ., **1** (2), (1919).
3. BJERKNES, J., and H. SOLBERG: Geofysiske Publ., **3** (1), (1922).
4. BJERKNES, V.: Geofysiske Publ., **2** (4), (1923).
5. EXNER, F. M.: "Dynamische Meteorologie," 2d ed., Springer, Vienna, 1925.
6. JEFFREYS, H.: Quart. J. Roy. Meteorolog. Soc., **52**: 85 (1926).
7. STEWART, H. J.: Proc. Nat. Acad. Sci., **26** (10), (1940).
8. STEWART, H. J.: Quart. Appl. Math., **1** (3), (1943).
9. ROSSBY, C.-G.: Proc. 5th Int. Cong. Appl. Mech. 373 (1938).
10. ROSSBY, C.-G.: J. Mar. Research, **2** (1), (1939).

11. BJERKNES, V., J. BJERKNES, H. SOLBERG, and T. BERGERON: "Physikalische Hydrodynamik mit Anwendung auf die dynamische Meteorologie," Springer, Berlin, 1933.
12. BERGERON, T.: *Geofysiske Publ.*, **5** (6), (1928).
13. PETTERSSEN, S.: *Geofysiske Publ.*, **11** (6), (1935).
14. PETTERSSEN, S.: *Geofysiske Publ.*, **10** (2), (1933).
15. LAMB, H.: "Hydrodynamics," 6th ed., Cambridge, London, 1932.
16. STEWART, H. J.: *Proc. Nat. Acad. Sci.*, **28** (5): 161 (1942).
17. BJERKNES, V.: *Vid.-Selsk. Skrifter*, Christiania, 1918.
18. STARR, V. P., and M. NEIBURGER: *J. Mar. Research*, **3**: 202 (1940).
19. BRUNT, D., and C. M. K. DOUGLAS: *Mém. Roy. Meteorolog. Soc.*, **3** (22): 29 (1928)
20. VAN MIEGHEM, J.: *Acad. Roy. Belgique Cl. Sci. Mém. Coll.*, **19** (3): 65 (1941).
21. DINES, W. H.: *Nature*, **99**: 24 (1917).
22. VON KÁRMÁN, T.: *Bull. Am. Math. Soc.*, **46**: 8, 615 (1940).
23. GRINDLEY, J. H., and A. H. GIBSON: *Proc. Roy. Soc. (London)*, A, **80**: 114 (1907).
24. VON HELMHOLTZ, H.: *Wiener Sitzb.*, **40**: 607 (1860).
25. NAVIER, C. L. M. H.: *Mém. Acad. Sciences*, **6**: 389 (1822).
26. STOKES, SIR G. G.: *Cambridge Trans.*, **8**: 287 (1845).
27. REYNOLDS, O.: *Phil. Trans. Roy. Soc. (London)*, (1883).
28. EKMAN, V. W.: *Arkiv. Mat. Astron. Fysik*, **6** (12), (1910).
29. DRYDEN, H. L., and A. M. KUETHE: *Nat. Advisory Comm. Aeronautics Tech. Rept.* 320, (1929).
30. JEANS, SIR J.: "Kinetic Theory of Gases," Cambridge, London, 1940.
31. BOUSSINESQ, T. V.: *Mém. pres. par. div. Sav.*, **23**, Paris, (1877).
32. GRIMMINGER, G.: *Trans. Am. Geophys. U.*, 19th Ann. Meeting, (1938).
33. SCHMIDT, W.: "Der Massenaustausch in freier Luft und verwandte Erscheinungen," Henri Grand, Hamburg, 1925.
34. LETTAU, H.: "Atmosphärische Turbulenz," Akademische Verlagsgesellschaft m. b. H., Leipzig, 1939.
35. EKMAN, V. W.: *Nyt. Mag. Naturv.*, **40** (1), (1902).
36. ÅKERBLOM, F. A.: *Nova Acta*, Ser. 4, **2** (2), (1908).
37. DOBSON, G. M. B.: *Quart. J. Roy. Meteorolog. Soc.*, **40**: 123 (1914).
38. MILDNER, P.: *Beitr. Phys. freien Atmos.*, **19** (1932).
39. PRANDTL, L.: *Z. angew. Math. Mech.*, **5** (2): 136 (1925).
40. VON KÁRMÁN, T.: *Nachr. Ges. Wiss. Göttingen, Math.-Phys. Klasse*, 58 (1930).
41. VON KÁRMÁN, T.: *J. Aeronaut. Science*, **1**: 1 (1934).
42. ROSSBY, C.-G.: *Papers Phys. Oceanogr. Meteorolog.*, **4** (3), (1936).
43. PRANDTL, L.: *Beitr. Phys. freien Atmos.*, Bjerknes-Festschrift, **19**: 188 (1932).
44. ROSSBY, C.-G., and R. B. MONTGOMERY: *Papers Phys. Oceanogr. Meteorolog.*, **3** (3), (1935).
45. HELLMANN, G.: *Meteorolog. Z.*, **32**: 1 (1919).
46. SHAW, N.: "Manual of Meteorology," Vol. 4, Cambridge, London, 1931.
47. WÜST, G.: *Veröffentlich. Inst. Meereskunde Universität Berlin*, Neue FolgeA (6): 73 (1920).
48. PAESCHKE, W.: *Beitr. Phys. freien Atmos.*, **24**: 163 (1937).
49. TAYLOR, G. I.: *Proc. Roy. Soc. (London)*, A, **92**: 196 (1916).
50. EXNER, F. M.: *Sitzb. Akad. Wien.*, Abt. 2a, **136**: 453 (1927).
51. SVERDRUP, H. U.: *Proc. 5th Int. Cong. Appl. Mech.*, 369 (1938).
52. ROSSBY, C.-G.: *Papers Phys. Oceanogr. Meteorolog.*, **1** (4), (1932).
53. JEFFREYS, H.: "Cartesian Tensors," Cambridge, London, 1931.
54. SVERDRUP, H. U.: *Geofysiske Publ.*, **11** (7), (1936).
55. TAYLOR, G. I.: *Phil. Trans. Roy. Soc. (London)*, A, **215**: 1 (1915).
56. JAHNKE, E., and F. EMDE: "Funktionentafeln mit Formeln und Kurven," B. G. Teubner, Berlin, 1909.
57. JEFFREYS, H.: *Phil. Mag.*, **35**: 270 (1918).
58. SUTTON, O. G.: *Proc. Roy. Soc. (London)*, A, **146**, (1934).
59. THORNTHWAITE, C. W., and B. HOLTZMAN: *Monthly Weather Rev.*, **67**: 4 (1939).

60. MONTGOMERY, R. B.: *Papers Phys. Oceanogr. Meteorolog.*, **7** (4), 1940.
61. SCHMIDT, W.: *Sitzb. Akad. Wiss. Wien.*, **127**: 1889 (1918).
62. SCHWERDTFEGER, W.: *Veröffentlich. Geophys. Inst. Leipzig*, 2d ser., **4**: 253 (1931).
63. HAURWITZ, B.: *Trans. Roy. Soc. Canada*, 3d ser., Sec. III, **30**: 1 (1936).
64. GRIMMINGER, G.: *op. cit.*, p. 163.
65. PARR, A. E.: *J. Conseil*, **11** (3), Copenhagen, 1936.
66. ROSSBY, C.-G., and collaborators: *Bull. Am. Meteorolog. Soc.*, **18**: 201 (1937).
67. ROSSBY, C.-G.: *Proc. 5th Int. Cong. Appl. Mech.*, 373, 379 (1938).
68. DEFANT, A.: *Geog. Ann.*, **3**: 209 (1921).
69. TAYLOR, G. I.: *Proc. Roy. Soc. (London)*, A, **151**: 421 (1935); **156**: 307 (1936).
70. VON KÁRMÁN, T.: *J. Aeronaut. Science*, **4**: 131 (1937).
71. DRYDEN, H. L.: *Quart. Appl. Math.*, **1**: 7 (1943).
72. NORMAND, C. W. B.: *Quart. J. Roy. Meteorolog. Soc.*, **64**: 71 (1938).
73. MARGULES, M.: *Meteorolog. Z.*, **23**: 481 (1906).
74. KOSCHMIEDER, H.: "Dynamische Meteorologie," p. 336, Akademische Verlagsgesellschaft m.b.H. Leipzig, 1933.
75. LITTWIN, W. V.: *Monthly Weather Rev.*, **64**: 397 (1936).
76. EKMAN, V. W.: "On Dead-Water," Scientific Results of the Norwegian Polar Expedition, Nansen ed., Christiania, 1906.
77. WEGENER, A.: *Beitr. Phys. freien Atmos.*, **2**: 55 (1906–1908).
78. HAURWITZ, B.: *Gerlands Beitr. Geophys.*, **34**: 213 (1931); **37**: 16 (1932).
79. BJERKNES, V., J. BJERKNES, H. SOLBERG, and T. BERGERON: *op. cit.*, p. 574.
80. ROSENHEAD, L.: *Proc. Roy. Soc. (London)*, A, **134**: 170 (1932).
81. LAMB, H.: *Proc. Roy. Soc. (London)*, A, **74**: 551 (1911).
82. COPSON, E. T.: "Functions of a Complex Variable," Oxford, New York, 1935.
83. TAYLOR, G. I.: *Proc. Roy. Soc. (London)*, A, **126**: 169, 728 (1929).
84. PEKERIS, C. L.: *Proc. Roy. Soc. (London)*, A, **158**: 650 (1937).
85. MARGULES, M.: *Wiener Sitzb.*, **101**: 597 (1892); **102**: 11 (1895).
86. CHAPMAN, S.: *Quart. J. Roy. Meteorolog. Soc.*, **44**: 271 (1918).
87. KELVIN, Lord: *Proc. Roy. Soc. (Edinburgh)*, **11** (1892).
88. CHAPMAN, S.: *Quart. J. Roy. Meteorolog. Soc.*, **50**: 165 (1923).
89. ROSSBY, C.-G.: *J. Mar. Research*, **2**: 38 (1939).
90. ELLIOT, R. D.: Weather Bureau Symposium on Wave Phenomena in the Atmosphere, Washington, D.C., 1944.
91. HAURWITZ, B.: *J. Mar. Research*, **3**: 254 (1940).
92. HAURWITZ, B.: *Trans. Am. Geophys. U.*, (2): 263 (1940).
93. STARR, V. P.: "Basic Principles of Weather Forecasting," Harper, New York, 1942.
94. GREY, A., G. B. MATHEWS, and T. M. MACROBERT: "Bessel Functions," Macmillan & Company, Ltd., London, 1931.

SECTION VII

THE SCIENTIFIC BASIS OF MODERN METEOROLOGY

By C.-G. Rossby

CONTENTS

SECTION VII

THE SCIENTIFIC BASIS OF MODERN METEOROLOGY*

By C.-G. Rossby

The science of meteorology does not yet have a universally accepted, coherent picture of the mechanics of the general circulation of the atmosphere. This is partly because observational data from the upper atmosphere still are very incomplete, but at least as much because our theoretical tools for the analysis of atmospheric motions are inadequate. Meteorology, like physics, is a natural science and may indeed be regarded as a branch of the latter; but, whereas in ordinary laboratory physics it is always possible to set up an experiment, vary one factor at a time, and study the consequences, meteorologists have to contend with such variations as nature may offer, and these variations are seldom so clean-cut as to permit the establishment of well-defined relationships between cause and effect.

Under these conditions, theoretical considerations become more important than ever. The atmosphere may be considered as a turbulent fluid subjected to strong thermal influences and moving over a rough, rotating surface. As yet no fully satisfactory theoretical or experimental technique exists for the study of such fluid motions; yet it is safe to say that, until the proper theoretical tools are available, no adequate progress will be made either with the problem of long-range forecasting or with the interpretation of past climatic fluctuations.

Certain phases of the admittedly oversimplified analysis of the circulation of the atmosphere presented here may be traced as far back as 1888, to the German physicist von Helmholtz, but other parts are the result of recent research and should, to some extent, be considered as the author's personal view. The combination of these various elements into a still fairly crude bird's-eye view of the atmosphere and its circulation was undertaken during the last few years as a by-product of an intensive study of Northern Hemisphere weather, with which the author had the good fortune to be associated. This study was conducted as a cooperative project between the U.S. Department of Agriculture and the Massachusetts Institute of Technology and was to a large extent supported by Bankhead-Jones funds.

CONVECTIVE CIRCULATION

The energy that drives the atmosphere is obtained from the sun's radiation. Just outside the atmosphere an area exposed at right angles to the sun's rays would receive radiant energy at the rate of about 2 gram-cal per sq cm per min. The earth is approximately spherical, and its surface area is therefore four times as large as the cross-sectional area intercepting the sun's rays; it follows that on an average each square centimeter at the outer boundary of the atmosphere receives about $\frac{1}{2}$ gram-cal per min. Part of this radiation is reflected back to space from the upper surface of clouds in the atmosphere, and part is lost through diffuse scattering of the solar radiation by air molecules and dust. It is estimated that a total of about 40 per cent on

* This material is reprinted from "Climate and Man," Yearbook of Agriculture for 1941, U.S. Government Printing Office, Washington, D.C.

502

an average is lost through these processes. (It is this reflection that determines the whiteness, or albedo, of the earth as a planet, and thus the earth is said to have an albedo of about 40 per cent.) The remaining radiation passes through the atmosphere without much absorption and finally reaches the surface of the earth, where it is absorbed, generally without much loss through reflection. Snow surfaces furnish an important exception, since they may reflect as much as 80 per cent of the incident solar radiation.

Since the mean temperature of the earth does not change appreciably, it follows that the heat gained from the sun must be sent back to space. On the surface of the earth, the solar radiation is transformed into heat, and this heat is returned as "long" (infrared) radiation toward space. The rate at which the ground sends out such radiation is very nearly proportional to the fourth power of the absolute temperature. Thus, at 10°C, the heat loss of the ground through radiation is about 15 per cent greater than at freezing, and it follows that the surface temperature of the earth must increase if the incident solar radiation increases, to maintain equilibrium between radiation income and loss. The ground is very nearly a perfect radiator, *i.e.*, it sends back to the atmosphere and to space the maximum amount obtainable from any surface at a particular temperature.

All the invisible radiation from the ground cannot escape to space; the larger part of it is absorbed in the atmosphere, principally by the water vapor in the lower layers but also to some extent by small amounts of ozone present in the upper atmosphere. The atmosphere in turn emits infrared (heat) radiation upward and downward. The radiation emitted in either direction increases with the mean temperature of the atmospheric column but is always less than the radiation emitted by a perfect radiator of the same temperature. The more perfectly the atmospheric column absorbs the ground radiation from below, the more perfectly it radiates. It is evident that, if radiative processes alone controlled the behavior of the atmosphere, it would have to emit as much as it absorbs. Since it emits in two directions but absorbs appreciably only from the ascending ground radiation, it follows that the mean temperature of the atmosphere must be lower than that of the ground. Also, since the ground receives not only solar radiation but also infrared radiation from the atmosphere, the ground temperature must be higher than might be expected from the intensity of the incident solar radiation alone.

The preceding analysis shows that the atmosphere serves as a protective covering that raises the mean temperature of the surface of the earth. This is often referred to as the "greenhouse" effect of the atmosphere. The analysis also shows that the atmosphere as a whole must be colder than the ground. The reasoning may be refined by considering the atmosphere as consisting of a number of superimposed horizontal layers, and it may then be shown that the mean temperature must decrease upward from layer to layer, fairly rapidly near the ground, then more and more slowly. At great heights (above 10 km, or 6 miles), the temperature becomes very nearly constant. This layer is called the *stratosphere*.

Up to this point, our analysis has been based on the assumption that no appreciable direct absorption of solar radiation occurs in the atmosphere. This is approximately correct as far as the lowest 20 km (12 miles) of the atmosphere are concerned. However, between approximately 20 and 50 km (12 and 30 miles) above the ground, the atmosphere contains a certain amount of ozone, increasing from the equator to high latitudes, and this ozone is capable of directly absorbing some of the radiation emitted by the sun and also, to a somewhat lesser extent, part of the long-wave radiation from below. As a result, in the upper portions of this ozone layer, the air temperature appears to exceed the mean air temperature next to the ground. It is this ozone layer which protects us from the extreme ultraviolet rays in the sun's radiation.

At still higher levels, the oxygen and the nitrogen in the earth's atmosphere are capable of intense direct absorption of solar radiation. For this and other reasons, it is now generally believed that the temperature again rises to values which may be as high as 500 to 1000°A above 150 km (93 miles) above sea level.

This uppermost region of high temperature is located at a height where the air density is so small that no appreciable direct dynamic effect on air circulation in the lower layers of the atmosphere can be expected. It is necessary to admit, however, that temperature variations and resulting circulation at the ozone level may be significant so far as weather and wind in the lowest strata are concerned. No existing theory suggests a definite mechanism for control of sea-level circulation and weather by an ozone layer. On the other hand, circulation in the lowest 20 km of the atmosphere has definitely been shown to produce redistributions of the ozone aloft, and hence atmospheric ozone has of late become an element of decided interest even to practical meteorologists, as a means of tracing air movements in the stratosphere.

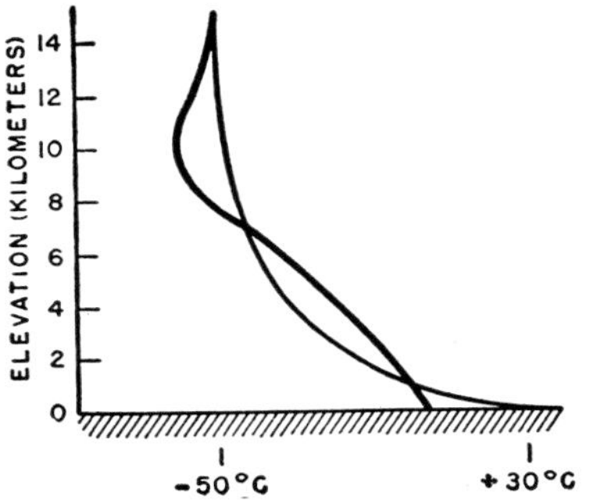

FIG. 1.—The thin line shows how the temperature would drop with height if radiation alone controlled the state of the atmosphere. This is an unstable arrangement resulting in violent overturning. The heavy line shows the result—cooling next to the ground, a moderate temperature drop with elevation in the lower atmosphere (troposphere), then a marked temperature minimum (tropopause), and above that an almost constant temperature.

Until our knowledge of the absorption and emission of radiation by water vapor increases, it is not possible to state what the final decrease of temperature with elevation would be in case of purely radiative equilibrium, but it is probable that in the lowest portion of the atmosphere it would be far steeper than the decrease actually observed (about 6°C in 1 km, or 17°F in 1 mile). As a result of evaporation, the lowest layers would be very nearly saturated with water vapor. It can be demonstrated that a saturated atmosphere in which the decrease in temperature with elevation is more rapid than the value just indicated must be mechanically unstable or, allowing for its compressibility, top-heavy, and thus must tend to turn over. Violent vertical currents (convection) would result, which would carry water vapor and heat from the earth's surface to higher levels.

At these upper levels, the water vapor would condense as the result of expansion cooling, and from there the heat realized through the condensation processes would be sent out to space through infrared radiation. In this way, the free atmosphere would actually give off more heat by radiation than it would absorb, and the loss would be balanced by convective transport of latent heat upward. Thus an atmosphere heated uniformly from below in all latitudes and longitudes, i.e., the atmosphere of a uniformly heated nonrotating globe, would show no sign of organized circulation between different latitudes but would be characterized by violent convection somewhat like that in a kettle of water that is being heated from below.

Rising bodies of air expand with decreasing air pressure and cool as the result of the expansion. In the convectively unstable atmosphere here described, the ascending currents would acquire their momentum in the overheated layers next to the ground and rise beyond the level where they reach temperature equilibrium with the environment. As a result of this overshooting, a layer of minimum temperature would be created at the top of the lower, convective portion of the atmosphere (the troposphere). It follows that troposphere and stratosphere would be separated by a narrow transition zone (tropopause) of rapid temperature increase upward. The effect of convection on an unstable temperature distribution assumed to have been established by radiation is shown in Fig. 1. Figure 2A shows the convective circulation of the troposphere on a uniformly heated nonrotating globe, as described.

MERIDIONAL (NORTH-AND-SOUTH) CIRCULATION

Into this chaotic state, a certain order is brought through the fact that the incoming solar radiation is far from uniformly distributed over the surface of the earth. Because of the low angle of incidence of solar radiation in polar regions, a given horizontal area in high latitudes receives far less solar radiation than an equal area closer to the equator. To determine the consequences of this concentration of heat income in equatorial regions, it is advisable for a moment again to disregard the rotation of the earth and investigate what would happen if the earth stood still but the sun followed its normal path across our sky. In response to the greater heat income in low latitudes, the temperature there would rise until the increased temperature of

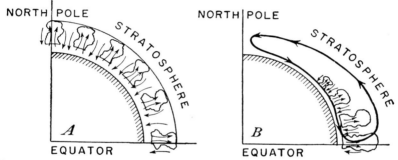

Fig. 2.—Schematic diagrams illustrating: *A*, the heavy, irregular convective activity, accompanied by cumulus and thunderstorm clouds, that would characterize the atmosphere if the sun's heat were applied uniformly everywhere and the earth did not rotate; *B*, concentration of convection in the vicinity of the Equator, north winds near the ground and south winds aloft, which would result on a nonrotating earth with the sun's heat applied mainly in low latitudes, as it actually is on the rotating earth.

the atmosphere in this region would result in increased infrared radiation toward space, capable of reestablishing complete balance with the heat received from the sun. Such an equilibrium corresponds to a far greater increase from pole toward equator in surface temperature or in the temperature of the lower atmosphere than is actually observed. As a result of this heating of the atmosphere in low latitudes, the air there would expand vertically, while the cooling in high latitudes would result in a vertical shrinking.

Thus, at a fixed level of, say, 5 km (3 miles) above sea level, a greater portion of the total atmospheric air column would be found overhead near the equator than near the poles. Since the air pressure always measures the weight of the superimposed air column, it follows that higher pressure would prevail at the 5-km level near the equator than near the poles. Air, like any fluid, tends to move from high to low pressure, and thus the upper atmosphere would be set in motion from the equator poleward. This motion obviously would raise the sea-level pressure near the poles and reduce it near the equator, and as a result the surface air would move from poles toward equator.

Considering the Northern Hemisphere only, it is thus evident that the inequality in heat income between latitudes produces, on a nonrotating globe, south winds aloft and north winds below.   This circulation scheme is illustrated in Fig. 2*B*.   Relatively warm air is carried northward aloft, and relatively cold air is carried southward near the ground.   As a result, it is no longer necessary for the ground and the sea surface in low latitudes to reradiate all the local radiation income to space, but part of this heat is used up in evaporation and is transported northward and upward in the form of latent and realized heat.   It is finally returned to space from higher latitudes or higher elevations through infrared radiation.

Thus a vertical column of air in high latitudes no longer absorbs as much radiation as it emits but is continually suffering a net loss of heat through the combined effects of the various radiative processes.   This loss is balanced by a gain of heat (realized or latent) resulting from the exchange of air with more southerly latitudes and by a gain resulting from realization of latent heat through condensation.   Even though our present knowledge of the radiative processes in the atmosphere is still very incomplete, it is becoming increasingly probable that everywhere in high and middle latitudes the free atmosphere above a shallow layer of air next to the ground is constantly losing heat by radiation.   The significance of water vapor as a carrier of heat (latent) poleward and upward is, for the same reason, becoming more and more appreciated.

As a result of the net transport of heat poleward, the temperature contrast between pole and equator is reduced.

The poleward flow of warm air aloft and the transport toward the equator of relatively cold air next to the ground has the further effect of bringing about a reduction in the vertical temperature decrease.   Thus instability is cut down and vertical convection reduced in middle and high latitudes.

The picture thus obtained, that of an atmosphere heated from below and rising near the equator, chilled and sinking in higher latitudes, requires a few additional comments to eliminate possible misunderstandings.   Even in high latitudes the temperature decreases upward.   Thus, if the air in these regions is steadily sinking, the individual air particles must be getting warmer, which would hardly seem to be in accord with the idea that the air in this region is losing heat through radiation.   It must be remembered, however, that air is highly compressible.   As the air sinks, it comes under the influence of higher pressure.   Owing to the compression, its temperature rises, just as the temperature of a gas in a cylinder rises with compression.   In the atmosphere, the rate at which the temperature of a sinking particle would rise as a result of compression is about 10°C in 1 km.   Thus, if the temperature of a sinking air particle rises only 6°C in 1 km, it means that the particle has given off heat corresponding to a temperature drop of about 4°C.   Dry air, rising through the atmosphere, cools through expansion at the same rate, 10°C in 1 km.   Thus, if a rising current shows a temperature drop of only 6°C per km, heat must have been added, corresponding roughly to a temperature increase of 4°C per km.   If this additional amount of heat is not available, the temperature at fixed upper levels would obviously drop.

In saturated air, rising through the lower atmosphere, a certain amount of heat is made available through the condensation of the water vapor carried along by the current.   Thus the expansion cooling of saturated air is less intense than the expansion cooling of dry air.   The actual rate of cooling varies, but it amounts to about 6°C per km in the lower atmosphere for temperatures around freezing.

The circulation described above between a heat source in low latitudes and at low elevations and a cold source in middle and higher latitudes but distributed over all elevations works on the same principle as a simple heat engine.   In such an engine, the difference between the heat received at the heat source and the heat given off at the cold source is converted into work.   Here it appears as kinetic energy of the wind system.   Unless brakes are applied, the wind must constantly increase in speed.

Such brakes are provided by the friction between the winds and the ground (or sea surface). Since no appreciable changes are wrought in the surface of the earth, the energy of the winds is steadily converted into heat, which ultimately must be radiated back to space. Thus, in the final analysis, the total amounts of radiation received and given off by the earth must equal each other.

The scheme just outlined differs sharply from the true situation observed in the atmosphere, and this discrepancy depends mainly on two factors completely neglected up to the present time. The first of these is the rotation of the earth; the second is the distribution of continents and oceans. The effect of the rotation will be discussed first, the earth's surface still being considered as uniform.

INFLUENCE OF THE EARTH'S ROTATION

The earth does not appreciably change its speed of rotation, and thus it may be assumed that on an average it neither receives momentum from nor gives off momentum to the atmosphere. The rotation, however, does profoundly affect the character of the flow patterns observed in the atmosphere.

To understand the influence of the earth's rotation, consider first this simple experiment: If a marble attached to a piece of string is placed on top of a smooth table and swung around the free end of the string and if then the length of the string is shortened, it will be found that the speed of the marble increases. If the string is shortened to one-half its original length, the speed of the marble is doubled; if the string is reduced to one-third its original length, the speed of the marble is tripled. Thus the product of speed and radius of rotation remains constant during the experiment. This product is usually referred to as the angular momentum (per unit mass) of the marble.

A ring of air extending around the earth at the equator, at rest relative to the surface of the earth, spins around the polar axis with a speed equal to that of the earth itself at the equator. If somehow this ring is pushed northward over the surface of the earth, its radius is correspondingly reduced; and it follows from the principle set forth that the absolute speed of the ring from west to east increases. Since the speed of the surface of the earth itself from west to east decreases northward, it follows that the moving ring, in addition to its northward velocity, must acquire a rapidly increasing speed from west to east relative to the earth's surface.

The principle itself may be illustrated very simply by an extreme and absurd case. The eastward speed of the earth itself at the equator is about 465 m (509 yd) per sec. A ring of air displaced from this latitude to latitude 60, where the distance from the axis of the earth is only half that at the equator, would appear in its new position with double the original absolute velocity, or 930 m (1,017 yd) per sec. Since the speed of the earth itself at this latitude is only half what it is at the equator, or 232 m (254 yd) per sec, it follows that a ring of air thus displaced would move eastward over the surface of the earth with a relative speed of 698 m per sec (about 1,560 mph). Obviously such wind speeds never occur in the atmosphere, one reason being the effect of frictional forces, another the fact that large-scale atmospheric displacements never are symmetric around the earth's axis or as large as those indicated.

It is apparent, however, that this tendency toward the establishment of west winds in northward-moving rings of air and of east winds in southward-moving rings must modify the previously described meridional (north-south) circulation scheme considerably. This is best seen if one assumes that the meridional circulation scheme characteristic of a nonrotating earth (Fig. 2*B*) is suddenly set in operation on a rotating earth in which the atmosphere previously was at rest relative to the ground. The moment the circulation begins, west winds (relative to the earth) would begin to develop in the upper atmosphere, with a slight component northward, and east winds in the lower atmosphere, with a slight component southward.

In this scheme, ground friction plays a basic role, since it prevents the development of excessive east winds in the surface layers. The upper atmosphere, in which west winds prevail, is not in direct contact with the earth's surface; however, mixing of air between the upper and lower strata must reduce the west winds aloft as well as the east winds below. Since the momentum of the east winds also is reduced from below, through the effect of ground friction, it is apparent that the mass of the west winds aloft would far exceed the mass of the easterlies near the surface. Figure 3*A* illustrates what the velocity distribution would be a short time after the rotation began.

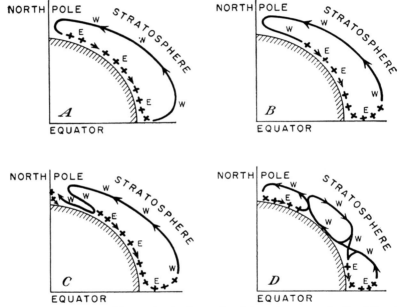

Fig. 3.—Why the earth's rotation leads to a breakdown into several cells of the simple meridional (north-south) circulation represented in Fig. 2*B*. *A*, A short time after the meridional circulation indicated by the arrows is set in operation, west winds appear aloft, east winds below. *B*, Gradually the upper west winds are brought down to the ground near the Pole, and the east winds rise near the Equator. *C*, The west winds are retarded by friction and seek their way northward, but cooling and sinking continue next to the Pole. Finally, *D*, a complete three-cell circulation system develops.

Certain features of this picture agree well with observed conditions. Above 4 to 5 km (2½ to 3 miles), westerly winds prevail in all latitudes. At sea level, easterly wind components are normally observed between latitude 30°N and 30°S. Other belts of easterly wind components are observed in the polar regions, north of latitude 60°N and south of latitude 60°S. Unexplained, however, is the fact that in each hemisphere westerly winds prevail also at sea level within a broad belt between latitudes 30 and 60°, approximately.

It is fairly easy to see that the theoretical model in Fig. 3*A*, characterized by east winds everywhere in the surface layers, is physically impossible as a steady state. If east winds prevailed in all latitudes, friction between the atmosphere and the solid earth would constantly tend to reduce the rotation of the earth. On the other hand, the atmosphere would constantly gain momentum from the earth. Sooner or later a state of equilibrium would be established in which the atmosphere would neither gain nor lose momentum through contact with the earth—an equilibrium that is known to prevail, since the rotation of the earth for practical purposes can be regarded as constant. Such an equilibrium requires that the retarding influence of the east winds must be offset by the accelerating influence of a belt or belts of west winds, also in the surface layers. It is clear, however, that this argument is incapable of determining the number, width, and strength of the required west-wind belts. It is the purpose of the four diagrams in Fig. 3 to explain why the initial meridional circulation, under the influence of the earth's rotation, necessarily must break down into at least three separate cells on each hemisphere.

In order to understand the successive stages of development indicated in Fig. 3, it is first necessary to discuss, in some detail, the effect of the rotation of the earth on the relative motion of air over its surface. Wherever a ring of air parallel to a latitude circle is rotating more rapidly than the surface of the earth itself, it is acted upon by an excess of centrifugal force which tends to throw the ring away from the axis of the earth, which in the Northern Hemisphere means southward.* If the ring rotates with the same speed as the earth (*i.e.*, if it is at rest relative to the surface of the earth), this excess of centrifugal force vanishes. If this were not the case, any object resting on the surface of the earth would be thrown toward the equator. A ring of air rotating more slowly than the earth itself, and hence appearing as an east wind relative to the earth, suffers from a deficiency in centrifugal force and tends to move toward the axis of the earth, *i.e.*, in this hemisphere, toward the north. To keep a west-wind belt from being thrown southward, the atmospheric pressure must be higher to the south than to the north of the ring (in the Northern Hemisphere), thus producing a force directed northward and capable of balancing the excess centrifugal force. If no such pressure force (gradient) is available, the ring will be displaced slightly southward until enough air has piled up on its south side to bring about the required cross-current pressure rise to the south and equilibrium. The total displacement needed for this purpose is usually quite small as compared with the width of the current.

To keep an east-wind belt in equilibrium, the atmospheric pressure must be higher on the north side than on the south side (in the Northern Hemisphere), so that the resulting pressure force balances the deficiency in centrifugal force acting on the ring. It has already been brought out that, in the Northern Hemisphere, air moving northward tends to acquire a velocity eastward, while air moving southward tends to acquire velocity westward. To offset this tendency toward deflection eastward,† a northbound current of limited width piles up air to the east and creates higher pressure to the east than to the west, while the reverse applies to a southbound current.

All these results may be generalized so as to apply to any wind direction. It is thus found that in the Northern Hemisphere steady winds always blow in such a fashion that the air pressure drops from right to left across the current for an observer facing downstream. The stronger the current flows, the steeper the drop in cross-current pressure. If, in any horizontal plane, lines of constant air pressure (isobars) are drawn, it may be seen that the air follows the isobars and moves counterclockwise

---

* That part of the centrifugal force (per unit mass) which corresponds to the earth's own rotation is balanced by a component of the earth's gravitational attraction. The resultant of this component of the centrifugal force and of the earth's total true gravitational attraction is perpendicular to the earth's surface and constitutes what is normally referred to as gravity.

† The method of compensation here described is obviously impossible for circumpolar rings of air. Hence rings of air displaced northward acquire west-wind tendencies, south-bound rings east-wind tendencies. as brought out previously.

around regions of low pressure (cyclones) and clockwise around regions of high pressure (anticyclones). In the Southern Hemisphere, the direction of motion around highs and lows is reversed.

It is apparent from the preceding reasoning that the relationship between wind and horizontal pressure distribution is truly mutual; a prescribed pressure distribution will gradually set the air in motion in accordance with the law set forth; likewise, if somehow a system of horizontal currents has been set up in the atmosphere, the individual current branches will very quickly be displaced slightly to the right (in the Northern Hemisphere) until everywhere the proper cross-current pressure drop from right to left has been established. Owing to the ease with which the atmosphere thus builds up the cross-current pressure drop required for equilibrium flow, the reasoning just outlined merely helps in understanding why the pressure in the Northern Hemisphere always rises from left to right for an observer looking downstream but does not by itself indicate that one current pattern is more likely to be established than another. To establish the character of the current patterns, either the pressure distribution must be known, or additional physical principles must be utilized.

It is now possible to return to a discussion of the circulation development in Fig. 3. In an axially symmetric atmosphere, such as the one here discussed, the absolute angular momentum of individual parcels of air does not change except through the influence of frictional forces. Under these conditions it is evident that the meridional (north-south) movements indicated in Fig. 3A must gradually redistribute the absolute angular momentum so as to create west winds next to the ground in the polar regions, and east winds aloft over the equator. This is the state illustrated in Fig. 3B. If the meridional circulation is slow, the pressure distribution in the atmosphere must constantly adjust itself fairly closely to the prevailing zonal winds. Thus, in Fig. 3B, there would be a sea-level-pressure maximum at the transition point between the east winds in low latitudes and the west winds farther north. This latter belt of west winds can continue its southward displacement as long as it is acted upon by an excess of centrifugal force. However, part of the air in this west-wind belt must steadily lose momentum through frictional contact with the ground. Under the influence of the resulting deficiency in centrifugal force, this shallow portion of the belt next to the earth's surface must seek its way northward, as indicated in Fig. 3C. Since air continues to cool and sink next to the pole, it follows that the retarded west winds, for purely dynamic reasons, are forced to escape aloft some distance from the pole. Finally a cellular state develops, as indicated in Fig. 3D.

The Three Hemispheric Circulation Cells. Up to this point the breakdown of the original simple meridional circulation scheme has been treated as a purely dynamic effect. It now becomes necessary to investigate whether the final mean circulation scheme is compatible with the thermal processes of the atmosphere. For this purpose we must fall back upon our as yet very incomplete knowledge of radiative processes in the atmosphere.

It was stated previously that, practically everywhere above a shallow layer next to the ground, the free atmosphere suffers heat losses through the combined effects of the various radiative processes to which it is subjected. In middle latitudes at 2 or 3 km above sea level, these losses would produce a cooling at fixed levels of the order of magnitude of perhaps 1 or 2°C per day.

Thus, with the possible exception of equatorial regions, the free atmosphere everywhere serves as a cold source (condenser) for the circulation engine. Heat sources are located at the earth's surface and above the surface layers in those regions where latent heat is released through condensation. The release of latent heat through scattered, unorganized convective action is of little consequence for the atmospheric heat engine, but those regions where there is organized ascending mass motion with attending condensation and release of latent heat become important heat sources capable of driving the atmospheric circulation. It follows that the atmosphere itself to a considerable extent has the power to regulate the distribution of its heat sources and also, through dynamically produced temperature changes, to modify the intensity of its cold sources.* The latter effect follows from the fact that a temperature change modifies the emission, but not the absorption, of a given parcel of air.

It follows from the preceding discussion that the air ascending in the equatorial belt and spreading poleward at upper levels must lose heat fairly quickly and that parts of it must reach ground again when it is in the horse latitudes (in the neighborhood of 30°N or S). A branch of the descending air spreads poleward; another branch equatorward. The poleward branch will appear as a west or south-west wind, and must eventually meet the cold air seeping equatorward from the pole. Forced ascent results, requiring a heat source which is provided through the release of latent heat in the ascending air.

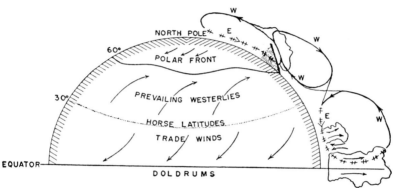

Fig. 4.—The final cellular meridional circulation on a rotating earth: Convection near the Equator, a clear zone of descending air motion north of it (about latitude 30° N), and heavy slanting cloud masses with accompanying precipitation in the polar-front zone (55 to 60° N).

Thus the original single meridional circulation cell characteristic of each hemisphere in the original scheme breaks up in such a fashion that one cell extends from the equator to the horse latitudes, another from about latitude 60° poleward. The resulting scheme of circulation is illustrated in Fig. 4. In both extreme cells, the heat sources are found at low levels, the cold sources well distributed along the vertical. Looking eastward at a meridional vertical section through the Northern Hemisphere, one would observe counterclockwise circulation in each of these extreme cells. These circulations are direct in the sense that they carry heat from heat source to cold source, all the while transforming a small fraction of the heat energy received into kinetic energy. The direct cell to the south may be called the trade-wind cell, since the southward-moving lower branch of this cell is responsible for the steady northeast trade winds just north of the equator. For reasons that will appear, the northern cell will be referred to under the name *polar-front cell*.

It now seems possible to offer an explanation also for the circulation in middle latitudes. In the two direct circulation cells to the north and to the south, strong westerly winds are continually being created at high levels. Along their boundaries with the middle cell, these strong westerly winds generate eddies with approximately

* This would become significant in an atmosphere free from water vapor.

vertical axes.   Through the action of these eddies, the momentum of the westerlies in the upper branches of the two direct cells is diffused toward middle latitudes, and the upper air in these regions is thus dragged along eastward.   The westerlies observed in middle latitudes are thus frictionally driven by the surrounding direct cells.   The excess of centrifugal force acting on these upper west winds of middle latitudes forces the air southward, but equilibrium is never reached, since the air still farther to the south, instead of piling up and thus permitting the establishment of an adequate cross-current pressure drop, cools through radiation and sinks to lower levels.

It has already been pointed out that the air that sinks in the horse latitudes spreads both poleward and equatorward.   The poleward branch must obviously appear as a west wind accompanied by a cross-current pressure drop northward.   It continues to move northward, since the retarding influence of ground friction continually keeps the surface westerlies below the intensity required for equilibrium.   Aloft the situation is reversed, since there the frictionally driven winds always remain slightly in excess of the value required to balance the cross-current pressure drop.   The result is a motion northward near the surface, a slight motion southward aloft.

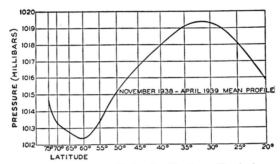

Fig. 5.—Pressure profile for the Northern Hemisphere.

Thus, to an observer looking eastward, the meridional circulation in middle latitudes is clockwise and opposite to the direct counterclockwise circulations to the north and south.   This middle cell serves as a necessary brake on the general circulation driven by the direct-working heat engines farther to the north and to the south. It has already been stressed that part of the (relative) momentum eastward generated aloft in the direct cells to the north and south spills over into the middle cell (through large-scale horizontal friction), where it is destroyed through slow southward displacement.   The surface westerlies established in this middle cell serve the additional purpose of balancing the retarding force exerted on the earth itself by the easterlies farther south and north.

It follows from the previously established rule for the relation between wind and pressure that the sea-level pressure must drop from the pole southward to about 60°N, then rise to about 30°N, and finally drop from there on toward the equator.   At higher levels, where the wind is everywhere westerly, the pressure rises steadily from the pole toward the equator.   Thus at sea level a trough of low pressure is established in latitude 60° and a ridge of high pressure in the vicinity of 30°N.   The observed mean pressure as a function of latitude for the winter 1938–1939, shown in Fig. 5, is in good agreement with the result of the previous analysis.

Climatic Zones.   By this time it becomes possible to talk of climatic zones.   The ascending motion in the equatorial region will obviously be attended by a great deal of convective activity, violent because of the extreme instability of the vertical temperature drop. Owing to the absence of horizontal contrasts, this convective activity will follow the sun with a great deal of regularity and produce the heavy afternoon showers characteristic of this climatic zone. The air descending in the horse latitudes will have lost a great deal of its moisture, and the descending motion leads to warming by compression at intermediate levels. Thus, in spite of relatively high surface temperatures, the vertical temperature drop is fairly weak and the air itself so dry as to prevent convection. This region, then, will be characterized by an arid, or semiarid, climate.

The region of ascending motion around 55 or 60°N will obviously be characterized by a great deal of precipitation from the ascending air. It also is evident that this precipitation will be of an entirely different nature from that observed in the tropics. Because of the southward movement of cold polar air along the ground and the northward movement of warm and relatively moist subtropical air aloft, the vertical temperature drop in this region will be too weak to permit violent convection, and the precipitation must be associated with the orderly ascent of moist, warm air over the wedgelike tongues of polar air that extend southward. In this region, cold polar air and moist subtropical air converge next to the earth's surface. This, then, must be a region in which the surface isotherms, or lines of equal temperature, are constantly being crowded together and where abrupt transitions from subtropical to polar air conditions may be observed.

This transition zone, incessantly regenerated and incessantly destroyed, is in modern meteorology referred to as the *polar-front zone.* Here cold and warm air masses are in constant battle. This battle expresses itself through the formation of quasi-horizontal waves that normally progress from west to east along the front. The length of individual waves varies between 1,000 and 5,000 km (about 600 and 3,000 miles). Because of the constant battle of air masses, the polar-front region is characterized by strong temperature contrasts and a rapid succession of dry and wet spells.

The polar regions, characterized in the main by the descent of air that has lost most of its moisture in the ascent over the polar front, are characterized by cold arid or semiarid climate.

PLANETARY FLOW PATTERNS AND THE STABILITY OF ZONAL CIRCULATION

The preceding analysis indicates that the polar front tends to occupy a mean position parallel to a latitude circle. The Southern Hemisphere, with its practically uniform water cover, is probably to a large extent characterized by such a zonal arrangement of the different wind belts and of the polar front. Thus it is fairly well established that the storms (polar-front waves) of high southerly latitudes move with far greater regularity from west to east than the storms of the Northern Hemisphere. This difference in behavior results from the influence of the nonzonal distribution of oceans and continents in the Northern Hemisphere, which leads to a breakdown of the polar front. The breakdown is reflected also in the sea-level-pressure distribution.

Figure 6 shows the practically zonal normal pressure distribution observed in the Southern Hemisphere, and Fig. 7 shows the extremely asymmetric normal pressure distribution characteristic of the Northern Hemisphere. Both figures refer to winter conditions. It is evident that, particularly in the Northern Hemisphere, the belts of high and low pressure have broken down into separate closed centers of high and low pressure. These centers are usually referred to by the somewhat misleading name *centers of action.* During the winter season, at least five such centers may be observed in our hemisphere—the Icelandic and Aleutian lows, the Pacific high, the Bermuda

or Azores high, and the Asiatic high.   The daily synoptic weather charts for the
Northern Hemisphere show a great many more moving high- and low-pressure systems,
associated with the battle between cold and warm air masses along the polar front.
The construction of mean charts permits the elimination of these moving disturbances
and brings out clearly the quasi-permanent character of the centers of action listed
above.

Of late such mean pressure charts for periods of a week, a month, or a season have
received a great deal of attention.   When a sequence of weekly mean charts is studied

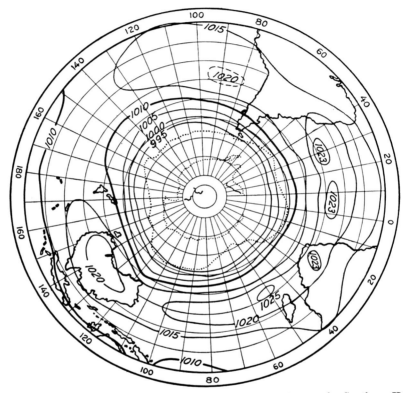

Fig. 6.—Normal sea-level-pressure distribution (millibars) over the Southern Hemisphere in July.   Compare this with Fig. 7 and note the great regularity and symmetry of
the pressure distribution in the Southern Hemisphere, due, presumably, to the absence of
large land masses.   (*After Shaw.*)

it is found that the centers of action may move very slowly, eastward or westward,
several weeks in succession.   Frequently one or several of these centers of action may
break up into several parts.   With the breaking up of the zonal pressure distribution
into separate centers of action or cells goes a breaking up of the mean polar front into
two, three, or four separate portions, usually extending from southwest to northeast.
The position of these separate mean fronts is determined by the size, development, and
position of the individual centers of action.   Since, on the other hand, most of the
storms that control weather in our latitudes move along the frontal zones thus established, it is easy to see why it is imperative to understand the factors that lead to the

breakdown of the ideal zonal circulation into individual centers of action. As a first step, it is necessary to discuss briefly the possible types of nonzonal steady flow patterns that can exist on the earth.

It is a well-known fact in mechanics that a rotating rigid body will not change its rate of rotation unless it is subjected to a force that produces a moment (torque) around the axis of rotation. If the body does not rotate initially, a torque is needed to set it in rotation. This simple principle can be applied to vertical columns of air as they move over the surface of the earth, but in so doing one must keep in mind that

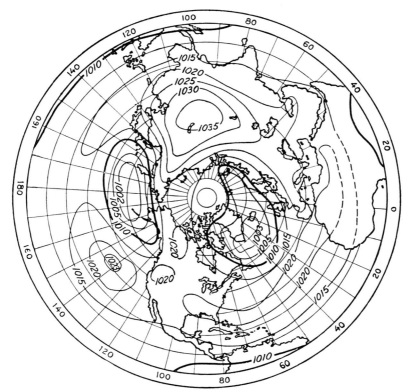

Fig. 7.—Normal sea-level-pressure distribution (millibars) over the Northerm Hemisphere in January. (*After Shaw*).

Note the great irregularity of the pressure distribution in the Northern Hemisphere as compared with that in the Southern (Fig. 6), the two deep cyclonic (low-pressure) centers over the northernmost oceans, and the well-developed anticyclone (high-pressure center) over Asia.

the earth itself is everywhere in a state of rotation around the vertical. To demonstrate this rotation, it is sufficient to refer to the case of an ordinary freely suspended pendulum, which swings back and forth in a vertical plane. If a pointer is attached to the pendulum weight and permitted to trace the path of the pendulum in a sand bed just below, it will be found that the plane of the pendulum slowly turns clockwise (in the Northern Hemisphere). At the pole the plane of oscillation would make one complete turn (360 deg) in 24 hr; in latitude 30° it turns more slowly, making one complete turn in 48 hr.   It is well known that the time of rotation of the pendulum plane may be obtained by dividing 24 by the sine of the latitude.

It is obvious that this rotation of the plane of oscillation simply means that below our feet the earth rotates counterclockwise (clockwise in the Southern Hemisphere), rapidly at the poles and more and more slowly as we approach the equator. Vertical air columns that move from one latitude to another tend to take their rotation with them. Thus a current of air originating in high northerly latitudes, where the cyclonic (counterclockwise) rotation of the earth is strong, and moving southward to a latitude where the cyclonic rotation of the earth is weak, will possess an excess cyclonic rotation around the vertical over that of the earth itself when it arrives at its destination. This excess rotation can express itself in two different ways or in a combination of both, as illustrated in Fig. 8.

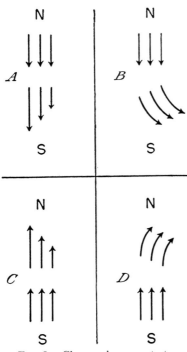

The current can follow a straight path but develop a shear so that the right edge of the current (looking downstream) moves faster than the left edge (Fig. 8*A*). In the atmosphere, there are several influences at work that normally prevent the development of strong shear zones and lead to the establishment of currents of fairly uniform velocity cross stream. In that case, the current is forced to bend around cyclonically, as indicated in Fig. 8*B*. Figures 8*C* and 8*D* illustrate the corresponding effects in a northbound current developing anticyclonic (clockwise) rotation.

Fig. 8.—Changes in current structure resulting from southward and northward movements in the Northern Hemisphere: *A*, A narrow current moving southward would, if forced to follow a straight path, acquire a strong cyclonic (counterclockwise) shear; *B*, if free to seek its own path, it would curve around cyclonically; *C*, a narrow current moving northward would, if forced to follow a straight path, acquire a strong anticyclonic (clockwise) shear; *D*, if free to seek its own path, it would curve around anticyclonically. In all cases the current would pile up air on the right-hand side looking downstream, but the slight deflections needed to bring this about would not materially affect the flow patterns illustrated.

When the mean zonal circulation was discussed, a uniform seeping southward of cold air from high latitudes was assumed. In that case, there is of course no possibility for the establishment of cyclonically curved streamlines, and the excess of cyclonic rotation in the southward-moving belt of cold air should therefore lead to strong cyclonic shear in the northern belt of easterlies. Likewise, in the free atmosphere, where rings of air are displaced northward toward regions where the earth itself rotates more rapidly counterclockwise (cyclonically) around the vertical, the resulting deficiency in rotation of the displaced air columns must express itself in the form of anticyclonic

shear.   However, in both cases it can be said that the statistical mean northward and southward velocities associated with the general circulation between latitudes are so weak that frictional forces of all kinds have ample time to prevent the establishment of sharp shear zones. The situation is different in the case of air currents that for some reason or other are definitely deflected from their east-west motion. In these currents, latitude changes occur so quickly that the frictional forces have inadequate time to act. In such currents, the excesses or deficiencies in rotation have a tendency to produce curved flow patterns. Initially straight currents from the north curve around cyclonically; currents from the south curve around anticyclonically.

It is particularly interesting to apply these results to a study of the stability of west-wind or east-wind belts of limited width. If a current from the west at some point in its path is subjected to a cyclonic torque that gives it a slight deflection northward, it follows that the current from then on will head toward higher latitudes, where the rotation of the earth itself around a vertical becomes stronger and stronger. Thus the relative cyclonic rotation (curvature) which the current acquired at the initial point of deflection will decrease and eventually, after sufficient displacement northward, change into an anticyclonic curvature. The current will then finally bend back toward its equilibrium position. As it moves southward, it will pick up cyclonic relative rotation (curvature), and the net effect will be a sinusoidal, or wavelike, oscillation of the west-wind belt around a certain mean latitude, such that the current will form a series of standing waves downstream from the point at which it was disturbed initially.

The amplitude of these waves depends upon the intensity of the initial disturbance, but the wave length depends principally on the strength of the west-wind belt. The stronger the wind, the longer the wave length. For the wind velocities prevailing in the upper troposphere in our middle latitudes in wintertime, this wave length is of the order of magnitude of 5,000 km (3,000 miles). These "resonance" waves give us a length scale for the large semipermanent centers of action into which the previously described symmetric zonal circulation actually breaks up.

If the same type of reasoning is now applied to a narrow current from the east, which at a given point in its path is given a slight cyclonic rotation (cyclonic curvature), it follows that the current from then on moves slightly southward. However, the farther south it moves, the stronger will be the cyclonic rotation of the current, since it is constantly moving toward latitudes where the earth's own cyclonic rotation around the vertical becomes weaker and weaker. Thus the current is deflected farther and farther away from its equilibrium latitude. Finally it will have turned around completely and will then appear as a west wind, but with sufficiently strong cyclonic curvature to return to its original path.

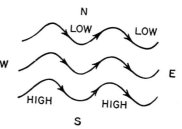

Fig. 9.—Diagram of waves on a broad west-wind belt, showing troughs of low pressure and ridges of high pressure.

If the current had originally been deflected northward, it would describe a complete anticyclonic circuit. This may be of importance in connection with the breaking up of the high-pressure belt around latitude 30°. It is fairly evident that the cold easterly winds to the north, because of the large body of cold air over the arctic, are constrained to break up into cyclonic vortices.

The analysis shows that easterly winds are unstable and tend to break up into large cyclonic or anticyclonic eddies. The dimensions of the eddies thus formed increase with the velocity of the east wind itself, and they agree reasonably well with the dimensions of the cyclonic centers of action referred to above. Figure 9 illustrates a steady resonance wave on a broad west-wind belt, and Fig. 10 the trajectory of a deflected east-wind belt of narrow width.

To understand completely the behavior of cold currents from the north a further reference should be made to the spinning-marble experiment discussed earlier. It was pointed out that, by a shortening of the string, the marble could be made to spin faster and by a lengthening of the string to spin more slowly. In the same way, the outer edge, or periphery, of a rotating column of air that is stretched vertically and shrunk horizontally will spin around more rapidly; if the column shrinks vertically and stretches horizontally, it will spin more slowly. In applying this result to vertical columns in the atmosphere, it is necessary to consider the absolute rotation. It thus follows that air columns that stretch vertically must acquire an excess (cyclonic) rotation relative to the earth, while air columns that shrink vertically acquire a deficient (anticyclonic) rotation relative to the surface of the earth.

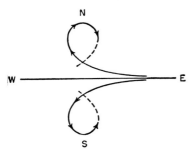

FIG. 10.—The upper curve represents the path of a narrow east-wind belt that has received a small initial deflection northward. The lower curve represents the path of a current that has received an initial deflection southward. Since a current in a state of steady motion cannot intersect itself, the analysis suggests that narrow eastwind belts are unstable, resulting in the intermittent formation of large vortices.

It is a well-established fact that cold currents from the north gradually sink and spread out next to the surface of the earth. This sinking is most marked along the right-hand edge of the cold current (for an observer facing downstream). Thus the left-hand branch of the current curves around cyclonically as a result of the decrease in latitude, but the right-hand branch, in which strong sinking occurs, curves around anticyclonically. As a result, the deflected cold current spreads south in a fanlike fashion.

If for some reason the westerlies of middle latitudes are disturbed and a quasistationary wave pattern is established, the pressure will drop wherever a wave trough is being established. Cold air from the north will be pulled into these low-pressure troughs. Hence intermittent outbreaks of cold polar air from the previously undisturbed east-wind belt to the north are likely to occur wherever there is a trough toward the south in the westerlies. Corresponding outbreaks of warm and moist air from the southern belt of easterlies tend to occur where the westerlies are deflected northward. One obtains in this way the pattern of flow illustrated in Fig. 11, which in several respects well describes the observed flow pattern in the atmosphere. It is obvious that this pattern leads to a breaking up of the polar front into separate portions that run roughly from southwest to northeast. Along these individual polar fronts, waves form that move northeastward and gradually die out in the central low-pressure areas to the north.

Figure 12 represents a modification of Fig. 11 obtained by assuming that the crests and valleys of the quasi-permanent waves in the west-wind zone extend from southwest to northeast, parallel to three of the four principal boundaries between oceans and continents in the Northern Hemisphere. This theoretical picture agrees remarkably well with the observed mean pressure and frontal distribution in any one of the principal frontal zones of the Northern Hemisphere.

To each speed of the westerlies corresponds one definite resonance wave length, and this wave length increases with increasing strength of the westerlies. Thus it appears that the number of points where polar air is simultaneously injected into the

west-wind zone depends upon the strength of the westerlies and decreases as the westerlies increase. In the Southern Hemisphere, where land masses are relatively insignificant or, in the case of the antarctic continent, symmetrically distributed, these
injection points may occur in any longitude, and thus the mean wind and pressure

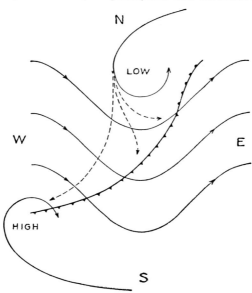

'Fig. 11.—Breakdown of the zonal polar front and establishment of a typical branch
front through the injection of polar air into a trough in the westerlies. Subsiding, under-
cutting branches of the polar air take on anticyclonic curvature (broken lines); nonsub-
siding branches will curve around cyclonically and form the left edge of the cold wave.
The polar front is indicated by a barbed line.

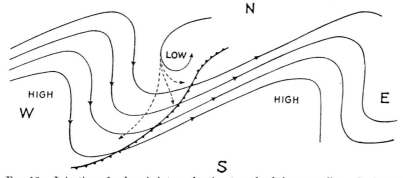

Fig. 12.—Injection of polar air into a slanting trough of the westerlies. Such troughs
may easily form as a result of thermal perturbations of the westerlies set up along the
slanting eastern coast lines of Asia and North America-Greenland.

distributions should appear fairly uniform around the globe. In the Northern
Hemisphere, on the other hand, preferential injection points are established, and the
final mean-pressure distribution is thus far from symmetric around the axis of the
earth.

INTENSITY FLUCTUATIONS IN ZONAL CIRCULATION
AND THE CIRCULATION INDEX

In view of the fact that the dimensions and positions of the circulation patterns associated with strong and weak circulation differ in a characteristic fashion, it is evidently important to develop a simple index to the intensity of the zonal circulation. It has been pointed out previously that the surface westerlies in middle latitudes form a part of a reverse meridional circulation cell which is driven by the direct cells to the north (polar-front cell) and to the south (trade-wind cell). Because of this frictional drive, it is reasonable to assume that these surface west winds, at least qualitatively, furnish a good measure for the variations in the general zonal circulation of the atmosphere. A simple measure of the intensity of these sea-level westerly winds may be obtained by taking the difference between the mean pressure observed in latitude 35°N, near the center of the subtropical high, and the mean pressure observed in latitude 55°N, just south of the pressure trough normally prevailing in the vicinity of latitude 60°N. This mean-pressure difference is very nearly proportional to the mean-wind component from the west prevailing within this zone, if surface frictional forces are disregarded. In view of the variations in the difference between the mean temperature of an air column in latitude 35°N and that of another

Fig. 13.—Zonal circulation variations during the period November, 1936 to May, 1938. This index to the zonal circulation is computed weekly as the difference between the mean pressure at latitude 35° and the mean pressure at latitude 55°. The greater this index is, the stronger the prevailing west wind in middle latitudes. Note the large amplitude of fluctuations in winter, particularly the occurrence of long trends downward or upward through several weeks in succession.

air column in latitude 55°N, it is, unfortunately, impossible to use the same mean-pressure difference as a quantitative measure of the mean west-wind component at higher levels within the same zone, but it should at least serve as a qualitative index to the variations in circulation intensity at higher levels.

Mean weekly values of this circulation index have been computed since 1936 and show amazingly strong fluctuations from week to week in the circulation intensity. During the winter season, these fluctuations range from an index value of about 15 mb to one of about −5 mb, the latter value indicating an actual reversal of flow in the surface layers (east wind). These fluctuations are well illustrated by the curve in Fig. 13. The variations in circulation intensity are obviously fairly irregular, but it is also evident that trends persisting through 3 or 4 weeks are fairly common during the winter. The irregularity of the variations in circulation intensity is sufficiently pronounced to eliminate effectively all possibility of long-range forecasting on the basis of periodicities, but the persistence tendency in the index curve is sufficiently high to permit judicious extension (extrapolation) of a trend for a week at a time. During the summer, the fluctuations in circulation intensity are somewhat smaller and also more irregular. It is hard to see how adequate short-term long-range weather forecasts can be developed until there is an adequate understanding of these amazing fluctuations in the zonal circulation intensity.

It has already been mentioned that the energy of the westerlies depends upon the meridional circulation between heat sources and cold sources in the two direct meridional cells, to the north and to the south of the westerlies. It would therefore seem probable that a satisfactory understanding of the variations in the zonal circulation intensity cannot be reached until fairly complete temperature and humidity data from the upper atmosphere over the entire Northern Hemisphere are available in the form of daily routine measurements. No adequate physical theory is available at the present time from which the fluctuations in circulation intensity may be computed, but recent studies suggest that these fluctuations may be associated with the intermittent establishment of a direct inflow of deep moist air from the equatorial trade-wind belt into the westerlies of middle latitudes. There is good reason to hope that the problem of the circulation fluctuations will be brought much closer to its solution within the next few years.

INFLUENCE OF LAND MASSES AND OCEANS ON THE CIRCULATION PATTERN

In older writings, it is sometimes stated that the difference between land climate and sea climate is caused by the difference in specific heat between water and solid rock. It is more correct to emphasize that the upper layers of the ocean are nearly always in a state of violent stirring whereby heat losses or heat gains occurring at the sea surface are distributed throughout large volumes of water. This mixing process sharply reduces the temperature contrasts between day and night and between winter and summer.

In the ground, there is no turbulent redistribution of heat, and the effect of molecular heat conduction is very slight. Thus violent contrasts between seasons and between day and night are created in the interior of continents. During the winter, the snow cover that extends over large portions of the northern continents reflects back toward space a large part of the sparse incident solar radiation. For these various reasons, the northern continents serve as efficient manufacturing plants for dry polar air. The polar air cap is no longer symmetric but is displaced far to the south, particularly over the interior of Asia. This in turn means that the mean polar-front zone is deformed and tends to follow the boundaries of the northern continents, extending northeastward along the Pacific coast of Asia, then southeastward along the Rockies, and finally northeastward along our Atlantic coast toward Iceland. Our knowledge of the upper westerlies at high levels is still very incomplete, but it appears that they, too, are displaced southward over the Asiatic continent.

The polar air that is being steadily manufactured over the interior of Asia generates a polar front that in the main follows the Pacific coast line of that continent. Just as pure easterly winds are established behind the mean polar front on a symmetric globe, northeasterly winds will be established to the north and west of the Asiatic polar front and southwesterly winds south and east of the same front. This arrangement of currents is, however, highly unstable. It has already been brought out that currents from the north tend to assume cyclonic curvature unless they have an opportunity to sink and spread out. The cold currents from the north behind the Asiatic polar front cannot spread out toward the interior where still deeper masses of cold air are stored. Hence they must stream south over the Pacific, and these intermittent outbursts help to maintain a deep cyclonic vortex off the Pacific coast of Asia.

The air masses that are found south and east of the Asiatic frontal zone and at higher levels stream toward the Aleutian Islands. As they move northward, they must gradually curve around anticyclonically (clockwise). It has already been pointed out that the wave length of such an oscillating west-wind belt increases with increasing wind velocity. Hence, if the southwest or west-southwest winds off the coast of Asia are sufficiently strong, they will follow the boundaries of the North Pacific Ocean, curving around anticyclonically as a result of their northward displacement. In this case, a single frontal zone is established, along which storms move rapidly in an eastward direction, crossing the Pacific coast of North America in fairly high latitudes (British Columbia, Washington, Oregon).

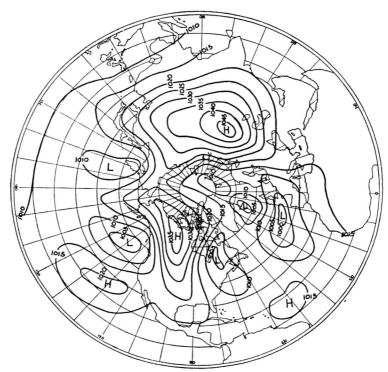

Fig. 14.—Mean sea-level-pressure distribution (millibars) during a week of slow zonal circulation, Nov. 14 to 20, 1937. Note two separate low-pressure centers in the Pacific, a well-developed continental high in North America, the split character of the Icelandic low, and the displacement toward Europe of the Asiatic high.

On the other hand, as the southwesterly winds off the coast of China grow weaker, they tend to curve around anticyclonically much more sharply. A trough of low pressure may then be created in the middle or eastern part of the Pacific, and thus another injection point for polar air may be established. A new polar front, extending across the mid-Pacific from southwest to northeast, may thus be established by purely dynamic means, whenever the general circulation slows down. Storms (waves) traveling northeastward along this polar front bring moist southerly winds to California. These moist air masses are trapped between the Pacific polar front on the one hand and the mountains and the cold air masses over the continent on the other. They are therefore forced to ascend, and in so doing they probably produce a large portion of the winter rains in southern and central California.

As a result of the production of polar air over Greenland and the North American continent, another polar-front zone is established over the eastern states, extending from the lower Mississippi Valley over New England, Newfoundland, and the northern North Atlantic toward northern Norway. When the circulation in middle latitudes is very weak, a second Atlantic polar front may be established over western Europe. This doubling of the Atlantic polar front appears to be a more infrequent phenomenon than the doubling of the Pacific polar front.

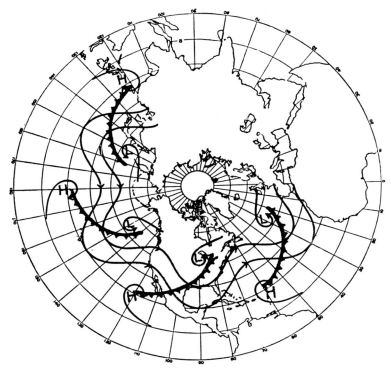

Fig. 15.—An example of the theoretical planetary flow pattern for weak zonal circulation. Compare this diagram with Fig. 14 and note the presence in both of a split Aleutian low and a split Icelandic low with one branch centered over eastern North America. In comparing the two diagrams it should be remembered that the wave pattern of the westerlies indicated here is in reality Fig. 14 obscured by shallow cold-air anticyclones and that it would appear clearly on a corresponding chart for the 3-km level.

Figures 14 and 15 represent an attempt at a comparison of the observed pressure distribution during a period of very weak zonal movement in middle latitudes with the computed air trajectories during a period of weak circulation. The observed pressure distribution (Fig. 14) shows that the Aleutian low has split into two separate centers, one off Kamchatka and one in the Gulf of Alaska. Likewise, the Icelandic low has split up, with one center over Labrador and a second center in the form of a long trough extending southwestward from Spitsbergen to a point off Ireland. The center positions of these observed cyclonic whirls agree fairly well with the computed circulation centers in Fig. 15.

In computing this last circulation diagram, it was assumed that the easterlies to the north have a mean velocity of 8.9 m per sec, the westerlies in middle latitudes a mean velocity of 15.5 m per sec, and the easterlies still farther to the south a mean velocity of about 13.3 m per sec. These velocities were chosen so as to give a proper wave length for the westerlies and the proper dimensions for the cyclonic and anti-cyclonic eddies to the north and south. Thus the only claim that can be made for this theoretical analysis is that, with reasonable values for the prevailing zonal winds,

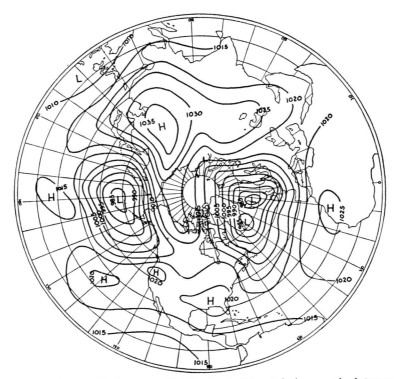

Fig. 16.—Mean sea-level-pressure distribution (millibars) during a week of strong zonal circulation (Jan. 9 to 15, 1938). Note the presence of a single strong Aleutian low, a single Icelandic low, and the displacement toward the Pacific of the Asiatic high.

it leads to flow patterns of a high degree of verisimilitude. In view of the disturbing influence on the mean zonal pressure distribution of the relatively shallow, cold anti-cyclones over Asia and North America, it is impossible to start the analysis from the observed zonal pressure distribution.

It is fairly apparent that both the theoretical and the observed circulation imply the existence of double polar fronts both in the Pacific and in the Atlantic. The positions of these mean fronts have been indicated by broken lines in Fig. 15.

During periods of strong circulation, the theoretical resonance wave length analyzed previously becomes too large for the development of two frontal zones either in the Pacific or in the Atlantic. The resonance wave pattern is no longer free to develop, and the circulation pattern is probably mainly a function of the distribution of continents and oceans.

Figure 16 is typical of the mean pressure distribution during periods of strong circulation. A comparison of Figs. 14 and 16 reveals certain marked contrasts that appear to be typical.

During periods of strong circulation both the Aleutian and the Icelandic lows are characterized by single well-developed centers and by large dimensions. The Aleutian low is then normally located near the Alaskan peninsula and the Icelandic low in the vicinity of Iceland or even east and north of it.

During periods of weak circulation, one or the other or even both of these centers split into two separate cells of smaller dimensions than normal. One part of the Aleutian low may be found near Kamchatka; one in the Gulf of Alaska. At the same time the Icelandic low is frequently displaced westward and southward, or it may, as in Fig. 14, split into two cells.

During periods of strong circulation, a fairly well developed high-pressure area, often referred to as the Great Basin high, is usually found over the southern portion of the Rocky Mountain states. This is really a part of the subtropical (warm) high-pressure area. North of this high-pressure area there is a rapid inflow of relatively mild Pacific air masses over the United States, and rapid air motion eastward prevails over the northern states. There are very few indications of the development of a cold continental anticyclone over the interior of this continent.

During such periods of strong circulation, maritime inflow from the southwest characterizes weather conditions in northwestern Europe. Both our Pacific Northwest and northwestern Europe are then dominated by a rapid succession of wave cyclones—warm, moist, subtropical air masses alternating with relatively mild, maritime polar air masses moving in from the west or northwest. Rainfall on the Pacific coast occurs mainly far to the north, in British Columbia or Washington and Oregon.

Finally, during such periods of strong circulation, the Asiatic high appears to be displaced toward the Pacific side of Eurasia.

On the other hand, as the zonal circulation of middle latitudes weakens and finally reaches a minimum value, there is a marked tendency for the Great Basin anticyclone to disappear and for a strong continental anticyclone to develop over the interior of North America. The center of this anticyclone is located far to the north, in Canada, and a wedge of high pressure extends southward into the United States. Thus, in the surface layers, there is very little air movement from west to east across North America. At the same time, the Asiatic anticyclone is usually displaced westward, toward Europe. The effects of these pressure changes on weather conditions are profound. There will now be an outflow of cold continental air from the east over Alaska and even over British Columbia, and another outflow from the southeast of extremely cold continental air from Asia over northwestern Europe.

The polar-front cyclones, which move up over the Pacific toward that portion of the Aleutian low which during periods of weak circulation is located in the Gulf of Alaska, are quite likely to bring with them warm moist air masses from the southwest. It has already been brought out that these moist currents are frequently trapped between the polar air masses that come down over the Pacific to the west and the mountains and continental air masses to the east. Thus forced to ascend, the moist air yields heavy rainfall fairly far south on the Pacific coast.

During periods of weak circulation, the theoretical sea-level-pressure distribution is so disturbed through the development of continental anticyclones that it may become unrecognizable. For this reason, it is of some interest to look at the pressure distribution at higher levels, say, 3 km (about 2 miles), as determined with the aid of upper-air data now available daily from a number of stations in the United States (Figs. 17 and 18). In Fig. 18, corresponding to a period of weak circulation, there are good indications of low-pressure troughs off the Pacific coast and east of the

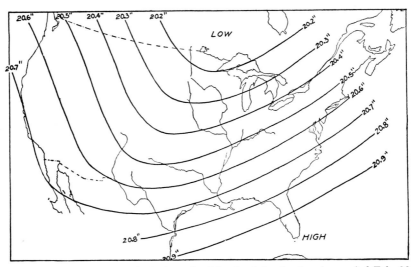

Fig. 17.—Mean pressure (inches) at the 3-km level for the five-day period Feb. 26 to Mar. 2, 1939. This map corresponds to a period of strong circulation. Notice the general trend of the isobars from west-southwest to east-northeast in the eastern half of the country, indicating a general west-southwestern wind. This type of wind distribution aloft in the eastern part of the United States corresponds to temperatures well above normal.

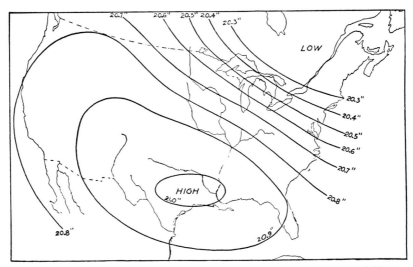

Fig. 18.—Mean pressure (inches) at the 3-km level for the five-day period Mar. 19 to 23, 1939. This map corresponds to a period of weak circulation. Notice the west-north-west winds over the Middle West and the East, corresponding to temperature well below normal in these sections. This pressure distribution suggests two troughs, one over the Atlantic coast and another over the Pacific just off the California coast.

Atlantic coast.   Figure 17, corresponding to a case of strong circulation, shows a single well-marked trough, probably an extension of the Icelandic low, extending southwestward through the Mississippi Valley.

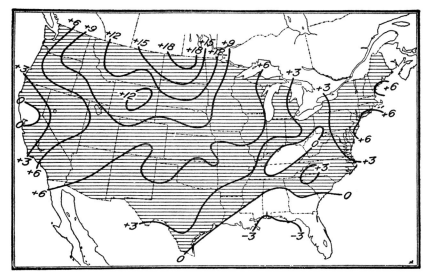

FIG. 19.—Temperature departure from normal, in °F, for a period of maximum circulation. The index at this time was 13.2 millibars.

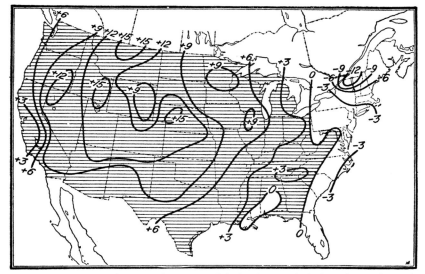

FIG. 20.—Temperature departure from normal, in °F, for a period of maximum circulation. The index at this time was 12.5 millibars.

During periods of strong circulation, characterized by strong west-to-east movements in middle latitudes, the belt of westerlies is usually displaced somewhat to the north of its normal position.   Because of the prevailing strong winds, intense lateral mixing and turbulence develop in middle latitudes.   This mixing process transports heat northward and creates positive temperature anomalies in middle latitudes and

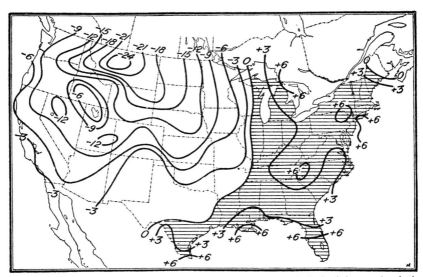

Fig. 21.—Temperature departure from normal, in °F, for a period of minimum circulation. The index at this time was −1.4 millibars.

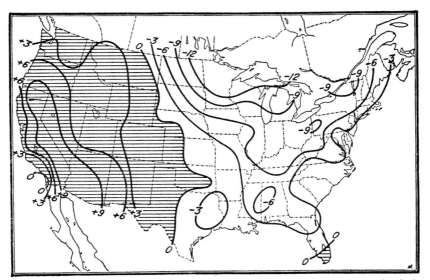

Fig. 22.—Temperature departure from normal, in °F, for a period of minimum circulation. The index at this time was 0 millibars.

presumably negative ones farther south.   During periods of weak circulation, the west-east components decrease in intensity, and there is a pronounced tendency toward the development of large-scale north-south current systems. At such times, regions of positive and negative temperature anomalies will appear side by side, but their position will not always be the same, as may be inferred from the previous discussion of the relation between the size of the stationary flow patterns and the prevailing zonal-circulation intensity. The four anomaly charts, Figs. 19 to 22, are intended to bring out these differences between periods of strong and weak circulation.

The value of the relationship described above between the observed circulation pattern and the intensity of the zonal circulation lies in the fact that it reduces the number of variables to be considered in any discussion of weather types by establishing two idealized world pressure patterns, for periods of maximum and minimum circulation intensity. Given the values of the zonal circulation index during a few consecutive weeks (during the winter season) it is probably possible, from these values alone, to give a description of the mean pressure distribution at the end of the period that will be decidedly better than a pure guess, although definitely subject to a considerable margin of uncertainty. Until it becomes possible to predict the fluctuations in the zonal-circulation index, it is of course impossible to utilize this knowledge with full effectiveness in forecasting.

The preceding discussion suggests another important application. It should be possible to establish relationships corresponding to (but not necessarily identical with) the ones described above, between mean zonal-index values and mean circulation patterns for longer periods (months or years). The establishment of such climatic patterns, having a physical background, should be of great value in the analysis of past climatic fluctuations and should serve to emphasize the need for restraint in this field of research, by bringing out the self-evident fact that the sequence of past climatic events can vary only very slightly from point to point. Hence, the geographic distribution of the climates assumed to have prevailed during a certain geological period must follow a pattern that is compatible with accepted physical and meteorological principles.

SECTION VIII

METEOROLOGICAL INSTRUMENTS

By Louvan E. Wood

CONTENTS

SECTION VIII

METEOROLOGICAL INSTRUMENTS

By Louvan E. Wood

INTRODUCTION

Meteorological instruments may be divided into three general groups, those in routine use for surface observations, *i.e.*, measurements of pressure, temperature, humidity, sunshine, etc., at or near the earth's surface; those in routine use for upper-air observations, *i.e.*, pilot balloons, associated equipment, and radiosondes; and last, a large variety of instruments that either are used in very limited numbers, applicable only for special problems, or are connected with closely related sciences.

Space permits a brief description of only the first two groups.

BAROMETERS

Barometers may be divided into two basic types, mercurial and aneroid. In general, mercurial barometers are characterized by a high degree of accuracy, permanence, and fragility as compared with aneroids, which are small, light, highly portable, and relatively rugged.

A mercurial barometer can be designed to measure pressure accurately without reference to another standard (although they are seldom so designed in practice), but all aneroids must be compared with a standard to establish calibration.

Mercurial Barometers. A mercurial barometer consists in principle of a column of mercury balanced against the weight of the atmosphere. A vertical column of mercury 30 in. long exerts approximately the same pressure per unit area as the atmosphere at sea level. The height of the mercury column is proportional to the pressure (see Fig. 1).

The basic relationship for the measurement of pressure by a mercurial barometer is

$$P_d = \rho_0 g(1 - \beta t)h$$

where P_d = pressure, dynes/cm^2

ρ_0 = density of mercury at 0°C, 13.5951 grams/cm^3

g = acceleration of gravity at sea level and 45°lat., 980.621 cm/sec^2

β = coefficient of thermal expansion of mercury, 0.0001818

t = temperature or mercury, °C

h = height of column of mercury, cm.

One millibar = 1,000 dynes/cm^2; therefore pressure in millibars is given by

$$P = \frac{\rho_0 g(1 - \beta t)h}{1,000}$$

Pressure measurement with a mercurial barometer depends on three principal factors.

1. Measurement of the height of the mercury column.
2. Determination of ρ, the density of the mercury.
3. The value of the acceleration of gravity.

There are two common types of mercurial barometers, the fixed cistern and the adjustable cistern.   In the simple U tube shown in Fig. 1, the mercury level changes equally in both sides with a change in pressure.   Since it is inconvenient to measure both levels, the open side of the barometer is made much larger than the closed side and is known as the *cistern*.   The changes in mercury level in the cistern are less than those in the closed tube in the same ratio as the area of the cistern is greater than that of the closed tube.

In an adjustable-cistern barometer, the mercury level is manually adjusted to a reference point prior to determining h, the height of the mercury column. The height h is always measured from this reference point.

In the fixed-cistern barometer, a fixed amount of mercury is used, and the level of the mercury in the cistern is allowed to vary with changes in pressure.

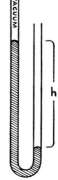

The variation in level in the cistern bears the same relationship to the variation in level in the tube as the cross-sectional area of the tube does to the area of mercury surface in the cistern, *i.e.*, if the ratio of the area of the mercury surface in the cistern to the area of the tube is 50 to 1, the range of fluctuation of mercury level in the cistern will be one-fiftieth of that in the tube.   Since the pressure is determined by measuring the level in the tube, allowance must be made for the change in cistern level.   This is done by contracting the measuring scale in the above case by $^{50}/_{51}$.   The simple U tube may be considered as a fixed-cistern barometer having a unity area ratio, in which case $\frac{1}{2}$ in. on the scale represents 1 in. pressure.   With a fixed cistern, it is essential that the cistern and tube have constant cross-sectional areas throughout the ranges of mercury movement in order to have a linear scale.   This is not necessary with an adjustable cistern.

Fig. 1. Principle of mercurial barometer.

In Fig. 2a is shown a cross section through a type of fixed-cistern barometer, while 2b shows a *Fortin* cistern, a common type of adjustable cistern.

With an adjustable-cistern instrument, h is measured by means of a vernier-equipped scale based on the tip of the ivory cone as a starting point.   With a fixed cistern, h is measured by means of a vernier-equipped scale based on the level of the mercury in the cistern at zero pressure as a starting point.

Barometer scales are ordinarily made of brass.   Consequently, for accurate measurement of h, it is necessary to correct for the temperature effect on the scales.

The density of mercury changes with temperature.   With an adjustable cistern, the density of mercury is important only inasmuch as it affects the length h through the change in weight per unit length of the column.   The change in volume is of itself of no importance.   With a fixed cistern, there are two effects, that changing the length of the column and that due to the change in volume of the mercury.   The latter is a function of the physical proportions of the barometer and is independent of the pressure.

In practice, the temperature correction for an adjustable-cistern barometer consists of a single composite correction for the scale and mercury errors, whereas that for a fixed-cistern barometer consists of the same correction plus a relatively small correction for the particular design of instrument in question.   In some barometers, that correction is so small as to be negligible under normal conditions of pressure and temperature.

Barometers reading in inches normally have scales cut to be correct at 62°F, and the mercury is referred to 32°F.   This combination gives zero temperature correction at 28.5°F.   Metric barometers have zero temperature correction at 0°C, both the scale and the mercury being referred to 0°C.

A thermometer is mounted on a mercurial barometer with its bulb exposed so as to indicate a temperature as representative as possible of that of the entire barometer. It is important to keep a barometer at as uniform a temperature as possible.

The acceleration of gravity varies with both latitude and altitude above sea level. A correction must be applied to all mercurial-barometer readings for the local value of g. This is given by $C = p(g_1 - g_0)/g_0$, where C is the correction and p the pressure, both in the same units. g_1 is the local gravity acceleration, and g_0 is the standard value. In practice, p_n, the normal pressure for the station in question, may be used.

Barometer readings are made in terms of the length of the mercury column measured to the top of the convex meniscus, which gives too large a value, the excess amount depending on the size of the tube. A large tube has very little meniscus error,

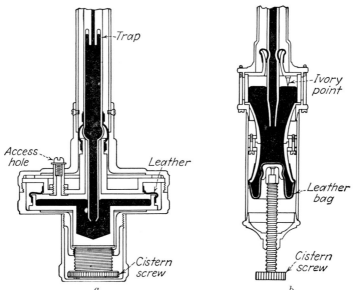

Fig. 2.—Barometer cisterns: *a*, fixed, and *b*, Fortin.

and a small tube has a relatively large amount. In practice, the meniscus error and any residual error in the location of the scale are grouped in an instrumental error peculiar to and constant for any given barometer. Many barometers are adjusted at manufacture to make this correction zero.

In Fig. 3a is shown a fixed-cistern marine barometer. When the clamp is loosened, the barometer swings out from the wall and hangs freely from the gimbal ring attached to the end of the hinged support arm. The tube is constricted through part of its length, as indicated in Fig. 2a, to reduce the volume of mercury and lessen pumping when the barometer is used on shipboard. To read the barometer, note the thermometer reading, tap the barometer lightly, set the vernier so that the bottom edge is tangent to the top of the meniscus, and read the vernier and apply the necessary corrections.

The cistern screw, when unscrewed as shown, lets the steel cistern rest on the outer case and is the normal position. For shipping, the screw is turned up, forcing the mercury up in the tube. The leather joint is impervious to mercury.

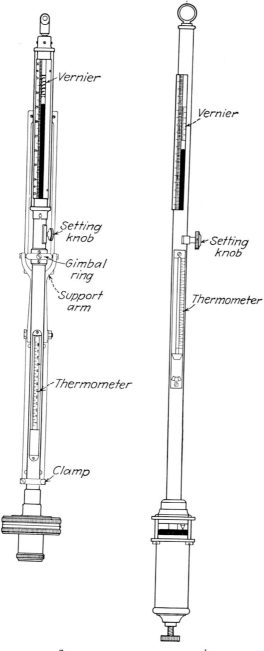

a   *b*
Fig. 3.—Mercurial barometers: *a*, fixed, and *b*, Fortin.

To read the Fortin barometer shown in Fig. 3*b*, note the thermometer reading, set the mercury level in the cistern even with the tip of the ivory point, tap the tube lightly, set the lower edge of the vernier tangent to the top of the meniscus, read the vernier, and apply the mercury corrections.

The cistern screw of a Fortin instrument is screwed up enough to fill the tube and cistern nearly full when it is desired to move the barometer.

FIG. 4.—Principle of aneroid barometer.

Adjustable-cistern barometers are generally more accurate than the fixed-cistern type. However, the latter type is simpler to use, more rugged, and better adapted to marine use.

A barometer should be mounted on a solid support where it will not be exposed to direct sunlight. It must hang vertically and be at a height convenient to read. Mercurial barometers are fragile and must be carefully handled.

Aneroid Barometers. The principle of an aneroid barometer is shown in Fig. 4, which represents a cross section through a simple aneroid mechanism. Figure 5 shows the mechanism of a precision aneroid. The pressure-sensitive element is a beryllium copper diaphragm which contracts and expands with changes in atmospheric pressure. The rate of a diaphragm, *i.e.*, the movement per unit change in pressure, may be controlled within reasonable limits by the thickness and temper of the metal and the shape. Well-designed diaphragms are substantially linear over wide pressure ranges. The movement of the diaphragm is transmitted to a pointer through a lever

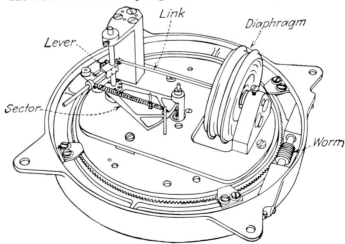

FIG. 5.—Aneroid mechanism.

system and sector and pinion. Aneroid instruments are subject to errors due to temperature, drift, and hysteresis.

The temperature error is due to the change in elasticity of the diaphragm with temperature. It is compensated for by incompletely evacuating the diaphragm or by use of a bimetallic compensator or both. A bimetallic compensator is a bimetal lever or shaft arranged to react to a temperature change, in the opposite direction to the diaphragm.

The stiffness of beryllium copper, or phosphor bronze, materials commonly used for diaphragms, increases with decreasing temperature. Thus, when the temperature of the instrument drops, the diaphragm will expand slightly, indicating a nonexistent drop in pressure. When a little air is left in the diaphragm, it will contract with a drop in temperature, thus decreasing the pressure inside the diaphragm and offsetting the increase in stiffness of the metal. Bimetallic compensation employs a lever or shaft of bimetal so placed in the lever system as to offset the changes in the diaphragm. In general, compensation methods in simple form are exact at only one pressure and temperature, but in practice they give good results over the ranges required.

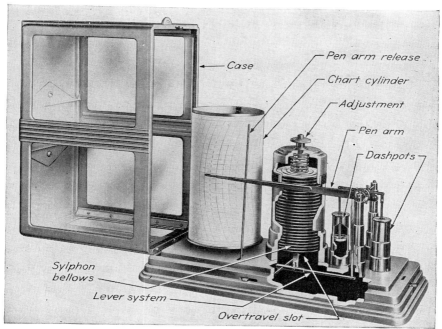

Fig. 6.—Microbarograph.

Hysteresis in aneroids that are well designed and constructed is very small, as is also drift with time.

In high-quality instruments, jeweled pivots are used and the mechanism is balanced so that position errors are negligible. Knife-edged pointers are used in conjunction with mirror scales.

Aneroid barometers are easily portable, convenient to use, and when checked against mercurial barometers at reasonable intervals are reliable and accurate. An adjustment is always provided for setting an aneroid to indicate the correct pressure. In the mechanism illustrated, the worm when turned rotates the entire mechanism while the dial remains stationary.

BAROGRAPHS

Conventional barographs employ aneroid mechanisms sufficiently powerful to operate recording pens. Mercurial barographs have been made but are relatively complicated and not in widespread use.

Figure 6 illustrates a microbarograph. The prefix *micro-* is added to the name because of the open scale, 2½ in. on the chart corresponding to 1 in. of mercury. Many barographs have 1 in. on the chart representing 1 in. of mercury.

The pressure-sensitive element is a sylphon bellows with an internal spring. This element, through a suitable lever system, moves the pen arm. Temperature compensation is accomplished with a small amount of air in the sylphon and a bimetal in the lever system.

The pressure range covered is 2½ in. of mercury, but the instrument can be adjusted to operate over that range at any reasonable altitude.

Dashpots are provided to prevent vibration from causing a ragged record.

The scale is linear, and the pen can be set to the correct pressure by means of the adjusting screw that moves the sylphon up and down.

The chart cylinder is driven by a jeweled clock movement mounted inside the cylinder. Weekly charts are generally used, but different time scales can be obtained by changing gears in the chart cylinder drive.

Chart drive clocks require the same care and attention normally given high-grade clocks. They should be cleaned, oiled, and repaired only by competent clock repairmen.

The pen release arm moves the pen arm away to make it easy to change the chart.

THERMOGRAPHS

A thermograph is shown in Fig. **7**. The sensitive element is a Bourdon tube. This is a metal tube of approximately elliptical cross section bent in a curve, the

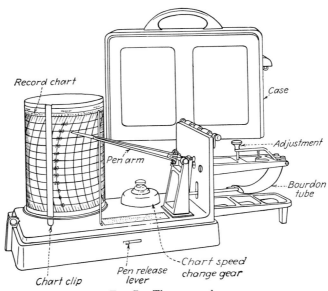

FIG. **7**.—Thermograph.

major axis of the cross section being parallel to the axis about which it is curved. For temperature measurement, the tube is filled with alcohol. An increase in temperature results in an increase in volume of the alcohol which results in straightening of the tube.

The scale is linear, and the pen can be set to record correctly by means of the adjusting screw.

The pen arm is supported in such a way that it bears against the chart with a fixed pressure due to its own weight. This arrangement for controlling pen pressure is also used on other recording instruments.

Different time scales are possible, although weekly charts are generally used.

A thermograph clock is exposed to extremes of temperature and consequently in cold climates will require special lubrication. Low-temperature oils are used, and in extreme cases all oil is removed. This may be done by means of a suitable solvent such as pure gasoline.

The temperature element on the instrument shown in Fig. 8 is a finned bimetal. The fins are made of copper and greatly reduce the time lag, making a very responsive instrument.

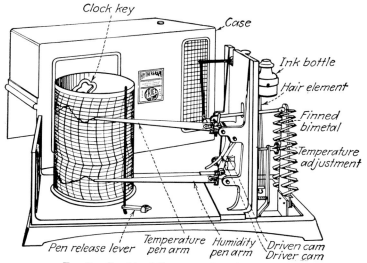

Fig. 8.—Combined thermograph and hygrograph.

For measurement of soil and water temperatures, thermographs are used in which the sensitive elements consist of liquid-filled bulbs connected to the recording mechanisms by capillary tubing. Three types of remote recording thermographs use (1) mercury in a steel bulb and capillary, (2) an organic liquid in a nonferrous bulb and capillary, and (3) a liquid- and vapor-filled system in which the vapor pressure varies with the temperature. The sensitive element connected to the pen may be either a Bourdon tube, a diaphragm, or a sylphon bellows. The recording mechanism is essentially the same as in the ordinary thermograph. Frequently two elements are employed so that temperatures of the air and soil, air and water, or other combinations may be recorded on the same chart. For recording ocean temperatures at sea, the bulb may be mounted in the condenser water intake of a ship.

HYGROGRAPHS

Figure 9 shows a hygrograph, or relative-humidity recorder, employing human hairs as a sensitive element. In Fig. 8 is shown a combined humidity and temperature recorder. In both instruments, the humidity mechanism operates in the same manner. Human hair is the only humidity-sensitive element in widespread use at meteorological observatories for recording relative humidity.

Human hairs, after they are treated to remove the natural oils, are characterized by elongation and shortening with changes in relative humidity. This change in length is nonlinear, being greater the lower the humidity. In the instruments illustrated, a pair of cams rolling against each other have a nonlinear motion that counteracts the nonlinear action of the hairs, thus giving the chart a linear scale.

Hair elements are essentially free from temperature effects, except for becoming very sluggish in their action at subfreezing temperatures.

The clock, pen release, etc., are the same as for barographs and thermographs. An adjustment is provided for setting the pen to record correctly.

The hairs should be kept as free from dust as possible and particularly from fumes or dirt that would form a nonhygroscopic film on the hairs. The hairs can be cleaned with a camel's-hair brush and distilled water when necessary.

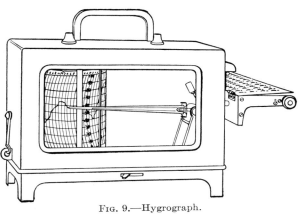

Fig. 9.—Hygrograph.

Barographs, thermographs, and hygrographs are ordinarily operated in conjunction with and checked by eye observations of indicating instruments, which are generally more accurate. The recorder charts are corrected to agree with the eye observations. Recording instruments are accurate and reliable but require checking against standard instruments.

In general, instruments depending on the deformation of an elastic system are subject to drift with time and to hysteresis. In well-designed instruments, both are small. An adjustment is always provided for shifting the pen arm to the correct point on the chart. The lever system always magnifies greatly the motion of the sensitive element. An adjustment in the lever system allows this magnification to be varied in amount so that the motion of the pen can be matched to the chart. This adjustment can be made only when the instrument is checked at two or more widely separated points in its measurement range, whereas shifting the pen to the correct point on the chart can be done whenever necessary without disturbing the calibration.

THERMOMETERS

The thermometer ordinarily used in meteorological work is the liquid-in-glass type, which depends on the differential expansion of glass and the liquid for its action. Mercury, which expands approximately ten times as much as glass, is used in most thermometers that measure temperatures above the freezing point of mercury. For extremely low temperatures, alcohol or other liquids with low freezing points are used. Figure 10 shows a simple mercury thermometer.

The relationship between the cross-sectional area of the capillary and the volume of the bulb determines the scale length per degree. Meteorological thermometers are usually accurate to within a few tenths of a degree centigrade and have as open scales as the accuracy justifies. Cylindrical bulbs, because of their greater surface area for the same volume, have less time lag than spherical bulbs. Consequently, when time lag is a factor, cylindrical bulbs are employed. When properly made of well-seasoned glass and accurately calibrated, thermometers retain their characteristic over long periods of time.

The maximum thermometer shown in Fig. 11 uses mercury and has a constriction near the point where the capillary joins the bulb. This constriction is of such a size that, when the temperature rises, the mercury will force its way through. When the temperature drops, the mercury column will break at the constriction. The mercury in the bulb can then contract while the mercury column remains above the constriction. The maximum thermometer is mounted in a special support (see Fig. 12) that holds it with the bulb end slightly below the other end. A catch is provided on the support so that the thermometer can be released and turned to a vertical position. The mercury column will then drop against the constriction, and the upper end will show the maximum temperature since the thermometer was last set. It can be reset by whirling it rapidly so that centrifugal force drives the mercury past the constriction.

The minimum thermometer shown in Fig. 11 is filled with alcohol and has a small glass index in the capillary inside the alcohol column where it is held by surface tension. When the temperature drops, the alcohol column diminishes in length and pulls the index along with it. If the temperature rises, the alcohol flows freely past the index, leaving it in the lowest position reached since the previous setting. The minimum thermometer may be read by noting the scale reading opposite the end of the index farthest from the bulb. The thermometer can be reset by turning its bulb end up, in which position the index will slide down to the end of the alcohol column.

A remote-indicating thermometer, known as a *telethermoscope*, is used to determine the outside temperature from indoors. This instrument usually employs a resistance coil made of platinum or nickel wire which is mounted in the instrument shelter. The indicator is a form of Wheatstone bridge with the dial calibrated in degrees of temperature. The circuit of such an instrument is shown in Fig. 13. The four arms of the bridge are represented by T, the temperature element, $R + V$, $S + b + S_1$, and $V_1 + a$. The two

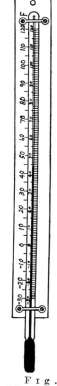

Fig. 10.—Mercurial thermometer.

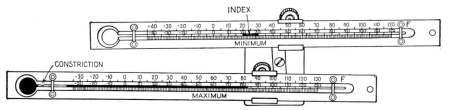

Fig. 11.—Maximum and minimum thermometers.

slide wires are mechanically connected and so proportioned that $S + b + S_1 = V_1 + a$. When the bridge is balanced, $T = R + V$. All moving contacts are in the galvanometer and battery circuits. Three wires are run from the indicator to the sensitive resistance element so that equal lengths of wire are connected in each side of the bridge and thus balance out any effect of temperature on the connections. To determine the temperature, it is necessary only to balance the bridge and read the temperature directly from the dial.[4]

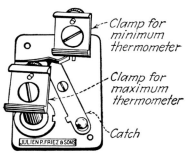

Fig. 12.—Support for maximum and minimum thermometers.

Figure 14 shows a bimetal. This consists of two metals securely bonded together and having different coefficients of expansion. The invar is essentially unaffected by temperature changes, whereas the brass expands and contracts, thus changing the shape of the bimetal. Bimetals are frequently straight, as illustrated, or coiled in the form of a helix. They are used extensively in thermographs and for temperature compensation of pressure instruments. Different combinations of metals are used, depending on the characteristics desired. The amount of movement of

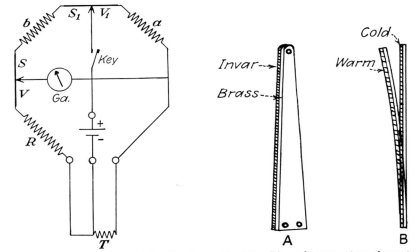

Fig. 13.—Telethermoscope circuit—Leeds and Northrup design.

Fig. 14.—Bimetal temperature element.

a straight bimetal as illustrated is proportional to the temperature change, the square of the length, and the reciprocal of the thickness.

PSYCHROMETERS

Meteorological humidity measurements are generally made with a wet- and dry-bulb psychrometer, consisting of two thermometers with carefully matched scales, the bulb of one being covered with a muslin wick. When this wick is wet and the psychrometer is exposed to the air, evaporation takes place from the wet bulb wick, cooling the thermometer an amount dependent on the vapor pressure. *An air flow of at least 15 ft per sec over the bulbs is needed.*

The amount of cooling of the wet bulb taken in conjunction with the dry-bulb indication, *i.e.*, air temperature, gives a reliable measure of water-vapor pressure and consequently of relative humidity and dew point.

Wet- and dry-bulb psychrometers are made in several forms, namely, sling, rotor, hand-aspirated, and motor-aspirated. A sling psychrometer is shown in Fig. 15. This instrument is whirled so that the bulbs travel with a circular motion around the hand of the operator. A form of hand-aspirated psychrometer uses a rubber bulb as a pump to cause a flow of air over the thermometer bulbs. In a motor-aspirated psychrometer, a small electric motor drives a blower that draws air over the bulbs (see Fig. 16). In making psychrometer measurements, it is important that the instrument have no effect on the humidity being measured. Consequently all fans, blowers, etc., must draw air over the thermometers rather than blow it on them to avoid heating the air.

Tables giving relative humidity and dew point as a function of the wet- and dry-bulb temperatures are used to facilitate obtaining the desired data. Various nomograms and slide rules are also in use.

In order to make accurate humidity measurements, certain precautions are necessary. It is vitally important that the two thermometers are well matched, since in many cases the difference between the wet- and dry-bulb temperatures is very small and must be known accurately. The wick must be clean and free from deposits from the evaporation of hard water. It must be completely saturated prior

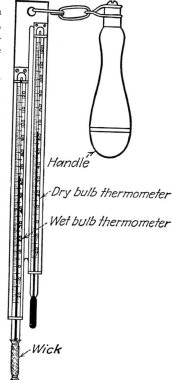

Fig. 15.—Sling psychrometer.

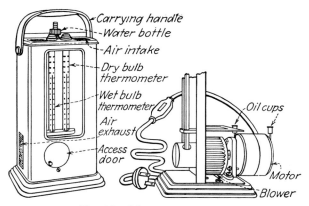

Fig. 16.—Motor psychrometer.

to use with water at the prevailing air temperature or else given extra time for thermal equilibrium to be reached.

In freezing weather, the water must be allowed to freeze and the resulting ice allowed to cool to the air temperature. Sublimation of the ice will then cool the thermometer. Wet-bulb readings must be taken at the lowest indication reached by the thermometer.

WIND INSTRUMENTS

Wind velocity is measured by two instruments, an anemometer that gives the movement and a wind vane that gives the direction. These are often combined in one unit.

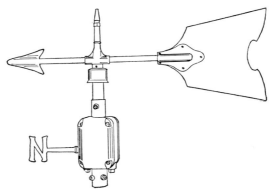

Fig. 17.—Wind vane.

Wind Vanes. A wind vane of the conventional type is shown in Fig. **17.** Well-designed wind vanes have the following characteristics:

1. Light weight so that the vane will have a low moment of inertia and will respond well to quick changes in wind direction without excessive overswing.
2. Accurate balance that will prevent side thrust on the bearings and will prevent a tendency to point in one direction if the axis of rotation is not exactly vertical.
3. Good bearings so that the vane will turn freely with light winds.
4. Sufficient size and correct shape to give an adequate turning moment in light winds.

The indications from wind vanes are commonly transmitted by one of three methods:

1. A wind vane may be directly connected to a rod extending downward through the roof to an indicator or recorder. Universal joints are provided to allow for slight misalignments. Because of its mechanical limitations, this method is seldom used any more.
2. The simplest common method is to have the wind vane turn a cam inside a ring of electrical contacts. By means of 8 contacts corresponding to the cardinal and intercardinal points of the compass, 16 directions can be indicated by means of a cam overlapping two contacts so that for intermediate directions two contacts are made. These contacts are connected by suitable wire to indicating and recording instruments inside the office. This system is in widespread use.
3. The best means for transmitting wind-direction indications is by means of a self-synchronous transmitting motor connected directly to the wind vane. This method gives reliable, continuous indications.

TABLE 1.—WIND-SPEED INSTRUMENTS

Wind-sensitive element	Transmitting mechanism in anemometer	Indicator	Recorder	Power supply	Remarks
	Gearing	Counting dials register wind movement	None	No external power required	Dials located in anemometer
	Mile contacts, gear-driven and closed for every mile of wind passing cup wheel. 9th- and 10th- mile contacts bridged to facilitate counting	None	Pen moves sideways on chart for duration of each contact	Schoenmehl or storage cells	Speed determined from movement recorded and chart time scale
	$\frac{1}{60}$-mile contacts, gear-driven and closed for each $\frac{1}{60}$ mile of wind passing cup wheel	Buzzer, connected to buzz when contacts close	None	Batteries, or transformer connected to 115-volt 60-cycle a-c	Speed determined by counting buzzes during a measured time interval. Buzzes per minute equal miles per hour
		Light, connected to flash when contacts close	None	Batteries or transformer connected to 115-volt 60-cycle a-c	Speed determined by counting flashes during a measured time interval. Flashes per minute equal miles per hour
Cup anemometer measures wind movement by rotating at rate proportional to wind speed		Capacitor discharge circuit with meter	None	115-volt 60-cycle a-c	A capacitor is discharged through a highly damped meter circuit at closing of each contact. Indicates speed on microammeter movement. Sluggish because of infrequent contacts
	Small permanent magnet generator directly connected to cup wheel	Voltmeter calibrated in terms of wind speed	Recording voltmeter	No external power required	Voltage generated proportional to wind speed
	Magnetic drag tachometer	Dial mounted on tachometer mechanism	None	No external power required	Used on portable or hand anemometer
	Self-synchronous motor driven from anemometer spindle through reduction gearing	Disk and roller tachometer mechanism	Self-synchronous motor recorder used in conjunction with torque amplifier	115 volt 60 cycle a-c	Indicates speed directly. Large dials, pointers, and angular deflections used as contrasted with meter movements

TABLE 1.—WIND-SPEED INSTRUMENTS.—(*Continued*)

Wind-sensitive element	Transmitting mechanism in anemometer	Indicator	Recorder	Power supply	Remarks
	Electronic pick-up unit using vaned disk rotating past inductance	Milliammeter calibrated in terms of wind speed	Recording milliammeter	115-volt 60-cycle a-c	Uses electronic frequency meter to convert frequency from pick-up coil to current
Bridled anemometer. Angular deflection of rotor proportional to wind speed	Self-synchronous motor gear-driven from rotor	Self-synchronous motor with dial and pointer	Self-synchronous motor recorder	115-volt 60-cycle a-c	Indications proportional to square root of air density.
Anemograph Pitot tube system	Pressure tube directed into wind by vane. Static pressure openings around vane axis	None	Slack diaphragm operated by pressure differential moves pen on chart. Linear record	No external power required	Indications proportional to square root of air density. Comparatively difficult to install owing to piping required. Responsive to gusts

Wind-speed Instruments. There are two common types of wind-speed instruments, those employing freely rotating cup wheels and those operating by virtue of the air pressure produced by the wind on a restrained system. Bridled and pressure tube anemometers are examples of the latter type.

Wind recorders are of two general types, those registering movement along a time axis and those making a graph of speed against time. The former type is simple, reliable, and in widespread use. A record of wind movement on a time scale is very useful for climatological purposes since total movement and averages, means, and extremes are easily measured.

Recorders plotting speed against time are usually used in conjunction with indicating instruments. When an instant knowledge of wind speed is needed, the indicator will show it, and the recorder will keep a record of it.

Table 1 outlines the more common types of wind-speed instruments.

Cup Anemometers. Wind movement may be measured in a number of ways, the most common of which is the rotating-cup anemometer. The cup anemometer in its simplest form measures the movement of the wind, *i.e.*, each rotation of the cup wheel corresponds to a definite distance traveled by the wind. Therefore, the number of turns the cup wheel makes in a given time interval corresponds to the distance the wind traveled in that interval. The wind speed can be determined by dividing the distance traveled by the time taken.

Cup anemometers have been made in a variety of sizes and styles but usually have three or four cups either hemispherical or conical in shape. A good cup anemometer such as that shown in Fig. 18 is substantially linear throughout the range of wind speeds commonly encountered, as shown by the graph, and is independent of air density. This anemometer has three conical-shaped cups with beaded edges. The cup wheel is light in weight and mounted on a spindle turning in high-grade ball bearings. Light weight makes it respond to rapid fluctuations in wind speed, and low friction on the bearings allows it to start turning in an extremely gentle wind. Deviations of the wind stream from the plane of rotation of the cup wheel have little effect

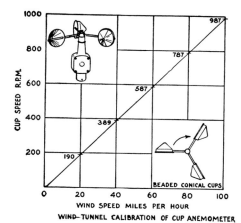

WIND–TUNNEL CALIBRATION OF CUP ANEMOMETER

FIG. 18.—Cup anemometer and calibration curve.

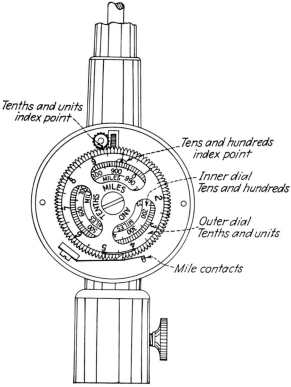

FIG. 19.—Anemometer dials.

on its rate of rotation.   From the graph, it will be noted that the rate of rotation in a wind of 100 mph is less than 1,000 rpm, and consequently the anemometer is a slow speed device not subject to rapid wear or deterioration.   Cup anemometers are

Fɪɢ. 20.—Wind indicator.

equipped with counting dials, electric contacts, self-synchronous motors, magnetic drag tachometers, and magnetos or electrical pick-up devices.   By the above means, wind movement and/or speed may be determined by observing the anemometer

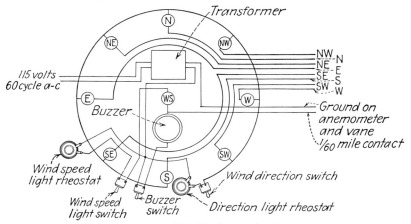

Fɪɢ. 21.—Wind-indicator circuit.

itself or by indicating or recording devices operating at a point remote from the anemometer.

Anemometer Dials.   Figure 19 shows a dial and contact mechanism.   The dials are geared to the cup-wheel spindle.   The two dials mesh with the same pinion.

The inner dial has 99 teeth, and the outer dial has 100 teeth. Thus, for every turn of the outer dial the inner dial gains one tooth. This system counts to 990 miles and then repeats. There is a contact pin set in the outer dial at each mile point and the ninth and tenth miles are bridged by a long contact. The dials illustrated read 919.4 miles, and the long contact between the ninth and tenth miles is closed. This long contact is physically located between the fourth- and fifth-mile points, since the contact shoe is diametrically opposite the index point. Contacts that close for each $\frac{1}{60}$ mile are located behind the dials. Conventional mechanical counters are also used in anemometers.

The mile contacts connect to a recorder in which a pen makes a jog in a straight line on a clock-driven chart for each contact made. Wind movement is thus plotted on a time scale, and determination of speed is simple.

Wind Indicator. Figure 20 shows an indicator for wall mounting. This is wired to an eight-contact wind vane and the one-sixtieth-mile contacts on an anemometer. The indicating elements are nine small light bulbs and a buzzer. The eight outer bulbs corresponding to directions are wired to the appropriate contacts in the vane head (see Fig. 21). The center light and buzzer are wired to the one-sixtieth-mile contacts so that either visual or aural indication of speed may be had. Rheostats for varying the light intensity are provided. To determine direction, the direction switch is snapped on, and the lights are noted. If two are on, the intermediate point is taken. To measure speed, the buzzer or center light is turned on and the buzzes or flashes are counted for 1 min. The number equals the miles per hour.

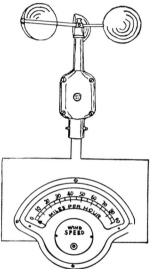

FIG. 22.—Magneto anemometer and indicator.

Magneto Anemometer. The cup wheel is directly connected to a small permanent magnet generator. This generator, or magneto, generates a d-c voltage proportional to the rate of rotation and consequently to the wind speed. The unit shown in Fig. 22 employs a Weston magneto in the anemometer and a Weston voltmeter calibrated in miles per hour as an indicator. This system requires no external power and is simple to install, requiring only two wires suitable for low voltage.

Wind-velocity Transmitter. The concentrically mounted vane and anemometer illustrated in Fig. 23 is representative of the type of construction used in wind instruments. The vane and anemometer both turn in ball bearings. Each is geared to a self-synchronous motor. The direction motor is geared one to one to the vane. The speed motor turns much slower than the anemometer spindle.

Self-synchronous motors are used for transmitting angular position or rotation from one point to another, as from a wind vane to a direction indicator. Figure 23a shows the circuit for two motors, a transmitter, and a receiver. The two motors are identical and have single-phase primary windings and three-phase secondary windings. Slip rings are provided to connect to the rotors. Either primary or secondary may be the rotor, depending on design. When the transmitter rotor is turned, the receiver rotor turns a like amount. The effort necessary to turn the transmitter rotor need be only large enough to overcome the brush and bearing friction of the two motors plus the load, if any, imposed on the receiving motors.

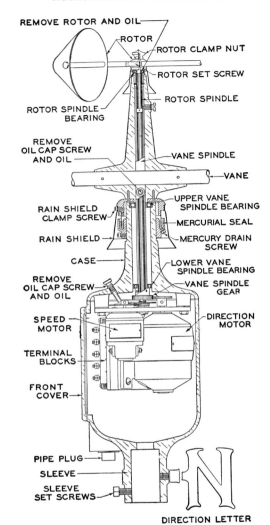

REMOVE ROTOR AND OIL

ROTOR

ROTOR CLAMP NUT

ROTOR SET SCREW

ROTOR SPINDLE

ROTOR SPINDLE BEARING

REMOVE OIL CAP SCREW AND OIL

VANE SPINDLE

VANE

RAIN SHIELD CLAMP SCREW

UPPER VANE SPINDLE BEARING

MERCURIAL SEAL

RAIN SHIELD

MERCURY DRAIN SCREW

CASE

LOWER VANE SPINDLE BEARING

REMOVE OIL CAP SCREW AND OIL

VANE SPINDLE GEAR

SPEED MOTOR

DIRECTION MOTOR

TERMINAL BLOCKS

FRONT COVER

PIPE PLUG

SLEEVE

SLEEVE SET SCREWS

DIRECTION LETTER

FIG. 23.—Wind-velocity transmitter.

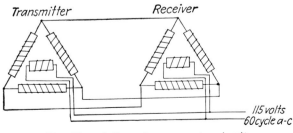

Transmitter

Receiver

115 volts
60 cycle a-c

FIG. 23a.—Self-synchronous motor circuit.

Disk and Roller Wind-speed Indicator Mechanism. Since the self-synchronous speed motor in Fig. 23 turns continuously, the corresponding motor on the receiving end will do the same. In order to get wind speed, it is therefore necessary to have some device for converting the rate of rotation of the receiving self-synchronous motor into an indication of speed. This may be done by means of the disk and roller mechanism shown in Fig. 24. In this mechanism, the disk-drive motor turns two polished hardened-steel disks at a constant speed in opposite directions. These disks bear against a steel roller. The axis of the roller is perpendicular to the axis of rotation of the disks. On an extension of the roller axis is a circular rack and a screw or worm. This worm is connected by a gear train to the self-synchronous motor driven from the anemometer. A pointer is mounted on the shaft of the pinion meshing with the

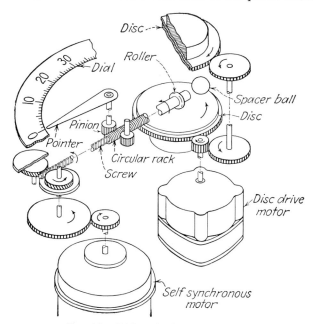

FIG. 24.—Disk and roller mechanism.

circular rack. When the anemometer cups are not turning, the roller is in the center of the disks and consequently does not turn. With the wind blowing and the cup wheel turning, the worm will pull the roller away from the center of the disks, and it will turn at a rate proportional to its distance from the center.

As soon as the roller has moved far enough to be driving the worm at the same rate that the anemometer tends to, equilibrium is reached, and the pointer will indicate the wind speed since it is moved by the sliding of the circular rack. The gears and parts making up the mechanism are so proportioned that, for the maximum wind speed for which the equipment is designed, the roller will be at the outer edge of the disk. The time factor for determining speed from wind movement is supplied by the synchronous disk-drive motor.

Wind-velocity Torque Amplifiers. When it is desired to operate several wind speed and direction indicators from one anemometer and vane, a means of torque amplification is necessary. Figure 25 shows a wind-direction torque amplifier

Essentially the same mechanism is used for speed. The wind vane, through the self-synchronous motor, turns the leader contact arm. This leader contact fits between the follower contacts with very little clearance. When the leader contacts move, connection is made to a follower contact that energizes the power motor, driving the follower contacts in the direction to open the circuit. The power motor also drives a self-synchronous motor, which is thus kept in step with the vane. This motor can

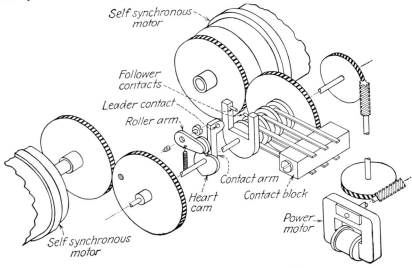

Fig. 25.—Wind-direction torque amplifier.

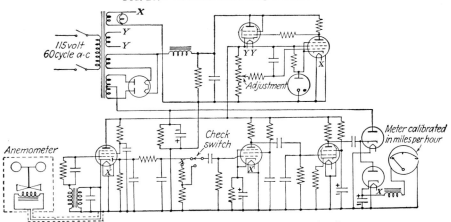

Fig. 26.—Electronic wind-speed indicator circuit.

operate several indicators, the power to drive them coming from the power motor. The leader contact arm is driven through a roller and heart cam. This is a safety device whereby, if the vane moves more rapidly than the follow-up mechanism can respond, the roller will ride up on the high part of the cam until the rest of the mechanism can catch up. The average wind direction is indicated, rapid fluctuations being smoothed out.

A torque amplifier is more necessary for indicating wind speed than for indicating direction, the power available from the anemometer being much less than that from the vane. Many wind systems use no amplification for direction. For speed, the leader contacts may be positioned by the disk and roller mechanism, the contact arm replacing the pointer. The indicators to operate with the torque amplifier mechanism

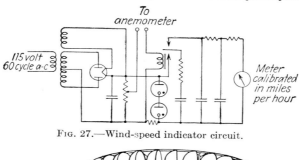

Fɪɢ. 27.—Wind-speed indicator circuit.

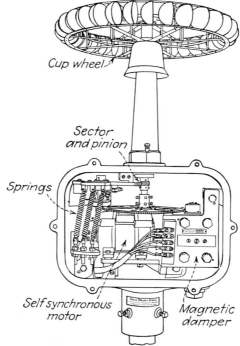

Fɪɢ. 28.—Bridled anemometer.

consist of self-synchronous motors mounted in suitable cases with pointers and dials. The recorder shown in Fig. 29 works with self-synchronous motor wind instruments.

Electronic Wind-speed Indicator. Most wind-speed instruments place a small load on the anemometer. The electronic type, a typical circuit of which is shown in Fig. 26, avoids this. Attached to the cup-wheel spindle is a light disk bearing copper vanes that rotate past the pole pieces of a coil, the Q of which is changed at a rate proportional to the rate of rotation. This coil is connected in such a manner that it modulates an oscillator which through a suitable circuit controls the current flowing through a milliammeter movement. Recording may be accomplished by means of a recording milliammeter.

Simple Electrical Wind-speed Indicator. To determine accurate speed from the indicator shown in Fig. 20 takes at least 30 sec. The indicator, the circuit for which is shown in Fig. 27, operates from the same one-sixtieth-mile contacts, and the speed may be read directly from a meter scale. Each time the anemometer closes the one-sixtieth-mile contacts, a capacitor charged to a fixed voltage is discharged through the meter circuit. Since with light winds appreciable time intervals elapse between contacts, the meter circuit is highly damped, and the meter shows average speed.

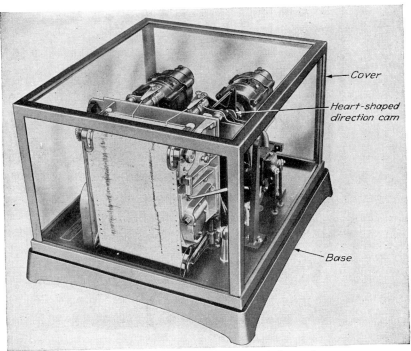

Fig. 29.—Self-synchronous motor wind recorder.

Hand Anemometer. Portable anemometers are made in several forms. One type employs a conventional cup wheel connected to a magnetic drag tachometer mechanism. The observer holds the anemometer with the cup wheel above his head and reads the speed from the dial.

Bridled Anemometer. Figure 28 shows an anemometer with a multicup rotor that is restrained by springs from turning freely. Since the wind pressure is proportional to the square of the speed, the springs are so arranged that their restraining force is proportional to the square of the angular deflection, thus linearizing the anemometer. The indications of all anemometers depending on air pressure are affected by air density.

A magnetic damper is used to prevent excessive oscillation of the rotor near the zero point. The rotor is geared to a self-synchronous motor, the angular position of which is a direct indication of speed.

Anemograph. The anemograph records both wind direction and speed on a clock-driven chart. Wind direction is recorded by means of a rod running from the vane through the roof and connecting to the recorder. Speed is measured by the pressure difference between static-pressure holes concentric with the vane axis and a pressure orifice directed into the wind by the vane. Static and pressure lines run to the recorder and connect to opposite sides of a diaphragm that moves a pen. The direction rod turns a cam that also moves a pen. Speed indications are sensitive to air density. This instrument requires no external power but is limited in its application, owing to the direct-connected vane.

PRECIPITATION GAUGES

A simple form of rain and snow gauge is shown in Fig. 30. The collector is 8 in. in diameter and therefore catches the rain falling on the area of a circle of that diameter. The water collected runs into a tube having a cross-sectional area one-tenth that of the collector. Consequently, the depth of water in this measuring tube is ten times the actual rainfall. The amount is measured by means of a wooden stick graduated in tenths of an inch, each graduation being equivalent to $\frac{1}{100}$ in. of precipitation. Under weather conditions conducive to rapid evaporation, measurement should be made immediately after precipitation stops. In freezing weather, the collector and measuring tube are removed, and snow is caught directly in the main body of the gauge. It is then melted and measured in the tube.

The tipping bucket gauge shown in Fig. 31 is one of the most widely used recording gauges. The tipping bucket is divided into two compartments and is so balanced that, when $\frac{1}{100}$ in. of rain has accumulated, it will tip, emptying itself, making an electrical contact, and exposing the other side of the bucket. It thus tips back and forth, making a momentary contact for each tip. These contacts are recorded by the quadruple recorder described

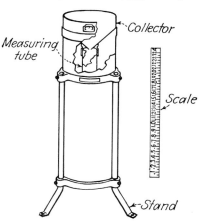

Fig. 30.—Simple rain gauge.

on page 558. The tipping bucket gauge has a faucet on the bottom by means of which the water can be drained into a measuring tube and measured by stick to check the recorded amount. This type of gauge is accurate for low and moderate rates of rainfall but has an error that increases with the rate of rainfall, owing to the rain lost while the bucket is tipping, the time required for tipping being constant regardless of the rate of rainfall. During freezing weather, the tipping bucket must be removed to prevent damage.

Figure 32 illustrates the Fergusson type of weighing rain and snow gauge. Precipitation is collected in a receiver mounted on a spring balance that positions the pen on a record chart. The chart drum is driven by clockwork. To give adequate openness of scale, the lever system is designed to make the pen traverse the record three times. This makes possible $1\frac{1}{2}$ in. on the chart for each inch of precipitation with a total capacity of 9 in. This gauge is particularly suitable when a graphic record is desired without the need of external power or frequent visits to the equipment. The record is accumulative to the maximum capacity of the gauge, the slope shows the rate of rainfall, and loss through evaporation is evident.

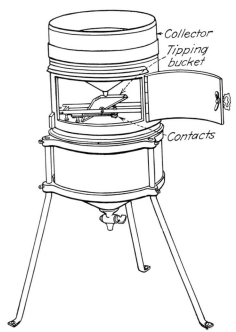

Fig. 31.—Tipping bucket rain gauge.

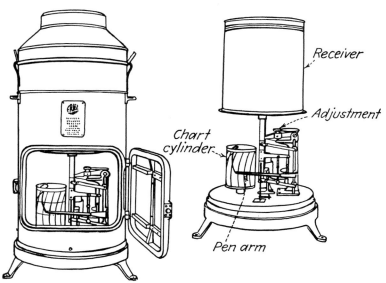

Fig. 32.—Fergusson rain gauge.

The accurate measurement of precipitation is seriously interfered with by the wind. In general, precipitation gauges catch too little in windy weather. In a steady wind, the direction of movement of the rain or snow will be the resultant of vertical and horizontal components and will strike the collector of the gauge at an angle. Thus the stronger the wind is the less the effective area of the gauge collector is. Also, in a gusty wind, the precipitation may be blown over and around the gauge. The Alter flexible shield in Fig. 33 breaks the force of the wind and thus diminishes its effect on the gauge.

SUNSHINE-DURATION TRANSMITTER

In Fig. 34 is illustrated a type of differential thermometer used to record the duration of sunshine. The construction of the black bulb is shown in the inset. When the sun shines, the black bulb is heated and the air expands, driving the mercury up past the contacts and thus closing a circuit to the quadruple recorder. When the

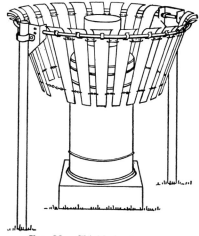

Fig. 33.—Shielded rain gauge.

sun is obscured by clouds, the temperatures of the clear and black bulbs tend to equalize, and the mercury moves below the contacts. This instrument must be set at such an angle that the sunshine falls as nearly perpendicular to the axis of the tube

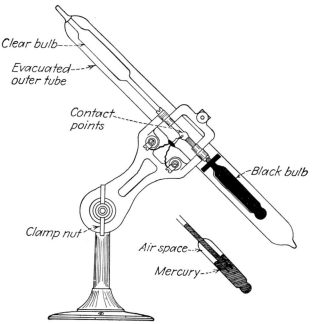

Fig. 34.—Sunshine tube.

as possible. It should be readjusted for the different seasons. It is essentially an on-off device and will show if the sun is shining except in the morning and evening, when twilight corrections must be applied, since even with a clear sky the radiation is then too weak to heat the black bulb.[2]

Another type of sunshine recorder which, in addition to showing whether or not the sun is shining, gives an indication of intensity of the sunshine is the Campbell-Stokes design. This consists of a glass ball approximately 3 in. in diameter with a heavy paper chart mounted underneath it. When the sun strikes the ball, the rays are focused on the paper chart and burn a trace in it as the earth rotates. In this recorder, the position of the chart must be varied to suit the season of the year.

QUADRUPLE RECORDER

This instrument, also known as a *station meteorograph*, is in widespread use for recording wind movement and direction, sunshine duration and rainfall. Figure 35

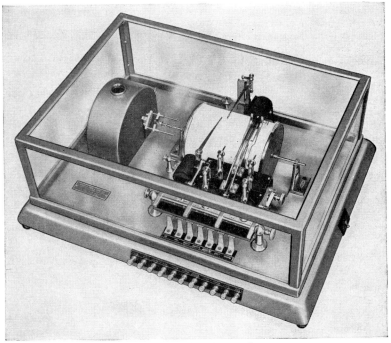

Fig. 35.—Quadruple recorder.

shows the general form and construction Fig. 36 shows the circuit. The record is made on a daily chart that turns once in 6 hr and moves axially as it turns, being driven by a clock movement. Wind movement is shown by sideways movement of a pen caused by the anemometer contacts closing the circuit. The wind-direction record is made by four dotting pens, one for each cardinal point. Intercardinal points are shown by two simultaneous dots. The circuit to the vane is closed once each minute by the clock, and the direction is therefore recorded at 1-min intervals. The sunshine-duration circuit is also closed once each minute by the clock. The sunshine pen is moved by a cam back and forth in a series of steps when the sun is shining.

It draws a straight line if the sun is not shining. The rain-gauge circuit is connected to actuate the pen used for sunshine. The record is similar to that made for sunshine except that, instead of the pen's moving each minute, it moves whenever the tipping bucket empties, thus making a record readily identified. The recorder is usually equipped with a buzzer connected to the one-sixtieth-mile contacts on the anemometer and with a switch for manually closing the direction circuit so that wind direction and movement may be readily determined without the need of interpreting the record or

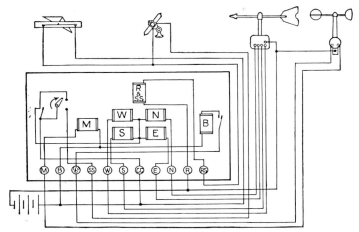

Fig. 36.—Quadruple-recorder circuit.

waiting for clock contacts to close. Recorders of this general type are used for recording wind movement and direction and in many cases just movement alone.

DETERMINATION OF UPPER WIND BY PILOT-BALLOON OBSERVATIONS

When a properly inflated spherical balloon is released, it rises at a substantially constant rate and drifts freely with the winds through which it passes. From a determination of the balloon's position in space at regular time intervals, with reference to its point of release as an origin, the speed and direction of the wind at various levels can be derived.

There are two principal methods used for determining the position of the balloon in space, the single-theodolite method, which is relatively simple and in widespread use; and the double-theodolite method, most accurate but complicated and seldom used except for experimental purposes. Only the single-theodolite method is described here.

To fix the balloon in space with one theodolite, it is necessary to measure the horizontal, *i.e.*, azimuth angle, from a reference point (true north), the vertical, *i.e.*, elevation angle, above the horizon, and to obtain the height from the ascension rate and time in the air.

In Fig. 37, P_1, P_2, and P_3 represent the positions of a balloon at the ends of the first, second, and third minutes, respectively, after release from 0. The lines D_1, D_2, and D_3 represent the horizontal distances of the balloon, and h_1, h_2, and h_3 represent the corresponding heights. The angles α_1, α_2, and α_3 are the corresponding elevation angles measured above the horizon. Therefore $D_1 = h_1 \cot \alpha_1$ and $D_2 = h_2 \cot \alpha_2$, etc. The height of the balloon, which is a function of time, and the elevation angle specify the location of the balloon insofar as distance from the origin is concerned. The azimuth angles β_1, β_2, and β_3 correspond to the first, second, and third minutes of the balloon's path, respectively. The line O, P_1, P_2, and P_3, etc., represents the path of the balloon insofar as it can be determined by 1-min intervals. More frequent observations would define the path more accurately, and less frequent observations would define it less accurately. However, standard practice requires an observation every minute.

In practice, the horizontal projection of the balloon is drawn to scale on a plotting board, using the horizontal distances and the azimuth angles. This will give the broken line O, a, b, c. To determine the wind speed at point b, the distance on the plotting board is measured from point a directly to point c. This distance is divided

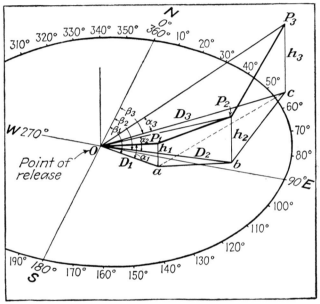

Fig. 37.—Diagram of pilot-balloon observation.

by the time involved, which in this case is 2 min, and the result will be the wind speed representative of the region around b. The wind speed for any other point may be determined in a similar manner. For instance, that for point a is determined by drawing a line from O to b, which in the case of the first minute coincides with the horizontal distance. Special scales are ordinarily used for evaluating speeds from the plotting board. These scales must be suitably proportioned to match the scale of the plotting board, which is normally 1 cm to 200 m. When dealing with high winds, it may be necessary to double this scale and plot on the basis of 1 cm to 400 m, taking care to make the necessary allowances in scaling wind speeds.

The wind direction at the point b is denoted by a line through O parallel to the line ac. An angle measured clockwise from north to this line, which shows the direction from which the wind came, is the direction desired. In routine pilot-balloon observations, special equipment is used for the plotting and determination of upper-wind conditions.

As previously stated, the ascension rate of the balloon is assumed to be constant, except for the first 5 min of ascent. The ascension rate can be calculated by the following formula:

$$V = 72 \left(\frac{l}{L^{2/3}}\right)^{5/6}$$

in which V is the speed in meters per minute, L the total lift of the balloon in grams; l the free lift of the balloon in grams which is equal to the total lift minus the weight of the balloon.

Fig. 38.—Pilot-balloon theodolite. (*Courtesy David White Company.*)

Hydrogen or helium is used to inflate the balloons, helium being preferred because of its freedom from fire hazard. For regular observations, balloons are inflated to give a normal ascension rate of 200 mpm, to which is added 20 per cent during the first minute, 10 per cent during the second and third minutes, and 5 per cent during the fourth and fifth. Twelve-inch balloons giving approximately twice the ascension rate are used when fast ascents are desired to save time. The use of standard ascension rates makes possible tables giving the horizontal distance of the balloon as a function of time and elevation angles.

Plotting boards are made with different mechanical arrangements and to different scales, but all use the basic method outlined above. The surfaces of the plotting boards are prepared so they can be marked on with a soft lead pencil and the marks erased when the work is completed.

A type of theodolite used for pilot-balloon observations is shown in Fig. 38.   This instrument has a telescope of approximately 20 power arranged in such a manner that it can be pointed in any direction above the horizon.   This is accomplished by a prism in the center of the telescope which causes a right-angle bend in the light rays, allowing the observer to sight always along the horizontal axis of the instrument. The theodolite is equipped with a leveling head, vertical and horizontal circles, tangent screws, and verniers similar to those used on conventional surveyors' instruments.

Pilot balloons are nominally of 30 or 100 grams weight, 6 or 12 in. in diameter prior to inflation and are used in several colors.   Uncolored balloons are used on clear days because they are readily observed against a bright sky.   On cloudy days, black balloons are most easily visible, and red, yellow, or orange balloons work best with broken clouds and intermediate conditions.   The balloons can be inflated with a reasonable degree of accuracy simply by using a definite free lift and assuming an average weight for the balloon itself, or they can be accurately inflated by means of an inflation balance, making allowances for individual differences in the balloons.   An inflation balance consists of an ordinary beam balance with special scales and a device for holding the balloon and measuring its lift while it is being inflated.

Other accessories used in connection with balloon observations are telephone systems for connecting the theodolite observer with the man plotting the data inside, and a timing device for signaling at the end of each minute when the theodolite angles are to be read.

For a pilot-balloon ascent at night, a small paper lantern containing a candle or a small battery-operated light bulb is tied to the balloon.   The theodolite is then sighted on the light.   The theodolite cross hairs and scales are provided with suitable illumination.

CEILING-HEIGHT INSTRUMENTS

Ceiling-height measurements in cloudy weather are made in the daytime by means of small 10-gram rubber balloons inflated to ascend at a definite rate.   The time

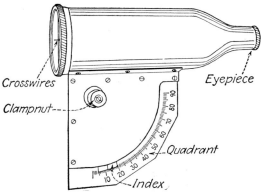

Fig. 39.—Clinometer.

interval between release and disappearance into the clouds multiplied by the ascension rate gives the ceiling height.

At night, a searchlight directed vertically will throw a spot of light on the cloud base.   By means of a known base line and measurement of the elevation angle of the light spot, the height can be determined from the relation $h = b \tan a$.   In Fig. 39 is shown a clinometer for measuring the angle of elevation.   In use, the observer stands at one end of the base line and sights through the eyepiece at the spot of light directly above the other end of the base line.   The spot of light is centered on the cross wires, and the index is clamped in place.   The angle can then be read, and by means of tables the ceiling height can be determined.   The light uses a 420-watt bulb and projects a concentrated beam of light.

A recent development for the measurement of ceiling heights, both day and night, is the photoelectric clinometer (Nephohypsometer) that, operating in conjunction with a modulated beam searchlight, determines ceiling heights on the same principle as the conventional clinometer and ceiling light.   The modulated light beam is reflected from the clouds and is detected by the photoelectric unit which responds only to the modulation frequency.[11]

AEROMETEOROGRAPH

The aerometeorograph is designed to be attached to an airplane and records the pressure, temperature, and relative humidity of the air through which it is carried.

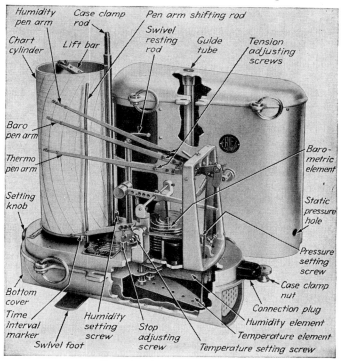

FIG. 40.—Aerometeorograph.

Prior to the development of the radiosonde, this instrument was used extensively for upper-air soundings (see Fig. 40).

RADIOSONDE EQUIPMENT

There are a number of radiosonde systems, but space permits a description only of the modulated audio-frequency system in widespread use in the United States.[8]

The principal parts of a radiosonde are shown in Fig. 41, and a schematic diagram is shown in Fig. 42.   The complete radiosonde measures about 8½ by 8 by 4 in. and, less its battery, weighs about 700 grams.   It is designed for use with a balloon giving an ascension rate of approximately 200 m per min.   The balloon weighs about 350

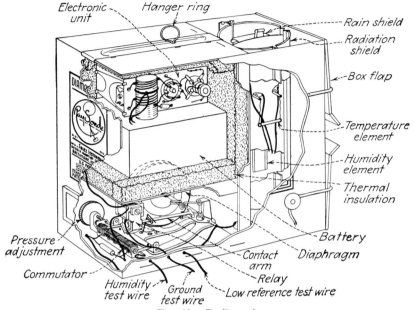

Fig. 41.—Radiosonde.

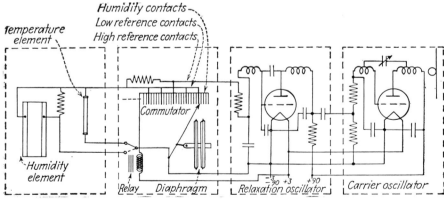

Fig. 42.—Radiosonde circuit.

grams and is capable, under favorable conditions, of reaching an altitude of 10 to 12 miles with normal load.   Inflation with helium or hydrogen to a free lift of 500 grams gives the requisite ascension rate.   A paper or cloth parachute is tied about 6 ft below the balloon, and the instrument is tied about 50 ft below the parachute. In gusty winds, when launching is difficult, the instrument will frequently be coupled close to the balloon to facilitate ground handling. The total load is supported directly by the neck of the balloon.

The instrument case is made of double-faced corrugated cardboard and covered with a waterproof surface that will reflect solar radiation.

The electronic unit combines in one assembly a radio-frequency oscillator operating at 72.2 mc and a modulating oscillator of the relaxation type operating over a nominal audio range of 10 to 200 cycles per sec. Changes in the rate of relaxation are caused by changes in temperature and humidity, thus varying the frequency in relation to atmospheric conditions. The power supply consists of a dry battery that has a 3-volt filament section and a 90-volt plate section. Two sizes of batteries of equal electrical characteristics but different shelf life are used. One with a shelf life of 6 months weighs approximately 20 oz, and the other with 3 months' shelf life weighs only 12 oz.

The pressure unit has aneroid capsules that move a contact arm over a commutator that has alternate conducting and insulating segments. The position of the contact arm is a measure of the pressure. In addition to measuring pressure, the unit acts as a switch to control the temperature, humidity, and reference circuits. The temperature element is a resistor having a large negative temperature coefficient of resistance. Temperature measurements depend on the ratio of the resistances at different temperatures, rather than on the actual resistance value.

The humidity element employs a chemical film, the resistance of which varies with humidity and temperature. Allowance is readily made for the temperature effect in evaluating the record.

The temperature and humidity elements are mounted in a cylindrical radiation shield to protect them as much as possible from direct exposure to solar radiation. The shield is arranged so that a free flow of air takes place both inside and outside.

The relay controlled by the pressure unit connects the temperature and humidity units in the circuit.

Circuit. The modulating oscillator has its rate of relaxation controlled by a resistance network connected in the grid circuit. The plate circuit of the modulating oscillator is capacitively coupled to the grid circuit of the radio-frequency oscillator, thus modulating it at the relaxation frequency.

Referring to Fig. 42, when the pressure contact arm is on an insulating spacer, the temperature element is in the circuit; when the contact arm is on a humidity contact, the relay is energized, thus disconnecting the temperature element and connecting the humidity element. When the contact arm strikes a low reference contact, a signal is emitted that has nominally a frequency of 190 cycles per sec, or 95 divisions on a 100-division recorder scale commonly used in radiosonde work. This signal is used as an upper limit for the measuring range of the instrument and is arbitrarily adjusted to 95 recorder divisions for evaluation of records, regardless of its true value. The adjustment of the low reference frequency to 95 is an expansion or contraction of the measuring scale of the particular radiosonde in question to make it fit a standard calibration.

When the contact arm touches a high reference contact, the signal is nominally 97 divisions in value and is used for identifying the signal sequence, each fifteenth contact being a high reference.

In the radiosonde illustrated, there are 80 contacts, each fifth being a low reference contact except for the fifteenth, thirtieth, etc., which are high reference up to and including the sixtieth. Above the sixtieth, each fifth is a high reference, and intermediate contacts are connected to give low reference signals. Near sea level, one contact and its associated insulating spacer represent approximately a 500-ft change in altitude.

The scale range for temperature measurement extends from 80 divisions for high temperatures to 5 for low. The humidity scale ranges from approximately 93 to 5 divisions.

Figure 43 gives the temperature versus resistance curve of the temperature element and the recorder divisions (*i.e.*, frequency) versus resistance curve of the relaxation oscillator.

Temperature elements all have curves of the same shape but differ slightly from each other in absolute value of resistance. To evaluate the temperature from a radiosonde record, it is necessary to find the resistance corresponding to the recorder reading in question and then find the temperature corresponding to that resistance. In practice, the resistance, being common to both curves, is eliminated, and the

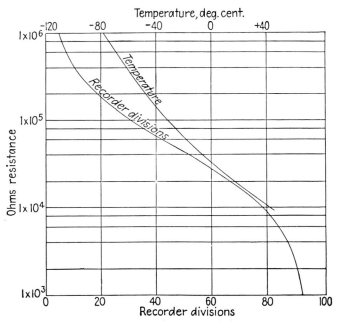

Fig. 43.—Radiosonde-characteristic curves.

temperature is read opposite recorder divisions on a special slide rule, the setting of which is done for the particular instrument in question prior to release.

Humidity readings are evaluated by means of a graph that gives humidity as a function of recorder divisions and temperature.

Ground Equipment. The major components of a radiosonde receiving and recording equipment are an antenna system, a receiver, an electronic frequency meter, and a recorder. Figure 45 is a schematic diagram from which the power-supply circuits have been omitted for simplicity.

A vertical half-wave dipole antenna is used and is connected to the receiver by a gas-filled or solid coaxial transmission line.

In Fig. 44, the top unit is the receiver, the next is the frequency meter, and beneath that is the recorder. The receiver is of the superregenerative type and is equipped with a tuning meter, a loud-speaker, and a special audio amplifier the output of which connects to the frequency meter input. The frequency meter has one stage of amplification that controls two Thyratron tubes. These Thyratrons are connected in push-pull and alternately discharge capacitors through a diode rectifier and indicating and recording meter circuit. The frequency meter input switch allows the input circuit to be grounded or connected to 60 cycles, thus permitting checking the zero point and a point corresponding to 60 cycles on the recording scale.

The recorder meter, while physically located in the recorder, is connected in series with the panel meter and operates in unison with it. An equalizing adjustment

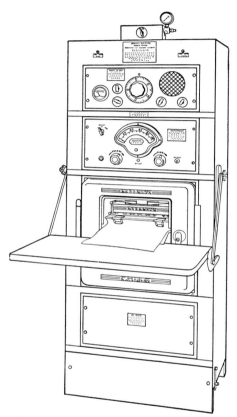

Fig. 44.—Radiosonde ground equipment.

enables the two meters to be set to agree exactly at the reference frequency. Rheostats, one a coarse adjustment and one a fine adjustment, provide for expansion and contraction of the scale to fit individual radiosondes.

The recorder mechanism is shown diagrammatically in Fig. 46. The recorder meter is scanned once every 2 sec by a light beam revolving concentrically with the meter axis. This beam of light, as it revolves, is interrupted by the wide meter pointer. After each interruption, the light strikes the photocell suddenly, causing its resistance to drop and through the medium of the amplifier to energize the tapper bar magnets, causing the tapper bar to press the ribbon and chart paper against a spiral ridge, the position of which is related to that of the meter pointer. The gear ratios are such that, in scanning the full meter scale, the spiral drum makes one turn, thus expanding the meter scale over a chart 10 in. wide. The paper is fed by a constant-speed motor while another motor drives the ribbon through an automatic reversing mechanism.

The power supplies for the frequency meter and amplifier are well regulated to prevent variations in supply voltage from affecting the operation.

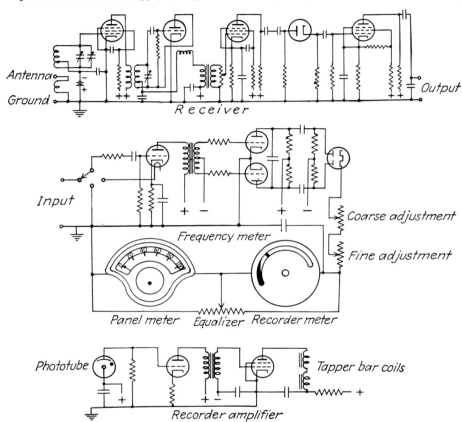

Fig. 45.—Ground-equipment schematic diagram.

Superheterodyne receivers are also used for radiosonde work. This type is characterized by greater selectivity and freedom from local interference than is the superregenerative receiver.

An alternate type of recorder for radiosonde signals, the Leeds and Northrup Company Speedomax, is a null-method recording potentiometer. In this recorder, the record is made by an ink pen moved back and forth across the chart. The record and pen are completely visible, eliminating the necessity of the indicating meter on the frequency meter panel.

Prior to release, a radiosonde is placed in an instrument shelter and connected to a motor-driven switch. This switch alternately connects the different circuits, causing the instrument to send temperature, humidity, and reference signals. These signals are recorded and checked against the actual temperature and humidity in the shelter. The instrument is then released, and as it rises the pressure-operated switch performs the same function in the air that the test switch in the shelter performs on the ground. The radiosonde record is in the form of a graph with temperature, humidity, and

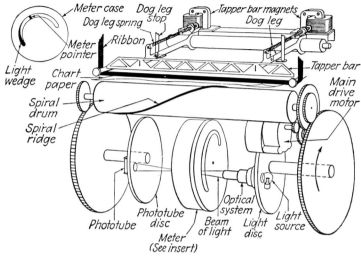

Fig. 46.—Recorder diagram.

reference signals plotted in terms of recorder divisions against time, while the pressure is from the contact sequence.

EXPOSURE OF INSTRUMENTS

The value of meteorological observations depends to a large extent on the exposure of the instruments. The importance of proper exposure can hardly be overemphasized.

Thermometers and thermographs are normally mounted in an instrument shelter such as that shown in Fig. 47. This has a double roof and louvered sides. The double roof helps prevent radiation from heating the interior, and the louvered sides allow free air circulation over the instruments. Shelters are painted white to reflect radiation. Instruments should be placed near the center, since the walls will be heated above the true air temperature at times. A shelter should be oriented so that direct sunlight will never strike the interior when the door is open. The best location is in an open space over sod with the floor of the shelter at least 4 ft above the ground. Shelters, because of lack of a better place, must often be mounted on roofs. A shelter may also be mounted on the shady side of a building with an air space between. Adequate exposure of instruments on shipboard is difficult. In general, the shelter should be placed where it will be uninfluenced by heat generated by or radiation reflected from the ship, and exposed to as free air circulation as possible.

Maximum and minimum thermometers are mounted on a crosspiece near the center of the shelter. When a rotor psychrometer is installed, care must be taken to allow sufficient room for it to revolve safely.

Wind instruments must be placed where the wind is undisturbed by surrounding objects. In general, wind speed increases with elevation above the ground. In the interest of comparable wind measurements, the instruments at different stations

should be exposed under standard conditions at a uniform height.   Since this is difficult, the vane and anemometer are usually placed as high as possible.   The wind vane
must be oriented correctly so that true directions will be indicated.

A rain gauge should preferably be placed on the ground in an open space.   Near-by
low objects that act as windbreaks may assist in obtaining accurate measurements.

A sunshine-duration transmitter must be located where it will be exposed to the
solar radiation throughout the entire day.

Successful pilot-balloon runs require a theodolite location that allows clear vision
in all directions to within a few degrees of the horizon.   The theodolite must be
oriented to true north.

The only part of radiosonde equipment critical with respect to location is the receiving antenna, which should preferably be located much as a theodolite, *i.e.*, so placed

Fig. 47.—Instrument shelter.

as to allow line-of-sight transmission from all directions.   The antennas should also
be placed as near the receiver as possible to avoid excessive length of transmission
line.   From a convenience standpoint, the instrument shelter where the radiosondes
are ground checked, the balloon-inflation station, and the ground equipment should
be as close to each other as practicable.

In locating both theodolites and radiosonde antennas, consideration should be
given to the prevailing winds, and the most favorable locations available should be
used.

Needless to say, the ideal exposure of all instruments is generally impossible at
any one station.   Care should be taken to use the best locations available, consideration being given to the function and relative importance of each instrument at the
particular station in question.

AUTOMATIC WEATHER STATIONS

Automatic weather stations have been developed for taking scheduled surface
observations at isolated locations and transmitting the data to the points where used.
An automatic station consists of an assemblage of measuring instruments, a mechanism for converting measurements into a form suitable for transmission, a radio

transmitter, program and control equipment, and the necessary power plant, house, and accessories.   Designs are such as to facilitate transportation to an installation at isolated places.

What is probably the most common type operates on the same principle as the audio modulated radiosonde.[10]   Other types have been built using various methods of converting the meteorological data into a form suitable for transmission by radio.

CALIBRATION AND TEST EQUIPMENT

In order to calibrate and/or test instruments, it is generally necessary to subject them to the range of conditions that they are designed to measure.

Pressure measuring instruments are tested in chambers in which the pressure may be varied over the needed range and accurately measured, usually with a mercury manometer.   For precise work, the manometers are maintained at a constant temperature or corrections applied for temperature effects.

Control of temperature in pressure chambers is necessary for checking the temperature effect on the instruments being tested and for adjusting the compensating means.   Temperatures are usually read from liquid in glass thermometers visible through windows.

Pressure changes are obtained by vacuum and pressure pumps, and temperature changes by electric heaters and mechanical refrigeration or dry ice.   The chambers are thermally insulated and of sufficient strength to withstand the pressure differentials encountered.

Testing of aneroids, barographs, aerometeorographs, etc., is usually done under static conditions, that is, the pressure and temperature are brought to the desired values and allowed to stabilize before readings are taken.   Air circulation in the chamber is necessary to prevent large temperature gradients.

A specialized test known as a *flight similitude test* is employed for testing radiosondes.   The pressure and temperature are decreased at rates approximating those encountered in actual radiosonde flights.   Equipment for such tests requires careful design to eliminate pressure and temperature gradients since measurements are made under constantly changing conditions.   Approximately forty readings of pressure and temperature are taken during a test and then compared with a record made by the radiosonde being tested.

A typical flight similitude chamber consists of an aluminum or copper can large enough to hold a radiosonde and provide room for air circulation around and through its radiation shield.   This can is mounted in a well-insulated cabinet.   Cooling is obtained by placing dry ice on the can or by surrounding the can with a liquid cooled with dry ice.

Satisfactory control of rates of evacuation and cooling are obtained by a valve in the suction line from the vacuum pump and the careful application of the dry ice in small quantities.   A temperature of $-60°C$ can be reached in this way in approximately 75 min. at a substantially linear rate.   Mechanical refrigeration is used to a limited extent for this work.   A manometer covering the full range of atmospheric pressures and a thermocouple potentiometer outfit for accurate temperature measurement are used as standards to check the radiosonde.

Humidity tests require a chamber in which the humidity can be varied and for some types of tests, the temperature also.   A humidity chamber should be lined with a nonhygroscopic material, equipped with an air-circulating fan, and wet- and dry-bulb thermometers.   Variations in relative humidity are usually obtained by one of two methods.   Various chemical salt solutions will maintain fairly constant and fixed values of humidity in a chamber by absorbing moisture from or giving it up to the air.   Eight different solutions quite adequately cover the relative humidity range.

The other method frequently used employs, for low humidities, the condensation of the moisture in the air by a cooling unit and reheating of the air, while for high humidities moisture is supplied by an atomizing spray.

Thermometers and any temperature instrument that can be safely placed in a liquid are best checked in a liquid bath which can be accurately controlled, and in which gradients are substantially eliminated by stirring and careful design.

Wind instruments require wind tunnel tests for checking over-all performance and obtaining calibrations for new designs. Routine checks of various types of rotation anemometers can be made by removing the rotor and driving the spindle at a constant known speed by means of a synchronous motor with suitable gear reduction, or other means. The indicated speed should agree within the allowable tolerance with the wind tunnel speed known to produce that rate of spindle rotation.

Radiosonde ground equipment for the audio-modulated-frequency system is checked for recording accuracy by means of a frequency generator, which supplies accurate frequencies to the frequency meter input. These frequencies are generated by a tuning fork controlled multivibrator or small permanent magnet alternators driven at a constant speed.

Bibliography

1. Barometers and the Measurement of Atmospheric Pressure, *U.S. Weather Bur. Circ. F.*
2. Care and Management of Electrical Sunshine Recorders, *U.S. Weather Bur. Circ. G.*
3. Instructions for the Installation and Maintenance of Wind Measuring and Recording Apparatus, *U.S. Weather Bur. Circ. D.*
4. Instructions for Installation and Use of Telethermoscopes, U.S. Weather Bureau.
5. Instructions for Making Pilot Balloon Observations, *U.S. Weather Bur. Circ. O.*
6. KLEINSCHMIDT, E.: "Handbuch der meteorologischen Instrumente," Springer, Berlin, 1935.
7. MIDDLETON, W. E. KNOWLES: "Meteorological Instruments," University of Toronto Press, Toronto, 1941.
8. DIAMOND, HARRY, WILBER S. HINMAN, JR., FRANCIS W. DUNMORE, and EVAN G. LAPHAM: Upper-air Weather Soundings by Radio, *A.I.E.E., Tech. Paper* 40–47, December, 1939.
9. FERGUSSON, STERLING P.: Experimental Studies of Anemometers, Harvard Meteorological Studies No. 4.
10. DIAMOND, HARRY, and WILBER S. HINMAN, JR.: An Automatic Weather Station, *Nat. Bur. of Standards Research Paper* RP 1318.
11. LAUFER, MAURICE K., and LAURENCE W. FOSKETT: The Daytime Photoelectric Measurement of Cloud Heights, *J. Inst. Aeronaut. Sciences*, **8** (5): 183, March, 1941.

SECTION IX

TRANSMISSION AND PLOTTING OF METEOROLOGICAL DATA

By George R. Jenkins

CONTENTS

573

SECTION IX

TRANSMISSION AND PLOTTING OF METEOROLOGICAL DATA

By George R. Jenkins

METEOROLOGICAL CODES*

A major problem of synoptic meteorology is the collection and dissemination of reports. The sheer bulk of data required to present a clear picture of weather over a large enough area to be suitable for forecasting has necessitated the use of many media of communication and the reduction of the size of messages by the use of codes of various kinds. A world-wide survey of all meteorological codes and means of transmitting weather information in use is not feasible in this volume; the currently used codes are far too numerous. There are many reasons for the complexity in weather codes and reports; the diversity of elements to be observed, the various techniques of observation, variation in the information desired by analysts in different parts of the world, and lack of uniformity in the codes adopted by separate political units are some of the reasons. Therefore, it is intended here to outline the basic principles upon which the transmission of meteorological data by coded reports is based. Details can be found in the references cited.

The International Codes Organization

Successive agreements and adoptions by various countries through the agency of the International Meteorological Congress Committees have resulted in a considerable degree of standardization of weather codes. Approval of a proposed code is indicated by its being assigned an appropriate International designator. This designator is in the form of F followed by a one-, two-, or three-digit number. The first of the digits is the general-category indicator: 1 indicates a form for surface reports from land stations; 2 is for surface reports from ships; 3 is for nephoscope and upper-air reports from both land stations and ships; 4 is for forecasts; and 5 is for analyses. Second and third digits are used to indicate the separate forms in use within these categories.

Although many political units use International codes, others use portions of these codes or have devised forms of their own. Details of the many code forms and their areas of utilization can be obtained in most complete form from the Hydrographic Office, U.S. Navy, in the publication H.O. 206, Radio Weather Aids to Navigation. In the following discussions, only the most widely used or more representative forms of codes are given; the choices are based on prewar and current usages as much as security considerations permit.

International forms are indicated by the International designators; United States forms are indicated by USWB. Definitions of the letter designators are given on successive pages under "Explanations of Code Symbols," and, where appropriate, code numerals are tabulated as to meaning under "Code Tables."

---

* Appreciation for valuable contributions and suggestions is due to Captain H. T. Orville, U.S.N., and to Commander Peter Lackner, U.S.N.

In addition to the numeral codes, mention should be made of the symbolic weather reports in use as basic data for airways forecasting in the United States. They are used to transmit over Schedule A hourly weather observations, route forecasts, and other aids to air navigation. Explanation of the code used in Schedule A hourly reports is given separately in Table 4, page 588 to 593.

Surface Synoptic Reports

Land Stations. The F 1 code, consisting of only four groups, is not used alone to any extent, but it constitutes the first part of most codes. Wide variation of practice occurs with additional groups. F 11 is the fundamental code form for middle-latitude stations; it provides for the reporting of high clouds and pressure tendencies. In turn, there are four extensions of the F 11 according to the location of stations. F 111 and F 112 are for inland stations where maximum and minimum temperatures are significant, along with state of the ground; they are used as a pair, the choice being determined by the time of reporting, which, in turn, determines whether maximum (F 111) or minimum (F 112) temperature is reported. Similarly, F 113 and F 114 are used as a pair for coastal stations, and the time of reporting determines whether the state of the sea, the visibility to seaward, and the state of the ground, or the difference between air and sea temperatures, the swell, and the direction of swell are to be reported.

F 12 and its elaborations are for tropical stations; pressure tendencies are of less significance than details of past weather, especially time of commencement of "weather," and time, direction, and force of maximum winds.

The USWB code, with possible use of 10 groups, is the most elaborate of those cited. Its ninth group takes cognizance of the problems of reporting pressures in the western highlands, and its tenth group permits special reports on state of sea and swell from coastal stations (thus eliminating the use of separate codes, as in the International) as well as many other special phenomena.

F 1 $IIIC_LC_M \ wwVhN_h \ DDFWN \ PPPTT$
F 11 F 1   plus UC_Happ
F 111 F 11 plus RRT_xT_xE
F 112 F 11 plus RRT_nT_nE
F 113 F 11 plus $RRSV_sE$
F 114 F 11 plus RRT_dKD_k
F 12 F 1   plus $UURRt_w$
F 123 F 12 plus $D_LC_HD_{H/M}T_xT_x \ DDF_xGG$
F 124 F 12 plus $D_LC_HD_{H/M}T_nT_n \ DDF_xGG$
$USWB$ $IIIC_LC_M \ wwVhN_h \ DDFWN$ (00FF0 or 55FF5) $PPPTT \ UC_Happ \ T_sT_sD_c$-$T_{x/n}T_{x/n} \ 7h_ch_cVV \ 8R_sR_tRR$ (000R'R') $9P_mP_mP_ma_3 \ 0S_PS_PS_pS_p$.

Ships. As with land-station reports, ship reports are standardized as to their initial groups; F 2 is the basic code. F 232 was adopted for standard middle-latitude use; F 242 differs from it in reporting height of clouds in place of form of high clouds and was introduced for use in regions traversed by air lines. F 261 simplifies cloud and pressure-tendency data, while F 291 accents sea conditions.

F 2   ship $YQLLL \ lllGG \ DDFww \ PPVTT$
F 232 F 2 plus $3C_LC_MC_HN$ $T_dKD_kWN_h \ d_sv_sapp$
F 242 F 2 plus $4C_LC_MhN$ $T_dKD_kWN_h \ d_sv_sapp$
F 261 F 2 plus $6KD_kCN$ $T_dd_sAWC_H$
F 291 F 2 plus $9SKD_kW$

Upper-air Reports

Pilot-balloon Observations. Two basic forms of code are in use; their differences are in the methods of reporting heights and wind velocities. F 331 and F 341 are designed to report primarily significant wind levels as determined by changes in the direction or velocity; the USWB form is distinctive in that it provides for reporting winds for standard levels. In $F331$ and $F341$, two numbers are needed to identify the elevation, and velocity is reported by only one number; this introduces some complexity to velocity tables. In the case of standard-level reporting, the levels are easily identified by the use of one number, and two numbers are available for reporting velocity. In both forms, direction is reported by two numbers.

F 331 pilot $IIIGG$ $HHddv_5$ $HHddv_5$ . . . etc. $(C_LC_M$ $HHM)$ $(44444$ $m_1m_2m_3m_4m_5$ $m_6m_7m_8xx)$
F 341 pilot ship $YQLLL$ $lllGG$ $HHddv_5$ $HHddv_5$ etc.
$USWB$ $II(I)GG$ $0ddvv$ $(H_a)ddvv$ $(H_a)ddvv$ etc.

Temperature and Humidity Soundings. As with pilot-balloon soundings, upper-air observations of temperature and humidity may be reported either in terms of significant levels or in terms of standard levels, which, in turn, may be standard pressure levels or standard elevations. The International forms cited report significant levels and standard pressure levels; the USWB forms recently and currently in use have accented significant levels and have shifted in secondary accent from standard elevations to standard pressure levels. $USWB(1)$ is the form used up to Feb. 1, 1945; $USWB(2)$ is the form used from Feb. 1, to May 1, 1945; $USWB(3)$ is the form from May 1, to July 1, 1945; $USWB(4)$ is the form used from July 1, 1945 to Jan. 1, 1946· and $USWB(5)$ is the form proposed for use in 1946.

F 36 temp $IIIGG$ $P_1P_1P_1$ $TTTUU$ $P_2P_2P_2$ $TTTUU$ etc. (for significant levels)
F 363 temp $IIIGG$ $H_dH_dP_1P_1P_1$ $TTTUU$ $H_dH_dP_2P_2P_2$ $TTTUU$ etc. (for significant levels) 00000 $H_dH_dH_dH_dH_d$ $TTTUU$ etc. (for standard pressures, hundreds of millibars)
F 37 temp ship $YQLLL$ $lllGG$ $P_1P_1P_1$ $TTTUU$ etc. (as $F36$)
F 373 temp ship $YQLLL$ $lllGG$ $H_dH_dP_1P_1P_1$ $TTTUU$ etc. 00000 $H_dH_dH_dH_dH_d$ $TTTUU$ (as F 363)

$USWB(1)$ (first transmission) $IIIGG$ $PPPTT$ $UUH_5H_{10}H_{20}$ $PPTTU$ $PPTTU$ etc. 10171 $Lm_rm_rm_rm_r$ $Lm_rm_rm_rm_r$ . . . etc. . . . $(50)HHH$; (second transmission) $IIIG_{50}G_{50}$ $YYH_{33}H_{43}H_{53}$ $PPTTU$ $PPTTU$ etc. $(101A_{df}A_{df})$ . . . etc. 10177 $L_xP_{px}P_{px}P_{sx}P_{sx}$ $L_xZ_{fx}Z_{fx}d_{wx}d_{wx}$ $L_xd_{ix}d_{ix}S_{ix}S_{ix}$ $L_yP_{py}P_{py}P_{sy}P_{sy}$ $L_yZ_{fy}Z_{fy}d_{wy}d_{wy}$ $L_yd_{iy}d_{iy}S_{iy}S_{iy}$ $L_zP_{pz}P_{pz}P_{sz}P_{sz}$ $L_zZ_{fz}Z_{fz}d_{wz}d_{wz}$ $L_zd_{iz}d_{iz}S_{iz}S_{iz}$ $101A_{df}A_{df}$ etc.

$USWB(2)$ (first transmission) same as $USWB(1)$; (second transmission) $IIIG_{50}G_{50}$ $YYH_{33}H_{43}H_{53}$ $PPTTU$ $PPTTU$ etc. . . . $(101A_{df}A_{df},$ etc.$)$ 10177 $L_hP_{px}P_{px}P_{sx}P_{sx}$ $L_tP_{py}P_{py}P_{sy}P_{sy}$ $L_uZ_{fx}Z_{fx}Z_{fy}Z_{fy}$.

$USWB(3)$ (first transmission) same as $USWB(1)$; (second transmission) $IIIG_{50}G_{50}$ $YYH_{33}H_{43}H_{53}$ $PPTTU$ $PPTTU$, etc. . . . 10197 $OOTTU$ $HHHm_rm_r$ $85TTU$ $HHHm_rm_r$ etc. (for the 1,000-, 850-, 700-, 500-, and 300-mb levels in that order) $(101A_{df}A_{df},$ etc.$)$.

$USWB(4)$ (first transmission) $IIIGG$ $PPPTT$ $UUee$ $PPTTU$ $PPTTU$ etc. . . . 10171 $Lm_rm_rm_rm_r$ etc. . . . $OOHHH$ $85HHH$ $70HHH$ $50HHH$; (second transmission) $IIIG_{50}G_{50}$ $YYee$ $PPTTU$ $PPTTU$ etc. . . . $30HHH$ $20HHH$ $10HHH$ 10198 $PPTTU$ $PPTTU$ etc. . . . 10171 $Lm_rm_rm_rm_r$ etc. (these mixing ratios referring in order to the 5,000-, 10,000-, 15,000-, and 20,000-ft levels only) $101A_{df}A_{df}$ etc.$)$.

$USWB(5)$ (first transmission) same as $USWB(4)$; (second transmission) $IIIG_{50}G_{50}$ $YYee$ $PPTTU$ $PPTTU$ etc. . . . $30HHH$ $20HHH$ $10HHH$ $(101A_{df}A_{df}$ etc.$)$.

Miscellaneous

There are many aspects to the reporting of meteorological data other than those covered. Several are noted briefly below, selected according to their general usefulness in synoptic work.

Aircraft Reports. These have been used for some time as a part of the airways reports in the United States; they are identified as "PIREPS" and utilize mixed symbolic code and abbreviated adaptation of plain language. More formal methods of utilizing aircraft reports were developed before the war; meanwhile they have been considerably modified. Although this is an important step in meteorological reporting, it is not thought appropriate to include an extended account of the reports in this discussion; many current developments are considered to be confidential, and it is further likely that current usages may be changed during the war and after it. The most complete and widely used aircraft reporting system is that developed by the United Nations; it has been published by the U. S. Weather Bureau under the title "Combined Aircraft Weather Report Form (CAW-C)."[3]

U.S. Teletype Sequence Indicators. Teletype transmissions are organized as to nature of data and area of its coverage into sequences, each of which is identifiable by its indicator at the head of the sequence. The general form of these indicators for Schedule C (those for Schedule A are only slightly different) is as follows:

$$(PDW)\ RI\ SI\ YYGGggZ.$$

If PDW appears, it indicates that the sequence is out of its normal order and constitutes delayed weather data. RI is here used to symbolize "Report Indicator"; the indicators used, with their meanings, are as follows:

AG North Atlantic reports
AL Alaskan reports
CA Central and South American reports
CR Crop region reports
FA Forecasts, airways
FD Forecasts, 5 day
FR Frontal Passage résumé
FS Forecasts, states
IR Icing résumé
MA Master analysis
MB Bermuda reports
MC Miscellaneous Canadian reports
MD General reports
ME Eastern Canadian reports
MF AACS Canadian reports
MG Eastern Atlantic reports
MJ Pacific reports
MK Mexican reports
MM Miscellaneous United States reports

MN Mexican, Central American, and western and eastern South American reports
MO Caribbean Islands and Eastern South American reports
MP Primary United States reports
MR Great Lakes reports
MS Supplementary United States reports
MT Primary United States reports
MW Western Canadian reports
MX South Atlantic reports
PA Prognostic map analysis
PB Pilot-balloon observations
RR Precipitation bulletins
RS Radiosonde and aerograph soundings
TA 5,000 and 10,000-ft weather-map analysis
TX Temperature extremes bulletin
$3M$ 3-hourly map reports

SI is here used to symbolize "Sequence Indicator," always a two-digit number. For details of the stations included, points of relay, and numbers of sequences, see the Federal Airways Manual of Operations.[4]

The rest of the indicator has meaning according to the definitions given under Explanations of Code Symbols, page 573; briefly, it supplies the day of month and the time of reports in hours and minutes.

Forecasts, Analyses, Advices, Etc. The general F 4 group of International codes is not given here. In the United States, forecasts are issued in plain language or an abbreviated adaptation. The general technique and structure of a code for transmitting an analysis has been developed to a high degree in the "Combined Analysis Code (CAC)." The form is completely general; any type of analysis (surface, upper-air constant level, upper-air constant pressure, prognostic surface or upper-air, etc.) can be transmitted with ample specification of details to permit very satisfactory rendition of the analysis. Complexities of the code are too great to permit its being presented in this volume; it has been published in useful form by the U.S. Weather Bureau.[4]

Explanations of Code Symbols

Most of the letter symbols of the codes given are purely symbolic abbreviations for the meteorological elements. In coded reports, they are replaced by numbers. Often, primarily when two numbers are available to describe an element, it is necessary to give only the definition, the range of elements, and the practices in force, in order for the coded reports to be interpreted. At other times, primarily when a single number is assigned to the element, the total range of values is too great to be reported by a single number without reduction of the range to no more than 10 categories. Such cases are defined below and in succeeding code tables.

Some letters and numbers in the symbolic forms of the codes appear in coded reports as given. These are indicators that may serve to identify either the code form itself or the content of individual groups within the reports. In the explanations, such indicators are identified by a single asterisk (*) for those used in International codes and by a double asterisk (**) for those used with USWB codes.

A Amount and character of barometric tendency (3-hr), expressed by a single figure (Table 1).

$A_{df}A_{df}$ Last two digits of "additional data designator group" $(A_dA_dA_dA_{df}A_{df})$ of USWB radiosonde code. $A_dA_dA_d$ is always transmitted as 101. $A_{df}A_{df}$ may range from 01 to 96, with some figures not used. 33 to 64, inclusive, give reasons for no report or for termination of record; 65 to 96 inclusive require additional groups to follow the designator. Most commonly encountered are 77 (isentropic data follow) and 71 (mixing-ratio data follow). Recent important additions are: 97 (constant pressure data in the form $OOTTU\ HHHm_rm_r\ 85TTU$ $HHHm_rm_r$ etc. follow); and 98 (constant level data for 5,000-, 10,000-, 15,000-, and 20,000-ft and 10-, 13-, and 16-km levels in form $PPTTU\ PPTTU$ etc. . . . 10171 $Lm_rm_rm_rm_r$, etc. follow).

a Characteristic of the barometric tendency for the 3-hr period preceding the time of report (see also pp). (Table 1.)

a_3 Characteristic of the barometric tendency for the 3-hr period ending 3 hr before the time of report (Table 1).

C Form of predominating cloud (used when only one form is reported). (Table 1.)

C_L Form of low cloud (Table 1).

C_M Form of middle cloud (Table 1).

C_H Form of high cloud (Table 1).

$CORR\ (RECTIF)$*   Corrected observations. CQN** same as $CORR$.

DD True direction of the wind near the ground. It is reported from 00 (calm) to 32 (north), hence to 32 points of the compass. Commonly only even values are used. With the International code, 33 is added to the coded value of DD when unusual squalliness or gustiness has occurred during the hour preceding observation; 67 is added if there has been a line squall.

dd Direction of wind in the upper air, or cloud movement, on scale 01 to 36, *i.e.*, to nearest 10 with the 0 omitted; 00 denotes calm. With USWB Pibal code and International forms similar to it, 50 is added to dd if the wind velocity exceeds 99 units (see vv); with F 331 and F 341, 50 is added to dd when wind velocities are between 30 and 60 mph or between 92 and 122 mph (see v_5).

D_K Direction from which swell comes. (9 denotes all directions or no definite direction.) (Table 1.)

D_C Direction from which clouds move. Refers to C_M if reported; if there is no C_M, refers to C_H; lastly, refers to C_L (9 denotes direction unknown). (Table 1.)

$D_{H/M}$ Direction from which C_H move when reported; if no C_H, refers to C_M (9 is not used). (Table 1.)

D_L Direction from which C_L move when reported (9 not used). (Table 1.)

d_s Direction in which ship is moving (Table 1).

$d_{wx}d_{wx}$, $d_{wy}d_{wy}$, $d_{wz}d_{wz}$ Direction of wind at the isentropic surface x, y, or z, whichever is indicated by the subscript. It is coded the same as dd.

$d_{ix}d_{ix}$, $d_{iy}d_{iy}$, $d_{iz}d_{iz}$ Direction of isobars at the isentropic surface x, y, or z, whichever is indicated by the subscript. It is coded the same as dd.

eee   Elevation of the reference level to which the initial or surface pressure included in the radiosonde report is reduced; in North America it is reported in tens of feet; elsewhere it may be reported in tens of feet or in whole meters.

E   State of the ground (Table 1).

F   Wind velocity expressed in Beaufort force. International forms report winds above force 9 as 9, and the group is followed by words indicating the force.

FF   Appears in added group of USWB surface code, following the usual third group when wind velocity exceeds force 9. The form 00*FF*0 is used when the wind has been measured; the form 55*FF*5 when the wind is estimated. The following table applies:

FF	Mph	*FF*	Mph	*FF*	Mph	*FF*	Mph
10	55–64	12	74–82	14	93–103	16	115–125
11	65–73	13	83–92	15	104–114	17	126–136

F_x Maximum wind force (Beaufort), usually for the 6 hr preceding the time of observation (Table 1).

GG   Hour of the day, GCT nearest to the time of surface observations, or of commencing pilot-balloon and radiosonde ascents. Numbers from 01 to 24 correspond to the 24-hr clock.

$G_{50}G_{50}$ Hour of the day, GCT, with 50 added. Used in the USWB radiosonde code to indicate that the following report is the second transmission.

gg   Minutes of time.

h   Height of base of cloud. Refers to base of lowest type of cloud if the base is below 8000 ft. If there are fragments of detached cloud lower than that of the predominating type, the height of the fragments may be sent in plain language in addition to the regular report (Table 1).

h_ch_c Height of ceiling in hundreds of feet.

H   Elevation in thousands of feet of the level for which data on wind velocity apply. Used in USWB pilot-balloon reports. Only alternate groups have such level designators, except for levels above 25,000 ft (Table 1).

H_5, H_{10}, H_{20} The numbers of the three groups in the first transmission of the USWB radiosonde code containing the data for the 5,000-, 10,000-, and 20,000-ft mandatory levels. These groups are further identifiable by each being followed by a period. 15,000 ft data are also followed by a period.

H_{33}, H_{43}, H_{53} The same as the preceding, but appearing in the second transmission and indicating the positions of the groups giving the data of the 10-, 13-, and 16-km levels. Periods also follow these data groups.

HH   Height in hectometers above sea level to which upper-air wind data refer. Mandatory levels are 10 and 20 hectometers; the lowest level of each stratum of distinctive wind directions of velocity is also transmitted.

H_dH_d Heights in hectometers for which data of temperature and humidity are transmitted.

HHH   Height above sea level of a mandatory pressure level, reported in tens of feet or in tens of meters (according to the practices of the reporting agencies).

III, *II*   Station designators, either index numbers or letters.

K   State of swell in open sea (Table 2).

LLL   Latitude in degrees and tenths, the latter being obtained by dividing the number of minutes by 6 and neglecting the remainder.

lll   Longitude in degrees (tens, units, and tenths: if the longitude exceeds 99°, its value is determined from the quadrant of the globe; see explanation of Q).

L Designator of levels for which mixing-ratio data are sent. It is used in the first transmission of the USWB radiosonde code. Significant levels are numbered in order from the surface (1) on up. The mixing-ratio groups are composed of the number of each succeeding odd-numbered level, the mixing ratio of that level in units and tenths, and the mixing ratio of the next even-numbered level in units and tenths. When mixing ratio exceeds 9.9 grams per kg, only the units and tenths are transmitted; the decoder supplies the additional figure by noting the relative humidity, temperature, and pressure, the combination of which indicate the mixing ratio intended.

L_x, L_y, L_z Group designators of isentropic data: x, y, and z denote the last digits of the values of the potential temperatures of the three isentropic surfaces for which data are sent.

L_h, L_t, L_u Group designators of isentropic data used with $USWB(2)$ radiosonde code. The h, t, and u denote the hundred, ten, and unit digits of the value of the lower of the two reported isentropic levels.

M Reason for ceasing upper-wind observations (1 and 5, clouds; 3, haze; 7, rain).

$m_1m_2m_3m_4m_5$ See explanation of v_5.

m_rm_r Mixing ratio in grams per kilogram (see explanation of L).

N Amount of sky covered by cloud, in tenths (Table 2).

N_h Amount of cloud of which height is reported by h. In ship reports where height of cloud is not given, N_h indicates only the amount of low cloud, C_L (use table for N, Table 2).

PPP   Pressure in millibars (tens, units, and tenths, initial 9 or 10 omitted), reduced to sea level.

$P_mP_mP_m$ Pressure in millibars (tens, units, and tenths, initial 7 or 8 omitted), reduced to 5,000 ft above sea level. Used by high-level stations in the United States, transmitted in the ninth group of the surface synoptic reports.

PP   Pressure in millibars (tens and units, hundreds omitted). Used to report pressures of upper-air reports.

$P_1P_1P_1, P_2P_2P_2$, etc.   Pressure in hundreds, tens, and units of millibars at transmitted levels in upper-air reports.

pp   Amount of pressure change in 3 hr immediately preceding time of observation. Usually transmitted in fifths or halves of millibars. In the USWB code, if the pressure change exceeds 99 fifths (19.8 mb), the total amount may be indicated by following the fifth group with an additional group in the form 00*ppp* where *ppp* is the total pressure change (in fifths of millibars).

$P_{px}P_{px}, P_{py}P_{py}, P_{pz}P_{pz}$ Pressures in hundreds and tens (units omitted) of millibars at the isentropic surfaces x, y, and z.

$P_{sx}P_{sx}, P_{sy}P_{sy}, P_{sz}P_{sz}$ Condensation pressures (hence measures of the humidity) in hundreds and tens (units omitted) of millibars at the isentropic surfaces x, y, and z.

PILOT*   Information following gives upper-air wind data.

Q Octant of the globe (Table 2).

RR   In International forms, the amount of rainfall (at 0700 for the preceding 13 hr, and at 1800 for the preceding 11 hr). Expressed in whole millimeters with the following exceptions:

Code No.	RR, mm	Code No.	RR, mm	Code No.	RR
91	0.1	94	0.4	97	Trace
92	0.2	95	0.5	98	Over 90 mm
93	0.3	96	0.6	99	Unknown

RR In USWB surface code, the amount of precipitation in the 6-hr period preceding the observation (if R_t is sent as /, RR is for the preceding 12-hr period), in hundredths of inches, except that coded 99 denotes trace. When precipitation exceeds 0.99 in., RR is encoded with the hundredths in excess of whole inches, and the eighth group is followed by another group of the form $000R'R'$ where $R'R'$ gives the whole inches.

R_s Depth of snow on the ground at the time of observation, in inches except that code figure 9 denotes less than 0.5 in.

R_t Time precipitation began or ended, in hours before the time of observation (Table 2). R_t gives time precipitation began if ww indicates precipitation, as current or within the last hour.

S State of the sea (Table 2).

$S_P S_P$ Special phenomena, general description (refer to USWB weather code for details of the group).

$s_p s_p$ Special phenomena, detailed description (refer to USWB weather code for details of the group).

$S_{ix}S_{ix}$, $S_{iy}S_{iy}$, $S_{iz}S_{iz}$ Angular spacing of isobars in degrees of latitude, at the isentropic surfaces x, y, and z.

Ship (Navire, Schiff, Navios)*   Identify ship reports.

SYN*   Sometimes precedes synoptic reports from land stations in International collective messages.

TTT   Temperature of the upper air in degrees and tenths, Fahrenheit or centigrade. Negative temperatures are indicated by adding 500 to the numerical value when encoded.

TT Temperature of the air in whole degrees, Fahrenheit or centigrade. When centigrade is used, negative temperatures are indicated by adding 50 to the numerical value of the temperature; in Fahrenheit, negative temperatures are added algebraically to 100 when encoded.

$T_x T_x$ Maximum temperature in previous period of length determined by the time of observation. Negative temperatures reported as explained under TT.

$T_n T_n$ Minimum temperature in previous period of length determined by the time of observation. Negative temperatures reported as explained under TT.

temp*   Following information gives humidity and temperature observations of upper air.

$T_s T_s$ Temperature of the dew point in whole degrees. Negative temperatures reported as explained under TT.

T_d Difference between sea and air temperatures (Table 2).

T_w Time of commencement of weather reported by W (Table 2).

U Relative humidity reported as a single number (Table 2).

UU Relative humidity in whole percentages.

vv Velocity of the wind in the upper air; it is used in the USWB code and reported in miles per hour. In International forms, it is usually designated as $v_1 v_1$ and reported in kilometers per hour. When the velocity exceeds 99 units, 50 is added to dd, and the velocity is recorded with the hundreds omitted.

v_5 (and $m_1 m_2 m_3 m_4$ etc.)   Velocity of upper winds at height HH above sea level. Two tables apply:

    a. When no qualifying groups (44444 $m_1 m_2 m_3 m_4$ etc.) are sent, v_5 represents wind velocity according to subtable A under v_5 of Table 2.

    b. When the qualifying groups are sent, v_5 represents wind velocity for level x according to the value of m with subscript number corresponding to the level x. When $m_x = 0$, wind velocity for the level x is obtained from subtable A; when $m_x = 1$, wind speed for the level x is obtained from subtable B under v_5 of Table 2.

v_s Speed of ship in knots (Table 2).

V Horizontal visibility (Table 2).

V_s Horizontal visibility toward the sea (Table 2).

VV Horizontal visibility in units and tenths of miles for code figures 00 through 90 (9 miles); thereafter, 91 = 10 miles, 92 = 20 miles, etc., and 99 = over 90 miles.

(*Continued on page 593.*)

TABLE 1.—CODE TABLES

Transmission and plotting of data—Meteorological codes

CODE NO.	A		a AND a_3	C	C_L	C_M
	CHARACTER OF CHANGE	3-HOUR AMOUNT OF CHANGE 0.2 MB/0.5 MB	PLOTTING SYMBOL DESCRIPTION	PLOTTING SYMBOL DESCRIPTION	PLOTTING SYMBOL DESCRIPTION	PLOTTING SYMBOL DESCRIPTION
0	STEADY	0-3 / 0 OR 1	RISING, THEN FALLING	STRATUS OR FRACTOSTRATUS	NO LOWER CLOUDS	NO MIDDLE CLOUDS
1	RISING SLOWLY	4-8 / 2 OR 3	RISING, THEN STEADY; OR RISING, THEN RISING MORE SLOWLY	CIRRUS	CUMULUS OF FINE WEATHER	TYPICAL ALTOSTRATUS, THIN
2	RISING	9-18 / 4-7	UNSTEADY	CIRRO-STRATUS	CUMULUS HEAVY AND SWELLING, WITHOUT ANVIL TOP	TYPICAL ALTOSTRATUS, THICK (OR NIMBOSTRATUS)
3	RISING QUICKLY	19-30 / 8-12	STEADY OR RISING	CIRRO-CUMULUS	CUMULONIMBUS	ALTOCUMULUS, OR HIGH STRATO-CUMULUS, SHEET AT ONE LEVEL ONLY
4	RISING VERY RAPIDLY	>30 / >12	FALLING OR STEADY, THEN RISING; OR RISING, THEN RISING MORE QUICKLY	ALTO-CUMULUS	STRATOCUMULUS FORMED BY THE FLATTENING OF CUMULUS CLOUDS	ALTOCUMULUS IN SMALL ISOLATED PATCHES; INDIVIDUAL CLOUDS OFTEN SHOW SIGNS OF EVAPORATION AND ARE MORE OR LESS LENTICULAR IN SHAPE
5	FALLING SLOWLY	4-8 / 2 OR 3	FALLING, THEN RISING	ALTO-STRATUS	LAYER OF STRATUS OR STRATOCUMULUS	ALTOCUMULUS ARRANGED IN MORE OR LESS PARALLEL BANDS OR AN ORDERED LAYER ADVANCING OVER THE SKY
6	FALLING	9-18 / 4-7	FALLING, THEN STEADY; OR FALLING, THEN FALLING MORE SLOWLY	STRATO-CUMULUS	LOW BROKEN UP CLOUDS OF BAD WEATHER	ALTOCUMULUS FORMED BY A SPREADING OUT OF THE TOPS OF CUMULUS
7	FALLING QUICKLY	19-30 / 8-12	UNSTEADY	NIMBO-STRATUS	CUMULUS OF FINE WEATHER AND STRATOCUMULUS	ALTOCUMULUS ASSOCIATED WITH ALTOSTRATUS OR ALTOSTRATUS WITH A PARTIALLY ALTOCUMULUS CHARACTER
8	FALLING VERY RAPIDLY	>30 / >12	FALLING	CUMULUS OR FRACTOCUMULUS	HEAVY OR SWELLING CUMULUS, OR CUMULONIMBUS AND STRATOCUMULUS	ALTOCUMULUS CASTELLATUS, OR SCATTERED CUMULIFORM TUFTS
9	NOT USED		STEADY OR RISING, THEN FALLING; OR FALLING, THEN FALLING MORE QUICKLY	CUMULO-NIMBUS	HEAVY OR SWELLING CUMULUS (OR CUMULONIMBUS) AND LOW RAGGED CLOUDS OF BAD WEATHER	ALTOCUMULUS IN SEVERAL SHEETS AT DIFFERENT LEVELS, GENERALLY ASSOCIATED WITH THICK FIBROUS VEILS OF CLOUD AND A CHAOTIC APPEARANCE OF THE SKY

TABLE 1.—CODE TABLES (*Continued*)
Transmission and plotting of data—Meteorological codes

CODE NO.	C_H PLOTTING SYMBOL / DESCRIPTION	D_C, $D_{H/M}$, D_K, D_L, d_s	E PLOTTING SYMBOL / DESCRIPTION	F_x (IN BEAUFORT FORCE)	h METERS / FEET	H_a (FEET)
0	NO HIGH CLOUDS	NO MOTION	DRY	10	0–50 / 0–163	SURFACE 10000 20000 30000 ETC.
1	CIRRUS, DELICATE, NOT INCREASING, SCATTERED AND ISOLATED MASSES	NE	WET	11	50–100 / 164–327	NOT USED
2	CIRRUS, DELICATE, NOT INCREASING, ABUNDANT BUT NOT FORMING A CONTINUOUS LAYER	E	FLOODED	12	100–200 / 328–655	2000 12000
3	CIRRUS OF ANVIL CLOUDS, USUALLY DENSE	SE	FROZEN, HARD AND DRY	0–3	200–300 / 656–983	NOT USED
4	CIRRUS, INCREASING, GENERALLY IN THE FORM OF HOOKS ENDING IN A POINT OR IN A SMALL TUFT	S	PARTLY COVERED WITH SNOW OR HAIL	4	300–600 / 984–1967	4000 14000
5	CIRRUS (OFTEN IN POLAR BANDS) OR CIRROSTRATUS ADVANCING OVER THE SKY BUT NOT MORE THAN 45 ABOVE THE HORIZON	SW	COVERED WITH ICE OR GLAZED FROST	5	600–1000 / 1968–3280	25000 35000 45000 ETC.
6	CIRRUS (OFTEN IN POLAR HANDS) OR CIRROSTRATUS ADVANCING OVER THE SKY AND MORE THAN 45 ABOVE THE HORIZON	W	COVERED WITH THAWING SNOW	6	1000–1500 / 3281–4920	6000
7	VEIL OF CIRROSTRATUS COVERING THE WHOLE SKY	NW	COVERED WITH SNOW <6"; NOT FROZEN	7	1500–2000 / 4921–6561	NOT USED
8	CIRROSTRATUS NOT INCREASING AND NOT COVERING THE WHOLE SKY	N	COVERED WITH SNOW < 6"; FROZEN	8	2000–2500 / 6562–8202	8000
9	CIRROCUMULUS PREDOMINATING, ASSOCIATED WITH A SMALL QUANTITY OF CIRRUS	NOT USED WITH d_s, D_L, $D_{H/M}$; WITH D_K AND D_C, UNKNOWN OR ALL DIRECTIONS	COVERED WITH SNOW > 6"	9	NO C_L	NOT USED

TABLE 2.—CODE TABLES

Transmission and plotting of data—Meteorological codes

CODE NO.	K	N / N_h	Q	R_+ (HOURS SINCE PRECIPITATION BEGAN OR ENDED)	S	t_d °C / °F
0	NONE	(circle) ABSOLUTELY NO CLOUDS IN SKY	0–90° W	NO PRECIPITATION	FLAT; NO RIPPLE; OILY SEA	5.0 / 9
1	SHORT OR AVERAGE LENGTH; LOW: h<2m.	LESS THAN ONE TENTH	90–180° W	<1	CALM, RIPPLED	3.1– 5.0 / 6– 9
2	LONG; LOW: h<2m.	ONE TENTH	180–90° E	1–2	SMOOTH; WAVELETS	1.6– 3.0 / 3– 6
3	SHORT; MODERATE HEIGHT: h=2 TO 4m.	TWO OR THREE TENTHS	90–0° E	2–3	SLIGHT; ROCKS BUOY OR SMALL BOAT	0.6– 1.5 / 1– 3
4	AVERAGE LENGTH; MODERATE HEIGHT: h=2 TO 4m.	FOUR, FIVE OR SIX TENTHS	NOT USED	3–4	MODERATE; FURROWED	0.0– 0.5 / 0– 1
5	LONG; MODERATE HEIGHT: h=2 TO 4m.	SEVEN OR EIGHT TENTHS	0–90° W	4–5	ROUGH; MUCH DISTURBED	0.1– 0.5 / 0– 1
6	SHORT; HEAVY: h>4m.	NINE TENTHS	90–180° W	5–6	VERY ROUGH; DEEPLY FURROWED	0.6– 1.5 / 1– 3
7	AVERAGE LENGTH; HEAVY: h>4m.	MORE THAN NINE TENTHS BUT WITH OPENINGS	180–90° E	6–12	HIGH; ROLLERS WITH STEEP FRONTS	1.6– 3.0 / 3– 6
8	LONG; HEAVY: h>4m.	SKY COMPLETELY COVERED WITH CLOUDS	90–0° E	>12	VERY HIGH; LARGE ROLLERS WITH STEEP FRONTS	3.1– 5.0 / 6– 9
9	CONFUSED	SKY OBSCURED BY FOG, DUSTSTORM, OR OTHER PHENOMENON	NOT USED	UNKNOWN	ENORMOUS; SUCH AS IN CENTER OF HURRICAINE	5.0 / 9

Q column: codes 0–4 labelled NORTHERN HEMISPHERE; codes 5–8 labelled SOUTHERN HEMISPHERE.

t_d column: codes 1–4 labelled $T_{AIR} > T_{SEA}$; codes 5–8 labelled $T_{SEA} > T_{AIR}$.

TABLE 2.—CODE TABLES (*Continued*)
Transmission and plotting of data—Meteorological codes

CODE NO.	t_w (HOURS SINCE START OF "WEATHER")	U (%)	SUBTABLE A	v_s	SUBTABLE B	v_s (KNOTS)	V (METERS)
			MPH	dd = 01–32	MPH		V AND v_s (MILES)
			MPH	dd = 51–86	MPH		(UPPER LIMITS)
0	NOTHING TO REPORT	95–100	0–1 / 30–32		61–63 / 92–94	0	50 / 1/32
1	0–1	00–19 OR TOO LOW TO BE RECORDED	2–4 / 33–35		64–66 / 95–97	1–3	200 / 1/8
2	1–2	20–29	5–7 / 36–38		67–69 / 98–100	4–6	500 / 5/16
3	2–3	30–39	8–10 / 39–41		70–73 / 101–104	7–9	1000 / 5/8
4	3–4	40–49	11–13 / 42–45		74–76 / 105–107	10–12	2000 / 1¼
5	4–5	50–59	14–17 / 46–48		77–79 / 108–110	13–15	4000 / 2½
6	5–6	60–69	18–20 / 49–51		80–82 / 111–113	16–18	10000 / 6
7	6–8	70–79	21–23 / 52–54		83–85 / 114–116	19–21	20000 / 12
8	8–10	80–89	24–26 / 55–57		86–88 / 117–119	22–24	50000 / 30
9	>10	90–94	27–29 / 58–60		89–91 / 120–122	>24	>50000 / >30

TABLE 3.—WEATHER CODE FIGURES AND SYMBOLS
Transmission and plotting of data—Meteorological codes

WW (CODED FROM 00 TO 99)

W — PLOTTING SYMBOL — DESCRIPTION	0	1	2	3	4	5	6	7	8	9
0 — CLEAR OR SCATTERED CLOUDS	**0** CLOUDLESS (FROM NO CLOUDS UP TO BUT NOT INCLUDING ONE-TENTH)	**1** PARTLY CLOUDY. (FROM EXACTLY ONE TENTH TO EXACTLY FIVE TENTHS)	**2** CLOUDY. (OVER 5 TENTHS UP TO AND INCLUDING EXACTLY 9 TENTHS)	**3** OVERCAST. (OVER NINE TENTHS)	**4** LOW FOG, WHETHER ON GROUND OR AT SEA	**5** HAZE, (VISIBILITY PLUS 1000 M., 1100 YDS.)	**6** DUST DEVILS SEEN	**7** DISTANT LIGHTNING	**8** LIGHT FOG, (VISIBILITY 2000 M., 1100 TO 2200 YDS.)	**9** FOG AT DISTANCE, NOT AT STATION
1 — PARTLY CLOUDY. OR VARIABLE SKY	**10** PRECIPITATION WITHIN SIGHT	**11** THUNDER WITHOUT PRECIPITATION AT STATION	**12** DUST STORM WITHIN SIGHT, BUT NOT AT STATION	**13**	**14** SQUALLY WEATHER	**15** HEAVY SQUALLS... IN LAST 3 HOURS	**16** WATERSPOUTS SEEN IN LAST 3 HOURS	**17** UGLY, THREATENING SKY	**18** DUST STORM, VISIBILITY PLUS 1100 YDS.	**19** SIGNS OF TROPICAL STORM (HURRICANE)
2 — CLOUDY OR OVERCAST	**20** PRECIPITATION IN LAST HOUR BUT NOT AT TIME OF OBSERVATION	**21** DRIZZLE, OTHER THAN SHOWERS, IN LAST HOUR BUT NOT AT TIME OF OBSERVATION	**22** RAIN, OTHER THAN SHOWERS, IN LAST HOUR BUT NOT AT TIME OF OBSERVATION	**23** SNOW, OTHER THAN SHOWERS, IN LAST HOUR BUT NOT AT TIME OF OBSERVATION	**24** RAIN AND SNOW MIXED IN LAST HOUR BUT NOT AT TIME OF OBSERVATION	**25** RAIN SHOWERS IN LAST HOUR BUT NOT AT TIME OF OBSERVATION	**26** SNOW SHOWERS IN LAST HOUR BUT NOT AT TIME OF OBSERVATION	**27** HAIL OR RAIN AND HAIL SHOWERS IN LAST HOUR BUT NOT AT TIME OF OBSERVATION	**28** SLIGHT THUNDERSTORM IN LAST HOUR BUT NOT AT TIME OF OBSERVATION	**29** HEAVY THUNDERSTORM IN LAST HOUR BUT NOT AT TIME OF OBSERVATION
3 — SAND STORM, DUST STORM, OR BLOWING SNOW	**30** DUST OR SAND STORM	**31** DUST OR SAND STORM HAS DECREASED	**32** DUST OR SAND STORM, NO APPRECIABLE CHANGE	**33** DUST OR SAND STORM HAS INCREASED	**34** LINE OF DUST STORMS	**35** STORM OF DRIFTING SNOW	**36** SLIGHT STORM OF DRIFTING SNOW- GENERALLY LOW	**37** HEAVY STORM OF DRIFTING SNOW- GENERALLY LOW	**38** SLIGHT STORM OF DRIFTING SNOW- GENERALLY HIGH	**39** HEAVY STORM OF DRIFTING SNOW GENERALLY HIGH

Table 3.—Weather Code Figures and Symbols (Continued)
Transmission and plotting of data—Meteorological codes

4	FOG OR THICK DUST HAZE V<1100 YDS.
40	FOG
41	MODERATE FOG IN LAST HOUR, BUT NOT AT TIME OF OBSERVATION
42	THICK FOG IN LAST HOUR, BUT NOT AT TIME OF OBSERVATION
43	FOG, SKY DISCERNIBLE—HAS BECOME THINNER DURING LAST HOUR
44	FOG, SKY NOT DISCERNIBLE—HAS BECOME THINNER DURING LAST HOUR
45	FOG, SKY DISCERNIBLE, NO APPRECIABLE CHANGE DURING LAST HOUR
46	FOG, SKY NOT DISCERNIBLE, NO APPRECIABLE CHANGE DURING LAST HOUR
47	FOG, SKY DISCERNIBLE—HAS BECOME THICKER DURING LAST HOUR
48	FOG, SKY NOT DISCERNIBLE—HAS BECOME THICKER DURING LAST HOUR
49	FOG IN PATCHES
5	DRIZZLE
50	DRIZZLE
51	INTERMITTENT SLIGHT DRIZZLE
52	CONTINUOUS SLIGHT DRIZZLE
53	INTERMITTENT-- MODERATE DRIZZLE
54	CONTINUOUS MODERATE DRIZZLE
55	INTERMITTENT-- THICK DRIZZLE
56	CONTINUOUS THICK DRIZZLE
57	DRIZZLE AND FOG
58	SLIGHT OR MODERATE DRIZZLE AND RAIN
59	THICK DRIZZLE AND RAIN
6	RAIN
60	RAIN
61	INTERMITTENT SLIGHT RAIN
62	CONTINUOUS SLIGHT RAIN
63	INTERMITTENT MODERATE RAIN
64	CONTINUOUS MODERATE RAIN
65	INTERMITTENT HEAVY RAIN
66	CONTINUOUS HEAVY RAIN
67	RAIN AND FOG
68	SLIGHT OR MODERATE RAIN AND SNOW, MIXED
69	HEAVY RAIN AND SNOW, MIXED
7	SNOW OR SLEET
70	SNOW (OR SNOW AND RAIN, MIXED)
71	INTERMITTENT SLIGHT SNOW IN FLAKES
72	CONTINUOUS SLIGHT SNOW IN FLAKES
73	INTERMITTENT MODERATE SNOW IN FLAKES
74	CONTINUOUS MODERATE SNOW IN FLAKES
75	INTERMITTENT HEAVY SNOW IN FLAKES
76	CONTINUOUS HEAVY SNOW IN FLAKES
77	SNOW AND FOG
78	GRAINS OF SNOW (FROZEN DRIZZLE)
79	ICE CRYSTALS (OR FROZEN RAINDROPS (SLEET)
8	SHOWERS
80	SHOWERS
81	SHOWERS OF SLIGHT OR MODERATE RAIN
82	SHOWERS OF HEAVY RAIN
83	SHOWERS OF SLIGHT OR MODERATE SNOW
84	SHOWERS OF HEAVY SNOW
85	SHOWERS OF SLIGHT OR MODERATE RAIN AND SNOW
86	SHOWERS OF HEAVY RAIN AND SNOW
87	SHOWERS OF SLIGHT SNOW PELLETS (SOFT HAIL)
88	SHOWERS OF SLIGHT HAIL, OR RAIN AND HAIL
89	SHOWERS OF HEAVY HAIL, OR RAIN AND HAIL
9	THUNDERSTORM
90	THUNDERSTORM
91	RAIN AT TIME, THUNDERSTORM DURING LAST HOUR BUT NOT AT TIME OF OBSERVATION
92	SNOW OR RAIN AND SNOW MIXED AT TIME, THUNDERSTORM DURING LAST HOUR BUT NOT AT TIME OF OBSERVATION
93	THUNDERSTORM, SLIGHT WITHOUT HAIL, BUT WITH RAIN, (OR SNOW), AT TIME OF OBSERVATION
94	THUNDERSTORM, SLIGHT WITH HAIL, AT TIME OF OBSERVATION
95	THUNDERSTORM, MODERATE WITHOUT HAIL, (OR SNOW), AT TIME OF OBSERVATION
96	THUNDERSTORM MODERATE WITH HAIL AT TIME OF OBSERVATION
97	THUNDERSTORM HEAVY WITHOUT HAIL, BUT WITH RAIN (OR SNOW) AT TIME OF OBSERVATION
98	THUNDERSTORM COMBINED WITH DUST STORM AT THE TIME OF OBSERVATION
99	THUNDERSTORM HEAVY WITH HAIL AT TIME OF OBSERVATION

TABLE 4.—USWB SYMBOLIC WEATHER CODE FOR AIRWAYS
HOURLY TELETYPE REPORTS

EXPLANATION OF SYMBOL WEATHER REPORTS

(*based on instructions contained in Weather Bureau Circular N 1941
and later modifications*)

To illustrate the method used in transmission and deciphering of symbol weather reports, the following example of such a report is given. Each element of the report is connected by a line with a description of all symbols and conditions which might be used in that particular phase of the report. Elements of observations are always transmitted in the same order; therefore all symbol weather reports may be deciphered by reference to this chart.

WA	N	SPL	$231028E$	$W7V$	$\oplus \; \oplus$	$\tfrac{3}{4}V$
Station	Classification of report	Type of report	Date and time of report	Ceiling	Sky	Visibility
$L-$	$F-$	$068/$	$48/$	47	$\rightarrow 8$	$/963/$
Weather	Obstructions to vision	Barometric pressure	Temperature	Dew point	Wind	Altimeter setting

$CIG \; VRBL \; 5 \; to \; 8 \; W13 \oplus \; \oplus V \oplus \; 2\oplus \; VSBY \; VRBL \; \tfrac{1}{2} \; to \; 1\tfrac{1}{4}$

Remarks

For example, the report given above would be deciphered as follows: Washington—instrument; special report on 23rd at 10:28 a. m., Eastern War Time; indefinite ceiling at 700 feet, variable; overcast, lower broken clouds; visibility ¾ mile, variable; light drizzle; light fog; barometric pressure 1006.8 millibars; temperature 48°F; dew point 47°F; wind west 8 miles per hour; altimeter setting, 29.63 inches; ceiling variable 500 to 800 feet, overcast layer at indefinite 1,300 feet, broken clouds variable to scattered, scattered clouds at 200 feet, visibility variable from ½ to 1¼ miles.

	Detailed explanations
Station............	Lists of station names and their representative call letters are posted on Weather Bureau airport station bulletin boards for the information of all concerned.
Classification of report.	The symbols C or N are used immediately following, after 1 space, the station letters to classify the weather conditions at airports specifically designated as control airports. C (contact): Ceiling 1,000 feet or more and visibility 3 miles or more. N (Instrument): Below class C minimums. If no classification letter is used the station is not located at a control airport.
Type of report.....	"SPL," when it appears in this position in a weather report, indicates a "special" observation; "LCL" similarly indicates a "local extra" observation.
Date and time of report.	This group consists of 6 figures and one letter, the first two figures referring to the month, the remaining figures indicating the time of day based on the 24-hour clock, and the letter showing the time zone. For example, 150307P means 3:07 a. m., Pacific War Time of the 15th day of month; 181338C means 1:38 p. m., Central War Time 18th day of the month. This group is transmitted in all reports other than those transmitted in regular sequences in which the date and time appear in the sequence heading.
Ceiling............	The absence of a "ceiling" group indicates an "unlimited" ceiling (above 9,750 feet).

TABLE 4.—USWB SYMBOLIC WEATHER CODE FOR AIRWAYS
HOURLY TELETYPE REPORTS.—(*Continued*)

	Detailed explanations

Ceiling (*Cont.*)	Figures representing the number of *hundreds* of feet which apply are used to indicate the height of the ceiling between 51 and .9,750 feet, inclusive, above the station, *e.g.*, "35" indicates 3,500 feet, "3" indicates 300 feet, etc.
	The figure naught (0) is used when the ceiling is zero (below 51 feet).
	Ceiling values are always (except as noted in the following paragraph) preceded by one of the following ceiling classification letters:
	  M Measured ceiling.
	  E Estimated ceiling.
	  W Indefinite ceiling.
	  P Precipitation ceiling.
	  B Balloon ceiling.
	  A Ceiling reported from aircraft.
	A plus sign (+) is used (instead of a ceiling classification letter) whenever the balloon was blown from sight, at the height represented by the figures, before reaching the clouds.
	The letter *V*, immediately following the figure(s) for ceiling, indicates that the height of the ceiling is changeable.

Sky...............	The absence of a symbol for sky indicates that precipitation or obstructions to vision are present which reduce the ceiling to zero and/or the visibility to less than one-fourth mile and make the sky unobservable.
	The sky condition is indicated by the following symbols unless the condition given above is present.
	  ○ Clear.
	  ◑ Scattered clouds.
	  ◔ Broken clouds.
	  ⊕ Overcast.
	  ◑/High scattered.
	  ◔/High broken.
	  ⊕/High overcast.
	  ⊕◔ Overcast, lower broken.
	  ⊕◑ Overcast, lower scattered.
	  ◔◔ Broken, lower broken.
	  ◔◑ Broken, lower scattered.
	  ◑◔ Scattered, lower broken.
	  ◑◑ Scattered, lower scattered.
	  ⊕/◔ High overcast, lower broken.
	  ⊕/◑ High overcast, lower scattered.
	  ○/◔ High broken, lower broken.
	  ◔/◑ High broken, lower scattered.
	  ◔/◔ High scattered, lower broken.
	  ◑/◑ High scattered, lower scattered.
	The plus (+) or minus (−) sign *preceding* the cloudiness symbol indicates "dark" and "thin," respectively.
	Height of lower scattered clouds is indicated by the entry of a figure, representing the hundreds of feet applying, immediately preceding the scattered clouds symbol.
	Only two layers of clouds are reported in the body of the report. Additional layers are reported in "Remarks."

Visibility.........	The absence of a figure for visibility indicates that the visibility is 10 miles or more.

TABLE 4.—USWB Symbolic Weather Code for Airways
Hourly Teletype Reports.—(*Continued*)

	Detailed explanations
Visibility (*Cont.*)	The value of the visibility below 10 miles is indicated by figures representing the number of miles and/or fractions of miles. The letter *V*, immediately following the figure for visibility, indicates a fluctuating visibility.
Weather..........	The "weather" element of the report is indicated, when appropriate, by the following symbols: *R* −Light rain. *R* Moderate rain. *R* +Heavy rain. *S* −Light snow. *S* Moderate snow. *S* +Heavy snow. *ZR* −Light freezing rain. *ZR* Moderate freezing rain. *ZR* +Heavy freezing rain. *L* −Light drizzle. *L* Moderate drizzle. *L* +Heavy drizzle. *ZL* −Light freezing drizzle. *ZL* Moderate freezing drizzle. *ZL* +Heavy freezing drizzle. *E* −Light sleet. *E* Moderate sleet. *E* +Heavy sleet. *A* −Light hail. *A* Moderate hail. *A* +Heavy hail. *AP* −Light small hail. *AP* Moderate small hail. *AP* +Heavy small hail. *SP* −Light snow pellets. *SP* Moderate snow pellets. *SP* +Heavy snow pellets. *SQ* −Light snow squall. *SQ* Moderate snow squall. *SQ* +Heavy snow squall. *RQ* −Light rain squall. *RQ* Moderate rain squall. *RQ* +Heavy rain squall. *T* Thunderstorm. *T* +Heavy thunderstorm. *SW* −Light snow showers. *SW* Moderate snow showers. *SW* +Heavy snow showers. *RW* −Light rain showers. *RW* Moderate rain showers. *RW* +Heavy rain showers. TORNADO *W* (always written out in full and followed by a letter showing direction from station).
Obstructions to vision.	The "obstructions to vision" element of the report is indicated, when appropriate, by the following symbols: *F* − −Damp haze. *F* − Light fog. *F* Moderate fog.

TABLE 4.—USWB SYMBOLIC WEATHER CODE FOR AIRWAYS
HOURLY TELETYPE REPORTS.—(*Continued*)

	Detailed explanations
Obstructions to vision (*Cont.*)	$F+$ Heavy fog. $GF-$ Light ground fog. GF Moderate ground fog. $GF+$ Heavy ground fog. $IF-$ Light ice fog. IF Moderate ice fog. $IF+$ Heavy ice fog. H Hazy. $K-$ Light smoke. K Moderate smoke. $K+$ Heavy smoke. $D-$ Light dust. D Moderate dust. $D+$ Heavy dust. $BS-$ Light blowing snow. BS Moderate blowing snow. $BS+$ Heavy blowing snow. $GS-$ Light drifting snow. GS Moderate drifting snow. $GS+$ Heavy drifting snow. $BD-$ Light blowing dust. BD Moderate blowing dust. $BD+$ Heavy blowing dust. $BN-$ Light blowing sand. BN Moderate blowing sand. $BN+$ Heavy blowing sand.
Barometric pressure	The barometric pressure is indicated by a group of 3 figures; tens, units, and tenths of millibars involved. Thus a pressure of 1015.2 millibars would be written as "152"; 962.8 as "628"; 1025.7 as "257"; etc. Sent only by stations equipped with mercurial barometers.
Temperature.......	Temperature is indicated by figures giving its value to the nearest degree Fahrenheit. Values below zero Fahrenheit are indicated by the entry of a minus sign ($-$) immediately preceding the figures for temperature. Zero is entered as "0."
Dew point.........	Dew point is indicated by figures giving its value to the nearest degree Fahrenheit. Values below zero Fahrenheit are indicated by the entry of a minus sign ($-$) immediately preceding the figures for dew point.

Wind.............	The wind *direction* is indicated by arrows, as follows:
	  ↓ North.
	  ↓↙ North-northeast.
	  ↙ Northeast.
	  ←↙ East-northeast.
	  ← East.
	  ←↖ East-southeast.
	  ↖ Southeast.
	  ↑↖ South-southeast.
	  ↑ South.
	  ↑↗ South-southwest.

TABLE 4.—USWB SYMBOLIC WEATHER CODE FOR AIRWAYS
HOURLY TELETYPE REPORTS.—(*Continued*)

	Detailed explanations
Wind (*Cont.*)	↗ Southwest. →↗ West-southwest. → West. →↘ West-northwest. ↘ Northwest. ↓↘ North-northwest. C Calm. The *velocity* is indicated by figures representing its value in miles per hour. If wind is calm no entry is made for velocity. If estimated, this is indicated by the entry of the letter *E* immediately following the velocity figures. The *character* of the wind is indicated, when appropriate, by entry, following the velocity, of a minus sign (−) for "fresh gusts" and a plus sign (+) for "strong gusts." No symbol for character means the wind is steady. *Wind shifts* which have occurred at the reporting station are indicated, immediately following the other wind data, by an arrow showing the direction (to 16 points) from which the wind was blowing prior to the shift, followed by the local time, on the 24-hour clock, at which the shift occurred, with following letter showing the time zone. The intensity of the shift is indicated by the minus sign (−) for "light," the absence of a sign for "moderate," and the plus sign (+) for "heavy," the signs being entered immediately following the standard-of-time letter.
Altimeter setting....	Indicated by a group of 3 figures representing the inch and hundredths of an inch of pressure involved. Thus, 30.00 would be written as "000"; 29.98 as "998"; etc. Sent only by designated stations equipped with mercurial barometers.
Remarks..........	Remarks are transmitted in authorized English abbreviations and teletype symbols. Lists of the abbreviations are available for inspection at all the Weather Bureau Airport Stations. The teletype symbols used are shown on this chart.
Special data........	Special data comprising pressure change and characteristic, 5,000-foot pressure at selected stations, cloud, thunderstorm, and snow depth data, special data from selected stations, etc., are entered in code at certain times by the stations designated to do this, as separate groups, immediately following the report proper.
Missing data.......	Elements normally sent, but for some reason missing from the transmission, will be indicated by the letter *M* entered in the report in place of the missing data.
Nonmeteorological data............	Nonmeteorological remarks are occasionally added by the CAA Communications personnel concerning field conditions and the operation of radio facilities.
Supplementary data	Every three hours supplementary (additive) groups are appended by selected stations. These facilitate the preparation of 3-hourly synoptic maps for forecasting, intermediate to the 6-hourly reports of Schedule *C*; besides appearing on Schedule *A*, some of these reports are also "converted" for transmission

TABLE 4.—USWB Symbolic Weather Code for Airways
Hourly Teletype Reports.—(*Continued*)

	Detailed explanations
Supplementary data (*Cont.*)	in the surface reports code of Schedule C in the 6-hourly reports and in the "$3M$" sequences. The symbolic form of the hourly reports with additive groups is as follows: (hourly report)/$app\ C_LC_MC_HDch_+M_+$ (or $C_LC_MC_HD_C$) $9P_mP_m$-P_{ma_3} $(T_{n/x}T_{n/x}OS_PS_PS_PS_p)$: the longer form, including maximum or minimum temperature and special phenomena, is used at the times that correspond with the regular 6-hourly reports; intermediate 3-hourly reports do not include this information. The symbols have the same meanings and code tables as given on pp. 578–593 for the USWB surface reports code with the addition of h_+ and M_+ which have the following meanings: h_+ Height above the surface of base of cloud layer which predominates in amount between 10,000 and 20,000 ft (Code 0 = none or 10,000 ft; Code 1 = 11,000 ft; . . . etc. . . . ; Code 9 = 19,000 ft; Code = 20,000 ft or more). M_+ Method by which h_+ is determined (Code 0 = missing; Code 1 = measured; Code 2 = estimated; Code 3 = from Pilot Report).

W Past weather, generally for the period 6 hr preceding the time of observation. If precipitation has occurred in the last hour or is current at the time of observation, W is coded as the most characteristic weather of the 5-hr period ending 1 hr before the observation. (Table 3.)

ww Present weather (Table 3).

w The initial figure of code ww, thus indicating the general state of the weather (Table 3).

YY Day of the month.

Y_mY_m Same as YY.

Y Day of the week (1 denotes Sunday, etc.).

$Z_{fx}Z_{fx}$, $Z_{fy}Z_{fy}$, $Z_{fz}Z_{fz}$ Value of the stream function in units and tenths (hundreds and thousands omitted and to be supplied by the decoder) of millions of ergs per gram. The thousands and hundreds are virtually always equal to the first two figures of the potential temperature of the isentropic surface.

Z^{**} Time given in GCT.

00, 85, 70, 50, 30, 20, 10 Indicators of data groups in the USWB radiosonde code that refer to the 1,000-, 850-, 700-, etc. mb levels.

00000* Following information gives upper-air temperatures and humidities for fixed pressures at intervals of 100 mb (1,000, 900, 800, etc.).

00FF0** See FF.

3*, 4*, 6*, 9* Internal indicators designating the code form being used for surface reports from ships. These appear as the first figure in the fifth group of F 232, F 242, F 261, and F 291, respectively.

44444* Indicator group for wind velocity. See $m_1m_2m_3m_4m_5$ and v_5.

(50) Indicator of group giving height of the 500-mb level in the first transmission of the USWB radiosonde code. It appears only when the reporting station is not on a circuit and its report is relayed.

55FF5** See FF.

7**, 8**, 9**, 0** Internal indicators used to identify the optional groups in USWB surface reports.

- (hyphen, - . . . . . -), X, / (slant), 99 Data missing or code spaces have no meaning.

Bibliography

1. Admiralty Weather Manual, Hydrographic Department, Admiralty, London, 1941.
2. Aerographer's Manual, U.S. Navy Department, Washington, D.C.
3. Circular Letters, U.S. Weather Bureau, Washington, D.C., various dates. Particularly important are:
     a. No. 86–44, "Amendment to 1943 Radiosonde Code; Isentropic Data," Nov. 23, 1944.
     b. No. 21–45, "Amendments and Interpretations; 1942 Weather Code," Mar. 2, 1945.
     c. No. 27–45, "Combined Aircraft Weather Report Form (CAW-C)," Mar. 28, 1945.
     d. No. 28–45, "Amendment to 1943 Radiosonde Code," Mar. 31, 1945.
4. Code for Weather Analysis Transmissions, W.B. No. 1345 (revised effective Sept. 1, 1944). U.S. Weather Bureau, Washington, D.C., August, 1944.
5. Fasicule I, Manuel des Codes Internationaux, 2d ed., Secretariat de l'Organisation Météorologique Internationale, No. 9, Paris, 1936.
6. Federal Airways Manual of Operations, vol. 3, Communications Division, Chap. B, U.S. Department of Commerce, CAA, Washington, D.C., 1944.
7. Instructions for Airway Meteorological Service, *U.S. Weather Bur. Circ. N*, 1941.
8. Numeral Weather Code, U.S. Weather Bureau, Washington, D.C., 1942.
9. Radio Weather Aids to Navigation, H.O. 206, Hydrographic Office, U.S. Navy Department, 1942, 1943.
10. 1943 Radiosonde Code, U.S. Weather Bureau, Washington, D.C.

PLOTTING WEATHER CHARTS

Surface Synoptic Charts. The chart on which plotting is to be done has most of the stations indicated by three-number designators. In North America, the first

STATION MODELS

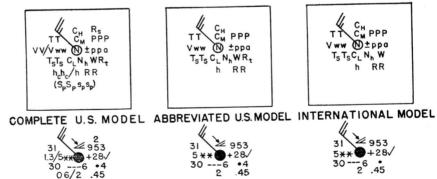

COMPLETE U.S. MODEL   ABBREVIATED U.S. MODEL   INTERNATIONAL MODEL

Fig. 1.—Plotting models, surface reports.

digit of the designators usually increases in value with increasing latitude; the second digit increases in value from east to west. The third digit varies at random within the rough quadrangles determined by the variation of each of the first two digits.

The data to be plotted are obtained from the 6-hourly teletype sequences (schedule C in the United States). Several systems of plotting the elements around the station circle are shown in Fig. 1.

Constant-level Charts. Data for constant-level charts are transmitted in convenient form for the 5,000-, 10,000-, and 20,000-ft pressures in the first transmission of radiosonde observations. Data for 10, 13, and 16 km are sent in the second transmission. The first transmission sends not only pressure, temperature, and relative humidity, but also mixing ratio; the last item is not sent in the second transmission for the highest three levels. Additional data are obtained from the pilot-balloon observations for the level being plotted. When no wind report is available for this level, but a report is available for the adjacent lower level, the latter report should be plotted, properly noted. The order of arrangement of data around the station circle, with an illustration of the use of a lower-level wind direction, is given in Fig. 2a and b.

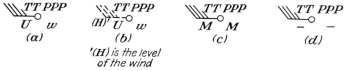

'(H) is the level
of the wind

FIG. 2.—Plotting models, constant-level charts.

Distinction should be made between data that are missing for any cause, best connoted by recording M in the proper place (Fig. 2c), and data (humidity elements) that are "unknown" because they are too low to record, best connoted by dashes (Fig. 2d).

The 10,000-ft chart may be extended from the continent over the adjoining ocean areas by use of reports of surface pressures from ships and island stations, provided that the following conditions prevail in order to justify an assumption that the lapse rate between the surface and 10,000 ft is nearly moist-adiabatic: (1) isobars are cyclonically curved in the vicinity of the report; (2) fresh to strong surface winds; (3) convective clouds or precipitation present at the time of report; and (4) absence of stability phenomena (fog, stratus, drizzle, haze). Under such circumstances, the Graph for Extrapolating the Pressure at the 10,000-ft Level (Fig. 49, Sec. I) may be used to obtain satisfactory approximations. In latitudes below 25°, an assumed mean lapse rate between the moist- and dry-adiabatic is more suitable than a moist-adiabatic.

TABLE 5

Months	Potential temperatures		
	L_x	L_y	L_z
January to February	290	296	302
March to April	296	302	308
May to June	302	308	314
July to August	308	314	320
September to October	302	308	314
November to December	296	302	308

Isentropic Charts. Data for isentropic charts are derived from the second transmission. The indicator group 10177 indicates "isentropic data follow." There are six data elements transmitted for each of three isentropic surfaces. The six elements are pressure, condensation pressure (moisture), stream function, wind direction,

direction of isobars, and angular spacing of isobars.* The three isentropic surfaces are spaced at 6° of potential temperature apart and are changed every 2 months as indicated in Table 5. For the United States, L_y, the middle surface, is generally most suitable; in very detailed analysis, all three may be plotted.

When a particular level to be plotted has been chosen, its data can be identified in the second transmission following the indicator group by noting the three groups, each of which begins with the number that is the last figure in the potential-temperature value of the surface. The decoding of these data can be determined from the explanation of codes.

$$10177\ L_x P_{px} P_{px} S_x S_{sx}\ L_x Z_{tx} Z_{tx} d_{wx} d_{wx}\ L_x d_{tx} d_{tx} S_{tx} S_{tx}\quad L_y P_{py} P_{py} S_{sy} S_{sy}\quad L_y Z_{fy} Z_{fy} d_{wy} d_{wy}\ L_y d_{ty} d_{ty} S_{ty} S_{ty}\ etc$$

$$10177\ 6\ 7\ 9\ 5\ 6\quad 6\ 8\ 2\ 3\ 1\quad 6\ 2\ 2\ 1\ 8\quad 2\ 7\ 2\ 4\ 8\quad 2\ 3\ 4\ 3\ 2\quad 2\ 3\ 0\ 1\ 6\ etc$$

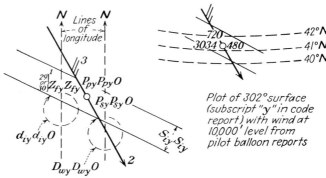

Fig. 3.—Plotting model, isentropic charts.

Wind speeds and additional wind-direction data may be obtained from pilot-balloon observations. When $P_{py}P_{py}0$ has been plotted, the elevation of the surface may be estimated from Table 6. With the elevation thus estimated, wind reports

TABLE 6

Pressure, mb	Height, 1,000 ft	Pressure, mb	Height, 1,000 ft
1,000	1	600	14
950	2	550	16
900	3	500	18
850	4–5	450	20
800	6	400	24
750	8	350	27
700	10	300	30
650	12		

from pilot-balloon observations can be plotted where winds are given at the proper level to be in or near the isentropic surface. When winds are 1,000 or 2,000 ft below the isentropic surface, they should be entered in dotted form. If these observations do not quite agree with the wind directions given in the isentropic data, it should be recalled that the two observations are not exactly synoptic.

Plotting of the station model is as shown in Fig. 3.

* From Feb. 1, to May 1, 1945, data for only two isentropic surfaces were transmitted, these being February, 290 and 295; March and April, 295 and 305, moreover, data included only pressure at the surface, condensation pressure, and stream function for each surface. After May 1, 1945, no further transmission of isentropic data is scheduled.

Cross Sections.   Cross-section charts are usually constructed with coordinates of horizontal distance as abscissa and elevation as ordinate.   An approximate pressure scale in millibars may be superimposed using the average pressure-height equivalent of Table 6 (or may be found more accurately in Sec. I).

The track of the cross section is chosen either from the latest surface map for the purpose of contributing most to the analysis, or from a proposed flight plan.   The topography along the track should be traced, either from a template or from the elevations of stations in conjunction with a relief map.   Simultaneously, positions of reporting stations should be labeled and marked with vertical lines.

The data entered on the cross section are of two kinds: radiosonde reports (first transmission) and pilot-balloon reports.   Radiosonde observations should be plotted for every available station as follows: for each reported level as determined by the pressures, plot the temperature, relative humidity, and specific humidity; in addition, enter a pseudoadiabatic chart with the pressure and temperature, and determine the potential temperature.   Pilot-balloon observations of wind velocities are plotted by referring to the height scale to determine the position of reported levels.   Directions are plotted with north to the top of the cross section.   Usually winds need to be

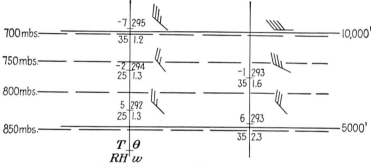

Fig. 4.—Plotting model, cross sections.

plotted only for each 2,000 ft.   If the station has no radiosonde report, the pilot-balloon data are plotted on the station ordinate; if the station does have radiosonde data, it is usually best to offset the wind data to right or left far enough to prevent overlap of wind arrows on the radiosonde data.

An example of plotted cross-section data showing both radiosonde and pilot-balloon observations is given in Fig. 4.

Pseudoadiabatic Diagrams.   The particular form of thermodynamic diagram used by the Weather Bureau and other operating agencies in the United States is the pseudoadiabatic diagram.   Pressure in millibars (scaled as $p^{0.288}$) is the ordinate, and temperature in degrees centigrade is the abscissa; the basic pressure-temperature grid is usually printed in brown.   Also in brown are adiabats, straight lines that slope in general from upper left to lower right.   Superimposed in red are pseudoadiabats, rendered as curving broken lines, and saturation mixing-ratio lines, rendered as fine unbroken curves.

A sounding is plotted by entering the chart with temperature and pressure as arguments for each level sent.   The points so established are connected by straight lines; through each point there is also drawn a horizontal line to designate the level clearly for purposes of facilitating analysis of the sounding.   Relative humidity and mixing ratio are entered to left and right of the significant points.   Examples of plotted soundings may be found at several places in this volume.

Upper-wind Charts.   Upper-air wind velocities are transmitted in the pilot-balloon observations code with directions to the nearest 10 deg with the 0 omitted, and speeds in miles per hour.   The velocities are plotted, however, with directions to the nearest one of 16 points of the compass and with speeds indicated in Beaufort force.   Conversions of the data should be done by the decoder (refer to the code tables and Beaufort force tables elsewhere in this volume to get equivalents).

The charts issued for use by the U.S. Weather Bureau and other agencies usually indicate that winds are to be plotted for the surface and at successive 2,000-ft levels *above sea level* up to 10,000 ft; and by 5,000-ft intervals thereafter to 20,000.   It is important to note that wind levels other than surface are indicated in thousands of feet above sea level, not station level.   For purposes of having upper-wind data in most useful form for plotting constant-level charts, 5,000-ft winds may be plotted in place of those for the 6,000-ft level.   Calm at a particular level is indicated by a C over the station circle; when weather conditions prevent the making of observations, appropriate weather symbols should be plotted at the station circles of the surface-wind chart.

Constant-pressure Charts.   The last group of the first transmission of radiosonde observations (HHH) gives the elevation of the 500-mb level above each station. The plotter or analyst must supply the first and last figures of the elevation in the form $1HHH0$; the resultant elevation is in feet to the nearest 10.   Although the contour lines of the pressure surface are also virtually streamlines, corroboratory data of wind velocities may be plotted from the pilot-balloon observations wherever these observations give winds reasonably close to the elevation of the 500-mb surface.

When radiosonde data are sent in the forms $USWB(3)$, $USWB(4)$, and $USWB(5)$, the plotting of constant-pressure charts will be a major plotting procedure in place of the plotting of constant-level charts.   The station plotting model for constant-pressure charts is the same as that for constant-level charts, HHH being substituted for PPP. In addition, changes of height in particular time intervals may be entered above the HHH figures.   With present forms of reports for upper-air winds, there is no alternative but to plot winds for the nearest approximate elevation to that of the pressure height.

METEOROLOGICAL MAPS

Maps used in plotting data for purposes of weather forecasting must represent relatively large portions of the earth.   It is therefore necessary to consider the kind of projection to be used.   Since no flat map can have all the desirable properties of the globe, the projection chosen must achieve accuracy in some dominant property at the expense of others.   In general, a map projection for meteorological plotting should have the following properties as closely as possible:

1. Conformality; *i.e.*, the property that the angle between any two intersecting curves on the map is the same as the angle between the curves on the earth represented by the curves on the map.   Thus, parallels are normal to meridians at their points of intersection.   This property has obvious advantages for the plotting of wind directions.

2. Minimum of variation in scale; *i.e.*, distances on the map should be as true as possible.   The plotting of trajectories, the use of gradient scales of various kinds, and the forecasting of displacements are obvious cases in which map scale is important. It is noteworthy that recent developments of gradient scales include measurements by degrees of latitude.   This eliminates scale problems when proper regard is given to the central latitude over which any distance is measured.

3. Equality of area; *i.e.*, a given area on the map bears the same ratio to the area that it represents that any other area on the map bears to the area it represents.

This property ensures that weather systems on the map are comparable in areal extent wherever they may occur.

4. Rendition of great circles on the globe as straight lines on the map. This property is desirable in considering radii of curvature and, consequently, the application of gradient-wind monograms and allied problems. It is also desirable for navigation purposes, and flight-planning maps are usually on projections having this property or closely approximating it; weather maps may appropriately be on similar projections.

The maps actually adopted are generally conformal. Because plotting charts do not endeavor to present the entire globe, other desirable properties are approached satisfactorily. The projections recommended for use in synoptic meteorology by resolution of the International Meteorological Committee are all conformal; the particular projections are designated according to the part of the earth for which the weather analysis is to be done. For polar regions, the polar stereographic projection with standard parallel at 60° is recommended. The U.S. Weather Bureau has used stereographic projection for its Northern Hemisphere maps. The recommended projection for middle latitudes is the Lambert conformal conic type with two standard parallels (recommended to be at 30 and 60°, but varied in practice according to the area being projected). Besides being conformal, this projection, when the latitudinal spread of the map is not too great, reduces scale errors and inequality of area to negligible amounts; in addition, over most of the map, straight lines closely approximate great-circle courses. It is particularly suited to rendition of areas of wide longitudinal extent, such as are needed for middle-latitude synoptic work. Consequently, this projection has been widely adopted. North American maps have also been produced on polyconic projection; this projection is suitable to North America because of the continent's wide latitudinal range; the maps have highly desirable properties along the central meridian but become increasingly less satisfactory for meteorological purposes with increasing longitudinal distance from the center. Recommended projection for equatorial regions is the Mercator with true scale at $22\frac{1}{2}°$. For obvious reasons, the Mercator projection should be confined to use in low latitudes, but it is very good for that purpose.

Climatological data requiring world maps present a quite different problem. Care should be exercised in selecting a projection suitable to the kind of data to be shown and the weight to be given to particular regions. Winds in general should be plotted on conformal maps, whereas air-mass source regions might well be presented on an equal-area map. In practice, no world map can be best suited to all kinds of data, and some compromise projection is often adopted in order to gain consistency of presentation without entirely sacrificing suitability of projection.

TIME

Concepts of time are associated with the two motions of the earth, the year with the revolution of the earth around the sun, the day with the rotation of the earth on its axis. Unfortunately, the two motions are not precisely synchronous, *i.e.*, the period of revolution is not an integer multiple of the period of rotation. Moreover, the period of rotation is a constant only in relation to a fixed celestial body; in reference to the sun, it changes in response to the variable distance between earth and sun, the consequent variation in the speed of the earth in its orbit, and the effect of the moon. As a result, there are several systems of measuring time.

The Sidereal System. The sidereal day is defined as the interval between two successive upper transits of the first point of Aries across any given meridian. It is thus the exact period of rotation and is constant in length, but it has the inconvenience that a particular time notion, such as noon, varies through all hours of the clock. Similarly, the sidereal year is the time taken by the earth to complete one revolution around the sun from a given star to the same star.   Its length is 365 days, 6 hr, 9 min, 9 sec.*   Sidereal time is used primarily for astronomical purposes.   It is of importance to meteorology only in problems involving expressions of the angular rotation of the earth.

Solar Time.   The *apparent solar day* is defined as the interval between two consecutive transits of the sun across any given meridian.   This interval is not constant (its annual range is approximately 31 min) and clock mechanisms cannot of course be constructed to keep this time.   The *mean solar day* was instituted to gain the advantage of regularity while retaining the convenience of correlation between hour numbers and the daily solar succession.   Mean time assumes a fictitious sun that moves in the plane of the equator with a uniform eastward velocity that equals the mean velocity of the true sun in the ecliptic.   The equation of time is the difference between apparent and mean time; it is zero at four times of the year, with maximum values of slightly over 16 min.

Corresponding to the solar day, *the tropical year* is the time interval between two successive passages of the vernal equinox by the sun.   It is 20 min 23 sec shorter than the sidereal year or 365 days 5 hr 48 min 46 sec.*   Since the tropical year is dependent upon the apparent motions of the sun and therefore can be correlated with the seasons, it is taken as the year of civil reckoning.   Actually, it is shorter by 26 sec than the Gregorian calendar year, but this is insignificant, amounting to only 1 day in about 3,300 years.

The division of the year into months originally stemmed from changes of the moon and is still so based in the Mohammedan lunar calendar.   It was changed in Europe with the establishment of the Julian calendar (45 B.C.); the monthly structure of the Gregorian calendar is essentially like that of the Julian.

For practical purposes, even mean solar time on a local meridional basis is unsatisfactory.   As a consequence, "standard time" has been adopted in some form in virtually all parts of the world.   Its fundamental concept is that, starting from a standard or initial meridian (Greenwich or 0° meridian), there are designated successive standard meridians at intervals of 15° of longitude; the entire 15° zone bisected by a standard meridian keeps the same time as the mean solar time for the meridian.   Such a purely geometric division is followed over the oceans, but, on large land areas and island groups, the standard-time-zone boundaries depart from meridians in order to accommodate political boundaries, railroad facilities, and the like.   Some political units choose to keep time that is based on mean solar time for a local meridian.   Details of the standard time zones are given by Time Zone Chart of the World, H.O. 5192, published by the Hydrographic Office, U.S. Navy.

Each time zone is given a "zone description" in hours from 1 to 12; the description is given a sign (+ for zones in west longitude, − for zones in east longitude); and the magnitude of the description increases with longitude.   The twelfth zone is divided medially by the 180° meridian.   This meridian, modified by departures for political boundaries, is also the International Date Line; when the meridian is crossed from east to west, 1 day is added to the date, when it is crossed from west to east, 1 day is subtracted from the date.

Meteorological reports are commonly referred to Greenwich civil time.   Local standard time, for purposes of considering diurnal effects, can be derived from GCT by algebraically adding the "zone description" to the GCT.   If "local mean time" or "local apparent time" are desired, reference should be made to the Nautical or Air Almanacs to apply longitude factors and the equation of time.   "Daylight saving time" and "war time" amount simply to adding −1 algebraically to the zone description.

* All time intervals are given in mean solor time units.

SECTION X

SYNOPTIC METEOROLOGY AND WEATHER FORECASTING

By F. A. Berry, Jr., E. Bollay, J. F. O'Connor, Norman R. Beers,
George R. Jenkins, R. A. Baumgartner, Civilian Staff,
Institute of Tropical Meteorology, Rio Piedras,
Puerto Rico, Vincent J. Oliver,
and Mildred B. Oliver

CONTENTS

601

SECTION X

SYNOPTIC METEOROLOGY AND WEATHER FORECASTING

By F. A. Berry, Jr., E. Bollay, J. F. O'Connor, Norman R. Beers,
George R. Jenkins, R. A. Baumgartner, Civilian Staff,
Institute of Tropical Meteorology, Rio Piedras,
Puerto Rico, Vincent J. Oliver.
and Mildred B. Oliver

HISTORICAL BACKGROUND

The role played by weather conditions in the struggles of mankind and the fortunes of war has been recorded in history. It was realized early that, since we cannot control the weather, the alternative was to forecast weather conditions.

Despite its importance to human welfare, economy, and warfare, little progress was made in weather forecasting until the development of extensive telegraphic networks during the nineteenth century. The advent of radio communication permitted the collection of weather reports from ships at sea and from outlying observation stations. The rapid collection of simultaneous weather observations has made synoptic meteorology possible.

During the First World War, detailed forecasts were put to tactical use by both sides. In the Second World War, weather forecasts have played an important part in the successful prosecution of the war. Strategic and tactical use of weather information is being made by all sides.

The greatest advancements in daily weather forecasting were made in the intermission between the First and Second World Wars. It was during this period that V. Bjerknes and his collaborators introduced the theory of the air-mass- and frontal-analysis method of forecasting, which is now in common usage in middle-latitude countries. C.-G. Rossby and his collaborators were notably active in upper-air analysis applied to short-term forecasting, and more recently they, I. P. Krick, and others have attacked the extended-period forecasting problem.

It should be noted that the introduction of the polar-front theory and our concept of air masses as well as the methods of upper-air analysis developed during the last 30 years have made it possible to train en masse the great number of qualified forecasters necessary to supply the military and naval needs in the present global conflict.

The rapid development of air transportation will undoubtedly continue to stimulate synoptic meteorology and may ultimately lead to techniques resulting in successful long-range forecasts. Owing to the complexity of the problem, it is questionable whether the accuracy found in the exact sciences will ever be achieved. Synoptic meteorology will always be a science necessitating the use of sound judgment based on experience and training. New techniques and a better physical background should ultimately lead to a forecast accuracy comparable with the prognostic accuracy now found, say, in the medical profession.

AIR MASSES

By F. A. Berry, Jr., and E. Bollay

An air mass is generally defined as an extensive portion of the atmosphere which is approximately homogeneous in its horizontal distribution of temperature, humidity, and lapse rate. These properties are acquired by air remaining over a large portion of the earth's surface having similar temperature and humidity conditions until equilibrium is reached by convection and radiation processes between the surface and the lower layers of the atmosphere.

Regions where air masses originate are referred to as *source regions*. Snow- or ice-covered polar regions, tropical seas, and desert areas are generally sufficiently uniform to act as source regions. The middle latitudes, owing to the steep latitudinal temperature gradient and the nonuniform temperature conditions established by the distribution of land and sea areas, are not sufficiently homogeneous to act as source regions. These portions of the earth serve as transition zones.

When air finally leaves its source region, it has the properties characteristic of that portion of the earth's surface. However, these properties are subject to modifications as the air travels over surfaces having different temperature and moisture conditions.

The weather conditions encountered in an air mass depend on the characteristic properties acquired at the source and on the extent to which modifications have taken place. For successful forecasting, it is therefore essential to follow the life history of the various air masses encountered on the synoptic chart.

It has been observed that the lower layers of air in the troposphere are subject to rapid modifications depending largely on the nature of the surface over which they are moving. In order to ascertain what modifications are taking place, it is essential to watch the trajectory of the various air masses to determine what physical and dynamical processes occur. The properties of an air mass may be modified by any one or all, or any combination, of the following processes:

1. Radiation (gain or loss of heat).
2. Evaporation or condensation.
3. Convergent or divergent flow.

Special attention should be given to whether the trajectory is principally cyclonic or anticyclonic. With a cyclonic trajectory, it has been found that the lower layers will have a predominantly steep lapse rate accompanied by associated instability phenomena. An anticyclonic trajectory results in a pronounced stable stratification and subsidence inversions at intermediate levels, which tend to suppress any vertical developments that may be initiated in the surface layers. Weather conditions in the two cases are generally quite different.

SOURCE REGIONS

The principal source regions of air masses are indicated in Figs. 1 and 2 for summer and winter conditions.

The Arctic and Polar Continental Source. The arctic and polar continental source comprises the arctic fields of snow and ice and the snow-covered portions of the continents where the wind system is predominantly of anticyclonic nature. In the

604

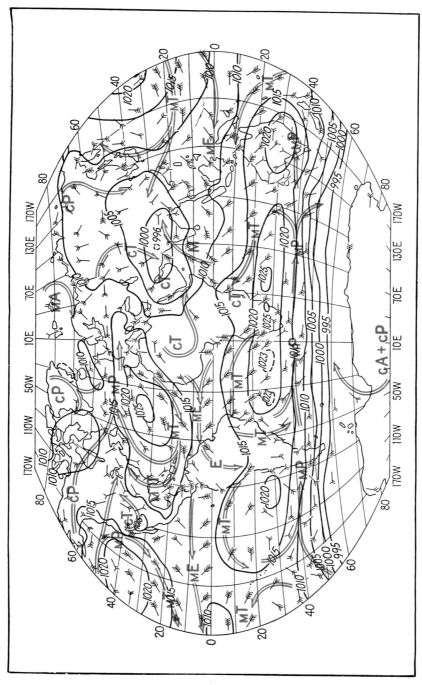

FIG. 1.—Air-mass sources in July.

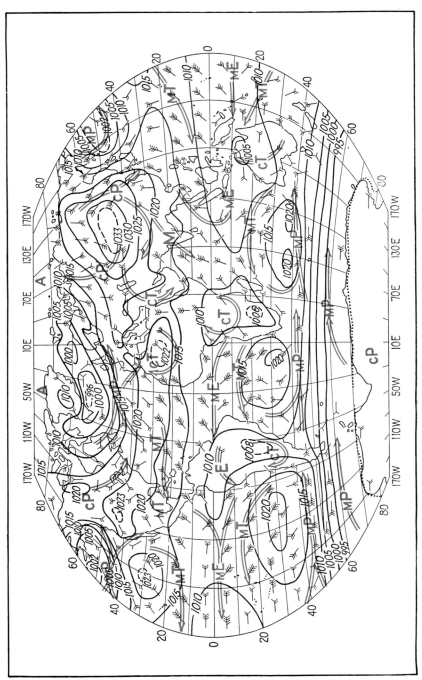

Fig. 2.—Air-mass sources in January.

Southern Hemisphere, this source is confined to the snow- and ice-covered portions of the antarctic continent.

The Tropical Maritime Source. The tropical maritime source comprises the regions on the equatorial side of the subtropical high-pressure belt over the oceans in both hemispheres.

The Polar Maritime Source. The polar maritime source is in reality an oceanic zone of transition separating the tropical maritime source from the polar and arctic continental source. Cyclonic flow conditions and pronounced west-east surface temperature gradients exist in the higher latitudes, while anticyclonic flow conditions and less marked zonal-temperature gradients prevail in the lower latitudes of this oceanic zone of transition. In the Southern Hemisphere, the air mass produced in this transition zone is virtually the only polar air reaching the continental areas. Polar air from a continental source is practically never observed beyond the antarctic continent.

The Tropical Continental Source. The tropical continental source is found in the Northern Hemisphere over the large land masses of North Africa, Asia, southwestern United States, and southern Europe in the summertime, and over North Africa in winter. In the Southern Hemisphere, a tropical continental source exists over the interior of south Africa, Australia, and over a limited area in the interior of South America.

The Equatorial Source. The equatorial source occupies the equatorial belt between the subtropical anticyclones of the Northern and Southern Hemispheres. This region is mainly oceanic and has an extremely uniform horizontal temperature distribution.

Source of Monsoon Air. The monsoon circulation predominates in the southern and southeastern parts of Asia. In winter, cP air originating in the interior of Asia flows across the Himalayan mountain ranges toward the equator. This air is relatively dry and is warmed adiabatically as it descends the southern slopes of The Himalaya. In summer, air from the equatorial source flows up over the mountains toward the interior of Asia. This air over India resembles unmodified equatorial air. Weather conditions encountered with the winter monsoon are in general fair. The summer monsoon produces a hot, damp climate with excessive rainfall.

Source of Superior Air. Superior air is essentially a high-level air mass, although it may appear at the surface. It is believed to be created largely by subsidence. This air is most frequently observed over the southwestern part of North America.

CLASSIFICATION OF AIR MASSES

Bergeron introduced the concept of air masses in forecasting practice and has laid down the basic principles of air-mass analysis. The classification established by Bergeron has come into almost universal usage, since it classifies air masses from the geographical as well as the thermodynamic point of view.

This classification separates air masses into four basic types:

1. Equatorial air masses (E), typical of the source region for equatorial air.
2. Tropical air masses (T), typical of the source region for tropical air.
3. Polar air masses (P), typical of the source region for polar air.
4. Arctic air masses (A), typical of the source region for arctic air.

In addition, each of the four principal groups may be subdivided into maritime (m) or continental (c) masses according to whether the source region is oceanic or continental.

The thermodynamic feature of Bergeron's classification consists of two main types:

1. Cold air masses (k), indicating air that is colder than the underlying surface.
2. Warm air masses (w), indicating air that is warmer than the underlying surface.

The thermodynamic classification indicates stability or instability conditions in the lower layers of the air mass, and the geographical classification indicates the source region of the air mass.

DIFFERENTIAL CLASSIFICATION OF AIR MASSES

Designator	Description
cAk	Continental arctic air that is colder than the surface over which it lies
cAw	Continental arctic air that is warmer than the surface over which it lies
mAk	Maritime arctic air that is colder than the surface over which it lies
cPw	Continental polar air that is warmer than the surface over which it is moving
cPk	Continental polar air that is colder than the surface over which it is moving
mPw	Maritime polar air that is warmer than the surface over which it is moving
mPk	Maritime polar air that is colder than the surface over which it is moving
mTw	Maritime tropical air that is warmer than the surface over which it is moving
mTk	Maritime tropical air that is colder than the surface over which it is moving
cTw	Continental tropical air that is warmer than the surface over which it is moving
cTk	Continental tropical air that is colder than the surface over which it is moving
mEk	Maritime equatorial air that is colder than the surface over which it is moving
mEw	Maritime equatorial air that is warmer than the surface over which it is moving
M	Monsoon air, found over southeastern Asia. In winter this air is transitional polar continental air being transformed into equatorial air. In summer this air is unmodified equatorial air
S	Superior air, found generally aloft over the southwestern United States, and occasionally at or near the surface

GENERAL CHARACTERISTICS OF AIR MASSES

The temperature and moisture content are the distinguishing properties of air masses. Since the temperature and moisture content of a parcel of air can be uniquely expressed in terms of the pseudo-wet-bulb potential temperature or the equivalent potential temperature, it has become customary to identify the various air masses by their equivalent potential or their pseudo-wet-bulb potential temperatures. These two temperatures are invariant for adiabatic changes (see Sec. V). They therefore offer a certain amount of conservatism not found in the simple observation of the free-air temperature and the usual expressions of moisture content.

For the synoptic analyst, it is relatively important to know the mean or typical values of the representative meteorological elements for the various air-mass types by regions and seasons. In the past, only the mean values of the different meteorological elements of the various air masses have been tabulated, and very few "typical" soundings have been published. This difficulty will no doubt be overcome when the rich accumulation of upper-air data now being collected throughout the world can be published.

CHARACTERISTIC AIR-MASS PROPERTIES OVER NORTH AMERICA

The characteristic air-mass properties of North America were first compiled in an organized form by Willett.[8] The data used were based primarily on kite observations. Showalter[5] published mean seasonal values of significant properties of North American air masses based on a short period of airplane observations. His results offer the most complete information to date and are therefore reproduced for use in identifying air masses (Table 1).

TABLE I-1.—MEAN SEASONAL VALUES OF SIGNIFICANT PROPERTIES OF NORTH AMERICAN AIR MASSES. (*After Showalter.*)

ORDER OF DATA

$\dfrac{\text{R.H.} \quad | \quad \text{T}}{q \quad \diagup\!\!\!\uparrow \quad \theta_E}$ — NO. OF OBS. (in circle)

CAW TYPE OF AIR MASS — MEAN SEASONAL VALUES OF SIGNIFICANT PROPERTIES — WINTER 1935-36

Station	SFC.	500	1000	1500	2000	2500	3000	4000	5000
SHREVE-PORT	65/-4.4 1.8/272 (7)	57/-5.9 1.5/274 (7)	50/-4.3 1.5/281 (4)	37/-2.5 1.4/287 (1)	32/0.4 1.6/296 (1)	38/0.4 2.0/303 (1)			
BILLINGS	57/-19.8 0.7/264 (31)			57/-16.6 1.0/27 (28)	54/-22.0 0.6/271 (14)	55/-22.6 0.5/273 (8)	36/-28.2 0.2/274 (2)		
LAKEHURST	60/-7.3 1.4/269 (22)	59/-9.7 1.2/270 (22)	59/-11.6 1.1/273 (21)	57/-13.6 1.0/275 (18)	55/-15.6 0.8/278 (14)	52/-16.9 0.8/282 (9)	63/-22.5 0.6/281 (4)	58/-30.6 0.4/283 (2)	
FARGO	75/-25.7 0.5/249 (50)	75/-23.6 0.5/253 (50)	70/-22.3 0.6/261 (46)	66/-21.7 0.6/266 (40)	62/-22.3 0.5/271 (34)	56/-24.4 0.4/275 (24)	53/-26.6 0.4/276 (13)		
OKLA. CITY	62/-9.4 1.3/269 (14)	62/-9.7 1.3/270 (14)	60/-11.5 1.1/273 (12)	48/-7.5 1.2/281 (3)	33/-7.6 0.8/286 (2)				
DAYTON	75/-13.1 1.2/264 (30)	71/-13.4 1.2/266 (29)	77/-13.6 1.2/273 (29)	67/-13.5 1.1/276 (24)	66/-15.6 0.9/278 (15)	67/-16.0 0.9/283 (9)	69/-16.1 1.1/289 (3)	57/-21.5 0.7/293 (1)	50/-28.2 0.4/295 (1)
OMAHA	71/-17.1 0.9/260 (49)	70/-16.6 1.0/262 (49)	69/-14.5 1.1/270 (45)	65/-13.2 1.2/276 (36)	58/-15.1 0.9/279 (17)	61/-16.7 1.0/282 (5)	64/-20.3 0.7/283 (1)		

TABLE I-2.

ORDER OF DATA

$\dfrac{\text{R.H.} \quad | \quad \text{T.}}{q \quad \diagup\!\!\!\uparrow \quad \theta_E}$ — NO. OF OBS. (in circle)

cPw TYPE OF AIR MASS — MEAN SEASONAL VALUES OF SIGNIFICANT PROPERTIES — WINTER 1935-36

Station	SFC.	500	1000	1500	2000	2500	3000	4000	5000
FARGO	87/-9.5 2.2/270 (22)	88/-8.5 2.0/273 (22)	89/-6.5 2.5/283 (13)	76/-8.2 2.0/284 (4)	73/-14.6 1.2/280 (2)	90/-13.4 1.7/289 (1)			
MURFREESBORO	79/-0.4 3.1/281 (16)	74/-0.1 3.0/285 (16)	70/-0.3 2.9/289 (13)	73/-3.0 2.7/291 (5)	68/-3.0 2.5/296 (1)				
OKLA. CITY	73/-4.9 2.1/276 (22)	71/-4.3 2.1/278 (22)	64/-3.8 2.1/283 (15)	62/-4.9 2.0/287 (9)	62/-3.9 2.2/294 (3)				
OMAHA	87/-3.4 2.7/279 (18)	84/-3.1 2.7/281 (16)	79/-3.9 2.5/285 (12)	50/-1.5 2.1/291 (3)	69/-4.8 2.4/294 (1)				
PENSACOLA	74/4.0 4.0/286 (18)	67/1.4 3.3/287 (11)	57/0.4 2.7/289 (4)	54/2.2 2.9/298 (1)					
DAYTON	80/-4.4 2.4/276 (10)	81/-4.8 2.4/279 (9)	81/-6.3 2.1/281 (8)	73/-7.4 2.0/285 (4)	66/-9.1 1.5/287 (3)				
MONTGOMERY	67/0.7 2.8/280 (15)	56/-0.1 2.4/283 (14)	51/0.1 2.3/288 (12)	53/-1.8 2.3/291 (5)	48/-5.6 1.8/290 (2)				

TABLE I-3.

S TYPE OF AIR MASS
MEAN SEASONAL VALUES OF SIGNIFICANT PROPERTIES
WINTER 1935-36

ORDER OF DATA:

$$\frac{R.H. \quad | \quad T}{q \quad / \quad \theta_E}$$

NO. OF OBS. (in circle)

Each cell: R.H. | T (top), obs. no. (circle), q | θ_E (bottom)

	SFC.	500	1000	1500	2000	2500	3000	4000	5000
PENSACOLA		48|14.6 (1) 5.2|306	42|10.1 (6) 3.9|303	32|7.4 (15) 2.5|302	28|6.0 (26) 2.1|304	27|4.0 (39) 1.9|307	28|1.2 (46) 1.7|308	23|-4.0 (52) 1.1|312	22|-10.3 (24) 0.7|314
SAN ANTONIO		38|20.0 (1) 5.8|309	27|15.0 (6) 3.2|306	25|12.1 (14) 2.6|306	24|9.6 (23) 2.2|308	24|6.2 (32) 1.9|309	22|3.0 (34) 1.5|310	20|-3.3 (40) 0.9|312	17|-10.3 (33) 0.5|314
OKLA. CITY	55|6.8 (2) 3.4|292	46|9.8 (2) 3.6|296	34|12.1 (3) 3.8|300	36|11.3 (9) 3.5|309	33|7.8 (15) 2.9|309	32|4.1 (18) 2.3|308	31|3.0 (21) 1.7|308	31|-6.5 (29) 1.1|309	29|-13.6 (28) 0.7|311
SPOKANE					29|-2.4 (1) 1.1|293	36|2.2 (5) 2.2|305	29|-0.8 (9) 1.5|306	27|-8.2 (11) 1.0|307	29|-15.8 (12) 0.6|308
SHREVEPORT	77|8.1 (6) 5.2|295	44|14.8 (6) 4.7|305	34|14.2 (8) 3.9|307	27|12.5 (9) 3.1|309	25|7.5 (14) 2.2|306	26|4.5 (20) 2.0|307	25|1.7 (23) 1.6|309	25|-4.5 (33) 1.2|311	24|-11.0 (30) 0.7|314
BILLINGS				38|6.3 (5) 2.7|301	35|4.3 (7) 2.3|303	32|0.7 (8) 1.8|303	32|-3.0 (8) 1.4|303	33|-0.1 (11) 1.0|305	31|-6.6 (11) 0.6|307
MONTGOMERY			30|12.0 (2) 2.9|302	27|9.5 (7) 2.5|303	25|7.0 (10) 2.0|305	24|4.3 (15) 1.8|307	26|0.0 (21) 1.4|306	24|-6.7 (29) 1.0|309	24|-13.7 (24) 0.6|311

TABLE I-4.

MPk TYPE OF AIR MASS
MEAN SEASONAL VALUES OF SIGNIFICANT PROPERTIES
WINTER 1935-36

ORDER OF DATA:

$$\frac{R.H. \quad | \quad T}{q \quad / \quad \theta_E}$$

NO. OF OBS. (in circle)

	SFC.	500	1000	1500	2000	2500	3000	4000	5000
SPOKANE	85|1.7 (22) 3.9|290		83|1.7 (22) 4.0|295	77|-1.3 (27) 3.3|295	74|-5.6 (34) 2.4|294	73|-9.1 (36) 2.0|294	68|-13.2 (35) 1.5|293	64|-20.7 (27) 0.9|294	57|-28.2 (24) 0.4|296
OKLA. CITY	71|8.3 (1) 5.1|299	72|7.3 (1) 4.9|298	68|1.5 (4) 3.4|290	63|-1.1 (5) 2.7|293	59|-4.2 (8) 2.2|294	56|-7.1 (11) 1.8|295	55|-10.5 (10) 1.4|296	58|-16.9 (9) 1.0|298	60|-23.2 (8) 0.7|301
FARGO			51|-0.6 (1) 2.3|289	64|-4.0 (7) 2.2|290	61|-7.7 (11) 1.7|290	56|-13.6 (20) 1.2|287	57|-17.6 (31) 0.9|287	52|-25.5 (46) 0.5|289	51|-32.2 (35) 0.3|292
SEATTLE	82|7.0 (6) 5.2|293	83|4.5 (6) 4.7|294	82|2.0 (6) 4.0|295	81|-1.1 (6) 3.4|295	78|-4.0 (6) 2.9|296	70|-7.4 (6) 2.2|296	66|-10.8 (6) 1.7|296	52|-17.8 (5) 0.8|297	45|-26.4 (3) 0.3|296
OMAHA	78|3.8 (2) 4.0|290	78|5.0 (2) 4.2|294	72|-2.3 (5) 2.6|286	69|-4.3 (15) 2.3|289	65|-8.7 (29) 1.7|288	61|-11.7 (37) 1.4|289	57|-14.7 (35) 1.1|290	53|-21.5 (30) 0.7|292	52|-28.4 (28) 0.4|295
CHEYENNE	71|-5.5 (23) 2.3|291				67|-4.9 (24) 2.3|293	64|-8.7 (27) 1.8|294	70|-11.1 (32) 1.5|294	69|-20.2 (30) 0.9|295	65|-28.9 (29) 0.5|295
BILLINGS	64|-0.5 (13) 2.7|290			54|0.2 (16) 2.5|295	57|-5.3 (28) 1.9|292	61|-9.8 (33) 1.6|292	67|-14.4 (41) 1.3|292	69|-22.5 (38) 0.8|292	62|-29.6 (28) 0.4|294

TABLE I-5.

mPw TYPE OF AIR MASS
MEAN SEASONAL VALUES OF SIGNIFICANT PROPERTIES
WINTER 1935-36

ORDER OF DATA legend: R.H. (top-left) | T (top-right); q ↗ θ_E (bottom); circle = NO. OF OBS.

Each cell: R.H. | T / (No. obs) / q | θ_E

Station	SFC.	500	1000	1500	2000	2500	3000	4000	5000
CHEYENNE	52 / 0.6 (15) 2.6 / 299	—	—	○	49 / 0.7 (24) 2.5 / 300	49 / -1.7 (30) 2.2 / 302	53 / -5.6 (36) 1.9 / 302	51 / -13.0 (40) 1.2 / 303	50 / -20.2 (45) 0.7 / 305
MONTGOMERY	72 / 3.7 (5) 3.9 / 286	49 / 9.3 (7) 3.8 / 296	41 / 6.3 (8) 3.0 / 296	42 / 3.6 (8) 2.8 / 298	38 / 1.1 (11) 2.0 / 299	38 / -1.5 (13) 1.8 / 301	35 / -3.1 (11) 1.5 / 304	34 / -8.9 (8) 1.1 / 307	34 / -13.9 (4) 0.8 / 312
FARGO	—	○	50 / -1.0 (2) 2.0 / 286	43 / -0.8 (8) 1.8 / 292	45 / -1.8 (11) 1.9 / 296	42 / -4.4 (12) 1.5 / 297	39 / -6.0 (13) 1.2 / 298	40 / -15.8 (14) 0.7 / 299	33 / -21.7 (13) 0.4 / 302
OKLA. CITY	70 / 4.1 (6) 3.7 / 290	60 / 7.4 (7) 3.7 / 294	44 / 8.4 (12) 3.4 / 300	43 / 5.2 (14) 2.8 / 300	47 / 0.5 (23) 2.4 / 300	48 / -2.5 (25) 2.1 / 300	47 / -5.3 (26) 1.8 / 302	47 / -11.0 (22) 1.3 / 305	46 / -17.1 (21) 0.9 / 308
SEATTLE	87 / 5.1 (5) 4.8 / 290	71 / 6.3 (5) 4.5 / 295	84 / 4.6 (2) 5.0 / 301	85 / 1.4 (2) 4.4 / 300	88 / -2.0 (2) 3.7 / 300	81 / -4.7 (2) 3.0 / 301	64 / -6.4 (2) 2.2 / 302	38 / -13.7 (1) 0.8 / 301	—
SPOKANE	89 / 0.8 (10) 3.9 / 289	—	89 / 1.9 (10) 4.4 / 295	76 / 1.9 (9) 4.0 / 299	72 / 0.6 (9) 3.4 / 301	68 / -3.3 (7) 2.8 / 302	65 / -6.9 (11) 2.2 / 302	61 / -13.4 (14) 1.4 / 303	55 / -20.6 (17) 0.8 / 304
OMAHA	85 / 0.1 (3) 3.4 / 284	72 / 3.1 (5) 3.6 / 290	50 / 7.0 (10) 3.6 / 300	50 / 3.8 (12) 2.9 / 300	51 / 0.8 (16) 2.5 / 301	51 / -3.2 (19) 2.4 / 300	58 / -7.5 (24) 1.8 / 300	61 / -14.4 (29) 1.3 / 302	57 / -20.8 (31) 0.8 / 305

TABLE I-6.

mT TYPE OF AIR MASS
MEAN SEASONAL VALUES OF SIGNIFICANT PROPERTIES
WINTER 1935-36

Station	SFC.	500	1000	1500	2000	2500	3000	4000	5000
SHREVEPORT	83 / 17.6 (3) 10.5 / 319	88 / 15.3 (4) 10.2 / 320	74 / 14.1 (4) 8.2 / 320	76 / 10.8 (3) 7.0 / 318	28 / 11.4 (2) 4.2 / 316	26 / 6.3 (1) 2.0 / 310	28 / 2.1 (1) 1.8 / 311	57 / -6.4 (2) 2.0 / 312	52 / -12.7 (2) 1.3 / 315
PENSACOLA	97 / 19.4 (7) 13.3 / 328	85 / 18.0 (8) 11.7 / 328	89 / 14.9 (10) 10.6 / 327	77 / 11.7 (11) 7.9 / 322	81 / 9.0 (9) 7.3 / 322	82 / 5.5 (9) 6.2 / 321	82 / 2.5 (10) 5.1 / 320	59 / -2.5 (8) 3.2 / 320	65 / -9.5 (5) 2.3 / 322
SAN ANTONIO	94 / 16.3 (3) 11.1 / 320	89 / 15.9 (4) 10.5 / 321	88 / 13.0 (5) 9.1 / 320	87 / 9.8 (6) 7.4 / 318	82 / 8.2 (6) 6.8 / 320	62 / 5.9 (3) 4.8 / 317	56 / 3.2 (2) 3.4 / 314	12 / -1.0 (1) 0.6 / 314	11 / -8.7 (1) 0.4 / 315
COCO SOLO (DEC,1936)	88 / 24.7 (28) 7.0 / 345	90 / 22.1 (28) 15.9 / 344	87 / 19.1 (28) 13.7 / 341	80 / 16.3 (28) 11.4 / 337	70 / 14.3 (27) 9.1 / 334	59 / 12.2 (27) 7.2 / 331	52 / 9.8 (27) 5.7 / 333	40 / 5.6 (26) 3.8 / 327	20 / 0.5 (25) 1.7 / 330
ST. THOMAS (JAN,1937)	84 / 23.4 14.9 / 337	91 / 21.4 15.2 / 341	88 / 18.6 13.0 / 338	87 / 15.8 11.5 / 336	80 / 13.6 9.7 / 334	58 / 11.7 6.6 / 328	47 / 9.5 4.9 / 327	33 / 4.2 2.7 / 326	26 / -1.8 1.5 / 326

TABLE I-7.

cAw TYPE OF AIR MASS
MEAN SEASONAL VALUES OF SIGNIFICANT PROPERTIES
SPRING 1936

ORDER OF DATA

$\dfrac{\text{R.H.} \quad | \quad \text{T}}{q \quad \nearrow | \quad \theta_E}$ NO. OF OBS.

	SFC.	500	1000	1500	2000	2500	3000	4000	5000
BILL-INGS	75 / -11.2 (10) 1.5 / 275			75 / -13.3 (10) 1.3 / 277	78 / -16.7 (9) 1.1 / 278	81 / -18.1 (5) 1.1 / 281	61 / -17.8 (1) 0.8 / 287		
CHEY-ENNE	84 / -12.3 (6) 1.7 / 283				82 / -12.8 (6) 1.9 / 283	77 / -14.2 (5) 1.4 / 287	86 / -16.2 (2) 1.4 / 290		
FARGO	81 / -5.3 (32) 2.5 / 275	73 / -5.5 (28) 2.3 / 277	64 / -9.1 (25) 1.8 / 278	55 / -10.8 (21) 1.3 / 280	56 / -14.9 (14) 0.9 / 279	56 / -18.4 (6) 0.8 / 280	40 / -21.7 (2) 0.5 / 282		
SPO-KANE	73 / -5.2 (6) 2.1 / 278		62 / -4.8 (5) 2.0 / 283	55 / -8.4 (4) 1.4 / 282	55 / -13.1 (3) 1.0 / 282	54 / -20.7 (1) 0.6 / 278			

TABLE I-8.

cPw TYPE OF AIR MASS
MEAN SEASONAL VALUES OF SIGNIFICANT PROPERTIES
SPRING 1936

ORDER OF DATA

$\dfrac{\text{R.H.} \quad | \quad \text{T}}{q \quad \nearrow | \quad \theta_E}$ NO. OF OBS.

	SFC.	500	1000	1500	2000	2500	3000	4000	5000
FARGO	83 / 5.1 (14) 4.9 / 293	74 / 4.4 (12) 4.3 / 293	53 / 3.0 (8) 2.6 / 292	51 / 0.5 (6) 2.2 / 294	46 / -1.1 (4) 2.0 / 296	(5)			
LAKE-HURST	71 / 5.0 (19) 4.0 / 288	61 / 4.2 (19) 3.6 / 290	57 / 2.8 (19) 3.3 / 293	49 / -2.4 (10) 1.9 / 289	56 / -4.3 (8) 2.0 / 293	52 / -5.0 (3) 1.9 / 297	24 / -6.7 (1) 0.8 / 298		
MUR-FREES-BORO	77 / 6.9 (22) 5.0 / 294	71 / 6.2 (15) 4.6 / 295	65 / 3.4 (12) 3.7 / 295	50 / 2.2 (10) 2.6 / 296					
OMAHA	69 / 7.2 (22) 4.9 / 296	63 / 7.5 (18) 4.7 / 298	63 / 5.4 (13) 4.2 / 299	60 / 4.1 (12) 3.7 / 302	62 / 2.6 (7) 3.7 / 305				
ST. LOUIS	75 / 5.6 (21) 4.6 / 291	59 / 6.2 (16) 3.8 / 294	54 / 2.4 (11) 2.9 / 292	36 / 2.9 (8) 2.2 / 296	33 / 0.0 (2) 1.6 / 296	23 / -2.4 (1) 1.0 / 298			
WASH-INGTON	71 / 7.2 (21) 4.9 / 292	46 / 8.1 (18) 3.5 / 294	46 / 5.0 (18) 3.0 / 295	43 / 2.6 (15) 2.6 / 297	55 / -1.7 (7) 2.4 / 297	78 / -7.4 (3) 2.4 / 296			

TABLE I-9.

S TYPE OF AIR MASS
MEAN SEASONAL VALUES OF SIGNIFICANT PROPERTIES
SPRING 1936

ORDER OF DATA: $\frac{\text{R.H.} \mid T}{q \mid \theta_E}$ (circled value = NO. OF OBS.)

Cell format: RH, T (n) q, θ_E

	SFC.	500	1000	1500	2000	2500	3000	4000	5000
EL PASO	24, 15.0 (34) 3.0, 308	○	○	26, 16.3 (36) 3.7, 315	25, 13.3 (38) 3.2, 315	26, 9.4 (39) 2.6, 315	27, 5.6 (39) 2.3, 315	27, -2.8 (36) 1.4, 314	25, -9.5 (36) 0.9, 316
SAN ANTONIO	○	○	○	32, 14.5 (8) 3.9, 313	28, 13.2 (16) 3.5, 316	27, 9.7 (22) 2.8, 315	26, 6.9 (26) 2.4, 316	26, 1.5 (31) 1.6, 317	25, -6.9 (33) 1.0, 318
MONT-GOMERY	○	○	32, 10.1 (4) 3.0, 301	27, 7.3 (7) 2.1, 301	27, 8.4 (16) 2.5, 308	25, 4.4 (22) 1.9, 308	25, 2.7 (27) 1.8, 311	22, -2.9 (33) 1.3, 314	21, -8.8 (35) 0.9, 317
WASH-INGTON	○	○	31, 14.8 (3) 4.0, 308	32, 11.3 (9) 3.5, 308	28, 6.4 (17) 2.5, 306	22, 3.7 (19) 1.7, 305	20, 1.0 (22) 1.3, 306	19, -4.8 (23) 1.0, 311	20, -9.8 (26) 0.7, 315
ST. LOUIS	○	○	37, 16.4 (3) 4.9, 313	38, 11.7 (6) 3.8, 320	31, 8.8 (12) 2.9, 309	29, 6.7 (17) 2.5, 311	26, 3.5 (27) 1.9, 312	25, -3.3 (32) 1.2, 313	26, -9.7 (32) 0.9, 316
MUR-FREES-BORO	○	○	○	34, 8.6 (3) 2.9, 305	33, 9.1 (11) 2.9, 310	32, 6.1 (18) 2.6, 311	34, 3.7 (28) 2.4, 313	30, -2.1 (36) 1.6, 315	29, -8.7 (39) 1.0, 317
PENSA-COLA	○	○	32, 14.5 (8) 3.5, 307	29, 12.4 (19) 3.0, 308	28, 9.0 (25) 2.5, 309	25, 6.3 (27) 2.0, 310	26, 4.3 (32) 2.0, 313	22, -1.3 (34) 1.4, 316	21, -6.5 (30) 0.9, 319

TABLE I-10.

MPk TYPE OF AIR MASS
MEAN SEASONAL VALUES OF SIGNIFICANT PROPERTIES
SPRING 1936

ORDER OF DATA: $\frac{\text{R.H.} \mid T}{q \mid \theta_E}$ (circled value = NO. OF OBS.)

Cell format: RH, T (n) q, θ_E

	SFC.	500	1000	1500	2000	2500	3000	4000	5000
BILL-INGS	56, 1.8 (13) 2.7, 292	○	○	51, 6.7 (13) 2.4, 294	50, -3.2 (13) 1.9, 294	58, -8.5 (15) 1.6, 293	69, -14.2 (19) 1.3, 291	67, -20.9 (20) 0.8, 294	61, -28.8 (15) 0.4, 295
CHEY-ENNE	67, -5.8 (7) 2.2, 290	○	○	○	65, -5.7 (7) 2.2, 291	66, -8.9 (7) 1.8, 293	66, -13.2 (10) 1.4, 292	74, -21.1 (11) 0.9, 293	66, -28.0 (11) 0.5, 296
FARGO	84, 2.4 (12) 4.0, 288	59, 3.9 (13) 3.2, 290	59, 0.7 (14) 2.7, 290	64, -2.3 (17) 2.6, 292	55, -6.0 (21) 1.9, 291	49, -11.4 (25) 1.1, 288	46, -15.2 (27) 0.8, 289	47, -21.3 (23) 0.6, 293	50, -28.2 (20) 0.3, 296
OMAHA	65, 3.0 (9) 3.2, 285	60, 3.7 (9) 3.1, 289	57, 2.1 (9) 2.9, 291	56, -1.3 (10) 2.3, 292	48, -4.2 (10) 1.7, 292	44, -8.9 (11) 1.2, 291	46, -13.6 (11) 0.9, 291	48, -21.5 (9) 0.5, 292	47, -28.3 (8) 0.3, 296
SPO-KANE	76, 3.3 (28) 3.9, 292	○	64, 3.4 (21) 3.5, 295	62, 0.1 (18) 2.9, 295	63, -4.6 (16) 2.3, 294	62, -9.3 (17) 1.7, 293	61, -14.4 (16) 1.2, 291	59, -22.7 (15) 0.6, 292	54, -31.1 (13) 0.3, 293
	○	○	○	○	○	○	○	○	○
	○	○	○	○	○	○	○	○	○

Table I-11.

mPw TYPE OF AIR MASS
MEAN SEASONAL VALUES OF SIGNIFICANT PROPERTIES
SPRING 1936

Order of data per cell: R.H. ⌐ T / q ↗ θ_E, with circled NO. OF OBS.

Station	SFC.	500	1000	1500	2000	2500	3000	4000	5000
BILLINGS	57⌐8.8 / 4.6 (34) 305			49⌐9.0 / 4.2 (29) 308	55⌐4.9 / 3.7 (38) 308	56⌐1.8 / 3.2 (38) 308	59⌐-1.7 / 2.8 (37) 309	63⌐-9.0 / 2.0 (30) 309	54⌐-16.7 / 1.1 (31) 309
CHEYENNE	62⌐4.6 / 4.1 (44) 307				59⌐4.4 / 3.8 (34) 308	52⌐2.9 / 3.3 (36) 310	55⌐-1.0 / 2.8 (35) 310	59⌐-8.7 / 1.9 (40) 310	57⌐-16.0 / 1.1 (40) 310
FARGO	82⌐11.0 / 7.2 (9) 306	67⌐13.2 / 6.9 (11) 308	62⌐10.5 / 5.6 (16) 309	60⌐7.3 / 4.6 (18) 308	58⌐4.3 / 3.9 (21) 308	58⌐3.2 / 3.1 (25) 307	59⌐-3.0 / 2.7 (27) 307	59⌐9.8 / 1.9 (30) 308	58⌐-5.5 / 1.3 (27) 311
OMAHA	73⌐10.0 / 5.9 (14) 301	62⌐10.5 / 5.1 (15) 302	48⌐9.4 / 4.1 (17) 303	44⌐6.5 / 3.3 (20) 303	44⌐3.1 / 2.7 (23) 303	50⌐-0.5 / 2.6 (25) 305	54⌐-3.7 / 2.4 (27) 306	59⌐-10.6 / 1.7 (26) 307	57⌐-17.7 / 1.0 (27) 308
SPOKANE	72⌐9.4 / 5.6 (35) 303		63⌐10.9 / 5.8 (37) 309	58⌐8.8 / 4.9 (36) 310	59⌐5.2 / 4.1 (38) 309	59⌐1.8 / 3.5 (38) 309	64⌐-1.9 / 3.0 (37) 310	65⌐-8.7 / 2.1 (33) 310	58⌐-15.9 / 1.2 (33) 310
DAYTON	79⌐9.4 / 6.2 (14) 301	63⌐10.9 / 5.4 (16) 303	52⌐9.6 / 4.4 (20) 304	52⌐6.4 / 3.9 (20) 304	53⌐3.0 / 3.3 (23) 304	55⌐0.1 / 2.9 (22) 305	57⌐-3.1 / 2.5 (20) 306	59⌐-9.1 / 1.8 (17) 309	59⌐-15.8 / 1.2 (8) 310
OKLA. CITY	50⌐10.8 / 4.3 (13) 299	52⌐11.4 / 4.7 (17) 301	48⌐10.5 / 4.2 (17) 304	47⌐7.9 / 3.8 (16) 305	48⌐5.7 / 3.5 (20) 308	48⌐2.3 / 3.0 (23) 308	51⌐-0.5 / 2.8 (22) 310	49⌐-6.8 / 1.8 (16) 311	40⌐-11.6 / 1.0 (4) 312

Table I-12.

mT TYPE OF AIR MASS
MEAN SEASONAL VALUES OF SIGNIFICANT PROPERTIES
SPRING 1936

Station	SFC.	500	1000	1500	2000	2500	3000	4000	5000
SHREVEPORT	87⌐9.7 / 12.5 (25) 327	71⌐20.0 / 10.8 (30) 328	72⌐17.3 / 9.9 (30) 328	72⌐14.2 / 8.4 (31) 326	69⌐11.1 / 7.1 (27) 324	66⌐8.1 / 5.8 (24) 322	69⌐4.7 / 5.2 (22) 322	66⌐-1.4 / 3.6 (23) 322	50⌐-9.2 / 1.7 (3) 319
EL PASO	56⌐18.4 / 8.3 (14) 327			56⌐18.0 / 8.1 (17) 329	53⌐15.1 / 7.0 (20) 328	55⌐11.9 / 6.2 (21) 327	56⌐8.4 / 5.3 (20) 327	60⌐0.2 / 3.7 (28) 324	64⌐-7.5 / 2.6 (27) 323
SAN ANTONIO	93⌐19.0 / 13.1 (25) 329	87⌐19.0 / 12.5 (34) 331	79⌐17.3 / 10.7 (37) 330	68⌐15.9 / 8.8 (34) 328	68⌐12.4 / 7.4 (32) 326	63⌐9.9 / 6.2 (30) 325	62⌐6.0 / 5.0 (30) 323	62⌐-4.2 / 3.7 (26) 323	58⌐-7.4 / 2.3 (21) 322
MONTGOMERY	86⌐18.8 / 11.5 (21) 323	63⌐21.1 / 10.2 (25) 327	62⌐18.7 / 9.1 (26) 327	67⌐15.0 / 8.3 (28) 326	62⌐12.2 / 6.8 (23) 324	67⌐8.6 / 6.3 (22) 324	63⌐5.5 / 5.3 (19) 323	58⌐0.0 / 3.6 (15) 323	50⌐-6.0 / 2.3 (11) 324
MURFREESBORO	85⌐18.3 / 11.2 (8) 322	63⌐21.1 / 10.1 (24) 326	64⌐18.4 / 9.3 (29) 327	71⌐14.6 / 8.3 (30) 325	74⌐10.8 / 7.4 (32) 324	68⌐7.9 / 6.0 (29) 322	69⌐4.3 / 5.0 (23) 321	62⌐-2.2 / 3.4 (22) 320	58⌐-7.9 / 2.3 (20) 322
OKLA. CITY	82⌐19.3 / 11.9 (18) 329	71⌐19.1 / 10.3 (26) 325	59⌐19.0 / 8.8 (26) 327	60⌐16.0 / 8.0 (23) 326	61⌐13.1 / 7.1 (21) 322	60⌐9.9 / 5.9 (20) 325	59⌐5.9 / 4.8 (15) 322	63⌐-1.8 / 3.4 (15) 320	64⌐-6.9 / 2.7 (6) 325
PENSACOLA	91⌐20.7 / 13.9 (34) 331	82⌐19.8 / 12.4 (36) 331	79⌐17.1 / 10.7 (38) 329	83⌐13.9 / 9.8 (32) 329	80⌐11.2 / 8.3 (31) 327	77⌐8.6 / 7.2 (29) 327	72⌐5.7 / 5.9 (27) 325	66⌐0.0 / 4.0 (25) 325	63⌐-5.7 / 3.1 (22) 327

TABLE I-13.

ORDER OF DATA
R.H. | T
q / θE
NO. OF OBS.

cAw TYPE OF AIR MASS
MEAN SEASONAL VALUES OF SIGNIFICANT PROPERTIES
AUTUMN 1936

	SFC.	500	1000	1500	2000	2500	3000	4000	5000
FARGO	73|-2.5 (46) 2.5|279	68|-1.0 (41) 2.7|283	66|-2.5 (38) 2.5|286	57|-3.0 (34) 2.2|290	52|-4.9 (25) 1.9|292	51|-7.7 (19) 1.5|293	42|-11.4 (10) 1.0|293	◯	—
BOSTON	68|2.2 (26) 3.4|283	61|0.8 (24) 2.9|285	58|-1.7 (23) 2.3|286	50|-2.8 (22) 2.0|289	58|-4.7 (21) 1.5|290	44|-7.8 (13) 1.3|292			
OMAHA	78|1.3 (41) 3.5|285	68|2.3 (34) 3.3|288	66|-2.2 (30) 3.0|290	58|-1.7 (22) 2.3|291	54|-3.5 (16) 2.0|294	59|-6.9 (9) 1.9|295	54|-8.4 (4) 1.4|298		
DAY-TON	87|1.1 (25) 3.8|285	76|1.4 (22) 3.4|287	70|-0.4 (20) 2.8|289	53|-1.3 (18) 2.1|291	52|-4.6 (10) 1.8|292	38|-5.4 (5) 1.2|295			
SAULT STE. MARIE	78|-0.8 (43) 3.4|283	74|-2.2 (40) 3.1|284	75|-3.7 (38) 3.3|285	61|-5.4 (35) 1.9|287	52|-7.2 (27) 1.5|289	43|-9.7 (13) 1.1|290	34|-12.9 (6) 0.6|291	23|-12.8 (1) 0.5|302	
	◯	◯	◯	◯	◯	◯	◯	◯	◯
	◯	◯	◯	◯	◯	◯	◯	◯	◯

TABLE I-14.

ORDER OF DATA
R.H. | T
q / θE
NO. OF OBS.

cP TYPE OF AIR MASS
MEAN SEASONAL VALUES OF SIGNIFICANT PROPERTIES
AUTUMN 1936

	SFC.	500	1000	1500	2000	2500	3000	4000	5000
EL PASO	58|9.4 (21) 4.9|307	◯	◯	53|11.5 (17) 5.1|313	55|8.9 (11) 4.8|314	47|7.5 (9) 4.1|316	54|5.5 (5) 4.2|320	◯	
FARGO	76|14.9 (11) 8.7|314	58|17.2 (12) 7.8|316	54|15.4 (10) 6.8|317	46|13.0 (10) 5.2|316	43|9.7 (9) 4.0|314	38|7.9 (8) 3.5|316	42|4.5 (6) 3.2|316	35|-1.2 (4) 2.0|318	—
OMAHA	77|12.2 (12) 6.9|306	64|13.5 (14) 6.5|308	55|13.8 (11) 7.8|312	58|10.9 (12) 5.5|313	48|8.5 (11) 4.3|312	47|7.6 (7) 4.0|315	60|4.8 (4) 4.5|320	◯	◯
DE-TROIT	87|13.8 (9) 9.0|312	65|15.5 (13) 7.7|314	56|13.7 (14) 6.3|314	48|11.5 (14) 5.0|313	46|8.6 (12) 4.1|312	44|5.7 (7) 3.4|313	52|1.8 (6) 3.3|314	◯	◯
SAULT STE. MARIE	89|11.2 (11) 7.7|307	73|12.2 (11) 6.9|308	63|11.9 (11) 6.3|312	56|10.2 (11) 5.4|313	48|7.3 (12) 4.1|311	58|3.5 (13) 3.9|315	60|0.8 (15) 3.5|314	57|-6.2 (10) 2.3|313	48|-11.4 (10) 1.5|316
WASH-INGTON	85|16.8 (11) 10.2|316	62|15.7 (14) 7.3|312	58|13.6 (15) 6.2|312	59|9.9 (12) 5.3|312	65|6.6 (11) 4.9|312	58|5.2 (7) 4.3|315	53|2.5 (6) 3.5|314	59|-1.3 (4) 3.2|320	
	◯	◯	◯	◯	◯	◯	◯	◯	◯

TABLE I-15.

S TYPE OF AIR MASS

MEAN SEASONAL VALUES OF SIGNIFICANT PROPERTIES

AUTUMN 1936

Order of data per cell: R.H. | T (top), q / θ_E (bottom), with circled NO. OF OBS.

	SFC	500	1000	1500	2000	2500	3000	4000	5000
SPOKANE			45 17.2 / 6.1 (4) 317	33 16.9 / 4.9 (10) 319	30 16.0 / 4.2 (10) 321	26 11.0 / 3.0 (14) 317	27 6.8 / 2.5 (18) 317	27 -0.2 / 1.8 (24) 318	26 -7.1 / 1.1 (27) 319
MIAMI					39 13.2 / 4.0 (10) 317	29 11.9 / 3.3 (16) 319	27 9.6 / 2.7 (21) 320	24 3.8 / 1.9 (27) 321	23 -2.4 / 1.2 (31) 324
EL PASO				31 16.7 / 4.5 (8) 317	28 15.5 / 3.9 (12) 320	28 11.3 / 3.2 (14) 318	26 7.3 / 2.5 (17) 317	21 1.4 / 1.5 (21) 319	18 -4.0 / 1.0 (33) 322
PENSACOLA				23 11.9 / 2.6 (7) 306	22 12.1 / 2.5 (17) 311	20 10.3 / 2.1 (23) 314	19 7.3 / 1.7 (28) 315	18 2.1 / 1.3 (37) 319	15 -3.0 / 0.8 (38) 323
OMAHA			32 21.0 / 5.5 (4) 320	30 16.5 / 4.3 (5) 317	30 14.0 / 3.9 (11) 318	32 8.8 / 2.9 (17) 315	32 5.2 / 2.4 (22) 315	28 -2.0 / 1.5 (32) 315	26 -8.9 / 1.1 (34) 317
OAKLAND		34 20.3 / 5.5 (3) 314	30 18.6 / 4.5 (14) 313	27 15.3 / 3.6 (19) 313	26 11.2 / 2.7 (25) 311	24 8.4 / 2.2 (25) 312	23 5.2 / 1.8 (26) 313	21 -0.9 / 1.2 (26) 315	22 -7.8 / 0.8 (25) 317
MUR-FREES-BORO				28 9.6 / 2.4 (7) 303	27 8.3 / 2.4 (22) 307	23 6.7 / 1.9 (29) 309	23 4.4 / 1.7 (37) 311	21 -1.1 / 1.3 (45) 315	20 -6.5 / 0.8 (54) 318

TABLE I-16.

MPk TYPE OF AIR MASS

MEAN SEASONAL VALUES OF SIGNIFICANT PROPERTIES

AUTUMN 1936

	SFC	500	1000	1500	2000	2500	3000	4000	5000
SPOKANE	80 6.5 / 5.1 (30) 298		57 10.3 / 4.7 (29) 306	51 7.3 / 3.9 (29) 305	50 4.9 / 3.3 (26) 306	58 0.5 / 3.0 (22) 306	61 -2.5 / 2.6 (19) 307	52 -10.1 / 1.4 (14) 307	50 -18.0 / 0.8 (11) 307
FARGO	62 9.0 / 4.6 (7) 297	64 8.3 / 4.6 (9) 298	50 7.5 / 3.6 (10) 300	51 3.2 / 2.9 (10) 299	49 0.3 / 2.4 (15) 300	47 -2.2 / 2.1 (17) 301	50 -5.9 / 1.8 (25) 302	48 -14.1 / 1.0 (25) 301	45 -21.8 / 0.6 (17) 302
BILLINGS	59 6.2 / 3.8 (28) 299			56 6.9 / 4.0 (20) 305	58 3.2 / 3.4 (17) 304	61 -0.7 / 2.9 (15) 304	61 -4.8 / 2.3 (15) 303	64 -12.3 / 1.6 (11) 304	58 -20.4 / 0.8 (8) 304
CHEYENNE	67 6.8 / 3.3 (24) 300				62 1.7 / 3.4 (21) 302	62 0.5 / 3.2 (16) 306	66 -3.3 / 2.7 (15) 306	55 -12.0 / 1.4 (13) 304	50 -19.9 / 0.7 (10) 304
SALT LAKE CITY	75 2.0 / 3.7 (19) 296			56 5.1 / 3.6 (23) 302	51 3.7 / 3.2 (15) 304	51 -0.2 / 2.6 (15) 304	56 -4.4 / 2.3 (11) 304	60 -13.4 / 1.4 (8) 302	47 -20.2 / 0.7 (6) 304

TABLE I-17.

Key (Order of Data): R.H. | T / q ↗T θE / NO. OF OBS.

Each cell lists: R.H. | T, (No. of Obs.), q / θE

MPw TYPE OF AIR MASS — MEAN SEASONAL VALUES OF SIGNIFICANT PROPERTIES — AUTUMN 1936

Station	SFC.	500	1000	1500	2000	2500	3000	4000	5000
SPO-KANE	77|13.3 (11) 7.7/313	—	59|13.9 (22) 6.5/314	53|12.6 (25) 5.7/316	57|9.5 (29) 5.3/317	60|6.9 (30) 4.9/318	59|3.9 (27) 4.2/319	54|-2.6 (22) 2.7/318	56|-9.4 (21) 1.9/319
FARGO	83|14.1 (2) 8.6/313	57|19.8 (2) 8.5/322	44|18.8 (4) 5.9/319	45|13.0 (8) 5.0/316	48|8.4 (11) 4.0/313	50|5.5 (13) 3.6/314	54|1.8 (12) 3.3/315	59|-5.9 (18) 2.3/314	60|-12.7 (24) 1.6/316
BILL-INGS	55|12.7 (14) 5.8/312	—	—	46|13.4 (29) 5.2/315	43|11.3 (29) 4.5/317	46|8.1 (29) 4.1/317	53|3.8 (32) 3.7/317	56|-2.3 (28) 2.8/318	55|-8.8 (27) 1.9/320
OMAHA	87|18.0 (2) 11.5/325	63|10.6 (2) 9.7/326	54|16.4 (6) 6.9/318	46|14.1 (9) 5.4/317	45|9.2 (11) 4.0/313	49|5.2 (12) 3.6/315	48|2.4 (19) 3.1/314	50|-5.7 (21) 2.7/313	50|-13.1 (21) 1.2/313
SALT LAKE CITY	64|8.8 (10) 5.2/308	—	—	53|11.3 (14) 5.2/313	49|8.7 (19) 4.3/313	52|5.2 (19) 3.8/313	54|1.1 (18) 3.2/313	58|-6.2 (20) 2.2/313	52|-12.8 (18) 1.3/314
OKLA CITY	85|10.6 (3) 6.9/305	72|11.7 (6) 6.8/308	57|13.6 (6) 6.2/313	60|10.1 (8) 5.4/312	53|8.7 (13) 4.7/314	51|6.5 (13) 4.2/315	51|-3.6 (13) 3.5/316	55|-3.5 (13) 2.6/317	49|-8.7 (12) 1.7/320

TABLE I-18.

Key (Order of Data): R.H. | T / q ↗T θE / NO. OF OBS.

MT TYPE OF AIR MASS — MEAN SEASONAL VALUES OF SIGNIFICANT PROPERTIES — AUTUMN 1936

Station	SFC.	500	1000	1500	2000	2500	3000	4000	5000
MIAMI	92|23.2 (64) 16.4/341	81|23.6 (65) 15.6/345	78|20.4 (65) 12.9/339	74|17.3 (64) 10.9/336	75|14.4 (60) 9.7/334	67|12.4 (57) 8.1/333	67|9.7 (52) 7.0/333	65|4.0 (46) 5.3/333	61|-1.8 (42) 3.6/332
OKLA. CITY	74|24.0 (40) 14.1/341	70|24.3 (22) 14.0/342	62|23.8 (22) 12.7/343	66|19.6 (24) 11.0/340	65|16.6 (24) 9.7/338	64|13.3 (24) 8.2/335	65|10.1 (24) 7.1/334	63|3.5 (22) 5.0/332	58|-2.3 (22) 3.4/332
PENSA-COLA	92|21.8 (40) 15.1/336	79|22.4 (42) 14.0/339	78|19.4 (45) 12.2/336	79|16.5 (44) 10.9/335	73|13.9 (42) 9.2/333	69|11.5 (39) 7.8/332	67|8.9 (37) 6.7/331	68|2.9 (32) 5.1/331	65|-2.7 (28) 3.8/333
EL PASO	68|19.7 (26) 11.0/336	—	—	64|20.7 (26) 14.0/342	68|18.3 (27) 11.0/343	73|14.4 (29) 10.0/342	70|11.3 (30) 8.4/339	74|4.2 6.1/335	70|-1.7 (22) 4.4/335
SAN AN-TONIO	96|22.3 (33) 16.5/342	88|22.5 (25) 15.7/344	76|20.5 (27) 12.6/339	71|18.1 (28) 10.7/336	69|15.2 (33) 9.1/334	68|12.7 8.1/334	67|9.7 7.0/333	65|4.1 (23) 5.4/332	68|-2.4 (16) 3.9/332
SHREVE-PORT	87|23.4 (24) 15.6/338	75|23.8 (25) 14.6/342	73|21.7 (24) 13.0/341	75|18.5 (24) 11.8/340	74|15.0 (25) 9.8/336	66|12.0 7.7/332	60|9.5 (25) 6.2/330	68|3.4 (23) 5.1/330	45|-0.1 (1) 3.0/332
MONT-GOMERY	91|21.7 (30) 14.8/335	70|22.8 (31) 12.9/337	70|20.5 (33) 11.8/336	72|17.3 (33) 10.5/334	64|14.9 (30) 8.5/324	60|11.9 (30) 6.9/330	63|8.8 (28) 6.3/330	65|3.0 (24) 4.8/330	63|-2.6 (15) 3.5/331

TABLE I-19.

ORDER OF DATA R.H. | T / q / θE / NO. OF OBS.	cP TYPE OF AIR MASS — MEAN SEASONAL VALUES OF SIGNIFICANT PROPERTIES — SUMMER 1936							
SFC.	500	1000	1500	2000	2500	3000	4000	5000
FARGO 71|15.3 / 7.8 (29) 312	55|19.0 / 7.7 (27) 318	53|17.1 / 6.9 (25) 319	52|14.2 / 6.1 (26) 319	51|11.4 / 5.3 (17) 319	51|8.0 / 4.4 (10) 318	70|2.2 / 4.3 (6) 316	73|-2.1 / 3.8 (1) 322	
BOSTON 77|16.6 / 9.0 (25) 313	56|17.8 / 7.4 (25) 316	57|15.1 / 6.8 (24) 316	62|11.9 / 6.2 (21) 317	60|8.9 / 5.3 (17) 316	59|6.8 / 4.9 (5) 316	66|5.3 / 5.2 (3) 324		
OMAHA 69|17.4 / 8.9 (28) 317	56|19.8 / 8.4 (27) 321	52|17.6 / 7.1 (23) 320	50|14.7 / 6.1 (18) 319	51|11.9 / 5.3 (9) 319	49|9.5 / 4.6 (5) 320	55|8.6 / 5.4 (1) 327		
ST. LOUIS 74|17.6 / 9.6 (21) 317	56|19.7 / 8.6 (18) 321	49|12.4 / 6.8 (15) 321	49|14.0 / 5.7 (12) 317	52|9.8 / 5.2 (8) 316	56|8.5 / 5.2 (3) 322	44|5.2 / 3.4 (1) 318		
DAYTON 77|18.5 / 10.6 (29) 322	68|20.3 / 10.7 (30) 328	66|17.3 / 9.1 (17) 326	66|14.6 / 8.1 (15) 325	71|10.1 / 7.7 (10) 324	65|6.5 / 5.4 (8) 319	81|4.8 / 6.3 (3) 325		
WASHINGTON 83|18.7 / 11.5 (38) 322	51|20.0 / 7.8 (38) 319	49|17.1 / 6.7 (31) 318	55|13.5 / 6.2 (29) 318	57|9.9 / 5.5 (22) 318	63|7.1 / 5.4 (14) 320	72|6.1 / 6.1 (3) 326		
DETROIT 81|15.6 / 9.2 (45) 315	60|18.2 / 8.0 (48) 318	54|16.4 / 7.0 (41) 318	52|13.0 / 5.8 (36) 317	52|9.2 / 4.7 (25) 315	53|5.6 / 4.0 (18) 314	70|1.6 / 4.4 (6) 317	38|-7.3 / 1.5 (2) 310	

TABLE I-20.

ORDER OF DATA R.H. | T / q / θE / NO. OF OBS.	mP TYPE OF AIR MASS — MEAN SEASONAL VALUES OF SIGNIFICANT PROPERTIES — SUMMER 1936							
SFC.	500	1000	1500	2000	2500	3000	4000	5000
FARGO 73|19.8 / 11.0 (3) 326	39|24.9 / 8.2 (4) 326	39|20.5 / 6.4 (8) 322	45|16.8 / 6.3 (9) 323	51|12.3 / 6.9 (10) 322	57|9.3 / 5.5 (9) 323	61|3.6 / 4.2 (10) 318	59|-2.3 / 3.1 (11) 320	54|-9.0 / 1.9 (11) 319
SPOKANE 66|15.7 / 7.7 (43) 316		54|19.8 / 8.5 (42) 326	52|19.7 / 7.4 (41) 326	51|13.4 / 6.2 (34) 324	57|9.2 / 5.4 (34) 322	59|5.9 / 4.6 (32) 322	55|-3.8 / 3.2 (28) 322	52|-7.5 / 2.1 (27) 322
BILLINGS 59|16.9 / 7.9 (32) 323			47|18.4 / 7.2 (30) 327	48|14.9 / 6.3 (30) 326	52|11.1 / 5.7 (29) 325	61|6.2 / 5.2 (30) 325	61|-0.1 / 3.7 (25) 324	52|-6.6 / 2.2 (18) 323
CHEYENNE 64|13.6 / 7.5 (30) 327				57|15.3 / 7.4 (26) 330	48|12.5 / 5.7 (16) 328	51|7.8 / 4.8 (11) 325	58|0.9 / 3.8 (9) 326	56|-7.4 / 2.2 (7) 322
OMAHA 68|23.4 / 12.7 (5) 335	57|24.0 / 11.4 (6) 334	40|23.0 / 7.5 (3) 326	34|17.7 / 5.1 (4) 320	46|14.4 / 5.1 (5) 320	59|8.0 / 5.0 (5) 320	55|4.7 / 4.1 (5) 319	57|-1.2 / 3.1 (6) 321	51|-7.6 / 2.0 (6) 321

TABLE I-21.

ORDER OF DATA R.H. ⌐ T q ⟋⊤ θE NO. OF OBS. SFC.	MT TYPE OF AIR MASS MEAN SEASONAL VALUES OF SIGNIFICANT PROPERTIES SUMMER 1936							
	500	1000	1500	2000	2500	3000	4000	5000
SPO-KANE 76/18.0 (2) 10.4/326	○	62/21.5 (8) 11.1/336	54/22.2 (14) 10.8/342	58/18.1 (21) 9.3/339	61/14.7 (21) 8.3/337	67/10.6 (21) 7.8/336	76/2.5 (24) 5.6/332	68/-4.4 (23) 3.4/329
BILL-INGS 55/21.6 (8) 9.9/334	○	○	40/22.7 (14) 8.1/335	41/19.8 (18) 7.3/335	46/16.1 (19) 6.8/334	51/12.6 (22) 6.5/335	65/5.0 (25) 5.5/335	74/-1.9 (25) 4.4/335
CHEY-ENNE 68/16.0 (20) 9.3/334	○	○	○	59/17.7 (20) 8.9/336	43/18.6 (34) 7.6/340	45/14.8 (36) 6.6/338	53/6.6 (38) 5.1/335	66/1.4 (41) 4.1/334
OMAHA 65/24.6 (29) 12.8/337	56/26.6 (29) 12.8/340	46/27.4 (32) 11.6/345	46/24.7 (33) 10.3/343	51/20.6 (36) 9.2/341	54/16.5 (40) 8.1/338	56/13.0 (38) 7.1/337	58/5.8 (34) 5.2/335	67/-2.2 (33) 3.9/333
ST. LOUIS 71/25.0 (33) 14.1/339	53/27.6 (43) 12.8/342	49/26.2 (43) 11.6/343	51/23.4 (40) 10.7/342	53/19.6 (41) 9.4/340	51/16.0 (40) 8.4/339	51/12.4 (40) 7.1/336	50/5.0 (35) 5.2/333	63/-1.8 (28) 3.7/332
DAY-TON 78/23.4 (20) 14.4/339	68/24.6 (27) 13.8/342	63/24.7 (38) 13.6/347	65/21.7 (39) 12.6/346	67/18.1 (41) 11.0/343	72/14.5 (39) 10.2/342	71/11.1 (40) 8.3/339	70/4.7 (39) 5.9/335	66/-1.4 (37) 4.1/334
WASH-INGTON 87/23.4 (18) 14.4/339	65/23.1 (19) 13.8/342	62/20.8 (23) 13.6/347	63/15.6 (24) 12.6/346	71/14.7 (24) 11.0/343	68/11.6 ( ) 10.2/342	75/8.3 (29) 8.3/339	74/2.4 (30) 5.9/335	73/-3.2 (29) 4.1/334

TABLE I-22.

ORDER OF DATA R.H. ⌐ T q ⟋⊤ θE NO. OF OBS. SFC.	MT TYPE OF AIR MASS MEAN SEASONAL VALUES OF SIGNIFICANT PROPERTIES SUMMER 1936							
	500	1000	1500	2000	2500	3000	4000	5000
EL PASO 56/22.9 (40) 10.7/339	○	○	53/24.1 (48) 11.5/346	53/22.0 (51) 10.8/347	56/18.3 (51) 9.6/345	60/14.3 (52) 8.4/343	69/6.3 (48) 6.5/339	75/-1.5 (46) 4.6/335
SAN AN-TONIO 91/22.7 (58) 16.0/342	86/23.9 (60) 16.6/348	73/22.5 (58) 13.8/344	73/19.7 (58) 12.3/343	67/18.8 (51) 10.1/339	63/14.5 (45) 8.4/337	60/11.6 (39) 7.0/335	60/5.1 (34) 5.1/333	62/-2.1 (25) 3.6/331
OKLA. CITY 64/24.8 (56) 12.8/338	60/26.2 (59) 13.2/342	53/26.0 (60) 12.2/344	53/23.0 (60) 10.8/343	53/19.7 (60) 9.3/340	52/16.2 (52) 7.8/337	54/12.5 (51) 6.7/335	58/5.1 (44) 4.9/332	55/-1.4 (36) 3.4/332
SHREVE-PORT 81/24.5 (50) 15.4/341	66/25.7 (51) 14.1/341	62/23.2 (51) 12.2/341	62/20.0 (49) 10.6/338	64/17.2 (45) 9.5/337	61/14.2 (39) 8.0/335	57/11.2 (36) 6.3/333	56/5.1 (32) 4.8/332	61/-1.2 (15) 3.7/333
PENSA-COLA 91/24.0 (65) 16.9/343	76/24.4 (67) 15.1/344	73/21.4 (66) 12.9/340	74/18.1 (64) 11.3/338	71/15.2 (60) 9.6/335	70/12.3 (56) 8.4/334	67/9.5 (54) 7.0/333	68/3.4 (50) 5.3/332	65/-2.2 (41) 3.8/333
MONT-GOMERY 87/23.6 (59) 16.0/341	68/25.1 (60) 14.0/342	64/22.7 (62) 12.3/340	67/19.3 (60) 11.0/338	65/16.0 (58) 9.3/335	64/12.9 (53) 7.9/333	61/9.9 (44) 6.5/331	61/4.0 (36) 4.9/331	53/-2.1 (28) 3.2/331
MUR-FREES-BORO 89/23.7 (43) 15.6/342	65/25.2 (56) 13.6/341	63/23.5 (58) 12.4/342	63/20.3 (58) 11.0/340	66/17.0 (58) 10.0/339	65/13.8 (53) 8.4/338	60/10.9 (52) 6.7/333	63/4.5 (46) 5.2/333	60/-1.6 (37) 3.6/333

TABLE I-23.

ORDER OF DATA R.H. ╱ T q ╱ θ_E NO. OF OBS.	S TYPE OF AIR MASS — MEAN SEASONAL VALUES OF SIGNIFICANT PROPERTIES — SUMMER 1936								
	SFC.	500	1000	1500	2000	2500	3000	4000	5000
FARGO	43⌐25.6 ⑨ 8.8⌐327	36⌐29.0 ⑪ 9.5⌐335	33⌐27.7 ⑫ 8.5⌐336	31⌐23.9 ⑬ 6.9⌐332	33⌐19.6 ⑮ 6.0⌐330	36⌐15.0 ⑰ 5.2⌐328	35⌐12.4 ⑰ 4.2⌐326	29⌐3.9 ⑰ 2.3⌐323	30⌐-3.3 ⑱ 1.6⌐324
CHEY-ENNE	48⌐17.2 ⑧ 7.3⌐330				43⌐20.1 ⑫ 7.4⌐335	31⌐21.1 ⑯ 6.6⌐339	33⌐17.0 ㉑ 5.6⌐338	35⌐8.7 ⑳ 4.0⌐334	35⌐-0.1 ⑲ 2.4⌐331
OMAHA	44⌐28.3 ⑦ 10.4⌐334	38⌐31.5 ⑦ 11.4⌐343	29⌐30.1 ⑪ 9.5⌐341	33⌐25.7 ⑭ 8.1⌐338	29⌐21.6 ⑱ 5.9⌐332	28⌐18.2 ⑯ 5.0⌐331	30⌐13.5 ㉑ 4.2⌐329	42⌐5.8 ㉖ 2.8⌐327	37⌐-1.9 ㉙ 1.7⌐327
WASH-INGTON		26⌐29.0 ② 7.2⌐338	29⌐25.1 ⑤ 6.5⌐327	37⌐19.1 ⑦ 6.1⌐324	35⌐14.5 ⑨ 4.6⌐320	28⌐11.1 ⑭ 3.1⌐318	24⌐7.7 ⑳ 2.3⌐317	26⌐1.9 ㉓ 1.8⌐320	23⌐-3.6 ㉗ 1.2⌐323
PENSA-COLA			28⌐23.0 ① 5.4⌐331	35⌐19.3 ③ 5.8⌐323	38⌐16.2 ⑧ 5.4⌐324	30⌐13.2 ⑬ 3.8⌐322	35⌐10.8 ⑯ 2.9⌐322	24⌐4.7 ㉗ 2.1⌐324	22⌐-1.1 ㉗ 1.4⌐326
MONT-GOMERY				50⌐18.1 ③ 7.7⌐327	39⌐15.9 ⑤ 5.6⌐325	35⌐13.3 ⑫ 4.4⌐323	32⌐11.0 ⑳ 3.8⌐325	30⌐4.6 ㉗ 2.6⌐325	27⌐-1.5 ㉚ 1.7⌐327
MUR-FREES-BORO			47⌐21.9 ① 9.1⌐330	38⌐19.9 ④ 6.6⌐326	27⌐16.4 ⑦ 3.8⌐320	29⌐13.2 ⑬ 3.6⌐321	25⌐10.7 ⑳ 2.8⌐321	25⌐4.4 ㉓ 2.1⌐323	23⌐-14 ㉗ 1.4⌐326

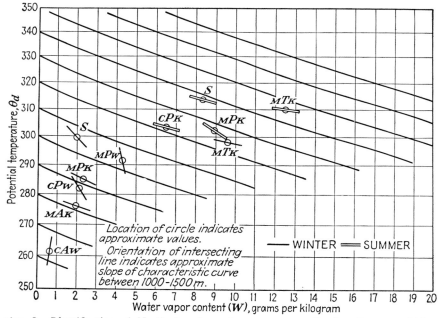

Fɪɢ. 3.—Identification of North American air masses on a Rossby diagram. (*After Showalter.*)

Figure 3 shows the approximate values of North American air masses on a Rossby diagram. This figure is exceedingly useful in identifying air masses whose trajectory or history is unknown or in comparing a particular air mass with the mean values.

WEATHER AND FLYING CONDITIONS IN NORTH AMERICAN AIR MASSES

cPk and cAk in Winter. The weather and flying conditions with *cPk* and *cAk* air over the United States depend primarily on the trajectory of the air after leaving the source region. Trajectories usually observed are shown in Fig. 4.

Types *A* and *B* in Fig. 4 are cyclonic trajectories and are usually indicative of strong as well as deep outbreaks of polar or arctic air. A steep lapse rate maintained by the cyclonic flow conditions is accompanied by active turbulence. The lapse rate often becomes absolutely unstable in the lower layers, owing to heating from

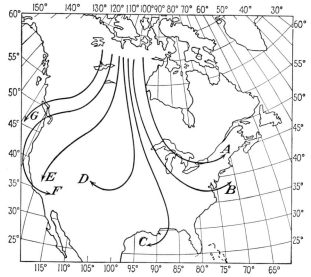

FIG. 4.—Trajectories of *cP* and *cA* air in winter. *A* and *B* are under cyclonic control; *C* and *D* are under anticyclonic control; and *E*, *F*, and *G* are infrequent Pacific coast trajectories.

below. The injection of moisture over the large areas of relatively warm water of the Great Lakes makes the air convectively unstable. The convective instability thus formed is released as the current of cold air flows over the highlands and ridges of the Appalachians.

On the lee side of the Great Lakes and on the windward side of the Appalachians, one may expect a rather low broken to overcast sky condition with frequent and widespread snow squalls. The tops of the clouds are on the order of 6,000 to 8,000 ft on the lee side of the Great Lakes and on the order of 12,000 to 15,000 ft over the Appalachians. Rather severe icing conditions may be expected over the mountains and light to moderate icing on the lee side of the lakes. Rough and gusty flying conditions are the rule as long as the outflow of cold air continues. East of the Appalachians, the sky condition is generally clear except for scattered fractocumulus clouds. Visibility is excellent, and surface temperatures are moderately high owing to turbulent mixing. As a rule, the upper-air temperatures are relatively low for *cPk* and *cAk*, owing to the steep lapse rate, which often extends to at least 3 km.

The weather conditions experienced over the central United States under the influence of trajectories similar to C and D (Fig. 4) are quite different. Unusually smooth flying conditions are found in this region, except near the surface where a turbulence layer results in a steep lapse rate and some bumpiness. Condensation forms are as a rule absent, except that some low stratus or stratocumulus may form at the top of the turbulence layer. As the cold air stagnates and subsides under the influence of the anticyclonic trajectory, marked haze layers develop indicating the subsidence inversions. The surface visibility also deteriorates as the air stagnates and subsides, owing to an accumulation of smoke and dust spreading laterally in the surface layers. This is especially noticeable during the early morning hours when the stability in the surface layers is most pronounced. In the afternoon, when surface heating has reached a maximum, the visibility will usually improve because of the steep lapse rate and resultant turbulence then present in the surface layers.

Movement of cPk and cAk westward over the Rocky Mountains to the Pacific coast is infrequent, owing to the stability of this air. However, when successive outbreaks of cold air build up a deep layer of cP air on the eastern slopes of the Rocky Mountains, relatively cold air can flow toward the Pacific coast unhindered.

If the trajectory of the cold air is similar to E in Fig. 4, rather mild temperatures and low humidities result on the Pacific coast because of the adiabatic warming the air undergoes as it flows down the mountain slopes.

Occasionally, however, the trajectory may pass out over the Pacific Ocean, as indicated in F and G in Fig. 4. The air then arrives over central and southern California as a cold current of convectively unstable air. This type is characterized by squalls and showers and even snow as far south as southern California.

cPk in Summer. This air mass has characteristics and properties quite different from those of its winter counterpart. Because of the long days and the higher altitude of the sun as well as the absence of a snow cover over the source region, this air is usually unstable in the surface layers, in contrast to the marked stability found in cP air at its source in winter. By the time this air reaches the United States, it can no longer be distinguished from air coming in from the North Pacific or from the Arctic Ocean.

Clear skies or scattered cumulus clouds with unlimited ceilings characterize this mass at its source region. Occasionally when this air arrives over the central and eastern portion of the United States it will be characterized by early-morning ground fogs or low stratus decks. Visibility is generally good except near sunrise when it is often reduced by haze or ground fog. Convective activity, usually observed during the daytime, ensures that no great amounts of smoke or dust accumulate in the surface layers. An exception to this is found under stagnant conditions near industrial areas, when visibility may be reduced in those areas day and night. Pronounced surface diurnal temperature variations are observed in cP air during summer.

The convective activity of this air is generally confined to the lower 7,000 to 10,000 ft, and flying conditions will therefore be smooth above approximately 10,000 ft except when local showers develop. Showers, when observed, usually develop in a modified type of cPk over the southeastern part of the country. The base of cumulus clouds that form in this air will usually be on the order of 4,000 ft because of the relative dryness of this mass, and they are therefore of ample height for the safe operation of aircraft.

mP in Winter—Pacific Origin. Maritime polar air from the Pacific dominates the weather conditions of the west coast during the winter months. In fact, with high-index circulation patterns, *i.e.*, rapid west-east motion, this air is often found to influence the weather over most of the United States.

Pacific coast weather, while under the influence of the same general air mass, varies considerably as a result of different trajectories of *mP* air over the Pacific. Thus knowledge of trajectories is of paramount importance in forecasting on the west coast.

Figure 5 shows various trajectories that *mP* air may take prior to reaching the coast. Occasionally the air takes a direct route from the Aleutians and the interior of Alaska across the Gulf of Alaska, as indicated in *A*, Fig. 5. In this case, when the air reaches the coast of British Columbia and Washington after 2 to 3 days over the water, it is found to be convectively unstable to about the 2-km level. This instability

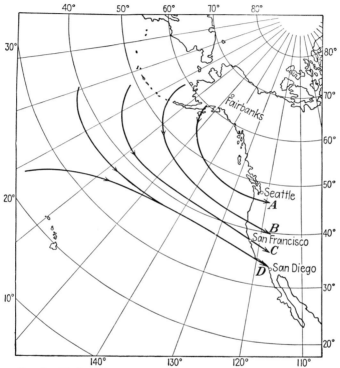

Fig. 5.—Trajectories of *mP* air observed over the Pacific coast.

is released when the air is lifted by the coastal mountain ranges. Showers and squalls are common with this condition. Ceilings are generally on the order of 1,000 to 3,000 ft along the coast and generally zero over the coastal mountain ranges. Cumulus and cumulonimbus are the predominating cloud types, and they generally extend to very high levels. Visibility is generally good because of turbulence and high winds commonly found with this trajectory. Of course, in areas of condensation and precipitation, the visibility will be low. Icing conditions, generally quite severe, are present in the clouds. After this *mP* air has been over land for several days, it has stabilized sufficiently to permit the safe operation of aircraft.

mP air following trajectories *B* and *C* in Fig. 5 undergoes a longer water trajectory, and the surface heating therefore extends to higher levels. Convective instability is usually present up to the 3-km level.   The degree of instability is less, however, than that found with trajectory *A*.   Showers in this air are generally not so intense as in the former case.   The total amount of precipitation is larger with this trajectory, particularly if lifting of the entire mass takes place.

mP air following trajectory *D* in Fig. 5 will arrive on the coast with an anticyclonic trajectory.   Whether this air originally came from the interior of Siberia or Alaska or whether it originally came around the western periphery of the Pacific anticyclone can usually not be determined and is not pertinent.   This trajectory usually is over water sufficiently long to permit modifications to reach equilibrium at all levels. When the air reaches the coast, it is a very stable current, usually evidencing one or two subsidence inversions.   Stratus or stratocumulus type clouds are frequently found. Ceilings are on the order of 500 to 1,500 ft, and the tops of clouds are generally less than 4,000 ft.   Icing conditions are seldom observed in air following trajectory *D*. Visibility is fair except during the early-morning hours, when, owing to the greater stability of the air, haze and smoke frequently reduce the visibility to less than 1 mile. This type of air is found over the entire Pacific coast at times.   It is at times referred to as *mT* air, since it follows the northern limb of the Pacific anticyclone.   Perhaps more appropriately, this air should be labeled *mP* air with an anticyclonic trajectory. True *mT* occasionally does reach the west coast.   Storm conditions of major intensity usually accompany an influx of *mT* air, and the weather conditions are in great contrast to those observed with *mP* air following trajectory *D*.

mP air gradually drifts eastward with the prevailing west-east circulation.   In crossing the Coastal Ranges and the Rocky Mountains, much of the moisture in the lower layers is condensed out, and the heat of condensation liberated is absorbed by the intermediate layers of air.   On the eastern slopes of the mountains, the air is warmed as it descends dry-adiabatically.   As it flows over the cold and quite often snow-covered land surface east of the mountains, the warm *mP* air becomes very stable in the lower layers.

The flying conditions in *mP* air east of the Rocky Mountains are in general the best that are experienced in winter.   Relatively large diurnal temperature ranges are observed.   Turbulence is almost absent, and visibility is good except for smoke and haze in industrial areas.   Ceilings are generally unlimited, since no or only a few high clouds are present.   This type of mild winter weather spreads eastward to the Atlantic coast at times.

When *mP* air crosses the Rocky Mountains and encounters a deep dome of *cP* air, which it is forced to overrun, storm conditions producing blizzards over the plains states result.

In winter, the Great Basin over the western United States is occasionally under the influence of an intense dynamic anticyclone.   Maritime polar air that is over this region at that time will be modified extensively by the subsidence effect from aloft. The characteristics of *mP* air are lost under such conditions, and the air is referred to as *polar basin* air by west coast meteorologists.

This air is very dry and is warmer than *cP* and typical *mP* air.   When it flows eastward over the Great Plains or westward to the Pacific Coast under the anticyclonic flow pattern, it brings fine weather with generally clear skies and mild temperatures with low relative humidity.   Flying conditions are excellent in this air.

mP in Summer—Pacific Origin.   With a fresh inflow of *mP* air over the Pacific coast, generally clear skies or a few scattered cumuli over the coastal mountains are observed.   As this air flows southward along the coast, a marked turbulence inversion reinforced by subsidence from aloft is observed.   Stratus or stratocumulus clouds generally form at the base of the inversion.   Ceilings are then generally on the order of 500 to 1,500 ft with tops of clouds seldom above 3,500 ft. The formation of the stratus condition along the coast of California is greatly enhanced by the presence of the upwelling of cold water along the coast. East of the Rocky Mountains, this air has the same properties as cP air.

mP in Winter—Atlantic Origin. Polar maritime air whose origin is in the Atlantic becomes significant at times along the east coast. It is not nearly so frequent over North America as the other types, owing to the normal west-east movement of all air masses. This type of air is observed over the east coast in the lower layers of the atmosphere whenever a cP anticyclone moves slowly off the coast of the Maritime Provinces and New England. This air, originally cP, undergoes less heating than its Pacific counterpart, because the water temperatures are colder and also because it spends less time over the water. This results in the instability being confined to the lower layers of this air. The intermediate layers of this air are very stable owing to the anticyclonic trajectory the air is subjected to. Showers are generally absent; however, light drizzle or snow and attendant low visibility are common. Ceilings are generally on the order of 700 to 1,500 ft with tops of the clouds near 3,000 ft. Marked subsidence above the inversion ensures that clouds due to penetrative convection will not exist above that level.

The synoptic weather condition favorable to mP air over the east coast is usually also ideal for the rapid development of a warm front with maritime tropical air to the south. Maritime tropical air will then overrun the mP air, and a thick cloud deck will form. Clouds extending from near the surface to at least 15,000 ft are then observed. Ceilings will then be near zero, and icing conditions will be severe in the cold air mass. Frequently freezing rain and sleet will be observed on the ground. Towering cumulus clouds prevail in the warm air and often produce thunderstorms.

Flying conditions are rather dangerous with mP air because of turbulence and icing conditions present near the surface. Poor visibility and low ceilings are additional hazards. The cloudiness associated with the mP current usually extends as far west as the Appalachians. *The rapidity of the above development is of great concern to all forecasters.*

When mT air overruns mP air, flying conditions become impossible over the east coast because of the thickness of clouds.

mP in Summer—Atlantic Origin. In spring and summer, this air is occasionally observed over the east coast. Marked drops in temperature that frequently bring relief from heat waves usually accompany the influx of this air. Just as in winter, there is a steep lapse rate in the lower 3,000 ft of this mass. Stratiform cloud forms usually mark the inversion. Ceilings are on the order of 500 to 1,500 ft and the tops are usually 500 to 1,000 ft higher. No precipitation occurs from these cloud types. Owing to turbulence in the surface layers, visibility is usually good. This air usually does not constitute a severe hazard to flying.

mT in Winter—Pacific Origin. This type of air is observed only infrequently on the Pacific coast, particularly near the surface. Any air flowing around the northern limb of the Pacific anticyclone is usually referred to as mP air. This air has the weather characteristics as well as the low temperature of mP air, having had a long trajectory over the water.

Occasionally the eastern cell of the Pacific anticyclone breaks down, and one portion moves far to the southward off the coast of Lower California. This portion of the anticyclone will then be able to produce an influx of mT air. Generally the influx of mT air is carried aloft by a rapidly occluding frontal system somewhere over southern California, producing the heaviest precipitation recorded in that area. Occasionally mT air is observed above the surface with pronounced storm developments over the Great Basin. Since large open warm sectors of mT air do not occur along the west coast, representative air-mass weather is not experienced. Flying conditions are generally restricted when this air is present, owing mainly to low frontal cloud systems and reduced visibility in precipitation areas.

mT in Summer—Pacific Origin. This type of air has no direct influence on the flying conditions and weather over the Pacific coast. During the summer season, the Pacific anticyclone has moved northward and dominates the Pacific Coast weather with *mP* air.

mT in Winter—Atlantic Origin. Maritime tropical air is in unmodified form when it reaches the southern portion of the United States. Temperature and moisture content are higher in this air mass than in any other American air mass in winter. In the southern states along the Atlantic coast and Gulf of Mexico, mild temperatures and high humidities with much cloudiness are found, especially during the night and early morning. This is the characteristic weather found in *mT* air in the absence of frontal conditions. The stratus- and stratocumulus-type clouds that form at night tend to dissipate during the middle of the day, and fair weather is then the rule. Visibility is generally poor when the cloudiness is present; however, it improves rapidly because of convective activity when the stratus clouds dissipate. The ceilings associated with the stratus condition are generally on the order of 500 to 1,500 ft, and the tops are usually not higher than 3,500 to 4,500 ft. The tops of the clouds usually coincide with the inversion of temperature and decrease in humidity found at those elevations. Precipitation does not occur in the absence of frontal action. With frontal activity, the convective instability inherent in this air is released, producing copious precipitation.

Occasionally when land has been cooled along the coastal area in winter, maritime tropical air flowing inland will produce an advection fog over extensive areas.

In general, flying conditions within this air mass are fair. Ceilings and visibilities are occasionally below safe operating limits; however, flying conditions are relatively smooth, and icing conditions are absent near the surface layers.

As the trajectory carries the *mT* current northward over progressively colder ground, the surface layers are cooling off and becoming saturated. This cooling is greatly accelerated if the surface is snow- or ice-covered or if the trajectory carries the air over a cold-water surface. Depending on the strength of the air current, fog with light winds or a low stratus deck with moderate to strong winds will form rapidly because of surface cooling. Occasionally fine drizzle will fall from this cloud form, and visibility even with moderate winds will then be very poor. Owing to the rapidity with which the low clouds and fog form within this air mass over very large areas, particularly during the night, flying conditions are hazardous whenever extensive surface cooling is indicated. Frontal lifting of *mT* air in winter, even after the surface layers have become stabilized, will release the convective instability and yield large amounts of precipitation in the form of rain or snow.

During the winter, air resembling *mT* is occasionally observed flowing inland over the Gulf and south Atlantic states. Generally this air is *cP* air that had a relatively short trajectory over the warm water off the southeast coast. Clear weather usually accompanies this type of air in contrast to cloudy weather accompanying a deep current of *mT* air. On surface synoptic charts, the apparent *mT* current can be distinguished from true *mT* currents by the surface dew-point-temperature value. True *mT* air will always have dew-point-temperature values in excess of 60°F. The highly modified *cP* air will usually have dew-point values between 50 and 60°F.

mT in Summer—Atlantic Origin. The weather in the eastern half of the United States is dominated by *mT* air in summer. As in winter, warmth and excessive amounts of moisture characterize this air. In summer, convective instability extends to higher levels, and there is also a tendency toward increasing instability since the air moves over a warmer land mass.   This is contrary to winter conditions.

Along the coastal area of the southern states, the development of stratocumulus clouds during the early morning is typical.   These clouds tend to dissipate during the middle of the morning and immediately reform in the shape of scattered cumulus. The continued development of these clouds leads to scattered showers and thunderstorms during the late afternoon.   Ceilings in the stratocumulus clouds are generally ample for the operation of aircraft, on the order of 700 to 1,500 ft.   Ceilings become unlimited with the development of the cumulus clouds.   Flying conditions are generally favorable despite the shower and thunderstorm conditions, since the convective activities are scattered and can be circumnavigated.   Visibility is usually good except near sunrise when the air is relatively stable over land.

When mT air moves slowly northward over the continent, ground fogs frequently form at night.   Sea fogs develop whenever this air flows over a relatively cold current off the east coast.   The fogs over the Grand Banks are usually formed by this process.

In late summer, the Bermuda anticyclone becomes intensified at times and seems to retrograde westward.   This results in a general flow of mT air over Texas, New Mexico, Arizona, Utah, Colorado and even southern and Lower California.   The mT air reaching these areas is very unstable because of the intense surface heating and orographic lifting it undergoes after leaving the source region in the Caribbean and Gulf of Mexico.   Shower and thundersorm conditions, frequently of cloudburst intensity, will then prevail over the southwestern states.   Locally this condition is termed *sonora* weather.

cT Air.   This type of air is found over the United States only in summer.   Its source region is the relatively small area over the northern portion of Mexico, western Texas, New Mexico, and eastern Arizona.   As might be suspected, high surface temperatures and low humidities are the main air-mass characteristics.   Large diurnal temperature ranges and absence of precipitation are additional properties of cT air. Flying conditions as pertaining to ceilings and visibilities are excellent.   Uncomfortable flying conditions prevail during the daytime in this air, owing to turbulence extending from the surface throughout the average flying levels.

Superior Air.   The origin of this air mass is not known.   It usually first appears in our field of observation over the southwestern states and is believed to be the result of strong subsiding motions.   Most frequently this air is observed only at higher and intermediate levels of the atmosphere, but it is not uncommon to find it at the surface particularly in the south central states.   In its lower layers, it is the warmest air mass observed in the United States.   Owing to a steep lapse rate, it becomes colder than mT at high levels.   Relative humidities are usually less than 40 per cent; quite often they are too low to measure with the present meteorographs.

Superior air is observed in both summer and winter.   Flying conditions are excellent in this air mass, since no cloud forms are present and visibilities are usually very good because of the dryness.

From a forecasting standpoint, this type is very important.   It frequently is observed with strong westerly winds that as a rule will overrun the northerly moving mT current from the Gulf of Mexico.   Unless the mT air mass is deep, the superior air will put a lid on all convective activity found in the maritime tropical current and thus prevent the formation of shower and thunderstorm conditions. If the maritime tropical current is deep and a current of S air is present at high levels, then the characteristic dryness and relative coldness of S air tend to steepen the lapse rate.   Convective activity within the mT mass will then be intensified and extend to higher levels.

CHARACTERISTIC AIR-MASS PROPERTIES OVER EUROPE

Although in general the characteristics of air masses over Europe are much the same as those found over North America, certain differences exist. One reason for this is the fact that an open ocean extends between Europe and North America toward the arctic. This allows an influx of mA to reach Europe. This type of air is not encountered over North America. The location of an extensive mountain range in an east-west direction across southern Europe is an additional influence not present over North America, where the prevailing range is oriented in a north-south direction.

If the trajectory of the air is observed carefully and the modifying influences of the underlying surface are known, little difficulty should be encountered in understanding the weather and flying conditions encountered in an air mass over any continent or ocean.

mA in Winter. This air is observed primarily over western Europe. Strong outbreaks of this air, originating in the arctic between Greenland and Spitsbergen, usually follow in the wake of a deep cyclonic system moving across Scandinavia. A deep cold anticyclone is generally associated with this invasion of arctic air.

By the time this air mass reaches the British Isles, it is very unstable. Cumulus and cumulonimbus are the typical clouds, and widespread showers and squalls are frequent. Visibility is generally good because of turbulence. Icing conditions in this air mass often affect aircraft operation.

TABLE 2.—ARCTIC AIR IN WEST EUROPE AFTER HAVING PASSED THE NORWEGIAN SEA (WINTER)*

Station	Elevation, gm	Press.	Temp., °C	W	θ	θ_{SW}
Nancy, Cologne, and Frankfurt	—	1,000	2	3.6	275	274
	800	900	−3	2.9	278	274
	1,700	800	−10	1.9	281	274
	2,700	700	−17	1.1	284	274
	3,850	600	−26	0.45	287	275
	5,100	500	−36	0.21	290	276
Trappes, Dijon, Lyon, and Châteauroux	—	1,000	4	3.6	277	275
	800	900	−2	2.6	279	275
	1,700	800	−8	2.0	282	275
	2,700	700	−16	1.2	285	275
	3,850	600	−25	0.52	288	276
	5,150	500	−32	0.25	294	278
Hamburg, Kiel, and Norderney	—	1,000	3	3.8	276	275
	800	900	−2	3.0	279	275
	1,700	800	−8	1.7	283	275
	2,750	700	−14	0.9	287	275
	3,900	600	−21	0.5	292	277
	5,150	500	−30	0.27	297	279
Lindenberg, Berlin, and Breslau	—	1,000	2	3.7	275	275
	800	900	−2	2.9	279	275
	1,700	800	−9	2.0	282	275
	2,700	700	−15	1.1	286	275
	3,850	600	−23	0.56	290	277
	5,150	500	−32	0.26	294	278

* After Petterssen.

With the presence of a secondary cyclonic system over France or Belgium, arctic air will occasionally sweep southward across France to the Mediterranean, giving rise to the severe mistral of the Rhone Valley and the Gulf of Lions. Heavy showers and thunderstorm conditions are typical of this situation.

mA in Summer. In summer, this type is so shallow that in moving southward from its source region it modifies to the extent that it can no longer be identified as a separate entity. It is then indicated as *mP*.

TABLE 3.—ARCTIC AIR IN SUMMER NEAR SOURCE REGION*

Station	Eleva- tion, gm	Press.	Temp., °C	W	θ	θ_{SW}
Murmansk and Archangel	0	1,000	9	5.1	282	279
	850	900	2	3.7	283	278
	1,750	800	−2	3.5	289	281
	2,800	700	−11	1.8	291	279
	3,850	600	−16	0.8	298	281
	5,150	500	−27	0.3	301	281

* After Petterssen.

mP in Winter. Maritime polar air observed over Europe usually originates in the form of *cP* air over North America. It reaches the west coast of Europe by vari-

TABLE 4.—POLAR MARITIME AIR OVER WEST EUROPE (WINTER)*

Station	Eleva- tion, gm	Press.	Temp., °C	W	θ	θ_{SW}
Trappes, Dijon, Lyon, and Châ- teauroux	—	1,000	5	4.5	278	276
	850	900	2	3.9	284	279
	1,750	800	−4	2.9	287	279
	2,750	700	−9	1.9	293	280
	3,950	600	−16	1.3	298	282
	5,250	500	−24	0.6	305	283
Hamburg, Kiel, and Norderney	—	1,000	4	4.4	277	276
	850	900	1	3.5	282	277
	1,750	800	−5	2.2	287	277
	2,750	700	−10	1.3	291	278
	3,950	600	−18	0.8	295	279
	5,250	500	−27	0.5	300	281
Nancy, Cologne, and Frankfurt	—	1,000	4	4.7	277	276
	850	900	2	4.1	283	278
	1,750	800	−5	2.8	286	278
	2,750	700	−11	1.6	290	278
	3,950	600	−19	0.9	295	279
	5,250	500	−30	0.5	297	280
Lindenberg, Berlin, and Breslau	—	1,000	2	4.1	275	275
	850	900	0	3.3	281	276
	1,750	800	−4	2.3	287	278
	2,750	700	−11	1.1	290	278
	3,950	600	−19	0.7	295	279
	5,250	500	−28	0.4	300	281

* After Petterssen.

ous trajectories and therefore is found in different stages of modification and produces weather similar to *mP* air over the west coast of North America.   Table 4 shows average conditions of this air over West Europe.

TABLE 5.—POLAR MARITIME AIR OVER WEST EUROPE (SUMMER)*

Station	Elevation, gm	Press.	Temp., °C	W	θ	θ_{SW}
Hamburg, Kiel,	—	1,000	15	8.5	288	286
N o r d e r n e y,	850	900	10	6.3	291	285
Swinemünde, and	1,800	800	4	4.7	295	285
Königsberg	2,850	700	−2	3.1	300	285
	4,050	600	−9	1.8	306	285
	5,400	500	−17	1.0	312	286
Helsingfors, Utti,	—	1,000	14	7.4	287	284
and Slutsk	850	900	11	6.0	293	285
	1,800	800	4	4.0	296	284
	2,850	700	−2	2.8	301	284
	3,950	600	−9	1.9	306	285
	5,300	500	−19	1.2	310	286

* After Petterssen.

TABLE 6.—POLAR CONTINENTAL AIR OVER RUSSIA AND CENTRAL EUROPE (WINTER)*

Stations	Elevation, gm	Press.	Temp., °C,	W	θ	θ_{SW}
Archangel	—	1,000	−26	0.36	247	247
	800	900	−19	0.85	262	260
	1,650	800	−21	0.80	269	265
	2,600	700	−27	0.50	273	267
	3,700	600	−36	0.25	275	268
	4,950	500	−42	0.16	282	272
Moscow and Smolensk	—	1,000	−17	1.0	256	256
	800	900	−14	1.0	267	264
	1,650	800	−16	1.0	272	269
	2,600	700	−23	0.56	269	270
	3,700	600	−32	0.39	267	271
	4,950	500	−41	0.20	263	272
Munich and Friedrichshafen	—	1,000				
	800	900	−11	1.4	270	267
	1,700	800	−17	1.0	273	268
	2,650	700	−21	0.60	279	271
	3,750	600	−26	0.38	282	275
	5,050	500	−36	0.16	290	276
Nancy, Cologne, and Frankfurt	—	1,000	−6	2.0	267	266
	800	900	−9	1.5	272	268
	1,700	800	−15	1.1	275	269
	2,650	700	−20	0.72	281	272
	3,800	600	−25	0.50	287	275
	5,100	500	−34	0.20	292	277

* After Petterssen.

mP in Summer. This air when observed over Europe is similar to *mP* air, as observed on the west coast of North America. The weather conditions associated with this air are generally good. Occasionally, because of surface heating, a shower or thunderstorm is observed in the daytime over land. Table 5 shows average conditions of this air over West Europe.

cA and cP Air in Winter. The source region for this air is over northern Russia, Finland, and Lapland. It usually is observed over Europe in connection with an anticyclone centered over northern Russia and Finland. Occasionally it reaches the British Isles, and also at times it extends southward to the Mediterranean.

Owing to the dryness of this air, clouds are usually absent over the continent. Fair-weather cumulus are the typical clouds when this air is observed over the British Isles. Over the Mediterranean, this air soon becomes unstable and gives rise to cumulus and cumulonimbus clouds with showers. Occasionally it initiates the development of deep cyclonic systems over the central Mediterranean.

Visibility is usually good; however, after this type becomes modified, haze layers form and reduce the visibility.

cA and cP Air in Summer. The source region for this air is the same as for its counterpart in winter. It is a predominantly dry mass and produces generally fair weather over the continent and the British Isles. The visibility is usually reduced because of haze and smoke in the surface layers. As this air streams southward, the lower layers become unstable, and eventually convective clouds and showers develop in the later stages of its life cycle.

TABLE 7.—POLAR CONTINENTAL AIR OVER RUSSIA AND EUROPE (SUMMER)*

Station	Eleva-tion, gm	Press.	Temp., °C	W	θ	θ_{SW}
Moscow, Smolensk, Cracow, and Kiev	—	1,000	17	7.4	290	287
	900	900	16	7.0	298	289
	1,850	800	9	5.1	301	288
	2,900	700	2	3.5	305	286
	4,100	600	−5	2.2	311	287
	5,500	500	−14	1.1	316	288
Murmansk and Archangel	—	1,000	14	8.6	287	285
	850	900	11	7.3	293	287
	1,800	800	5	5.6	295	286
	2,850	700	−1	4.0	301	286
	3,950	600	−5	2.4	311	287
	5,350	500	−15	1.9	315	288
Berlin, Lindenberg, Breslau, Swine-münde, and Kön-igsberg	—	1,000	17	9.6	290	288
	850	900	14	7.5	296	288
	1,850	800	8	5.7	300	287
	2,900	700	2	3.9	305	288
	4,050	600	−6	2.3	309	287
	5,450	500	−14	1.3	316	288
Helsingfors, Utti, and Slutsk	—	1,000	19	9.2	292	288
	900	900	14	7.3	296	288
	1,850	800	8	5.0	300	287
	2,900	700	1	3.5	304	287
	4,100	600	−7	2.2	308	286
	5,450	500	−17	1.1	313	287

* After Pettersen.

mT Air in Winter. Maritime tropical air that arrives over Europe usually originates over the southern portion of the North Atlantic under the influence of the Azores anticyclone. It is marked by pronounced stability in the lower layers and typical warm-mass cloud and weather conditions. Owing to the stability, it is relatively smooth even with moderately strong winds. Relatively high temperatures accompany the influx of this air, and the moisture content is greater than in any other air mass observed in the middle latitudes of Europe.

Visibility is as a rule reduced, owing to its stable stratification and the presence of fog and drizzle, which are frequently observed with an influx of mT air.

Maritime tropical air is most frequently observed in West Europe. By the time it reaches Russia, it is generally found aloft and only infrequently on the surface.

TABLE 8.—MARITIME TROPICAL AIR OVER EUROPE (WINTER)*

Station	Elevation, gm	Press.	Temp., °C	W	θ	θ_{SW}
Trappes, Dijon, Lyon, and Châteauroux	—	1,000	10	7.0	283	282
	850	900	6	5.4	288	282
	1,800	800	3	4.1	294	283
	2,850	700	−2	2.7	300	284
	4,050	600	−9	1.8	306	285
	5,400	500	−18	1.0	312	286
Nancy, Cologne, and Frankfurt	—	1,000	9	6.7	282	281
	850	900	7	5.2	289	283
	1,800	800	3	4.0	294	283
	2,850	700	−3	2.5	299	283
	4,000	600	−11	1.4	304	284
	5,400	500	−19	1.0	310	286
Hamburg, Kiel, and Norderney	—	1,000	9	6.2	282	281
	850	900	5	5.4	287	282
	1,800	800	2	3.7	293	283
	2,850	700	−4	2.2	298	283
	4,000	600	−12	1.4	303	284
	5,400	500	−19	0.8	311	286
Lindenberg, Berlin, and Breslau	—	1,000	6	5.5	279	278
	850	900	5	5.5	286	282
	1,800	800	1	4.2	293	283
	2,850	700	−6	2.7	297	283
	4,000	600	−11	1.8	304	284
	5,400	500	−19	1.1	311	286
Königsberg and Swinemünde		1,000				
		900	4	5.6	286	282
		800	1	3.4	292	282
		700	−6	2.1	296	281
		600	−14	1.5	301	282
		500	−18	1.3	311	286
Moscow and Slutsk		1,000				
		900				
		800	−3	3.4	288	280
		700	−10	2.0	292	279
		600	−15	1.4	299	282
		500	−23	0.8	305	284

* After Petterssen.

Table 8 gives average conditions in mT air as observed over Europe.

mT Air in Summer. In general, this air has the same properties as its counterpart in winter with the exception that it is less stable over land because of surface heating. This air mass loses its warm-mass characteristics soon after passing inland over the relatively warmer land surface during the summertime.

Over water, this mass is still a typical warm mass. Sea fogs frequently occur in the approaches to the English Channel during the spring and early summer in mT air.

Visibility in mT air is generally better in summer than in winter particularly over land where convection currents usually develop.

TABLE 9.—MARITIME TROPICAL AIR OVER EUROPE (SUMMER)*

Elevation, gm	Press.	Temp., °C	W	θ	θ_{SW}
—	1,000	20	8.9	293	288
900	900	14	7.5	296	288
1,850	800	9	6.2	301	289
2,900	700	3	4.6	306	289
4,100	600	−4	3.3	312	289
5,500	500	−11	2.1	320	290

* After Petterssen.

TABLE 10.—CONTINENTAL TROPICAL AIR OVER EUROPE (SUMMER)*

Stations	Elevation, gm	Press.	Temp., °C	W	θ	θ_{SW}
Königsberg and Swinemünde	—	1,000	21	11.6	294	291
	900	900	17	8.7	300	290
	1,850	800	11	6.6	302	289.5
	2,925	700	3	4.6	305	288.5
	4,125	600	−4	2.9	312	289
	5,500	500	−12	1.6	319	289
Berlin, Lindenberg, and Breslau	—	1,000	20	9.5	293	289
	900	900	18	8.5	300	270
	1,850	800	11	6.5	303	289
	2,925	700	4	4.6	308	289
	4,150	600	−3	2.7	313	289
	5,500	500	−14	1.6	317	288
Helsinki, Utti, and Slutsk	—	1,000	20	10.4	293	290
	900	900	17	8.0	299	289
	1,850	800	10	5.8	302	288
	2,900	700	4	4.1	307	288
	4,150	600	−4	2.6	312	288
	5,500	500	−12	1.3	319	289
Moscow, Smolensk, Kiev, and Cracow	—	1,000				
	900	900	23	9.3	305	293
	1,900	800	15	7.2	307	291
	3,000	700	6	4.5	309	289
	4,200	600	−1	3.0	315	290
	4,600	500	−10	1.6	321	290

* After Petterssen.

Maritime tropical air flowing over the Mediterranean in summer usually changes to a cold mass, since the water temperature of the Mediterranean is then slightly higher than that of the air. Weak convection currents prevail then, usually sufficiently strong to form cumulus-type clouds, but seldom sufficiently strong to produce showers. Table 9 gives average conditions in mT air over Europe in summer.

cT in Winter. The continental tropical air that arrives over Europe in winter originates over North Africa. By the time it reaches central Europe it differs little from mT air. In general, this air mass is much more prevalent over southern Europe than over central or western Europe.

Although the moisture content of this air is less than observed in mT air, the visibility is not much better than in mT air, owing primarily to the greater dust content.

This air constitutes the major source of heat for the development of the Mediterranean cyclonic storms, most common during the winter and spring months.

cT Air in Summer. Air in this category usually develops over North Africa, Asia Minor, and the southern Balkans. At its source region, the air is dry and warm as well as unstable. In its northward flow over southern Europe, cT air will absorb moisture and increase its convective instability. The summer showers and thunderstorms observed over southern Europe are often produced by a modified cT air mass.

This air mass is much more prevalent over southern Europe than is its winter counterpart. Table 10 shows the average condition of this air over Europe in summer.

CHARACTERISTIC AIR-MASS PROPERTIES OVER ASIA

The air masses commonly observed over Asia, particularly eastern Asia, are continental polar, maritime tropical, equatorial, and monsoon air. Maritime polar and continental tropical air play a minor part in the air-mass cycle of Asia.

TABLE 11.—POLAR CONTINENTAL AIR OVER SIBERIA AND CHINA (WINTER)*

Station	Elevation, gm	Press.	Temp., °C	W	θ	θ_{SW}
Khabarovsk	—	1,000	−25	0.40	248	247
	750	900	−27	0.39	254	253
	1,600	800	−28	0.37	261	259
	2,500	700	−32	0.30	267	263
	3,500	600	−38	0.20	273	266
	4,750	500	−44	0.13	280	271
Vladivostok	—	1,000	−19	0.55	254	253
	750	900	−22	0.50	259	257
	1,600	800	−25	0.45	265	261
	2,550	700	−32	0.22	267	263
	3,600	600	−39	0.16	271	265
		500				
Peiping	—		−9	0.4		
	500		−13	0.3		
	1,000		−18	0.2		
	1,500		−22	0.2		
	2,000		−27	0.1		

* After Petterssen.

cP Air. Continental polar air or arctic air as observed over the interior of Asia is the coldest air on record. This is brought about by the fact that the interior of Asia, made up of vast steppes, serves as an ideal source region. High mountain ranges across southern Asia aid in the production of *cP* air. They tend to keep the polar air over the source region for a long time and to block the inflow of tropical air over the higher latitudes.

The weather conditions over eastern Asia are governed by this air mass throughout the winter. Successive outbreaks of this air occur over Siberia, China, and the Japanese Islands and establish the winter weather regime. The weather conditions prevailing in this air are similar to those found in *cP* air over the eastern portion of North America.

The cold air that is forced southward over The Himalaya to India and Burma arrives in a highly modified form and is generally known as winter monsoon air.

Table 11 gives average conditions of *cP* air over Siberia and China.

mT Air. Maritime tropical air is usually observed along the coast of China and over the Japanese Islands during the summer. In structure it is almost identical with *mT* air as observed off the east coast of North America. The weather conditions encountered in this air are similar to those of its North American counterpart.

E Air. Equatorial air is observed over southeastern Asia. During the summer all of India and Burma are under the influence of *E* air, owing to the summer monsoon circulation. In the wintertime, when offshore winds prevail, *E* air is found not over the land masses but some distance offshore.

Equatorial air is an extremely warm and moist air mass. It has great vertical depth, often extending beyond 20,000 ft in height. This entire column is unstable, and either any slight lifting or a small amount of surface heating tends to release the instability and produce showers and squalls.

The equatorial air observed over India and Burma is almost identical in structure with *E* air found all along the equatorial zone over the entire earth.

Table 12 gives yearly average values of *E* air over Batavia, Java.

TABLE 12.—EQUATORIAL AIR OVER BATAVIA, JAVA*

Elevation, gm	Temp., °C	R.H., %	W, g/Kg
10	26	86	18.8
1,000	21	77	13.3
2,000	15	74	10.0
3,000	10	66	7.1
4,000	4	64	5.4
5,000	−2	60	3.8
6,000	−7	57	2.7
7,000	−13	53	1.9
8,000	−19	49	1.1

* Yearly average after A. Wagner.

Monsoon Air. Monsoon air as observed over India and Burma consists of unmodified equatorial air during the summer monsoon and of highly modified *cP* air during the winter season.

The weather conditions during the summer monsoon consist of cloudy weather with almost continuous rain and widespread shower activity. High temperatures and high humidities further add to the discomfort.

The weather conditions during the winter monsoon are dominated by the dry and adiabatically warmed polar air flowing equatorward. It is during this time that generally pleasant weather prevails over most of the area influenced by monsoon conditions.

Tables 13 and 14 give average conditions in monsoon air over India.

TABLE 13.—MONSOON AIR IN INDIA (WINTER)*

Stations	Eleva-tion, gm	Press.	Temp., °C	W	θ	θ_{SW}
Agra	—	996	19	4.3	292	283
	850	900	10	4.2	292	283
	1,800	800	5	3.5	296	283
	2,850	700	1	3.1	305	286
	4,050	600	−5	1.7	310	287
	5,400	500	−14	1.2	320	288
Poona	1,000	900	22	6.8	304	290
	2,000	800	14	4.6	306	288
	3,200	700	7	2.8	311	288
	4,300	600	0	1.4	317	288
	5,700	500	−9	0.6	322	289
Karachi	—	1,000	18	6.6	291	285
	900	900	15	5.0	297	286
	1,850	800	8	3.4	300	285
	2,900	700	4	3.4	307	287
	4,150	600	−3	2.6	313	291

* After Petterssen.

TABLE 14.—MONSOON AIR IN INDIA (SUMMER)*

Stations	Eleva-tion, gm	Press.	Temp., °C	W	θ	θ_{SW}
Madras	—	1,000	31	18.0	304	299
	950	900	25	17.0	307	298
	1,950	800	17	13.5	310	297
	3,050	700	10	10.0	314	296
	4,300	600	3	7.0	320	295
	5,700	500	−5	4.2	327	294
Agra	—	1,000				
	900	900	28	19.0	311	300
	1,900	800	21	15.0	314	299
	3,000	700	15	12.0	320	298
	4,250	600	6	7.5	342	296
	5,700	500	−1	5.0	332	296

* After Petterssen.

CHARACTERISTIC AIR-MASS PROPERTIES
IN THE SOUTHERN HEMISPHERE

Air masses encountered in the Southern Hemisphere differ little from their counterpart in the Northern Hemisphere. Since by far the greater portion of the Southern Hemisphere is oceanic, it is not surprising to find maritime climates predominating in that hemisphere.

Since the two large continents of the Southern Hemisphere, Africa and South America, both taper from the equatorial regions toward the South Pole and have only small land areas at high latitudes, cP air is unknown except over the antarctic continent. Maritime polar air is the coldest air mass observed over the middle latitudes of the Southern Hemisphere.

In the interior of Africa, South America, and Australia, cT air can be observed during the summer. Over the remainder of the Southern Hemisphere, the predominating air-mass types are mP, mT, and E air. The structure of these air-mass types is almost identical with those found in the Northern Hemisphere.

Bibliography

1. BERGERON, T.: Über die dreidimensional verknüpfende Wetteranalyse, *Geofysiske Publ.*, **5**(6).
2. SCHINZE, G.: Die praktische Wetteranalyse, *Archiv deut. Seewarte*, vol. 52, 1932.
3. KRICK, I. P.: Airline Meteorology (unpublished notes).
4. HUANG, H. C.: Aerological Soundings by Kite Flight at Peiping, China, National Research Institute of Meteorology, Academia Sinica, Nanking, China.
5. SHOWALTER, A. K.: Further Studies of American Air Masses, *Monthly Weather Rev.*, July, 1939.
6. PETTERSSEN, S.: "Weather Analysis and Forecasting," McGraw-Hill, New York, 1940.
7. TAYLOR, G. F.: "Aeronautical Meteorology," Pitman, New York, 1938.
8. WILLETT, H. C.: "American Air Mass Properties," *Physical Oceanog. Met.*, vol. II, No. 2, 1934.

FRONTS

By E. Bollay

The theories of atmospheric boundaries were presented by Helmholtz, Margules, and V. Bjerknes in various classical papers. J. Bjerknes and H. Solberg were first to apply these theoretical concepts to synoptic meteorology in the theory of frontal analysis and the formation of wave cyclones. T. Bergeron did much to refine the original concepts of fronts and laid the ground work of air-mass analysis.

Bjerknes and Solberg first called attention to the fact that the major weather changes occur along the boundaries between the different air masses. In reality these boundaries are transition zones that for practical purposes can be considered as discontinuities or frontal surfaces. Owing to the general circulation of the air, the frontal surfaces are in equilibrium whenever they form a small angle with the horizontal. A frontal surface may therefore be defined as an inclined narrow zone of transition in which the meteorological elements vary more or less abruptly. A front is defined as the line or zone of intersection between the frontal surface and the earth's surface or some other horizontal plane. The theories of atmospheric boundaries are discussed in Sec. VI.

CLASSIFICATION OF FRONTS

Fronts can be classified geographically and also according to the motion of the air masses involved.

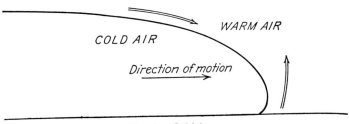

Fig. 1.—Cold front.

Geographical Classification. The formation of fronts occurs from time to time in widely different regions over the earth's surface. However, it follows from what has been said about frontogenesis in Sec. VI that the development of fronts occurs most frequently in certain geographical regions, particularly where deformation fields prevail with a suitable temperature gradient. Regions favorable to frontogenesis are usually found along the boundaries of the four major air masses. The climatological charts in Sec. XII show these major deformation fields and mean frontal zones in each hemisphere. The fronts along these boundaries are called arctic, polar, and intertropical front according to their mean positions. These positions and their annual variation are further discussed in Sec. XII.

Classification of Fronts According to Motion of Air. This classification was introduced by Bjerknes and Solberg. It is based primarily on the displacement of fronts and the resultant temperature changes.   Four basic types of fronts are provided by this system: (1) cold front, (2) warm front, (3) occlusion, and (4) stationary front.

A *cold front* is defined as a front along which cold air replaces warmer air at the surface (Fig. 1).

A *warm front* is defined as a front along which warm air replaces colder air at the surface (Fig. 2).

An *occlusion* is defined as a front resulting when a cold front overtakes a warm front.   Occlusions are subdivided into warm-front-type occlusions and cold-front-type occlusions (Figs. 3 and 4).

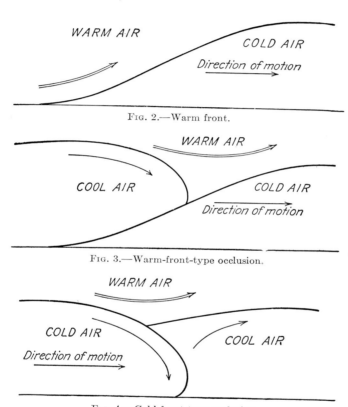

Fig. 2.—Warm front.

Fig. 3.—Warm-front-type occlusion.

Fig. 4.—Cold-front-type occlusion.

A *stationary front* is a front that has little or no horizontal motion.   On synoptic charts, these fronts are indicated as quasi-stationary fronts.

For forecasting purposes, it is not sufficient to know whether one is dealing with a cold front or a warm front.   In order to forecast the weather conditions accurately, it is further necessary to know the stability conditions in both air masses as well as the distribution of vertical velocities.   An estimate of the stability of the respective cold and warm air masses can usually be gained from upper-air soundings and cloud observations.   The determination of vertical currents on a large scale is not feasible so far; however, an estimate of the sign of the vertical velocity can usually be obtained from the distribution of hydrometeors and clouds.

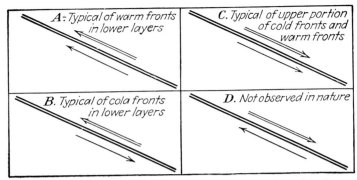

FIG. 5.—Possible distributions of vertical velocities at frontal surfaces.

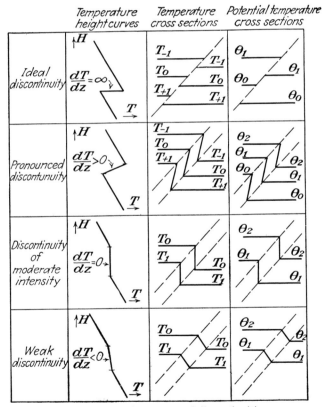

FIG. 6.—Thermal structure of discontinuities.

The various possible combinations of vertical velocities at frontal surfaces are shown in Fig. 5.   Types *A*, *B*, and *C* are observed with various frontal types in nature. With type *D* the frontal cloud system would be found in the ascending currents of the cold air, since no frontal cloud system could exist in the warm descending currents. From observation, it is known that frontal cloud systems are embedded in the warm air and that type *D* would therefore lead to conditions not consistent with observations.

Thermal Structure.   The thermal structure of discontinuities depends upon the width of the transition zone and the temperature difference across the zone.   The strength of a frontal zone is a function of the thermal structure.   The strength or intensity of fronts determines the ease with which discontinuities can be located. Strong fronts have marked temperature changes in both the horizontal and the vertical, whereas weak fronts are marked by only minor changes in temperature.   Figure 6 indicates the thermal structure of fronts or discontinuities of different intensity.

The intensity or strength of a frontal zone is not to be confused with frontal activity.   The activity of a front is more a function of frontal slope, vertical velocities, and stability conditions.

WARM FRONTS

A front along which warm air replaces colder air at the surface is called a warm front.   The velocity of the warm air normal to the front is usually greater than the

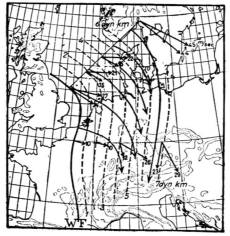

Fig. 7.—Streamlines of the geostrophic motion of tropical air relative to the moving warm-front surface, Feb. 15, 1935, 6 P.M.   (*After Bjerknes and Palmen.*)

velocity of the cold air retreating from the front.   This results in active upglide motion in the warm air.

The active upglide motion usually extends to the highest levels of the warm front in the vicinity of a cyclone.   In regions not under pronounced cyclonic flow conditions, the warm air recurves anticyclonically as it ascends.   In such regions, the vertical components decrease with altitude, and eventually descending motions develop.   A good example of this is shown in Fig. 7.   It can be seen that in this case the motion at the 6-km level is almost parallel to the contour lines.

Since the slope of warm fronts is about 1:150 on an average, it follows that the vertical components in the warm air are never large, even with relatively strong

winds in the warm air.   The typical weather conditions and flow patterns associated
with warm fronts are shown in Fig. 8.

Warm-front Cloud Systems and Precipitation.   Characteristic cloud systems are
usually observed with warm fronts, covering a wide area and often coinciding with the
prefrontal, falling isallobaric pattern.   The first clouds indicating the approach of
the warm front are usually observed 700 to 1,000 miles from the surface position of the
front and consist of cirrus.   As a rule, the cirrus do not coincide with the frontal
boundary but are embedded in an adjacent layer of air some distance above the front.
There is some disagreement as to their exact cause.   That these clouds are not formed
on the warm-front surface is inferred from the fact that the warm-front surfaces do
not reach the level at which these clouds are observed.

With the approach of the front, the cirrus increase and merge into cirrostratus.
As a rule, the boundary of the cirrostratus is found about 600 miles from the surface
front.   Gradually the cirrostratus thickens and merges into altostratus as the front
approaches.   The leading edge of the altostratus is found with a typical condition

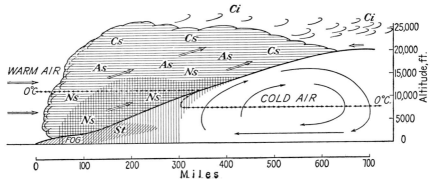

Fig. 8.—Typical weather conditions and flow patterns observed with warm fronts.

about 500 miles in advance of the surface front.   Precipitation is common in the
lowering altostratus deck up to 300 miles in advance of the warm front.   The cloud
type in the precipitation area is referred to as nimbostratus.

The cloud systems found with warm fronts coincide closely with the frontal surface
up to the cirrostratus level.   As a rule the frontal clouds are thickest over the surface
portion of the front and decrease steadily along the warm-front slope.   It is not
unusual to find the cirrostratus, altostratus, and nimbostratus arranged in layers
instead of a solid cloud mass resting on the frontal surface.

When the warm air is convectively unstable, showers and thunderstorm conditions are frequently superimposed on the steady warm-front-type precipitation.
Cumuliform clouds will then be embedded in the nimbostratus and altostratus clouds
and extend to high levels within the warm mass.

Warm-front precipitation increases and reaches its greatest intensity with the
approach of the steepest portion of the warm-front slope.   Little if any rain is observed
when the surface front passes.   However, it is not uncommon to find drizzle and
stratus-type clouds with the frontal passage and in the warm sector after the front
has passed.

Particularly common with warm fronts of rapidly developing systems is the
warm-front fog observed in the wedge of cold air adjacent to the surface front.   As a
rule, this fog will disappear only with the passage of the warm front.

In examining pilot-balloon observations, warm fronts will be found in regions of pronounced clockwise shifts in the wind direction with increasing elevation. When clouds are present in and above the warm-front surface, this veer in wind direction can of course not be observed; however, in such cases, the pilot-balloon data will terminate at the cloud base, thus giving indirect evidence of the height of the frontal zone. Along the upper portion of the warm-front surface, the winds will usually flow parallel to the frontal slope or may even flow downslope. Under such circumstances, wind data will not necessarily indicate the frontal position.

On radiosonde soundings, warm fronts can be located with ease whenever the front is active. That is particularly true in the lower portions of the frontal surface, where the vertical currents are at a maximum. At the upper levels, where downslope motions may be present, the frontal characteristics are masked to such a degree that the warm-front surface cannot be located with any degree of certainty or con-

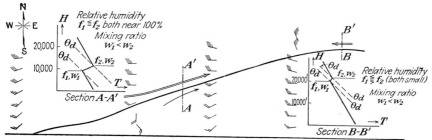

F<small>IG</small>. 9.—Wind distribution and characteristic properties through a warm-front surface oriented in a *NW—SE* direction.

fidence. Figure 9 shows the wind distribution and characteristic properties through a warm-front surface.

Warm-front Characteristics Observed on Surface Charts

Pressure Distribution. Active warm fronts can generally be located by pressure troughs on the surface charts. The troughs are seldom so pronounced as those observed with cold fronts, and other meteorological elements are therefore usually resorted to in locating warm fronts accurately.

Pressure Tendencies. Pressure usually falls for an appreciable length of time prior to the frontal passage. Advection of warmer air and divergence above the frontal surface, and the removal of colder air in the lower layers account for the falling tendency. Prior to the frontal passage, the tendency characteristics are ⌇ or ⟍. A warm-front passage is usually indicated by a ⟍ tendency characteristic. Extensive negative isallobaric patterns precede warm fronts and roughly coincide with the warm-front cloud shield. A marked decrease in the isallobaric gradient is observed in the warm sector except when rapid deepening of the associated cyclone is taking place.

Wind. The wind velocity usually increases in advance of warm fronts because of the general increase in the pressure gradient. Gustiness, which is noticeably absent in the underlying cold air, has been observed to increase below the warm-front altostratus deck, particularly during the cold season of the year over continents and during the warm season over the oceans. This is explained by the greater pressure gradient and by the destructive effect of the altostratus deck on the surface inversion. Radiation from the surface becomes negligible under cloudy conditions, and the surface inversion will then be destroyed by turbulent mixing.

The approach of the front is usually marked by increasing gustiness and stronger winds blowing nearly parallel to the surface front. A maximum in the wind velocity is reached just prior to the frontal passage. As the front passes, the wind will veer and generally diminish in intensity.

Cloud Forms. Warm fronts are nearly always well defined by typical cloud forms. A transition from cirrus to cirrostratus to altostratus to nimbostratus is the rule. The relative distance of these cloud types from the surface position of the front is indicated in Fig. 8.

Fractostratus and stratus and occasionally fog are observed in the immediate vicinity of the warm front. Within the warm mass away from the front, only typical air-mass clouds are found.

Cumulus-type clouds will form in the warm air being lifted if the air mass is convectively unstable. It is also under such circumstances that mammatocumulus are observed from the ground.

Precipitation Forms. The precipitation area of warm fronts extends about 300 miles in advance of the surface front. Especially typical hydrometeors of warm-front precipitation are the following states of weather:

76	74	72	70	69
CONTINUOUS HEAVY SNOW IN FLAKES	CONTINUOUS MODERATE SNOW IN FLAKES	CONTINUOUS SLIGHT SNOW IN FLAKES	SNOW (OR SNOW AND RAIN, MIXED)	HEAVY RAIN AND SNOW, MIXED
68	66	64	62	60
SLIGHT OR MODERATE RAIN AND SNOW, MIXED	CONTINUOUS HEAVY RAIN	CONTINUOUS MODERATE RAIN	CONTINUOUS SLIGHT RAIN	RAIN

Hydrometeors typical of the frontal zone are the states of weather as shown on page 645.

Temperature Changes. Temperature rises slowly with the first indications of the warm-front cloud system. When the altostratus shield approaches and the surface inversion is destroyed, rapid increases in the surface temperature occur, often leading to the apparent development of a fictitious warm front. In precipitation and fog areas, the temperature drops several degrees. When the front passes, a general and rapid rise in temperature is commonly observed.

Dew-point Temperature. The temperature of the dew point increases slowly with the approach of the warm front. In the precipitation area, there is a marked increase in the dew point. A further increase follows the frontal passage whenever the air in the warm sector is of maritime tropical origin.

Visibility. In advance of the front particularly underneath the altostratus deck, the visibility is unusually good, owing to the absence of a surface inversion. In the precipitation area, the visibility decreases and becomes quite low just prior to the frontal passage. After the warm-front passage, the visibility improves to the characteristic values of the air in the warm sector.

77 SNOW AND FOG	67 RAIN AND FOG	61 INTERMITTENT SLIGHT RAIN	59 THICK DRIZZLE AND RAIN	58 SLIGHT OR MODERATE DRIZZLE AND RAIN
57 DRIZZLE AND FOG	56 CONTINUOUS THICK DRIZZLE	55 INTERMITTENT THICK DRIZZLE	54 CONTINUOUS MODERATE DRIZZLE	53 INTERMITTENT-- MODERATE DRIZZLE
52 CONTINUOUS SLIGHT DRIZZLE	51 INTERMITTENT-- SLIGHT DRIZZLE	50 DRIZZLE	48 FOG, SKY NOT DISCERNIBLE--HAS BECOME THICKER DURING LAST HOUR	46 FOG, SKY NOT DISCERNIBLE, NO APPRECIABLE CHANGE DURING LAST HOUR
44 FOG, SKY NOT DISCERNIBLE--HAS BECOME THINNER DURING LAST HOUR	41 MODERATE FOG IN LAST HOUR, BUT NOT AT TIME OF OBSERVATION	40 FOG	24 RAIN AND SNOW MIXED IN LAST HOUR BUT NOT AT TIME OF OBSERVATION	23 SNOW, OTHER THAN SHOWERS, IN LAST HOUR BUT NOT AT TIME OF OBSERVATION

22 RAIN, OTHER THAN SHOWERS, IN LAST HOUR, BUT NOT AT TIME OF OBSERVATION	21 DRIZZLE, OTHER THAN SHOWERS, IN LAST HOUR BUT NOT AT TIME OF OBSERVATION	8 LIGHT FOG, (VISIBILITY 1000 TO 2000 M., 1100 TO 2200 YDS.)	5 HAZE, (VISIBILITY PLUS 1000 M., 1100 YDS.)

Ceilings. The cloud ceilings in advance of the warm front follow the slope of the frontal surface as long as no precipitation takes place. In the precipitation area, the ceilings are lower because of clouds forming within the cold air. Very low ceilings prevail along the surface front, and a rapid improvement follows the frontal passage.

COLD FRONTS

A front along which cold air replaces warmer air at the surface is defined as a cold front.

Marked pressure rises due primarily to the advection of colder air are associated with cold-front passages. As a cold front passes, the wind will veer sharply and the temperature and humidity decrease. The visibility improves markedly in the colder air, and the cloud ceilings increase almost immediately after the cold-front passage.

The distribution of cloudiness and precipitation depends to a large extent on the distribution of the vertical velocities of the warm air. Bergeron differentiated between two basic types of cold fronts on the basis of vertical velocities. Type I is defined as a cold front where the warm air is being lifted to high levels over the cold-front surface, owing primarily to the intrusion of cold air in the lower levels. Type II

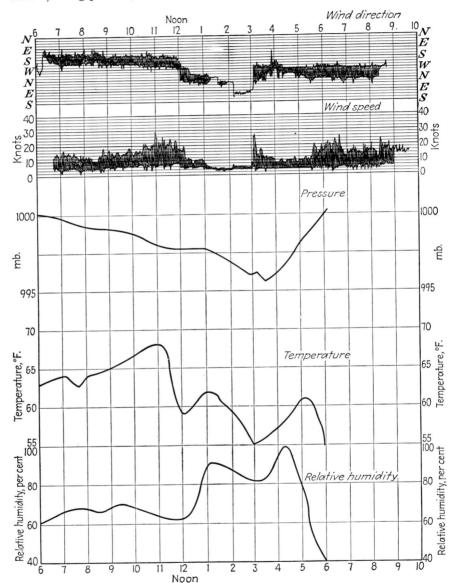

Fig. 10.—Autographic record of cold-front passage. Mar. 17, 1944, Annapolis, Md.

is defined as a cold front where the warm air is lifted only along the leading edge of the intruding wedge of cold air. At higher levels, the warm air is moving faster than the cold air, and therefore no ascending motions are observed in the warm air above the frontal surface. Frequently descending motions develop along the upper portion of the frontal surface with this type of fast-moving cold front.

Cold Front Type I

Cold fronts of this classification are observed outside the zone of cyclonic activity. They are generally slowly moving fronts or may even be quasi-stationary. There

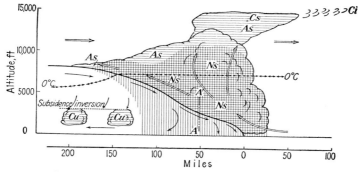

Fig. 11.—Type I cold front. (*After Bergeron, 1934.*)

is a general upglide of warm air over the entire frontal surface so that the cloud system resembles that of a warm front. An observer on the ground will of course observe the clouds in reverse order. Nimbostratus and rain extend for several hundred miles to the rear of the front and are followed by altostratus and cirrostratus. The slope of this type of front is on an average about 1:100; however, near the ground the slope is appreciably steeper owing to surface friction. Unlike the gradual and uniform upglide motion observed with warm fronts, the lifting along the lower portion of the cold front is pronounced. This lifting produces a high cloud bank often resembling a gigantic cumulonimbus cloud. Showers and thunderstorms and squalls are the typical conditions that persist for hundreds of miles along this type of front. The horizontal extent of the cloud and precipitation area is generally smaller than that observed with warm fronts. Since the motion in the cold air is generally descending, no cloudiness is present in the cold air some distance behind the front. If the cold air is unstable, convective cloudiness may develop behind the surface front

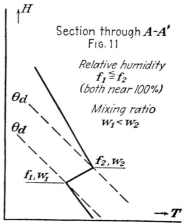

Fig. 12.—Characteristic properties through a type I cold front.

where the descending motions have become negligible. Figure 11 shows the flow conditions and cross section through a cold front of type I.

On vertical soundings, this type of front is usually well marked. The temperature-height curves will indicate pronounced inversions. The relative humidity is generally high in both the cold air and the warm air near the surface front owing to cloudiness and precipitation. The mixing ratio in the warmer air is larger than that in the cold air. Figure 12 shows the variation in characteristic properties through a type I frontal zone.

Cold Front Type II

This type of cold front is very important from a forecasting standpoint. They are observed within the zone of cyclonic activity and as a rule move very rapidly. Since the warm air above the frontal surface moves even faster, descending motion actually develops in the warm air along the frontal surface at the higher levels. The warm air near the surface is pushed vigorously and forcibly upward. As a rule, descending currents prevail in the cold air immediately behind the frontal surface.

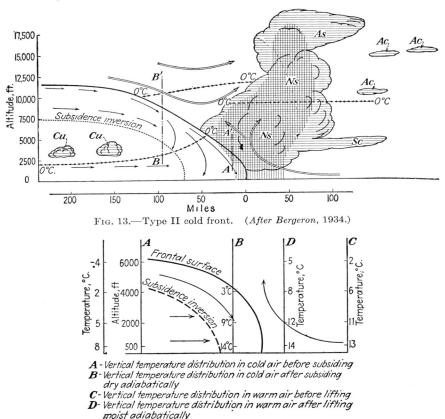

Fig. 13.—Type II cold front. (*After Bergeron, 1934.*)

A - Vertical temperature distribution in cold air before subsiding
B - Vertical temperature distribution in cold air after subsiding
  dry adiabatically
C - Vertical temperature distribution in warm air before lifting
D - Vertical temperature distribution in warm air after lifting
  moist adiabatically

Fig. 14.—The development of a temperature discontinuity within a cold air mass.

It is with this type that a squall line frequently extends along or slightly in advance of the surface front for many miles, induced by the ascending warm air and the rapidly protruding and overturning cold air near the surface layers. Because of high wind velocities associated with this type, steep frontal slopes can be maintained, ranging from 1:40 to 1:80.

The cloud system forming in the ascending warm air resembles a huge cumulonimbus cloud. Frequently stratus clouds grow laterally from the main cloud bank at some intermediate level, and nimbostratus and altostratus form after the ice-crystal level is reached. This cloud deck may drift ahead of the surface front by as much as 150 miles, where it generally breaks up into small isolated patches of lenticular altocumulus. The vertical extent of the cloud system in the warm air is determined by the inversion formed by the high-velocity current aloft and the predominantly ascending warm air. The trailing edge of the cloud system usually consists of stratocumulus opacus or altocumulus opacus and may be observed within a 50-mile zone behind the surface front.

Since descending motions are pronounced in the cold air behind the front, clouds are noticeably absent in this air. If the cold air is unstable, cumulus-type clouds will develop when the descending components are no longer significant.

Precipitation usually occurs at or slightly in advance of the surface position of the front in the form of heavy showers. Line squalls and thunderstorms should be anticipated if the warm air is sufficiently unstable and no pronounced inversions are present to block vertical convection.

The width of the precipitation area is frequently so small that it may not show up at all on synoptic charts. In such cases, past weather and the amounts of recorded precipitation are the only clues as to the activity of the front. Careful attention should be given these items.

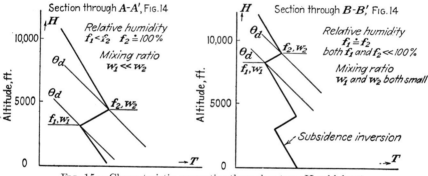

Fig. 15.—Characteristic properties through a type II cold front.

When the warm air is relatively dry, type II cold fronts will sweep through without producing any precipitation and sometimes even without producing cloudiness.

Owing to the sinking motion of the cold air behind the front and the resultant adiabatic warming, the temperature change across the front is often destroyed or may even be reversed. In such cases, the frontal passage is marked by only a wind shift, an increase in the visibility, and a decrease in the moisture content. A real temperature change is observed under these conditions as far as 50 to 100 miles behind the masked front and is often designated as a secondary cold front. This temperature discontinuity marks the boundary between the adiabatically warmed cold air and the horizontally moving current of cold air observed farther behind the front, as shown in Fig. 14. This temperature discontinuity may also be marked by a change in the wind velocity due to the absence of descending motions. No clouds or other frontal phenomena are observed with the temperature discontinuity.

Difficulties are often encountered in analyzing cross sections through the lower portion of type II cold fronts. The dry-adiabatically descending cold air is warmed more than the moist-adiabatically ascending warm air is cooled, thereby destroying the temperature contrast across the front or possibly even reversing the temperature gradient, as discussed in the preceding paragraphs. The subsidence inversion within **the cold air is often more marked than the frontal surface on a temperature-height** curve. This results frequently in some confusion as to where the frontal surface should be located. Figure 15 shows the characteristic properties through a type II cold front.

Cold-front Indications Observed on Surface Charts

Pressure Distribution. Cold fronts are located in well-defined pressure troughs whenever there is a marked density contrast between two air masses. A careful analysis of the isobars will in most cases indicate the correct position of the pressure trough that contains the front. This method of isobaric analysis is frequently the only possible means of locating fronts over ocean areas or regions of scanty surface reports.

Pressure Tendencies. In advance of cold fronts, the tendency characteristic is usually indicated by a ⤌ and ⟍. The isallobars of falling pressure in advance of the front usually form an elongated pattern approximately parallel to the front. Frontal passages are indicated by ⤍, ✓, and ⟍ tendency characteristics.

Type I cold fronts will indicate weak pressure rises behind the frontal passage. The isallobars of rising pressure behind the front are widespread and do not show strong isallobaric gradient.

Type II cold fronts frequently have sudden and strong rising-pressure tendencies following the frontal passage, and the isallobars of rising pressure are often confined to a narrow and oval-shaped area behind the front.

Wind. With the approach of the front, the wind will usually back until it is almost parallel to the front. The wind will be highest at the time of the frontal passage, particularly with type I fronts. In advance of type II fronts, the change in wind speed is not so pronounced as that observed with type I.

A clockwise shift occurs with the frontal passage. Very gusty winds and occasionally a line squall occur in the frontal zone. After the frontal passage, the winds decrease rapidly with type I fronts. Strong winds, frequently gusty and turbulent, are observed for long periods of time in the rear of type II cold fronts.

Cloud Forms. Type I cold fronts are preceded by the cloud forms typical of the warm air mass. Towering cumulus and cumulonimbus as well as stratocumulus and nimbostratus are observed with and immediately behind the surface front. Altostratus and cirrostratus usually trail far behind the surface front.

Type II cold fronts are generally preceded by lenticular altocumulus that gradually increases and changes to altostratus. With the approach of the front, the altostratus merges with nimbostratus and huge cumulonimbus in a cloud bank at the front. Stratocumulus opacus and altocumulus opacus are the predominant cloud types found a short distance behind type II cold fronts.

Precipitation Forms. Showers and sometimes thunderstorms will occur with the cold-front passage, and continuous precipitation will be observed for some hours after the frontal passage with cold fronts of type I flow conditions.

Shower and sometimes thunderstorm activity of short duration will occur with or slightly in advance of type II cold fronts. Rapidly clearing conditions are typical after the frontal passage.

Temperature. With type I cold fronts, there will be a pronounced drop in temperature with and following the frontal passage. Type II cold fronts will generally be preceded by a temperature fall in the warm air due to the prefrontal shower activity. Immediately behind the front, rapid clearing and quite often descending motions tend to keep the temperature in the cold air about the same as in the warm air. An abrupt temperature change is encountered under such circumstances some distance from the cold front.

Dew-point Temperature. The dew-point temperature will generally aid in locating the fronts. A drop in dew point will be observed with both type I and type II cold fronts. When the temperature effect across a type II cold front is masked by descending currents, the front can often be located by the change in dew-point temperature that will be pronounced under such circumstances.

Visibility. With the approach and passage of type I cold fronts, the visibility decreases. In the cold air a slight improvement in the visibility is observed. Since the precipitation continues with this type, no marked improvement takes place until the precipitation ends. Type II cold fronts are preceded by regions of poor visibility due to shower activity. A pronounced improvement will almost always follow this type of cold-front passage.

Ceilings. With type I fronts, the ceilings become very low at the time of the frontal passage and lift only gradually after the front has passed the station. Type II cold fronts will have low ceilings restricting aircraft operation only in the vicinity of the front. Unlimited ceilings usually are found in the cold air, and, far in advance of the cold-front, ceilings are ample for operation of aircraft.

INFLUENCE OF MOUNTAIN RANGES

Mountain ranges greatly influence the speed of fronts and frontal slopes, and the frontal activity. The orientation of the front relative to the mountain range and the

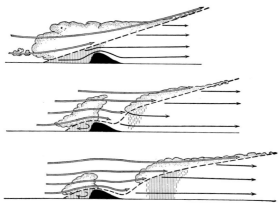

Fig. 16.—A warm front crossing a mountain. (*After J. Bjerknes and Solberg, 1921.*)

size of the mountains as well as the stability of the air masses involved determine the amount of modification that develops.

The retardation of fronts in crossing mountain ranges frequently results in the development of new cyclonic waves on the front. This effect is observed on the south coast of Greenland, along the south coast of Norway, and also along the southern section of the Alleghenies in the United States.

Warm Fronts. Figure 16 represents vertical sections through a warm-front surface passing a mountain range. As a rule, the slope of warm fronts is much less than the slope of mountain ranges. Therefore, when a warm front approaches a mountain range, the warm-front surface reaches the mountain ridge first and traps a wedge of cold air along the mountainside, as shown in Fig. 16. Since the cold mass cannot escape, the warm-front surface will accordingly become stationary along the leading slopes of the mountains. The precipitation area common to the lower portion of the warm front will then persist over the same region for a long time. The upper portion of the warm front will pass over the mountains and descend on the lee side of the mountain range. Precipitation and sometimes cloudiness are absent on this side owing to sinking motions. When the warm front arrives over flat country, it regains its characteristic cloud and precipitation pattern.

The effect of mountain ranges is therefore primarily to widen the precipitation area and to increase the duration as well as the intensity of the precipitation on the windward side of the mountains and to decrease the precipitation area and intensity on the lee side.

Cold Fronts. Figure 17 shows successive vertical sections through a cold front crossing a mountain range. The cold wedge is retarded on the windward side and accelerated on the lee side of the mountains.

If the air on the lee side of the mountains is colder than the advancing wedge of cold air, the cold front will merely travel across the top of the cold dome on the lee

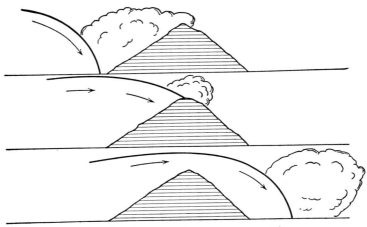

Fɪɢ. 17.—A cold front crossing a mountain.

side of the mountains as an upper-air front. This frequently happens during the winter east of the Rocky Mountains and also east of the Alleghenies during the early-morning hours. At that time, the stagnant air east of the mountains will be colder because of radiation processes, while the well-stirred air associated with the cold front will maintain higher temperatures and therefore overrun the stagnant mass on the lee side of the mountains. Frontal activity will be at a minimum under such conditions until surface heating has become active in reversing the temperature conditions. Thunderstorm conditions and squalls may develop when the upper-air front finally breaks through to the ground.

If the air on the lee side of the mountains is warmer than the advancing wedge of cold air, the cold front will sweep the warm air away. The frontal activity continues unhindered in this case.

The angle with which a cold front approaches and crosses a mountain range determines the increase in frontal activity on the windward side and the relative decrease on the lee side of the mountains. The maximum change in frontal activity occurs when the front lies parallel to the mountain range.

Occluded Fronts. Occluded fronts crossing mountain ranges are influenced in the same way as cold fronts and warm fronts   Cold-front-type occlusions behave like cold fronts, and warm-front-type occlusions cross mountain ranges like warm fronts.

Mountain ranges frequently tend to accelerate the occlusion process when a warm-sector cyclone approaches. Under such circumstances, the warm front is retarded by the mountain range while the cold front moves unhindered until it reaches the mountain range and the stalled warm front. This condition is observed often on the west coast of continents during winter.

OVERRUNNING COLD FRONTS

Overrunning cold fronts indicated in Fig. 18 are commonly associated with violent thunderstorms and tornadoes in the Gulf states and the Great Plains region. Although in general this type of frontal structure is indicative of great instability, it can be

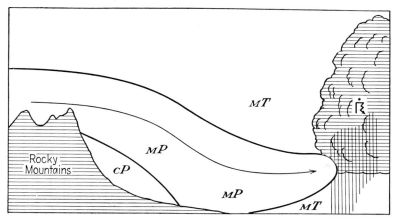

Fig. 18.—Overrunning cold front.

shown that this situation may under certain circumstances persist for periods of 12 to 24 hr.

Overrunning cold fronts are observed primarily with mP cold fronts invading mT air. Maritime polar air when observed east of the Rocky Mountains has approximately the same or sometimes less density than the maritime tropical air from the Gulf of Mexico and the Atlantic Ocean.

The horizontal velocity distribution favoring the development of overrunning cold fronts is such that strong westerly winds are found in the mP air and strong southerly winds prevail in the mT air. The synoptic situation necessary for this flow pattern consists of a deep cyclonic center located over the northern and central Great Plains. This cyclonic development is usually the result of pronounced outbreaks of cP air southward along the eastern slopes of the Rocky Mountains and the simultaneous influx of a warm mT current over the Gulf states. The mP air, accelerated by the deepening cyclonic system, will then flow rapidly eastward and impinge on the mT current in a jetlike effect.

The forced lifting of the mT current by the mP air will set off convective activity, producing heavy showers and thunderstorms. The mP air will occasionally be driven by the jetlike effect to levels where its density becomes greater that that of the underlying mT air. The boundary zone between the mP and mT then becomes autoconvectively unstable. This is one of the conditions under which tornadoes develop.

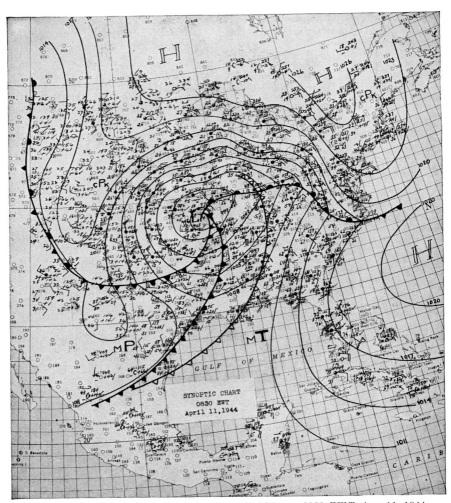

Fig. 19.—Weather map showing overrunning cold front, 0830 EWT, Apr. 11, 1944.

The distance to which the overrunning takes place has been observed to be up to 300 miles. Because of turbulence and convective activity, the overrunning nose eventually breaks down, and the front continues in the conventional fashion eastward.

Figures 19 and 20 show the synoptic conditions for the development of a typical overrunning cold front. Tornadoes were observed with this synoptic situation.

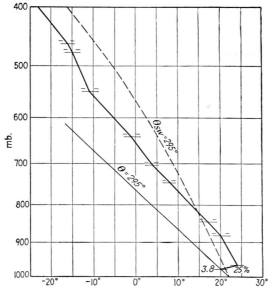

Fig. 20a.—San Antonio, Tex., sounding, 0000 *EWT*, Apr. 11, 1944

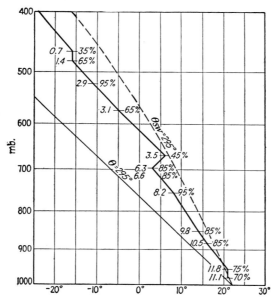

Fig. 20b.—Atlanta, Ga., sounding, 0000 *EWT*, Apr. 11, 1944.

OCCLUDED FRONTS

Warm-front-type Occlusions. In wave cyclones, where the advancing cold air is warmer than the cold air below the warm-front surface, a warm-front-type occlusion results if the cold front moves faster than the warm front. In this case, cool air will lift the warm air from the surface and gradually will flow up the warm-front surface. The cold air below the warm-front surface will move slowly under such circumstances. Figure 21 shows the cycle of events associated with this type of occlusion.

The warm-front-type occlusion has a typical warm-front-type cloud system that is gradually destroyed by the advancing wedge of cool air, *i.e.*, the upper-air cold front. Eventually only cirrus clouds mark the remnant of the original cloud system.

For forecasting purposes, the progress of the upper-air cold front is of greatest importance, since the zone of bad weather precedes and accompanies this front. The

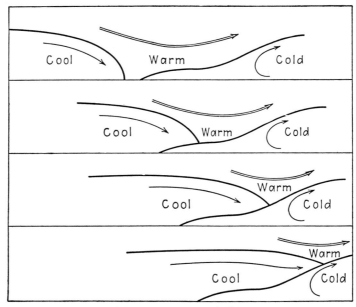

FIG. 21.—Warm-front-type occlusion process.

surface occluded front is of interest only in forecasting the wind direction and temperature changes near the ground. The position of the upper-air front can best be followed by noting the trailing edge of the prefrontal precipitation pattern and the pressure tendencies. A line of cumulonimbus clouds and thunderstorms often accompanies the upper-air cold front, provided that the overrunning cold air has not progressed too far up the warm-front slope.

Between the upper-air cold front and the surface occluded front, precipitation is generally absent unless induced by the topography.

Warm-front-type occlusions are particularly prevalent along west coasts of continents during the winter. The cool air flowing onshore is usually warmer than the cold continental air lying along the immediate costal region. This type of occlusion is also very common east of the Rockies in winter. Cold *cP* air over the Great Plains is invariably colder than *mP* air. Any cyclonic activity between these two air masses will produce a warm-front-type occlusion.

The weather conditions change rapidly with this type of occlusion. When the upper-air cold front is first established, moderate frontal activity should be expected. However, as the upper-air front progresses to higher levels and gradually eliminates the warm air over the warm-front surface, the frontal activity diminishes, and the front is gradually lost on the synoptic chart. This type of occlusion is therefore very important during its initial stages and of only minor interest in its later stages.

Cold-front-type Occlusions. When the air behind the cold-front surface is coldest, the occlusion that results when the cold front overtakes the warm front is called a cold-front-type occlusion.

In the early stages of the occlusion process, the occlusion is preceded by all the phenomena associated with warm fronts, and at the surface it acts in exactly the same

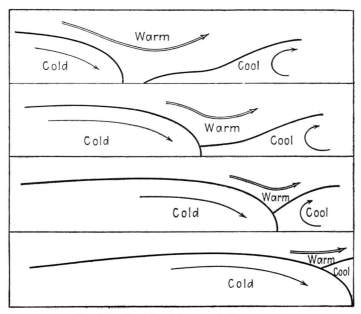

Fig. 22.—Cold-front-type occlusion process.

manner as a cold front. Figure 22 shows the cycle of events associated with this type of occlusion.

During the later stages of the occlusion process when the warm air undergoes no further lifting, the prefrontal cloud system disappears almost entirely. The occlusion, from then on, acts once more as a simple cold front and should be indicated as such on surface charts.

Cyclonic activity associated with outbreaks of *cP* or *cA* air usually results in the formation of cold-front-type occlusions. Occlusions produced over continents and along east coasts of continents are therefore generally of the cold-front type.

The zone of poor weather conditions occurs along and in advance of the occlusion. A marked improvement in the weather conditions, similar to that experienced with cold fronts, occurs after the occlusion passes.

CYCLONES AND ANTICYCLONES

By E. Bollay

CYCLONES

Experience in synoptic meteorology has shown that cyclonic and anticyclonic systems are primarily the result of the interactions between different air masses. This is especially true of cyclones, which have their origin on the fronts separating the major air masses.

In the initial stages, cyclones have the characteristics of waves. If the frontal wave is dynamically unstable, which is true in the majority of the cases, then the occlusion process begins and the disturbance assumes the characteristics of a vortex. The structure of the cyclone is therefore not fixed but in the course of its life cycle is undergoing constant changes, forming as a wave and ending as a vortex. The initial appearance of cyclones as waves not only is an observed fact but is also required by the theory of the development of cyclones.

Wave Action on Fronts, Stable and Unstable Waves. The question whether a frontal disturbance is dynamically stable or unstable depends on various circumstances that have so far been investigated only after stringent simplifying assumptions have been made. No complete and entirely satisfactory stability criteria, directly applicable to synoptic meteorology, have been found so far.

Theory indicates that the stability of a frontal surface depends on the wave length of the wave. Experience indicates that frontal disturbances with a wave length less than 600 miles are usually stable, and frontal disturbances having a greater wave length are usually unstable. Following theoretical considerations, based on certain simplifying assumptions, the critical frontal slope is found to be about 1:100. Slopes greater than this result in unstable waves. Observations in synoptic meteorology seem to indicate that 1:150 is more nearly the critical slope of frontal surfaces. The temperature gradient and wind discontinuity across the front are further factors influencing the dynamic stability of the frontal surface.

On the basis that the stability of the frontal surface decreases with increasing frontal slopes, Bergeron has offered the following useful conclusions:

1. Warm fronts are generally dynamically stable and therefore less subject to cyclogenesis.

2. The southern portion of cold fronts and quasi-stationary fronts are dynamically unstable and therefore subject to cyclogenesis.

3. Horizontal temperature inversions and the upper portions of frontal surfaces are dynamically stable and therefore not subject to cyclogenesis.

Experience also seems to indicate that frontal waves become more stable with increasing stability of the adjacent air masses. When dealing with a deformation field and a frontogenetical zone, the eastern sector is generally more subject to cyclogenesis than the western sector.

The Development of Frontal Cyclones. If a wave forming on a major air-mass boundary is dynamically unstable, its amplitude will increase as it travels along the front, and the associated low-pressure system will deepen. Figure 1 shows the life cycle of a cyclone as first presented by Bjerknes and Solberg in 1922. Stage *a* represents the initial undisturbed condition of the front; stage *b* shows the developing frontal wave with a converging flow pattern and a small precipitation area to the north of the warm front.   If the wave is dynamically stable, it will not develop further but will travel along the front in this stage of development.   If the wave is unstable,

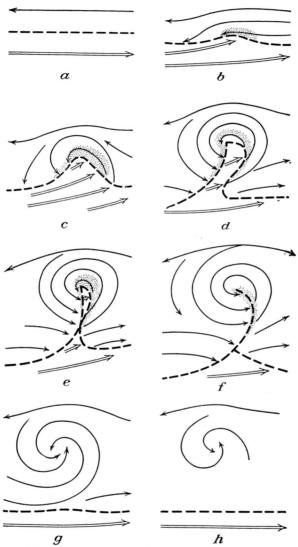

Fig. 1.—Life cycle of a cyclone.   (*After Bjerknes and Solberg*, 1922.)

stage *c* follows.   The wave development has now progressed sufficiently far that the disturbance can be called a *young wave cyclone*.   A well-formed warm sector and a steadily deepening low-pressure system forming at its peak are characteristic of this phase.

In stage *c*, the flow of air in the warm sector is such as to produce overrunning when encountering the warm front. A broad warm-front rain area is typical of this condition. The cold front also becomes more pronounced at this stage, owing to the increased cyclonic pressure field. Type II cold-front weather conditions are typical in stage *c*.

By this time the pressure at the cyclonic center is generally 10 mb lower than in its initial stage, indicating of course that air is being removed in the upper levels.

The direction of movement of the young wave cyclone is determined almost exclusively by the direction of the warm-sector isobars. The speed of the wave cyclone is usually greatest at this time. It will decrease as the cyclone gradually changes from its wave characteristics to those of a vortex.

Stages *e* and *f* indicate the occluded stage of the cyclone. The greatest intensity, the minimum central pressure, and the highest vertical extent are observed in these stages.

Stages *g* and *h* indicate the final dissolution of the cyclone. All frontal activity has disappeared, and only the characteristics of a dissipating vortex remain. The polar front has now been displaced southward, and new waves will form if the proper initiating forces are present.

Energy Considerations. The main source of kinetic energy found in cyclones is believed to be brought about by the transformation of potential energy of the horizontal distribution of mass into kinetic energy. The sinking of cold air and the simultaneous rising of warm air along the frontal surface creates a solenoid field favorable to the development of large amounts of kinetic energy.

The kinetic energy of the preexisting air currents on both sides of the front is another source of energy. This energy is necessary to bring about the initial vertical displacements that will release the large amounts of kinetic energy discussed above.

The potential energy of conditionally unstable stratifications is a further supply of energy. Refsdal has shown that cyclones with moist and conditionally unstable air increase more rapidly in kinetic energy. Likewise, cyclones that are already occluded may suddenly experience a redevelopment with the injection of conditionally unstable warm air. This is often observed in winter when an old cyclone passes out to sea. The injection of moist unstable air will then frequently produce a vigorous redevelopment.

Pressure Variations in Moving Cyclones. J. Bjerknes has shown that the local pressure change at some fixed point at a level *h* is brought about by

1. Horizontal divergence above *h*.
2. Vertical air transport through the level *h*.
3. Horizontal advection above level *h*.

The following tendency equation shows the combined effect of these factors:

$$\left(\frac{\partial p}{\partial t}\right)_h = -\int_h^\infty g\rho \operatorname{div}_2 \mathbf{v}\, dz + (g\rho v_z)_h - \int_h^\infty g\left(v_x \frac{\partial \rho}{\partial x} + v_y \frac{\partial \rho}{\partial y}\right) dz \qquad (1)$$

In examining the pressure variations experienced with a schematic pressure field shown in Fig. 2, Bjerknes concludes that in advance of a trough line the pressure falls because of removal of air by means of horizontal divergence in that part of the atmosphere above level *h*. The rear part of a trough is associated with rising pressure because of horizontal convergence above level *h*.

Applying this analysis to a young polar-front cyclone, Bjerknes explains the pressure variations observed at different levels in terms of the tendency equation.

Considering the pressure distribution in advance of the warm front, it is apparent that the trend of the isobars at the earth's surface is such that there will be divergence.

This will cause the pressure to fall in advance of the warm front. In addition, the motion is such that warm air replaces colder air. The advective pressure term would therefore also cause falling pressure. At the earth's surface, the vertical air transport through the base of the column is zero, and therefore this term of the tendency equation vanishes at the ground. Horizontal divergence and advection are therefore the only terms affecting the pressure variation in advance of the warm front at the ground.

In the warm sector at the ground, the last two terms of the tendency equation vanish, since no vertical velocities are present and the air is supposedly horizontally homogeneous. Any pressure change in the warm sector must therefore be due solely to horizontal divergence in the air column.

In the cold air behind the cold front, a marked pressure rise is observed on the ground. This is due principally to the horizontal advection of colder air. Con-

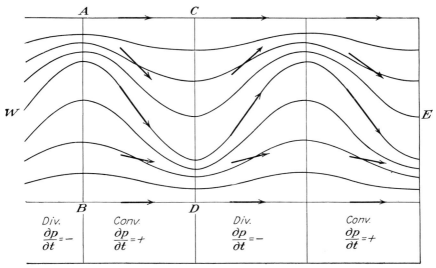

Fig. 2.—Schematic pressure field at a level h of the upper troposphere. (*After J. Bjerknes,* 1937.)

vergence within the cold mass also tends to cause a pressure rise. Above the dome of cold air, conditions are somewhat different. The isobars in the warm current have a trough at some distance behind the surface cold front. Divergence producing a pressure fall is found in the warm air between the surface cold front and the upper-air pressure trough. This pressure fall will reduce the surface pressure rise. This effect is slight with young wave cyclones but becomes more pronounced as the cyclone develops. After this upper-pressure trough passes the station, a more marked pressure rise will be observed at the surface. Convergence in both the cold and warm air as well as advection of colder air will then contribute positive pressure changes. The marked pressure rise associated with the upper pressure trough is frequently erroneously interpreted as a frontal trough or a bent-back occlusion.

The pressure variations associated with upper-air waves cannot be explained on the basis of the divergence term or the advection term and therefore must be largely brought about by the vertical air transport $(g\rho v_z)$. Since the warm air descends over the cold-front surface and ascends over the warm-front surface, this term will produce falling pressure over the cold-front surface and rising pressure over the warm-front surface at the levels through which this vertical motion takes place. These are the pressure changes necessary in increasing the amplitude of the upper-air pressure waves. As these waves develop, they will reflect their influence on the pressure distribution on the ground. According to Bjerknes, this pressure variation is the essential cause of the deepening of cyclones. Since this term is largely brought about by the flow over the frontal surfaces, it is apparent that the frontal wave is the primary source of the cyclone. Figure 3 shows the barograms to be expected on the basis of the tendency equation at various levels. These observations are in general accord with observational facts.

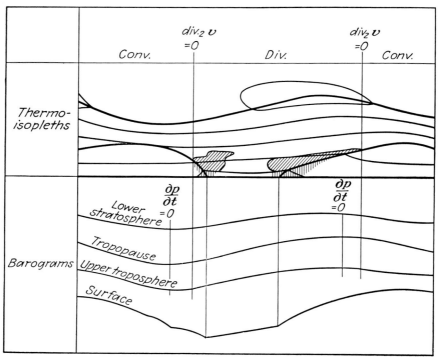

Fɪɢ. 3.—Thermo-isopleths and barograms of a young polar-front cyclone. (*After J. Bjerknes*, 1937.)

Structure of Cyclones. *The Structure of a Young Wave Cyclone.* The structure and typical weather conditions of a young wave cyclone in the Northern Hemisphere are shown schematically in Fig. 4. The streamlines show the general air flow in both the warm and cold air and are therefore representative of the isobars. The isobars in the warm sector are generally straight and orientated in a southwest-northeast direction. In the cold air, the isobars show strong cyclonic curvature except in advance of the warm front where the flow is divergent. During this stage of development, the central pressure of the depression is usually not more than 10 to 15 mb lower than in the undisturbed environment.

Two vertical cross sections show the structure and flow conditions north and south of the center. The lower cross section crosses the warm sector and indicates

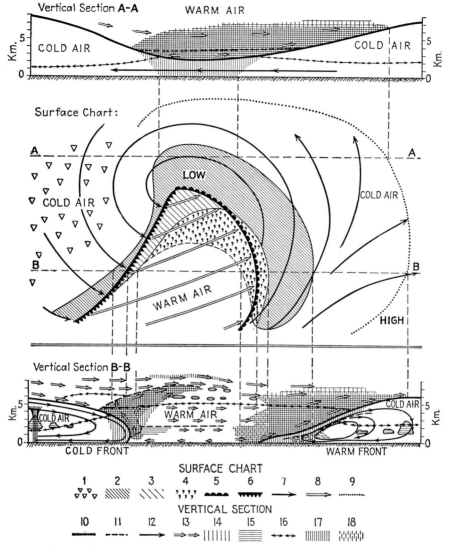

Fig. 4.—Model of a young cyclone. (*After Bergeron,* 1937.)   Air motion in the vertical
sections refers to a moving-coordinate system.

Surface Chart: 1, Shower; 2, rain area in cold air; 3, rain area in warm air; 4, drizzle
area; 5, warm front; 6, cold front; 7, streamlines of the cold air; 8, streamlines of the warm
air; 9, outer cirrostratus boundary.

Vertical Section: 10, frontal surface; 11, other discontinuity surface; 12, motion of the
cold air relative to the center; 13, motion of the warm air relative to the center; 14, falling
ice crystals; 15, suspended cloud particles; 16, lower ice crystal boundary; 17, rain or snow;
18, drizzle.

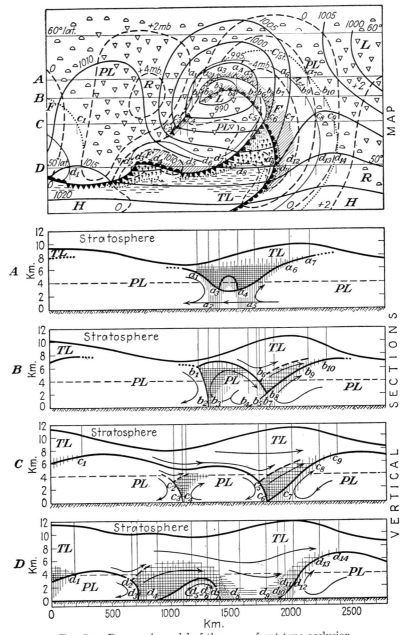

FIG. 5a.—Bergeron's model of the warm-front-type occlusion.

On the Map: 1, Showers △; 2, "upslide" rain ▨; 3, warm-sector rain ⫲;
4, stratus or fog ⹀; 5, warm front ●●●; 6, cold front ▲▲▲; 7, drizzle ,,,,.

Vertical Sections: 8, Cloud mass consisting of ice crystals ||||||||; 9, cloud mass con-
sisting of water droplets ≣; 10, ordinary precipitation |||||||||; 11, warm air ▨;
12, drizzle ▨.

the vertical motions of the warm air over the warm front and the typical warm-front cloud system as well as the motions of air associated with the cold front and its associated cloud system. The upper cross section indicates the frontal surface at some distance above the surface of the earth and the frontal precipitation that is falling through the cold air.

In Fig. 4, the weather conditions in the cold air in advance of the warm front are governed more or less by active convection as long as the higher clouds are lacking. The typical clouds are *Cu* and *Cb*. With the approach of the typical warm-front cloud system, the insolation is reduced considerably, and the convection is damped out. The convection-type clouds shrink and gradually disappear under the *Cs-As* shield.

Typical warm-front conditions will then set in and continue until the surface front passes. Sometimes the typical warm-front conditions discussed in the paragraphs on fronts are disrupted by shower and thunderstorm conditions. This can be expected if the overrunning warm air is convectively unstable. Maritime tropical air is usually convectively unstable if it had a short trajectory from the primary source region, and stable if it had a long trajectory, such as around the northern limb of a subtropical anticyclone. Thus maritime tropical air arriving over the southeastern United States is usually convectively unstable while the same air flowing around the Bermuda and Azores anticyclone is stable when it arrives over the British Isles and the continent.

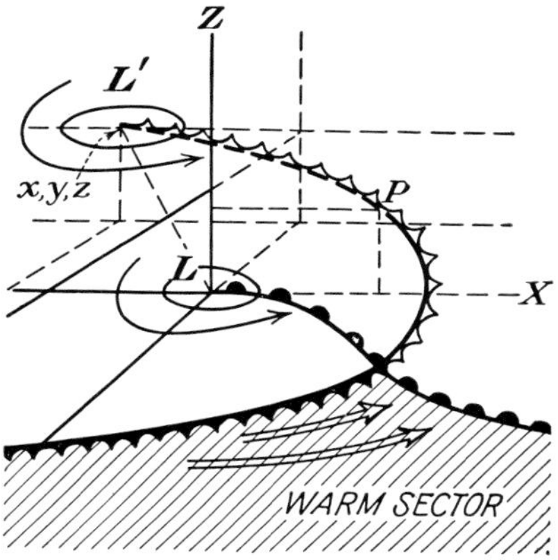

The weather conditions in the warm sector are a direct function of the stability. Convective-type cloudiness should be expected if the instability can be realized by surface heating. In general, the weather conditions in the warm sector are ideal except in the peak, where the air undergoes a certain amount of surface cooling and

FIG. 5b.—Three-dimensional picture of the warm-front-type occlusion. *L* is low-pressure center at the ground; *L'* is low-pressure center in the warm air aloft; *P* is the point where the upper-cold front pierces the *x, z*-plane.

convergence. An *St* or *Sc* cloud deck frequently forms in this zone, and occasionally it lowers to the surface and forms fog.

Sometimes thunderstorm conditions break out in the warm sector when an overrunning cold front impinges on the warm current. This condition is discussed in some details under Fronts, pages 653 to 655.

The weather conditions experienced in the rear of the cyclone after the cold-front-type weather has passed are primarily a function of the stability and the flow conditions in the cold air. Cyclonically curved isobars usually indicate convective activity, and anticyclonically curved isobars will usually stabilize the intermediate layers sufficiently to minimize the convective development. An *Sc* or *St* deck is then the typical cloud if there is sufficient moisture in the lower layers. Frequently clear skies prevail with anticyclonic flow conditions.

Structure of the Occluded Cyclone. The wave cyclone discussed thus far is only a transitional stage in the life cycle of the cyclone, lasting on an average 12 to 24 hr. The occluded stage in contrast maintains itself longer. Figure 5 shows Bergeron's model of the warm-front-type occlusion.

When the occlusion process starts, the pressure at the center of the depression is usually 15 to 20 mb lower than the environment. The cyclone is then represented by a system of closed isobars having a steep gradient.

One of the clearest indications of an occluded cyclone is the lack of warm air near the surface. A vortex of polar air exists on the surface. The warm air that has not been deflected to the right in ascending over the warm front also turns cyclonically and forms a vortex of warm air above the cyclone center. Mixing between these two masses takes place, and the warm air is gradually transformed into polar air.

The cloud and precipitation forms associated with this type of occlusion are shown in Fig. 5.

ANTICYCLONES

Anticyclones are characterized by small horizontal pressure gradients and light winds in the inner portion of the system. In the Northern Hemisphere, the air motion is clockwise and diverging in the lower layers. At the upper levels, there is a corresponding inflow of air that is particularly pronounced while the anticyclone is building up. The diverging air flow in the lower layers is brought about by the

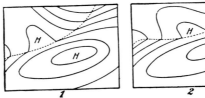

Fig. 6.—Migrating wedges of polar air north of the subtropical anticyclone. (*After Bjerknes and Bergeron.*)

general subsiding effect. This also is responsible for the formation of subsidence inversions. A general warming and drying effect accompanying the subsidence produces the clear and generally fine weather normally associated with anticyclones.

Anticyclones can be divided into four main types:

1. Rapidly Moving High-pressure Area between the Individual Cyclones of a Cyclone Family. Such systems are more like wedges than like closed anticyclones. They form exclusively within the cold air. Owing to divergence, no fronts are found within these systems despite the fact that a good temperature contrast may exist between the front side and the rear of the wedge. When these systems approach a quasi-stationary anticyclone such as a subtropical cell, the polar wedges will appear simply as wedges of the subtropical anticyclone, as shown in Fig. 6. Under such circumstances, a strong front will separate the polar air from the tropical air.

2. Cold Anticyclones, Which Follow the Last of a Series of Cyclones. These systems are similar to the first type in that they form wholly within the cold air. This type of anticyclone usually develops to moderate intensity and produces a number of closed isobars around the center. The areal extent of anticyclones of this type is comparable with that of the average well-occluded cyclone. As this anticyclone moves southward, it gradually slows and tends to become quasi-stationary. This usually happens in preferred zones where conditions are favorable for an anticyclonic circulation. One such region is the location of subtropical anticyclones. With the approach of a cold anticyclone, the subtropical cell usually retreats toward the equator and begins to break down. The cold anticyclone will then take its place and reform into a new subtropical cell, as shown in Fig. 7.

3. Persistent, Slowly Moving Anticyclones Composed of Polar or Arctic Air.
As a rule, this type of anticyclone forms as a result of successive outbreaks of cold
air. It maintains itself and quite often increases in intensity after reaching middle
latitudes. It covers large areas often the size of Europe or the eastern half of North
America. One reason why this type of anticyclone persists for periods of 5 to 10 days
in middle latitudes is the fact that it gradually changes to a warm or dynamic anticyclone. Since dynamic anticyclones maintain their anticyclonic circulation to
very high levels, the prevailing westerly circulation is interrupted and thus the
pressure centers tend to remain stationary with respect to west-east motion. The
formation of this type of anticyclone is frequently linked to tropopause waves and
their coupling effect on polar-front waves.

4. Subtropical Anticyclones. These are dynamic or warm anticyclones that are
maintained by the general circulation. They are frequently reinforced by the southward movement of the polar anticyclones. The popular concept partly established
by climatological data that these systems are stationary is erroneous. Portions of

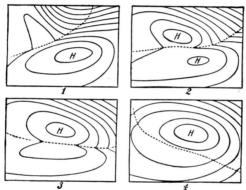

Fig. 7.—Transformation of subtropical anticyclone. (*After Bjerknes and Bergeron.*)

the Pacific or Atlantic anticyclones frequently break off and progress eastward.
This is particularly noticeable in advance of an eastward-moving cyclone at higher
latitudes.

Bibliography

BERGERON, T.: Die dreidimensional verknüpfende Wetteranalyse, Part II, 1934.
BERGERON, T.: "How the Weather Originates and How to Predict It," pp. 199–231,
    Ymer, Stockholm, 1937.
BERGERON, T.: Über die dreidimensional verknüpfende Wetteranalyse, Part I,
    Geofysiske Publ., **5** (6), 1928.
BJERKNES, J.: Theorie der ausertropischen Zyklonenbildung, *Meteorolog. Z.*, **54:**
    462–466 (1937).
BJERKNES, V., J. BJERKNES, H. SOLBERG and T. BERGERON: "Physikalische Hydro-
    dynamik mit Anwendung auf die dynamische Meteorologie," Springer, Berlin,
    1933.
BJERKNES, J., and E. PALMÉN: Investigations of Selected European Cyclones by
    Means of Serial Ascents, *Geofysiske Publ.*, **12** (2), 1937.
BJERKNES, J., and H. SOLBERG: Meteorological Conditions for the Formation of Rain,
    Geofysiske Publ., **2** (3), 1921.
BYERS, H. R., and V. P. STARR: *Monthly Weather Rev. Suppl.* 47, 1941.

CHROMOW, S.: "Einführung in die synoptische Wetteranalyse," Springer, Vienna, 1940.

LENNAHAN, C. M.: Summary of Aerological Observations, *Monthly Weather Rev. Supp.* 38, 1938.

NAMIAS, J.: Elements of Cyclonic Structure, *Bull. Am. Met. Soc.*, **16** (5), 1935.

PALMÉN, E.: Über die Temperaturverteilung in der Stratosphäre und ihren Einfluss auf die Dynamik des Wetters, *Meteorolog. Z.* **51**: 17–23 (1934).

PETTERSSEN, S.: "Weather Analysis and Forecasting," McGraw-Hill, New York, 1940.

SCHINZE, G.: Die praktische Wetteranalyse, *Archiv deut. Seewarte*, vol. 52, 1932.

WEXLER, H.: Some Aspects of Dynamic Anticyclogenesis, Misc. Rept. 8, Institute of Meteorology, University of Chicago, 1942.

DISPLACEMENT OF PRESSURE SYSTEMS

By J. F. O'Connor

The speed of individual isobars is given by the expression

$$U = \frac{bN}{\Delta p}$$

where U = speed of isobars in N units per 3 hr

b = 3-hr pressure change

N = distance between two isobars in arbitrary length units (degrees of latitude will be used here)

Δp = the pressure difference between two isobars that are N distance units apart

It can be shown[1] that the displacement of any symmetrical pressure system along a chosen axis is given by the average of the speeds of two isobars, equidistant from the trough or wedge line

$$U = \frac{1}{2}(U_A + U_B) = \frac{1}{2}\left(\frac{b_A N_A}{\Delta p_A} + \frac{b_B N_B}{\Delta p_B}\right)$$

where the subscripts A and B refer to the points at which the axis of computation intersects the isobars, in advance, and to the rear of the symmetry line, respectively.

Smoothed isallobars must be drawn through the pressure tendencies plotted on the map. The term b is then interpolated from these smoothed isallobars at the desired point. This term may be positive or negative, the isobar moving toward higher pressure when negative, and toward lower pressure when positive.

The term N, in combination with Δp, is an expression for the representative pressure gradient in the vicinity of the point where b is chosen. N is measured along the axis chosen for computation. Following Byers,[1] N is considered positive when the pressure decreases in the positive direction of U, and negative when the pressure increases in this direction.

Reliable results from the above expressions depend upon the proper choice of three things:

A. The representative points on the isobars at which the terms b are read.

B. The location and orientation of the axes of computation.

C. The choice of the distance terms N.

A. The choice of representative points on isobars should be governed by the following considerations:

1. For the computation of displacements of individual isobars, points should be chosen close to the isallobaric maximum or minimum, in regions where the isobars are closely packed (see Fig. 1).

2. For the computation of displacements of pressure systems, points must not be chosen such that either one will fall beyond the isallobaric maximum or minimum, measured from the center or center line of the pressure system. Since the chosen points must be equidistant from the center of the pressure system along the

axis of computation, the point closest to the center determines the distance of both points from the center (see Fig. 2).

NOTE: Point B is at the maximum distance from the center along the west-east axis. Thus point A is fixed such that distance AO equals distance OB.

B. The choice of the computation axis should be governed by the following:

1. For the computation of displacement of an individual isobar, the axis may be chosen in any direction, provided that the point at which the axis intersects the isobar in question fulfills condition 1 in paragraph A.

2. For the computation of displacement of a complete pressure system:

a. It is desirable to choose the axis normal to the trough or wedge line, when possible. However, the axis may frequently be chosen in any direction oblique to the trough or wedge line, as long as it intersects a well-defined isallobaric field (see axis CD, Fig. 2).

b. For approximately circular pressure systems, the axis may be chosen in any direction. Usually the largest component of motion is obtained along an axis connecting the pressure center with the isallobaric center in advance of it.

c. To determine the future position of individual pressure centers, it is advisable to compute the displacements along two convenient axes (see Fig. 3). Points are then located on both axes representing the future positions of the center in those directions. The resultant position of the center is then found at the intersection of perpendiculars dropped from these points.

NOTE: This is different from the vectorial sum or resultant, except for the case of two axes chosen perpendicular to each other.

C. The choice of distance N is governed by the following:

1. N should be sufficiently long to give a true representation of the pressure gradient in the vicinity of the point where its related pressure tendency b is read.

2. N should not be so long that it extends beyond the trough or wedge line or cuts the isobars of adjacent pressure systems.

3. N lengths in advance and in the rear of the center need not be equal. However, the proper pressure difference Δp corresponding to each must be used in the denominator of the displacement formula.

Problem. Compute the 12-hr displacement of the 996-mb isobar along a west-east axis (see Fig. 1). *Solution.* 3-hr displacement of isobar
where $b = -6.0$ mb per 3 hr
$\qquad N = -2.0°$ lat. in length
$\qquad \Delta p = 6.0$ mb (the pressure difference between the extremities of N)
NOTE: N is minus in this case, since U is considered positive to the east, and the pressure increases in this direction.

$$U = \frac{(-6.0)(-2.0)}{6.0} = 2.0° \text{ lat. per 3 hr}$$
$$= 8.0° \text{ lat. per 12 hr eastward}$$

Thus the 996-mb isobar will move eastward along the chosen axis to the vicinity of Albany, N.Y.

NOTE: Individual isobars extrapolated by means of the above expression will not give reliable results for more than 12 hr.

Problem. Compute the 24-hr displacement of the low-pressure center and occluded front in **Fig. 2.** *Solution.* A west-east axis is chosen through the low center. It appears that there will be no appreciable displacement of the low center in a north-south direction, since the isallobaric gradient in this direction is small. Points A and B are chosen on the axis equidistant from the center. Point B is chosen at the maximum distance from the center, *i.e.*, on the isallobaric ridge (B cannot be chosen to the left of this position). Thus the position of the point A is also fixed in advance of the

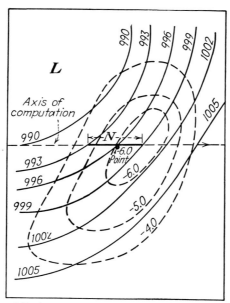

Fig. 1.—Computation of displacement of an isobar. The point chosen for computation is located at Lansing, Mich. This is a portion of Fig. 2.

center. Distances N are chosen for 9 mb of pressure difference in the vicinity of points A and B and are expressed in units of degrees of latitude.

$$U = \frac{1}{2}(U_A + U_B) = \frac{1}{2}\left(\frac{b_A N_A}{\Delta p_A} + \frac{b_B N_B}{\Delta p_B}\right)$$

where $b_A = -2.2$ mb per 3 hr
$\quad N_A = -3.8°$ lat. per 9 mb of Δp_A
$\quad b_B = +5.0$ mb per 3 hr
$\quad N_B = +2.6°$ lat. per 9 mb of Δp_B

$$U = \frac{1}{2}\left[\frac{(-2.2)(-3.8)}{9} + \frac{(+5.0)(+2.6)}{9}\right] = 1.18° \text{ lat per 3 hr}$$

$$= 9.4° \text{ lat. per 24 hr.}$$

Thus the low center will move eastward to the vicinity of Doucet, Canada. This will also mark the northern fix of the occluded front in 24 hr.

To determine the displacement of the southern portion of the occlusion in the next 24 hr, a west-east axis is chosen along the 38.8° parallel of latitude. This axis is chosen because it intersects a well-defined isallobaric field. Points C and D are chosen on this axis, as indicated in Fig. 2.

$$U = \frac{1}{2}[U_C + U_D]$$

where $b_C = -4.2$ mb per 3 hr
   $\quad N_C = -2.2°$ lat. per 3 mb of Δp_C
   $\quad b_D = +1.3$ mb per 3 hr
   $\quad N_D = +1.8°$ lat. per 3 mb of Δp_D

$$U = \frac{1}{2}\left[\frac{(-4.2)(-2.2)}{3} + \frac{(+1.3)(+1.8)}{3}\right] = 1.93° \text{ lat. per 3 hr}$$

$$= 15.4° \text{ lat. per 24 hr}$$

Thus, the occlusion will advance eastward along the southern west-east axis to approximately the 72d meridian.   *The actual position of the low center and occlusion as indicated*

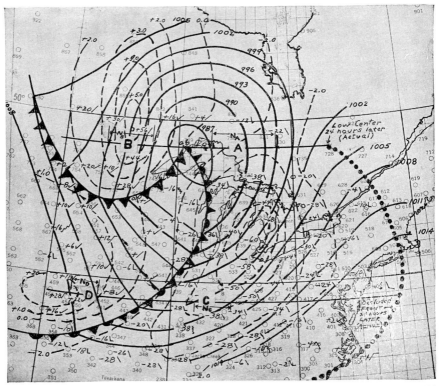

Fig. 2.—Computation of displacement of cyclonic center and its occluded front, 0830 EWT, Jan. 14, 1943.
Data plotted are pressure tendencies.   Isobars are the solid lines, isallobars are the dashed lines.   The -1.0 mb isallobar and the $+1.0$ mb isallobar north of the secondary cold front have not been drawn.

by the 24-*hr succeeding synoptic map has been indicated in Fig.* 2 *as a dotted curve, which agrees quite well with the computed positions.*

Problem.   Compute the 24-hr displacement of the anticyclonic center in Fig. 3. *Solution.*   Two axes for computation are chosen.   Points A and B are marked off on the west-east axis, C and D on the north-south axis. N_A and N_B are the distances chosen for 6 mb of pressure difference along the west-east axis, while N_C and N_D are chosen for 3 mb of pressure difference.   In each case, the N distances chosen are convenient and adequate to represent the true pressure gradients.

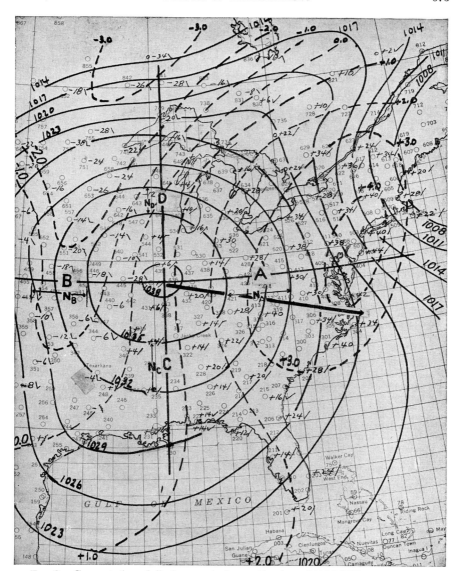

Fig. 3.—Computation of displacement of an anticyclone, 2030 EWT, Feb. 12, 1944.

Along the west-east axis

$$b_A = +3.2 \text{ mb per 3 hr}$$
$$N_A = +3.7° \text{ lat. per 6 mb of } \Delta p_A$$
$$b_B = -1.5 \text{ mb per 3 hr}$$
$$N_B = -3.2° \text{ lat. per 6 mb of } \Delta p_B$$
$$U_{WE} = \frac{1}{2}\left[\frac{(3.2)(3.7)}{6} + \frac{(-1.5)(-3.2)}{6} \right]$$
$$= 1.38° \text{ lat. per 3 hr}$$
$$= 11.0° \text{ lat. per 24 hr}$$

Along the north-south axis

$$b_C = +0.8 \text{ mb per 3 hr}$$
$$N_C = +2.9° \text{ lat. per 3 mb of } \Delta p_C$$
$$b_D = +0.1 \text{ mb per 3 hr}$$
$$N_D = -1.5° \text{ lat. per 3 mb of } \Delta p_D$$
$$U_{NS} = \frac{1}{2}\left[\frac{(0.8)(2.9)}{3} + \frac{(+0.1)(-1.5)}{3} \right]$$
$$= 0.36° \text{ lat. per 3 hr}$$
$$= 2.9° \text{ lat. per 24 hr}$$

The resultant (see *B,c*) of the north-south and west-east components is the displacement indicated by the solid arrow pointing east-southeastward from the center in Fig. 3. Thus the computed 24-hr displacement of the center places it about 60 miles northeast of Cape Hatteras. The average speed in that direction is about 28 knots.

NOTE: In these computations, the observed tendencies have not been corrected for diurnal variation. It was felt that the precision of both the method of computation and the accuracy of the analysis did not warrant such corrections. Also, the greatest distance used between two computation points along a chosen axis is about 600 miles. Over this distance, the variation of diurnal tendency is negligible compared with the magnitude of the isallobaric gradients involved.

Bibliography

1. BYERS, H. R.: A Simplification of the Formulas for Displacement of Pressure Systems, *Bull. Am. Meteorolog. Soc.*, January, 1943, p. 13.
2. PETTERSSEN, S.: "Weather Analysis and Forecasting," Chap. 9, McGraw-Hill, New York, 1940.

UPPER-AIR CONDITIONS

By E. Bollay

Vertical Temperature Distribution and Height of Tropopause. The distribution of temperature at sea level over the globe is presented in charts contained in the section on Climatology. In general these show that the isotherms deviate from the parallels of latitude on account of the thermal effects of the sea and land distribution.

The distribution of temperature above the surface of the earth indicated in Fig. 1 shows a systematic decrease in temperature with altitude up to the tropopause. An isothermal lapse rate or even an increase in temperature can be expected in the stratosphere.

The highest temperature of the troposphere is found near the equator on the earth's surface. In the upper air, too, up to an altitude of about 10 dyn km, the

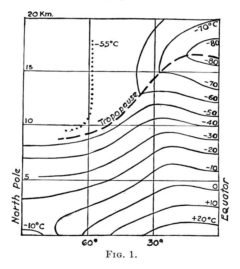

Fig. 1.

tropical regions are the warmest. However, at higher elevations the equatorial zone has the lowest temperature. In equatorial regions, the decrease in temperature continues up to about the 17 dyn km level, while at other latitudes the decrease in temperature ceases at lower levels. Thus the temperature distribution of the stratosphere is characterized in all seasons by lower temperatures over the tropics than over the middle and high latitudes.

More detailed information of the vertical temperature distribution in the lower 5 km during the winter months is given by the mean temperature-pressure curves of selected stations illustrated in Figs. 2 to 10.

The height of the tropopause lies somewhat higher in summer than in winter. Also, as indicated above, the tropopause is highest over the equatorial regions, and its elevation decreases poleward.

675

Upper-air Pressure. Charts showing the mean pressure at levels above the surface have been prepared by Shaw in his "Manual of Meteorology," vol. II. Owing to the lack of sufficient upper-air data, these charts are based primarily on extrapolated surface data. With upper-air data now being observed over a wider region of the earth's surface, these charts will probably be revised to fit observed conditions. Byers and Starr in *Supplement* 47 of the *Monthly Weather Review* have recently refined the 2- and 4-km normal January pressure charts of Shaw by means of new data made available from North America. These are shown in Figs. 11 and 12.

Winds in the Upper Troposphere and Lower Stratosphere over North America. Stevens compiled the resultant winds at 6,000, 8,000, 10,000, 12,000, and 14,000 m over the United States in *Supplement* 36 of the *Monthly Weather Review*. The resultant winds during various times of the year are shown in Figs. 13 to 17.

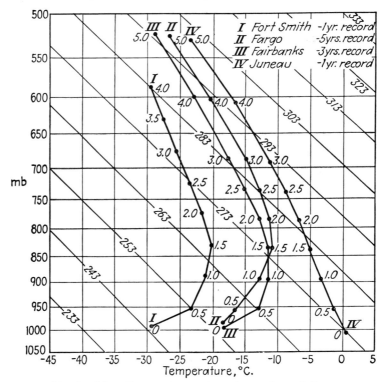

Fig. 2.—Mean January temperatures for northern stations.

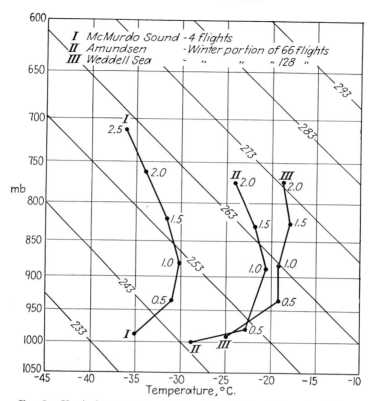

FIG. 3.—Vertical temperature distribution at polar stations in winter.

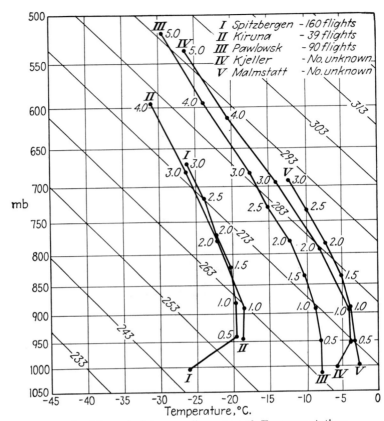

Fig. 4.—Mean of winter soundings at north European stations.

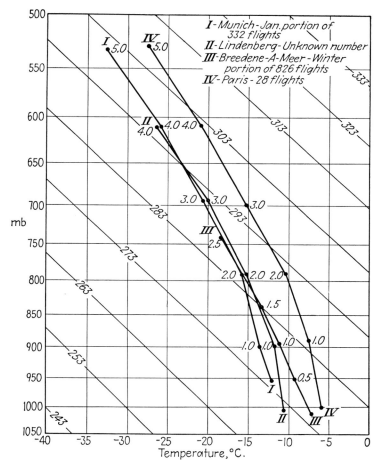

FIG. 5.—Mean winter soundings in western Europe.

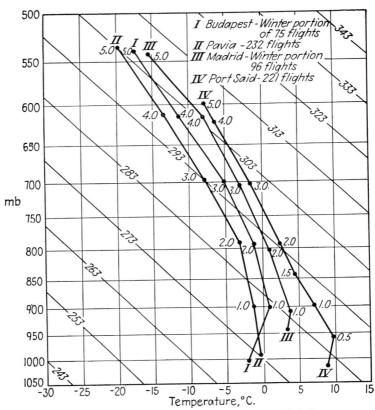

Fɪɢ. 6.—Mean of winter soundings in southern European and Mediterranean stations.

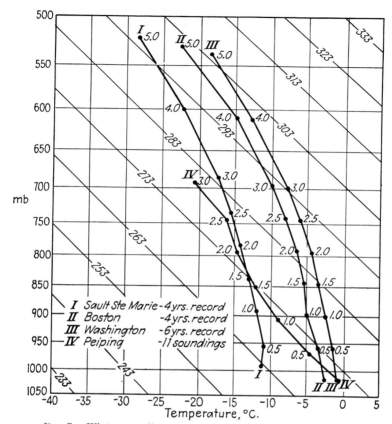

FIG. 7.—Winter soundings at northeastern continental stations.

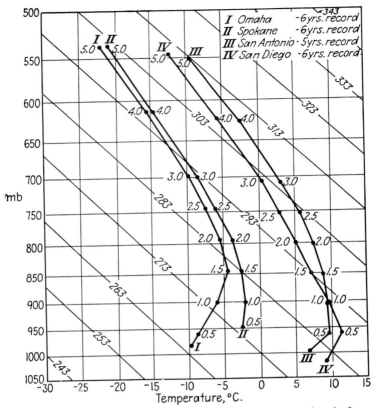

Fig. 8.—Mean free-air temperatures at selected United States stations in January.

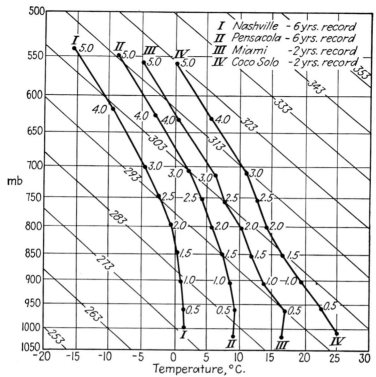

FIG. 9.—Mean free-air temperatures in January at low-latitude American stations.

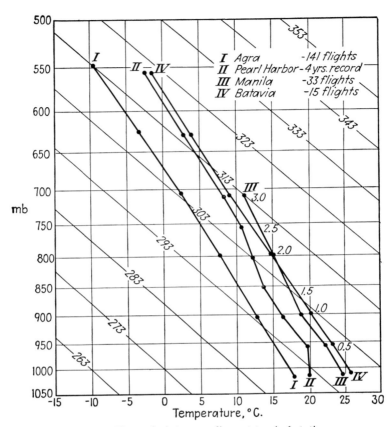

FIG. 10.—Mean of winter soundings at tropical stations.

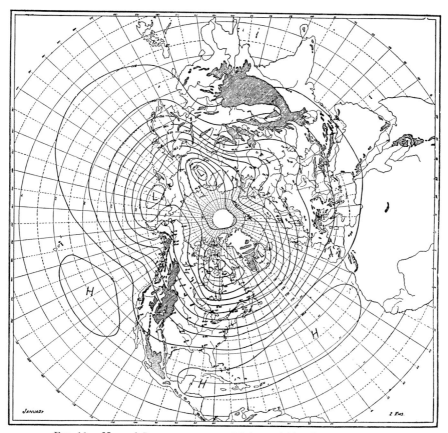

FIG. 11.—Normal January pressure at 2 km.   (*After Byers and Starr.*)

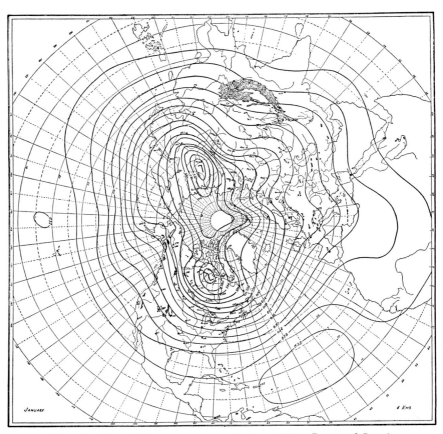

Fig. 12.—Normal January pressure at 4 km.   (*After Byers and Starr.*)

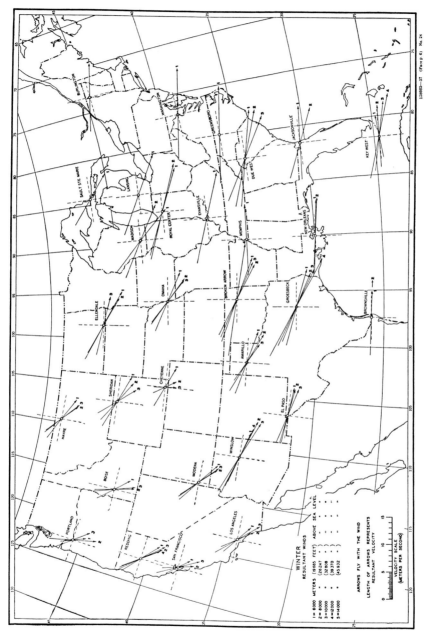

FIG. 13.—United States winter resultant winds. (*After Stevens.*)

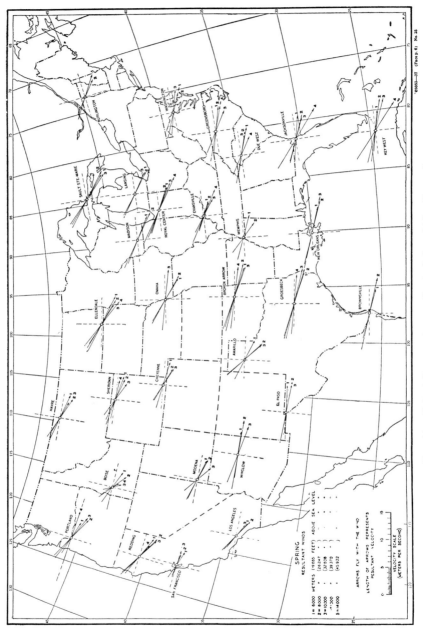

Fig. 14.—United States spring resultant winds. (*After Stevens.*)

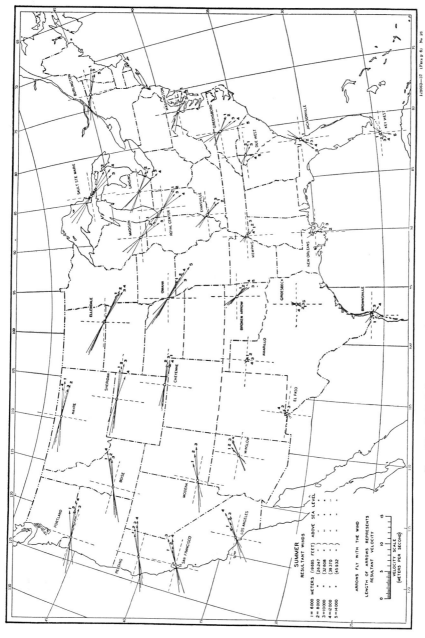

FIG. 15.—United States summer resultant winds.   (*After Stevens.*)

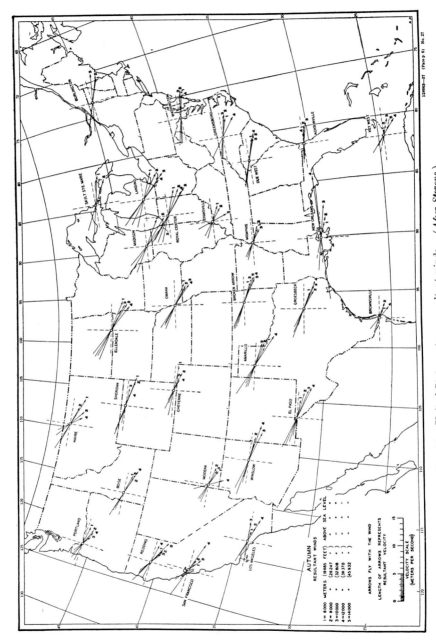

Fig. 16.—United States autumn resultant winds. (*After Stevens.*)

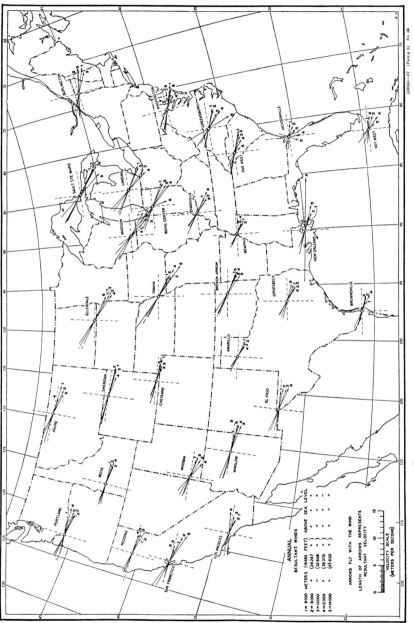

Fig. 17.—United States annual resultant winds. (*After Stevens.*)

ATMOSPHERIC STABILITY AND INSTABILITY

BY

NORMAN R. BEERS

GENERAL PRINCIPLES, SLICE METHOD (Ċ)

The physical significance of the saturated-adiabatic ascent of air through a dry-adiabatically descending environment was investigated by J. Bjerknes.[1] Bjerknes' paper gives a theoretical treatment of the cumulus convection in an atmosphere with a temperature lapse rate between the saturated-adiabatic and the dry-adiabatic (conditional instability). Two important conclusions are made: (1) cumulus convection does not convert heat into kinetic energy unless the ratio between the width of the cumulus towers and that of the cloudless intervals is below a certain critical value, and (2) the atmosphere is always less unstable with respect to a system of finite cloud towers than with respect to the infinitely small saturated particle of the so-called "parcel method." Since the cloud-tower method deals with conditions that are more real than those of the parcel method, Bjerknes concludes that "Its results are likely to be better suited for all applications to real nature."

S. Petterssen[2] applied the elements of Bjerknes' investigation with some important extensions to the practical problem of forecasting convective phenomena (with emphasis on thunderstorm analysis and forecast). The "slice method," as originally presented by Bjerknes and adapted by Petterssen, actually classifies the stability of a single level of the atmosphere of unit thickness. Although good results have been obtained generally using the techniques suggested by Petterssen, experience has indicated some desirable changes. In particular, it has been found essential to take account of the stability layer by layer through the atmosphere, especially through the column where cumulus activity may occur. This requires an extension of the theory and particular attention to the distribution of moisture in the column.

The criteria of stability developed here apply in the case of an initially barotropic atmosphere subject to perturbations of velocity from any source. The suggested techniques of analysis may be used with normal care to forecast cumulus activity triggered off by fronts, orographic barriers, or surface heating. The procedures can be learned and applied by rote; the forecast itself will in most cases require some judgment. After a moderate amount of experience, the forecaster can make a quick estimate of the situation with the slice method. This estimate appears to be more reliable than the result obtained by the parcel method. In addition to this, the slice or convective-cell method provides a quantitative measure of stability for the entire column. With this information, the forecaster has a real basis for statements on time and extent of development of the expected cumulus.

The quantitative measure of stability used is the *circulation acceleration*. The results follow from an application of the circulation theorem of V. Bjerknes. See a paper by E. Höiland[3] for a similar though less extensive independent treatment. The atmosphere is assumed to be stratified in layers, each layer being determined by significant levels of radiosonde data. The circulation acceleration is determined for each layer separately; the results are shown to be additive for the entire column. Account is taken (before the computation) of known changes in the sounding before

693

the time of the forecast period (*e.g.*, removal of night ground inversions by insolation). The cumulus activity resulting from velocity perturbations is forecast on the basis of the distribution of circulation acceleration through the entire column. The analyst must of course base his forecast on all available information. In other words, he will continue to analyze the surface synoptic chart, upper-air charts, etc. Techniques and a routine for analysis are given below after a discussion of the theory. The whole is illustrated by several selected examples.

State of Atmosphere. The thermodynamic state of the atmosphere over a particular station at a particular time is given with moderate accuracy by the report of a modern radiosonde or aerograph. The report is plotted on a thermodynamic diagram and called the *sounding.* Since the report does not give a continuous record of pressure, temperature, and humidity, it is generally assumed that straight lines drawn between significant points will adequately represent the state of the atmosphere between significant levels as well as at those levels. The assumption is warranted, provided that the levels have been chosen according to current instructions (see *U.S. Weather Bureau Circular P*).

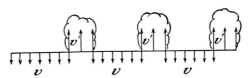

Fig. 1.—Region of cumulus clouds—schematic.

The following simplifications are introduced as assumptions:

1. Horizontal motion does not maintain any net inflow to or outflow from any stratum determined by the air between significant levels.
2. Conditions are barotropic initially.
3. All motions are adiabatic above the surface layers.

These assumptions are generally valid in circumstances where cumulus activity is expected because of surface heating. They may be modified by frontal or orographic lifting.

Consider simplification (1). The situation is represented schematically by Fig. 1. Attention is fixed on a horizontal area in the atmosphere large enough to contain several cumulus clouds, and small enough so that isobaric surfaces are also geopotential surfaces. Cumulus clouds indicate regions of ascending currents of air. Cloudless sky indicates regions of descending air. For the present, all velocities other than those indicated in Fig. 1 may be neglected. For simplicity, it is assumed that the ascending velocities are constant across any one region of ascending air. It is also assumed that this constant upward velocity persists throughout the thickness of a layer determined by consecutive significant levels. More rigorously, one may define average values of the velocity through the stratum and across the horizontal area in question. Similar assumptions are made for descending velocities. These conditions are made useful by writing the equation

$$M'v' + Mv = 0 \tag{1}$$

where M' and M are the ascending and descending masses, respectively, in one region and stratum. v' and v are the corresponding velocities. Equation (1) is the precise statement of simplification (1) above.

Assumed barotropy means simply that the isobaric and isothermal surfaces are coincident initially. These surfaces are also taken horizontal. If the atmosphere

is not barotropic in the case to be investigated, the solenoids computed for the barotropic field may be considered to be added to those already in existence before the
perturbation commenced. In practice, the slice method is used only when conditions
are almost or actually barotropic.

Simplification (3) is the basic one made in many thermodynamic analyses. All
evidence indicates it to be valid whenever the motions take place with moderate
speed away from the earth's surface. Since it is used in the analysis for air above the
convective condensation level generally, and for periods of a few hours at most, the
assumption is warranted.

Velocity Perturbations. It is desired to know what changes in the state of the
atmosphere will be produced by a certain type of velocity perturbation. Initially
the conditions are as described above. When vertical velocities have persisted for a
small interval of time, there will be changes. These changes depend upon the initial
state and upon the perturbations received. In particular, it is essential to determine
whether initially small perturbations will be intensified or dissipated by the distribution of temperature and moisture originally present. It is not necessary for
the time being to specify the perturbations more completely than was done above.
Their origin is also not important; they may be thought of as being due to impulsive
or continuous forces.

When only vertical velocities exist at a fixed point in space, the local changes in
temperature per unit time are given by (see Sec. VI)

$$\frac{\partial T'}{\partial t} = v'(\gamma - \gamma_d) \qquad \text{in ascending nonsaturated air} \qquad (2)$$

$$\frac{\partial T'}{\partial t} = v'(\gamma - \gamma_m) \qquad \text{in ascending saturated air} \qquad (3)$$

$$\frac{\partial T}{\partial t} = v(\gamma - \gamma_d) \qquad \text{in descending nonsaturated air} \qquad (4)$$

$$\frac{\partial T}{\partial t} = v(\gamma - \gamma_m) \qquad \text{in descending saturated air} \qquad (5)$$

where v' and v are the ascending and descending velocities, respectively, T' and T
are the temperatures in the ascending and descending currents, γ_d is the dry-adiabatic
lapse rate, γ_m is the saturated- or wet-adiabatic lapse rate, and γ is the actual lapse
rate in the atmosphere at the moment considered. The very small change in γ_d if
some but not 100 per cent moisture is present is neglected.

It may be noted in passing that the local changes in pressure in the atmosphere considered, under small perturbations of velocity, will be negligible. This is so because,
aside from dynamic effects, the pressure at a fixed point depends essentially upon the
weight of the air above the level in question. This amount of weight is changed so
little by small changes of velocity in relatively thin layers that to an excellent approximation geopotential surfaces remain isobaric surfaces throughout the motions.

Stability of a Layer. The stability of a layer of the atmosphere is investigated
using the circulation theorem of V. Bjerknes. This theorem gives the rate of change
of circulation around a closed path that moves with the fluid (see Sec. VI). On
neglecting frictional forces and the term $2\omega \, dF/dt$ where F is the projection of the
path on the equatorial plane, one obtains

$$\frac{dC}{dt} = \dot{C} = - \oint \frac{dp}{\rho} \qquad (6)$$

where C is the circulation. The rate of change of circulation is frequently called
(and is here called) the *circulation acceleration*. It will generally be designated by
$\dot{C}$ ("C dot") for brevity. The dimensions of $\dot{C}$ are the same as the dimensions of

energy per unit mass. Equation (6) is essentially a prognostic equation, since it gives $\dot{C}$ in terms of present values of the density (or of the specific volume) field. On introducing the equation of state for a perfect gas, Eq. (6) becomes

$$\dot{C} = -\oint RT \frac{dp}{p} \tag{6'}$$

where p is pressure, T absolute temperature, and R the gas constant for the air. The value of R depends on the moisture present. It may be noted in passing that the term $2\omega \, dF/dt$ is not always negligible at low latitudes. The effect of it is to increase $\dot{C}$ when the path of integration changes the enclosed area rapidly, *e.g., after Cb* clouds have formed.

Consider a portion of the horizontal region that contains one area of ascending and one area of descending air. Applying the concepts of the model atmosphere already described, consider the vertical currents to persist through layers of finite thickness. The scheme is shown in Fig. 2. The closed path 1-2-3-4 consists of two isobars and two vertical legs; the path is fixed in space, and the air flows across the

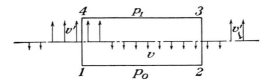

FIG. 2.—Schematic view of convective cell.

isobars. Let the temperature at elevation z in the stratum be T initially. After an interval of time Δt, the temperature at z in the two vertical legs is given by

$$T + \left(\frac{\partial T'}{\partial t}\right) \Delta t \qquad T + \left(\frac{\partial T}{\partial t}\right) \Delta t \tag{7}$$

for ascending and descending air, respectively.

Equation (6') can be applied to the path 1-2-3-4, since the layer is assumed to have a finite thickness, *i.e.*, after a small interval of time, the space average temperatures in the vertical legs of the circuit are very closely equal to the average temperatures in the path that travels with the fluid. The approximation becomes better the thicker the layer and the smaller the time interval. If in fact the perturbations were initiated by impulsive forces, no approximation is involved in the instant after the motion commenced. Since it is desired to apply the theorem only for layers of finite thickness and for small intervals of time, the error introduced by integrating around 1-2-3-4 is negligible. On performing the integration, one readily finds

$$\dot{C} = R \ln \frac{p_0}{p_1} (T_{41} - T_{23}) \tag{8}$$

where T_{23} and T_{41} are the average temperatures in the vertical legs of the circuit at the moment considered. The top and bottom pressures in the stratum are p_1 and p_0. The small variation in R through the layer due to the presence of water vapor is neglected, *i.e.*, R will be assumed to be constant through any one layer. A refinement could be made by using virtual temperature, but this is not generally necessary.

Since conditions are barotropic initially by assumption, the initial values of the average temperatures in the vertical legs are the same. Initially of course the velocities are zero. Initially then $\dot{C}$ is zero. When, however, the perturbing velocities have persisted for a time Δt, the average temperatures in the vertical legs are given by

$$\bar{T} + \left(\frac{\partial T'}{\partial t}\right)\Delta t \qquad \bar{T} + \left(\frac{\partial T}{\partial t}\right)\Delta t \tag{7'}$$

where the lapse rates that are used to evaluate $\partial T'/\partial t$ and $\partial T/\partial t$ are average values through the stratum. On substitution into Eq. (8), one finds

$$\dot{C} = R \ln \frac{p_0}{p_1}\left(\frac{\partial T'}{\partial t} - \frac{\partial T}{\partial t}\right)\Delta t \tag{9}$$

Equation (9) is the fundamental result of the slice or convective-cell method. The right-hand side of the equation can be evaluated from any sounding. On recalling the definition of the circulation, it is apparent that $\dot{C}$ is a measure of the change in kinetic energy through the interval of time Δt. In fact, it is clear from Eq. (8) that the difference on the right-hand side (*i.e.*, $\dot{C}$ in this special path of integration) is the difference in the geopotentials represented by the ascending and descending columns. Since energy is conserved in the system, this is also the difference in kinetic energy.

When the initial temperature and moisture distributions in the atmosphere are such that the right-hand side of Eq. (9) is positive (*i.e.*, $\dot{C}$ greater than zero), the perturbing velocities increase with time. In such a case, the layer is said to be in *unstable equilibrium.* The state of equilibrium is of course in reference to the type of perturbation assumed earlier, *i.e.*, to the vertical velocities v' and v where $M'v' + Mv = 0$. When, on the other hand, $\dot{C}$ is zero or negative, the layer is said to be in *neutral* or *stable equilibrium,* respectively. These cases produce either no change in or a diminution of the perturbating velocities.

Effect of Moisture Distribution. Equation (9) gives essentially different answers respecting stability according to the distribution of moisture in the atmosphere. Three possible cases occurring in nature must be considered. These cases and the results of appropriate substitutions of preceding equations are

I. *Dry Layers.* Both ascending and descending currents are initially and remain so dry that condensation does not occur. Then

$$\dot{C} = R \ln \frac{p_0}{p_1}\left[(\gamma - \gamma_d)\left(1 + \frac{M'}{M}\right)\right]v'\,\Delta t \tag{10}$$

II. *Liquid Water Present.* The layer contains enough liquid water so that both ascending and descending currents follow the saturated-adiabatic, *i.e.*, condensation occurs in the ascending current while evaporation occurs in the descending current. Then

$$\dot{C} = R \ln \frac{p_0}{p_1}\left[(\gamma - \gamma_m)\left(1 + \frac{M'}{M}\right)\right]v'\,\Delta t \tag{11}$$

III. *Wet Layers.* The ascending current is initially so nearly saturated that condensation occurs immediately the motion commences and continues throughout the interval considered. The descending current produces no evaporation. Thus parcels of air in the two currents follow the saturated- and the dry-adiabatic, respectively. Then

$$\dot{C} = R \ln \frac{p_0}{p_1}\left[(\gamma - \gamma_m) + \frac{M'}{M}(\gamma - \gamma_d)\right]v'\,\Delta t \tag{12}$$

It is clear from Eq. (10) above that, when the stratum is dry and when $\gamma < \gamma_d$, $\dot{C}$ is negative and the layer is question is stable with respect to all perturbations. When $\gamma > \gamma_d$ there is instability in the dry layer. In case II, the layer is stable or unstable according to the sign of the term $(\gamma - \gamma_m)$. Case III, which is the most interesting in nature, will be seen from Eq. (12) to depend on the magnitude of the ratio M'/M when $\gamma_m < \gamma < \gamma_d$. In this case the layer is said to be "selectively unstable." In other words, $\dot{C}$ will in certain cases be positive for a given sounding and M'/M but negative for the same sounding and different M'/M. The functional dependence of $\dot{C}$ on M'/M is best shown graphically (see Figs. 4 to 9).

Thickness of the Layer. Equations (10) to (12) show that the magnitude of $\dot{C}$ depends on the "thickness" of the layer considered. More precisely, $\dot{C}$ is proportional to the natural logarithm of the ratio of bottom to top pressure across the layer. This result suggests a convenient unit of measurement for $\dot{C}$. For the linear distance between two isobars on an Emagram is exactly proportional to $\ln p_0/p_1$. Also the linear distance between isobars on a pseudoadiabatic diagram is approximately proportional to this factor. The effect of the thickness of the layer considered on its stability is then very conveniently obtained by multiplication by the linear spread of top and bottom isobars for the layer on the thermodynamic diagram. In present practice, the linear scale of temperature on the pseudoadiabatic diagram is used.

Bjerknes' Method. Bjerknes[1] arrives at results similar to those given in Eqs. (9) and (12) by considering the "accumulation of heat per second" within a layer that is kept fixed in space, while the air passes through it upward and downward. The pressure at the level remains almost constant. Then the rate of change of heat (more precisely enthalpy) is given by

$$\frac{\partial Q}{\partial t} = c_p \left[M' \frac{\partial T'}{\partial t} + M \frac{\partial T}{\partial t} \right] \equiv c_p(M' + M) \frac{\partial T}{\partial t} + c_p M' \left(\frac{\partial T'}{\partial t} - \frac{\partial T}{\partial t} \right) \quad (13)$$

Small variations of the specific heat c_p with changes in moisture content are neglected. The right-hand side of Eq. (13) is divided up into two terms (1) the "uniform heating" of the whole unit layer at the rate $\partial T/\partial t$, and (2) the "surplus heating" (positive or negative) of the cloud parts of the layer at a rate proportional to $(\partial T'/\partial t) - (\partial T/\partial t)$.

The surplus heating in the layer for the three cases considered is found to be

(I) $$-c_p M'v' \left(1 + \frac{M'}{M} \right) \alpha$$

(II) $$c_p M'v' \left(1 + \frac{M'}{M} \right) \beta \qquad (14)$$

(III) $$c_p M'v' \left(\beta - \alpha \frac{M'}{M} \right)$$

where $\alpha \equiv \gamma_d - \gamma$ and $\beta \equiv \gamma - \gamma_m$. As Bjerknes remarks, "The surplus heating of the clouds is the only part of the heat of condensation available for the production or annihilation of kinetic energy." The uniform heating of the layer under ordinary circumstances consumes most of the energy put into the system by the heat of condensation. The uniform heating also involves a raising of the center of gravity of the whole air mass, *i.e.*, an increase of its potential energy.

Note on the Derivations. It will be noted that the only difference beween the expressions (14) for the surplus heating and the right-hand side of Eqs. (10) to (12) is the factor involving the thickness of the layer considered. The term $R \ln p_0/p_1 \, \Delta t$ has been replaced by $c_p M'$. Bjerknes' expressions apply to a layer of unit thickness. The equations derived for $\dot{C}$ apply to the stratum between the pressures p_0 and p_1. In each case, the sign of the result is the same for a given sounding; the difference lies in the magnitude of the "stability" to be assigned.

Intuitively it is not difficult to see that $\dot{C}$ is a measure of the change in kinetic energy. It may be noted, however, that no effort has so far been made to suggest precisely what system has this amount of energy. The refinement may be made using the circulation theorem, provided that some additional simplifications are introduced. Consider the following special case: At time t_0 when there is no motion, a perturbation of velocity v_0' is applied as an impulse. Assume the layer to be so thick that the ratio of the length of the ascending column to the length of the descending column is almost unity. This assumption is valid for even very thin layers if the interval of time is small enough.

Initially the circulation around a path consisting of two isobars and two verticals is zero. After the motion has commenced, it is $C = Lv' - Lv$ or $Lv'(1 + M'/M)$ where L is the thickness of the stratum in linear measure. For this simple path, the circulation acceleration is given by direct differentiation to be $\dot{C} = L\,dv'/dt\,(1 + M'/M)$. Let h be the elevation above an arbitrary base of the lower boundary of the layer in question. Then $dh/dt = v'$. The linear thickness of the layer is given in terms of the mean temperature through it to be $L = RT_m \ln p_0/p_1 \cdot 1/g$. On combining the two expressions for $\dot{C}$ and eliminating L, one obtains

$$\frac{d^2h}{dt^2} = \alpha^2 h$$

where

$$\alpha^2 = \frac{g\gamma_d\dot{C}/K}{T_m[1 + (M'/M)]}$$

$$K = R \ln \frac{p_0}{p_1} \cdot \gamma_d v' \, \Delta t$$

The solutions of the above differential equations for the cases where $\alpha^2 > 0$ and $\alpha^2 < 0$ are

$$h = \frac{v_0'}{\alpha} \sinh \alpha t \qquad v' = v_0' \cosh \alpha t \qquad \alpha^2 > 0$$

$$h = \frac{v_0'}{|\alpha|} \sin |\alpha| t \qquad v' = v_0' \cos |\alpha| t \qquad \alpha^2 < 0$$

with the boundary conditions that $v' = v_0'$ and $h = 0$ when $t = 0$. It will be noted that α^2 is determined by the sounding and choice of the ratio M'/M. The above form for α^2 was selected because it is later found convenient to plot $\dot{C}/K$ for use in practical analysis (Figs. 4 to 9).

In applying the above equations for h or v', it is essential to remember the conditions under which they are valid. The time t must be so small that the displacement h does not materially alter the original fields of temperature and pressure. It is reasonable, for example, to take t large enough to make h/L about 0.10.

It is seen that, when $\alpha^2 > 0$, the perturbing velocities are increased. When, on the other hand, $\alpha^2 < 0$, the result of the perturbation is an oscillation of simple harmonic motion about the original position. These two cases respond to unstable and stable stratification, respectively.

In nature, the effect of friction cannot be neglected. Without a complete analysis, one can still conclude qualitatively that friction will set an upper limit to the velocity that will be attained in the case of instability. Similarly, friction will damp out the oscillations in the case of a stable stratification.

A numerical example for the above solutions neglecting friction is as follows (for $M'/M = 0.2$):

Unstable Layer	Stable Layer
$\dfrac{\dot{C}}{K} = 0.40$ (from sounding)	$\dfrac{\dot{C}}{K} = -1.60$ (from sounding)
$T_m = 270°\mathrm{K}$	$T_m = 270°\mathrm{K}$
$\alpha^2 = 1.21 \times 10^{-4} \sec^{-2}$	$\alpha^2 = -4.84 \times 10^{-4} \sec^{-2}$
$v_0' = 2$ mps (assumed)	$v_0' = 2$ mps (assumed)
$h = 182 \sinh \alpha t$	Amplitude $= 91$ m
$v' = 2 \cosh \alpha t$	Period $= 4.75$ hr

From the velocities as given by the above equations, one can find the kinetic energy of the motion. In all that follows, attention is centered, however, on the quantity $\dot{C}$ rather than on the velocity in either the ascending or the descending air. This is done in order to avoid the additional assumptions made above. It will also be seen that $\dot{C}$ more readily than the velocities lends itself to a description of the stability of the column as a whole.

Stability of Stratified Atmosphere.—Inspection of two or more soundings taken within one air mass will show that the atmosphere is more or less stratified in parallel layers. In the following, it is assumed that this stratification is accurately enough represented by the layers between consecutive significant levels of radiosonde data for the station desired. Through each such layer, the thermodynamic proper-ties and the mixing ratio have either con-stant magnitudes or constant vertical gradi-ents. Thus lapse rate is about constant through one layer, as is the decrease (or increase) of mixing ratio. Where these quantities have sudden changes, new signifi-cant levels are inserted in the report.

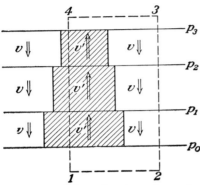

The assumed dynamical structure of the atmosphere during cumulus formation is shown schematically in Fig. 3. Each layer is assumed to have the simple structure used above in investigating the stability in one stratum. The hatched area represents the cloud area. It is not necessary to make any statement about precisely when velocity perturbations reach any particular layer, nor about how large the perturbations are. Such additional information would permit more detailed analysis and forecast but could be added in our present state of knowledge only as assumptions. Since the significant levels are isobars, the circu-lation theorem provides the value of $\dot{C}$ for the cumulus tower as a whole, *i.e.*

Fig. 3.—Schematic view of cumulus tower. Hatched portion represents cloud; clear portion represents clear sky.

$$\dot{C}(1\text{-}2\text{-}3\text{-}4) = \Sigma\dot{C} \text{ (each stratum)}$$

where the summation on the right-hand side will take account of the fact that indi-vidual layers may be either wet or dry.

For use in the above, it is convenient to rewrite Eqs. (10) to (12) as follows:

$$\frac{\dot{C}}{K} = \left(\frac{\gamma}{\gamma_d} - 1\right)\left(1 + \frac{M'}{M}\right) \qquad \text{dry ascent and descent} \qquad \text{(I)}$$

$$\frac{\dot{C}}{K} = \left(\frac{\gamma - \gamma_m}{\gamma_d}\right)\left(1 + \frac{M'}{M}\right) \qquad \text{wet ascent and descent} \qquad \text{(II)}$$

$$\frac{\dot{C}}{K} = \frac{\gamma}{\gamma_d}\left(1 + \frac{M'}{M}\right) - \left(\frac{\gamma_m}{\gamma_d} + \frac{M'}{M}\right) \qquad \text{wet ascent and dry descent} \qquad \text{(III)}$$

$$K = R\ln\frac{p_0}{p_1}\cdot\gamma_d v'\,\Delta t \qquad\qquad\qquad\qquad\qquad\qquad \text{(IV)}$$

Since the liquid water content of the strata is generally unknown (and in any case must be very small during early stages of formation), the case of wet-adiabatic ascent and descent is excluded by assumption. The circulation acceleration for the entire cumulus tower is then given by the sum of Eqs. (I) and (III) determined for the dry and wet strata, respectively.   Then

$$\dot{C} = \sum_{\substack{\text{dry} \\ \text{layers}}} K \left(\frac{\gamma}{\gamma_d} - 1\right) \left(1 + \frac{M'}{M}\right) + \sum_{\substack{\text{wet} \\ \text{layers}}} K \left[\frac{\gamma}{\gamma_d} \cdot \left(1 + \frac{M'}{M}\right) - \left(\frac{\gamma_m}{\gamma_d} + \frac{M'}{M}\right)\right] \quad (15)$$

It will be noted that both the factor K and the lapse rates used in Eq. (15) may vary from layer to layer.   As implied by the sketch, one may also consider M'/M to have different values for different strata.   Equation (15) is the general expression for the stability of the entire sounding.

Total Heating at Any Elevation.   It is of some interest to note, following Bjerknes[1] and Petterssen,[2] the total heating of the layers in the three cases considered.   By substitution into the appropriate part of Eq. (13), it is easily found that

(I) $$\frac{\partial Q}{\partial t} = 0$$

(II) $$\frac{\partial Q}{\partial t} = 0 \qquad\qquad\qquad (16)$$

(III) $$\frac{\partial Q}{\partial t} = c_p M' v' (\gamma_d - \gamma_m)$$

The rate of change of heat (enthalpy) at any level is then zero except in the case of wet-adiabatic ascent accompanied by dry-adiabatic descent.   This means that the temperature averaged over a large horizontal area is constant in time in case I and case II.   The temperature change within the ascending and descending streams is of course given by the appropriate one of Eqs. (2) to (5).   In the case of wet ascent and dry descent, however, the average temperature over a large horizontal area will increase with time.   This follows from (III) above since $\gamma_d > \gamma_m$ always.

The net effect of wet-adiabatic ascent and dry-adiabatic descent in parallel streams as visualized here is then to warm the levels concerned.   This warming over a horizontal area of appreciable extent will increase the stability of the warmed layers with respect to the air below and will decrease the stability with respect to the air above. When, for example, condensation occurs just under a dry layer, the warmed air beneath will become very unstable with respect to the dry air above.

Formation of Cumulus.   Physically the sequence of events that ends in the existence of cumulus clouds is not simple.   The fact of vertical gust velocities through the atmosphere is well established by observation.   These vertical velocities may be thought of as initiating in a great variety of ways.   The turbulent surface layer, for example, in flowing over any but a mirror smooth surface will provide vertical perturbations to the layers above.   When the circulation acceleration is positive in the layer over the source of the gust, greater velocities result.   When the turbulent layer is capped with a strong inversion (in which $\dot{C}$ becomes negative), perturbations from below are soon damped out.

The preceding analysis shows that the atmosphere is more unstable to perturbations with M'/M small.   The result is that, while initially there may be no "preferred value" for M'/M, yet the smaller towers will build up while those of greater cross-sectional area will be damped out.   Observation of the growth of cumulus shows, especially in the early stages, that the cloud area is small compared with the clear spaces.   After many clouds form, the entire sky may of course be covered because of wind shear in the vertical and consequent spreading out of the individual towers. There is also a physical limit to the smallness of M'/M.   For as the ascending current becomes very small in area, turbulent mixing in the horizontal will destroy cloud as soon as it forms.   It is difficult to visualize a real case in which M'/M would be less than about 0.10.

There seems to be no basis in fact for stating that M'/M will have any particular value in nature.   A complete statistical investigation may provide further information on what value should be used in practice.   Experience to date indicates that M'/M assumed to be from 0.20 to 0.30 in mid-latitudes will lead to a correct analysis in most cases.   It may be noted that the ratio M'/M is associated with cloud cover; cloud cover $= M'/(M' + M)$.   It is evident also that, as M'/M approaches zero, the slice method reduces to the parcel method (see Sec. V).

It is considered probable that the effects of impulses from below are transmitted upward through the atmosphere somewhat like a shock wave.   The rate of upward development of cumulus as observed shows definitely that this "wave" speed is relatively small (order of magnitude a few meters per second).   The impulses will be transmitted relatively rapidly through layers where $\dot{C} >> 0$ and will penetrate slowly if at all those layers where $\dot{C} < 0$.   The growth of velocity in the simplest case is of course given by the preceding equations.   These cannot be used to give an exact time of development of the complete cloud, for the equations are only approximately true after the displacements have a finite value.   They may, however, be used in a qualitative way as indicated.

If the layers are all completely dry ($w \equiv 0$), there will naturally be no cloud formation.   Whether a given layer should be considered "wet" or "dry" in the determination of $\dot{C}$ depends on the amount of moisture present.   If the permitted displacements produce condensation, the layer is wet and $\dot{C}$ is given by Eq. (12).   If no condensation results from the displacements, the layer is dry and $\dot{C}$ is given by Eq. (10).   It will be recalled that condensation may occur when relative humidities are below 100 per cent.   The rule that relative humidities are usually at 100 per cent in cumulus clouds does not apply in the analysis, for during the time of perturbations and small displacements the cloud is only beginning to form.   After the formation is complete, the simplifications introduced do not apply in any case.   In particular the atmosphere is not barotropic after cumulus have formed.

The "weather" resulting from cumulus formation depends on a number of factors beyond the scope of this stability investigation.   See Sec. III for the physics of condensation.   It is adequate here merely to remark that if instability is large through relatively thick layers around the 0-deg isotherm, thunderstorms, hail, etc., are bound to result when perturbations of velocity trigger off the instability.

Modification of the Real Atmosphere.   The application of Eq. (15) to a sounding gives a quantitative measure of the stability of the atmosphere as it was at the time of the radiosonde (or aerograph) flight.   The soundings used in forecasts are generally taken at night when very stable conditions exist in the lower layers (over land).   Unless the insolation of the early day is sufficient to remove the inversions and to steepen the lapse rate in the lower layers, cumulus formation is improbable.   The analyst must then take probable modifications of the sounding into consideration before making a forecast.

The major changes in the sounding will be due to one or more of the following: (1) surface heating, (2) lifting of the entire layer as by a front or an orographic barrier, (3) advection, and (4) large-scale divergent or convergent flow.   These changes may be taken into account before $\dot{C}$ is computed.   Minor changes in the atmospheric structure will also occur because of change in moisture by mixing aloft and evaporation at the surface.   Under certain circumstances, radiation from moist layers aloft will cause appreciable changes in lapse rate.   This is a cause of some nocturnal thunderstorms.   The latter modifications may generally be accounted for after $\dot{C}$ is computed for the sounding.

Evaluation of the Record. Given a sounding modified by the analyst to take account of surface heating, divergence or convergence, advection, and lifting expected, the circulation acceleration may be computed by Eq. (15). In practice, the ratio M'/M will be taken to be 0.20 or 0.30. Numerical results are given for v' and Δt equal to unity. The final $\dot{C}$ is given in arbitrary units resulting from using the linear spacing of isobars on a pseudoadiabatic diagram as "thickness factor" (see page 707).

The distribution of $\dot{C}$ for the sounding gives the stability of the atmosphere. In drawing conclusions from the result, one must remember the restrictions on it. Thus the stability is accurately indicated only for the atmosphere satisfying the original assumptions and for (1) relatively small displacements and (2) perturbations of velocity that fit the equation $M'v' + Mv = 0$. One may extrapolate the results qualitatively beyond these rigid boundaries, but the extrapolated values can be approximations to reality at best.

In drawing conclusions from the record for forecasting the weather, one must also have some knowledge of the physical structure of the clouds. The atmosphere may be very unstable (in the sense of having much turbulence) without resulting weather. The slice or convective-cell method like all others requires some judgment and physical knowledge to ensure accurate forecasts. The inherent advantage of the present method is that its simplifying assumptions are close approximations to reality. *What has been accomplished is a means of stating stability or instability in terms of the magnitude and distribution of one variable ($\dot{C}$), instead of the several variables previously employed.* It also appears that, in a good percentage of cases investigated, a correct forecast may be issued with relatively small reliance on general experience.

PROCEDURES AND TECHNIQUES OF ANALYSIS FOR CUMULUS ACTIVITY

The procedures and techniques suggested below are designed to meet two needs: (1) to give additional information to the experienced analyst through convenient use of the slice or convective-cell method and (2) to establish systematic routines by which the nonexpert may improve his analysis and forecast.

Surface Synoptic Chart. Examine the surface synoptic chart to answer the questions

1. What air mass is expected over the station? What modifications are expected in it during the forecast period? What will be the trajectory of the air arriving over the station? What changes may be expected in moisture content in lower layers?
2. What fronts are in the vicinity? What is the general flow pattern (convergent or divergent)?
3. What is the expected cloud cover? Will insolation be sufficient to remove night ground inversions?
4. Is the station in a region of a dynamic anticyclone?
5. What is the expected maximum temperature?
6. What surface winds will prevail through the forecast period?

In many cases, the analyst can easily make a correct forecast from the synoptic chart alone. If the answers to the above questions show that intense cumulus activity is possible, he should proceed to use other data. In passing, it is always wise to consult a reliable climatological atlas unless the climate is familiar (see Sec. XII).

Upper-air Charts. Examine the constant-level charts, the isentropic charts, the constant pressure charts, and the pilot-balloon charts to answer the questions.

1. Is the air over the station unusually wet (dry)? What changes are expected?
2. Will mixing along the isentropic surfaces increase (or decrease) the moisture content?

3. Is upslope (or downslope) motion expected along the isentropic surfaces?
4. Will the upper winds increase so rapidly with elevation that the tops of cumulus will be sheared off and dissipated by horizontal turbulence?
5. Is the flow pattern convergent (divergent) over the station? What levels are affected?
6. Will advection aloft bring in colder (warmer) air? What levels?

Radiosonde Reports. Examine radiosonde reports from the station and from one or more stations upwind to answer the questions

1. Is the atmosphere clearly so stable that cumulus activity will not occur? Such is generally the case under a strong subsidence inversion.
2. Is there sufficient moisture present to permit clouds?

The radiosonde reports plotted on a thermodynamic diagram will also permit a quick estimate of the situation by either the parcel method or the slice method. If the parcel method shows absolute stability for air carried above the convective condensation level (CCL), no cumulus will show any appreciable development. If the parcel method shows a positive area, the slice method should be applied. **A quick qualitative estimate may be made by inspection.** If the sounding shows near saturation everywhere and has a lapse rate close to or less than the wet-adiabatic, the atmosphere is stable. If the sounding is wet and lies appreciably to the left of the wet-adiabatics (greater than wet-adiabatic lapse rate), cumulus formation may be severe. If the air aloft is dry, the atmosphere is stable except when lifted frontally or orographically unless the sounding shows a lapse rate greater than the dry-adiabatic. Stability from the surface up through 10,000 to 15,000 ft will almost always stop extensive cumulus activity.

Slice Method Procedures. After analysis of the synoptic chart, modify the appropriate sounding according to what type of convective activity is expected.

1. *Cumulus Due to Surface Heating.* (a) Mark the convective condensation level (CCL) of the air in the surface layers. The CCL is located at the intersection of the sounding with the saturation mixing ratio line equal to the average mixing ratio in the lower layers. *Rule:* Take the average mixing ratio to be the arithmetical mean of the mixing ratios transmitted for the lowest three significant levels plus 2 grams per kg. The 2 grams added are to take account of the usual diurnal increase. Circumstances may indicate that a greater or lesser value should be added (*e.g.*, surface winds coming from wet or dry terrain). The addition usually lowers the CCL. (b) Draw the dry-adiabatic intersecting the CCL from the CCL to the surface isobar. The stratum between the CCL and the surface isobar is the source of velocity perturbations to the upper layers; this is called the *turbulent surface layer*. The temperature at the intersection of the surface isobar and the dry-adiabatic is the temperature to which the surface must be warmed before cumulus activity may become pronounced. At lesser temperatures, velocity perturbations will not exist at the CCL. Greater surface temperatures will increase the magnitude of the turbulent transfer and, consequently, the magnitude of the perturbing velocities.

2. *Cumulus Due to Frontal or Orographic Lifting.* If the initial lifting of the air leaves undisturbed its barotropic condition, the slice method may be applied. Some account should be taken of divergence and convergence in important cases. The general effects of lifting may be determined by computing $\dot{C}$ (circulation acceleration) after modification of the sounding as follows: (a) Move the surface level isobarically to the expected temperature before the frontal passage (or as the air approaches the orographic barrier). (b) Lift each significant point of the sounding (surface point after the above change) adiabatically a predetermined number of millibars. It is generally sufficient to lift each level the same amount in millibars, thus assuming

there is zero convergence. The amount of lift will depend on the slope of the front or the orographic barrier. In practice, it seems best to lift all points 100 mb. The analyst may then use his judgment to decide whether in this particular case a greater or less lift should have been employed.

3. Make a mental note or jot on the thermodynamic diagram what minor changes are expected in the sounding between the time of the sounding and the time of expected cumulus activity. Pay particular attention to changes expected in mixing ratio and temperature aloft. (See a recent paper by Martin[4] for discussion of changes due to advection.) When current upper winds are available, show their values on the thermodynamic diagram (see illustrations on succeeding pages).

4. Divide the sounding into layers. The lowest layer lies between the CCL and the first significant level above it. Other layers lie between consecutive significant levels above. Each layer is considered separately in the analysis. It is usually sufficient (and desirable) to proceed to the limits of the first "raob" transmission (400 mb). When aerograph soundings are used, this elevation is generally not reached.

5. Label each layer as "wet" or "dry." A layer is said to be wet when perturbing velocities produce condensation in the upward currents. A layer is said to be dry when there is no condensation in upward currents. *Rule:* Experience indicates that in mid-latitudes a layer may be assumed wet when the relative humidity is greater than about 70 per cent; otherwise the layer is taken to be dry.

NOTE: This step is clearly very important. For the sign of C will generally depend upon whether the layer is wet or dry. In case of doubt, use another sounding if available. It is sometimes desirable to make computation for a layer both wet and dry.

6. Determine the circulation acceleration for each wet layer separately. The equation for wet layers is

$$\frac{\dot{C}}{K} = \frac{\gamma}{\gamma_d}\left(1 + \frac{M'}{M}\right) - \left(\frac{\gamma_m}{\gamma_d} + \frac{M'}{M}\right) \tag{III}$$

where

$$K = R \ln \frac{p_0}{p_1} \cdot \gamma_d v' \, \Delta t \tag{IV}$$

and $\dot{C}$ = circulation acceleration (positive for instability)

M'/M = ratio of the masses of ascending and descending currents of air in a convection cell

$\gamma, \gamma_d, \gamma_m$ = actual lapse rate, dry-adiabatic lapse rate, and wet-adiabatic lapse rate, respectively

R = gas constant for the air

p_0 and p_1 = pressures at bottom and top of the layer

v' = magnitude of the ascending velocity in the cell

Δt = duration of the velocity perturbation

NOTE: The ratio M'/M may be interpreted in terms of the expected cloud cover at the start of the cumulus activity (in case clouds are due to surface heating). In fact

$$\text{Cloud cover} = \frac{M'/M}{1 + (M'/M)}$$

Figures 4 to 9 give for convenience the value of C/K for different ratios M'/M and for all likely lapse rates. To apply the above equation proceed as follows:

a. For each layer separately mark off the temperature change along the sounding, the dry-adiabatic, and the wet-adiabatic. The average values of the ratios of the lapse rates through the layer are given by

$$\frac{\gamma_m}{\gamma_d} = \frac{\Delta T_m}{\Delta T_d} \qquad \frac{\gamma}{\gamma_d} = \frac{\Delta T}{\Delta T_d} \tag{17}$$

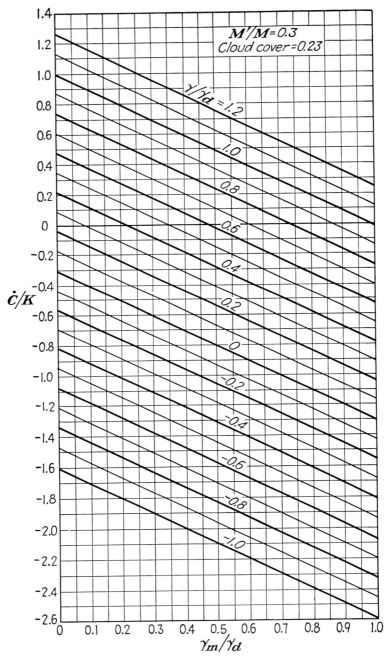

Fig. 4.—Nomogram for circulation acceleration. $(\dot{C}/K.)$

The procedure is illustrated in Fig. 10.   Alternatively look up the mean value of γ_m in Fig. 33 or Table 74 in Sec. I.   Care should be used in finding these values.

 b. Select a value of M'/M.   *Rule:* Experience indicates that in mid-latitudes M'/M will be from 0.20 to 0.30.   Reference to the figures shows that, if a layer is slightly unstable for a certain M'/M, it is more unstable for smaller M'/M and stable for larger M'/M.   Use $M'/M = 0.3$ first.

 c. Enter the appropriate figure (for selected M'/M) with the ratios γ_m/γ_d and γ/γ_d.   Read off the value of C/K (positive or negative).   Record this value on the thermodynamic diagram in the layer.

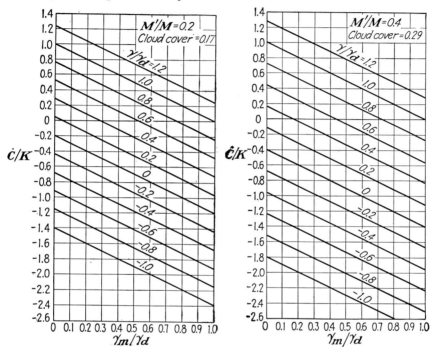

FIG. 5.—Nomogram for circulation acceleration.

FIG. 6.—Nomogram for circulation acceleration.

 d. Take the thickness of the layer into account.   Using dividers or marks on a piece of white paper, measure the linear distance between isobars for the layer.   Use as scale the linear scale for temperature at the base of the pseudoadiabatic diagram.   Thus a layer will be so many "degrees centigrade thick" (see Fig. 10).   This choice of unit is arbitrary but has the advantage of convenience.   Record this thickness in the layer.

 e. Multiply C/K by the thickness of the layer.   Record this product in the layer. *In the arbitrary units selected, this number is the stability of the layer.*

 NOTE: The value of C may be computed in specific energy units (*e.g.*, joules per gram) if desired [see Eq. (IV)].

 7. Determine the circulation acceleration for each dry layer separately.   The equation for dry layers is

$$\frac{\dot{C}}{K} = \left(\frac{\gamma}{\gamma_d} - 1\right)\left(1 + \frac{M'}{M}\right) \tag{I}$$

where the symbols have the meaning used in Eq. (IV) above. It will be noted that $\dot{C}/K$ is negative unless the sounding has a superadiabatic lapse rate. Since this is unlikely in the upper air save for very thin layers, the effect of dry layers is generally stabilizing to initial velocity perturbations.

NOTE: When a dry layer overlies a wet layer that becomes warmed because of cumulus formation, the lapse rate will increase appreciably, and instability will develop.

    a. Find γ/γ_d as in Eq. (17) above.

    b. Use same M'/M as in wet layers. Compute $\dot{C}/K$ by simple multiplication, as shown in Eq. (I).

    c. Multiply $\dot{C}/K$ by thickness of layer as in wet layers. Record final $\dot{C}$ in the arbitrary units explained above in the layers on thermodynamic diagram.

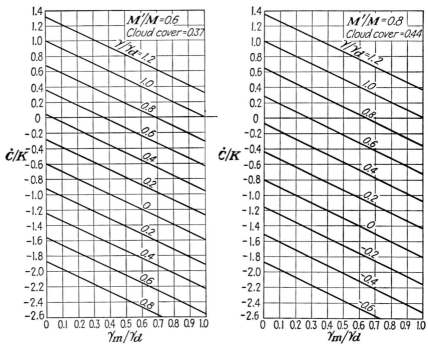

FIG. 7.—Nomogram for circulation acceleration

FIG. 8.—Nomogram for circulation acceleration.

Note on Computation. In the rather rare cases when lapse rate and moisture distribution are almost constant through layers thicker than those transmitted on the teletype sequence, it is correct to apply the above procedures to the thicker layer as if it were one. Particular care must be taken to obtain average values for the ratios of the lapse rates in such cases. In working up a sounding, read off lapse rates to the nearest $0.01°C$ per 100 m. Measure thickness of layers to the nearest $0.10°C$. Record final $\dot{C}$ to the nearest 0.1 arbitrary unit.

Distribution of $\dot{C}$. Figures 11, 12, and 13 show the order of magnitude of $\dot{C}$ in some general cases. Figure 11 is a skeleton pseudoadiabatic diagram [*N* Aer. 448(?)].

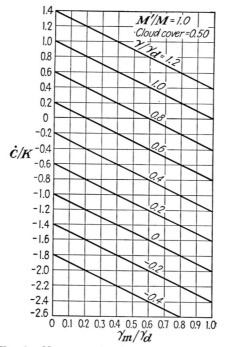

FIG. 9.—Nomogram for circulation acceleration.

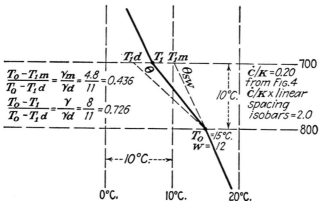

FIG. 10.—Sample computation for circulation acceleration for "wet layer." Ratio $M'/M = 0.3$.

The numbers in the rectangular blocks (each 100 mb by 10°C) show, reading from top to bottom;

   1. The average ratio of the wet-adiabatic lapse rate to the dry-adiabatic lapse rate in the block (a pure number).

   2. The value of $\dot{C}$ in the arbitrary units explained above for a sounding that is saturated along the dry-adiabatic through the middle of the block—a very unstable case.

   3. The value of $\dot{C}$ in same units when the sounding is saturated, $M'/M = 0.3$, and the lapse rate is the mean of the wet- and dry-adiabatic lapse rates for the block.

   4. The value of $\dot{C}$ in same units when the sounding is saturated, $M'/M = 0.3$, and the lapse rate is equal to the average wet-adiabatic lapse rate in the block.

KEY:
1. γ_m / γ_d
2. $\dot{C}(\gamma = \gamma_d)$
3. $\dot{C}\left(\gamma = \frac{\gamma_m + \gamma_d}{2}\right)$
4. $\dot{C}(\gamma = \gamma_m)$

$M'/M = 0.3$ for all $\dot{C}$

mb	−30 to −20	−20 to −10	−10 to 0	0 to 10	10 to 20	20 to 30	30 to 40
400–500	.810 / 2.80 / 1.96 / −0.55	.670 / 4.85 / 3.40 / −1.44	.560 / 6.46 / 4.52 / −1.91	.437 / 8.27 / 5.80 / −2.44			
500–600	(KEY)	.730 / 3.45 / 2.42 / −0.98	.585 / 5.25 / 3.68 / −1.57	.480 / 6.60 / 4.61 / −1.96			
600–700		.755 / 2.74 / 1.92 / −0.83	.640 / 4.03 / 2.82 / −1.22	.510 / 5.49 / 3.84 / −1.66	.408 / 6.62 / 4.62 / −2.00		
700–800		.780 / 2.20 / 1.54 / −0.66	.657 / 3.43 / 2.40 / −1.03	.531 / 4.69 / 3.28 / −1.41	.428 / 5.72 / 3.98 / −1.72	.364 / 6.46 / 4.52 / −1.91	
800–900		.830 / 1.57 / 1.10 / −0.47	.690 / 2.85 / 2.00 / −0.86	.560 / 4.05 / 2.84 / −1.23	.450 / 5.05 / 3.53 / −1.53	.375 / 5.75 / 4.03 / −1.74	.300 / 6.45 / 4.51 / −1.95
900–1000			.710 / 2.46 / 1.72 / −0.75	.584 / 3.54 / 2.48 / −1.07	.470 / 4.50 / 3.15 / −1.37	.390 / 5.19 / 3.63 / −1.57	.332 / 5.67 / 3.97 / −1.72

Temperature, °C.

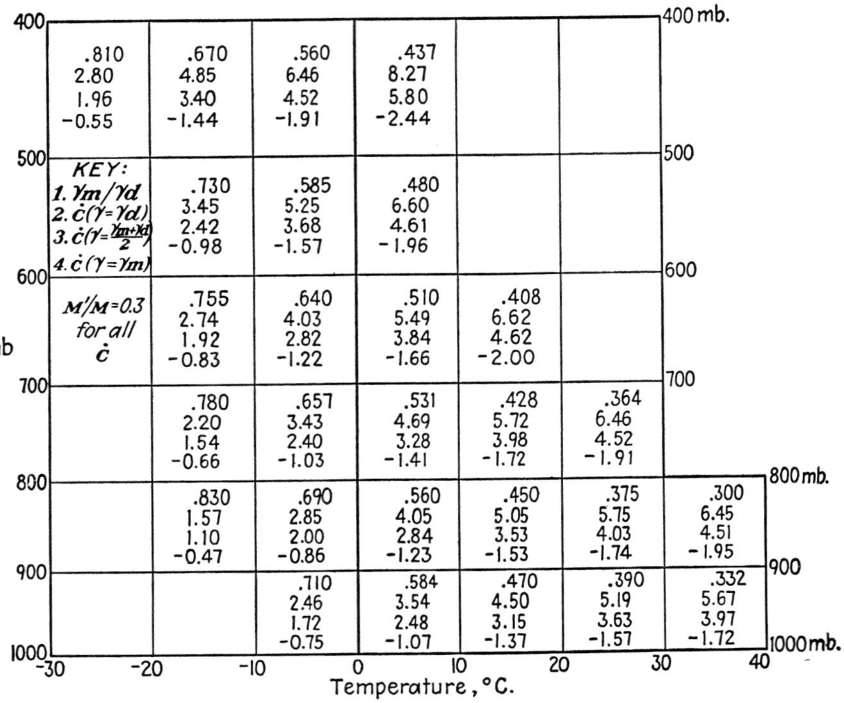

FIG. 11.—Average wet-adiabatic lapse rates and circulation acceleration for various lapse rates through 100 mb intervals. See *KEY* in figure.

Reference to Fig. 11 will prevent any gross errors in calculation. It will be noted that interpolation from block to block is not linear.

Figure 12 shows $\dot{C}$ computed for average values of mP air in summer in Europe. It will be noted that the atmosphere is unstable below 900 mb; above 900 mb there is considerable stability. Figure 13 shows $\dot{C}$ for cP air in summer in Europe. The air is very stable in its lowest 100 mb. From 900 to 800 mb there is some instability. Above 800 mb the sounding is stable or neutral (700 to 600 mb). All layers have been assumed "wet" and M'/M taken to be 0.3 in the calculation. Both figures are for average conditions.

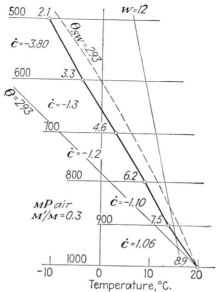

Fig. 12.—Circulation acceleration for mean conditions in *mP* air.

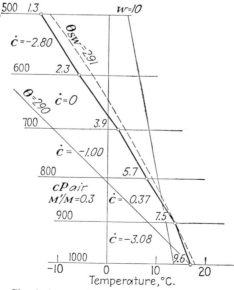

Fig. 13.—Circulation acceleration for mean conditions in *cP* air.

THUNDERSTORMS AND THUNDERSTORM FORECASTING

Three major considerations arise in a study of thunderstorms:

1. *Meteorological problem:* What are the conditions of the air mass(es) concerned? What lifting will occur? Where and when will thunderstorms occur? What will be their violence?
2. *Structure and life history:* What is the physical structure in the thunderstorm throughout its life history? What are the probable trajectories of various thunderstorms?
3. *Operations:* What instructions will be issued to personnel who will be in or in the vicinity of thunderstorms (aloft or at the surface)? What measures will be taken to safeguard equipment?

The meteorological problem must be solved by the meteorologist. He may be guided in his conclusions by pilot reports. The structure and life history of thunderstorms will be understood only through the wholehearted cooperation of meteorologist and pilot. Without this cooperation, forecasts will be issued in terms not understood by the pilot. The too often earth-bound aerologist must rely on the descriptions given by the man who has flown the weather. The question of what to do about the weather is the final responsibility of the operations officer. He cannot make the correct decisions without correct forecasts from the aerologist.

General Structure and Description of Thunderstorms. A thunderstorm consists of a core of uprushing air fed by an inrush of air from the front of the storm. Inertia carries the rising mass far above its equilibrium level. The top of the rising column finally falls over and pours down. Cold rain, hail, or snow may assist in cooling the downdrafts. These increase in violence and sometimes reach the surface as cold currents. A strong downdraft is usually found immediately behind the updraft, with the air farther to the rear settling more gradually.

The intensity of a thunderstorm is an indefinite term. One storm may provide strong turbulence and hail at a station while less than a mile away the storm is considered mild. Moreover, the severity of the storm may change greatly in an hour's time. At the right time and place, the conditions within a localized convective thunderstorm can be just as bad as they are at any spot in a long line squall. The classification of thunderstorms must be based largely on what starts them off. After they have commenced, the associated turbulence and precipitation will depend on what the structure of the atmosphere was when they started and upon how it modifies as they progress.

NOTE: No thunderstorm will be reported as "mild" by a pilot caught in its center.

There is no reliable method of recognizing in advance a thunderstorm that will produce large hailstones. It appears that large hailstones occur in convective thunderstorms as often as in line squalls. About one storm in 800 on the Denver–New York airway produces hail as large as walnuts. One in 5,000 produces hail as large as baseballs. Large hail appears more frequently when the base of the unstable region is below 7,000 ft and when lapse rates are $4\frac{1}{2}$°F per 1,000 ft or greater. The moist layer generally straddles the freezing level and is at least 15,000 ft deep. Hail sometimes occurs only in a very small column of the storm. Severe hail has been observed to fall apparently from the blue sky close in advance of and behind the storm. This has apparently formed in the chimney and been thrown out at the top.

It appears that lightning strikes on airplanes are more common in cumulus- or thunderstorm-type clouds and precipitation at a temperature within 6 deg either side of freezing. The majority of strikes are preceded by St. Elmo's fire (coronal discharge, visible at night). This begins when precipitation is encountered. Static in the radio and turbulence generally increase with the increase of the St. Elmo's fire. The entire cycle including the strike often takes place within less than ½ min. The strike may often be avoided by choosing a new flight level. Pilots should turn on the cockpit light and watch the instrument panel to guard against being blinded by flashes. Structural damage to the aircraft is unlikely. The radio and flight instruments may be permanently damaged by a strike; they will generally be temporarily out of commission at least.

Turbulence appears to occur largely at the boundary surfaces between vertical currents. A number of experiences have been reported in which the plane has been carried rapidly but smoothly within a strong vertical current. More frequently the effect is that of a "sharp edge gust." Accelerations up to 10 *g* have been recorded by aircraft. Visual evidence of turbulence is given by the scud roll observed at the forward edge of a typical thunderstorm between an updraft and the following downdraft. Strongest vertical currents generally occur in the lower two-thirds of the thundercloud.

Gusty winds and wind shifts generally accompanying the passage of a thunderstorm over the station may overturn grounded aircraft. Loose gear may be picked up and thrown against other objects, causing considerable damage. Meteorologists are expected to notify hangar personnel in advance of such winds.

Two things are necessary before a thunderstorm will occur: (1) a potentially unstable atmosphere with some of the air warmer than 0°C and (2) a trigger action to start the vertical motions. Thunderstorms are generally classified according to the manner in which the vertical motions are started. Once started, the potential instability of the atmosphere is turned into real instability that produces strong vertical velocities and the attendant thunderstorm.

Thunderstorm Types. Thunderstorms may be classified as follows:

1. Air-mass thunderstorms.
    a. Convective.
    b. Orographic.
    c. Nocturnal.
    d. Isentropic.
2. Frontal thunderstorms.
    a. Cold front.
    b. Prefrontal.
    c. Warm front.
    d. Stationary front.
    e. Upper front.

It should be noted that the thunderstorm occurs in the warm air mass in all cases. In air-mass thunderstorms, this is obvious, for only one air mass is in the vicinity. The statement is still true when cold and warm air masses are present and the thunderstorm is frontal in character. In this case, the cold air serves only as a means of lifting the warm air, thus triggering off its instability.

Convective thunderstorms occur with greater frequency than any other type and are common to most temperate latitudes during the summer months. Other names are (1) heat thunderstorm, (2) thundershower, (3) local thunderstorm, and (4) thundersquall. The triggering action is provided by perturbations of velocity, local in character, from the turbulent surface layers of the air after they are heated by the sun. Characteristics are (1) rapid development, (2) small area, (3) slow movement, (4) intense local precipitation, (5) strong downdrafts and local squalls, (6) risk of local hail, and (7) dissipate by midnight at the latest (over land). They are most common at the time of maximum temperature or shortly thereafter.[7]

Orographic thunderstorms occur when conditionally or convectively unstable air is lifted by the terrain. Very rapid development and rather large area covered are characteristic. The storm frequently remains stationary over the elevated land causing its start. It is fed by new supplies of moisture brought in by fresh air. "Cloudbursts" are generally caused by orographic thunderstorms. Contact flight should not be attempted beneath such storms, since they have a tendency to hug the peaks. Hail is a common feature in terrain such as the eastern slopes of the Rockies south of Montana. The orographic thunderstorm is frequently found along with frontal thunderstorms.

Nocturnal thunderstorms are common in the plains states between 2200 and 0600. They are generally scattered. They are frequently due to radiation from the top of the moist layer. Nocturnal thunderstorms are often associated with obscure overrunning conditions in which a weak warm front was indicated on the surface map. Factors that seem to precede these storms are (1) a weak warm front overrunning from the southwest, (2) a convectively unstable air mass, and (3) altocumulus or stratocumulus clouds, generally broken.

Isentropic thunderstorms occur as the result of isentropic vertical displacement of a convectively unstable air mass. These may be associated with warm-front or nocturnal thunderstorms in many cases.

Cold-front or "line-squall" thunderstorms are started by the lifting of convectively unstable air by the intrusion of the cold air along a cold front. They occur in the warm air. Greater energy is generally released when the front moves rapidly and/or when it has a steep slope. The thunderstorms on a cold front may persist along hundreds of miles of the front. Hence the name *line squall*. The storms are generally intensified in the afternoon and weakened at night. This indicates that convective action in the day is a large factor. In some cases, the contrast in the air masses is so great that very violent storms occur even at night. Topographical features distort the microscopic structure of the front and cause great variation in the strength of storms along it. Thunderheads at 12,000 ft or so are frequently separated by clear spaces. At other times, the cold front may present a solid wall rising above 40,000 ft.

Prefrontal thunderstorms, often without breaks, sometimes develop between 75 and 300 miles ahead of a cold front. These storms are most active from afternoon to midnight with a concentration around the end of the afternoon. During the period of greatest activity, the original cold front becomes greatly weakened but intensifies again after the prefrontal line squall dissipates. These storms are frequently accompanied by tornadoes in the United States. There are various explanations for the development of these squall lines (see page 653). No completely reliable procedure for forecasting all prefrontal line squalls is known.

Warm-front thunderstorms are relatively infrequent. In general, they occur at high altitude. They then give showers superimposed on the usual warm-front weather. High-level storms appear to be most active from later in the evening to sunrise and least active in the afternoon, the reverse of the cold-front type. This fact may be due to radiation from the top of the moist layer at night. When the instability of the warm air is great, the storm will occur after only slight lifting and may become violent at low levels.

Thunderstorms produced by *stationary or quasi-stationary fronts* may have warm-front or cold-front characteristics according to the slight movement of the front. Violent storms may result, especially if the front is near the edges of cold water in the spring.

Upper-cold-front thunderstorms are similar to the warm-front type but are usually more severe. They occur along the upper cold front where the lifting of the warm air ahead of the front is greatly increased. There is a layer of stable air beneath the storm base except where the upper front becomes a surface front over higher country.

Time of Occurrence of Thunderstorms. Unless the climate of the region is familiar, the forecaster should consult a reliable climatological atlas (see the section on Climatology, page 995). Convective thunderstorms are most likely to occur when the temperature of the surface minus the temperature of the air over the surface is a maximum. Such maxima occur over land in the afternoon and over the oceans at night. Frontal thunderstorms occur whenever the unstable air is lifted by the front. They are likely to be most violent when aided by convective activity (cold front) or radiation from the top (warm front).

Forecasting of Thunderstorms. Thunderstorm forecasting may be done a few hours ahead with a little practice by observation of the clouds. Flat cumulus that show no tendency to develop indicate stable conditions aloft and no possibility of thunder or showers. If towers show, precipitation is probable. If the clouds show signs of ice-crystal development, precipitation with thunder is certain. Radio static increasing is frequently a good sign of approaching thunderstorms.

From the pseudoadiabatic diagram it is possible to determine certain properties of the atmosphere as they existed at the time the sounding was made. Allowance must be made for modifications between the time of the sounding and the time of possible thunderstorm development. All information available from upper-air data and pilot reports should be used. Lapse rates in the atmosphere under thunderstorm conditions are generally between $3\frac{1}{2}$ and $5\frac{1}{2}$°F per 1,000 ft. In general, the steeper the lapse rate (the nearer to $5\frac{1}{2}$ deg), the more unstable is the atmosphere. Lapse rates by themselves are of little use since even a steep lapse rate is not serious if sufficient moisture, initial perturbations (lifting forces), and other factors are not available.

Early-morning inversions will generally be eliminated by early-day surface heating. Inversions at higher levels may limit the development to stratocumulus clouds. When strong differences of wind exist at different levels, air-mass thunderstorms fail to appear even when other conditions are right. This is probably due to shear and turbulence in the horizontal. Low-pressure areas and troughs generally favor thunderstorms. Convergence effects may be dominant even when fronts are not present. Isentropic upslope motion may produce thunderstorms over wide areas.

Thunderstorms in mid-latitudes in other than the summer season will not generally extend to great heights. Thunderstorms in the tropics at any season, and in summer in mid-latitudes, may extend to great heights. Pilots have reported the tops of cumulonimbus clouds well above 50,000 ft. The deeper the storm, in general, the more intense it is.

Detailed procedures and techniques for forecasting thunderstorms due to convective activity will be found on pages 703 to 711. The methods follow some of the original ideas of J. Bjerknes. The parcel method is discussed on pages 402 to 405. Frontal and orographic thunderstorms are forecast after the atmosphere is found to be unstable to lifting in layers. The effects of the lifting are discussed on pages 405 to 408. The effects of convergence and divergence are discussed on pages 407 to 408. A discussion of how to estimate the sign of divergence or convergence that will occur in a given flow pattern is given on pages 813 to 815. The physics of condensation and precipitation is given on pages 252 to 262. The thermodynamic approach and the general analysis of thermodynamic diagrams are treated in Sec. V.

The discussions referred to above are complete in themselves. A thorough understanding of their principles is requisite to correct forecasting in the wide variety of situations that may develop. It is impossible to illustrate all possible combinations of circumstances; the understanding that comes from a knowledge of the principles plus experience will ensure the best possible forecasts. From a combination of experience and principles, the individual will deduce his own set of forecasting rules.

Since a long set of such rules is unnecessary for the experienced man and is likely to provide unwarranted confidence in the beginner, no such set is given here.

It should be noted that the forecaster can hope to do no more than determine whether thunderstorms are due in the vicinity, their probable time of development, and their probable severity (expressed in terms of vertical development, turbulence, hail, etc.). Forecasts of exact time and place of occurrence are almost impossible except after long experience with the terrain in question.

EXAMPLES OF ANALYSIS AND FORECAST

After the routine work of finding the distribution of C along the sounding is done, the forecast is made. A few general rules for forecasting may be given, but emphasis

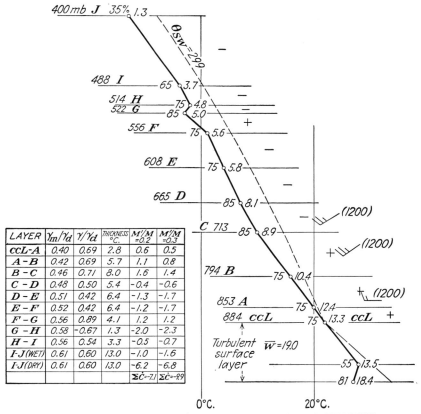

LAYER	γ_m/γ_d	γ/γ_d	THICKNESS °C.	M'/M =0.2	M'/M =0.3
CCL-A	0.40	0.69	2.8	0.6	0.5
A-B	0.42	0.69	5.7	1.1	0.8
B-C	0.46	0.71	8.0	1.6	1.4
C-D	0.48	0.50	5.4	-0.4	-0.6
D-E	0.51	0.42	6.4	-1.3	-1.7
E-F	0.52	0.42	6.4	-1.2	-1.7
F-G	0.56	0.89	4.1	1.2	1.2
G-H	0.58	-0.67	1.3	-2.0	-2.3
H-I	0.56	0.54	3.3	-0.5	-0.7
I-J(WET)	0.61	0.60	13.0	-1.0	-1.6
I-J(DRY)	0.61	0.60	13.0	-6.2	-6.8
				$\Sigma \dot{C}$=-7.1	$\Sigma \dot{C}$=-9.9

Fig. 14.—Sounding for Washington, D.C., June 28, 1943, 0000 EWT.

is placed on the physical meaning of the number $\dot{C}$ rather than on a long list of rules. Thus positive $\dot{C}$ in a layer means that when perturbing velocities reach it they will become larger; the layer is unstable. The larger $\dot{C}$ is in a layer, the more unstable the layer is. When, on the other hand, $\dot{C}$ is negative in a layer, perturbing velocities that reach it will become smaller; the layer is stable. According to the magnitude and distribution of $\dot{C}$ through the entire sounding, the analyst will forecast weather from clear to severe thunderstorms and hail. Experience and constant attention to all available synoptic data are required before correct forecasts can be made in some cases; in others the distribution of $\dot{C}$ is so striking that a correct forecast may be made by a beginner. The following statements may be used as guides:

1. The distribution of $\dot{C}$ worked up for a sounding is for the sounding as modified; the analyst must decide from synoptic data what other modifications will occur during the forecast period.

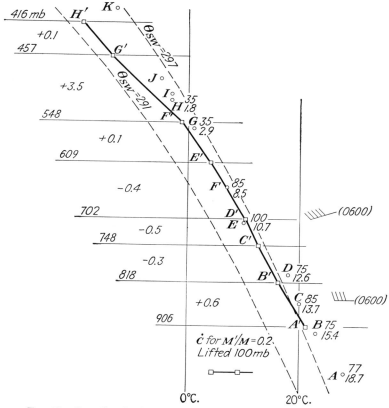

FIG. 15.—Sounding for Washington, D.C., June 29, 1943, 0000 EWT.

2. Dry layers are stable unless the lapse rate is superadiabatic.

3. Wet layers are stable unless the lapse rate is appreciably greater than the wet-adiabatic lapse rate.

4. The time factor may be considered qualitatively. Thus perturbations persisting for several hours in layers where $\dot{C}$ is positive will result in velocities large enough to penetrate thin stable layers above.

5. The magnitude of the initial perturbations may be considered qualitatively. Thus impulses from a thick turbulent surface layer in which the wind is strong will be greater than if light winds were present in a thin surface layer.

6. When $\dot{C}$ is positive for all layers separately, the instability will certainly result in cumulonimbus, thunderstorms, and hail.

7. When $\dot{C}$ is negative for all layers separately, the stability present will prevent any cumulus formation.

8. When $\dot{C}$ summed over all layers is positive and greater than a certain magnitude (depending on the scale used to account for thickness of layers), cumulus activity is certain.

9. When $\dot{C}$ summed over all layers is negative and its absolute magnitude is large enough, extensive cumulus activity is impossible.

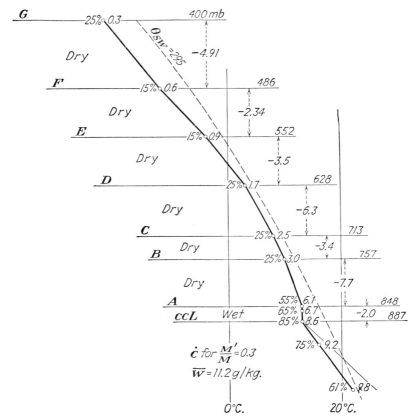

Fig. 16.—Sounding for Miami, Fla., May 4, 1944, 0000 EWT.

10. When $\dot{C}$ is alternately positive and negative up the sounding (the usual case in nature), the analyst must pay particular attention to the location and thickness of the layers:

a. If positive $\dot{C}$ occurs through several thick layers and clusters around the 0-deg isotherm, thunderstorm activity with precipitation is probable.

b. If positive $\dot{C}$ occurs only well below the 0-deg isotherm, cumulus clouds may form (extent depends on thickness of layers in which $\dot{C}$ is positive), but precipitation is improbable.

NOTE: Precipitation is not uncommon in the tropics even when the tops of the cumulus are well below the 0-deg isotherm.

Figure 14 is plotted from the teletype sequence for Washington, D.C., June 28, 1943, 0000 EWT. The synoptic situation was such that the early day was expected to be clear. The maximum temperature expected was 92°F. The average mixing ratio in the turbulent surface layer was assumed to be 19 grams per kg, since a considerable increase was expected during the day with southerly winds. The CCL was at about 884 mb. The parcel method showed a substantial positive area. Wind shear aloft was not expected to be excessive. Considerable moisture was found at all levels. On the basis of the parcel method, thunderstorms were forecast for the area.

The slice-method analysis gives positive $\dot{C}$ with considerable instability up to 713 mb at 10°C. Above this layer, there is great stability except for the relatively

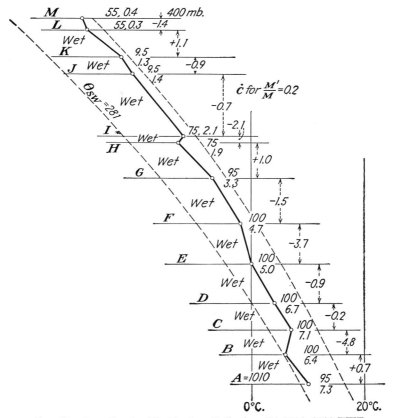

Fig. 17.—Sounding for Washington, D.C., Apr. 27, 1944, 0000 EWT.

thin layer *FG* which has positive $\dot{C}$. The table shows $\dot{C}$ computed for $M'/M = 0.2$ and 0.3. Since the instability is confined largely to layers below the 0-deg isotherm, and since $\Sigma\dot{C}$ is negative and large in magnitude, partly cloudy should be forecast on the basis of the slice method. Observation at Annapolis on June 28 showed about one-fourth cloud cover from 0800 EWT throughout the day; cumulus humilis were logged with bases estimated at 3,500 ft. Maximum temperature attained 90°F. Winds were about southerly, Beaufort 4, throughout the day.

Figure 15 is plotted from the teletype sequence for Washington, D.C., June 29, 1943, 0000 EWT. Circles and letters *A*, *B*, etc., represent the significant levels The parcel method again shows a positive area. The slice method again shows great cumulus activity improbable from surface heating. The synoptic chart showed, however, that a cold front of good depth and strength was expected to pass Annapolis about 1400. Therefore the $\dot{C}$ was computed for the "lifted sounding." The points in squares and with letters A', B', etc., represent the result of lifting the sounding point by point. Each level was lifted 100 mb. The lift is along the dry-adiabatic until saturation and then along the wet-adiabatic until the 100 mb have been passed. The

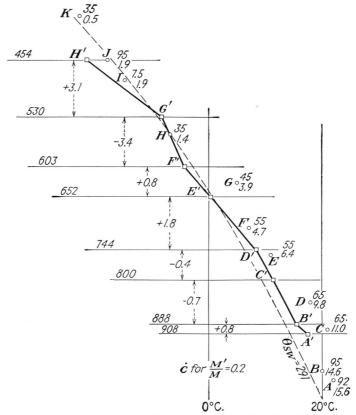

FIG. 18.—Sounding for Lake Charles, La., Mar. 27, 1944, 0000 EWT.

resulting $\dot{C}$ for the lifted sounding with $M'/M = 0.2$ is shown for each layer in Fig. 15. The pseudo-wet-bulb potential temperature decreases about 5°C through the sounding. The distribution of $\dot{C}$ shows instability through and below the 0-deg isotherm with great instability from F' to G'. Fair stability exists from B' to E'. There is instability in the lowest lifted layer. The sum of $\dot{C}$ is positive. On the basis of instability around the 0-deg isotherm, frontal passage expected during period of maximum insolation, and considerable depth to front, thunderstorms should be forecast. Hail is unlikely, since the total instability in lower layers is small. The thunderstorms should be "moderate." Observation at Annapolis on June 29 showed light thunderstorms and showers beginning at 1330. The cold front passed the station at about 1430.

Note: some allowance may also be made for convergence. Thus, if the flow is convergent in the layer *BC*, for example, one should lift level *C* more than *B*. When the flow is divergent, on the other hand, the final spread between the levels in millibars is less than it was originally.

Figures 16 and 17 show extreme stability. In the Miami sounding, the stability is produced by the effect of a dynamic high centered about 400 miles to the northeast. All layers aloft are dry. In the Washington sounding, the stability is produced by advection aloft of warm moist air. These examples are stable by any analysis.

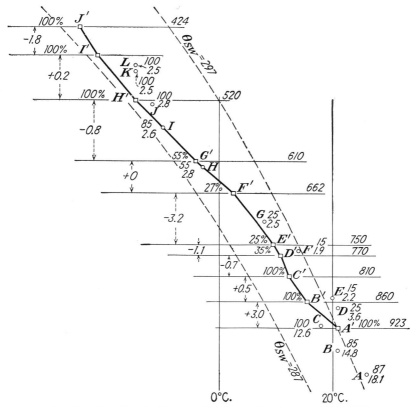

Fig. 19.—Sounding for Brownsville, Tex., May 4, 1944, 0000 EWT.

Figure 18 is for Lake Charles, La., Mar. 27, 1944, 0000 EWT. The sounding levels are shown by circles and letters *A*, *B*, etc. It is desired to know the result of lifting. Each level was lifted 100 mb; the result is shown by squares and letters *A'*, *B'*, etc. The distribution of $\check{C}$ shows great instability through and around the 0-deg isotherm. On the basis of the slice analysis, severe thunderstorms and hail should be forecast for this lifted air. Actually, when a strong cold front lifted the column sometime later large hailstones fell at Memphis and Nashville.

Figure 19 is for Brownsville, Tex., May 4, 1944, 0000 EWT. A cold front was expected to pass the station sometime after 1200. The lifted sounding (100 mb) and the distribution of $\check{C}$ is shown in the figure. There is strong instability in lower

layers and neutral equilibrium through the 0-deg isotherm. Since the front is expected
close to the time of maximum insolation, thunderstorms should be forecast. "Slight-
intensity" thunderstorms 4 hr ago were reported by Brownsville on the 1430 map
of May 4 with the cold front by the station. There was a trace of precipitation.
Present thunderstorms of moderate intensity were shown on the same map.

　Figure 20 shows another case of the effect of considerable dry air aloft and of the
time of frontal passage. The sounding as received shows moderate instability around

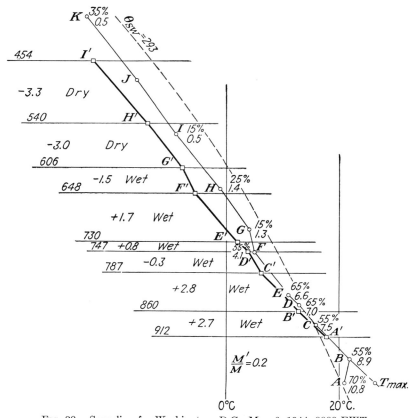

Fig. 20.—Sounding for Washington, D.C., May 6, 1944, 0000 EWT.

the 0-deg isotherm. Insolation was not expected to give convective thunderstorms.
A cold front was expected during the afternoon. The lifted sounding shows instability,
and thunderstorms were forecast. Actually the front did not pass Annapolis until
very late evening. A light shower was logged beginning at 2030. Showers continued throughout the night. No thunder was logged. Washington on the other
hand received moderate thunderstorms with showers in the late afternoon.

　Figure 21 shows a tropical sounding from the Carribbean area, early autumn,
0000 local time. The parcel method shows a tremendous positive area. On taking
account of the dryness of the atmosphere, the slice method shows stability. Using

a slightly different version of the present method, partly cloudy was forecast. Observations showed $\frac{1}{10}$ *Cu*, $\frac{7}{10}$ *As* from 0600 to 0900; $\frac{4}{10}$ *Cu*, $\frac{2}{10}$ *Ci* from 1200 to 1600. No convective clouds seem to have penetrated the stable dry layer *BCD*.

Figure 22 is a sounding taken at Henlow (Beds), England, 0600 GMT, June 25, 1935. The parcel method shows a moderately large positive area. The slice method shows all layers unstable up to 700 mb. Above 700 mb there is rather small stability. Since $\dot{C}$ is uniformly positive through all lower layers, severe thunderstorms should

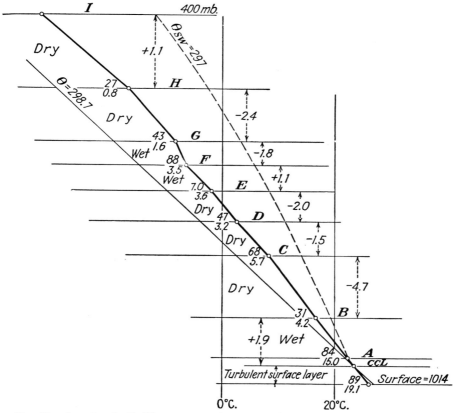

FIG. 21.—Sounding in Caribbean area, early autumn, 0000 Local Time. $M'/M = 0.3$.

be forecast. Observations showed clear sky in the early morning. Violent thunderstorms occurred all over southern England shortly after noon. Maximum temperature attained 81°F.

Figure 23 is a sounding taken at Pasadena, Calif. Cumulus clouds were already in evidence at the time of the flight. The distribution of $\dot{C}$ shows that clouds will carry well past the 0-deg isotherm with considerable velocity. Thunder and hail should be forecast, particularly since some orographic lifting will occur in the vicinity of Pasadena. Observation showed light thunderstorms and considerable hail and showers beginning at 1245.

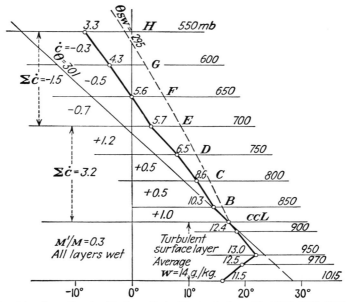

Fig. 22.—Sounding for Henlow, Beds, England, June 25, 1935, 0600 GMT.

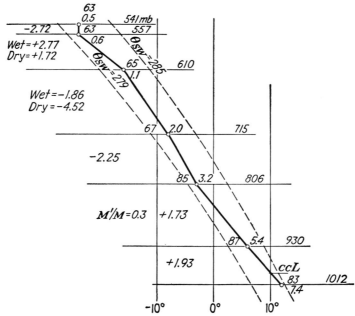

Fig. 23.—Sounding for Pasadena, Calif., Mar. 22, 1937, early morning.   Aircraft flight.

Bibliography

1. BJERKNES, J.: Saturated Ascent of Air through a Dry-adiabatically Descending Environment, *Quart. J. Roy. Meteorolog. Soc.*, vol. 65, 1938.   Also BJERKNES, V., J. BJERKNES, H. SOLBERG, and T. BERGERON: "Physikalische Hydrodynamik," paragraph 715, Springer, Berlin, 1933.
2. PETTERSSEN, S.: "Weather Analysis and Forecasting," McGraw-Hill, New York, 1940.
3. HÖILAND, E.: On the Interpretation and Application of the Circulation Theorem of V. Bjerknes, *Archiv Math. Naturvidenskab*, 42 (5), Oslo, 1939.
4. MARTIN, F. L.: Some Aspects of Stability Deduced from Vertical Wind Shear, *Bull. Am. Meteorolog. Soc.*, 25 (3), March, 1944.
5. MEANS, L. L.: The Nocturnal Maximum Occurrence of Thunderstorms in the Midwestern States, *Misc. Rept.* 16, Institute of Meteorology, University of Chicago, 1945.
6. CROWLEY, D. M.: Causes of Nighttime Thunderstorms over the Middle West, *J. Aeronaut. Science*, **11** (4), October, 1944.
7. BEERS, N. R.: Temperature and Turbulence in the Lower Atmosphere, *J. of Meteorology*, **1** (3-4), December, 1944.

FOG AND FOG FORECASTING

By J. F. O'Connor

Fog results from the condensation of atmospheric water vapor into water droplets that remain suspended in the air in sufficient concentration to reduce the surface visibility below 1,100 yd. Saturation of the air and the presence of sufficient nuclei of condensation are usually necessary for fog formation. In some cases, however, deliquescent nuclei may produce foggy air with relative humidity considerably below 100 per cent (see page 255). These nuclei are frequently found in industrial areas in the form of sulfur compounds resulting from combustion, and also in masses of air with large concentrations of hygroscopic salts due to previous trajectory over oceanic areas.

The principal physical processes that cause saturation are evaporation and cooling. The common types of fog resulting from each process are

1. Evaporation fogs.
   a. Frontal fog.
   b. Steam fog.
2. Cooling fogs.
   a. Advection fog.
   b. Radiation fog.
   c. Inversion fog (stratus).
   d. Upslope fog.

Frontal Fog and Stratus. When warm rain falls through cold air, *e.g.*, at a frontal surface, supersaturation develops owing to evaporation from the warm rain into the cooler air, and condensation on the nuclei proceeds, forming stratus cloud or fog. If the air through which the warmer rain falls is originally unsaturated, it will be cooled to its wet-bulb temperature by the evaporation process. Thus, wherever the rain temperature exceeds the T_{sw} of the cold air, stratus or fog will form. George[1] has shown that pre-warm-frontal fogs offer the greatest obstacle to operation of scheduled flying during winter months in the southern and eastern United States.

In the rear of fast-moving cold fronts, fog does not result from this process, owing principally to turbulent mixing (see page 399) of the cold air beneath the front. In general, cold air that is thoroughly mixed beneath a front because of high wind velocity is usually characterized by a stratus cloud with no fog.

NOTE: If the cold air mass is so much colder than the warm mass that, even with an adiabatic lapse rate established through mixing, its temperature at the ground is appreciably colder than the falling rain, then fog will form regardless of the intensity of turbulent mixing.

Following are given some vertical-temperature distributions through frontal surfaces, before rain starts to fall from the frontal cloud (hatched zone). The temperature of the falling rain T_1 is considered to be the temperature at the top of the frontal inversion AB. When the air below the frontal inversion has become saturated by evaporation from the falling rain, the original lapse rates (solid lines) will have cooled to the wet-bulb lapse rates (dashed lines) between those levels where the dashed lines fall to the left of the T_1 isotherms (line of circles).   Consider the four cases in Fig. 1.

Case I.   Rain at temperature T_1 falls through the cold air below, imparting the saturation vapor pressure at its temperature T_1 to the air.   Above the level C, this will be in excess of the saturation value at the air temperature, and condensation will occur.   A stratus cloud will form in stratum BC.   Below level C, the air temperature exceeds the rain temperature, and no condensation occurs.

Case II.   Frontal fog forms from the base of warm-front cloud B down to surface, since the rain is warmer than the cold air throughout.   Furthermore, the lapse rate after saturation will have changed to the dashed line (T_{sw} curve).   After the air reaches its wet-bulb temperature, the energy for further evaporation of the rain is provided by the enthalpy of the rain itself.   If the air column below the frontal surface is too deep, the rain will cool to the air temperature before it reaches the ground.   In such cases, the stratus does not build down to the surface.   Thus frontal fogs are usually found in the shallow wedge of cold air near the front.   This case

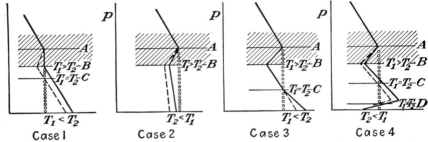

Fig. 1.—Vertical temperature distributions through precipitating frontal surfaces, together with strata developing fog or stratus.

represents negligible turbulent mixing in the cold air, as indicated by its stable lapse rate.

Case III.   Figure 1 represents case II with turbulent mixing having developed in the air below the frontal surface.   The lapse rate below the frontal inversion is therefore adiabatic.   The fog that existed at the ground in case II has been lifted to level C by turbulent mixing.   This has raised the surface temperature to T_2, which now exceeds T_1.   Thus the final lapse rate below level C (condensation level of surface air) is dry-adiabatic (solid line), moist-adiabatic from level C to B where the stratus exists, with the frontal inversion from B to A, where the base of the frontal nimbo-stratus is found.

Case IV.   The inversion at the surface is caused by cooling from below (cold air under the front moving northward over colder ground).   In this case, a stratus deck should develop in the stratum BC.   No condensation would occur in the stratum CD. Fog would also exist in a shallow layer between the surface and level D.   Increasing velocities would rapidly dissipate the surface fog, since the mixing would destroy the surface inversion.

Steam Fog.   Steam fog is an unstable fog type, produced by intense evaporation from a water surface into relatively cold air.   Rapid condensation then occurs, and, since this process involves heating from below as well as addition of moisture, a strong lid in the form of an inversion some distance above the surface must be present to prevent the fog particles from dissipating into the drier air above.   Steam fogs are observed in the middle latitudes in the vicinity of lakes and rivers in autumn when the water surfaces are still warm and the air cold.   For example, at night after stagnant air has developed a strong radiationally produced surface inversion, it may drift over the adjacent warm water.   The heating from below destroys the lower portion of the surface inversion, and the steam fog fills this stratum.   This is frequently a hazard to aviation, since many airports are located along rivers.   This process is shown diagrammatically in Fig. 2.   If the heating from below is sufficient, it may entirely wipe out the inversion, and then the fog dissipates owing to mixing with the dry air above.

Oceanic areas adjacent to source regions of arctic air in winter frequently experience this type of fog.   In such regions, *e.g.*, the Aleutians, it is known as *arctic sea smoke*, occasionally building up from the water surface to heights of 5,000 ft.[2]   In such cases, it takes the form of cumulus clouds with bases on the water, often with clear spaces between.   Flying is hazardous through this type because of severe icing and turbulence.   At other times, the "sea smoke" forms a continuous sheet over the water surface, and then it is much shallower.

Advection Fog.   Advection fog is produced by the transport of moist air over a colder surface, resulting in the cooling of the surface layers below their dew points,

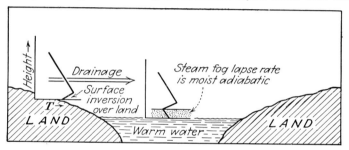

Fig. 2.—Formation of steam fog over a warm body of water, due to drainage of radiationally cooled air in the surface layers over the adjacent land.

with condensation taking place in the form of fog.   All sea fogs are of the advection type except the steam fog mentioned above.

Moist air whose dew point is in excess of water temperature, moving at gentle to moderate velocities over an underlying surface that becomes progressively colder downwind, gives the conditions favorable for its formation.   The maximum frequency of advection fog is observed with light to gentle winds.   Under such conditions, the fog is relatively shallow and stable, *i.e.*, it is characterized by surface inversion.   With lower velocities, the effective temperature difference between air and water decreases, and the almost complete absence of turbulent mixing permits only a very shallow layer at the surface to be cooled.

With winds in excess of moderate velocity, the effective temperature difference increases, and the change favors fog formation.   On the other hand, intense turbulent mixing develops, lifting the fog off the surface to a stratus deck.   However, some cases of sea fog have been reported with very high winds (tropical air fog), *e.g.*, in the North Pacific in summer.   In such cases, the fog is quite deep and turbulent, filling the entire area below the inversion.   Such a condition can exist only in regions where the sea surface is much colder than the air traveling over it.   For example, tropical Pacific air moving northward over the cold waters of the Bering Sea and the Oyashio current, is cooled in excess of 20°F, resulting in fog over large areas.

The Pacific Oyashio waters off Kamchatka and the Atlantic waters of the Labrador current off the Grand Banks of Newfoundland experience the maximum yearly fog frequency in the Northern Hemisphere, with the season of greatest frequency being summer (see the chart on page 994). High frequencies of advection fog also occur in summer along the west coasts of continents (*e.g.*, the San Francisco area) where low sea-surface temperatures exist because of upwelling (see page 1042). Prevailing onshore winds transport the warmer air from farther out to sea inland over the cold coastal waters, producing fog. However, the subtropical latitudes along west coasts experience a high frequency in summer of what is essentially "inversion fog" (see page 731).

Cold waters off the east coasts of continents frequently produce fog in summer, owing to warm moist air blowing seaward from the continent. Frequently the fog banks can be observed lying some distance offshore while a land breeze prevails, *e.g.*, along the New England coast. During late afternoons, these fog banks may be transported inland by the sea breeze. With clear skies, however, the fog dissipates because of convection over the warm land, while in the evening after the land has cooled off, the fog may be transported for a considerable distance inland and persist until convection is set up by the sun's heating next morning.

It should be noted that, when warm moist air is transported over a snow surface, although the initial rapid cooling may produce a fog, there is a marked tendency for dissipation inland. Petterssen[3] ascribes this condition to eddy flux of moisture downward on the snow, which has a constant temperature of 0°C (since the snow is melting because of warmer air above), thus tending to keep the humidity of the air below saturation. Subsequent condensation on the snow liberates latent heat, contributing further to the snow's ablation.

It is also observed that sea fog drifting in over coastal snow surfaces results in rapid dissipation of the snow and a reciprocal dissipation of the fog. Thus, advection fogs transported inland over snow surfaces can persist only for a certain distance, but, of course, as long as the fog continues to blow inland, the coastal stations remain befogged. Arctic summer fogs are predominantly advection fogs that eventually dissipate with sufficiently long trajectory over the melting ice.

Radiation Fog. Radiation fog, or "ground fog," is produced when stagnant moist air is in contact with ground that has become progressively cooler during the night because of an excess of outgoing radiation. The cooling from below produces a temperature inversion in the layers next to the ground. The depth of this inversion depends upon the amount of turbulent mixing. Under calm conditions, turbulent mixing is practically zero, and an extremely shallow inversion results in fog only a few feet thick, or simply in the formation of dew or frost. Frequently, shortly after sunrise, stirring develops, and the fog rapidly increases in depth. For the production of a ground fog sufficient to hamper terminal weather conditions on airways, a breeze of Beaufort force 1 or 2 is necessary. With velocities in excess of these values, turbulent heat redistribution within the mixed layer counteracts surface radiational cooling, and fog usually does not develop. Thus, for the formation of radiation fog, the most favorable conditions are

1. Air with an initial dew point high enough that the expected cooling during the night will lower the air temperature a few degrees below this value.
2. Initial gradient wind not in excess of gentle velocities, so that, when the air is stabilized by surface cooling, just enough turbulent stirring will remain to keep a fairly deep layer cooled below its dew point.
3. Constant or increasing mixing ratio with height in the surface layers, initially, so that the stirring will increase the dew point near the surface.
4. Cloudless or only high cirriform clouds in the sky, in order to promote the maximum heat loss from the ground to outer space by long-wave radiation.

Radiation fogs do not form at sea, owing to the negligible diurnal variation of sea-surface temperatures. Radiation water fogs do not develop over snow, owing to the ice-crystal effect[4] (see page 253). Briefly, at air temperatures below 0°C moisture sublimes on the snow at relative humidities less than 100 per cent, and saturation cannot occur. This is precisely the mechanism of desiccation that operates in the source regions of polar air. Figure 1, page 253, indicates the values of relative humidity existing in the air in excess of which sublimation occurs on the snow surface, at various temperatures below 0°C. If the air contains active sublimation nuclei, however, dense radiation ice fogs may then form. Such sublimation nuclei are active only at temperatures substantially below freezing. These fogs are common in continental areas at high latitudes in winter. It should be noted that water-droplet fogs have been observed at temperatures as low as −30°C. Simpson[5] described such an observation in the antarctic. However, these are probably of an advective nature.

It is usually difficult to distinguish true radiation fog, since frequently other processes are in operation. A frequent combination of processes resulting in fog formation is advection of warm moist air inland or northward over colder ground in winter, resulting in near saturation of the air, with subsequent radiational cooling of the surface layers, producing sufficient further cooling to cause fog formation.

Valleys are particularly subject to radiation fogs. Drainage of air cooled at higher elevations into the valleys prevents excessive cooling or fog formation on the slopes and ridges. This cooled air accumulates in the valleys, resulting in fogs of considerable intensity in low-lying terrain. This same mechanism, *i.e.*, radiation plus air drainage, accounts for the high frequency of frost conditions in low-lying places in certain seasons, while higher elevations are frequently frost-free.

Inversion Fog. Inversion fog is the name given to any type of fog or stratus cloud that initially develops at the top of a moist layer under a subsidence inversion. Cooling, both by turbulence and radiation at the top of the moist layer, accompanied by subsidence above the inversion, intensifies the latter and produces the stratus that may build down to the ground as a fog. The most persistent inversion fogs are those occurring on the subtropical west coasts of continents in summer, *e.g.*, southern California.[6] The trajectory of the onshore flow takes the air over isotherms of increasing temperature (since the minimum sea-surface temperatures are farther north); thus, the air approaching the coast both by day and by night is unstable below the inversion. Added to this effect, radiation to space from the top of the moist stratum cools the air at this level below its dew point, producing stratus. With sufficient radiational cooling, the cloud may build down to the surface.

Neiburger[13] recently suggested that adiabatic cooling of the top of the moist stratum, resulting from vertical motion, plays an important role. The land-sea breeze effect is advanced as the probable cause of this vertical motion. Briefly, off the coast the normal wind flow is assumed to remain unchanged. Near the coast during the day, the onshore component is augmented by the sea breeze. As a result, the maritime air stratum below the inversion is spread farther inland, becoming thinner vertically. The inversion thus lowers, warming adiabatically, and dissipates the stratus. At night, on the other hand, the maritime air stratum stretches vertically, with the inversion ascending and cooling adiabatically, causing the stratus to form below the inversion. After its formation, radiational cooling from the top of the stratus is suggested as playing a major role in the thickening of the stratus.

Inversion stratus is a frequent occurrence in winter over any locality where cold air flows anticyclonically over warm bodies of water. For example, polar air flowing anticyclonically southward over the Great Lakes in winter acquires heat and moisture from below, developing a convective stratus deck below the base of the subsidence inversion. Radiation from the top of the moist layer at night results in thickening of the stratus deck below the inversion, which may even cause precipitation in the form of granular snow. Analogous developments occur when polar air moves off the northeast coast of the United States and returns as an anticyclonic current to eastern United States coastal regions after a trajectory over the relatively warm ocean surface. Light precipitation in the form of drizzle-type snow may then occur along the coast. Similarly, cold air passing southward off the Gulf coast, experiencing an anticyclonic trajectory over the warm Gulf waters in winter, returns inland over the Gulf coastal regions embedded with stratus formations, from which precipitation frequently falls as an air-mass drizzle after radiational cooling from the top of the moist layer during the night has intensified the cooling from above. In such cases, however, surface temperatures rarely rise during the day to the values necessary for the dissipation of the stratus decks, except along the Gulf coast. Thus, the stratus is very persistent as long as the anticyclonic flow conditions continue, although some lifting of the cloud base during the day does occur.

Other occurrences of inversion fog over the United States occur in winter in west coast valleys with access to the sea, *e.g.*, the San Joaquin Valley in California. The synoptic conditions necessary are the presence of a Great Basin high, with its accompanying persistent subsidence inversion above the valley floor. Beneath this inversion exists stagnating moist maritime air. Because of the excessive outgoing radiation at night due to exceedingly dry air in the subsiding anticyclone, stratus frequently develops at the base of the inversion and builds downward to the floor of the valley. These fogs are remarkably persistent, sometimes continuing for as long as 2 weeks without breaking. Similar persistent high fogs occur with stagnating polar maritime anticyclones over central Europe. These fogs eventually dissipate with a change in the synoptic situation, *e.g.*, when an occlusion moves into this region terminating the subsidence. Sometimes these fogs build up from the ground, and as such, are not true inversion fogs, but rather advection-radiation type (maritime fog).[7]

Upslope Fog. Upslope fog is a stable-type fog resulting from the gradual orographic uplift of convectively stable air (see page 407). The air cools adiabatically, and the fog begins to form when it reaches an elevation where the air has cooled to saturation. This elevation may be determined on an adiabatic chart from the initial conditions and is its lifting condensation level (see page 383). The dew point must be initially sufficiently close to the air temperature that saturation can be attained before the ascending motion ceases. The wind velocities must not be so great that turbulent lifting of the fog away from the surface will result. If the air is initially convectively unstable, convective currents will develop on saturation.

In the United States, this type of fog is frequently observed on the eastern slopes of the Rockies, when easterly winds produce a general ascent of moist air from the low elevations of the plains to the mile-high foothills of the Continental Divide. "Cheyenne" fog is of this type.[8]

Frequent occurrences of dense fog that are reported by observers at high elevations are due to convective and other types of clouds engulfing the station. This condition is, of course, different from what is understood as upslope fog. When convective clouds engulf the observer, frequent clearings are experienced as the convective elements move past the station. Observers in the valleys would, of course, report this condition for what it truly is, *i.e.*, cumulus clouds.

Actually, there is no physical difference between a fog and cloud, except that clouds have a greater variety of forms. Whether a mass of water droplets restricting the visibility is reported as fog or cloud depends merely upon the observer's position with respect to it.

FOG FORECASTING

The forecasting of radiation fog depends upon

1. A determination of the amount of nocturnal cooling that will be necessary to lower the air temperature sufficiently below the dew point to produce a fog.
2. The determination that the temperature will or will not drop to that value during the night.

Fog-prediction diagrams that facilitate the determination of (1) are available (see page 395). Entering the diagram with the air and dew-point temperatures of the previous evening, the temperature necessary for fog formation can be obtained. The determination of (2) depends upon expected wind, sky, and soil conditions, *i.e.*, involves the general techniques of minimum-temperature forecasting. General rules for fog forecasting can, at best, serve only as a guide, since fog formation depends to a great extent upon local factors.

A statistical approach to the problem of fog forecasting has been found to give valuable results. For example, fog-prediction diagrams for individual localities may be constructed on the basis of past frequencies of fog formation related to the gradient wind, "spread" or dew-point depression, and visibility prevailing at some hour on the previous evening. A diagram for use on clear nights with light winds is given on page 396. This diagram relates the fog expectation under such conditions to the dew-point depression prevailing at various hours during the night. Other fog-expectation diagrams relate gradient wind as well as dew-point depression prevailing the previous evening to the time of morning when the visibility will decrease below certain limits.[1]

Even with the best fog-prediction diagrams available, attention must continually be paid to the other factors affecting fog formation. Topography, for example, is especially important, with its accompanying drainage effects mentioned above. Haze and smoke in industrial areas accelerate formation, other conditions being the same. The gradient-wind direction also affects fog formation. This is, however, a more important factor for coastal areas, where advection plays the most important role. A nonconvective middle or low overcast during the day restricting the sun's heating, with subsequent clearing at night, will permit the air to cool rapidly at night below its dew point and result in fog. This is usually a post-cold-frontal situation, or *mixing-radiation* fog after George.[9]

The forecasting of the development of inversion fog depends upon a knowledge of the amount of radiation that will emanate from the moist layer to space during the night, as well as a knowledge of air trajectories. An approach to the solution of this problem may be made through the use of the Elsasser radiation chart (see page 301). Since experience indicates that, under the conditions indicated previously, its development is probable, some procedures for its quantitative estimation may be mentioned.

The elevation at which the top of the stratus will be located is determined by the height of the base of the subsidence inversion. The latter is easily obtained from a temperature-height (T-H) curve and is assumed to be invariant within the forecast period, unless rapid changes in circulation are impending. The forecasting of the height to which stratus will build down after forming is apparently closely related to the wind velocity near the gradient level.[10] A relationship between these two factors can be generally determined for each locality.

To forecast the time of dissipation of such a fog,[11] a T-H curve extending at least up into the inversion stratum must be at hand in the early morning after the stratus has been formed. If an aerological sounding is at hand, this is sufficient. However, in its absence, a T-H curve may be constructed as follows on an adiabatic chart (see Fig. 3):

1. From the isobar corresponding to the surface pressure, draw a dry adiabat through the surface dry-bulb temperature T.

2. Terminate dry adiabat found in (1) at the elevation of the cloud base (level 2), which may be determined by ceiling measurement.

3. The w-line through the point where the dry adiabat intersects the level of the cloud base represents the "effective" specific humidity[12] of the turbulent layer. This specific humidity is usually smaller than the surface specific humidity, and the cloud base is therefore found to be higher than the LCL* of the surface air.

4. The cloud top must be known from pilot's reports or other sources.

5. The T-H curve from the cloud base (level 2) to the reported elevation of the cloud top (level 4) will be a moist adiabat continuous with the termination of the dry adiabat found in (1).

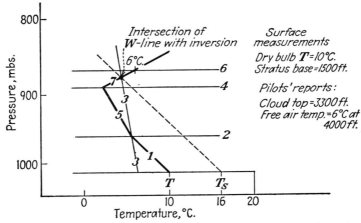

FIG. 3.—Determination of surface temperature T_s necessary to dissipate inversion stratus.

6. To construct the T-H curve of the inversion stratum, the temperature at some elevation above the cloud top must be known, either from a pilot's report or from some high-level observation station in the vicinity. Of course, the temperature determined in the latter fashion must be representative of the free air.

7. The inversion T-H curve is then determined by a line drawn from the termination of the moist adiabat located in (5) at the elevation of the cloud's top (level 4), through the free-air temperature determined above the cloud top at its proper elevation (level 6).

The T-H curve from the surface through the subsidence inversion may thus be synthesized from available data, and consists of a dry adiabat up to the cloud base, a moist adiabat through the cloud stratum, and an inversion above the cloud top. To determine the surface temperature T_s that will be necessary to dissipate the cloud, a dry adiabat is dropped to the surface isobar from the intersection of the w-line (3) and the inversion (7). The temperature at the intersection of this adiabat with the surface isobar (T_s) must be attained during the day to just dissipate the cloud. From the curve of the normal diurnal temperature variation that has been constructed statistically for the particular locality under previous conditions of stratus overcast for the various seasons, the time at which this temperature will be attained may be

* Lifting condensation level.

estimated.   Also, in this manner, the time at which the cloud will lift to any elevation
necessary for flight operations may be forecast.   It should be realized that the time
of dissipation of the stratus determined in this manner will be the earliest possible.
Actually, it may take longer, since under normal conditions there is a certain amount
of evaporation from the soil with diurnal heating, which raises the dew point, requiring
a consequent higher surface temperature for the dissolution of the stratus.   This
procedure may be used in any locality where radiationally produced stratus has

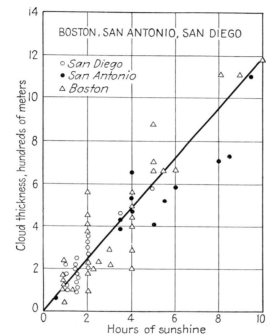

Fig. 4.—Plot of number of hours of sunshine necessary for breaking varying thicknesses of
cloud at indicated localities.   (*After Wood.*)

developed during the night, if the necessary data, such as height of cloud tops, free-air
temperature, etc., are known.

It may be seen from the above method that a general relationship must exist
between cloud thickness and time necessary for clearing.   Figure 4, after Wood,[12]
is a plot of the number of hours of sunshine required to produce clearing against
thickness of the stratus.   This plot is an average based on observations at Boston,
San Antonio, and San Diego.   It appears from the scatter that the amount of daylight to dissipate a given stratus thickness may vary by a few hours.   There does
appear to be an average linear relationship that may serve as a rough guide.

Bibliography

1. GEORGE, J. J.: Fog, Its Causes and Forecasting, with Particular Solutions for the
    Airports at Chattanooga, Tenn., *et al.*, Eastern Air Lines Meteorological Depart-
    ment, Atlanta, 1939.
2. TATOM, LT. CDR. J.: Summer Fogs and Winter Winds of the Aleutian Islands.

3. PETTERSSEN, S.: "Weather Analysis and Forecasting," McGraw-Hill, New York, 1940.
4. BERGERON, T.: On the Physics of Clouds, *Mem. Meteorolog. Assoc., Int. U. for Geodesy and Geophysics*, Lisbon, 1933.
5. BRUNT, D.: "Physical and Dynamical Meteorology," Cambridge, London, 1939.
6. PETTERSSEN, S.: On the Causes and the Forecasting of California Fog, *Bull. Am. Meterolog. Soc.*, 1938.
7. WILLETT, H. C.: Synoptic Studies in Fog, *M.I.T. Meteorolog. Papers* 1 (1), 1930.
8. STONE, R. G.: *Bull. Am. Meteorolog. Soc.*, 1938.
9. GEORGE, J. J.: Fog, Its Causes and Forecasting with Special Reference to the Eastern and Southern U.S., *Bull. Am. Meteorolog. Soc.*, April, 1940.
10. TAYLOR, G. F.: "Aernautical Meterology," Pitman, New York, 1938.
11. KRICK, I. P.: Forecasting the Dissipation of Fog and Stratus Clouds, *J. Aeronaut. Science*, 4 (9).
12. WOOD, F. B.: The Formation and Dissipation of Stratus Clouds beneath Turbulence Inversions, *M.I.T. Prof. Notes*, 10, Cambridge, Mass., 1937.
13. NEIBURGER, M.: Temperature Changes during Formation and Dissipation of West Coast Stratus, *J. Meteorology*, 1 (1–2), September, 1944.

EVAPORATION AND DISTRIBUTION OF WATER VAPOR IN THE ATMOSPHERE

By Norman R. Beers

One of the fundamental questions arising in synoptic meteorology is, "What is the moisture distribution?" A prominent scientist has remarked privately, "Once you know the distribution of moisture in the atmosphere, then you can attempt to forecast." The answer to this question is supplied as fully as circumstances warrant by surface and upper-air synoptic data. Although desired information is sometimes lacking because of paucity of reports or breakdown of facilities, yet in principle the moisture distribution at a given time may be determined with reasonable speed and accuracy. One of the fundamental elements leading to correct forecasts is then the answer to, "What will the moisture distribution become?" This is fundamental, because, if the analyst could accurately predict (1) the flow of the air, (2) the temperature field, and (3) the moisture distribution, he would not fail also to forecast accurately the "weather." The above three elements are of course not independent. Nor are they the only three that may be selected for attention. But whatever the choice of elements may be, one must be the moisture distribution.

Two aspects of the problem are recognized: (1) how does the water vapor get into the air and (2) what happens to it after it is in the air? The first of these aspects is the subject of *evaporation* (generally called *sublimation* if water vapor results from ice or snow directly). Evaporation is the name usually given to the physical process by which liquid water becomes water vapor, whether from a free-water surface, from a soil surface, or from transpiration from plants. C. W. Thornthwaite refers to evaporation from liquid water plus transpiration from plants as *evapotranspiration*. The question of the further distribution of the vapor after it is in the air is the subject of *diffusion*. Diffusion is taken to include molecular diffusion and, of far greater magnitude outside of laminar boundary layers, eddy diffusion or transfer of the water vapor. These are both, strictly speaking, three-dimensional problems unsteady in time. Solutions of an exact nature are available only in relatively simple cases. Approximate solutions are however generally available.

NOTE: Eventually, of course, the water vapor may condense and precipitation may result, thus completing the hydrologic cycle. The subjects of condensation and precipitation are treated on pages 252 to 263.

The necessity for close attention to all aspects of evaporation and diffusion is easily demonstrated. Byers[1] comments, for example, "In a moist, fertile region such as the eastern part of the United States in summer, the rate of addition of water vapor to the atmosphere by evaporation and the transpiration of plants is probably greater than the addition of moisture by an ocean surface." It is to be noted, of course, that the addition of water vapor to the lower surface layers is not enough to cause much "weather." The moisture must also be distributed through the vertical. Striking proof of this statement is evident on the synoptic chart when a dynamic high with strongly subsiding air aloft lies over the southeastern United States. In such a case, only radiation fog can prevent the logging of partly cloudy or clear skies.

Similarly, in the tropics, when the trade-wind inversion is low and strong, if dry air is aloft, no severe cumulus activity will occur.

Process and Factors of Evaporation. The fundamental mechanism of evaporation is a problem in kinetic theory. When a substance and its vapor are in equilibrium together, molecules of the substance are continually passing into the vapor phase, while at the same time an equal number of molecules (on the average) are passing from vapor into liquid (or solid). If the vapor density in the space surrounding the substance is lowered a little, as by the removal of some molecules to a separate space, there is a net passage of the liquid into the vapor phase. If, on the other hand, the density of the vapor is somewhat increased, there is a net flow of molecules into the liquid phase. What is usually called evaporation is a net flow of molecules from liquid to vapor. When the flow is in the other direction, "condensation" is occurring.

The prediction of precisely how much substance will evaporate under specified conditions is difficult enough in the laboratory. In the field of meteorology, where there is no control of the factors and where it is difficult to measure the conditions accurately, the prediction is approximate at best. A few qualitative statements may be made, however.

1. Other things remaining the same, the evaporation is proportional to the difference between the saturation vapor pressure at the temperature of the water and the actual vapor pressure of the air.

2. Evaporation will be continuous only if energy is continuously received from some outside source (*e.g.*, insolation).

3. Evaporation will be proportional to the rate at which vapor is carried away from the immediate vicinity of the evaporating substance. The evaporation rate into still air, for example, is much smaller than evaporation into a wind.

4. Evaporation is more rapid from fresh than from salt water.

Determination of Evaporation. Thornthwaite and Holzman[2] have published promising results of theoretical and observational procedures. Their method is to measure accurately the vertical distribution of moisture in the atmosphere, measurements being taken at two or more levels. Using theoretical developments of Prandtl, Rossby, Sverdrup, and others, they find the formula

$$E = \frac{17.1(e_1 - e_2)(u_2 - u_1)}{T + 459.4}$$

where E = evaporation, in. per hr, for an installation where the upper observations are taken at 28.6 ft and the lower at 2 ft above the ground

T = temperature, °F

e_2 and e_1 = vapor pressures, in. Hg at upper and lower levels

u_2 and u_1 = wind velocities, mph at upper and lower levels

The formula is rigorously correct for an adiabatic atmosphere. Some corrections are made for other conditions, but these are minor except when inversions are strong or when the adiabatic lapse rate is considerably exceeded.

With respect to evaporation studies generally, Thornwaite and Holzman comment:

In general, most researches on the problem of evaporation have been directed toward the determination of moisture losses from bodies of water. This has led to a deficiency in basic and quantitative information regarding actual moisture losses from watersheds or land surfaces. Studies of the evaporation from free water surfaces alone have proved wholly inadequate from the standpoint of soil conservation. What is required is the actual moisture losses from all types of geographic surfaces, or in other words, a quantitative estimate of the evaporative phase of the hydrologic cycle. . . . In an effort to supply this much needed information a method for determining the evaporation from either land or water areas has been presented in this preliminary report. The practicability of the technique has been completely demonstrated. It is hoped that with proper instrumental installation it will be possible to determine transpiration rates and moisture requirements of various field crops and forest trees, the effectiveness of various moisture conserving practices, and the relative importance of evaporation and transpiration in the hydrologic cycle.

Because of the fundamental nature of the methods used by Thornthwaite and Holzman,[2] much may be expected from measurements made along the lines suggested. A network of observation points over an area, with readings taken at several levels on each, would give invaluable information to all interested in this important question. Some rather brief case histories are given by the above authors.

Other Methods. A number of other methods have been applied to the study of evaporation. Among these may be mentioned

1. Formulas based on meteorological elements.
2. Heat balance.
3. Correlation with climatic records.
4. Direct measurement from pans, etc.

Formulas based on meteorological elements vary from the simple expression given by Dalton in 1802 $[E = c(e_s - e_d)]$ to the rather complex equations of Jeffreys, Sutton, et al. (see Brunt[3]). Dalton's equation with a factor $1 + ku$ added to account for wind is about as useful as any for the simpler applications. Here e_s is the saturation vapor pressure at the temperature of the water, e_d is the saturation vapor pressure at the dew-point temperature of the air, u is the wind velocity, and k is a constant. The more complex equations have necessarily been derived for certain special systems, e.g., evaporation from circular pans holding a free surface of water. They do not in any case give much immediately useful information to the synoptic meteorologist.

The study of evaporation from the point of view of heat balance has been used with considerable success in determinations of the evaporation from the oceans by Sverdrup.[4] His theoretical results agree well with those of Wüst based on observations (after a correction factor is applied to the results of Wüst). The method consists in equating the energy received from the sun to the energy used by evaporation, warming, back radiation, etc. The results would, of course, be exact, provided that exact measurements could be made of all the energies involved.

The correlation of climatic records offers another means of making exact determinations of evaporation from large areas, provided that the records are complete and accurate. Thus, if the total precipitation is known as well as the total runoff by surface and ground water and the change in storage over a given area, the difference will be the amount of total evaporation (soil, free-water surface, snow surface, and transpiration). The method has been applied with some success by the U.S. Geological Survey. See the article by C. H. Lee in Meinzer's "Hydrology."[5]

A new method of determining the amount of water transferred to the atmosphere from land surfaces through processes of evaporation and transpiration has recently been developed by Thornthwaite. A preliminary description of the method appears in the annual report for 1944 of the Committee on Transpiration and Evaporation of the American Geophysical Union.[8]

Thornthwaite obtained all of the available observations of water losses from land areas in different parts of the country. These include water requirements in various irrigation districts in the West and the data on evapotranspiration from different types of cover on the weighing lysimeters of the Soil Conservation Service in Coshocton, Ohio. They also include computations of water loss from small watersheds as a difference between precipitation and runoff for months when it was apparent that there was no reduction of evapotranspiration because of a moisture deficiency in the soil and where it was possible to estimate the lag in runoff and thus to assign the water available for runoff to the month when the rainfall occurred.

From these observations Thornthwaite derived a general equation by means of which daily, monthly, and annual evapotranspiration can be determined from records of temperature, length of day, and precipitation. Since runoff can then be computed as a difference between precipitation and evapotranspiration, the many runoff measurements of the Geological Survey and other agencies provide a rigorous test of the validity of the formula. A map of average annual runoff in eastern United States computed from Weather Bureau data of temperature and precipitation and published in the report of the American Geophysical Union Committee on transpiration and evaporation shows that the formula is trustworthy. Maps of average annual and average July evapotranspiration computed by Thornthwaite's method appear on pages 741 and 742. They are preliminary and subject to revision.

The direct measurement of evaporation from pans offers many advantages in some applications. A variety of pans and special surfaces has been devised. See Meinzer[5] for examples. Measurements must be corrected for size and shape of the pan. Moreover, any pan placed in the open is subject to various disturbances, some of which are difficult to predict and to account for. Finally the rate of evaporation from such a device is so dependent on the temperature and wind that very accurate measurements of these elements must be made before complete correlations are possible.

Quantitative Results of Measurement of Evaporation. Some quantitative results of Wüst and Sverdrup for the oceans are given in Table 1. According to Wüst, the figures for the oceans as a whole are as follows:

Total evaporation from oceans = 334,000 km^3 per year
Total precipitation on oceans = 297,000 km^3 per year
Amount supplied oceans by runoff = 37,000 km^3 per year

The figures for the land surfaces of the earth are as follows:

Total precipitation on land = 99,000 km^3 per year
Evaporation from land surfaces and inland waters = 62,000 km^3 per year

TABLE 1.—AVERAGE VALUES OF EVAPORATION E AND PRECIPITATION P, CM PER YEAR*

Latitude	Atlantic Ocean		Indian Ocean		Pacific Ocean	
	E	P	E	P	E	P
40°N	94	76			94	93
30°	121	54			116	65
20°	149	40	(125)	(74)	130	62
10°	132	101	(125)	(88)	123	127
0°	116	96	125	131	116	98
10°S	143	22	99	156	131	96
20°	132	30	143	59	121	70
30°	116	45	134	58	110	64
40°	81	9	83	73	81	84
50°	43	72	43	79	43	84

* After Sverdrup and Wüst.

Plant physiologists and foresters have accumulated a vast amount of interesting data on the amount of transpiration by plants. Plants were grown in pots or tanks and the amount of water utilized by them was carefully measured. The ratio of total water lost by a plant through transpiration in a given time interval to the weight of dry matter produced is called the water requirement. In the experiments conducted by Briggs and Shantz,[9] between 1911 and 1917, the water requirement of

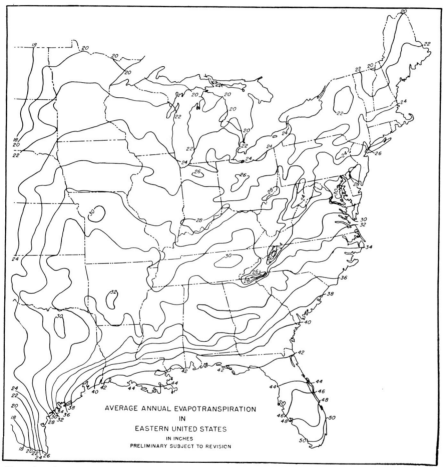

Fig. 1.—Average annual evapotranspiration in eastern United States in inches. Preliminary, subject to revision. (*Courtesy of C. W. Thornthwaite.*)

alfalfa grown at Akron, Colorado, averaged about 850, and ranged between 657 in 1912 and 1,068 in 1911. The growing season temperature was low in 1912 and high in 1911. They found that the water requirement varied through the season; it was much higher in midsummer than in spring or autumn. In an experiment conducted in 1912 the water requirement of the first crop of alfalfa (June 3 to July 26) was 615, of the second crop (July 26 to Sept. 7) was 975, and of the third crop (Sept. 7 to Nov. 4) was 479.[10] The average water requirement for a number of field crops studied at Akron ranged from about 250 to nearly 900. The water requirement of forest trees appears to be somewhat higher, *e.g.*, larch about 1,200. See Meinzer[5] for more complete data. These observations suggest that the variation in water efficiency (reciprocal of water requirement) of a single plant under the changing conditions of temperature and insolation during the growing season is as great as that found in different species. Thus it is virtually impossible to estimate the transpiration rates of these plants from the determinations of their water requirements.

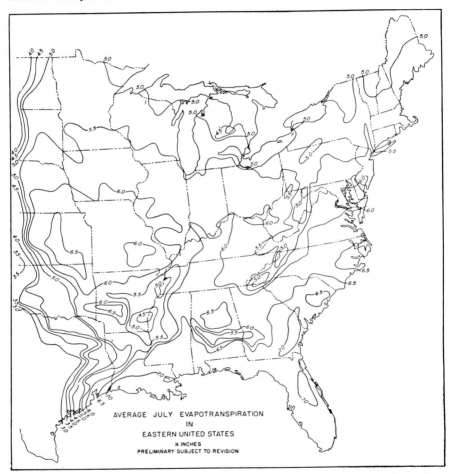

AVERAGE JULY EVAPOTRANSPIRATION
IN
EASTERN UNITED STATES
IN INCHES
PRELIMINARY SUBJECT TO REVISION

FIG. 2.—Average July evapotranspiration in eastern United States, in inches. Preliminary, subject to revision. (*Courtesy of C. W. Thornthwaite.*)

The recent studies of Thornthwaite suggest that transpiration is a function of incident radiation upon the plant surface and of the ambient temperature and is practically independent of the kind of vegetation involved. Although this conclusion is contrary to current notions concerning transpiration rates it is not unreasonable when we consider that transpiration, like evaporation, is related to the energy received from the sun, so long as there is no moisture deficiency in the plant.

According to Thornthwaite's preliminary map (Fig. 1), the average annual water transfer into the atmosphere in eastern United States is less than 20 in. along the Canadian border and more than 50 in. in southern Florida. In the Great Plains evapotranspiration drops off sharply in response to reduced precipitation. This is striking along the Gulf Coast of Texas and in the lower Rio Grande valley where almost every drop of rain returns directly to the atmosphere and practically none appears as runoff. An annual evapotranspiration of less than 18 in. there corresponds to more than 48 in. in the same latitude in Florida. Thus the loss to the atmosphere in south Texas would be at least 30 in. greater than it is if the water were available.

According to Thornthwaite's map, the range of evapotranspiration from north to south in July is quite small (see Fig. 2). It is about 5 in. along the Canadian border and less than 7 in. along most of the Gulf Coast. Only in southern Louisiana and in two small areas in Georgia and Florida does it rise above 7 in. In the major part of the area east of the 100th meridian the average loss to the atmosphere is between 5.5 and 6.5 in., about 6.0 in. This is approximately 0.2 in. per day.

Holzman[11] has computed the total average depth of water in the atmosphere for July, 1936. In eastern United States it ranged from less than 0.9 in. in the north to more than 1.7 in. on the Gulf Coast. If Holzman's computations for 1936 can be considered representative, then it is evident that the daily increment of moisture to the atmosphere is a considerable part of the total. From Thornthwaite's map it appears that the average daily evapotranspiration is about 0.16 in. in the north and 0.22 in. in the south. *Apparently, then, the moisture in the atmosphere can be increased daily by one-sixth to one-eighth.*

From Wüst's[12] data, the July evaporation from the surface of the Gulf of Mexico and the Caribbean is estimated at less than 5.5 in. Over most of the land area of eastern United States the evapotranspiration is greater than this amount.

The amount of evaporation from bare soil is a function of the depth of the water table. When the soil is saturated at the surface, evaporation is rapid. When the water table reaches a depth exceeding the limits of capillary action, evaporation ceases. For practical purposes, there is no evaporation after the water table is below 4 ft. Evaporation from sand, saturated at the surface, is slightly in excess of the evaporation from a free-water surface. Evaporation from average loamy soil is about 90 per cent of that from a free-water surface. Data for various bare well-drained soils for the year are given by Meinzer.[5] Some representative values follow:

Type of Soil	$\dfrac{\text{Average evaporation}}{\text{Average precipitation}} \times 100$
Fine sand	16
Clay	72
Loam	69
Sandy loam	69
Peat	35

The evaporation from soil itself when covered with vegetation is less than indicated above. The total evaporation, including transpiration, is usually larger, however.

Process of Diffusion. Water vapor that has been evaporated into the atmosphere will be diffused through the available space according to the magnitude of the eddy coefficients of diffusivity. These coefficients are functions of the type and velocity of flow. They are large, and large diffusion results when the air is turbulent; diffusion is small when the flow is laminar and winds are light. In the lower surfaces, when the lapse rate is close to the dry-adiabatic, diffusion along the vertical is considerable. Thus, within a period of a few hours, the mixing ratio will become almost constant through several thousand feet of turbulent surface layer, even when the moisture was originally largely confined to the lower few hundred feet or less. In the upper layers of the atmosphere when there is no pronounced convective activity and no cumulus clouds, diffusion is largely along isentropic surfaces.

In theoretical studies of diffusion, one investigates the transfer of the quantity mixing ratio (or sometimes specific humidity). Thus Grimminger[6] has made a study of the lateral mixing from isentropic charts. His results derive from an application of the fundamental equation of diffusion

$$\rho \frac{dq}{dt} = \frac{\partial}{\partial x}\left(A_x \frac{\partial q}{\partial x}\right) + \frac{\partial}{\partial y}\left(A_y \frac{\partial q}{\partial y}\right) + \frac{\partial}{\partial z}\left(\eta \frac{\partial q}{\partial z}\right)$$
$$= \rho\left[\frac{\partial q}{\partial t} + u\frac{\partial q}{\partial x} + v\frac{\partial q}{\partial y} + w\frac{\partial q}{\partial z}\right]$$

where A_x, A_y, η are the coefficients of eddy diffusivity in the x, y, z directions (z-axis up); ρ is the density; q is specific humidity; and u, v, w are x, y, z velocities. After making certain simplifying assumptions, Grimminger was able to evaluate the eddy coefficients for the isentropic sheet both for mean isentropic maps and for daily maps. Average values of the *kinematic coefficient of eddy diffusion* ($\nu = A_x/\rho$) run about 5×10^9 cm^2 sec^{-1}. The value of η was taken to lie between zero and 50 grams cm^{-1} sec^{-1}. It must be borne in mind that the A's will vary somewhat with lapse rate and with the wind speed. Extreme values of ν found by calculation were 0.3×10^9 and 40×10^9.

Important theoretical investigations of the lateral diffusion of currents have been based numerically on the results obtained from the study referred to above. See a second paper by Grimminger.[7] See also Sec. VI of this book.

Qualitative Estimates of Diffusion. No ready means exists at present for the calculation of diffusion in the atmosphere, and the analyst must in general rely on qualitative estimates. Some very general synoptic situations may be recognized. For example, the presence of considerable moisture in the lower 5,000 ft, say, of the air over Miami, Fla., does not necessarily mean that the atmosphere will be moist at 10,000 ft next day over the South Central states. Before the moisture can be carried aloft in quantity, some mechanism must exist whereby vertical diffusion is large. This vertical distribution is accomplished generally by one or more of the following: (1) fronts, (2) isentropic upslope, (3) orographic barriers, and (4) convective storms.

A frequent demonstration of the above is found in summer on the isentropic chart (or on the 10,000-ft map). Here, for example, it is not uncommon to find isolated patches of moisture at higher levels. It is generally easy to establish that the patch of moisture was caused by thunderstorm activity or orographic lifting local in character. Fronts or large-scale isentropic upslope will generally produce tongues of moisture on the upper-level charts.

It should also be noted that, just as certain trajectories of the air tend to increase the moisture present, other trajectories will decrease the moisture. Since mixing ratio is conservative (in the absence of condensation), moist tongues will become less moist as the water vapor is diffused into dry air. Similarly, moisture may be lost in the lower layers when the air mass travels over desert regions.

Bibliography

1. Byers, H. R.: "Synoptic and Aeronautical Meteorology," McGraw-Hill, New York, 1937.
2. Thornthwaite, C. W., and B. Holzman: The Determination of Evaporation from Land and Water Surfaces, *Monthly Weather Rev.*, January, 1939.
3. Brunt, D.: "Physical and Dynamical Meteorology," Cambridge, London, 1939.
4. Sverdrup, H.: "Oceanography for Meteorologists," Prentice-Hall, New York, 1942.

5. MEINZER, O. E., Editor: "Hydrology," (Physics of the Earth, IX), McGraw-Hill, New York, 1942.
6. GRIMMINGER, G.: Lateral Mixing from Isentropic Charts, *Bull. Amer. Meteorolog. Soc.*, May, 1941.
7. GRIMMINGER, G.: Lateral Diffusion of a Current, *J. Franklin Inst.*, November, 1943.
8. WILM, H. G., and others: Report of the Committee on Transpiration and Evaporation, 1943–1944, *Trans. Am. Geophys. Union*, in press.
9. SHANTZ, H. L., and LYDIA N. PIEMEISEL: The water requirements of plants at Akron, Colo., *J. Agr. Research*, **34** (12): 1–44, (1927).
10. BRIGGS, LYMAN J.: Effect of frequent cutting on the water requirement of alfalfa and its bearing on pasturage, *U.S. Dep. Agr.*, Paper 228, p. 3, 1915.
11. HOLZMAN, BENJAMIN: Sources of Moisture for Precipitation in the United States, *U.S. Dep. Agr. Tech. Bull.* 589, p. 21 (1937).
12. WÜST, G.: Verdunstung und Niederschlag auf der Erde, *Gesell. Erdkunde, Berlin, Ztschr.*, (1–2): 35–43, 1922.

DIURNAL VARIATION OF THE METEOROLOGICAL ELEMENTS

By George R. Jenkins

PRESSURE

In low and middle latitudes, mean hourly pressures are found to follow an approximately systematic daily succession which is usually characterized by two minima occurring near 0400 and 1600 and two maxima occurring near 1000 and 2200, local time.   Commonly, this total daily succession is called the *diurnal variation of pressure*. When the diurnal variation is subjected to harmonic analysis, it is found to have at least two important components and possibly four.[2,10]

The first harmonic is usually explained as being due to the alternate daily heating of the air by the sun and nocturnal cooling of the air by terrestial radiation.[6]   It varies with cloudiness; between oceanic and continental areas; with latitude; and, over the oceans, between the major subdivisions thereof[7]; over equatorial oceans (10°N to 10°S), it has an amplitude between 0.35 and 0.52 mb, being least over the Atlantic; between latitudes 10 and 20°N and S, the amplitude generally decreases to around 0.3 mb.

The second harmonic has maxima and minima closely corresponding to those of the diurnal succession.   It varies with latitude, the time of year, and location; over equatorial oceans, its amplitude is generally 1.1 to 1.2 mb; this decreases to 1.0 to 1.1 mb in latitudes 10 to 20°N and S.[7]   At Washington, D.C., it is very nearly 1.2 mb.[2]

The third harmonic, the terdiurnal variation, according to Albright,[2] is absent over the equator, reaches a maximum at 30° latitude, has opposite meridional phases between Northern and Southern Hemispheres, and shifts 180° between winter and summer.   A quatrodiurnal variation is also offered by Albright.

Not entirely satisfactory explanations have been developed for these several components of the normal diurnal variation of pressure, nor have the effects of various locations been analyzed.   Diurnal pressure changes in high latitudes and at high altitudes are quite different from those at low elevations in middle and low latitudes for which the harmonics have been mentioned.   As a consequence of the many variations that enter into the diurnal variations at different places, a thorough systematization for the world is not possible.   Examples of diurnal variations representative of different situations are given in Fig. 1.

Tropical stations *A*, *B*, *B'*, and *C* show remarkably uniform diurnal variations of pressure.   Total daily amplitude is 3 to 4 mb; and this total amplitude is equal to the daytime variation from 1000 to 1600.   The variation at Batavia is presented for the 2 months of minimum (June) and maximum (September) amplitudes.   As a generalization, tropical diurnal amplitudes are least at the solstices and largest at the equinoxes; at Batavia, a secondary maximum is observed in March and a secondary minimum in December.   At Singapore, the monthly successions of amplitude are essentially the same as at Batavia.[4]   The total difference in amplitude as shown by *B* and *B'* is not large, however; as a consequence, curve *C* seems generally representative for tropical regions, and this uniformity accounts for the general usefulness of table 1.   No examples of diurnal variation of pressure at tropical continental locations are available for inclusion in the data for purposes of noting the effect of continentality in low latitudes.

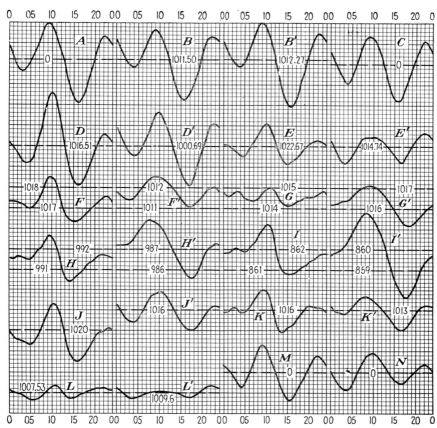

Fig. 1.—Representative examples of diurnal variation of pressure. Examples *A*, *C*,
M, and *N* are annual curves and, owing to the annual variation of pressure, are usually
derived as averages of hourly departures from monthly mean pressures; hence, the base line
is "zero departure." Examples *B*, *B'*, *D*, *D'*, *E*, *E'*, *L*, and *L'* are monthly curves computed as
the mean hourly pressures plotted from a base line of the monthly mean pressure usually
reduced to mean sea level. All other examples are plots of monthly averages of hourly
station pressures. In all cases, ordinate scale units are 0.2 mb, and abscissa units are
hours. An index of the curves, including latitude, longitude, elevation, time interval of
curve, and source, is as follows:

A, Singapore (1°N, 104°E), annual.[8]
B, Batavia (6°S, 107°E, 8 m), June; *B'*, September.[17]
C, Pacific Ocean (4½°, composite general curve), annual.[9]
D, Calcutta (22½°N, 88°E, 6.5 m), January; *D'*, July.[17]
E, Cape of Good Hope (34°S, 18½°E), July; *E'*, January.[17]
F, San Diego (33°N, 117°W, 28 ft), January; *F'*, July.[18]
G, Seattle (48°N, 122°W, 30 ft), January; *G'*, July.[18]
H, Kansas City (39°N, 96°W, 750 ft), January; *H'*, July.[18]
I, Grand Junction (39°N, 109°W, 4602 ft), January; *I'*, July.[18]
J, Jacksonville (30°N, 82°W, 31 ft), January; *J'*, July.[18]
K, Boston (42°N, 71°W, 29 ft), January; *K'*, July.[18]
L, Aberdeen (57°N, 2°W, 26.8 m), January; *L'*, July.[17]
M, 20° (composite general curve), annual.[9]
N, 33½° (composite general curve, oceanic), annual.[9]

In middle latitudes, continentality and elevation as well as latitude and season have significant effects on the diurnal variation of pressure. It is notable that, at most stations and seasons, the absolute maximum is around 1000, the absolute minimum around 1600. Seasonal variations are distinctive, but are not consistent; Calcutta, Cape of Good Hope, San Diego, Jacksonville, Boston, and Aberdeen show larger amplitude in winter than in summer; the reverse is true at Seattle, Kansas City, and Grand Junction, the last two showing the effects of continentality. The Pacific coast stations show considerable differences from the Atlantic coast stations of the United States; Jacksonville shows an expected larger amplitude than higher latitude San Diego, but the January amplitude at Boston is double that at Seattle; and in reversal of normal latitudinal variation, the July amplitude at Seattle is larger than that at Boston. Perhaps the dominant features of the middle-latitude curves are the damping out of the nocturnal range at Kansas City and Grand Junction. In January, the 2000 maximum is nearly absent and is displaced to 0200 or 0300, while the usual 0400 minimum, also nearly absent, occurs at 0500 at both stations; in July, the nocturnal range is entirely absent at Grand Junction and barely discernible at Kansas City. Especially prominent is the daytime range at Grand Junction in July; the maximum is 2 hr earlier than normal, the minimum is 1½ hr later than normal, and the total range exceeds slightly that of any presented for the tropics. The same kind of curve is characteristic for inner Asia in summer.[9] Continental effects are also noticeable when curve M is compared with those for Calcutta; the latter shows considerably larger amplitude than the average for the Pacific Ocean in about the same latitude. The coastal stations in the United States and Cape of Good Hope show less variation from the oceanic average amplitude in comparable latitudes; the ocean curve is notable, however, for its absolute minimum at 0400 instead of the usual 1600; it is matched in this respect only by Aberdeen in both months and Cape of Good Hope in January.

TABLE 1.—To Correct* Barometric Pressure for Diurnal Range at Sea
within the Tropics

Zone time	Northern Hemisphere				Southern Hemisphere			
	Spring	Summer	Autumn	Winter	Spring	Summer	Autumn	Winter
0400	+0.8	+0.7	+0.8	+0.3	+0.6	+0.7	+0.7	+0.5
0800	−1.1	−0.9	−0.9	−0.9	−1.0	−1.0	−0.8	−0.9
1200	−0.9	−0.6	−0.7	−0.6	−0.5	−0.4	−0.4	−0.4
1600	+1.3	+1.2	+1.3	+1.4	+1.4	+1.3	+1.1	+1.2
2000	+0.1	+0.1	−0.1	0.0	0.0	−0.1	−0.2	−0.2
2400	−0.4	−0.3	−0.3	−0.2	−0.5	−0.4	−0.4	−0.5

* The amounts given should be algebraically added to the observed pressure to extract the effect of the diurnal variation.

High latitudes show departures from the characteristics noted for low latitudes. The Aberdeen curves show that the total amplitude is small, especially in July, and the nocturnal range is as important as the daytime. Shaw[17] gives data showing the maximum ranges to be 0.5 mb in the arctic (February) and 0.68 in the antarctic (August); thus the maximum range occurs near the time of low sun in each case. The average amplitude in both the polar ranges is near 0.3 mb, and the diurnal range of pressure may properly be considered as negligible.

Interest in the use of pressure tendencies for extrapolating the motions of pressure systems and in single-station forecasting has accentuated the need for systematic knowledge of the diurnal variation of pressure. For the United States, this need has been largely met.[18]   In the tropics, the amplitude of normal pressure variation is so great that the 3-hr pressure tendencies have little significance, and 24-hr tendencies have more forecasting value (see pages 782). However, for purposes of aiding in the identification of developments of tropical storms, Table 1 is useful; it is suggested[1] that, when pressure tendencies are less (algebraically) than those of the table, and the corrected pressure is more than 3 mb below the normal for the time and place, it is likely that a tropical storm is forming or is near by.

WIND

Wind Velocity.   At the earth's surface, over land areas, the variation of wind during the day is a well-marked phenomenon. The time of maximum velocity is commonly in the early-afternoon hours, when thermal convection is at a maximum and provides a coupling effect between surface air and the usually more rapidly moving air aloft. The time of minimum is usually in the early-morning hours when the atmosphere is most stable. This variation obviously is dependent on the synoptic situation, but the situation is favorable often enough that the variation is clearly defined in experience and statistics.

The same coupling effects that cause the surface winds to increase during the day cause winds aloft above plains regions and surface winds at high-altitude stations to show an average minimum of velocity during midday hours. The decrease is not precisely the converse in time or amount of the increase in surface winds. This effect presumably is noticeable to the height to which convection penetrates. The shift in the nature of the diurnal variation takes place comparatively close to the earth's surface; Landsberg[12] suggests that the level is as low as 200 ft, while Brunt[5] cites a study by Hellman indicating that with light winds the shift is definable at 16 m. in winter and between 16 and 32 m. in summer.

There are considerable seasonal differences in the amplitude and frequency of well-marked diurnal variation in wind velocity in middle latitudes; moreover, there are wide differences between topographic situations. Along coasts and lake shores, the diurnal variation is not solely a function of vertical convection but is enhanced by the accompanying development of land and sea breezes. Since the latter are usually stronger than the land breezes, the resulting regime is not much different from that over land surfaces. In regions of rugged topography, the diurnal variation of wind is complicated by the nocturnal development of katabatic mountain breezes that are generally stronger than the daytime valley breezes and may result in a diurnal variation contrary to that of plains regions.

Diurnal variation of wind over the oceans has not been explored as thoroughly as has that over land. The small diurnal variation of sea-surface temperature has suggested[1] that there is no significant variation in wind. In middle latitudes, it is likely that this is substantially correct. Over the tropical oceans, however, the large amplitude of diurnal variation of pressure suggests that there may well be small but definable corollary variations in wind. Moreover, average nocturnal maximum of convection produced by the vertical distribution of nocturnal radiation implies a likelihood of maximum wind velocity at night. Brunt[5] cites a single reference to Gallé to the effect that there is a maximum velocity at night in the southeast trade winds over the Indian Ocean. If such variation is real and has the causation suggested, there might well be a shift in winds with sunrise, when convection is lessened by insolational warming of the air. Observational data are insufficient to establish the facts.

Wind Direction.   Diurnal variations in wind direction are strong where local conditions result in the domination of the winds by such effects as land and sea breezes (such as at some low-latitude coastal stations) and mountain and valley breezes. Other variations appear in statistics but are small in amplitude.   For example, associated with the increasing wind force as convection couples the surface and upper-air flows, there is a slight veer in the Northern Hemisphere (a back in the Southern Hemisphere).   Moreover, Humphreys[10] notes what may be called *heliotropic variations*, whereby the wind veers gradually during the day from east to west, following the sun; this, of course, is a statistical product and is not observed as an actual shift in the wind.   Landsberg[12] gives data from H. H. Clayton on time of maximum frequency of cloud drift from variation directions at Blue Hill Observatory, which show also a continued veer during the day and night from east at 0500 to west at 1700 to north at 2300 and return to east at 0500.   Humphreys offers the explanation that wind will tend to blow toward the region of highest heating.

TEMPERATURE

The fact of diurnal variation of temperature needs no amplification.   The amount of the daily range, however, varies widely with many factors.   Among these, cloudiness and humidity of the air, nature of the earth's surface (principally a function also of moisture as well as vegetative cover and other properties of the surface), the vertical lapse rate of temperature, wind, elevation, and latitude may be mentioned. The first is quite obvious in its influence on the penetration of insolation to the earth's surface by day and retardation of the net loss of heat by terrestial radiation at night. The second is clearly demonstrated by the very small diurnal variation of temperature over the oceans; on land, after heavy rains when soil is moist and water stands on the surface, temperature ranges are less than during dry weather, apart from the cloudiness.   In general, the average diurnal range of temperature increases with distance from water bodies.

In air masses with steep lapse rates, heating during the day is accompanied by deep convection whereby energy absorbed by air near the earth's surface (over land) is rapidly distributed through a thick layer of air.   Similarly, at night, steep lapse rates are often accompanied by considerable winds and turbulent mixing that keeps the lower layers warmer than in conditions of still, stable air in which ground inversions develop.   Stable air during the daytime damps convection and confines turbulent mixing to a small thickness of air.   As a consequence, then, air with steep lapse rates is likely to show smaller diurnal temperature ranges than stable air, provided, of course, that cloudiness does not overwhelmingly interfere.   These same conditions in general account for the fact that diurnal ranges of temperature are usually smaller at high-level stations than in near-by valleys; more prevalent windiness and mixing of the air at high levels is in contrast to air drainage at night and high heating by day in valleys.

There is no simple statement for latitude effects.   Daily ranges of temperature normally increase with latitude up to subtropical latitudes.   Maximum daily ranges have been recorded in subtropical deserts, where clear air, dry land surfaces, and weak pressure gradients combine to make conditions ideal for high maximum and minimum temperatures.   In these same latitudes, however, along foggy coasts paralleled by cool ocean currents, sea breezes effectively chop off the maximum temperatures, and fog often interferes with terrestial radiation at night, so that daily ranges are among the lowest in the world.   In middle latitudes, daily ranges vary less with latitude than with distance from the sea and season; maximum ranges usually occur in summer, minimum ranges in winter.   In very high latitudes, diurnal ranges decrease again, owing to the lessened effectiveness of the daily succession of sunlight and darkness.

HUMIDITY OF SURFACE AIR

There are many expressions for the moisture content of the air. Of these, relative humidity shows the largest and most regular diurnal variation, in response primarily to the diurnal variation in temperature. For synoptic purposes, more significance is found in dew point and mixing ratio; for practical purposes, these may be considered to vary together, and the following discussion is confined largely to mixing ratio.

Diurnal variation of mixing ratio depends upon the diurnal distribution of condensation, evaporation, diffusion of water vapor, and, to a negligible degree, air pressure. On the basis of average diurnal conditions, condensation is a night to early-morning phenomenon. Evaporation, on the other hand, increases during the day-time period of rising temperatures, although it is also strongly influenced by wind, relative humidity (both of these normally correlate with temperature), and nature of the surface. Diffusion of water vapor depends upon vertical convection and the vertical distribution of moisture. Obviously, the diurnal variation of mixing ratio must vary with different localities.

Over the oceans, small diurnal variation of temperature of the sea surface and of the lower levels of the atmosphere prohibits strong diurnal variation in mixing ratio. Along mid-latitude coasts, and in low latitudes, where daily temperature ranges are not large, the normal diurnal variation of mixing ratio consists of a late-night to early-morning minimum, and an afternoon maximum, with a small total amplitude on the order of 1 gram per kg. However, during the summer season, when transpiration and other surface conditions may yield as much or more water vapor to the air than do free-water surfaces,[13] the amplitude may reach as much as 4 grams per kg on some days. Such conditions are probably typical of continental interiors of the tropics with rain-forest vegetation.

In arid regions and continental interiors in middle latitudes in summer, the large diurnal temperature ranges initiate potentially large diurnal variations of moisture. However, air aloft is normally drier than low-level air; and, as convection develops, vertical diffusion of water vapor may offset evaporation processes and cause a secondary afternoon minimum of mixing ratio. This is shown in data presented by Shaw[17] for several stations (the data give diurnal variation of vapor pressure, which may be converted to an approximation of variation of mixing ratio from the expression $(dw/w) = (de/e) - [d(p - e)/p - e]$, the last term of which is negligible). In extreme conditions such as at Calcutta in March, the premonsoon hot dry season, the average daily variation of mixing ratio is near 2.5 grams per kg. Absolute daily minima of vapor pressure in the afternoon to early-evening hours are quite common apparently; they are reported by Shaw at Calcutta from October through April, at Paris from April through August, at Dar es Salaam, coast of Tanganyika, in June and July, at Stuart, central Australia, from November through March, and at Naha, Oshima Shoto, south of Japan, from July through September. Thus the minimum is not confined to arid regions; it is a summer phenomenon except at Dar es Salaam. In general, however, the magnitude of the variation is not large, the one given for Calcutta being the largest reported in the data.

In polar regions, the average vapor pressures and mixing ratios are so low that diurnal variation can necessarily be only minute.

CLOUDINESS

Diurnal variation in cloudiness appears where either radiative or convective processes of cloud formation are dominant. Thus, in tropical regions, at land stations, cloudiness shows a strong tendency to a maximum in the afternoon. Over tropical oceans, the periodicity is not so pronounced; there is a tendency toward a late-night or early-morning (0600) maximum of cloudiness, related to nocturnal radiation from upper levels and possibly to convergence in advance of the morning crest of the diurnal pressure variation (see the section on Oceanography, pages 1048–1049).

In middle latitudes, there is considerable seasonal variation in cloudiness. In general, summer shows afternoon maximum of cloudiness associated with convective cumulus. Winter, on the other hand, usually shows a sunrise maximum associated with radiational stratiform types; but the winter maximum is much less distinct than the summer one at most stations, owing to the smaller hourly range in cloudiness associated with high frequency of cyclonic systems.

Polar regions are reported by Landsberg[12] to have a midday to afternoon maximum of cloudiness; data presented by Shaw[17] for the arctic verify this in general; 8 months including both winter and summer periods, have slight maxima of cloudiness from 1200 to 2000, inclusive.

PRECIPITATION

Diurnal variation of rainfall has much in common with that of cloudiness, but it is less pronounced. There is also a distinction to be made between variation in frequency of rainfall and variation in quantity of precipitation. In reference to the latter, most inland stations at all latitudes, but particularly in the tropics and in the middle latitudes in summer, show an afternoon maximum in amount of rainfall per hour, associated with the commonly heavy falls from afternoon and early-evening convectional storms. On tropical islands, coasts, and oceans, the most distinctive diurnal patterns appear, with maximum of precipitation at night; this is related to upper-air convection released by radiation from cloud and moist layers. In middle latitudes, the total magnitude of the variation is usually quite small; the more cyclonically controlled the climate is, the less is the variation. Thus at New York, the amplitude of variation of intensity of precipitation in inches per hour for 3-hourly intervals of the day is only 0.009 in.[12] The same order of magnitude has been found for English stations.[16]

Time of maximum frequency of precipitation may show much or little relation to time of maximum intensity. In the tropics and in middle-latitude inland stations in summer, the correlation is high. But at middle-latitude coastal stations, and in the winter period in the middle latitudes generally, the correlation is slight. Some stations show quite marked periodicity in time of rainfall, however; *e.g.*, Valencia, Ireland,[16] has rain on 25 to 30 per cent of all days at 0800, and on only 15 per cent of days at 1200; it does not have well-defined periods of maximum intensity. Other stations in the British Isles do not show the marked diurnal variation that occurs at Valencia. At Barcelona, Spain, the time of maximum frequency of rainfall shows particularly low order of relation with time of maximum intensity, and both vary widely from month to month.[6]

Most land stations having thunderstorms show a strong maximum of incidence in the afternoon. A secondary maximum often appears between 0300 and 0400; and, at maritime locations (western and northwestern Europe), this secondary period is quite pronounced.[12] As noted in connection with cloudiness, over tropical oceans and coasts, convective activity reaches a maximum in the early-morning hours, while the afternoon hours show only a secondary maximum.

VISIBILITY

Diurnal variation of visibility is marked commonly by a maximum in the afternoon and a minimum in the morning in humid regions. The factors involved are several. Fog usually shows maximum frequency in the early morning; this is naturally reflected in the morning minimum of visibility, and that minimum varies with the seasons and consequent incidence of radiational fogs.   The afternoon maximum is associated with the average minimum of relative humidity and maximum convective turbulence that thins air-plankton by distributing it through maximum depth of air.   Consequently, the diurnal variation is usually greater in summer than in winter in middle latitudes. It is, of course, much greater over land than over the oceans.

In arid regions, the afternoon maximum of visibility is reduced or eliminated by the increasing quantity of dust that is picked up from the ground as winds increase to their afternoon maximum.   Similarly, near forest fires, accelerated rate of burning during afternoon hours produces more smoke and haze than at other times of the day.[14]

Bibliography

1. Admiralty Weather Manual, Hydrographic Department, Admiralty, London, 1941.
2. ALBRIGHT, JOHN G.: "Physical Meteorology," Prentice-Hall, New York, 1942.
3. ANGOT, ALFRED: La Nebulosité à Paris, *Ann. Bur. Cent. Met. de Paris*, 1891.
4. ANGOT, ALFRED: Etude sur la marche diurne du barometer, *Ann. Bur. Cent. Met. de Paris*, c.1883.
5. BRUNT, D.: "Physical and Dynamical Meteorology," Cambridge, London, 1939.
6. CAMPO, GABRIEL: Distribucio horària de la pluga a Barcelona, *Notes D'Estudi*, No. 65, Servei Meteorologic de Catalunya, Barcelona, 1936.
7. FARQUAHARSON, J. S.: The Diurnal Variation of Wind over Tropical Africa, *Quart. J. Roy. Meteorolog. Soc.*, vol. 65, 1939.
8. GEDDES, A. E. M.: "Meteorology," Blackie, London, 1921.
9. HANN-SURING: "Lehrbuch der Meteorologie," 4th ed., Tauchnitz, Leipzig, 1926.
10. HUMPHREYS, W. J.: "Physics of the Air," 3d ed., McGraw-Hill, New York, 1940.
11. JAMESON, H.: Summary of Investigations on the Diurnal Variation of Barometric Pressure in Tropical Seas, *Quart. J. Roy. Meteorolog. Soc.*, vol. 67, 1941.
12. LANDSBERG, H.: "Physical Climatology," The Pennsylvania State College, School of Mineral Industries.   1941.
13. MEINZER, O. E., Editor: "Hydrology" (Physics of the Earth, IX), McGraw-Hill, New York, 1942.
14. MIDDLETON, W.E.K.: "Visibility in Meteorology," 2d ed., University of Toronto Press, 1941.
15. PEKERIS, C. L.: Atmospheric Oscillations, *Proc. Roy. Soc. (London)*, Ser. A, **158** (895) February, 1937.
16. SCOTT, ROBERT H.: Diurnal Range of Rain at Seven Observatories, 1871–1890, *Gr. Brit. Met. Off.*, No. 143.
17. SHAW, SIR NAPIER: "Manual of Meteorology," vol. II, 2d ed., Cambridge, London, 1936.
18. 10-year Normals of Pressure Tendencies and Hourly Station Pressures for the United States, *U.S. Weather Bur.*, *Tech. Paper* 1, Washington, D.C., 1943.

ICING ON AIRCRAFT

By E. BOLLAY

Aircraft icing constitutes one of the chief hazards to the safe operation of aircraft. The condition favorable for icing will always require careful study and attention by the forecaster and pilot, despite the introduction of mechanical deicing devices. The present trend of going to higher flight levels, particularly on long flights, will greatly reduce the icing hazard.   Icing conditions at the terminals will then constitute the major problem.

Icing will result when liquid water is present at temperatures below freezing. It is well known that liquid water can exist at temperatures below 0°C.   The region where liquid water in the supercooled state is most frequently observed is between 0 and −18°C. *Any cloud area with temperatures between 0 and* −18°*C is therefore a potential icing zone.* Liquid water and icing have been observed with temperatures as low as −40°C.

Types of Icing. Icing as encountered by aircraft is divided into three main categories.

1. *Clear Ice.* Transparent ice formed by the freezing of large-sized droplets is known as clear ice. It presents the greatest hazard of the three types. Clear ice forms clear deposit of ice, deforming the contours of the airplane and adding much additional weight. The clear ice on the leading edge of the airfoils greatly changes the aerodynamic properties of the wing and results in decreased lift and increased drag.

2. *Rime Ice.* Opaque ice, formed in clouds by the rapid freezing of minute-sized droplets is known as rime ice. This type of ice consists of a rough opaque mass that

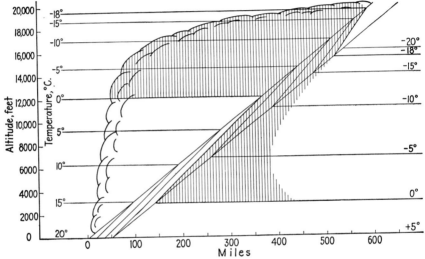

Fig. 1.—A warm-front icing situation. Shaded areas indicate zones of potential aircraft icing.

builds into the wind and will affect the operation of the aircraft sooner than clear ice. Mechanical deicing equipment removes this type of ice more effectively than it removes clear ice.

3. *Hoarfrost.* Ice crystals deposited directly on the aircraft from the water vapor in the air are known as hoarfrost. This presents far less difficulty to aircraft, particularly while in flight. During landing and take-off operations, the presence of hoarfrost presents hazards of which the pilot must be aware. With hoarfrost present, the normal relationships between stalling speed and wing attitude may change appreciably. Only power landings should be attempted with hoarfrost, rime ice, or clear ice present. No take-offs should be made with ice or hoarfrost covering the wings or control surfaces.

Wet snow also presents hazards to flying, particularly at low speeds. Under such circumstances, snow clings to the surfaces more readily and thus deforms the aerodynamic properties of the airfoils in a fashion quite similar to that of ice. Climbing to higher levels will usually remedy this situation.

Meteorological Conditions for Icing. Icing conditions will be found to exist in most cloud types with the proper temperature distribution. The intensity of icing is more pronounced if turbulence is present, and it also is believed to increase with the speed of the aircraft.   Rime ice is more common with little turbulence, and clear ice predominates when turbulence and vertical velocities are observed.   Occasionally clear ice and rime ice form simultaneously, indicating the presence of both small and large droplets of supercooled water.   Since vertical velocities are usually associated with icing conditions, they are most frequently observed with warm fronts, cold fronts, occlusions, and cumulus clouds, and over mountaineous terrain where vertical currents are orographically induced.

Warm Fronts.   Owing to the extensive cloud systems, both in the vertical and horizontal, associated with warm fronts, severe icing may be associated with them. The temperature distribution in both the cold and the warm air should be carefully studied, and regions in the 0 to −18°C zone should be delineated as icing zones when

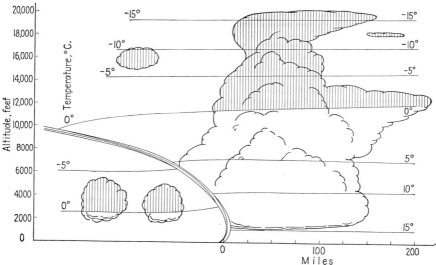

Fig. 2.—A cold-front icing situation.   Shaded areas indicate zones of potential aircraft icing.

clouds are present or suspected.   Particularly severe icing conditions result when warm rain falls through a very cold layer near the ground.   In such cases, the freezing rain will produce impossible flying conditions.

Cold Fronts.   With cold fronts, the primary cloud system is in advance of the front in the warm air.   This region is therefore the zone of potential icing, in cloud areas with temperatures between 0 and −18°C.   Strong vertical currents are frequently found in this zone, carrying large amounts of liquid water into the critical icing area. If warm rain is falling from the overlying warm air through the cold air, freezing rain may result if the cold-air temperatures are below freezing.

Cumulus Clouds.   Aside from the dangers from structural strains and control difficulties from the violent vertical currents experienced with certain cumulus-type clouds, these clouds present a great icing hazard.   Whenever possible, they should be circumnavigated.   Flying over the top may be impossible, and flying below them is often dangerous because of turbulence and strong descending and ascending currents.

Orographic Conditions.   Mountain ranges introduce severe icing hazards whenever an air mass with a suitable temperature range and high moisture content is forced to ascend. Under such circumstances, the clouds forming should be avoided. Icing conditions will be most severe where the vertical velocities are greatest. Figure 3 shows the distribution of the icing zone over mountains.

Carburetor Icing. Carburetor icing is encountered by aircraft when flying through humid air with temperatures generally above freezing. When the air passes through the expansion channels of carburetors, the reduction of the air temperature due to expansion and the high moisture content will produce a dangerous icing situation.

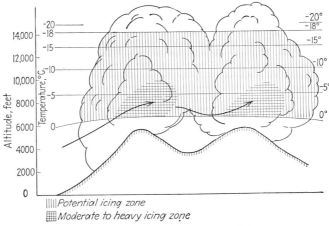

|||| *Potential icing zone*

▦ *Moderate to heavy icing zone*

Fig. 3.—Icing situation over mountains.

With present aircraft equipment, heating devices are available whereby the temperature in the carburetors can be regulated, thereby greatly reducing this icing hazard.

ATMOSPHERIC CROSS SECTIONS FOR FLIGHT OPERATIONS

By E. Bollay

Atmospheric cross sections are extremely useful in depicting weather conditions along a certain path at a given time. They are of great assistance in preparing a safe as well as efficient flight plan for extended flight operations. Atmospheric cross sections usually indicate the location of the frontal surfaces, cloud types with estimated heights of the tops and bases of clouds, precipitation areas, thunderstorm zones, regions of aircraft icing, the distribution of the upper-air winds along the flight path, and the distribution of the free-air temperature. Figure 1 is an example of such a cross section. The data plotted on atmospheric cross sections consist of the information received from radiosonde and airplane soundings and the pilot-balloon data as well as reports from other aircraft flying along the same route. The surface position of the fronts and current surface observations are plotted on the base of the cross section and are obtained from surface synoptic charts.

Method of Analysis. In analyzing cross sections, the frontal surfaces are located first. These should be located in accordance with the characteristics of frontal inversions and wind variations outlined in the section on Fronts, pages 643 to 650.

Care must be taken not to confuse all temperature inversions as frontal inversions. Inversions also develop because of such processes as cooling, turbulence, and subsidence. Often more than one of these processes is active in forming an inversion,

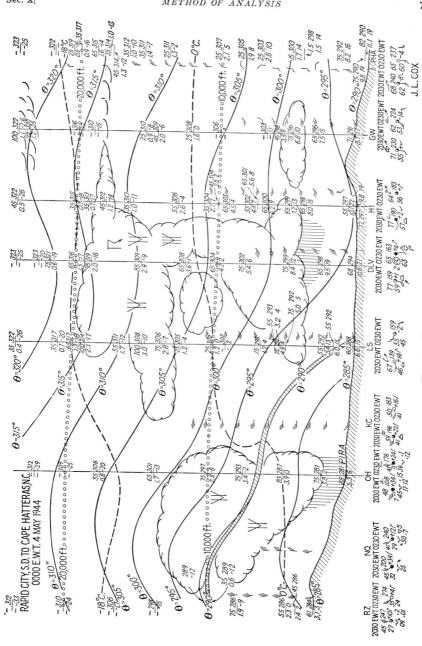

Fig. 1.—Cross section.

such as cooling from below and subsidence aloft, or cooling and turbulent mixing. Figure 2 indicates the various criteria of nonfrontal temperature inversions.

Nonfrontal temperature discontinuities are drawn on cross sections to indicate to the forecaster regions of great vertical stability. To the pilot, these layers appear as haze layers and regions of smooth flying conditions. Bumpiness is usually observed at the base of the inversion; therefore no difficulty will be had in recognizing these nonfrontal temperature discontinuities.

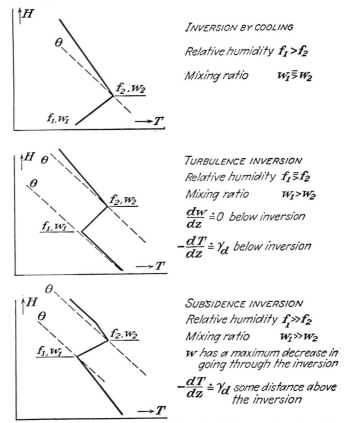

Inversion by cooling

Relative humidity $f_1 > f_2$

Mixing ratio $\quad w_1 \gtreqless w_2$

Turbulence inversion

Relative humidity $f_1 \gtreqless f_2$

Mixing ratio $\quad w_1 > w_2$

$\dfrac{dw}{dz} \doteq 0$ below inversion

$-\dfrac{dT}{dz} \doteq \gamma_d$ below inversion

Subsidence inversion

Relative humidity $f_1 \gg f_2$

Mixing ratio $\quad w_1 \gg w_2$

w has a maximum decrease in going through the inversion

$-\dfrac{dT}{dz} \doteq \gamma_d$ some distance above the inversion

Fig. 2.—Characteristic properties of nonfrontal temperature inversions.

Frequently the frontal slopes cannot be uniquely determined by frontal inversions or pilot-balloon data. In such cases, the slope of the potential-temperature surfaces should be utilized. It is generally conceded that a frontal zone consists of pronounced concentration of potential-temperature surfaces as shown in Fig. 3. Since these surfaces are nearly parallel, the frontal surface can be drawn parallel to the general slope of the potential-temperature surfaces.

The cloud types, precipitation areas, and thunderstorm zones are indicated on the basis of the surface observations, and on the data from the soundings and airplane observations. It should be remembered that, when indicating cloud areas, relative humidities may be less than 100 per cent, as explained in Sec. III, page 253.

Regions of aircraft icing are indicated in cloud areas between the temperatures of 0 and −18°C. Whenever vertical currents are suspected in these potential icing areas, moderate to heavy icing should be anticipated. For additional information on icing, see pages 753 to 756.

The distribution of the free-air temperature and the upper-air winds are included in the plotted data on cross sections and therefore become apparent at a glance. On long flights, these data are extremely useful since the free-air temperature and upper-air winds enter into the performance computations of aircraft power plants.

In performance computations of aircraft power plants, the free-air temperature is of considerable importance. The actual flight level may be determined by the free-air temperature in order to obtain the maximum range with the minimum fuel consumption. Winds aloft, unless strong, may play a secondary role in such cases. In actual practice, the determination of the flight level is of course a compromise

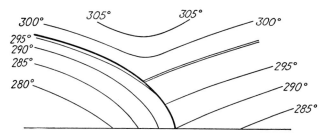

FIG. 3.—Potential temperature distribution through a frontal zone.

between the gain achieved by a suitable temperature distribution and the gain of speed due to winds, provided that other meteorological conditions are the same at the levels under consideration.

Atmospheric cross sections show the vertical structure of the atmosphere along a selected route as interpreted by the meteorologist from the surface synoptic chart and the upper-air data. Owing to the paucity of upper-air data, the cross sections are by necessity estimates of existing conditions. Since forecasts fall in the same category, the meteorologist should illustrate as realistically as possible the weather conditions along the route.

AIR TRAJECTORIES

BY R. A. BAUMGARTNER

The trajectory of a parcel of air is its track over the earth's surface. Assuming gradient-flow conditions, the isobars represent trajectories only when the pressure system is stationary and its structure is unchanging. Since pressure systems move rather than remain stationary, the isobars on synoptic maps will rarely indicate the actual path of a parcel of air for any appreciable length of time. It is desirable to foresee the actual paths for various air parcels in considering the movements within air masses for forecasting temperatures at fixed stations, for fog forecasting,[1] and for a general understanding of the existing distribution of meteorological elements.

A simple relationship has been determined[2] between the curvature of isobars and the curvature of trajectories as follows:

$$\frac{1}{r_t} = \frac{1}{r_i}\left(1 - \frac{c}{v}\cos\Psi\right)$$

where r_t = radius of curvature of the trajectory
    r_i = radius of curvature of the isobar
    c = speed of the pressure system
    Ψ = angle between the direction of motion of the system and direction of the gradient wind (v)

The r_i of a parcel is determined by first finding the center of curvature of the isobar in its vicinity. The center of curvature is found at the intersection of the perpendicular bisector of two chords of the isobar in the vicinity of the parcel.

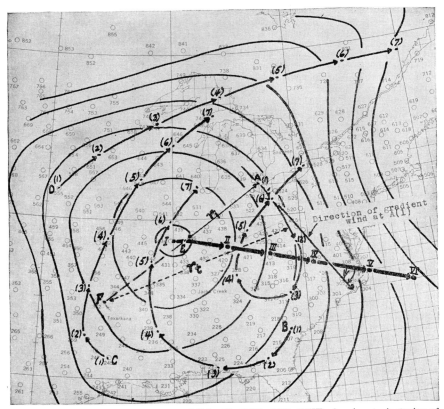

Fig. 1.—Synoptic weather map of Feb. 12, 1944, 2030 EWT, showing trajectories of air parcels A, B, C, and D in an anticyclone over Eastern United States. Arabic numerals locate successive six-hourly positions of the air parcels. Roman numerals refer to six-hourly positions of the center of the anticyclone.

The speed and direction of motion of the pressure system may be estimated from previous history or computed by means of the displacement formula on page 669. The speed of the gradient wind may be determined on the assumption that the wind blows along the isobars from a gradient-wind nomogram such as that found on page 54. The angle Ψ is considered as the angle, smaller than 180 deg, between the positive direction of motion of the system and the positive direction of the wind vector.

The trajectories of a few parcels of air in an anticyclone (see Fig. 1) were computed with the above equation. It was determined that the anticyclone moved east-southeast at approximately 28 knots. The following indicates the procedure used in obtaining the trajectory of parcel A:

At the initial point on trajectory A, which is located at position A (1) in Fig. 1, the following values were determined:

$$r_i = EA = 5.8° \text{ lat.}$$
$$c = 28 \text{ knots} = 2.8° \text{ lat. per 6 hr}$$
$$v = 43 \text{ knots} = 4.3° \text{ lat. per 6 hr}$$
$$\Psi = 40 \text{ deg}$$

Substituting in the formula

$$\frac{1}{r_t} = \frac{1}{5.8}\left(1 - \frac{2.8}{4.3} \cos 40 \text{ deg}\right)$$
$$r_t = 11.4° \text{ lat.}$$

Along EA, the radius of curvature of the isobar, mark off 11.4° latitude (AF). Using this radius, $AF = r_t$, construct an arc of a circle from A (1) equal to the computed 6-hr displacement (4.3° latitude). This will determine the position of point (2), which, to a first approximation, is its position on the earth's surface 6 hr from the current map. Point 2 is now located on the next 6-hourly map where the above procedure is followed in locating point 3. After repeating the above for successive 6-hr intervals, the complete, 36-hr trajectory of point A is determined as indicated in Fig. 1.

For forecast purposes, it may be assumed the structure of the system will not change appreciably in the next 24 to 36 hr. The trajectory of a parcel may then be computed from the current map for a given period of time. For example, point 2 of parcel A is located from the current map as described above. Since the pressure system will be centered at point II 6 hr later (from previous history or calculation), the parcel A is thus located in a different position relative to the system's center from the current map. To determine point 3, the above procedure is repeated for point 2 with the pressure system centered at its new position II. This is repeated for as many 6-hourly intervals as it may be assumed that the movement and structure of the system remain unchanged.

The following table gives an illustration of the air temperatures in degrees Fahrenheit as reported by the stations over which the various parcels were located (Fig. 1) at each successive 6 hr.

Time	Position	A	B	C	D
2030	1	14	33	43	15
0230	2	12	30	37	15
0830	3	15	32	31	11
1430	4	27	32	31	19
2030	5	26	32	26	14
0230	6	13	27	24	15
0830	7	15	24	25	15

It is of interest to note that temperatures of parcel A at 24-hr intervals are practically the same with the exception of points 4 and 5. Increased temperatures at these points are due to increased daytime heating. Temperatures at the end of 36 hr, under similar sky conditions, are considerably warmer for parcels B and C in comparison with parcel A, which is as one would expect from the trajectories of the air parcels.

Bibliography

1. GEORGE, J. J.: Fog: Its Causes and Forecasting with Special Reference to Eastern and Southern United States, *Bull. Am. Meteorolog. Soc.*, **21** (4), April, 1940.
2. PETTERSSEN, S.: "Weather Analysis and Forecasting," p. 225, McGraw-Hill, New York, 1940.

TROPICAL SYNOPTIC METEOROLOGY

By Civilian Staff, Institute of Tropical Meteorology,

Rio Piedras, Puerto Rico*

INTRODUCTION

Approximately half the troposphere lies above that portion of the earth's surface included between the Tropics of Cancer and Capricorn. On the whole, synoptic meteorologists have neglected this region, partly because reliable weather data are scanty, partly because an ill-founded tradition is current among high-latitude meteorologists, that tropical weather is determined in a simple manner and, except for occasional hurricanes in some parts, is unaffected by secondary circulations. In the last few years, however, practical necessity has demanded that synoptic analysis near the equator receive closer attention; military and civil meteorologists whose duty has taken them into equatorial and subequatorial regions have soon learned how ill-founded the standard accounts of those regions may be. The picture these accounts give would indicate that convection in the tropics is intense, chaotic, and unorganized, especially on and near the equator, that the wind systems are simple, and that the weather is dominated primarily by the diurnal and seasonal variation of insolation. This concept is derived from climatology, and the climatological means are based largely on records from land stations, many of them quite unrepresentative. Such a picture is of little or no use to the forecaster and may occasionally lead him seriously astray.

In their attempt to grapple with the difficult problem of the aperiodic variations of low-latitude weather, forecasters have naturally availed themselves of ideas developed in the temperate zone. It is unquestionable that organized systems of bad weather, bearing a striking resemblance to the fronts of the temperate zone, are frequently encountered by aircraft flying over the tropical oceans. It is also unquestionable that such organized systems may be observed to pass over well-exposed land stations, their passage being accompanied by rain, wind shifts, cloud sequences, and sometimes temperature oscillations so marked as to tempt the most conservative to recognize them as fronts. Moreover, it is often possible to draw the systems as fronts on one or a few synoptic maps. But these "fronts" will soon behave in a manner quite unpredictable by the usual rules for high-latitude forecasting: they will move at improbable speeds, and the anticipation of intensity changes will be mainly guesswork. Under these conditions, it is not possible to maintain continuity, and the forecasting problem becomes correspondingly difficult.

PRINCIPLES OF TROPICAL METEOROLOGY

Lack of Well-defined Air Masses. In temperate meteorology, great emphasis is placed upon the concept of the air mass, a huge body of air possessing properties that are more or less homogeneous in the horizontal. It is recognized, of course, that

* The Institute of Tropical Meteorology is located on the campus of the University of Puerto Rico and is operated jointly by the University of Chicago and the University of Puerto Rico. The initial organization of the teaching and research program was carried out by Mr. C. E. Palmer of the New Zealand Meteorological Service, who served as director of the Institute during the first year of its existence.

this horizontal homogeneity is only approximate—that convection, radiation, and other processes will always produce gradients of property within the air mass; but

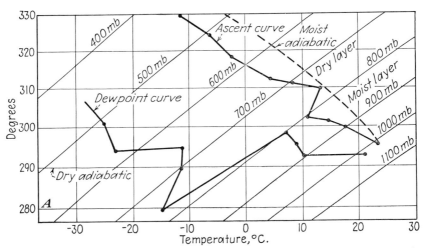

FIG. 1.—$T\varphi$ diagram for San Juan, Puerto Rico, at 2300 local time on Dec. 18, 1943 showing the "trade inversion" and the moist and dry layers.

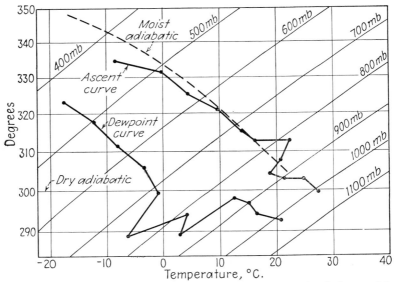

FIG. 2.—$T\varphi$ diagram for Henderson Field, Guadalcanal, at 1000 local time on Mar. 5, showing the separation of the so-called "monsoon air" into moist and dry layers.

such gradients are insignificant compared with those found at the *boundaries* of the air masses. In the tropics, however, the changes in property that occur across zones that could possibly be regarded as air-mass boundaries are insignificant compared with those that result from the action of convection and radiation upon the "air masses."

The Moist and Dry Layers. The horizontal gradients of air-mass properties, therefore, are of subordinate utility in tropical analysis. On the other hand, the vertical distribution of moisture is of greatest importance. A series of upper-air soundings obtained from a representative tropical station, especially in an area where trade winds prevail for a large part of the year, will show that for long periods the humid air supposedly characteristic of the tropics extends to only moderate heights (Figs. 1 and 2). At a comparatively low level, the moisture content falls off very rapidly with height; and the region of rapid change may be marked by an inversion of temperature or, failing this, a change of lapse rate. Above lies extremely dry air, the superior air. This inversion of temperature has long been known as the "trade-wind inversion," and much significance has been attached to it. However, it now seems advisable to recognize that a sharp zone of demarcation between the humid lower layer and the dry upper air almost always exists—even in the absence of any temperature inversion—and to name the layers the moist and dry layers, respectively. The distinction between these layers is evident not only in the trade-wind zones but also in regions that are traditionally supposed to be the seat of widespread chaotic convection, *e.g.*, the equatorial Pacific Ocean. It cannot be too strongly emphasized that the atmosphere in the tropics is not necessarily highly unstable, not even in the equatorial zone. In some regions, it retains a stable stratification for very long periods, showing well-marked moist and dry layers on the equator itself. Figures 1 and 2 illustrate this separation of the tropical atmosphere into moist and dry layers. Both show a marked inversion in the zone of separation. The sounding at Henderson Field (Fig. 2) is particularly important, since it was made in the "monsoon air" on the northern side of the equatorial front.

The forecaster thus encounters his first analytic problem: to determine the depth of the moist layer and its variations in space and time. The solution is not so simple as it appears, since, in the tropical current, areas of ascending air several miles wide alternate with clear areas where the air subsides. For this reason, radiosonde or airplane observations will report the depth of the moist layer differently according to whether the flight is made in an area of ascent or one of subsidence. Moreover, many soundings are made on islands or in mountainous terrain, where the local effects overshadow those of synoptic importance. The data from the sound-

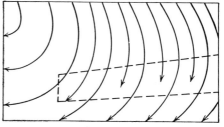

Fig. 3.—Streamlines disappearing in a region of convergence.

ings may then completely mislead the analyst. Accurate measurements or expert estimation of the height of the tops of convective clouds taken from aircraft flying over the open sea provide the best means of determining the depth of the moist layer. These observations have shown how widespread is the separation of the tropical atmosphere into well-marked moist and dry layers.

Representation of the Velocity Field. In addition to the depth of the moist layer, the tropical forecaster needs to determine the structure of the velocity field at all levels in the troposphere, and probably also in the lower stratosphere. Through the wartime extension of aircraft operations and the establishment of close pilot-balloon networks, wind observations are now available in some tropical regions in great quantity. Using them, some tropical meteorologists have attempted to develop streamline analysis, with special emphasis on locating areas of appearing and disappearing streamlines. Those areas where streamlines disappear are regions in which there is horizontal convergence in the velocity field, and those where new streamlines appear are regions of horizontal divergence (Figs. 3 and 4).

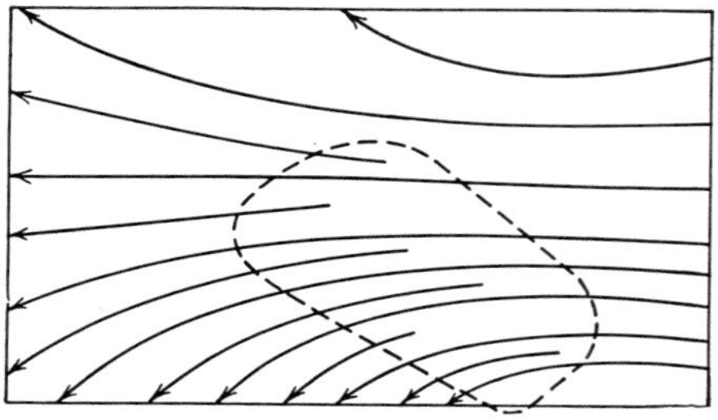

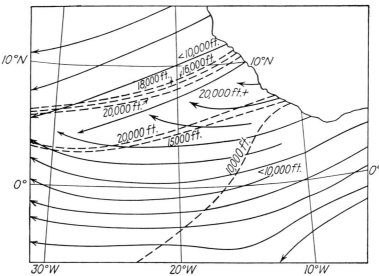

Fig. 4.—Streamlines spaced in inverse proportion to the wind speed. New streamlines a p p e a r i n an area of horizontal divergence.

In the low troposphere, convergence in a layer several thousand feet thick nearly always results in ascending motion because the air cannot move through the earth's surface, and divergence results in descending motion. In that part of the atmosphere to which on the average we find the moist layer restricted, then, the disappearance of streamlines implies a stretching of the air columns and deepening of the moist layer; and the appearance of streamlines is accompanied by subsidence and a contraction of the moist layer. Streamline maps that also contain lines of depth of the moist layer bring out this relationship clearly. In the regions in which streamlines disappear, the moist layer is deep; in regions in which they appear, there is evidence of subsidence. Such maps (Fig. 5) used in conjunction with the surface charts, constitute a powerful instrument in tropical analysis.

Fig. 5.—Streamline map of the tropical North Atlantic for the 9,000-ft level surface showing variations in the depth of the moist layer.

Basic Currents of the Tropics. If areas of horizontal convergence and divergence could be forecast expansion and contraction of the moist layer could be anticipated. Then the type, the amount, and the thickness of the convective cloud could be predicted. Neither the surface map nor the streamline analysis automatically makes this possible. It is necessary first to understand the processes governing the development and travel of the areas of convergence and divergence and the manner in which these areas appear in the pressure and velocity fields. The knowledge so far acquired may best be summarized in the form of synoptic models.

Before discussing the models as such, it is useful to consider the broad currents in which the different systems may lie. Three general wind regimes occur in the tropics:

1. Deep easterlies extending to great heights, at least to 6 km. This current is found usually between 15°N and 15°S and extends farther poleward in summer than in winter.
2. Equatorial westerlies overlain by easterlies at high levels. They also lie within the belt mentioned in (1) and are generally separated from the polar westerlies by the deep easterlies.
3. Shallow easterlies overlain by the polar westerlies of temperate latitudes. These occur on the poleward sides of the tropical zone and extend farthest equatorward in winter.

It is not correct to think that west winds are found above the easterlies along all meridians or in all tropical latitudes. Observations show quite clearly that the upper winds of the tropics are controlled by the anticyclonic cells centered at high levels at about 15 to 25°N and S. On the equatorward side of these cells are deep easterlies. Thus the clear-cut schematic diagram sometimes given, showing the surface easterlies everywhere overlain by westerlies, is an oversimplification. The height of the base of the westerlies increases very rapidly near the central parts of the subtropical highs. In a few degrees of latitude, it may rise from 1.5 to 8 km or more.

In addition, the equator is not necessarily the warmest region in the tropics, even on the average. Rather it may be said that within 20°N and S, at least over the open oceans, temperature gradients in the lower troposphere are weak or absent. A general poleward outflow from the equatorial zone aloft is rarely observed.

SYNOPTIC MODELS OF THE TROPICS

The models described below belong to two general categories: (1) those of the tropics proper, generally steered by the deep easterlies, though westerlies may characterize the lower layers, and (2) those representing modifications of temperate systems, steered by the polar westerlies, though easterlies may occur near the surface.

Waves in the Easterlies. Although the disturbances known as *waves in the easterlies* occur in many parts of the tropics, they have been primarily studied in the tropical North Atlantic and the Caribbean (Fig. 6). Dunn[1] has pointed out that during summer a series of isallobaric highs and lows moves from east to west over the region and that the intensity of convection within the current varies with the succession of these isallobaric centers. More insight into the mechanism of these systems has recently been gained by Riehl.[2] Figures 7 and 8 show a model wave in the easterlies in the Northern Hemisphere.

The deformation in the isobars is more pronounced near 10,000 ft than at the surface. Indeed, sometimes when a disturbance passes aloft, the surface-pressure change is less than 1 mb. Also, the wave appears best in the velocity field. Thus a streamline analysis at various levels outlines the wave better than the isobaric analysis. The onset of the wave is attended by a turning of the trade wind, both at the surface and aloft, toward the northeast in the Northern Hemisphere. When the axis of the trough has passed, the wind shifts fairly rapidly to the southeast.

Streamlines drawn to a sufficiently dense network of winds show that the forward part of the wave (west of the wave axis) is a region of divergence and descending motion; to the rear, however, there is usually convergence with ascending motion. The wave crest generally slopes eastward with elevation.

It is noteworthy that, with a northeast wind and falling pressure, the weather over the sea and even over the mountainous islands of the Caribbean is fine, the moist layer varying between 5,000 and 8,000 ft in depth; and the bad weather is concentrated not so much at the axis of the wave but some little distance behind. A deep moist

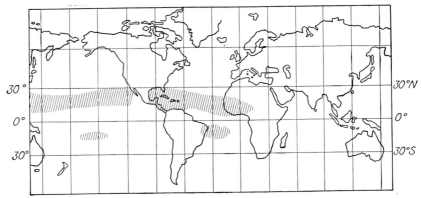

Fig. 6.—General locations in which waves in the easterlies have been observed or suspected.

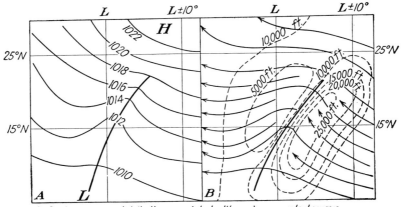

A = Surface pressure distribution associated with a strong easterly wave
B = Streamlines at 3000 ft. for the same wave, and isolines showing the depth of the moist layer

Fig. 7.—The easterly wave in the Northern Hemisphere.

layer and an extensive area of unsettled weather accompany the isallobaric high behind the wave axis.

There is evidence that many waves in the easterlies are generated near the west coast of North Africa, more especially when the low-pressure trough usually associated with the equatorial regions has moved far into the Northern Hemisphere; that is to say, in the northern summer. It seems that the waves develop inland rather than along the coast line and that they pass on to the ocean and then travel more or less uniformly toward the west. Easterly waves, of the type described, occur also in the South Pacific as far west as the northern Cook Islands.   There they can be traced in summer first passing across the Marquesas Islands and later over Penrhyn Island and Nassau.   Observations at Palmyra and the Caroline Islands in the North Pacific indicate that they pass these islands also.   At Natal, on the east coast of South America, weather and wind sequences suggest that the South Atlantic is also subject to easterly waves.   This distribution suggests that the waves develop mainly in those regions where the oceanic anticyclones are large, quasi-stationary, and very stable. In regions where the high-pressure belt is largely made up of a succession of migratory anticyclones, easterly waves are comparatively rare, so far as is known.   As an example, the western South Pacific west of the Cook Islands has few.   Occasionally, however, even in this region, a migratory high-pressure cell will slow up and become stationary, and then small easterly waves may develop along its northern border.

Tropical Modifications of Temperate Systems.   The waves in the easterlies discussed in the preceding section are comparatively simple in that there is little or no

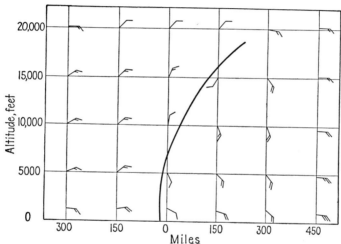

Fig. 8.—Vertical cross section along the 15° N parallel of latitude through the wave shown in Fig. 7.

temperature gradient in the lower troposphere when they occur (in summer), and hence advective changes are negligible.

Strong temperature gradients on the other hand are usually associated with mid-latitude waves in the westerlies.   Specifically, the air behind the trough is colder than that ahead of it.   This distribution means that the trough must slope upward toward the rear.

With the trough is, more often than not, a cold front that has, of course, the two essential characteristics of a front:

1. A gradient of temperature.
2. A cyclonic wind shear.

Quasi-periodic outbreaks of polar air into the tropics occur in the Northern Hemisphere almost exclusively in winter, but in the Southern Hemisphere the poleward parts of the tropics are affected by temperate systems throughout the year.

During such an outbreak, several developments may occur.

1. The upper trough may advance to the east of the cold front.

2. The density discontinuity across the front may disappear while the wind shear persists.

3. A new trough may develop by the splitting of a high on the equatorward side of a developing middle latitude cyclone.

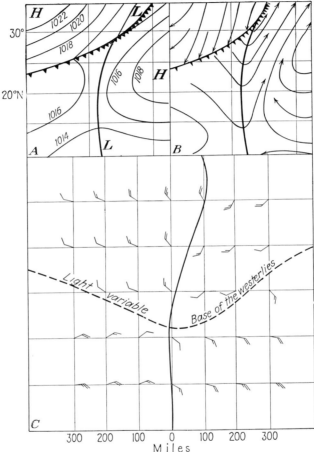

A = *The surface pressure pattern. The surface discontinuity is still strong but will frontolyse*
B = *The 5,000 ft. streamlines*
C = *Vertical E-W cross section through a polar trough*

Fig. 9.—The polar trough.

The structures that result from these processes are called (1) polar troughs, (2) shear lines, and (3) induced troughs.

The Polar Trough. In most cases when a cold front reaches lower latitudes, the accompanying wave in the upper westerlies advances far to the east of the surface front (Fig. 9). In the lower troposphere, the front soon becomes weak, because it lies in a frontolytic wind field. A temperature discontinuity at low levels may persist for a short time, but in most cases it becomes very diffuse. However, the wave in the upper westerlies continues to advance above the surface easterly winds. Since the wave is accompanied by a pronounced pressure trough aloft, the surface easterly isobars also show a cyclonic wave pattern. The surface-pressure trough, it must be emphasized, is a reflection of the trough in the higher westerlies. Since the latter is continuous with the surface trough of higher latitudes, the whole structure is termed the *polar trough.* Because of the presence of deep westerlies above the shallow layer of easterlies, the direction of movement of the polar trough is opposite to that of the waves in the deep easterlies and therefore cannot be understood without reference to these upper winds.

The convergence and divergence in the lower levels have the same distribution as about a wave in the deep easterlies. East of the trough line, there is usually horizontal convergence with bad weather. To its west, one finds divergence with a shal-

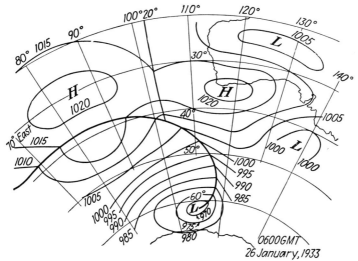

Fig. 10.—Polar front in the South Indian Ocean.

low moist layer and fine weather. The bad weather east of the trough line consists of convective clouds, either cumulus congestus or cumulonimbus. But sheets of altostratus associated with the cumulonimbi may link up and form a continuous deck, from which steady rain falls. Figure 9 represents a model polar trough at the stage when the surface front is still fairly strong though undergoing frontolysis. The streamlines are drawn for the 5,000-ft level. Figure 9C is a vertical cross section through the polar trough. The axis of the trough is drawn to slope eastward at high levels, a frequent but not necessary event. The contour of the base of the westerlies deserves close attention because of its importance in forecasting the upper winds and the steering current.

In the large regions between the subtropical highs, the speed of the polar troughs is relatively fast, and their intensity may be very persistent. Immediately to the west of the subtropical cells, on the other hand, the troughs are quasi-stationary. Here they present an equatorward extension of the mean positions of the oceanic polar fronts.

A "doubling" of the oceanic anticyclones is associated with the formation of an additional polar trough in mid-ocean. Under such circumstances, there are two complete polar-front systems in temperate latitudes and two pronounced quasi-stationary polar troughs in the tropics.

In general, the moving polar troughs form near the west coasts of the continents. They tend to dissipate as they approach the semipermanent troughs and to merge with the latter. For example, consider the occlusions that develop in the South Indian Ocean[3] and move toward Australia (Fig. 10). As they approach the continent, polar troughs form along the west coast well in advance of the occlusions and extend far equatorward.

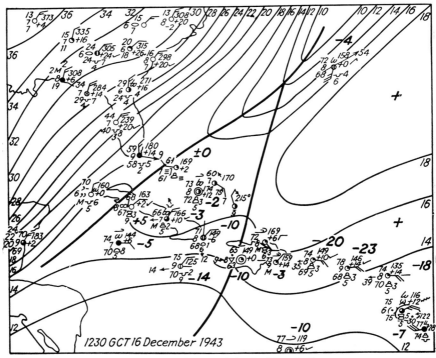

Fig. 11.—A polar trough in the West Indies.

In the subtropics and lower middle latitudes, a migratory anticyclone follows the occlusion, somewhat equatorward from the low center, and an inverted V-shaped depression appears on the surface map between this anticyclone and its predecessor. The trough in the upper westerlies lies far to the east of the weak surface front in the tropics, and it keeps this relative position during the whole of the passage toward the east, at least until it reaches the region of Tahiti (150°W) where it encounters the South Pacific cell.

Figures 11 to 13 show a polar trough over the West Indies. Significant features of the surface map (Fig. 11) are (1) the separation of a small anticyclonic circulation over the Bahamas, (2) the formation of the trough over Haiti, and (3) the persistence of the principal high farther to the east. The anticyclonic circulation over western Cuba and the trough over Haiti appear clearly in the upper winds (Fig 12). The similarity of air-mass structure on both sides of the trough is illustrated by the soundings in Fig. 13.   Notice that the air above San Juan is actually colder than that to the west.

The Shear Line.   The fast-moving troughs just described coincide with an eastward progression of migratory warm highs that frequently reinforce the semipermanent highs.

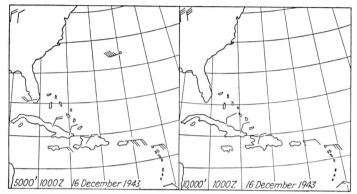

FIG. 12.—Winds aloft near a polar trough.

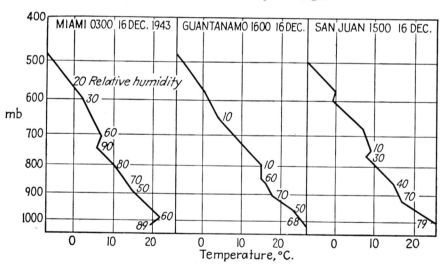

FIG. 13.—Upper air soundings near a polar trough.

The latter, however, may be replaced in another manner.   If a very strong polar outbreak occurs behind a deep depression on the poleward side of the quasi-permanent high, a marked cold front with a predominantly east-west orientation will advance into the subtropics.   The oceanic high then collapses, and it is replaced, not by a warm high proceeding from the west, but by the cold high associated with the polar outbreak, moving equatorward.   This high rapidly assumes warm characteristics, building up soon into the higher atmosphere.   Considerable subsidence in the cold air above a shallow surface layer coincides with the transformation of the high (Fig.

14). Therefore, the density discontinuity at the front quickly disappears above the surface layer.   But the zone of cyclonic wind shear not only persists but may actually strengthen, especially if subsidence on its poleward side is extreme.   This persistent velocity discontinuity is known as a *shear line.*

In the lower levels, the cyclonic shear exists between strong east winds on the poleward side of the line and weak east winds on the equatorward side.   At high levels,

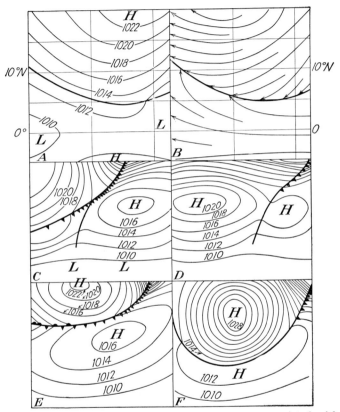

Fig. 14.—The "shear line."   *A,* surface pressure pattern associated with a "shear line" moving into very low latitudes.   *B,* streamlines at the 5,000-ft level corresponding with the pressure pattern in *A.*   *C* and *D,* replacement of an oceanic anticyclone by a migrating high moving from the west behind a polar trough.   *E* and *F,* replacement of the oceanic anticyclone by an intense high associated with a cold outbreak.   A "shear line" moves rapidly into low latitudes.

it is evidenced by a trough between two highs: (1) the old oceanic high still persisting aloft but decreasing in intensity, and (2) the new high replacing the former and building into the upper air.   While this new anticyclone continues to intensify and the old one to weaken, the shear line will move into lower latitudes, sometimes reaching the equatorial zone itself.   Its arrival produces a sudden freshening of the trade wind which has long been known in the tropics as the *surge of the trades* or *reestablishment of the trades.*

Severe weather, including several lines of cumulonimbus and thunderstorms, accompanies the shear line. At some distance to the rear of the shear line, the weather usually becomes very fine with a shallow moist layer and very dry air aloft; near the top of this moist layer, radiation stratus or stratocumulus may form.

The Induced Trough. The formation of an induced trough can be expected when a front, while moving into fairly low latitudes, preserves its extratropical characteristics and becomes quasi-stationary there. The front will be oriented nearly east-west, giving rise to frontal waves. The increase in the westerlies above the frontal surface in connection with such a wave development may result in the formation of a new closed anticyclonic eddy on its equatorward side. The dynamic trough between this eddy and the principal high farthest east is called the induced trough (Fig. 15).

The distribution of weather at the induced trough is similar to that about the polar trough, provided that the induced trough moves to the east. A deep moist layer with bad weather is located on the eastward side, and to the west there is good weather.

If the frontal depression responsible for the formation of the trough moves toward the east, the induced trough will move with it but will die quickly.

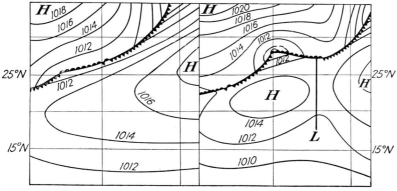

Fig. 15.—Two surface maps showing the development of the induced trough.

If, however, the frontal depression remains stationary, the induced trough will also remain stationary, and it may intensify with bad weather on both sides.

The Equatorial Trough and the Equatorial Front. Several concepts of the synoptic systems in the equatorial zone have been advanced. Brooks and Braby[4] described a synoptic structure which they called the *equatorial front.* Basing their discussion chiefly on the climatological records of certain islands in the equatorial western Pacific, they came to the conclusion that in that region the trade winds from the two hemispheres converge along a narrow zone and that the Southern Hemisphere air of the southeast trades ascends over the Northern Hemisphere air, forming a front rather similar to the warm front of extratropical latitudes. East of the 180th meridian, they stated, the two trade systems do not meet at a "sharp angle"; consequently there is less convergence and cloud.

Later the Norwegian meteorologists adopted the idea of the equatorial front (Bergeron[5]) and extended it. According to their concept, the trade winds of the hemispheres meet along a zone of convergence that completely encircles the earth. This zone they call the *intertropical front,* similar to the polar front, with one of the air masses ascending over the other.

At the same time, a considerable group of meteorologists have continued to use the old concept of the doldrum zone, in which the trade winds of the two hemispheres are thought to merge not along a narrow band, but in a fairly broad belt of light and variable winds and unorganized, heavy convective showers. This concept is based on climatology, and it should be emphasized that climatological records are inadequate as a basis for synoptic forecasting.

The Equatorial Trough.   The high-pressure belts of the two hemispheres are separated by a trough of low pressure, which in general coincides with the equator in

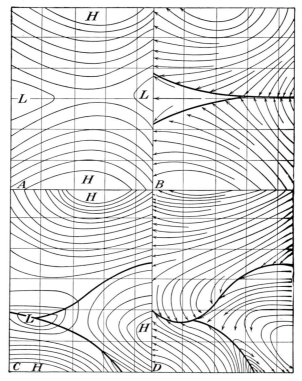

Fig. 16.—The equatorial trough and the equatorial front.   *A*, the equatorial trough near 0°.   *B*, streamlines for *A*.   *C*, the equatorial front south of 0° reinforced by a shear line.   *D*, streamlines for *C*.

oceanic regions but which fluctuates with the seasons, especially in the neighborhood of the great land masses. This is the "equatorial trough" (Fig. 16*A*, *B*). When the trough lies along the equator or within 3° latitude N or S of it, the winds on both sides are predominantly easterly in direction. When it departs widely from the equator, however, the trough will frequently contain a closed depression with predominantly westerly winds on the equatorward side and easterlies on the poleward side only. As an approximate rule, we may say that the equatorial trough must depart more than 5° away from the equator for a closed system to exist within it.

Equatorial westerlies, however, may be found also in certain seasons when the trough is located very close to the equator, especially above a shallow layer of surface easterlies. Thus they appear along the equatorial coast of Brazil during the wet season. The difference may best be stated as follows:

1. When the trough departs more than 5° latitude away from the equator, the appearance of equatorial westerlies can be counted upon in nearly all instances.
2. When the trough lies near the equator, the westerlies will be found only in specific synoptic situations favorable for their formation.

Observations show definitely that along only a portion of its length is the equatorial trough a zone of convergence. Rather it is a zone in which regions of convergence alternate with regions of divergence. Where there is an east-west gradient of pressure along the equator, divergence occurs in the equatorial trough to the west of a line joining the center of the subtropical anticyclones; to the east of this line there is convergence.

Such a distribution of convergence and divergence may also be deduced from the cell theory of the general circulation: the east winds of subequatorial regions cannot everywhere have a component toward the equator. Rather, there will be alternating regions round the globe in the first of which the east wind has a slight component toward the equator and in the next of which it has a slight component away from the equator. In view of the fact that deep easterlies prevail predominantly in the subequatorial oceanic regions, this velocity distribution will exist through the larger part of the troposphere.

These alternate regions of convergence and divergence are abundantly confirmed by the climatological means of precipitation, cloudiness, and thunderstorm activity along the equator and also by the daily synoptic charts. The regions of horizontal divergence are dry zones, and fine weather prevails for long periods. A strong inversion usually exists at about 8,000 ft with very dry air above. As a result, the clouds are mainly cumulus humilis or trade cumulus. There is very little precipitation. In the equatorial dry zone of the Pacific, many islands, *e.g.*, Malden, Christmas, and Canton Islands are practically devoid of vegetation. In the western parts of the equatorial Atlantic, somewhat similar, though less extreme, conditions prevail over a considerable part of the year.

The Equatorial Front. When the equatorial trough is more than 5° latitude away from the equator, convergence between the winds on the polar side of the trough and those on the equatorward side takes place along a narrow elongated zone, called the equatorial front (Fig. 16C, D). The streamline analysis shows that there is considerable ascent of air at the front but that both Northern and Southern Hemisphere air masses ascend; there is little, if any, evidence of overrunning. The result is the formation of a band of cumulonimbi with bases reaching almost to the surface and the tops frequently to above 40,000 ft. An extensive sheet of altostratus is usually associated with the cumulonimbus. The height and thickness of this altostratus may differ in the two currents. Fighter aircraft ascending through the altostratus sometimes report as many as six separate sheets of this cloud. At high levels, a broad deck of cirrostratus spreads out on either side of the front. Heavy squalls, water spouts, and thunderstorms may accompany an intense equatorial front, and frequently the convective clouds seem to be oriented in two or three parallel bands separated by comparatively clear spaces approximately 50 miles wide. *

* This phenomenon can be linked to the structure of tropical clouds over the open sea. Far from land, convective clouds in the equatorial zone tend to orient themselves in long lines (cumulus *en échelon*) lying parallel to the direction of the wind. With fine-weather cumulus and trade cumulus, the lines are close together, but as the depth of the cloud increases so also does the distance between the cloud lines. The same tendency of the equatorial air to form longitudinal vortices parallel to the wind probably accounts for the double and triple structure that is sometimes associated with a strong equatorial front.

The equatorial front does not always have the structure outlined above. Its intensity varies considerably from day to day, and a front of the intensity just described may in 2 days be reduced to a narrow line of cumulus congestus.

The equatorial front furthermore is subject to rapid dissipation after a marked poleward advance. When this occurs, a new front will develop, usually in the mean seasonal position. This sequence of events is associated with the development of depressions on the front, some of which attain hurricane intensity.

Tropical Depressions. * The depression characteristic of the tropics is that known in different parts of the world as the typhoon, hurricane, or tropical cyclone. The extension of the observational network into low latitudes, however, has shown quite clearly that many more low-pressure centers are found in the tropics than was previously suspected and that only a small proportion of them attain hurricane intensity. Indeed, the hurricane or typhoon is a rather rare and extreme form of a fairly common development.

Tropical depressions can form in any of the systems discussed in the previous sections and not only in the wet season, as has previously been supposed, but in the winter months as well. While the systems previously discussed are difficult to detect by the analysis of the surface synoptic chart alone, tropical lows are picked up fairly easily with an adequate network.

The following classification is in use by the U.S. Weather Bureau and has been adopted by the Institute of Tropical Meteorology, Rio Piedras, Puerto Rico:

1. Tropical disturbance: circulation slight or absent on the surface, possibly more marked aloft, but no closed surface isobars.
2. Tropical depression: one or more closed surface isobars; wind force equal or less than Beaufort 6.
3. Tropical storm: closed surface isobars, wind force more than Beaufort 6.
4. Hurricane: wind force Beaufort 12 (75 mph) or more.

The life history of tropical lows also may be divided into four stages:

1. The formative stage, which begins when the circulation develops along the equatorial front, in an easterly wave or in a polar trough, and ends when the circulation reaches hurricane intensity.
2. The stage of immaturity during which the circulation reaches its maximum intensity, is most symmetrical, and covers a relatively small area.
3. The stage of maturity, when the isobars associated with the storm are gradually spreading out over a wide area and there is no further deepending. The intensity may be slowly decreasing, although the area covered by the cyclonic wind system may be larger than at any other period during the life of the storm.
4. The stage of decay, when the circulation either assumes extratropical characteristics or dissipates.

Hurricanes sometimes go through all four stages entirely within the tropics, but winter storms may recurve and move so quickly that they reach the polar front quite early, perhaps in stage 1.

The exact causes of the formation of tropical depressions are unknown. In spite of this, however, it is possible to divide them into different types according to the synoptic systems in which they originate. We may distinguish the following:

1. Easterly wave depressions: any wave in the easterlies in midsummer and early fall is likely to give rise to a closed circulation which in many cases will attain hurricane intensity. It is characteristic of this type of depression that it is very small and violent in the stage of immaturity and that it may develop with great suddenness. It usually develops in that part of the wave known as the *crest*, that is to say, the region

* Most of the material on hurricanes is from an unpublished summary by G. E. Dunn.

where the surface isobars show the greatest amplitude. Many of the most destructive storms in the Carribbean and Gulf of Mexico belong to this category. Not all easterly wave depressions attain hurricane intensity, however. In the fall and spring, there is a great tendency for such depressions to be feeble, to move fairly rapidly away from the parent wave, and to recurve into a polar trough over the United States or in the western Atlantic. In this case, they will induce on the polar front wave depressions that may become very strong.

2. Equatorial front depressions: the equatorial front gives rise to series of depressions of various intensities. When this occurs, the front will usually be well marked and will lie along the greater or east-west axis of the depressions, which in the earliest stage will show little movement but increasing intensity. From the beginning, the region of low pressure is extensive, sometimes covering 30° longitude and 10° latitude within the central isobar. Inside this "envelope" there are light variable winds, with easterly directions predominating on the poleward side of the equatorial front and westerly directions on the equatorward side. Invariably, however, in this early stage the winds have a marked component toward the front.

Eventually one or two smaller depressions will develop within the flat low-pressure area, but the rate of development is usually much slower than that characteristic of easterly wave depressions. Soon after such a small intensifying depression is established, it will begin to move toward the west. At the same time, the equatorial front will move farther away from the equator, at least on the eastern side of the low. Once this happens, deepening in the central depression is usually rapid, and a true hurricane may develop. In all cases when the equatorial front shows a pronounced poleward excursion to the east of the major low-pressure center, a smaller depression of at least storm intensity may be suspected within the outer flat "envelope."

3. In winter, strong polar troughs in the tropics show a pronounced tendency to produce small depressions, which will appear to develop some considerable distance on the equatorward side of the frontal system. The upper-air analysis shows that they form in the polar trough. They seldom attain any great intensity in the tropics but move rapidly poleward, meet the front associated with the temperate part of the polar trough, and induce a frontal wave in middle latitudes. Some of the violent storms off the east coast of the United States may be traced to such small depressions developing near Cuba or Hispañola. The Kona storms of Hawaii also appear to belong to this category.

4. Depressions may develop in the induced trough but only if the polar front wave that has induced the trough remains stationary. Such depressions, like those forming in the polar trough, are likely to produce wave cyclones on the polar front in higher latitudes.

In their formative stage, equatorial-front depressions may show a pronounced asymmetry in the distribution of weather and cloud, whereas most easterly wave storms are symmetrical about the center. In all cases, there is a marked tendency for the weather on the left of the track in the Northern Hemisphere to be particularly fine. In some cases, strong anticyclogenesis on the equatorward side of the center takes place, and a small high appears even on the surface map. This is especially the case in the South Pacific. Soundings in the area of anticyclogenesis show that the moist layer there is very shallow and is topped by a strong inversion. It has long been known that, in many hurricanes, both before and during recurve, there is a tendency for the bad weather to be concentrated in the poleward portions and ahead of the moving depression. It now appears that this applies also to small and weak depressions, such as those developing in the polar trough.

It should be mentioned that the equatorial front can be detected for long periods during the development of the equatorial-front hurricane. In the central region of strong winds, the bad weather is more or less uniformly distributed, but outside this area a definite discontinuity in wind and weather can often be found long after the storm has reached full intensity.

It is noteworthy that in all types of depressions the cyclonic circulation can frequently be observed at 8,000 to 10,000 ft before it reaches the surface. This fact is often useful in forecasting. When a well-developed storm or hurricane moves through a sparse network of observing stations, the circulation aloft is often more extensive than at the surface, and the 10,000-ft wind chart will therefore show a circulation that is too small to pick up accurately on the surface chart.

TROPICAL ANALYSIS

As in higher latitudes, an analyst in the tropics must consider

1. The region he is analyzing. The variations of weather and climate from place to place are considerable.
2. The seasonal variation of the weather. In some regions, the primary circulation changes little with the season whereas in others it undergoes large variations.
3. Migratory synoptic systems.
4. Local effects. The changes in the meteorological elements due to the diurnal variation of insolation and to the effects of orography are in general far greater than those due to all except the most intense secondary circulations. Consequently it is necessary to separate the diurnal and orographic variations from those attributable to the synoptic systems.

An important difference between high and low latitudes is the more complicated relationship between wind and pressure in the tropics. The gradient wind is approximated in only the higher tropical latitudes; and even there it is often a very poor approximation. Hence new methods of analysis of the velocity field are essential.

Methods of Analysis

The charts and cross sections on which tropical analysis currently is based are

1. For the general analysis:
   a. The surface charts.
   b. Upper-level charts—mostly of wind data.
   c. Time sections showing the variation of the weather elements with time at one station.
2. For the special analysis based on aircraft reports:
   a. Aircraft spot reports—a method of representing aircraft reports on maps.
   b. Flight cross sections—a section along the track of aircraft showing in detail the vertical distribution of clouds and weather.
   c. Weather-distribution charts—a method of collecting a large number of visual observations into a graphic, usable form, especially one that can be used by nonmeteorological personnel.
3. For microanalysis:
   a. Application of all items listed under 1 and 2 to a small area.
   b. Station-circle reports—a graphic method of representing the weather visible from observations at one station.

The Surface Map. *Surface Temperature.* The surface temperatures at land stations in the tropics are entirely unrepresentative for synoptic analysis. Those taken on very small, flat islands far removed from large land masses are somewhat more useful.

1. The diurnal far exceeds the seasonal range of temperature. At Penrhyn Island, an atoll of the South Pacific (9°S, 158°E), the mean diurnal range is 8°F and the mean seasonal range only 2°F. At stations in the interior of continents, the diurnal range is much greater. An an example, a portion of the thermogram at Medellín, Colombia, is given in Fig. 17. This trace is typical also of stations on large mountainous islands, especially on the lee side, where a diurnal range of 20°F is not uncommon. The foehn effect tends to give these stations a continental character.

2. The diurnal variation of temperature is most marked on clear days. Clouds, even in small amounts, will lower the surface temperature in daytime. Precipitation, however, including small orographic showers, has the greatest effect, since the evaporation from falling rain quickly cools the air to the wet-bulb temperature. If the showers occur during what is ordinarily the hottest part of the day, the temperature will fall so suddenly and markedly as to resemble the typical temperature change with mid-latitude cold fronts. The coincidence of temperature change and rainfall often has led to faulty frontal analyses.

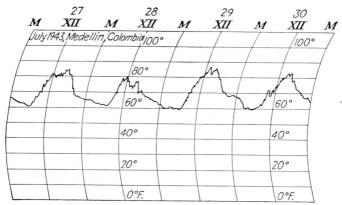

FIG. 17.—Thermograph for Medellín, Colombia.

So closely does the temperature at land stations follow the trend of cloudiness and rainfall that under no circumstances should it be used except for microanalysis.

The Dew Point. The dew point is affected in the tropics by two processes:

1. Evaporation from falling rain or from surface water.
2. Mixing processes that bring dry air from aloft to the surface.

These effects far exceed those which are due to difference in origin of the air masses; consequently, the dew point is not used in tropical as in temperate analysis. It serves to detect subsidence inversions at comparatively low levels and hence to give evidence of descending motion. The dew point should not be used when the station recording it is reporting rain or has reported rain within the last hour. In addition, it becomes unrepresentative on the lee sides of mountains, where often it shows a considerable diurnal variation. On low tropical islands, however, this diurnal variation is very small, and marked changes can in most cases be traced to vertical mixing.

Pressure. Pressure readings are representative except for dynamic effects in mountainous areas. The gradient of pressure due to dynamic causes across a chain of mountainous islands may be as much as 3 mb. This can lead to a fictitious picture of pressure troughs between individual islands. For the drawing of isobars, it should always be remembered that the pressure-wind relationship cannot be relied upon as an aid. It is best to carry out the isobaric analysis with the use of the pressures alone.

Pressure Changes. The diurnal variation of pressure in the tropics is very great and becomes greater as one approaches the equator in oceanic regions or proceeds from the shores inland in continental regions. The 3-hr pressure changes which are still being recorded at most tropical stations are without value in the analysis. On the other hand, the 24-hr tendencies are very valuable, so much so that, where possible, they should be computed hour by hour.

Surface Wind. The surface wind at tropical land stations is notoriously unrepresentative. At coastal stations, the land and sea breeze is more marked near the equator than anywhere else. Orographic distortions of the wind flow are very great in mountainous country. An analyst should be constantly on his guard against using wind shifts due to the land and sea breeze to identify the passage of synoptic systems. Observations over the open sea far from large land masses, however, are representative.

Hydrometeors. Clouds and precipitation at land stations are largely determined by the orography. Even islands of small elevation tend to become convection points during the daylight hours and to be free of cloud at night. The intensity of convection on the other hand depends on the synoptic situation. Thus it is possible to use the height of the orographic cloud and the time of commencement of orographic precipitation as an index of stability.

Upper-level Charts. Since the radiosonde network in the tropics has not yet attained sufficient density to permit accurate drawing of upper isobars, upper-level charts serve mainly for analysis of the wind field, as given by pilot balloons and aircraft reports. Since the wind changes and the zones of convergence and divergence are usually more pronounced aloft than at the surface, such analyses outline the synoptic situation quite clearly, especially if carried out for several levels, *e.g.*, 5,000, 10,000, and 20,000 ft.

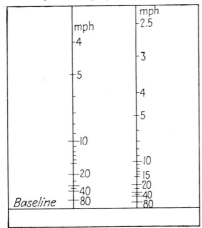

Fig. 18.—Reciprocal scales of the type used for streamline analysis.

Streamline analysis should be used if the reporting network is dense enough. Its value is the same as that of the representation of the flow by isobaric analysis in middle and high latitudes, in that it permits an estimate of vorticity and divergence and locates discontinuities in the wind field.

The objects of streamline analysis are to represent the direction and speed of the flow and to show the vertical motion. The streamline is parallel everywhere to the wind direction. The distance between adjacent streamlines is drawn inversely proportional to the wind speeds,* *e.g.*, a speed of 5 mph is represented by a spacing twice as great as that for 10 mph, a speed of 10 mph by a spacing twice as great as that for 20 mph, etc. (Fig. 18). Usually the streamlines are not everywhere continuous but begin and end in certain regions. A beginning streamline shows divergence in the velocity field and an ending streamline convergence (Fig. 19).

Because of the direct relationship of the streamline pattern to weather, the analysis should not be carried out without reference to the other data, especially the weather-

* Some meteorologists use other methods of representing wind speed.

distribution chart. Such a procedure is less necessary when the network of winds is very dense and accurate. If the streamlines are drawn for a level near the surface, downward motion from that level is not likely, and the ending of a streamline will represent an upward motion of the air at that point. If it is necessary to start a new streamline at any point of the same low-level chart, that point will lie within an area of subsidence.

For higher levels, it must be remembered that at any level, the vertical motion depends on the total divergence between that level and the earth's surface. If there

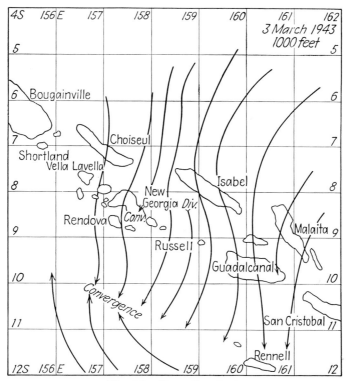

Fig. 19.—Streamline chart.

is a net divergence between the surface and 10,000 ft, the wind velocity at 10,000 ft will have a downward component, even though there may be convergence at the 10,000-ft level itself. At a higher level, then, it is possible to have either upward or downward motion, or even no vertical motion at all, regardless of whether or not convergence, divergence, or neither is occurring at that particular level.

Therefore, streamline analysis for one upper-level chart, *e.g.*, 10,000 ft, does not necessarily indicate the sign of the vertical motion. A simultaneous analysis of several charts between the surface and 10,000 ft gives the information necessary. For instance, if the lines disappear at 5,000 and 10,000 ft in the same area, upward motion can nearly always be inferred. However, if streamlines disappear at 10,000 ft and appear at 5,000 ft, subsidence between 5,000 and 10,000 ft is probable. For an accurate determination of vertical motion at a particular level, all heights between the surface and that level must be considered.

Spot Reports from Aircraft. In order to acquaint the forecaster as quickly as possible with observations from aircraft, a number of codes for the transmission of in-flight reports by radio have been devised.

A brief summary of the elements that are most desirable in such observations is presented here. However, it should be emphasized that spot reports are not considered to be so satisfactory as a plain-language report by the pilot or observer. They should never be used as a substitute for briefing the pilot after the flight.

Standard elements in aircraft position reports are (1) the time of report, usually Greenwich civil time, (2) the position of the aircraft in degrees and tenths of latitude and longitude, and (3) the flight altitude in thousands of feet. The positions of successive observations should be close enough together to have overlapping visibility fields. This usually requires that observations be made each quarter hour, with additional observations where special phenomena or rapid changes occur. Winds at flight altitude are readily obtained by measurement of drift with the drift meter. The double-drift method is recommended, although other methods using the drift meter are reasonably accurate. Estimated winds should never be reported, because they are often wrong and may be misleading.

All different types and layers of clouds should be reported separately, each with amount in tenths, and heights of tops and bases. Since several different types of low cloud are often observed at the same time, it is desirable to allow for three or four. It makes little difference if several types of high cloud are present except in special cases. Most of the weather in the tropics is shower-type precipitation, and for forecasting purposes it is useful to expand the classification of showers according to frequency and intensity.

The following such classification is suggested:

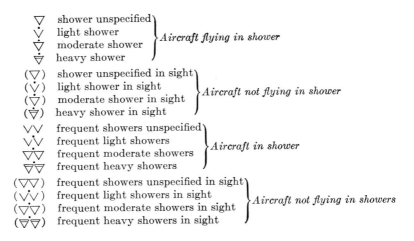

Other weather may be described using the International classification. Visibility is seldom seriously restricted in the tropics, except in clouds or precipitation. Haze, however, is useful as an index to stability conditions, and visibility is therefore included in aircraft reports. Below is a model of an aircraft report:

$$N_H\ C_H\ \frac{T}{B}$$

$$30 \longleftarrow TT\ \ N_M C_M\ \frac{T}{B}\ \ GG{:}gg$$

$$VV_{\underline{ww}}\ \bigodot N_h \quad H_A$$

$$N_L\ C_L\ \frac{T}{B}\ \ W$$

$$N_L\ C_L\ \frac{T}{B}$$

$$N_L\ C_L\ \frac{T}{B}$$

1. Surface wind is given by a barbed arrow, 10 knots for a full barb and 5 knots for a half barb.
2. Wind at flight level is given by an arrow with the velocity in knots written in.
3. N_h is the total amount of low clouds.
4. N_L is the amount of low clouds of one type. C_L is the type of low clouds.
5. T/B is the height of the tops and the height of the bases in thousands of feet.
6. $N_M C_M$ is the amount and type of middle clouds.
7. $N_H C_H$ is the amount and type of high clouds.
8. TT is the temperature in degrees centigrade.
9. VV is the visibility in miles.
10. ww is the symbol for weather.
11. $GG{:}gg$ is Greenwich civil time.
12. H_A is the height of the aircraft.
13. W is the weather encountered since the last observation.

It is recommended that aircraft reports be plotted on a base map of large scale.

Flight Cross Section. Drawn by the pilot or observer while in flight, the flight cross section shows, in pictorial form, all the weather encountered by the airplane. It is intended to be as true as possible a representation of cloud patterns, distribution of haze, fog, and precipitation, together with notes on wind direction, amount of sky cover and of different cloud decks, and other weather phenomena (Fig. 20).

The vertical scale is in thousands of feet, and the horizontal scale may be of either time or distance. Use of a time scale has the advantage of leaving the observer free of navigation problems until the end of the flight when the position of the airplane at various times can be worked out from air speed, wind directions, and headings. Furthermore, the time scale permits the observer to change the horizontal scale at will. For speeds of near 200 mph, a scale of 15 min of flight to 1 in. of section gives a very satisfactory spread of clouds. Where weather conditions change rapidly, as in a thunderstorm area, a scale of 10 or even 5 min of flight per inch of section is desirable. The flight path is drawn to aid in entering the cloud heights and to show the chosen flight elevation in the various weather situations. Pertinent navigational data should be put on the section or in the notes. If a time scale is used, it suffices to check the exact times when the heading is changed and to mark the new heading along the base of the section together with either the air speed or the ground speed.

Clouds are sketched as they appear during the flight, except for the distortion due to the exaggeration of the vertical scale. They are outlined in blue pencil; and, in cases of large cumulonimbi or thick stratus, the body of the cloud is hatched in or shaded in blue. Vertical green hatching denotes showers or rain between the cloud and the surface, and a dashed line gives the height of the probable inversions. All indications of shear in the wind field should be carefully noted. At the base of the section are entered appropriate symbols for weather phenomena, such as thunderstorms, haze, virga, squall lines, and halos, and also wind observations both at the flight level and at the surface.

The observer should write remarks in blank spaces of the section, covering such subjects as air temperature, weather observed on either side of the flight path, and positions of the airplane relative to landmarks and, more particularly, to topographic features and land and water areas.

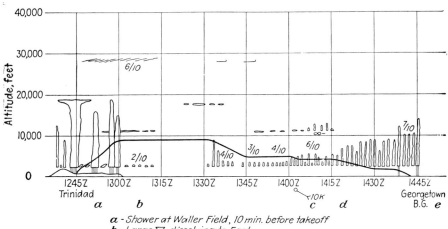

a - *Shower at Waller Field, 10 min. before takeoff*
b - *Large ☒ dissolving to East*
c - *Headed inland*
d - *In and out of clouds*
e - *Started to rain after landing*

Fig. 20.—Flight cross section.

Weather-distribution Chart. To provide a simple and complete method of representing cloud and weather conditions, many forecasters have found the weather-distribution chart useful. This is a graphic chart upon which weather data are entered as appropriate symbols (Fig. 21). It is used as a supplement to the standard synoptic map. It gives the location of the cloud systems with respect to coast lines, islands, mountains, as well as synoptic pressure patterns. Aircraft and ship reports, as well as representative land reports, are used in the preparation of the chart (Fig. 22). Nevertheless, personal interviews with pilots who patrol the areas furnish the best information concerning the weather on and surrounding the flight path. These data cannot be expressed completely on the aircraft weather-report form.

Time Sections. The use of time sections, a standard tool of meteorologists for many years, is a special aid in the tropics, where stations are widely spaced. Several of the most applicable types of time sections are listed below.

1. Surface pressure: Graphs of the 24-hr pressure change are particularly helpful since they represent the synoptic changes.

2. Upper pressures: The 24-hr change is used in the same manner as, and in conjunction with, the surface-pressure variations.

The upper-pressure changes can also be represented in the form of altimeter corrections, as developed by John C. Bellamy. Centers of high positive values of altim-

SYMBOLS
CLOUD AND WEATHER DISTRIBUTION

LOW CLOUD IN BLUE HEIGHTS IN FEET (TOPS/BASES)		WEATHER PRECIPITATION IN GREEN FOG IN YELLOW	
Scattered small clouds (Cumulus, Fracto-cumulus)	3500 / 2000	Rain (or drizzle)	Solid green
Large heap clouds (Towering cumulus)	8000 / 2500	Scattered or intermittent rain (or drizzle), if distinguished from continuous rain	
Tall shower clouds with anvil tops (Cumulonimbus)	18000 / 2000	Moving showers (of rain)	▽ ▽ ▽
Broken layer cloud (Stratocumulus, stratus)	2500 / 1000	Snow	* * * *
Continuous layer cloud (Stratocumulus, stratus, nimbostratus)	4000 / 1000	Moving showers (of snow or sleet)	
Sharp edge of cloud sheet or clear area marked by boundary line	3000 / 1500	Drizzle, if distinguished from rain	, , , ,
Area where cloud cover effectively unbroken marked by diagonal hatching		Fog (yellow) Or, area of yellow shading	

UPPER CLOUD (OPTIONAL) IN BLACK		SPECIAL PHENOMENA IN RED	
Middle cloud in patches, globules, etc. (Altocumulus)	10000 / 8000	Thunderstorm	ℝ
Middle cloud in layer, continuous, (Altostratus)	15000 / 10000	Hail	△
		Squall	∧
High cloud, scattered or broken (Cirrus, Cirrucumulus)	30000	Gale at surface	⟶
		Freezing level	FREEZING 8000
		Icing level (Base)	ICING 8000
High cloud in continuous sheet (Cirrostratus)	20000	Visibility in miles or yards (Purple)	V=15 / V=500 YDS.
Combinations of symbols may be used; relative spacing indicates amount of cloud cover, precise amount may be entered in tenths if desired. Relative size of symbols indicates relative size of individual clouds (not amount) or of showers, thunderstorms, etc.	8/10 / 5/10 / 400 / V=½-1 / 300 / 1000	Upper wind in knots or m.p.h. at standard level, or level on shaft	15⟶ 5000' 25
		Cold front (Blue line) Warm front (Red line) Occluded front (Purple line)	
		Equatorial front or other wind discontinuity (Orange line)	

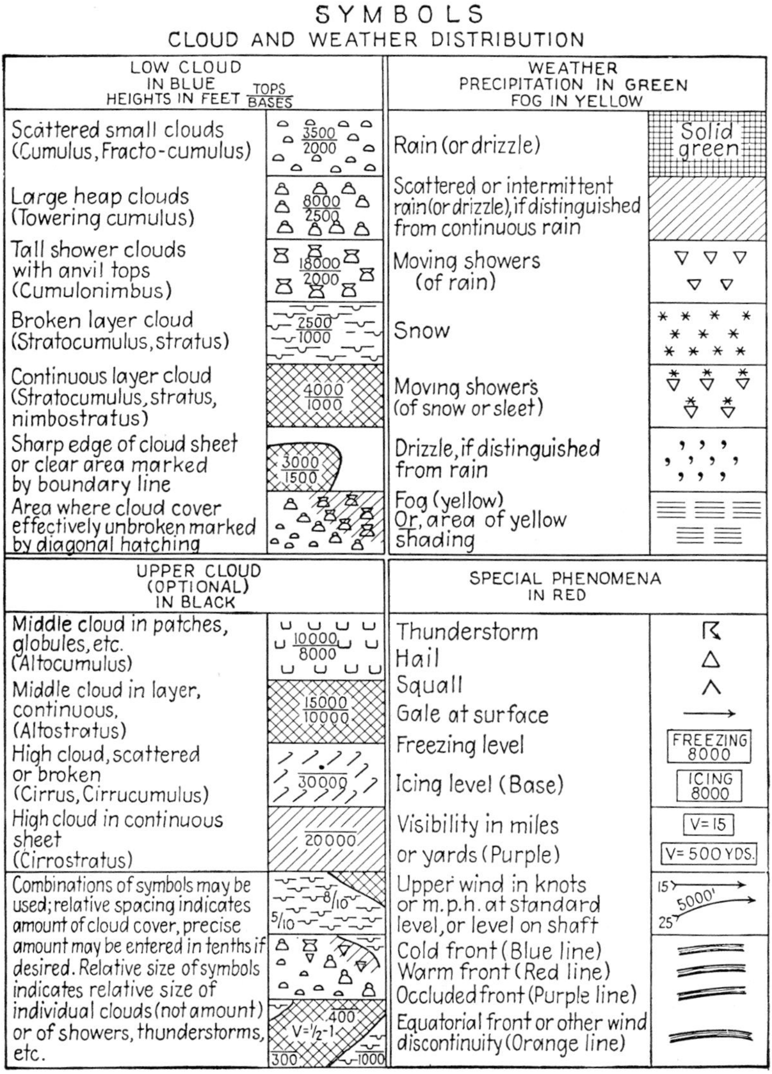

FIG. 21.—Symbols: cloud and weather distribution.

eter correction are also centers of high pressure relative to their surroundings in the horizontal and vertical, and vice versa. Thus the altitude of the greatest intensity of the centers may be determined immediately. Greatest variation of the altimeter corrections is found above 6 km, but, for operational use, a running plot of the daily soundings should be kept for all heights (Fig. 23).

3. Height of the moist layer: Preferably from aircraft reports. A cross section of θ_e may be kept in connection with that of the height of the moist layer, or it is sometimes kept separate (Fig. 24).

4. Upper winds: Together with the 24-hr pressure changes at the surface, this cross section is currently the most important one for the forecaster. It commonly consists of arrows plotted for each 2,000-ft interval. It is of especial importance to

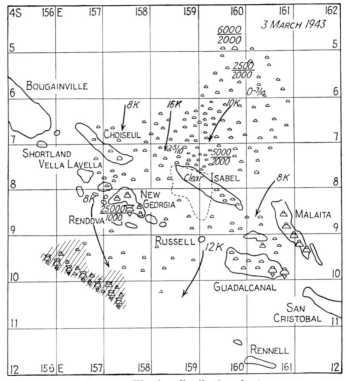

Fig. 22.—Weather distribution chart.

determine the base of the polar westerlies during the winter season, but that level is often impossible to locate because of the small number of pilot balloons reaching it.

Like all analyses in the tropics, that of the time section should not confuse the local and the synoptic effects. The weather at a station may be affected by its leeward or windward position and not show a change representative for its position in a synoptic system.

The various types of time section just described are usually combined in a single form. One illustration of such a form is contained in Fig. 25.

Microanalysis. For military air operations in the tropics, forecasts of a general nature indicating roughly the presence of cloud and rain in an area are insufficient. A forecast for operational use requires information of specific and detailed nature, *e.g.*, whether a particular target or landing strip will be in a shower or under cloud at a

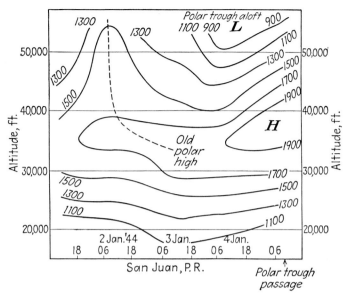

Fig. 23.—Time section of altimeter correction showing southward movement of old polar high followed by passage of a polar trough. Notice that both are preceded by distinctive patterns only in the higher troposphere and lower stratosphere.

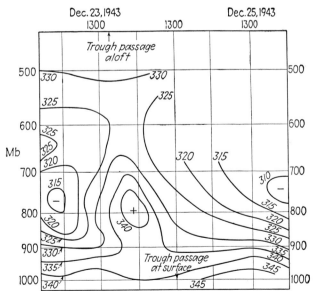

Fig. 24.—Time section of θ_E showing strong mixing preceding passage of a polar trough and relative position of the trough aloft and at the surface.

given time, or exactly when and where fighters will leave a cloud cover. The preparation of such a forecast in a region where local and diurnal effects are as marked as synoptic changes requires something more than the usual large-scale analysis. The technique developed to facilitate such detailed forecasts is called microanalysis.

Fundamental to the microanalytic approach is specialization to fit the particular problem of the area. In the Panama region, for instance, this problem is primarily

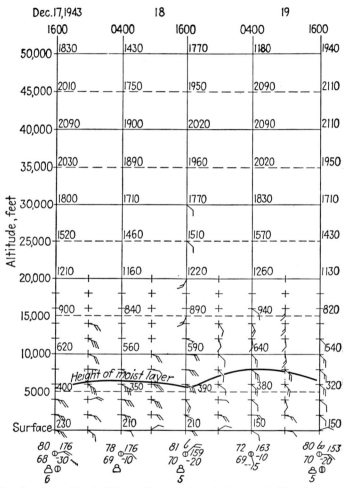

Fig. 25.—San Juan, Puerto Rico. Cross section of winds and altimeter corrections.

to provide the best possible defense of the Canal Zone. To do this it is necessary that interceptor aircraft be able to strike at maximum range as rapidly as possible. When they return from such a mission, fuel supplies will usually be insufficient to enable the aircraft to wait out a shower before landing or to reach a weather-free field unless warned in advance. The weather officer is able to supply the necessary information readily because a dense network of weather-reporting stations makes possible the charting of all major cloud groups in the area.   He follows the hourly development and movement of these clouds and showers and can make satisfactory short-term forecasts.   In this region, the primary concern is not whether there will be showers, but rather when they will form and where they will move.

The weather-distribution chart and the station-circle plot form the basic tools for microanalysis.

Weather-distribution Chart (Fig. 22).   The scale of the weather-distribution chart is large enough so that individual large clouds may be located on it.   A scale of 1° latitude per inch will satisfy most purposes.   Each individual cumulonimbus cloud

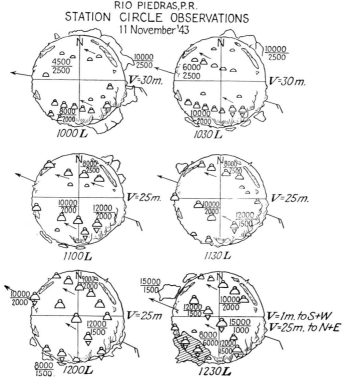

Fig. 26.—Station-circle observations.

is drawn on the map in its exact position at the observation time.   Towering cumuli are entered individually if prominent enough, but where many cumuliform clouds are present over a considerable area, scattered symbols suffice without specifying individual clouds.   In the latter case, the density of symbols is proportional to the number of tenths of clouds of the particular type, and the size of the symbols is proportional to the size of the cloud.   Hatched areas or a variation of the L_5 symbol of the International cloud code represent stratiform clouds.   The heights of the tops and bases of all clouds are entered in feet besides the cloud symbols.   Green shading indicates areas of continuous precipitation and areas of rain from a stationary cumulus or cumulonimbus.   Green shower symbols designate moving showers, with a distinction between light and heavy showers.

Wind directions may be entered for particular levels using red arrows, but the chart is generally used in conjunction with a separate streamline map.

In operations, a prognostic weather-distribution chart has far more meaning to a pilot than the same forecast in writing. A glance locates the areas of good and of bad weather and gives a definite idea of the intensity and extent of convective activity and of cloud heights throughout the area. Another great advantage of the weather-distribution chart is that it makes the weather forecast popular with the using personnel. This popularity makes it well worth the added labor of its construction.

Station-circle Reports. At stations where the weather depends largely on the distribution of sea and land masses and on topography, detailed visual observations of the location of weather with respect to these features aid the forecaster considerably. The station circle (Fig. 26) is designed to record such observations against prominent hills, mountain ridges, coast lines, and other reference points. The outline of clouds in the distance is drawn around the periphery of the circle. Those clouds lying within the circle of vision are located as accurately as possible within the circumference, using the same symbols as described for the weather-distribution chart. Blue symbols represent clouds; green symbols and shading delineate shower and rain areas. Wind arrows with barbs show the direction and force of the surface wind.

Beside the station circle, the observer should make notes covering the amount of each cloud type, the heights of tops and bases, the visibility, and other phenomena. Since the direction of clouds is of major importance for forecasting, the observer must determine it with great accuracy.

Observations of this type must be frequent to be of value. Use of successive reports makes it possible to trace across the sky the path of each prominent cloud or shower from the time it appears in the field of vision until it passes out of sight or dissipates. For developments near the station, observations should be made every 15 min, certainly at no longer intervals than $\frac{1}{2}$ hr.

Stability Analysis. Stability of the atmosphere may be determined instrumentally or visually. A general, but very useful, visual classification is

1. *Extreme stability throughout atmosphere.* Poor visibility, haze layers, stratiform clouds only or none at all.
2. *Extreme stability in one layer.* One stable layer such as at the top of a layer of cold-air advection, flat-topped cumulus humilis, haze layer, perhaps altostratus or altocumulus, sometimes stratocumulus cumulogenitus.
3. *Slight stability.* Haze layers, trade cumulus over the ocean, limited tops to the cumulus and often stratocumulus.
4. *Slight instability.* Good visibility, no haze layer, cumulus congestus and chimney cumulus, cumulonimbi over the mountains, occasional light showers.
5. *Extreme instability.* Organized cumulonimbus systems with associated cirrus, altostratus, and stratocumulus decks, turbulence, frequent moderate to heavy showers, often with steady light rain between showers.

Clouds. Clouds in the tropics form primarily as a result of vertical motion. In fact, the cumulonimbus is the parent cloud of low latitudes, *i.e.*, the stratiform and cirriform types usually observed are outgrowths from cumulonimbus. Some of these outgrowths may resemble the warm-front sequence of cirrus, cirrostratus, and altostratus observed in middle latitudes; but they do not indicate overrunning. One important type of cloud—stratus due to radiation and turbulence—does not form because of vertical motion.

Clouds Due to Vertical Motion. The factors that control the development of cumuliform clouds in the tropics are (1) the depth of the moist layer, (2) orography, (3) horizontal convergence in the wind field, and (4) to a lesser extent, stability conditions.

Depending on the intensity and distribution of these factors, various types of cumuli-form clouds develop.

1. *Cumulus humilis* is far more frequent in the tropics than was formerly supposed. It often has a rough, tattered appearance caused by turbulence (then it is frequently called *fractocumulus*). It indicates a shallow moist layer and negligible convergence (Fig. 27a).

2. *Trade cumulus* in general has a "blocklike" appearance since it ends abruptly at the trade inversion. A group of these clouds shows considerable symmetry because their bases are all at the same level and they all have nearly the same vertical extent (Fig. 27b).

3. *Chimney clouds* have greater vertical than horizontal extent. Frequently they take the form of long "necks" protruding above the trade inversion from the tops of the lower cumuli. Chimney clouds indicate a moist layer of moderate height. The observer should note whether or not the necks are dissipating or developing with time. The "necks" may lean with height, thus indicating the vertical wind shear (Fig. 27c).

Fig. 27.—Cumiliform clouds.

4. *Cumulus congestus* forms when cumulus builds to a considerable height and takes on a massive, mountainous appearance. At the edges of the main body of this cloud are smaller developments, called *outriders*, which give the cloud a very broad base. Showers fall from cumulus congestus when the cloud has reached a thickness of at least 6,000 to 8,000 ft, regardless of whether or not it has reached the ice-crystal level. These showers can be heavy (Fig. 27d).

5. *Cumulonimbus* forms when a cumulus congestus builds far into the freezing layer. The height of individual cumulonimbi varies from 20,000 to 50,000 ft. The cloud has many outriders, and sheets of altostratus and cirrostratus formed by the lateral spreading of the developing cloud. Since these sheets frequently spread far from the parent cloud, they have often been mistaken for frontal surfaces. After the parent cloud disintegrates, some of these layers may remain. Thick anvil cirrus persisting under such circumstances is called *cirrus nothus* (Fig. 28).

Array of Clouds. Frequently cumuli are observed in ordered files with definite clear areas between. Theoretically, the orientation of such files should depend upon the direction and magnitude of the vertical wind shear. With small shear, the files should be oriented normal to the direction of shear; with great shear, the files should be parallel to this direction.   However, too few observations have been taken to check the validity of this rule in practice.

Clouds in Relation to the Distribution of Land and Sea.   It is generally supposed that convection over land tends to reach a maximum during the afternoon; while that over the sea reaches a maximum near sunrise.   However, there are many exceptions to this simple rule.   On mountainous islands, nocturnal showers occur in some localities, and afternoon showers occur in others.   In still others, there are two maxima of convective activity, one at night and one in the afternoon.   Also, during the night and again during the forenoon near such islands, a line of cumulus congestus or cumulonimbus may form parallel to the coast and at a distance of 30 to 50 miles from it.

Radiation-turbulence Stratus.   Stratus forming at the base of an inversion is an exception to the predominance of cumuliform clouds in the tropics.   The causes of

Fig. 28.—Cumulonimbus with associated stratus.

this cloud are (1) moisture transport aloft by vertical mixing due to turbulence and (2) cooling by radiation of the top of the moist layer.   This type of stratus occurs in many parts of the tropics.   Some of the principal tropical regions are the west coast of South America from a few degrees south of the equator to northern Chile, the coast of Southwest Africa off Angola, the coast of Northwest Africa, off Río de Oro, the coast of Brazil between São Luiz and Natal and the coast of Lower California.   Radiation clouds may occur also in many parts of the tropics in modified polar air behind a cold front or shear line when this air is undergoing subsidence.

Over wide areas, both the height of the base and that of the top of the stratus are nearly uniform.   The cloud reaches its maximum development in the early morning near sunrise.   During the forenoon, it weakens, especially near coasts, and it may dissipate there entirely.   The first stage in the breaking up of the stratus is its transformation to stratocumulus.   However, in many places a sheet of this cloud persists for weeks at a time, particularly over oceanic regions.

Summary of Analysis Rules.   This section will present a survey of methods of analysis for the synoptic systems of the tropics.   The analysis is based on the weather elements and charts previously described.

Waves in the Easterlies. Once formed, waves in the easterlies can be recognized by

1. 24-hr pressure falls, generally less than 3 mb, ahead of, and corresponding rises behind the waves. Because of the considerable time interval over which the pressure changes is computed, the wave crest itself will usually be located in the region of greatest 24-hr falls.

2. A cyclonic deformation of the surface isobars.

3. The presence of easterlies to at least 20,000 ft.

4. A turning of the upper winds, especially between 5,000 and 15,000 ft to northeast ahead of the wave crest (Northern Hemisphere) and a shift to southeast behind.

5. Unusually fine weather ahead of the wave crest and broad convection areas with showers to its rear.

Intensity. The intensity of a wave is proportional (1) to the amount of deformation in the flow pattern and (2) to the magnitude of the 24-hr pressure changes. Increases of intensity of the waves will be discussed under tropical storms.

A decrease of the intensity is indicated by

1. A decrease of the magnitude of the 24-hr pressure variations.

2. A decrease of the deformation in the upper wind field.

3. A lessening in the changes of the height of the moist layer across the wave.

4. A decrease of the convection behind the wave; and a maintenance of the trade cumuli ahead of it.

The Equatorial Trough. The equatorial trough can simply be located by the surface isobaric analysis near the equator.

The Equatorial Front. The analyst may suspect the presence of an equatorial front when the equatorial trough is about 5° latitude away from the equator. It can readily be located between a current of deep easterlies on the poleward side, and equatorial westerlies on its equatorward side. These westerlies are found not always at the surface but frequently above a shallow layer of easterlies, 1,000 to 2,000 ft thick. It is important to ascertain the depth of the westerlies, which may range from only a few thousand feet (notably in the East Atlantic Ocean) to 20,000 ft and more (Dutch East Indies). Long bands of cumulonimbus oriented mainly east-west serve as a further indication.

The 24-hr pressure change is not a reliable indicator, since it depends entirely on the type of transformation the equatorial front is undergoing. A stationary equatorial front may have no change at all.

The Polar Trough. A polar trough moving eastward in the tropics can be recognized by

1. 24-hr pressure falls, rarely in excess of 5 mb, ahead of the trough with usually smaller rises to its rear.

2. A cyclonic deformation of the surface isobars.

3. A low base of the polar westerlies; determination of its heights is of the greatest importance.

4. Southeast winds ahead of the trough below the base of the westerlies.

TROPICAL FORECASTING

Two kinds of short-term forecast are usually demanded (1) those that require prediction of only a few hours and (2) those that extend over a period of 1 to 2 days. The latter can be prepared only after analyses of weather maps covering a wide region. The technique of forecasting only a few hours ahead is usually much simpler. The two kinds will be considered separately here. Since forecasting in the tropics is still in the development stage, some of the forecasting principles indicated must necessarily remain more vague than would be desired.

Forecasts for Periods Not Exceeding 12 Hr

Short-term forecasts for periods not exceeding 12 hr can be subdivided into

1. Terminal, patrol, and operational forecasts for coastal areas and islands.
2. Forecasts for the open sea.
3. Route forecasts for aircraft in transit.

The forecasts are prepared from (1) microanalysis and (2) modifications imposed by the larger scale synoptic situation.

1. For forecasts over coastal areas and islands the dependency of the daily weather sequence on the local orography and the land and water distribution is of primary importance. Synoptic systems will modify this diurnal cycle, but only the strongest will suppress it altogether.

The forecast from microanalysis is dependent largely on the wind in the lower 10,000 ft in addition to the topography of the forecasting area. Since the winds are very steady in most parts of the tropics for considerable periods of the year, one should determine, as a preparation to forecasting during periods of the prevailing wind:

For the Land and Sea Breeze.

1. Duration, depth, and extent.
2. Time of beginning and ending.

For Daytime Convection over Land.

1. A stability scale on which the expected amount of build-up can be based.
2. Time of beginning and end of convection.
3. Favored sites for development of convective clouds.
4. Sites where the clouds break off from areas of origin.
5. Principal lines of drift of showers.
6. Time of day at which clouds begin to move out over the open sea.
7. Areas in which clouds move out over the open sea.

For Nocturnal Showers along the Coast.

1. Location of principal build-ups.
2. Time of beginning and dissipation of coastal clouds.
3. Periods of principal shower activity along the coast.
4. Distances that showers extend inland.
5. Duration of relatively clear periods between the diurnal convection maxima.

The following rules serve to guide in the forecasting:

1. The best estimate of stability conditions can be made a short time after convection has started.
2. Usually the clouds can be forecast to move with their steering current as represented by wind direction and velocity between 5,000 and 8,000 ft. Stationary clouds, however, are frequently found over the favored sites of convection.
3. Strong wind shear with height is detrimental to the development of large cumuli and showers.

Different synoptic systems will produce variations from normal of wind and stability conditions. Because both the rate of movement of these secondary circulations and the weather associated with them change but gradually in the tropics, extrapolation of these systems assuming constant velocity and weather pattern will usually be successful.*

* Exceptions are an equatorial front moving actively poleward, and a tropical storm deepening rapidly or recurving.

The abnormal wind velocities necessitate entirely different forecasts from micro-analysis. Thus, for instance, the site of origin and the time of formation of clouds will change, as well as their track and time of dissipation. Forecasters should, for the more frequent abnormal wind direction, make the same study outlined for the normal. The effect of changed stability conditions is much easier to estimate, since it depends on the intensity of the synoptic system that produces it and on the position of the forecasting area relative to the system.

2. Forecasts over the open sea usually need to consider the secondary circulations only. However, microanalysis with subsequent forecast of the areas of formation and tracks of large individual clouds will still be of value in many types of operation. For carrier-based aircraft, the problem is similar to that for fighter aircraft based in the Panama Canal Zone, outlined under microanalysis. Upon return from missions, these airplanes also are generally short of fuel and need to land almost immediately. Prediction of the movement of individual clouds and squall lines will permit steering of the carrier into the clear zones between clouds. A similar type of forecast of squall lines can be made for terminal conditions at seaports.

Route forecasts can be based primarily on the extrapolation of synoptic systems and their attendant areas of good and bad weather. The position of individual clouds is of little importance, unless the pilot is expected to encounter a solid wall of cumuionimbus.

Forecasts for Periods of 24 to 48 Hr

As stated previously, a forecaster needs to analyze maps covering a wide region on all sides of his forecasting area in order to prepare forecasts 1 or 2 days ahead. For

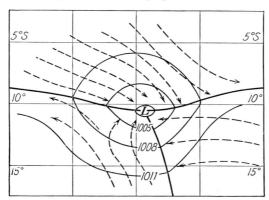

FIG. 29.—Cyclogenesis at a triple point on the equatorial front.

instance, in order to understand the sequence of events in the Antilles, maps for North America and the North Atlantic Ocean as well as the equatorial zone have to be drawn. Events taking place at the Azores or Seattle may profoundly affect weather over the Antilles 2 days later.

The forecasting methods are

1. Extrapolation for shear lines and troughs.
2. Path method for centers.
3. Special rules for formation, intensification, and dissolution.

The subject may be conveniently divided into (1) summer (wet-season) forecasting and (2) winter (dry-season) forecasting.

Summer (Wet-season) Forecasting. *The Equatorial Front.* In those regions of the world where the equatorial front undergoes large seasonal departures from the equator, it furnishes the principal forecasting problem. This front differs from other systems in that it moves toward the north or south rather than toward the east and west, and in that, while other systems may move over the oceans for 1,000 miles or more without changes in intensity, the equatorial front necessarily intensifies while moving actively poleward. In addition, it usually does not move equatorward over considerable distances.

The equatorial front will persist in a steady state when

1. It lies near its mean position for the season.
2. It undergoes only minor oscillations north and south.
3. The weather at the front is light or moderate.
4. Winds normal to it on either side do not increase.

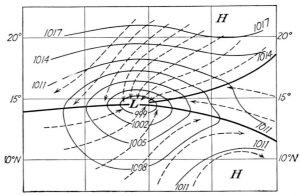

Fig. 30.—Cyclogenesis on the equatorial front due to strengthening of the trades.

Increasing activity is to be expected when the current on either side of the front strengthens and the wind component normal to it increases. Such an intensification is primarily derived from

1. Formation of a triple point between the front and a polar trough on its poleward side (Fig. 29).
2. Strengthening of the deep easterlies, generally due to the approach of a shear line from the poleward side (Fig. 30).
3. Strengthening of the equatorial westerlies, primarily due to a shorter trajectory of the air flowing across the equator (Figs. 31 and 32).

In case of a triple-point formation, bad weather can be expected to develop fairly uniformly within the three currents involved; in the other cases, the current whose activity causes deepening at the equatorial front will contain the worst weather. In all three situations described, there will form on the equatorial front a tropical disturbance, which may develop to hurricane intensity. This tropical disturbance need not form in the region where the equatorial front at any given instant has its greatest poleward bulge; and the orientation of the front need not change for some time after the center has begun to move slowly along it, generally westward.

Deepening disturbances can be recognized by their tendency to move poleward, while continued movement along the front, especially where the front bends equatorward, leads to gradual dissipation of the center. If deepening occurs, that part of the equatorial front which lies to the east of the forming depression can be expected to move poleward with the center to about latitude 15 to 20°; while on the equatorward side of the whole system very fine weather will develop.   After reaching latitude 15 to 20°, the front can be forecast to break down and to reform simultaneously near its mean seasonal position.   As mentioned before, equatorward movement of the front, other than the small oscillations during steady state, should not be forecast.

The exact time of re-formation of the front is difficult to estimate.   Indications will usually be furnished by a break of the streamline pattern near the seasonal mean

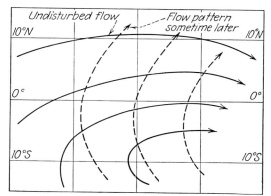

FIG. 31.—Strengthening of the southwest monsoon due to shortening of the trajectory across the equator.

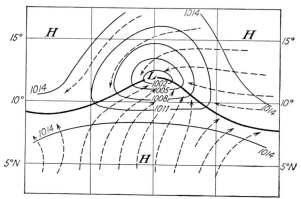

FIG. 32.—Cyclogenesis on the equatorial front due to strengthening of the southwest monsoon.

position of the front.   As soon as this occurs, rapid reestablishment of a zone of moderately bad weather can be forecast at once.

Waves in the Easterlies.   Since waves in the easterlies are embedded in a deep easterly steering current, usually of more than 1,000 miles' extent, they often advance steadily for very long distances over the tropical oceans.   A forecast of regular progression is therefore usually accurate.   The wave moves with constant speed of about 10 to 15 mph, the direction of movement being about normal to the trough line and parallel to the steering current.   If 3-hr tendencies are available, application of the kinematic methods of Petterssen[6] should yield excellent results.

Forecasters should watch primarily the equatorward eastern side of the subtropical highs for formation of waves in the easterlies, especially if a marked semipermanent trough is located east of the high. Weakening and often dissipation can be expected in the wide area to the west of the oceanic highs.

Intensification of the 24-hr pressure falls and strong increases of the wind velocities, especially near the 5,000-ft level, together with a marked increase of the meridional wind components, may be taken as a warning that a closed surface center is forming. Deceleration should be forecast immediately, together with dissolution of that part of the wave in which the deepening is taking place. The part well on the equatorward side of the forming center, however, will continue its movement westward as an independent unit.

Tropical Storms. Intensification. The forecasting principles regarding the formation of tropical storms, as far as now known, were presented in the two preceding sections. For forecasting intensification, the following rules have been found useful in the West Atlantic:

1. A tropical storm in the formative stage moving in excess of 20 mph will not intensify.
2. A rate of movement of 15 mph or less is favorable for intensification.
3. Storms developing coincident with or following the intrusion of polar or superior air into the tropics will not become more intense.
4. If a depression has attained winds of Beaufort force 6 at the surface, it will usually increase to hurricane intensity during the main hurricane season, provided that (1) no polar or superior air enters the system, (2) it does not pass inland over a large land mass or reach a high mountain range on an island, and (3) the depression stays below 25° latitude.

The following rule will also hold in general: A tropical disturbance will intensify strongly if it is passed by a polar trough moving eastward on its poleward side, while deep easterlies of at least 20,000 ft of thickness maintain themselves between the two systems.

Movement. While the storm is in the tropics, the deep easterlies are nearly always the steering current. In determining this steering current, care must be taken not to make use of stations that are affected by the hurricane circulation itself. For immature storms, the average wind between 8,000 and 12,000 ft will be most representative; for mature storms, higher elevations to the cirrus level must be included. If a strong turning of winds with height exists, in particular if the base of the polar westerlies lies at a relatively low level, the steering current is weak or nonexistent. The problem of recurvature then confronts the forecaster.

Steady progression can be forecast for all storms that are moving westward or west-northwestward under the influence of the trade wind as in the case of waves in the easterlies. The storms will move slightly to the right of the steering current (Northern Hemisphere), about 20 to 30 deg, and, according to recent studies, will be displaced with 70 to 80 per cent of its speed.

Recurvature. It was believed by Mitchell[7] that

. . . all tropical storms apparently seek to move northward (in the northern hemisphere) at the first favorable opportunity. Any tropical storm will recurve into a trough of relatively low pressure that may exist when the tropical storm arrives in the same region. No storm will break through and recurve until it reaches a region where south or southwest winds prevail aloft and relatively low pressure to the northward is shown on the weather map.

Since, as was just brought out, the upper winds play a vital role in determining the recurvature, and since the surface map does not necessarily show the flow aloft, a forecast of recurvature from the surface map alone should not be attempted.   Many hurricanes have by-passed troughs and low-pressure centers on their poleward side and continued their westward movement under circumstances that presented a picture on the surface map similar to other situations in which recurvature did take place.   A decision as to whether recurvature will occur, then, has to be made more than once in the lifetime of many tropical disturbances.

The following rules, as yet quite imperfect, may be taken as guiding principles:

1. Eastward movement of polar troughs in excess of 25 mph is unfavorable for recurvature.
2. If a polar trough is followed by a strong secondary cold front at some distance, the probability of recurvature is greatly increased.
3. Tropical storms will not recurve eastward away from the polar trough into which they are moving.
4. The farther the polar westerlies intrude into the tropics and the lower their base west of the hurricane, the more likely recurvature will be.   Mitchell's statement concerning the upper flow is entirely applicable, but as yet it has not been determined to what level the base of the westerlies must lower to warrant a forecast of recurvature.

During recurvature, very slow movement, often 6 to 10 mph, can be forecast. After completion of the recurvature, when the storm has come fully under the influence of the upper south or southwest current, considerable acceleration will often take place.   The rate of acceleration depends on the strength of that upper current, and movements of as much as 15° latitude in 24 hr or more have been observed.

The storm can be expected to pick up the polar front or to develop fronts of its own while moving into the polar zone.   Its isobars will tend to become V-shaped or oval, and the rainfall and wind distribution will lose the symmetry of the tropics.

Dissipation without Reaching High Latitudes.   Dissipation in the tropics can be forecast if a tropical storm moves on a tropical continent, such as Central America or Indo-China.   Dissipation can also be expected, when the storm reaches a subtropical continent, such as the southern United States or southern China, and is blocked from moving poleward by a large warm high over the continent.   Hardly ever does dissipation over a tropical ocean occur.

Winter (Dry-season) Forecasting.   Especially in the poleward parts of the tropics, winter forecasting is complicated by the fact that a uniform steering current does not exist.   During the quasi-periodic outbreaks of cold air, the base of the polar westerlies extends far into the tropics; while, in the periods between outbreaks, the deep easterlies tend to reestablish themselves.   The forecasting of the height of the base of the polar westerlies thus assumes great importance, both for the upper-wind forecast and for determining the type and direction of motion of disturbances to be expected in the tropics.   During the period when the base of the westerlies is low, motion with a component toward the east can be expected; and the disturbances will be polar troughs, frontolyzing cold fronts and wintertime tropical disturbances.   The reestablishment of deep easterlies coincides with the equatorward advance of very active shear lines.

Polar Troughs.   Polar troughs will not change speed, unless undergoing changes of intensity.   Again, the kinematical methods of Petterssen[6] should yield good results, if the 3-hourly tendencies are available.

Steady progression of polar troughs will occur primarily in the broad regions between the semipermanent subtropical cells and can be forecast with assurance if the migratory anticyclones of the subtropics are expected to move steadily from west to east, without much meridional displacement, and if no large-scale outbreak of arctic air is predicted.

As a migratory polar trough approaches a semipermanent cell from the west, it can be forecast to merge with the quasi-stationary polar trough found there. The latter will move slowly westward at the approach of a migratory trough; after the two systems have combined, it will return to its previous position.

Weakening of polar troughs other than that just described is indicated by a general rise of the base of the polar westerlies, a lessening of the upper thermal gradient across the trough and a decrease of the meridional wind components. If strengthening of the trades occurs ahead of a polar trough, the latter should be forecast to decelerate and weaken rapidly. It may even reverse its direction of movement while in the process of dissolution.

Cold Fronts. Forecasting the displacement of the cold front can best be done by the standard methods. After the thermal gradient across the front loses its importance, the forecaster will depend mainly on a correct judgment of the intensity and displacement of the upper anticyclone forming to the rear and over the front. Since this anticyclogenesis is associated with strong subsidence in the middle troposphere, high convective clouds will rapidly disappear except at the forward edge of the front, and there will remain only a low deck of stratocumulus due to turbulence in the old polar air.

Wintertime Tropical Disturbances. The forecasting of displacement of tropical storms in winter is quite analogous to that for the summertime storms. The steering current usually flows from the south or even west of south at the time of the formation of the disturbance. Therefore, immediate movement northward or east of north can be predicted. The speed will generally be more rapid than in summer, corresponding to the stronger circulation of winter. Wintertime storms can rarely be forecast to acquire the intensity of the summer storms, and they will reach the polar front usually within 2 or 3 days after their formation.

As in summer, the disturbances at times lose connection with the polar trough in which they are moving poleward. This separation is followed by westward recurvature and deepening of the disturbances. Since a breakaway from the fast-moving troughs of winter takes place more readily than from the slow-moving troughs of summer, the forecaster must be constantly on the alert for such a separation during the colder season.

Maintained Shear Lines. The displacement of equatorward- and also westward-moving shear lines in connection with the reestablishment of the deep easterlies is more difficult to forecast and is part of the general problem of determining the direction of the steering current. This problem is closely connected with the expected changes of the zonal indices. The index of the subtropical easterlies generally increases about 5 days after the index of the polar westerlies strengthens. This development follows a strong arctic outbreak and the simultaneous equatorward movement over the oceans of a large polar high, which rapidly is transformed into a dynamic high as it progresses into the subtropics.

If this sequence is observed on the weather map, a rapid rise—and often disappearance—of the base of the polar westerlies in the tropics can be forecast. At the forward edge of the deepening easterlies, a strong shear line with an elongated belt of severe weather and a rapid increase of the easterlies to its rear can be expected to move equatorward. This shear line will frequently reach the equatorial front or the equatorial trough and cause deepening there.

Bibliography

1. Dunn, G. E.: Cyclogenesis in the Tropical Atlantic, *Bull. Am. Meteorolog Soc.*, **21**: 125 (June, 1940).

2. RIEHL, H.: Waves in the Easterlies and the Polar Front in the Tropics, *Misc. Rept.* 17, Department of Meteorology, University of Chicago, January, 1945.
3. PALMER, C. E.: Synoptic Analysis over the Southern Oceans, New Zealand Meteorological Office, *Professional Note*, No. 1, Wellington, 1942.
4. BROOKS, C. E. P., and H. W. BRABY: The Clash of the Trades in the Pacific, *Quart. J. Roy. Meteorolog. Soc.*, **47**: 1 (January, 1921).
5. BERGERON, T.: Über die dreidimensional verknüpfende Wetteranalyse, Part I, *Geofysiske Publ.*, **5** (6), 1928.   Also Richtlinien einer dynamischen Klimatologie, *Meteorolog. Z.*, p. 246, 1930.
6. PETTERSSEN, S.: "Weather Analysis and Forecasting," McGraw-Hill, New York, 1940.
7. MITCHELL, C. L.: West Indian Hurricanes and Other Tropical Cyclones of the North Atlantic Ocean, *Monthly Weather Rev. Suppl.*, 1924.

SOUTHERN HEMISPHERE SYNOPTIC METEOROLOGY

By Civilian Staff, Institute of Tropical Meteorology,

Rio Piedras, Puerto Rico*

INTRODUCTION

To the synoptic meteorologist who has acquired his training and experience in the Northern Hemisphere, weather maps of any portion of the Southern Hemisphere present some unfamiliar features. There are, of course, the well-known differences that depend upon the relation of wind and pressure south of the equator; the sense of the wind circulation about anticyclones and depressions is opposite to that to which he is accustomed; vorticity relationships at frontal boundaries seem strange, and frontal waves and occlusions have unfamiliar shapes. But over and above these differences, which are after all not unexpected, there are those that are more difficult to analyze, that are apparent in the development and movement of high- and low-pressure systems, in the properties of the air masses and in the nature of the disturbances of such frontal systems as are found in the hemisphere. This chapter is concerned with the description and elucidation of the latter more subtle differences. It is also concerned with the mechanical difficulties that, arising out of the nature of the weather-reporting networks, beset the northern meteorologist in his first attempts at the construction of a southern analysis.

All these differences and difficulties can be traced to the prime geographical feature in which the Southern Hemisphere differs from the Northern. The great land masses of the earth are concentrated north of the equator; for most practical purposes, the Southern Hemisphere is a water hemisphere, and its meteorology is oceanic meteorology. This is particularly apparent in the higher latitudes. It is true that the antarctic continental icecap surrounds the pole and extends to the antarctic circle, but, as will be seen later, there are reasons for suspecting that its importance as a source of a continental air mass is not so great as has been supposed. Apart from the narrow southern part of South America, the stretch between 45 and 65°S is unbroken ocean right around the hemisphere.

THE GENERAL CIRCULATION

The mean annual pressure map of the Southern Hemisphere shows the following salient features: three large quasi-permanent anticyclones are located in middle latitudes over the great oceans, one in the eastern South Pacific extending from the west coast of South America to longitude 140°W, the second almost completely covering the South Atlantic Ocean, and the third more or less centrally placed in the South Indian Ocean. Between the highs, the axes of which run approximately along 30°S latitude, lie indifferent regions, or cols. South of the anticyclonic belt, with its quasi-permanent centers and intervening cols, the mean isobars run almost parallel to the latitude circles. Southward pressure decreases rapidly to about the 65th circle of

---

* The Institute of Tropical Meteorology is located on the campus of the University of Puerto Rico and is operated jointly by the University of Chicago and the University of Puerto Rico. The initial organization of the teaching and research program was carried out by Mr. C. E. Palmer of the New Zealand Meteorological Service, who served as director of the Institute during the first year of its existence.

latitude, where in the mean is found, not separate low-pressure cells, as in the Northern Hemisphere, but a continuous pressure trough that encircles the antarctic continent (the so-called "antarctic trough"). South of the trough, pressure again rises under the influence of a continental anticyclone that is coextensive with the antarctic continent. North of the anticyclonic belt, also, pressure decreases toward the equator, and here we find the so-called "equatorial trough," which separates the high-pressure belt of one hemisphere from that of the other.

As a whole, the pressure pattern in the Southern Hemisphere shows far less seasonal variation than that of the Northern Hemisphere. Moreover, the major seasonal changes affect the tropics, and this is to be expected, since the major part of the southern land masses lies in low latitudes. In summer, "heat lows" appear south of the Amazon Basin in South America, over the whole of the eastern part of South Africa and over northern Australia. The largest of these low-pressure areas is that which covers northern Australia, the Dutch East Indies, and Central Indian Ocean. These "heat lows" are to be regarded as more extensive sections of the equatorial trough, which, under the increased surface heating of summer, moves southward from the equator on and near the land masses. In winter, although pressure is relatively high over the land, separate subtropical anticyclones can still be distinguished. The quasi-permanent high that, according to the mean-pressure maps, appears over Australia in the winter months has misled many northern meteorologists, who have regarded it as a continental anticyclone similar to that found over Greenland. Synoptic charts, however, show clearly that Australia is only rarely, even in the wintertime, the seat of a stationary anticyclone. It must be emphasized that, so far as it is known, the only pronounced seasonal variation of pressure in middle and high latitudes in the Southern Hemisphere is the shift of the axis of the subtropical highs. Both these and the tracks of the migratory anticyclones presently to be described move northward during summer and southward in winter. The excursion, however, is only over a very few degrees of latitude, much less, indeed, than the somewhat similar movement in the Northern Hemisphere.

Before we pass to a consideration of the air masses and frontal configuration of the Southern Hemisphere, it will be well to compare the picture given by the mean annual and seasonal pressure maps with that revealed by experience with daily maps. The latter show that the quasi-permanent oceanic highs have indeed a considerable permanence on the day-to-day synoptic charts. However, the regions between the highs, which appear as cols in the mean-pressure pattern, are occupied on the synoptic charts by a train of warm migratory anticyclones. These are best developed in the Australian and western South Pacific regions but also appear on South African and South American maps. As an example, we may take the Australian region. Between the permanent anticyclone of the South Pacific and that in the Indian Ocean, there are usually two or three migratory highs moving steadily at about 10° longitude per day from west to east. Between these anticyclones protrude northward the upper parts of inverted V-shaped depressions which, of course, move with the anticyclones from west to east. The few analyses that have been found possible over the great southern ocean in high latitudes show that the V-shaped depression is but the northern part of a great closed cyclonic system somewhat similar to the stationary Icelandic or Aleutian low of the Northern Hemisphere but, unlike those depressions, moving from west to east with the anticyclones to the north.

The synoptic charts also show that, along the western border of the great South Pacific, South Indian Ocean and South Atlantic cells there is a quasi-stationary low-pressure trough, which not only extends through the belt of the westerlies but, at least in the upper air, also runs into the tropics. Even the mean-wind charts of the oceans show very clearly the deformation fields associated with these permanent polar troughs. It has already been pointed out that the daily maps show that the so-called "continental anticyclone" of Australian mean-pressure maps is not a quasi-stationary cold high but is to be attributed to the fact that the migratory anticyclones moving from the west are usually most intense in winter over the continent and lose their vigor after they pass eastward into the Pacific. In summer, the huge low-pressure area over and about northern Australia is seen on synoptic maps to be composed of several separate centers. Some of them over the neighboring open sea attain hurricane intensity and recurve toward the south and east. A similar low-pressure area, in which equatorial-front hurricanes develop, lies north and northeast of Madagascar and is continuous with the extensive summer low-pressure area over South Africa. On the other hand, the ocean to the east of the tropical parts of the South American continent only rarely seems to be an area of very low pressure, since in the Atlantic the equatorial trough remains in the Northern Hemisphere both summer and winter.

AIR MASSES

We are now in a position to inquire if frontal zones exist in the Southern Hemisphere. From the mean-pressure maps and the synoptic charts, we should expect to find continental polar air (antarctic air) over the antarctic continent. We should also expect to find that the quasi-permanent anticyclones over the great oceans are source regions for the production of tropical air. In both these regions, of course, observations are scanty, but those that exist confirm this expectation. The antarctic air has properties similar to those of the arctic air found over Greenland. Similarly, the tropical air that originates in the great anticyclones resembles the air that moves out of the northern subtropical highs, *e.g.*, the Azores anticyclone. The few high-latitude synoptic charts available show that antarctic air rarely, if ever, moves out from the continent in a great outburst, accompanying a cold polar anticyclone moving from antarctic to lower latitudes. It seems rather as if this continental air, moving northward to the west of the great traveling depressions already mentioned, becomes very rapidly transformed over the ocean into maritime polar air. There is very little evidence that it advances as a body behind an "antarctic front," but rather evidence seems to indicate that it mixes with the air over the polar oceans. Thus the chief polar air mass of the Southern Hemisphere appears to be maritime polar air. The history and properties of this air mass differ from those expected on the classical air-mass theory. In the mean, the west winds of middle and high latitudes blow parallel to the surface isotherms of the polar seas. Thus, though they are of considerable strength, the moving air undergoes little transformation by heating or cooling from below. In these zones, then, the maritime polar air mass moves rapidly and shows a considerable gradient of property (particularly of temperature) along the meridians. It is not homogeneous in the horizontal nor does it originate in an anticyclonic region.

The great permanent subtropical anticyclones are source regions for a tropical maritime air mass entirely similar to that found in similar situations north of the equator. As already pointed out, a quasi-permanent polar trough lies along the western border of each subtropical oceanic high. This provides a good wind field along which frontogenesis can take place. On the western and southern sides of the high, then, we expect to find, and do find, a polar-front system following the classical model of the Northern Hemisphere. It is far otherwise with the regions in which migratory anticyclones are found. Here a more or less uniform gradient of temperature extends from the center of the anticyclone into high latitudes. The air-mass contrast is not so much between the air masses originating in the moving anticyclones and the polar westerlies to the south, but rather between the air which moves southward along the western border of the anticyclone and that which moves northward along the eastern border of its successor.

An air mass which appears to have considerable importance in some synoptic situations is that which originates in the dry central regions of South Africa and Australia. This has all the properties associated with continental tropical air. At its source it is dry and cloudless and possesses characteristically a steep lapse rate of temperature. When it moves from its source onto the neighboring oceans, a strong inversion usually forms in the lower layers (owing to cooling and turbulence above the sea surface). The moisture added from the sea is then restricted to the lower layers. Sheets of stratocumulus and stratus are characteristic of continental tropical air when it passes on to the ocean.

It has already been pointed out in a preceding chapter that, in the tropics, especially in oceanic regions, the concept of an air mass is of only limited application. We shall therefore not describe the hypothetical equatorial air mass that appears in the rather confused accounts of equatorial meteorology in the standard texts.

FRONTS

It is apparent from the preceding account that, on the average, we should find three major frontal systems in the Southern Hemisphere (1) that which runs from a point a little south of Tahiti southeastward along the southern boundary of the quasi-stationary subtropical anticyclone of the South Pacific, (2) that which crosses South Africa somewhere between 30 and 40°S latitude and extends southeastward on the poleward side of the South Indian anticyclone, and (3) that which lies in the vicinity of the mouth of the Plata River and stretches across the South Atlantic Ocean and far to the south of South Africa. The three major fronts are subject to wave disturbances entirely similar to those of the polar fronts of the North Atlantic Ocean and of the North Pacific. As an example, we may take a wave originating on the South Pacific polar front to the south of Tahiti. It moves rapidly toward the southeast as an unstable wave, which finally occludes off the southern tip of South America. By the time it reaches that position, the depression associated with the wave has become a formidable cyclone, and this passes south of the south Orkney Islands into the Weddell Sea; here it usually remains stationary and fills up. Such depressions, of course, are the great storms for which the passage around Cape Horn was famous in sailing-ship days. Similar sequences are found in the South Atlantic and South Indian Oceans. In the South Indian Ocean, however, the great depressions that form as a result of wave activity on the South Indian polar front do not remain stationary but continue to move eastward south of Australia and New Zealand. It should be mentioned that the young wave cyclones on the South Atlantic and South Indian polar fronts can frequently be detected in their early stages over the continents of South America and South Africa, respectively. In that condition, they seem to be in every way similar to waves on the polar fronts in the Northern Hemisphere, and, as in the Northern Hemisphere, they rarely form singly on the polar front but form instead in families of from two to six, usually in rapid succession. The wave sequence on the polar front ends when the final members of the series run together and occlude in a great central cyclone to the southeast of the quasi-permanent anticyclone. The great mass of maritime polar air that sweeps around the western and northern parts of this cyclone then carries the fronts away eastward.

As already mentioned, both the mean-pressure and wind charts and the daily synoptic maps show us that the whole of the southern atmosphere is organized into regions that alternate along the latitude circles. The first of these regions corresponds roughly to the area where the great subtropical anticyclones are located, where the zonal circulation is strong and is interrupted only occasionally by the passage of large wave depressions on the polar front. But between these regions lie others in which the meridional circulation, though intermittent, is more intense. These are the regions in which migratory anticyclones are found, and between them are the intense troughs that end in the great migratory depressions centered at about 60°S latitude. A comparable region in the Northern Hemisphere is the United States, but here the picture is considerably complicated by the presence in wintertime of continental polar air. The anticyclones under these circumstances are at first cold and have a considerable equatorward component in their trajectories.

The V-shaped depression that lies between two migratory anticyclones is in most cases occupied by a strong front of the cold-front type. Recent work has shown that this cold front can be derived historically from an occlusion that originates as a result of wave activity on the polar front associated with the stationary subtropical anticyclone lying to the west of the region of migratory highs. In most cases, this front is not oriented northwest-southeast except in its northern subtropical parts; rather it sweeps in the form of a great arc with a meridian as chord. In the South Pacific, it has been termed the *meridional front*. When the procession of migratory anticyclones from west to east is regular, the meridional fronts will move from western Australia across New Zealand and into the South Pacific without showing the slightest sign of the instability that results in frontal-wave formation. This sequence, called *normal progression*, of undeformed meridional fronts and migratory anticyclones, though basic to an understanding of the meteorology of the area, is not very common. More often the east coast of Australia acts as a region of disturbance, so that meridional fronts passing from the west into the region will undergo wave deformation there. The waves thus formed will move across to the north or to the south of New Zealand, giving rise to storms that sometimes reach formidable dimensions and intensity.

As they approach the region of the quasi-permanent anticyclone of the South Pacific, the migratory highs and their associated meridional fronts may have one of two fates. The most usual sequence is for the migratory high to the east of the meridional front to collapse and disappear. The approaching meridional front then tends to become weak and to move into the quasi-permanent polar trough along the western border of the subtropical high, where it reinforces the already existent polar front. The second sequence is for the anticyclone following the meridional front to become very intense, bringing up much cold air from southern latitudes. The meridional front then becomes the new quasi-stationary polar front of the South Pacific. When this happens, the migratory high that preceded it replaces the old subtropical anticyclone, and the old quasi-permanent polar trough moves into the eastern South Pacific; here under most circumstances it quickly becomes reduced to a weak trough between the two high-pressure cells. Occasionally, however, it may preserve its intensity, and in this case we have doubled cells in the South Pacific anticyclone, and two polar fronts.

Both Northern and Southern Hemisphere synoptic meteorology, but more especially the latter, show us that the quasi-permanent oceanic high-pressure cells can be replaced in two ways, depending on the strength of the zonal index. The first has just been described; it is the replacement of the high by a warm migratory cell of great vertical extent that moves in from the west behind a meridional or some similar front. The second is perhaps better known and has been described by the Norwegian meteorologists. Periodically, the great migratory cyclones centered at 60°S latitude, after their establishment at the eastern end of the polar-front system, will not move away eastward but will remain stationary. In this case, the anticyclone forming in the polar maritime air to the west of the cyclone will not migrate eastward and become a warm moving high, but will remain stationary and intensify *in situ*. As it intensifies, the old subtropical high to the north rapidly loses intensity and is completely replaced by the new high as the latter builds up into the higher atmosphere. Future research should show that the type of replacement of the subtropical cells depends on the zonal index.   When the index is high, the replacement should be from the west.   When the index is low, the eastward progress of the systems slows down, and replacement comes from the south.

TROPICAL METEOROLOGY

The more theoretical aspects of tropical meteorology and those principles that are of wide application have already been discussed in a previous chapter.   Here it suffices to treat the local peculiarities of tropical meteorology in the Southern Hemisphere.

In spring (October, November, and December), the equatorial trough migrates from the Northern into the Southern Hemisphere in two regions, the western Pacific from about 160°W to the Timor Sea, and the western Indian Ocean.

The equatorial front reaches the peak of its development and its farthest south position in late February and early March.   It is then found in the Pacific crossing the equator west of the 160th west meridian, from which position it passes between the Solomon Islands and the New Hebrides across the Coral Sea on to northern Australia. It leaves the western Australian coast in the neighborhood of Broome, but its position in the Timor Sea varies greatly from year to year.   In the middle part of the Indian Ocean, it usually runs south of the equator (at about 5°S latitude).   There is, however, a strong probability that at about 80°E longitude an equatorial dry zone is intermittently established, but there are few observations to confirm this.   The Indian equatorial front runs from about 65°E longitude and 5°S latitude southwestward on to the coast of South Africa near the northern tip of Madagascar.   So far as is known, the equatorial front is not active in the South Atlantic Ocean.   Similarly, in the eastern Pacific east of 160°W, the equatorial front appears to lie 3° or more north of the equator throughout the year.   A pronounced equatorial dry zone lies in the central Pacific from about 140°W, 5°S to about 170 or 180°E.   Islands like Canton Island, Malden Island, Howland Island, and Christmas Island, which have a dry climate, lie wholly within the dry zone, and the great variability of rainfall on such islands as Penrhyn and Ocean Islands which are on its southern and western borders, respectively, depends upon a shift in the position of the dry zone.

Hurricanes are found, as would be expected, in those regions where the equatorial front is active, that is to say, the western South Pacific from about 170°W to the coast of Australia; occasionally in the Gulf of Carpenteria; in the Timor Sea; in the western South Indian Ocean and about Madagascar.   At the height of the hurricane season, storms may form near the northwest coast of Australia and move entirely across the Indian Ocean to recurve in the neighborhood of Mauritius and Madagascar, just as in the Northern Hemisphere hurricanes are known to form near the Cape Verde Islands, move across the Atlantic and pass into the Caribbean and the Gulf of Mexico. So far as it is known, hurricanes do not occur in the far eastern South Pacific nor do they occur in the South Atlantic.   In forecasting for southern hurricane areas, it must be remembered that the storms will recurve rapidly into the quasi-permanent polar trough should they form in the neighborhood of the western boundaries of the quasi-permanent anticyclones.   Thus, some hurricane tracks show very little east-to-west movement, the storms apparently passing straight into the polar trough after formation.   Another point to remember is that the polar trough between two migratory anticyclones is in most cases just as effective as a channel for the passage of the hurricane into high latitudes.   On the other hand, in the South Pacific many cases are known of hurricanes moving southward to about 35°S meeting a large stable migratory anticyclone and then becoming stationary and filling up.

In winter, the equatorial low-pressure trough lies wholly in the Northern Hemisphere; consequently those narrow zones of convergence associated with it and known as the *equatorial front* are not found south of the equator. The southeast trades then extend into the Northern Hemisphere, turning to become the southwest monsoon in those regions where the equatorial trough is far enough removed from the equator. However, in winter, strong "shear lines" are common in low southern latitudes, and these may be mistaken for the equatorial front. This is particularly the case in the eastern Pacific south of Galápagos Islands and in the eastern Atlantic in the vicinity of Ascension Island.

The troughs associated with the quasi-permanent polar-frontal systems, as already pointed out, are of considerable strength and permanence. Moreover, they extend in the upper westerlies in the tropics to a very low latitude. This is especially true of the polar trough of the South Pacific, which extends almost to the equator in wintertime. Naturally, it is subject to considerable changes of intensity with variation in the zonal index and the consequent movement and development of the migratory anticyclones to the west of its position. Easterly waves are known to occur in the eastern South Pacific. They form somewhere east of the Marquesas Islands, passing over these islands and over Penrhyn and Danger Islands to lose their identity in the quasi-permanent trough in the South Pacific. There seems to be evidence that they exist also in the South Atlantic, but this is a matter for future research. Nothing is known of their appearance in the Indian Ocean, but observations at Mauritius suggest that they will be found here also.

ANALYTICAL PECULIARITIES OF THE SOUTHERN HEMISPHERE

The technique of synoptic analysis in the Southern Hemisphere does not differ in principle from that used in the north. But the emphasis on certain aspects that do not, perhaps, receive very great attention north of the equator is considerable. Since the Southern Hemisphere is largely a water hemisphere, oceanic analysis assumes the greatest importance. The principles of oceanic analysis in middle and high latitudes may be summarized as follows:

1. Most weight should be given to observations made on ships and aircraft. This particularly applies to the wind and (in the case of ships) the pressure observations. Every attempt should be made to space the isobars over oceanic regions by using these representative winds in conjunction with a gradient-wind scale. The isobars can then be boldly extrapolated toward neighboring land masses. The cloud and precipitation observations also receive great weight.

2. Observations from representative island stations should receive the next greatest weight. To do this correctly, however, requires very accurate knowledge of the topographic character of the islands and of the situation of the stations upon them. It is advisable to classify the islands according to their topography and to weight them in the following order:

a. Low-lying coral islands or atolls serve best as observing stations. The observations are probably as representative as those taken on board ship.

b. Small islands of moderate elevation, such as the phosphatic islands of the South Pacific make excellent stations, especially if the observation point can be located on the windward side with respect to the prevailing wind. Orographic effects are usually slight enough to be neglected (*e.g.*, Ocean Island, Chatham Islands).

c. Small islands of considerable elevation will show marked orographic effects, usually on the cloud and wind. Stations on the windward side should receive the most weight (*e.g.*, Lord Howe Island).

d. Stations situated on large mountainous islands yield data that must be handled with great caution. The observations taken from stations on the windward side of an island may be fairly representative as regards wind, but precipitation and cloud will be quite unrepresentative. Lee-side stations usually show marked orographic effects of the foehn type and may be quite misleading if used uncritically in the analysis.

3. Over the continents of Australia, South Africa, and South America, the principles applicable to continental analysis should be used. However, two facts must be kept in mind:

a. The interior of South Africa is a plateau of considerable elevation, and the continent of South America, more especially, the Andes, offers a tremendous obstruction to the air flow of all latitudes. To a minor extent, the islands of New Zealand constitute a similar obstruction to the high-latitude flow.

The distortion produced by high mountain chains in middle and high latitudes is well known in the Northern Hemisphere. In such regions in the north, however, there is usually a large continental space to the east of the chain, so that there is no difficulty in drawing isobars in the lee side of the mountains. South America and, to a lesser degree, New Zealand are so narrow in middle and high latitudes that we have virtually a mountain chain rising from the sea. The observations on both sides of the chain are considerably influenced; indeed, in many situations they appear quite unrepresentative. When the wind blows almost at right angles to the chain, there is considerable local increase of pressure on the windward side and a corresponding decrease in pressure on the leeward side of the obstacle. This results in a great distortion of the surface isobars and, of course, the wind flow. A ridge of high pressure is built up on the windward side of the land mass with a compensating low-pressure trough on the leeward side. To one inexperienced in mountain meteorology, this considerable fall of pressure across the chain is very puzzling. It may lead in the analysis to a fictitious front, entered on the lee side of the mountain. In the actual drawing of the isobars over a narrow mountainous country like New Zealand, it is best to sketch in tentatively isobars of the undeformed flow (which can be estimated by smoothing the observed pressures across the mountain chain) and then later to correct them to agree with the actual pressures on the coast. If this is not done, an exaggerated curvature will appear in the isobars over the ocean, and this in turn will affect the location on the map of fronts moving toward the land mass.

In both the analysis and the forecast, southern meteorologists should remember that orographic distortion of the isobars is greatest in air masses that are stable. However, unstable flow, if at all strong, will almost invariably show some orographical distortion, provided that it strikes the mountain chain at an angle greater than 35 deg.

When a large number of stations exist on a small mountainous island, orographic effects similar to the one just described can be detected with a strong wind flow. The effect must be allowed for in the isobaric analysis.

Isobaric analysis over South Africa presents considerable difficulty, especially in the summertime. In that season, the central plateau is very strongly heated. The surface isobars in the interior then give very little information that can aid in locating fronts; it is often advisable to draw merely for the pressures recorded by stations situated along the coastal plain. Analysis of the interior is best based upon pilot-balloon observations.

b. In summer the interior of South Africa, Australia, and to a lesser degree the tropical parts of South America are covered by quasi-permanent low-pressure areas due to prolonged heating. To analysts who are not accustomed to dealing with such regions, so-called "heat lows" present very difficult problems, particularly since the air flow aloft is so markedly different from that of the surface.

It should always be remembered that "heat lows" may be fictitious, that is to say, the low sea-level pressure may be due to faulty temperature correction of barometers situated on strongly heated high plateaus. Nevertheless, even when this effect is allowed for, there will often remain low-pressure areas that must be directly attributed to the effects of prolonged heating on the density of the lower atmosphere. If the low is sufficiently far removed from the equator, a cyclonic wind circulation and marked horizontal convergence will develop in the lower atmosphere over the heated area. It is characteristic of such systems, however, that they are replaced at no great height (usually below 10,000 ft) by relatively high pressure, anticyclonic circulation, and horizontal divergence. Because "heat lows" usually develop only over deserts and semiarid regions, the steep lapse rates of the lower atmosphere are not accompanied by cloud or precipitation, but dust storms, dust devils, and other instability phenomena characteristic of hot arid conditions are common.

Occasionally a front will move into a region that is the seat of a "heat low." High-level thunderstorms may then develop. These are not persistent, and the front will quickly dissipate. A new heat low is likely to develop in the region where the front has disappeared, and this may tempt the inexperienced to retain the front on the map and attribute the low to a frontal wave disturbance. Examination of the pilot-balloon and radiosonde observations from the area should quickly reveal the fallacy in such analyses.

4. Since the data, even over the continents, are so scanty, special devices must be used, more particularly at the representative island stations, to follow closely changes in the meteorological elements with time. It is advisable to choose certain key stations from the island groups, together with a few representative coastal stations from the continents, and to maintain for these stations time cross sections or meteorograms. Wherever possible, each meteorogram should show in the form of time sequences the following meteorological elements: surface pressure; 10,000-ft pressure and 20,000-ft pressure; surface wind; upper winds plotted in the so-called "ladder form"; temperature at the surface, 10,000 and 20,000 ft; surface dew point; specific humidity at 5,000, 10,000, 15,000, and 20,000 ft; the cloud forms and precipitation.

5. Certain key pilot-balloon stations should be chosen; for these hodographs should be drawn and the methods of single-station forecasting as developed at the University of Chicago should be applied.

6. Since southern lands lie to a great extent in low latitudes, great attention must be paid to the relation between the secondary circulations of the tropical and temperate zones. In particular, the tropics should be analyzed very carefully at all times in order to trace the development and travel of tropical depressions, which may recurve into higher latitudes. In certain areas, it is most important in the summer to keep a close watch on old hurricanes, since these are likely to accelerate considerably after recurving into the temperate zone.

FORECASTING THE WEATHER WITH THE AID OF UPPER-AIR DATA*

By Vincent J. Oliver and Mildred B. Oliver

FORECASTING THE WEATHER IN MIDDLE LATITUDES FROM THE DIRECTION AND CHANGE OF DIRECTION OF THE WINDS AT 10,000 FT

The flow pattern of the air in the free atmosphere above the layer of frictional influence next to the earth is indicative of the type of weather that should occur at the surface of the earth. Meteorologists long ago noted that a current of air moving from south to north (in the Northern Hemisphere) will usually turn anticyclonically to the east and that a current of air moving from north to south will turn cyclonically to the east. More recently, investigations have been made to determine the exact paths of these currents and under what circumstances they deviate from these normal paths. Rossby† has demonstrated that, if a current of air conserves its absolute vorticity as it moves across latitude lines, it must change its vorticity relative to the earth. This vorticity will show up as a curvature of its path or as a shear in the current. Any current moving across latitude lines toward the south will acquire increased cyclonic vorticity (Northern Hemisphere) with respect to the earth. In more familiar terms, this means that the current will turn cyclonically or will acquire a cyclonic shear (the wind will decrease from the right side of the current toward the left side). The change in vorticity usually appears as a change in curvature of the streamlines rather than shear.

---

* At the time of preparation of this manuscript, constant-level charts were being drawn throughout the United States; hence all of the illustrations and discussions here refer directly to these constant-level charts. Since at the time of publication constant-pressure charts have replaced constant-level charts, it might be supposed that this discussion is no longer of current interest. It is the purpose of this introduction to show that all that has been said about the 10,000-ft chart is equally true of the 700-mb chart.

Contour lines of the 700-mb chart are drawn parallel to the winds and spaced inversely as the strength of the wind just as on the 10,000-ft chart. Therefore all the discussion in the following pages that deals with the spacing or curvature or direction of the isobars at 10,000 ft holds equally well for the constant-pressure chart for 700 mb.

Isotherms are drawn through equal values of temperature on both charts and their motion and orientation with respect to the isobars or contour lines are of the same forecasting importance on both charts. The isotherms on the constant-pressure chart have the additional significance of being lines of constant potential temperature and constant density, which the isotherms at 10,000 ft only approximate.

The moisture lines are drawn and used identically on either chart. On the constant-pressure chart the mixing ratio lines have the additional significance of being equivalent-potential temperature lines and condensation temperature lines for particles at 700 mb.

The overlay or intersection method of drawing mean isotherms as indicated on page 829 can also be done with constant-pressure charts. The surface map should be drawn for intervals of 4 mb and the 700-mb contours for intervals of 100 ft in order to have the lines drawn through the intersections approximate the mean temperatures. The intersection method will give lines of exact mean values when two constant-pressure maps are used—such as between 1,000 and 700 mb.

In all other respects the 700-mb and the 10,000-ft charts are the same and all rules for their use as an aid to forecasting the weather are identical for both charts.

† The principles discussed here may be stated in terms of the following equation:

$$\frac{f + \zeta}{\Delta p} = \text{const}$$

In this equation f is the Coriolis parameter, ζ is the vorticity relative to the earth, and Δp is the pressure difference between the bottom and the top of an individual fluid column. The absolute vorticity is the quantity $f + \zeta$.

Consider now a deep current of air from near the surface to 10,000 ft which continues to move in a straight line toward the north for some time. If this current were moving with a constant absolute vorticity, it should be turning anticyclonically to the right. Since it is observed to be moving in a straight line instead, we know that there must be something acting to change the absolute vorticity of the stream. The only process by which the absolute vorticity of a current in the free atmosphere may be changed is horizontal convergence (causing cyclonic vorticity) or divergence (causing anticyclonic vorticity). In this case, since the vorticity of the stream with respect to the earth is less anticyclonic than would be expected if the absolute vorticity remained constant, we see that horizontal convergence must be occurring. From this we can make the rule that follows:

Any extensive current of air moving across latitude lines from the south, which is moving in a straight or a cyclonically curved path, must be converging horizontally.

Since in this discussion we are referring to a current of air extending from near the surface up to at least 10,000 ft, the gradual convergence in the above-mentioned current will produce slow upward motion in the air mass. This will usually cause shower-type precipitation if the air mass is convectively unstable or steady precipitation and stratus-type clouds if the air mass is convectively stable. If this latter type of weather occurs in the warm sector of a cyclone, it will often be misinterpreted by the analyst as an indication that overrunning is occurring and that there is a warm front to the south of the region of steady rain and altostratus clouds. In reality, the clouds and rain are due to the gradual upward motion in the warm air as the result of horizontal convergence.

It is evident from the above discussion that the more sharply the current of air from the south turns cyclonically, the more rapid must be the convergence, and therefore the more pronounced the vertical motions and weather.

Consider now a current of air moving from the north. Owing to the change in the vorticity of the earth with latitude, this current should acquire a cyclonic curvature and turn to the east if the absolute vorticity were to remain constant. We find that the usual motion of such a current is not in accord with the assumption of constant absolute vorticity. The central portion of the stream will travel in a straight line while the extreme right edge will turn anticyclonically. We know that horizontal divergence must be occurring in the right side of the current, since it is acquiring a greater anticyclonic vorticity relative to the earth than it had when it started. Therefore, we can deduce that this portion of the current must be subsiding. It follows that the portion of the air mass which is turning cyclonically will have a steeper lapse rate than the portion of the air mass which is moving in a straight line from the north or turning anticyclonically, for the subsidence in the latter two portions of the current would be stabilizing the air mass. In the portion turning cyclonically, more instability and shower-type precipitation occurs than in the other parts. A quantitative determination of the amount of convergence and divergence taking place can be made by comparing the observed trajectories of the air stream with the trajectory the stream would have if it maintained its original absolute vorticity. A method of computing this latter trajectory has been developed by Rossby.[1]

The above principles can be used in practical forecasting as follows:

1. Cloudiness and precipitation are prevalent under cyclonically curved isobars aloft (about 10,000 ft), regardless of the presence or absence of fronts on the surface-weather map.

2. In a cold air mass the instability showers, cumulus, and stratocumulus clouds will be found only in the portion of the air mass moving in a cyclonically curved path.

3. In a warm air mass moving with a component from the south, cloudiness and precipitation will be abundant under a current turning cyclonically or even moving in a

straight line. Only where the current turns sharply anticyclonically will there be no clouds and precipitation.

Rule 1 will hold best in mountainous regions, over the oceans, and in the warm sectors of cyclones. The mountains will help release the instability of the air mass in the region of cyclonic curvature; the oceans will supply sufficient moisture so that, wherever instability is indicated, precipitation will occur; and warm-sector air will usually be moist enough so that small amounts of convergence will bring about the cloudiness and precipitation as indicated. In the summer in warm sectors containing mT air, thunderstorms and occasionally tornadoes will be observed if the air flow aloft is cyclonic.

4. Clear skies will occur wherever a current of air is moving from the north in a straight line or in an anticyclonically curved path. Clear skies will also be observed in a current of air moving from the south if it is turning sharply anticyclonically.

Examples of the above four rules will be found in Figs. 10 to 23. On nearly all these maps, the precipitation is under the regions of cyclonic curvature, and clear or scattered skies occur under the regions of anticyclonic curvature. These rules hold regardless of the proximity of surface frontal systems.

The method used by the authors in these figures to illustrate the relation between the weather and the upper-air curvature is to superimpose the upper-air map on the surface map. This technique is quite fruitful in bringing to light the relationships between the two levels and is suggested for forecasters as a standard procedure to be followed when analyzing the two maps.

5. Elongated V-shaped troughs will have cloudiness and precipitation in the southerly current in advance of the trough with clearing at the trough line and behind it.

An illustration of this rule will be found on the map for Sept. 27, 1942, shown in Fig. 12. This figure shows a deep 10,000-ft trough in the central United States with overcast skies and precipitation in the entire area covered by the southwesterly flow. The trough line at 10,000 ft marks the change to clear skies. The entire region under the strong flow from the northwest has clear skies, even though there is a frontal system in that current as well as in the southerly current.

FORECASTING FRONTAL WEATHER FROM WINDS ALOFT

When the component of the wind perpendicular to a cold front increases with height through the front, the front will produce no weather, since the front will not be lifting the air mass above it. Rather, the upper air mass will be flowing downslope along the frontal surface. The weather at the front will be due to convergence along the surface trough line, and not due to any lifting by the front. Under these circumstances, the general weather distribution is as follows: Convergence produces a certain amount of weather in the warm sector. This weather increases toward the cold front where the effects of the convergence become greatest, but stops abruptly after the front has passed.

Frequently, with the above wind distribution, the weather at the cold front is absent, while the weather some distance ahead of the front is still present. The zone of weather ahead of the front is often called a *prefrontal squall line*. This condition is to be explained as follows:

If the air above the cold front is moving faster than the front, that air will be flowing downslope and will be warmed adiabatically (see Fig. 1), and the relative humidity will decrease. This drier air over and just ahead of the surface cold front will prevent much cloudiness or precipitation from occurring there. Farther ahead of the front, the warm air will be converging into the trough, rising as it does so. As it reaches higher elevations where the winds are stronger, this air will move away from the front and carry the prefrontal clouds with it. The prefrontal squall lines and weather will occur in the portion of the warm sector where the upward motion

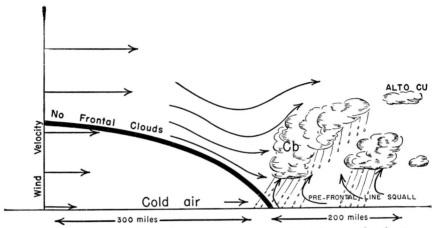

Fig. 1.—Inactive cold front. Wind speed increases with height normal to front.

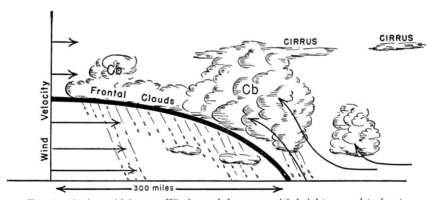

Fig. 2.—Active cold front. Wind speed decreases with height normal to front.

takes place. The distance of this weather ahead of the cold front will depend upon the rate of increase of the winds with height, for that determines the distance ahead of the front at which the dry subsiding air will be found.

These types of cold fronts are referred to as *inactive* cold fronts and usually move rapidly. *Active* cold fronts are those in which the component of the wind perpendicular to the front decreases with height through the fronts.

Active cold fronts (see Fig. 2) have widespread frontal cloudiness and precipitation at and behind them, since the air above the front is actually being lifted in this case. Relative to the front, the air above the front moves backward up along the front.

With warm fronts, an increase of the wind component with height perpendicular to the front gives an active front, one with pronounced prefrontal clouds and precipitation. Inactive fronts, characterized by broken cirrus and altocumulus, are produced by a decrease with height of the wind component perpendicular to the front. However, the activity of a warm front depends as much on the curvature of the isobars at 10,000 ft above it as upon the direction of these isobars. If the wind is turning cyclonically as it crosses the front, the front is far more active than if the flow is anticyclonic above the front. Where the wind is blowing anticyclonically over the warm front, the cold air mass will usually be subsiding so rapidly that the apparent upslope motion of the warm air over the cold air will be completely neutralized by the subsidence, and no upslope motion will actually be occurring. Besides this, if the warm air mass is turning anticyclonically, the relative humidity will usually be much lower than if it is turning cyclonically, and greater lifting will therefore be required to bring the air to saturation.

The activity of a cold or warm front may be recognized by a comparison between the 10,000-ft wind flow and the frontal positions as follows:

6. If the 10,000-ft flow is perpendicular to the cold front, the front will be inactive, because the winds normally increase with height.

7. If the 10,000-ft flow is parallel to the cold front, the front will be active. The cloudiness and precipitation will extend as far behind the front as the winds at 10,000-ft are parallel to the front.

8. Warm-front cloudiness and precipitation will occur where the 10,000-ft wind flow is across the warm front from the warm to the cold side and turning cyclonically or moving in a straight line.

9. The 10,000-ft ridge line is usually the forward limit of the prewarm-frontal cloudiness. The sharper the ridge, the more accurate the rule is.

10. Warm fronts are accompanied by no weather and few clouds if the 10,000-ft flow above them is anticyclonic. Frontolysis of such fronts is usual.

FORECASTING THE MOTION OF SURFACE SYSTEMS AND THE WEATHER ASSOCIATED WITH THEM FROM THE DIRECTION OF UPPER-AIR FLOW

11. Centers of rising and falling pressure on the surface map move parallel to the wind flow at 10,000 ft across the surface isallobaric center, with a speed proportional to that of the 10,000-ft wind.

Hence, most lows move parallel to the 10,000-ft winds with a velocity proportional to that of these winds since most surface lows move with the surface isallobaric centers. Surface highs are for the most part larger than surface lows and will often intensify or merely bulge in the direction of the surface isallobaric maxima.

A surface low with the flow at 10,000 ft across it is a warm low and is low only because it is warm. It will move with the flow at 10,000 ft since the warm air will do so. Surface lows that are merely the reflection of a closed upper-level center are usually cold lows and will move in the direction of and with the speed of the upper-level center.

Petterssen[2] has pointed out that wave cyclones move in the direction of the warm-sector isobars. This rule is identical with the above statement whenever the temperatures in the warm sector are homogeneous or whenever the isotherms are parallel to the warm-sector isobars. Whenever there is some other temperature distribution in the warm sector, the surface isobars in the warm sector are not parallel to the winds aloft. In such a case, the surface low will tend to move in the direction of the upper flow rather than parallel to the warm-sector isobars.

(The above guiding of surface systems by the upper-level flow has been referred to as the *steering* principle.) The following statements refer to the principal steering patterns observed in North America:

12. During periods of continued strong westerly flow aloft (high index) over North America, surface fronts move rapidly eastward with little air-mass contrast across them and practically no frontal precipitation or cloudiness south of the belt of strongest westerlies aloft.

East of the Rockies, such fronts are accompanied by cirrus and altocumulus or clear skies with little or no precipitation, but on the west coast the westerly flow up the mountains may produce copious rain. For an example of the former effect, see Fig. 19.

13. During periods of northwest flow at 10,000 ft from western Canada to the eastern United States, surface lows move rapidly from northwest to southeast bringing cold air outbreaks southward east of the Continental Divide.

The amount of precipitation occurring ahead of these surface lows depends upon the position of the ridge aloft. With the 10,000-ft ridge centered on the west coast and the main trough aloft over the western Great Lakes, cyclonic curvature aloft will exist between the Continental Divide and the Great Lakes region. This favors widespread cloudiness and precipitation in advance of each new outbreak of cold air from the northwest. For an example, see the 0130 maps for Sept. 25 and 26, 1942 (Figs. 10 and 11). Note particularly the low that comes in north of Montana and moves to the border of South Dakota and Minnesota.

If the 10,000-ft ridge is centered over the Continental Divide, anticyclonic curvature aloft will prevail through the central United States, and precipitation will be localized in the northeastern portion of the country. In general, the cold-air outbreaks will move into the country farther east. See the 0130 map for Nov. 13, 1942 (Fig. 16).

14. When anticyclonic flow aloft prevails over the Mississippi Basin, fair weather occurs throughout the region under the anticyclonic flow, and storms move around the periphery of the high.

These large high-pressure areas frequently stall and prevent cold air from entering the United States except in New England. They stagnate more often in summer than in winter. See the maps for June 19 and 20, 1943 (Figs. 20 and 21).

15. When southwest or southerly flow aloft prevails over the United States, storms move toward the north, giving widespread bad weather.

16. During periods of southerly flow at 10,000 ft along the east coast, secondary storms frequently develop in the vicinity of Cape Hatteras.

With such a flow aloft, primary disturbances form in the Gulf or lower Mississippi Valley and move northeastward just west of the Appalachians. These lows are generally occluded by the time they reach the Great Lakes region. Then secondaries form along the Atlantic coast where the flow aloft has the greatest cyclonic curvature. The energy for these storms is concentrated where the tropical air over the Gulf Stream comes into contact with cold air trapped between the Appalachians and the coast.

The motion of these secondaries, once they have formed, can be forecast from the 10,000-ft pressure pattern. With southwesterly flow aloft, the storms move parallel to the Atlantic coast, producing northeasterly gales and heavy precipitation along the coast. See the maps for Mar. 6 and 7, 1943 (Figs. 14 and 15). If the flow at 10,000 ft is southerly or southeasterly, the storms move inland. Heavy snow or rain and strong east winds occur along the coast, but west of the low the northerly winds are very light.

FORECASTING DISPLACEMENT OF SURFACE SYSTEMS BY USE OF PRESSURE CHANGES ALOFT

The rule that follows is based on statistical studies made by Bice and Stephens.[3] Previously it had been suggested by German meteorologists, but no supporting statistical evidence of its validity for the United States was given by them.

17. The 24-hr surface isallobaric minimum will move parallel to the upper wind flow to the current position of the 20,000-ft 24-hr isallobaric ridge line in 24 hr. Pressure changes aloft should be 5 mb or larger for this rule to apply.

The above empirical rule may be explained physically and geometrically as follows: Pressure rises at 20,000 ft are usually caused by rising pressure in the stratosphere and upper troposphere. These pressure rises accompany tropical air in the stratosphere and upper troposphere as the air moves in aloft from its source region. It has been observed that this inflow of tropical air in the stratosphere occurs approximately 24 hr before the tropical air reaches the same region in the lower levels. The forward edge of the inflow of tropical air at low levels is the region of greatest surface-pressure falls, for it is there that the lighter air is replacing heavier air most rapidly. In this manner, the 24-hr isallobaric maximum at 20,000 ft is related to the 24-hr surface isallobaric minimum.

Geometrically, the above rule may be explained as follows: Consider a surface wave in Texas with the 20,000-ft flow approximately west to east with slight cyclonic curvature over the wave. Suppose now that, during the next 24 hr, large pressure rises occur at 20,000 ft in New England, and during this same period the surface low moves from Texas to Tennessee parallel to the 20,000-ft wind flow. The large pressure rises aloft in New England will have changed the isobar direction at 20,000 ft from nearly west-east to more southwest-northeast. The surface low during the next 24 hr will follow this modified steering pattern up to the northeast. We see, then, in order for the storm to move from Texas to New England, it was necessary for the pressure to rise aloft in the New England region. This reasoning may be applied to other types of systems with similar results.

FORECASTING THE DEEPENING AND FILLING OF WAVES BY USE OF THE 10,000-FT CHART

18. A wave will be unstable and deepen if the 10,000-ft wind field over it possesses cyclonic relative vorticity (*i.e.*, cyclonic curvature of the isobars, or cyclonic shear). A wave will be stable if the 10,000-ft wind over it possesses anticyclonic vorticity.

In applying this rule, it is necessary to consider the shear as well as the curvature of the streamlines. Thus the relative vorticity of a current is given by the following expression:

$$\text{Vorticity} = \frac{c}{r} + \frac{\partial c}{\partial r}$$

where c is the wind speed and r is the radius of curvature of the streamlines. It is clear that a current may have vorticity even though it is straight (r infinite) if it has shear. The shear is taken as positive (cyclonic) if the velocity increases from left to right (Northern Hemisphere) when looking downstream.

We know from observation that in all intense cyclones the cyclonic rotation is present at all levels. If we assume that the development of an incipient wave into a large storm is produced by the release and redistribution of the potential energy in the air masses whose boundaries form the wave, we can reason as follows: Whenever the circulation over an incipient wave cyclone is cyclonic to high levels, there is very little release or readjustment of energy needed to produce a large storm except in the lower levels, since the circulation in the higher levels already corresponds to that observed over all large storms.   On the other hand, whenever the circulation over an incipient wave is anticyclonic at high levels, this portion of the circulation must be completely changed before the wave can become an intense storm.   Since the amount of change necessary for the development of the wave is much less when the upper atmosphere already possesses cyclonic vorticity, the wave will be more likely to develop when it is under a current possessing cyclonic vorticity than when it is under a current possessing anticyclonic vorticity.

19. If there are several waves along a front, the one with the most intense cyclonic vorticity aloft will develop at the expense of the others.   This is usually the one nearest the axis of the trough aloft.

Minor local influences frequently produce irregularities in the shape of a front. When the winds aloft are perpendicular to the front, these irregularities are usually unimportant, since they produce no weather and are transitory in character.   They are usually disregarded on 6-hourly surface maps.   But when the winds aloft are parallel to the front, even a slight bend in the front will produce upslope motion. The winds aloft will blow upslope over the forward edge of the bend and produce clouds and/or precipitation.   Thus, energy is released that is used to cause further development of the wave.   The winds aloft blowing down over the cold air to the rear of the bend pull the cold air along and make the bend persist so that there is no tendency for the wave to flatten out once it has formed.

20. Waves develop along fronts when the 10,000-ft flow is parallel to the front, or nearly so.

On the 0130 map of Mar. 12, 1943 (Fig. 17), there is a good example of this type of activity.

FORECASTING WITH THE AID OF MEAN TEMPERATURES AND MEAN-TEMPERATURE CHANGES

21. Cloudiness and precipitation will occur where the 10,000-ft winds blow across the mean isotherms from the warm side to the cold side; clear skies will occur where the 10,000-ft winds blow across mean isotherms from the cold side to the warm side.

When the winds aloft blow across the mean isotherms from warm to cold air, they are blowing up over the more slowly moving cold dome.   Thus the air aloft is lifted. This results in cloudiness and precipitation.   When the winds aloft blow from cold to warm air, they blow downslope over the cold dome, and weather is at a minimum.

22. If the mean isotherms in the cold air are packed along a front, it is a strong front and will tend to produce weather.

Where the mean isotherms are closely packed, there is considerable potential energy.   When a storm enters this region, the potential energy is released, and the storm develops.   If there is the same packing of mean isotherms on both sides of a frontal zone, there is no marked front as yet.   A marked front can develop only along the warm side of the closely packed isotherms.   This follows from the definition of a front.   See maps for Sept. 25 to 27, 1942 (Figs. 10 to 12).

23. If the mean isotherms are perpendicular to a front, the front is weak, will produce little weather, and is subject to frontalysis.

If the mean isotherms are perpendicular to the front, there is no temperature or density contrast across the front.   Hence, by definition it can hardly be called a front, but is merely a trough line. The following rules were derived from research being conducted by the authors on mean temperatures. The research is by no means completed, and the following results should therefore be regarded as tentative.

With warm lows, the mean isotherms show the highest temperature directly over the surface low, which is about halfway between the 10,000-ft trough and ridge line, *i.e.*, the mean isotherms and 10,000-ft isobars are 90 deg out of phase. Since warm lows move with the mean speed of the warm air over them, they will be rapidly moving systems.

If, on the other hand, the highest mean temperatures occur under the 10,000-ft ridge (mean isotherms and 10,000-ft isobars in phase), the ridge itself is warm while the low is a cold and therefore slowly moving low. From these relationships, we obtain the following rules:

24. If the mean-temperature chart shows the maximum temperature along a latitude line to be about halfway between the 10,000-ft high and the next low, the surface low beneath will move rapidly. If the maximum temperatures occur in the high aloft, the surface low will tend to stagnate, with weak fronts moving rapidly around it.

25. A change in the type of advection behind a slowly moving cold front is a sign of a new frontal wave forming.

Instead of uniform cold-air advection behind the front, there will be a decrease in cold-air advection or even slight warm-air advection behind one portion of the front with continued strong cold-air advection on both sides of this region. This change in advection may be detected on mean-temperature charts or by inspection of pilot-balloon observations at stations behind the front.

26. The first indication of prewarm-frontal clouds over a cold high is a layer characterized by advection of warm air at some height between 10,000 and 20,000 ft.

This layer of warm-air advection will show up as a veering of the wind with elevation at those heights.

FORECASTING THE 10,000-FT CHART

The following are several general methods for forecasting 10,000-ft charts:

1. Extrapolation. The easiest method of forecasting the 10,000-ft pattern is extrapolation. Either trough lines and ridge lines or 24-hr isallobaric centers may be extrapolated. Since the patterns at 10,000 ft are more persistent than at the surface, extrapolation gives better results at 10,000 ft.

In using pressure changes aloft, the procedure is to extrapolate the isallobaric centers and add the changes to the present pressures in order to get the prognostic map. Changes in the shape of the 10,000-ft pattern may be forecast by this method, whereas pure extrapolation would not indicate these variations.

2. Isotherm-isobar Relationships. The 10,000-ft pressure pattern may be forecast from the isobars and isotherms at that level. From the kinematics of wave motion, assuming that temperature is a conservative element and that there is no convergence or divergence, Rossby[4] derived the following equation:

$$\frac{U}{U-c} = \frac{A_T}{A_p}$$

where c is the speed of a system, U is the mean speed of the westerlies, A_T is the amplitude of the isotherms, and A_p is the amplitude of the isobars.

The authors have found that for practical purposes the above equation indicates that, wherever cold air moves into a system, the pressures at 10,000 ft will fall; where warm air moves in, the pressures at 10,000 ft will rise.   This can be demonstrated as follows:

Let us first consider the case where $A_T < A_p$.   From the above equation, it is seen that $U - c$ must be larger than U, i.e., that c must be negative.   This means that the system should retrograde, which in turn means that pressures should fall

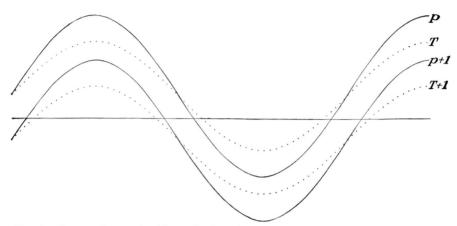

Fig. 3.—Retrograde trough with amplitude of isotherms smaller than amplitude of isobars.

to the west of the system and rise to the east.   The preceding diagram (Fig. 3) showing $A_T < A_p$ indicates that colder air is being brought into the western end of the trough and warmer air into the eastern portion, since the winds will move the isotherms in that manner.   Thus, where the isobars indicate that temperatures will fall, the equation indicates that the pressures should fall; and similarly, where the rising temperatures are expected, rising pressures are indicated.   Any indications of retrograde motion from this formula should ordinarily be interpreted as deepening of the system rather than as retrogression.

Consider now the case when $A_p < A_T$ (Fig. 4).   From the above temperature and pressure distribution, it is clear that temperatures would be increasing to the rear of the trough and decreasing in advance of it.   At the same time, the above equation shows that the trough should move eastward.   In other words, pressures will fall in front of the trough and rise behind it.   Again we see that the region of temperature fall is the region of pressure fall, and likewise the regions of temperature and pressure rise are coincident.   Moreover, the formula shows that the larger the amplitude of the isotherms relative to the amplitude of the isobars, i.e., the more nearly the winds blow perpendicular to the isotherms, the faster the trough will move.   This means that the greater the temperature change at a given point at 10,000 ft, the greater the pressure change is.

If the isotherms and isobars are parallel, $U = U - c$, hence $c = 0$.   Such a system will have no temperature advection around it and will stagnate without appreciable deepening or filling.   For example, see Sept. 25, 1942 (Fig. 10).

If A_T/A_p is negative, the temperature and pressure changes would be of the opposite sign.   However, it is the opinion of the authors that such lows seldom occur at 10,000 ft.

If we assume that the pressure changes at 10,000 ft are due to horizontal advection of air in the stratosphere, we can explain the above relationship between temperature and pressure changes as follows:

Cold, heavy air in the stratosphere is most frequently found in tropical regions where the 10,000-ft temperatures are highest. As the cold air in the stratosphere moves northward, the warm air at 10,000 ft accompanies it. If, then, we can predict the place to which the warm air at 10,000 ft will move and can assume that the cold

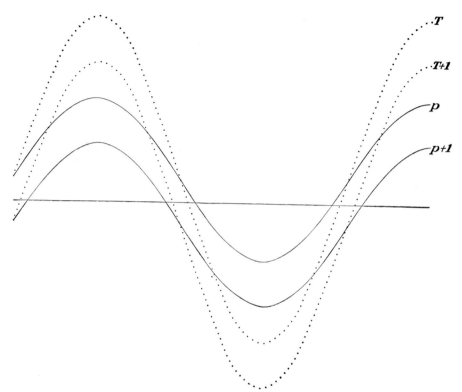

Fig. 4.—Slowly progressing cold trough with amplitude of isotherms greater than amplitude of isobars.

air in the stratosphere will move with it, we can forecast that pressure rises will occur there at 10,000 ft because of the inroad of cold dense air in the stratosphere above it.

Conversely, cold air at 10,000 ft is associated with warm light air in the stratosphere. Hence, in regions where cold air moves in at 10,000 ft, there will be pressure falls, since above 10,000 ft the stratospheric air is becoming less dense and exerts less pressure.

3. Solenoids. If we examine the temperature and pressure field for solenoids, we see that they, too, indicate rising pressure where the temperature is rising, and falling pressure where the temperature is falling. Wherever warm-air advection is indicated, the horizontal solenoidal field indicates anticyclonic circulation acceleration. This increasing anticyclonic circulation may be interpreted as indicating rising pressure, for the two usually occur together. On the other hand, wherever the iso-therm-isobar intersections indicate cold-air advection, the solenoidal field indicates cyclonic circulation intensification and therefore falling pressures. We see then that the rule that temperature falls indicate pressure falls and temperature rises indicate pressure rises is indicated by all three types of reasoning: first, the formula based on kinematics alone; second, the stratospheric changes that should accompany the temperature changes; and third, the solenoidal field.

In forecasting the 10,000-ft pattern, it is necessary to use the above relationships quantitatively. Empirical results show that *the temperature change in degrees centigrade is in the mean approximately equal to the pressure change in millibars during the same period of time.**

The procedure in forecasting is to move the temperature at one point on the 10,000-ft chart along the trajectory of that point determined by the present and future winds at 10,000 ft. Ordinarily this trajectory will not be coincident with the isobars, nor will it be a straight line. Having moved the point in question, subtract its temperature from that observed at its terminus to determine the temperature change in the forecast period. Add this change algebraically to the present pressure at the terminus to determine the prognostic pressure. The future winds at 10,000 ft may be arrived at by simple approximation of air trajectories or, more accurately, by Rossby's method outlined below.

4. Rossby's Trajectory Method. This method of forecasting the future trajectories of air particles, and hence the future streamline pattern, is based on the assumption that free-air flow proceeds with constant absolute vorticity.

Particles are selected where the vorticity relative to the earth is zero, *i.e.*, where the isobars are straight and there is no shear. If these particles change latitude but conserve their absolute vorticity, they acquire a vorticity relative to the earth and travel in a curved path. The curve they follow is periodic, with constant amplitude, wave length, and frequency as far as the particle is followed. This obviously does not take into account the loss of energy, and hence momentum, due to lateral mixing or changes in vorticity due to convergence and divergence.

After many particles have been followed, a map can be constructed with isobars drawn from the predicted wind directions and velocities. Bellamy has made some simple charts that facilitate and extend the use of this method, making it unnecessary to use only the inflection points.

Forecasting the 10,000-ft map, when extrapolation fails because the pattern is changing, may be done adequately by the temperature-change method, by computing trajectories, or preferably by a combination of both. Two examples of these methods will be given below.

Consider first the case of a deep, cold trough in eastern North America (Fig. 5) with a surface low in the region marked *A*. We see, from the pressure and temperature distribution indicated on this map, that large temperature rises will occur in the northern portion of the trough in the region marked "Rising" where the strong westerly flow would bring warm air into the trough, while temperature falls will occur in the southern part of the trough, marked "Falling." By the temperature-change method, we should therefore forecast pressure rises in the north and falls in the south. These pressure changes would gradually alter the map until it looks like Fig. 6, with the northern portion of the trough filling up and the southern portion becoming a more pronounced low. This also means that the northern portion of the trough will be displaced eastward more rapidly than the southern portion, and the trough line will tilt from northeast to southwest.

---

* This principle was brought to the attention of the authors by C. L. Mitchell of the U.S. Weather Bureau.

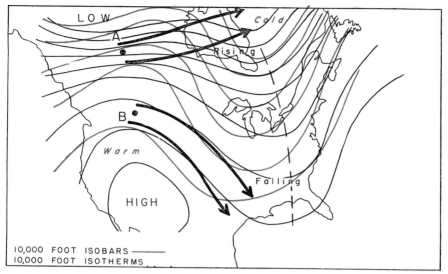

FIG. 5.

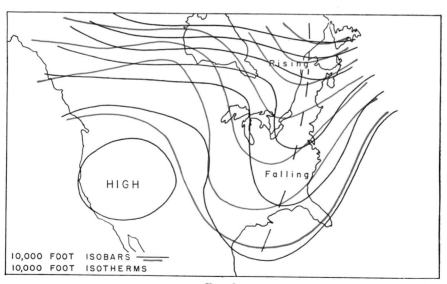

FIG. 6.

As these changes continue, a closed low will develop in the southern part of the trough, as is indicated in Fig. 7. Not only has the low in the south been completely disconnected from the rest of the trough in the north, but the belt of westerlies has also moved north of it so that it is likely to stagnate.

Now let us consider the problem of forecasting the changes in the same pressure and temperature distribution by Rossby's trajectory method. In the region marked *A* (Fig. 5), over the intense surface low, trajectories would have a long wave length and small amplitude, as shown by the arrows starting from that region. Farther to the south, south of *A*, where the wind velocities are lighter and the isobars have more anticyclonic curvature, the trajectories would have a shorter wave length and follow the arrows indicated there.

Consider now the effect of these future trajectories on the present pressure distribution, keeping in mind the principle that, when a strong current of air enters a region in which the isobars do not balance the Coriolis force, as near the regions marked "Rising" and "Falling," pressures will be built up to the right of the current and will fall to the left, until the pressure gradient does balance the Coriolis force and the isobars are parallel to the winds. As is seen from the diagram, "Rising" is to the right of the current there and "Falling" is to the left, so that rising pressures should occur at "Rising" and falling pressures at "Falling." This is in agreement with the results obtained above by the temperature method.

For a second example, consider a stagnant deep low in the eastern United States. The forecasting problem here is to determine when the low will start to fill up and move. As long as the isotherms and isobars remain parallel as in Fig. 8, no temperature changes should occur. Therefore no pressure changes should occur, and the low should remain stagnant. As the isobars and isotherms are parallel, the trajectories would be the same as the isobars. They, too, would indicate no motion.

Let us now consider the effect on this stagnant low of the temperature and pressure distribution associated with a surface low pressure at *A* in Fig. 8. The temperature and pressure distribution is now such that marked temperature rises should occur in the region marked "Rising." This will produce rising pressures there, which, in turn, will produce more westerly flow to the north of the region marked "Rising." Subsequently, more warm-air advection will occur north of this region with the rising pressures gradually extending northward and eastward. It is apparent from this that the low-pressure center will be displaced rapidly northeastward ahead of the rising pressures.

Associated with the surface disturbance at *A* in Fig. 8, the 10,000-ft chart shows stronger winds with less anticyclonic vorticity at that point. The trajectories of air particles from the vicinity of *A* are indicated by the heavy lines that pass through the eastern low. As before, pressures will rise to the right of this current and fall to the left, so that the low in the east will move northeastward. Again the trajectories and temperature changes indicate the same motion of the pressure system.

5. Use of Stratospheric Conditions for Forecasting the 10,000-ft Pattern. The 13-km chart seems to be the most useful high-level chart for forecasting changes that will occur below at 10,000 ft. On this chart, pressure lines and potential-temperature lines are drawn. At this level, lines of constant potential temperature approximate lines of constant density, since potential temperature is a function of both temperature and pressure. High temperatures and low pressures give high potential temperatures and low density, while low temperatures and high pressures give low potential temperatures and high densities. We can recognize regions at 13 km where the air is very dense by noting the regions of low potential temperatures and regions where the air is rare by high potential temperatures If we assume that the changes that occur at these high levels are due primarily to advection, it follows that, wherever

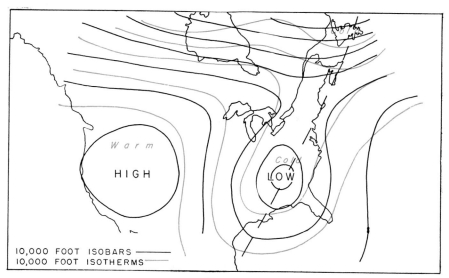

FIG. 7.

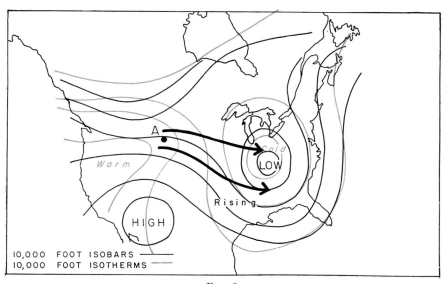

FIG. 8.

the potential temperature at 13 km rises, the pressures beneath 13 km will fall, since less dense air has been moving in at 13 km. The 13-km potential temperatures will move with the 13-km winds, which are indicated by the isobars at that level.

The technique of forecasting from this chart is to examine the 13-km potential temperature and pressure pattern and determine where the potential temperatures at 13 km will rise, and fall, in the forecast period. The pressures at 10,000 ft should fall wherever the 13-km potential temperature rises and should rise where the potential temperature decreases at 13 km. The magnitude of the pressure changes at 10,000 ft will be roughly proportional to the magnitude of the potential-temperature changes at 13 km.

If this technique were applied, not just at 13 km, but at all the upper levels and if the changes at all levels were integrated, a much more accurate prognosis of 10,000-ft pressure changes should be obtained. More detailed information on this use of stratospheric conditions may be obtained in a paper by Wulf and Obloy.[5]

A chart indicating the height of the tropopause can be used in a manner similar to that using the 13-km potential temperature. The tropopause will be low wherever the 13-km potential temperature is high, and the tropopause will be high where the 13-km potential temperature is low. Wherever the tropopause becomes lower, the pressure will fall at 10,000 ft. A high tropopause will give high pressure at 10,000 ft. The motion of the tropopause can best be forecast by the winds at the level of the tropopause. Since this level is not constant, the isobars at one constant level will not give this motion. The above discussion assumes that change in height of the tropopause is due mainly to horizontal motion. The authors are not prepared to make a statement on the validity of this assumption.

At present the upper-air data as it comes in over the teletype is too sparse to give a unique solution for any of these high-level charts. Therefore, for current data, forecasting methods utilizing the radiosonde data below 400 mb only will be more accurate than those relying upon the second transmission of radiosonde data. Moreover, winds up to 10,000 ft are many times more plentiful and accurate than those above. Therefore, pressures at 10,000 ft may be checked by the pilot-balloon observations, whereas errors at the higher levels may be overlooked.

6. Use of the Surface Map and Mean Isotherms. First, forecast the motion of the surface low by use of a combination of upper-air steering and deepening principles and surface continuity. Then, from the present distribution of mean temperature and pressure, forecast the future distribution of mean temperature. A combination of the surface map and mean temperatures automatically determines the 10,000-ft map. An easy method of determining this graphically was devised by G.T. Stephens at the University of Chicago. This method will be discussed later in this chapter.

The 0130 map of Sept. 25, 1942 (Fig. 10) shows how this method of forecasting may be applied. The surface low-pressure center located just north of Montana is forecast to move to the border of South Dakota and Minnesota and deepen, since strong cyclonic shear and curvature exist over it at 10,000 ft. The mean temperatures in advance of the low are slightly lower than those behind the low, and only slight warming should take place during the next 24 hr, since very little advection is indicated by the change of winds with height. The surface low in 24 hr will then have approximately the same temperatures on both sides of it, and the 10,000-ft low should therefore be directly over the surface low. The map of Sept. 26 (Fig. 11) shows that this occurred.

7. Use of Current Mean-temperature Distribution. Pressure rises at 10,000 ft will occur in 24 hr over regions of lowest mean temperature, and pressure falls will occur in 24 hr over regions of highest mean temperature.[6] This rule, like the rules on the use of 24-hr isallobars aloft, frequently holds but like all empirical rules has many exceptions.

8. Use of Wave Length. At 10,000 ft, major troughs and ridges of long wave length tend to persist. Minor troughs will deepen when they move through the major troughs and will fill when they move over major ridges. This is one of the principles used by the 5-day forecast group in Washington.[7] They determine where the 5-day mean troughs will be and forecast that individual troughs will deepen as they move into the mean troughs.

9. Effect of Cold Water Surface. Upper-level highs tend to stagnate and intensify over the Great Lakes region and off the coast of New England during the season when the water is colder than the air.[8] When the air flows over colder water, a surface inversion is produced that prevents ground friction from affecting the air above the inversion. The frictional outflow from the high is thereby reduced, and the high tends to persist.

10. Various Equations Developed for Forecasting the Displacement of the 10,000-ft Systems. With the 3-hr pressure changes as computed by the method of Miller and Thompson,[9] Petterssen's trough and ridge formulas may be used at 10,000 ft.

Rossby and Wexler have developed the following equation:

$$C = U - \frac{\beta L^2}{4\pi^2}$$

where C is the velocity of a trough eastward, U is the mean west-east wind speed, L is the wave length, and the other quantities are constant for a given latitude. The main difficulty with this formula is the determination of the wave length. The best plan seems to be to use the large-scale troughs and ridges to determine the wave length, and to ignore minor troughs superimposed on those. With a map only as large as the United States, an accurate determination of the wave length is always problematical.

METHOD FOR THE CONSTRUCTION OF MEAN ISOTHERMS BETWEEN THE SEA LEVEL AND 10,000 FT

The mean isotherms for the layer between sea level and 10,000 ft may be constructed graphically as follows: Draw the 10,000-ft isobars for intervals of 4 mb and the surface map for intervals of 3 mb on the same size map base. Now select the intersection of any one of the 10,000-ft isobars and a surface isobar. From this point, draw a line to the intersection of the next higher 10,000-ft isobar and the surface isobar 6 mb higher. From this second point, continue on to the next higher 10,000-ft isobar and the second higher surface isobar, and so forth. Finally, when one 10,000-ft isobar does not intersect the corresponding surface isobar, continue the line through another intersection of the same two isobars and proceed to the intersection of the next lower 10,000-ft isobar and the surface isobar 6 mb lower. This line which has been constructed is a mean isotherm. It will be noted that the value of the surface isobars that pass through all the intersections through which the mean isotherm has been drawn was even if the value of the first surface isobar was even, or odd if the value of the first surface isobar was odd. After this first isotherm has been drawn, draw the rest of the mean isotherms through the other intersections of the 10,000-ft isobars and the corresponding surface isobars. Through all the intersections on the map that are used, either odd- or even-valued surface isobars will pass. This may be used as a check on the construction. In drawing these isotherms, the following rules must be followed:

1. Always cross an intersection from the higher pressure side of *both* the surface and 10,000-ft isobars to the lower pressure side of *both*, or from the lower pressure side of *both* to the higher pressure side.

2. Never cross a 10,000-ft pressure line unless it intersects one of the surface isobars which is being used (*i.e.*, either odd- or even-valued surface isobars).

3. Never cross a surface isobar, odd or even depending upon which is used, except where it intersects the corresponding 10,000-ft isobar.

In weather stations, these isotherms may be drawn easily on the 10,000-ft map by superimposing it on the surface map on a light table. It is easier to draw mean iso-

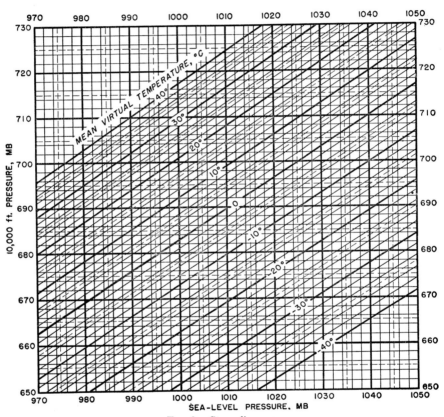

Fig. 9.—Starr diagram.

therms if the isotherms are first drawn near a low where there are numerous intersections and a definite isotherm pattern.

The isotherms drawn by this method are for intervals of approximately 4°C up to +14°C and for intervals of approximately 5°C for higher temperatures. Hence, when one line is labeled by the use of the Starr diagram (Fig. 9), the others may be labeled immediately with no further recourse to the diagram.

Illustrations of these mean isotherms may be found in the maps in other sections of the paper.

SYNOPTIC EXAMPLES

In the following discussion, we shall attempt to show how the rules we have given can be used in forecasting the motion and intensity of surface highs, lows, and frontal systems, as well as the formation and spread of cloudiness and precipitation that accompany these storms.

Map for Sept. 25, 1942. On the map for Sept. 25, 1942 (Fig. 10), our first problem is to forecast the motion and change in intensity of the two waves on the front just east of the Rocky Mountain region. Using the 10,000-ft flow to guide the two lows, we see that the low in eastern Colorado should move straight east, while the one just north of Montana should move southeast. The velocity of these two lows should be approximately equal to the velocity of the wind at 10,000 ft over them. We see that at the present the northern wave has much stronger winds over it than the southern one. The former is moving at about 35 mph, while the one to the south is moving between 15 and 20 mph. Before deciding whether these values will be constant for the next 24 hr, the forecaster must forecast the change in the 10,000-ft winds over the path of the low. This change will become evident in this case when we consider more carefully the intensity of the two lows.

The low in eastern Colorado is under a current that has only a little cyclonic curvature but has anticyclonic shear along its path. Note how much stronger the 10,000-ft winds are to the left of the approximate future path of this low than they are to the right of this path. This anticyclonic shear indicates that the low will not develop as it moves to the east. On the other hand, the low just north of Colorado should move to the southeast toward Iowa at about 35 mph and deepen as it moves. The deepening is indicated by the cyclonic curvature and cyclonic shear at 10,000 ft over and ahead of the surface low. The cyclonic shear is very evident in this case. Note that to the left of the path of the low the 10,000-ft winds are only about 10 to 15 mph. To the right and over the path of the low, the winds are much stronger. Note in particular the strong winds at Bismarck, Moline, and Chicago as compared with those at Winnipeg, La Crosse, and Milwaukee.

Knowing then that the surface low will move to Iowa and deepen as it does so, we can use this knowledge to help with the forecast of the 10,000-ft pattern. We know that the 10,000-ft low will be displaced toward the colder air from the surface low. If we examine now the temperatures in advance of the surface low, we see that the air mass is extremely cold for this time of the year. In fact, it was in some sections of Illinois and Iowa the coldest air that had ever been observed there that early in the fall. Knowing this, we can say that the air mass that will move in behind the low will not be much, if any, colder than the air mass ahead of the low. We see then that we are forecasting that there will be a deep low in Iowa and that the air mass behind it will be about the same temperature or perhaps even slightly warmer than the one preceding it. We can say, therefore, that the 10,000-ft low will be located nearly directly over the surface low and will be closed or nearly closed, as is the case with all intense surface lows that have cold air surrounding them.

Let us now consider what the effects of this 10,000-ft low will be on the motion of the surface system. First, since it will be over the surface low, the surface low will not have so strong a current over it as it has on this map. Therefore, instead of moving it 35 mph for the entire 24 hr, we must slow it down toward the end of the period when the closed low forms over it. Twenty-five to thirty miles per hour would be a better forecast for its average speed. If now we consider the effects of a closed low aloft over Iowa on the 10,000-ft winds in Kansas and Missouri, we see that the winds should be much stronger than they are now, because in general the strongest winds aloft are found just to the south of a low center aloft. We must now revise our

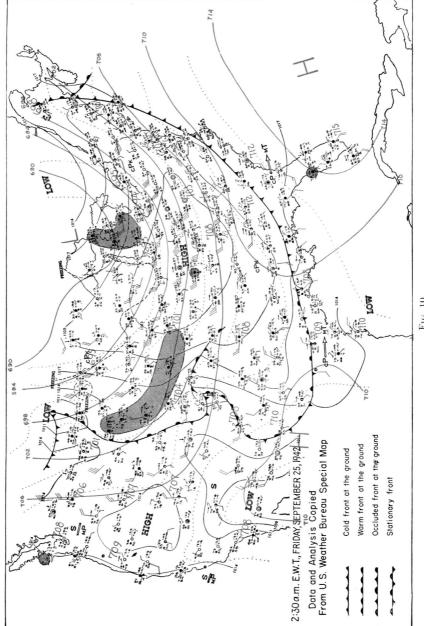

2:30 a.m. E.W.T., FRIDAY, SEPTEMBER 25, 1942

Data and Analysis Copied
From U.S. Weather Bureau Special Map

Cold front at the ground

Warm front at the ground

Occluded front at the ground

Stationary front

FIG. 10.

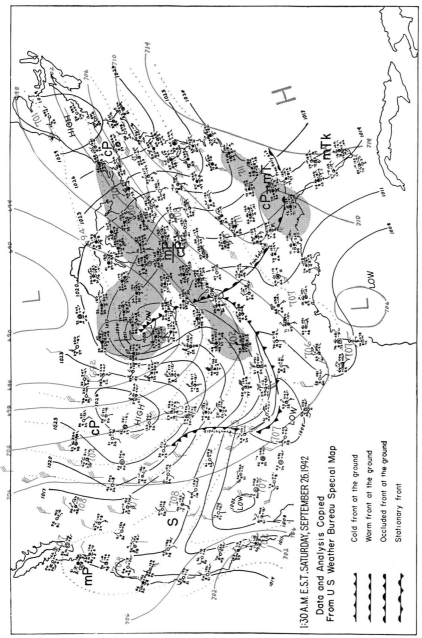

1:30 A.M. E.S.T. SATURDAY, SEPTEMBER 26, 1942

Data and Analysis Copied
From U S Weather Bureau Special Map

Cold front at the ground

Warm front at the ground

Occluded front at the ground

Stationary front

FIG. 11.

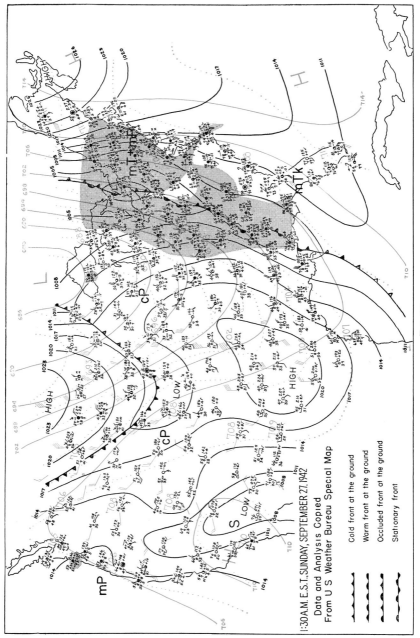

1:30A.M. E.S.T. SUNDAY, SEPTEMBER 27, 1942

Data and Analysis Copied
From U S Weather Bureau Special Map

Cold front at the ground

Warm front at the ground

Occluded front at the ground

Stationary front

Fig. 12.

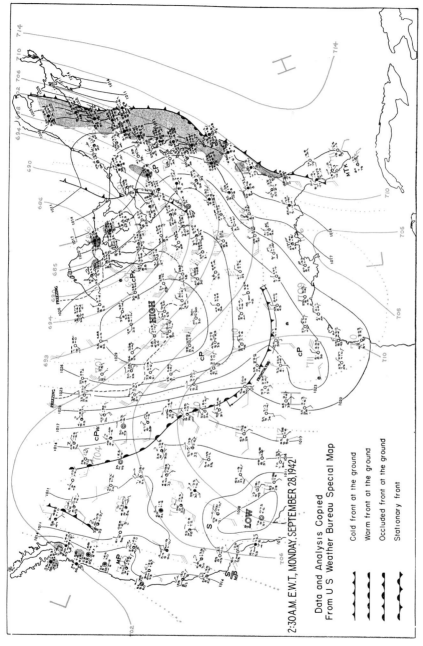

2:30 A.M. E.W.T, MONDAY, SEPTEMBER 28,1942

Data and Analysis Copied
From U S Weather Bureau Special Map

Cold front at the ground

Warm front at the ground

Occluded front at the ground

Stationary front

Fig. 13.

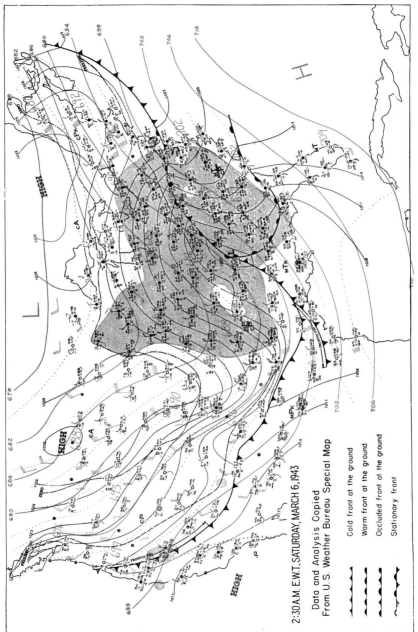

2:30 A.M. E.W.T., SATURDAY, MARCH 6, 1943

Data and Analysis Copied
From U.S. Weather Bureau Special Map

Cold front at the ground
Warm front at the ground
Occluded front at the ground
Stationary front

FIG. 14.

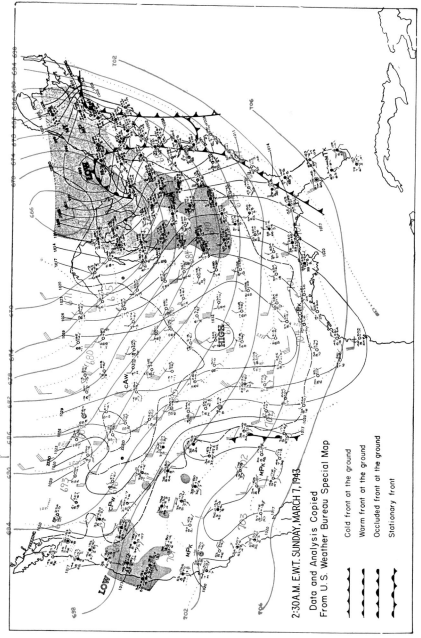

2:30 A.M. E.W.T. SUNDAY, MARCH 7, 1943

Data and Analysis Copied
From U.S. Weather Bureau Special Map

Cold front at the ground

Warm front at the ground

Occluded front at the ground

Stationary front

Fig. 15.

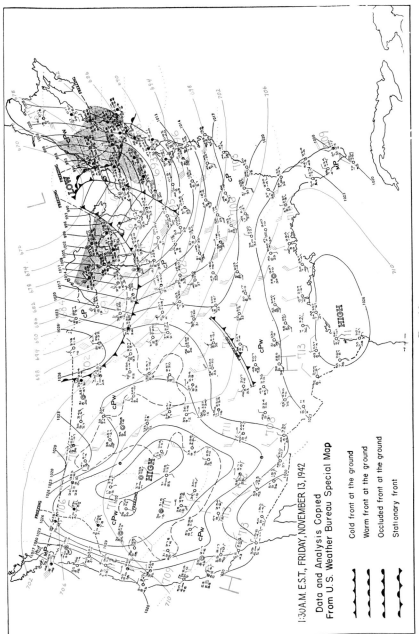

1:30A.M. E.S.T., FRIDAY, NOVEMBER 13, 1942

Data and Analysis Copied
From U.S. Weather Bureau Special Map

Cold front at the ground

Warm front at the ground

Occluded front at the ground

Stationary front

Fig. 16.

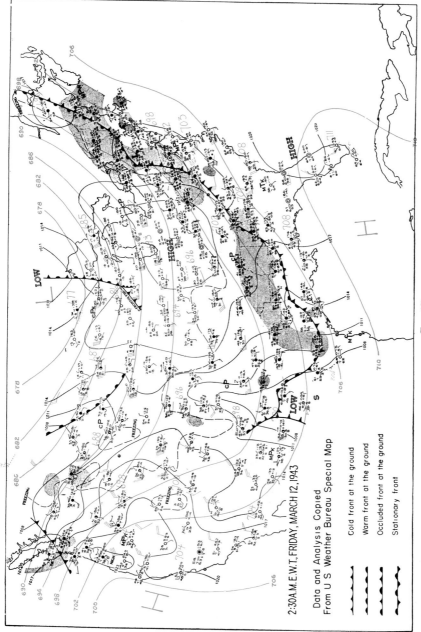

2:30 A.M. E.W.T., FRIDAY, MARCH 12, 1943

Data and Analysis Copied
From U S Weather Bureau Special Map

	Cold front at the ground
	Warm front at the ground
	Occluded front at the ground
	Stationary front

Fig. 17.

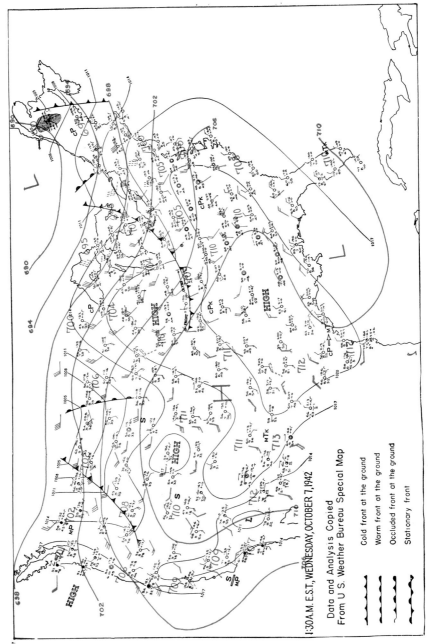

1:30 A.M. E.S.T., WEDNESDAY, OCTOBER 7, 1942

Data and Analysis Copied
From U.S. Weather Bureau Special Map

Cold front at the ground

Warm front at the ground

Occluded front at the ground

Stationary front

Fig. 18.

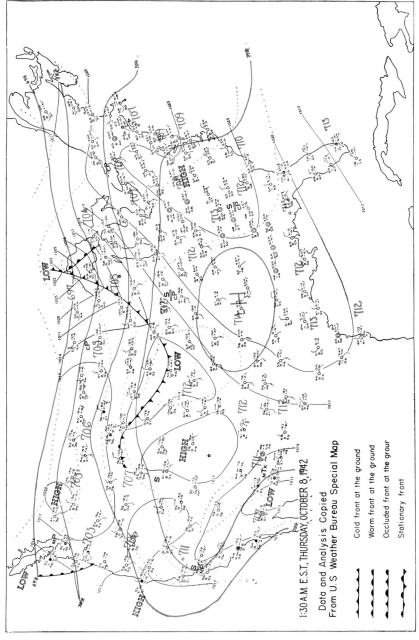

1:30 A.M. E.S.T., THURSDAY, OCTOBER 8, 1942

Data and Analysis Copied
From U.S Weather Bureau Special Map

Cold front at the ground
Warm front at the ground
Occluded front at the grour
Stationary front

Fig. 19.

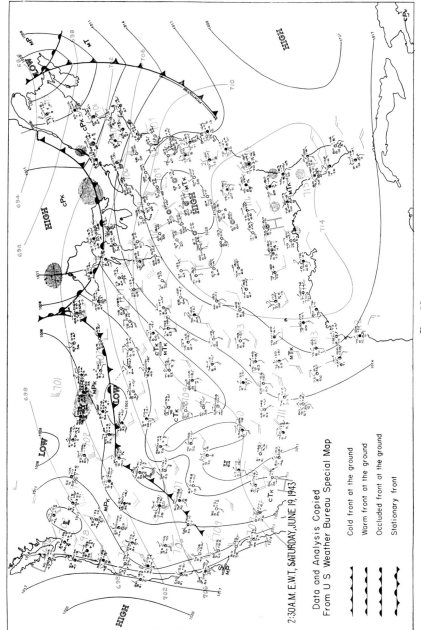

2:30 A.M. E.W.T., SATURDAY, JUNE 19, 1943

Data and Analysis Copied
From U S Weather Bureau Special Map

Cold front at the ground

Warm front at the ground

Occluded front at the ground

Stationary front

Fig. 20.

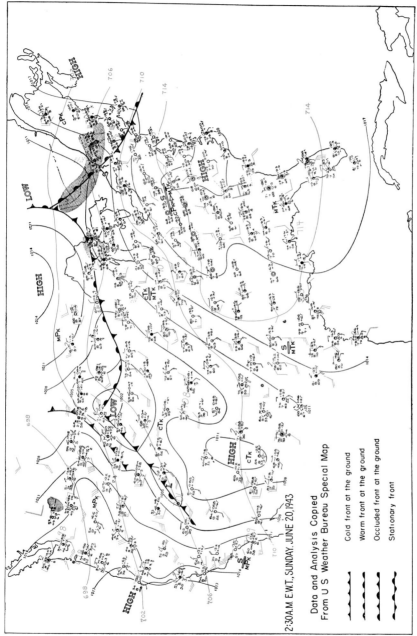

2:30 A.M. E.W.T., SUNDAY, JUNE 20, 1943

Data and Analysis Copied
From U S Weather Bureau Special Map

Cold front at the ground

Warm front at the ground

Occluded front at the ground

Stationary front

FIG. 21.

forecast movement for the low in Colorado. Instead of moving it 15 to 20 mph, we should move it more rapidly. Twenty-five or thirty miles per hour would be more in line with the results of our forecast of the northern system.

So far we have attempted to make a forecast of the motion of the two surface lows using the 10,000-ft winds and then using this motion and the mean-temperature distribution to forecast the 10,000-ft pattern. We then made certain changes in motion of our systems to make them consistent with the changes in the 10,000-ft pattern.

The next problem is forecasting the distribution of clouds and precipitation. We see that all of Oklahoma, Kansas, and Nebraska and North and South Dakota is covered with clouds. Most of these clouds are middle clouds and are due to mP air overrunning the cP air in that region. It is obvious that none of the cloudiness is from mT air, for the winds aloft show that none is moving into these regions. Consider now the region to which this mP air will move in the next 24 hr. If in 24 hr there will be a closed low at 10,000 ft over Iowa, we see that the winds will be from the south or southwest from Iowa eastward. The mP air that is producing the rain and clouds will be moving to the northeast. We see at the present from the mean isotherms that the coldest air is in the Great Lakes region. The very light winds observed there now indicate that this cold air will still be there 24 hr later. If we still have a dome of cold air over the Great Lakes and have south or southwest winds aloft going up over the dome, we see that upslope motion will be occurring there. The air that will be going upslope is the air that is now producing the clouds and precipitation farther west. Since this air is already saturated, the upslope motion should cause precipitation to occur over as much of the Great Lakes region as this current of air overruns. The mean isotherms, by locating the domes of cold air, facilitate the forecasting of clouds and precipitation caused by overrunning.

Let us now consider the likelihood of frontal weather in North Dakota, South Dakota, Nebraska, and Kansas, behind the front that should extend from the low in northern Iowa south and southwest to the mountains. It is apparent from the fact that we have placed our 10,000-ft low directly over the surface low that, no matter in what direction the front extends from the center of the low, it will run perpendicular to the 10,000-ft winds for at least several hundred miles from the center of the low. It follows, therefore, that there should be little or no frontal clouds or precipitation behind the cold front in these regions. However, both clouds and precipitation should occur throughout the region of convergence under the closed 10,000-ft center.

The cold front along the southeast coast is under a region of anticyclonic curvature of the 10,000-ft isobars. It should therefore gradually frontolyze. To be consistent with our forecast of a deep 10,000-ft trough extending from Iowa southward, we should forecast the 10,000-ft winds over this region of frontolysis to become south or southwest. In either case, there would be active overrunning of tropical air north of this region. We should thus forecast increasing cloudiness and precipitation there, even though the front is weakening and should continue to weaken. The extremely low velocities at 10,000 ft at present would indicate that this overrunning must be quite slow until the low from the northwest deepens considerably.

The disappearance of the 10,000-ft trough in Michigan was quite plainly indicated on this map. Upon examination of the change in wind direction with height in the region just to the south and west of the 10,000-ft low, we see that very little advection of either warmer or colder air is indicated. It follows from this that the surface pressure changes are the result of pressure changes occurring at 10,000 ft. Since we know the 10,000-ft pressure is rising in and to the rear of this trough, we should expect it to move and fill during the following period.

Map for Sept. 26, 1942. This situation (Fig. 11) has several interesting features.

The closed surface and 10,000-ft lows in southern Minnesota show the result of the strong cyclonic shear in that region on the previous day.   Since this low is vertical, the steering principle cannot be used to forecast its motion.   The motion of lows of this type can be forecast by using either Petterssen's formulas or the upper-air rules for the motion of upper centers, because the upper and lower centers will both move along together.   In this case, we see that the centers should move to the northeast and deepen as they do so.   It should be noted that surface pressures are falling all through northern North Dakota, Minnesota, and Wisconsin, although there is little advection of warm or cold air there.   If these falls are not due to warm-air advection, as is indicated by the mean isotherms and isobars, they must indicate that the pressure is falling at 10,000 ft.   We should therefore forecast a deepening of the 10,000-ft low as it moves to the northeast.   The falling pressures and slight warm-air advection in Montana indicate the approach of another frontal system from the north, since all such disturbances are preceded by pressure falls and temperature rises.

The motion of the cold front in Oklahoma and Texas is very important in forecasting the general distribution of weather.   We note in this region that the surface winds are considerably stronger than the 10,000-ft winds.   This frequently happens in this region, and it is the writers' opinion that it is caused by the topography.   The cold air moving from the north is prevented from spreading out to the west by the mountains there, and wherever the center of the low is south of the Canadian border there seems to be a funnel effect that increases the wind velocities between the mountains and the low trough.   This is aided in the south by the Ozarks.   The cold front should move south and southeast with the speed of the surface winds behind it. Since it is moving faster than the 10,000-ft winds, it should be active even though the 10,000-ft winds are perpendicular to it.   The surface winds behind it are between 30 and 40 mph, and the front should therefore move about 35 mph to the south. Its eastward motion should be less than this but should increase because these strong winds will become northwest as they move farther south.

The large-scale flow of cold air to the Gulf should be accompanied by falling pressures at 10,000 ft.   From this and the deepening indicated in the north, we should forecast a well-developed trough at 10,000 ft extending from about Lake Superior to the Gulf.   East of this trough, the 10,000-ft winds should be from the southwest, and west of it, they should be from the northwest.   The cold front should be considerably east of this trough line, since it was moving quite fast and should be oriented northeast to southwest.   We should forecast, therefore, that this front will be quite active because it will be parallel to the 10,000-ft winds over it.   On the other hand, the new cold front, expected from Canada, could not be parallel to the northwest winds expected there unless it were moving from the northeast, which would be impossible with the present winds aloft.   We therefore should expect no weather with it.

Map for Sept. 27, 1942.   This is a very good example of an active cold front in the east (Fig. 12).   The cloud and precipitation area extends as far behind it as the winds at 10,000 ft remain parallel to the front.   There are several small waves along the front; but, since there is no cyclonic shear at 10,000 ft, they will not develop.   The cold front in the northwest, on the other hand, is perpendicular to the 10,000-ft winds and therefore has no waves or weather associated with it.

East of the eastern cold front, the 10,000-ft isobars indicate a current of air moving from the south in a straight line or turning slightly cyclonically.   This, then, is a region of convergence, and we should expect the cloudiness and precipitation throughout this region to continue.   The mean isotherms show very little change in temperature from Florida to New York, indicating that a warm front in this region would be impossible.   This is the type of situation in which many analysts will indicate a warm front somewhere along the eastern seaboard to account for the rain and altostratus clouds, even though the mean temperature may be higher on the cold side of the "warm front" than in the warm sector.

This is also an example of an elongated or V-shaped 10,000-ft trough. Such troughs will have clouds and precipitation quite generally in the region east of them and mostly clear skies to the west, regardless of the position of the surface fronts.

Map for Sept. 28, 1942. Along the east coast, this map (Fig. 13) again illustrates a cold front with small waves forming along it in the region where the 10,000-ft winds are parallel to the front. The important change indicated on this chart is the influx of air from the west over the Rockies. It is important to note that this influx of air is taking place more rapidly in the south than it is in the north. When this occurs, pressures rise most rapidly in the southwest quadrant of the 10,000-ft low, and the low consequently fills. This filling of the southwest quadrant of the low is also indicated by the surface pressure tendencies. We see, by comparing the surface isobars with the 10,000-ft isobars in Kansas and Nebraska, that warm-air advection is occurring there. This would tend to cause falling pressures at the surface. The surface pressures are observed to be rising in this region. From this, we conclude that the 10,000-ft pressures are rising even more rapidly than the surface ones, for they must compensate for the lighter air coming in between the surface and 10,000 ft.

This map also shows a good example of an inactive warm front. The warm front east of the Continental Divide is a pronounced front in that there is a large temperature contrast across it and yet there is no weather with it. This will nearly always be the case when the 10,000-ft isobars turn anticyclonically over and on both sides of the front. Compare the 10,000-ft curvature on this map with that on the map of Sept. 25, and note the difference in the weather.

The map for the following day is not included in this report. It shows that the warm front moved eastward and partly frontolyzed. Clear skies still prevailed over and east of the front. The 10,000-ft trough in the Great Lakes region filled very rapidly, giving nearly straight west winds over the eastern half of the country. This then increased the speed of the frontal system along the east coast and changed it from an active front with waves and weather behind it to an inactive one, because the 10,000-ft winds became perpendicular to it when the 10,000-ft trough filled up.

Map for Mar. 6, 1943. This map (Fig. 14) is included to illustrate the motion and development of the large low center in Tennessee and the subsequent development of a secondary center along the east coast. From the steering principle, we see the low in Tennessee should move rapidly to the northeast. The 10,000-ft isobars are cyclonically curved over the low and show pronounced cyclonic shear. The low should therefore deepen as it moves. This deepening is occurring at the time of the map, as is indicated by the surface pressure changes. If we forecast this low to move to the northeast and deepen, we should expect the 10,000-ft trough to deepen in the northeast also and be centered west of the surface low, because the mean isotherms show the air to be much colder to the west of the low. This position of the 10,000-ft trough would indicate southwest winds along the east coast of the United States. We should expect a secondary low to develop along the front that now crosses the east coast. This coastal storm should intensify quite rapidly as the storm in Tennessee occludes and moves northeastward west of the Appalachians. This east coast development occurs whenever the 10,000-ft winds become parallel to the coast, the 10,000-ft isobars show cyclonic curvature or cyclonic shear with straight flow, and a cold air mass is present over the coastal plain. This map is another example of how the precipitation area extends from the front as far to the northwest as the winds remain parallel to the front. The trough line at 10,000 ft marks the edge of this region.

Map for Mar. 7, 1943. This map (Fig. 15) illustrates the development of the secondary storm along the east coast where the winds aloft are parallel to the coast and are cyclonically curved. Note the pressure tendencies in Maine. This map also contains an illustration of an inactive cold front. The front extending from Montreal to Florida now is nearly perpendicular to the 10,000-ft winds. There is no frontal precipitation or frontal cloudiness behind this front. The cyclonic curvature at 10,000 ft behind the front indicates that the air mass is quite unstable. Note the instability showers and stratocumulus clouds that extend throughout the Great Lakes region and from the Mississippi River to the mountains. These instability phenomena are increased by the heating effect of the Great Lakes and the lifting by the mountains. The snows occurring as far south as Nashville and Evansville are too far south to be due to the Lakes and too far west to be due to the mountains. They are instability air-mass showers set off by turbulence under the 10,000-ft trough, because the air is most unstable there. The downslope effect east of the mountains is sufficient to prevent the showers and to dissipate most of the clouds.

Map for Nov. 13, 1942. This map (Fig. 16) is included to give an example of a polar outbreak over the Great Lakes region. The 10,000-ft ridge and the accompanying west winds over Montana are effective in preventing the cold air from flowing southeastward just east of the Continental Divide. The stronger the circulation from the west around this ridge and the farther east the ridge is located, the farther east will the cold-air outbreak occur.

Map for Mar. 12, 1943. This map (Fig. 17) affords another illustration of a cold front which is parallel to the 10,000-ft wind flow. It is therefore an active cold front and has several waves forming on it. The frontal cloud sheet extends as far to the rear of the front as the flow is parallel to the front. Note also, that there is considerable cloudiness and precipitation ahead of the front. This again illustrates the principle that, whenever a current of air moves from the south in a straight line or in a cyclonically curved path, convergence is occurring, and clouds and precipitation should be expected. In this case, the clouds and precipitation are occurring only where the 10,000-ft isobars indicate this type of flow is present. Note the clear skies in Florida where the 10,000-ft flow is anticyclonic. Of all the waves on the front, the one in Texas is most likely to develop because it is closest to the 10,000-ft trough line. There may be considerable doubt as to the correctness of the frontal system in Texas, but there seems to be little doubt as to the presence of a wave of some sort there, and since it is closest to the 10,000-ft trough line it should deepen and move to the east-northeast, as is indicated by the 10,000-ft wind flow. Since the 10,000-ft winds over this wave average about 35 mph, it should move at this speed.

The front in Wisconsin is perpendicular to the 10,000-ft winds and therefore has no weather with it.

Maps for Oct. 7 and 8, 1942. These two maps (Figs. 18 and 19) afford a good illustration of fair-weather situations over the entire United States. Note the large region of anticyclonic flow at 10,000 ft in the west and central part of the country and the anticyclonic shear over the entire country. Whenever the strongest westerlies occur along or north of the Canadian border, the anticyclonic shear thus indicated from the border southward is effective in weakening all the fronts and preventing the formation of any extended areas of clouds or precipitation. Subsidence is quite general throughout these regions of anticyclonic curvature of the 10,000-ft isobars, and the *S* air produced will help maintain the cloudless skies that occur under these conditions. The strong westerly winds aloft usually cause copious precipitation along the west coast; but, if they are also turning anticyclonically, as must be the case on these two days, even this region will be without precipitation. The fronts that reach the central United States usually are accompanied by some cirrus and altocumulus clouds, but in this case even these were not observed.

Maps for June 19 and 20, 1943. These two summer maps (Figs. 20 and 21) were chosen to illustrate a rather frequent summer situation—a large 10,000-ft high south of the Great Lakes region and frontal systems moving around the periphery of the high giving frequent frontal passages in New England where the 10,000-ft winds are from the northwest, causing recurrent pushes of cold air from Canada to move to the southeast in this region. The map of June 19 shows that such an outbreak has just passed New England behind the cold front off the coast. Another should be expected during the next 6 hr behind the front just to the northwest of New England. This front should move southeastward over New England at about 35 mph, since that is the speed of the 10,000-ft winds there. The map of June 20 shows that this cold air has completely covered New England and that another outbreak is approaching from central Canada. This system should be past New England in 24 hr, because it has winds at 10,000 ft over it of about 30 mph and therefore should move at about 30 mph. There is a small trough indicated by the 10,000-ft winds extending from Sault Ste Marie to between Chicago and St. Paul. Note the northwest wind at St. Paul and the southwest wind at Chicago. As this trough moves eastward, we should expect showers and thunderstorms to occur under it in the warm sector of the low to the north. These would be much more frequent and severe during the afternoon than at night, because the air would be most unstable then. The thunderstorms that are occurring in northern Michigan at the time of the map are probably associated with this small 10,000-ft trough rather than with the frontal system to the north. The wave on the front farther to the west should move parallel to the 10,000-ft wind flow and have a history similar to the previous two just discussed.

THE CONSTRUCTION AND USE OF ISENTROPIC CHARTS

An isentropic chart is a chart depicting meteorological elements on a surface of constant potential temperature. Except for nonadiabatic processes, particles move along an isentropic surface. The main purposes of the chart are (1) to aid the forecasting of atmospheric circulations and the weather produced by them, and (2) to study the flow patterns in the free atmosphere. In order to do the latter, it is assumed that the mixing ratio of a particle of air is conservative and that the motions of moist and dry air are followed on the chart.

Data Used on Isentropic Charts. Ordinarily pressure lines, moisture lines, and streamlines are drawn on the isentropic chart.

The pressure lines are isobars, drawn for intervals of 50 mb. Their spacing and direction at each radiosonde station are sent over the teletype wherever they may be determined. This is another form of the information previously sent over the teletype as the "shear-stability ratio vector."

The moisture lines are either lines of constant mixing ratio or lines of constant condensation pressure. The latter are drawn for 50- or 100-mb intervals and have a certain advantage over mixing ratio lines. The proximity of a condensation pressure line and its corresponding actual pressure line shows at a glance the degree of saturation of the air.

The streamlines bear the same relationship to winds that isobars do on a constant-level chart. Values of the isentropic stream function are sent and are used to determine the spacing of streamlines, with higher values of the stream function to the right of the wind.

Other useful data that should be plotted on isentropic charts are the 24-hr changes in pressure and condensation pressure at each station. If the pressure and condensation pressure are approaching each other, the air is nearing its condensation point. This may be useful in forecasting.

In determining areas of condensation, it is not necessary for the actual and condensation pressures at a point to be equal. Because of instrumental errors in radiosonde observations, a condensation pressure 10 or 20 mb less than the actual pressure may denote saturation.

On isentropic charts, regions of saturation are shaded. The condensation pressure lines are not drawn through such regions, since they would have to coincide with the actual pressure lines or show supersaturation. Furthermore, the assumption of conservatism of potential temperature is not fulfilled in such regions, and the interpretation of the lines if drawn would not be the same as elsewhere.

The winds at the radiosonde stations should be plotted on the isentropic chart. Winds at other stations are helpful for very accurate isentropic analysis. These are determined by sketching in the isobars lightly and labeling them as to elevation, assuming a standard atmosphere. The wind is picked from the pilot-balloon run at the elevation indicated by the isobar running through the station. This wind is then plotted on the isentropic chart.

Choice of Isentropic Level. The data for three different isentropic levels are sent daily over the teletype. These levels vary with the season, for in order to be of practical use they must be such that (1) they are above the layer of frictional influence, and (2) they are sufficiently low to have a large range of mixing-ratio values. The following values have been used:

Winter... 290–296°A
Spring and fall....................................... 296–302°A
Summer... 308–320°A

Ordinarily the forecaster should select the intermediate level sent and use it on consecutive days for continuity. However, changes must be made occasionally when unusual weather occurs.

Relationships of Meteorological Elements on Isentropic Charts. The following relationships between meteorological elements on isentropic charts are useful in the interpretation of such charts for forecasting:

1. Pressure lines are lines of constant temperature
2. Pressure lines are lines of constant density.
3. Pressure lines are approximately lines of constant height.
4. The elevation of the potential-temperature surface is an indication of the mean temperature of the air column below the surface, a high elevation implying low mean temperature, and a low elevation implying high mean temperature.
5. Condensation-pressure lines are lines of constant equivalent-potential temperature and lines of constant wet-bulb potential temperature.
6. The wind velocity at a given latitude is inversely proportional to the spacing of the streamlines.
7. The isobars at any point on the isentropic surface are approximately parallel to the vertical shear of the geostrophic wind (*i.e.*, the thermal wind) at that point.

This fact can be used as an aid in constructing the isentropic isobar field, since the directions of the thermal winds at various points on the given isentropic surface can be determined with sufficient accuracy by estimating from pilot-balloon reports the actual wind shear in the vicinities of these points. The direction and horizontal spacing of the isobars for the isentropic surface are included in the data transmitted over the teletype whenever the winds can be measured to sufficient heights.

Very small or very large values of isobar spacing, as sent in on the teletype, are not useful, since such values will usually apply to only a limited area in the immediate vicinity of the observing station. Light thermal winds are frequently inaccurate because of the coding of direction of winds as sent on the teletype to the nearest

10 deg and because of errors of observation due to vertical motions. With a very steep lapse rate, the turbulence in the region will make the assumed ascension rate of the balloon fallacious and will therefore give incorrect results for the computed isobar spacing and direction.

Principles for Drawing Isentropic Charts. On isentropic charts, the data are sufficiently widely scattered so that a unique pattern may not be drawn on the chart. Judgment must be exercised in using the data so that the resultant chart will be reasonable. Since the surface map is usually more definite, the isentropic chart should, first of all, be consistent with it and, secondly, should be consistent with other charts and data, such as the 10,000-ft chart.

To ensure such consistency, the first thing to do in isentropic analysis is to draw the surface fronts on the isentropic chart. Centers of surface highs and lows should also be marked. Then, with the 10,000-ft chart at hand, the analysis of the isentropic chart is begun.

In analyzing isentropic charts, the streamlines should be drawn first, after all the winds have been plotted as outlined above. The advantage of drawing them first is that the analyst will then be conscious of the main flow pattern. Since air particles move with the winds, he will be deterred, for example, from moving a dry or moist tongue in from the east in regions where the streamlines show strong west winds. In general, drawing the streamlines first makes for a more reasonable isentropic chart.

Once the general flow pattern is determined, the pressure and condensation-pressure lines are drawn for intervals of 50 mb. In the past, it has been customary to space these lines as evenly as possible, but from a physical point of view such spacing is highly improbable. Pressure lines should be closely packed behind strong cold fronts or ahead of strong warm fronts. They should be parallel to these fronts. If the data show that pressure lines cross a front or are widely spaced on the cold side of the front, frontolysis or a weakening of the front is indicated. In warm sectors, the air is fairly homogeneous so that pressure lines should have wide spacing and should be oriented east-west, in general, since latitudinal differences in heating produce the main temperature gradients in warm sectors. Thus, it appears that the practice of drawing evenly spaced isobars on isentropic charts should be avoided in favor of lines drawn to fit actual surface and upper-air conditions.

Where pressure lines are drawn to fit surface fronts, cold tongues should push in from the direction shown by the streamlines behind cold fronts. Warm tongues develop ahead of and over cold fronts with the maximum temperatures immediately in advance of the cold front. Warm tongues branch when a surface system occludes, with one branch turning anticyclonically and the other turning cyclonically back along the occlusion. Ordinarily both warm and cold tongues tend to have an anticyclonically curved axis and frequently have closed isobars at the forward end of the tongue.

Condensation-pressure lines ordinarily have the same pattern as the actual pressure lines, so that ordinarily warm tongues are moist and cold tongues dry.

In drawing these tongues, besides taking into consideration their direction as determined by streamlines, the analyst must consider the source regions. He must see that dry and moist tongues originate in appropriate regions. This means that in general the isobars are open to the north in dry tongues and to the south in moist tongues.

Summer isentropic patterns are generally persistent, except for diurnal changes. However, they are complicated by the fact that convection during the day often reaches to considerable heights, often up through the isentropic surface. This frequently occurs in the southeastern United States. The source region then is below the isentropic level, and moist tongues frequently are not open toward the south. This is so because in summer the Gulf of Mexico is often relatively cool and has less convection over it than there is over the hot land. In summer, the Pacific Ocean is not a source of warm air because of the relatively cool coastal waters. Moist air in the southwest in summer originates at low levels and penetrates into the isentropic surface by convection in the vicinity of Arizona and New Mexico, the ultimate source of moisture being the Gulf of Mexico.

Up to the present time, moist and dry tongues have been shaded on isentropic charts to bring out the contrasts between air masses. However, the authors feel that a better picture of the temperature and moisture distribution could be obtained if the chart were shaded in red toward the higher moisture values starting with some moisture line that divides the chart approximately in half, and if the chart were shaded in blue on the low-pressure side of the isobar having the same numerical value as the above-mentioned moisture line. This method of shading accentuates both the moisture and the temperature distribution and to some extent the distribution of relative humidity.

Arrows are drawn in dry and moist tongues with the shaft of the arrow parallel to the axis of the tongue and the arrowhead pointing in the direction of the near-by streamlines.

After analyzing the isentropic chart, the analyst should recheck his work with the surface map and winds aloft. Without some sort of radio direction finder, pilot-balloon observations are impossible to make through cloud layers. Hence, saturated areas should be devoid of winds. Saturated areas should appear where there are clouds or precipitation reported at the surface. Such clouds and precipitation will appear on some isentropic surfaces. The elevations given for the clouds should be checked to see if they are near the isentropic surface drawn. Finally, a check for continuity from the previous isentropic charts, allowing for diurnal changes, should be made.

For an example of an isentropic chart, see Fig. 22.

The Use of Isentropic Charts in Forecasting. Isentropic charts may be used (1) in forecasting the development and deepening of cyclones and anticyclones, (2) in forecasting frontogenesis and frontolysis, and (3) in forecasting cloudiness and precipitation.

Forecasting the Development of Cyclones and Anticyclones. Cyclones and anticyclones develop, in general, in regions where the isentropic chart shows a crowding of isobars. Usually such regions have considerable potential energy concentrated in them. If, however, the lapse rate is very steep in either air mass, but not in both, the close packing of the isobars will be due to this difference in lapse rate. The air mass where the steep lapse rate exists will appear colder than the other but may have nearly the same surface temperature. The forecaster should check this point by use of the radiosonde observations in the region.

If on successive isentropic charts we observe that a cold dome has built up over one portion of the country, we should realize that this dome may represent a large concentration of potential energy, and we should be on the lookout for the development of a storm on the eastern side of the dome if the isobars become crowded there. To release the potential energy, the wind there should blow from the cold into the warm air. The cold air will then start spreading out, with some of the cold air turning cyclonically and some anticyclonically. Hence, a low will form to the east or northeast with an anticyclone toward the west or southwest.

It has been found that 36 hr usually elapse from the time when the cold air starts spreading out until the time when the storm has deepened completely.*

* This statement is in agreement with conclusions reached by Lt. G. P. Cressman from an unpublished study of a number of weather situations made at the University of Chicago.

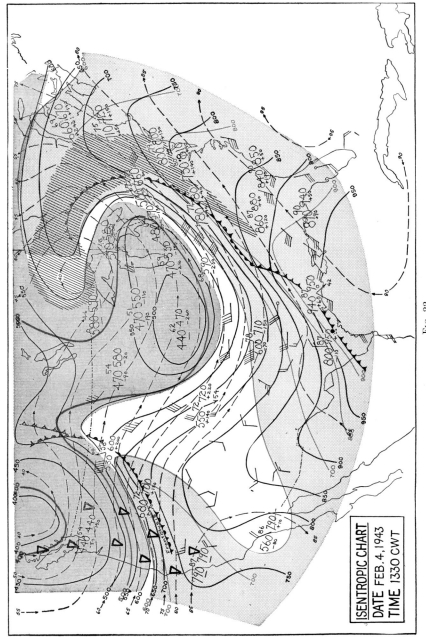

ISENTROPIC CHART
DATE FEB. 4, 1943
TIME 1330 CWT

Fig. 22.

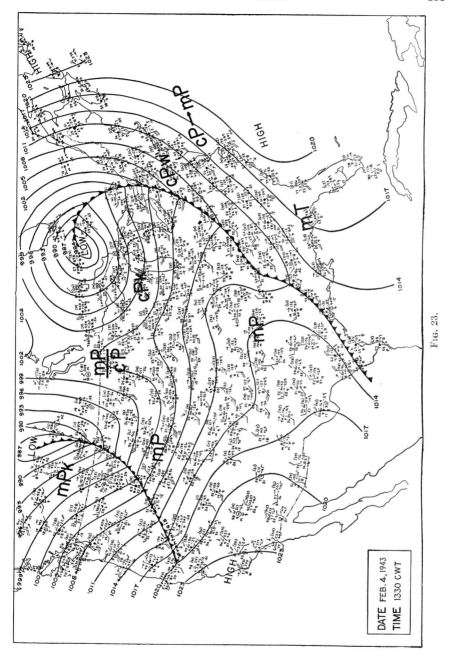

FIG. 23.

The isentropic chart can also be used in forecasting by noting cold pushes from the north. These appear as cold tongues. In almost all cases when the cold tongue is pronounced, cyclogenesis will take place to the left of the tongue, and anticyclogenesis will take place to the right. The same thing occurs when there is a pronounced warm tongue from the south.

The development of cyclones may be forecast by the consideration of moist tongues on isentropic charts. When a moist tongue *starts* to split into two branches, one curving cyclonically and the other anticyclonically, cyclogenesis is likely to occur beneath the branching. Moist tongues continue to branch for some time after a cyclone is fully developed and is no longer deepening. Hence it is important in using this pattern for forecasting cyclogenesis to make such a forecast when the moist tongue first starts to branch, not after it has been branching for considerable time. Care must be taken to be sure that the moist tongue is branching. Frequently some moist air will come in from another direction and will make a moist tongue appear to branch. In order to be useful in forecasting, the isentropic chart must be drawn with careful regard for continuity.

Forecasting Frontolysis and Frontogenesis. When the isobars on an isentropic chart begin to spread out in the region of a surface front, they indicate that a decreasing amount of potential energy is available at the front. We can then forecast that the front will become less pronounced.

If the isobars on an isentropic chart actually cross the surface front at right angles, or nearly so, then the energy of the front will be dissipated through lateral mixing. As soon as this condition exists, frontolysis should be forecast.

In this connection, it might be of use to mention some effects of lateral mixing noted by Namias.[10] Lateral mixing generally takes place along isentropic surfaces. Hence, frontal surfaces parallel to isentropic surfaces will easily persist, whereas fronts that intersect the isentropic surfaces will frontolyze unless they are in highly frontogenetical regions.

When precipitation occurs, the equivalent potential temperature, rather than the potential temperature, is conservative, so that lateral mixing will occur along surfaces of constant equivalent potential temperature instead of along isentropic surfaces. Under these circumstances, isentropic surfaces may cross frontal surfaces in the vicinity. Near cold fronts, widespread precipitation is rare, so that in general the isentropic surfaces are parallel to cold fronts and frontolysis is therefore not so likely as in the case of warm fronts.

A packing of isobars on an isentropic chart in a region where no surface fronts exist indicates the possibility of frontogenesis. This is particularly useful in forecasting the development of warm fronts in the eastern United States ahead of a surface low-pressure system moving in from the west or northwest.

Forecasting Cloudiness and Precipitation. Most cloudiness and precipitation occurs in regions of moist tongues on isentropic charts. In summer, in particular, moist tongues are useful for forecasting, since air-mass and frontal thunderstorms are concentrated where these tongues exist. At any time of year, cloudiness and precipitation are more prevalent toward the left and forward sides of moist tongues than elsewhere.

In general, when the winds, as shown by the streamlines, blow across the isobars, upslope or downslope motion results, since winds generally increase with height. Of course, if there were no changes in wind velocity with height, there would be no such upslope or downslope motion. In order to determine this motion, recourse may be had to relative-motion charts. On these charts, the isobars at only one level are drawn. The central isentropic level is used for economy of effort. Because it is assumed that all three isentropic surfaces have approximately the same pressure pattern, it is unimportant which particular surface is used. The difference between the stream function at the top and bottom isentropic surfaces is plotted at each station. Then lines are drawn for equal values of this difference. These will be relative streamlines. These lines indicate the direction of flow at the upper level with respect to the lower level. Assuming that the pressure pattern moves with the speed of the lowest level, these relative streamlines will indicate where upslope and downslope motion exists on the isentropic surface.

Where the relative wind, as indicated by the relative streamlines, blows across the isobars from higher to lower pressure, upslope motion exists. In regions where this occurs, cloudiness and precipitation are prevalent; while in regions of downslope motion, where the relative winds blow from lower to higher pressure, cloudiness is curtailed, precipitation infrequent, and frontolysis likely. In general, the larger the angle between the isobars and streamlines on an isentropic chart, the greater the upslope or downslope motion is and the more pronounced are the effects of such motion. Where the streamlines are very widely spaced, however, the winds are too light to produce any marked effects.

In forecasting cloudiness from upslope motion, the forecaster should take into account the moisture content of the air. The nearer the air is to saturation, the sooner cloudiness and precipitation should be forecast. In New Mexico, for instance, upslope motion often exists, but the air is far too dry to produce cloudiness.

There are a few generalities that can be made concerning upslope motion. In general, air moving toward the north moves upslope, since it is generally colder to the north. Air usually moves downslope toward the south. Therefore, increasing cloudiness and precipitation can be expected with northward-moving air currents and increasing fair weather as the air moves toward the south.

Use of Isentropic Weight Charts in Forecasting. Isentropic weight charts are constructed by plotting the difference between values of actual pressures at two isentropic surfaces on a map and superimposing the streamlines from one isentropic surface. The difference of pressure between the two levels is a measure of the difference in height between them. The farther apart two isentropic surfaces are, the steeper must be the lapse rate between them. Thus at a glance the isentropic weight chart shows the relative lapse rates in different air masses.

To be useful for forecasting the patterns of pressure, differences on the isentropic weight chart must move more slowly than the wind between the isentropic levels. Suppose, now, that the isentropic weight chart ("thick-thin" chart) shows, by the streamlines, that air is moving from a region where the isentropic surfaces are close together to one where they are far apart, *i.e.*, the air blows from "thin" to "thick." The current of air moves along potential-temperature lines with the top at one potential temperature and the bottom at the other. As the column of air moves from the "thin" to the "thick" region, it will stretch vertically as the isentropic surfaces bounding the top and bottom of the column are farther apart in the "thick" region. This vertical stretching must be compensated for by horizontal convergence. This, in turn, leads to cyclogenesis. Conversely, when air moves from "thick" to "thin," the possibility of anticyclogenesis is enhanced.

The use of isentropic weight charts, and of relative motion charts as well, is usually not worth the time necessary to construct them. Most of the time the patterns do not give definite indications one way or another. But occasionally, particularly in forecasting intense cyclogenesis, or anticyclogenesis, good results have been obtained.

Limitations of Isentropic Charts. There are two types of limitation to isentropic charts, the first physical, the second practical. In following the motion of particles on isentropic charts, it is assumed that the potential temperature is constant, a condition that is not fulfilled in the following nonadiabatic processes common in the atmosphere:

1. Radiational cooling and heating.
2. Evaporation and condensation.
3. Convective activity.

It is further assumed that the moisture content of air is conservative, and this, too, varies.

Other limitations are practical. In the course of the above discussion, isobars and moisture lines on isentropic charts have been treated together for the most part. In the winter, the moist tongues almost always coincide with the pressure troughs in the isentropic surface, while dry tongues are coincident with regions where the isentropic surface is at a high level. Hence the only information added by the moisture lines on the isentropic chart is the proximity of the air to saturation.

The reason that the *position* of moist and dry tongues is not useful in winter depends upon the fact that the vertical gradient of moisture is far larger than the horizontal. This is true in all seasons, but in winter the isentropic surface varies so greatly in elevation that the moisture lines merely reflect the changes in elevation, rather than delineate the horizontal moisture differences. In summer, on the other hand, the isobaric pattern is frequently flat, and the moist tongues are really representative of horizontal moisture gradients.

Another practical limitation of isentropic charts is the time element. At the present time, the isentropic data come in on the second radiosonde transmission. For instance, the isentropic data for the midnight sounding come in about 0800. This means that the isentropic chart is usually not plotted and analyzed until 1100 or noon, not until *after* the main forecasts for the day must be made. The consideration of time has led various meteorologists to look for substitutes for the isentropic chart that can be used as soon as the first radiosonde transmission comes in at about 0230.

Proposed Substitutes for Isentropic Charts. As far as forecasting the intensity of fronts, cyclogenesis, and anticyclogenesis is concerned, the mean isotherms between sea level and 10,000 ft may be used in the same way that isobars on isentropic charts are used. Packing or spreading out of mean isotherms has the same significance as a similar configuration of isobars on isentropic charts.

As a substitute for condensation pressure lines on an isentropic chart, G. T. Stephens of the U.S. Weather Bureau has been using condensation temperatures at 10,000 ft, assuming a constant pressure of 700 mb. The proximity of a condensation-temperature (or dew-point) line and its corresponding temperature line gives essentially the degree of saturation of the air, and thus moist tongues will be delineated.

The authors feel that this is more satisfactory than the isentropic chart but that even better results could be obtained by dealing with the mean moisture in a layer and mean temperatures, rather than with the moisture at one surface, whether isentropic or constant level. Research is being done on this subject at present.

Bibliography

1. STARR, V. P.: "Basic Principles of Weather Forecasting," Appendix, Harper, New York, 1942.
2. PETTERSSEN, S.: "Weather Analysis and Forecasting," Chap. IX, McGraw-Hill, New York, 1940.
3. BICE E. G. and G. T. STEPHENS: Correlation of Isallobaric Patterns in the High Atmosphere with Those at the Surface," *Misc. Rept.* 12, Institute of Meteorology, University of Chicago, 1944.

4. Rossby, C.-G.: Kinematic and Hydrostatic Properties of Certain Long Waves in the Westerlies, *Misc. Rept.* 5, Institute of Meteorology, University of Chicago, August, 1942.

5. Wulf, O. R., and S. J. Obloy: The Utilization of the Entire Course of Radiosonde Flights in Weather Diagnosis, *Misc. Rept.* 10, Institute of Meteorology, University of Chicago, March, 1944.

6. Starr: *op. cit.*, p. 23.

7. Namias, J.: "Methods of Extended Forecasting," p. 23, U.S. Weather Bureau, Washington, D.C., 1943.

8. Wexler, H.: Some Aspects of Dynamic Anticyclogenesis, *Misc. Rept.* 8, Institute of Meteorology, University of Chicago, 1943.

9. Miller, C. H., and W. L. Thompson: A Proposed Method for the Computation of the 10,000-foot Tendency Field, *Misc. Rept.* 1, Institute of Meteorology, University of Chicago, 1942.

10. Namias, J., *et al.:* An Introduction to the Study of Air Mass and Isentropic Analysis, p. 141, American Meteorological Society, Milton, Mass., 1940.

WEATHER ANALYSIS FROM SINGLE-STATION DATA

By Vincent J. Oliver and Mildred B. Oliver

INTRODUCTION

During the last few years, wartime conditions have made it necessary for isolated combat units to issue forecasts in regions where no network of meteorological stations could be made available. Frequently the data from several stations or from reconnaissance planes are available, but in some regions the forecaster must rely only on surface and upper-air observations made at his own station. This is particularly true in the case of ships at sea. Hence, it is important to develop proficiency in extracting information from limited aerological data. The technique outlined below was developed for working with data from a single station, but if the data from two or more stations are available, the analysis will be much more accurate. The same technique is then applied to each station reporting weather conditions.

After the forecaster has analyzed all the available material, he reconstructs the surface and 10,000-ft maps to fit all the data and then forecasts the weather from a combination of these charts according to the methods described in the section on the use of upper-air data in forecasting (pages 813–857). In such a reconstruction the meteorologist must be thoroughly familiar with the characteristics of the normally observed patterns of surface and upper-air circulation. He must further be able to distinguish between solutions which are internally consistent, and those which are inconsistent and therefore to be discarded. The mean-temperature chart is drawn to check the consistency between the sea-level and the 10,000-ft pressure patterns. The procedures outlined below apply primarily to regions in middle latitudes but may be adapted for other localities.

OUTLINE OF GENERAL PROCEDURE IN FORECASTING FROM LOCAL AEROLOGICAL DATA

I. Plot pilot-balloon observation on a polar diagram.
II. Plot pressure-temperature and characteristic (θw) curves for the entire radiosonde flight on a tephigram or other energy diagram. The analysis is facilitated if both curves are on the same diagram.
III. Plot a graph of the variation with time of as many of the following elements as are available:
    A. Sea level, 10,000 ft, and 10 km pressures.
    B. Surface temperature and dew point.
    C. Mean temperature between sea level and 10,000 ft.
    D. 13-km potential temperature.
    E. Wind direction, sky conditions, and clouds, which should be plotted at the bottom of the graph at 6-hr intervals.
    F. Frontal passages.
IV. Determine from a careful investigation of all the data the most probable circulation pattern, both aloft and at the ground.
    A. To determine the circulation pattern at 10,000 ft:
        1. Note the air masses present at the surface and aloft. The type of air mass gives some indication of the general circulation if the source regions of the various types of air mass are known. For example, tropical air found at a

station in a northerly latitude would indicate that the circulation pattern aloft was one with ridges and troughs of large amplitude, since such a pattern is required for tropical air to be transported far to the north.

2. Determine whether there is a warm- or cold-type low- or high-pressure area in the vicinity of the station. Warm lows in middle latitudes with warm highs to the south usually exist with strong west to east winds at 10,000 ft over the lows, and weak 10,000-ft circulation over the highs (high-index conditions). Cold highs and lows are associated with 10,000-ft ridges and troughs of large amplitude (low-index conditions).

3. Examine the temperature and height of the tropopause. A cold high tropopause indicates tropical air, and a warm low tropopause indicates arctic type air. Either of these indicates a circulation pattern of waves with large amplitude.

4. Note the wind direction and velocity at high levels. Large northerly or southerly components occur only with waves of large amplitude.

5. From the wind direction and velocity and 10,000 ft pressure, determine the location of the 10,000-ft ridges and troughs with respect to the station. Since high pressures lie to the right of the wind and low pressures to the left (Northern Hemisphere), the general picture of the 10,000-ft pattern near the station may be obtained from the wind alone. The 10,000 ft pressure should be compared with the normal 10,000 ft pressure for the station for the season of the year. Any large departures from normal indicate a trough or ridge of considerable intensity, usually of large amplitude.

6. From the wind direction and velocity at 10,000 ft, determine the spacing and direction of isobars at that level.

B. To determine the sea-level circulation pattern:

1. Use the gradient wind direction and changes in it during the past 24 hr to determine the location of sea-level pressure systems with respect to the station, just as the 10,000-ft wind was used to determine the circulation at that level.

2. Locate any fronts that are in the vicinity of the station as follows:

   a. Determine from the surface reports the hours at which fronts passed the station, and displace them from the station at a speed determined from the pilot-balloon observations.

   b. Examine the radiosonde observation to see if there are any fronts approaching the station.

   c. Examine the pilot-balloon observation to see if any fronts are approaching the station and to ascertain how rapidly they are moving.

   d. Determine the orientation of near-by fronts by use of the mean isotherms, orientation of clouds, the change in gradient wind direction, or by a stability analysis.

   e. Determine the spacing and direction of isobars at the surface.

V. Construct the sea-level and 10,000-ft isobars on one chart. Draw them to fit all the deductions made from the data in the previous analysis. Check the internal consistency of the patterns by use of the mean-temperature chart.

VI. Shade in precipitation and cloud areas on the map in accordance with the principles discussed in the section on the use of upper-air data in forecasting.

VII. Forecast from the reconstructed map, taking into account deepening or filling of systems determined from the pilot-balloon and radiosonde data.

TECHNIQUE OF UTILIZING THE PILOT-BALLOON DATA IN SINGLE-STATION ANALYSIS

Of all the measurements made in the atmosphere, the pilot-balloon run is the most important in single-station work. In the following paragraphs, we shall discuss the different ways we can use the pilot-balloon run, after it is plotted on a polar diagram.

On the polar diagram, a point for the wind speed and direction is plotted for every thousand feet, as in the example below. The line joining these points constitutes a hodograph (Fig. 1).

1. The Gradient Wind. In order to determine the direction and spacing of the sea-level isobars from the balloon run, it is necessary to determine the gradient wind. The effect of surface friction shows up on the hodograph as a turning of the wind to the right (Northern Hemisphere) and an increase in velocity with height throughout the

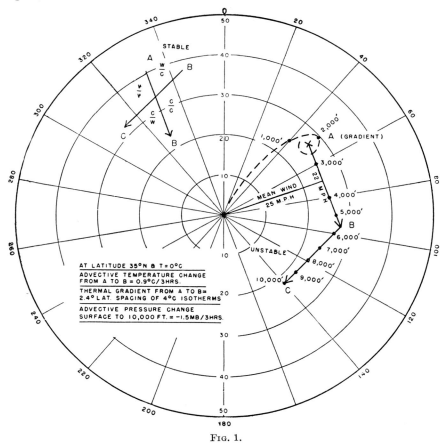

FIG. 1.

frictional layer. Under certain assumed conditions, Ekman has shown that this change in direction and velocity with height will appear on the hodograph as a spiral extending from the center of the hodograph. The gradient wind would be determined by a vector from the center of the hodograph to the center of the spiral. Actually, since Ekman's conditions rarely exist in the atmosphere, a clear-cut spiral does not usually appear on the hodograph. Hence, in most cases the spiral cannot be used to determine the gradient wind. In such cases, the forecaster will obtain comparable results from successive pilot-balloon runs if he assumes that the gradient wind is approximately 20 to 30 deg to the right of the surface wind in direction and has a velocity approximately equal to the wind velocity at about 1,500 or 2,000 ft above the ground. Surface winds that are unrepresentative because of some local factor should not be used. These approximations depend upon topography and the type of air mass involved. In making these approximations, the forecaster should decide what factors are distorting the Ekman spiral and should take all of them into account.

At any level above the gradient level, the direction and spacing of isobars may be determined directly from the pilot-balloon run, since the winds will be nearly parallel to the isobars.

2. Horizontal Temperature Distribution. The horizontal temperature distribution is deduced from the hodograph by using the change of wind direction and velocity with height. A vector from a point on the hodograph to some higher point, *e.g.*, from A to B (Fig. 1), will be parallel to the mean isotherms in the layer between the two points with cold air to the left and warm air to the right (Northern Hemisphere). The length of the vector is proportional to the magnitude of the horizontal temperature gradient.

The mean isotherms in a cold air mass will be parallel to any strong front within 200 or 300 miles of the station. They will also be closely packed in that region. We can therefore tell from the direction and spacing of the isotherms as indicated by the hodograph whether there is a front near by, and if there is what orientation it has. Small shears on the hodograph indicate weak horizontal temperature gradients. In view of the fact that such gradients occur in warm sectors and in the midst of large highs where no fronts exist, small shears should not be used to determine the orientation of fronts.

Besides giving the orientation of fronts, the mean isotherms indicate on which side of a dome of cold air the station is located, because the dome of cold air is always to the left of the mean shear vector. Knowing where the dome is located and knowing the direction of the upper-air winds, the forecaster can tell where upslope and downslope weather is probable. This type of analysis should be used to determine where precipitation and cloud areas should be entered on the reconstructed map.

The spacing of the mean isotherms can be used in deducing the mean temperatures in the regions surrounding the station. The surface temperature may be arrived at by taking the lapse rate at each station into consideration.

Since the direction of upper-air winds is sent to every 10 deg and since the accuracy of the velocities depends on the assumption of an ascensional rate for the pilot balloon, which is invalidated by upward or downward motion, indicated shears that could be due to these inaccuracies in the actual pilot-balloon observations should not be used to determine the orientation of fronts, etc. As a general rule, any shear produced by a change in the wind direction of 10 deg or less or a change in velocity of 10 mph or less should not be relied upon.

3. Advection. Once the reliable shears have been drawn on a hodograph, it is a simple matter to determine how much warm or cold air is being brought in over the station. Consider the shear AB in Fig. 1. The perpendicular to the shear AB from the center of the diagram gives the mean wind in the layer AB perpendicular to the horizontal temperature gradient, and therefore the direction and rate of motion of the mean isotherms. The value of the resulting temperature change can be determined by using the Horizontal Temperature and Pressure Gradient Scale which was devised by John C. Bellamy of the University of Chicago for this purpose, or it can be computed with the aid of the thermal wind equation. In the example referred to (Fig. 1) the shear indicates that warmer air lies to the west-southwest of the station. The mean wind perpendicular to these isotherms is likewise from the west-southwest, 25 mph. This indicates that the warm air to the west-southwest of the station is approaching at the rate of 25 mph.

By a continuation of this reasoning, it is clear that warm-air advection will always show up as a veering of winds with height, and cold-air advection as a backing of winds with height (Northern Hemisphere).

The most important use of advection is to determine whether a front is approaching or receding. All types of fronts are preceded by warm-air advection near the ground, and with warm and occluded fronts the warm advection extends to high levels. In the case of cold fronts, the warm-air advection in the lower levels is usually accompanied by slight cold-air advection in a layer somewhere between 6,000 and 15,000 ft as much as 200 miles ahead of the front. This cold-air advection aloft in advance of the front is often the only indication that a cold front is within 200 miles of the station.

Cold-air advection in the lower levels indicates that the cold air mass is getting deeper and that hence the cold front is moving away from the station. When this cold advection ceases, the center of the cold dome has been reached. Beyond the center of the cold dome, warm-air advection begins. Therefore, if the type of advection is noted for several consecutive periods, the location of the station with respect to the center of cold domes may be readily ascertained.

The distribution of advection with height may be used for obtaining the degree of activity (i.e., the amount of cloudiness and precipitation) of fronts. If upon approach of a warm front, the balloon run indicates very little advection in the lower 3,000 to 5,000 ft and marked warm-air advection above, the approaching front will be active, because the warm air aloft will be moving in more rapidly than the cold air in the lower levels is retreating. The approach of an active warm front is indicated by a rapid increase of wind velocity with height even before it shows up as the above distribution of advection. On the other hand, if the strongest warm-air advection occurs near the surface, or if the winds decrease in velocity with elevation, an approaching warm front will be inactive and will have an attenuated cloud system.

Cold fronts are inactive when the winds aloft are perpendicular to them and increasing with elevation. It follows that, in periods of strong west or northwest wind aloft, all cold fronts approaching the station from a northwesterly direction will be inactive. With active cold fronts, the winds aloft are parallel to the front or blow across the front from the warm to the cold side. The precipitation and cloudiness behind the front persist as long as the wind distribution indicates presence of activity. This is usually until the trough line aloft has been reached. If the forecaster knows the approximate orientation of an approaching cold front, he can infer the activity of the front from the winds aloft ahead of the front. The approximate orientation may be obtained from a knowledge of the location of source regions of cold air with respect to the station. For example, in the central United States, most cold fronts are oriented northeast-southwest so that northwest winds accompany inactive fronts, whereas in Texas and Oklahoma most cold fronts are oriented east-west so that northerly winds aloft would be associated with inactive fronts and westerly winds with active ones.

4. Stability. The horizontal distribution of stability also may be determined from the hodograph. Consider two shears AB and BC, as indicated in Fig. 1. Since in the lower layer (A to B) the warm air lies to the south and west of the station with cold air to the north and east, while in the upper layer (B to C) the warm air lies to the west and cold air to the east as indicated above, there is relatively cold air over relatively warm air to the south of the station. Hence in this quadrant the lapse rate will be steeper than in the other quadrants. To the north, where relatively warm air overlies cold air, the stability will be greatest. Instability usually shows up in the direction of cyclonically curved isobars near the ground and aloft. Since cold fronts are usually associated with the major troughs, the instability usually shows up toward a cold front.

If there is no change of the isotherm direction with height, a change of shear may still exist.   Consider the case where there is a shear from west to east from the ground to 6,000 ft and no shear above that level.   This indicates that in the lower 6,000 ft it is warmer to the south and colder to the north, while above 6,000 ft the temperature is uniform in all directions.   The lapse rate, then, would be steepest where the surface temperature is highest, *i.e.*, toward the south.

   5. Advective-pressure Change.   The pressure tendency at levels above the ground can be computed with the aid of the hodograph.   The surface-pressure tendency can be considered as being equal to the tendency at 10,000 ft (or any other level) plus the rate of change of weight of the air between the surface and 10,000 ft. The rate of change of weight between the surface and 10,000 ft will be referred to as the advective-pressure change and may be computed with the aid of the following equation:

$$\frac{\partial \, \Delta p}{\partial t} = 2.16 \times 10^4 \bar{\rho} f C_n C_T$$

where $\partial \, \Delta p / \partial t$ is the advective pressure change in millibars per 3 hr, $\bar{\rho}$ is the mean density in the layer considered (about 1.1×10^{-3}), f is the Coriolis parameter at the latitude in question, C_n is the wind velocity perpendicular to the shear, and C_T is the shear.   It is assumed that velocities are measured in miles per hour.   On a hodograph, a radius vector from the center gives the wind at each level.   The above formula states that the advective-pressure change in a given layer is proportional to the area of a triangle bounded by the two vectors representing the gradient and the 10,000-ft winds and the shear line between the ends of the two vectors.   The 10,000-ft-pressure tendency is equal to the surface-pressure tendency (corrected for diurnal change) minus the advective-pressure change.   For the hodograph in Fig. 1, the advective-pressure change is -1.5 mb per 3 hr.

   The 10,000-ft-pressure change is used to determine the motion or changes in intensity of the 10,000-ft troughs and ridges.   For instance, a southwest wind at 10,000 ft in the central United States with rising pressures would indicate that the trough to the west of the station was either filling up or moving westward and that the high to the east of the station was intensifying or moving westward.   Since retrograde systems are rare, we can assume that the trough at 10,000 ft to the west is decreasing in intensity.   This indicates to the forecaster that the frontal system associated with the trough is becoming less active, since the winds aloft will be becoming more nearly west-east.

   At a station in the central United States with northwest winds at 10,000 ft and a rising tendency at that level, we should know that there is a ridge to the west of the station that is approaching the station or intensifying.   Either of these would produce the same type of weather in the environs of the station.

   With a northwest wind and falling pressures at 10,000 ft, we should know that the high to the west is diminishing in intensity or the low to the east intensifying. Both of these suggest a new outbreak of cold air approaching from the north or northwest.

   Other wind directions and pressure changes at 10,000 ft may be interpreted by similar reasoning.

TECHNIQUE OF UTILIZING RADIOSONDE OR AIRPLANE OBSERVATIONS IN SINGLE-STATION ANALYSIS

   In analyzing a sounding, one of the first steps should be to determine the air masses present aloft over the station.   Seasonal normals of equivalent potential temperature, temperature, and moisture content such as Showalter (see page 609)has prepared for various air masses may be used to help in the identification of air masses. The trajectory of the air mass may cause such pronounced modifications that individual air masses do not compare well with the normals. Probably characteristics of the lapse rates, such as multiple inversions in cP or cA air, and the upper-air pressures and tropopause characteristics are more useful for identification than any other normals that have been compiled up to the present time.

Wulf and Obloy[1] in their work with upper-air data in the vicinity of the tropopause have found that changes there are a good indication of the type of air mass in the vicinity of the station. The 10-km level is near the level where, with change of air mass, the largest pressure and temperature changes occur. Very low pressures at 10 km indicate that air of arctic origin is over the station, and very high pressures indicate that air of tropical origin is over the station. Hence, in single-station forecasting, the pressure changes at 10 km are quite useful. Rises of pressure in the stratosphere indicate the influx of more tropical air aloft and the subsequent influx of more tropical air near the surface, since the changes aloft precede those on the ground. Similarly, falls at 10 km indicate air of more arctic origin aloft and the subsequent inflow of a colder air mass at the surface. Table 1 shows the average pressure for each month of the year at Omaha, Neb., and Oklahoma City, Okla. It should be noted that there is a very large variation from summer to winter. On any one day, the observed 10-km pressure may be compared with the mean from the month to determine whether the air present has come from a more northerly or more southerly latitude than is normally the case. The normal flow for any region may be determined from the monthly normal upper-level maps published by the U.S. Weather Bureau 5-day forecasting section.

TABLE 1.—NORMAL 10-KM PRESSURE

	Jan.	Feb.	Mar.	Apr.	May	June	July	Aug.	Sept.	Oct.	Nov.	Dec.
Omaha...	257	259	260	267	273	282	286	286	283	277	267	264
Oklahoma City...	264	265	267	271	279	284	287	288	286	280	274	270

It might be pointed out that there exists a positive correlation between the pressures at 10 km at a given station and the elevation of the tropopause, so that a high value of the pressure is usually accompanied by a high tropopause. It is thus apparent that the elevation of the tropopause may be used in the same manner as the 10-km pressures.

Since the winds at upper levels are usually stronger than at the surface, these tropopause characteristics will often precede the surface air mass. One of the best ways of forecasting the approach of a cold front is to watch the tropopause and pressures aloft. When a low, warm tropopause appears and the stratospheric pressures start to fall, a cold front may be expected soon. The arctic-type tropopause may precede the surface front up to approximately 300 miles. Frequently when this occurs there is a double tropopause due to the incoming arctic-type air and the remaining more tropic air present ahead of the front.[1]

Double tropopauses have another interpretation useful for forecasting. Suppose a cold front has passed with a deep low to the northeast and the cold air has become quite deep with a characteristic arctic-type tropopause. If then a second tropopause appears above the first, it may indicate that warm air is coming in all the way around the low to the northeast bringing with it a colder, higher tropopause. Here a double tropopause indicates the type of low with a bent-back occlusion, or trough hanging back to the rear.

After a cold front passes a station, new cold outbreaks frequently follow with no clouds to presage them. Then falls in stratospheric pressures, with northwest winds aloft, indicate that the storm is deepening and that a new outbreak of cold air is approaching. Large 12-hr pressure changes aloft may be used similarly.

Small 12-hr pressure changes aloft are not very useful, since there is considerable diurnal variation, the magnitude and cause of which are not well known. This problem is being studied by O. R. Wulf of the U.S. Weather Bureau.

After the air masses have been determined and various conclusions drawn from the stratospheric pressures, the forecaster should investigate the sounding for evidences of fronts, regions of clouds, and stable layers. All inversions should be studied to find fronts, regions of subsidence, etc. The height at which the base of cumulus or cumulonimbus clouds could form during the day and the height to which they could extend should be investigated. The sounding should be tested for conditional and convective instability so that the forecaster can forecast clouds and showers after he determines frontal patterns in the vicinity. This type of analysis is the same as that usually made in any kind of forecast work involving soundings (see pages 371, 649, 758).

When the heights of frontal surfaces are determined by the forecaster, he has some idea of the distance of the front from the station. Normally warm fronts have a slope of about 1:100 or 1:200, cold fronts 1:150 close to the front and occlusions 1:100. Another method of computing the distance to a front is by use of Margules' formula

$$\tan \alpha = \frac{f}{g\gamma} \frac{T_m(u_1 - u_2)}{(T_1 - T_2)}$$

where f is the Coriolis parameter, u_1 and u_2 are the wind components parallel to the front in the two air masses (determined from the pilot-balloon observation), T_1 and T_2 are the temperatures in the two air masses on either side of the front, and T_m is the mean of these temperatures. The main difficulty in computing the distance of a front by means of the slope is that the slope usually changes with distance from the surface position. Hence another method of spacing systems must be used. This will be treated later. The proximity of the front may often be arrived at fairly accurately by noting changes in the stratosphere, discussed above.

Besides being useful in the identification of air masses and fronts, the soundings also indicate what type of index condition exists. Here normals come into play. If the 10,000-ft pressure in the United States is far above or below the monthly normal, low-index conditions are indicated. Approximately normal 10,000-ft pressures indicate high index. Similarly above or below normal mean temperatures between the sea-level and 10,000 ft, as determined from the sounding, indicate an abnormal flow from the north or south and hence low index. Normal mean temperatures persisting for several days indicate high index. With either high or low index, normal pressure at 10,000 ft may occur for a brief time between troughs and ridges, but, with low index, the mean temperature will usually deviate from the normal. For this reason, both 10,000-ft pressures and mean temperatures should be utilized in determining the index.

The mean temperature in the lowest layers is often an indication of the trajectory of the air. For instance, below normal mean temperatures throughout an entire low, especially in the warm sector, would rule out the possibility of an mT warm sector and in North America would suggest northwest steering with a deep low. In the Great Lakes region, very high mean temperatures in a warm sector would suggest a trajectory from the Gulf. The surface and 10,000-ft isobars should be drawn to be consistent with these indicated trajectories.

The principle of the guidance of surface systems by the upper-air flow (usually at 10,000 ft) has been used successfully in single-station forecasting. All systems except deep cold lows or warm highs move in the direction of the upper flow. Conversely, if the direction of motion of surface systems may be determined from surface observations, the 10,000-ft wind direction may be inferred.

TECHNIQUE OF USING SURFACE OBSERVATIONS IN SINGLE-STATION FORECASTING

In reconstructing a weather map for the region near the station, the past surface observations should be plotted in positions displaced from the station in accordance with the steering principle mentioned above. Changes due to the motion of the 10,000-ft systems, generally toward the east, must be considered. Hence, past surface observations should be moved in a direction halfway between that in which they would be moved by the 10,000-ft wind, and due east. The velocity with which these reports are displaced should be the velocity of the surface system, which may usually be determined by the gradient wind behind a cold front. If other data are not available, the forecaster should use the seasonal mean velocity for fronts near his station. Usually past reports for 6-hr intervals for 24 or 36 hr are plotted. These may then be drawn for, unless there is intense deepening or filling, and they are a great asset in determining the surface map east of the station. Past fronts and centers of highs and lows should also be displaced in a similar fashion and should be drawn for unless the forecaster has good indications that a change in velocity has occurred.

The sea-level pressures are useful in determining the spacing of systems, which is one of the most difficult problems facing the forecaster. The surface high is located approximately halfway between consecutive lows. Hence, after a surface low passes a station, the forecaster should note the pressure carefully and also note the time between the passage of the low and the arrival of the surface ridge. This time multiplied by the speed at which the past low was moving gives the distance between the low and the high. The next low should be an equal distance to the west. The fronts may be placed in this second low with results more reliable than those derived from computations of the slope of fronts.

In summer, the actual pressure is an indication of the likelihood of frontal passages. For instance, no pronounced cold fronts pass Chicago with the pressure 1,014 mb or higher. Such empirical rules may be devised for any locality if sufficient data are available.

The cloud sequences show whether or not a warm front or occlusion will be active. This is particularly useful in the absence of pilot-balloon observations. If a sequence of cirrus, cirrostratus, and thick altostratus appears, the isobars aloft are straight or cyclonically curved over the warm front, and the winds increase in velocity with height. There would also be warm-air advection where the clouds occur. From the surface wind and from cloud directions, the direction of the 10,000-ft isobars and of the mean isotherms may be inferred. The lapse of time between the appearance of cirrostratus and the beginning of rain is an indication of the velocity of the winds aloft and may help in determining the speed of the system.

Cloud and haze observations are also useful in constructing a sounding when an actual radiosonde observation is not available. Visibility near the ground indicates the presence or absence of a surface inversion. If cumulus clouds are present, a dry-adiabatic lapse rate may be assumed to the base, and a moist-adiabatic lapse rate may be assumed through the cloud layer. Inversions aloft show up as haze layers or as the spreading out of the tops or sides of clouds. When pilot-balloon observations and radiosonde observations are not available, cloud observations are extremely useful. Hodographs and soundings can be constructed and the 10,000-ft wind direction and pressure computed. Since clouds have a different appearance and significance in different parts of the world, the forecaster must make a careful study of cloud sequences in his region preceding and following weather disturbances in order to use them to best advantage in making single-station forecasts.

The forecaster must know the effects of local topographic influences. He can determine these from a careful study of past records. A knowledge of the location of the station with respect to the principal storm tracks and air-mass sources is likewise essential. Typing of maps for each locality can be done and is often of considerable help, provided that allowances are made in individual cases for departures from the normal types.

When the map is complete, the forecaster may forecast from it as he would from a regular weather map.

In conclusion, we shall touch upon the usefulness of single-station forecasting. That such forecasting is important for military units forced to operate where weather-station networks are not feasible has been demonstrated in actual combat conditions. Furthermore, the application of the method to analysis from several stations along the periphery of wide water surfaces has been made in the North Pacific and Atlantic. Frequently, in northerly latitudes where weather reports are transmitted by radio and where atmospheric conditions often interfere with radio reception, the analysis must be made from the data at only a few stations.

Aside from these practical uses of single-station forecasting, there are great benefits to be gained by a meteorologist who studies the method. Single-station forecasting forces the forecaster to make use of *all* the data. This is usually not possible with the profusion of information ordinarily available. The method forces the forecaster to see the interrelationship and dependence of the weather in one region upon that in another region. It forces him to consider the atmosphere in three dimensions and hence gives him a complete picture of the structure and changes in the atmosphere. This makes for improved forecasting.

EXAMPLE OF ANALYSIS AND FORECAST

An Example of a Single-station Analysis and Forecast. An analysis of the data taken during a 24-hr period at Chicago, Ill., is given in Tables 2 and 3 and Figs. 2 to 6, which cover the period from 1200 CWT Oct. 20, to 1330 CWT Oct. 21, 1942. The hourly sequences were taken at the Chicago Municipal Airport. The radiosonde and pilot-balloon observations were taken at the Joliet Airport, which is about 30 miles southwest of Chicago.

Discussion of Data for First 12-*hr Period.* The surface observations during this 12-hr period show that the station is on the western side of a high-pressure area. Since the temperatures are quite high, it must be a modified *mP* or a very old *cP* anticyclone. The dew points are low enough to exclude the possibility of *mT* air being present. The scattered high clouds, disappearing after dark, indicate that there is no active warm front approaching the station. Also, the temperature is so high that it would be unlikely that a warmer air mass is approaching. The falling surface pressures indicate the approach of a trough; and, since most troughs occur at fronts, we should expect the approach of a cold front. Since the pressure is falling rapidly at the close of the period, and since it has been falling for at least 12 hr, we should conclude that the front is fairly close, *i.e.*, less than 400 miles.

The clear skies suggest that, if there is a front approaching, it is not within 100

Table 2.—Radiosonde Observations, Chicago, Ill., Oct. 20, 1942, 2300 CWT

Elevation, km	Press., mb	Temp., °C	R.H.%	Mixing ratio, grams/kg
Surface	986	11	79	6.4
0.4	966	16	72	8.7
0.5	950	16	64	7.8
1.2	874	11	70	6.4
1.5	846	8	72	5.5
1.6	833	8	69	5.4
2.9	708	0	53	3.0
3.8	635	−7	54	3.3
4.9	552	−14	57	1.3
6.1	466	−22	57	0.8
7.0	416	−27	57	0.6
7.2	400	−28	60	0.4
8.0	360	−33	55	0.3
10.1	265	−50		
11.2	224	−58		
13.0	166	−58		
14.1	140	−58		
15.0	122	−60		
15.7	108	−59		

```
5,000-ft pressure.....................................................  840 mb
10,000-ft pressure....................................................  696 mb
15,000-ft pressure....................................................  573 mb
20,000-ft pressure....................................................  467 mb
10-km pressure........................................................  268 mb
13-km pressure........................................................  167 mb
```

Hourly Sequence Data for Chicago, Ill., Oct. 20 and 21, 1942

```
√ 1200C-①/139/66/40 ⟋14/992
R 1230C-①/129/68/36→ ⟋12/990
√ 1300C-①/129/68/36→ ⟋12/990
R 1330C-①/115/70/37 ⟋12/986/912    0407    44
√ 1400C-①/115/70/37 ⟋14/986
R 1430C-①/105/72/36 ↑ ⟋16/983
√ 1500C-①/105/72/36 ↑ ⟋16/983
R 1530C-①/098/71/36 ↑ ⟋15/981
√ 1600C-①/098/71/36 ↑ ⟋15/980
R 1630C-①/091/71/36 ↑ ⟋17/980/810    0505
√ 1700C-①/091/71/36 ↑ ⟋11/979
R 1730C-①/88/65/43 ↑ ⟋9/978
√ 1800C-①/088/65/43 ↑ ⟋7/976
R 1830C-①/5K-085/63/40 ↑ ⟋9/976
√ 1900C ○6K-085/63/40 ↑ ⟋9/976
R 1930C ○6K-081/60/40 ↑ ⟋8/975/606    0409    73
√ 2000C ○6K-081/60/40 ↑ ⟋8/975
R 2030C ○7 081/58/42 ↑ ⟋11/975
√ 2100C ○8 081/58/42 ↑ 10/974
R 2130C ○8 075/56/41 ↑ ⟋10/973
√ 2200C ○9 075/56/41 ↑ 8/973
R 2230C ○9 068/55/43 ↑ 10/972/906
√ 2300C ○9 068/55/43 ↑ 8/971
R 2330C ○9 061/53/44 ↑ 7/907
√ 0000C ○8 061/53/44 ↑ 9/969
R 0030C ○8 054/53/45 ↑ 10/968
√ 0100C ○8 054/53/45 ↑ 10/968
R 0130C ○8 044/53/45 ↑ ⟋8/966/810    73
```

miles of the station, because in this region most fronts are preceded by some clouds at least 100 miles in advance of them.

The pilot-balloon observation taken at 1700 CWT (Fig. 2) indicates that the surface isobars extend from southwest to northeast and the 10,000-ft isobars extend from west to east. This change in direction with height indicates that warm-air advection is occurring. Warm-air advection occurs in advance of all types of fronts, so that the sounding indicates that the station is located in advance of some type of

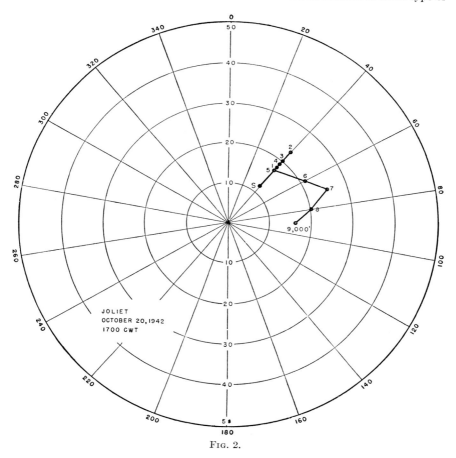

Fig. 2.

front. The mean isotherms extend from north to south with the warmer air to the west. In view of the fact that normally cold air is found to the north and warm air is found to the south, this indicated departure of the temperature field from normal suggests that the approaching disturbance is well developed, since small disturbances do not distort the normal temperature distribution so much.

The change in the shear with elevation indicates that the most unstable air is to the south. This shows that the isobars should be more cyclonically curved to the south, which in turn suggests some type of disturbance to the south of the station. In this case, it turned out to be the remnant of a system that was undergoing frontolysis in the southeastern United States.) The uniformity of the wind direction and velocity up to 5,000 ft should be interpreted as indicating an adiabatic lapse rate and complete mixing up to 5,000 ft. The advective-pressure change is less than the surface fall. The computation gives a pressure tendency of -0.7 mb per 3 hr at 9,000 ft. This, accompanying a west wind at 9,000 ft, indicates that a trough is approaching from the northwest and deepening, as is normally the case with nearly all upper-air troughs moving southward.

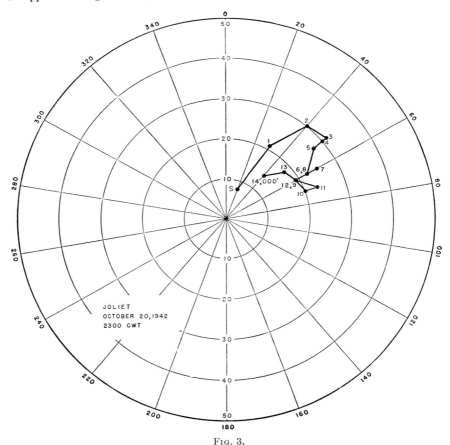

JOLIET
OCTOBER 20, 1942
2300 CWT

Fig. 3.

The pilot-balloon observation taken at 2300 CWT (Fig. 3) shows very little change as compared with the earlier one. The surface frictional layer stands out more clearly now that the air has become more stable next to the ground. The 10,000-ft winds have become more nearly southwest, indicating that the trough is approaching. The change in direction with height up to 10,000 ft still indicates that warm-air advection is occurring. From 10,000 to 14,000 ft, the backing of the winds indicates that it is getting cooler at those levels. This cold-air advection occurring above a layer of warm-air advection is typical of the region extending for about 200 miles in advance of the cold front. This, then, is the first indication that the cold front is within 200 miles of the station. The mean isotherms between the gradient level and 10,000 ft are still oriented north to south with the colder air to the east. From 10,000 to 14,000 ft, the mean isotherms extend nearly east to west with the warmer air to the north. This again indicates the presence of a well-developed low-pressure system, because this temperature distribution is quite different from the mean. The decrease of wind velocity from the gradient level up to 14,000 ft suggests that the cold front will be an active one, since the strong winds at low levels cause the cold front to move faster than the air mass aloft. Computations from the hodograph again indicate that the 10,000 ft pressure is falling. This again, in view of the fact that the 10,000-ft winds are west-southwest and weak, indicates deepening of the 10,000-ft trough as well as motion to the southeast. This deepening should result in stronger winds at the

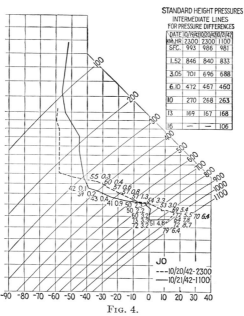

STANDARD HEIGHT PRESSURES INTERMEDIATE LINES FOR PRESSURE DIFFERENCES			
DATE	10/19/42	10/20/42	10/21/42
KM.HR.	2300	2300	1100
SFC.	993	986	981
1.52	846	840	833
3.05	701	696	688
6.10	472	467	460
10	270	268	263
13	169	167	168
16	—	—	106

JO
---10/20/42-2300
———10/21/42-1100

-90 -80 -70 -60 -50 -40 -30 -20 -10 0 10 20 30 40

FIG. 4.

upper levels and, therefore, less frontal cloudiness or precipitation behind the front. The change in the shear with height indicates that the most unstable conditions are to the southwest of the station. This also points to the fact that there is a cold front to the west of the station and that the greatest curvature of the isobars ahead of the front is to the southwest.

The west-east component of the velocity of the air between the surface and 10,000 ft is 20 mph. Assuming the front to extend north to south, we can use this velocity of the wind perpendicular to the front for the velocity of the front. This value will be too large if the front is inactive and too small if the front is active, but it will be close enough to use as an approximation. We have found so far that the distance of the front is less than 200 miles but more than 100 miles away. If we assume that it is moving 20 mph and accelerating (due to the deepening of the low), we should expect it to pass the station in 5 to 10 hr (closer to 5 if the storm deepens much).

The radiosonde observation for 2300 CWT (Fig. 4) shows clearly the nocturnal inversion next to the ground produced by surface cooling. If the maximum tempera-

TABLE 3.—RADIOSONDE OBSERVATIONS, CHICAGO, ILL., OCT. 21, 1942, 1100 CWT

Elevation, km	Press., mb	Temp., °C	R.H., %	Mixing ratio, grams/kg
Surface	981	12	51	4.8
1.4	841	0	72	3.2
1.6	827	0	73	3.3
2.3	760	1	60	3.2
2.9	703	−2	50	2.2
3.0	694	−2	50	2.3
4.6	560	−16	41	0.0
5.9	474	−25	43	0.4
7.1	400	−33	39	0.2
7.9	356	−39	42	0.1
9.0	302	−42		
11.3	216	−45		
12.4	182	−45		
14.9	124	−54		
16.6	96	−56		
19.6	60	−52		

5,000-ft pressure..	833 mb
10,000-ft pressure...	688 mb
15,000-ft pressure...	565 mb
20,000-ft pressure...	460 mb
10-km pressure..	263 mb
13-km pressure..	168 mb
16-km pressure..	106 mb

HOURLY SEQUENCE DATA FOR CHICAGO, ILL., OCT. 21, 1942

CWT

√ 0200C ○8044/53/45 ↑ ↗8/966
R 0230C①/8034/52/45 ↑ ↗8/962
√ 0300C①/8034/52/45 ↑ ↗8/962
R 0330C①/8024/53/46 ↑ ↗8/959
√ 0400C①/8024/53/46 ↑ ↗8/959
R 0430C①/8017/53/45 ↑ ↗8/957/812   0309
√ 0500C①/7017/53/45 ↑ ↗8/957
R 0530C80①6K-014/52/47→ ↗6/956
√ 0600C80 ⓟ6K-014/52/47→ ↗6/956
R 0630CE70 ⊕8014/52/44→↘14→ ↗060Sc/956/Binovc
√ 0700CE70 ⊕7014/52/44→↘14/956
R/s 0730CE70 ⊕5R-K-010/52/40↘7/955/60399   0706   51
√ 0800C ⊕ /60①5R-K-010/52/40→↘12/95S
R/s 0830C ⊕ /60①3K-014/50/42→12/956 Binovc
√ 0900C-ⅅ /4K-014/50/42→12/956
R 0930C-ⓓ /4K020/50/40→↘15/957
√ 1000C-①/6K-020/50/40→↘16/957
R 1030C ○017/51/38→↘18/957/103
√ 1100C30①017/51/38→↘18/957
R 1130C35①017/53/37→18/957
√ 1200C35①017/53/37→16/957
R 1230CE35ⓓ014/53/36→16/956
√ 1300CE35 ⊕014/53/36→16/956/Binovc
R 1330CE35 ⊕003/52/35→↘16/953/Binovc/90799   5006   50
   PIREPS TOVC 45 MSL TMP AT 60 MSL
   37 TEMP IN CLDS 33 LGT TURBC

ture for the day (73°F) is plotted on this diagram, it is apparent that during the afternoon the lapse rate was equal to the dry-adiabatic rate from the surface up to 5,000 ft. This is in agreement with the deduction we made from the pilot-balloon run for 1700 CWT. The tropopause shows up distinctly at 11 km. This indicates that the air mass is maritime polar, because tropical air has a much higher tropopause, and arctic or polar continental has a lower tropopause. There are no frontal inversions indicated by the sounding, substantiating our conclusion that there was no warm front or occlusion approaching but only a cold front. The relative humidities are high enough so that a forecast of cloudiness and precipitation with the passage of the front is indicated. During the last 24 hr, pressure has fallen at all levels, indicating a more arctic type of stratospheric air moving in above 13 km. This is an indication of the approach of a cold front, because stratospheric-pressure falls usually precede a surface front.

From the surface observations, winds aloft, and radiosonde data taken during this 12-hr period, we have found that an active cold front is approaching and should pass within the next 5 to 8 hr. We also know that there is some sort of disturbance to the southeast, because the first balloon run indicated cyclonically curved isobars in that direction. We think that the main system is deepening, since the upper-level pressure falls with weak westerly winds would indicate this. The map should therefore be drawn showing a low to the northwest, a high over the mountains west of the front, and another high along the east coast east of the disturbance to the southeast of Chicago.

We shall now examine the data for the second 12-hr period and see what further developments are indicated. The surface reports indicate a frontal passage at 0630 followed by rain and then rapid clearing. Since the temperature remained nearly constant from 0630 to 1330, we can infer that a cold-front or cold-type occlusion passed, because otherwise the temperature would have risen during the day. The dew point fell slowly from the time of the frontal passage. Because only scattered high clouds preceded the front, a cold front rather than an occlusion is indicated. The clearing behind the front was quickly followed by the formation by turbulence of a stratocumulus cloud deck indicating that the air mass must be unstable at least in the lower levels. The pressure continued to fall for 2 hr after the frontal passage, then rose very slowly, and then started falling again. This could indicate rapid deepening of the low accompanying the cold front or the approach of a secondary cold front. There is insufficient information in the surface data for us to decide which alternative is correct. We know from experience, however, that, if a secondary front is found to the rear of a low, the low will usually deepen. We should therefore interpret these falling pressures as an indication of the approach of a secondary cold front as well as the deepening of the low, since these two frequently occur together.

The pilot-balloon run taken at 0500 (Fig. 5) indicates quite marked cold-air advection above 2,000 ft. This sounding was taken over 1 hr before the front passed. We should therefore interpret the cold-air advection that is occurring at levels so much lower than is usually the case in advance of cold fronts as an indication that the front is very close to the station, less than 50 miles. Since the wind still decreases in velocity with height above 7,000 ft, we should expect the front to be active, because it should be moving faster than the portion of the air mass ahead of it which lies above 7,000 ft. The frontal lifting therefore should be most pronounced above 7,000 ft. The surface observations at and shortly after the passage of the front substantiate this, in view of the fact that the clouds were reported at about 7,000 ft.

A computation from this balloon run indicates that the pressure is falling at the rate of 3.7 mb per 3 hr at 9,000 ft. This very rapid drop in the upper-level pressure accompanying such weak upper winds indicates rapid deepening of the upper-level low-pressure trough. We know that deepening of an upper-level low occurs when cold air enters the rear of the trough (see pages 822–823). We should therefore conclude that cold air is entering the rear of the upper-level trough. The weak winds at the surface and 1,000 ft and the large change in wind direction in this layer indicate that the stable layer produced by nighttime radiation is still present.

The balloon run at 1100 (Fig. 6) indicates that the front has passed, and we see that pronounced cold-air advection is now occurring. The mean isotherms, as

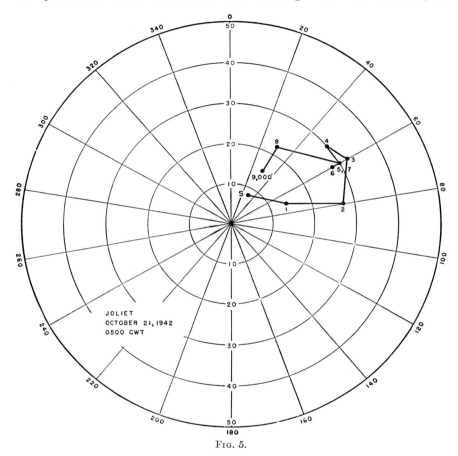

Fig. 5.

indicated by the shear vector, are oriented from south to north with the cold air to the west. Since this report is taken just a few hours after the front has passed and since the shear is quite pronounced, it should be used to indicate the orientation of the cold front. Since the mean isotherms are oriented from north to south, we conclude that the cold front is oriented from north to south. The velocity of the cold front is equal to the velocity of the wind perpendicular to the front. In this case, we see that the front should move to the east at about 25 mph. As we mentioned before, there must be an adiabatic lapse rate up to the base of the clouds and considerable turbulence throughout this layer. We should therefore realize that the effects of surface friction will be present up to about 2,500 or 4,000 ft, and we should choose that level as the gradient level, remembering that the directions and velocities below that level may be considerably in error because strong turbulence invalidates the assumption of constant ascension rate of the balloon.

If, on the other hand, we assume that the report is completely reliable in the lower levels, we should conclude that only slight warm-air advection is occurring below 4,000 ft and marked cold-air advection is occurring above that level. From this,

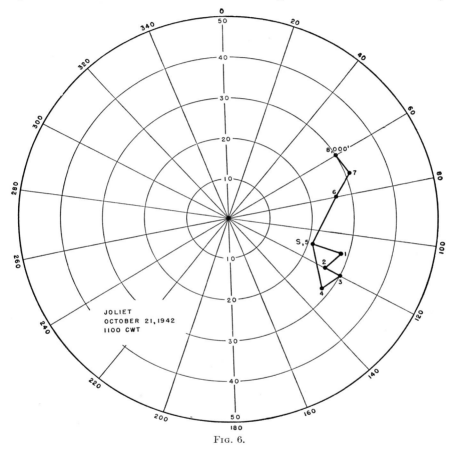

Fig. 6.

we should forecast a rapid steepening of the lapse rate and therefore increasing air-mass instability and resulting cloudiness and weather.

The computation of the pressure tendency at 8,000 ft again indicates that the pressures aloft are falling very rapidly. This again indicates that deepening aloft is occurring, and a secondary front should be expected to the rear of the trough. The deepening of the upper-level trough indicates that more air-mass instability should occur, and we should therefore expect that the low scattered clouds observed forming at the time of the balloon run will increase.

The radiosonde observation taken at 1100 CWT (Fig. 4) indicates that considerable cooling has occurred between 10,000 ft and the surface. This must be due

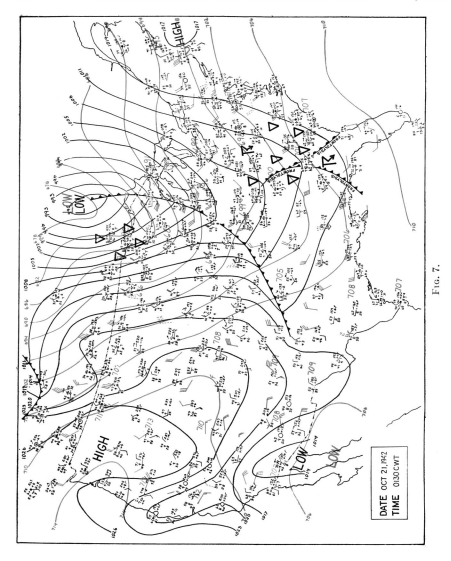

FIG. 7.

DATE OCT 21, 1942
TIME 0130 CWT

to the passage of the cold front. Some cooling has occurred between 10,000 ft and 10 km, also. This may be due to lifting of the upper air mass by the front, or it may be that cooler air has moved in aloft over the front. We see, though, that by far the most pronounced change in temperature has occurred in the stratosphere. The tropopause has dropped from 11 to 8 km while the temperature in the lower portion of the stratosphere has increased from -58 to $-43°C$, a rise in temperature of 15°C. This is much greater than the change produced at the surface by the passage of the cold front, and its effect on the surface pressure is more pronounced than the effect of the cold air. It is the influx of this very warm and light air in the stratosphere that is responsible for the large falls in pressure observed at all levels below 13 km. The slight rise at 13 km is less than the diurnal rise at that level, but it does indicate that the air above 13 km was not responsible for any of the deepening that occurred below.

It is apparent from this sounding that, if the surface air were carried by turbulence up as far as the lapse rate is adiabatic, stratocumulus should form at about 4,000 ft.

We shall now attempt to describe how a map could be drawn for 1330 CWT, Oct. 21, 1942, and a forecast made for the following 24 hr from the results of the above analysis of the data.

We must first place the cold front that passed at 0630 CWT. Since the pilot-balloon observation showed that it was moving at 25 mph and since it passed 7 hr before 1330, we should place it 175 miles to the east of Chicago. This would be just to the west of Toledo, Ohio. Since the winds on the ground and aloft had a westerly component both before and after the passage of the front, we know that the center of the low passed to the north of Chicago. Just how far to the north we cannot tell from these data. We know also that the 10,000-ft trough is still to the west of Chicago and deepening rapidly. Since the 10,000-ft pressure is already 688 mb, which is 14 mb below the October normal pressure of 702 mb, we can conclude that the trough line is not very far to the west of Chicago, not more than 100 or 200 miles. We should therefore draw the 10,000-ft map showing a deep trough in the center of the United States with a well-developed ridge along both coast lines. The two ridges cannot be drawn closer to the trough for the following reason: In order that the 10,000 ft pressure should average somewhere near normal over the country as a whole, the two ridges on either side of this deep low must have somewhat above the normal pressure for these regions. We cannot draw ridges with central pressures above normal closer to the trough without making the pressure gradients between them and the 10,000-ft trough to the west of Chicago much greater than are ever observed in these regions. Furthermore, since the uppermost winds observed in this trough have been quite weak, it would be unwise to draw a map showing extremely strong gradients anywhere around this trough.

The secondary front indicated by the data must be somewhere to the northwest of the station, because it could not be preceded by falling pressures aloft (10 km) if it were coming from any other direction. We cannot tell yet just how far away it is except that it is over 200 miles away and therefore will not pass in the next 8 hr. We should watch the surface wind direction closely to see when the ridge between the two surface fronts passes. When this ridge is reached, Chicago should be halfway between the two fronts. Knowing then how far away the first front is, we can say how far away the second one is. Since this ridge line has not as yet been reached, we know that the secondary front is farther to the west than the first front is to the east. The wind direction rather than the pressure should be used to define the ridge line, since with a deepening system the pressures will fall long before the ridge line has arrived.

We should expect a surface high to be located to the east (displaced toward colder air) of each of the 10,000-ft ridge lines. This would place one of them over the Great Basin region and the other off the east coast. The instability that was indicated on

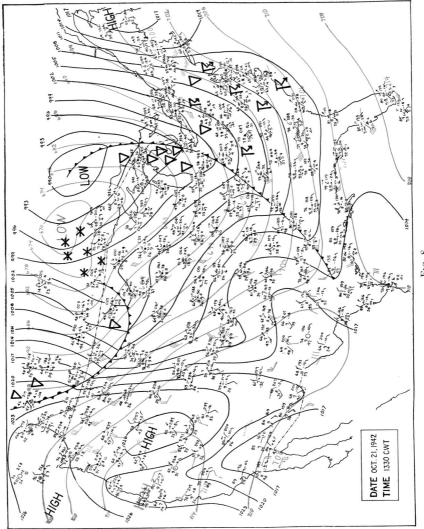

DATE OCT. 21, 1942
TIME 1330 CWT

FIG. 8.

the first balloon run to the southeast of Chicago should be effective in producing showers or thunderstorms over the mountain regions of Kentucky, Tennessee, and Virginia. The secondary front to the northwest should be accompanied with low clouds and showers, for the air mass should be quite unstable under the 10,000-ft trough.

The forecast for Chicago, based on the above reconstructed picture would be for continued instability clouds during the afternoon, diminishing at night. We must assume that the secondary front will pass at some time between 8 and 24 hr from 1330, because, if it were any closer, the ridge mentioned previously would have passed. We should therefore pick some time between these two limits for its passage and make our forecast accordingly. We should forecast the surface wind to shift to southwest or south halfway between the forecasted time of the front passage and the time of passage of the previous one (0630). Since the 10,000-ft trough is deepening rapidly, we should expect considerable instability-type precipitation both before and after the passage of the secondary front.

The observed data showed that the ridge line passed Chicago at 1900 on the current day (Oct. 21). This is about 13 hr after the passage of the first front. From this time until 0800 on the following day, the wind remained from the southwest. At 0800, the secondary front passed, 13 hr after the ridge line. Precipitation began at 0400 and lasted intermittently until 1600. The ceiling averaged between 1,000 and 1,500 ft during this period, fluctuating with each shower.

It was the purpose of this discussion to show the method used in single-station analysis and forecasting. Each case presents a new set of problems that must be considered. There are numerous local effects that must be taken into account in different regions, but the general technique of investigating the data is basically the same. Such assumptions as the one used above, that the ridge is halfway between the two troughs, of course do not always hold as in this case, yet the results from such an assumption will give in nearly all cases a closer approximation to the distance of the next disturbance than any other method such as assuming a low every 3 days, or that all systems are some constant distance apart.

Bibliography

1. WULF, O. R., and S. J. OBLOY: The Utilization of the Entire Course of Radiosonde Flights in Weather Diagnosis, *Misc. Rept.* 10, Institute of Meteorology, University of Chicago, March, 1944.
2. WOLF and OBLOY: *op. cit.*

SECTION XI

CLOUDS AND STATES OF THE SKY

By P. L. Schereschewsky

CONTENTS

881

SECTION XI

CLOUDS AND STATES OF THE SKY

By P. L. Schereschewsky

HISTORY AND INTRODUCTION

Clouds have been observed and used for short-period weather forecasting from time immemorial. Empirical knowledge acquired by sailors and peasants was first expressed in familiar sayings. When more scientific studies of weather began in the nineteenth century, emphasis was first placed on winds, pressures, and temperatures rather than on clouds. Abercromby and Hildebrandsson recognized this emphasis when they published the first international classification of clouds. To quote the present International Atlas, "In the opinion of the authors, the foremost application of cloud observations was the determination of the direction of winds at different altitudes."

Clouds are now considered essential and accurate tools for weather forecasting. Every feature of the air masses (discontinuity, subsidence, instability and stability, etc.) is reflected by the shape, amount, and structure of the clouds. Thus close scrutiny of clouds will assist the analyst in identifying and analyzing air masses. Relationships among the details of cloud structure, synoptic weather analysis, and structure of the atmosphere are being investigated and progressively determined by meteorologists in all countries. Progress along this line makes advisable more and more detailed observations and eventually more detailed classifications. Synoptic studies of clouds (*nephanalysis*) indicate the possibility of a visual weather analysis based on cloud observations. Such an analysis is of great help in areas where a dense network of ground stations cannot be operated (*e.g.*, in the oceans or polar areas).

Actual scientific studies of clouds began in 1802 and 1803 when Lamarck in France and Howard in England published their classifications. Their crude drawings of clouds were inadequate, since small structural details may have a great significance. Both men were hampered by lack of proper instruments to keep accurate records of observations. No instruments were available to measure air currents, height of clouds, temperature, and humidity in the free atmosphere. Furthermore it was not believed that cloud forms were the same all over the world.

Further scientific progress was made in the era beginning in 1879 when H. Hildebrandsson published an atlas of 16 cloud photographs. In 1887, Abercromby and Hildebrandsson published a classification of clouds in which great importance was attached to height as a criterion. The present International classification was the direct offspring of this work. The first International Atlas was published in 1896. It contained 27 sketches and photographs. The classification laid down in the Atlas soon became official and was used generally in almost all countries.

Another period of progress, greatly accelerated by aviation, commenced with the First World War. Pilots had to live among the clouds, to use them, and to avoid their dangers. Forecasters began to utilize information derived from synoptic studies of clouds. A new and much more extensive atlas was prepared. The classification of 1896 was followed quite closely. In addition, the states of the sky were classified on the basis of the work emerging from studies of depressions by the Norwegian and

TABLE 1.—CLOUD NAMES*

Cloud genera

A. *High clouds*	Cirrus	Ci
	Cirrocumulus	Cc
	Cirrostratus	Cs
B. *Middle clouds*	Altocumulus	Ac
	Altostratus	As
	Stratocumulus	Sc
C. *Low clouds*	Stratus	St
	Nimbostratus	Ns
D. *Clouds with vertical development*	Cumulus	Cu
	Cumulonimbus	Cb

Cloud species most frequently observed

Cirrus	Ci filosus	(Ci fil.)
	Ci uncinus	(Ci unc.)
	Ci densus	(Ci den.)
	Ci nothus	(Ci not.)
Cirrostratus	Cs nebulosus	(Cs neb.)
	Cs filosus	(Cs fil.)
Altocumulus	Ac translucidus	(Ac tra.)
	Ac opacus	(Ac op.)
	Ac cumulogenitus	(Ac cug.)
Altostratus	As translucidus	(As tra.)
	As opacus	(As op.)
	As praecipitans	(As prae.)
Stratocumulus	Sc translucidus	(Sc tra.)
	Sc opacus	(Sc op.)
	Sc vesperalis	(Sc vesp.)
Cumulus	Cu humilis	(Cu hum.)
	Cu congestus	(Cu con.)
Cumulonimbus	Cb calvus	(Cb cal.)
	Cb capillatus	(Cb cap.)

The principal varieties are

	Abbreviation	Symbol
Fumulus	(Fum.)	$\underset{\tilde{}}{X}$
Cumuliformis floccus	(Cuf. flocc.)	$\widehat{X}$
Cumuliformis castellatus	(Cuf. cast.)	$\overset{n}{X}$
Lenticularis	(Lent.)	$\overset{\frown}{X}$
Mammatus	(Mam.)	$\underset{U}{X}$
Undulatus	(Und.)	$\overset{\frown}{X}$
Radiatus	(Rad.)	$\overset{*}{X}$

Special features may occur

Virga		$\overset{''}{X}$
Pileus		$\overset{\frown}{X}$
Incus		$\overset{\triangledown}{X}$
Arcus		$\underset{\frown}{X}$

Development of clouds

On the horizon	(X)
Increasing	X|
Decreasing	|X
Intermittent	|X|
Has ceased during the last hour	X]
Has begun during the last hour	[X
Chronological connection	X-X
Spatial connection	$\overset{X}{\underset{X}{\cdot}}$

The above symbols may be used to facilitate the making of entries (X representing the cloud designation).

* Taken from International Atlas, Paris, 1932.

French schools, and the title of the table in the Atlas was altered to Atlas of Clouds and States of the Sky. The plates in the 1932 edition of the Atlas include 101 photographs taken from the ground, 22 photographs taken from aircraft, and 51 "states of the sky."

Clouds play a much more important part in meteorology today than ever before. Aircraft are flying in progressively more dangerous weather, and the pilots must therefore have a better knowledge of the clouds they will encounter, their structure and hazards. The familiar features of dangerous thunderclouds are often hidden behind other clouds of secondary importance; but, even in such cases, if pilots pay careful attention to structure, they can generally identify and avoid danger zones. From the analyst's point of view, clouds should be considered as accurate tools for forecasting. The detailed structure of clouds, together with their diurnal evolution and association with other clouds, or the succession of clouds at a particular station, are refined and directly visible evidence of the structure of the atmosphere. This fact is being used more and more in synoptic meteorology. It is not necessary to enumerate in detail here all that is being done at present to further understanding of all the physical processes involved in the formation and evolution of clouds. Progress has again been accelerated by military necessity, and much more knowledge of clouds will become available when peacetime security permits publication of all data.

INTERNATIONAL CLASSIFICATION OF THE CLOUDS AND STATES OF THE SKY

Necessity of Detailed Description of the Clouds. Clouds are divided into four *families:* (1) high clouds, (2) middle clouds, (3) low clouds, and (4) clouds with vertical development. Each family is divided into *genera.* There are 10 genera. Each of the families contains two or three genera corresponding to clouds that appear (*a*) isolated, (*b*) in sheets, or (*c*) in more or less continuous sheets. Genera are subdivided into *species.* Species are again subdivided into *varieties.* The International Atlas mentions explicitly 20 species for 7 out of 10 genera, and 7 general varieties that can apply to many of the species. For example, *altocumulus translucidus cumuliformis floccus* refers to the genus altocumulus, the species translucidus, the variety cumuliformis, and the subvariety floccus. In addition, special features may be mentioned whenever feasible. Thus it is helpful to know whether the cloudiness is increasing, decreasing, has begun during last hour, etc. Also details of amount, density, distribution in the sky, optical phenomena, and direction of motion are helpful in analysis. Table 1 indicates the classification of cloud forms fixed by international agreement.

CLOUDS*
Table of Cloud Classification

At nearly all levels clouds may appear under the following forms:

a. Isolated, heap clouds with vertical development during their formation, and a spreading out when they are dissolving.

b. Sheet clouds which are divided up into filaments, scales, or rounded masses, and which are often stable or in process of disintegration.

c. More or less continuous cloud sheets, often in process of formation or growth.

CLASSIFICATION INTO FAMILIES AND GENERA

Family A: HIGH CLOUDS (mean lower level 6,000 m.)[1]

Form *b* { 1. Genus cirrus.
          { 2. Genus cirrocumulus.

Form *c* { 3. Genus cirrostratus.

---

* From International Atlas, Paris, 1932.

[1] It should be noted that the heights given are for temperate latitudes, and refer, not to sea level, but to the general level of the land in the region. It should be noted that in certain cases there may be large departures from the given mean heights, especially as regards cirrus, which may be found as low as 3000 metres in temperate latitudes, and in polar regions even almost as low as the surface.

Family B: MIDDLE CLOUDS (mean upper level 6,000 m., mean lower level 2,000 m.)

Form a
Form b } 4. Genus altocumulus.[1]

Form c { 5. Genus altostratus.

Family C: LOW CLOUDS (mean upper level 2,000 m., mean lower level close to the ground)

Form a
Form b } 6. Genus stratocumulus.[1]

Form c { 7. Genus stratus.
8. Genus nimbostratus.

Family D: CLOUDS WITH VERTICAL DEVELOPMENT (mean upper level that of the cirrus, mean lower level 500 m.)

Form a { 9. Genus cumulus.
10. Genus cumulonimbus.

DEFINITIONS AND DESCRIPTIONS OF THE FORMS OF CLOUDS

I. Cirrus (Ci) [*Howard* 1803[2]]

A. Definition. Detached clouds of delicate and fibrous appearance, without shading, generally white in colour, often of a silky appearance.

Cirrus appears in the most varied forms, such as isolated tufts, lines drawn across a blue sky, branching feather-like plumes, curved lines ending in tufts, etc.; they are often arranged in bands which cross the sky like meridian lines, and which, owing to the effect of perspective, converge to a point on the horizon, or to two opposite points (cirrostratus and cirrocumulus often take part in the formation of these bands).

B. Explanatory Remarks. Cirrus clouds are always composed of ice crystals, and their transparent character is due to the state of division of the crystals.

As a rule these clouds cross the sun's disc without dimming its light. But when they are exceptionally thick they may veil its light and obliterate its contour. This would also be the case with patches of altostratus, but cirrus is distinguished by its dazzling whiteness and silky edges.

Halos[3] are rather rare in cirrus.

Sometimes isolated wisps of snow are seen against the blue sky, and resemble cirrus; they are of a less pure white and less silky than cirrus; wisps of rain are definitely grey, and a rainbow, should one be visible, shows their nature at once, for this cannot be produced in cirrus.

Before sunrise and after sunset, cirrus is sometimes coloured bright yellow or red. These clouds are lit up long before other clouds and fade out much later; sometimes after sunset they become grey. At all hours of the day cirrus near the horizon is often of a yellowish colour; this is due to distance and to the great thickness of air traversed by the rays of light.

Cirrus, being in general more or less inclined to the horizontal, tends less than other clouds to become parallel to the horizon, under the effect of perspective, as the horizon is approached; often on the contrary it seems to converge to a point on the horizon.

C. Species. Amongst the more remarkable forms one may note:

1. Cirrus filosus [*Clayton* 1896[4]]. More or less straight or irregularly curved filaments (neither tufts nor little points) and without any of the parts being fused together.

2. Cirrus uncinus [*Maze* 1889[5]]. Cirrus in the shape of a comma, the upper part ending in a little tuft or point.

3. Cirrus densus [*Besson* 1921[6]]. Cirrus clouds with such thickness that without care an observer might mistake them for middle or low clouds.

4. Cirrus nothus [*C.E.N.* 1926[7]] (Hybrid cirrus). Cirrus proceeding from a cumulonimbus and composed of the debris of the upper frozen parts of these clouds.

D. Varieties. Ordinary cirrus may appear in many very different forms. One may particularly note the forms floccus and vertebratus which are really aspects of the varieties cumuliformis and undulatus radiatus respectively.

[1] Most altocumulus and stratocumulus clouds come under category b; but the varieties cumuliformis and particularly castellatus belong to category a.

[2] "On the Modification of Clouds" reprinted in: Neudrucke von Schriften und Karten über Meteorologie und Erdmagnetismus. Tome III, Berlin 1894, p. 6.

[3] Cf. p. 886, note 5.

[4] "Discussion of the Cloud Observations" made at the Blue Hill Observatory of Harvard College, Annals of the Astronomical Observatory of Harvard College, Cambridge 1896, Vol. XXX; part IV, p. 347.

[5] Sur la Classification des Nuages, Mémoires du Congrès Météorologique International de Paris 1889, p. 32.

[6] "La Classification détaillée des Nuages en usage à l'Observatoire de Montsouris," Annales der Services Techniques d'Hygiène de la Ville de Paris, tome I, Paris 1921, p. 304.

[7] "Procès-verbaux de la Commission Internationale pour l'Etude des Nuages—Session de Paris, Avril 1926," Circulaire no. 47 de la C.E.N, p. 37.

II. Cirrocumulus (Cc) [*Howard* 1803,[1] *Renou* 1855[2]]

A. Definition.   A cirriform layer or patch composed of small white flakes or of very small globular masses, without shadows, which are arranged in groups or lines, or more often in ripples resembling those of the sand on the sea shore.

B. Explanatory Remarks.   In general cirrocumulus represents a degraded state of cirrus and cirrostratus both of which may change into it.   In this case the changing patches often retain some fibrous structure in places.

Real cirrocumulus is uncommon.   It must not be confused with small altocumulus patches on the edges of altocumulus sheets.   There are in fact all states of transition between cirrocumulus and altocumulus proper; this is only to be expected as the process of formation is the same.   In the absence of any other criterion the term cirrocumulus should only be used when:

1. There is evident connection with cirrus or cirrostratus.
2. The cloud observed results from a change in cirrus or cirrostratus.
3. The cloud observed shows some of the characteristics of ice crystal clouds which will be found enumerated under cirrus (p. 885).

Clear rifts are often seen in a sheet of cirrocumulus.

III. Cirrostratus (Cs) [*Howard* 1803,[3] *Renou* 1855[4]]

A. Definition.   A thin whitish veil, which does not blur the outlines of the sun or moon, but gives rise to halos.   Sometimes it is quite diffuse and merely gives the sky a milky look; sometimes it more or less distinctly shows a fibrous structure with disordered filaments.

B. Explanatory Remarks.   A sheet of cirrostratus which is very extensive, though in places it may be interrupted by rifts, nearly always ends by covering the whole sky.   The border of the sheet may be straight edged and clear-cut but more often it is ragged or cut up.

During the day, when the sun is sufficiently high above the horizon, the sheet is never thick enough to mask the shadows of objects on the ground.

A milky veil of fog is distinguished from a veil of cirrostratus of a similar appearance by the halo phenomena[5] which the sun or the moon nearly always produce in a layer of cirrostratus.

What has been said above of the transparent character and colours of cirrus is true to a great extent of cirrostratus.

C. Species.   Cirrostratus has two principal aspects which correspond to the two following species:

1. Cirrostratus nebulosus [*Clayden* 1905[6]].   A very uniform light nebulous veil, sometimes very thin and hardly visible sometimes relatively dense but always without definite details and with halo phenomena.

2. Cirrostratus filosus [*C.E.N.* 1930[7]].   A white fibrous veil, where the strands are more or less definite, often resembling a sheet of cirrus densus from which indeed it may originate.

IV. Altocumulus (Ac) [*Renou* 1870[8]]

A. Definition.   A layer, or patches composed of lamina or rather flattened globular masses, the smallest elements of the regularly arranged layer being fairly small and thin, with or without shading. These elements are arranged in groups, in lines, or waves, following one or two directions, and are sometimes so close together that their edges join.

The thin and semitransparent edges of the elements often show *irisations* which are rather characteristic of this class of cloud.

From the definition it follows that altocumulus comprises the sub-genera:

1. Altocumulus translucidus [*C.E.N.* 1930[9]].   Altocumulus formed of elements whose colour—from dazzling white to dark grey—and whose thickness vary much from one example to another, or even in the same layer; the elements are more or less regularly arranged and distinct.   In the definition of the elements it is the variation in the transparency of the layer, variable from one point to another, that plays the essential part.   There appears in the interstices either the blue of the sky, or at least a marked lightening of the layer of cloud due to a thinning out.

[1] Ibidem, p. 8.

[2] First to be found with the present meaning in: "Instructions Météorologiques," Annuaire de la Société Météorologique de France, tome III, Paris 1855, p. 143.

[3] Ibidem, p. 9.

[4] First to be found with the present meaning in Ibidem, p. 142.

[5] The following are the principal halo phenomena:—I. A circle of 22° radius round the sun or moon; this is roughly the angle subtended by the hand placed at right angles to the arm when the latter is extended; this halo is sometimes, but rarely, accompanied by one of 46° radius.   2. Parhelia, Paraselenae (mock suns or mock moons) luminous patches, often showing prismatic colours, a little over 22° from the sun or moon and at the same elevation.   3. A luminous column (sun pillar) placed vertically above and sometimes below, the sun.

Often only small fragments of these appearances are visible but they are none the less characteristic of high clouds.

[6] "Cloud Studies," Londres 1905, p. 45.

[7] "Atlas International des Nuages et des Etats du Ciel—Extrait a l'usage des Observateurs," Paris 1930, p. 6.

[8] Bulletin de l'Observatoire de Montsouris, 13 Février 1870.   See also "Rapport sur la Classification des Nuages," Mémoires du Congres Météorologique International de Paris 1889, p. 15.

[9] "Atlas International des Nuages et des Etats du Ciel—Extrait a l'usage des Observateurs," Paris 1930, p. 7.

2. Altocumulus opacus [*C.E.N.* 1930¹]. Altocumulus sheet which is continuous, at least over the greater part of the layer, and consisting of dark and more or less irregular elements, in the definition of which transparency does not play a great part, owing to the thickness and density of the layer; but the elements show in real relief on the lower surface of the cloud sheet.

B. Explanatory Remarks. The limits within which altocumulus is met are very wide.

At the greatest heights, altocumulus, made up of small elements, resembles cirrocumulus; altocumulus however is distinguished by not possessing any of the following characters of cirrocumulus:

1. Connection with cirrus or cirrostratus.
2. An evolution from cirrus or cirrostratus.
3. Properties due to physical structure (ice crystals) enumerated under cirrus.

At lower levels, where altocumulus may be derived from a spreading out of the tops of cumulus clouds, it may easily be mistaken for stratocumulus; the convention is that the cloud is altocumulus if the smallest, well defined, and regularly arranged elements which are observed in the layer (leaving out the detached elements which are generally seen on the edges) are not greater than ten solar diameters in their smallest diameters.

When the edge of a thin semitransparent patch of altocumulus passes in front of the sun or moon a *corona* appears close up to them; this is a coloured ring with red outside and green inside; the colours may be repeated more than once. This phenomenon is infrequent in the case of cirrocumulus and only the higher forms of stratocumulus can show it.

Irisation, mentioned above, is a phenomenon of the same type as the corona; it is a sure mark of altocumulus as distinguished from cirrocumulus or stratocumulus.

Altocumulus clouds often appear at different levels at one and the same time. Often too they are associated with other types of cloud.

The atmosphere is often hazy just below altocumulus clouds.

When the elements of a sheet of altocumulus fuse together and make a continuous layer, altostratus or nimbostratus is the result. On the other hand a sheet of altostratus can change into altocumulus. It may happen that these two aspects of cloud sheet may alternate with each other during the whole course of a day. It is also not rare to have a layer of altocumulus coexisting with a veil resembling altostratus at a height very little less than the altocumulus (*altocumulus duplicatus* de Quervain).

It is interesting to note that one may often observe filiform descending trails to which the name *virga* has been given.

C. Species. Among the most remarkable species one may note:

Altocumulus cumulogenitus (*Cumulo-Stratus* de Quervain). [*C.E.N.* 1926.²] This is an altocumulus cloud formed by the spreading out of the tops of cumulus, the lower parts of the cumulus clouds having melted away; the layer has the appearance of altocumulus opacus in the first stages of its growth.

D. Varieties. An important variety of altocumulus should be noted, namely, *altocumulus cumuliformis* which has two different aspects:

Altocumulus floccus [*Vincent* 1903³]. Tufts resembling small cumulus clouds without a base, and more or less ragged.

Altocumulus castellatus [*Clement Ley* 1894⁴]. Cumuliform masses with more or less vertical development, arranged in a line, and resting on a common horizontal base, which gives the cloud a crenellated appearance.

The hoods or veils which form above a cumulus by the uplift of a damp layer, and which may be pierced by the tops of the cumulus, are considered as a detail of cumulus, and denoted by the term *pileus* attached to the name cumulus; but in reality they are aberrant forms of altocumulus translucidus. Moreover, similar clouds, independent of cumulus, can be formed by the same process by the effect of a rising current caused by a mountain or an obstacle. They are then named altocumulus, and they are classed, on account of their form, with the variety *lenticularis*.

V. Altostratus (As) [*Renou* 1877⁵]

A. Definition. Striated or fibrous veil, more or less grey or bluish in colour. This cloud is like thick cirrostratus but without halo phenomena; the sun or moon shows vaguely, with a gleam, as though through ground glass. Sometimes the sheet is thin with forms intermediate with cirrostratus (altostratus translucidus). Sometimes it is very thick and dark (altostratus opacus), sometimes even completely hiding the sun or moon. In this case differences of thickness may cause relatively light patches between very dark parts; but the surface never shows real relief, and the striated or fibrous structure is always seen in places in the body of the cloud.

Every form is observed between high altostratus and cirrostratus on the one hand, and low altostratus and nimbostratus on the other.

Rain or snow may fall from altostratus (altostratus praecipitans), but when the rain is heavy the cloud layer will have grown thicker and lower, becoming nimbostratus; but heavy snow may fall from a layer that is definitely altostratus.

¹ Ibidem, p. 7.
² Circulaire C. E. N, no. 47, p. 40.
³ "Les Variétés de l'Alto-Cumulus, Annales de l'Observatoire Royal de Belgique," Nouvelle série, tome VI, Bruxelles 1903, p. 14.
⁴ Stratus castellatus "Cloudland," London 1894, p. 56.
⁵ "Rapport sur la Classification des Nuages," Mémoires du Congrès Météorologique International de Paris 1889, p. 15.

From the definition of altostratus it follows that there are three sub-genera:

1. Altostratus translucidus [*C.E.N.* 1926[1]]. A sheet of altostratus resembling thick cirrostratus; the sun and the moon show as through ground glass.

2. Altostratus opacus [*Besson* 1921[2]]. An opaque layer of altostratus of variable thickness which may entirely hide the sun, at any rate in parts, but showing a fibrous structure in some parts.

3. Altostratus praecipitans [*Clement Ley* 1894[3]]. A layer of opaque altostratus which has not yet lost its fibrous character, and from which there are light falls of rain or snow, either continuous or intermittent. This precipitation may not reach the ground in which case it forms *virga*.

B. Explanatory Remarks. The limits between which altostratus may be met are fairly wide (about 5000 to 2000 metres).

A sheet of high altostratus is distinguished from a rather similar sheet of cirrostratus by the convention that halo phenomena are not seen in altostratus, nor are the shadows of objects on the ground visible.

A sheet of low altostratus may be distinguished from a somewhat similar sheet of nimbostratus by the following characters: Nimbostratus is of a much darker and more uniform grey, and shows nowhere any whitish gleam or fibrous structure; one cannot definitely see the limit of its undersurface which has a "soft" look, due to the rain, which may not reach the ground.

The convention is also made that nimbostratus always hides the sun and moon in every part of it, while altostratus only hides them in places behind its darker portions, but they reappear through the lighter parts.

Careful observation may often detect *virga* hanging from altostratus, and these may even reach the ground causing slight rain. If the sheet still has the character of altostratus it will then be called altostratus praecipitans, but not if it has become nimbostratus.

A sheet of altostratus, even if it has rifts in places, has a general fibrous character. A cloud layer, even a continuous one, which has no fibrous structure, and in which rounded cloud masses may be seen, is classed as altocumulus or stratocumulus, according to circumstances.

Altostratus may result from a transformation of a sheet of altocumulus, and on the other hand altostratus may often break up into altocumulus.

C. Varieties. There are many varieties; some of them may be distinguished by adding one of the general qualifying adjectives to the name of the sub-genera (*e.g.*, altostratus opacus undulatus).

VI. Stratocumulus (Sc) [*Kaemiz* 1841[4]]

A. Definitions. A layer or patches composed of lamina or globular masses; the smallest of the regularly arranged elements are fairly large; they are soft and grey, with darker parts.

These elements are arranged in groups, in lines, or in waves, aligned in one or in two directions. Very often the rolls are so close that their edges join together; when they cover the whole sky, as on the continent, especially in winter, they have a wavy appearance.

From the definition it follows that stratocumulus comprises two sub-genera:

1. Stratocumulus translucidus [*C.E.N.* 1930[5]]. A not very thick layer; in the interstices between its elements either the blue sky appears, or at any rate there are much lighter parts of the cloud sheet, which here is thinned to its upper surface.

2. Stratocumulus opacus [*C.E.N.* 1930[6]]. A very thick layer made up of a continuous sheet of large, dark rounded masses; their shape is seen not by a difference in transparency, but they stand out in real relief from the under surface of the cloud layer.

There are transitional forms between stratocumulus and altocumulus on the one hand and between stratocumulus and stratus on the other.

B. Explanatory Remarks. The difference between stratocumulus and altocumulus is given under the latter (page 886).

It should also be noted that the cloud sheet called altocumulus by an observer at a small height would appear as stratocumulus to an observer at a sufficient height.

It often happens that stratocumulus is not associated with any clouds of the B or C families; but it fairly often coexists with clouds of the D family.

The elements of thick stratocumulus (stratocumulus opacus) often tend to fuse together completely, and the layer can, in certain cases, change into nimbostratus. The cloud is called nimbostratus when the cloud elements of stratocumulus have completely disappeared and when, owing to the trails of falling precipitation, the lower surface has no longer a clear-cut boundary.

Stratocumulus can change into stratus and vice versa. The stratus, being lower, the elements appear very large and very soft, so that the structure of regularly arranged globular masses and wave disappears as far as the observer can see. The cloud will be called stratocumulus as long as the structure remains visible.

C. Species. Among the more remarkable species one may note:

1. Stratocumulus vesperalis [*Besson* 1921[7]]. This name is given to flat elongated clouds which are often seen to form about sunset as the final product of the diurnal changes of cumulus.

---

[1] Circulaire C.E.N., no. 47, p. 40.

[2] Ibidem, p. 309.

[3] Stratus praecipitans, Ibidem, p. 61.

[4] "Vorlesungen über Meteorologie," Halle 1841, p. 151.

[5] "Atlas International des Nuages et des Etats du Ciel, Extrait à l'usage des Observateurs," Paris 1930, p. 12.

[6] Ibidem, p. 12.

[7] Ibidem, p. 312.

2. Stratocumulus cumulogenitus [*C.E.N.* 1930¹]. Stratocumulus formed by the spreading out of the tops of cumulus clouds, which latter have disappeared; the layer in the early stages of its formation looks like stratocumulus opacus.

D. Varieties. The cloud called *roll cumulus* in England and Germany is designated *stratocumulus undulatus;* its wave system is in one direction only. It must not be confused with flat cumulus clouds ranged in line. Stratocumulus often has a *mammatus* (festooned) character, that is to say there is a high relief on the lower surface where pendant rounded masses or corruptions are observed, and at times these look as though they would become detached from the cloud. Care must be taken not to confuse this cloud with some kinds of altostratus opacus whose under surface may appear to be slightly corrugated; the latter is distinguished by its fibrous structure.

VII. Stratus (St) [*Howard* 1803² *Hildebrandsson and Abercromby* 1887³]

A. Definition. A uniform layer of cloud, resembling fog, but not resting on the ground. When this very low layer is broken up into irregular shreds, it is designated fractostratus (Frst).

B. Explanatory Remarks. A veil of true stratus generally gives the sky a hazy appearance which is very characteristic, but which in certain cases may cause confusion with nimbostratus. When there is precipitation the difference is manifest: Nimbostratus gives continuous rain (sometimes snow), precipitation composed of drops which may be small and sparse, or else large (at least some of them) and close together, while stratus only gives a drizzle, that is to say small drops very close together.

When there is no precipitation a dark and uniform layer of stratus can easily be mistaken for nimbostratus. The lower surface of nimbostratus however has always a soft appearance (widespread trailing precipitation, virga); it is quite uniform and it is not possible to make out definite detail; stratus on the other hand has a "dryer" appearance, and however uniform it may be it shows some contrasts and some lighter transparent parts, that is, places less dark where the cloud is thinner, corresponding to the interstices between the rolls and globular masses of stratocumulus, but considerably larger, while nimbostratus seems only to be feebly illuminated, and as though lit up from within.

Stratus is often a local cloud, and when it breaks up the blue sky is seen.

Fractostratus sometimes originates from the breaking up of a layer of stratus; sometimes it forms independently and develops till it forms a layer below altostratus or nimbostratus, which latter may be seen in the interstices.

A layer of fractostratus may be distinguished from nimbostratus by its darker appearance, and by being broken up into cloud elements. If these elements have a cumuliform appearance in places the cloud layer is called fractocumulus and not fractostratus.⁴

VIII. Nimbostratus (Ns) [*C.E.N.* 1930⁵]

A. Definition. A low, amorphous, and rainy layer, of a dark grey colour and nearly uniform; feebly illuminated seemingly from inside. When it gives precipitation it is in the form of continuous rain or snow.

But precipitation alone is not a sufficient criterion to distinguish the cloud which should be called nimbostratus even when no rain or snow falls from it.

There is often precipitation which does not reach the ground; in this case the base of the cloud is always diffuse and looks "wet" on account of the general trailing precipitation, *virga,* so that it is not possible to determine the limit of its lower surface.

B. Explanatory Remarks. The usual evolution is as follows: a layer of altostratus grows thicker and lower until it becomes a layer of nimbostratus. Beneath the latter there is generally a progressive development of very low ragged clouds, isolated at first, then fusing together into an almost continuous layer, in the interstices of which however the nimbostratus can generally be seen. These very low clouds are called fractocumulus or fractostratus according as to whether they appear more or less cumuliform or stratiform.

Generally the rain only falls after the formation of these very low clouds, which are then hidden by the precipitation or may even melt away under its action. The vertical visibility then becomes very bad.

¹ Ibidem, p. 13.

² Ibidem, p. 8.

³ First to be found with the present meaning in: "Quarterly Journal of the Royal Meteorological Society," April 1887, p. 148.

⁴ Such fractostratus or fractocumulus clouds may be called nimbus.

⁵ The introduction of nimbostratus is considered indispensable by the President of the International Committee for the study of clouds, on account of criticisms made on the definition of nimbus in the Provisional Atlas.

The following are the reasons for this new definition:—the definition of nimbus in the Atlas of 1910 led to some confusion; in fact, in different countries the name nimbus was given (*a*) to a low amorphous rainy, layer, originating directly by change from a descending layer of altostratus, (*b*) to very low, dark, ragged clouds isolated at first though not later, which form very often below altostratus or under the rainy layer described above, (*a*)

In the present Atlas it was intended to give the cloud (*a*) the new name of nimbostratus, which is a better name than nimbus for a continuous layer which is formed by evolution from altostratus. As to clouds (*b*) they are classed as fractocumulus or fractostratus (according as they are more or less cumuliform) from which they are not distinguished either in shape, or in their mode of formation (turbulence); when however they look dark in colour, owing to special lighting (*e.g.,* the presence of a higher cloud sheet) and so have a very different colour from ordinary fractocumulus or fractostratus, they may if thought necessary, be called nimbus.

In certain cases the precipitation may precede the formation of fractocumulus or fractostratus, or it may happen that these clouds do not form at all.

Rather rarely a sheet of nimbostratus may form by an evolution from a stratocumulus.

IX. Cumulus (Cu) [*Howard* 1803[1]]

A. Definition. Thick clouds with vertical development, the upper surface is dome shaped and exhibits protuberances, while the base is nearly horizontal.

When the cloud is opposite to the sun the surfaces normal to the observer are brighter than the edges of the protuberances. When the light comes from the side, the clouds exhibit strong contrasts of light and shade; against the sun, on the other hand, they look dark with a bright edge.

True cumulus is definitely limited above and below, its surface often appears hard and clear cut. But one may also observe a cloud resembling ragged cumulus in which the different parts show constant change. This cloud is designated fractocumulus (Fcu).[2]

B. Explanatory Remarks. Typical cumulus develops on days of clear skies, and is due to the currents of diurnal convection; on the continent it appears in the morning, grows, and then more or less dissolves again towards the evening.

Cumulus whose base is generally of a grey colour has a uniform structure, that is to say it is composed of rounded parts right up to its summit, with no fibrous structure. Even when highly developed, cumulus can only produce light precipitation.

Cumulus, when it reaches the altocumulus level, is sometimes capped with a light, diffuse, and white veil of more or less lenticular shape, with a delicate striated or flaky structure on its edges; it is generally shaped like a bow which may cover several domes of the cumulus, and finally be pierced by them. This cloud which does not constitute a species is given the name of *pileus*, a cap or hood.

The clouds which form below altostratus or nimbostratus and which can develop into a complete layer, through whose interstices the altostratus or nimbostratus is generally seen, are usually fractostratus; but if they have a cumuliform appearance they should be classed as fractocumulus. They rarely have this appearance during or soon after rain; on the other hand it is frequent at the beginning of the formation of the low cloud, and when it breaks up.

C. Species. Among the more remarkable species one may note:

1. Cumulus humilis [*Vincent* 1907[3]]. Cumulus with little vertical development, and seemingly flattened. These clouds are generally seen in fine weather.

2. Cumulus congestus [*Maze* 1889[4]]. Very distended and swollen cumulus, whose domes have a cauliflower appearance.

X. Cumulonimbus (Cb) [*Weilbach* 1880[5]]

A. Definition. Heavy masses of cloud, with great vertical development, whose cumuliform summits rise in the form of mountains or towers, the upper parts having a fibrous texture and often spreading out in the shape of an anvil.

The base resembles nimbostratus, and one generally notices *virga*. This base has often a layer of very low ragged clouds below it (fractostratus, fractocumulus, cf. p. 889).

Cumulonimbus clouds generally produce showers of rain or snow and sometimes of hail or soft hail, and often thunderstorms as well.

If the whole of the cloud cannot be seen the fall of a real shower is enough to characterise the cloud as a cumulonimbus.

B. Explanatory Remarks. Even if a cumulonimbus were not distinguished by its shape from a strongly developed cumulus its essential character is evident in the difference of structure of its upper parts, when these are visible (fibrous structure and cumuliform structure). Masses of cumulus, however heavy they may be, and however great their vertical development, should never be classed as cumulonimbus unless the whole or a part of their tops is transformed or is in process of transformation into a cirrus mass.

Although the upper cirriform parts of a cumulonimbus may take on very varied shapes, yet in certain cases they spread out into the form of an anvil. To this interesting variety the name *incus* is given.

In certain types of cumulonimbus, which are especially common in spring in moderately high latitudes, the fibrous structure extends to nearly the whole cloud mass, so that the cumuliform parts almost wholly disappear; the cloud is reduced to a mass of cirrus and of *virga*.

The veil cloud *pileus* is seen with cumulonimbus clouds as with cumulus.

When a cumulonimbus covers nearly all the sky the base alone is visible, and resembles nimbostratus, with or without fractostratus or fractocumulus below. The difference between the base of a cumulonimbus and a nimbostratus is often rather difficult to make out. If the cloud mass does not cover all the sky, or if even small portions of the upper parts of the cumulonimbus appear, the difference is evident. If not it can only be made out if the preceding evolution of the clouds has been followed, or if precipitation occurs; its character is violent and intermittent (showers) in the case of cumulonimbus, as opposed to the relatively gentle and continuous precipitation of a nimbostratus.

The front of a thunder cloud of great extent is sometimes accompanied by a roll cloud of a dark colour in the shape of an arch, of a frayed out appearance, and circumscribing a part of the sky of a

[1] Ibidem, p. 7.

[2] Poey 1863 "Sur deux nouveaux types de nuages observés à La Havane," C. R. Académie des Sciences de Paris 1863, tome 56, p. 361.

[3] "Atlas des Nuages," Bruxelles 1907, p. 8.

[4] Ibidem, p. 35.

[5] "Formes de Nuages dans l'Europe Septentrionale," Annales du B. C. M. de France, année 1880, tome I, partie B, p. 27.

lighter grey. This cloud is named *arcus* and is nothing more or less than a particular case of fracto-cumulus or fractostratus.

Fairly often *mammatus* structure appears in cumulonimbus, either at the base, or on the lower surface of the lateral parts of the anvil.

When a layer of menacing cloud covers the sky and *virga* and *mammatus* structure are both seen it is a sure sign that the cloud is the base of a cumulonimbus, even in the absence of all other signs.

Cumulonimbus is a real *factory of clouds;* it is responsible in great measure for the clouds in the rear of disturbances. By the spreading out of the more or less high parts and the melting away of the under-lying parts, cumulonimbus can produce either more or less thick sheets of altocumulus or stratocumulus (spreading out of the cumuliform parts) or dense cirrus (spreading out of the cirriform parts).

C. Species. Amongst the remarkable species may be noted:

1. Cumulonimbus calvus [*C.E.N.* 1926[1]]. Cumulonimbus characterised by the thunderstorm or the shower that it causes, or by *virga*, but in which at first no cirriform parts can be made out. Never-theless the freezing of the upper parts has already begun; the tops are beginning to lose their cumulus structure, that is to say their rounded outlines and clear-cut contours; the hard and "cauliflower" swellings soon become confused and melt away so that nothing can be seen in the white mass but more or less vertical fibres. The freezing, accompanied by the change into a fibrous structure, often goes on very rapidly.

2. Cumulonimbus capillatus [*C.E.N.* 1926[2]]. Cumulonimbus which displays distinct cirriform parts, having sometimes, but not always, the shape of an anvil.

Instructions for the Observation of Clouds

Determination of the Cloud Form, Varieties and Casual Details

At the time of each observation it is essential to determine the family to which the clouds belong (high cloud, middle cloud, low cloud, or cloud with vertical development).

The observer will then specify and enter in the observation book:

1. The genus of the cloud described by the international code number used in the Atlas. It must be remembered that typical forms are relatively rare; it is generally forms more or less intermediate that are observed; consequently one must determine the typical form which the cloud observed most closely resembles.

2. The species (the particular kind of cloud belonging to the genus already determined) making use of the definitions, and the names given in the Atlas under the cloud in question.

3. The variety (that is, one of the particular forms common to different genera) using the definitions and abreviations given below.

4. The casual details, which do not characterise either species or varieties, according to the defini-tions given below.

If the cloud is in process of evolution the observer should note both the present and the preceding forms.

The general and principal species have been described on pages 884–891. The most important varieties and special features are defined below.

A. Principal Varieties. The chief varieties common to different genera are as follows:

1. Fumulus (Fum.) [*Ritter* 1880[3]]. At all levels, from cirrus to stratus, a very thin veil may form, so delicate that it may be almost invisible. These veils seem to be most frequent on hot days, and in low latitudes. Occasionally they may be observed to thicken rapidly, forming clouds easily visible, espe-cially cirrus and cumulus. The clouds thus produced seem unstable however, and usually melt away soon after their formation.

Cirrus fumulus must not be confused with cirrostratus nebulosus. The latter is much more stable and does not show the phenomenon of the formation and subsequent rapid disappearance of cirrus clouds.

2. Lenticularis (Lent.) [*Clement Ley* 1894[4]]. Clouds of an ovoid shape, with clean cut edges, and sometimes irisations, especially common on days of föhn, sirocco, and mistral winds. This form exists at all levels from cirrostratus to stratus.

3. Cumuliformis (Cuf.) [*Atlas* 1896[5]]. The rounded form resembling cumulus which the upper parts of other clouds may sometimes assume. This may be seen at all levels from cirrus to stratus.

4. Mammatus (Mam.) [*Clement Ley* 1894[6]]. This description is given to all clouds whose lower surfaces form pockets or festoons. This form is found especially in stratocumulus and in cumulonimbus, either at the base, or even more often on the lower surface of anvil projections. It is also found, though rarely, in cirrus clouds, probably when they have originated in the anvil of a dispersing cumulonimbus.

5. Undulatus (Und.) [*Clayton* 1896[7]]. This term is applied to clouds composed of elongated and parallel elements, like waves of the sea. It is well to note the orientation of these lines or waves. When there is an appearance of two distinct systems, as when the cloud is divided up into rounded masses by undulations in two directions, the observer will note the orientation of the two systems. Observation should be made on lines as near the zenith as possible to avoid errors due to perspective. If the *undu-latus* variety is very pronounced, the orientation of the bands should be determined and recorded.

[1] Circulaire C. E. N. n° 47, p. 44.
[2] Ibidem, p. 44.
[3] Annuaire de la Société Météorologique de France, tome XXVIII, Paris 1880, p. 109.
[4] Stratus lenticularis, Ibidem, p. 49.
[5] Atlas International des Nuages, Paris 1896, p. 8.
[6] Cumulo-Stratus et Cumulo-Nimbus mammatus, Ibidem, pp. 84, 104.
[7] Ibidem, p. 346.

6. Radiatus (Rad.) [*C.E.N.* 1926[1]]. This term is applied to clouds in parallel bands (polar bands), which owing to perspective seem to converge to a point on the horizon, or to two opposite points if the bands cross the whole sky. The point is called the radiant point, or vanishing point, and its position on the horizon should be noted (N, NNE, etc.). The point on the horizon should be noted where the bands, if prolonged, would converge, if they do not actually reach the horizon.

B. Casual Varieties. The chief casual varieties are the following:

1. *Virga,* wisps or falling trails of precipitation; applied principally to altocumulus and altostratus.
2. *Pileus,* a cap or hood; applied principally to cumulus or cumulonimbus.
3. *Incus,* anvil; applied to cumulonimbus.
4. *Arcus,* arch cloud; applied to cumulonimbus.

THE NECESSITY OF NOTING THE STATE OF THE SKY AS A WHOLE AND OF FOLLOWING THE EVOLUTION OF THE CLOUDS

A. The Necessity of Noting the State of the Sky as a Whole. It is evident from the specifications of the cloud code that to describe the sky at the station at a particular time logically and completely it is not enough to know the genera or even the species of the clouds present; for example, altocumulus appears in seven specifications of the code, and cirrus in nine. In reality each specification of the code, as the explanations show, is not so much a dry enumeration of the genera or species of clouds in the sky, as a general indicaton of the structure, the organization, and the evolution of the cloud complex which makes up the state of the sky. Some specifications only refer to the general structure; for example $C_M = 9$ is a thundery sky; everyone knows that in thundery conditions degenerate cloud forms are met with which are very difficult to classify, while the thundery look of the whole sky is apparent immediately and without any doubt.

Each specification of the code corresponds to a state of the sky, lower, middle, or high. The observer should have at his fingers' ends the commentaries accompanying the definitions; he should consider as a whole the lower, middle, and high clouds there described, and try to make a considered judgment of the observed sky as a whole, so that he can directly apply to it a number of the code.

The detailed analysis of the individual clouds should follow and not precede this recognition of the state of the sky as a whole. If the observer gets used to this course he will find in a short time that the different states of the sky, lower, middle and high, corresponding with the code, will seem just as "live" as the typical cloud forms, and it will be just as easy to identify a state of the sky as the form of a cloud.

B. Necessity of Following the Evolution of the Sky. The aspect of the sky is continually changing, and many transitional forms exist between the different types of cloud described. It is relatively rare for the observer to see typical clouds of one genus which float past, or persist in the sky for any considerable time; in most cases he will find that he has difficulties at the time of observation if he has not taken the trouble to watch the sky since the last observation. If however he has taken this precaution he will often be able to refer a confusing state of the sky, or a particular cloud, to a previous state which was typical and easy to identify. Moreover most of the specifications of the cloud code take into account the evolution of the clouds. A single isolated observation is insufficient.

As regards evolution, the recognition of the state of the sky as a whole, recommended in the previous paragraph, is better than the identification of clouds considered by themselves, for as a matter of fact the evolution of the state of the sky can be followed indefinitely at one station, while the evolution of a cloud, if, as is usual, it is a "migrant" can only be observed during the relatively short time that it takes to cross the sky.

CLOUD CODE

Lower Clouds

0. No lower clouds
1. Cumulus of fine weather
2. Cumulus, heavy and swelling, without anvil top
3. Cumulonimbus
4. Stratocumulus formed by the flattening of cumulus clouds
5. Layer of stratus or stratocumulus
6. Low broken-up clouds of bad weather
7. Cumulus of fine weather and stratocumulus
8. Heavy or swelling cumulus, or cumulonimbus, and stratocumulus
9. Heavy or swelling cumulus (or cumulonimbus) and low ragged clouds of bad weather

Middle Clouds

0. No middle clouds
1. Typical altostratus, thin
2. Typical altostratus, thick
3. Altocumulus, or high stratocumulus, sheet at one level only
4. Altocumulus in small isolated patches, individual clouds often showing signs of evaporation and being more or less lenticular in shape

[1] Circulaire C. E. N., n° 47, p 36.

5. Altocumulus arranged in more or less parallel bands, or an ordered layer advancing over the sky
6. Altocumulus formed by a spreading out of the tops of cumulus
7. Altocumulus associated with altostratus, or altostratus with a partly altocumulus character
8. Altocumulus castellatus, or scattered cumuliform tufts
9. Altocumulus in several sheets at different levels, generally associated with thick fibrous veils of clouds, and a chaotic appearance of the sky

Upper Clouds

0. No upper clouds
1. Cirrus, delicate, not increasing, scattered and isolated masses
2. Cirrus, delicate, not increasing, abundant but not forming a continuous layer
3. Cirrus of anvil clouds, usually dense
4. Cirrus, increasing, generally in the form of hooks ending in a point or in a small tuft
5. Cirrus (often in polar bands), or cirrostratus advancing over the sky, but not more than 45 deg above the horizon
6. Cirrus (often in polar bands) or cirrostratus advancing over the sky, and more than 45 deg above the horizon
7. Veil of cirrostratus covering the whole sky
8. Cirrostratus not increasing, and not covering the whole sky
9. Cirrocumulus predominating, associated with a small quantity of cirrus

STATES OF THE SKY*

CLASSIFICATION OF STATES OF THE SKY[1]

This classification is based not only upon knowledge of the physical processes involved in the formation of clouds but also upon the synoptic study of disturbances.

The enumeration of the kinds of cloud which cover the sky at a given moment is not a sufficient characterization of the "state of the sky." What really characterises it is the aggregate of the individual clouds and their organization.

A distinction will be drawn between the "upper and middle sky" and the "lower sky." (The divisions of the space which may be occupied by clouds, or rather the cloud complexes contained in them, will be referred to repeatedly in the following pages so that it was necessary to adopt abbreviated terms.)

The state of the upper and middle sky is determined by thermodynamical processes associated with big atmospheric disturbances.

It is most often a question of adiabatic expansion which accompanies a general fall of pressure. Turbulent processes also are always to be found to a greater or lesser extent and many aspects peculiar to the middle sky are due to turbulence. Even convection may sometimes have effect up to these altitudes. But in general these states of the sky depend almost entirely upon the synoptic situation; they correspond to definite sectors of big atmospheric disturbances and are like them migrant or durable. They appear most clearly in middle latitudes, for in low latitudes, apart from tropical cyclones, variations of pressure are often feeble and local while in northern regions the upper and middle sky is often masked by a very overcast lower sky.

The state of the lower sky on the contrary is determined by the meteorological conditions existing in the low layer which is directly influenced by the surface of the earth. Cloud formations in that part are essentially caused by the processes of convection or turbulence and sometimes by cooling by radiation. These states of the sky are therefore understood to depend largely upon the nature of the surface of the earth whether sea or land, upon the relief of the land, its heating, etc. They are particularly abundant and heavy in maritime and northern climates.

I. Upper and Middle Skies

These states of the sky are one of the principal manifestations of the life of large normal atmospheric disturbances, and characterise the various zones: front, central, rear and lateral.

In some of them it is advisable to distinguish between the case of a feeble disturbance and the case of a typical, well developed disturbance. Further, in the latter case the same attenuated aspects are to be found in the south lateral section.

Finally it is expedient to treat separately the big thundery disturbances of middle latitudes which exhibit characteristics peculiar to themselves and a degree of complexity unknown in ordinary disturbances. In such cases no useful purpose is served by distinguishing more than the front zone on the one hand and the central and rear zone on the other.

* International Atlas, Paris, 1932.
[1] This classification has been based upon experience gained in Europe by studying extra-tropical depressions.

There are, therefore, in short the following states of the sky:
1. Emissary sky.
2. Front zone sky, typical.
3. Front zone sky, weak.
4. Lateral sky.
5. Central sky, typical.
6. Central sky, weak.
7. Rear sky.[1]
8. Pre-thunderstorm sky.
9. Thundery sky.

II. Lower Skies

A. Simple Skies. *a. "Convection" Skies.*[2]   The extent and intensity of convection depend upon the vertical instability of the atmosphere which results from the heating of the surface layer of a mass of air which is cold above.   It also varies with the water content of the upper layers.   Three states of the sky are to be distinguished according to the intensity of this convection:
1. Cumuliform sky of fine weather.
2. Disturbed cumuliform sky (without cumulonimbus).
3. Disturbed cumuliform sky (with cumulonimbus).
   b. "Turbulence" Skies.[2]   When air is slightly disturbed over a large surface on a small scale, horizontal layers of cloud are produced.   The following distinctions are drawn according to whether the area of disturbance is clearly bounded above by a much more stable layer, or whether it has no distinctly defined limits:
4. Stratiform sky (stratus or stratocumulus).[3]
5. Amorphous sky (fractostratus or fractocumulus of bad weather).[4]
   B. Mixed Skies.   Some combinations of two of these skies are evidently impossible.   For example, the amorphous sky cannot exist at the same time as the cumuliform of fine weather because there is then not enough humidity; the association of a stratiform sky with that of cumulonimbus is rare, the marked inversion of the former being hardly compatible with the powerful convection of the latter.
   The following combinations are conceivable:
6. Cumuliform sky of fine weather (1) and stratiform sky (4).
7. Disturbed cumuliform sky (2 or 3) and stratiform sky (4).
8. Disturbed cumuliform sky (2 or 3) and amorphous sky (5).

III. Combinations of Lower Skies and Upper and Middle Skies

The various states of the lower sky on the one hand and of the upper and middle on the other hand can naturally exist separately.

It must, however, be noted that the amorphous sky associated with turbulence can only exist beneath a layer of middle cloud (central zone) or under a vast cumulonimbus (rear zone).

On the other hand, the rear zone which, in its upper and middle parts contains some fragments of clouds (fragments from middle layers or cirrus and altocumulus cumulogenitus), is necessarily favourable to very vigorous convection which extends to high altitudes producing disturbed cumulus or even cumulonimbus.   A disturbed cumuliform sky or one with cumulonimbus is therefore necessarily associated with the upper and middle part of a rear zone and itself constitutes the essential characteristic of the rear sector of a disturbance.

Many combinations of lower skies with upper and middle skies are possible, but not all, however. Thus, for example, beneath the middle layers there very often exists a marked subsidence of the air which tends to weaken convection so that this state of the sky is incompatible with a disturbed cumuliform or cumulonimbus sky.

DEFINITION AND DESCRIPTION OF THE STATES OF SKY

A. Upper and Middle Skies

I. Emissary Sky.   Sky of cirrus.   Either isolated or forming little groups at some distance from each other.

The types of sky which may be directly replaced by an emissary sky are: the cumuliform of fine weather, the stratiform, the rear zone (which it follows fairly often when there is overlapping of two disturbances) or the thundery sky.

These clouds are the first indications of the presence at a distance of a disturbance, from which the cirrus clouds emanate.

When the observer is on the extreme edge of the disturbance, with only these emissary cirrus in view, he cannot know whether he is in front or at the side of the depression.

This state of the sky is distinguishable from the beginning of the front zone sky by the fact that the cirrus have no tendency to coalesce.

Compatible lower skies.   All lower skies may be accompanied by emissary cirrus above.

---

[1] The more or less accentuated characteristics of this section of the sky are determined by the nature of its lower part.

[2] This term is not to be understood in its precise physical sense and does not presuppose the exact process of cloud formation.

[3] Stratus and stratocumulus do not differ essentially except in height.

[4] These clouds may also be called nimbus.

II. Typical Front Zone Sky.   Sky of cirrus in fine filaments, regularly arranged, passing progressively into a transparent veil of cirrostratus or altostratus.

The upper and middle skies which may be directly followed by a typical front zone sky are: the emissary sky, the rear zone sky (superimposition possible) and the thundery sky.

The typical front zone sky begins with cirrus sometimes accompanied by cirrocumulus.   The cirrus are generally of the fibrous type in long transparent bands, never of the foamy or irregular type which is encountered in thundery disturbances.

The most rapidly moving cirrus are to be found in this state of the sky.   In Europe the direction of the current in which they travel is generally between W.S.W. and N.W.   The cirrocumulus are generally in the classical form (little white globular masses rather transparent and close together).

The cirrus are soon replaced by an immense cirriform sheet which is very transparent at first.   The most beautiful solar phenomena are to be observed with this type of cirrostratus; the complete halo of 22° is an almost regular occurrence.

At the time when the first clouds of the front zone sky appear, it not infrequently happens that the last cloud elements of the preceding disturbance have not yet disappeared; with the result that the overlapping of a rear zone or thundery sky complex with that of a front zone may be observed.

Compatible lower skies: various convection and stratiform types.

Under the cirrus there is often at first a cumuliform type below (only over the land and provided the time of day is favourable).   But as the veil of cirrostratus develops, these cumulus clouds "subside," flatten out and generally end by disappearing.

If a stratiform sky has prevailed before the arrival of the front zone, the low clouds can generally be seen to break into fragments and disappear but there are exceptions to this rule.   It may happen that the stratiform sky hides the whole evolution of the front zone sky.

III. Weak Front Zone Sky.   Sky composed of middle clouds (more rarely high), regularly arranged, often in parallel bands.   Banks of altocumulus not very thick and with clear outlines, uniting gradually into a continuous layer.

The states of the sky which may be directly followed by a weak front zone sky are: the emissary sky, the rear zone sky (superimposition possible) and the thundery sky.

The weak front zone sky begins with altocumulus which is as likely as not to be accompanied by cirrus.   (There might sometimes be a tendency to confuse the altocumulus with cirrocumulus.)

There is not much to be said about the cirrus.   It is generally of the type described above under the typical front zone heading but it is generally less abundant and less rapid.

The altocumulus clouds belong to two different varieties but often exist together.   The one kind covers the sky with an immense patchwork; the rectilinear fissures which divide them call to mind the ornamentation of a crocodile's skin; their matter is often diaphanous and the blue of the sky which is seen through their mass blends with them at the edges in an almost imperceptible fashion.   On the edge of these sheets or of the fragments which become detached, the cloudlets are much smaller and, in places, very similar to cirrocumulus.   The other kind, on the contrary, congregate and form large, elongated bands the front edge of which crosses the sky, clear-cut and rectilinear in shape; their matter is white and opaque while the cloudlets constituting the bands are in the shape of rolls or ovoid stones.   To this variety belong usually the clouds which on some summer evenings hide the sunset with a horizontal violet barrier and which are sometimes called in error "stratus of the horizon."   The characteristic optical phenomenon occasioned by these clouds is that of the solar or lunar corona.

Altocumulus with "scaly" edges of so frequent occurrence in thundery conditions are not found with this type.

Compatible lower skies.   The lower types which may exist below a weak front zone sky are the same as in the case of a typical front zone sky, that is to say, convection or stratiform skies and there is the same tendency for low clouds to disappear.

IV. Lateral Sky.   Little isolated banks of high and middle cloud, often lenticular, irregularly arranged and in perpetual transformation.[1]

The states of the sky which may be directly followed by the lateral type are: cumuliform sky of fine weather, stratiform sky and emissary sky.

The lateral sky almost always follows the isolated cirrus, the distant forerunners of the disturbance, on the flanks as well as in the front.

After these cirrus clouds, which in some instances, however, may be wanting, the altocumulus appear; these middle clouds should not be confused with the "shreds" of the rear zone sky nor with the regularly arranged masses in the front zone sky.   Here they appear in isolated banks of altocumulus, arranged irregularly.   Lenticular altocumulus, of pearly whiteness, is of frequent occurrence in this type.   In the general way, these banks of altocumulus are remarkably unstable in form.   They are incessantly dispersing or developing afresh; if any one looks at the sky after having ceased to do so for a few minutes he will find it completely changed.

The wind generally remains light since the place of observation is far removed from the centre of the disturbance.

The sky is never completely overcast; the amount of altocumulus reaches a maximum, then diminishes again and the fine weather sky reappears without any more serious effect of the disturbance having been experienced.

Compatible lower skies.   These are of the convection (excepting disturbed cumuliform with cumulonimbus) and stratiform types.

[1] This description of a lateral sky only applies to the lateral zone on the warm side of a disturbance (southern side in Europe).   On the cold side (northern side) the lateral part in a typical case presents exactly the same aspect as the front zone (cirrus, then complete veil of cirrostratus or altocumulus).   Fine weather, however may follow immediately instead of the rainy weather which follows in the front zone.

Beneath the lateral sky a cumulus type can persist without subsiding much. Nevertheless large cumulus congestus are rarely found there (except in mountainous regions). A stratiform sky is frequently associated with a lateral zone sky especially in maritime climates.

V. Typical Central Sky. Low ceiling, opaque veil of altostratus or of nimbostratus becoming covered by fractostratus or fractocumulus and perhaps giving continuous rain.

The only upper and middle state of the sky which can precede a typical central zone sky is the typical front zone sky (sometimes hidden by a stratiform sky).

The veil of cirrostratus in the front zone sky is soon followed by a thicker veil; the halo disappears and the sun only remains visible as a silvery spot of ill-defined contour and fading brightness. This is the typical altostratus appearance. Altocumulus and cirrocumulus, or rather clouds looking like the latter but being perhaps only altocumulus, are still to be seen but instead of appearing white against the blue of the sky, they stand out dark grey against the pearly grey of the altostratus veil. Often they are masses of little round spots; more rarely they extend in close parallel lines reminiscent of the undulations which the sea forms on the sand of the seashore.

It is at this moment that a characteristic phenomenon may be seen: the transparency of the air has become perfect. When the veil of altostratus advances, the horizontal screen of cumulus disappears, as also the mist which shrouds distant objects; beneath the high vault of the sky the atmosphere becomes immense and limpid.

But the sun soon disappears and the ceiling of altostratus darkens as it becomes gradually lower and passes into nimbostratus. Rain, generally of the continuous kind, begins and lasts for several hours. In some conditions the rainfall may be prolonged and last the whole day. Rainfall accompanying altostratus or nimbostratus is generally more intense than the light rain from altocumulus in a weak central sky type but it is not as abundant as in the downpours from the typical rear zone sky which follows. The atmosphere is very humid, another point of difference between central rain and rear zone showers. Visibility is rather bad. In general the sheet of altostratus or nimbostratus terminates shortly after the rain stops, its rear edge being generally fairly clear-cut, rectilinear in shape and of fibrous structure. In fewer cases it may be possible to witness a progressive thinning of the altostratus or nimbostratus for some hours after the rain.

Compatible lower skies. The amorphous sky is necessarily associated with the typical central zone sky.

The complete passage of an altostratus system is always accompanied by low clouds. If the lower part of the sky is clear of clouds at the time of the arrival of the altostratus system, low clouds are seen to appear in the form of fractostratus, fractocumulus or stratocumulus; but in the last case also the amorphous condition (fractostratus) dominates in the end. Even if there are still some cumuliform clouds when the central zone conditions set in, they are gradually transformed into fractostratus. The latter sometimes form eventually a continuous layer which entirely hides the altostratus or nimbostratus. When the rain has nearly ceased falling, however, the fractostratus is often transformed into fractocumulus.

VI. Weak Central Sky. Low ceiling. Opaque veil of altostratus formed from altocumulus which have coalesced, sometimes but not always becoming covered by fractostratus or by fractocumulus and being capable of giving more or less continuous though slight rainfall. Misty sky. Wind generally rather light.

The only upper and middle state of the sky which can be directly followed by a weak central zone sky is the weak front zone type.

The banks of altocumulus in the weak front zone sky are followed by a sheet of altocumulus of the second variety described on page 887 but much more extensive than those banks. On the front edges altocumulus without shadows are clearly seen. But almost immediately shadows begin to form on the altocumulus and they develop into a uniform sheet, true altostratus, the altocumulus structure being almost indiscernible; mist intervenes, an essential factor in the weak central zone sky and one which contributes towards the uniform appearance of the altostratus in such a fashion that if the edges had not been observed it would be impossible to distinguish it from an altostratus of cirriform structure.

Where the mist is not too thick to be pierced by the sun, it diffuses a warm and gilded hue over the whiteness of the altocumulus. In all other parts a greyish-mauve veil spreads over everything. Visibility deteriorates rapidly, sometimes to only a few hundred metres.

Fractostratus intermingles with the mist in small shreds of little density. These clouds, which in a typical central zone sky are almost always associated with rain, often pass without giving rain in this type. When rain does fall, it is fine and resembles at times a heavy drizzle. Altostratus of this type does not exhibit the same continuity as in typical central zone cases. Gaps or thin places may appear. Sometimes even a rift may divide it into two sheets; the mist then becomes momentarily invisible; only the whitish colour of the sky reveals it.

The rear edge of the sheet is formed of altocumulus without shadow, shiny white and very opaque. It stands out clearly against the sky and is often broken by well-marked fissures through which the blue sky is seen.

Compatible lower skies: amorphous sky. The lower sky is analogous to the corresponding one in a typical central sky. The quantity of low clouds is always less (in exceptional cases there is none at all) and the fractostratus is seldom sufficient to hide the sky above completely.

VII. Rear Zone Sky. Unstable weather, characterised by rapid alternations of pronounced improvements with exceptional visibility and menacing, disturbed skies which may be accompanied by showers and squalls. The upper and middle parts of the sky contain remains of dense and relatively low, high or middle clouds.

In the case of the rear zone sky the lower sky is inseparable from the upper and middle sky. It is characterised by more or less intense convection. That gives the dominating note to the scene and the typical or "weak" nature of the disturbance is manifested by the degree of intensity of convection.

If the lower sky contains cumulonimbus (heavy showers, squalls) the disturbance is a typical one. If there are only disturbed cumuliform clouds (showers and feeble gusts of wind) it is weak.

A description of a typical rear zone sky will be followed by that of a weak one but in both, the lower part of the sky plays a large part.

First it should be noted that:

The rear zone sky generally follows the central zone type.[1]

a. Typical rear zone sky. It has just been shown that the dominating characteristic of a central zone sky is its uniformity. It is just the opposite in the rear zone.

It is not an uncommon occurrence for rain in a depression to be falling simultaneously over an area equal to half of France. In the rear-zone, however, the most violent shower almost always leaves one corner of the horizon illuminated by the sun which is soon to reappear. Diversity obtains; an observer flying at a great height would see beneath him, following upon the impenetrable mantle of the central sheet, a heterogeneous collection of cumulus, fractocumulus and cumulonimbus and beneath this host of clouds, the earth still wet, variegated with shade and light. Squall clouds and extraordinary transparency of the atmosphere are the two essential characteristics of the diverse aspects which the rear-zone sky may assume.

The typical silhouette of cumulonimbus is not generally to be seen owing to the presence of cumulus or other low clouds (fractocumulus, fractostratus) which hide the cumulonimbus clouds at the moment when they are distant enough to be seen in profile.

Electrical manifestations are feeble in this zone (a few peals of thunder, occasional lightning), but they do occur sometimes and even in winter the fall of temperature does not preclude their occurrence completely. Another characteristic of squalls associated with depressions is the regularity of their travel over large areas.[2]

The cumulus which surround cumulonimbus are distinguishable from the cumulus of fine weather (cumulus humilis) by the ragged appearance of their lower edge and the bulging of their upper part. This tendency for cumulus to bulge always exists in rear zone skies.

In cumulus humilis the ratio of the vertical height to the horizontal length is of the order of $\frac{1}{10}$ but in the cumulus discussed above it may reach or surpass $\frac{1}{2}$. Further, the usual typical rear zone skies do not exhibit the most beautiful examples of dome formation which are more clearly shown by the misty cumulus of the weak rear zone sky and the marble-like piles of showery cumulus.

In the clearances in the rear zone, visibility is very good and sometimes exceptional. The sky assumes a characteristically deep blue colour, very different from the pale blue of anticyclonic fine weather which always makes one think of a vivid colour seen through a light mist as though through a thin veil of dust. The colour of the sky is paler at the horizon and when the cirrus in advance of a new depression does not cause a red sunset, the sun sets below the horizon on a background of shining straw-coloured yellow.

It goes without saying that the heterogeneous character of the rear zone sky is observable there too and if one glances at the points on the horizon which are dominated by the squall clouds, clumps of very low clouds may be seen floating in their shadow.

Finally the last clouds withdraw and fine weather is established.[3] When the passage of a depression is completed by the end of the day, the night air is very pure and favourable to the formation of fog. If it passes during the night the distant views are slightly veiled and the blue of the sky is less vivid.

The rear zone phase may last a long time (several days) owing to the persistance of a current of cold, unstable air in which cumulonimbus are formed. This is exemplified in the showery weather typical of maritime polar air, on the coasts of north-west Europe.

b. Weak rear zone sky. The clouds to be found in the weak rear zone sky are nominally the same as those of the typical rear zone but their characteristics are different.

Although the lower sky generally does not degenerate beyond the disturbed cumuliform stage, cumulonimbus are not entirely lacking; although the tops bulge considerably they do not present any danger and frequently appear lost in the mist. Electrical discharges from them are practically unnoticeable and the gusts of wind which accompany them are not at all strong. They frequently pass without a shower and they never cause heavy rainfall. In weak disturbances it might almost be said that mist takes the place of rain.

This attenuation of virulence corresponds with the substitution of sheets of middle clouds for those of high ones. Further the harmless cumulonimbus belonging to the passage of the type in question are generally associated with sheets of altocumulus, while those of thunderstorms and typical rear zone squalls are associated with cirrostratus and false cirrus. Moreover, isolated fragments of altocumulus may be seen trailing across the sky, quite independent of any cumulonimbus. This is of much rarer occurrence in the typical rear zone sky but may be seen in the thundery sky. The aggregate of these fragments at the same height and following at the central sheet and following at some distance in its wake gives a very peculiar appearance to the weak rear zone sky. It is also hardly possible to confuse these altocumulus clouds with those of the weak front zone sky. They are much more opaque, more brilliant and their irregularly arranged banks have silhouettes which are absurdly easy to recognise.

Finally in the place occupied, as will be seen later, by stratocumulus in the rear of a thundery disturbance there are generally found with the type in question, layers which may be called by the same

---

[1] In some cases, unstable conditions which give rise to rear zone clouds may develop after fine weather without the previous appearance of the central zone type. On the other hand, in the rear of a central sky the rear zone sky may be lacking and be replaced by a stratiform type.

[2] In thundery conditions on the contrary the thunderstorms seem to come to birth, to rage and disappear again within a small area so that they are seldom known to maintain their force for more than a few hours.

[3] Provided that the front zone complex of a new disturbance is not too near.

name but which exhibit a peculiar structure. The layer looks just like altocumulus in vast rolls or in patchwork formation, but the cloudlets are infinitely greater than in true layers of altocumulus, less geometrical and obviously lower. They present certain analogies with stratocumulus vesperalis derived from the diurnal evolution of cumulus.

The weak rear zone sky is likewise the type which exhibits in its clearer parts certain well-developed cumulus which are sometimes designated "silver-edged" cumulus. These are huge clouds of grey colour, tinged with violet, which when situated in front of the sun remain altogether dark except for a thin streak of silver or gold which silhouettes them. In the atmosphere, always a little misty at such times, the sun's rays which are intercepted by these clouds form an immense pencil of diverging rays which are clearly visible owing to the mist and have a very fine, aesthetic effect.

Eventually fine weather returns, provided that the front zone cloud complex of a new disturbance is not very near.

VIII. Pre-thunderstorm Sky. Dense cirrus in very soft outlines. Partial veil of thick cirrostratus, domed altocumulus (cumuliformis) in separate tufts—since the beginning of the approach of the storm no thunder clouds in the strict sense of the word have yet arrived—very light wind.

The types of sky which may be immediately followed by the pre-thunderstorm sky are: cumuliform fine weather sky and the rear zone sky.

The pre-thunderstorm sky begins with cirrus, generally of a very characteristic type. The forms are diverse and most of the varieties which experts have distinguished in the family of cirrus may be said to belong to thundery disturbances. But one characteristic common to almost all varieties is their opacity, an opacity purely relative and especially so in comparison with the remarkable transparency of the fibrous cirrus which moreover is often in accompaniment. This opacity may make them resemble fragments of fractocumulus. Among their varieties of form may be quoted at random: foamy tufts, fern leaves, hooks or rings. In Europe their direction is from between west and south.

These cirrus are generally derived from the anvils of cumulonimbus clouds whose cumuliform parts have disappeared.

They are accompanied or preceded (sometimes by as much as 24 hours) by isolated altocumulus cumuliformis in a fine weather sky which also indicates a high temperature gradient in the upper layers.

Compatible lower skies. The only lower type which can be associated with the pre-thunderstorm sky is the cumuliform fine weather sky which characterises warm periods in summer and which may develop considerably in mountainous regions.

IX. Thundery Sky. Sky characterised by its overcast aspect, chaotic, heavy and as it were stagnant. The lower layers of cloud often having a mammatus appearance. A general absence of wind except during the storms or thundery showers. The upper and middle sky is composed chiefly of dense cirrostratus and diverse forms of altocumulus.

The only state of the sky which can immediately precede the thundery sky is the pre-thunderstorm sky.

As in the case of the rear zone sky of an ordinary disturbance the lower sky is inseparable from the upper and middle sky. Strong convection (cumulonimbus) is an essential characteristic of thundery skies, except in the extreme rear where layers of stratocumulus are often to be found.

It is as well to draw attention to the fact that in the case of thunderstorm systems, the central and rear zones are very difficult to distinguish, convection playing an essential part in both. Also the expression "thundery sky" applies to both the central and rear parts in the passage of a thunderstorm system.

The central sheet often produces no rain. It is sometimes difficult to identify and in many cases one is tempted to say that it is altogether lacking, since a thunderstorm system truly consists essentially of a multitude of cumulonimbus. Some thundery disturbances bring abundant and lasting rain which falls from a vast altostratus but strangely enough these altostratus clouds do not precede the cumulonimbus: they follow or are intermingled with them. The name of "thundery rains" is kept for these characteristic rains.

Nevertheless, the precursory cirrus which form the pre-thunderstorm sky are separated from the cumulonimbus by clouds of the cirrus or altocumulus families which do not necessarily form a sheet but take its place.

In the cirrus family are to be included the extensive sheets of fibrous structure, often fairly transparent and assuming at times the air of altostratus. A rather curious veil of cloud is to be seen in these conditions; it is so vast that it seems to cover the sky and so light that its existence is conjectured rather than perceived. This immense, almost imperceptible gauze is chiefly revealed by the irregularities of its structure and by the effect it has upon the appearance of the stars.

Of the altocumulus family "altocumulus cumuliformis" is to be seen, a characteristic type in thundery systems. It generally appears in the form of large tufts with ragged edges, without shadows, and set somewhat apart (altocumulus floccus). The colour of the sky at the time of the passage of thundery altocumulus is remarkable; the blue becomes mingled with a violet hue as it loses its vividness. Sometimes the dome formations are accentuated and develop considerably (altocumulus castellatus). These clouds are often white, sometimes grey and frequently merge into a greyish cirrostratus.

The thundery sky, especially near the end of its passage, is essentially composed of cumulonimbus with their attendant train of fractocumulus and fractostratus. In Europe, the most violent thunderstorms occur in this phase. The direction from whence they arrive deviates little from S.W. and hardly ever touches S. or W. Their velocity is rather variable, of the order of about 50 km./hr. A curious point about these clouds is their rapid development. If efforts are made to follow minutely the course of a thunder squall by telephone it is surprising to observe its birth, sudden growth, maintenance of its force over a limited course, of the order of 150 km., and its death without any evident reason. The cumulus clouds show clearly the tendency to form domes. Visibility in the clear intervals admits of the same observations as in the rear zone skies of ordinary disturbances. The peculiar trait in the thundery type is the presence of two clouds, the mammatocumulus and a type of stratocumulus which does not appear with absolute regularity in other cases. The mammatocumulus is in a wellknown form but often somewhat different from its classic form: in place of rounded protuberances it often exhibits rolls or species of festoons.[1]

The thundery stratocumulus is formed of elongated clouds, dull in colour, of a dark bluish grey. It appears in the rear of thunderstorms and particularly in the following circumstances: when a thunderstorm finishes in the evening and there seems some justification for expecting clear weather the next morning it often happens that when the sun rises the sky is clouded with an extensive bank of this stratocumulus. It seldom lasts, however, beyond the first few hours of the day.

In conclusion, isolated banks of cirrostratus may be seen at the same time as the full cumulonimbus and lightning sometimes flashes in them during the night. These clouds are probably the cirrus anvils remaining from cumulonimbus clouds.

B. Lower Skies

I. *"Convection" Skies*

I. Cumuliform Fine Weather Sky. Moderate "convection" sky, characterised by the presence of cumulus with horizontal bases, more or less dome-shaped but without excessive vertical development and exhibiting a clear diurnal variation which may result in altocumulus or stratocumulus.

This type of sky has a definite diurnal variation, especially on the Continent. The evolution of "settled fine" weather may often be observed in the morning from the more or less dome-shaped, fine weather cumulus which last during the day but disperse in the evening. In the tropics as well as in the mountainous regions of the temperate zone, fine weather cumulus may grow until they develop into cumulonimbus which give heat thunderstorms. If the state of the sky reaches this stage it should be classed according to circumstances either as "thundery sky" or as disturbed cumuliform sky with cumulonimbus.

Isolated cumulus sky may develop and assume very varying aspects according to circumstances. It may happen that in the course of their vertical development the cumulus reach a stable layer which prevents them from developing to any greater height. The result is a spreading out of the tops of the cumulus which form a sheet resembling altocumulus or stratocumulus. If subsequently the convection currents weaken, the cumuliform part of the clouds disperses first while the altocumulus may survive. Thus a fine weather cumulus sky may be transformed, especially towards evening, into a sky characterised by banks of low altocumulus (altocumulus cumulogenitus).

If the stable layer is only a little above the level of condensation, the cumulus become extended quite near to their base and may coalesce into an almost continuous layer. A rather analogous phenomenon may frequently be observed in the evening when dome-shaped cumulus subside and spread at the level of their bases (stratocumulus vesperalis).

II. Disturbed Cumuliform Sky without Cumulonimbus. Intense convection sky, characterised by thick cumulus, jagged and seething but with no cirriform structure at their summits.

When convection takes place in a sufficiently strong current the forms of dome-shaped cumulus are not so regular and they tend to become cumulus congestus. The weather is then more disturbed with rapid alternations of threatening skies and bright periods.

This motley sky of changing cumulus congestus is characteristic of cold, unstable currents which are warmed by contact with warm earth or sea.

Cumulus of this kind—especially those which are formed over the sea—show much less diurnal variation than fine weather cumulus.

These conditions are especially likely to occur after the passage of a disturbance. In the latter case a cumulus congestus sky is also associated with fragments or broken sheets of middle cloud of altocumulus structure.

III. Disturbed Cumuliform Sky with Cumulonimbus. Intense convection sky, characterised by the presence of cumulonimbus.

In the same meteorological conditions convection may increase sufficiently for some of the cumulus congestus to become cumulonimbus. The sky then assumes the typical form associated with showers and bright periods. Sometimes there are thunderstorms too but they are of short duration and of little importance compared with those of thundery systems.

II. *"Turbulence" Skies*

IV. Stratiform Sky. Rather low ceiling layer of stratocumulus, stratus or fog; weather not rainy (excepting fine drizzle) but misty.

In those parts of the continents which are within the temperate zone, this state of the sky is mainly observed in winter in anticyclonic conditions; it is very rare in summer. In the intermediate seasons, autumn and spring, it exhibits a clear diurnal variation. First of all there is low cloud, then fog or stratus is formed by cooling during the night and it is not dispersed until after the sun rises the next day. In winter insolation is often too small to dissipate the fog or stratus which has formed during the night and the sky may remain overcast for several days in succession—above the stratus there is generally a clear sky, as may be seen from mountains of no great height.

The same skies of fog, stratus or stratocumulus are observable at sea as a result of warm air currents and they may thus extend to regions which are not anticyclonic.

Western Europe is often overrun by this maritime variety of stratus weather. On the northern edge of subtropical anticyclones there is often a humid current from west or south-west in which the stratus and stratocumulus often constitutes a lateral connecting zone between two successive disturb-

---

[1] Mammatocumulus often precedes a thunderstorm; in America it often appears in front of a tornado

ances. It should be noted that stratocumulus or stratus of this type do not have a definite diurnal variation; they sometimes spread over the sky in the middle of the day.

Stratus does not generally give any precipitation; but if it is very dense it may give drizzle, which is accentuated if orographical effects are also present.

V. Amorphous Sky. More or less continuous layer of fractostratus or of fractocumulus of bad weather.

This state of the sky somewhat resembles the preceding one. Nevertheless it is distinguishable by its more ragged appearance and complete absence of regular undulations. On the other hand low clouds are never the only ones in this type; above them there is always either a sheet of altostratus (possibly of altocumulus structure) or cumulonimbus from which more or less continuous precipitation is falling.

Fractostratus or fractocumulus appear at first in isolated masses beneath the veil of middle cloud but they multiply rapidly and form before long a more or less continuous layer.

III. Mixed Skies

VI. Stratiform and Cumuliform of Fine Weather. A mixed stratiform and cumuliform sky may develop when a sheet of stratus or stratocumulus is dispersing and fractostratus is being transformed into fractocumulus. Cumulus is thus formed at the expense of stratus and at the same level where the stratus previously existed.

This type of sky is always transitory, for in the long run one of the two tendencies always triumphs over the other and forms a simple cumuliform or stratiform sky.

Another and much more lasting variety of a mixed stratiform and cumuliform sky may develop when cumulus are formed beneath the cloud sheet (which in this instance consists of stratocumulus, stratus being too low for cumulus to exist beneath it). In the case considered the cumulus are of the fine weather type, remain small and do not reach the sheet of stratocumulus.

VII. Stratiform and Disturbed Cumuliform Sky. If convection is very strong and if the cumulus clouds assume a disturbed aspect, their summits generally penetrate into or even right through the cloud layer above them, a phenomenon which cannot be seen from the ground but may be seen by an aviator. In humid climates this type of sky may give showers.

VIII. Amorphous and Disturbed Cumuliform Sky. This type only exists in showery weather when fractostratus or fractocumulus may be formed beneath cumulonimbus (sometimes even beneath heavy, disturbed cumulus). In humid climates, fractostratus or fractocumulus may form a layer which fills the spaces between the cumulonimbus. In this case the individual cumulonimbus clouds cannot be recognised in their typical form; nevertheless their passage is indicated by a temporary darkening of the sky and showers.

REMARKS ON TAKING OBSERVATIONS OF THE STATE OF THE SKY

I. Distribution of Cloud Forms in Different States of the Sky. A knowledge of the genera or even of the species of the individual clouds present in the sky at any given moment is not enough to identify the state prevailing at the station in question. In reality the same cloud genera or groups of genera may be found in different states of the sky. Inversely, supposing the state of the sky to be known, it is not possible to deduce with certainty the genera of individual clouds of which it is composed because a given state of the sky does not always contain the same genera.

Some interesting conclusions, although mainly of negative value, can be drawn from the distribution of cloud forms in the various states of the sky: certain cloud genera and, still more, certain groups of cloud genera are incompatible with certain states of the sky. From an analysis of the clouds present, certain states of the sky can in some cases be rejected. Inversely, if the state of the sky in question is known, it is also a foregone conclusion that certain clouds and associations of clouds cannot exist. Thus, in both cases, the process of elimination of various solutions simplifies the diagnosis of the remainder.

II. Applications of Knowledge of State of the Sky. 1. It facilitates the identification of cloud genera present in the sky.

2. In some cases, it makes up, in part at least, for indecision as to cloud genera.

This is frequently the case in thundery situations; degenerate cloud forms then occur which are very difficult if not impossible to name according to the international cloud classification, while the thundery nature of the sky as a whole is immediately manifest beyond all doubt.

Thus if an observer cannot name the cloud forms with certainty, he can sometimes know what section of a cloud system is prevailing and it will be granted that that is an important piece of information with regard to the clouds.

The identification of a sky as a whole compared with that of the individual clouds is further greatly facilitated by the fact that the observer can make use of the preceding history of the sky. The evolution of the sky at one station can be followed indefinitely, while the evolution of a cloud—if it is a "migrant" as is usually the case—can only be observed during the relatively short time that it takes to cross the sky.

3. Even if the individual cloud forms are identifiable, clouds of the same genera assembled in the same quantities may correspond to two states of the sky equally characteristic but altogether distinct— and therefore to two different situations.

For example, cumulus and altocumulus may be associated in the same proportions in totally different states of the sky: (1) preceding rain (front zone), (2) following rain (rear zone), (3) on the border of a rain area (lateral zone).

Therefore, even after the cloud forms present in the sky have been identified accurately it may still be necessary to identify the aggregate condition in order to distinguish between two sky complexes, the composition of which is qualitatively and quantitatively the same as regards cloud genera but which correspond to totally different situations.

For these various reasons the classification of states of the sky is indispensable and it is necessary to know how to proceed with the direct and indirect identification of the state of the sky. In principle the observer's diagnosis of the state of the sky ought always to be accompanied by an analysis of the individual clouds present at the time.

DEFINITION OF HYDROMETEORS*

(With German, French, and English Names.)

Regen, Pluie, RAIN: Somewhat uniform precipitation of rather large drops (ordinary rain, steady, widespread rain) from a continuous cloud covering. The sky is either covered with a real layer of rain cloud which has developed from a succession of layer clouds or with a uniform, grey but comparatively high layer of cloud, generally with shapeless cloud masses below, which may in fact be present in such numbers that the layer clouds are completely hidden.

Schnee, Neige, SNOW: Somewhat uniform precipitation of hexagonal stars from a continuous cloud covering (appearance of the sky as in the case of rain).

Regenschnee, Pluie et neige, SLEET: Somewhat uniform precipitation from melting snow or snow mixed with rain.

Reifgraupeln Neige roulée, GRANULAR SNOW: White opaque pellets, 1 to 5 mm. in diameter, of snow-like structure. They are brittle and easily compressed. If they fall on a hard surface, they rebound and then break easily. They chiefly occur with temperature about 0°C. and mainly over the land, often before or at the same time as ordinary snow.

Frostgraupeln, Grésil, SOFT HAIL: Semi-transparent, round, seldom conical pellets, about 2 to 5 mm. in diameter. They often have a centre of granular snow covered by a thin layer of ice. Even when they fall on a hard surface, they frequently remain lying on it without breaking; they are not easily compressed or brittle. Since they generally fall with temperatures about 0°C., often at the same time as rain, they are wet.

Hagel, Grêle, HAIL: Irregular little pieces of ice varying in size from that of peas to that of a man's fist. They are either completely transparent or formed of alternating clear and opaque (snow-like) layers. They fall almost exclusively during severe or protracted thunderstorms and never with temperature below freezing point.

Eiskörnchen, Grains de glace, PELLETS OF ICE: Transparent pellets, hard as glass, 1 to 4 mm. in diameter. If they fall on a hard surface they rebound. They develop from raindrops which fall through a cold surface layer and thus freeze.

Eisnadeln, Aiguilles de glace, FLOATING ICE CRYSTALS: Very little sticks, pellets or scales of ice, which appear to hover in the air. They are especially visible when they glitter in the sunshine and may give rise to sun pillars or other halo phenomena. They occur in stable winter conditions mostly with intense cold in polar winter or in the higher layers of the free atmosphere.

Nieseln, Bruine, DRIZZLE: Somewhat uniform precipitation of numerous tiny droplets (diameter generally less than ½ mm.) which almost seem to hover in the air and share its slightest movements. Drizzle falls from a continuous, dense, low layer of stratus. In particular along coastlines and on mountain ranges drizzle can sometimes yield considerable amounts of precipitation (at any rate up to 20 mm. in 24 hours).

Nebel, Brouillard, FOG: Microscopical little drops of water, which hover in the air and cause a wet and cold feeling. If watched carefully the water droplets may in some circumstances be almost seen to pass in front of one's eyes. On the whole fog appears white except near industrial regions where it becomes dirty grey or yellow. With a real fog which is not dispersing, the horizontal visibility should, according to international agreement, be less than 1 km., at least in one direction.

Leichter Nebel, Brouillard léger, MIST: Light fog, in which visibility must be greater than 1 km. It does not cause a raw or damp feeling because the water droplets in mist are too small and far apart; it often has a greyish appearance in distinction from actual fog.

Dunst, Brume, HAZE: Little particles of dust from dry regions or of salt, which are dry and so extraordinarily small that they are neither felt nor perceived with the naked eye but which give the air a characteristically smoky appearance. Haze casts a completely uniform veil over the landscape and dulls its colours. Against a dark background this veil has a bluish colour ("blue distance") but a dirty yellow or reddish yellow against a light background (*e.g.*, clouds on the horizon, snow-covered summit, sun). Haze may be distinguished by this characteristic from greyish mist which it sometimes equals in intensity.

Reine, durchsichtige Luft, Visibilité exceptionnelle, UNUSUAL VISIBILITY: Even when there is no particular obscuration of the atmosphere, it is possible to observe the same blue veil as in hazy conditions against a sufficiently remote, dark background. At sea level (in the lower layers of air) this veil which is caused by the pure air itself completely hides the outlines of mountains at a distance of over 500 km. even in the most favourable circumstances and in general does so at less than 100 km. This may also occur at twilight or when visibility in any direction is reduced, for example, by a shower, provided that the air is otherwise clear and transparent. If there is no range of hills at a sufficiently great distance, the purity of the air can be recognised by the fact that the colours and details of the landscape within 5 or 10 km. are clear and distinct (no semblance of being veiled), that ranges of hills to a distance of about 30 km. are sharply defined in cloudy weather and appear almost pure dark blue against the sky, while details are clearly visible (at least with a telescope) when the sun shines upon them.

* International Atlas, 1932.

Schauer, Averse, SHOWER: Of the above mentioned hydrometeors, rain, snow, sleet, granular snow, soft hail, hail and grains of ice may fall in showers. A shower is not only characterised by the quick onset and cessation of the precipitation or its rapid changes of intensity but also and more especially by the appearance of the sky. Showery weather is characterised by a rapid alternation of dark, threatening shower-clouds and bright periods of short duration but often with deep blue skies (April weather). If there is no definite clearance between the showers it is due either to a layer of high cloud (which is often the forerunner of further rainfall) or to the fact that the space between the shower-clouds is filled with low though lighter clouds. It may even happen that the precipitation never completely stops, the arrival of a shower being marked in such cases by sudden darkness.

Schneegestöber, Tourmente de neige, DRIFT-SNOW *(a):* Snow fall in squally weather.

Schneetreiben, Chasse-neige (1), DRIFT-SNOW *(b* 1): No actual precipitation. The snow is whirled so high from the ground that visibility is considerably decreased both horizontally along the ground and vertically. The condition of the sky is not recognisable.

Schneefegen, Chasse-neige (2), DRIFT-SNOW *(b* 2): Snow, torn from the ground by the wind, is driven along it low down so that vertical visibility is not materially decreased. The motion is almost in a straight line.

Sandsturm, Tempête de sable, SANDSTORM: No actual precipitation but sand or dust is whirled into the air so that visibility at eye level is less than 1000 m.

Tau, Rosée, DEW: Little drops of water which in consequence of direct condensation from the neighbouring layer of air are deposited on surfaces which have been cooled by nocturnal radiation.

Reif, Gelée blanche, HOARFROST: Ice crystals which are precipitated in the same way as dew.

Rauhreif, Givre (1), RIME (1): Ice crystals which are deposited when fog and frost occur together mainly upon vertical surfaces—in particular upon the points and corners of objects. To windward they may grow to considerable layers of hoarfrost-like structure. The process is in some ways analogous to the formation of granular snow.

Rauhfrost, Givre (2), RIME (2): Masses of ice, which are deposited in an analogous way to rime but which occur with wet fog or supercooled drizzle so that their structure is analogous to that of soft hail. Both forms are specially apt to occur in mountainous districts.

Glatteis, Verglas, GLAZED FROST: Homogeneous transparent layers of ice formed on horizontal and vertical surfaces from super-cooled drizzle or rain.

Beschlag, Dépôt de buée: Deposition of water vapour from the lowest layers of the atmosphere which have been cooled below the dew point by outward radiation from the ground.

Gewitter, Orage, THUNDERSTORM: Both lightning and thunder are observed.

Wetterleuchten, Eclairs, SHEET LIGHTNING: Lightning without thunder.

Donner, Tonnerre, THUNDER: Distant thunder is heard, but no lightning is seen. The direction in which thunder and lightning are observed should be stated if possible.

REMARKS ON THE INTERNATIONAL CLASSIFICATION

The number of internationally accepted varieties of clouds will increase in the future. Even at present, more is known about the significance of the varieties than can adequately be expressed by the International classification. It is consequently the observer's duty to use the existing classification as extensively as possible. He should always express species and variety as well as genus whenever he is sure of the observation. The reason for this is that the synoptic significance of the cloud usually lies in the variety and detailed structure. The necessity of detailed observation applies not only to the individual clouds, but also to the successive aspects of the same layer of clouds during the day. The behavior of a layer of cumulus throughout the day, for example, indicates the synoptic situation as well as or better than does the shape of the individual clouds at a given time.

A good cloud observation helps in the analysis of the synoptic map. Such an observation may be discussed under three headings: (1) correctness, (2) detail, and (3) continuity.

1. Correctness. A doubtful observation must be reported as doubtful. If it is intended for a teletype message in which doubt cannot be mentioned, the observation should be omitted.

Example. White lines close to the horizon may be cirrus, cirrostratus, alto-cumulus, haze, or simple stratus. Each of these clouds has significance in the analysis, stratus being the most harmless, altocumulus being the most significant. Generally speaking, clouds close to the horizon (below 10 to 15 deg elevation) should be examined with extra care, and doubt as to their name should be clearly indicated. An incorrect report is worse than no report. In the case of a teletype message, for example, an observer should remember that he is cooperating in a network with many other observers. Clouds that appear low on his horizon may be described easily from the next station. Little is lost by omission, but much harm may be done by sending a doubtful message that conflicts with the correct report from the neighboring station. The analyst is not always able to decide which report is correct from the data available.

2. Details. Certain clouds are found under so many different circumstances that mention of the genus alone does little good in synoptic analysis. Even mention of species as well may not suffice. The variety, the most elaborate term in the present classification, is more typical, and there is an obvious need for subvariety observations in order to keep pace with the progress in other fields of meteorology. Similarly, great detail is very useful in understanding the chronological evolution of a given cloud layer. For best results, frequent inspections of the sky (hourly at least) and many stations (50-mile grid) are necessary.

3. Continuity. The successive aspects of the clouds observed at a given station vary continuously. So do the clouds observed on a cross-country flight. This continuity is related to the continuity in the slopes and shapes of frontal surfaces and the elements that control the physics of the clouds. Important here are (1) surfaces of discontinuity, (2) inversions, (3) subsidence layers, (4) ascending currents, (5) descending currents, (6) distribution of moist and dry layers, and (7) the distribution and varieties of the vertical temperature lapse rates.

SKY COVER

The cloudiness or sky coverage is usually given in fractions of $\frac{1}{10}$ or $\frac{1}{4}$. The total cloudiness refers to the amount of the sky covered by all genera; the partial cloudiness refers to each genus observed separately. Close to the horizon, perspective reduces the amount of blue sky that can be seen, causing cloudiness to appear greater than it really is. It is consequently advisable to ignore that portion of the sky when estimating the cloudiness. There are no rules that automatically give a correct estimate of total or partial cloudiness. Clear or overcast skies are easy to identify. In other cases, the following indications usually give satisfactory results: The observer should ask himself the questions (1) is the amount of clouds (or blue sky) almost negligible? and, if necessary, (2) is the amount of blue sky obviously greater than the amount of clouds? If cloudiness is almost negligible, the report should be $\frac{1}{10}$. If blue sky is almost negligible, the report should be $\frac{9}{10}$. If cloudiness and blue sky appear equal, the report is $\frac{1}{2}$. If the blue sky predominates but cloudiness is greater than $\frac{1}{10}$, the report should be $\frac{1}{4}$. If cloudiness predominates but the blue sky is greater than $\frac{1}{10}$, the report should be $\frac{3}{4}$.

Experience shows that the above method leads to estimates that are independent of the observer. From the viewpoint of the synoptic analyst, this is a valuable achievement. The scale of cloudiness reads then $0-\frac{1}{10}-\frac{1}{4}-\frac{1}{2}-\frac{3}{4}-\frac{9}{10}-1\frac{0}{10}$. In the writer's opinion, closer steps are really not necessary and probably would not be estimated identically by different observers.

NEPHANALYSIS

The study of synoptic charts on which only clouds and weather are plotted demonstrates three facts that can be labeled (1) individuality, (2) typicalness, and (3) continuity. Cloud forms and cloud species over large areas form synoptic entities (called *nephsystems*) usually surrounded by fair-weather areas of clear sky, cumulus humilis, or cirriform clouds. These systems usually correspond to groups of frontal lines, the most typical of which is the V-shaped group consisting of one warm front and one cold front (*i.e.*, a cyclone). Other nephsystems are not associated with fronts. Here, for example, we have the thundery nephsystems frequently found within the tropical air in the southern United States.

Nephsystems can be divided into various sectors that are characterized by various kinds of weather. Subspecies (varieties) of clouds yield valuable data concerning the location of the station with respect to the sectors of the nephsystem and, in consequence, with respect to the fronts. If an itinerary is traced within the nephsystem, all cloud forms successively encountered constitute a succession that is continuous. Through these continuous changes, cloud genera and species, although appearing quite diverse (such as altostratus and cumulus, or stratocumulus duplicatus and cumulonimbus), are proved to be related by a series of intermediate forms. Thus

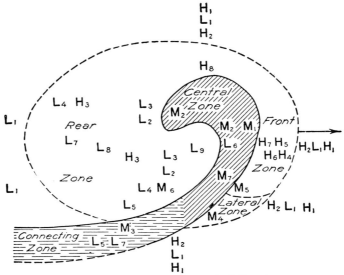

Fig. 1.—Sky and clouds in a typical disturbance.

NOTE: 1. This diagram corresponds to a typical disturbance—to be precise, the first of a family—arriving from the west in western Europe.

2. It sometimes happens that the rear zone is much more extensive and that it persists for several days over the same region.

3. Fractocumulus may occur practically anywhere in the rear zone.

4. The specifications M_8 and M_9 correspond, respectively, with the front or lateral sector and with the central or rear sector of a *thunderstorm* disturbance. They can find no place in this diagram, which is that of a *normal* disturbance.

5. Observers should not take the form of cloud corresponding to the position in this diagram as necessarily determining the form of cloud to be reported.

the clouds are found to be part of a large synoptic continuum, of an organized "society." This entity is the nephsystem.

The geographical relationship and organized association of numerous different genera, species, and varieties of clouds is the most characteristic datum in the concept of the nephsystem. Various types of nephsystems have been mentioned above (*i.e.*, cyclonic and thundery types). Other types may also be discovered when air-mass weather, tropical, and arctic weather have been more thoroughly studied. The major significance of the concept of the nephsystem is attached, not to a given geographic relationship of clouds, but to the existence of a patterned association of clouds.

Within the nephsystem, certain discontinuities exist. These discontinuities correspond to the surfaces of discontinuity in frontal analysis. They are encountered only in the itineraries that cross them, at the point of traverse. The basic cloud of the nephsystem is altostratus.   Since altostratus is usually encountered in connection with a surface of discontinuity (namely, a warm front), its position is fundamental in weather analysis.   Almost all other forms of cloud can be related to the altostratus by a continuous succession of intermediate forms such as have been described elsewhere.   Such successions are called *sky* or *cloud sequences*.   A study of cloud sequences obviously yields practical advice on the use of clouds in forecasting.

A general sketch of a nephsystem is given in Fig. 1.   It is important to explain how the figure should be interpreted.   Forecasters already familiar with the simple models of fronts and cyclones used in frontal analysis, and with the complex shapes of actual fronts and cyclones as found on the weather map, will appreciate the difference between the idealized nephsystem and those found in practice.   In other words, the simple model of a nephsystem should not be described or interpreted with a dogmatic mind.   The model is merely a guide for the interpretation of a neph-analysis map, comparable with the cyclone models used in frontal analysis.   Furthermore, the simple nephsystem model is based largely on observations gathered in temperate climates and especially in western Europe, as were the early models used in frontal analysis.   Every analyst will then be able to adapt the simple model to the weather conditions as found in his region, whether it be in the tropics, in the arctic, or in remote parts of the temperate zone.   Three essential facts are (1) individuality of nephsystems, (2) typicalness of cloud varieties, and (3) continuity of cloud sequences.

Observations plotted on maps for nephanalysis include clouds, weather, height and amount of clouds, ceilings, and precipitation.   The observations are presented by means of curves called *nephcurves*.   Nephcurves are comparable to isobars or fronts on the usual synoptic map, in that they delineate certain areas by sharp lines. In practice, nephanalysis consists in identifying the various nephsystems and the external zones separating them.   The most useful nephcurves are (1) the clear sky curve, (2) the border line of the nephsystem, (3) the curves of precipitation, and (4) various auxiliary curves such as ceiling, cloudiness, instability, etc.   The nephcurves enable the analyst ultimately to trace the various nephsystems and to locate their sectors.   Predicting the motions and transformations of the systems is the next and most important step.

Most of the nephcurves need no definition since their names are self-explanatory, *e.g.*, the clear sky curve, the precipitation curves, ceiling curves, and cloudiness curves. The border line and the lines of weakness are defined as follows: As can be seen in Fig. 1, the typical borderline curve is the altocumulus, with exception of the area designated as connecting zone.   The border line consequently separates the stations reporting altocumulus from those reporting none.   In the connecting-zone area, which is characterized by stratiform and convective-stratiform states of the sky, the stratocumulus is the best index of the border line.   If typical, it belongs to the system. False or nontypical stratocumulus, however, may appear, especially on the midday charts.   They are merely dense cumulus and belong to the external zone.   This is also true of stratocumulus of local origin such as the orographic type.

Lines of weakness are used to mark the separation of two nephsystems that are otherwise not entirely separated by an area of fair weather.   The two systems may at first appear as one.   The latter occurs, for instance, when small waves cause the frontal line to bend.   Weakness in the cloud formations can show in various ways: (1) the total sky coverage may decrease; (2) the sky may remain overcast while the ceiling rises, (3) the distance between the border lines may decrease, etc.

PRACTICE OF OBSERVATIONS OF CLOUDS AND OF STATES OF THE SKY

Some errors may be committed in spite of a good knowledge of the international definitions.   The following will be helpful in deciding which name to select when (as often happens) a cloud form is intermediate between two or even three typical forms.

1. Cirrus. Vague white layers are not necessarily composed of cirrus. Confusion is especially easy in hazy weather. When cumulus begin to form in the morning, condensing vapor sometimes appears as a transparent veil, which may be mistaken for cirrus. Usually, however, these veils may be identified as low clouds because of their motion parallel to the fumulus and small cumulus that appear at the same time. They may also be identified by their angular speed, which, because of their low altitude, is far greater than that of cirrus. When white patches of altocumulus or high stratocumulus appear on the horizon, their wavy or cellular structure is difficult to see. They may also be mistaken for cirrus or cirrostratus, a mistake that would confuse the central analyst. This is the reason why the observer must be very careful when recording clouds less than 20 deg above the horizon.

2. Altocumulus. When the early-morning sky is overcast with stratocumulus or stratus, the layer often progressively dissolves thereafter and may altogether disappear before noon. When sky cover during this period falls below $\frac{2}{10}$, remnants of the layer may resemble altocumulus. In places, the clouds appear as white opaque rolls or look like thin transparent paving blocks. Were it not for the previous history of the sky, the clouds might be mistaken for altocumulus. Altocumulus, however, are middle clouds and have nothing to do with this synoptic situation. The correct observation is "dissolving stratus or dissolving stratocumulus." From the synoptic analyst's point of view, any mention of altocumulus should be avoided unless they are undoubtedly present; for, when altocumulus are located on the edge of a nephsystem, they serve to delineate the system. They mark the boundary between the system and the external zone where only cirrus and cumulus are found. The incorrect mention of altocumulus may consequently lead to the mistaken belief that a system is approaching. The remedy clearly lies in closely following the history of the sky.

3. Altostratus. Since altostratus is the backbone and the most important cloud of the system, it should never be mentioned unless it is certainly present. Its dimensions are characteristically large, covering hundreds of miles in either direction. Thus descriptions of a sky whose cloudiness is only $\frac{1}{10}$ or $\frac{2}{10}$, coupled with reports of altostratus, would therefore puzzle the central forecaster. Actually, the reported altostratus is sometimes a thick cirrostratus, the anvil of a cumulonimbus, an isolated anvil left behind by a series of thunderstorms, or even an altocumulus lenticularis (which is completely developed and has lost its altocumulus structure). The international definition of altostratus reads, "striated or fibrous veil more or less bluish in color." For practical purposes, it is best to stay strictly within the limits of this definition. This advice agrees with the spirit of the modern classification. Whenever a cloud is observed that is of an intermediate kind and may be called either altostratus, cirrostratus, anvil, etc., the name altostratus should be avoided and the alternate name applied.

CLOUD OBSERVATIONS IN THE TROPICS

Orographic and continental effects and the effects of diurnal variations in the tropics are much greater than in temperate zones. The net effect is a large increase in the number and size of cumulus congestus clouds. These often prevent observation of middle and high clouds. Consequently the observer is unable to report the presence (or absence) of these important types. Similarly the congestus often prevent proper reporting of dangerous cumulonimbus. The disturbing effect of the cumulus congestus in the tropics compares with that of stratocumulus in temperate zones. The disturbance takes place even more frequently. Unless caution is used, therefore, observers over large areas will report nothing but cumulus congestus, thereby critically lessening the data available for analysis and forecast.

Numerous observation points are located near the seashore. Here many clouds observed in the direction of the land are under strong continental influences, whereas clouds over the water generally indicate more accurately the undisturbed structure of the tropical air mass. In addition, there may be coast-line orographic effects. The significance of a typical congestus is then entirely different in the middle of the day depending on whether it is observed over water or over land. Over the sea it shows structural instability of the atmosphere; over a hilly plain, it simply results from diurnal heating of the ground. To avoid misunderstanding, it is advisable to report cloud conditions over land and over sea separately. Such a double description is in line with the thought of the Atlas (which provides for a circle of horizon outlining cloud conditions in all directions), and with the great necessity for description from below and above when reporting from aircraft.

Cumulus. The observer should avoid a tendency to limit his vocabulary to cumulus humilis, cumulus congestus, and cumulonimbus. All known forms should be mentioned when viewed in order to differentiate the various aspects of connections and their various significances. In particular, the name congestus should be used only when no other is applicable. In extended tropical areas, especially south of large anticyclonic masses, there is a natural tendency to stability in the lower part of the middle troposphere (probably due to subsidence), usually between 5,000 and 20,000 feet. Such stability is reflected in the shape of convection clouds and should be noted.

Altostratus. The altostratus plays a great part in the structure of the tropical hurricane and of the squall lines connected with them. The present tendency to reduce the name altostratus to a layer connected with a large phenomenon of synoptic significance applies, consequently, especially well to the tropics.

Example. Caribbean hurricane Aug. 19 to 23, 1943. On a flight from San Juan Puerto Rico to Trinidad, cirrus and cirrostratus were observed halfway, progressively thickening into altostratus covering the whole sky, with little or no low clouds for about 150 miles, until a line of cumulonimbus was reached. A hurricane was detected soon after the flight, and its center located next day (Aug. 20) about 200 miles east of Guadeloupe Island. The hurricane moved west northwest. Altostratus was connected with it and with one of its squall lines for 3 days. The altostratus extended from south to north up to the center of the hurricane. The layer of clouds was 60 miles wide from west to east, and several hundred miles long from south to north.

The thundery variety of altostratus opacus seems to be the most frequent under hurricane conditions. Its typical features are as follows: Cirrus and cirrostratus ahead of it have a more turbulent and foamy structure than the altostratus of temperate zones. Its surface is not so smooth. It seems to be lined with other clouds of the altocumulus genus, sometimes altocumulus cumuliformis and floccus. It looks in some respects like the thundery altostratus that appears within the tropical air mass in the southern United States. Heavy rains and thunderstorms occur.

Cumulonimbus. Tropical cumulonimbus are much higher than in temperate zones (up to 40,000 ft). They can consequently be seen from much greater distances. Even in temperate zones, cumulonimbus towering from behind the horizon have been photographed by Rudaux with special lenses (focal length 120 cm) at distances up to 150 miles. In the tropics, especially, a misleading impression is conveyed by the observer (especially the aerial observer) to the analyst and forecaster unless he specifies whether the clouds are low or high on the horizon. A complete description will also state whether the clouds are seen in all directions or in only one sector of the horizon.

Degenerated Cumulonimbus. Degenerated or weak types of cumulonimbus are observed in the tropics, *e.g.*, in some troughs of the easterlies. Both from the synoptic point of view and for the safety of flying personnel it is necessary to distinguish these from their more dangerous cousins the true cumulonimbus. The anvils of the degenerate type are low and often have no defined cirriform structure. The clouds are perhaps best compared in appearance with high altocumulus or even strato-cumulus. In the International classification, they fall under the species of strato- or altocumulus cumulogenitus. These names do not clearly express the relationship between the degenerate cumulonimbus and the cumulonimbus; but the difference should be kept clearly in mind by the observers.

Whenever feasible, observers should avoid mentioning cumulonimbus. Really dangerous spots will thus be clearly delineated on the maps, and safe areas without cumulonimbus or with weak varieties of cumulonimbus will be indicated more easily to airplane pilots. For additional information regarding clouds in the tropics, see Sec. X, pages 792–794.

CLOUD OBSERVATIONS IN TEMPERATE LATITUDES

The synoptic significance of cumulus clouds in temperate zones is less important than that of middle clouds. However, it is not to be neglected; careful observation of whatever clouds may form may always be associated with the synoptic map.

Diurnal Evolution of Fair-weather Cumulus. Over land, the cumulus of fair weather is a day cloud. Over the sea, it may be observed at any time. The land cumulus starts in the morning; small white tufts without shadows appear simultaneously at one or more places in the clear sky and start growing. In the first stage, the cloud does not really fit the definition of the standard cumulus, for it shows neither vertical development nor base. It is advisable to describe such clouds as "fumulus," this being the name from the International classification that best fits the case. Later both the base and vertical development appear. When fair weather is accompanied by morning fog, the fog may dissolve completely before the fumulus appear, or the fractostratus in which the fog breaks up may be transformed directly into fractocumulus and cumulus.

If the vertical development is especially small, the cloud is called *cumulus humilis*. If it is especially distended and swollen, it is called *cumulus congestus*. The humilis variety is found in the interval between two nephsystems in the "moderate convective sky," at the edge of the cumulus area, close to the anticyclone or the anticyclonic ridge. The congestus variety is found within the nephsystem in its rear zone, near the cold front, or in the warm sector.

The International classification does not explicitly distinguish other species of cumulus. An extension is implicitly favored when it is stated that the humilis and congestus are "among the more remarkable species." There are in fact several intermediate forms of cumulus between the typical humilis and the typical congestus. For the time being, the general *varieties* may be used for distinguishing the types that have other synoptic significance and must consequently not be confused with humilis or congestus.

Cumulus with normal vertical development come under the variety *cumulus cumuliformis*. In the tropics, an interesting variety corresponds to the relatively undisturbed flow of the trade winds and is called by some meteorologists *trade-wind cumulus*. Another interesting variety encountered in the tropics is the *chimney cumulus*, so called because of its unusual vertical development and very small base. In this cloud, a large number of narrow vertical currents or "chimneys" characterize the formation. This is in contrast to the congestus, in which the ascending current occupies a greater horizontal area. The chimney cumulus gives little if any precipitation.

The "table" cumulus (German *Tafelwolke*) is a more or less swollen cumulus, the top of which is remarkably flat. All upward bulges seem to be stopped by the same horizontal plane. The indication, of course, is that a strongly stable layer lies just above the top of the cloud. The utility of the classification is obvious.

A simple rule for establishing a border line between cumulus cumuliformis and cumulus congestus is that the height of the cumuliformis is moderate and does not exceed appreciably the dimension of its base, which is also small. The height is rarely above 3,000 ft. The base and height of the congestus often measure 5,000 to 7,000 ft. The thickness of humilis is below 1,500 ft, and the base is always somewhat larger. The altitude of the base varies, with 4,500 ft a normal figure. The altitude will vary during the day, increasing with ground temperature. An increase of a thousand feet between 1000 and 1400 local time is not unusual.

When the sun goes down, the fair-weather cumulus progressively shrinks and disappears, but the process of dissolution is not the reverse of the process of growth. The typical cloud of dissolution is the stratocumulus vesperalis. The dome-shaped top flattens, and the base spreads out. The empty spaces between several individual clouds may disappear, thus creating at places one single large and flat cloud. At the same time, the structure of the base changes. Undulations, or cells, appear in the previously uniform gray base. Similarities between these clouds and altocumulus or stratocumulus will be noted. The final name *stratocumulus vesperalis* is given, since the formation occurs in the evening. This is the typical cloud of fair-weather sunset. After sunset, the layer becomes thinner, breaks up, and dissolves completely.

Abnormal Evolution of Fair-weather Cumulus. The diurnal evolution described above occurs when the station remains for a sufficient time (12 hr or longer) far enough from a nephsystem where congestus and fractocumulus appear, and also far enough away from an anticyclone where the weather is fair but no cumulus appear. This normal evolution frequently does not occur. According to the departures from it, the forecaster may draw conclusions regarding the displacement of anticyclones and nephsystems toward or away from the station. This is of course especially helpful for single-station analysis. Anomalies are of numerous types.

Example. 1. Observations from the ground and from a reconnaissance aircraft at Chicago, Ill., Aug. 18, 1943. Swelling fractocumulus were observed in the early morning following the passage of a cold front. The clouds flattened rapidly instead of bulging according to normal evolution, and at 1100 only fair-weather cumulus were observed. These showed no increase in number or size up to 1200. A decrease was noted shortly after 1200. All these observations conflict with the known normal evolution of clouds. No stratocumulus formed, and at 1600 (long before sunset) the cumulus began to disappear. At Dubuque, west of Chicago, the cumulus disappeared at least an hour earlier than at Chicago. Then the observations indicated quite clearly that, even though the cold front was still fairly close to the station to the east, an anticyclonic ridge had reached the area from the west and dominated the cloud formations.

2. Observations at Fort Bragg, North Carolina, Apr. 5 and 6, 1944. A strong cPk air mass has invaded the area. On Apr. 5, a typical evolution of fair weather occurs. Fumulus appear at 1100 EWT. Stratocumulus vesperalis last until after sunset. On Apr. 6, the sky is clear in the morning. Fumulus do not appear until 1200. They develop into cumulus humilis about 1400 and a little later once again look like fumulus. No vesperalis forms. The fumulus disappear about 1700. The difference in cumulus conditions on the two successive days was striking. The atmosphere was already stable the first day, as was shown by the poor vertical development of the humilis. The stability was strongly increased the second day. At

upper levels, temperatures increased markedly. See Figs. 2 and 3 for a 24-hr sequence of changes in the air mass. Subsidence aloft was so strong on Apr. 6 that descending currents were observed by light aircraft making frequent ascents.

Cumulus Congestus. Cumulus congestus are distended and swollen cumulus whose domes have a cauliflower appearance. No special varieties are mentioned in the Atlas of 1932. The great expansion of air travel since that time (especially in tropical areas where swollen cumulus are frequent), and the synoptic relation between these clouds and the more dangerous cumulonimbus make it desirable to go into greater detail in the observations and reports. Whenever possible, mention of congestus should be avoided; thus one should apply the term "table," "trade wind," "chimney," cumuliformis, etc., whenever possible. Even when the species name

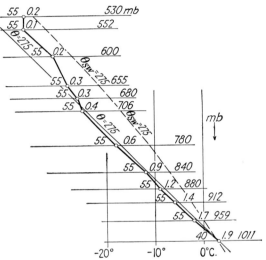

Fig. 2.—Normal evolution of fair-weather cumulus. Sounding taken at Norfolk, Va., Apr. 5, 1944, 1200 EWT. Visual observations made at Fort Bragg, N.C., showed a typical evolution of fair-weather cumulus. The sounding is seen to lie close to the dry adiabatic up to 706 mb, which agrees with the formation of the cumulus. Stable layers above 706 mb prevent the mushrooming of the clouds.

congestus applies, there is one variety that should always be distinguished. This is the "silver-lined" congestus that can be seen in Plate 78 of the International Atlas. The silver-lined appearance and the divergent sunrays that usually accompany the cloud are due to the hazy atmosphere. It will be noted that cumulonimbus found in the vicinity of the silver-lined congestus do not give strong squalls or heavy showers or thunderstorms.

Thermal Structure of the Atmosphere. Kochansky in Lwow, Poland, has carefully studied the thermal structure of the atmosphere by means of aircraft soundings (one to three per day). The soundings show a stable layer with isothermal or slightly inverted temperature distribution above the condensation level in the case of cumulus humilis. This accounts for the absence of bulges. Near midday, the lapse rate between condensation level and ground is close to the dry-adiabatic, with even superadiabatic lapse rates near the ground. Diurnal variations of temperature (within the same air mass) decrease with elevation above ground. The change at condensation level is practically negligible. The conclusion is that the diurnal evolution of the fair-weather cumulus is therefore due to changes in the vertical currents at the cloud level.   Changes in general air-mass conditions, such as subsidence and advection, will of course produce changes aloft.   These changes are reflected in the nature of the cloud formations.   Turbulent conditions occur frequently in the adiabatic layer below the condensation level, especially near the condensation level.   Relative humidity increases from the ground up to the condensation level.   Above the tops of

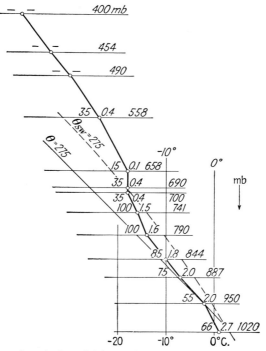

Fig. 3.—Abnormal evolution of fair-weather cumulus.   Sounding taken at Norfolk, Va., Apr. 6, 1944, 1200 EWT.   Visual observations made at Fort Bragg, N.C., showed abnormal evolution of fair-weather cumulus.   The cumulus were flat and lasted less than 2 hr, being succeeded by a few fumulus.   Although the sounding had not changed much since the previous day from the surface up to about 1,500 m, marked subsidence and warming had occurred aloft.   Airplane soundings at 10,000 ft showed an increase in temperature of about 9.5°F (5.3°C) between 0930 and 1345 EWT.   The greatest warming aloft occurred at Nashville, Tenn., and Louisville, Ky.   The striking change in the behavior of the cumulus on the successive days is the conspicuous sign of the subsidence warming aloft.   The subsidence was associated with the passage of a surface anticyclone moving from northwest to southeast.

cumulus humilis, relative humidity usually drops suddenly to a low figure, as low as 15 to 20 per cent.   See Sec. V for thermodynamics of the atmosphere.   General stability and instability conditions are treated in Secs. V and X.

COLOR OF THE SKY (CYANOMETRY)—HAZE LAYERS

When the sky is cloudless or contains on the horizon only low clouds whose characteristics cannot be determined, observation must not be suspended.   The shade of blue is significant; it is related to the origin of the air mass.   The haze levels and their colors are also significant.

Cyanometry. A logarithmic scale proposed by Linke (1927) comprises 14 shades of blue from deep ultramarine to pure white. Color samples are compared with the bluest spot in the sky, which is taken to be about 90 deg to one side if the observer faces away from the sun. Comparison takes place after a 30-sec period of accommodation.

A simpler scale has been proposed and tried by Neuberger; its six shades can readily be identified by an observer without reference to a prepared color scale.

In the eastern states of the United States, the shades observed usually range from Nos. 4 to 8 of the Linke scale. Elsewhere, *e.g.*, in Morocco, a far wider range can be found.

Table 2 is an example of the relationship between color measurements of the sky and the origin of air masses in French Morocco.

TABLE 2

Blue shade No. (Linke)	Vapor pressure, mm	Origin of air mass
14	10	Polar maritime or continental
13	11	
12	13	
11	12	Tropical maritime
10	14	
9	16	
8	18	
7	17	Tropical continental
6		

Haze. The International definition reads, "Little particles of dust from dry regions or of salt which are dry and so extraordinarily small that they are neither felt nor perceived with the naked eye but which give the air a characteristically smoky appearance. . . . Against a dark background this veil has a bluish color, (blue distance), but a dirty yellow or reddish yellow against a light background (*e.g.*, clouds on the horizon, snow covered summit, sun). Haze may be distinguished by this characteristic from greyish mist which it sometimes equals in intensity."

There is no sharp division between thin clouds and thick haze. Haze often has a sharply defined top below an inversion, which is called the *haze level*. It can be determined with a precision of approximately 100 ft in altitude. The light of the sun is reflected by the haze particles and by the top of the haze. As a result, the haze layer looks white to an observer in an airplane who looks toward the sun; it looks purple-brownish when he looks in the opposite direction. After the plane has reached and passed the haze level, the sky right above the haze is clear, and clouds very low on the horizon appear with sharply delineated features. Light white streaks are seen floating on top of the haze. There are often several haze levels, especially in anticyclonic areas. Sometimes the beginning of a haze layer can be noticed by the fact that it appears on only one-half of the horizon. This undoubtedly has a synoptic significance.

Study of synoptic distribution of haze layers and of their use for air-mass analysis is still in its initial stage but is expected to yield valuable information. In fair-weather areas, it is the most important element for visual analysis.

CLOUD PHOTOGRAPHY

Cloud photography is an essential part of weather observation that has no substitute. No matter how detailed the International classifications become in the future, visual observations and written descriptions cannot replace photographs.

1. Pilots can assist greatly in improving weather forecasting by photographing cloud formations. This will also help the forecaster to explain the weather to the pilots; the forecaster has to describe how cloud formations look when seen aloft. Photographs are helpful in this respect.

The cost of photographing is small compared with the other operating expenses of an airplane or of a weather station. Photographs provide an efficient method of keeping a record of interesting flight or weather conditions. They should be taken as regularly as possible.

2. Weather-history Photographs. Interpretation of photographs taken aloft is much easier if photographs of clouds in the same area are taken simultaneously from the ground. If such ground photographs are taken every day at about the hours of international observation, whether photographs have or have not been taken aloft, they constitute a real "history of the sky." Three photographs a day (morning, noon, and afternoon) are usually sufficient for that purpose. This history can be used as a background to interpret all photographs taken aloft and is therefore very valuable to the forecaster. If an interesting event takes place between international observation hours, it is recommended that additional photographs be taken.

Example. If the sky breaks up, thus showing the structure of the overcast, or if a mammatus variety forms, photographs taken between the regular hours will make the sequence of events clear.

3. Time. Best hours are those of international observations—0730Z, 1330Z, 1830Z, and 0030Z; but the time is of secondary importance, especially for photographs taken aloft.

4. Size. Small photographs are of little use, as a rule. Of course, they can be enlarged in order to show details more clearly; but in the field, scratches, stains, etc., on the original cannot entirely be avoided and are enlarged as well as the details. Best results are obtained from cameras that use large film sizes. The minimum size recommended is 4 by 5 in. (K20 camera, for instance). Photographs of this size can easily be enlarged to 8 by 10 in., which is a convenient size for studying photographs or showing them to individual pilots. Photographs for filing should be kept attached to a cardboard of the usual commercial size, 8 by 11 in.

5. Marginal Data. If a photograph is to be used for improving weather forecasts, it should be compared with the map that corresponds to the area taken, and it must therefore bear the following essential data:

1. Filing number.
2. Location (*e.g.*, aloft 55 miles, northwest of Birmingham, Ala., or Mitchel Field, New York).
3. Altitude (of airplane and of cloud layers traversed below).
4. Date (use standard abbreviations).
5. Time (indicate time and also time zone; *e.g.*, 1230 EWT).
6. Orientation, *e.g.*, facing north, south, etc.

A photograph without the above data is almost useless for the purpose of synoptic studies. The data are so important that it is advisable to write them directly on the original film, preferably in a part of the landscape and not on the clouds. They will appear automatically on each print.

The following data are interesting but optional:

7. Subject.
8. Film used.
9. Filter used.
10. Length of exposure and aperture.

It is advisable to keep a record of the preceding data on sheets or in a book for all pictures taken.

6. Choice of Subject. Photographs should preferably be taken according to the directions of an experienced photographer and forecaster. If none is available, the following suggestions may be useful:

1. Since clouds are widely disseminated throughout the sky, two photographs in opposite or different directions should be taken if possible.

2. If there are several levels of clouds, shoot a part of the sky where clouds appear at all levels, or at least at two levels. If there is a third level, use the second shot to show it.

3. If there is a break in a large layer of clouds, place it in the picture. The structure of cloud layers appears more clearly on the edges of the break.

4. Low clouds are less interesting than medium and high clouds; emphasis should therefore be placed on medium and high clouds.

7. Framing. The horizon should also be clearly shown, preferably including a tree or building, because it indicates the size of the clouds. If clouds lie relatively close to the zenith, the horizon cannot be shown; a place in the picture, a treetop, telephone pole, roof of a building, tower, or chimneytop will then be a useful guide. These, like the horizon, help to orient the photograph and provide a relative scale for its details.

Many photographs are hard to use because the horizon or an object of reference is not shown; altocumulus clouds, for instance, may be confused with cirrocumulus or stratocumulus.

8. Filters. Since most of the synoptically interesting clouds are found in a relatively gray sky of uniform brightness, their characteristics should be artificially rendered more conspicuous in order that their structure may be better understood. Film, moreover, is not so sensitive as the human eye to contrast in shades of light gray and blue. This heightening or dramatizing of photographs is a practical necessity. The quality of the results obtained from them therefore greatly depends on the proper use of filters.

Red and yellow filters are used most commonly. Red filters are preferable when gray shades predominate. Yellow filters are recommended when the blue of the sky appears in large areas of the subject. In case of doubt, use a red filter.

9. Length of Exposure. Cloud photographs should be taken under all conditions, including the most adverse, and the length of the exposure consequently varies widely. Even professional photographers make mistakes if they rely on guesswork in choosing the proper period of exposure. A good exposure meter is strongly recommended.

It should be borne in mind that the exposure time indicated on commercial meters refers to a subject on the ground, containing an average amount of bright spots and dark shades. Clouds, on the other hand, are a different kind of subject. They never contain shadows comparable with those of a ground subject, and the exposure time indicated by meters must consequently be lessened, usually divided by two or even four. Overexposure is a common shortcoming of cloud photographs. If the exposure is correct for the clouds, the landscape is usually underexposed, but this does not matter.

10. Aperture. Small apertures are recommended on photographs taken from the ground because clouds are usually lucent. Photographs taken from airplanes are sharper if exposure is short, which implies larger apertures.

11. Stereoscopic Photographs. Stereoscopic views of clouds are much more easily understood than are simple photographs. They avoid errors due to perspective effects and other causes. The first photographs of clouds that could be used to a certain extent for stereoscopic views were taken in order to measure cloud altitudes; the two stations were about 1 mile apart. Stereophotographs are especially easy to take from airplanes that traverse a long distance within a few seconds. Two photographs are taken successively from the same airplane, the camera being aimed laterally at the same cloud formation. The elapsed time between the two exposures is several seconds, varying with the speed of the plane and the estimated distance of the subject. Since the average distance between the pupils of the eyes is $2\frac{1}{2}$ in. and the normal range of strong stereoscopic sight is less than a yard (1 : 10 approximately), calculations for stereoscopic photographs of a cloud group 10 miles off, for instance, can be reckoned accordingly. A 1-mile base may be acceptable for low clouds, but a much greater distance is necessary for high altocumulus and cirrocumulus.

12. Duo photographs. Photographs of the same cloud layer may be taken from the ground and from above in quick succession or simultaneously. Few, unfortunately, have been made, but their value is great, and as many of these air-ground photographs as possible should be taken. They enable forecasters to explain to pilots how weather will appear to them when seen from above. Conversely, they enable pilots to describe what they have seen to the forecaster; thus in both ways they contribute to improved forecasts.

13. Prints. Glossy paper brings out details better than mat paper and also lends itself better to reproduction of the photographs from the print, if photographs are to be used for publication when the original negative is not available.

Contrasting paper is usually recommended in order to bring out all details of the clouds.

When the print is developed, emphasis should be placed on cloud details. As with overexposure, the landscape comes out gray and dull, but this is unimportant.

U.S. WEATHER BUREAU CLOUD CODE CHART

Introductory Remarks. Figures 4 to 37 inclusive are reprinted by permission of the U.S. Weather Bureau. With the exception of Figs. 5, 8, 12, 19, and 28 these are from the Weather Bureau's *Cloud Code Chart*. The exceptions were made in order to obtain glossy prints for the cuts; the additional photographs were kindly supplied by the Librarian of the Weather Bureau. The *Cloud Code Chart* was "prepared with a view to aiding in the interpretation of cloud reports in the international figure code." The size of the photographs does not permit detailed descriptions of the cloud structures that characterize their significance from the viewpoint of synoptic meteorology.

Fig. 4.—Cumulus humilis.

C_L 2

Fɪɢ. 5.—Cumulus congestus.

C_L 3

Fɪɢ. 6.—Cumulonimbus calvus.

C_L 3

Fɪɢ. 7.—Cumulonimbus incus.

C_L 4

FIG. 8.—Stratocumulus vesperalis.

C_L 5

FIG. 9.—Stratocumulus translucidus.

C_L 5

FIG. 10.—Stratocumulus opacus.

C_L 5

FIG. 11.—Stratus.

C_L 6

FIG. 12.—Fractocumulus of bad weather.

C_L 7

FIG. 13.—Cumulus humilis and stratocumulus.

C_L 8

FIG. 14.—Cumulus congestus and stratocumulus.

C_L 9

FIG. 15.—Cumulonimbus and fractocumulus

C_M 1

FIG. 16.—Altostratus translucidus.

C_M 2

Fig. 17.—Altostratus opacus.

C_M 3

Fig. 18.—Altocumulus translucidus.

C_M 4

Fig. 19.—Altocumulus lenticularis.

C_M 4

Fig. 20.—Altocumulus patches (lenticularis).

C_M 5

Fig. 21.—Altocumulus translucidus undulatus.

C_M 6

Fig. 22.—Altocumulus cumulogenitus.

Cм 7

Fɪɢ. 23.—Altocumulus duplicatus.

Cм 7

Fɪɢ. 24.—Altocumulus opacus.

Cм 8

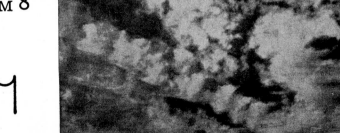

Fɪɢ. 25.—Altocumulus floccus.

C_M 8

Fig. 26.—Altocumulus castellatus.

C_M 9

Fig. 27.—Thundery sky.

C_H 1

Fig. 28.—Cirrus filosus, scarce.

C~H~ 2

FIG. 29.—Cirrus filosus.

C~H~ 3

FIG. 30.—Cirrus nothus.

C~H~ 3

FIG. 31.—Cirrus densus.

C_H 4

FIG. 32.—Cirrus uncinus.

C_H 5

FIG. 33.—Cirrus below 45 deg.

C_H 6

FIG. 34.—Cirrus above 45 deg.

C_H 7

FIG. 35.—Cirrostratus filosus.

C_H 8

FIG. 36.—Cirrostratus.

C_H 9

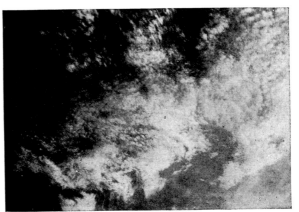

FIG. 37.—Cirrocumulus.

SECTION XII

CLIMATOLOGY

By H. Landsberg

CONTENTS

SECTION XII

CLIMATOLOGY*

By H. Landsberg

INTRODUCTION

Climate is the collective state of the atmosphere at a given place or over a given area within a specified period of time.

The individual weather events and the meteorological elements characterizing them form at any place a group with properties peculiar to the particular locality. A group of observed values can be described by statistical measures such as the mean, the range, the frequency of various intervals, the sequence of events and their periodic repetition. In a narrower sense, many consider climate only as the mean local condition of the atmosphere. As such, it can be abstracted by the "mean values" (a term often used as equivalent to "normal values") of the various meteorological elements, namely, temperature, pressure, humidity, wind direction and speed, cloudiness, etc. A characterization of the collective "weather" in this fashion permits one to establish distinctions between atmospheric conditions at various localities. One should, however, not lose sight of the fact that an abstraction of this type serves only to simplify the scientific digest of a tremendous number of individual meteorological observations.

By analogy, let the individual citizen of a country correspond to a single weather event; collectively the individuals form a nation, and this collective corresponds to climate. From the traits of the individuals, one can obtain statistically an "average citizen" of a nation. This fictitious character cannot be identified with any individual in the population; yet he may serve as a symbolic contrast to the "average citizen" of another nation. The same advantages and shortcomings that can be attributed to the concept of an "average citizen" are inherent in that of the "normal climate" at a given place. In considering climate, it is therefore necessary to remember always that the climate is a result of single weather events. Hence the same basic factors that cause weather as an instantaneous manifestation in the atmosphere bring about the climate. This means that the climatic conditions can be derived from a study of the general atmospheric circulation and its local modifications.

The circulation in the atmosphere is determined by a multitude of processes converting energies in the atmosphere. Some of these processes are well known, others barely studied. Their influence, interrelation, and interaction are very complex. Therefore no analytical and quantitative treatment of all the causes underlying climate can be given at the present time. Rather, climatic conditions are usually described by the empirically observed quantities that characterize the instantaneous state of the atmosphere. The observed facts and the figures derived therefrom can then be fitted into the scheme of general circulation that is currently accepted and based on a more or less qualitative appraisal of the many factors involved in atmospheric dynamics.

* The author is indebted to his assistant, Miss Odette Neuman, for considerable help in plotting the world charts of climatic elements presented in this section.

PLANETARY (RADIATION) CLIMATE

Even if the earth had no atmosphere, one could speak about a climate because its surface would show a temperature different from that of other celestial bodies. This temperature would also show variations in time and space, particularly as a function of latitude. These differences are caused by position and motion of the earth in the solar system and also by changes in the primary source of energy, the sun. The sun, a star with a surface temperature of about 6000°C, radiates energy into space. On a surface exposed normal to the sun's rays at the mean distance of the earth from the sun, an energy of 1.94 cal per cm² per min is received on an average. There are slight variations around this mean value, probably amounting to not more than 1 per cent either way from the mean. The fluctuations of this value are of shorter and longer duration, days and years. They are intimately related to solar activity, but their influence on climatic conditions is as yet insufficiently explored.

Because it was originally assumed that the solar radiation within an era long compared with human history was constant, this energy amount of 1.94 cal per cm² per min was called the *solar constant*. Even though its fluctuations are now an accepted fact, the term *solar constant* is still used to designate the intensity of the solar energy received at a mean distance from the earth to the sun on a surface normal to the rays. The energy actually received at the boundary of the earth's atmosphere is variable not only because of the solar fluctuations but also because of the variation of distance between sun and earth during the course of the year. For a certain day of the year, this distance is approximately given by the formula

$$D = a \left(1 - e \cos \frac{2\pi d}{Y} \right)$$

where D = distance from sun to earth
$\quad a$ = mean distance from sun to earth (92,900,000 miles)
$\quad e$ = eccentricity of orbit (0.0167330)
$\quad Y$ = length of the year in days (365.2563 days)
$\quad d$ = number of days since Jan. 1 (approximate least distance)
$\quad 2\pi$ = 360 deg

During the current era, the earth is closest to the sun on about Jan. 1 (91,342,000 miles), farthest away on about July 2 (94,452,000 miles). The mean intensity of the solar radiation received on these days at the boundary of the atmosphere is 2.007 and 1.877 cal per cm² per min, respectively. The mean value for any individual day can be obtained if one remembers that the radiation intensity varies inversely with the square of the distance.

A further climatic influence inherent in the position of the earth in the universe is the inclination of the axis of the earth with respect to the orbit. At the present time, the axis is inclined 66°33′ against the plane of the orbit. This inclination is essentially the reason for the existence of seasons. Only at the time of equinox (Mar. 21, Sept. 23) does the dividing line of the lighted and the dark half of the earth parallel a meridian and pass through the poles. Between Mar. 21 and Sept. 23, the North Pole is tilted toward the sun, and from Sept. 23 to Mar. 21, the South Pole is tilted toward the sun. This causes the variable length of the day in all latitudes except the equator. The length of the astronomical day as a function of latitude can be computed or represented by a nomographic chart (Fig. 1).

The existence of an atmosphere around the earth complicates matters further. Even the length of daylight is determined not by the astronomically possible duration of direct solar illumination but by additional light received before sunrise and after sunset through the scattering of the sun's light in the atmosphere. The scattering effect that produces the twilight is variable because of atmospheric refraction that changes the apparent position of the sun, particularly in polar regions. The amount of refraction is a function of the temperature distribution in the atmosphere, but no fixed rules for this effect can be given here. However, an approximate estimate of the length of twilight at either end of the daylight period can be obtained by setting

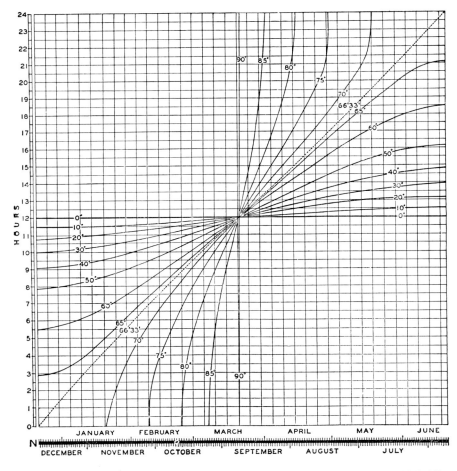

Fig. 1.—Duration of day (sunrise to sunset) in various latitudes. Use scale of dates marked "N" for Northern Hemisphere, scale "S" for Southern Hemisphere. Select date on horizontal scale reading upper portion on either "N" or "S" scale from left to right, lower portion from right to left. Place ruler vertically to scale through date. Obtain intersection with appropriate latitude. Follow horizontally across to ordinate scale of hours on left of diagram and read time interval, in hours, between sunrise and sunset. Time of sunrise or sunset can be approximately obtained by dividing this interval by 2 and subtracting it from or adding it to time of local noon.

an arbitrary limit of solar depression below the horizon at which light normally becomes so dim that outdoor activities cannot be carried on without the benefit of artificial illumination.   This limit is usually fixed at a solar depression of 6 deg below the horizon, and the time interval between this point and sunrise or sunset is called *civil twilight*.   The length of the civil twilight can also be nomographically represented (Fig. 2).

The scattering of solar radiation in the atmosphere is of climatic importance even during the daylight hours, because it causes a separation of the solar radiation into

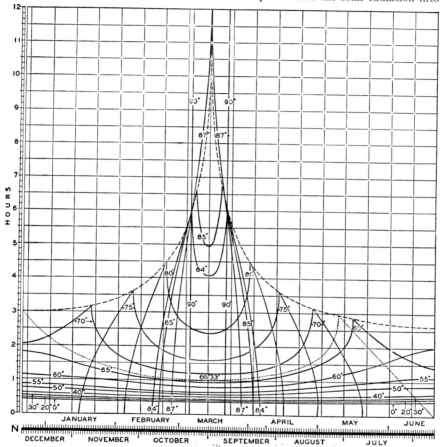

Fig. 2.—Duration of civil twilight in various latitudes.   Use scale of date marked "N" for Northern Hemisphere, scale "S" for Southern Hemisphere, reading upper portion of either "N" or "S" scale from left to right, lower portion from right to left.   Place ruler vertically to scale through date.   Obtain intersection with appropriate latitude line. Follow horizontally across to ordinate scale of hours and read duration of civil twilight at either end of day.   To obtain time of beginning or end of civil twilight subtract interval from time of sunrise or add to time of sunset.

direct radiation from the sun and diffuse sky radiation. The total amount of solar and sky radiation that should theoretically be received on a horizontal surface, *e.g.*, the surface of the earth, can be calculated. If the atmosphere had no absorbing capacity, this would be only a function of the solar elevation above the horizon, *i.e.*, the angle of incidence of the sun's rays. This angle of elevation would also govern the ratio between the direct and the diffuse radiation. Actual observations show that, on clear days at solar elevations of 70 deg, about 85 per cent of the total radiation comes from the sun, 15 per cent from the sky. At a solar elevation of 10 deg, only 60 per cent of the total radiation comes from the sun and 40 per cent from the sky. Diffuse scattering is also responsible for a certain amount of loss of incoming radiation to the terrestrial heat balance, because scattering is uniform in all directions, and hence part of the incoming radiation goes back into the universe.

On and beyond scattering, the absorbing properties of the atmosphere are of importance in radiative processes. Before ever hitting the surface of the earth, part of the radiation is absorbed by gases in the atmosphere that show selective absorption. In the high atmosphere, there are two absorbing constituents, ozone and nitrogen pentoxide. The exact amounts absorbed by these gases are as yet insufficiently explored, and matters are particularly complicated because these components probably owe their existence to radiative processes. In the lower atmosphere, water vapor, carbon dioxide, and suspended dust act as absorbing agents. Most important for the heat budget of the atmosphere is the behavior of the water in it. As vapor, it absorbs radiation of long wave length, and, in suspended liquid form (clouds), it reflects considerable portions of the incoming radiation. The reflection from cloud surfaces is probably the largest single factor in the atmospheric heat budget. All estimates of the heat transactions in the earth's atmosphere (and at present all values are at best rough estimates) have to be based on assumptions about cloudiness, because reliable cloud observations are lacking in many parts of the globe. It is known, however, that the surfaces of clouds reflect from 60 to 90 per cent of the incoming radiation. According to Mosby, the mean incoming radiation received at the surface of the earth on a horizontal plane is given by the following empirical formula:

$$Q = k \cdot h(1 - 0.0071C) \quad \text{cal/cm}^2\text{/min}$$

where k = transparency of atmosphere (varying from 0.023 at equator to 0.027 in 70° lat.)
h = mean elevation of sun above horizon
C = mean cloudiness of locality in percentage of sky cover

After the solar radiation reaches the surface of the earth, further transformations take place. First of all, a certain amount of energy is again reflected. The amount of reflection (albedo) depends on the type of surface. For solid surfaces, the albedo in percentage of the incoming radiation is shown in Table 1.

TABLE 1.—ALBEDO OF SOLID SURFACES

Type of surface	Cultivated soil and vegetation	Sand areas	Fresh snow	Old snow and sea ice
Per cent reflection	7–9	13–18	80–90	50–60

For a smooth sea surface, the values for reflection of solar and sky radiation on a clear day as a function of solar elevation are given in Table 2, according to Sverdrup.

TABLE 2.—ALBEDO OF SMOOTH SEA SURFACE

Solar elevation, deg	5	10	20	30	40	50–90
Per cent reflection	40	25	12	6	4	3

For overcast sky, the reflection of a smooth sea surface is 8 per cent. The over-all energy reflection of the earth, including that of clouds, has been estimated to be between 37 and 45 per cent.

The observed amount of heat received on a horizontal surface on the average clear day (*i.e.*, mean of all clear days of the year) with respect to latitude on the Northern Hemisphere is shown in Table 3, according to Perl.

TABLE 3.—MEAN DAILY HEAT SUM RECEIVED ON CLEAR DAYS IN VARIOUS LATITUDES OF THE NORTHERN HEMISPHERE

Latitude, °	0	15	30	45	60	75
Heat sum, cal/cm²/day	510	510	470	380	300	220

Because of cloudiness, the values actually received are in many spots of the earth only one-half to two-thirds of the indicated quantities.

At the surface of the earth, interactions between the surface and the atmosphere, which are of fundamental importance for the energy budget of the atmosphere, take place. In this respect, it is important to note that the total heat budget remains essentially constant within the current era. Hence as much energy as is received is again lost. The minute amounts of heat flowing from the interior of the earth outward and the small amounts stored by photochemical reactions in plants are negligible. Very influential, however, is the change that takes place in transforming the incoming radiation of short wave length from sun and sky (maximum energy at wave length of 0.5μ) to an outgoing radiation of long wave length (maximum energy at about 10μ). Each point on the surface of the earth radiates energy according to the law of Stefan-Boltzmann, *i.e.*, heat loss is proportional to the fourth power of the absolute temperature of the surface. The heat loss from the surface of the earth would far exceed the amount received from sun and sky were it not for the interception of considerable amounts of energy by the water in the atmosphere.

Clouds and water vapor will absorb a large amount of the dark outgoing radiation and radiate it back to the earth's surface. Thus the heat loss that, for a surface of 287° absolute temperature (mean temperature of surface of the earth as a whole 14°C = 57°F = 287° K), would be about 0.54 cal per cm² per min is actually only about 0.13 cal per cm² per min. This last quantity is called the *effective radiation* of the earth and is, of course, a mean value only. Observed effective radiations vary greatly throughout the year and from place to place. For the earth as a whole and the average day of the year, the mean heat budget at the surface of the earth is shown in Table 4, according to Conrad.

TABLE 4.—HEAT BUDGET OF THE EARTH AS A WHOLE FOR AVERAGE DAY OF THE YEAR

Heat received by	Cal/cm²/day	Heat lost by	Cal/cm²/day
Direct solar radiation	170	Outgoing radiation	790
Diffuse radiation	140	Evaporation	120
Back radiation from atmosphere	600		
Total	910	Total	910

G. C. Simpson has calculated the annual mean value of the net heat balance for all latitudinal zones with considerations of all factors involved. His results are shown in Table 5.

TABLE 5.—NET RADIATION BALANCE OF VARIOUS ZONES

Latitude, °	0–10	10–20	20–30	30–40	40–50	50–60	60–70	70–80	80–90
Net radiation balance, cal/cm²/min:									
Northern Hemisphere	+0.060	+0.055	+0.043	+0.013	−0.034	−0.077	−0.103	−0.163	−0.168
Southern Hemisphere	+0.059	+0.053	+0.035	0.00	−0.053	−0.096	−0.126	−0.163	−0.169

On the whole, the earth is losing heat at all latitudes higher than 35° and gaining heat between 35°N and S. This again represents, however, only a mean picture for the year. There is a considerable annual variation, as can be noted from Figs. 3 and 4, which show the radiation balance of the earth in January and July. It is noteworthy that the hemisphere that has summer has a positive radiation balance to latitudes as high as 60°. The irregularities of the lines of equal radiation balance are caused mainly by differences in the temperature of the radiating land and water masses as well as by the differences in cloudiness.

From data such as these, Milankovich has calculated a theoretical temperature that should be expected on the surface of the earth if the atmosphere were completely at rest and the distribution of land and sea were even over the surface of the earth. This mean planetary temperature of various zones produced by solar radiation and its modifications is shown in Table 6. For comparison, the temperatures actually observed on the two hemispheres are also given.

TABLE 6.—THEORETICAL PLANETARY AND OBSERVED TEMPERATURES FOR VARIOUS ZONES

Latitude, °	0	10	20	30	40	50	60	70	80	90
Planetary temperature, °F	91	89	83	72	57	37	12	−11	−26	−31
Observed Northern Hemisphere, °F	79	80	78	69	57	42	30	14	2	?
Observed Southern Hemisphere, °F	79	78	73	65	54	42	28	11	−4	?

The planetary temperature that is a result of the net radiation effects agrees with the observed values only in the latitudes of about 40°. From there toward the equator, actually observed values are lower; toward the poles observed values are higher than the theoretical ones. This observation is a consequence of the *general circulation*. The temperature differences between pole and equator caused by radiation produce pressure gradients resulting in transport of air from one place to another. This general circulation is a most important influence upon climatic conditions. Table 5 also indicates that the observed temperature values on the two hemispheres are different. That fact is caused by the uneven *distribution of land and sea* between the two hemispheres. This differentiation of the surface of the earth into continental and oceanic areas is another control of climate that produces some of the most pronounced climatic differences observed on this planet. These differences are mostly a result of the diverse heat transactions taking place in various component parts of the earth's surface. Three types of surface, widely different in their properties, are of major importance: solid ground, water, ice and snow.

In solid ground, only a very shallow layer participates in the heat transactions. The molecular heat transfer is almost solely responsible for transport of heat from the surface to greater depths. For example, the annual variation of temperature usually does not penetrate deeper into the ground than 30 to 50 ft.

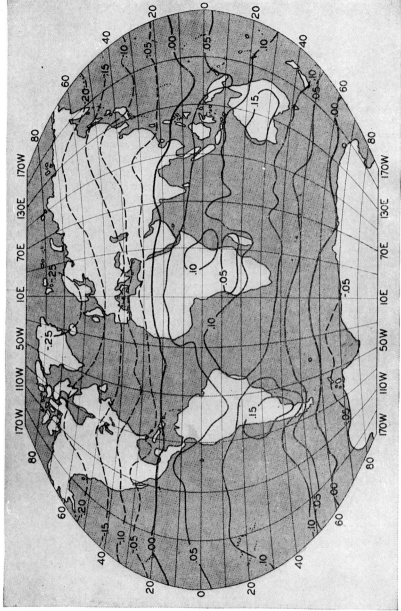

Fig. 3.—Distribution of net radiation gains or losses over the earth, January. (*After G. C. Simpson.*) Units: cal per cm² per min. Positive values indicated by solid lines (areas of net heat gain). Negative values indicated by broken lines (areas of net heat loss).

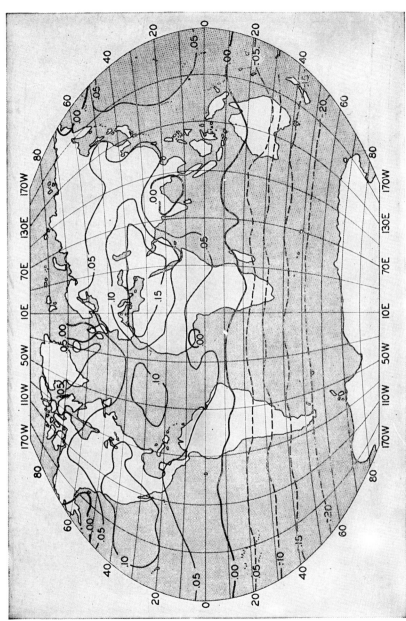

Fig. 4.—Distribution of net radiation gains or losses over the earth, July. (*After G. C. Simpson.*) Units: cal per cm² per min. Positive values indicated by solid lines (areas of net heat gain). Negative values indicated by broken lines (areas of net heat loss).

In water, this variation penetrates to at least 600 ft, mainly because of turbulent heat transfer. Moreover, to raise the temperature of a unit volume of water by one unit takes about two and one-half times as much heat as for an equal volume of soil. Hence a surface of solid ground will respond very rapidly to changes of radiation, *i.e.*, it will heat and cool quickly. In contrast, a water surface will warm or cool only very slowly. A further consequence is that air in contact with the soil will follow the temperature changes of the soil, and, by upward- or downward-directed convection, the higher layers of the atmosphere will participate in these heat transfers. Over water, air will show only very slow temperature changes. From water surfaces some heat will also be lost by evaporation. This heat will eventually benefit the air when condensation occurs, but this may be at a considerable distance from the point where the evaporation took place.

Surfaces of ice and snow have entirely different properties. They will reflect considerable amounts of radiation, part of which will be absorbed in the air. If the temperature of the ice or snow surface is close to the freezing point, part of the heat received at the surface will be used for melting. In addition, especially in fresh snow, besides a very high reflection of energy, little heat will penetrate into the ground because of poor conductivity. At night, when snow radiates just about as a black body, the temperature of the air in contact with the snow surface will be lowered considerably. If the air is calm, daily temperature extremes over snow will be accentuated.

GENERAL CIRCULATION OF THE ATMOSPHERE

The radiative processes on the earth will create a priori at the surface of the planet a zone in the region of the equator that is warmer than zones at the poles. In addition the mosaic of continents interspersed between oceans establishes areas that will follow radiation changes rapidly while the oceans themselves will react only very sluggishly to such changes. The influence of temperature changes in the oceanic part of the surface will be widespread partly because the oceans cover five-sevenths of the total surface area of the earth and partly because the currents in the ocean can "transport" climate far from the region of origin. In a medium as mobile as air, temperature differences that are created by radiation will immediately give rise to currents that tend to equalize them.

On an imaginary earth that had a homogeneous surface and was nonrotating, the differences in radiation would lead at the surface to sources of cold air at the poles and a source of warm air at the equator. Ascending motion would prevail at the equator, subsidence at the poles. A pressure gradient would exist aloft from equator to the poles resulting in a transport of air from equator to the poles aloft. As a consequence, at the surface a pressure gradient from the poles to the equator would be established, producing transport of air from the poles to the equator.

This simple cycle is, of course, immediately destroyed by the rotation of the earth from west to east. Air originating in equatorial latitudes has the high angular momentum existing at this great distance from the axis of rotation. Air originating near the poles has a small angular momentum because of proximity to the axis of rotation. The conservation of momentum will cause a mass of air moving poleward to advance faster eastward than a point on the surface at higher latitude. Hence, with respect to the surface of the earth, a poleward-directed motion will acquire a west-to-east component. That means that the winds from the equator to the pole tend to become west winds and, vice versa, that a flow directed toward the equator will, with respect to a point on the surface, acquire an east-to-west component because it moves from a region of small angular momentum to one with higher angular momentum. Thus a wind from the pole to the equator tends to become an easterly current.

The deformation of the single cell envisaged for a nonrotating earth will lead on the rotating earth to westerly winds at high elevations in the poleward-directed currents as actually observed. The lower winds moving toward the equator become easterly winds, a condition also actually observed from the poles to latitudes of about 60°. The upper air moving from the equatorial region (as a westerly current) with a slight northward- or southward-directed component cools because of high radiative heat losses aloft and will gradually sink to the surface. This subsiding process actually takes place in latitudes of 30°. With a pressure gradient at the surface existing toward the equator, part of this air will return toward the equator, now acquiring an easterly component because during the slow subsidence process it will acquire the angular rotation momentum of the 30° region. In the polar regions, the very cold air existing at the surface will create a pressure gradient toward a region of minimum surface pressure between 50 and 60° latitude. However, in the high atmosphere, pressure stays highest at the equator and lowest at the poles. The surface trough of low pressure in 50 to 60° latitude will cause an inflow of air from both south and north, which will tend to reinforce the easterly winds on the poleward side. On

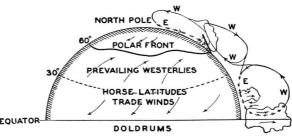

Fig. 5.—Scheme of general circulation over a rotating earth with uniform surface. (*After C.-G. Rossby.*)

the equatorial side of the trough, an inflow of air from regions closer to the equator will yield westerly winds.

In the Northern Hemisphere, this circulation will result in planetary winds, as shown in a schematic diagram given by Rossby (Fig. 5). The Southern Hemisphere would represent a mirror image of this picture, with the equator as symmetry axis. As can be seen, near the equator winds are light (doldrums), and ascending air currents produce convective cloudiness with precipitation. North and south of the equator to about 30° latitude, the trade winds are prevalent at the surface (northeast in the Northern, southeast in the Southern Hemisphere). The trade winds from both hemispheres often meet in the equatorial belt and create a zone of convergence that is called the *equatorial front*. In the regions of 30° latitude (horse latitudes), subsiding air leads to clear skies. In the latitudes of 50 to 60°, between the southwesterly winds originating in the subtropics and the northeasterly winds of the polar regions, a discontinuity, the *polar front*, is in existence. Here the warmer southern air masses ascend over the colder polar air masses, creating another zone of cloudiness and precipitation.

The meridional triple-cell arrangement in each hemisphere is an idealized state that would determine the climate of various zones, provided that the surface of the earth were homogeneous. The entirely different conditions of radiation and heat exchange over continents and oceans distort this theoretical pattern very badly, at least at the surface. This can readily be seen from a set of pressure and wind maps, giving the actually observed values (Figs. 6 to 11). Evidently the pattern with three

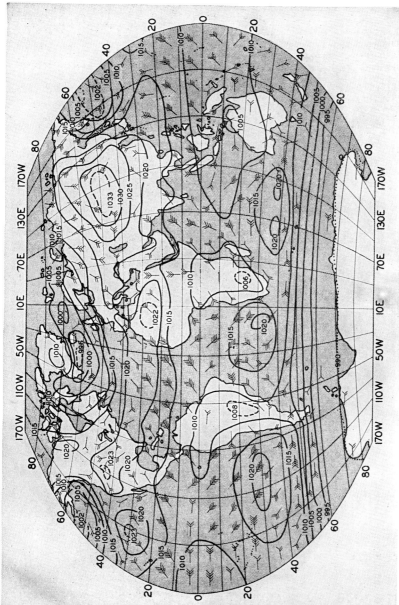

FIG. 6.—Mean isobars at sea level (in millibars), month of January. Arrows indicate directions of the most frequently observed surface winds. Barbs mark the frequency of the winds in percentage of the observations, each barb representing 10 per cent. Thus: ⇗ south wind 20 per cent of observations; ⇙ northeast wind, 80 per cent of observations.

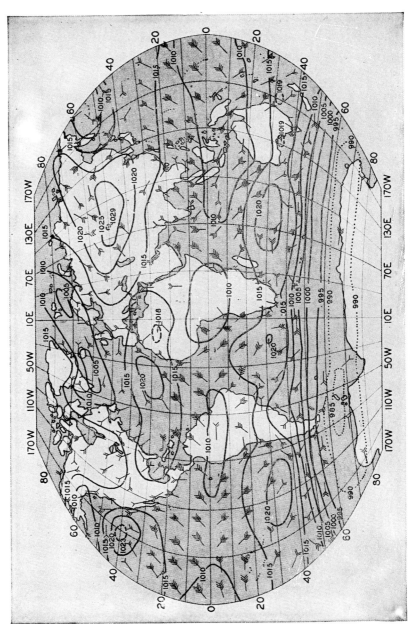

Fig. 7.—Mean isobars at sea level (in millibars), month of March. (Legend, see Fig. 6.)

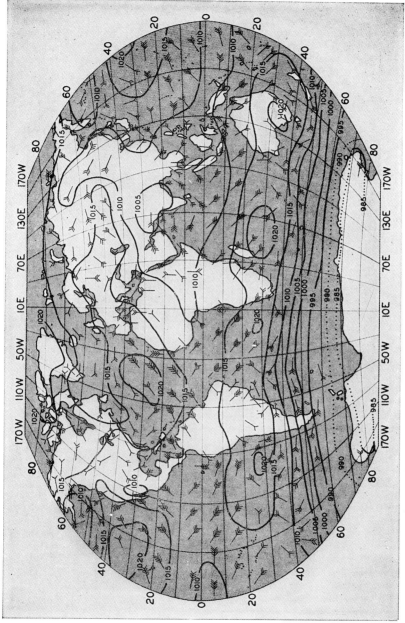

Fig. 8.—Mean isobars at sea level (in millibars), month of May. (Legend, see Fig. 6.)

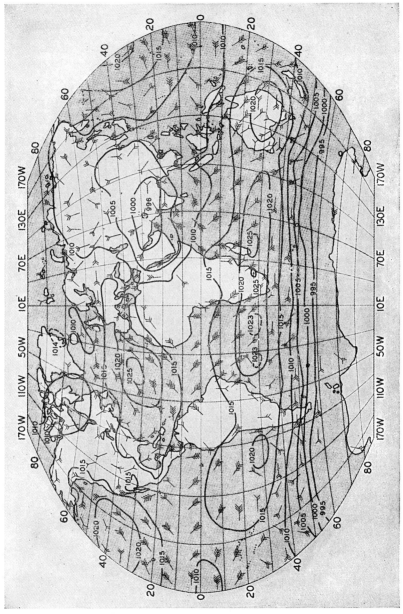

Fig. 9.—Mean isobars at sea level (in millibars), month of July. (Legend, see Fig. 6.)

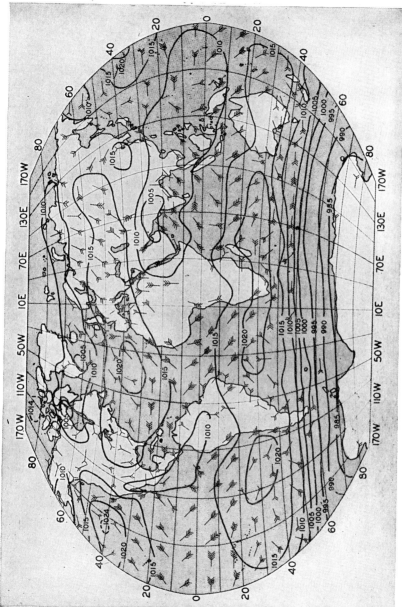

Fig. 10.—Mean isobars at sea level (in millibars), month of September.   (Legend, see Fig. 6.)

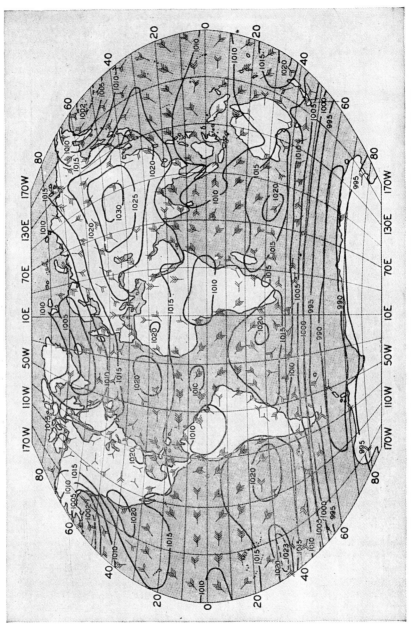

Fig. 11.—Mean isobars at sea level (in millibars), month of November. (Legend, see Fig. 6.)

cells of circulation between the poles and the equator is better preserved in the Southern Hemisphere (land area 19.2 per cent, ocean area 80.8 per cent) than in the Northern Hemisphere (land area 39.2 per cent, ocean area 60.8 per cent). In general, this simple circulation scheme is more recognizable over the large oceans than over the land masses. In the region of the Asiatic-Australian continental area in many months of the year, just a single high- and low-pressure system dominates the circulation from 60°N to 30°S latitude.

There is, of course, in addition to the distortions introduced by the differentiation of land and water surfaces, a seasonal migration of the circulation belts. This can perhaps be best illustrated by the mean positions of the two polar fronts and the equatorial front in the months in which these fronts reach farthest to the north and to the south (Fig. 12). In the Northern Hemisphere, the polar front has its extreme northern position in July, its extreme southern position in January. During this

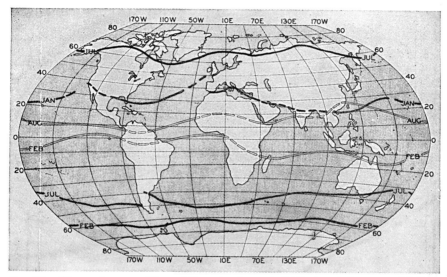

Fig. 12.—Mean extreme positions of the polar fronts (solid line) and the equatorial front (double line). Broken lines indicate highly variable or doubtful positions.

last month, frequent wave development leading to cyclonic activity along the front and the distortion by the continents make it quite difficult to assign a mean position to the front in many parts of the hemisphere. The distance between the positions in January and July is quite considerable, again caused by continental influence. In the Southern Hemisphere, the migration of the polar front is much less than in the Northern Hemisphere. The southern polar front is closest to the pole in February, during the latter part of the southern summer, when the water masses reach their highest temperature. The equatorial front reaches its extreme positions in February and August. The distance covered by this migration is greatest in the Asiatic-Australian region and least over the ocean areas of the South Atlantic and the Southeast Pacific.

CONTINENTAL AND OCEANIC INFLUENCES ON CLIMATE

Next to the general circulation, as is plainly evident from the preceding paragraphs, the simple planetary climate is mostly influenced by the distribution of continents and

oceans. The major wind systems of the general circulation are largely responsible for
a circulation in the oceans. Although temperature distribution in the water masses
would set up a circulation in the water even if no winds were acting, nevertheless the
persistent wind systems are the mainspring of the huge transport of water masses
across the length and width of the oceans. More or less stable currents carry their
temperature characteristics to far-distant parts of the oceans, moving warm water
poleward and cold waters toward the equator. Examples of warm currents are
the Gulf Stream and its extension, the North Atlantic Drift, in the Atlantic, the
Kuroshio and the North Pacific Current in the Pacific. Moving in a general northeast direction from the coasts of southeastern North America and Southeast Asia
they create, coupled with the prevailing westerly winds, temperature conditions on
the coasts of Northwest America and Northwest Europe that are much warmer than
could be expected in latitudes close to the arctic circle. The contrast between the
temperatures of the western and eastern oceans in the Northern Hemisphere is further
reinforced by southward-moving cold currents in the western portion of the oceans,
such as the Labrador Current in the North Atlantic and the Oyashio in the North
Pacific. The influence of these cold currents on the temperature of the surroundings
and on the formation of fog stamps the climate of these regions.

The climate of some coasts, where the prevailing winds blow the surface water
away from the coast, is influenced by the temperature of water welling up from greater
depth to replace the surface water. This upwelling water usually has a rather low
temperature, and the phenomenon has a conspicuous effect on cloudiness of the coasts
of Southwest Africa, Peru, and California.

Aside from the effects of the currents, the oceans dampen and distort temperature
variations. The daily amplitude of air temperature is much less over the ocean than
over the land. The same is true for the annual range of temperature (difference
between mean temperatures of warmest and coldest months). Over the ocean, the
lag of temperature extremes behind the time of maximum and minimum radiation
intensity is longer than over land. In the Northern Hemisphere, except for polar
regions, the warmest month inland is ordinarily July, the coldest January. Over the
oceans and on some coasts, the warmest month is August, the coldest February, i.e.,
about a month later than inland. In polar regions, because of the peculiar radiation
conditions, the maximum may be in either July or August, but the minimum temperature is usually not reached until late February or early March, just before the single
great sunrise of the arctic.

The daily range of air temperature over the open ocean is not much over 2 to 3°F;
coastal fringes may experience mean daily ranges of 8 to 10°F, but 300 miles inland
the mean daily range is ordinarily not less than 15°F.

In continental areas under a regime of prevailing onshore winds, the oceanic
influence is carried far inland; while, in oceanic regions under a regime of offshore
winds, continental influences are felt out at sea. The latter effect is usually not
nearly so pronounced as the former. The contrast of continental and oceanic climates
can be described numerically by indices of "continentality." An index of that type
which is most adapted to modern meteorological concepts is the ratio of the frequencies
of air masses of continental (C') and maritime (M') origin. At the present time,
analyzed weather maps for a sufficient number of years, which would enable one to
determine this ratio, are available for only a few regions of the world, mainly Europe
and North America. From these, the C'/M' ratio for the eastern United States for
the year as a whole turns out to be about 1.2. This indicates that the area is slightly
more under the influence of continental than of maritime air masses. In western
Europe, the ratio C'/M' is about 0.3, denoting that maritime air masses are about
three times as prevalent as continental air masses.

A less dynamic, although very descriptive, approach is the use of the annual temperature range for the calculation of a factor of continentality. Johansson defines an index of continentality based on this element as follows:

$$K = \frac{0.9A}{\sin \varphi} - 14$$

where K = index of continentality
A = annual temperature range, °F
φ = geographic latitude

A purely oceanic climate would have an index value of 0, an extremely continental climate one of 100. The constants of the equation are obtained by assigning the values 0 and 100 to two extreme stations. Johansson used Thorshavn (Faro Islands) and Verkhoyansk (Siberia) as the two extreme stations. Use of this formula will yield, for example, for Winnepeg (Manitoba) an index of 51, for San Francisco the value 3.

Defining "oceanicity" rather than continentality, F. Kerner makes use of the fact that in maritime climates the spring months are much colder than the corresponding autumn months. Kerner calls his index the *thermoisodromic ratio*, which is given by the formula

$$O = 100 \frac{(t_O - t_A)}{A}$$

where O = oceanicity
t_O = mean temperature of October
t_A = mean temperature of April
A = annual temperature range

The use of this formula yields values of 4 for Verkhoyansk, 5 for Winnipeg, and 44 for San Francisco.

On the whole, the annual temperature range can well serve to illustrate the contrast between continental and oceanic areas (Fig. 13) although it is also in part a function of latitude. Except for the immediate neighborhood of the equator, the values of annual temperature range increase from the shores inland. The lines of equal range follow roughly the outlines of the continents. Where prevailing wind currents transport maritime air masses toward the continent, as in the case of the westerlies over Eurasia, the core of maximum range is displaced from the center of the continent toward the lee. In this case, the annual ranges increase only very slowly from the Atlantic Ocean into Eurasia. On the Asiatic shores of that continent, it is noticeable that the continental influences do not reach far offshore, for the values of the range decrease very rapidly toward the Pacific.

TABLE 7.—RELATION BETWEEN INTERDIURNAL TEMPERATURE VARIATION AND CONTINENTALITY

Area	Mean interdiurnal temp. variation, °F	Index of continentality of Johansson
West coast of North America........	2.7	3
Central North America............	6.3	60
East coast of North America........	5.2	56
Western Europe..................	3.2	27
Western Russia..................	4.7	45
Central Siberia..................	5.8	86

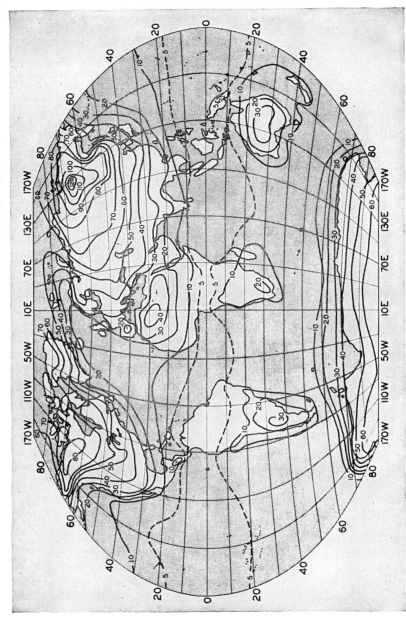

Fig. 13.—Lines of equal annual range of temperature (difference between mean temperature of warmest and coldest months) in degrees Fahrenheit.

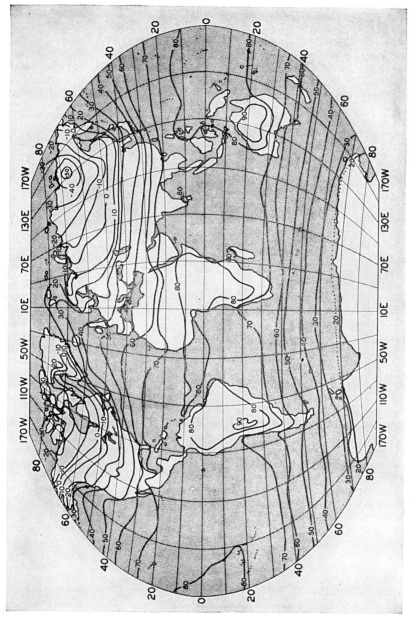

FIG. 14.—Mean isotherms (in degrees Fahrenheit) at sea level, month of January.

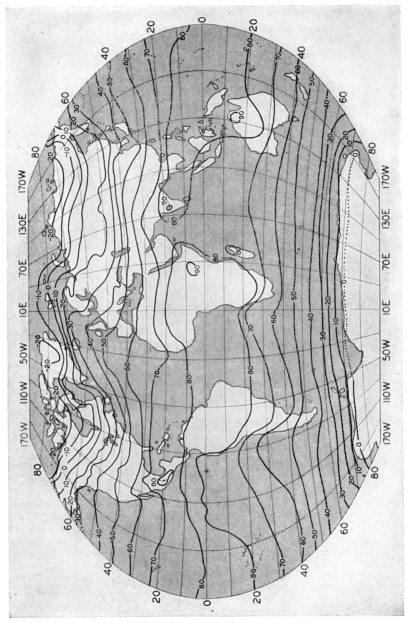

Fig. 15.—Mean isotherms (in degrees Fahrenheit) at sea level, month of March.

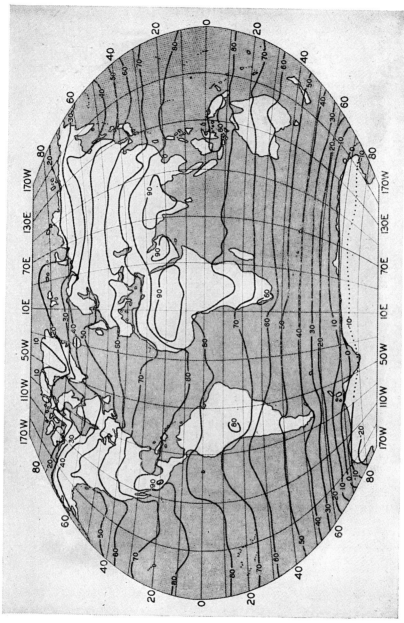

Fig. 16.—Mean isotherms (in degrees Fahrenheit) at sea level, month of May.

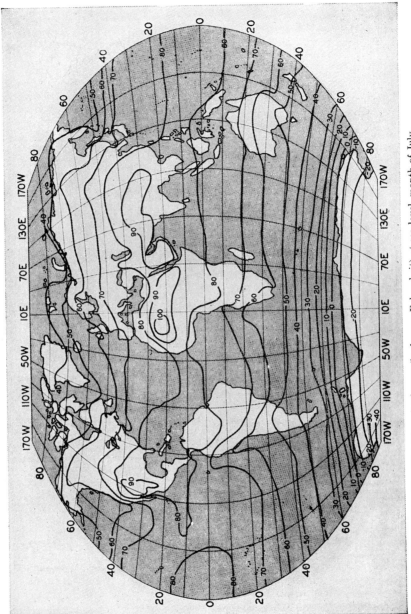

FIG. 17.—Mean isotherms (in degrees Fahrenheit) at sea level, month of July.

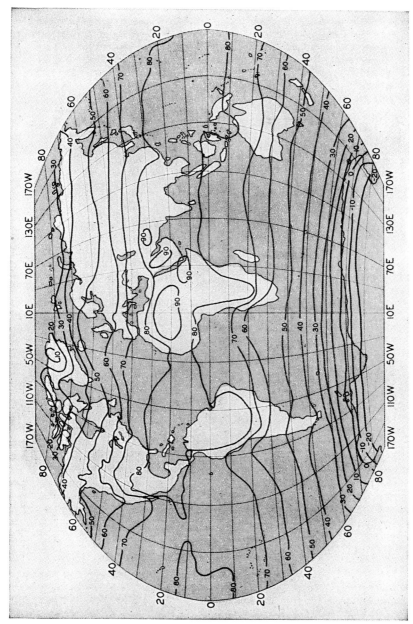

Fɪɢ. 18.—Mean isotherms (in degrees Fahrenheit) at sea level, month of September.

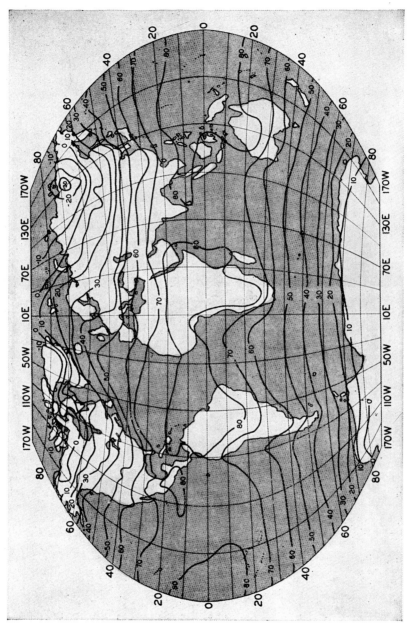

Fig. 19.—Mean isotherms (in degrees Fahrenheit) at sea level, month of November.

The leveling influence of the ocean is also effective in reducing the temperature changes from day to day (interdiurnal temperature variation) caused by alternating of various air masses especially in the regions of polar-front cyclonic activity. Air masses passing over the ocean lose rapidly their original characteristic properties near the surface and acquire those of the immediate environment. As a result, the mean interdiurnal temperature variation is less over the oceans than over land. There exists considerable correlation between the index of continentality and the mean interdiurnal temperature variation, as shown in Table 7.

The general effect of continents and oceans on the temperature distribution can be well visualized from Figs. 14 to 19, which represent world-wide isotherms at sea level. Especially the maps for January (Fig. 14) and July (Fig. 17) will make the

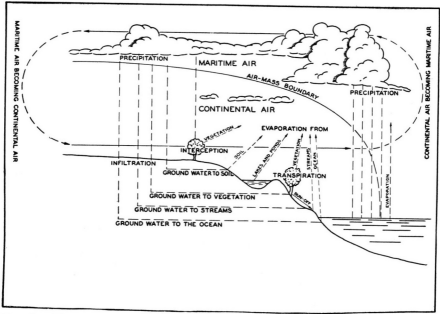

Fig. 20.—The hydrologic cycle of evaporation, precipitation, and runoff between oceans and continents. (*After B. Holzman.*)

following axiomatic statements clear: In the summer season the sea-level isotherms over the oceans bulge toward the equator, over the continents toward the poles. Continental areas around the tropics show centers of maximum heat in summer. Cold oceanic currents are distinctly marked in the summer isotherms. In contrast, during winter, the sea-level isotherms over the oceans bulge toward the pole, over the continents toward the equator. Continental areas near or beyond the arctic circle show centers of maximum cold. Warm oceanic currents are distinctly marked in the winter isotherms.

It is, of course, plainly evident that the influence of continents and oceans is not restricted to thermal conditions. All the phenomena connected with the water in the atmosphere are related to the distribution of land and water: evaporation and hence the amount of water vapor in the atmosphere, condensation and hence cloudiness and sunshine, and ultimately precipitation. It is not possible to exclude the influence of dynamic effects such as frontal activity and subsidence or those of mountain ranges on the water cycle in the atmosphere. Nevertheless a few general statements are valid, namely, cloudiness decreases, sunshine increases inland; humidity and precipitation, as well as wind speeds, decrease inland.

Evaporation over the ocean is considerable and, according to estimates by Wüst, amounts to 81,000 cubic miles of water yearly. In comparison, from inland water bodies and wet soil only 15,000 cubic miles evaporate. Most of this inland evaporation occurs in dry air masses, is transported as water vapor out to sea, and is precipitated. Of the water evaporated from the ocean a considerable portion is carried by maritime air masses to the continents, where total precipitation amounts to about 24,000 cubic miles annually. In the moderate latitudes this precipitation is mostly caused by frontal activity. In those areas the hydrologic cycle follows essentially the scheme indicated in Fig. 20. To make up for the excess of precipitation over evaporation inland, 9,000 cubic miles of water returns to the sea by runoff.

Climatically very important are the superposition of air currents caused by thermal contrasts between land and ocean upon the general circulation. It is a well-known fact that in daytime at the surface a thermal gradient will exist from land to sea and at night from sea to land. This will lead to a low-level pressure distribution that is often sufficient to affect the flow of air, particularly if the general pressure gradients are weak. At the surface, this results in a sea breeze during the day and a land breeze at night. As usual, the oceanic effect is more pronounced. The sea breeze is generally stronger than the land breeze and is felt up to 30 or 40 miles from the coast. The land breeze is usually weak and reaches at best a few miles out to sea. Both land and sea breeze are shallow phenomena, restricted to the lower 1,500 ft of the atmosphere. For some coastal areas, especially in tropical regions, the more or less regular land and sea breeze is of considerable influence upon the climate.

Infinitely more spectacular than this small-scale phenomenon is a circulation superimposed upon the general circulation by the seasonal thermal differences between continent and ocean. In winter when the land is cold and hence surface pressures are high, an outflow of air toward the ocean takes place that may reinforce or weaken currents set up by the planetary atmospheric circulation. Similarly, in summer the land is warm, surface pressures are lowered, and a tendency for an inflow of air from the ocean inland is the result. Again this gradient is superimposed upon the general circulation. These seasonal land and sea winds are called _monsoons_. They affect all continents, yet the degree of influence varies from slight deflection of winds to absolute dominance. The difference in appearance is caused by the fact that in some areas pressure gradients of the general circulation are too strong or the continents too small to produce more than an additional wind component, which is insufficient to cause a complete seasonal reversal of winds. In other cases, mainly that of the largest continent, Asia, the pressure pattern of the general circulation is sufficiently distorted to produce opposite wind directions in winter and summer.

TABLE 8.—MEAN VALUES OF SELECTED CLIMATIC ELEMENTS AT TYPICAL STATIONS UNDER INFLUENCE OF LAND AND SEA MONSOONS

Station	Month	Mon-soon	Frequency of winds, %									Mean precip., in.	No. days precip.	Mean cloud., %	Mean rel. hum., %
			N	NE	E	SE	S	SW	W	NW	C				
Shanghai (China)	Jan.	Land	27	15	9	8	4	3	8	24	2	2.0	10	61	78
	June	Sea	5	8	18	33	19	7	5	4	1	7.3	14	77	84
Bombay (India)	Feb.	Land	26	41	24	4	1	0	1	3	0	0.1	0.2	13	71
	July	Sea	0	0	1	4	4	29	59	3	0	24.3	21.0	88	87

The position and intensity of the subtropical high-pressure cells on both hemispheres, hence dynamic effects, are distinctly linked with the purely thermal effects of continents and oceans. The classical Indian and east Asiatic monsoons are composite effects of both influences. The climatic consequences of fully developed monsoons are prominent. In a typical monsoon climate, the following conditions coincide: During the winter (land) monsoon, the prevailing wind is offshore; precipitation, cloudiness, and humidity show minima. During the summer (sea) monsoon, the prevailing wind is onshore; precipitation, cloudiness, and humidity show maxima. Examples for two stations, one in East Asia, the other in India, will illustrate these statements. All the pertinent elements are shown in Table 8 for the peak of the respective monsoon seasons.

INFLUENCE OF TOPOGRAPHY ON CLIMATE

Large areas of ocean territory have quite uniform climatic conditions. Over land, however, climates may be widely differentiated even within very short distances because of land forms and the variations in elevation. A change of all climatic values is observed as a function of elevation. This is true for the free atmosphere as well as for different levels of land. The changes of temperature with height above sea level are well explored. The observations show that, on an average, the temperature decreases with elevation from the surface to the height of the tropopause. The heat absorbed by the surface of the earth leads to convection, and the winds caused by the general circulation produce turbulence. Both these factors in conjunction with fronts and topographic obstacles lead to vertical motions in the atmosphere. Upon ascending, the air cools adiabatically, and if it were completely dry a lapse rate (decrease of temperature per unit height) of 0.55°F per 100 ft would result up to an elevation at which convective motions normally cease, *i.e.*, the tropopause. The lapse rates that are actually observed are frequently less because of the existence of inversions and because condensation processes add heat to the ascending air masses. In the lowest layers, inversions produced by radiative loss of heat at the surface decrease the mean lapse rates markedly in winter.

For the whole earth the annual mean lapse rate is approximately 0.27°F per 100 ft. This value is valid up to elevations of about 12,000 ft. In regions with a distinct winter season, the mean lapse rate of that season is only 0.22°F per 100 ft, and even less in the lowest 1,000 ft. Those regions have in summer a mean lapse rate of 0.36°F per 100 ft, a value that prevails in the tropics throughout the year. The use of mean lapse rates makes it possible to draw isothermal maps on which all observations are reduced to sea level (such as those shown in Figs. 14 to 19). Sea-level values are obtained by adding the appropriate amount to the mean temperature values observed at stations above sea level. If this reduction is not performed, an isotherm map will mainly reflect the topography. From a map showing the sea-level isotherms, one can, of course, by reversing the procedure obtain the mean temperature of a station if its elevation is known, as follows:

$$t_h = t_0 - \gamma h$$

where t_h = mean temperature at height h
t_0 = sea-level temperature
γ = mean lapse rate (for unit height)
h = height above sea level in same units of height as chosen for γ

On maps with sea-level temperatures, interpolation between the isolines is permissible; whereas, on maps showing true isotherms, interpolation is a rather unsatisfactory procedure, particularly in mountainous areas.

Mountains have a powerful influence on climates even beyond the mere effect of altitude, which is also present in the free atmosphere. Especially the long, high mountain chains act as climatic divides. Their local influence is comparable with that of ocean currents. The high obstacles placed in the path of the currents of the general circulation deflect the tracks of low-pressure centers and block the passage of air masses in the lower levels. Even if the pressure gradients are strong enough to force air masses across the mountains, the forced ascent and descent on the slopes modifies the air masses profoundly. The direction of the mountain chains may block access of certain air masses to the lee side. For example, the Alps and the Himalayas extending from west to east will prevent fresh polar air masses from advancing southward. Hence the climates of northern Italy and India are warmer than corresponds to their latitude. The chains of the Andes in South America and the Coastal Ranges in North America running parallel to the coast from north to south prevent the unmodified passage of maritime air masses from the ocean and cause a lack of precipitation to the lee.

The distribution of precipitation is probably most noticeably affected by mountains. On both the windward and the lee slopes, there is an increase of precipitation with elevation. Yet the values observed on equal levels are much higher on the windward than on the lee side. In the moderate climates, the increase of precipitation along slopes can be represented in first approximation for annual mean values by the following formula:

$$r_h = r_0 + 0.072h$$

where r_h = precipitation in inches at level h

r_0 = precipitation in inches at foot of mountain

h = height above foot of mountain, ft

In the region of westerly circulation, the amounts of precipitation increase fairly uniformly up to the crest heights of the mountains, as observations in the Alps up to elevations of 10,000 ft show. In the subtropical trade-wind zone, as, for example, in the Hawaiian Islands, precipitation amounts increase only to about 3,000-ft elevation and then decrease gradually, although even at 6,000 ft more rain is received than at sea level.

TABLE 9.—DIFFERENTIAL OF CLIMATE ON WINDWARD AND LEE SIDE OF MOUNTAINS

Element	Period	Wind-ward side	Lee side	Observed difference	
				Tacoma excess	Yakima excess
Mean temperature.........	Winter	Higher	Lower	9.7°F	
Mean temperature.........	Summer	Lower	Higher		6.8°F
Temperature range........	Daily	Smaller	Larger		9.8°F
Temperature range........	Annual	Smaller	Larger		19.6°F
Temperature range........	Absolute	Smaller	Larger		44°F
Precipitation............	Annual	Larger	Smaller	31 in.	
Relative humidity........	Afternoon	Higher	Lower	25%	
Sunshine duration.........	Annual	Shorter	Longer		25% of possible
Clear days..............	Annual	Less	More		66 days
Cloudy days.............	Annual	More	Less	56 days	
Wind speed.............	Winter	Higher	Lower	3.7 mph	
Wind speed.............	Summer	Nearly equal		0.7 mph	

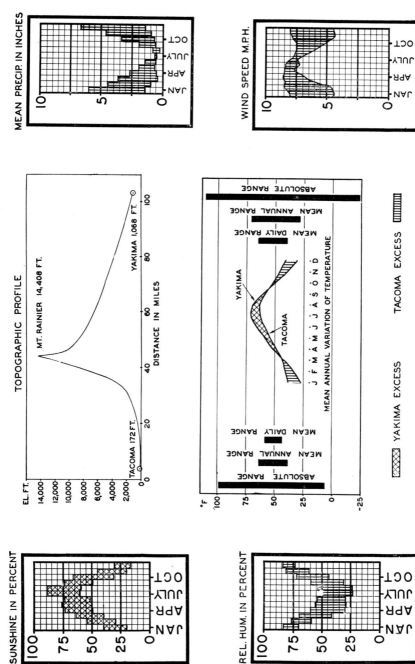

Fig. 21.—Climatic differences between windward and lee side of mountain, illustrated by the example of the influence of Mount Rainier, located between the stations Tacoma and Yakima.

The over-all effect of a mountain on the climate in the zone of westerlies can be illustrated best by an example. Conditions on the windward and lee sides of Mount Rainier (Washington), elevation 14,408 ft, present a most interesting and instructive picture. At the foot of the mountain on the windward side is the station Tacoma (172 ft); 100 miles from Tacoma to the lee of the mountain is Yakima at an elevation of 1,068 ft. Some of the important climatic elements and their annual variation are shown in Fig. 21.

From these observations, a number of general rules can be derived. They are presented in Table 9 with the corresponding values for Tacoma and Yakima given as examples.

All the observed differences between windward and lee sides of mountains can be explained by the difference in cloudiness. The ascending air currents on the windward side cause condensation and precipitation. The subsidence on the lee side leads to dissolution of clouds. Under clear sky conditions, the effects of radiation on temperature are accentuated. Thus all ranges of temperature are increased on the lee side. In winter, marked inversions are the result, and hence the wind speeds in the cold air that clings to the ground are reduced. In summer, the lee-side lapse rates are steep under the influence of radiation, turbulence is increased, and the wind speeds near the surface are virtually the same as those on the windward side.

The increased temperature range on the lee side shows that the continentality is raised markedly. This poses the question if the inland position of Yakima might not be responsible for the observed effects. Fortunately there exists a pair of stations with comparable distance between them and with the same distance of each one from the ocean but without intervening mountains. These are Hamburg and Berlin, about 100 miles apart, in the northwest European plain. The observations there show that Berlin receives 78 per cent of the annual rainfall that is noted for Hamburg. By comparison Yakima receives only 21 per cent of the amount falling at Tacoma. Berlin has 114 per cent of the sunshine recorded at Hamburg, whereas Yakima receives 164 per cent of that observed at Tacoma. The annual range of temperature at Berlin is only 3.7°F larger than that of Hamburg, but Yakima's annual range tops that of Tacoma by 19.6°F. In order to find a correspondingly large increase in annual temperature range in the northwest European plain, one has to go about 1,000 miles inland from Hamburg to the region near Vyazma in Russia. This shows conclusively how much the mountain influence outweighs the factor of increased distance from the ocean.

Among other effects of mountains on climate is the development of a diurnal wind system in the lowest layers, particularly on clear days. This wind system, the mountain and valley breeze, corresponds to the land and sea breeze observed on shore lines. In daytime, an upslope wind blows (valley breeze); at night, a downslope wind blows (mountain breeze). Aided by the effect of upward-directed convection in daytime, a temperature gradient develops directed from the heated slopes of the mountains toward the same levels in the free atmosphere over the lowlands, which are cooler. This results in a pressure gradient directed toward the mountain and a wind in that direction. At night, aided by cold gravitational slope flow of air from higher to lower levels, a gradient in the opposite direction is created aloft, and the mountain breeze results.

Other winds of mountainous areas, the so-called "katabatic" or downslope winds, are of considerable climatic importance. Among these are the foehn, or Chinook-type, winds. Their effect is already implicitly contained in the differences noted for windward and lee sides of mountains. In those cases, a moisture-burdened air, ascending the windward side of a mountain, cools for a considerable portion of the ascent at the moist-adiabatic lapse rate, loses water by precipitation, and descends on the lee side, heating at the dry-adiabatic lapse rate.   The consequence is that a relatively cool, moist air mass after crossing the mountain arrives at the foot to the lee as a warm dry air mass.

Not all descending air masses, however, act as warm air masses.   In some areas where inland plateaus are adjacent to coastal regions, the air over the coast, especially in winter, may have superadiabatic lapse rates because of the continuous heat supply from a warm water body.   If a pressure gradient forces air to cascade down the slopes of the plateau toward the sea, this air will heat only adiabatically and hence will arrive cooler at the foot than the air it replaces.   These cold katabatic winds are called *bora* winds from their prototype observed at the Dalmatian coast of the Adriatic Sea.   They can be found on all coasts where cold highlands are adjacent to warm seas.

All the foregoing are the large-scale, or macroclimatic, effects of topography. Locally, however, the variation of land forms creates an infinite variety of smaller climatic differences called *microclimates*.   *A discussion of all these local variations of climate would fill a book.*   Here only a table showing some of the most important qualitative differences between concave (valley) and convex (crest) land forms will be presented (Table 10).

TABLE 10.—DIFFERENCE OF CLIMATES OF CONCAVE AND CONVEX LAND FORMS

Element	Period	Concave land form	Convex land forms
Minimum temperature.....	Daily	Much lower	Higher
Range of temperature......	Daily and annual	Larger	Smaller
Frost.....................	Night	More frequent	Less frequent
Wind speed...............	Night	Lower	Higher
Low visibility and fog......	Early morning	Much more frequent	Less frequent

INFLUENCE OF LAKES ON CLIMATE

The degree of influence of lakes on climate is not so large as that of a high mountain range, but in the case of large lakes it exceeds the influence of local topography. Large lakes in regions with pronounced prevailing wind direction show considerable differences between the windward and the lee shore.   The air masses that are passing across the lakes are modified in the lowest layer by the lake.   This influence can be well illustrated by observations from Lake Michigan, which is located in the region of prevailing westerly winds.   Stations on the west shore of the lake have a more continental climate than those on the east shore.   The main characteristics of the differences between windward and lee shores of large lakes are given in Table 11 with the actually observed differences between Milwaukee (west shore of Lake Michigan) and Grand Haven (east shore of Lake Michigan).   The two stations are separated by a stretch of 85 miles of water surface.

The effect of an open water body is most pronounced in winter in a generally continental climate.   Air masses that move to the lee shore pick up moisture over the lake.   Therefore cloudiness, precipitation amounts and frequency, and relative humidities increase, and the temperatures rise.   The free sweep over the relatively smooth lake surface will increase the wind speed at the surface.   In summer, when the lake is colder than the environment, air masses moving to the lee shore are stabilized, and hence the tendency toward instability precipitation is reduced.

Inland lakes that freeze in winter distort the annual temperature variation in their vicinity.   In autumn, the temperature of these lakes stays above freezing while

TABLE 11.—DIFFERENCE IN CLIMATE BETWEEN WINDWARD AND LEE SHORES OF
LARGE LAKES

Element	Period	Windward shore	Lee shore	Example		
				Period	Milwaukee excess	Grand Haven excess
Mean temperature	Midwinter	Lower	Higher	Jan.		3.6°F
Mean temperature	Midsummer	Higher	Lower	Aug.	2.0°F	
Temperature range	Annual	Larger	Smaller	Year	4.9°F	
Temperature range	Diurnal	Larger	Smaller	Dec.-Feb.	2.4°F	
Precipitation amounts	Winter	Smaller	Larger	Dec.-Feb.		2.09 in.
Precipitation amounts	Summer	Larger	Smaller	June-Aug.	0.55 in.	
Days with precip. ($\leqq$ 0.01 in.)	Annual	Less frequent	More frequent	Year		22 days
Days with snowfall ($\leqq T$)	Annual	Less frequent	More frequent	Year		44 days
Relative humidity	Winter	Lower	Higher	Dec.-Feb.	1%	5%
Relative humidity	Summer	About the same		June-Aug.		
Sunshine	Winter	More	Less	Dec.-Feb.	20% of possible	
Sunshine	Summer	About the same		June-Aug.		2% of possible
Wind speed	Winter	Less	More	Dec.-Feb.		1.4 mph
Wind speed	Summer	About the same		June-Aug.	0.1 mph	

the air temperature of the general region sinks below it. Stations immediately on the
lake shores usually benefit from the heat surplus of the lake and stay a little warmer
than the general surroundings. Steam fogs may develop at the same time over the
lake. In spring, however, the lake may still be frozen while the air temperature
has generally risen above the freezing point. The shore stations may show, for a
few weeks, considerably lower temperatures than the general environment. This
condition lasts until the ice breaks and the water has had a chance to warm.

On clear days with weak general pressure gradient, the shores of all great lakes
show lake breezes that are identical in origin with the sea breezes on the oceanic coasts.
The main differences between lake and sea breezes are the dimensions. Lake breezes
rarely exceed 500 to 800 ft in height and usually do not penetrate more than about
1 or 2 miles inland. They affect particularly the maximum temperatures on the
shores of the lake, which remain 4 to 5°F lower on days with lake breeze than would
be expected from the normal diurnal temperature variation.

INFLUENCES OF VEGETATION AND HUMAN ACTIVITY ON CLIMATE

In addition to the microclimatic effects of various land forms, other climatic
influences, usually of a subordinate nature, are introduced by presence or absence of
vegetation and by human activities.

Among the influences of vegetation, those on the hydrologic cycle (see Fig. 20)
are the most important ones. Trees intercept falling precipitation, which will be
evaporated before reaching the ground. Evaporation will also be increased by the
transpiration of plants. On the other hand, precipitation that reaches the soil will
not readily evaporate nor will it run off easily, because the soil of forests has a spongy
structure that can absorb and store considerable quantities of water. Snow covers
in forests, because of protection from direct radiation and from falling warm rain,
may stay on the ground for a period considerably longer than in open spaces. Both
the retardation of snow meltage and the water storage in the soil will diminish runoff
from wooded slopes and reduce flood hazards. Inside forests, temperature maxima
are lower and minima higher than over open land. The wind speed is sharply reduced
at the surface, and relative humidities in the forest are raised.

Tillage of ground, which under cultivation may lie barren for a considerable part of the year, will increase water loss from the surface by evaporation. It may also lead to lower minimum temperatures at the surface because of the poor heat conduc-

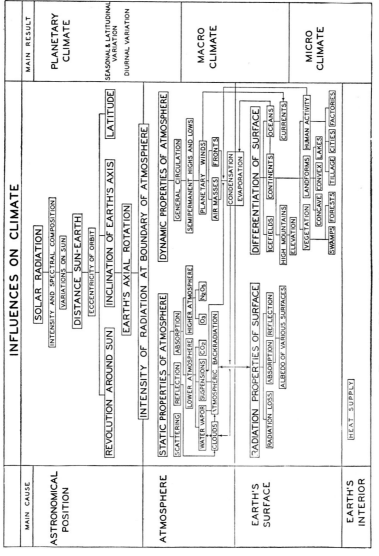

FIG. 22.—Scheme of influences on climate.

tivity of loose soil, which will radiate heat without being able to draw on the reservoir of heat in great depths. Plowed ground will also lead to increased soil blowing and, at high wind speeds, to dust storms.

Industrial activity and large human settlements have measurable influence upon climate. By increasing the atmospheric pollution, they change the radiation balance in the vicinity of the sources of pollution. Dust and smoke deplete particularly the ultraviolet part of the solar spectrum. The considerable amount of heat produced in cities serves to raise the air temperature. Mainly affected are the daily minimum and the winter temperatures. These are from 1 to 3°F higher than those of the environment. In summer, increased convection causes higher cloudiness over cities, coupled with somewhat higher rain frequencies and amounts. Because of the good artificial drainage, however, evaporation is reduced and the mean relative humidities are lower in cities than in the surrounding countryside. The last fact may be somewhat surprising because the urban pollution with its hygroscopic condensation nuclei leads to an increased number of fogs. Visibilities are usually reduced, if not by fog, by smoke and haze. In built-up sections, the roughness of the ground, hence friction, is increased and the wind speeds at the surface are decreased and the number of calms rises. The stagnation of the air will accentuate the effects of pollution.

SCHEME OF CLIMATIC CONTROLS

As has been said initially, the various factors acting as controls of climate are an interwoven plexus with many obscure relationships of cause and effect. A look at Fig. 22, which indicates schematically the numerous basic influences, will make it clear why climatology is still largely a descriptive and observational science and why progress toward analytical treatment has been very slow.

CLIMATIC VARIATIONS

The major features of the climate of a place have certain attributes of constancy. In many areas, year after year weather events repeat themselves within a normal range and, during the course of a human lifetime, climatic conditions seem to conform in general to an established pattern. Single years may be abnormal, but usually a return to normalcy is again achieved after a short interruption. If one considers intervals longer than the span of a human life, climatic features may undergo rather radical changes. During geologic ages, arctic climates have prevailed where moderate climates rule now, or tropical rain conditions may have dominated areas that are now desert. It is likely that some of these large-scale changes were caused by the cyclical variations in the elements of the earth's orbit, by the changes in the inclination of the earth's axis, and by a different distribution of land and sea on the surface of the earth.

Even during the relatively short period of recorded history as preserved in human writings and in natural documents as, for example, tree rings, certain areas have undergone climatic changes. These have led to established migrations of man, fauna, and flora. Climatic history documented by reliable and uninterrupted instrumental records is very brief. Less than 300 years comprise the longest series, and for vast spaces of the earth such as portions of the interior of Africa, the Arctic and Antarctic zones, Central Asia, Central Greenland, remote parts of the oceans, 2 to 5 years may be the best available data, and a few isolated observations from expeditions may be the only available meteorological information. Any analysis of current climatic trends is seriously restricted by the shortness of the series of observations. Yet even in the longest records there is no conclusive evidence of one-sided change in climate. There seem to be only longer or shorter intervals in which the climatic elements vary through a considerable range. Years of higher temperature or precipitation are followed by sequences that are colder or drier. These variations are rather irregular in occurrence; and, in spite of many efforts, no persistent regular cycles of appreciable amplitude have been discovered in the now available data, with the sole exception of the diurnal and annual variation.

There are, however, certain rhythms in the series of climatic observations. These are irregular in phase and amplitude. Often several basic rhythms seem to be hidden in the observational series. Sometimes these reinforce, at other times they weaken each other. No physical reasons can be given as causes for many of the apparent cycles, and hence their reality has the support only of statistical criteria. Probably those rhythms associated with variations in solar radiation will have the best chance of eventual substantiation. Among them are correlations to the irregular sunspot rhythm, which has a mean value of 11 years. The effects of this solar variation seem to be most noticeable in the observations of tropical stations, for both temperature and precipitation.

For precipitation, the changes are usually noted only in records of lake levels and river stages. These integrate rainfall effects over wide areas and are, of course, a statistically more reliable sample than that cut out by an 8-in. gauge which is to represent supposedly the rainfall over many square miles. But, even if one accepts the variations paralleling the sunspot variations as real, they are very small. The differences in mean temperature values between the extreme years of the sunspot rhythm are not much over 1°F. Another basic solar cycle, according to Abbot, 23 years in length, is noted in the climatic record preserved by tree rings and also in the shorter series of instrumental observations. A rhythm, which is apparently the result of several superimposed basic fluctuations, was found by Brückner in precipitation and lake-level data. It was originally assumed to be a cycle of 35 years' duration. Actually it varies in length between 30 and 40 years. During the driest and wettest period of the rhythm, the mean deviation of precipitation from normal is only 6 per cent. A look at almost any series of meteorological data will show that fluctuations from year to year are much larger than this value. Therefore, the incidental variations at most places obscure whatever cyclical or rhythmical elements may be present in the record.

It is likely that the aperiodic variations in climatic values from year to year, which are of great practical importance, for example, in agriculture, are caused by shifts in the general circulation. These shifts are indicated by displacement of the semipermanent high- and low-pressure cells (centers of action) from their normal position and by weakening or strengthening of the pressure gradients prevailing between them. The effects of changes in the general circulation may lag behind the basic shifts, in part because it takes some time for climatologically important basic elements, as ocean currents, to adjust themselves to the new circulation. Attempts have been made to exploit such time lags to establish correlations between meteorological events in distant regions. One of the most remarkable attempts was that by Sir Gilbert Walker and his associates and successors to use correlations between conditions on the Indian and Pacific Oceans to predict the intensity of the Indian summer monsoon. Yet even here, if one accepts the correlations as cause and effect relationships, not much more than 30 per cent of the observed variations are accounted for in this fashion.

The changes in the general circulation of the atmosphere over longer periods of time and the causes behind them are the fundamental but as yet unsolved problems in climatology. Of course, after a fairly long series of climatic observations has been secured, the range of variation of the individual elements is given, and statistical measures for the variability around the mean value can be established. Thus, for places with long records, the effect of climate on human enterprises can be entered as a calculable risk into economic considerations. In case of agricultural staple crops, for example, precipitation is probably the most important single climatic factor. The absolute variability of precipitation is a function of the amounts. The greater the rainfall is, the greater are the absolute changes from year to year. A map on

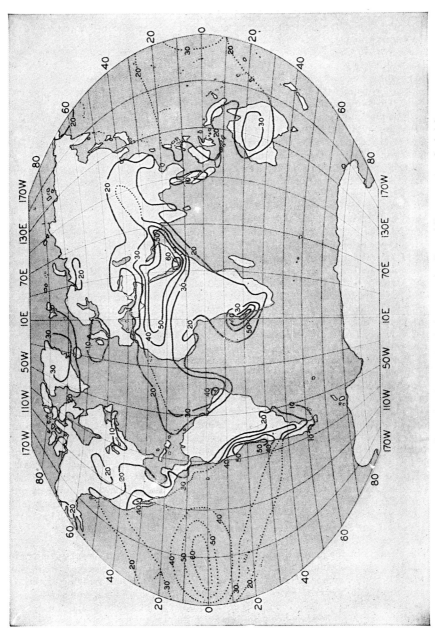

Fig. 23.—Relative variability of rainfall, in per cent. Dotted lines are very doubtful.

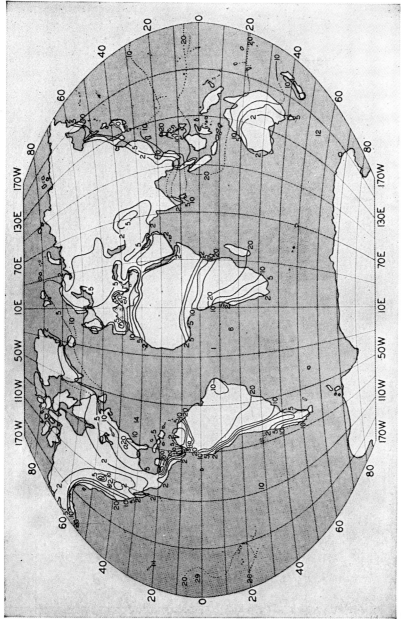

FIG. 24.—Lines of equal mean precipitation, in inches, for season December to February. Data over oceans are too sparse to draw lines. Values for some islands are given as number placed over approximate position of station.

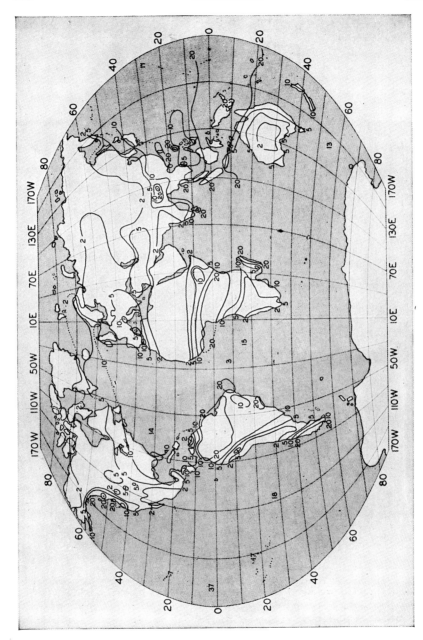

FIG. 25.—Lines of equal mean precipitation, in inches, for season March to May. (Legend, see Fig. 24)

FIG. 26.—Lines of equal mean precipitation, in inches, for season June to August.   (Legend, see Fig. 24.)

Fig. 27.—Lines of equal mean precipitation, in inches, for season September to November. (Legend. see Fig. 24.)

Fig. 28.—Lines of equal annual precipitation, in inches. (Legend, see Fig. 24.)

which absolute variability of precipitation is shown would indicate a pattern similar to that of an isohyetal map. Biel and Conrad have used the relative variability as a factor that will show more clearly those areas where changes in the circulation will produce the greatest relative changes in rainfall. The relative variation is given as

$$v_r = 100 \frac{\sum_1^n |\epsilon_i|}{n\bar{p}} \qquad \epsilon_i = p_i - \bar{p}$$

where v_r = relative variability, per cent
ϵ_i = departure of individual year from normal
$\bar{p}$ = normal (mean) precipitation
p_i = precipitation in individual year
n = number of years in series

Figure 23 shows lines of equal relative variability of rainfall for the world. Over the oceans, these lines are very doubtful and based only on some island stations. The great variability in the Central Pacific is caused by shifts in the ocean currents, which, in turn, follow changes in the general circulation. Over the continents, the greatest relative variabilities are encountered in arid areas. This pattern of variability should be kept in mind when viewing maps showing mean values of precipitation (Figs. 24 to 28).

Some of the boundaries within which climatic elements have varied over the earth during the time since instrumental observations started are shown in Table 12. For comparison, the extreme values observed in the United States are also given. With many portions of the earth as yet very incompletely covered by meteorological stations, the values shown in this table may yet be exceeded by a few units in each

TABLE 12.—EXTREMES OF SOME CLIMATIC ELEMENTS, OBSERVED ON EARTH*

Element	Value	Location	In the United States	
			Value	Location
Absolute maximum temperature	136°F	Aziziya, Tripolitania	134°F	Greenland Ranch, Death Valley, Calif.
Absolute minimum temperature	−108°F	Oimekon, Siberia, USSR.	−66°F	Riverside, Yellowstone Park, Wyo.
Absolute maximum pressure...	1,078 mb	Western Siberia	1,060 mb	Lander, Wyo.
Absolute minimum pressure...	887 mb	Pacific Ocean, off Phillippines	892 mb	Lower Matecumbe Key, Fla.
Absolute maximum rainfall in 24 hr...................	46 in.	Baguio, Phillipines	25.83 in.	Camp Leroy, Calif.
Absolute maximum rainfall in 1 month................	241 in.	Cherrapunji, India		
Highest mean annual rainfall..	476 in.	Mt. Waialeale, T.H.†	150 in.	Wynoochee Oxbow, Wash.
Absolute maximum rainfall measured in 1 min.........	....	....	1.02 in	Opid's Camp, Calif.
Absolute maximum wind speed, accurately measured	....	....	225 mph	Mount Washington, N.H.
Maximum mean wind speed, accurately measured for 5 min	‡	‡	188 mph	Mount Washington, N.H.
Maximum mean wind speed, accurately measured for 1 hr	....	....	173 mph	Mount Washington, N.H.
Maximum mean wind speed, accurately measured for 24 hr	....	....	129 mph	Mount Washington, N.H.

* The values given do not include measurements in the free atmosphere.
† Value based on 7 years of interrupted record. A computation by using the normal value at a base station with long record (formula on p. 958) yields 462 in. Both values are definitely above the long-record value of 428 in. at Cherrapunji, India.
‡ For winds the values given for Mount Washington in the column for the U.S. extremes are the highest that are accurately measured at stations on the earth's surface. They are known to have been exceeded in the free atmosphere.

case. Yet the order of magnitude of climatic range during the current era is probably reasonably well established by these data.

CLASSIFICATION OF CLIMATES

Within the rather wide range of atmospheric conditions shown in Table 12 an infinite variety of combinations of climatic values, their frequencies, and their diurnal and annual variation appear. It is therefore natural to attempt grouping of kindred climates and obtaining a classification that will permit the establishment of regional boundaries between areas of uniform climatic conditions. But to set up climatic categories is by no means easy. Two places may show the same temperature throughout the year but have different amounts and annual variation of rainfall. Others may agree in these two elements but show discrepancies in fog occurrence or in wind speeds. Furthermore, at a considerable number of places, only temperature and precipitation are observed, and other elements cannot be considered for the purpose of classifying the climate. The best that can be achieved is a classification of climate for a specific purpose rather than a climatic taxonomy comparable with that of animals and plants. The purpose of a climatic classification may be, for example, to establish the limits of areas suitable for a given crop plant, or areas with comfortable climates for human settlement, or climates where weather conditions are favorable for flying, etc. In each case, another set of limiting conditions will govern, and hence the climatic classification of a place will change with the objective toward which the classification is directed. Some of the most widely known climatic classifications are actually an attempt to relate the extent and type of natural vegetation on the surface of the earth to climatic conditions. Among these are the classifications of Koeppen and Thornthwaite.

In Koeppen's classification, five main climatic classes with a total of eleven climatic provinces are given. One of the classes, the "desert climate" is distinguished from the others mainly by a precipitation limit, while the others are separated from each other by characteristic temperature limits. The limiting values for the climatic provinces of Koeppen are shown in Table 13.

It was soon noted that for many purposes the Koeppen classification was too coarse, and an attempt was made to remedy the situation by adding another 20 limiting factors. This permits further subclassification but renders the classification almost as unwieldy as the original mass of data.

While the main emphasis of Koeppen's classification is on temperature limits, Thornthwaite's climatic classes are based on the effectiveness of precipitation. This factor bears a close relation to plant growth and hence permits more refined analysis of climatic problems related to vegetation and agriculture than does the Koeppen scheme. The effective precipitation is the ratio between precipitation and evaporation at a given place. Evaporation is an element that is difficult to measure and for which records are available at a few localities only. Thornthwaite circumvents the difficulty by using a statistical relation between temperature-precipitation values and precipitation-evaporation values. This defines the precipitation-evaporation ratio as

$$\frac{p}{e} = 11.5 \left(\frac{p}{t - 10} \right)^{10/9}$$

where p = monthly precipitation amounts, in.

  e = monthly evaporation amounts, in.

  t = mean monthly temperature, °F

From the sum of the 12 monthly p/e ratios the precipitation effectiveness (p-e) index is derived.

$$p\text{-}e \text{ index} = \sum_1^{12} 115 \left(\frac{p}{t - 10} \right)^{10/9}_n$$

TABLE 13.—MAIN GROUPS OF KOEPPEN'S CLIMATIC CLASSIFICATION

Main type	Climatic province	Symbol	Mean temperature		Precipitation	
			Coldest month	Warmest month	Main season	Amounts
Tropical rain climate	Rain forest	Af	$> 64.4°F$		All year	Driest month > 2.4 in. Annual > 40 in.
	Savanna	Aw	$> 64.4°F$		Winter	Driest month < 2.4 in. Annual > 40 in.
Dry climate	Steppe	BS			Winter Summer Even	Annual $< 0.44t\text{-}14$* Annual $< 0.44t\text{-}3$ Annual $< 0.44t\text{-}8.5$
	Desert	BW			Winter Summer Even	Annual $< 0.22t\text{-}7$ Annual $< 0.22t\text{-}1.5$ Annual $< 0.22t\text{-}4.25$
Humid mesothermal climate	Moderate, oceanic Moderate, winter dry Moderate, summer dry	Cf Cw Cs	$< 64.4°F, > 26.6°F$ $< 64.4°F, > 26.6°F$ $< 64.4°F, > 26.6°F$		Even Summer Winter	Annual $> 0.44t\text{-}8.5$ Annual $> 0.44t\text{-}3$ Annual $> 0.44t\text{-}14$
Humid microthermal climate	Oceanic boreal Continental boreal	Df Dw	$< 26.6°F$ $< 26.6°F$	$> 50°F$ $> 50°F$	Even Summer	Annual $> 0.44t\text{-}8.5$ Annual $> 0.44t\text{-}3$
Polar climate	Tundra Eternal frost	ET EF		$< 50°F, > 32°F$ $< 32°F$		

*t is mean temperature in degrees Fahrenheit.

Thornthwaite establishes five climatic provinces that correspond closely to natural plant covers. The climatic zones are shown in Table 14 with the respective limits of the p-e index.

TABLE 14.—THORNTHWAITE'S CLIMATIC CLASSIFICATION

Climatic province	Type of vegetation	PE index
Wet.............................	Rain forest	≥ 128
Humid............................	Forest	64–127
Subhumid.........................	Grassland	32–63
Semiarid.........................	Steppe	16–31
Arid.............................	Desert	< 16

The undeniable importance of plant cover and agriculture to human culture has resulted in widespread use of these climatic classifications by geographers. Yet they show no simple relation to the reactions of the human body to climate. The human body has to lose a certain amount of heat per unit of time. It amounts to about 1 mcal per cm² of surface per second at rest and to 7 mcal per cm² per sec when working hard. If less than this amount is lost, the body overheats, and discomfort results; if more is lost, the body has to be protected by clothing, or discomfort and freezing set in. The heat loss is essentially regulated by the conditions of the surrounding atmosphere. The temperature, humidity, radiation intensity, and wind speed all enter prominently into the problems of human climatic comfort. Because radiation observations are not generally available, comfort conditions are usually expressed in terms of two or more of the other elements. The American Society of Heating and Ventilating Engineers has made extensive determinations of conditions of comfort, and the classification diagram shown in Fig. 29 was the result. An instantaneous atmospheric condition or a set of mean values can be entered in this diagram. Either dry-bulb and wet-bulb temperatures may be used or dry-bulb temperature and relative humidity (slanting lines). The comfort conditions indicated on the diagram are essentially applicable for indoor atmospheres, but they permit also to distinguish, in first approximation, between various climatic regions.

More applicable to outdoor atmospheric conditions is the use of the "cooling power," a factor first introduced into human bioclimatology by L. Hill and based originally on measurements with the katathermometer. These gave the heat loss of a physical body in terms of millicalories per centimeter squared per second under the influence of the atmospheric environment. Actual measurements were made at a few places only, but it was found that the measured values could be approximated with an accuracy sufficient for climatic purposes by the use of the normally observed climatic elements. Two formulas are given below, one for the dry cooling power, essentially applicable when perspiration enters little into the physiological heat process, and the other for wet cooling power, which is more satisfactory because it incorporates the cooling associated with perspiration also.

The dry cooling power is represented by

$$C_d = (0.14 + 0.12v^{0.62})(98 - t) \qquad \text{mcal/cm}^2\text{/sec}$$

and the wet cooling power by

$$C_w = (0.21 + 0.17v^{0.62})(98 - t') \qquad \text{mcal/cm}^2\text{/sec}$$

where v = wind speed, mph
t = air temperature, °F
t' = temperature of wet-bulb thermometer, °F

Several attempts have been made to correlate the values of cooling power to subjective sensations. It is, of course, readily seen that persons of different age, status of nutrition and health, and stages of acclimatization will react differently. Therefore the classifications are at best orders of magnitude and mainly convenient as a relative

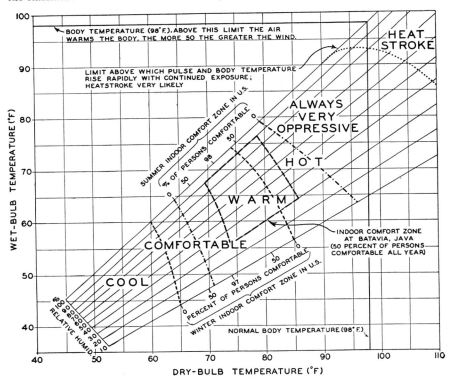

Fig. 29.—Comfort zones in terms of humidity and temperature.

distinction between climatic zones. With this reservation, the sensation scale of Schmid for the clothed body is given in Table 15.

TABLE 15.—CLIMATIC CLASSIFICATION ACCORDING TO COOLING POWER

Climatic comfort zone	Cooling power, mcal/cm²/sec	
	Dry	Wet
Bitterly cold	> 55	> 90
Very cold	41–55	76–90
Cold	31–40	59–75
Cool	22–30	48–58
Mild	15–21	39–47
Warm	10–14	30–38
Sultry	< 10	< 30

According to Siple, the undressed human body at rest in the shade will experience a sensation of "warm" at dry cooling powers below 6 mcal per cm^2 per sec and of "cold" at values above 23 mcal per cm^2 per sec.

Indices such as the cooling power can be used to obtain estimates on the clothing required in different regions at various seasons. A very simplified scheme of a "clothing map" is shown in Fig. 30. Only three zones are distinguished. One is the tropical zone (area covered by dots on map). There no distinct change between summer and winter exists, and light clothing is appropriate for the whole year. North and south of this zone are two belts (without hatching on map) of moderate climates where a distinct change between summer and winter occurs and where different types of clothing are required for each season. In these zones closer toward the equator, the

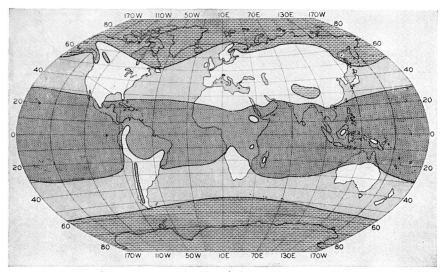

FIG. 30.—Major clothing zones on the earth. Dotted area: tropical zone, warm throughout year. Blank area: moderate zones, regular change between warm and cold season. Dashed area: Polar zones, cold throughout year.

warm season is much longer than the cold season and, vice versa, toward the poles in this zone the cold season lasts much longer than the warm season. Around the poles, there are zones of frigid polar climate (dashed areas on map) where cold-weather clothing will be required the year round. It would be, of course, more appropriate to draw such maps for each month of the year, but lack of space prohibits the presentation of such maps here.

REGIONAL CLIMATES AS RELATED TO THE DYNAMICS OF THE ATMOSPHERE

To the meteorologist, the relation of climatic conditions to the general circulation offers a basis for systematization. Various climates blend into each other without sharp boundaries. This is caused by the fact that no single dynamic feature rules exclusively in any area. Moreover, upon the major dynamic controls local conditions superimpose themselves. The paramount dynamic factors of circulation in the atmosphere are (1) the polar-front circulation, (2) the anticyclonic circulation of the subtropical anticyclones, (3) the trade winds, (4) the equatorial convergence, (5)

NORTH AMERICAN AREA

MIXED POLAR FRONT-CYCLONIC AND ANTICYCLONIC CIRCULATION
WITH MARITIME INFLUENCE

SAN FRANCISCO, CAL

SAN DIEGO, CAL.

SAN ANTONIO, TEXAS

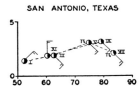

KINGSTON, JAMAICA

KEY WEST, FLORIDA

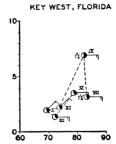

MEXICO CITY, MEXICO

FIG. 31.—Climographs for selected stations.   Ordinates, precipitation in inches, abscissas, temperatures in degrees Fahrenheit.   Bimonthly values plotted for cloudiness, weather conditions, prevailing wind, mean wind force.   For full explanation, see pages 979, 980.

NORTH AMERICAN AREA

POLAR FRONT-CYCLONIC CIRCULATION WITH CONTINENTAL INFLUENCE

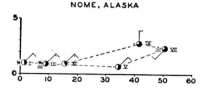

NOME, ALASKA

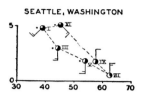

SEATTLE, WASHINGTON

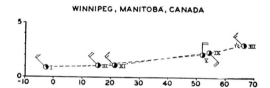

WINNIPEG, MANITOBA, CANADA

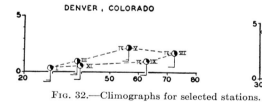

DENVER, COLORADO

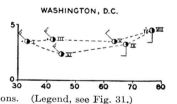

WASHINGTON, D.C.

Fig. 32.—Climographs for selected stations.   (Legend, see Fig. 31.)

the large monsoonal circulations.   These factors are modified by the relative oceanicity or continentality, *i.e.*, the frequency of oceanic and continental air masses.

Figures 31 to 40 show climographs for 50 selected stations.   Graphs of this type were introduced into climatology by Griffith Taylor.   The scheme employed for plotting is as follows: The diagram is a temperature-precipitation coordinate system. Ordinates are precipitation (in inches), abscissas are temperatures (in degrees Fahrenheit).   The mean values of the elements for a given month determine a point in this diagram.   One point can be plotted for every month of the year, but in the figures presented here only the odd-numbered months were chosen for reasons of space. This point, monthly mean temperature versus monthly normal precipitation, serves as center for a station circle.   In the circle, the mean cloudiness of the month is entered in tenths of sky cover.   The most frequent wind direction is entered in the same fashion as on weather maps as a pointer (north wind upward, east wind toward

WEST EUROPEAN AREA

POLAR FRONT-CYCLONIC CIRCULATION WITH MARITIME INFLUENCE

VALENTIA IS., IRELAND

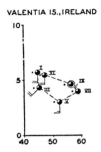

BERGEN, NORWAY

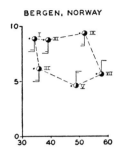

CENTRAL EUROPEAN AREA

POLAR FRONT-CYCLONIC CIRCULATION WITH MIXED MARITIME AND CONTINENTAL INFLUENCE

BERLIN, GERMANY

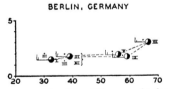

BUCHAREST, ROMANIA

Fig. 33.—Climographs for selected stations. (Legend, see Fig. 31.)

the right, south wind downward, west wind toward the left). The barbs on these pointers indicate the mean wind speed for the month in Beaufort scale divisions ($\frac{1}{2}$ barb length = 1 scale division). To the left of the circle, weather symbols are noted as follows: a dot, in case the month has an average of 10 or more days with precipitation; a star, if the month has 10 or more days with snowfall; a thunderstorm symbol for 5 or more days with thunderstorms; and a fog symbol for 5 or more days with fog during the particular month. The month is indicated in Roman numerals to the right of the station circle. Dashed lines link consecutive months.

On these climographs, maritime climates are noted by the close clustering of the individual points. Continental climates are marked by a considerable horizontal extent of the diagrams. Monsoonal climates have a marked vertical dimension on the graph, usually starting near the zero precipitation line. Trade-wind climates also often show considerable vertical extent on the climographs but with an appreciable amount of rainfall during the driest month. The changes in precipitation from month to month there are essentially a function of the steadiness of the trade wind.

RUSSIAN AREA

POLAR FRONT-CYCLONIC CIRCULATION WITH CONTINENTAL INFLUENCE

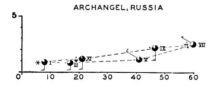

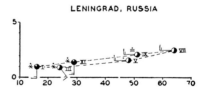

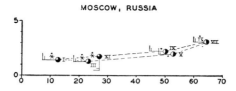

WITH EXTREME CONTINENTAL INFLUENCE

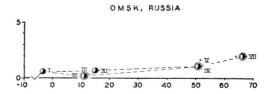

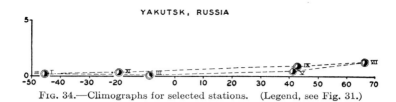

FIG. 34.—Climographs for selected stations.   (Legend, see Fig. 31.)

SOUTH EUROPEAN AREA

MIXED POLAR FRONT-CYCLONIC AND ANTICYCLONIC CIRCULATION
WITH MIXED MARITIME AND CONTINENTAL INFLUENCE

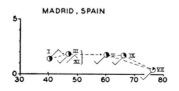

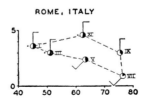

AFRICAN AREA

ANTICYCLONIC AND TRADE CIRCULATION

CONTINENTAL INFLUENCE   MARITIME INFLUENCE

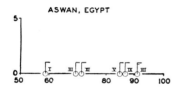

MIXED POLAR FRONT-CYCLONIC AND ANTICYCLONIC CIRCULATION
WITH MARITIME INFLUENCE

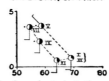

FIG. 35.—Climographs for selected stations. (Legend, see Fig. 31.)

AFRICAN AREA   SOUTH AMERICAN AREA

EQUATORIAL CONVERGENCE CIRCULATION WITH MONSOONAL INFLUENCE

DAR-ES-SALAAM, E.AFRICA

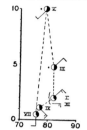

MANAOS, BRAZIL

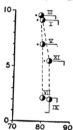

SOUTH AMERICAN AREA

ANTICYCLONIC CIRCULATION WITH MARITIME INFLUENCE

RIO DE JANEIRO, BRAZIL

MIXED POLAR FRONT-CYCLONIC AND ANTICYCLONIC CIRCULATION WITH MARITIME INFLUENCE

SANTIAGO, CHILE

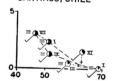

BUENOS AIRES, ARGENTINA

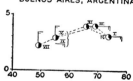

FIG. 36.—Climographs for selected stations.   (Legend, see Fig. 31.)

SOUTH AMERICAN AREA   NEW ZEALAND AREA
POLAR FRONT-CYCLONIC CIRCULATION WITH MARITIME INFLUENCE

PUNTA ARENAS, CHILE

WELLINGTON, NEW ZEALAND

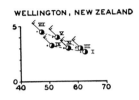

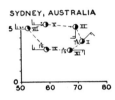

AUSTRALIAN AREA
MIXED POLAR FRONT-CYCLONIC AND ANTICYCLONIC CIRCULATION
WITH MARITIME INFLUENCE

SYDNEY, AUSTRALIA

MELBOURNE, AUSTRALIA

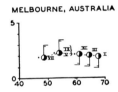

PERTH, AUSTRALIA

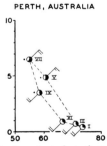

FIG. 37.—Climographs for selected stations.   (Legend, see Fig. 31.)

AUSTRALO-ASIATIC MONSOON AREA

MONSOONAL CIRCULATION AND COMBINED MONSOON-TRADE CIRCULATION

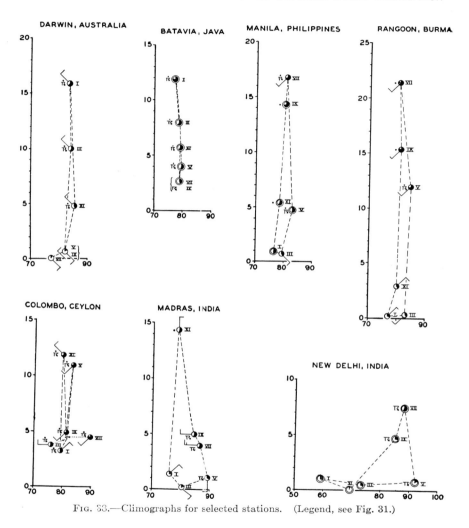

FIG. 33.—Climographs for selected stations.   (Legend, see Fig. 31.)

EAST ASIATIC AREA

POLAR FRONT-CYCLONIC CIRCULATION WITH MONSOON INFLUENCE

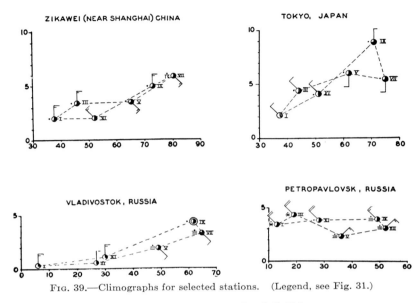

Fig. 39.—Climographs for selected stations.   (Legend, see Fig. 31.)

PACIFIC ISLAND AREA

TRADE CIRCULATION WITH MARITIME INFLUENCE

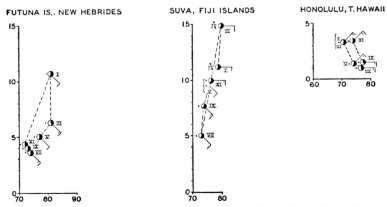

Fig. 40.—Climographs for selected stations.   (Legend, see Fig. 31.)

The main circulation features creating various climograph patterns are indicated in the headings of Figs. 31 to 40. The coordinates of the stations represented in these figures are shown in Table 16 with brief notes on environmental topography where this factor has a strong influence upon the climate.

TABLE 16.—COORDINATES OF STATIONS USED IN FIGS. 31 TO 40

Region	Station	Latitude	Longitude	Elevation, ft	Topographic influences
North America	San Francisco, Calif.	37°47′N	122°25′W	123	
	San Diego, Calif.	32°43′N	117°10′W	87	Mountains to E
	San Antonio, Tex.	29°27′N	98°28′W	693	
	Kingston, Jamaica	18°01′N	76°48′W	24	Rising terrain to NW
	Key West, Fla.	24°33′N	81°48′W	5	Mountains to N and NE
	Mexico City, Mexico	19°30′N	99°08′W	7,572	On high plateau
	Nome, Alaska	64°30′N	165°24′W	22	
	Seattle, Wash.	47°38′N	122°20′W	125	
	Winnipeg, Manitoba	49°53′N	97°07′W	760	
	Denver, Colo.	39°45′N	105°00′W	5,292	On high plateau, high ranges to W
	Washington, D.C.	38°54′N	77°03′W	112	
West Europe	Valencia Island, Ireland	51°56′N	10°15′W	39	
	Bergen, Norway	60°23′N	5°21′E	72	Steep mountains to E
Central Europe	Berlin, Germany	52°33′N	13°21′E	115	
	Bucharest, Romania	44°25′N	26°06′E	269	
Russia	Archangel	64°35′N	40°36′E	22	
	Leningrad	59°56′N	30°16′E	15	
	Moscow	55°50′N	37°33′E	543	
	Omsk	54°48′N	73°20′E	290	
	Yakutsk	62°01′N	129°43′E	335	
South Europe	Madrid, Spain	40°24′N	3°41′W	2,188	On plateau, mountains to NW
	Rome	41°54′N	12°29′E	164	
Africa	Aswan, Egypt	24°02′N	32°53′E	328	
	Lugar, Cape Verde Islands	16°54′N	25°04′E	49	
	Swakopmund, Southwest Africa	22°42′S	14°32′E	26	
	Capetown, South Africa	33°56′S	18°29′E	40	
	Dar es Salaam, East Africa	6°49′S	39°18′E	26	
South America	Manaus, Brazil	3°08′S	60°01′W	147	
	Rio de Janeiro, Brazil	22°54′S	43°10′W	200	Hills to S
	Santiago, Chile	33°27′S	70°42′W	1,706	Mountains to N and E
	Buenos Aires, Argentina	34°36′S	58°22′W	82	
	Puntarenas, Chile	53°10′S	70°54′W	39	
New Zealand	Wellington	41°16′S	174°46′E	10	
Australia	Sydney	32°52′S	151°13′E	138	
	Melbourne	37°49′S	144°58′E	115	
	Perth	31°57′S	115°50′E	197	
	Darwin	12°28′S	130°51′E	97	
Dutch East Indies	Batavia, Java	6°11′S	106°50′E	23	
South and East Asia	Manila, Philippines	14°35′N	120°59′E	46	
	Rangoon, Burma	16°47′N	96°13′E	16	
	Colombo, Ceylon	6°54′N	79°53′E	24	
	Madras, India	13°04′N	80°14′E	22	
	New Delhi, India	28°35′N	77°20′E	718	
	Zi-Ka-Wei (Shanghai), China	31°12′N	121°26′E	23	
	Tokyo, Japan	35°41′N	139°56′E	19	Mountains to W, NW, N
	Vladivostok, U.S.S.R.	43°06′N	131°56′E	443	
Pacific Islands	Petropavlovsk, U.S.S.R.	52°53′N	158°43′E	335	
	Futuna Islands, New Hebrides	19°30′S	170°13′E	98	
	Suva, Fiji Islands	18°08′S	178°26′E	44	
	Honolulu, T.H.	21°19′N	157°52′W	12	

SUPPLEMENTARY CLIMATIC CHARTS

For some modern human activities, such as aviation, the elements of pressure, temperature, and precipitation shown in preceding figures are of subordinate interest. Quantities most important to the aviator are ceiling heights, visibilities, icing conditions, and upper winds. At present, records of these elements are short and regionally

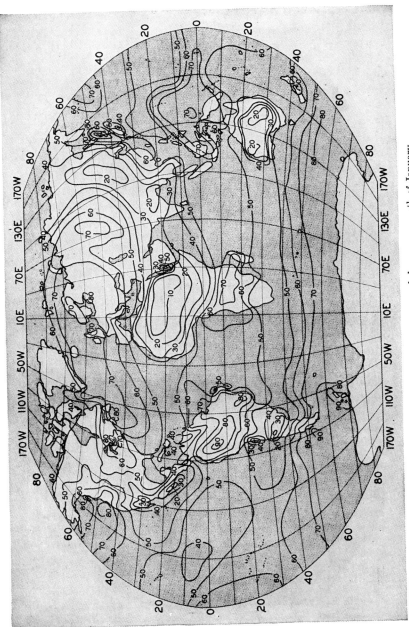

FIG. 41.—Mean cloudiness in percentage of sky cover, month of January.

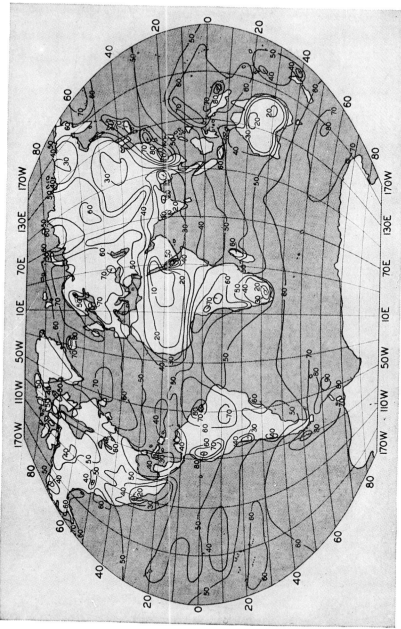

Fig. 42.—Mean cloudiness in percentage of sky cover, month of March.

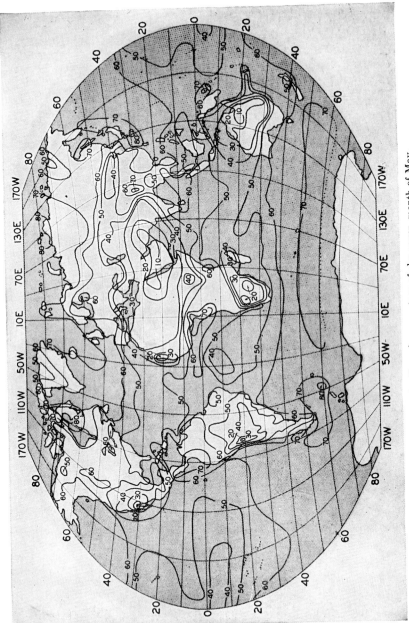

Fig. 43.—Mean cloudiness in percentage of sky cover, month of May.

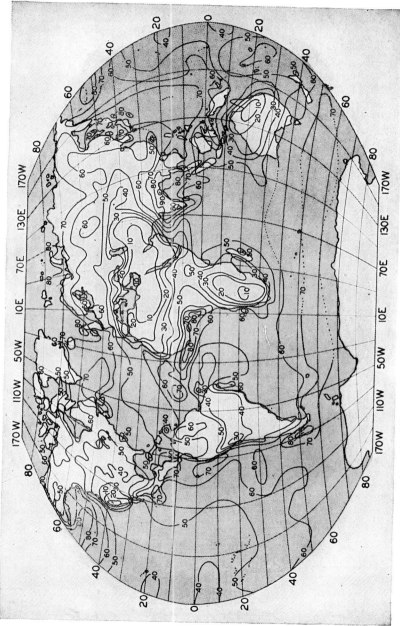

Fig. 44.—Mean cloudiness in percentage of sky cover, month of July.

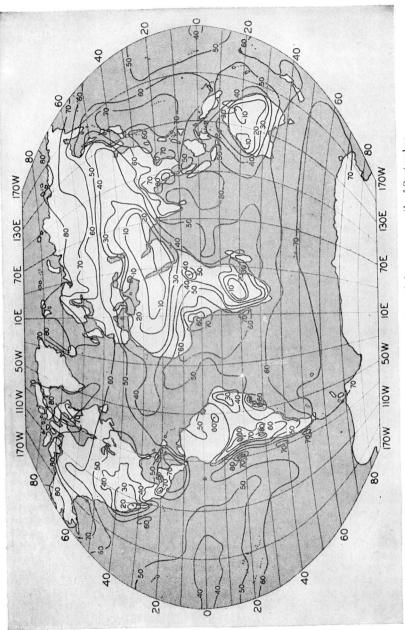

FIG. 45.—Mean cloudiness in percentage of sky cover, month of September.

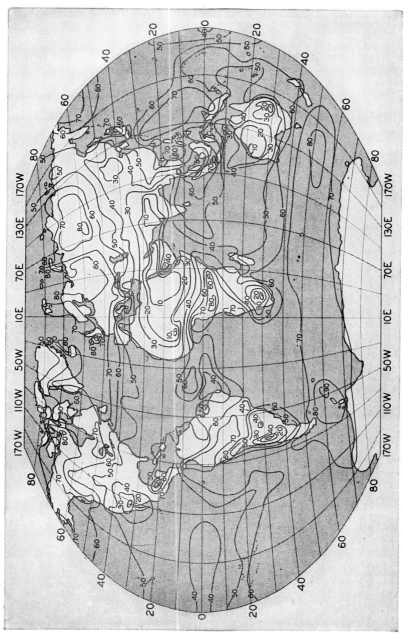

Fig. 46.—Mean cloudiness in percentage of sky cover, month of November.

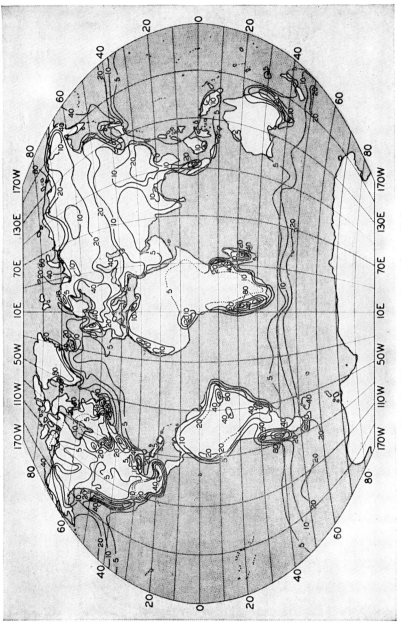

Fig. 47.—Mean annual number of days with fog.

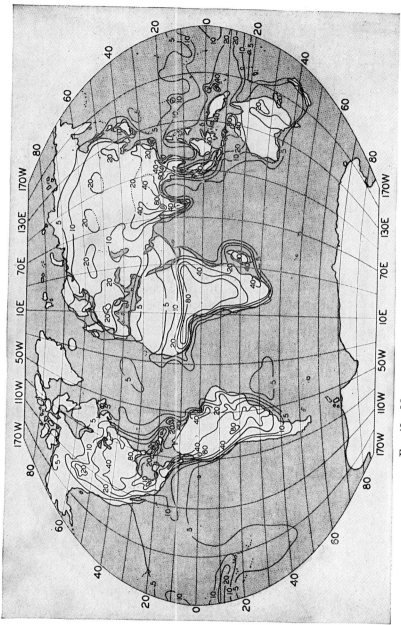

FIG. 48.—Mean annual number of days with thunderstorms.

very incomplete, and presentation of them in world maps would be premature. The same holds for observations of upper level pressures, temperatures, humidities, and winds. Such data are as yet available from a very few stations for any length of time. From vast regions of the world, especially the oceans, only the scantiest observational material for the upper air has been published. But for such other data of interest in aerological climatology as cloudiness, fog, and thunderstorm occurrence sufficient material for world charts is available (Figs. 41 to 48).

In trying to make practical use of maps showing mean cloudiness, one should keep the peculiar frequency distribution of cloudiness in mind. In the case of most climatological elements, the mean value is close to the center of a frequency distribution that has its maximum near the mean value and tapers off toward the extreme values. In case of cloudiness, however, the frequency distribution in many regions has, instead of the normal bell shape, a distinct U shape. The two extreme values, clear and overcast, occur more frequently than the intermediate values. The number of observations near the center of the distribution, 50 per cent cloudiness, are distinctly rare, especially in regions under the influence of polar-front cyclonic circulation. Only in the tropics are intermediate cloud steps more frequent than the extremes. Yet values of "mean cloudiness" as shown on the maps, even though they are least frequent in the extratropical regions, have a distinct meaning. Studies have shown that statistical relations exist between mean cloudiness and the number of clear days (mean cloudiness less than 20 per cent) and cloudy days (mean cloudiness more than 80 per cent) in a month. As a matter of fact, mean cloudiness is a nearly linear function of the difference between the number of cloudy and clear days in a month. In first approximation, even the clear and cloudy days separately show simple functional relationships to mean cloudiness. From data collected at many United States stations, the relations between mean cloudiness and the mean monthly numbers of clear and cloudy days are indicated in Table 17.

TABLE 17.—RELATION BETWEEN MEAN CLOUDINESS AND THE NUMBER OF CLEAR AND CLOUDY DAYS (FROM UNITED STATES OBSERVATIONS)

Mean cloudiness, %	January		July	
	Clear	Cloudy	Clear	Cloudy
10			28	1
20	22	3	23	2
30	18	4	18	3
40	15	6	14	5
50	12	10	10	8
60	9	13	6	10
70	6	17	4	12
80	3	21		

From observations gathered elsewhere, it seems likely that the values for all regions poleward from latitudes of 20°N and 20°S are not greatly different from those presented in the table. For the extratropical regions, these data may serve as a guide for the approximate interpretation of mean cloudiness values.

Bibliography

ABBOT, C. G.: Solar Radiation and Weather Studies, *Smithsonian Inst. Misc. Collections* **94** (10) 1935.

BIEL, E. R.: Veränderlichkeit der Jahressumme der Niederschläge auf der Erde, (mit Weltkarte); *Geog. Jahrb. a. Oesterr.*, XIV, XV, Vienna, 1929.

BLAIR, T. H.: "Climatology, General and Regional," Prentice-Hall, New York, 1942.

BROOKS, C. F., A. J. CONNOR, and OTHERS: "Climatic Maps of North America," Harvard University Press, Cambridge, Mass., 1936.

CLAYTON, H. H.: "World Weather Records," *Smithsonian Inst. Misc. Collections,* **79** (1927); **90** (1934).

Climate and Man, "Yearbook of Agriculture," U.S. Department of Agriculture, 1941.

Climatic Summary of the United States by Sections, *U.S. Weather Bur. Bull. W,* 1932–1936.

CONRAD, V.: The Variability of Precipitation, *Monthly Weather Rev.* **69**: 5–11 (1941).

CONRAD, V.: "Fundamentals of Physical Climatology," Harvard University, Blue Hill Observatory, Milton, Mass., 1942.

ECKARD, W. R.: "Grundzüge einer Physioklimatologie der Festländer," Berlin, 1922.

GORCZYNSKI, W.: "Nouvelles isothermes de la Pologne, de l'Europe et du globe terrestre," Warsaw, 1918.

HANN, J.: "Handbuch der Klimatologie," J. Engelhorn, Stuttgart, 1908.

HETTNER, A.: "Die Klimate der Erde," Heft 5, Geografische Schriften, Leipzig and Berlin, 1930.

JOHANSSON, O. V.: Die Hauptcharakteristika des jährlichen Temperaturganges, *Gerl. Beitr. Geophysik.,* **33**, 406–428, 1931.

KAMINSKY, A., and E. RUBINSTEIN: "Klima der U.S.S.R." Part I, Leningrad, 1927; Part II, Leningrad, 1932.

KENDREW, W. G.: "Climate of the Continents," 3d ed., Oxford University Press, Oxford, 1937.

KENDREW, W. G.: "Climate," 2d ed., Oxford University Press, Oxford, 1938.

KERNER, F.: "Thermoisodromen; Versuch einer kartographischen Darstellung des jährlichen Ganges der Lufttemperatur," K. K. Geogr. Ges. Vienna, Bd. VI, 1905. Nr. 3.

KOEPPEN, W.: "Grundriss der Klimakunde," 2d ed., W. de Gruyter, Berlin and Leipzig, 1931.

KOEPPEN, W., and GEIGER, R.: "Handbuch der Klimatologie," Borntraeger, Berlin, 1932–1938.

LANDSBERG, H.: "Physical Climatology," 3d print., The Pennsylvania State College, School of Mineral Industries, 1943.

McDONALD, W. F.: "Atlas of Climatic Charts of the Oceans; U.S. Weather Bureau, No. 1247, 1938.

MOSBY, H.: Sunshine and Radiation; Norwegian North Polar Expedition with the *Maud* 1918–1925, *Scientific Results,* **1**, 17.

PERL, G.: Zur Kenntnis der wahren Sonnenstrahlung in verschiedenen geographischen Breiten, *Met. Zeits.,* **52**, 85–89, 1935.

SHAW, SIR NAPIER: "Manual of Meteorology," Part II, 2d ed., Cambridge University Press, Cambridge, 1936.

SIMPSON, G. C.: The Distribution of Terrestrial Radiation, *Mem. Roy. Meterolog. Soc.,* **3**(23) 53–78 (1929).

SVERDRUP, H. U.: "Oceanography for Meteorologists," Prentice-Hall, New York, 1942.

TANNEHILL, I. R.: "Weather around the World," Princeton, 1943.

TAYLOR, GRIFFITH: Control of Settlement by Temperature, *Australia Commonwealth Bur. Meteor. Bull.* 14, Melbourne, 1916.

THORNTHWAITE, C. W.: The Climate of North America According to a New Classification, *Geog. Rev.,* **21**: 631–655 (1931).

THORNTHWAITE, C. W.: Atlas of Climatic Types in the U.S., 1900–1939, *Soil Con. Serv. Misc. Publ.* 421, Washington, D.C., 1941.

TREWARTHA, G. T.: "Introduction to Weather and Climate," 2d ed., McGraw-Hill, New York, 1943.

WARD, R. DeC.: "Climate Considered Especially in Relation to Man," Putnam, New York.

WARD, R. DeC.: "The Climates of the United States," Ginn, New York, 1935.

SECTION XIII

HYDROMETEOROLOGY

By A. S. Fry and A. K. Showalter

CONTENTS

SECTION XIII

HYDROMETEOROLOGY

By A. S. Fry and A. K. Showalter

WATER PROJECT SYSTEM OPERATIONS

By A. S. Fry

INTRODUCTION

Definition. Hydrology as defined by the American Geophysical Union "deals with the distribution and disposal of precipitation on the land areas of the Earth, and as such is a borderline science of interest to agronomists, Engineers, foresters, meteorologists, soils technicians, geologists, and others."

Hydrometeorology is a combination of specialized meteorology and hydrology that has application in the solution of problems and the understanding of natural phenomena for which neither science is in itself sufficient. Hydrometeorology functions not primarily as a separate science but rather as a medium for the integration of hydrology and meteorology.

Applications. Hydrometeorology is largely concerned with projects that in some way involve the control, conservation, and use of water. Such projects include chiefly those designed to accomplish either the separate or combined purposes of flood control, hydroelectric power development, navigation, irrigation, drainage, domestic and industrial water supply, and recreation. The control of water on the land by changes in land use and management involves complex applications of hydrometeorology.

One of the more important applications of hydrometeorology is the determination of maximum storms that can occur over an area. This is the subject of another section of this handbook. Such determinations are invaluable to the hydraulic engineer and hydrologist engaged on large water-project planning and operation.

In many of these projects, applications of hydrometeorology begin with the planning, carry through the design and construction, and continue into the operation phases. For example, in the development of the water resources of a region, one or more dams may be planned to create a water-storage reservoir or a system of such reservoirs. Meteorology supplies the essential and basic information regarding the maximum and minimum annual, seasonal, and storm rainfall that may be expected over the region in the future. Predicated upon this information, in whole or in part, engineers using hydrologic methods determine the volumes of water that are available for useful purposes and the volumes of water that must be considered in designing the reservoirs to provide adequate flood control and to ensure the safety of the structures in times of maximum floods. Similarly, in the planning of bridges across large streams, the maximum flood flow to be used in determining the area of waterway required for safe passage of flood waters may be estimated by hydrometeorologic methods.

During the construction of engineering works in river channels, such as dams, bridges, or waterworks intakes, current storms are important. Meteorological forecasts of such storms aid the engineers in predicting flood heights and issuing advance

1000

warnings of floods so that construction operations may be protected from loss or damage.

In constructing earth embankments for dams, levees, and high road fills, modern methods require close control of moisture in the material used for the embankment. Forecasts of rainfall and weather conditions are useful in the control of moisture content that is necessary on such work.

Probably the most important application of hydrometeorology in water-control and utilization projects comes after the storage reservoirs and other works have been completed and have passed into the operations phase. This phase continues throughout the life of the project, which may be for centuries.

During the past decade, there has been a notable increase in the construction of large water-control and utilization projects. Among others, these include the Tennessee Valley Authority, reservoirs on the Ohio River watershed above Pittsburgh, the Central Valley project in California, and reservoirs in the Columbia River Basin in the northwest. A definite national trend is indicated toward the further and continued development of the water resources of the United States. The trend is also toward the development of entire river systems by a number of unified projects rather than by isolated projects and by multipurpose rather than single-purpose projects. Hydrometeorology is coming to be recognized as essential to the most efficient operation of such projects in achieving maximum control, conservation, and use of water.

Quantitative Forecasts. Advances in the science of meteorology during the past quarter of a century have given meteorologists the necessary tools for making quantitative rather than generalized forecasts of rain that may be expected over particular drainage basins. Admittedly, these tools are still in a state of development, but the start that has been made and the acceptance of the forecasts in water-project operation indicate the need for this type of work. Operators of water projects have always sought information concerning the weather and the probability of storm rainfall. But, until recent years, only forecasts in general terms have been available. These forecasts have been helpful, but they could not be evaluated into definite amounts of runoff resulting from predicted storm rainfall.

Although the availability of quantitative forecasts is comparatively recent, the value of this type of forecast is apparent when the problems of water control and utilization are considered. Floods must be controlled and water must be stored and conserved in order that it may be gainfully used for power, navigation, irrigation, or water supply. The need for water is frequently so great that the utmost degree of conservation is demanded. In order that control may be effective and use efficient, as exact knowledge as possible is needed at all times of the amounts of water that must be controlled and that are available for use. Once precipitation has fallen, engineers can estimate the amount of the resulting runoff. Equally important is the runoff from rain that has not fallen but is expected in a future period of 2 or 3 days. To estimate this rainfall quantitatively is the function and responsibility of the meteorologist. On the basis of dependable quantitative forecasts, the engineers can plan ahead for the runoff from future precipitation and thereby accomplish better control of floods and a greater measure of conservation and utilization of water. For example, water already in storage may be used for power production or other uses and replaced from runoff from future precipitation. At the same time, reservoir storage space would be provided for regulating that runoff, thus making the most efficient use of available reservoir storage.

U.S. Weather Bureau and TVA Agreement. Recognition of the value of quantitative precipitation forecasts is evidenced by a joint agreement in the field of hydrometeorology between the TVA and the U.S. Weather Bureau made in 1940. This provides for utilizing both meteorology and hydrology in the forecasting of precipitation and weather conditions and the determination of runoff and stream flow as a basis for river and reservoir stage and volume predictions and for dispatching water through the TVA reservoir system. This agreement between the two Federal agencies illustrates the practical value of hydrometeorology in the operation of water projects. The TVA is the first large water-control project to make definite use of hydrometeorology in its operating program.

The watershed of the Tennessee River and its tributaries includes an area of 40,910 sq miles. Within this watershed are 28 reservoirs, 21 of which are units in the TVA system of multipurpose reservoirs operating primarily for flood control, power development, and navigation. All the reservoirs must be considered in operating the TVA system. In such a large reservoir system, the dispatching of water from reservoir to reservoir is an intricate and complex process.

Under the joint agreement, techniques have been developed for the work that meteorologists on the one hand and hydrologists on the other contribute to the operation of the water-project system. These procedures have been proved through daily use for several years and may therefore be considered applicable to the operation of other water projects.

METEOROLOGY IN WATER PROJECT SYSTEM OPERATIONS

Forecasts. Water in streams and in reservoirs, like the weather, is seldom static but is constantly in motion. Water projects are in continuous operation, and the engineers must control and use water 24 hr every day whether stream flows are high or low. The meteorologist, by supplying regular quantitative weather forecasts, can give important and valuable assistance to the engineer in this work of water control. Continuous contact with weather and water is necessary for both meteorologist and hydrological engineer concerned with water-project operation. The meteorologist must keep the hydrologist advised of the state of the weather and the expected precipitation. The hydrologist must keep informed regarding the quantities of water that are flowing in all parts of the watershed and particularly through the reservoir system. The meteorologist should prepare forecasts of expected precipitation and weather conditions on a regular twice-daily schedule with additional forecasts during the season of flood expectancy and during other periods when protracted rainfall or exceptional conditions demand. Teletype communication between the offices of the meteorologist and the engineer is desirable for transmission of forecasts.

Preliminary Forecasts. In order to give the hydrologic engineer engaged on stream and reservoir forecasting and water dispatching an estimate of expected rainfall, the meteorologist should prepare what may be called a *Preliminary Forecast* for transmission to the hydrologic engineer early each morning. This forecast is based on the 1:30 A.M. EST data disseminated by teletype throughout the Weather Bureau offices. The purpose of this forecast is to furnish the engineers with information early enough in the day so that it may be considered in making necessary adjustments to the schedule for the day's use of water throughout the system. An example of such a forecast is the following:

PRELIMINARY FORECAST FOR MARCH 28 AND 29, 1944 TODAY AND TONIGHT

MOSTLY OVERCAST CLOUDINESS WITH INTERMITTENT RAIN TODAY AND TONIGHT WITH INTERMITTENT RAIN, SHOWERS, AND THUNDERSTORMS SLIGHTLY COOLER TODAY, DUE TO THE CLOUDINESS AND PRECIPITATION, AND BECOMING COLDER OVER THE WESTERN THIRD OF THE AREA TONIGHT GENTLE TO FRESH WINDS MAXIMUM TEMPERATURES TODAY RANGING FROM NEAR 55 DEGREES ALONG THE NORTHERN EDGE OF THE AREA UP TO

65 DEGREES ALONG THE SOUTHERN EDGE OF THE AREA MIN TEMPS **TONITE** RANGING FROM 36 DEGREES IN THE EXTREME NORTHWEST UP TO 60 **DEGREES** IN THE EXTREME SOUTHEAST LIGHT INTENSITY TODAY 3 TO 4

WEDNESDAY

BROKEN TO OVERCAST CLOUDINESS, BECOMING **BROKEN** TO SCTD OVER THE WESTERN HALF OF THE AREA AND MOSTLY BROKEN OVER THE EASTERN HALF OF THE AREA BY THE CLOSE OF THE PERIOD SHOWERS AND THUNDERSTORMS ENDING OVER THE EASTERN HALF OF THE AREA, BECOMING COLDER MODERATE TO FRESH WINDS LIGHT INTENSITY 3

PRECIP AMOUNTS DURING THIS FORECAST PERIOD ARE EXPECTED TO AVERAGE FROM 1.20 INCHES TO 1.60 INCHES WITH LOCAL HIGH SPOTS UP TO 3.50 INCHES

GENERALLY FAIR AND COLDER WEATHER IS INDICATED FOR THURSDAY

USWB

Complete Forecast. The *Preliminary Forecast* should be followed by a *Complete Forecast* based on the information from the Weather Bureau teletype system for 7:30 A.M. EST. This forecast is based on a more complete analysis of the data particularly with regard to upper-air conditions. It refines and brings up to date the Preliminary Forecast. The estimated amounts of rainfall that are expected during the forecast period are given in inches for the subdivisions of the main watershed that are important in estimating runoff.

The daily schedule for water-control operations must be prepared by the engineers and based on information received prior to about 9:00 A.M. CWT each day. The Complete Forecast is not prepared by that time but should be submitted by about 11:00 A.M. CWT each day, which is prior to the issuance of daily river and reservoir predictions in bulletin form. After receiving the Complete Forecast, the engineers check their previously prepared predictions for rivers and reservoirs and revise these if such revision is warranted. Usually the Complete Forecast confirms rather than changes the Preliminary Forecast.

The following is an example of a Complete Forecast:

COMPLETE FORECAST FOR MARCH 28 AND 29, 1944

TODAY

MOSTLY CLOUDY WITH RAIN AND SCTD THUNDERSTORMS LIGHT INTENSITY MOSTLY 4 MODERATE WINDS

TONIGHT

CLOUDY WITH RAIN AND SCTD THUNDERSTORMS OVER THE CENTRAL AND EASTERN PORTIONS COLDER OVER THE WESTERN PORTION LITTLE TEMPERATURE CHANGE OVER THE CENTRAL AND EASTERN PORTIONS MODERATE WINDS

WEDNESDAY

MOSTLY CLOUDY AND COLDER WITH OCNL RAIN OVER THE EASTERN PORTIONS, ENDING DURING THE PERIOD LIGHT INTENSITY 4 TO 3 MODERATE TO FRESH WINDS

OUTLOOK FAIR AND MODERATELY COLD THURSDAY

PRECIP AMOUNTS DURING THIS FORECAST PERIOD ARE EXPECTED TO AVERAGE NEAR 2.00 INCHES OVER THE SOUTHERN LIMITS TAPERING DOWNWARD TO NEAR 1.20 INCHES

OVER THE NORTHERN LIMITS, EXCEPT OVER THE WESTERN SECTION, WHERE AMOUNTS
ARE EXPECTED TO AVERAGE BETWEEN .50 AND .80 OF AN INCH

SOME LOCAL HIGH SPOTS UP TO 3.00 INCHES MAY BE EXPECTED OVER THE SOUTHERN
SECTIONS

CHATT 54–58 KX 52–58 MEM 41–52 NASH 43–41 KY 36–49 FONT 51–58

USWB

Light intensities are included in both the Preliminary and Complete Forecasts.
These have an influence on the utilization of water, since the amount of power needed
to supply lighting varies with the intensity of light and affects the quantity of water
that is used for power production. Where power is produced by steam plants, light
intensities are equally important. Because there is no standard scale for light-
intensity designation, the following scale has been adopted for use in the Tennessee
Valley:

Light intensity 1.......... Artificial lighting unnecessary
Light intensity 2.......... Artificial lighting necessary only in dark rooms
Light intensity 3.......... Artificial lighting necessary except in best lighted rooms
Light intensity 4.......... Artificial lighting necessary in all rooms

In the field of light intensity, there is a need for the adoption of uniform standards of
measurement and for the collection of continuous records of actual intensities.

Temperature predictions are made in both Preliminary and Complete Forecasts.
Temperatures are important in estimating runoff and, during the winter season, are
useful for protection of construction work from freezing damage. Temperatures
also influence electric power loads where electricity is used for air-conditioning cooling
in summer and for heating in winter. In the Complete Forecast, temperatures for
four large cities and two dam-construction projects are given just above the signature.
The first figure is the predicted minimum for the coming night, and the second figure
is the maximum for the following day.

The operation and maintenance of overhead lines for power distribution and communications are affected by winds, sleet, and snow. Navigation and recreation may
also be affected to some extent by these factors. The forecasts should therefore
include this type of information.

In mountainous regions, the observed and predicted elevations of the freezing line
are valuable in estimating runoff from snow.

Supplementary Flood-season Forecasts. During the season of the year when
past records show that floods are most likely to occur, it is necessary to maintain
closer contact with weather and river and reservoir conditions than during other
months of the year. During the flood season, the amount of rainfall, the continuance
or stopping of the rain, and the areal location of the rain may require changes in the
plans for controlling flood water. At such times, these plans are subject to continuous
check and possible change. In the Tennessee Valley, the flood season extends from
about Dec. 1 to May 1, and during that time the meteorologists regularly make a
night forecast, designated as a *Supplementary Forecast*. This is transmitted about
8:00 P.M. CWT and is based on the teletype data for 1:30 P.M. EST. These forecasts
may also be required during summer flood periods such as occurred in August, 1940,
in the Tennessee Valley when a tropical hurricane in August produced runoff of flood
magnitude.

The form of the Supplementary Forecast follows generally that of the *Complete
Forecast* but omits light intensities. A typical Supplementary Forecast is as follows:

SUPPLEMENTARY FORECAST FOR APRIL 14–15, 1944

TONIGHT

INCREASING CLOUDINESS AND WARMER WITH SHOWERS AND SCATTERED THUNDER-
STORMS SPREADING OVER MOST OF AREA BY MORNING.   MIN TEMPS 52 TO 62.   MDT
TO OCNLY STRONG WINDS.

SATURDAY

MOSTLY CLOUDY AND CONTINUED MILD, EXCEPT BECOMING COOLER WESTERN PORTIONS.
SHOWERS AND SCATTERED THUNDERSTORMS.   MAX TEMPS 68 TO 78.   MDT TO OCNLY
STRONG WINDS

SATURDAY NIGHT

CONSIDERABLE CLOUDINESS WITH SHOWERS AND SCATTERED THUNDERSTORMS OVER
EASTERN TWO-THIRDS OF AREA.   COOLER WESTERN HALF AND BECOMING COOLER
OVER MOST OF EASTERN HALF BY MORNING.   GENTLE TO FRESH WINDS.

PRECIP AMOUNTS ARE EXPECTED TO BE EXTREMELY VARIABLE DUE TO THUNDERSTORM
ACTIVITY.   LOCAL HIGH SPOTS OF 2 TO 3 INCHES ARE LIKELY.

PRESENT ESTIMATE OF AVERAGE AMOUNTS DURING NEXT 36 HOURS 1.00 TO 1.25 INCHES
OVER AREA, EXCEPT UNDER 1.00 NORTHEASTERN SECTION AND 1.25 TO 1.75 SOUTH-
EASTERN SECTION AND SOUTHERN PORTION OF EAST CENTRAL SECTION.   MOST OF THIS
PRECIP IS EXPECTED TO OCCUR DURING THE NEXT 24 HOURS.

INDICATIONS ARE THAT THE OHIO VALLEY WILL ALSO RECEIVE SOME SUBSTANTIAL
FALLS DURING THE NEXT 36 HOURS

USWB

In a water-project system as large as that of the TVA where tremendous volumes of water must be controlled in flood times, these Supplementary Forecasts are essential to satisfactory water control.   The flood situation may change almost hourly depending on rainfall and the outlook for rainfall.   At such times, the meteorologist has a heavy responsibility to keep the engineers advised regarding the state of the weather and probable quantitative rainfall.   To the engineer, it is important to know how much additional rainfall is to be expected at any time and to know when the storm will end.   If he can receive reliable meteorological information, then he can revise his estimates of flood-water volumes in accordance with the latest forecasts of expected rainfall, and the actual flood waters may be safely dispatched through the reservoir system with a minimum of damage.   Any uncertainty in the probable amount of rainfall requires that a corresponding reserve in reservoir storage space be maintained and may result in not making full use of the storage space to reduce the flood crest. An example of this occurred in December, 1942, when the volume of water discharged at Chickamauga Reservoir near Chattanooga was increased in expectancy of additional heavy rain that did not materialize.

Extended Forecasts.   The regular forecasts ordinarily cover periods of 36 hr in advance.   Knowledge of expected rainfall for longer periods is much to be desired, but the science of meteorology does not as yet provide a satisfactory basis for long-range quantitative forecasts.   It is, however, helpful, even if quantities may not be stated, to know what the general prospects are for weather conditions for several days in advance.   To fulfill this need, extended forecasts for a 5-day period are made. An example of such a forecast is the following:

EXTENDED FORECAST FOR APRIL 15 TO 19, INCLUSIVE

PRECIP IS EXPECTED TO BE HEAVY AND TEMPS ARE EXPECTED TO AVERAGE NEAR NORMAL

THE NORMAL MEAN TEMP FOR THIS PERIOD FOR YOUR AREA IS ABOUT 58 DEGREES

USWB

The Complete Forecasts contain, in addition to quantitative forecasts for 36 hr, an outlook forecast for the day following the forecast period. The outlook forecasts are generalized synoptic summaries and are not specific as to quantity. They do, however, furnish information on conditions beyond that for which quantitative forecasts are made. Illustrating such a forecast is the following:

OUTLOOK A DISTURBANCE NOW CENTERED OVER WYOMING IS EXPECTED TO MOVE EASTWARD WITH CHANCE OF LIGHT PRECIPITATION DEVELOPING OVER THE WESTERN PORTION THURSDAY NIGHT OR FRIDAY PRESENT INDICATIONS ARE THAT THIS DISTURBANCE WILL MOVE EASTWARD OVER THE NORTHERN PORTION OF THE COUNTRY AND PRODUCE ONLY LIGHT AMOUNTS OVER YOUR AREA.

Weather Maps. It is desirable that the water-control engineers be furnished a copy of the weather map for 7.30 A.M. EST upon which the Complete Forecast is made. On a large project, where the offices of the meteorologists and engineers should necessarily be near one another, this presents no obstacles. On projects remote from the meteorological-forecasting office, the furnishing of the map may not be feasible. Having the weather map enables the engineer to interpret the forecast better and to visualize storm movements. He thereby gains a conception of the speed of passage of storms which is desirable in estimating the time of occurrence of stream flow.

Personal Conferences. In addition to the forecasts that are transmitted by teletype, personal conferences between the meteorologists and the engineers responsible for water control are desirable, particularly during storm periods. The benefits from such conferences are mutual to the meteorologist and engineer. No matter how carefully worded or in how much detail the forecasts may be, personal interpretation by the meteorologist of prevailing weather conditions has a material value in conveying to the engineers the storm possibilities that the weather maps and charts indicate. Only through personal contact can the degree of certainty felt by the forecaster be conveyed to the hydrologist, thereby expanding or narrowing his limit of operations.

The meteorologist gains an understanding of the uses to which his forecasts are put and the value of these in the operation of the water project. He learns the importance of accurate and courageous forecasts and sees the difficulties that arise when forecasts are inaccurate and rainfall amounts have been over- or underestimated. As a means of integrating meteorology and hydrology into the operation of water projects and of developing confidence in each other by meteorologists and engineers, personal conferences are invaluable.

Language of Forecasts. Care should be taken in the wording of forecasts so that these may be correctly understood. Language should be simple, direct, and easy of interpretation. It should be sufficiently descriptive to give a clear picture of existing weather conditions, storm movements, and the developments that are expected to occur. Meaningless words and phrases or indefinite generalities should be omitted. Brevity is desirable but not at a sacrifice of clear understanding. A forecaster should cultivate an interesting style of writing forecasts that the reader can follow and understand.

HYDROLOGY IN WATER PROJECT SYSTEM OPERATIONS

It is beyond the scope of this section to treat in detail the hydrological phases of hydrometeorology. However, some understanding of this is desirable by meteorologists as background for forecasts for projects utilizing hydrometeorology. Hydrology also includes collection of meteorological as well as other records for the project area.

Rainfall and Stream Gauges. The engineers must gather, at least daily, sufficiently complete information regarding rainfall and stream flow throughout the project drainage basin so that they can estimate with considerable exactness the volume of water that is flowing and will be flowing for several days in all the principal streams throughout the watershed. To accomplish this requires a comprehensive network of rain and stream gauge stations.

This network is basic in the success of the hydrological phases of water-project operation. The extent of the network of rain gauges will vary with the size and topography of the drainage area. In general, in mountainous regions where variations in rainfall are most pronounced, the number of stations per unit area must be greater than in flatter country where orographic influences are not so important. In water-project operations, as in other hydrologic investigations, the smaller the area of interest, the larger the number of rain gauges required per unit area.

In the entire Tennessee Valley, the density of rain gauges is one to each 80 sq miles. In the mountainous eastern sections, the square miles per gauge approximate half of the average; and, in the rolling western portion of the basin at lesser elevation, the density is about 130 sq miles per gauge. The network includes sufficient recording rain gauges so that intensity and duration of storm rainfall are known. Gauges are located with respect to topography so that rainfall at high as well as low altitudes is measured.

Stream-gauging stations are located at a sufficient number of places throughout the drainage area to measure the volumes of actual runoff from main watersheds and from subwatersheds. Stream heights are recorded by automatic continuous recorders so that heights may be read from the charts by local observers at whatever time intervals may be desired.

In an area of considerable size, such as that of the Tennessee Valley, the complete network of rainfall and stream gauges necessarily contains so many stations that time does not permit for consideration of data from all of them each day during current forecasting and prediction work. Consequently, on that project, daily reports are obtained from selected and representative key rainfall- and stream-gauge stations that provide sufficient data on amounts and intensities of actual rainfall and the heights of important streams for current day-to-day predictions. Criteria are established, based on amounts of rainfall or heights of stream, whereby additional stations report in the event of heavy rainfall and during floods, thus providing ample coverage to give complete data on rainfall and stream flows at all times. Periodic checks of storms as they occur are made to determine whether key stations should be changed, eliminated, or increased in number.

The complete network of rainfall and stream-flow gauges provides information for the preparation and keeping up to date of an office manual of charts and diagrams covering hydrologic factors that the engineers use in river and reservoir forecasting. The preparation of this manual and improving the charts and diagrams as additional data become available require constant study of storms and floods that occur over the watershed. From these studies are developed basic hydrological relations among rainfall, surface and ground-water runoff, and water travel for each particular main and subwatershed which enable the hydrologist to estimate both the amount and the concentration of runoff at desired points from both observed and predicted rainfall.

Communication of Reports. The reporting of rainfall and streamflow observations from a large area, daily or more often, is essential, for stream and reservoir prediction depends upon this information and can be made only when this information is received. Reports should be sent by telephone or telegraph either to a central forecasting office or to intermediate offices. Observers in the vicinity of the central office report data direct to that office. Observers in the more remote portions of the area report to intermediate offices that relay the information to the central office. This system facilitates reporting and is efficient and economical.

Radio Gauges. Communication systems that depend upon overhead wires are likely to be interrupted during severe storms that produce flood conditions. There are also areas where there are no telephone or telegraph facilities. To obtain reports from these localities and to provide information during storms when wires may be down, automatic radio rain gauges and radio stream gauges have been developed by Tennessee Valley Authority engineers. The radio rain gauge collects rainfall in accordance with standard practice, and at intervals the amount of rainfall in inches that has been caught by the gauge is broadcast by a modified Morse-code signal. Similarly, radio stream gauges transmit heights of important streams. The broadcast signals are received at a forecasting office usually within 50 miles of each gauge. Reception is either manual or automatic by a tape recorder. A broadcast interval of 2 hr has been found to be convenient and satisfactory, for this supplies sufficient data and permits several gauges in one watershed to broadcast using the same radio frequency.

Forecasting River Discharge and Reservoir Stages. The reporting of rainfall and stream flow from key stations each morning is followed by immediate analysis of the data by experienced engineers with specialized training in hydrology. The volume of water that is expected to run off from each watershed or contributing area is estimated on the basis of reports of rain that has actually fallen. Runoff is also estimated for rain that has not yet fallen but is expected on the basis of forecasts by the meteorologist. The estimation of runoff from both rain that has fallen and that which is to be expected is translated into stream flow throughout the watershed and becomes the basis for determination of how the reservoir system shall be operated to accommodate the runoff of water actually on the ground and that which is yet in the clouds. Integrated with this in the daily schedule for water control and dispatching must be the requirements of the system for water for electric power, navigation, or other use.

Daily Cooperative Bulletins. The culmination of the daily meteorological forecasts and the hydrological predictions of river and reservoir stages and flows are expressed for the Tennessee River Basin by two bulletins issued daily under the joint name of the U.S. Weather Bureau and the Tennessee Valley Authority. Figures 1 and 2 show the front and back of one of these bulletins, which is designated the *Daily River Bulletin*. This bulletin is distributed to those directly concerned with the operation of the TVA reservoir system and with the fluctuations of river and reservoir levels. This includes TVA operating personnel, industrialists, and others having plants or other interests along the reservoirs.

The second bulletin is a large-size post card designated the *Daily Navigation Bulletin*. This is distributed generally to the public and is particularly aimed to supply navigation interests with river information and predictions.

Operation of Reservoir System. An illustrative example of the coordination of meteorology and hydrology in the operation of the TVA reservoir system for reduction of flood heights is that of the storm of mid-August, 1940. In this case, a tropical storm of hurricane intensity moved inland in the vicinity of Charleston,

KNOXVILLE, TENNESSEE   COOPERATIVE FORECAST SERVICE   SATURDAY
U. S. WEATHER BUREAU   Tennessee River Basin   APRIL 15, 1944

                              DAILY RIVER BULLETIN   TENNESSEE VALLEY AUTHORITY
                     REPORTED RIVER ELEVATIONS AND RAINFALL

	STATION	Miles Above Mouth	Top of Gates or Flood Stage	12 P.M.-12 P.M. Discharge	Elevation	Δ Change 24 hrs.	6 A.M. RAINFALL 24 hrs.	Month to date
	TENNESSEE RIVER							
1	Knoxville	648	818		814.0	-0.2	.30	2.51
2	FORT LOUDOUN RESERVOIR	602	815	14,300	813.81	-0.17	.48	2.39
	Fort Loudoun Lower Lock		758		742.4	-1.5		
3	WATTS BAR RESERVOIR	530	745	47,200	741.18	-0.36	.76	3.27
4	CHICKAMAUGA RESERVOIR	471	685	66,300	680.66	-0.22	.62	3.21
5	Chattanooga	464	651		636.1	-3.8	.56	3.81
6	HALES BAR RESERVOIR	431	629 (e)	77,000	631.93	-1.65	.43	3.02
7	GUNTERSVILLE RESERVOIR	349	595	89,200	594.78	+0.28	.20	3.25
8	Decatur †	305	559		556.3	-0.2	.28	2.34
9	WHEELER RESERVOIR	275	556	131,300	555.86	-0.10	.31	2.23
10	WILSON RESERVOIR	259	508	131,300	507.72	+0.20	.08	2.60
11	Florence †	257	419		-			1.49
12	PICKWICK RESERVOIR	207	418	144,800	414.35	-0.37	.20	2.16
	Pickwick Lower Lock		370		375.4	-1.5		
13	Savannah	190	370	144,000	368.6	-0.7	.03	3.21
14	Perryville	135	357		355.0	+0.9	.02	1.67
15	Johnsonville	96	352	166,000	344.3	+1.6	.47	3.12
16	Danville †	78	342		-			
17	KENTUCKY RESERVOIR	22	375		329.4	+3.0	.72	3.72
	Gilbertsville		316	148,000	323.8	+2.8		
18	Paducah (USED) †	0	325		-			3.94
	Average Rain Above Chatt.						.40	2.31
	CLINCH AND HOLSTON							
19	Tazewell †	160	1079	5,480	1066.2	-2.1		
20	Arthur	65	1058	4,000	1050.9	-2.5	.53	2.45
21	NORRIS RESERVOIR	80	1020	4,860	1023.00	+0.60	.49	2.29
22	Gate City †	9	1205	2,100	1201.9	-0.9		
23	Kingsport	6	1191	6,800	1188.0	-1.3	.24	1.40
24	CHEROKEE RESERVOIR	52	1075	4,752	1069.88	+0.59	.29	1.86
	FRENCH BROAD AND LITTLE TENNESSEE							
25	Bent Creek †	14	1030	3,980	1021.2	-1.2		
26	Newport	78	1018	4,800	1015.3	-0.3	.33	.85
27	DOUGLAS RESERVOIR	32	1002	3,519	996.45	+0.67	.30	2.08
28	GLENVILLE RESERVOIR	60	3100	0	3081.09	+0.28	.46	2.68
29	NANTAHALA RESERVOIR	22	2890	0	2880.31	+0.69	.51	3.00
30	SANTEETLAH RESERVOIR	10	1817	593	1816.14	+0.11	.65	3.02
31	FONTANA RESERVOIR	61		6,500	1305.15	+4.44	.50	2.98
32	CHEOAH RESERVOIR	51	1154	6,429	1153.7	+0.1	.42	3.32
33	CALDERWOOD RESERVOIR	44	965	6,934	965.0	+0.7	.38	2.57
	HIWASSEE AND OCOEE							
34	NOTTELY RESERVOIR	24	1780	680	1777.73	-0.04	.45	2.01
35	CHATUGE RESERVOIR	121	1928	680	1925.77	-0.02	.53	2.24
36	HIWASSEE RESERVOIR	76	1526	2,137	1525.20	+0.43	.43	2.84
37	APALACHIA RESERVOIR	66	1280	1,990	1279.65	+0.39	.56	2.24
38	BLUE RIDGE RESERVOIR	53	1690	1,307	1687.86	-0.26	.49	2.49
39	OCOEE NO. 3 RESERVOIR	29	1435	2,322	1435.16	+1.01	.35	1.60
40	OCOEE NO. 1 RESERVOIR	12	831	2,755	827.1	-0.5	.36	1.69
	ELK, DUCK, AND CANEY FORK							
41	Prospect †	42	587	7,200	573.5	-1.9	.50	-
42	Hurricane Mills †	20	379	11,400	372.0	-3.0		
43	GREAT FALLS RESERVOIR	91	805	7,175	805.92	+0.02	.34	1.55

Δ River observations are for 12:00 p. m. C. W. T. at the end of the previous day except as noted.
† Elevation and discharge are at about 6:00 a. m. C. W. T. today.   (e) Estimated.

FIG. 1.

PREDICTED RIVER FLOWS AND ELEVATIONS

SATURDAY
APRIL 15, 1944

Predicted elevations and discharges are based on stream flows and rainfall reported up to 6:00 a. m. of the date of issue, taking into account current water control operations. Elevation forecasts are for 12:00 p. m. C. W. T. at the end of the day and flows are average 12 p. m. to 12 p. m. in thousand cubic feet per second.

	STATION	April 15 Inflow	Discharge	Elevation	April 16 Inflow	Discharge	Elevation	April 17 Inflow	Discharge	Elevation
1	Knoxville		8.5	813.6		4.0	813.0		4.0	813.0
2	FORT LOUDOUN R.	9.0	11.5	813.5	5.0	9.0	813.0	5.5	5.5	813.0
3	WATTS BAR RES.	35.5	38.0	741.1	35.5	40.0	740.8	33.0	35.0	740.7
4	CHICKAM'GA RES.	51.0	52.0	680.6	52.5	55.0	680.4	44.0	50.0	680.1
5	Chattanooga			635.8			635.8			634.3
6	HALES BAR RES.	55.0	56.0	631.6	58.0	58.0	631.6	52.0	54.0	631.0
7	GUNT'RSV'LE RES.	66.0	75.0	594.8	65.0	65.0	594.7	59.0	60.0	594.7
8	Decatur			556.2			556.0			556.0
9	WHEELER RES.	99.0	108.0	555.9	83.0	90.0	555.7	75.0	75.0	555.7
10	WILSON RES.	110.0	110.0	507.5	92.0	92.0	507.5	76.0	76.0	507.5
11	Florence			416.0			415.4			413.5
12	PICKWICK RES.	113.0	123.0	413.9	94.0	115.0	412.9	78.0	95.0	412.0
13	Savannah		138.0	365.8		128.0	364.6		119.0	360.3
15	Johnsonville		164.0	344.3		152.0	343.5		137.0	342.8
16	Danville			340.1			340.4			340.2
17	KENTUCKY RES.			331.2			332.1			332.5
	Gilbertsville, Ky.		156.0	325.8		158.0	327.2		150.0	328.1
18	Paducah (USED)			324.3			325.7			326.6
21	NORRIS RES.	10.0	8.4	1023.1	7.6	12.0	1022.8	6.0	12.0	1022.5
24	CHEROKEE RES.	10.0	0	1070.6	8.2	0	1071.1	7.1	3.4	1071.4
27	DOUGLAS RES.	11.4	0	997.2	11.5	0	998.0	10.9	3.9	998.5
31	FONTANA RES.		6.7	1307.8		6.2	1306.0		5.5	1305.0
34	NOTTELY RES.	0.75	0.67	1777.8	0.70	0.71	1777.8	0.62	0.68	1777.7
35	CHATUGE RES.	0.80	0.61	1925.8	0.75	0.72	1925.8	0.70	0.73	1925.8
36	HIWASSEE RES.	3.40	1.40	1425.8	3.30	1.60	1426.4	3.20	3.00	1426.4
37	APALACHIA RES.	1.50	2.20	1278.4	1.70	1.10	1279.5	3.10	3.00	1279.6
38	BLUE RIDGE RES.	1.20	0.40	1688.4	1.10	0	1689.0	1.00	1.00	1689.0
39	OCOEE NO. 3 RES.	2.50	2.85	1434.0	1.60	1.60	1434.0	2.00	2.00	1434.0
40	OCOEE NO. 1 RES.	3.00	2.65	827.5	1.70	2.40	826.8	2.05	2.20	826.6
43	GREAT FALLS RES.	5.30	5.30	805.0	4.55	4.55	805.0	4.30	4.30	805.0

RIVER NOTICE--The release at Pickwick Dam is expected to be reduced from day to day. Barring heavy rains, probable crests at Lower Tennessee River stations are:

Johnsonville   - 344.3 today
Danville   - 340.5 April 17
Kentucky Res. - 332.5 April 18

WEATHER FORECAST FOR TENNESSEE VALLEY

Variable cloudiness this afternoon, tonight, and Sunday. Becoming cooler over the western half of the area tonight and over the eastern half of the area Sunday.

U. S. WEATHER BUREAU

FIG. 2.

S.C., and produced floods of major proportions during the period Aug. 8 to 15 in the mountainous streams of eastern Tennessee, western North Carolina, and southern Virginia, as the decaying storm center moved through that region.   Centers of high rainfall reached from 13 to 15 in. in the mountainous portions of the Tennessee Valley, the rain falling within a period of from 40 to 50 hr.   Predictions of rainfall for this storm were made by the U. S. Weather Bureau for the TVA.   As the rainfall amounts and upstream river stages were reported to the central river forecasting office, estimates of flow were made.   The flood originated in the tributary streams to the Tennessee River.   Determinations were made of the volume and time when the flood water would reach Chickamauga Reservoir immediately upstream from the city of Chattanooga, Tenn.   With this advance knowledge, the engineers were able to increase the volume of water discharges from Chickamauga Reservoir about 4 days ahead of the time that the flood reached the reservoir.   The discharge was restricted to a quantity that would cause only minor damage through and below the city of Chattanooga.   Thus storage space was provided in the reservoir for the flood water when it finally arrived.   As a result of this coordinated operation, the flood height at Chattanooga was reduced 7 ft from what it would have been under natural conditions.

In a similar operation, the same flood was carried through reservoirs in the system downstream from Chattanooga and regulated so that the flood height along the lower 200 miles of the Tennessee River was reduced by nearly 12 ft below that which would have occurred without the regulation.   Since the flood occurred during the crop season, this flood-height reduction prevented a considerable crop loss that would otherwise have resulted.

In late December, 1942, a flood occurred that was general over the Tennessee Valley.   This flood was forecast by the Weather Bureau, and the forecasts were taken into account by the hydraulic engineers in the operation of the TVA reservoirs. As a result of the regulation, the stages in Chattanooga were reduced by about 4 ft from what they would have been under natural conditions, and damages estimated at approximately $1,000,000 were prevented.   This illustrates the importance and the monetary value that results from water-control operations.

CRITERIA FOR WATER PROJECT METEOROLOGICAL FORECASTS

Area Forecast Concept.   The meteorologist making forecasts for an operating water-control project must approach the problem with the entire drainage basin in mind.   Forecasts must be made, not for point locations, but for areas comprising significant subdivisions of the drainage basin.   In converting predicted rainfall to stream flow, the hydrologist needs expected rainfall amounts for each subdrainage area tributary to each dam under construction or in operation.   The size of the areas will be influenced by the topography and drainage pattern of the forecast region and by the location of the water-storage reservoirs or places where reservoir or river-stage forecasts are to be made by the hydrological engineers.   The areas may vary from a few hundred to several thousand miles.   If a dam is under construction, the drainage area above that dam is the logical forecast area.   If many reservoirs are being operated for water control and utilization, as in the Tennessee Valley, then the entire watershed must be divided into a number of areas, the runoff from each of which is significant in the prediction of inflow into the reservoir system.   The meteorological forecasts should refer to the subdivisions, giving the rainfall forecast for each.

Illustrating the division of a large area into smaller areas, experience has shown that the Tennessee River Basin should be divided into seven areas approximating 6,000 sq miles each for the purpose of forecasting areal distribution of rainfall. Topographical features and the location of the several areas cause rainfall to develop most heavily in certain areas, depending on the direction and path of travel of the

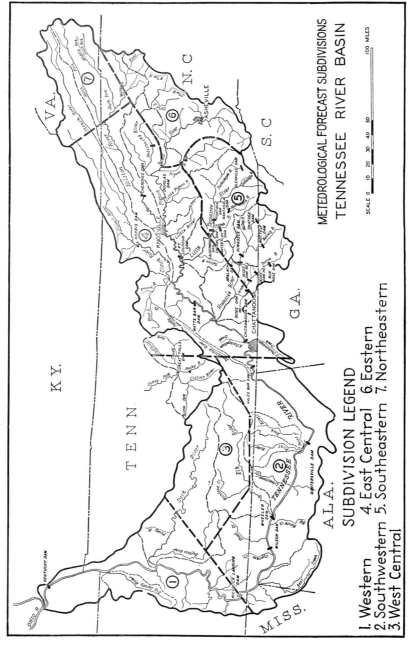

FIG. 3.

storm. Since these same features control the drainage system to some extent, the subdivisions selected to conform to the usual rainfall distribution conform approximately to the logical areas for stream-flow forecasting. The "western section" is subject to rainfall from storms that sometimes give very little rainfall in other sections of the basin. The "southwestern section" is subject to rainfall from storms approaching from the south and southwest, which often diminish rapidly before reaching other sections. The "southeastern section" is subject to heavy rainfall from storms approaching from the south and southwest, which lose intensity before reaching sections farther north and east. The "east central section" is frequently in the path of storms traveling up the main valley. The "eastern and northeastern sections" are generally shielded from storms approaching from the south and southwest but are subject to storms overreaching from the Atlantic.

Knowledge of Forecast Area. An intimate knowledge of the forecast region is a necessary asset to the meteorologist who makes quantitative forecasts for water projects. Particularly is this true where the terrain is mountainous and precipitation is subject to definite orographic influences. The greater the knowledge of the meteorologist as to the topography and its influence on storms as learned from actual experiences, the greater will be the accuracy and value of his forecasts.

Along with knowledge of the region goes knowledge of the storm types that may occur over the forecast areas. These will include seasonal considerations and variations in storm types depending upon locations and areas of interest. For example, in the western part of the Tennessee Valley, summer thunderstorms are of little areal significance. But in the eastern part of the valley where the rugged mountainous terrain reaches elevations above 6,000 ft, storms of this type may cover several hundred square miles and produce storm runoff in volumes that affect the TVA system operations.

Perhaps the day may come when meteorology will be such an exact science that any meteorologist using the techniques of his profession will be able to make satisfactory quantitative forecasts for any location. But that day has not yet come, and until it does, there is no substitute for the lessons of experience gained on the ground in the forecast region. As these lessons accumulate through the years, the value and worth of the individual meteorologist on such regional projects as that of the Tennessee Valley will likewise continue to become greater and greater.

Accuracy of Forecasts. Meteorological forecasts, to be of most significant value hydrologically, must express rainfall expectancies in definite quantitative figures, *i.e.*, inches of rainfall. Because of deficiencies in data and the present status of knowledge, rainfall amounts cannot be forecast with absolute accuracy. However, the meteorologist should have an appreciation of the value and importance of quantitative evaluation and should continually strive to develop judgment, skill, and techniques in order to improve the accuracy of his quantitative estimates.

The degree of accuracy or uncertainty in each particular forecast should be recognized by the meteorologist and made known to the engineer using the forecast so that the latter may take this into account in planning operations. However, the forecast should be as accurate and definite as the known facts permit. Otherwise, forecasts would tend to become broad generalizations of minimum value.

A tendency toward too conservative an outlook on quantitative rainfall may result in difficult situations for the engineer in controlling water through a system of reservoirs or even in a single reservoir. Precipitation of greater amounts than those forecast results in greater quantities of runoff than those for which the control operations are planned. In that case, it might be that sufficient storage volume for the actual runoff that does occur would not have been provided. The increased runoff would then have to be cared for on an emergency basis with the possibility that the maximum degree of flood control would not be accomplished, and damages might result that could otherwise have been prevented. Conversely, if the amount of rainfall is overestimated, large volumes of water may be released from a reservoir or a system of reservoirs in advance of flood water anticipated on the basis of predicted rainfall. If the rain fails to materialize and the flood-water volume is less than that estimated, valuable water may have been wasted, and damages may result that need not have happened.

Courage is required actually to forecast large amounts of rain even though the calculations and physical conditions indicate the probability of such amounts. On May 10, 1943, the U.S. Weather Bureau meteorologist made the following forecast for rain in the Tennessee Valley near Paducah, Kentucky:

HIGH SPOTS UP TO 6.00 INCHES OVER THIS AREA WITH THE PASSAGE OF THE DISTURBANCE

Courage was required to predict that much rain, but the actual rainfall at Paducah was 5.65 in., which justified the prediction.

Timing. Timing as well as quantity of rainfall is important hydrologically. The direction and rate of storm travel are reflected in the runoff pattern that results from the storm. Particularly valuable is the prediction of the cessation of rain. On small drainage areas, the peak storm flows occur soon after the end of heavy rainfall, and on large areas the timing of stream flows is influenced by the timing of the storm rainfall. During a flood with rains continuing and the flood developing, in order to provide for future runoff, water must be dispatched through storage reservoirs in volumes that are directly related to the volume of water yet to be expected. Forecasting with reasonable accuracy the time that the rain will end permits flood water to be dispatched to better advantage through a reservoir system and prevents unnecessary waste of water.

Field for Future Development. The present state of knowledge in both meteorology and hydrology provides ample opportunity for future developments. In meteorology, increase in accuracy of current forecasts is desirable. Also important is the development of long-range forecasting. The 5-day generalized forecasts now being made by the U.S. Weather Bureau have considerable value to engineers operating water projects as indicators of storm conditions that lie ahead. However, plans must be made for operating a large system or even one reservoir for weeks and months in advance. These plans must be sufficiently flexible to be adjustable according to current water conditions and changes. The more nearly that these water conditions can be predicted, the more efficient will be the utilization and conservation of water. Long-range weather forecasts are necessary if these improvements in planning the operation of water projects are to be accomplished.

Personnel. The field of hydrometeorology, as applied to water project operation, does not admit of mediocrity. The tremendous forces of nature that must be dealt with by both meteorologists and hydrologists are such as to require personnel with specialized technical education supplemented by experience. Both meteorological forecasting and hydrologic predictions must be made rapidly and accurately. Time does not permit mistakes to be made and corrected after thorough checking, as is the case with historical meteorological and hydrological studies. Satisfactory personnel must have keen alert minds, must be able to perform work rapidly and accurately under pressure, and must have courage and confidence in forecasts and predictions. They should be forward-thinking and capable of research for improving techniques and procedures. Experience has shown that these qualifications are essential to success in the field of hydrometeorology. Scientific personnel of considerable ability may fail in hydrometeorology because of not meeting the requirements stated.

QUANTITATIVE DETERMINATION OF MAXIMUM RAINFALL

By A. K. Showalter

The methods discussed here have been developed over the past 6 years, largely by the Hydrometeorological Section of the U.S. Weather Bureau. The investigations are being conducted for and under an allotment of funds from the Engineer Department, Corps of Engineers, War Department.

The problem of determining the maximum possible rainfall over any selected river basin logically divides into four major steps: (1) historic, (2) analytic, (3) geometric, and (4) dynamic.

HISTORIC BACKGROUND

Hydrologic. The natural beginning of any survey of flood possibility is the organization of all available information regarding flood occurrences. In most areas, it is possible to find rather long records of river stages for periods throughout which the hydrologic character of the river basin has remained unchanged. In recent years, construction of flood-control works, erosion control, land improvements, and changes in percentage of cultivated areas have resulted in changing the effects of rainfall on subsequent river stages. If the investigator is fully cognizant of such variations, he can establish comparative flood stages for various years of record and then determine the seasonal and annual frequency of stages above a datum flood level. The next important step is to establish the dates of all major floods and classify them as summer or winter floods, floods in which snow melt played an important part, those which occurred on saturated soil, or those which may have occurred with frozen soil.

Hydroclimatic. This includes a determination of rainfall frequencies for selected durations and a classification thereof by seasons. There follows the organization of data for all major rainstorms, snowfalls, snow-depth accumulations, and snow-melt periods.

ANALYTIC

Meteorologic. *Preliminary Isobaric Analysis.* The first step is to prepare a preliminary survey of isobaric patterns favorable for heavy rains of the local and general type in summer and in winter. With these patterns as a guide, a selection is made of storm periods to be subjected to more detailed analysis. In the final list of this type, it is necessary to include all the major rainstorms and also recent storms for which adequate upper-air data are available.

Identification of Major Storm Types. After analyses are completed for 10 or more of the major storms, it generally becomes obvious that there are no more than four or five basic types of synoptic patterns that can produce general heavy rains over the project basin. Once these major types are identified, it is advisable to reexamine the preliminary analyses for all recent storms and to select several of each type for detailed upper-air and surface analyses. It is desirable to select such storms from the last 2 or 3 years, for which period there is available a dense network of autographic rainfall stations.

Controlling Meteorological Phenomena. Then follows a detailed analysis of each of the important controlling meteorological phenomena. First in importance is the determination of the prevailing source direction of moisture for each of the major types. The rates of inflow of the moist air and the variation of the rates of inflow throughout the storm period are then evaluated. The next obvious step is to compute the moisture charge carried by the inflow winds. All synoptic types are then re-examined in order to determine the synoptic patterns or combinations of patterns that are most favorable for high rates of inflow, high moisture charge, and prolonged durations.

Hydrologic. The hydrologic analyses must be coordinated with the meteorologic analyses of the storms.

Preliminary Isohyetal Maps. The first step is the preparation of preliminary isohyetal maps for the areas of outstanding rainfall. Basic data for this purpose can usually be obtained from printed publications of the Weather Bureau.

Mass Curves of Rainfall. For the storms that are to be subjected to detailed meteorological analyses, a more accurate determination of the time distribution of rainfall throughout the entire storm area is required. During most of the period of record, comparatively few recording rain gauges have existed; and it is therefore necessary, by coordinating the hydrologic and meteorologic analyses, to identify the synoptic features most directly responsible for the heavy rainfall and to construct mass curves of rainfall for all nonrecording gauges in the area. A mass curve of rainfall is merely an accumulative plot of rainfall against time. The original data from cooperative observers establish points on these curves at 24-hr intervals; and, if beginnings and endings have been reported, at least one or two additional points may be established each 24 hr. If the meteorological analysts can identify various bursts of rain as being associated with frontal or wave activity and also establish the movement and timing of these events within the vicinity of the cooperative stations, they can make a good estimate of the most probable trend of the mass curves of rainfall during the periods between observations.

Time-area-depth Studies. After the mass curves of rainfall are prepared, isohyetal maps for the total storm or for increment periods of the storm are then plotted and planimetered. It is then possible to prepare area-depth curves and duration-depth curves for selected subbasins or unit areas.

GEOMETRIC

Concentration Time of Basin. This phase of the study is a simple and direct approach toward determining the importance of the relative dimensions of the basin and their effect upon concentrations of rainfall and assembling of flood waters from the tributary channels. Obviously, a symmetrical basin with very steep tributaries of approximately equal area, slope, and length will assemble flood waters in the main stream more rapidly than an elongated basin of relatively flat slope. It is very important for the meteorological analysts to know the approximate concentration time of the basin in order to fix the critical duration of rainfall and in order to distribute the rainfall through time and over area.

Relative Orientation of Basin Axis and Storm Axes. The second phase of the geometric problem is the determination of the relative orientation of the major axis of the river basin with respect to the axis of the moist tongue or tongues associated with each storm type. The synoptic analyses have already determined which source directions are most favorable for heavy rains in that general locality, but it may be necessary to test several storm types in order to determine the synoptic type that is capable of making the most efficient rainfall contribution to the specific basin.

Development of Rainfall Formula. *Rectangle of Best Fit.* After the most efficient synoptic storm type has been determined, the use of a theoretical rectangle that closely approximates the shape, size, and orientation of the basin will simplify the computation of theoretical rainfall over the area.

Determination of Basin Constant. Once the rectangle has been selected, the approach is straightforward. The basin constant becomes part of the formula for the theoretical computation of rainfall. The formula is simple and can be stated as follows:

$$R_t = \frac{1}{Y} \int_0^t V W_E \, dt$$

where R_t = average depth of rainfall in time t over area XY, in.
t = duration of storm, hr
V = velocity of inflow, mph
W_E = depth of water released from each unit column, in.
Y = length of basin parallel to direction of inflow

DYNAMIC

Physical Upper Limits. The synoptic analyses identify the important controlling meteorological phenomena, and each of these must be subjected to rigid statistical and theoretical analysis in order to determine the physical upper limits.

The Moisture Charge. The moisture charge of a column of air can be determined in terms of mixing ratio, absolute humidity, or precipitable water, the latter being most useful in computations of storm rainfall. Precipitable water is defined as the equivalent linear depth of water in the column of air if all water vapor is condensed. It is a function of the mixing ratio w and can be determined by graphical or tabular methods.

$$Wp = \int_{p_o}^{p} w \, dp$$

In tropical maritime air, the precipitable water generally varies between 1 and 3 in The amount of moisture that a saturated column of air can contain is conditioned by the temperature of the air by which the water vapor is surrounded, and it is therefore customary to speak of air of a given temperature as having a definite water-vapor capacity. The most important contribution of moisture to the air is by evaporation from the tropical ocean surface, but it is not possible to evaporate moisture into the air if the dew point of the air has a higher temperature than the ocean surface. In other words, if the vapor pressure of the air exceeded that of the ocean surface, condensation would occur on the ocean surface. The temperature of tropical oceans adjacent to the United States rarely exceeds 82°F, which automatically places an upper limit on the amount of moisture that can be added to maritime tropical air masses. The limiting moisture charges are therefore usually described in terms of maximum possible dew points. The maximum possible dew points for the Gulf coast stations may exceed 80°F; but, for stations farther inland and at higher altitudes, the effects of elevation and horizontal mixing demand that lower dew points be considered for the maximum possible.

If the surface maximum dew point is determined, the limiting moisture charge for the column above can be evaluated by assuming that under extreme conditions the column of air might be saturated with a pseudoadiabatic lapse rate from the surface to very high levels. It would not be logical to assume a more stable lapse rate since convection would be resisted, and any steeper lapse rate would decrease the possible moisture charge. For a project basin, therefore, a detailed statistical analysis is generally made in order to determine the logical upper limit for the surface dew point for various storm durations.

Temperature and Temperature Gradients. Intimately connected with the assumption regarding the maximum possible dew point is the maximum surface temperature to be expected in the warm sector of the flood-producing cyclone. This temperature will generally be near the assumed maximum dew point, especially for storms of long duration. The possible temperature gradients at the surface are largely controlled by the synoptic pattern selected, and a special study must be made of each type.

For temperature gradients in the free air, it is usually satisfactory to evaluate the possible gradients of temperature throughout the storm area at the 5-km level, which represents mid-troposphere conditions. The highest temperatures at 5 km will occur above the warm moist air and can be directly evaluated on a pseudoadiabatic chart by extrapolating the surface wet-bulb temperature upward along a pseudoadiabat. This temperature will generally be slightly lower than the maximum 5-km temperature of record for the area, since extremely high temperatures at this level are associated with subsidence in subtropical currents. The lowest temperatures at 5 km occur in maritime polar air masses, and they can also be evaluated by pseudoadiabatic extrapolations upward of the surface temperature of the air or sea over the most northerly source region of such an air mass. For example, the lowest possible sea-surface temperature can be assumed as $0°C$ and, with relative humidities of 50 to 75 per cent, a particle rising adiabatically from the sea surface to the condensation level and pseudoadiabatically from there to 5 km would attain a temperature of approximately $-45°C$. This temperature could subsequently fall by radiational loss of heat to underlying or overlying air masses, but there have been no occurrences of 5-km temperatures below $-50°C$ over the United States. Data on extreme minimum 5-km temperatures are now available for a period of 10 years or more and furnish a good estimate of the minimum possible temperatures to be expected for any area and season at that level.

Pressure Gradient. The lowest possible surface pressure to be expected is a characteristic of the most critical storm type and may vary for different localities. For example, a maximum storm in the Gulf coast area might be associated with a tropical hurricane and therefore extremely low pressures. Pacific coast storms are associated with major occluding systems and may also have extremely low pressures. However, the synoptic pattern associated with summer flash floods over interior sections of eastern United States will probably never have pressures less than 1,000 mb. Available data on maximum surface pressures are of no significance in connection with maximum storms, since all important rain activity is far removed from the centers of outstanding anticyclones. It is therefore necessary to make direct surveys of maximum pressure gradients and maximum wind velocities in the project area in order to evaluate surface rates of inflow.

Interdependence of Phenomena. *Free-air Pressure and Temperature Gradients.* The variations of pressure and pressure gradients with height are conditioned by gradients of density and pressure at the surface and by the temperature lapse rates throughout the storm area. The selection of limiting surface temperatures, pressures, densities, and their gradients is a process of mutual adjustment for consistency with the synoptic pattern most critical for the area. For example, a broad moist tongue associated with a quasi-stationary front in a summer storm might have fairly uniform horizontal temperature distribution and a not very steep pressure gradient. The lack of temperature gradient within this moist tongue would automatically demand that the pressure gradient and probably the wind velocity should decrease with height. However, in the cold cyclone associated with the Pacific coast occlusions, there could be steep gradients of both temperature and pressure from the center of the disturbance out to the heavy rain area, and the wind velocities and possibly the pressure gradient could increase with height. (It will be noted, by reference to the Hydrometeorological Section's *Report* 3 on the Sacramento Basin, that this possibility was investigated with the result that the limiting assumption for a Sacramento Basin storm was a pressure gradient constant with height.)

Another important effect of the interrelations between pressure and temperature is the relative orientation of isobars and isotherms and the resultant changes in wind direction as well as velocity with height. For large-area storms, the most critical conditions are that a northward-moving warm moist tongue have to its west, or left, an air mass that is colder at the same latitude or possesses a steeper lapse rate. This causes the wind to back with height and so distorts the upper-air circulation as to impose a cyclonic curvature on the warm moist air, therefore causing it to undergo horizontal convergence with resultant lift and realization of its potential instability. Heavy rates of precipitation are never associated with potentially stable air, and accumulations of great depths of precipitation require localized lifting and convergence.

Temperature Gradients and Moisture Charge. Rather elementary considerations make it obvious that the maximum rates of inflow and the proper turning of winds with height can best be attained in currents of air possessing marked gradients of temperature along the horizontal as well as along the vertical. Because of the fact that mixing ratio decreases with height, it is desirable to have a compensating effect of wind velocity increasing with height in order to bring into the area the maximum quantity of moisture. However, the rate at which wind velocity can increase with height is conditioned by the decrease in density and the increase in pressure gradient. The rate of decrease of density with height is approximately constant, and therefore this contribution to increasing wind velocity is only slightly variable. The more important contribution is produced by a pressure gradient that increases with height, but this requires that the pressure gradients and the density gradients always be oppositely directed and, hence, that the pressure gradients and temperature gradients be oppositely directed, since high pressure tends to cause high density whereas high temperature tends to cause low density. However, if the moist tongue is characterized by marked horizontal gradients of temperature, a limitation is placed on the potential moisture charge, since, for maximum moisture charge, it would be desirable to have a broad moist tongue characterized by the maximum possible wet-bulb potential temperature throughout.

For proper interpretation of the rates of inflow of moisture, it is necessary to integrate the product of mixing ratio times wind velocity through height. However, in most rain-producing cyclones of major flood importance, there is usually a relatively broad moist tongue with slight horizontal temperature gradient. In such an air mass, the wind velocity decreases with height in direct ratio to the decrease in absolute temperature. If the air mass is quite warm and possesses a pseudo-adiabatic lapse rate, the decrease in gradient wind from the surface to 10,000 ft would not be appreciable. For example, the absolute temperature at the surface would be of the order of 300°A; and, if the lapse rate is approximately half the dry-adiabatic, the decrease in temperature to the 3-km level would be only 15°C, enough to cause a 5 per cent decrease in wind velocity. For a northward-moving current, this will be the order of magnitude of the decrease of wind velocity on the east, or right-hand, side of the moist tongue, while on the west side the temperature contrast between the warm and cold air will tend to increase with height, therefore causing an increase of wind velocity with height. The gradient-wind velocity in the moist tongue will remain constant with height or show a slight increase if the moist tongue is part of a rapidly occluding and developing system. Only a slight error is involved, and the computations are much simplified by assuming wind velocity constant with height, and the surface gradient-wind velocity can then be multiplied directly by the precipitable water in the inflow layer.

The maximum increase in wind velocity with height is probably that consistent with a pressure gradient remaining constant with height, since it is difficult to justify

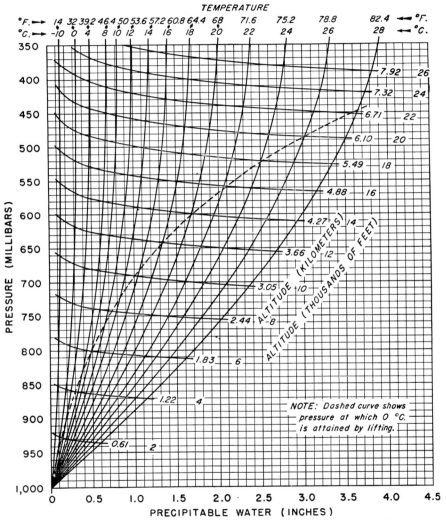

Fig. 1.—Depths of precipitable water in a column of air of given height above 1,000 mb, assuming saturation with a pseudoadiabatic lapse rate for the indicated surface temperatures.

pressure and temperature gradients at the surface that would produce increasing pressure gradients with height. This condition was fully discussed in the Sacramento report previously referred to.

Available Moisture Released as Precipitation. *Effective Precipitable Water.* This is the most difficult phase of the computation, since the quantities of moisture carried by the inflow layer are usually greater than the quantities of moisture that are eventually measured as rainfall at the surface.

A simple and direct approach for straight, parallel flow is to attempt to compute the rates of moisture outflow on the opposite end of the basin, but this is not necessarily reliable if extreme convective activity occurs within the basin. For the maximum possible storm, the safest approach seems to be that which attempts to calculate the greatest possible percentage of moisture that may be released from each unit column

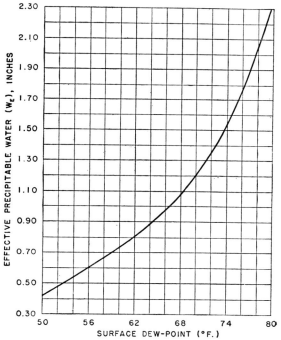

Fɪɢ. 2.—Effective precipitable water as a function of surface dew point-reduced pseudo-adiabatically to 1,000 mb—in a saturated column of air.

of air entering the basin. This was discussed rather fully in *Hydrometeorological Report* 2 for the Ohio River Basin above Pittsburgh. In that report, the concept of effective precipitable water was developed. Briefly, the method involved the assumption that each unit column of air would be subjected to lifting in an ideal convective cell. The moisture charge, depth of inflow layer, and maximum height of convection or lift were assumed to increase with increased surface dew point, and the amount of moisture made available for precipitation was then computed. The depth of inflow was assumed to be one-third of the maximum height of convection, and the layer was assumed to be lifted vertically a distance equivalent to twice its original depth, then undergoing outflow through a depth equivalent to the depth of inflow. By the use of thermodynamic and adapted charts showing precipitable water content (Fig. 1), it was possible to develop a curve expressing effective precipitable water as a function of surface dew point (Fig. 2). The latter chart attempts to represent the maximum amount of moisture that can be realized as precipitation through a convective process.

Loss to Evaporational Cooling. Near active centers of convection, an air mass frequently appears that has a wet-bulb potential temperature lower than any value that can be accounted for by surface advection. In other words, the cold air mass observed at the surface is brought down from aloft and cooled by evaporation of precipitation falling through the air mass. This is the typical cold squall wind associated with local thunderstorms. If this process occurs within an active convective cell, not all the moisture released by lifting of the warm moist air may be realized as precipitation, because some may be reevaporated into the initially unsaturated air mass from upper levels as it descends to form the cold squall wind and in turn helps to force more warm moist air upward. There is need for further research on this point, but the possibility seems to be indicated that any assumption regarding moisture released in ideal convective cells should include the subtraction of moisture through a reverse cell in which moisture is added to the descending and cooling air mass. If the reverse cell is of the same order of magnitude as the ascending portion, the amount of moisture subtracted can be estimated by using the effective precipitable-water curve of Fig. 2.

For example, if the surface wet-bulb temperature is 80°F and saturation exists, Fig. 2 indicates that 2.3 in. of moisture can be released from each unit column of air through convection. However, if to initiate and sustain such convection it is necessary to have superimposed an air mass of lower wet-bulb potential temperature that may absorb moisture from rainfall, cool, and descend, then there could be a net loss of rainfall equivalent to the effective precipitable water specified by the wet-bulb potential temperature of the colder air mass. If the descending air mass had a final surface wet bulb of 65°F, its effective precipitable water would be 0.9 in. This would be roughly the amount necessary to bring it to the ground in a saturated condition. It should be decreased, however, by the initial moisture content of the air, and a further subtraction should be made because the mass of descending air in the convective system may be less than the mass of ascending air and, in addition, all descending air may not reach the ground in a saturated condition. Neglecting these corrections, in a process in which convective instability is realized and the energy for lifting comes through evaporating rainfall into the cooling, descending air, the final quantity of moisture available for precipitation would then be not 2.3 in. but 2.3 minus 0.9 in., or only 1.4 in. available for precipitation.

The lower figure of 1.4 may represent the maximum for local convective storms, whereas in major cyclonic systems, in which the lifting force is a dynamic one of convergence, frontal activity, or orography, the higher figure of 2.3 might be more appropriate. For summer storms in which the source of moisture is maritime tropical air, the limiting wet-bulb potential temperatures in the surface layers will be in the range between 24 and 28°C, and convective instability will be necessary. The limiting lower values of the overlying air mass may range between 14 and 20°C. However, an air mass with wet-bulb potential of 14°C will probably never be found above one containing a wet-bulb potential temperature of 28°C. For specific areas, it is therefore necessary to make detailed studies of the typical seasonal variations of air-mass properties at various heights in order to establish reasonable limitations for the degree of convective instability to be assumed.

Intensity-duration Studies. After the rainfall equation has been established and the limiting values for all factors determined, it is necessary to establish intensity-duration curves for all of the important factors. It is difficult to determine these from purely theoretical considerations, and, up to the present, the technique of determining the duration has been largely through the use of statistical evidence of past behavior.

Wind Velocity and Pressure Gradient. A study of the ratio between maximum 5-min wind velocity and maximum 1- and 24-hr wind movement at stations of long periods of record furnishes reliable information for the expected possible durations of any given wind velocity at a point. Similar studies for adjacent regions and studies of a like nature for pressure gradients must also be made.

Dew Point. For the surface dew point, the curve of dew-point variations through time can also be determined by statistical studies. Reliable information can be obtained by the use of frequency studies based on hourly dew-point data. For example, for the month of June at Shreveport, La., a dew point in excess of 40°F may occur, on the average, over 700 hr, whereas a dew point of 80°F may occur, on the average, less than 1 hr per month. Preliminary studies to date seem to indicate that, in actual storm situations, the variation of dew point through time may be of the same order of magnitude as typical frequency curves for the area.

The latitudinal variations in dew point are significantly related to the latitudinal variations of storm rainfall intensities. For example, high dew points of 80°F have almost the same frequency in northern Iowa as in southern Texas, and intense rates of rainfall over small areas for short periods are therefore possible in both sections. However, the dew points of intermediate values show a much lower frequency and steeper duration curves in northern Iowa than in Texas, and the duration curves of intense rainfall over small areas will therefore be much steeper in Iowa.

The Maximum Possible Storm. *Comparison with Enveloping Values for Immediate Region.* After the theoretical storm has been computed for any selected river basin, it must be compared with the enveloping duration-depth curve from all major storms that have occurred over the basin or within the immediate region of meteorological homogeneity. It has also been the practice to compare the maximum observed dew points in all major storms with their maximum possible values and then to increase the storm rainfall by the ratio of the W_E values appropriate to the dew points. The final duration-depth curve envelops the adjusted values.

There is little justification for, and considerable danger in, the extrapolation of the dynamic features of major storms to their physical upper limits. For example, the maximum temperature gradients at 5 km and also the minimum 5-km temperatures have actually been produced or very closely approached by the major storms of record. Furthermore, any slight change in the temperature or pressure patterns can radically affect the behavior and movement of the parent cyclone. Present investigations suggest that the most important differences in magnitudes of rainfall in the major storms are caused by moisture-charge differences. Since all storms are tested for greater possible moisture charge, there is little security in attempting to test them for greater possible wind movement. That would merely be repeating the computations of the theoretical maximum possible storm.

Comparison with Record United States Storms. As a further check on theoretical computations, it is generally advisable to compare the storm with enveloping isohyets of the major storms of record over the United States, shown as Figs. 3 and 4, two charts expressing the maximum observed values of rainfall for all areas up to 10,000 sq miles and all durations up to 4 days as a family of enveloping isohyets. Figure 3 is based on data for storms caused by or directly associated with tropical hurricanes, and Fig. 4 includes all intense cyclonic storms or families of thunderstorms. Many of the data on Fig. 4 were obtained from unofficial surveys of major flooded areas, and the values in some cases may be greater than actually occurred; but the data have all been carefully checked, and sound engineering practice does not permit ignoring their possible occurrence.

Comparison with Mean Seasonal Values. Another approach to estimating the maximum possible rainfall is through use of enveloping curves of maximum observed percentages of mean seasonal rainfall. Such a curve for point rainfall is shown in Fig. 5. This method appears to furnish a reliable estimate in regions where the rainfall regime is largely seasonal and orographically controlled. In such areas,

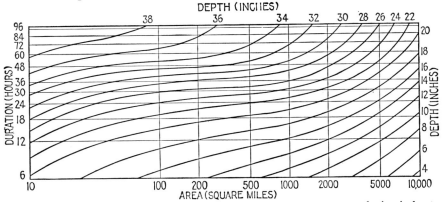

FIG. 3.—Maximum rainfall associated with tropical disturbances: enveloping isohyets of maximum observed depths for indicated areas and durations, Gulf of Mexico and Atlantic coastal regions of the United States.

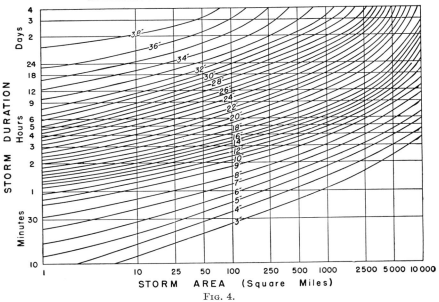

FIG. 4.

individual storms tend to produce isohyetal patterns similar to the mean seasonal patterns, and the percentage of mean seasonal rainfall shows less variation over the basin than any other single phenomenon directly measurable. The percentage of mean seasonal rainfall possible in 24 hr appears to decrease with increasing mean seasonal rainfall and also decreases with increased area. The latter effect is more difficult to evaluate but is the subject of considerable investigation by the United States Engineers in Pacific coastal areas.

Use of Frequency Studies. A number of attempts have been made to evaluate the maximum possible storm for a given area by the use of frequency studies that assume that the 1,000-year storm or some higher frequency is an approach to the maximum possible. However, this method is unreliable, because good records of rainfall for any area do not cover periods greater than 50 years, and at present there is no reliable statistical technique that can predict 1,000 years of behavior by use of data for only 50 years. Furthermore, most frequency studies are based on point-rainfall values, and the frequencies over areas of any selected magnitude may be quite different from the frequency for any point within the area.

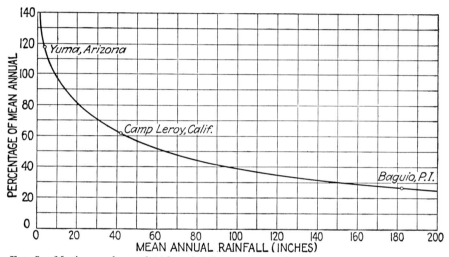

Fig. 5.—Maximum observed 24-hr rainfall, percentage of mean annual—tentative curve.

In areas that have pronounced seasonal-orographic regimes, the combination of factors necessary to produce the maximum possible may be approached more frequently than in areas subjected to cyclonic rainfall. For example, in the Sacramento River Basin of California, the major lifting mechanism is the Sierra Nevada Range, whose characteristics and position remain fixed. Whenever a strong southwest wind blows up these mountains, the basin can experience a storm approaching the maximum possible if the moisture charge of the air is near the maximum possible value. The source region from which the moisture comes is a very large ocean area whose annual and seasonal temperature variations are slight, and it is therefore not possible for any storm to produce air masses with abnormally high moisture charges. For these reasons, in the Sacramento Basin there are likely to be many occurrences exceeding 50 per cent of the maximum possible values within a period of 1,000 years.

On the other hand, in an area such as the Republican River in western Nebraska and Kansas, the occurrence of intense rainfall depends upon the fortuitous combination of a number of phenomena. First of all there must be strong cyclonic activity ideally situated with respect to the area. The moist air must also be of unusually high moisture content, and the surrounding circulation patterns must be such as to produce less than the normal rate of displacement of the cyclonic system.   Furthermore, the frontal activity or source of convergence producing the lift must be exactly and symmetrically superimposed on the river basin.   There was one unusual occurrence of rainfall over the Republican River on May 30–31, 1935, which could not have been anticipated by any frequency studies up to that time.   But even then the rainfall pattern was not strategically symmetrical with the basin's configuration. Furthermore, although this was an amazing storm and may not be approached in magnitude for many years, it still must be considered appreciably less than the maximum possible storm except over the very smallest areas.   Any statistical approach to the determination of the maximum possible storm over the Republican River, therefore, would require the use of a much longer period of record and the selection of a much longer recurrence interval than would similar studies over the Sacramento Basin.

When extremely long records of precipitation from dense networks become available and well organized, it is likely that a more reliable approach to the maximum possible can be obtained through the use of frequency studies.   However, if such studies are ever to be made, it would seem more logical to investigate the recurrence interval of such factors as dew point, moisture charge, rates of inflow, degree of convective instability, magnitudes of convergence, etc.   When the frequency of selected magnitudes of each of these elements has been determined and their interdependence evaluated, a recombination of values of selected frequency should provide a better approach.

Rainfall-computation Formulas.   *Storage—Outflow Negligible.*

$$R_t = \frac{V W_E t}{Y}$$

where symbols are as previously defined.

Storage—Inflow Minus Outflow.

$$R_t = \frac{(V_1 - V_2) W_E t}{Y}$$

where subscript 1 is for inflow, subscript 2 for outflow, and other symbols are as previously defined.

Radial Inflow.

$$R_t = \frac{2 V W_E t}{r}$$

where r is radius, V is velocity normal to circumference, and R is average depth in inches over area πr^2.

Point-rainfall Intensity.

$$I = \frac{V_Z \, \rho_0 (w_0 - w_1)}{7}$$

where I = intensity of rainfall over unit area, in./hr
V_Z = vertical velocity at condensation level, mps
ρ_0 = air density at condensation level, grams/m^3
w_0 = mixing ratio at condensation level, grams/gram
w_1 = mixing ratio at top of convective updraft, grams/gram

Bibliography

The basic theories underlying most of the techniques discussed, with practical illustrations of their application to specific basins, are fully discussed in the published reports of the Hydrometeorological Section of the U.S. Weather Bureau.

1. Maximum Possible Precipitation over the Ompompanoosuc Basin above Union Village, Vermont, March, 1940.
2. Maximum Possible Precipitation over the Ohio River Basin above Pittsburgh, June, 1941.
3. Maximum Possible Precipitation over the Sacramento Basin of California, May, 1942.

Three papers growing out of the section's work are also suggested:

1. SOLOT, S. B.: Computation of Depth of Precipitable Water in a Column of Air, *Monthly Weather Rev.*, **67** (April, 1939).
2. SHOWALTER, A. K., and S. B. SOLOT: Computation of Maximum Possible Precipitation, *Trans. Am. Geophys. U.*, 1942.
3. BERNARD, MERRILL: Primary Role of Meteorology in Flood Flow Estimating, *Proc. A.S.C.E.*, January, 1943.

SECTION XIV

OCEANOGRAPHY

By. H. U. Sverdrup

CONTENTS

SECTION XIV

OCEANOGRAPHY

By H. U. Sverdrup

PHYSICAL PROPERTIES OF SEA WATER

Specific Gravity. The specific gravity of sea water (the ratio between the density of a water sample and the density of distilled water of temperature 39.2°F at pressure of 1 atmosphere) depends upon three variables—temperature, salinity, and pressure. The *salinity* is conveniently defined as the ratio between the weight of the dissolved salts and the weight of the sample of sea water, although a more elaborate definition

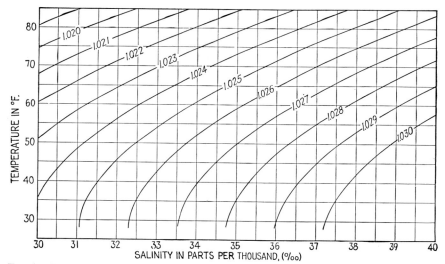

FIG. 1.—Density of sea water at atmospheric pressure as a function of temperature and salinity.

is used in oceanographic literature. The salinity is always expressed as parts per thousand or per mille, for which the symbol °/oo is used.

In the surface layers of the open oceans, the salinity varies between 32°/oo and 37.5°/oo, and the temperature between 29 and 88°F. To considerable distances off the mouths of rivers the surface water may be practically fresh. High surface salinities are found in partly isolated areas in low latitudes where excessive evaporation takes place. Thus, surface salinities up to 41°/oo are observed in the Red Sea.

The specific gravity ρ of the surface water of the open oceans varies between the limits 1.022 and 1.028. The relation between specific gravity, temperature, and salinity is shown in Fig. 1. In oceanography, the specific gravity at atmospheric pressure is commonly expressed as $\sigma_t = (\rho - 1)1,000$. Thus σ_t of the surface waters varies between the limits 22.0 and 28.0.

When dealing with wind waves or tides, the specific gravity of sea water can be taken as unity, but small differences are important when considering horizontal currents and when dealing with convection currents in the surface layers. Observations of sea-surface salinities and temperatures should therefore be accurate to within 0.1 per mille and 0.2°F, and for the deeper waters the limits of tolerance are 0.02°/$_{\circ\circ}$ and 0.02°F, respectively.

Vapor Pressure over Sea Water. The vapor pressure over sea water is somewhat lower than that over fresh water. The lowering is proportional to the salinity.

$$e_w = e_d(1 - 0.00053S)$$

where e_w is the vapor pressure of sea water of salinity $S°/_{\circ\circ}$, and e_d is the vapor pressure of distilled water at the same temperature. When dealing with the open ocean, one can employ the simple rule that the vapor pressure of sea water is 2 per cent lower than that of distilled water of the same temperature.

Freezing Point. The freezing point of sea water is lower than that of fresh water, and the lowering depends upon the salinity. With sufficient accuracy

$$\vartheta_f(°F) = 32 - 0.095S$$

where ϑ_f represents the freezing point of water of salinity $S°/_{\circ\circ}$.

Maximum Density. Fresh water reaches its maximum density at a temperature of 39.2°F, but for sea water the temperature of maximum density is lower, depending upon the salinity. With sufficient accuracy

$$\vartheta_m(°F) = 39.2 - 0.387S$$

where ϑ_m is the temperature of maximum density of water of salinity $S°/_{\circ\circ}$.

The temperature of maximum density coincides with the freezing point at a salinity of 24.70°/$_{\circ\circ}$ (a temperature of 29.6°F). The relation between temperature of maximum density and freezing point is of importance to the formation of ice in the sea (see page 1033).

Character of Sea Water in Respect to Radiation. The oceans receive energy by short-wave radiation from the sun and the sky and lose energy by outgoing long-wave radiation (see page 1044). Of the diffuse radiation from the sky and from clouds only 8 per cent is reflected from the sea surface. The amount of direct radiation from the sun that is reflected from the sea surface depends upon the height of the sun above the horizon: it is 2 per cent with the height of the sun 60 deg or more, 6 per cent with the height of the sun 30 deg, and 35 per cent with the height of the sun 10 deg. With a low sun, a considerable fraction of the incoming radiation is diffuse, and the reflection loss of the total income is therefore less than that of the direct radiation. On a clear day, the following values apply:

Altitude of the sun, deg	5	10	20	30	50–90
Percentage reflected	40	25	12	6	3

Thus, on an overcast day, 92 per cent of the incoming radiation penetrates the sea surface and on a clear day 60 to 97 per cent, depending upon the altitude of the sun, provided that the sun is 5 deg or more above the horizon.

The sea surface radiates nearly as a black body of the same temperature. The net loss of heat by radiation (the nocturnal radiation) depends therefore upon the temperature of the sea surface and upon the incoming long-wave radiation from the atmosphere and can be expressed approximately as a function of the temperature of the sea surface and the vapor pressure (relative humidity) of the air as observed on board ship. With a clear sky, the values range between 200 and 250 gcal per cm^2 per day. With an overcast sky, the loss is reduced.

$$Q = Q_0(1 - 0.083C)$$

where Q_0 represents the loss with a clear sky and where C is the cloudiness on the scale 0–10. The above equation applies to average conditions. If the sky is completely covered by cirrus, altostratus, or stratocumulus clouds, the radiation loss is about $0.75Q_0$, $0.4Q_0$, or $0.1Q_0$, respectively.

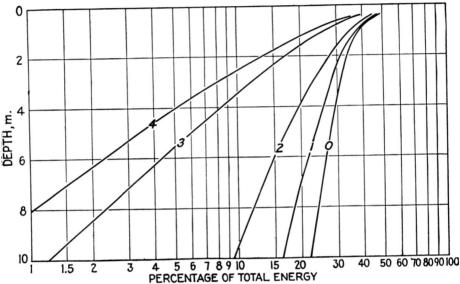

Fig. 2.—Percentages of total energy reaching different depths in pure water, clear oceanic, average oceanic, average coastal, and turbid coastal sea water—curves 0, 1, 2, 3, and 4 respectively.

Sea water free from suspended matter absorbs radiation in the same manner as distilled water, which is transparent for visible radiation only. The infrared part of solar radiation is completely absorbed in a layer that is a few tenths of an inch thick, and long-wave radiation is absorbed in a layer that has a thickness of a very small fraction of an inch, whereas blue light passes through several feet of water without appreciable reduction. However, sea water always contains suspended matter, particularly in coastal regions, and this greatly reduces the transparency in the visible portion of the spectrum. Figure 2 shows the percentage of the total energy of incoming radiation which after having penetrated the sea surface reaches to different depths between the surface and 10 m in oceanic and coastal waters. It is seen that even in the clearest oceanic water only 15 per cent reaches 10 m, and in the most turbid water less than 1 per cent reaches that depth. Thus, by far the greater fraction of incoming radiation is absorbed near the sea surface.

Eddy Viscosity, Conductivity, Diffusivity. When dealing with problems of friction. heat conduction, or diffusion of water vapor in the atmosphere, it is necessary to replace the ordinary coefficients by corresponding eddy coefficients that are related to the turbulent character of the wind. Similarly, when dealing with such problems in the sea, eddy coefficients have to be introduced because currents are generally present, and the flow is turbulent. The effect of the eddy viscosity is of particular importance when dealing with the currents produced directly by wind, and the effect of eddy conductivity is of importance to the diurnal variation of temperature in the surface layer.

Sea Ice. When the water in a fresh-water lake is cooled by loss of heat from the surface, the density of the surface layer is increased, provided that the temperature lies above 39.2°F, the temperature of maximum density of fresh water. Convection currents develop and continue until the entire body of water in the lake has been cooled to 39.2°F. With continued cooling, the convection ceases, the surface temperature drops rapidly to 32°F, and freezing of ice begins. Sea water of a uniform salinity higher than 24.7°/₀₀ behaves differently because the maximum density is reached only if the water is undercooled. The convection currents continue until the entire body of water is cooled to freezing point, and further cooling leads to formation of ice at the surface. In general, the salinity is uniform in a surface layer of thickness 10 to 20 fathoms, and this entire layer has to be cooled to freezing point before ice can form. Where fresh water spreads over sea water, conditions corresponding to those in fresh-water lakes are encountered, and this is an added reason that ice forms more readily in landlocked parts of the seas such as fiords and harbors that receive a substantial runoff from land and in which there are no strong currents.

When ice begins to form, elongated crystals of pure ice are produced. These crystals soon grow into a matrix in which sea water becomes mechanically trapped, in amounts that depend upon the rapidity of the freezing. When the temperature of this sea ice is lowered, part of the trapped sea water freezes to pure ice, the cells containing the liquid brine become smaller, and the salt concentration in the brine becomes greater. Thus, sea ice consists of pure ice in which are enclosed numerous small cells containing brine, the concentration of which depends upon the temperature of the ice. If the freezing takes place at a very low temperature, solid salts crystallize out on the surface of the ice, which at a slight rise in temperature becomes covered by concentrated brine. Sea ice may therefore appear wet in places at temperatures of 30°F or more below zero. When walking over the ice, such wet spots should be avoided because the brine penetrates through the best leather.

When the temperature of sea ice rises, the ice surrounding the cells melts, and the cells grow in size. When the temperature approaches zero, the cells will have grown to such a size that the ice becomes porous and the brine trickles down.

Drifting sea ice that is subject to the action of variable winds will be torn apart or pressed together. Where the ice is pressed together, pressure ridges and hummocks of ice are formed. When such hummocked ice is subject to melting during the summer, the upper portions become free from salt, and the remaining ice renders potable water. During winter, fresh water can be obtained by melting ice from old rounded hummocks that have been exposed to melting during the preceding summer. The melt water that forms pools in late summer at the base of hummocks is generally usable for drinking and cooking purposes.

The specific gravity of pure ice is 0.92, but the specific gravity of sea ice varies between 0.88 and 0.93, depending upon the salt content and the number of air bubbles in the ice.

Freezing of ice radically changes the thermal characteristics of the surface because of the high reflective power of the ice. Sea ice that is covered by rime or fresh snow reflects as much as 80 per cent of the incoming radiation, and old and grayish sea ice reflects 50 to 60 per cent, whereas the reflection from a water surface is much lower.

CURRENTS IN THE OCEAN

General Remarks. In oceanography, as in navigation, the direction of a current is given as the direction *toward which* the current flows. The velocity is given in centimeters per second, in knots, or in nautical miles in 24 hr.

Our knowledge of the sea-surface currents is based primarily on ships' observations of differences between positions according to fixes and dead reckoning. For several regions, charts have been prepared showing the streamlines and the steadiness of the average currents; but, on most charts used in navigation, the average direction of the currents is shown by arrows and the variability by current roses.

These average surface currents are composed of the large-scale, deep, and more or less permanent ocean currents that are related to the distribution of mass in the sea and upon which are superimposed the currents that are maintained directly by the prevailing winds. By far the greater part of the variability in the currents is directly related to the variability of the winds.

Besides these two types of currents, there are also present tidal currents and currents related to interval waves in the sea, but these are periodic and do not cause any net transport of water over a period of 24 hr. They are therefore not included in the ships' observations and are of no significance to meteorological problems over the open ocean.

Currents Related to Distribution of Mass. The large-scale ocean currents are related to the distribution of mass in the sea in a manner similar to that in which the winds are related to the distribution of mass in the atmosphere. The distribution of mass is reflected in the distribution of pressure from which in the atmosphere the geostrophic or the gradient wind can be computed. In dealing with the atmosphere, the distribution of pressure is commonly represented by isobars at sea level and at other standard level surfaces. The distribution of pressure could equally well be represented by the topography of standard isobaric surfaces, say, the 1,000-, 900-mb surfaces, and so on. If this presentation is used, the horizontal pressure gradient is conveniently replaced by the component of gravity acting along the sloping isobaric surfaces. In oceanography, it is customary to use the latter presentation of the distribution of pressure, *i.e.*, to show the topography of isobaric surfaces. The sea surface can be considered an isobaric surface, and the currents related to the distribution of mass can be computed from the slope of the sea surface if this can be established. Conversely, one can draw conclusions from observations of permanent currents as to the slope of the true sea surface. Assuming that the surface current can be computed by the equation corresponding to the equation for the geostrophic wind, one has

$$v = g \frac{\tan \alpha_p}{2\omega \sin \varphi}$$

where v is the velocity of the current along the contour lines of the sloping sea surface, g is the acceleration of gravity, α_p is the inclination of the sea, ω is the angular velocity of the earth, and φ is the latitude. By means of the above relation one finds, for instance, that in the Gulf Stream off Cape Hatteras in latitude 35°N, where current velocities up to 3 knots have been observed, the slope of the sea surface must equal 1.5×10^{-5}. The current flows toward the northeast, and the surface therefore rises toward the southeast, the rise being 0.11 in. in 1 nautical mile.

If it were possible to determine the true shape of the sea surface and of all other isobaric surfaces in the sea, a complete picture of the ocean currents related to the distribution of mass could be derived, assuming no acceleration, and neglecting friction. It is impossible, however, to determine directly the true shape of the sea surface; but, from observations of temperature and salinity, the density of the sea water can be found, and relative slopes of isobaric surfaces can be computed. If the currents can be shown to be negligible at certain depths, the true slopes are found and true currents are computed. Our knowledge of the large-scale ocean currents is based upon this type of observation and computation, which in several instances has been verified by direct measurements.

Where the density of the water is low, the distances between isobaric surfaces is great, and vice versa. Consequently, the sea surface slopes down from a region where the water is of low density to a region where the water is of high density or, since the density is more closely related to the temperature than to the salinity, it can be stated that the sea surface slopes down from a region of warm water to a region of cold water. The current that flows at right angles to the slope of the sea surface is directed such that in the Northern Hemisphere the warm water is to the right of an observer looking in the direction of flow, and in the Southern Hemisphere the warm water is to the left.

The best example of the application to practical problems of this method of computing currents is rendered by the work of the International Ice Patrol, which is maintained by the U.S. Coast Guard. The International Ice Patrol was established after the *Titanic* disaster in 1911 to safeguard shipping by reporting the presence of icebergs. For a number of years, the drifts of observed icebergs have been predicted satisfactorily from currents computed from rapid oceanographic surveys.

The computation of ocean currents is based upon the concept that a given distribution of mass can remain stationary only in the presence of corresponding currents or, conversely, that, where a permanent current exists, permanent slopes of isobaric surfaces must be maintained by a corresponding distribution of mass. No causal relationship is involved. When dealing with the causes of the ocean currents, it is necessary to examine processes that maintain the distribution of mass, such as heating and cooling, and to study the effects of the prevailing winds.

Thermal Circulation in the Sea. The atmosphere is a fairly efficient thermodynamic machine because heating takes place at high pressure (low levels), and cooling takes place at low pressure (high levels). In the oceans, on the other hand, heating and cooling take place at nearly the same level, the sea surface, and these processes can therefore not maintain any large-scale circulation. They are of the greatest importance to the development of convection currents in the surface layer and to the exchange of energy with the atmosphere, and they are also important to the slow deep-water circulation, but they do not contribute directly to the development of such currents as the Gulf Stream or the Kuroshio.

Wind Currents. The wind exerts a frictional stress (drag) on the sea surface. The stress is directed along the vector that represents the difference between wind and surface current, but, in general, the latter is so weak that the direction of the stress coincides with the wind direction. At low wind velocities, the sea surface has the character of a hydrodynamically smooth surface; but, at wind velocities above 10 to 12 knots, the sea surface can be considered hydrodynamically rough and characterized by a constant roughness length of 0.24 in. At low wind velocities, the stress is relatively small, and the relation to the wind velocity is somewhat complicated; but, at higher wind velocities, the stress is much greater and is proportional to the density of the air and the square of the wind velocity. The factor of proportionality depends upon the height at which the wind is measured or to which the estimated wind velocity applies. Assuming that the wind force as observed on board ships refers to a height of about 30 ft, the stress τ is obtained from the equation

$$\tau = 2.6 \times 10^{-3} \rho' v^2$$

where ρ' is the density of the air and v is the wind velocity. The factor 2.6×10^{-3} is a pure number and is therefore independent of the units in which stress, density, and velocity are measured, provided that all quantities are measured in the same units. The equation applies at wind velocities greater than 12 knots, and at lower wind velocities the stress is roughly one-third of the values found by the above equation.

The stress of the wind on the sea surface can lead either to a piling up of the water in the direction of the wind or to the development of a pure wind current, or to a combination of these two effects.

Piling up of water is frequently observed in bays. Where no wind currents are present, the stress of the wind is balanced by the component of gravity acting upon the entire body of water. The slope of the sea surface that can be maintained by a given wind is therefore greater in shallow water than in deep water.

For winds exceeding 12 knots, the slope of the sea surface in shallow water is obtained from the formula

$$\alpha_p = \frac{\tau}{\rho g d}$$

where ρ is the density of sea water, g is the acceleration of gravity, and d is the depth to the bottom. With wind velocity in knots and depths in feet, the slope is

$$\alpha_p = 4.7 \times 10^{-8} \frac{v^2}{d}$$

When applylng this formula, it is found that a very high wind velocity is required to bring about an appreciable change in water level. Where depths of 10 fathoms extend to a distance of 20 miles from the coast, a 100-knot wind is needed to raise sea level on the coast by 6 ft. Where a hurricane reaches a coast, the great observed rise of the water may therefore be due partly to the transport of water toward the coast by waves (see page 1052), but the effect of this transport cannot as yet be estimated.

Piling up of water also takes place over large ocean areas. Oceanographic observations indicate that, as an effect of the trade winds, the sea surface slopes upward from east to west across the equatorial portions of the oceans, the slope being of the order of 4×10^{-8}. A result of this piling up is that in the Gulf of Mexico the sea level is about 8 in. higher than at St. Augustine, Fla., so that the Florida Current flows "downhill." The difference in sea level is maintained by the trade winds; and, therefore, the velocity of the Florida Current must be influenced by changes in the trade winds. Another result is that, in the equatorial calm belt where the slope of the sea surface is not balanced by a corresponding wind stress, a current flows "downhill" from west to east as a *countercurrent* embedded between the equatorial currents. This equatorial countercurrent is particularly well developed in the Pacific.

The *pure wind current* develops in the open ocean where the flow is not impeded by coasts. Where a pure wind current is present, the stress of the wind is balanced by the Coriolis force acting upon the entire body of water that is set in motion. Since this Coriolis force must be directed against the stress, it follows that the total transport of water by pure wind currents is directed at right angles to the wind, in the Northern Hemisphere to the right, in the Southern Hemisphere to the left. The mass transport depends only upon the stress of the wind and the latitude

$$M = \int_{-\infty}^{0} \rho v \, dz = \frac{\tau}{2\omega \sin \varphi}$$

and is independent of the viscosity of the water.

In order to examine the details of the wind current, it is necessary to make certain assumptions as to the viscosity of the water.   Observations show that the ordinary viscosity has to be replaced by an eddy viscosity that is many thousand times greater. Ekman assumed a constant eddy viscosity and arrived at the simple results that in the Northern Hemisphere the wind current at the surface is directed 45 deg to the right of the wind, and that with increasing depth the current turns linearly to the right and its velocity decreases exponentially.   When projected on a horizontal plane, the end points of vectors representing the wind current at different depths lie on a logarithmic spiral (see Fig. 3).   In the Southern Hemisphere, the surface current is directed to the left and turns to the left.   A more complicated theory by Rossby and Montgomery leads to essentially similar results, although these authors find that the angle between wind and surface currents changes somewhat with latitude and

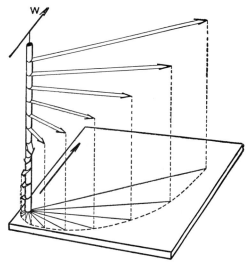

Fig. 3.—Schematic representation of a wind current in deep water, showing the decrease in velocity and change of direction at regular intervals of depth—the Ekman spiral. *W* indicates direction of wind.

wind velocity, and that the change of the currents with depth is somewhat more complicated.

The ratio between the velocity of the surface current v_0 and the wind velocity v depends upon the latitude.   Ekman has derived the semiempirical relationship

$$\frac{v_0}{v} = \frac{0.0127}{\sqrt{\sin \varphi}}$$

This relationship is based on ships' observations and applies therefore to the velocity of the current not at the very surface but at a depth corresponding to the average draft of ships, *i.e.*, at a depth of about 15 ft.   According to Rossby and Montgomery, the velocity is two to three times higher in the upper few feet of water, and this feature, which needs confirmation, may have considerable bearing on the drift of bodies of small draft such as life rafts and rubber boats.

It is probable that in a given latitude the ratio between surface current and wind varies with the season and is less in winter than in summer, but this question has not been examined sufficiently.

The wind current provides for an effective mechanism for stirring the surface waters and for establishing and maintaining a layer of nearly uniform temperature and salinity near the sea surface. In summer, the heating at the sea surface counteracts the development of such a homogeneous layer, and the effect of the wind does not penetrate so deeply as in winter when, in latitudes 45 to 50°N, a homogeneous layer may be found to a depth of 50 fathoms or more. In the trade-wind regions, the permanent winds provide for thorough mixing, and over large ocean areas a homogeneous surface layer is present of a thickness of 40 to 60 fathoms. Through this mixing, the energy that the surface layers received by absorption of radiation becomes distributed over a thick layer of water.

Secondary Effect of Wind in Producing Ocean Currents. In the open ocean, the total transport by a pure wind current is directed at right angles to the wind

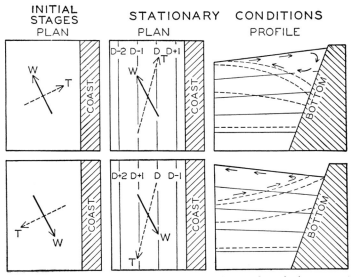

Fig. 4.—Schematic representation of effect of wind toward producing currents parallel to a coast in the Northern Hemisphere and vertical circulation. *W* shows wind direction and *T* shows direction of transport. Contours of sea surface shown by lines marked *D*, *D* + 1 . . . . Top figures show sinking near the coast; bottom figures show upwelling.

and depends only upon the stress of the wind and the latitude. This transport of surface water plays a prominent part in the generation and maintenance of large-scale ocean currents, because converging and diverging transports lead to accumulation of light surface waters in some regions, and to a rise of denser water from subsurface depths in other regions. Thus, the wind currents maintain a certain distribution of mass with which certain currents must be associated. These processes will be illustrated by a few examples.

Consider in the Northern Hemisphere a coast line along which a wind blows in such a manner that the coast is on the right-hand side of an observer who looks in the direction of the wind (Fig. 4). At some distance from the coast, the surface of the water will be transported to the right of the wind, *i.e.*, toward the coast, but at the coast all motion must be parallel to the coast line. Consequently, convergence takes place off the coast, leading to an accumulation of light water along the coast. This accumulation again establishes a downward slope of the sea surface away from the coast and a current running in the direction of the wind with the lighter water to the right.

Consider next a wind in the Northern Hemisphere that blows parallel to the coast with the coast on the left-hand side. In this case, the light surface water will be transported away from the coast and, owing to the continuity of the system, must be replaced near the coast by heavier subsurface water. Again the distribution of density is changed, but now in such a manner that the sea surface slopes upward when departing from the coast, and, corresponding to this slope, a current is established that again runs parallel to the coast in the direction of the wind. This particular process by which subsurface water is drawn to the surface is known as *upwelling* and is a conspicuous phenomenon in several localities. The upwelled water has a low temperature, and the upwelling will therefore establish and maintain low temperatures in the immediate vicinity of the coast. This temperature distribution is of great importance to the formation of fog (see page 1048). The upwelled water rises from depths not exceeding 100 to 150 fathoms, and the process therefore represents primarily an overturn of the upper layers.

A great deal of evidence shows that the upwelling is caused by the prevailing winds, but the process is so complicated that so far no quantitative relationships have been established between the wind velocities and the intensity of upwelling. Upwelling is counteracted by heating of the surface layers and formation of eddies. If the surface layers are heated, a current in the Northern Hemisphere will be deflected to the left, and this may explain why, in regions of upwelling, a tonguelike distribution of temperature is often present, regions of intense upwelling being separated by regions in which offshore water is carried toward the coast. This tonguelike distribution has considerable bearing on the cloudiness off a coast where upwelling takes place (page 1049).

Consider finally the effect of a permanent anticyclonic wind system in the Northern Hemisphere. The winds will transport the light surface water to the right, *i.e.*, toward the center of the anticyclone where the surface water will accumulate. The distribution of mass is altered in such a manner that the topography of the sea surface will show a high near the center of the anticyclone and, associated with this high, an anticyclonic current system must exist. A similar reasoning shows that a cyclonic current system develops within a permanent system of cyclonic winds.

This reasoning leads to the conclusion that the distribution of mass with which the large-scale currents of the ocean are associated is maintained by the prevailing winds in conjunction with the climatic factors that control the density of the surface layers. The circulation of the oceans is therefore closely related to the circulation of the atmosphere but displays features that are related to the boundaries of the oceans. In the Northern Hemisphere, the most conspicuous of these features are that the large water masses, which are transported from east to west by the equatorial currents, are squeezed together off the eastern coasts of the continents as they turn north and form the Gulf Stream of the Atlantic and the Kuroshio of the Pacific. A similar effect is not found off the western coasts, mainly because the warm waters when flowing north and east are cooled in the direction of flow and are therefore deflected to the right. By this process, the easterly currents in middle latitudes are gradually turned south, and no squeezing together of large water masses takes place.

Currents of the Oceans. On the basis of the preceding discussions of the manner in which the large-scale ocean currents are maintained, it is possible to give a brief general description of the major current systems. In lower and middle latitudes, anticyclonic currents prevail. The currents are wide and diffuse except off the eastern coasts of the continents where, in the North Atlantic and the North Pacific, the Gulf Stream and the Kuroshio are the outstanding branches, and, in the Southern Hemi-

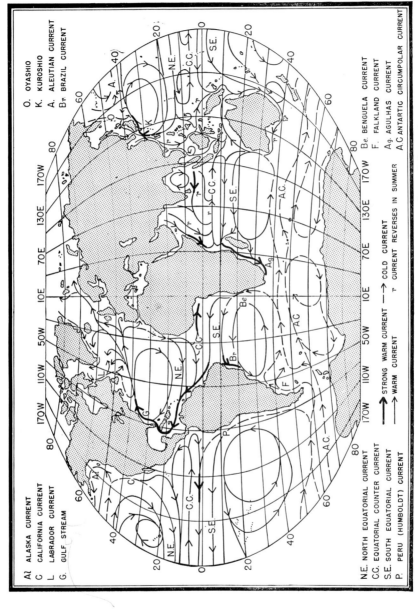

The prevailing ocean currents in Northern winter (December to February).

sphere, the Brazil Current and the Agulhas Current (off Southeast Africa) are weaker counterparts.   In the South Pacific, no narrow warm current is found.   In the higher latitudes of the Northern Hemisphere, cyclonic current systems prevail characterized by a warm branch off the western coasts of continents or islands and a cold branch off the eastern coasts.   Typical examples of these circulations are found in the Norwegian Sea where the last branch of the Gulf Stream carries warm water along the coast of Norway and the East Greenland Current carries cold water south along the coast of eastern Greenland, and in the Sea of Japan where the Tsushima Current carries warm water along the west coast of the islands of Japan and the Liman Current carries cold water along the Asiatic coast.   In the Southern Hemisphere, the Antarctic Circumpolar Current flows toward the east in the direction of the prevailing winds, sending one cold branch, the Falkland Current, to the north along the east coast of Argentina.   For details, it is necessary to consult special descriptions, or charts of currents such as those issued by the Hydrographic Office, U.S. Navy.

Drift of Life Rafts.   The drift of a life raft depends upon (1) the drift of the raft through the water (the leeway of the raft), (2) the effect of the average current, (3) the effect of a temporary wind current, and (4) the effect of the forward motion in waves.

The leeway of a raft cannot be predicted with certainty unless rafts are provided with sea anchors so that they drift with the wind, because without a sea anchor a raft may drift up to 30 deg to either side of the downwind direction.

The *average* current to be considered is that which is obtained from ships' observations and is composed of the permanent current and the average wind current.   This average current, as well as the average wind, should be known for every ocean area for which rescue operations are anticipated.

The *temporary* wind current can be computed if the present wind is known from synoptic charts and the average wind is known from climatological charts.   The temporary wind current is obtained from the deviation of the wind from the climatological average, applying the rules that (1) the speed of the wind current is

$$v_c = \frac{0.02}{\sqrt{\sin \varphi}}\, v$$

where v means the deviation of the wind speed from the average and that (2) the wind current is directed 45 deg from the wind deviation (to the right in the Northern Hemisphere, to the left in the Southern), provided that the wind has blown steadily for 24 hr or more and that (3) the angle between current and wind is proportionally smaller if the duration of the wind is less than 24 hr.   The factor used above to compute the wind current is somewhat larger than that given on page 1037 because of the shallow draft of a life raft.

The forward motion due to wind waves need not be considered separately because it is included in the empirical results as to the leeway of rafts presented in H.O. 235. This publication contains rules for determining the course of a raft which differ somewhat in details from the above.

SEA-SURFACE TEMPERATURES

Distribution.   The distribution of sea-surface temperature is primarily controlled by three factors, the latitude, the season, and the character of the ocean currents. In every month of the year, the effect of latitude generally dominates, as is evident from the general decrease of temperature from lower to higher latitudes; but in certain seasons and regions, the effect of currents may be great enough to change the direction of the sea-surface isotherms from an east-westerly to a north-southerly.

The regional distribution is greatly influenced by the character of the currents, as is evident from a comparison of charts of currents and of sea-surface temperatures. In lower and middle latitudes where the anticyclonic circulation dominates, the sea-surface isotherms bend away from the equator off the eastern coasts of continents and toward the equator off the western coasts. The latter feature is so greatly accentuated where upwelling takes place that the sea-surface isotherms run nearly north-south off the west coasts of parts of South America, Southwest Africa and Northwest Africa, and in spring and summer off the west coast of the United States. In higher latitudes where cyclonic circulation prevails, the contrast between the east and west is reversed, and the isotherms bend toward the equator off the east coasts and away from the equator off the west coasts. These features are particularly conspicuous in the Northern Hemisphere, but in the Southern Hemisphere the latitude effect is by far more noticeable. In the Northern Hemisphere, in about latitude 40°N, regions are found off the east coasts in which the sea-surface temperature changes very rapidly in a horizontal direction. In the Atlantic Ocean steep horizontal gradients are encountered in the boundary region between the warm North Atlantic Current (the continuation of the Gulf Stream) and the cold Labrador Current, and in the Pacific Ocean between the warm Kuroshio Extension and the cold Oyashio.

Along the equator, low surface temperatures are found in most months, particularly in the Pacific Ocean. These low surface temperatures are associated with a divergence along the equator and occur where the equatorial countercurrent is best developed. For these and other details, it is necessary to consult special descriptions or the charts of mean monthly sea-surface temperatures issued by the Hydrographic Office, U.S. Navy (H.O. 10,877). Also see section on Climatology.

Annual Variation. The annual variation of the sea-surface temperature is greatest in middle latitudes. It is very large off the eastern coasts of the continents of the Northern Hemisphere but small in the regions of upwelling. Thus the annual range of sea-surface temperature in 40°N is somewhat above 40°F on the east coast of Japan and less than 6°F on the coast of California. Over large tropical areas and in the arctic and antarctic, the annual range is less than 4°F.

Diurnal Variation. The diurnal variation in sea-surface temperature is small throughout. In the region of the trade winds, the diurnal range is only 0.4 to 0.6°F. In middle and higher latitudes, there is an annual variation in the range, but even in midsummer the average value is not more than 0.6 to 0.9°F. The range depends upon the wind velocity and the cloudiness and may on a calm day with a clear sky reach 4°F. Even this range is so small that it can in general be neglected when dealing with meteorological problems.

Difference of Sea Surface Minus Air Temperature. On an average, the sea surface is slightly warmer than the air at a height of 20 to 30 ft. In the tropics, the difference is constant throughout the year and, according to very careful observations on oceanographic expeditions, amounts to about 1.4°F. Routine observations on commercial vessels render smaller values, because during the daytime the readings on shipboard give somewhat too high air temperatures, since the thermometers are rarely protected sufficiently against the ship's heat.

In middle latitudes, large seasonal and regional differences occur. In winter, the sea surface is much warmer, on an average, than the air off the eastern coasts of the continents of the Northern Hemisphere (6 to 8°F), but off the western coasts the differences are small. In spring and summer, the sea surface is in many regions colder than the air. Thus, over the Grand Banks of Newfoundland, the sea surface in spring is 2 to 3°F colder on an average than the air, and off the coast of California the sea surface is also colder than the air in spring and summer.

The differences are related to the circulation of the atmosphere and of the oceans.

Off the eastern coasts, the large differences in winter occur because the prevailing westerly winds bring cold air from the continents, whereas the currents carry warm water toward the northeast. Temperature differences up to 15 or 20°F are found during intense outbreaks of polar continental air. The differences of opposite sign that are found over the Grand Banks in spring are related on the other hand to inflow of warm air over the cold Labrador Current.

ICE IN THE SEA

Two types of ice are encountered in the oceans, *sea ice*, which is ice frozen from sea water, and *icebergs*, which are broken-off pieces of glaciers. The numerous forms of sea ice depend upon the conditions of freezing, upon packing and piling up due to wind, and upon processes of disintegration and melting. A number of classifications and several codes for reporting sea ice have been proposed, and one code has been adopted by the International Meteorological Committee. Similarly, icebergs differ widely in appearance depending upon the character of the glaciers from which they originate and upon the processes of destruction to which they have been subjected. Icebergs can also be reported by code.

Sea ice is called *fast ice* where it forms a solid immobile cover. In winter, fast ice is encountered in bays and fiords, such as Bristol Bay (Bering Sea) and in the fiords of Greenland and Spitsbergen. In general, sea ice is broken up by winds and currents and is called *drift ice*. Drift ice that has been jammed together by wind and forms a more or less continuous cover of rugged ice is often described as *pack ice*.

In all seasons of the year, the Antarctic Continent is surrounded by drift ice. At the end of the southern winter, in October, the average limit of the drift ice lies in about 55°S in the Atlantic and Indian Ocean sectors and in about 63°S in the Pacific sector. At the end of the summer, in March, the coast of Graham Land is, on an average, free from ice, and otherwise the drift ice extends to short distances from land. The coast of Enderby Land in longitude 55°W and the entrance to Ross Sea in longitude 180° may be free from ice.

The icebergs of the Antarctic originate from the antarctic icecap. In general, the icebergs are flat; they may be 10 to 20 miles wide and up to 50 miles long and may rise more than 250 ft out of the water, corresponding to a total thickness of about 2,400 ft. The average northern limit of the antarctic icebergs lies about 15° latitude to the north of the winter limit of the drift ice. In the Atlantic Ocean, a piece of floating ice was reported on Apr. 30 1894, in 26°30′W.

In the Arctic, the Polar Sea remains ice-covered throughout the year, but in summer numerous irregular openings are present because such openings are formed under the action of the wind and do not freeze over, since the temperature remains at or slightly above freezing point for 2 months or longer.

In summer, lanes develop along the north coast of Siberia, and commercial shipping is possible if the merchant vessels are accompanied by ice breakers that can assist them where passage is difficult. In general, Bering Strait becomes ice-free in the beginning of July, but ships can rarely reach Point Barrow, Alaska, until August. Small vessels have made the Northwest Passage between the islands to the north of Canada, but the sounds between these islands are shallow and full of dangers. The west coast of Spitsbergen is free from drift ice, and to the north of Spitsbergen vessels can frequently go in open water to latitude 81° and have reached 82°. In all seasons, masses of drift ice are carried south along the east coast of Greenland. Even in summer, the belt of ice is so dense that in general no vessel can pass through it north of latitude 64°.

In late winter, the drift ice extends much farther south. In Bering Sea, it may reach to less than 50 miles from the north coast of the Aleutian Islands and is carried south along Kamchatka and the Kuril Islands to latitude 42°N. In European waters, the north coast of the Murmansk peninsula and the north coast of Norway remain free of ice, but Spitsbergen is surrounded by drift ice. A belt of drift ice is present around Greenland; but, even in midwinter, openings in the ice are found off the southwest coast. On the American side, the drift ice extends south along the coasts of Labrador and Newfoundland, and in February it may reach to latitude 42°N.

The icebergs of the Northern Hemisphere come primarily from the glaciers on Greenland. They drift south with the Labrador Current and represent in spring and early summer a menace to shipping in the North Atlantic (see page 1035). A few small icebergs are discharged from glaciers on Franz Josef Land and Spitsbergen. These first drift to the west and are later carried south by the East Greenland Current so that no icebergs are found over the large part of the Polar Sea. Some of the Alaskan glaciers in about latitude 58°N produce icebergs that may block fiords and sounds.

Ice limits are shown on the charts of mean monthly sea-surface temperatures issued by the Hydrographic Office, U.S. Navy (H.O. 10,877).

EVAPORATION AND HEAT EXCHANGE

On an average for all oceans, the amount of radiation received from the sun and the sky is greater than the loss by nocturnal radiation. During one year, this radiation surplus Q_r must be given off to the atmosphere as sensible heat Q_h or used for evaporation Q_e, because the mean temperature of the oceans can be considered constant: $Q_r = Q_h + Q_e$. It follows that the mean annual evaporation E can be expressed by the equation

$$E = \frac{Q_r}{L(1 + R)}$$

where E is the annual evaporation, Q_r is the radiation surplus, L is the latent heat of vaporization, and R is the ratio between the heat given off from the oceans and the heat used for evaporation. The radiation surplus has been determined from observations, and the ratio R has been estimated at about 0.1. The corresponding value of the average evaporation from all oceans between latitudes 70°N and 70°S is 40 in. per year, a value that is in accord with the probable value of the average precipitation.

In order to examine the evaporation in different seasons and regions, it is necessary to consider the process by which evaporation takes place. When a steady state has been established near the sea surface so that the vapor content of the very lowest layer of air remains constant, then the vertical flux of water vapor through this layer must be constant. Where the specific humidity decreases upward, this flux equals the evaporation

$$E = -F \frac{dq}{dz}$$

where F is the eddy transfer of water vapor and q is specific humidity. The gradient of the specific humidity can be replaced by the difference in vapor pressure at the sea surface e_w and in the air at a height of 20 to 30 ft, e. At moderate and high wind velocities, the eddy transfer is proportional to the wind velocity, so that the evaporation can be expressed as

$$E = k(e_w - e)v$$

where the numerical value of the factor k depends upon the height at which e and v are measured and upon the unit used. If the vapor pressure and the wind velocities are measured at a height of 20 to 30 ft and are expressed in millibars and knots, respectively, the evaporation is

$$E = 3.5 \times 10^{-3}(e_w - e)v \qquad \text{in./24 hr}$$

provided that the wind velocity exceeds 12 knots.   When applied to average monthly values, the factor becomes 3×10^{-3}.

From the above equation, it is evident that, at a given wind velocity, the evaporation increases with increasing values of the difference $(e_w - e)$.   Over the open ocean, the vapor pressure at the sea surface, e_w, depends on the temperature of the sea surface and equals $0.98e_d$ when e_d is the vapor pressure over distilled water (page 70). The difference $(e_w - e)$ is therefore always positive, even with the relative humidity at 100 per cent, when the sea surface is 0.6°F or more warmer than the air, and it is also positive if the sea surface is slightly colder, provided that the relative humidity is sufficiently below 100 per cent.   In the tropics, the sea surface is always a little warmer than the air which, at a height of about 30 ft, has a relative humidity of 75 to 80 per cent.   There evaporation takes place throughout the year, and by far the greater part of the radiation surplus is used for evaporation.   In middle latitudes, the difference sea surface minus air temperature varies from season to season and from region to region.   On the whole, the difference is much greater in winter than in summer, so that the greatest evaporation takes place in winter.   In the Northern Hemisphere, very large differences are found off the eastern coasts of the American and Asiatic continents where the warm currents, the Gulf Stream and the Kuroshio, flow north and where cold polar continental air flows out over the oceans.   On an average, the daily evaporation from the Gulf Stream region, in 35 to 40°W, is about 0.45 in. in the 3 months December, January, and February, and from the Kuroshio it is about 0.32 in.   During outbreaks of polar continental air, the daily evaporation may be five times the average winter maximum.

In the same latitude but in the eastern parts of the oceans, the daily evaporation in winter is about 0.10 in. off the coast of Spain and 0.07 in. off the coast of California. In summer, the daily evaporation in these latitudes nowhere exceeds 0.1 in., and in the eastern parts of the oceans it is less than 0.05 in.

The vertical transport of water vapor changes direction if the vapor pressure in the air is greater than that at the sea surface $(e > e_w)$.   This happens when warm moist air flows over colder water, in which case the air in contact with the water becomes saturated, and condensation takes place on the sea surface (dew forms).

The heat exchange between the ocean and the atmosphere follows a pattern similar to the evaporation.   Where the sea surface is warmer, heat is transferred from the sea to the air and is transported to considerable heights by eddy conductivity and by convection currents in the air.   In the tropics, the heat transfer is very small because the radiation surplus is mainly used for evaporation.   In middle and higher latitudes of the Northern Hemisphere, large regional and seasonal differences occur. It is noteworthy that, in the regions of the Gulf Stream and the Kuroshio, 40 per cent of the energy given off from the ocean to the atmosphere in winter is given off as heat and only 60 per cent is given off as latent heat of vaporization.   During outbreaks of polar continental air, the proportion may be even greater.   Because of this great heat transfer, the air temperature rises so rapidly that, in spite of intense evaporation, the relative humidity at the sea surface probably remains at about 70 to 80 per cent.

INTERACTION BETWEEN THE OCEAN AND THE ATMOSPHERE

General.   In the tropics, the radiation surplus received by the oceans (on an average, about 250 gcal per cm² per day) is primarily used for evaporation, which supplies the air in the trade-wind belts with huge quantities of moisture.   This moisture falls out as rain along the equatorial front or over the land masses that in the tropics represent the western boundaries of the oceans.

In middle and higher latitudes of the Northern Hemisphere, the influence of the oceans varies greatly from summer to winter.   In summer when the oceans receive the greatest radiation surplus, the air is generally as warm as or slightly warmer than the sea surface.   No heat is transferred from the ocean to the atmosphere, the evaporation is small, and by far the greater part of the radiation surplus is used for heating the surface layers of the sea.   In winter, this heat is given off or used for evaporation, particularly off the eastern coasts of the continents where cold continental air flows out over the warm waters of the Gulf Stream or the Kuroshio.   Thus the oceans function as a hot-water heating system.   The water is heated in summer, and in winter it gives off its heat in localized regions.

The ocean currents that play an important part in the exchange of heat between the ocean and the atmosphere are maintained by the prevailing winds and are therefore altered by changes in the circulation of the atmosphere.   On the other hand, the heat given off from the oceans is of importance to the circulation of the atmosphere, and changes in the currents must therefore be accompanied by changes in the prevailing winds.   The interaction is so complicated that in many instances it is impossible to separate cause and effect.

The following example illustrates some of the many considerations that enter into the interpretation of observed conditions.   Where the surface layers of a current are cooled rapidly, they are deflected to the right, in the Northern Hemisphere, and must be replaced by water that rises from some greater depth or is drawn in on the left-hand side of the current.   This implies that, when the Gulf Stream waters are cooled in winter, the surface layers are deflected to the east and south and are replaced by colder water from below or from the north.   The decrease in the surface temperature along the axis of the current is therefore not a result only of cooling but is partly due to replacement of the original surface water by colder water.   Consequently, the sea-surface temperature on the northwest coast of Europe must depend not only upon the temperature of the Gulf Stream water when it passed the east coast of the United States at some earlier date but also upon the outbreaks of polar continental air that led to a deflection of the surface layers.   For this reason, no close correlation can be expected between variations in the surface temperature of the Gulf Stream off the east coast of the United States and the surface temperature off northwestern Europe at some later date.   Since the North Pacific Ocean is much wider than the North Atlantic and the circulation is more complicated, it appears even more improbable that any correlations exist there between variations in sea-surface temperatures off the western and eastern shores.

Transformation of Polar Continental to Polar Maritime Air.   It has repeatedly been stated that polar continental air that in winter flows out over the ocean rapidly takes up heat and moisture.   The rapidity of these processes is best illustrated by an example.

On Dec. 21, 1935, a ship in latitude 35°71′W reported at 1300: Wind WNW 7, Cloudiness 10, Air temperature 41°F, Sea-surface temperature, 66°F.   The synoptic chart showed that the ship was in a region where an intense outflow of polar continental air took place and that the air which passed the ship at 1300 had left the coast, 400 miles away, about 12 hr earlier.   The vertical distributions of temperature and specific humidity at the time when the air left the coast could be reconstructed by interpolation between soundings at the U.S. Weather Bureau stations and are shown in Fig. 5.   Assuming that the relative humidity at the ship was 80 per cent and that the air above the condensation level was saturated, the corresponding vertical distributions at the ship could also be constructed, as shown in the same figure.   It is seen that the convection currents reached to a pressure of about 580 mb, *i.e.*, to a height of about 13,000 ft.   The increase in the water vapor of an air column of this height and cross section 1 cm² was 0.82 grams, and the increase in heat content was 945 gcal.   Part of the increase in heat content, however, was due to condensation of water vapor. Taking this into account, it is found that, during the 12-hr passage over the water, the average evaporation was 1.3 cm (0.51 in.), and the average heat transfer was 520 gcal per cm². These values agree well with values that can be computed from the wind velocity and the differences in vapor pressure and temperature at the sea surface.

It follows from this example that the transformation of the air takes place on a short distance and in a short time. It is also obvious that convection currents develop as soon as the air flows out over the water and that these convection currents reach to greater and greater height as the temperature of the lowest layer increases. Clouds form, first stratocumulus or fractostratus, later large cumuli accompanied by showers and perhaps cumulonimbus.

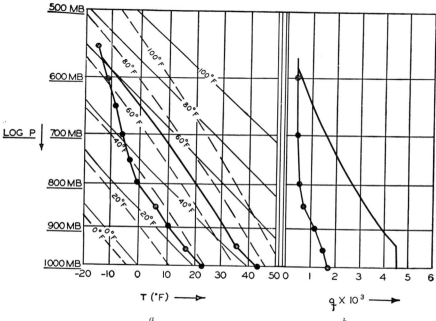

a *b*

FIG. 5.—*a*, temperature distribution in polar air on leaving the coast and over the ocean at 400 miles off the coast. *b*, corresponding distribution of specific humidity.

It seems probable that, when more is known about the rapidity with which the air temperature approaches the sea-surface temperature and when soundings on the coast are available, it will be possible to forecast the probable height of cloud tops at different distances from the coast and the distance at which showers can be expected.

Clouds in Air Masses. In general, the air in contact with the sea surface has a high relative humidity, about 80 per cent. Where the sea surface is warmer than the air, convection currents and clouds associated with these are present, but where the sea surface is colder, the sky may be clear or fog may form. In the trade-wind regions where the sea surface is warmer throughout the year, convection currents are generally present reaching to the trade-wind inversion. Below the inversion, cumulus or stratocumulus clouds are regularly present. In winter, in middle and higher latitudes, the sea surface is much warmer than the air, particularly off the east coasts of the continents of the Northern Hemisphere. In general, convection currents develop, but the polar air masses are sufficiently stable aloft to prevent the convection from reaching great heights. In this case, the convection clouds will appear as a nearly complete cloud cover which an observer on board ship will describe as stratus, strato-cumulus, or fractostratus. If the stability aloft is small or if an air mass of stable stratification flows out over very warm water, cumulonimbus develops accompanied by showers. In all these cases, the clouds occur in the air mass and are not a result of vertical motion near fronts, but such vertical motion may produce similar clouds.

In spring and summer, in middle and higher latitudes, the sea surface is often colder than the air. Where the temperature difference is small, a clear sky is frequently observed, but where the temperature difference is great, fog prevails.

Where upwelling occurs, the sea-surface temperature often shows a tonguelike distribution, tongues of warm water alternating with tongues of cold water. Air that flows across a tongue of cold water will ordinarily be warmer than the water, and if the air is nearly saturated fog may form. Offshore the temperature difference is generally too small to produce fog but sufficient to create a stable stratification near the sea surface such that a belt of clear sky is found over the tongue of cold water.

Diurnal Variation of Cloudiness. In the trade-wind and subtropic regions, the cloudiness shows a double diurnal variation with maximum amounts of clouds at about 0600 and 1800 (local time) and minimum amounts at about 1300 and 2200.

The double diurnal variation has not been explained, but it seems possible that it is related to the double diurnal variation of the atmospheric pressure. This variation has the character of a wave that travels around the earth and is accompanied by converging winds in front of the region of maximum pressure and diverging winds in the rear. In any given locality, maximum convergence occurs twice daily, at about 0600 and 1800 local time, *i.e.*, at the hours when the cloudiness of the tropical oceans reaches a maximum. This suggests a relationship that deserves further examination.

Of the two maxima, the morning maximum at 0600 is the greater, and showers and thunderstorms are most frequent in the morning. It seems probable that the shower frequency and the large maximum of cloudiness in the early morning is an effect of radiation from the upper surfaces of clouds which during the night leads to an increase of the instability of the lowest layer. So far, the shower frequency in the early morning is the only direct evidence of the effect of radiation on the cloudiness at sea, but as yet the effect has not been sufficiently examined.

In higher latitudes, a possible diurnal variation in cloudiness seems to be masked by the large unperiodic variations. It may be observed that the amplitude of the pressure variation decreases rapidly with latitude, and a variation in the cloudiness related to the pressure variation should therefore be expected to be small in higher latitudes.

Fog over the Oceans. When warm moist air flows in over colder water, the air is cooled from below and may be cooled to the dew point, so that condensation takes place and *advection fog* forms within the temperature inversion that is established at the sea surface.

With a weak wind, an inversion begins at the very sea surface, and the fog may extend nearly to the top of the inversion. With a strong wind, mechanical mixing leads to the establishment of a positive lapse rate in the lowest one hundred or few hundred feet, and within this layer water vapor is transported upward so that the fog lifts, leaving a clear space below a dense stratus cover.

The advection fog is by far the commonest type of fog over the open ocean areas. Regions known for their frequent fogs are the region of the Grand Banks of Newfoundland (maximum frequency in spring), the regions of the Kurils and the Aleutians, and the outskirts of the arctic and antarctic regions. Over the Polar Sea, advection fog is very common in summer.

In the regions of upwelling, coastal fogs are related to the belt of low temperature along the coast (see page 994). When moist maritime air flows in over this belt, it is cooled, and where the cooling begins a belt of clear sky may be found. As the cooling proceeds, fog forms and rolls in over the coast but is dissipated over land. In the seasons of upwelling, this type of advection fog is common on the coasts of California, northern Chile and southern Peru, and the coast of southwest Africa.

Steam fog, which is formed when very cold air flows over warm water, does not occur over the open ocean because the air is heated so rapidly that it does not become saturated in spite of intense evaporation. Steam fog does, however, form in protected coastal waters such as the fiords of southern Alaska and Norway where it can be a serious menace to navigation.

Radiation fog may perhaps be formed at sea in calm weather if a layer of moist air underlies very dry air. In this case, the cooling that leads to a formation of the fog is partly due to loss of heat from the sea surface by nocturnal radiation and partly due to loss of heat by radiation from the moist air.

Smoke Screens over the Ocean. A smoke screen that will "stick" to the surface can be laid only if the air is warmer than the sea surface. If the air is colder, convection currents are present and the smoke screen lifts in a manner that depends upon the magnitude of the temperature difference and the wind velocity. The convection currents often show a definite pattern, and the smoke will rise in regions of ascending currents but may remain at the sea surface in regions of descending currents.

With stable conditions (the air warmer), the wind velocity increases rapidly with increasing height and, consequently, the upper portion of a smoke screen will move faster than the lower part and may form a wedge-shaped cloud that reaches to the greatest height downwind.

Forecasting Cloudiness and Fog at Sea. Over the ocean, frontal clouds and fogs are forecast on the basis of considerations similar to those used over land. Forecasting of clouds and fog in the air masses and of the behavior of smoke screens, on the other hand, is partly a matter of forecasting the temperature difference between the air and the sea surface.

The air-sea temperature difference determines the stability of the very lowest layer and indicates whether or not convection currents can develop. For forecasting of clouds and fog, knowledge of the humidity is needed, and observations of the humidity should therefore be made on board ship and should be reported regularly. When observations are lacking, the forecaster has to make assumptions, based upon his experience. In the forecasting of clouds, it is necessary to estimate the height to which convection currents may reach, and for this purpose the stability aloft should be known. Soundings should be available, and it should be possible to evaluate the effects of subsidence and radiation. As yet, adequate rules for forecasting of clouds and fog cannot be formulated, but in the meantime useful information can be obtained from the expected stability near the sea surface, as indicated by the air-sea temperature difference. In order to facilitate forecasts of this temperature difference, weather base maps have been prepared for all oceans and all months on which mean sea-surface isotherms are printed as weak blue lines.

SURFACE WAVES

Waves of Very Small Height. Wind waves are progressive surface waves which when formed move under the influence of gravity. In a progressive surface wave, the crest advances one wave length L during one period T and the velocity of progress is therefore

$$C = \frac{L}{T}$$

The theory of waves deals with waves of very small height whose form can be represented by a sine curve. By waves of small height are understood waves for which the ratio of height to length is $\frac{1}{100}$ or less. In water of constant depth d, such waves travel with the velocity

$$C = \sqrt{g \frac{L}{2\pi} \tanh 2\pi \frac{d}{L}}$$

where g is the acceleration due to gravity.

If d/L is *large, i.e., if the wave length is small compared to the depth,* $\tanh 2\pi d/L$ approaches unity, and one obtains

$$C = \sqrt{g \frac{L}{2\pi}}$$

These waves are called *deep-water waves.*

If d/L is small, *i.e.,* if the wave length is large compared with the depth, $\tanh 2\pi d/L$ approaches $2\pi d/L$, and one obtains

$$C = \sqrt{gd}$$

These waves are called *shallow-water waves.*

In general, waves have the character of deep-water waves when the depth to the bottom is greater than one-half the wave length ($d > L/2$). However, for shallow-water waves, the depth must be less than one twenty-fifth of the wave length in deep water ($d < L/25$).

In low deep-water waves, the water particles move in circles. At any depth z below the surface, the radius of the circular path followed by a particle is

$$r = \frac{H}{2} e^{-2\pi \frac{z}{L}}$$

In this circle, the velocity is

$$\frac{2\pi r}{T} = \pi \frac{H}{T} e^{-2\pi \frac{z}{L}} = \pi \frac{H}{L} C e^{-2\pi \frac{z}{L}}$$

because the particles complete one revolution in the time T (see Fig. 6).

A water particle at the sea surface remains at the surface throughout its orbit. A water particle at a given average depth below the sea surface is farthest from the surface when it moves in the direction of wave progress.

In a low shallow-water wave, the vertical motion of the particles is negligible, and the horizontal motion is independent of depth. The particles move back and forth, following straight lines.

By *energy of the wave* is always understood the average energy over one wave length. The energy is in part potential, E_p, associated witht he displacement of the water particles above or below the level of equilibrium, and in part it is kinetic, E_k, associated with the motion of the particles. In surface waves, half the energy is present as kinetic and half as potential. The total average energy per square foot is $E = \frac{1}{8}g\rho H^2$, where g is the acceleration of gravity and ρ is the density of the water. Since g and ρ can be considered constant, the energy per unit area in a wave depends only upon the wave height and is proportional to the square of the wave height. For

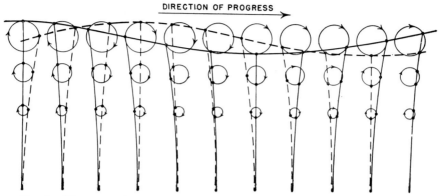

DIRECTION OF PROGRESS

Fig. 6.—Movement of water particles in a deep-water wave of very small height. The circles show the paths in which the water particles move. The wave profiles and the positions of a series of water particles are shown at two instants, which are one quarter of a period apart. The full-drawn nearly vertical lines indicate the relative positions of water particles that lie exactly on vertical lines when the crest or the trough of the wave pass, and the dashed lines show the relative positions of the same particles one quarter of a period later.

the total energy per unit width along a wave length, it is necessary to multiply the energy per unit area by the wave length.

In a deep-water wave, only *half the energy advances with wave velocity;* whereas, in a shallow-water wave, all the energy advances with wave velocity. The reason for this difference is that, in a deep-water wave, only the potential energy varies periodically and advances with the wave form; whereas, in a shallow-water wave, both potential and kinetic energy vary periodically, and both advance with the wave form. These laws can also be stated by saying that the energy advances at a rate which, in a deep-water wave, equals half the product of energy and wave velocity; whereas, in a shallow-water wave, it equals the product of energy and wave velocity.

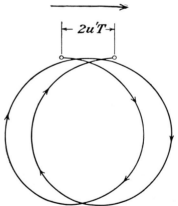

Fig. 7.—Orbital motion during two wave periods of a water particle in a deep-water wave of moderate or great height.

Deep-water Waves of Moderate and Great Height. By waves of moderate and great height are understood waves for which the ratio height to length (H/L) is from $\frac{1}{100}$ to $\frac{1}{25}$ and from $\frac{1}{25}$ to $\frac{1}{7}$, respectively. The form of these waves cannot be represented by a sine curve. For waves of moderate height, the form closely approaches the trochoid, *i.e.*, the curve that is described by a point on a disk that rolls below a flat surface. Waves of great height deviate from the trochoid; the troughs are wider and flatter and the crests narrower and steeper. The wave form becomes unstable when the ratio H/L equals $\frac{1}{7}$.

The wave velocity increases with increasing steepness (increasing values of H/L), but the increase of velocity never exceeds 12 per cent.

The water particles move approximately in circles the radii of which decrease rapidly with depth. The particle velocity is not uniform but is greatest when the particles are near the top of their orbit (moving in the direction of wave progress), with the result that the particles upon completion of each nearly circular motion have advanced a short distance in the direction of progress of the wave (Fig. 7). Consequently, there is a mass transport in the direction of progress of the wave. The *mass-transport velocity* (v_0') at the sea surface is expressed by the formula

$$v_0' = \left(\pi \frac{H}{L}\right)^2 C = \left(\pi \frac{H}{L}\right)^2 \frac{C}{v} v$$

where v is the wind velocity.

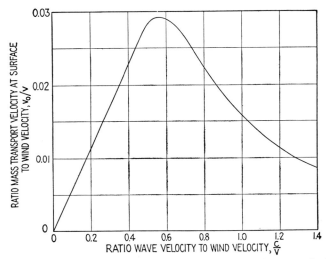

Fig. 8.—Ratio between mass transport velocity by a growing wind wave and wind velocity as a function of the ratio between wave velocity and wind velocity.

The velocity is appreciable for high, steep waves but is very small for low waves of long period. Mass transport in waves has received little attention in previous work because in most practical applications it is sufficient to consider the water particles as moving in circles regardless of the wave height. In order to understand the growth of waves through wind action, however, it is necessary to take the mass-transport velocity into account.

It will be shown later that, for growing waves, the ratio H/L depends upon the ratio between wave velocity and wind velocity, C/v. Therefore the ratio between mass-transport velocity at the sea surface and wind velocity v_0'/v depends only upon C/v. The relationship is shown in Fig. 8. For growing waves, the ratio C/v generally lies between 0.6 and 1.0, and in this range the mass-transport velocity varies between 2.9 and 1.6 per cent of the wind velocity, *i.e.*, at the sea surface, the mass-transport velocity is of the same order of magnitude as the velocity of the pure wind current (see page 1037).

Interference of Waves; Short-crested Waves; White Caps. When waves of different heights and lengths are present simultaneously, the appearance of the free surface becomes very complicated. At some points, the waves are opposite in phase and therefore tend to eliminate each other; whereas, at other points, they coincide in phase and reinforce each other.

As a simple case, consider two trains of waves that have the same height and nearly the same velocity of progress. Owing to interference, groups of waves are formed with wave heights roughly twice those in the component wave trains. Be-

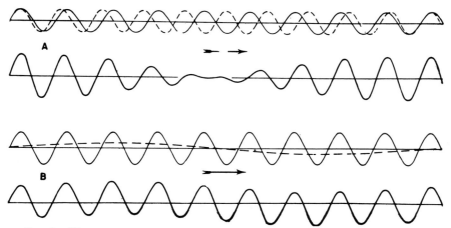

Fɪɢ. 9.—Wave patterns resulting from interference. *A*, interference of two waves of equal height and nearly equal length, forming wave groups. *B*, interference between short wind waves and long swell.

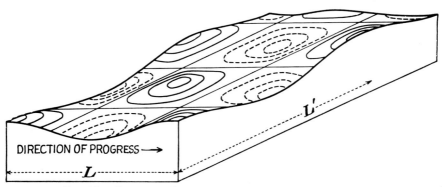

DIRECTION OF PROGRESS →

Fɪɢ- 10.—Short-crested waves. *L*, wave length, *L'*, crest length.

tween the wave groups are regions in which the waves nearly disappear (Fig. 9). Analysis shows that these *groups* advance with a velocity that is nearly equal to *one-half* the average velocity of the two trains.

As another example, consider the simultaneous presence of long, low swell and short but high wind waves. The resultant pattern is illustrated in Fig. 9*B*, from which it is evident that the short, high waves dominate to such an extent that the presence of the swell is obscured.

So far, the discussion has dealt only with long-crested waves, *i.e.*, waves with very long straight crests and troughs. Waves, however, can also have short, irregular crests and troughs. In the presence of such *short-crested waves*, the free surface shows a series of alternating "highs" and "lows," as indicated in Fig. 10. This figure illustrates the topography of the sea surface, "highs" being shown with full-drawn lines and "lows" with dashed lines.

White caps are formed by the breaking of relatively short waves, which often appear as "riders" on longer waves (Fig. 9B). Such short waves may grow so rapidly that their steepness reaches the critical value $H/L = \frac{1}{7}$ and they break. If interference occurs, long waves may attain this steepness and may break.

EMPIRICAL KNOWLEDGE OF WIND WAVES AND SWELL

Measurements of Waves and Swell. *Wind waves* are defined as waves that are growing in height under the influence of the wind.

Swell consists of wind-generated waves that have advanced into regions of weaker winds and are decreasing in height.

So far, the discussion of surface waves has dealt mainly with waves that appear as rhythmic and regular deformations of the surface. Because of interference, the formation of short-crested waves, and the breaking of waves, there is little regularity, however, in the appearance of the sea surface, particularly when a strong wind blows. Although individual waves can be recognized and their heights, periods, lengths, and velocities measured, such measurements are extremely difficult and comparatively inaccurate. The lengths of most waves and the heights of low waves are likely to be underestimated, while the heights of big waves are generally overestimated. Wave heights above 55 ft are extremely rare, yet the literature contains many reports of waves exceeding 80 ft in height. The reasons for such errors are probably due to the complexity of the sea surface and the movement of the ships from which measurements are made.

Reliable measurements of *wave height H* are so difficult that in general the reported values represent crude estimates. The height of a large wave is estimated as the eye height of the observer above the water line when the ship is on even keel in the trough of the wave, provided that the observer sees the crest of the wave coincide with the horizon. The height of a small wave is estimated directly, using the dimensions of the ship for comparison. On board a small ship, the height of waves that are more than twice as long as the ship can be recorded by a microbarograph.

The *wave period T* can be measured by recording the time interval between successive appearances (on a wave crest) of a well-defined patch of foam at a considerable distance from the ship. In order to obtain a reliable value, observations should be made for several minutes and averaged.

The *wave length L* can be estimated by comparing the ship's length to the distance between two successive crests. This procedure leads to uncertain results, however, because it is often difficult to locate both crests relative to the ship and because of disturbance caused by the movement of the ship.

The *velocity of the wave C* can be found by recording the time needed for the wave to run a measured distance along the side of the ship and by applying correction for the ship's speed.

Comparison of Measured and Computed Values. Theory indicates that for deep-water waves velocity, length, and period are interrelated by the formulas

$$C = \frac{L}{T} = \sqrt{g\frac{L}{2\pi}} \qquad L = \frac{2\pi}{g}C^2 = \frac{g}{2\pi}T^2 \qquad T = \sqrt{\frac{2\pi}{g}L} = \frac{2\pi}{g}C$$

With C in knots, L in feet, and T in seconds

$$C = 1.34 \sqrt{L} = 3.03T$$
$$L = 0.555C^2 = 5.12T^2$$
$$T = 0.422 \sqrt{L} = 0.33C$$

Thus, if one characteristic is measured, the other two can be computed, and if two or three are measured, the correctness of the theory as applied to ocean waves can be checked. Comparisons of measured and computed values have given satisfactory results, indicating that wind waves and swell in deep water do have the characteristics described above. In general, the conclusion that the ratio H/L always remains less than $\frac{1}{7}$ is also confirmed by observations, since waves of this or greater steepness are rarely reported.

Empirical Relationships between Wind and Waves. Observations of waves have not led to clear-cut conclusions on the empirical relationships between the wind and waves. The following nine approximate relationships have been proposed by various workers:

1. *Maximum Wave Height and Fetch.* For a given wind velocity, the wave height becomes greater the longer the stretch of water (fetch) over which the wind has blown. Even with a very strong wind, the wave height for a given fetch does not exceed a certain maximum value. For fetches larger than 10 nautical miles, it has been observed that

$$H_{max} = 1.5 \sqrt{F}$$

where H_{max} represents the maximum probable wave height in feet for *very strong* winds and F is the fetch in nautical miles.

2. *Wave Velocity and Fetch.* At a given wind velocity, the wave velocity increases with increasing fetch.

3. *Wave Height and Wind Velocity.* The height in feet of the greatest waves at low wind velocities has been observed to be about 0.8 of the wind velocity in knots. Over the entire range of wind velocities, the observed data conform to

$$H = 0.026v^2$$

where v represents the wind velocity in knots.

4. *Wave Velocity and Wind Velocity.* Although the ratio of wave velocity to wind velocity has been observed to vary from less than 0.1 to nearly 2.0, the average maximum wave velocity apparently slightly exceeds the wind velocity when the latter is less than about 25 knots and is somewhat less than the wind velocity at higher wind speeds.

5. *Wave Height and Duration of Wind.* The time required to develop waves of maximum height corresponding to a given wind increases with increasing wind velocity. Observations show that with strong winds high waves will develop in less than 12 hr.

6. *Wave Velocity and Duration of Wind.* Although observational data are inadequate, it is known that, for a given fetch and wind velocity, the wave velocity increases rapidly with time.

7. *Wave Steepness.* No well-established relationship exists between wind velocity and wave steepness, *i.e.*, the ratio of wave height to length. This is probably due to the fact that wave steepness is not directly related to the wind velocity but depends upon the stage of development of the wave. The stage of development, or *age of the wave*, can be conveniently expressed by the ratio of wave velocity to wind velocity

(C/v), because during the early stages of their formation the waves are short and travel with a velocity much less than that of the wind, while at later stages the wave velocity may exceed the wind velocity. In order to establish the probable relation between wave steepness and wave age, all wave observations were examined which appeared to be consistent with certain basic requirements and for which values of H, L (or C or T), and v were recorded. When corresponding values of H/L and C/v are plotted in a diagram, there appears to be a definite relationship between the steepness and the age of the wave, the scattering of the values being no greater than would be expected, considering the great errors of measurements.

8. *Decrease of Height of Swell.* The height of swell decreases as the swell advances. Roughly, the waves lose one-third of their height each time they travel a distance in miles equal to their length in feet.

9. *Increase of Period of Swell.* Some authors claim that the period of the swell remains unaltered when the swell advances, whereas others claim that the period increases. The greater amount of evidence at the present time indicates that the period of the swell increases as the swell advances from the generating area.

Sea and Swell. Where the wind blows, there are present long-crested and short-crested waves of varying length and steepness, some of which break and form white caps or streaks of foam. The irregular broken appearance of the sea surface is described by the term *the state of the sea*. By "swell" is understood, on the other hand, regular waves of moderate or great length and small steepness. These waves have been formed in a region of high winds where they accumulated their energy. Together with many other waves of the wind region, they advanced into regions of calm, but the shorter waves died out quickly, and only the longest and highest waves continued as swell to great distances from the wind region. In many instances, swell from distant wind areas crosses the waves related to local winds.

The frequency and direction of different sea states in certain parts of the oceans, as well as the frequency and direction of swell, are given on H.O. charts 10 and 712*A*, *C*, and *E*.

INDEX

For author names see bibliographies at end of individual sections

A

Abscissa, 127
Absolute stability, 404
Absolute temperature, 26, 335
Absolute value, 125
Absorption coefficient, 290
  generalized, 300
  pressure effect, 285, 300
Absorption spectrum, 298
Acceleration, 226, 229
  centripetal, 428
  Coriolis, 428
  of gravity, 90, 93, 429
  pressure systems, 423
  relative, 428
Acceleration potential, 401
Accuracy of forecast, 603, 1013
Action, centers of, 513
Active cold front, 816
Active warm front, 817
Adiabatic, 321, 348
  dry, 348
  equation, 431
  equilibrium, 373
  lapse rate, 373, 376
  wet, 359
Advection fog, 729, 732, 1048
Advective pressure tendency, 660
  nomogram, 118
  single-station, 861, 863
Aerometeorograph, 563
Air, composition of, 351
Air density, table, 116
Air masses, 604–637
  Asiatic, 634
  characteristics of, 608
  classification of, 607
  conservative properties, 400
  European, 628–634
  formation of, 306
  horizontal gradients of property, 764
  North American, 608–621
  seasonal values, 609–620
  Southern Hemisphere, 636, 806
  transformation, 306, 604, 1046
  in tropics, 763
  weather in, 621–627
Air molecules (visibility), 242
Air trajectories, 759–762
Aircraft, 756
  cross sections for, 756, 785
  effects of lightning on, 277
  icing, 753
  reports code, 577
Airways weather code, 558–593
Albedo, 296
  clouds, 932
  sea surface, 933, 1031
  solid surfaces, 932
Aleutian Islands fog, 729
Alignment charts, 136–138
Altimeter, 374
  Dalton computer, 374
  Kollsman scales, 96, 97
  setting, 88
Altitude time tables, 68, 69
Altocumulus, 886
Altostratus, 887
Ampere, 25
Amplitude, isobar-isotherm, 821

Analyses, codes for, 577
Analysis, 603–867
  air mass, 604–636
  cloud, 903
  constant-level, 813
  cross section, 756, 785
  fog, 727
  frontal, 638
  icing, 753
  isentropic, 850
  isobaric, 813n
  kinematic, 419
  single-station, 858
  soundings, 371
  Southern Hemisphere, 804
  thunderstorms, 703
  tropical, 780
  upper-air, 813
Anemograph, 555
Anemometers, 544–555
Aneroid barometers, 88, 536–537
Angle, 138, 139, 148
  bisector equation of, 162
  cross-isobar, 460
  dihedral, 141
  direction, 170
  multiple, 150, 151
  solid, 142
  between two curves, 178
  between two planes, 171
Angular velocity, 31, 427
Anticyclogenesis, 851
Anticyclone, 666–667
  displacement of, 672
  dynamic, 624
  eddies, 468
Anticyclonic, 437
  flow and stratus, 731
  isobars, 814
  shear, 516
  trajectory, 621
Antiderivative, 193
Anvil, 890
Apparent brightness, 242
Apparent color, 245
Apparent stresses, 452
Approximation, 134
  derivative, 177
  differential, 178, 200
  integral, 197
  interpolation, 134, 234
  notational error, 129
  series, 184
  solution of equations, 159, 160, 217
Arc, 138
Arctic sea smoke, 729
Arctic source, 604
Arcus, 892
Area, 140–142, 204
  bounded by a curve, 191
  equal, transformations, 363
  hodograph, 863
  solid of revolution, 192
  of surfaces, 207
  on thermodynamic diagrams, 363
  triangle, 161
Area forecast, 1011
Argand diagram, 153
Argument, 153
Arithmetic mean, 125, 236
Ascensional rates, balloon, 64–67

Knot, 27, 143
Kollsman number, 88, 96
Kurtosis, 236

L

Lagrange formula, 232
Lagrange's identity, 125
Lagrangian method, 412
Lake breeze, 433, 962
Lambert's law, 287
Lamellar vector field, 223
Laminar flow, 451
Land breeze, 433, 956
Land masses, influence of, 521
Laplace's equation, 218
Lapse rate, 371
  adiabatic, 373, 376
  any sounding, 381
  autoconvective, 372
  average in layer, 705
  and fog, 728
  normal, 676–684, 957
  polytropic, 377
  potential temperature, 381
  values, wet, 77, 79, 80
Latent heat, 340
  tables, 37, 70
Lateral sky, 895
Latitude, 142, 170
  barometer correction, 84
Latus rectum, 164, 168
Least squares, 232
Least upper bound, 125
Leeway, of raft, 1041
Leibnitz' rule, 176
Length, 27, 28, 99
  of day, 930
  unit, 669
Lenticularis, 891
Life cycle of cyclone, 659
Life rafts, drift of, 1041
Light intensity, 1004
Lightning, 264–282
  effects, 277, 712
  mechanism, 264
  protection, 278
Limit, 173, 174, 191
Line, 161, 169, 170
  contour, 200
  equations of, 162, 171
  families of, 162
  of flow, 200
  parametric equations of, 167
  in polar coordinates, 168
Line-element, 217
Line integrals, 208
Line squall, 815
Linear expansion, table, 36
Linear velocity field, 414
Local data, 858–879
Local influences, frontal, 820
Logarithms, 7, 18, 132
Longitude, 143, 170
Long-wave radiation, 297
Low clouds, 883
Lower skies, 894
Loxodrome, 143

M

Maclaurin series, 182–184, 202
Macroclimate, 963
Magnetic link, 269
Magneto anemometer, 549
Mammatus, 891
Maps, 143, 598
Marginal distributions, 238
Margules, 474
Marine barometer, 534
Mass, 210
  center of, 211
  distribution of, ocean, 1034
  fluid, 211
  transport, 1036, 1052

Mathematical tables, 3–24
Maxima, 179, 183, 202
Maximum possible storms, 1023–1026
Maximum thermometers, 541
Maxwell relations, 329
Mean, 125, 236
Mean isotherms, 829
Mean pressure, 526, 805
Mean soundings, 676–684
Mean temperatures, 609–620, 820
Mean-value theorem, 180, 192, 202, 205
Mechanical similarity, 456
Mechanics, 225
Median, 236
Melting point, 36
Mercurial barometers, 532–536
Mercurial thermometers, 540–541
Meridian, 113, 142
Meridional circulation, 505–507
Meteorological temperatures, 381–390
Meteorology, scientific basis of, 502–529
Meter, 25
Metric system, 25
Microanalysis, 788
Microclimate, 961, 963
Middle clouds, 883
Middle skies, 893
Midpoint, 161
Mile, 27
Millibar, 25
Minima, 179, 183, 202
Minimum thermometers, 541
Mixed skies, 893
Mixing of air masses, 395
  fog, 727
Mixing length, 456
Mixing ratio, 353, 358
  diurnal variation, 751
  table, 78
Mixtures, of gases, 348
Mode, 236
Modification, of air masses, 604, 1046
  of sounding, 702
Modulus, 136
  complex number, 153
  precision, 237
  real number, 125
  scale, 136
  vector, 219
Moist air, 352
Moist layer, 765
Moisture, 743
  charge, 1017
  daily increase, 743
  distribution of, 765
  and stability, 697
Moisture losses, 738
Molecular densities in air, 113
Moment of inertia, 211
Momentum, 227, 228
  angular, 228, 229
  moment of, 228, 229
Monsoon, 956, 957
Monsoon source, 607
Motion of systems, 669, 817
Motor psychrometer, 543
Mountain breeze, 960
Mountain influences, frontal, 651
M.S.L. pressure, 87
Multiple integrals, 203–205
Multiplicity, 147
Multipressure hygrometric chart, 392

N

Nautical mile, 27, 143
Navier-Stokes equations, 450
Necessary condition, 128
Nephanalysis, 903
Nephohypsometer, 563
New England Coastal fog, 739
Newfoundland fog, 729
Newton-Gregory formula, 234
Newton's law, 144
Newton's method, 160